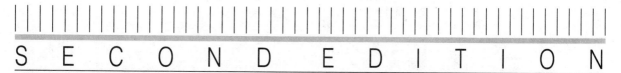

SECOND EDITION

STATISTICS FOR THE ENGINEERING AND COMPUTER SCIENCES

SECOND EDITION

STATISTICS FOR THE ENGINEERING AND COMPUTER SCIENCES

WILLIAM MENDENHALL   **TERRY SINCICH**

DELLEN PUBLISHING COMPANY
San Francisco, California

divisions of Macmillan, Inc.

COLLIER MACMILLAN PUBLISHERS
London

On the cover: "At the Garden Gate," by Robert Zupancic, 1987; mixed media on paper, 38 × 50 inches. Robert Zupancic is interested in natural science and technology. He received his graduate degree from San Francisco Art Institute. His work can be seen at Fuller Goldeen Gallery in San Francisco, California, and the Eugene Binder Gallery in Dallas, Texas.

Permissions: Dellen Publishing Company
400 Pacific Avenue
San Francisco, California 94133

Orders: Dellen Publishing Company
c/o Macmillan Publishing Company
Front and Brown Streets
Riverside, New Jersey 08075

Collier Macmillan Canada, Inc.

LIBRARY OF CONGRESS CATALOGING-IN-PUBLICATION DATA

Mendenhall, William.
 Statistics for the engineering and computer sciences/William Mendenhall,
  Terry Sincich.—2nd ed.
   Includes index.
   ISBN 0-02-380460-2
    1. Mathematical statistics. 2. Mathematics—Data processing.
   3. Engineering—Statistical methods. I. Sincich, Terry.
   II. Title.
   QA276.M428 1988
   519.5′024′62—dc19   87-33203
                                                                      CIP

Printing: 1 2 3 4 5 6 7 8   Year: 8 9 0 1 2

ISBN 0-02-380460-2

C O N T E N T S

P R E F A C E

The second edition of *Statistics for the Engineering and Computer Sciences* is a text for a two-semester introductory course in statistics for students majoring in engineering, the physical sciences, or computer science. Inevitably, once these students graduate and are employed, they will be involved in the analysis of data and will be required to make inferences from their analyses. Consequently, they need to acquire knowledge of the basic concepts of statistical inference and familiarity with some of the statistical methods that they will be required to use in their employment.

Chapters 1–7 identify the objectives of statistics, explain how we can describe data sets, and present the basic concepts of probability. Chapters 8 and 9 describe the two methods for making inferences about population parameters: estimation and testing hypotheses. These notions are extended to the problems of estimating the parameters of a general linear model, estimating the mean value of y, and predicting some future value of y for given values of a set of independent predictor variables, $x_1, x_2, \ldots, x_k$. The remaining chapters cover other topics that are useful in analyzing engineering and scientific data, including model building, the analysis of variance, the analysis of enumerative data, nonparametric statistics, and applications to the analysis of product reliability and quality control.

The assumed mathematical background is a two-semester sequence in calculus—that is, the course could be taught to students of average mathematical talent and a basic understanding of the principles of differential and integral calculus. Presentation requires the ability to perform one-variable differentiation and integration, but examples involving topics from multivariable calculus are included as optional topics. Thus, the theoretical concepts are sketched and presented in a one-variable context, but it is easy for the instructor to delve deeper into the theoretical and mathematical aspects of statistics using the optional topics, examples, and exercises.

Specific features of the text are the following:

1. **Blend of Theory and Applications.** The basic theoretical concepts of mathematical statistics are integrated with a two-semester presentation of statistical methodology. Thus, the instructor has the opportunity to present a course with either of two characteristics—a course stressing basic concepts and applied statistics or a course that, while still tilted toward application, presents a modest introduction to the theory underlying statistical inference.

2. **Computer Applications with Instructions on How to Use the Computer.** The instructor and the student have the option of using a computer to perform many of the statistical calculations. Chapter 2 explains how to enter data into a computer so that it can be accessed by any of four major statistical software packages: BMDP, Minitab, SAS, and SPSS[X].* As we proceed through the text,

---

*See the references to these statistical computer program packages at the end of Chapter 2.

we indicate how to call specific programs (for example, generation of random numbers, multiple regression analysis, or analysis of variance) from each package. No prior data-processing experience is needed. The instructions on how to use the computer in the analysis of statistical data apply to both large mainframe computers and personal computers (PCs).

3. **Coverage of Topics.** The text provides a good coverage of topics useful in analyzing engineering and other scientific data. The material often refers to theoretical material covered in earlier chapters but the presentation is oriented toward applications.

4. **Applied Exercises.** The text contains a large number of applied exercises designed to motivate a student and suggest future uses for the methodology. Many of these exercises require the student to analyze "real" data extracted from professional journals in the engineering and computer sciences.

5. **Theoretical Exercises.** Where appropriate, theoretical exercises are provided to motivate those students who have a stronger desire to understand the mathematical theory that forms an underpinning for the applications. These exercises are labeled "optional" since they require greater mathematical skill for their solution.

6. **Key Concepts Highlighted.** Definitions, theorems, and major concepts are boxed to enable the student to assimilate easily the most important facts in a chapter.

7. **Real Data Sets.** Explanations of basic statistical concepts and methodology are based on and motivated by the use of real scientific data sets. Three data sets are provided in the appendices for use as instructional vehicles:

> **Appendix III.** Length, weight, and DDT measurements for various species of fish collected from the Tennessee River by the U.S. Army Corps of Engineers.
>
> **Appendix IV.** The central processing unit (CPU) times of 1,000 computer jobs run by a small statistical consulting firm.
>
> **Appendix V.** Percentage iron content for 1.5-kilogram specimens of iron ore selected from a 20,000-ton consignment of Canadian ore.

These data sets can be entered into computer storage and then accessed by students for sampling and statistical inference. For example, the data sets can be used by the instructor to illustrate the concept of a sampling distribution and the concepts of estimation and tests of hypotheses.

Although the scope of coverage remains the same, the second edition contains several substantial changes, additions, and enhancements:

1. **Chapter 1: Some Basic Concepts.** We have expanded our discussion of data description to include several exploratory data analysis (EDA) techniques, including stem and leaf displays (Section 1.4) and box plots (Section 1.9).

2. **Chapter 7: Sampling Distributions.** A new section on the normal approximation to the binomial distribution (Section 7.5) has been added.

3. **Chapter 8: Estimation.** The entire chapter has been reorganized by population parameter. That is, each section (beginning with Section 8.5) now contains the estimation techniques corresponding to a particular parameter. Also, we have added a section on comparing the population means using matched pairs (Section 8.7) and have expanded our discussion on point estimation (Section 8.3) to include the method of moments, jackknife estimators, robust estimators, and Bayes estimators, in addition to the method of maximum likelihood.

4. **Chapter 9: Tests of Hypotheses.** As in Chapter 8, this chapter has been reorganized and now gives the test statistics appropriate for each population parameter in a separate section (beginning in Section 9.4), including a test of hypothesis for a matched-pairs experiment (Section 9.7). We have also added a new section on how to run tests for population means on the computer (Section 9.13).

5. **Chapter 10: Simple Linear Regression Analysis.** A test for correlation has been incorporated into Section 10.7 and a new section on computer printouts for a simple linear regression analysis (Section 10.11) has been added.

6. **Chapter 11: Multiple Regression Analysis.** The matrix algebra sections (Sections 11.4–11.6 in the first edition) have been moved to Appendix I. Our discussion of multicollinearity in Section 11.12 has been expanded to include methods for both detecting and solving the problem. We have also included a new section on the analysis of residuals (Section 11.13).

7. **Chapter 12: Introduction to Model Building.** Interpretations of the β parameters are provided for each model discussed in Sections 12.3, 12.4, and 12.6. Also, a new section on coding quantitative variables (Section 12.5) has been added.

8. **Chapter 13: The Analysis of Variance for Designed Experiments.** This chapter has been almost entirely rewritten to follow the more traditional approach to ANOVA. The relationship between ANOVA and regression, including the models for each type of design, is covered in a new section (Section 13.8).

9. **Chapter 16: Nonparametric Statistics.** We have added a new section on the sign test for a population median (Section 16.2).

10. **Chapter 18: Applications: Quality Control.** New sections on conducting a runs analysis for trend (Section 18.4) and establishing tolerance limits (Section 18.7) have been added. Throughout the entire chapter, our discussion of control charts has been embellished to give it more of a Shewhart-flavor.

11. **More Exercises with Real Data.** Many new "real-world" scientific exercises have been added to each chapter. These exercises are extracted from news articles, magazines, and professional journals to give students the opportunity to apply their knowledge of statistics to current practical problems in the engineering and computer sciences.

12. **Solutions Manual.** The second edition is accompanied by a student's exercise solutions manual, which presents the complete solutions to selected exercises contained in the text.

Numerous less obvious changes in details have been made throughout the text in response to suggestions by current users and reviewers of the first edition.

We want to thank those people who contributed to the development of this text. Chief among these was Susan Reiland, who provided a line-by-line review of the text and managed the production of both the first and second editions. E. Jacquelin Dietz, David Robinson, and Dennis Wackerly were helpful in reviewing the first edition and in helping us decide on the level of the text. The reviewers of the second edition suggested many of the changes listed above; they include Lydia Gans (California State Polytechnic University), Chand Midha (University of Akron), Carol O'Connor (University of Louisville), Donald L. Woods (Texas A&M University), and especially Herbert B. Eisenberg (West Virginia College of Graduate Studies). Finally, we thank Brenda Dobson, who typed the original manuscript, and Carol Springer, who typed the manuscript for the second edition.

C H A P T E R 1

OBJECTIVE

To identify the applications of statistics in data analysis and to present some graphical and numerical methods for describing data.

CONTENTS

SOME BASIC CONCEPTS

STATISTICS: AN INFORMATION SCIENCE

The science of statistics and statistical methodology are concerned with two types of problems:

1. The description of large data sets
2. The use of sample data to infer the nature of the data set from which the sample was selected

As an illustration of the descriptive applications of statistics, consider the United States census, which involves the collection of a data set that purports to characterize the socioeconomic characteristics of the 226,000,000 people living in the United States. Managing this enormous mass of data is a problem for the computer scientist, and describing the data utilizes the methods of statistics. Similarly, an engineer uses statistics to describe the data set consisting of the daily emissions of sulfur oxides of an industrial plant recorded for each of 365 days last year.

Sometimes the phenomenon of interest is characterized by a data set that is either physically unobtainable, or too costly or too time-consuming to obtain. In such situations, we sample the data set and use the sample information to infer its nature. To illustrate, suppose the phenomenon of interest is the waiting time for a data-processing job to be completed. You might expect the waiting time to depend on such factors as the size of the job, the computer utilization factor, etc. In fact, if you were to run the same job over and over again on the computer, the waiting times would vary, even for the same computer utilization factor. Thus, the phenomenon "waiting time before job processing" is charac-terized by a large data set that exists only conceptually (in our minds). To determine the nature of this data set, we *sample* it—i.e., we process the job a number n of times, record the waiting time for each run, and then use this sample of n waiting times to infer the nature of the large conceptual data set of interest. The branch of statistics used to solve this problem is called **inferential statistics**.

In statistical terminology, the data set that we want to describe, the one that characterizes a phenomenon of interest to us, is called a **population**. A **sample** is a subset of data selected from a population. Sometimes the words *population* and *sample* are used to represent the objects upon which the measurements are taken. In a particular situation, the meaning attached to these terms will be clear by the context in which they are used.

DEFINITION 1.1

A **population** is a data set that is the target of our interest.

DEFINITION 1.2

A **sample** is a subset of data selected from a population.

THE OBJECTIVE OF INFERENTIAL STATISTICS

The **objective of inferential statistics** is to make inferences about a population based on information contained in a sample.

CASE STUDY 1.1

CONTAMINATION OF FISH
IN THE TENNESSEE
RIVER

Chemical and manufacturing plants often discharge toxic waste materials into nearby rivers and streams. These toxicants have a detrimental effect on the plant and animal life inhabiting the river and the river's bank. One type of pollutant, commonly known as DDT, is especially harmful to fish and, indirectly, to people. The Food and Drug Administration sets the limit for DDT content in individual fish at 5 parts per million (ppm). Fish with DDT content exceeding this limit are considered potentially hazardous to people if consumed. A study was recently undertaken to examine the DDT content of fish inhabiting the Tennessee River (in Alabama) and its tributaries.

The Tennessee River flows in a west–east direction across the northern part of the state of Alabama, through Wheeler Reservoir, a national wildlife refuge. Ecologists fear that contaminated fish migrating from the mouth of the river to the reservoir could endanger other wildlife that prey on the fish. This concern is more than academic. A manufacturing plant was once located along Indian Creek, which enters the Tennessee River 321 miles upstream from the mouth. Although the plant has been inactive for over 10 years, there is evidence that the plant discharged toxic materials into the creek, contaminating all the fish in the immediate area. Have the fish in the Tennessee River and its tributary creeks also been contaminated? And if so, how far upstream have the contaminated fish migrated? In order to answer these and other questions, members of the U.S. Army Corps of Engineers in the summer of 1980 collected fish specimens at different locations along the Tennessee River and three tributary creeks: Flint Creek (which enters the river 309 miles upstream from the river's mouth), Limestone Creek (310 miles upstream), and Spring Creek (282 miles upstream). Each fish was first weighed (in grams) and measured (length in centimeters), then the fillet of the fish was extracted and the DDT concentration (in parts per million) in the fillet was measured.

Appendix III contains the length, weight, and DDT measurements for a total of 144 fish specimens.* Obviously, not all the fish in the Tennessee River and its tributaries were captured. Consequently, the data are based on a sample collected from the population of all fish inhabiting the Tennessee River. Here, the words *population* and *sample* are used to describe the objects upon which the measurements are taken, i.e., the fish. We could also use the terms to represent data sets. For example, the 144 DDT measurements represent a sample collected from the population consisting of DDT measurements for all fish inhabiting the river.

*Source: U.S. Army Corps of Engineers, Mobile District, Alabama.

Notice that the data set also contains information on the location (i.e., where the fish were captured) and species of the fish. Three species of fish were examined: channel catfish, large-mouth bass, and small-mouth buffalo. The different symbols for location are interpreted as follows. The first two characters represent the river or creek and the remaining characters represent the distance (in miles) from the mouth of the river or creek. For example, FCM5 indicates that the fish was captured in Flint Creek (FC), 5 miles upstream from the mouth of the creek (M5). Similarly, TRM380 denotes a fish sample collected from the Tennessee River (TR), 380 miles upstream from the river's mouth (M380). In this and subsequent chapters we will use the data in Appendix III to compare the DDT contents of fish at different locations and among the different species and to determine the relationship (if any) of length and weight to DDT content.

Although we will present some useful methods for describing data sets, the major emphasis in this text and in modern statistics is in the area of inferential statistics. In particular, we will see how probability enables us to utilize information in a sample to infer the nature of a sampled population.

EXERCISES 1.1–1.5

1.1 Checking all manufactured items coming off an assembly line for defectives would be a costly and time-consuming procedure. One effective and economical method of checking for defectives involves the selection and examination of a portion of the items by a quality control engineer. The percentage of examined items that are defective is computed and then used to estimate the percentage of all items manufactured on the line that are defective. Identify the population, the sample, and a type of statistical inference to be made for this problem.

1.2 Research engineers with the University of Kentucky Transportation Research Program have collected data on accidents occurring at intersections in Lexington, Kentucky, over a period of 5 years. One of the goals of the study was to compare the average number of left-turn accidents at locations with and without left-turn-only lanes, in order to develop numerical warrants (or guidelines) for the installation of left-turn lanes.
a. What is the population of interest?
b. What is the sample?
c. How can the sample information be used to attain the researchers' goal?

1.3 The reliability of a computer system is measured in terms of the lifelength of a specified hardware component (for example, the disk drive). In order to estimate the reliability of a particular system, 100 computer components are tested until they fail, and their lifelengths are recorded.
a. What is the population of interest? What is the sample?
b. How could the sample information be used to estimate the reliability of the computer system?

1.4 Researchers have developed a new precooling method for preparing Florida vegetables for market. The system employs an air and water mixture designed to yield effective cooling with a much lower water flow than conventional hydrocooling. In an effort to compare the effectiveness of the two cooling systems, 20 batches of green tomatoes were divided into two groups; one group was precooled with the new method, and the other with the conventional method. The total water flow (in gallons) required to effectively cool each batch was recorded.

a. Identify the population, the samples, and the type of statistical inference to be made for this problem.

b. How could the sample data be used to compare the cooling effectiveness of the two systems?

1.5 Computer tomography (CT) scanners are highly sensitive, visual computer systems designed to aid a physician's diagnosis by generating radiographlike images of inner organs and physiological functions. Suppose you want to estimate the average *scan time*—that is, the average time required for a CT scanner to project an image. Describe how you could collect the sample data necessary to make the desired inference. What is the population of interest?

SECTION 1.2

TYPES OF DATA

Data can be one of two types, quantitative or qualitative. **Quantitative data** are those that represent the quantity or amount of something. For example, the waiting time before a job begins processing is a quantitative variable, as is the size of the job. In contrast, **qualitative** (or **categorical**) data possess no quantitative interpretation. They can only be classified. The set of *n* occupations corresponding to a group of *n* engineering graduates is a qualitative data set. A list of the manufacturers of *n* minicomputers owned by *n* small businesses is a set of qualitative data.

DEFINITION 1.3

Quantitative data are those that represent the quantity or amount of something.

DEFINITION 1.4

Qualitative data are those that have no quantitative interpretation, i.e., they can only be classified into categories.

EXAMPLE 1.1

Refer to the data set in Appendix III (see Case Study 1.1). Classify each of the five variables in the data set, location, species, length, weight, and DDT concentration, as quantitative or qualitative.

SOLUTION

Length (in centimeters), weight (in grams), and DDT concentration (in parts per million) are all measured on a numerical scale; thus, they represent quantitative data. In contrast, location and species cannot be measured on a quantitative scale; they can only be classified (e.g., channel catfish, large-mouth bass, and small-mouth buffalo for species). Consequently, data on location and species are qualitative. ∎

**GRAPHICAL METHODS
FOR DESCRIBING
QUALITATIVE DATA**

When describing qualitative observations, we define the categories in such a way that each observation can fall in one and only one category. The data set is then described by giving the number of observations, or the proportion of the total number of observations, that fall in each of the categories.

DEFINITION 1.5

The **category frequency** for a given category is the number of observations that fall in that category.

DEFINITION 1.6

The **category relative frequency** for a given category is the proportion of the total number of observations that fall in that category.

Graphical descriptions of qualitative data sets are usually achieved using bar graphs or pie charts, and these figures are often constructed by a computer. **Bar graphs** give the frequency (or relative frequency) corresponding to each category, with the height or length of the bar proportional to the category frequency (or relative frequency). **Pie charts** divide a complete circle (a pie) into slices, one corresponding to each category, with the central angle of the slice proportional to the category relative frequency. Examples of these familiar graphical methods are shown in Figures 1.1 and 1.2.

Figure 1.1 is a vertical bar graph that shows the number of large nuclear reactors in each of nine countries. (Bar graphs can be vertical or horizontal.) Each bar corresponds to one of the nine countries, and the height of the bar is proportional to the number of large nuclear reactors in that country. Figure 1.2 shows the percentages of computer maintenance time that can be attributed to each of four categories: requirement specification errors, design errors, coding errors, and all others. The pie chart not only enables you to determine the exact percentage of maintenance time attributable to a given source, but it also provides a rapid visual comparison of the relative percentages (sizes of the slices) assignable to the various types of error.

EXAMPLE 1.2

Refer to the fish sample data in Appendix III. Construct a horizontal bar graph for the proportion of fish specimens captured of each of the three species: channel catfish, large-mouth bass, and small-mouth buffalo.

SOLUTION

A count of the 144 fish in the sample reveals that 36 are small-mouth buffalo, 12 are large-mouth bass, and the remaining 96 are channel catfish. The corresponding proportions (or relative frequencies) are $\frac{36}{144} = .25$, $\frac{12}{144} = .083$, and $\frac{96}{144} = .667$. A horizontal bar graph showing these relative frequencies is portrayed in Figure 1.3 (page 8).

FIGURE 1.1
Vertical Bar Graph
Showing the Number of
Large Nuclear Reactors
for Nine Countries

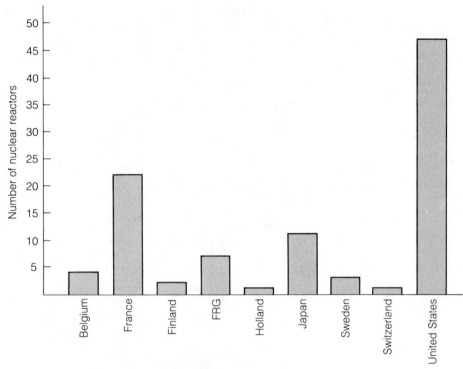

Source: Plews, M. J., Wakerly, M. W., and Winyard, R. A. "Comparing PWR Exposures Worldwide." *Nuclear Engineering International*, Vol. 31, No. 381, Apr. 1986, p. 46 (Table 1).

FIGURE 1.2
Pie Chart Showing the
Percentages of Computer
Maintenance Time
Attributable to Various
Sources of Error

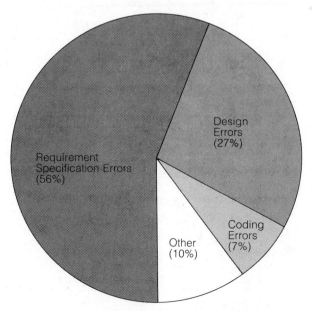

Source: *Data Communications*, Jan. 1982, Vol. 11, No. 1, p. 99. Copyright 1982 McGraw-Hill, Inc. All rights reserved.

FIGURE 1.3

Horizontal Bar Graph
Showing the Proportions
of Captured Fish for the
Three Species

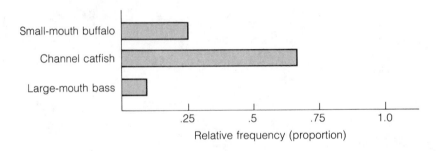

EXERCISES 1.6–1.13

1.6 State whether each of the following data sets is quantitative or qualitative.
  a. Arrival times of 16 reflected seismic waves
  b. Types of computer software used in a data base management system
  c. Brands of calculator used by 100 engineering students on campus
  d. Ash contents in pieces of coal from three different mines
  e. Mileages attained by 12 automobiles powered by alcohol
  f. Numbers of print characters per line of computer output for 20 line printers
  g. Shift supervisors in charge of computer operations at an airline company
  h. Accident rates at 46 machine shops

1.7 Many U.S. industries are considering switching to surface mounting of electronic components. Currently, the computer, telecommunications, and automotive industries are the only large assemblers using surface-mounted devices (SMDs). The annual sales (in millions of units) of several varieties of SMDs are shown in the table for 1985, 1987 (projected), and 1989 (projected).

SMD TYPE	ANNUAL SALES		
	1985	1987	1989
Plastic DIP	7,679	7,999	9,209
Cerdip	1,032	1,272	1,567
Ceramic DIP	176	221	277
Flatpack	73	64	55
Leaded chip carrier	49	164	553
Leadless chip carrier	35	113	369
Plastic leaded chip carrier	97	1,164	2,292
Quad flatpack	4	23	133
Small-outline transistor	485	2,209	3,687
Pin grid away	3	7	18
Other	70	139	277
TOTALS	9,703	13,375	18,437

Source: Lyman, J. "Inside Technology: What's Holding Back Surface Mounting." *Electronics*, Vol. 59, No. 6, Feb. 10, 1986, p. 26. Copyright © 1986 McGraw-Hill, Inc. All rights reserved.

  a. Construct bar charts for types of SMD sold in each of the three years, 1985, 1987, and 1989.
  b. Examine the bar charts of part **a**. Which SMDs are expected to grow the fastest in sales over the 4-year period?

1.8 Shore-based marine traffic systems have been proposed to improve the safety and efficiency of marine traffic. Prior to the installation of the traffic system, one study was conducted to assess the current level of risk of collision to vessels operating in European waters. Data on large-vessel collisions over the 5-year period 1978–1982 are presented in the table.

COLLISIONS BY LOCATION		COLLISIONS AT SEA BY ENCOUNTER ASPECT	
Location	Number of Ships	Aspect	Number of Ships
At sea	376	Meeting	131
Restricted waters	273	Overtaking	29
In port	478	Crossing	73
TOTAL	1,127	Unknown	143
		TOTAL	376

Source: Kemp, J. F. and Goodwin, E. M. "Risk Analysis Within the Cost 301 Project." *The Dock and Harbour Authority*, Vol. 66, No. 775, Dec./Jan. 1985–1986.

a. Construct a bar graph for the locations of large-vessel collisions in European waters. Interpret the graph.

b. Construct a bar graph for the encounter aspects of large-vessel collisions at sea in European waters. Interpret the graph.

1.9 In the United States, dyes are used in coloration products such as textiles, paper, leather, and foodstuffs, and are required by law to be in gasoline to indicate the presence of lead. The long-term effects of dyes and their degradation products, however, are unknown. In order to monitor environmental contamination, analytical methods must be developed to identify and quantify these dyes. In one study, thermospray high-performance liquid chromatography/mass spectrometry was used to characterize dyes in wastewater and gasoline. The accompanying table gives the relative abundance (relative frequency of occurrence) of commercial Diazo Red dye components in gasoline. Describe the relative abundance of red dye compounds with a bar graph. Interpret the graph.

RED DYE COMPOUND	RELATIVE ABUNDANCE	RED DYE COMPOUND	RELATIVE ABUNDANCE
H	.021	C_8H_{17}	.127
CH_3	.210	C_9H_{19}	.118
C_2H_5	.354	$C_{10}H_{21}$	.025
C_3H_7	.072	Others	.019
C_7H_{15}	.054		

Source: Voyksner, R. D. "Characterization of Dyes in Environmental Samples by Thermospray High-performance Liquid Chromatography/Mass Spectrometry." *Analytical Chemistry*, Vol. 57, No. 13, Nov. 1985, p. 2601 (Table I). Reprinted with permission. Copyright 1985 American Chemical Society.

1.10 The personal computer (PC) industry is growing at a phenomenal rate. The *Wall Street Journal* (Sept. 16, 1985) conducted a survey of businesses that use or are planning to use PCs in the office. The table gives a breakdown of the brand of PC purchased by the surveyed businesses in 1985.

PC MANUFACTURER	PERCENTAGE OF BUSINESSES USING THIS PC BRAND
Apple	8%
AT&T	4
Compaq	6
IBM	70
NEC	7
Other	5
TOTAL	100%

a. Suppose 1,000 businesses took part in the survey. Construct a bar graph for the data.

b. Which PC manufacturer dominated the business PC market in 1985?

1.11 Stainless steels are frequently used in chemical plants to handle corrosive fluids. However, in certain environments these steels are especially susceptible to stress corrosion cracking. One study identified stress corrosion cracking as the greatest single cause of steel alloy failure in Japanese chemical plants. The accompanying table lists the various modes of failure and their corresponding percentages of the total for 295 cases of alloy failures that occurred in oil refineries and petrochemical plants in Japan over the last 10 years.

CAUSE OF FAILURE	PERCENTAGE
Wet environment	
General corrosion	12.5%
Localized corrosion	15.9
Stress corrosion cracking	39.9
Miscellaneous	3.8
Dry environment	
Corrosion	8.2
Cracking	10.9
Decrease of mechanical properties	1.7
Miscellaneous	1.7
Material defects	2.0
Welding defects	3.4

Source: Yamamoto, K. and Kagawa, N. "Ferritic Stainless Steels Have Improved Resistance to SCC in Chemical Plant Environments." *Materials Performance*, June 1981, Vol. 20, No. 6, pp. 32–35.

a. Construct a bar graph for the causes of steel alloy failure in Japanese chemical plants.

b. Does the bar graph of part **a** support the claim made by the researchers?

1.12 The process of identifying and estimating the amount of different minerals present in a grain of rock is called *modal analysis*. In modal analysis, the area occupied by the section of the grain in a plane intersecting it is viewed under a powerful electron microscope. Points are selected within the area and the type of mineral in which each point falls is identified. The proportion of points identified as a particular mineral can then be used to estimate the volume of that mineral in the grain. Suppose a geophysical engineer selects 1,000 such points and identifies five minerals, with the results listed in the table.

MINERAL	NUMBER OF POINTS
A	196
B	82
C	400
D	12
E	310

a. Construct a pie chart showing the percentage of points identified for each of the five minerals.

b. Which minerals appear to have the largest volume in the rock grain?

1.13 A marketing research study of consulting engineering services to industrial firms in the Midwest was recently conducted. The main goal of the study was to gather information that will enable consulting engineers to effectively market their services to industrial firms. Of the 70 firms surveyed, 40 indicated that they have no need for outside consulting engineering services. The accompanying table gives the primary reasons cited by the "non-needers" and corresponding breakdown in percentages for both the large and small industrial firms in the survey.

REASON	LARGE FIRMS	SMALL FIRMS
Assistance obtained from corporate headquarters	62%	30%
No wastes, therefore, no need to improve	0	32
No improvements planned	0	24
Assistance obtained from staff engineers	19	6
Unfamiliar with consulting	10	2
Waiting for regulations	9	0
Other reasons	0	6
TOTALS	100%	100%

Source: Carey, R. J. and Brunner, J. A. "A Study of Marketing of Consulting Engineering Services to Industrial Firms." *Journal of the Boston Society of Civil Engineers Section*, American Society of Civil Engineers, Vol. 71, No. 1 and 2, 1985, p. 152.

a. Construct a pie chart that describes the reasons cited for not needing consulting engineering services at large industrial firms.

b. Repeat part **a** for small industrial firms.

c. Compare the two pie charts in parts **a** and **b**. Do you detect major differences in the reasons cited by large and small firms?

| | | | | | | | | | | | | |

S E C T I O N 1.4

GRAPHICAL METHODS FOR DESCRIBING QUANTITATIVE DATA

The most popular and traditional graphical method for describing quantitative data is the **relative frequency histogram** (often called a **relative frequency distribution**). For example, suppose we want to describe graphically the data shown in Table 1.1. The *CPU times* given in the table are the amounts of time (in seconds) 25 jobs were in control of a large mainframe computer's central processing unit (CPU). These 25 values represent a sample selected from the 1,000 CPU times listed in Appendix IV. The steps to follow in constructing a relative frequency histogram are listed in the box. The results for the data of Table 1.1 are shown in Table 1.2.

STEPS TO FOLLOW IN CONSTRUCTING A RELATIVE FREQUENCY HISTOGRAM

Step 1. Calculate the **range** of the data:

Range = Largest observation − Smallest observation

The range for the data of Table 1.1 is

Range = 4.75 − 0.02 = 4.73

Step 2. Divide the range into between five and twenty **classes** of equal width. The number of classes is arbitrary, but you will obtain a better graphical description if you use a small number of classes for a small amount of data and a larger number of classes for large data sets. The lowest (or first) class boundary should be located below the smallest measurement, and the class width should be chosen so that no observation can fall on a class boundary. We will use seven classes for the data of Table 1.1. Then the approximate class width is

$$\text{Approximate class width} = \frac{\text{Range}}{7} = \frac{4.73}{7} = .676$$

We will round this width upward and use a class width of .7. The first class will be located at .015, just below the smallest CPU time. The resulting seven class intervals are shown in Table 1.2.

Step 3. For each class, count the number of observations that fall in that class. This number is called the **class frequency**.

Step 4. Calculate each **class relative frequency**:

$$\text{Class relative frequency} = \frac{\text{Class frequency}}{\text{Total number of measurements}}$$

The class frequencies and relative frequencies for the data of Table 1.1 are shown in columns 4 and 5, respectively, of Table 1.2.

Step 5. The **relative frequency histogram** is essentially a bar graph in which the categories are the classes. The relative frequency histogram for the data of Table 1.1 is shown in Figure 1.4.

TABLE 1.1

A Sample of $n = 25$ Job CPU Times (in seconds) Selected from Appendix IV

1.17	1.61	1.16	1.38	3.53
1.23	3.76	1.94	0.96	4.75
0.15	2.41	0.71	0.02	1.59
0.19	0.82	0.47	2.16	2.01
0.92	0.75	2.59	3.07	1.40

TABLE 1.2

Calculating Class Relative Frequencies for the Data of Table 1.1

CLASS	CLASS INTERVAL	DATA TABULATION	CLASS FREQUENCY	CLASS RELATIVE FREQUENCY
1	.015–.715	ⱶⱶ	5	.20
2	.715–1.415	ⱶⱶ ||||	9	.36
3	1.415–2.115	||||	4	.16
4	2.115–2.815	|||	3	.12
5	2.815–3.515	|	1	.04
6	3.515–4.215	||	2	.08
7	4.215–4.915	|	1	.04
		Totals $n = 25$		1.00

Since the bars in a relative frequency histogram are of equal width, the area of a particular bar is proportional to the corresponding class relative frequency. If we let the total area of the bars equal 1, then the area of a particular bar is *equal* to its corresponding class relative frequency. Furthermore, if we select one observation from among the $n = 25$ observations in Table 1.1, then the probability that the observation will fall in a particular class is equal to the relative frequency of that class. The probability that the observation will fall in one of two or more specific classes is equal to the sum of their respective relative frequencies and is proportional to the total area of the bars corresponding to those classes. For example, the probability that the observation will be a CPU time less than 2.115 seconds is equal to .72, the sum of the relative frequencies for classes 1, 2, and 3 of Table 1.2. This probability is proportional to the shaded area of Figure 1.4 on page 14.

[Note: Most graphical descriptions of large data sets, whether qualitative or quantitative, are performed on a computer by one of several easy-to-use statistical computer program packages. In Chapter 2 we will show you how to enter data into a computer for use with each of four such packages: BMDP, Minitab, SAS, and SPSS[x]. In Sections 2.6 and 2.7, we will show you how to use the programs to construct a relative frequency histogram.]

Another graphical method for describing quantitative data is the **stem and leaf display**, which is widely used in exploratory data analysis when the data set is small. The next box contains the steps to follow in constructing a stem and leaf display for the 25 CPU times of Table 1.1.

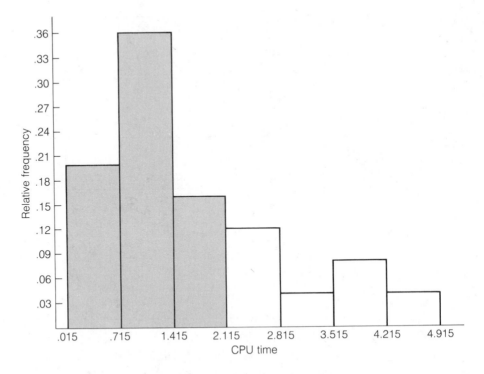

STEPS TO FOLLOW IN CONSTRUCTING A STEM AND LEAF DISPLAY

Step 1. Divide each observation in the data set into two parts, the **stem** and the **leaf**. We will designate the first digit of the CPU time (i.e., the digit to the left of the decimal point) as its stem; we will call the last two digits its leaf. For example, the stem and leaf of the CPU time 2.41 are 2 and 41, respectively:

STEM	LEAF
2	41

Although this assignment is arbitrary, the stem and leaf display will yield more information if you define the stems and leaves so that you obtain a smaller number of stems for a small amount of data and a larger number of stems for large data sets. Using the assignment specified here, we obtain a total of five stems: 0, 1, 2, 3, and 4.

Step 2. List the stems in order in a column, starting with the smallest stem and ending with the largest (see Figure 1.5).

Step 3. Proceed through the data set, placing the leaf for each observation in the appropriate stem row. The completed **stem and leaf display** for the data of Table 1.1 is shown in Figure 1.5.

Notice that if you rotate the stem and leaf display on its side, you obtain the same type of bar graph as provided by the (relative) frequency distribution. The stem and leaf display of Figure 1.5 partitions the data set into five classes corresponding to the five stems. The number of leaves in each class gives the class frequency.

FIGURE 1.5

Stem and Leaf Display for the $n = 25$ CPU Times of Table 1.1

STEMS	LEAVES	FREQUENCY	RELATIVE FREQUENCY
0	15 19 92 82 75 71 47 96 02	9	.36
1	17 23 61 16 94 38 59 40	8	.32
2	41 59 16 01	4	.16
3	76 07 53	3	.12
4	75	1	.04
	TOTALS $n = 25$		1.00

One advantage of a stem and leaf display over a frequency distribution is that the original data are preserved. That is, you can look at the display and resurrect the exact values of the data. A stem and leaf display also arranges the data in an orderly fashion and makes it easy to determine certain numerical characteristics to be discussed in the following sections. The third advantage is that the classes and the numbers falling in them are quickly determined once we have selected the digits that we want to use for the stems and leaves.

A disadvantage of the stem and leaf display is that there is sometimes not much flexibility in choosing the stems. For the data of Table 1.1, two stem and leaf options are possible. We could define the stems and leaves as shown in Figure 1.5. Or, we could let the first two digits represent the stem, in which case the number 2.41 would have the stem 24 and the leaf 1:

STEM	LEAF
24	1

The associated stem and leaf display for the data of Table 1.1 would then contain a total of 48 stems, ranging from 00 to 47, and most of these stem rows would contain no leaves. Clearly, this choice of stems and leaves would not provide as much information about the data as does the display of Figure 1.5. Consequently, we are left with the option of using a stem and leaf display that produces five stems (and thus, five classes for the frequency distribution) or one that produces 48 stems.*

Frequency distributions or relative frequency distributions are most often used in scientific publications to describe quantitative data sets. They are better suited to the description of large data sets and they permit a greater flexibility in the choice of class width.

---

*By sacrificing some of the simplicity of our procedure, we could define the stems and leaves so that the number of stems falls between 5 and 48. We omit discussion of this topic.

| | | | | | | | | | | | |

EXERCISES 1.14–1.19

1.14 Bacteria are the most important component of microbial ecosystems in sewage treatment plants. Water management engineers must know the percentage of active bacteria at each stage of the sewage treatment. The accompanying data represent the percentages of respiring bacteria in 25 raw sewage samples collected from a sewage plant. (Measurements were obtained by passing the samples through a membrane filter and staining with a chlorination solution.)

42.3	50.6	41.8	36.5	28.6
40.7	48.1	48.0	45.7	39.9
31.3	30.7	40.5	40.9	51.2
38.6	35.6	22.9	33.4	46.5
41.5	43.5	41.1	38.5	44.4

a. Construct a relative frequency distribution for the data.

b. Construct a stem and leaf display for the data.

c. Compare the two graphs of parts **a** and **b**.

1.15 Refer to the results of the U.S. Army Corps of Engineers analysis of fish samples (Appendix III). Consider the 50 DDT measurements corresponding to fish specimens identified on the data set by observations numbered 51–100.

a. Construct a relative frequency histogram for the 50 DDT values. Use 10 classes to span the range.

b. Repeat part **a**, but use only three classes to span the range. Compare the result with the relative frequency histogram you constructed in part **a**. Which is more informative? Why does an inadequate number of classes limit the information conveyed by a relative frequency histogram?

c. Repeat part **a**, but use 25 classes. Comment on the information provided by this histogram, and compare with the result of part **a**.

1.16 The National Research Council's Committee on Underground Coal Mine Safety was established to determine "the factors that distinguish the safest from the most dangerous mines." In order to evaluate differences between the safest and most dangerous mines, the committee collected data on 19 of the largest underground coal companies.

COMPANY	INJURY RATE	COMPANY	INJURY RATE
Old Ben	2.72	American Electric Power	5.11
Bethlehem	2.89	Rochester & Pittsburgh	5.12
Island Creek	2.97	Pittston	5.39
Consolidation	2.98	Ziegler	6.19
Mapco	3.17	Freeman United	6.83
U.S. Steel	3.58	Republic	6.84
Alabama By-Product	3.88	North American	7.47
Eastern Assoc.	4.66	West Moreland	7.68
Peabody	4.81	Valley Camp	8.71
Jones & Laughlin	4.87		

Source: Spokes, E. M. "New Look at Underground Coal Mine Safety." *Mining Engineering,* Vol. 38, No. 4, Apr. 1986, p. 267 (Table 1).

The "intermediate injury" rate (i.e., the number of disabling injuries resulting from falls of roof and sides, haulage, machinery, and explosive accidents per 200,000 worker hours) for each of the 19 companies is recorded in the table. Construct a stem and leaf display for the data.

1.17 *Scram* is the term used by nuclear engineers to describe a rapid emergency shutdown of a nuclear reactor. The nuclear industry has made a concerted effort to significantly reduce the number of unplanned scrams. The table gives the number of scrams at each of 56 U.S. nuclear reactor units in 1984. Summarize the data with a graphical descriptive technique.

				NUMBER OF SCRAMS					
1	0	3	1	4	2	10	6	5	2
0	3	1	5	4	2	7	12	0	3
8	2	0	9	3	3	4	7	2	4
5	3	2	7	13	4	2	3	3	7
0	9	4	3	5	2	7	8	5	2
4	3	4	0	1	7				

Source: Visner, S. "Reducing the Frequency of Scrams in U.S. Nuclear Power Plants." *Transactions of the American Nuclear Society*, Vol. 50, 1985, p. 504 (Figure 2). Copyright 1985 by the American Nuclear Society, LaGrange Park, Illinois.

1.18 A Harris Corporation/University of Florida study was undertaken to determine whether a manufacturing process performed at a remote location can be established locally. Test devices (pilots) were set up at both the old and new locations and voltage readings on the process were obtained. A "good process" was considered to be one with voltage readings of at least 9.2 volts (with larger readings being better than smaller readings). The table contains voltage readings for 30 production runs at each location.

OLD LOCATION			NEW LOCATION		
9.98	10.12	9.84	9.19	10.01	8.82
10.26	10.05	10.15	9.63	8.82	8.65
10.05	9.80	10.02	10.10	9.43	8.51
10.29	10.15	9.80	9.70	10.03	9.14
10.03	10.00	9.73	10.09	9.85	9.75
8.05	9.87	10.01	9.60	9.27	8.78
10.55	9.55	9.98	10.05	8.83	9.35
10.26	9.95	8.72	10.12	9.39	9.54
9.97	9.70	8.80	9.49	9.48	9.36
9.87	8.72	9.84	9.37	9.64	8.68

a. Construct a relative frequency histogram for the voltage readings of the old process.
b. Construct a stem and leaf display for the voltage readings of the old process. Which of the two graphs in parts **a** and **b** is more informative?
c. Construct a relative frequency histogram for the voltage readings of the new process.

d. Compare the two graphs in parts **a** and **c**. Does it appear that the manufacturing process can be established locally (i.e., is the new process as good as or better than the old)?

1.19 Given below are the monthly wages for 35 operations analysts. Construct a relative frequency distribution for the data.

$2,470	$2,790	$1,730	$2,090	$2,290	$2,000	$2,130
2,060	1,910	2,130	2,470	2,190	1,920	2,300
2,400	2,480	2,400	2,200	2,120	2,070	2,090
2,610	2,200	2,430	2,150	2,200	2,290	2,270
2,720	2,880	2,130	2,310	2,400	2,230	2,130

| | | | | | | | | | | | |

S E C T I O N 1.5

NUMERICAL METHODS FOR DESCRIBING QUANTITATIVE DATA

Numerical descriptive measures are numbers computed from a data set to help us create a mental image of its relative frequency distribution. The measures that we will present fall into three categories: (1) those that help to locate the *center* of the relative frequency distribution, (2) those that measure its *spread*, and (3) those that describe the *relative position* of an observation within the data set. These categories are called, respectively, **measures of central tendency, measures of variation**, and **measures of relative standing**. In the definitions and formulas that follow, we will denote the *variable* observed to create a data set by the symbol y and the n measurements of a data set by $y_1, y_2, \ldots, y_n$.

| | | | | | | | | | | | |

S E C T I O N 1.6

MEASURES OF CENTRAL TENDENCY

The three most common measures of central tendency are the **arithmetic mean**, the **median**, and the **mode**. Of the three, the arithmetic mean (or **mean**, as it is commonly called) is used most frequently in practice.

DEFINITION 1.7

The **arithmetic mean** $\bar{y}$ of a sample of n measurements, $y_1, y_2, \ldots, y_n$, is the average of the measurements:

$$\bar{y} = \frac{\sum_{i=1}^{n} y_i}{n}$$

EXAMPLE 1.3

Calculate the arithmetic mean for the set of $n = 5$ measurements: 4, 6, 1, 2, 3.

SOLUTION

Substitution into the formula for $\bar{y}$ yields

$$\bar{y} = \frac{\sum_{i=1}^{n} y_i}{n} = \frac{4 + 6 + 1 + 2 + 3}{5} = 3.2$$

∎

DEFINITION 1.8

The **median** m of a sample of n measurements, $y_1, y_2, \ldots, y_n$, is the middle number when the measurements are arranged in ascending (or descending) order, i.e., the value of y located so that half the area under the relative frequency histogram lies to its left and half the area lies to its right.

If the number of measurements in a data set is odd, the median is the measurement that falls in the middle when the measurements are arranged in increasing order. For example, the median of the $n = 5$ measurements of Example 1.3 is $m = 3$. If the number of measurements is even, the median is defined to be the mean of the two middle measurements when the measurements are arranged in increasing order. For example, the median of the $n = 6$ measurements, 1, 4, 5, 8, 10, 11, is

$$m = \frac{5 + 8}{2} = 6.5$$

DEFINITION 1.9

The **mode** of a sample of n measurements, $y_1, y_2, \ldots, y_n$, is the value of y that occurs with the greatest frequency.

If the outline of a relative frequency histogram were cut from a piece of plywood, it would be perfectly balanced over the point that locates its mean, as illustrated in Figure 1.6a. As noted in Definition 1.8, half the area under the relative frequency distribution will lie to the left of the median, and half will lie to the right, as shown in Figure 1.6b. The mode will locate the point at which the greatest frequency occurs, i.e., the peak of the relative frequency distribution, as shown in Figure 1.6c.

FIGURE 1.6 Interpretations of the Mean, Median, and Mode for a Relative Frequency Distribution

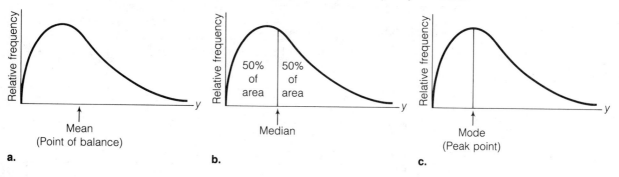

Although the mean is the preferred measure of central tendency for most data sets, it is sensitive to very large or very small observations. Consequently, the mean will shift toward the direction of **skewness** (i.e., the tail of the distribution) and may be misleading in some situations. For example, if a data set consists of the first-year starting salaries of computer science graduates, the high starting salaries of a few graduates will influence the mean more than the median. The median, then, would better represent the typical starting salary that a computer science graduate could expect. Alternatively, if the relative frequency of occurrence of y is of interest, then the mode would be preferred over the mean or median. For example, a supplier of carpenter's materials would be interested in the modal length (in inches) of nails he sells.

In summary, the best measure of central tendency for a data set depends on the type of descriptive information you want. For most data sets encountered in the engineering and computing sciences, this will be the mean.

EXERCISES 1.20–1.24

1.20 Find the mean, median, and mode for the data set consisting of the five measurements: 4, 3, 10, 8, 5.

1.21 Find the mean, median, and mode for the data set consisting of the eight measurements: 9, 6, 12, 4, 4, 2, 5, 6.

1.22 Find the mean, median, and mode for the data set in Exercise 1.15.

1.23 Find the mean, median, and mode for the data set in Exercise 1.17.

1.24 Find the mean, median, and mode for the data set in Exercise 1.19.

SECTION 1.7

MEASURES OF VARIATION

The most commonly used measures of data variation are the **range**, the **variance**, and the **standard deviation**.

DEFINITION 1.10

The **range** is equal to the difference between the largest and the smallest measurements in a data set:

Range = Largest measurement − Smallest measurement

DEFINITION 1.11

The **variance** of a sample of n measurements, $y_1, y_2, \ldots, y_n$, is defined to be

$$s^2 = \frac{\sum_{i=1}^{n}(y_i - \bar{y})^2}{n-1} = \frac{\sum_{i=1}^{n}y_i^2 - \frac{\left(\sum_{i=1}^{n}y_i\right)^2}{n}}{n-1}$$

DEFINITION 1.12

The **standard deviation** of a sample of n measurements is equal to the square root of the variance:

$$s = \sqrt{s^2} = \sqrt{\frac{\sum_{i=1}^{n} (y_i - \bar{y})^2}{n - 1}}$$

EXAMPLE 1.4

Find the variance and standard deviation for the $n = 5$ observations: 1, 3, 2, 2, 4.

SOLUTION

We must first calculate $\sum_{i=1}^{n} y_i$ and $\sum_{i=1}^{n} y_i^2$:

$$\sum_{i=1}^{n} y_i = 1 + 3 + 2 + 2 + 4 = 12$$

$$\sum_{i=1}^{n} y_i^2 = (1)^2 + (3)^2 + (2)^2 + (2)^2 + (4)^2 = 34$$

Then the variance is

$$s^2 = \frac{\sum_{i=1}^{n} (y_i - \bar{y})^2}{n - 1} = \frac{\sum_{i=1}^{n} y_i^2 - \frac{\left(\sum_{i=1}^{n} y_i\right)^2}{n}}{n - 1} = \frac{34 - \frac{(12)^2}{5}}{4} = 1.3$$

and the standard deviation is

$$s = \sqrt{s^2} = \sqrt{1.3} = 1.1402$$

■

It is possible that two different data sets could possess the same range, but differ greatly in the amount of variation in the data. Consequently, the range is a relatively insensitive measure of data variation. It is used primarily in industrial quality control where the inferential procedures are based on small samples (i.e., small values of n). The variance* is primarily of theoretical significance, but the standard deviation as a measure of data variation is easily interpreted by means of Tchebysheff's theorem and a rule of thumb known as the Empirical Rule.

*The variance of a data set is sometimes defined to be the mean of the squares of the deviations of the y values about their mean, i.e., with a divisor of n instead of $(n - 1)$. The definition that we give is, in fact, the formula for an unbiased estimator of a population variance (see Section 8.3 for the definition of unbiased). Since, for moderate to large values of n, there is very little numerical difference between the values of s^2 computed from the two definitions and since most statistical methods use s^2, we will find it convenient to use our definition for the sample variance.

Tchebysheff's theorem, which derives its name from the Russian mathematician P. L. Tchebysheff (1821–1894), is stated in the box without proof.

TCHEBYSHEFF'S THEOREM*

For $k \geq 1$, at least $(1 - 1/k^2)$ of a set of n measurements will lie within k standard deviations of their mean.

EXAMPLE 1.5

Use Tchebysheff's theorem to describe a data set with mean $\bar{y} = 16.2$ and standard deviation $s = 1.1$.

SOLUTION

In constructing intervals of the form $\bar{y} \pm ks$, we must select values of k such that $k \geq 1$. For convenience, we will let k take on the integer values 2 and 3. The results of the computations are shown in Table 1.3.

TABLE 1.3

k	$\bar{y} + ks$	$1 - 1/k^2$
2	$16.2 \pm 2(1.1)$ or 14.0 to 18.4	$\frac{3}{4}$
3	$16.2 \pm 3(1.1)$ or 12.9 to 19.5	$\frac{8}{9}$

From the table, you can see that at least $\frac{3}{4}$ of the measurements in the data set will lie in the interval $\bar{y} \pm 2s$, i.e., 14.0 to 18.4; at least $\frac{8}{9}$ of the measurements will lie in the interval $\bar{y} \pm 3s$, i.e., 12.9 to 19.5. Note the words *at least* in the statement of Tchebysheff's theorem. The fraction $(1 - 1/k^2)$ is a *lower bound* for the proportion of the total number of measurements in the interval $\bar{y} \pm ks$. The proportion of measurements in the interval will usually be much larger than $(1 - 1/k^2)$.[†]   ■

The Empirical Rule is not a theorem. Rather, it is the result of the practical experience of researchers in many fields who have observed many different types of real-life data sets.

EXAMPLE 1.6

Suppose a data set has a mound-shaped relative frequency distribution with mean $\bar{y} = 100$ and standard deviation $s = 6$. Use the Empirical Rule to describe the distribution.

---

*The proof of Tchebysheff's theorem is based on a variance defined as the mean of the squares of the deviations of the y values about their mean. Since this quantity is less than our defined value of s^2, it follows that at least $(1 - 1/k^2)$ of a set of n measurements will lie within ks of $\bar{y}$.

[†]The conservative nature of Tchebysheff's theorem is best illustrated for $k = 1$. The theorem states that "at least 0" of the n measurements will lie within 1 standard deviation of their mean. Thus, Tchebysheff's theorem gives no information about the proportion of measurements between $\bar{y} - s$ and $\bar{y} + s$.

THE EMPIRICAL RULE

If a data set has a mound-shaped relative frequency distribution, then the following rules of thumb may be used to describe the data set:

1. Approximately 68% of the measurements will lie within 1 standard deviation of their mean, i.e., within the interval $\bar{y} \pm s$.
2. Approximately 95% of the measurements will lie within 2 standard deviations of their mean, i.e., within the interval $\bar{y} \pm 2s$.
3. Almost all the measurements will lie within 3 standard deviations of their mean, i.e., within the interval $\bar{y} \pm 3s$.

SOLUTION

The intervals $\bar{y} \pm s$, $\bar{y} \pm 2s$, and $\bar{y} \pm 3s$, together with the proportions of the total number of measurements that we would expect to find in these intervals according to the Empirical Rule, are shown in Table 1.4. Visualize a mound-shaped distribution centered over $\bar{y} = 100$ with most (95%) of the distribution lying within the interval $\bar{y} \pm 2s$, i.e., 88 to 112.

TABLE 1.4

k	$\bar{y} \pm ks$	APPROXIMATE PROPORTION OF MEASUREMENTS IN THE INTERVAL
1	94 to 106	.68
2	88 to 112	.95
3	82 to 118	Almost 1.00

The percentages given in the Empirical Rule are approximate, particularly for the interval $\bar{y} \pm s$. The percentage of the total number of measurements that fall within 2 standard deviations of their mean will usually be quite close to 95%. For example, you can verify that the mean and standard deviation for the $n = 25$ CPU times of Table 1.1 are

$$\bar{y} = 1.63 \quad \text{and} \quad s = 1.19$$

The intervals $\bar{y} \pm ks$ and the proportions of the $n = 25$ measurements that fall in each of these intervals are shown in Table 1.5, for values of $k = 1, 2, 3$. You can see that, for each of the three intervals, the actual proportion of the $n = 25$ CPU times that lie in the interval is very close to that specified by the Empirical Rule.

TABLE 1.5

k	$\bar{y} \pm ks$	NUMBER OF MEASUREMENTS IN THE INTERVAL	PROPORTION OF MEASUREMENTS IN THE INTERVAL
1	.44 to 2.82	18	.72
2	−.75 to 4.01	24	.96
3	−1.94 to 5.20	25	1.00

1.25 Find the range, the variance, and the standard deviation of the following $n = 25$ measurements:

9	10	3	3	10
6	5	8	8	9
12	16	2	4	3
4	2	6	7	7
4	4	11	3	13

1.26 Refer to the data of Exercise 1.25. Construct the intervals $\bar{y} \pm s$, $\bar{y} \pm 2s$, and $\bar{y} \pm 3s$. Count the number of observations that fall within each interval and find the corresponding proportions. Compare your results to Tchebysheff's theorem and the Empirical Rule.

1.27 Sixty-six bulk specimens of Chilean lumpy iron ore (95% particle size, 150 millimeters) were randomly sampled from a 35,325-long-ton shipload of ore and the percentage of iron in each ore specimen was determined. The data are shown in the accompanying table.

ORE SPECIMEN	PERCENTAGE IRON	ORE SPECIMEN	PERCENTAGE IRON	ORE SPECIMEN	PERCENTAGE IRON
1	62.66	23	61.82	45	62.24
2	62.87	24	63.01	46	63.43
3	63.22	25	63.01	47	62.87
4	63.01	26	62.80	48	63.64
5	62.10	27	62.80	49	63.92
6	63.43	28	63.01	50	63.71
7	63.22	29	62.10	51	63.64
8	63.57	30	63.29	52	64.06
9	61.75	31	63.37	53	62.73
10	63.15	32	61.75	54	62.52
11	63.08	33	63.29	55	62.10
12	63.22	34	62.38	56	63.29
13	63.22	35	62.59	57	63.01
14	63.08	36	63.92	58	63.36
15	62.87	37	63.29	59	63.08
16	61.68	38	63.57	60	62.03
17	62.45	39	62.80	61	64.34
18	62.10	40	62.31	62	64.06
19	62.87	41	63.01	63	62.87
20	62.87	42	62.94	64	63.50
21	62.94	43	63.08	65	63.78
22	62.38	44	63.43	66	62.10

Source: Sato, T., Ito, K., Chujo, S., and Takahashi, U. "Example of Experiments on Systematic Sampling of Iron Ore." *Reports of Statistical Application Research*, Union of Japanese Scientists and Engineers. Vol. 18, No. 1, 1971.

a. Describe the population from which the sample was selected.

b. Give one possible objective of this sampling procedure.

c. Construct a relative frequency histogram for the data.

d. Calculate $\bar{y}$ and s.

e. Find the percentage of the total number ($n = 66$) of observations that lie in the interval $\bar{y} \pm 2s$. Does this percentage agree with the Empirical Rule?

1.28 The means and standard deviations of the length, weight, and DDT measurements for the data in Appendix III are shown in the accompanying table.

VARIABLE	MEAN	STANDARD DEVIATION
Length (cm)	42.81	6.88
Weight (g)	1,049.72	376.55
DDT (ppm)	24.36	98.38

a. Use the tabled information, in conjunction with the Empirical Rule, to describe the relative frequency distributions of the measurements for the three variables.

b. Use the tabled information, in conjunction with Tchebysheff's theorem, to describe the relative frequency distributions of the measurements for the three variables.

c. Which of the two rules, Tchebysheff's theorem or the Empirical Rule, appears to describe better the distribution of DDT measurements?

1.29 Refer to the scram data of Exercise 1.17. Construct the intervals $\bar{y} \pm s$, $\bar{y} \pm 2s$, and $\bar{y} \pm 3s$. Count the number of observations that fall within each interval and compare your results to Tchebysheff's theorem and the Empirical Rule.

1.30 Research has indicated that the inhabitants of some older cities in the United States may ingest small, but potentially harmful, amounts of lead that is introduced into their drinking water by the use of lead-lined pipes installed in some of the early city water systems. The data reported in the table are the mean lead, copper, and iron contents (milligrams per liter) for samples of water collected on each of 23 days from the Boston water system. The data were collected in 1977 after a sodium hydroxide water treatment system was installed. Each mean is based on approximately 40 measurements taken at different sites in the Boston water system where lead pipe was still being used.

LEAD			COPPER			IRON		
.035	.073	.030	.12	.07	.10	.20	.23	.17
.060	.047	.019	.18	.07	.04	.33	.18	.13
.055	.031	.021	.10	.08	.08	.22	.25	.15
.035	.016	.036	.07	.07	.05	.17	.14	.13
.031	.015	.016	.08	.04	.05	.15	.14	.14
.039	.015	.010	.09	.04	.04	.19	.12	.11
.038	.022	.020	.16	.05	.04	.17	.12	.11
.049	.043		.14	.07		.17	.16	

Source: Karalekas, P. C. Jr., Ryan, C. R., and Taylor, F. B. "Control of Lead, Copper, and Iron Pipe Corrosion in Boston." *American Water Works Journal*, Feb. 1983, pp. 92–95. Reprinted by permission. Copyright © 1983, American Water Works Association.

  a. Construct a relative frequency histogram for the 23 daily mean lead concentration measurements.
  b. Calculate the mean and standard deviation for the sample of part **a**.
  c. What percentage of the 23 daily mean lead concentrations lie in the interval $\bar{y} \pm 2s$?

1.31 Repeat Exercise 1.30, using the copper concentration means.

1.32 Repeat Exercise 1.30, using the iron concentration means.

| | | | | | | | | | | | | |

S E C T I O N   1.8

**MEASURES OF
RELATIVE STANDING**

Test scores and some types of sociological and health data are often reported in a manner that describes the location of an observation *relative* to the other scores in the distribution. Two measures of the relative standing of an observation are percentiles and **z-scores**.

DEFINITION 1.13

The **100pth percentile** of a data set is a value of y located so that $100p\%$ of the area under the relative frequency distribution for the data lies to the left of the 100pth percentile and $100(1 - p)\%$ of the area lies to its right.

EXAMPLE 1.7

If your grade in an industrial engineering class was located at the 84th percentile, describe the location of your grade relative to the other grades in your class.

SOLUTION

Eighty-four percent of the grades were lower than your grade and 16% were higher.   ■

    The median is the 50th percentile. The 25th percentile, the median, and the 75th percentile are called the **lower quartile**, the **midquartile**, and the **upper quartile**, respectively, for a data set.

DEFINITION 1.14

The **lower quartile**, Q_L, for a data set is the 25th percentile.

DEFINITION 1.15

The **midquartile** (or median), m, for a data set is the 50th percentile.

DEFINITION 1.16

The **upper quartile**, Q_U, for a data set is the 75th percentile.

For large data sets (e.g., populations), quartiles are found by locating the corresponding areas under the curve (relative frequency distribution). However, when the data set of interest is small it may be impossible to find a measurement in the data set that exceeds, say, *exactly* 25% of the remaining measurements. Consequently, the 25th percentile (or lower quartile) for the data set is not well defined. The box contains a few rules for finding quartiles with small data sets.

FINDING QUARTILES WITH SMALL DATA SETS

Step 1. Rank the measurements in the data set in increasing order of magnitude. Let $y_{(1)}, y_{(2)}, ..., y_{(n)}$ represent the ranked measurements.

Step 2. Calculate the quantity $\frac{1}{4}(n + 1)$ and round to the nearest integer. The measurement with this rank represents the *lower quartile* or 25th percentile. [*Note:* If $\frac{1}{4}(n + 1)$ falls halfway between the two integers, round up.]

Step 3. Calculate the quantity $\frac{3}{4}(n + 1)$ and round to the nearest integer. The measurement with this rank represents the *upper quartile* or 75th percentile. [*Note:* If $\frac{3}{4}(n + 1)$ falls halfway between the two integers, round down.]

DEFINITION 1.17

The **z-score** for a value y of a data set is the distance that y lies above or below the sample mean $\bar{y}$, measured in units of the standard deviation s:

$$z = \frac{y - \bar{y}}{s}$$

By definition, the z-score describes the location of an observation y relative to the sample mean $\bar{y}$. Negative z-scores indicate that the observation lies to the left of the mean; positive z-scores indicate that the observation lies to the right of the mean. Also, we know from Tchebysheff's theorem and the Empirical Rule that most of the observations in a data set will be less than 2 standard deviations from the mean (i.e., will have z-scores less than 2 in absolute value) and almost all will be within 3 standard deviations of the mean (i.e., will have z-scores less than 3 in absolute value).

EXAMPLE 1.8

The mean and standard deviation for the sample of $n = 25$ CPU times in Table 1.1 are $\bar{y} = 1.63$ and $s = 1.19$, respectively. Use the values to find and interpret the z-score for the CPU time of 3.76 seconds.

SOLUTION

Substituting $y = 3.76$, $\bar{y} = 1.63$, and $s = 1.19$ into the formula for z, we obtain

$$z = \frac{y - \bar{y}}{s} = \frac{3.76 - 1.63}{1.19} = 1.79$$

Since the z-score is positive, we conclude that the CPU time of 3.76 seconds lies a distance of 1.79 standard deviations above (to the right of) the sample mean of 1.63 seconds.   ■

1.33 Find the 25th, 50th, and 75th percentiles for the data set in Exercise 1.25.

1.34 Refer to Exercise 1.19.
   a. The 80th percentile for the monthly salaries of the 35 operations analysts is $2,470. Interpret this value.
   b. Find the upper and lower quartiles for the 35 salaries.
   c. Find the z-score for the monthly salary of $1,920. Interpret this value.

1.35 The relative frequency distribution of the 144 sample DDT measurements (ppm) (Appendix III) is summarized by the following numerical descriptive measures:

$$\bar{y} = 24.36 \qquad s = 98.38 \qquad m = 7.15$$

   a. Find the 50th percentile for the sample DDT measurements.
   b. Use Tchebysheff's theorem to find an upper bound for the 75th percentile of the sample of DDT measurements.
   c. Find the z-score for a fish specimen with a DDT measurement of 500 ppm. Interpret this value.

1.36 A simulation study was conducted to investigate the rounding accuracies of several new algorithms for functions in the FORTRAN computer program library (*IBM Journal of Research and Development*, Mar. 1986). For each new FORTRAN function, the rounding error for each of 10,000 trials was calculated.
   a. The 99th percentile of the rounding errors for one particular FORTRAN function was found to be .53. Interpret this value.
   b. For the FORTRAN function of part **a**, the mean rounding error is $\bar{y} = .22$ and the standard deviation (estimated) is $s = .07$. Use this information to calculate the z-score for the 99th percentile value of .53. Interpret the z-score.

1.37 Refer to the scram data given in Exercise 1.17.
   a. Find the upper quartile for the data set. Interpret this value.
   b. Find the z-score for a nuclear reactor unit with 9 scrams in 1984. Interpret this value.

Sometimes inconsistent observations are included in a data set. For example, when we discuss starting salaries for college graduates with bachelor's degrees, we generally think of traditional college graduates—those near 22 years of age with 4 years of college education. But suppose one of the graduates is a 34-year-old Ph.D. chemical engineer who has returned to the university to obtain a bachelor's degree in metallurgy. Clearly, the starting salary for this graduate could be much larger than the other starting salaries because of the graduate's additional education and experience, and we probably would not want to include it in the data set. Such an errant observation, which lies outside the range of the data values that we want to describe, is called an **outlier**.

DEFINITION 1.18

An observation y that is unusually large or small relative to the other values in a data set is called an **outlier**.

The most obvious method for determining whether an observation is an outlier is to calculate its z-score (Section 1.8). For example, if the z-score for a y value is -4.13, we know that it lies more than 4 standard deviations below the mean of the data set. Both the Empirical Rule and Tchebysheff's theorem tell us that almost all the observations in a data set will have z-scores less than 3 in absolute value. Thus, a z-score as small as -4.13 is highly improbable and points to the possibility of an errant observation.

Another procedure for detecting outliers is to construct a **box plot** of the data. With this method, we construct intervals similar to the $\bar{y} \pm 2s$ and $\bar{y} \pm 3s$ intervals of the Empirical Rule; however, the intervals are based on a quantity called the **interquartile range** instead of the standard deviation s.

DEFINITION 1.19

The **interquartile range**, IQR, is the distance between the upper and lower quartiles:

$$IQR = Q_U - Q_L$$

The procedure is especially easy to use for small data sets because the quartiles and interquartile range can be quickly determined. The steps to follow in constructing a box plot are given in the box on page 30.

A box plot for the 25 CPU times in Table 1.1 is shown in Figure 1.7. From the plot you can see that $Q_L = .82$, $m = 1.38$, $Q_U = 2.16$, and

$$IQR = Q_U - Q_L = 2.16 - .82 = 1.34$$

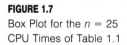

FIGURE 1.7

Box Plot for the $n = 25$ CPU Times of Table 1.1

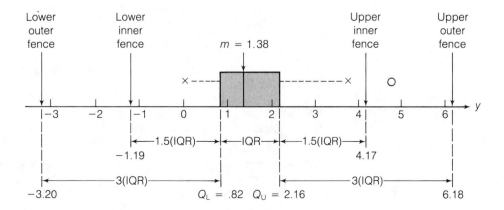

The inner and outer fences are located a distance of $1.5(IQR) = 1.5(1.34) = 2.01$ and $3(IQR) = 3(1.34) = 4.02$ below Q_L and above Q_U, respectively. Note that the data set contains only one suspect outlier, indicated by the small circle between the upper inner and outer fences. This is the CPU time of 4.75—the largest value in the data set.

The z-score and box plot methods both establish rule-of-thumb limits outside of which a y value is deemed to be an outlier. Usually the two methods produce similar results. However, the presence of one or more outliers in a data set can inflate the value of s used to calculate the z-score. Consequently, it will be less likely that an errant observation would have a z-score larger than 3 in absolute value. In contrast, the values of the quartiles used to calculate the fences for a box plot are not affected by the presence of outliers.

STEPS TO FOLLOW IN CONSTRUCTING A BOX PLOT

Step 1. Calculate the median, m, lower and upper quartiles, Q_L and Q_U, and the interquartile range, IQR, for the y values in a data set.

Step 2. Construct a box on the y-axis with Q_L and Q_U located at the lower corners (see Figure 1.7). The base width will then be equal to IQR. Draw a vertical line inside the box to locate the median, m.

Step 3. Construct two sets of limits on the box plot. **Inner fences** are located a distance of $1.5(IQR)$ below Q_L and above Q_U; **outer fences** are located a distance of $3(IQR)$ below Q_L and above Q_U (see Figure 1.7).

Step 4. Observations that fall between the inner and outer fences are called **suspect outliers**. Locate the suspect outliers on the box plot using small circles. Observations that fall outside the outer fences are called **highly suspect outliers**. Use solid dots to locate highly suspect outliers.

Step 5. To further highlight extreme values, **whiskers** are added to the box plot. Mark the y value in the region between Q_L and the lower inner fence that is closest to the inner fence with an $\times$, and join the $\times$ to the box with a dashed line—a whisker (see Figure 1.7). Similarly, use an $\times$ and attached whisker to locate the most extreme value between Q_U and the upper inner fence.

EXERCISES 1.38–1.41

1.38 Refer to the scram data given in Exercise 1.17.
   a. Construct a box plot for the data. Do you detect any outliers?
   b. Use the method of z-scores to detect outliers.

1.39 Refer to the voltage reading data for the old location, supplied in Exercise 1.18.
   a. Construct a box plot for the data. Do you detect any outliers?
   b. Use the method of z-scores to detect outliers.

1.40 Refer to the lead concentration data of Exercise 1.30.
   a. Construct a box plot for the data. Do you detect any outliers?
   b. Use the method of z-scores to detect outliers.

1.41 Refer to the 144 DDT measurements in Appendix III.
   a. Construct a box plot for the data. Do you detect any outliers?
   b. Use the method of z-scores to detect outliers.

S E C T I O N 1.10 **SAMPLE STATISTICS AND POPULATION PARAMETERS**	Numerical descriptive measures computed from sample data are often called **statistics**. Thus, the sample mean $\bar{y}$ and standard deviation s are examples of statistics. In contrast, numerical descriptive measures of the population are called **parameters**. Their values are typically unknown and can only be represented by symbols. For example, the population mean and standard deviation are commonly represented by the Greek letters μ and σ, respectively; the population variance is σ^2. Although we *could* calculate the values of these parameters if we actually had access to the entire population, we generally wish to avoid doing so, for economic or other reasons. Thus, as you will subsequently see, we will *sample* the population and then use the sample statistics to infer, or make decisions about, the values of population parameters.

DEFINITION 1.20

A **statistic** is a numerical descriptive measure computed from sample data.

DEFINITION 1.21

A **parameter** is a numerical descriptive measure of a population.

S E C T I O N 1.11 **THE ROLE OF STATISTICS IN THE SCIENCES**	Experimental scientific research, in engineering, information sciences, and other disciplines, involves the use of experimental data—a sample—to infer the nature of some conceptual population that characterizes a phenomenon of interest to the experimenter. This inferential process is an integral part of the scientific method. Inference based on experimental data is first employed to develop a theory about some phenomenon. Then the theory is tested against additional sample data.

How does the science of statistics contribute to this process? To answer this question, we must note that inferences based on sample data will almost always be subject to error, because a sample will not provide an exact image of the population. The nature of the information provided by a sample depends on the particular sample chosen, and thus will change from sample to sample. For example, suppose you manufacture computers and you want to estimate the proportion of all potential customers who would buy your product. You contract with a sample survey organization to sample 100 potential customers and find that 17 plan to buy your product. Does this mean that 17% of all potential customers plan to buy your product? Of course, the answer is "no." Suppose that, unknown to you, the true percentage of potential customers who would purchase your product is 14%. One sample of 100 potential customers might yield 17 in favor of your product, while another sample of 100 might yield only 12 potential purchasers. Thus, an inference based on sampling is always subject to *uncertainty*.

The theory of statistics uses *probability* to measure the uncertainty associated with an inference. It enables us to calculate the probabilities of observing specific samples, under specific assumptions about the population. The statistician then uses these probabilities to evaluate the uncertainties associated with sample inferences. Thus, the major contribution of statistics is that it enables us to make inferences—estimates of and decisions about population parameters—with a known measure of uncertainty. It enables us to evaluate the *reliability* of inferences based on sample data.

SECTION 1.12

SUMMARY

Statistics is concerned with two types of problems: (1) describing large masses of data, and (2) using **sample** data to make inferences about a large set of data— a **population**—from which the sample has been selected. Data description plays a role in **inferential statistics** because we must know how to describe a population before we can make inferences about it.

Data sets can be described using either **graphical** or **numerical methods**. **Pie charts** and **bar graphs** are most commonly used to describe **qualitative** data. **Quantitative** data sets are graphically described using either **relative frequency histograms (distributions)** or **stem and leaf displays**. Numerical descriptive measures are numbers that enable us to create a mental image of these graphical descriptions. For example, in conjunction with **Tchebysheff's theorem** and the **Empirical Rule**, the **mean** and **standard deviation** of a data set enable us to visualize both the approximate location and spread of the relative frequency distribution for the data set. **Percentiles** and *z*-scores are numbers that measure the relative standing of a value within a data set. Errant observations or highly unusual values, called **outliers**, can be detected using either *z*-scores or **box plots**.

As you will subsequently see, we will employ sample numerical descriptive measures—**statistics**—to make inferences about the values of their population equivalents—the **parameters** of a population relative frequency distribution. The most important contribution of statistics is that our methodology will enable us to assess the uncertainty associated with an inference and, thereby, to evaluate its reliability.

Since many statistical calculations are performed by computers, we will digress from our discussion of statistics to present in Chapter 2 the procedure for entering data into a computer for eventual use with one of four popular statistical program packages. Then as we proceed through the text, we will explain how to use various portions of a package to perform specific statistical calculations.

In Chapter 3, we will turn to a study of probability, the vehicle that is used both in making inferences and in evaluating their reliabilities. Subsequent chapters will show how probability is used in statistical inference and will present some useful statistical methods.

| | | | | | | | | | | | |

SUPPLEMENTARY EXERCISES 1.42–1.65

1.42 An assembly line that produces Christmas lights is considered to be operating successfully if less than 2% of the lights manufactured per day are defective. If 2% or more of the lights are defective, the line must be shut down and proper adjustments made. Since a halt in the production process for even as little as 5 minutes could cost the manufacturer millions of dollars in lost revenue, the line should be shut down only if the true percentage of defective lights is 2% or larger. However, to check each of the lights individually would be both impractical and cost-ineffective. As a solution, quality control inspectors select 100 lights from the day's production and inspect them for defects. The decision on whether to shut down the assmbly line is then made according to the percentage of defectives in the 100 lights tested.
a. Describe the population of interest to the manufacturer.
b. Describe the sample.
c. If the sample percentage of defectives is larger than 2%, is it necessarily true that the actual percentage of defective lights produced per day is larger than 2%? Explain.

1.43 A preliminary study was conducted to obtain information on the background levels of the toxic substance polychlorinated biphenyl (PCB) in soil samples in the United Kingdom. Such information could then be used as a benchmark against which PCB levels at waste disposal facilities in the United Kingdom can be compared. The accompanying table contains the measured PCB levels of soil samples taken at 14 rural and 15 urban locations in the United Kingdom. (PCB concentration is measured in .0001 gram per kilogram of soil.) From these preliminary results, the researchers reported "a significant difference between (the PCB levels) for rural areas . . . and for urban areas."

RURAL				URBAN			
3.5	1.0	1.6	12.0	24.0	11.0	107.0	18.0
8.1	5.3	23.0	8.2	29.0	49.0	94.0	12.0
1.8	9.8	1.5	9.7	16.0	22.0	141.0	18.0
9.0	15.0			21.0	13.0	11.0	

Source: Badsha, K. and Eduljee, G. "PCB in the U.K. Environment—A Preliminary Survey." *Chemosphere*, Vol. 15, No. 2, Feb. 1986, p. 213 (Table 1). Reprinted with permission. Copyright 1986, Pergamon Press, Ltd.

a. Construct a stem and leaf display for the PCB levels of rural soil samples.
b. Construct a stem and leaf display for the PCB levels of urban soil samples.
c. Combine the data for rural and urban soil samples and construct a stem and leaf display. Identify each of the urban PCB levels on the display with a circle. Does the graph support the researchers' conclusions?

1.44 Many worldwide manufacturers currently use, or are planning to use, newly designed computer-operated robots to perform certain assembly line tasks that may require as many as ten people to complete. Information on the estimated number of industrial robots currently being used in each of 11 countries is given in the table.

COUNTRY	NUMBER OF INDUSTRIAL ROBOTS	COUNTRY	NUMBER OF INDUSTRIAL ROBOTS
Finland	130	Poland	360
France	200	Sweden	600
Great Britain	185	United States	3,000
Italy	500	U.S.S.R.	25
Japan	10,000	West Germany	850
Norway	200		

Source: Adapted from *Time*, Dec. 8, 1980. Copyright 1980, Time Inc. All rights reserved. Reprinted by permission from *Time*.

a. Construct a bar graph to describe the distribution of industrial robots in the eleven countries.
b. Interpret the bar graph in part **a**.

1.45 Refer to Exercise 1.44. The number of industrial robots in the United States is expected to increase tenfold by the year 1990 (*Gainesville Sun*, Dec. 8, 1985). The pie chart shown here describes how the projected 33,600 robots in the work force will be used at U.S. industrial plants. Interpret the pie chart.

U.S. Robot Applications in 1990 (33,600 units)

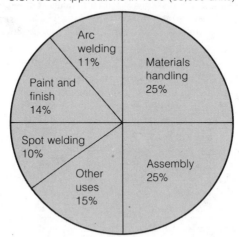

Source: Info Graphics, News America Syndicate, 1985.

1.46 In an effort to keep up with the most recent developments in computer software innovations, a large company created a task force to select the software package that would best meet the company's needs. The task force narrowed its selection to three vendors and then rated each software package in nine categories (4 = Excellent, 3 = Good, 2 = Fair, 1 = Poor). The results are shown in the table.

	VENDOR 1	VENDOR 2	VENDOR 3
Reliability	3.5	3.6	3.2
Efficiency	3.4	3.3	3.6
Ease of installation	3.1	3.2	3.3
Ease of use	3.5	3.2	3.4
Vendor's technical support			
Trouble shooting	2.9	3.0	2.2
Documentation	2.9	2.8	2.6
User education	2.7	2.8	2.6
Vendor's maintenance	3.1	2.8	2.6
Overall satisfaction	3.3	3.2	2.8

a. Compute the mean, median, and mode of the ratings for each of the three vendors.
b. Compute the range, the variance, and the standard deviation of the ratings for each of the three vendors.
c. Vendor 1 advertises that its software package is superior to those of the competitors because it has a higher overall average rating. Do you think this claim is misleading? Explain.

1.47 The Electric Power Research Institute (EPRI) is a nonprofit organization that was formed in the early 1970's to meet the research needs of the growing electric utility industry. Its major role is to manage and provide funding for research and development projects. The accompanying table gives the annual percentage of research dollars granted by EPRI to various types of contractors. Construct a bar chart for the data. What percentage of total research dollars does EPRI grant annually to universities or government laboratories?

CONTRACTOR	PERCENTAGE
Industrial, commercial	62
Consultant	13
University	9
Utility	7
Architect, engineer	6
Government laboratories	3

Source: Starr, C. "The Electric Power Research Institute." *Science*, Mar. 11, 1983, p. 1193.

1.48 The nuclear mishap on Three Mile Island near Harrisburg, Pennsylvania, on March 28, 1979, forced many local residents to evacuate their homes—some temporarily, others permanently. In order to assess the impact of the accident on the area population, a questionnaire was designed and mailed to a sample of 150 households within 2 weeks after the accident occurred. Two questions asked of the sampled residents were: (1) When did you learn about the accident? and (2) How did you learn about the accident? The responses to the two questions are illustrated in the frequency distributions shown on page 36.

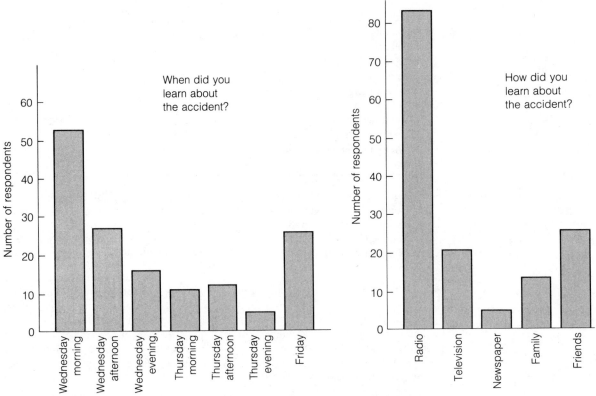

Source: Browń, S. et al. "Final Report on a Survey of Three Mile Island Area Residents."
Department of Geography, Michigan State University, Aug. 1979.

Based on these graphical descriptions, find each of the following percentages:

a. The percentage of the 150 respondents who learned about the accident on Wednesday afternoon

b. The percentage of the 150 respondents who learned about the accident on Friday

c. The percentage of the 150 respondents who learned about the accident from a radio report

d. The percentage of the 150 respondents who learned about the accident from television

1.49 A chemical manufacturing plant, under investigation by the Environmental Protection Agency (EPA) for possible violation of air pollution standards, reports that the distribution of the amount of dangerous chemicals in the plant's air emissions is approximately mound-shaped, with a mean of 5 parts per million and a standard deviation of 1.5 parts per million.

a. Use the Empirical Rule to describe the distribution of the amount of dangerous chemicals in the plant's air emissions.

b. The EPA sets a limit of 10 parts per million on dangerous chemicals in plant air emissions. Plants found exceeding this standard more than once per year are required to install air pollution control devices. Based on your description of the air emissions distribution of part **a**, do you believe that the chemical plant is in violation of air quality standards? Explain.

1.50 A study was conducted to determine the effect of the presence or absence of a company safety program on the number of work-hours lost due to work-related accidents. Fifty manufacturing companies were selected, and the number of employees and the number of hours lost due to work-related accidents were recorded for each. A ratio was computed to determine the number of hours lost due to work-related accidents per employee. The results follow (an asterisk indicates the company had a work safety program):

.0186	.0091	.0204	.0106*	.0128*	.0233	.0094	.0166
.0072*	.0116*	.0169	.0133	.0223	.0070*	.0095*	.0202
.0217	.0083	.0189*	.0089*	.0223	.0179	.0145	.0092*
.0138*	.0195	.0153	.0133	.0106*	.0099*	.0189	.0230
.0143*	.0164*	.0245*	.0234	.0116	.0111*	.0142*	.0198*
.0169	.0183	.0085*	.0213*	.0155*	.0211	.0173	.0153*
.0029*	.0123*						

a. Construct a relative frequency histogram for each of the following data sets:
  1. The complete set of measurements on lost work-hours per employee
  2. The set associated with those companies having no work safety program
  3. The set associated with those companies having a work safety program
b. Describe any differences you detect between the histograms of parts a2 and a3.
c. What inference can you make concerning the effectiveness of a company safety program?

1.51 Prior to 1980, Egypt had no access to electronic databases, either at home or abroad. In order to provide Egyptians with access to existing U.S. databases, the Georgia Institute of Technology designed and tested the "delayed online database search service" (which operated for 38 months until a direct online service became available).

DATABASE	FREQUENCY OF ACCESS	RELATIVE FREQUENCY
Medline	3,616	.343
Agricola	1,572	.149
Biosis	1,292	.123
CAB	1,113	.106
CA	742	.070
Compendex	239	.023
ERIC	213	.020
Food Science	201	.019
INSPEC	197	.019
PsychInfo	141	.013
All other databases	1,204	.114
TOTALS	10,530	.999

Source: Slamecka, V., El-Shishiny, N., and Bassit, A. A. "A Longitudinal Profile of a National Database Search Service." *Information Processing & Management*, Vol. 22, No. 3, 1986, p. 208. Reprinted with permission. Copyright 1986, Pergamon Press, Ltd.

As part of a performance study of the search service, the number of times each of the over 100 databases in the system was accessed during the 38-month period was recorded. The frequency of access for the ten most requested databases is reported in the table. Summarize the data in the table using a graphical descriptive method. Interpret the graph.

1.52 Refer to Exercise 1.51. Because of the large size of some databases in the "delayed online search service," users split some of them into "files"; therefore, in some cases, searching a database calls for accessing two or more files. The total number of times each of the 164 files in the system was accessed by Egyptian users over the 38-month period is recorded below.

213	2	3	201	1	10	1	481	5	3	112	
6	1	6	3	4	11	13	1,505	4	213	2	
5	3	14	1	11	7	1	1,603	1	2	2	
8	34	3	504	4	1	4	14	10	1	1	
575	1	12	7	86	7	8	6	2	3	2	
52	2	17	2	12	3	13	6	3	1	1	
15	2	14	4	4	1	9	1	12	19	5	
239	2	1	12	2	4	1	1	4	2	1	
1	2	6	12	64	1	1	2	1	1	1	
905	5	1	12	1	43	1	1	32	1	1	
141	2	1	5	1	8	1	1	11	1	1	
90	4	18	2	6	1	3	1	1	66	1	
107	65	2	8	3	1	3	1	3	138	4	
32	2	6	4	9	1	5	15	1	188	55	
36	9	1,113	42	1	667	6	8	1	241		

Source: Slamecka, V., El-Shishiny, N., and Bassit, A. A. "A Longitudinal Profile of a National Database Search Service." *Information Processing & Management*, Vol. 22, No. 3, 1986, p. 209. Reprinted with permission. Copyright 1986, Pergamon Press, Ltd.

a. Summarize the data using a graphical descriptive method.
b. Calculate the mean, median, and mode for the data set.
c. Calculate the range, variance, and standard deviation for the data set.
d. Calculate the lower quartile, upper quartile, and interquartile range for the data set.
e. Construct a box plot for the data and identify any outliers.

1.53 The projected 30-day storage charges for 35 data sets stored on disk at a university computing center are listed here:

$6.34	$1.36	$1.56	$3.09	$2.89	$ .98	$2.13
1.56	7.68	.98	.79	.79	5.38	1.17
1.56	3.28	1.36	6.34	1.75	5.38	.98
1.56	3.85	.79	1.36	6.34	13.03	9.78
.79	3.47	1.94	.79	.98	12.07	4.42

a. Construct a stem and leaf display for the projected charges.
b. Compute $\bar{y}$, s^2, and s.

c. Calculate the intervals $\bar{y} \pm s$, $\bar{y} \pm 2s$, and $\bar{y} \pm 3s$, and count the number of observations that fall within each interval. Compare your results to the Empirical Rule.

1.54 Find the mean, median, mode, range, variance, and standard deviation for each of the following data sets:
a. 50, 40, 100, 60, 100
b. 70, 90, 80, 60, 50
c. 70, 70, 70, 70, 70
d. 10, 100, 90, 95, 75

1.55 The Airborne Toxic Elements and Organic Substances (ATEOS) project was designed to measure atmospheric levels of more than 50 toxic and carcinogenic chemicals within three urban centers and one rural area in New Jersey. The pie charts on page 40 show the estimated contributions of various sources to the total inhalable particulate matter (IPM, measured in number of particles) at each of the three urban sites—Newark, Elizabeth, and Camden, New Jersey.
a. Which source contributes the most to the total IPM at each urban site?
b. What percentage of the total IPM in Newark is due to industry, oil burning, or motor vehicles? In Elizabeth? In Camden?

1.56 Industrial engineers periodically conduct "work measurement" analyses to determine the time required to produce a single unit of output. At a large processing plant, the number of total worker-hours required per day to perform a certain task was recorded for 50 days. This information will be used in a work measurement analysis.
a. What is the population of interest?
b. What is the sample?

1.57 Refer to Exercise 1.56. The total worker-hours required for each of the 50 days are listed below.

128	119	95	97	124	128	142	98	108	120
113	109	124	132	97	138	133	136	120	112
146	128	103	135	114	109	100	111	131	113
124	131	133	131	88	118	116	98	112	138
100	112	111	150	117	122	97	116	92	122

a. Compute the mean, median, and mode of the data set.
b. Find the range, variance, and standard deviation of the data set.
c. Construct the intervals $\bar{y} \pm s$, $\bar{y} \pm 2s$, and $\bar{y} \pm 3s$. Count the number of observations that fall within each interval and find the corresponding proportions. Compare the results to the Empirical Rule. Do you detect any outliers?
d. Construct a box plot for the data. Do you detect any outliers?
e. Find the 70th percentile for the data on total daily worker-hours. Interpret its value.

1.58 In his essay "Making Things Right," W. Edwards Deming considered the role of statistics in the quality control of industrial products.* In one example, Deming examined the quality control process for a manufacturer of steel rods. Rods produced

*From Tanur, J. et al., eds. *Statistics: A Guide to the Unknown.* San Francisco: Holden-Day, 1978, pp. 279–281.

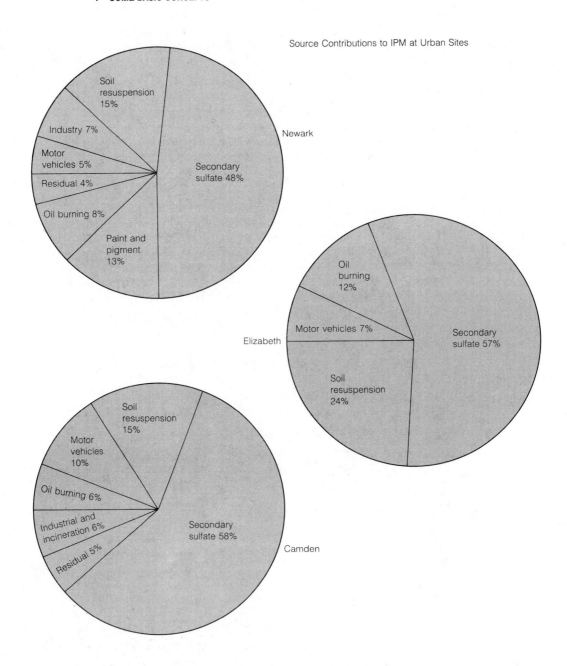

Source Contributions to IPM at Urban Sites

Source: Lioy, P. J. and Daisey, J. M. "Airborne Toxic Elements and Organic Substances."
Environmental Science & Technology, Vol. 20, No. 1, Jan. 1986, p. 12. Reprinted with permission.
Copyright 1986 American Chemical Society.

with diameters smaller than 1 centimeter fit too loosely in their bearings and ultimately must be rejected (thrown out). To determine if the diameter setting of the machine that produces the rods is correct, 500 rods are selected from the day's production and their diameters are recorded. The distribution of the 500 diameters for one day's production is shown in the accompanying figure. Note that the symbol LSL in the figure represents the 1-centimeter lower specification limit of the steel rod diameters.

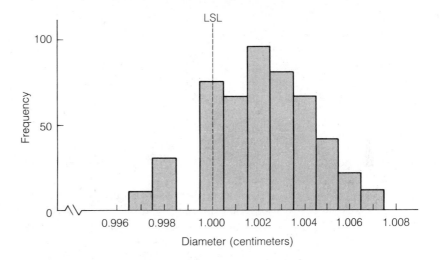

a. What type of data, quantitative or qualitative, does the figure portray?

b. What type of graphical method is being used to describe the data?

c. Use the figure to estimate the proportion of rods with diameters between 1.0025 and 1.0045 centimeters.

d. There has been speculation that some of the inspectors are unaware of the trouble that an undersized rod diameter would cause later on in the manufacturing process. Consequently, these inspectors may be passing rods with diameters that were barely below the lower specification limit, and recording them in the interval centered at 1.000 centimeter. According to the figure, is there any evidence to support this claim? Explain.

1.59 A major utilities company processes all work and service orders on a computer. Each business and residential customer has an account that includes current information on type of service received, corresponding billing cost, and customer payment records. Inherent in this computerized record and billing system is the potential for error (e.g., incorrect reports of equipment installed, errors in coding, and keypunch errors). The company executives want to determine the percentage of accounts that contain errors. Unfortunately, it is too time-consuming and costly to verify every account. Thus, the computer will be used to select 200 accounts from the master file of accounts and a task force will then examine these 200 records to establish an error percentage.

a. What is the population of interest? What is the sample?

b. If the company wants to determine the percentage of all residential customer accounts that contain errors, what is the desired statistical inference?

1.60 Many experts are predicting that energy, particularly in the form of liquid fuel, will be much more costly and in limited supply in the near future. Consumers are therefore being urged to conserve energy while new domestic supplies of liquid fuel are being developed. One seemingly popular form of energy conservation is carpooling. However, statistics compiled by the Bureau of the Census indicate that American motorists drive without anyone else in the car on 52% of their trips, as shown in the table.

NUMBER OF CAR OCCUPANTS	RELATIVE FREQUENCY OF CAR TRIPS
1	.52
2	.28
3	.10
4	.06
5 or more	.04

a. Use one of the graphical methods described in this chapter to summarize the data in the table.

b. What percentage of all automobile trips are made with no more than two people in the car?

1.61 Paper and lumber companies pay for timber by the weight per truckload of 16-foot logs. Consequently, an investor in forest land needs to estimate the total weight of logs that can be produced by a property. This is done by "cruising the property" and counting the total number of trees capable of producing 16-foot logs. A sample of trees (usually 10% of the total number) is selected from this group, and the diameter at chest height and the number of logs per tree (a visual guess) are recorded for each. A forester can then use the diameter and logs-per-tree measurements to calculate the *approximate* weight for each tree. The data shown in the table describe the chest height diameters and estimated average weights for 117 trees cruised in a 4-acre tract of short-leaf pine timber.

DIAMETER AT CHEST HEIGHT (inches)	ESTIMATED (AVERAGE) WEIGHT PER TREE (pounds)	NUMBER OF TREES
10	580	38
11	750	34
12	1,100	21
13	1,800	15
14	2,000	5
15	2,660	3
16	3,000	1
	Total	117

Source: Timber data and information courtesy of Delton F. Price, Fort Smith, Arkansas.

a. Describe the population of measurements from which the sample of tree-weight measurements was selected.

b. Construct a bar graph showing the number of trees cruised in each grouping of chest height diameter.

c. Compute the mean of the estimated weights for the sample of 117 trees. [Hint: Use the fact that $n_i\bar{y}_i = \sum_{j=1}^{n_i} \bar{y}_{ij}$, where n_i is the number of trees in group i, $\bar{y}_i$ is the average weight of group i, and y_{ij} is the weight of tree j in group i.]

1.62 Engineers have a term for unaided human acts of lifting, lowering, pushing, pulling, carrying, or holding and releasing an object—*manual materials handling activities (MMHA)*. M. M. Ayoub et al. (1980) have attempted to develop strength and capacity guidelines for MMHA. The authors point out that a clear distinction between strength and capacity must be made: "Strength implies what a person can do in a single attempt, whereas capacity implies what a person can do for an extended period of time. Lifting strength, for example, determines the amount that can be lifted at frequent intervals." The accompanying table presents a portion of the recommendations of Ayoub et al. for the lifting capacities of males and females. It gives the means and standard deviations of the maximum weight (in kilograms) of a box 30 centimeters wide that can be safely lifted from the floor to knuckle height at two different lift rates—1 lift per minute and 4 lifts per minute.

SEX	LIFTS/MINUTE	MEAN	STANDARD DEVIATION
Male	1	30.25	8.56
	4	23.83	6.70
Female	1	19.79	3.11
	4	15.82	3.23

Source: Ayoub, M. M., Mital, A., Bakken, G. M., Asfour, S. S., and Bethea, N. J. "Development of Strength Capacity Norms for Manual Materials Handling Activities: The State of the Art." *Human Factors*, June 1980, 22, pp. 271–283. Copyright 1980 by the Human Factors Society, Inc. and reproduced by permission.

a. Roughly sketch the relative frequency distribution of maximum recommended weight of lift for each of the four sex/lifts-per-minute combinations. The Empirical Rule will help you do this.

b. Construct the interval $\bar{y} \pm 2s$ for each of the four data sets and give the approximate proportion of measurements that fall within the interval.

c. Assuming the MMHA recommendations of Ayoub et al. are reasonable, would you expect that an average male could safely lift a box (30 centimeters wide) weighing 25 kilograms from the floor to knuckle height at a rate of 4 lifts per minute? An average female? Explain.

1.63 Toxic chemical leaks at Union Carbide plants in West Virginia and Bhopal, India, in the mid-1980's spurred the Environmental Protection Agency to tighten regulations on manufacturers of toxic substances and improve emergency plans. The table (page 44) lists the states with the most toxic chemical accidents from January 1983 to March 1985.

a. Construct a bar graph for the data. What information does the graph provide?

b. Construct a stem and leaf display for the data. What information does it provide?

STATE	NUMBER OF TOXIC CHEMICAL ACCIDENTS
California	244
Illinois	403
Indiana	117
Louisiana	289
Michigan	158
Ohio	321
Pennsylvania	172
Texas	451
Virginia	113
West Virginia	144

Source: United States Environmental Protection Agency.

1.64 According to the National Science Foundation, an estimated 740,000 engineers were employed in manufacturing industries in 1985. The pie chart shows the percentage breakdown of the 740,000 engineers according to occupation.

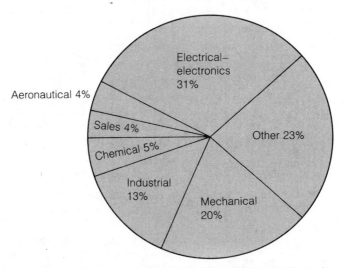

Source: "The Engineer at Large." *IEEE Spectrum*, Vol. 23, No. 4, Apr. 1986, p. 24. © 1986 IEEE.

a. Estimate the number of electrical engineers employed in sales in 1985.
b. What percentage of the 740,000 engineers were employed as electrical, mechanical, or industrial engineers?

1.65 The data in Appendix V are measurements of iron content (percentage of iron) for 390 1.5-kilogram specimens of iron ore selected from a 20,000-ton consignment of Canadian ore (*Reports of Statistical Application Research*, Vol. 18, No. 1, 1971).
a. Describe the population of interest.
b. Give a possible objective of the sampling experiment.

c. Suppose we regard the first 30 measurements as a sample from the 390 measurements. Construct a relative frequency histogram for the 30 measurements.

d. Calculate $\bar{y}$ and s for the sample of 30 measurements.

e. Find the percentage of the 30 measurements that fall in the interval $\bar{y} \pm 2s$. Does this result agree with the Empirical Rule?

f. Construct a box plot for the sample of 30 measurements. Do you detect any outliers?

g. The mean of all 390 measurements is $\bar{y} = 65.745$. Does the mean of $n = 30$ sample measurements from part **d** provide a good estimate of the mean of the 390 measurements? (We will discuss the reliability of this estimate in Chapter 8.)

REFERENCES

Devore, J. *Probability & Statistics for Engineering and the Sciences*. Monterey, California: Brooks/Cole, 1982.

Freedman, D., Pisani, R., and Purves, R. *Statistics*. New York: W. W. Norton and Co., 1978.

Huff, D. *How to Lie with Statistics*. New York: W. W. Norton and Co., 1954.

Koopmans, L. H. *An Introduction to Contemporary Statistics*. Boston: Duxbury Press, 1981.

McClave, J. T. and Dietrich, F. H. II. *Statistics*, 3rd ed. San Francisco: Dellen Publishing Co., 1985.

Mendenhall, W. *Introduction to Probability and Statistics*, 7th ed. Boston: Duxbury Press, 1987.

Sincich, T. *Statistics by Example*, 3rd ed. San Francisco: Dellen Publishing Co., 1987.

Tukey, J. W. *Exploratory Data Analysis*. Reading, Mass.: Addison-Wesley, 1977.

C H A P T E R 2

OBJECTIVE

To show you how to enter data into a computer for subsequent analysis by any of four popular statistical computer program packages; to show how to generate a relative frequency histogram and compute numerical descriptive measures for a data set.

CONTENTS

USING THE COMPUTER (OPTIONAL)

SECTION 2.1

INTRODUCTION

According to *Time* magazine, the "Age of Computers" is upon us.* Never before has the popularity of computers been so high. Many experts are predicting that personal computers for use at home will become as numerous as telephones and televisions in the near future. But what exactly is a computer and how does it work? In simplest terms, a computer is an extremely intricate electronic machine designed to receive, store, and analyze data at a very high rate of speed. But in order for the computer to work properly, you must present instructions to the machine in a language that it can understand—that is, in the form of a **computer program**.

All computer programs consist of a number of detailed, step-by-step commands that tell the computer exactly what it is to do. Many computer programs are specifically geared toward statistical analyses. An entire set of such computer programs is called a **statistical computer program package** (or **statistical software package**).

Various statistical program packages—such as the SAS System, Statistical Package for the Social Sciences (SPSS^X), Minitab, and BMDP—are available at most university computing centers. These systems provide "canned" programs for many of the statistical methods discussed in this text. However, before you attempt to use these packages, you must learn how to enter data into your computer (either mainframe, minicomputer, or personal computer) so that the program package will be able to read and analyze it.

There are several ways in which data may be entered into a computer: disk, magnetic tape, punched cards, or a computer terminal for mainframe (or mini-) computers; and floppy diskette, hard disk, or computer keyboard for personal computers. The easiest and most common way to enter a small amount of data into a computer is through a computer terminal (or keyboard). The computer monitor (or screen) permits you to enter up to 80 characters (or numbers) on a single line of type. Large data sets are typically stored on disk (or floppy diskettes).

The commands given in this chapter are appropriate for entering your data through a computer terminal or keyboard.[†] If your data are stored on disk, diskette, or tape, consult the references given at the end of this chapter.

The instructions to the computer and the manner in which data are entered vary depending on the statistical software package that you use. However, all the computer packages discussed in this chapter (SAS, SPSS^X, Minitab, and BMDP) utilize the following three basic types of instructions:[††]

---

*See "Machine of the Year: The Computer Moves In," *Time*, Jan. 3, 1983.

[†]These commands are also appropriate for entering your data through punched cards, where each command line on a terminal is equivalent to a punched computer card.

[††]When using these statistical software packages on a mainframe computer, additional instructions are required. The basic function of these statements, called **job control language (JCL) instructions**, is to inform the computer which statistical program package to execute. The appropriate JCL varies from one computing center to another. Your instructor will inform you of the appropriate JCL statements to use at your institution.

1. Data entry instructions
2. Statistical analysis instructions
3. Input data values

Instructions to the computer on how to enter your data are called **data entry instructions**. These commands are usually followed by **statistical analysis instructions**, which command the computer to perform various graphical and numerical analyses of the **input data values**.

In this chapter we will explain how to use the data entry instructions of each of the four packages, SAS, SPSS[X], Minitab, and BMDP. We will then apply this knowledge to generate a relative frequency histogram and numerical descriptive measures for the data recorded in Appendix IV. These data represent actual CPU times for 1,000 jobs submitted by a data management and statistical consulting firm. [Recall from Section 1.4 that *CPU time* is the amount of time that a submitted job is in control of the computer's central processing unit (CPU).]

SECTION 2.2

ENTERING DATA INTO THE COMPUTER: SAS

The **SAS System** is probably the most versatile of the four computer packages that we will discuss. All SAS computer programs consist of at least two steps: (1) DATA steps—used to create data sets (data entry instructions), and (2) PROC (procedure) steps—used to analyze these data sets (statistical analysis instructions).*

Let us now create a SAS data set consisting of the five CPU times listed in the first column of Table 1.1. The required statements are shown in Program 2.1.

Each of the eight statements in Program 2.1 is entered as a line of type on a computer terminal or keyboard, beginning in column 1.

PROGRAM 2.1

SAS Statements for Creating a Data Set Consisting of Five CPU Times

Command line	Column 1 ↓	
1	DATA CPU;	⎫
2	INPUT TIME;	⎬ Data entry instructions
3	CARDS;	⎭
4	1.17	⎫
5	1.23	⎪
6	0.15	⎬ Data values (one observation per line)
7	0.19	⎪
8	0.92	⎭

Command line 1 The first command in a SAS program is usually the DATA statement. In Program 2.1, the DATA statement asks SAS to create a data set named CPU. In a DATA statement, the word DATA must be followed by a blank

*When using a mainframe computer, the necessary job control language (JCL) statements precede all the SAS data entry and statistical analysis instructions. For example, the JCL instruction

```
// EXEC SAS
```

commands the computer to execute the SAS software package. Your instructor will give you the JCL instructions required at your computing center.

and the name of the data set that you wish to create. The name CPU was chosen arbitrarily for our example. SAS, as with most statistical program packages, will accept any data set or variable name with a maximum of eight characters (letters or numbers), as long as the characters are *not* separated by blanks and the name does *not* begin with a number. For example, CPUTIMES is a legitimate SAS name, but CPU TIME, 7ELEVEN, and ENGINEERS are not. The name CPU TIME includes a blank, 7ELEVEN begins with a number, and ENGINEERS includes more than eight characters.

Command line 2   The INPUT statement follows the DATA statement and describes the variables that are measured to form the data set. The variable names must follow the word INPUT and must be separated by blanks. In this example, the data set CPU will contain measurements on the single variable called TIME, which again was arbitrarily named.

Command line 3   The third command in a typical SAS program contains only a single word, CARDS. This statement signals the computer that the input data values follow immediately.* The computer reads the value of the variable TIME on each of the five data lines that follow, thereby creating a data set with a total of five observations.

Command lines 4–8   The data lines contain the actual raw data—in this case, the five CPU times listed in the first column of Table 1.1. Each data line includes only a single observation—that is, a single value of TIME.
  Note that all SAS statements, with the exception of the input data values, end with a semicolon. In general, each SAS statement *must* end in a semicolon. The *only* exceptions to this rule are the input data values (in this example, command lines 4–8).

EXAMPLE 2.1

a. Write the appropriate SAS statements to create a data set that contains information on the sex and quiz scores for the three students described in Table 2.1.

TABLE 2.1

STUDENT	SEX	QUIZ1	QUIZ2	QUIZ3
1	Male	87	91	90
2	Male	63	59	86
3	Female	100	85	82

b. Write the appropriate SAS statements that will compute the sum of the three quiz scores and add this new variable to the data set.

*The CARDS statement is required if the data are entered from punched cards *or* from a computer terminal.

SOLUTION

a. The six statements shown in Program 2.2 will create the desired data set. The data set, which we have called SCORES, has three observations (students), with four variables—SEX, QUIZ1, QUIZ2, and QUIZ3—per observation. Notice that a dollar sign ($) follows the variable named SEX on the INPUT statement (line 2). This is because SEX is a **character** variable—that is, a qualitative variable—whose values (M or F) are characters, not numbers. In contrast, QUIZ1, QUIZ2, and QUIZ3 are read as **numeric** variables, since their values are quantitative. Whenever you want to include a character (nonnumeric) variable in a SAS data set, a dollar sign must follow the variable name in the INPUT statement. This informs the computer that the values of this variable to be read on the input data lines are observations on a qualitative variable.

PROGRAM 2.2

SAS Statements for Creating the Test Scores Data Set

```
Command    Column 1
  line        ↓

    1        DATA SCORES;                          ⎫ Data entry
    2        INPUT SEX $ QUIZ1 QUIZ2 QUIZ3;        ⎬ instructions
    3        CARDS;                                ⎭
    4        M   87 91 90 ⎫ Data values
    5        M   63 59 86 ⎬ (one observation per line)
    6        F  100 85 82 ⎭
```

Note that the order in which the names of the variables appear in the INPUT statement must be the same as the order in which the values of the variables appear on the input data lines (lines 4–6). Thus, the value of SEX appears first on the data line, followed by the values of QUIZ1, QUIZ2, and QUIZ3, respectively. As long as these values are separated by at least one blank and appear in the proper order, they may begin in any of the 80 columns of the screen. For example, command lines 4–6 could have been typed as follows:

```
Command    Column 1
  line        ↓

    4           M   87    91      90
    5        M        63    59          86
    6              F   100    85     82
```

b. The SAS statements shown in Program 2.3 expand the data set of part **a** by creating an additional variable, named QUIZTOT, which represents the sum of the scores on the three quizzes. The only additional statement needed is that shown on command line 3:

```
QUIZTOT = QUIZ1 + QUIZ2 + QUIZ3;
```

SAS will now automatically compute QUIZTOT for each of the three observations (students) and include this variable on the data set.

PROGRAM 2.3

SAS Statements for Creating
the Test Scores Data Set

Command line	Column 1 ↓	
1	DATA SCORES;	⎫ Data entry
2	INPUT SEX $ QUIZ1 QUIZ2 QUIZ3;	⎬ instructions
3	QUIZTOT = QUIZ1 + QUIZ2 + QUIZ3;	⎭
4	CARDS;	
5	M 87 91 90	⎫
6	M 63 59 86	⎬ Data values
7	F 100 85 82	⎭

The complete SAS data set, which we have named SCORES, is stored in the computer as shown in Table 2.2.

TABLE 2.2

OBSERVATION	SEX	QUIZ1	QUIZ2	QUIZ3	QUIZTOT
1	M	87	91	90	268
2	M	63	59	86	208
3	F	100	85	82	267

Any SAS statements that are used to create variables in addition to those on the INPUT statement should follow the INPUT statement, but precede the CARDS statement. When creating variables in SAS or in any of the other computer packages we discuss in this text, you should use the standard arithmetic operation symbols, $+$, $-$, $*$, and $/$, for addition, subtraction, multiplication, and division, respectively. For example, if you want to double the value of QUIZ1 and then average the resulting scores in Example 2.1, insert the following statements after command line 3 in Program 2.3:

```
QUIZ1DBL = 2*QUIZ1;
QUIZAVE = (QUIZ1DBL + QUIZ2 + QUIZ3)/3;
```

Once a SAS data set has been created, you may begin analyzing it using PROC statements. These statistical analysis instructions will follow the input data values in the SAS program. Thus, the three basic types of commands (discussed in Section 2.1) appear in SAS in the following order:

1. Data entry instructions
2. Input data values
3. Statistical analysis instructions

We will illustrate the use of PROC statements in Section 2.6.

We end this section with a word of caution: Unlike some computer programming languages, a SAS statement may begin in *any* of the 80 columns of the computer screen. Also, it is not necessary to begin each SAS statement on a new line. SAS statements can run continuously on a line, as long as each of the separate commands ends in a semicolon. However, we strongly recommend that beginning users of SAS follow the approach of Programs 2.1–2.3: (1) Start typing in column 1, and (2) Begin each new SAS command on a new line. Our experience has indicated that those who follow this simple approach will make fewer programming errors and have an easier time "debugging" their programs.

| | | | | | | | | | | |
EXERCISES 2.1–2.5

2.1 Which of the following are legitimate SAS variable names?
  a. VOLTAGE   **b.** DDT   **c.** TENSILE STRENGTH   **d.** STATISTICS
  e. IRONORE   **f.** 16YEARS   **g.** MONTH8

2.2 Which of the following are legitimate SAS data set names?
  a. ALLSCORES   **b.** ALL SCORES   **c.** GRADES   **d.** TEST184
  e. 184   **f.** TEST 184

2.3 Explain why each of the following SAS commands is incorrectly coded:
  a. DATA DDT IN FISH;
  b. INPUT SPECIES $ WEIGHT
  c. PCT = DDT÷WEIGHT;
  d. CARDS;
     18; 42; 21; 20;

2.4 A company that services two brands of microcomputers wants to investigate the factors that affect the amount of time it takes service personnel to perform preventive maintenance on each brand. One variable believed to affect maintenance time is the service person's number of months of experience in preventive maintenance. The company obtained the information shown in the table for a sample of ten service people.

MAINTENANCE TIME Hours	BRAND	EXPERIENCE Months
2.0	A	2
1.8	A	4
.8	B	12
1.1	A	12
1.0	B	8
1.5	B	2 ·
1.7	A	6
1.2	B	5
1.4	A	9
1.2	B	7

  a. Write a SAS program to create a data set that contains the sample information.
  b. The cost of maintenance is $200 per hour. Write the SAS statements that will add the cost of maintenance to the data set.

2.5 The information shown in the table was obtained at the end of the academic year for a sample of 20 college freshmen. Write a SAS program that will enter the sample data into the computer. Include in the data set a new variable that represents the sum of the college entrance exam scores.

COLLEGE ENTRANCE EXAM SCORES		GRADE POINT AVERAGE	MAJOR[a]
Verbal, %	Mathematics, %		
81	87	3.49	COMPSCI
68	99	2.89	MATH
57	86	2.73	BUSADM
100	49	1.54	EDUC
54	83	2.56	ENG
82	86	3.43	ENG
75	74	3.59	MATH
58	98	2.86	MATH
55	54	1.46	BUSADM
49	81	2.11	EDUC
64	76	2.69	ENG
66	59	2.16	BUSADM
80	61	2.60	MATH
100	85	3.30	EDUC
83	76	3.75	COMPSCI
64	66	2.70	EDUC
83	72	3.15	ENG
93	54	2.28	BUSADM
74	59	2.92	EDUC
51	75	2.48	MATH

[a]COMPSCI = Computer Science, MATH = Mathematics, BUSADM = Business Administration, ENG = Engineering, EDUC = Education.

SECTION 2.3

ENTERING DATA INTO THE COMPUTER: SPSS[X]

The **Statistical Package for the Social Sciences (SPSS[X])** is similar to the SAS package in the sense that its commands or control statements may be classified into one of two groups: **data creation commands** and **statistical procedure commands**. Data creation commands are data entry instructions used to create a data set, and statistical procedure commands define the statistical procedure to be used to analyze the data set.* Unlike SAS instructions, the instructions on an SPSS[X] command line must begin in column 1 and all continuation lines must be indented at least one column.

Consider the SPSS[X] statements of Program 2.4, which create a data set containing the sex, first three quiz scores, and total quiz score of the three students of Example 2.1. Each SPSS[X] command begins with a **command keyword** (which may contain more than one word) in column 1. Although a few commands (such as BEGIN DATA and END DATA) are complete in themselves, most require

---

*The appropriate JCL instructions for a mainframe computer must precede all the SPSS[X] control statements. For example, the JCL statement

```
// EXEC SPSSX
```

commands the computer to begin executing the SPSS[X] program package. See your instructor for the JCL statements required at your institution.

specifications. Specifications are made up of variable names, subcommands, numbers, and other keywords required to complete an SPSS^X command. Unlike SAS statements, SPSS^X statements do not end with a semicolon.

PROGRAM 2.4 SPSS^x
Statements for Creating
the Test Scores Data Set

Command line	Column 1 ↓	
1	`DATA LIST FREE/QUIZ1 QUIZ2 QUIZ3 * SEX (A1)`	Data entry instructions
2	`COMPUTE   QUIZTOT=QUIZ1+QUIZ2+QUIZ3`	
3	`BEGIN DATA`	
4	`  87 91 90 M`	Data values
5	`  63 59 86 M`	
6	`100 85 82 F`	
7	`END DATA`	

Command line 1 The first SPSS^X statement, the DATA LIST command, identifies the variables to be read from the input data values. The command DATA LIST must appear as shown in columns 1–9. Specifications for this command include formats and names for the variables in the data set.

SPSS^X offers various options for arranging your data on the input data lines. We have typed the SPSS^X keyword FREE after the DATA LIST command. The FREE option enables you to enter the data anywhere on the input data line, as long as the variable values are separated by at least one blank and the order in which the values appear is consistent with the order given in the variable list (described in the following paragraph). Since the freefield-format mode of data entry is the simplest to describe and use, we will illustrate this method of entry in all our SPSS^X examples. If you want to use another method of data entry, you will need to consult the SPSS^X references given at the end of this chapter.

The names of the variables in the data set are listed following the slash (/) in the DATA LIST statement. These variable names (each up to eight characters long) must be separated by blanks. As in Example 2.1, we have named the variables SEX, QUIZ1, QUIZ2, and QUIZ3. Note that an asterisk (*) separates the numeric variables (QUIZ1, QUIZ2, and QUIZ3) from the character variable SEX and that the numeric variables are listed first. This is necessary whenever you use the freefield format for data entry. In addition, an alphanumeric format for the character variable must be specified, in parentheses, after the variable name. In Program 2.4, the format A1 implies that values of the variable SEX will occupy a single column on the input data lines.*

Command line 2 The COMPUTE command is used to create a new variable, arbitrarily called QUIZTOT, by summing the values of QUIZ1, QUIZ2, and QUIZ3. The algebraic expression specified in the COMPUTE statement may begin in any column following the COMPUTE command.†

*Similarly, the format A5 would be used for a character variable with a maximum length of five columns.

†The COMPUTE command is an optional SPSS^X statement. It is necessary only when you wish to create additional variables or transform existing variables.

As with SAS, an SPSS^X COMPUTE statement uses the standard arithmetic operation symbols, $+$, $-$, $*$, and $/$, for addition, subtraction, multiplication, and division, respectively. For example, to double the first quiz score and then average the three scores, use the following COMPUTE commands:

Column 1
↓

```
COMPUTE       QUIZ1DBL = 2*QUIZ1
COMPUTE       QUIZAVE = (QUIZ1DBL+QUIZ2+QUIZ3)/3
```

Command line 3   After compiling the necessary SPSS^X data entry instructions, the computer is ready to begin reading the input data. The BEGIN DATA statement instructs SPSS^X to start this process and signals the computer that the input data values (lines 4–6 of Program 2.4) follow immediately.

Command lines 4–6   The input data values follow the BEGIN DATA statement. Each data line represents a single observation or case. The only variables whose values must appear on the input data lines are those named on the DATA LIST command. In our example, you do *not* need to enter the value of QUIZTOT— the sum of the three quiz scores—on the input data line. The COMPUTE statement guarantees that SPSS^X will automatically calculate QUIZTOT for each observation and include it in the data set.

Since the freefield format was specified in command line 1, the values of the variables can appear anywhere on the input data lines, as long as they are separated by at least one blank and the order of appearance is consistent with the order specified in the DATA LIST command. Thus, the first value on the input data line is the value of QUIZ1, the second is the value of QUIZ2, etc.

Command line 7   To alert SPSS^X that all the input data values have been read, insert an END DATA statement after the last input data line.

The SPSS^X data file that results from Program 2.4 is stored in the computer as indicated in Table 2.3. Now that we have created the SPSS^X data file, we are ready to begin analyzing it using SPSS^X statistical procedure commands. These analysis instructions will be inserted *after* the data entry instructions (lines 1–2 of Program 2.4), but *before* the BEGIN DATA command. We will discuss some SPSS^X analysis instructions in Section 2.7.

TABLE 2.3

CASE	SEX	QUIZ1	QUIZ2	QUIZ3	QUIZTOT
1	M	87	91	90	268
2	M	63	59	86	208
3	F	100	85	82	267

To review, the three basic types of instructions (discussed in Section 2.1) appear in SPSS^X programs in the following order:

1. Data entry instructions
2. Statistical analysis instructions
3. Input data values

EXERCISES 2.6–2.9

2.6 Which of the following are legitimate SPSS^X variable names?
  a. VOLTAGE   **b.** DDT   **c.** TENSILE STRENGTH   **d.** STATISTICS
  e. IRONORE   **f.** 16YEARS   **g.** MONTH8

2.7 Explain why each of the following SPSS^X commands is incorrectly coded.

Column 1
↓

```
a.  DATA LIST            FREE/LENGTH WEIGHT DDT SPECIES (A10)
b.  COMPUTE              RATIO=LENGTH÷WEIGHT
c.  READ DATA
d.  60   125   100CATFISH
e.      END DATA
```

2.8 Enter the data of Exercise 2.4 into the computer using the SPSS^X computer package. Include the cost of maintenance in the data set.

2.9 Enter the data of Exercise 2.5 into the computer using the SPSS^X computer package. Include the sum of the college entrance exam scores in the data set.

SECTION 2.4

ENTERING DATA INTO THE COMPUTER: MINITAB

Minitab is another easy-to-use statistical program package that has been designed especially for those who have no previous experience with computers. Minitab commands can be given in English; that is, the commands can be expressed in nearly the same language you would use to instruct someone to do the computations, tasks, or analysis by hand. Also, Minitab commands may begin or end in any column of the command line of your terminal. We illustrate with Example 2.2.

EXAMPLE 2.2

Refer to Example 2.1. Write the Minitab commands that will create the quiz scores data set, which is reproduced in Table 2.4. Include the total of the three scores on the data set.

TABLE 2.4

STUDENT	SEX	QUIZ1	QUIZ2	QUIZ3
1	Male	87	91	90
2	Male	63	59	86
3	Female	100	85	82

SOLUTION

The appropriate Minitab statements are given in Program 2.5.

PROGRAM 2.5
Minitab Statements
for Creating the Test
Scores Data Set

Command line	Column 1 ↓
1	READ C1 C2 C3 C4} Data entry instruction
2	1 87 91 90 ⎫
3	1 63 59 86 ⎬ Data values (one observation per line)
4	0 100 85 82 ⎭
5	ADD C2 C3 C4 PUT INTO C5 ⎱ Data entry
6	STOP ⎰ instruction

Command line 1 Minitab stores data in a "worksheet" that it maintains inside the computer. The initial Minitab statement instructs the computer to read the data on lines 2–4 and to put the values of the variables into specific "columns" of the worksheet. (In contrast to the columns of a computer terminal screen, the columns of a Minitab worksheet may contain values with more than one digit—for example, 90, 128, 10772.) The columns are designated by the letter C and a corresponding number. Instead of referring to a variable by its name, we refer to it in Minitab by the column into which it is placed (for example, C1, C2, . . .). In this example, the computer reads the value of the first variable (SEX) and places it into column 1 (C1) of the worksheet. Likewise, the value of the second variable (QUIZ1) is read and placed into column 2 (C2); the value of the third variable (QUIZ2) goes into column 3 (C3); and the value of the fourth variable (QUIZ3) goes into column 4 (C4). In Minitab, the input data values always follow the READ command.

Command lines 2–4 The input data values contain the actual values of the variables read in the worksheet columns. As is the case with SAS and SPSS^X, these values must be separated by at least one blank. Notice, however, that the values of the qualitative variable SEX in C1 are entered as numbers (1 or 0) rather than as letters (M or F). Minitab requires that all data used in statistical analyses be numerical. Thus, if we want to use the qualitative variable SEX in a statistical procedure, we need to convert its possible values to numbers. In this example, we arbitrarily let 1 represent male and 0 represent female.

Command line 5 In contrast to the SAS and SPSS^X computer packages, Minitab does not, in general, recognize variable names (for example, SEX or QUIZ1) when requested to analyze or manipulate data. Thus, if you wish to add the values of the three quiz scores, you must refer to the variables by their column numbers—C2, C3, and C4—rather than by their names. The Minitab command ADD in line 5 of Program 2.5 instructs the computer to add the values of the variables stored in columns 2, 3, and 4 and to place the sum in column 5 (a new column in the worksheet that represents the variable called QUIZTOT in the previous sections).

Minitab commands for the other usual arithmetic operations (subtraction, multiplication, and division) are illustrated below:

Column 1
↓

```
SUBTRACT    C2 FROM C3 PUT INTO C6
MULTIPLY    C2 BY 2 PUT INTO C7
DIVIDE      C4 BY 10 PUT INTO C8
```

Note that the words FROM, BY, PUT, and INTO may be omitted from the Minitab commands for arithmetic operations.

Command line 6 All Minitab programs terminate with the STOP command.

The Minitab worksheet, stored in the computer as shown in Table 2.5, is now ready for data analysis.

TABLE 2.5

OBSERVATION	C1	C2	C3	C4	C5
1	1	87	91	90	268
2	1	63	59	86	208
3	0	100	85	82	267

The statistical analysis commands of Minitab follow the data input values. Thus, the three types of instructions (discussed in Section 2.1) generally appear in Minitab programs in the following order:

1. Data entry instructions
2. Input data values
3. Statistical analysis instructions

We will show you some Minitab statistical analysis commands in Section 2.7.

Note: The commands of Program 2.5 were written using the minimal amount of coded text required by Minitab. However, you may insert key words within each command to help you follow the logic of the program. For example, line 1 of Program 2.5 could be written as follows:

```
READ SEX IN C1, QUIZ SCORES IN C2, C3 AND C4
```

Minitab ignores the extraneous words (SEX, QUIZ SCORES, etc.), and reads only the command name (READ) and its arguments (C1, C2, C3, C4).

| | | | | | | | | | | | |

EXERCISES 2.10–2.13

2.10 Explain why each of the following Minitab commands is incorrectly coded.
  a. READ SUBJECT IN COLUMN 1, IQ IN C2, AGE IN C3
  b. MARY 120
     DAVE 110
  c. ADD IQ TO AGE, PUT IN C4
  d. END

2.11 Rewrite the Minitab program shown here using the minimal amount of coded text.

Command line	Column 1 ↓
1	READ LOT IN C1, MEASUREMENT A IN C2, MEASUREMENT B IN C3
2	1 100 250
3	2 50 60
4	3 75 150
5	4 200 500
6	5 150 300
7	MULTIPLY C2 BY 2, PUT INTO C4
8	SUBTRACT C4 FROM C3, PUT INTO C5
9	STOP

2.12 Enter the data of Exercise 2.4 into the computer using the Minitab computer package. Include the cost of maintenance in the data set. [*Hint:* The Minitab command for multiplication is MULTIPLY C1 BY 200.]

2.13 Enter the data of Exercise 2.5 into the computer using the Minitab computer package. Include the sum of the college entrance exam scores in the data set.

S E C T I O N   2.5

ENTERING DATA INTO THE COMPUTER: BMDP

The last of the statistical computer packages to be discussed in this chapter was originally developed for use in biomedical analyses and is called **BMDP Statistical Software**. BMDP employs three different types of commands: program identification (analysis) commands, data entry instructions, and input data values.

BMDP **program identification** statements are commands that identify the BMDP statistical analysis program you want to run. For example, the program identification statement

```
BMDP 1D
```

commands a personal computer to execute the BMDP program called 1D (which generates the simple descriptive statistics discussed in Chapter 1).*

The BMDP package differs from SAS, SPSS[X], and Minitab in that you cannot enter the data into the computer without, at the same time, telling the computer the type of statistical analysis that you wish to perform. Therefore, to enter data into the computer using the BMDP package, you must precede the data entry instructions and data values (to be explained in the following paragraphs) by the appropriate program identification commands (statistical analysis instructions). The appropriate BMDP program identification statements for generating a histogram and simple descriptive statistics are given in Section 2.7; other BMDP program identification and statistical analysis instructions will be presented as the need arises in subsequent chapters.

*On the mainframe computer, BMDP program identification commands are given as JCL statements, e.g.,

```
// EXEC BIMED,PROG=BMDP1D
```

See your instructor for the appropriate JCL statements required at your institution.

EXAMPLE 2.3   Refer to Examples 2.1 and 2.2. Write the BMDP commands that will create the quiz scores data set. Include the sum of the three quiz scores on the data set.

SOLUTION   The correct BMDP data entry instructions and input data values are shown in Program 2.6. BMDP data entry instructions are written in sentences (commands) that are grouped into paragraphs. Each paragraph begins with a slash (/) in column 1, followed by a blank space and an identifying paragraph name. (The paragraph name must be the first word in the paragraph.) Several paragraphs are common to all BMDP programs. In Program 2.6, the BMDP instructions consist of three paragraphs: the INPUT paragraph, the VARIABLE paragraph, and the TRANS-FORM paragraph. Note that each sentence within a paragraph ends with a period. Although we have written each sentence on a new line beginning in a fixed (but arbitrary) column (column 16), the instructions can be typed anywhere in columns 1–80. Sentences and paragraphs may be typed continuously as long as paragraphs are separated by a slash.

PROGRAM 2.6 BMDP Statements for Creating the Test Scores Data Set

```
Command     Column 1            Column 16
line          ↓                   ↓

   1        /   INPUT           VARIABLES ARE 4.                                      ⎫
   2                            FORMAT IS FREE.                                       ⎬  Data entry
   3        /   VARIABLE        NAMES ARE SEX, QUIZ1, QUIZ2, QUIZ3, QUIZTOT.  ⎬  instructions
   4                            ADD = NEW.                                            ⎭
   5        /   TRANSFORM       QUIZTOT = QUIZ1 + QUIZ2 + QUIZ3.
   6        /   END
   7        1       87    91    90 ⎫
   8        1       63    59    86 ⎬  Data values
   9        0      100    85    82 ⎭  (one observation per line)
  10        /END
```

Command lines 1–2   The INPUT paragraph is the first required paragraph of any BMDP program. When the data are read from lines or entered from a terminal, as will most likely be the case, the INPUT paragraph must specify the number of VARIABLES to be read and their FORMAT. In our example, the sentence

```
VARIABLES ARE 4.
```

informs the computer that four variables are to be read onto a data set. The FORMAT sentence instructs the computer to read the variables in a FREE format. The FREE format method of reading data allows you to enter the data anywhere on the input data lines, as long as the data values on each line are separated by blanks.

Command lines 3–4   The VARIABLE paragraph describes the variables on the data set. Generally, the VARIABLE paragraph is used to name the variables and to indicate whether new variables are to be created by data transformations. In our example, the sentence

```
NAMES ARE SEX, QUIZ1, QUIZ2, QUIZ3, QUIZTOT.
```

specifies the names of the four variables to be read from the input data lines. SEX is the name of the first variable to be read, QUIZ1 is the name of the second variable to be read, etc. As with SAS and SPSS^X, BMDP variable names are restricted to a maximum of eight characters. Note, however, that the values for SEX are numbers. BMDP, like Minitab, requires that all data be numerical. Notice also that five variable names are given in the NAMES sentence, even though the INPUT paragraph specifies that only four variables are to be read. This is because we will create a new variable in the next paragraph by summing the three quiz scores. Any variables created by data transformations (addition, subtraction, multiplication, division, etc.) must be named in the NAMES sentence. Thus, the number of names in the VARIABLE paragraph may be greater than the number of variables specified to be read in the INPUT paragraph. However, if additional variables are to be created by data transformations, the ADD sentence is also required. Since we want to create a new variable (QUIZTOT), we include the sentence

```
ADD = NEW,
```

Command line 5   The TRANSFORM paragraph provides options for computing new variables from existing ones, correcting miscoded data values, selecting only certain observations for analysis, and generating random samples. The most frequent use of the TRANSFORM paragraph is to create new variables. In this example, the sentence

```
QUIZTOT = QUIZ1 + QUIZ2 + QUIZ3,
```

creates an additional variable, QUIZTOT, which is the sum of the first three quiz scores of each student.

The TRANSFORM paragraph uses the standard arithmetic operation symbols, $+$, $-$, $*$, and $/$, for addition, subtraction, multiplication, and division, respectively. The following is another example of a typical TRANSFORM paragraph.

```
/ TRANSFORM QUIZ1DBL = 2*QUIZ1,
           QUIZAVE = (QUIZ1 + QUIZ2 + QUIZ3)/3,
           QUIZDIFF = QUIZ3 - QUIZ1,
```

Command line 6   BMDP data entry instructions are terminated with an END paragraph.

Command lines 7–9   The input data values follow the END paragraph. Note that each line of data represents an observation (student) and includes the values of the first four variables named (in their order of appearance) in the VARIABLE paragraph. These values can appear anywhere on the input data lines as long as they are separated by at least one blank.

Command line 10 The last data record must be followed by /END. Note that there should be no space (blank) after the slash when the /END card follows data.

The BMDP data entry instructions and input data values have created and stored in the computer a data set that appears as shown in Table 2.6. The data are now in the appropriate form to be analyzed using program (analysis) identification statements.

TABLE 2.6

CASE	SEX	QUIZ1	QUIZ2	QUIZ3	QUIZTOT
1	M	87	91	90	268
2	M	63	59	86	208
3	F	100	85	82	267

To review, the three types of instructions (described in Section 2.1) appear in a BMDP program in the following order:

1. Statistical analysis instructions
2. Data entry instructions
3. Input data values

EXERCISES 2.14–2.17

2.14 Which of the following are legitimate BMDP variable names?
a. VOLTAGE **b.** DDT **c.** TENSILE STRENGTH **d.** STATISTICS
e. IRONORE **f.** 16YEARS **g.** MONTH8

2.15 Explain why each of the following BMDP commands is incorrectly coded.
```
a. INPUT        VARIABLES ARE 2,
b.              FORMAT IS FREEFIELD,
c. / VARIABLE   NAMES ARE FISH SPECIES, LENGTH, DDT,
d.              ADD = NEW
e. / TRANFORM   RATIO = LENGTH÷DDT,
f. END
g. BASS 50 2,5
   CATFISH 10 7,1
h. / STOP
```

2.16 Enter the data of Exercise 2.4 into the computer using the BMDP computer package. Include the cost of maintenance in the data set.

2.17 Enter the data of Exercise 2.5 into the computer using the BMDP computer package. Include the sum of the college entrance exam scores in the data set.

SECTION 2.6

GENERATING A RELATIVE FREQUENCY HISTOGRAM AND COMPUTING NUMERICAL DESCRIPTIVE MEASURES: SAS

In this section we will show how to generate a relative frequency histogram and how to produce numerical descriptive measures using the SAS System. The large data set that we are interested in describing is given in Appendix IV. It is the collection of CPU times (in seconds) for 1,000 computer jobs submitted by a statistical consulting firm.

A careful examination of the data reveals that the largest CPU time is 20.79 and the smallest is .06. Thus, the range of the data set is

Range = 20.79 − .06 = 20.73

Suppose we want to construct a histogram with 14 class intervals;* then the width of each interval must be

$$\frac{\text{Range}}{14} = 1.48 \approx 1.5$$

Since no CPU time falls below .06 second, and we want no measurement to fall on a class boundary, we will define the first class interval to be .005–1.505 seconds. The subsequent class intervals are then set up as follows: 1.505–3.005, 3.005–4.505, 4.505–6.005, . . . , 18.005–19.505, and 19.505–21.005.

The appropriate SAS statements that will produce the histogram we have described are given in Program 2.7.

PROGRAM 2.7 SAS Statements That Produce a Histogram for the 1,000 CPU Times of Appendix IV

Command line

```
1       DATA CPUTIMES;  ⎫
2       INPUT TIME;      ⎬  Data entry instructions: Create the data set of 1,000 CPU times
3       CARDS;           ⎭
4       1.86   ⎫
5       3.49   ⎪
6       2.63   ⎪
7       3.49   ⎬  Data values
 .       .     ⎪  (one observation per line)
 .       .     ⎪
 .       .     ⎪
1,003   2.05   ⎭
1,004   PROC CHART;                                            Statistical analysis instructions:
1,005   VBAR TIME/TYPE=PERCENT MIDPOINTS=.755 TO 20.255 BY 1.5;  Generate the histogram
```

Command lines 1–1,003 The first 1,003 lines comprise the DATA step of the SAS program—that is, the data entry instructions and input data values. A data set called CPUTIMES is to be created. It will contain 1,000 observations on a single variable called TIME.

Command lines 1,004–1,005 The last two lines of the SAS program comprise the PROC (procedure) step, specifying the statistical analysis to be performed. PROC CHART commands SAS to generate either a bar graph or pie chart for the newly created data set. VBAR TIME requests that a vertical bar graph—that is, a histogram—be constructed for the variable TIME. The keywords TYPE and MIDPOINTS after the slash (/) in card 1,005 are SAS options available to the

*The number of class intervals was arbitrarily chosen. Any number of class intervals from 10 to 20 will adequately describe the data set.

CHART procedure.* TYPE=PERCENT requests that percentages or relative frequencies be plotted along the vertical axis, thus generating a *relative* frequency histogram. (If TYPE is omitted, a frequency histogram is constructed.) The MIDPOINTS option enables the user to select the midpoints of the class intervals. Since we want our intervals to be .005–1.505, 1.505–3.005, . . . , 19.505–21.005, the respective class interval midpoints are .755, 2.255, . . . , 20.255. (If the MIDPOINTS option is omitted, SAS will automatically choose classes with nicely rounded midpoints.)

The output for this SAS program is shown in Figure 2.1. Notice that approximately 65% of the 1,000 CPU times fall between .005 and 4.505 seconds and that the relative frequency histogram is skewed to the right.

FIGURE 2.1 Relative Frequency Histogram for the 1,000 CPU Times Generated by SAS

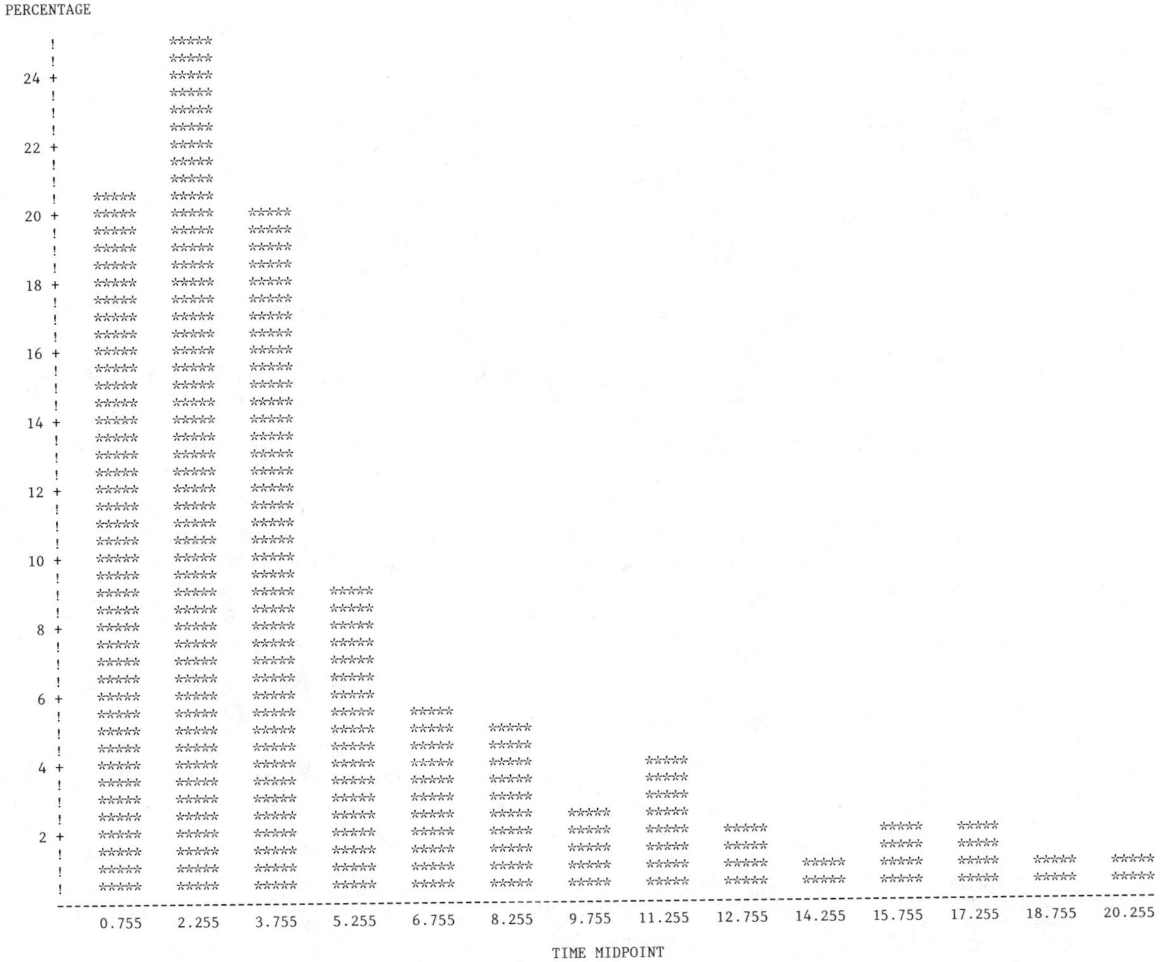

```
PERCENTAGE

   !                       *****
   !                       *****
24 +                       *****
   !                       *****
   !                       *****
   !                       *****
22 +                       *****
   !                       *****
   !                       *****
   !       *****           *****
20 +       *****           *****   *****
   !       *****           *****   *****
   !       *****           *****   *****
   !       *****           *****   *****
18 +       *****           *****   *****
   !       *****           *****   *****
   !       *****           *****   *****
   !       *****           *****   *****
16 +       *****           *****   *****
   !       *****           *****   *****
   !       *****           *****   *****
   !       *****           *****   *****
14 +       *****           *****   *****
   !       *****           *****   *****
   !       *****           *****   *****
   !       *****           *****   *****
12 +       *****           *****   *****
   !       *****           *****   *****
   !       *****           *****   *****
   !       *****           *****   *****
10 +       *****           *****   *****
   !       *****           *****   *****
   !       *****           *****   *****   *****
   !       *****           *****   *****   *****
 8 +       *****           *****   *****   *****
   !       *****           *****   *****   *****
   !       *****           *****   *****   *****
   !       *****           *****   *****   *****
 6 +       *****           *****   *****   *****
   !       *****           *****   *****   *****   *****
   !       *****           *****   *****   *****   *****   *****
   !       *****           *****   *****   *****   *****   *****
 4 +       *****           *****   *****   *****   *****   *****           *****
   !       *****           *****   *****   *****   *****   *****           *****
   !       *****           *****   *****   *****   *****   *****           *****
   !       *****           *****   *****   *****   *****   *****   *****   *****
 2 +       *****           *****   *****   *****   *****   *****   *****   *****   *****           *****   *****
   !       *****           *****   *****   *****   *****   *****   *****   *****   *****           *****   *****
   !       *****           *****   *****   *****   *****   *****   *****   *****   *****   *****   *****   *****   *****   *****
   !       *****           *****   *****   *****   *****   *****   *****   *****   *****   *****   *****   *****   *****   *****
   -------------------------------------------------------------------------------------------------------------------
        0.755   2.255   3.755   5.255   6.755   8.255   9.755  11.255  12.755  14.255  15.755  17.255  18.755  20.255

                                            TIME MIDPOINT
```

*TYPE and MIDPOINTS are optional SAS commands and may be omitted. If they are omitted, do not include the slash (/) in the VBAR command. Simply insert a semicolon after the VBAR TIME request.

In its present form, Program 2.7 is extremely long; it requires that the programmer type over 1,000 lines on a terminal—a tedious chore, to say the least. This is because the data set is large and we are allowing SAS to read only one observation (CPU time) per line. The program can be shortened considerably if we permit SAS to read multiple observations per input data line. This is accomplished by altering the form of the INPUT command (line 2), as shown in Program 2.8. The symbols @@ placed at the end of the INPUT statement, but before the semicolon, allow multiple observations to be read on a single input data line as long as the values of the variable are separated by blanks. The number of observations on a single line is limited only by the available space (80 columns). By permitting more than one observation per line to be read (we entered ten observations on each line), we have reduced the SAS program from 1,005 lines in Program 2.7 to a manageable 105 lines in Program 2.8.

PROGRAM 2.8   Reading Multiple Observations per Data Line in SAS

Command
line

1	DATA CPUTIMES; ⎫
2	INPUT TIME @@; ⎬ Data entry instructions
3	CARDS; ⎭
4	1.86 3.49 2.63 3.49 1.69 1.83 0.81 4.70 0.85 4.24 ⎫
5	3.49 2.75 1.65 0.92 0.62 0.41 3.23 4.13 3.23 1.89 ⎪ Data values
.	. . . . ⎬ (ten observations
.	. . . . ⎪ per line)
.	. . . . ⎪
103	2.02 2.82 2.19 5.13 2.43 0.23 0.43 6.22 3.98 2.05 ⎭
104	PROC CHART; ⎫ Statistical analysis
105	VBAR TIME/TYPE=PERCENT MIDPOINTS=.755 TO 20.255 BY 1.5; ⎬ instructions

You may also want to compute several numerical descriptive measures of the large data set, e.g., the mean, median, mode, range, variance, standard deviation, and percentiles. The additional SAS commands for producing simple descriptive statistics for the 1,000 CPU times are given below (these lines should follow command line 105 in Program 2.8):

Command
line

106	PROC UNIVARIATE;
107	VAR TIME;

In Section 2.7, we will list the program statements necessary to produce a histogram and numerical descriptive measures in SPSS^X, Minitab, and BMDP.

EXERCISES 2.18–2.19

2.18 Use the SAS computer package to generate a relative frequency histogram and produce numerical descriptive measures for the data of Exercise 1.12.

2.19 Use the SAS computer package to generate a relative frequency histogram and produce numerical descriptive measures for the data of Exercise 1.15.

| | | | | | | | | | | |

SECTION 2.7

GENERATING A RELATIVE FREQUENCY HISTOGRAM AND COMPUTING NUMERICAL DESCRIPTIVE MEASURES: SPSSˣ, MINITAB, AND BMDP

The program statements required to generate a histogram and simple descriptive statistics for the data of Appendix IV using SPSSˣ, Minitab, and BMDP are shown in Programs 2.9, 2.10, and 2.11, respectively. Notice that each program allows multiple observations to be read per input data line.* (We have arbitrarily chosen to read in 10 observations per line, but you may read in as many observations as will fit on the 80 columns of the computer terminal screen.) Also, some of the statements shown in Programs 2.9–2.11 are optional statements included solely for the purpose of labeling the printouts of the histograms, and thus may be omitted. (Consult the references at the end of the chapter for details on optional commands and their purpose.) The histograms produced by the SPSSˣ, Minitab, and BMDP programs are shown in Figures 2.2, 2.3, and 2.4 (pages 68–69), respectively.

PROGRAM 2.9

SPSSˣ Statements That Produce a Histogram and Descriptive Statistics for the 1,000 CPU Times of Appendix IV

Command line	Column 1 ↓									
1	DATA LIST FREE/TIME} Data entry instruction									
2	FREQUENCIES VARIABLES=TIME/									
3	HISTOGRAM=PERCENT INCREMENT(1.5)									
4	STATISTICS=ALL									
5	BEGIN DATA									
6	1.86	3.49	2.63	3.49	1.69	1.83	0.81	4.70	0.85	4.24
.	.	.	.	.	.	.	.	.	.	.
.	.	.	.	.	.	.	.	.	.	.
.	.	.	.	.	.	.	.	.	.	.
105	2.02	2.82	2.19	5.13	2.43	0.23	0.43	6.22	3.98	2.05
106	END DATA									

Lines 2–4: Statistical analysis instructions: Generate the histogram and descriptive statistics

Lines 6–105: Data values (ten observations per line)

PROGRAM 2.10

Minitab Statements That Produce a Histogram and Descriptive Statistics for the 1,000 CPU Times of Appendix IV

Command line										
1	SET CPU TIMES IN C1} Data entry instruction									
2	1.86	3.49	2.63	3.49	1.69	1.83	0.81	4.70	0.85	4.24
.	.	.	.	.	.	.	.	.	.	.
.	.	.	.	.	.	.	.	.	.	.
.	.	.	.	.	.	.	.	.	.	.
101	2.02	2.82	2.19	5.13	2.43	0.23	0.43	6.22	3.98	2.05
102	NAME C1 = 'TIME'} Data entry instruction									
103	HISTOGRAM OF C1, FIRST MIDPOINT AT .755, INTERVAL WIDTH 1.5									
104	DESCRIBE CPU TIMES IN C1									
105	STOP									

Lines 2–101: Data values (ten observations per line)

Lines 103–104: Statistical analysis instructions

*The FREE format in SPSSˣ, the SET command in Minitab, and the STREAM and SLASH formats in BMDP permit multiple observations to be read per line. Consult the references at the end of the chapter for further information on these commands.

PROGRAM 2.11 (a)   BMDP Statements That Produce a Histogram for the 1,000 CPU Times of Appendix IV

Command line	Column 1 ↓		
1	BMDP 5D} Statistical analysis instruction (program ID command): Generate a histogram		
2	/ PROBLEM	TITLE IS 'HISTOGRAM OF CPU TIMES',	
3	/ INPUT	VARIABLE IS 1.	
4		FORMAT IS STREAM.	
5	/ VARIABLE	NAME IS TIME.	
6	/ GROUP	CUTPOINTS(TIME) = 1.505 TO 19.505 BY 1.5.	Data entry
7		NAMES(TIME) = '1.505', '3.005', '4.505',	instructions
8		'6.005', '7.505', '9.005', '10.505',	
9		'12.005', '13.505', '15.005',	
10		'16.505', '18.005', '19.505',	
11		'21.005'.	
12	/ PLOT	VARIABLE IS TIME.	
13		SCALE = 0,4.	
14	/ END		
15	1.86 3.49 2.63 3.49 1.69 1.83 0.81 4.70 0.85 4.24		
·	· · · · · · · · · ·		Data values
·	· · · · · · · · · ·		(ten observations
·	· · · · · · · · · ·		per line)
114	2.02 2.82 2.19 5.13 2.43 0.23 0.43 6.22 3.98 2.05		
115	/END		

(b)   BMDP Statements That Produce Descriptive Statistics for the 1,000 CPU Times of Appendix IV

Command line	Column 1 ↓		
1	BMDP 1D} Statistical analysis instruction (Program ID command): Descriptive statistics		
2	/ PROBLEM	TITLE IS 'DESCRIPTIVE STATISTICS FOR CPU TIMES'.	Data
3	/ INPUT	VARIABLE IS 1.	entry
4		FORMAT IS STREAM	instructions
5	/ VARIABLE	NAME IS TIME.	
6	/ END		
7	1.86 3.49 2.63 3.49 1.69 1.83 0.81 4.70 0.85 4.24		
·	· · · · · · · · · ·		Data values
·	· · · · · ` · · · · ·		(ten observations
·	· · · · · · · · · ·		per line)
106	2.02 2.82 2.19 5.13 2.43 0.23 0.43 6.22 3.98 2.05		
107	/END		

FIGURE 2.2

Histogram for the 1,000
CPU Times of Appendix
IV, Generated by SPSS[x]

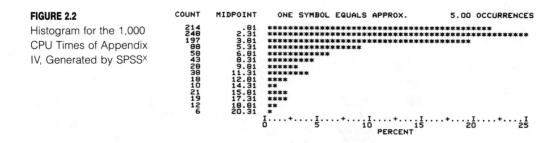

FIGURE 2.3

Histogram for the 1,000 CPU Times of Appendix IV, Generated by Minitab

```
TIME
EACH * REPRESENTS  10 OBSERVATIONS
MIDDLE OF    NUMBER OF
INTERVAL     OBSERVATIONS
   0.75         203     *********************
   2.26         252     **************************
   3.76         199     ********************
   5.26          91     **********
   6.76          53     ******
   8.26          49     *****
   9.76          26     ***
  11.26          40     ****
  12.76          19     **
  14.26          10     *
  15.76          21     ***
  17.25          19     **
  18.75           9     *
  20.25           9     *
```

FIGURE 2.4

Histogram for the 1,000 CPU Times of Appendix IV, Generated by BMDP

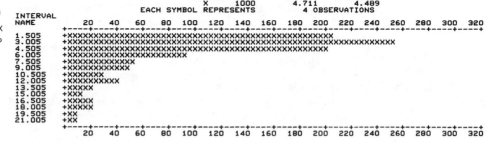

EXERCISES 2.20–2.25

2.20 Use the SPSS[X] computer package to generate a relative frequency histogram and numerical descriptive measures for the data of Exercise 1.12.

2.21 Use the SPSS[X] computer package to generate a relative frequency histogram and numerical descriptive measures for the data of Exercise 1.15.

2.22 Use the Minitab computer package to generate a relative frequency histogram and numerical descriptive measures for the data of Exercise 1.12.

2.23 Use the Minitab computer package to generate a relative frequency histogram and numerical descriptive measures for the data of Exercise 1.15.

2.24 Use the BMDP computer package to generate a relative frequency histogram and numerical descriptive measures for the data of Exercise 1.12.

2.25 Use the BMDP computer package to generate a relative frequency histogram and numerical descriptive measures for the data of Exercise 1.15.

S E C T I O N   2.8

SUMMARY

This chapter introduces four statistical computer program packages that are widely used at university computing centers, either on mainframe computers or personal computers (PCs): **SAS**, **SPSS**[X], **Minitab**, and **BMDP**. Each of the packages utilizes **data entry instructions**, **input data values**, and **statistical analysis instructions**. The data entry control statements for each package are presented in such a manner that students with little or no computer experience can, with the help of their instructor, enter and store data onto the computer.

In Sections 2.6 and 2.7 we give the **statistical analysis commands** that will generate a relative frequency histogram and produce simple descriptive statistics for each of the computer packages.

The four computer program packages can perform many statistical procedures not described in this text and, in addition, they are periodically revised. Consequently, if you use a particular software package, you may want to refer to the manual that explains how to use it. The manuals are readily available and are listed in the references.

REFERENCES

BMDP User's Digest, 2nd ed. MaryAnn Hill (ed.). Los Angeles: BMDP Statistical Software, 1982.

BMDPC: User's Guide to BMDP on the IBM PC. Los Angeles: BMDP Statistical Software.

Dixon, W. J., Brown, M. B., Engelman, L., Frane, J. W., Hill, M. A., Jennrich, R. I., and Toporek, J. D. *BMDP Statistical Software*, 1985 ed. Berkeley: University of California Press.

Norusis, M. J. *The SPSS Guide to Data Analysis*. 1986 ed. SPSS, Inc., Suite 3000, 444 N. Michigan Avenue, Chicago, Ill. 60611.

Norusis, M. J. *SPSS/PC+: SPSS for the IBM PC/XT/AT*. 1986 ed. SPSS, Inc., Suite 3000, 444 N. Michigan Avenue, Chicago, Ill. 60611.

Ryan, T. A., Joiner, B. L., and Ryan, B. F. *Minitab Reference Manual*. Minitab Project, University Park, Pa., 1985.

Ryan, T. A., Joiner, B. L., and Ryan, B. F. *Minitab Student Handbook*, 2nd ed. Boston: Duxbury, 1985.

SAS Procedures Guide for Personal Computers, Version 6 ed. 1986. SAS Institute, Inc., Box 8000, Cary, N.C. 27511.

SAS User's Guide: Basics. Version 5 ed. 1985. SAS Institute, Inc., Box 8000, Cary, N.C. 27511.

SAS User's Guide: Statistics. Version 5 ed. 1985. SAS Institute, Inc., Box 8000, Cary, N.C. 27511.

SPSS^X User's Guide. 1983 ed. SPSS, Inc., Suite 3000, 444 N. Michigan Ave., Chicago, Ill. 60611.

C H A P T E R 3

OBJECTIVE

To present an introduction to the theory of probability and to suggest the role that probability will play in statistical inference.

CONTENTS

PROBABILITY

| | | | | | | | | | | |

SECTION 3.1

THE ROLE OF PROBABILITY IN STATISTICS

You will recall that statistics is concerned with decisions about a population based on sample information. Understanding how this is accomplished will be easier if you understand the relationship between population and sample. This understanding is enhanced by reversing the statistical procedure of making inferences from sample to population. In this chapter we assume the population *known* and calculate the chances of obtaining various samples from the population. Thus, probability is the "reverse" of statistics: In probability we use the population information to infer the probable nature of the sample.

Probability plays an important role in decision-making. To illustrate, suppose you have an opportunity to invest in an oil exploration company. Past records show that for ten out of ten previous oil drillings (a sample of the company's experiences), all ten resulted in dry wells. What do you conclude? Do you think the chances are better than 50–50 that the company will hit a producing well? Should you invest in this company? We think your answer to these questions will be an emphatic "No." If the company's exploratory prowess is sufficient to hit a producing well 50% of the time, a record of ten dry wells out of ten drilled is an event that is just too improbable. Do you agree?

As another illustration, suppose you are playing poker with what your opponents assure you is a well-shuffled deck of cards. In three consecutive 5-card hands, the person on your right is dealt 4 aces. Based on this sample of three deals, do you think the cards are being adequately shuffled? Again, we think your answer will be "No" and that you will reach this conclusion because dealing three hands of 4 aces is just too improbable assuming that the cards were properly shuffled.

Note that the decisions concerning the potential success of the oil drilling company and the decision concerning the card shuffling were both based on probabilities, namely the probabilities of certain sample results. Both situations were contrived so that you could easily conclude that the probabilities of the sample results were small. Unfortunately, the probabilities of many observed sample results are not so easy to evaluate intuitively. For these cases we will need the assistance of a theory of probability.

| | | | | | | | | | | |

SECTION 3.2

EVENTS, SAMPLE SPACES, AND PROBABILITY

We will begin the discussion of probability with simple examples that are easily described, thus eliminating any discussion that could be distracting. With the aid of simple examples, important definitions are introduced and the notion of probability is more easily developed.

Suppose a coin is tossed once and the up face of the coin is recorded. This is an **observation**, or **measurement**. Any process of making an observation is called an **experiment**. Our definition of experiment is broader than that used in the physical sciences, where you would picture test tubes, microscopes, etc. Other practical examples of statistical experiments are recording whether a customer prefers one of two brands of electronic calculators, recording a voter's opinion on an important political issue, measuring the amount of dissolved oxygen in a polluted river, observing the closing price of a stock, counting the number of errors in an inventory, and observing the fraction of insects killed by a new insecticide. This list of statistical experiments could be continued ad infinitum, but the point is that our definition of experiment is very broad.

DEFINITION 3.1

An **experiment** is the process of making an observation or taking a measurement.

Consider another simple experiment consisting of tossing a die and observing the number on the up face of the die. The six basic possible outcomes to this experiment are:

1. Observe a 1
2. Observe a 2
3. Observe a 3
4. Observe a 4
5. Observe a 5
6. Observe a 6

Note that if this experiment is conducted once, *you can observe one and only one of these six basic outcomes*. The distinguishing feature of these outcomes is that these possibilities *cannot be decomposed* into any other outcomes. These very basic possible outcomes to an experiment are called **simple events**.

DEFINITION 3.2

A **simple event** is an outcome of an experiment that cannot be decomposed.

EXAMPLE 3.1

Two coins are tossed and the up faces of both coins are recorded. List all the simple events for this experiment.

SOLUTION

Even for a seemingly trivial experiment, we must be careful when listing the simple events. At first glance the basic outcomes seem to be Observe two heads, Observe two tails, Observe one head and one tail. However, further reflection reveals that the last of these, Observe one head and one tail, can be decomposed into Head on coin 1, Tail on coin 2 and Tail on coin 1, Head on coin 2.* Thus, the simple events are as follows:

1. Observe *HH*
2. Observe *HT*
3. Observe *TH*
4. Observe *TT*

(where *H* in the first position means "Head on coin 1," *H* in the second position means "Head on coin 2," etc.). ∎

---

*Even if the coins are identical in appearance, there are, in fact, two distinct coins. Thus, the designation of one coin as "coin 1" and the other as "coin 2" is legitimate in any case.

We will often wish to refer to the collection of all the simple events of an experiment. This collection will be called the **sample space** of the experiment. For example, there are six simple events in the sample space associated with the die tossing experiment. The sample spaces for the experiments discussed thus far are shown in Table 3.1.

DEFINITION 3.3

The **sample space** of an experiment is the collection of all its simple events.

TABLE 3.1
Experiments and Their
Sample Spaces

Experiment: Observe the up face on a coin.

Sample space: 1. Observe a head
          2. Observe a tail

This sample space can be represented in set notation as a set containing two simple events

   S: {H, T}

where H represents the simple event Observe a head and T represents the simple event Observe a tail.

Experiment: Observe the up face on a die.

Sample space: 1. Observe a 1
          2. Observe a 2
          3. Observe a 3
          4. Observe a 4
          5. Observe a 5
          6. Observe a 6

This sample space can be represented in set notation as a set of six simple events

   S: {1, 2, 3, 4, 5, 6}

Experiment: Observe the up faces on two coins.

Sample space: 1. Observe HH
          2. Observe HT
          3. Observe TH
          4. Observe TT

This sample space can be represented in set notation as a set of four simple events

   S: {HH, HT, TH, TT}

Just as graphs are useful in describing sets of data, a pictorial method for presenting the sample space and its simple events will often be useful. Figure 3.1 shows such a representation for each of the experiments in Table 3.1. In each case, the sample space is shown as a closed figure, labeled S, containing a set of points, called **sample points**, with each point representing one simple event. Note that the number of sample points in a sample space S is equal to the number of simple events associated with the respective experiment: two for the coin toss, six for the die toss, and four for the two-coin toss. These graphical representations are called **Venn diagrams**.

FIGURE 3.1

Venn Diagrams for the Three Experiments from Table 3.1

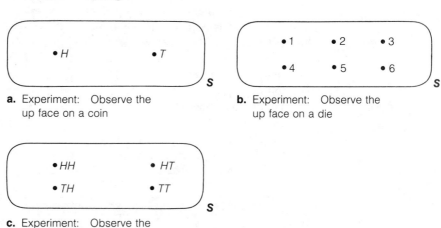

a. Experiment: Observe the up face on a coin

b. Experiment: Observe the up face on a die

c. Experiment: Observe the up faces on two coins

Now that we have defined simple events as the basic outcomes of the experiment and the sample space as the collection of all the simple events, we are prepared to discuss the probabilities of simple events. You have undoubtedly used the term *probability* and have some intuitive idea about its meaning. Probability is generally used synonymously with "chance," "odds," and similar concepts. We will begin our treatment of probability using these informal concepts and then solidify what we mean later. For example, if a fair coin is tossed, we might reason that both the simple events, Observe a head and Observe a tail, have the same chance of occurring. Thus, we might state that "the probability of observing a head is 50%," or "the odds of seeing a head are 50–50." Both these statements are based on an informal knowledge of probability.

DEFINITION 3.4

The **probability** of a simple event is a number that measures the likelihood that the event will occur when the experiment is performed. For a simple event E, we denote the probability of E as P(E).

The probability of a simple event is usually taken to be the relative frequency of the occurrence of a simple event in a very long series of repetitions of the experiment. Or, when this information is not available, we select the number based on experience. For example, if we are assigning probabilities to the two simple events in the coin toss experiment (Observe a head and Observe a tail), we might reason that if we toss a balanced coin a very large number of times, the simple events Observe a head (H) and Observe a tail (T) will occur with the same relative frequency of .5. Thus, the probability of each simple event is .5, i.e., $P(H) = P(T) = .5$.

In other cases we may choose the probability based on general information about the experiment. For example, if the experiment is observing whether a venture succeeds or fails (the simple events), we may assess the probability of success by considering the personnel managing the venture, the success of similar ventures, and any other information deemed pertinent. If we finally decide that the venture has an 80% chance of succeeding, we assign a probability of .8 to the simple event Success. We hope that .8 is a reasonably accurate measure of the likelihood of the occurrence of the simple event Success. If it is not, we may be misled on any decisions based on this probability or based on any calculations in which it appears.

No matter how you assign probabilities to the simple events of an experiment, the probabilities assigned must obey the two rules given in the box.

RULES FOR ASSIGNING PROBABILITIES TO SIMPLE EVENTS

Let $E_1, E_2, \ldots, E_k$ be the simple events in a sample space.

1. All simple event probabilities *must* lie between 0 and 1:

$$0 \leq P(E_i) \leq 1 \quad \text{for } i = 1, 2, \ldots, k$$

2. The sum of the probabilities of all the simple events within a sample space must be equal to 1:

$$\sum_{i=1}^{k} P(E_i) = 1$$

Sometimes we are interested in the occurrence of any one of a collection of simple events. For example, in the die tossing experiment of Table 3.1, we may be interested in observing an odd number on the die. This will occur if any one of the following three simple events occurs:

1. Observe a 1
2. Observe a 3
3. Observe a 5

In fact, the event Observe an odd number is clearly defined if we specify the collection of simple events that imply its occurrence. Such specific collections of simple events are called **events**.

DEFINITION 3.5

An **event** is a specific collection of simple events.

The probability of an event is computed by summing the probabilities of the simple events that comprise it. This rule agrees with the relative frequency concept of probability, as Example 3.2 illustrates.

THE PROBABILITY OF AN EVENT
The **probability of an event** A is equal to the sum of the probabilities of the simple events in event A.

EXAMPLE 3.2

Consider the experiment of tossing two coins. Suppose the coins are *not* balanced and the correct probabilities associated with the simple events are as follows:

SIMPLE EVENT	PROBABILITY
HH	$\frac{4}{9}$
HT	$\frac{2}{9}$
TH	$\frac{2}{9}$
TT	$\frac{1}{9}$

Define the following events:

A: {Observe exactly one head}

B: {Observe at least one head}

Calculate the probability of A and the probability of B.

SOLUTION

Examine the probabilities assigned to the simple events. These probabilities imply that, if the coin tossing experiment is repeated a very large number of times, approximately $\frac{4}{9}$ of the outcomes will result in HH, $\frac{2}{9}$ in HT, $\frac{2}{9}$ in TH, and $\frac{1}{9}$ in TT. Since the event A: {Observe exactly one head} will occur if either of the two simple events HT or TH occurs (see Figure 3.2), then approximately $\frac{2}{9} + \frac{2}{9} = \frac{4}{9}$ of the large number of experiments will result in event A. This additivity of the relative frequencies of simple events is consistent with our rule for finding $P(A)$:

$$P(A) = P(HT) + P(TH) = \frac{2}{9} + \frac{2}{9} = \frac{4}{9}$$

FIGURE 3.2
Coin Tossing Experiment Showing Events A and B as Collections of Simple Events

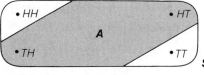

a. Event A

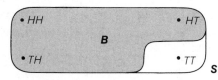

b. Event B

Applying this rule to find $P(B)$, we note that event B contains the simple events HH, HT, and TH—that is, B will occur if any one of these three simple events occurs. Therefore,

$$P(B) = P(HH) + P(HT) + P(TH) = \frac{4}{9} + \frac{2}{9} + \frac{2}{9} = \frac{8}{9}$$

We can now summarize the steps for calculating the probability of any event:*

STEPS FOR CALCULATING PROBABILITIES OF EVENTS

1. Define the experiment, i.e., describe the process used to make an observation and the type of observation that will be recorded.
2. List the simple events.
3. Assign probabilities to the simple events.
4. Determine the collection of simple events contained in the event of interest.
5. Sum the simple event probabilities to get the event probability.

EXAMPLE 3.3

A quality control engineer must decide whether an assembly line that produces manufactured items is "out of control"—that is, producing defective items at a higher rate than usual. At this stage of our study, we do not have the tools to solve this problem, but we can say that one of the important factors affecting the solution is the proportion of defectives manufactured by the line. To illustrate, what is the probability that an item manufactured by the line will be defective? What is the probability that the next two items produced by the line will be defective? What is the probability for the general case of k items? Explain how you might solve this problem.

SOLUTION

STEP 1 Define the experiment. The experiment corresponding to the inspection of a single item is identical in underlying structure to the coin tossing experiment illustrated in Figure 3.1a. An item, either a nondefective (call this a head) or a defective (call this a tail), is observed and its operating status is recorded.

Experiment: Observe the operating status of a single manufactured item.

STEP 2 List the simple events. There are only two possible outcomes of the experiment. These simple events are:

Simple events: 1. N: {Item is nondefective}
2. D: {Item is defective}

STEP 3 Assign probabilities to the simple events. The difference between this problem and the coin tossing problem becomes apparent when we attempt to assign probabilities to the two simple events. What probability should we assign to the simple event D? Some people might say .5, as for the coin tossing experiment, but you can see that finding $P(D)$, the probability of simple event D, is not so easy. Suppose that when the assembly line is in control, 10% of the items

*A thorough treatment of this topic can be found in Feller (1968).

produced will be defective. Then, at first glance, it would appear that $P(D)$ is .10. But this may not be correct, because the line may be out of control, producing defectives at a higher rate. So, the important point to note is that this is a case where equal probabilities are not assigned to the simple events. How can we find these probabilities? A good procedure might be to monitor the assembly line for a period of time, and record the number of defective and nondefective items produced. Then the proportions of the two types of items could be used to approximate the probabilities of the two simple events.

We could then continue with steps 4 and 5 to calculate any probability of interest for this experiment with two simple events.

The experiment, assessing the operating status of two items, is identical to the experiment of Example 3.2, tossing two coins, except that the probabilities of the simple events are not the same. We will learn how to find the probabilities of the simple events for this experiment, or for the general case of k items, in Section 3.7. ∎

For the experiments discussed thus far, listing the simple events has been easy. For more complex experiments, the number of simple events may be so large that listing them is impractical. In solving probability problems for experiments with many simple events, we employ the same principles as for experiments with few simple events. The only difference is that we need **counting rules** for determining the number of simple events without actually enumerating all of them. In Section 3.3, we present several of the more useful counting rules.

EXAMPLE 3.4

A computer programmer must select three jobs from among five jobs awaiting the programmer's attention. If, unknown to the programmer, the jobs vary in the length of programming time required, what is the probability that:

a. The programmer selects the two jobs that require the least amount of time?
b. The programmer selects the three jobs that require the most time?

SOLUTION

STEP 1 The experiment consists of selecting three jobs from among the five that are available.

STEP 2 We will denote the available jobs by the symbols $J_1, J_2, \ldots, J_5$, where J_1 is the shortest job and J_5 is the longest. The notation $J_i J_j$ will denote the selection of jobs J_i and J_j. For example, $J_1 J_3$ denotes the selection of jobs J_1 and J_3. Then the ten simple events associated with the experiment are as follows:

SIMPLE EVENT	PROBABILITY	SIMPLE EVENT	PROBABILITY
$J_1 J_2 J_3$	$\frac{1}{10}$	$J_1 J_4 J_5$	$\frac{1}{10}$
$J_1 J_2 J_4$	$\frac{1}{10}$	$J_2 J_3 J_4$	$\frac{1}{10}$
$J_1 J_2 J_5$	$\frac{1}{10}$	$J_2 J_3 J_5$	$\frac{1}{10}$
$J_1 J_3 J_4$	$\frac{1}{10}$	$J_2 J_4 J_5$	$\frac{1}{10}$
$J_1 J_3 J_5$	$\frac{1}{10}$	$J_3 J_4 J_5$	$\frac{1}{10}$

STEP 3 If we assume that the selection of any set of three jobs is as likely as any other, then the probability of each of the ten simple events is $\frac{1}{10}$.

STEP 4 Define the events A and B as follows:

A: {The programmer selects the two jobs that require the least amount of time}

B: {The programmer selects the three jobs that require the most time}

Event A will occur for any simple events in which jobs J_1 and J_2 are selected—namely, the three simple events $J_1J_2J_3$, $J_1J_2J_4$, and $J_1J_2J_5$. Similarly, the event B is comprised of the single event $J_3J_4J_5$.

STEP 5 We now sum the probabilities of the simple events in A and B to obtain:

$$P(A) = P(J_1J_2J_3) + P(J_1J_2J_4) + P(J_1J_2J_5)$$
$$= \frac{1}{10} + \frac{1}{10} + \frac{1}{10} = \frac{3}{10}$$

and

$$P(B) = P(J_3J_4J_5) = \frac{1}{10}$$ ∎

EXERCISES 3.1–3.5

3.1 Under "Operation Greenback," United States customs officials are using computers to help trace large sums of money from the illegal narcotics trade to the kingpins—those people who do not actually handle the drugs but who realize profits from drug sales. The computer monitors cash transactions at state and federal banks and flags all unusual transactions. Because some transactions are completely legitimate, investigators cross-reference the list of possible offenders with a list of suspected or convicted criminals. When a "hit" or match is made, the name is investigated by federal law enforcement agencies.

Suppose the computer detects four unusual cash transactions at a Miami bank. These four names are then cross-referenced with a list of suspected Miami drug dealers. Assume that the simple events in the sample space are equally likely.
a. What is the probability of obtaining exactly two matches?
b. What is the probability of obtaining at least one match?
c. Do you think that the assumption of equally likely outcomes is reasonable? Explain.

3.2 The Commission of the European Communities initiated a research program to determine the influence of traffic noise on sleep, subjective assessment, and psychomotor performance. For one portion of the study, a team of German acoustical engineers monitored the sleep of ten couples (one male and one female per couple) during twelve consecutive nights. All ten couples slept under usual conditions on seven of the nights. (This represents the *control phase* of the study.) For the other five nights (the *experimental phase*), the ten couples were divided into two groups of equal size. One group slept with the windows open and the other group slept with earplugs. The experimental setup is described in the table.

NIGHTS	CONTROL PHASE 1 2 3 4 5	EXPERIMENTAL PHASE 6 7 8 9 10	CONTROL PHASE 11 12
Earplugs group (couples 1, 4, 6, 9, 10)	Without earplugs Windows open	With earplugs Windows open	Without earplugs Windows open
Windows group (couples 2, 3, 5, 7, 8)	Windows closed No earplugs	Windows open No earplugs	Windows closed No earplugs

Source: Griefahn, B. and Gros, E. "Noise and Sleep at Home, a Field Study on Primary and After-effects." *Journal of Sound and Vibration*, Vol. 105, No. 3, Mar. 1986, p. 376 (Figure 2).

Suppose we randomly select one couple from the experiment on one randomly selected night and note whether or not the couple is wearing earplugs and whether the windows are open or closed.

a. List the simple events for the experiment.

b. Assign probabilities to the simple events.

c. What is the probability that the couple is wearing earplugs?

d. What is the probability that the windows are closed?

3.3 A manufacturer of industrial products has decided to computerize its operations and must select two of the seven vendors available to supply packaged software for the firm's computer system. Unknown to the firm, three of the seven software vendors will be unable to provide reliable service and support for the system. Find the probability that:

a. Both vendors selected will be able to provide service and support.

b. Neither vendor selected will be able to provide service and support.

c. At least one vendor selected will be able to provide service and support.

3.4 An improved method for measuring the electrical resistivity of concrete has been developed which eliminates difficulties due to polarization effects and capacitive resistance (*Magazine of Concrete Research*, Dec. 1985). The method was tested on concrete specimens with different water–cement mixes. Three different water weight ratios (40%, 45%, and 50%) and three different mixes of cement (300, 350, and 400 kilograms per cubic meter) were examined.

a. List all possible outcomes (water–cement mixes) for this experiment.

b. Suppose we determine the water–cement mix that yields the highest electrical resistivity. Before the experiment is performed, should equal probabilities be assigned to the simple events? Why or why not?

3.5 The YES/MVS (Yorktown Expert System/MVS Manager) is an experimental expert system designed to exert active control over a computer system and provide advice to computer operators. YES/MVS is designed with a knowledge base consisting of 548 rules that are triggered in response to messages or queries from the computer operator. The table on page 82 gives the number of rules allocated to different subdomains of the operator's actions. Periodically, the rules in the YES/MVS knowledge base are tested and adjusted, if necessary. Suppose a rule is selected at random for testing and its type (operator action/query) noted.

a. List the simple events for this experiment.

OPERATOR'S ACTION/QUERY	NUMBER OF RULES
Batch scheduling	139
JES queue space	104
C-to-C links	68
Hardware errors	87
SMF management	25
Quiesce and IPL	52
Performance	41
Background monitor	32
TOTAL	548

Source: Ennis, R. L. et al., "A Continuous Real-time Expert System for Computer Operations." *IBM Journal of Research and Development*, Vol. 30, No. 1, Jan. 1986, p. 19. Copyright 1986 by International Business Machines Corporation; reprinted with permission.

b. Assign probabilities to the simple events based on the information contained in the table.
c. What is the probability the rule is a C-to-C link or hardware error rule?
d. What is the probability the rule is not a performance rule?

SECTION 3.3

SOME COUNTING RULES

In Section 3.2 we pointed out that experiments sometimes have so many simple events that it is impractical to list them all. However, many of these experiments possess simple events with identical characteristics. If you can develop a **counting rule** to count the number of simple events for such an experiment, it can be used to aid in the solution of the problems.

EXAMPLE 3.5

A product (e.g., hardware for a computer system) can be shipped by four different airlines and each airline can ship via three different routes. How many distinct ways exist to ship the product?

SOLUTION

A pictorial representation of the different ways to ship the product will aid in counting them. This representation, called a **decision tree**, is shown in Figure 3.3. At the starting point (stage 1), there are four choices—the different airlines—to begin the journey. Once we have chosen an airline (stage 2), there are three choices—the different routes—to complete the shipment and reach the final destination. Thus, the decision tree clearly shows that there are $(4)(3) = 12$ distinct ways to ship the product.

The method of solving Example 3.5 can be generalized to any number of stages with sets of different elements. The framework is provided by the **multiplicative rule**.

FIGURE 3.3

Decision Tree for
Shipping Problem

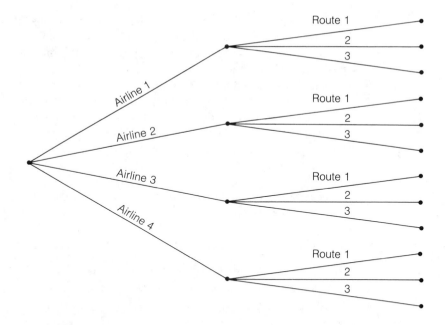

Route 1
2
3

Airline 1

Route 1
2
3

Airline 2

Airline 3

Route 1
2
3

Airline 4

Route 1
2
3

■

THEOREM 3.1

The Multiplicative Rule   You have k sets of elements, n_1, in the first set, n_2 in the second set, . . . , and n_k in the kth set. Suppose you want to form a sample of k elements by taking one element from each of the k sets. The number of different samples that can be formed is the product

$$n_1 n_2 n_3 \cdot \cdot \cdot \cdot \cdot n_k$$

Outline of proof of Theorem 3.1   The proof of Theorem 3.1 can be obtained most easily by examining Table 3.2. Each of the pairs that can be formed from two sets of elements—$a_1, a_2, \ldots, a_{n_1}$ and $b_1, b_2, \ldots, b_{n_2}$—corresponds to a cell of Table 3.2.

TABLE 3.2

Pairings of $a_1, a_2, \ldots, a_{n_1}$
and $b_1, b_2, \ldots, b_{n_2}$

	b_1	b_2	b_3	$\cdots$	b_{n_2}
a_1	$a_1 b_1$	$a_1 b_2$	$a_1 b_3$	$\cdots$	$a_1 b_{n_2}$
a_2	$a_2 b_1$	$\cdots$	$\cdots$	$\cdots$	$\cdots$
a_3	$a_3 b_1$	$\cdots$	$\cdots$	$\cdots$	$\cdots$
$\vdots$	$\vdots$	$\vdots$	$\vdots$	$\vdots$	$\vdots$
a_{n_1}	$a_{n_1} b_1$	$\cdots$	$\cdots$	$\cdots$	$a_{n_1} b_{n_2}$

Since the table contains n_1 rows and n_2 columns, there will be n_1n_2 pairs corresponding to each of the n_1n_2 cells of the table. To extend the proof to the case in which $k = 3$, note that the number of triplets that can be formed from three sets of elements—$a_1, a_2, \ldots, a_{n_1}$; $b_1, b_2, \ldots, b_{n_2}$; and $c_1, c_2, \ldots, c_{n_3}$— is equal to the number of pairs that can be formed by associating one of the a_ib_j pairs with one of the c elements. Since there are (n_1, n_2) of the a_ib_j pairs and n_3 of the c elements, we can form $(n_1n_2)n_3 = n_1n_2n_3$ triplets consisting of one a element, one b element, and one c element. The proof of the multiplicative rule for any number, say k, of sets is obtained by mathematical induction. We leave this proof as an exercise for the student.

EXAMPLE 3.6

There are 20 candidates for three different mechanical engineer positions, E_1, E_2, and E_3. How may different ways could you fill the positions?

SOLUTION

For this example, there are $k = 3$ sets of elements corresponding to:

Set 1:   Candidates available to fill position E_1
Set 2:   Candidates remaining (after filling E_1) that are available to fill E_2
Set 3:   Candidates remaining (after filling E_1 and E_2) that are available to fill E_3

The numbers of elements in the sets are $n_1 = 20$, $n_2 = 19$, $n_3 = 18$. Therefore, the number of different ways of filling the three positions is

$$n_1n_2n_3 = (20)(19)(18) = 6{,}840 \qquad \blacksquare$$

EXAMPLE 3.7

Consider an experiment that consists of tossing a coin ten times. Show that there are $2^{10} = 1{,}024$ simple events for this experiment.

SOLUTION

There are $k = 10$ sets of elements for this experiment. Each set contains two elements, a head and a tail. Thus, there are

$$(2)(2)(2)(2)(2)(2)(2)(2)(2)(2) = 2^{10} = 1{,}024$$

different outcomes (simple events) of this experiment. $\blacksquare$

EXAMPLE 3.8

Suppose there are five different space flights scheduled, each requiring one astronaut. Assuming that no astronaut can go on more than one space flight, in how many different ways can five of the country's top 100 astronauts be assigned to the five space flights?

SOLUTION

We can solve this problem by using the multiplicative rule. The entire set of 100 astronauts is available for the first flight, and after the selection of one astronaut for that flight, 99 are available for the second flight, etc. Thus, the total number of different ways of choosing five astronauts for the five space flights is

$$n_1n_2n_3n_4n_5 = (100)(99)(98)(97)(96) = 9{,}034{,}502{,}400 \qquad \blacksquare$$

The arrangement of elements in a distinct order is called a **permutation**. Thus, from Example 3.8, we see that there are more than 9 billion different *permutations* of five elements (astronauts) drawn from a set of 100 elements!

THEOREM 3.2

Permutations Rule   Given a single set of N distinctly different elements, you wish to select n elements from the N and arrange them within n positions. The number of different permutations of the N elements taken n at a time is denoted by P_n^N and is equal to

$$P_n^N = N(N - 1)(N - 2) \cdots \cdot (N - n + 1) = \frac{N!}{(N - n)!}$$

where $n! = n(n - 1)(n - 2) \cdots \cdot (3)(2)(1)$ and is called n **factorial**. (Thus, for example, $5! = 5 \cdot 4 \cdot 3 \cdot 2 \cdot 1 = 120$.) The quantity $0!$ is defined to be equal to 1.

Proof of Theorem 3.2   The proof of Theorem 3.2 is a generalization of the solution to Example 3.8. There are N ways of filling the first position. After it is filled, there are $N - 1$ ways of filling the second, $N - 2$ ways of filling the third, . . . , and $(N - n + 1)$ ways of filling the nth position. We apply the multiplicative rule to obtain

$$P_n^N = (N)(N - 1)(N - 2) \cdots \cdot (N - n + 1) = \frac{N!}{(N - n)!}$$

EXAMPLE 3.9

Consider the following transportation engineering problem: You want to drive, in sequence, from a starting point to each of five cities, and you want to compare the distances and average speeds of the different routings. How many different routings would have to be compared?

SOLUTION

Denote the cities as $C_1, C_2, \ldots, C_5$. Then a route moving from the starting point to C_2 to C_1 to C_3 to C_4 to C_5 would be represented as $C_2C_1C_3C_4C_5$. The total number of routings would equal the number of ways you could rearrange the $N = 5$ cities in $n = 5$ positions. This number is

$$P_n^N = P_5^5 = \frac{5!}{(5 - 5)!} = \frac{5!}{0!} = \frac{5 \cdot 4 \cdot 3 \cdot 2 \cdot 1}{1} = 120$$

(recall that $0! = 1$).   ■

EXAMPLE 3.10

There are four system analysts and you must assign three to job 1 and one to job 2. In how many different ways can you make this assignment?

SOLUTION

To begin, suppose that each system analyst is to be assigned to a distinct job. Then, using the multiplicative rule, we obtain $(4)(3)(2)(1) = 24$ ways of assigning the system analysts to four distinct jobs. The 24 ways are listed in four groups in Table 3.3 (where ABCD indicates that system analyst A was assigned the first job; system analyst B, the second; etc.).

TABLE 3.3

Ways of Assigning System Analysts to Four Distinct Jobs

GROUP 1	GROUP 2	GROUP 3	GROUP 4
ABCD	ABDC	ACDB	BCDA
ACBD	ADBC	ADCB	BDCA
BACD	BADC	CADB	CBDA
BCAD	BDAC	CDAB	CDBA
CABD	DABC	DACB	DBCA
CBAD	DBAC	DCAB	DCBA

Now, suppose the first three positions represent job 1 and the last position represents job 2. We can now see that all the listings in group 1 represent the same outcome of the experiment of interest. That is, system analysts A, B, and C are assigned to job 1 and system analyst D is assigned to job 2. Similarly, group 2 listings are equivalent, as are group 3 and group 4 listings. Thus, there are only four different assignments of four system analysts to the two jobs. These are shown in Table 3.4. ∎

TABLE 3.4

Ways to Assign Three System Analysts to Job 1 and One System Analyst to Job 2

JOB 1	JOB 2
ABC	D
ABD	C
ACD	B
BCD	A

To generalize the result obtained in Example 3.10, we point out that the final result can be found by

$$\frac{(4)(3)(2)(1)}{(3)(2)(1)(1)} = 4$$

The $(4)(3)(2)(1)$ is the number of different ways (*permutations*) the system analysts could be assigned four distinct jobs. The division by $(3)(2)(1)$ is to remove the duplicated permutations resulting from the fact that three system analysts are assigned the same jobs. And the division by (1) is associated with the system analyst assigned to job 2.

THEOREM 3.3

The Partitions Rule There exists a single set of N distinctly different elements and you want to partition them into k sets, the first set containing n_1 elements, the second containing n_2 elements, . . . , and the kth set containing n_k elements. The number of different partitions is

$$\frac{N!}{n_1! n_2! \cdots \cdot n_k!} \quad \text{where } n_1 + n_2 + n_3 + \cdots + n_k = N$$

Proof of Theorem 3.3 Let A equal the number of ways that you can partition N distinctly different elements into k sets. We want to show that

$$A = \frac{N!}{n_1! n_2! \cdots \cdot n_k!}$$

We will find A by writing an expression for arranging N distinctly different elements in N positions. By Theorem 3.2, the number of ways this can be done is

$$P_N^N = \frac{N!}{(N-N)!} = \frac{N!}{0!} = N!$$

But, by Theorem 3.1, P_N^N is also equal to the product

$$P_N^N = N! = (A)\,(n_1!)\,(n_2!)\,\cdots\,(n_k!)$$

where A is the number of ways of partitioning N elements into k groups of $n_1, n_2, \ldots, n_k$ elements, respectively; $n_1!$ is the number of ways of arranging the n_1 elements in group 1; $n_2!$ is the number of ways of arranging the n_2 elements in group 2; $\ldots$; and $n_k!$ is the number of ways of arranging the n_k elements in group k. We obtain the desired result by solving for A:

$$A = \frac{N!}{n_1!n_2!\,\cdots\cdots\,n_k!}$$

EXAMPLE 3.11

You have 12 system analysts and you want to assign three to job 1, four to job 2, and five to job 3. In how many different ways can you make this assignment?

SOLUTION

For this example, $k = 3$ (corresponding to the $k = 3$ different jobs), $N = 12$, $n_1 = 3$, $n_2 = 4$, and $n_3 = 5$. Then the number of different ways to assign the system analysts to the jobs is

$$\frac{N!}{n_1!n_2!n_3!} = \frac{12!}{3!4!5!} = \frac{12 \cdot 11 \cdot 10 \cdot\cdots\cdot 3 \cdot 2 \cdot 1}{(3 \cdot 2 \cdot 1)(4 \cdot 3 \cdot 2 \cdot 1)(5 \cdot 4 \cdot 3 \cdot 2 \cdot 1)} = 27{,}720 \quad \blacksquare$$

EXAMPLE 3.12

How many samples of 4 tin-lead solder joints can be selected from a lot of 25 tin-lead solder joints available for strength tests?

SOLUTION

For this example, $k = 2$ (corresponding to the $n_1 = 4$ solder joints you *do* choose and the $n_2 = 21$ solder joints you *do not* choose) and $N = 25$. Then, the number of different ways to choose the 4 solder joints from 25 is

$$\frac{N!}{n_1!n_2!} = \frac{25!}{(4!)(21!)} = \frac{25 \cdot 24 \cdot 23 \cdot\cdots\cdot 3 \cdot 2 \cdot 1}{(4 \cdot 3 \cdot 2 \cdot 1)(21 \cdot 20 \cdot\cdots\cdot 2 \cdot 1)} = 12{,}650 \quad \blacksquare$$

The special application of the partitions rule illustrated by Example 3.12—partitioning a set of N elements into $k = 2$ groups (the elements that appear in a sample and those that do not)—is very common. Therefore, we give a different name to the rule for counting the number of different ways of partitioning a set of elements into two parts—the **combinations rule**.

THEOREM 3.4

The Combinations Rule A sample of n elements is to be chosen from a set of N elements. Then the number of different samples of n elements that can be selected from N is denoted by $\binom{N}{n}$ and is equal to

$$\binom{N}{n} = \frac{N!}{n!(N-n)!}$$

Note that the order in which the n elements are drawn is not important.

Proof of Theorem 3.4   The proof of Theorem 3.4 follows directly from Theorem 3.3. Selecting a sample of n elements from a set of N elements is equivalent to partitioning the N elements into $k = 2$ groups—the n that are selected for the sample and the remaining $(N - n)$ that are not selected. Therefore, by applying Theorem 3.3 we obtain

$$\binom{N}{n} = \frac{N!}{n!(N-n)!}$$

EXAMPLE 3.13

Five sales engineers will be hired from a group of 100 applicants. In how many ways (*combinations*) can groups of five sales engineers be selected?

SOLUTION

This is equivalent to sampling $n = 5$ elements from a set of $N = 100$ elements. Thus, the number of ways is the number of possible combinations of five applicants selected from 100, or

$$\binom{100}{5} = \frac{100!}{(5!)(95!)} = \frac{100 \cdot 99 \cdot 98 \cdot 97 \cdot 96 \cdot 95 \cdot 94 \cdots \cdot 2 \cdot 1}{(5 \cdot 4 \cdot 3 \cdot 2 \cdot 1)(95 \cdot 94 \cdots \cdot 2 \cdot 1)}$$

$$= \frac{100 \cdot 99 \cdot 98 \cdot 97 \cdot 96}{5 \cdot 4 \cdot 3 \cdot 2 \cdot 1} = 75{,}287{,}520$$

Compare this result with that of Example 3.8, where we found that the number of permutations of 5 elements drawn from 100 was more than 9 billion. *Because the order of the elements does not affect combinations, there are fewer combinations than permutations.* ∎

When working a probability problem, you should carefully examine the experiment to determine whether you can use one or more of the rules we have discussed in this section. We will illustrate in Examples 3.14 and 3.15 how these rules can help solve a probability problem.

EXAMPLE 3.14

A computer rating service is commissioned to rank the top three brands of intelligent terminals. A total of ten brands are to be included in the study.

a. In how many different ways can the computer rating service arrive at the final ranking?

SUMMARY OF COUNTING RULES

1. *Multiplicative rule.* If you are drawing one element from each of k sets of elements, with the sizes of the sets $n_1, n_2, \ldots, n_k$, the number of different results is

$$n_1 n_2 n_3 \cdots \cdots n_k$$

2. *Permutations rule.* If you are drawing n elements from a set of N elements and arranging the n elements in a distinct order, the number of different results is

$$P_n^N = \frac{N!}{(N - n)!}$$

3. *Partitions rule.* If you are partitioning the elements of a set of N elements into k groups consisting of $n_1, n_2, \ldots, n_k$ elements ($n_1 + n_2 + \cdots + n_k = N$), the number of different results is

$$\frac{N!}{n_1! n_2! \cdots \cdots n_k!}$$

4. *Combinations rule.* If you are drawing n elements from a set of N elements without regard to the order of the n elements, the number of different results is

$$\binom{N}{n} = \frac{N!}{n!(N - n)!}$$

[*Note:* The combinations rule is a special case of the partitions rule when $k = 2$.]

---

b. If the rating service can distinguish no difference among the brands and there-fore arrives at the final ranking by chance, what is the probability that company Z's brand is ranked first? In the top three?

SOLUTION

a. Since the rating service is drawing three elements (brands) from a set of ten elements and arranging the three elements in a distinct order, we use the permutations rule to find the number of different results:

$$P_3^{10} = \frac{10!}{(10 - 3)!} = 10 \cdot 9 \cdot 8 = 720$$

b. The steps for calculating the probability of interest are as follows:

Step 1 The experiment is to select and rank three brands of intelligent terminals from ten brands.

Step 2 There are too many simple events to list. However, we know from part **a** that there are 720 different outcomes (i.e., simple events) of this experiment.

Step 3 If we assume the rating service determines the rankings by chance, each of the 720 simple events should have an equal probability of occurrence. Thus,

$$P(\text{Each simple event}) = \frac{1}{720}$$

Step 4 One event of interest to company Z is that its brand receives top ranking. We will call this event A. The list of simple events that result in the occurrence of event A is long, but the *number* of simple events contained in event A is determined by breaking event A into two parts:

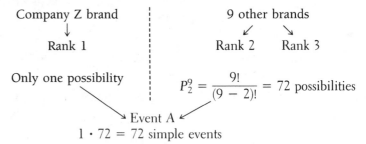

Company Z brand   9 other brands
        ↓   ↙   ↘
     Rank 1   Rank 2   Rank 3

Only one possibility $P_2^9 = \dfrac{9!}{(9-2)!} = 72$ possibilities

Event A
$1 \cdot 72 = 72$ simple events

Thus, event A can occur in 72 different ways.

Now define B as the event that company Z's brand is ranked in the top three. Since event B specifies only that brand Z appear in the top three, we repeat the calculations above, fixing brand Z in position 2 and then in position 3. We conclude that the number of simple events contained in event B is $3(72) = 216$.

Step 5 The final step is to calculate the probabilities of events A and B. Since the 720 simple events are equally likely to occur, we find

$$P(A) = \frac{\text{Number of simple events in } A}{\text{Total number of simple events}} = \frac{72}{720} = \frac{1}{10}$$

Similarly,

$$P(B) = \frac{216}{720} = \frac{3}{10}$$

∎

EXAMPLE 3.15

Refer to Example 3.14. Suppose the computer rating service is to choose the top three intelligent terminals from the group of ten, but is *not to rank the three*.

a. In how many different ways can the rating service choose the three to be designated as top-of-the-line terminals?

b. Assuming that the rating service makes its choice by chance and that company X has two brands in the group of ten, what is the probability that exactly one of the company X brands is selected in the top three? At least one?

SOLUTION

a. The rating service is selecting three elements (brands) from a set of ten elements *without regard to order*, so we can apply the combinations rule to determine the number of different results:

$$\binom{10}{3} = \frac{10!}{3!(10-3)!} = \frac{10 \cdot 9 \cdot 8}{3 \cdot 2 \cdot 1} = 120$$

b. We will follow the five-step procedure.

Step 1 The experiment is to select (but *not rank*) three brands from ten.

Step 2 There are 120 simple events for this experiment.

Step 3 Since the selection is made by chance, each simple event is equally likely:

$$P(\text{Each simple event}) = \frac{1}{120}$$

Step 4 Define events A and B as follows:

   A: {Exactly one company X brand is selected}

   B: {At least one company X brand is selected}

Since each of the simple events is equally likely to occur, we need to know only the number of simple events in A and B to determine their probabilities.

In order for event A to occur, exactly one company X brand must be selected, along with two of the remaining eight brands. We thus break A into two parts:

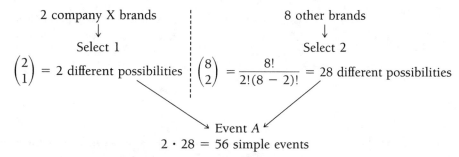

Note that the one company X brand can be selected in 2 ways, while the two other brands can be selected in 28 ways (we use the combinations rule because the order of selection is not important). Then, we use the multiplicative rule to combine one of the 2 ways to select a company X brand with one of the 28 ways to select two other brands, yielding a total of 56 simple events for event A.

The simple events in event B would include all simple events containing either one or two company X brands. We already know that the number containing exactly one company X brand is 56, the number of elements in event A. The number containing exactly two company X brands is equal to the product of the number of ways of selecting two company X brands out of a possible 2 and the number of ways of selecting the third brand from the remaining 8:

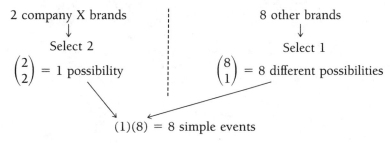

Then the number of simple events that imply the selection of either one *or* two company X brands is

$$\begin{pmatrix}\text{Number containing}\\\text{one X brand}\end{pmatrix} + \begin{pmatrix}\text{Number containing}\\\text{two X brands}\end{pmatrix}$$

or

$$56 + 8 = 64$$

Step 5 Since all the simple events are equally likely, we have

$$P(A) = \frac{\text{Number of simple events in } A}{\text{Total number of simple events}} = \frac{56}{120} = \frac{7}{15}$$

and

$$P(B) = \frac{\text{Number of simple events in } B}{\text{Total number of simple events}} = \frac{64}{120} = \frac{8}{15}$$

∎

Learning how to decide whether a particular counting rule applies to an experiment takes patience and practice. If you want to develop this skill, use the rules to solve the following exercises and some of the supplementary exercises given at the end of this chapter.

EXERCISES 3.6–3.14

3.6 An experiment consists of tossing a balanced die three times. Show that there are 216 simple events for this experiment.

3.7 A company specializing in data-communications hardware markets a computing system with two types of hard disk drives, four types of display stations, and two types of interfacing. How many systems would the company have to distribute if it received one order for each possible combination of hard disk drive, display station, and interfacing?

3.8 A security alarm system is activated and deactivated by correctly entering the appropriate three-digit numerical code in the proper sequence on a digital panel.
a. Compute the total number of possible code combinations if no digit may be used twice.
b. Compute the total number of possible code combinations if digits may be used more than once.

3.9 A study was conducted by Union Carbide to identify the optimal catalyst preparation conditions in the conversion of monoethanolamine (MEA) to ethylenediamine (EDA), a substance used commercially in soaps.* The initial experimental plan was chosen to screen four metals (Fe, Co, Ni, Cu) and four catalyst support classes (low acidity, high acidity, porous, and high surface area).
a. How many metal–support combinations are possible for this experiment?

*Hansen, J. L. and Best, D. C. "How to Pick a Winner." Paper presented at Joint Statistical Meetings, American Statistical Association and Biometric Society, Aug. 1986, Chicago, Ill.

b. From each catalyst support class a typical support was selected and all four tested in random order with one of the metals. How many different orderings of the four supports are possible with each metal?

3.10 Suppose you need to replace 5 gaskets in a nuclear-powered device. If you have a box of 20 gaskets from which to make the selection, how many different choices are possible; i.e., how many different samples of 5 gaskets can be selected from the 20?

3.11 A full-scale reinforced concrete building was designed and tested under simulated earthquake loading conditions (*Journal of Structural Engineering*, Jan. 1986). After completion of the experiments, several design engineers were administered a questionnaire in which they were asked to evaluate two building parameters (size and reinforcement) for each of three parts (shear wall, columns, and girders). For each parameter–part combination, the design engineers were asked to choose one of the following three responses: too heavy, about right, and too light.
a. How many different responses are possible on the questionnaire?
b. Suppose the design engineers are also asked to select the three parameter–part combinations with the overall highest ratings and rank them from 1 to 3. How many different rankings are possible?

3.12 In order to evaluate the traffic control systems of four facilities relying on computer-based equipment, the Federal Aviation Administration (FAA) formed a 16-member task force. If the FAA wants to assign 4 task force members to each facility, how many different assignments are possible?

OPTIONAL EXERCISES

3.13 Blackjack, a favorite game of gamblers, is played by a dealer and at least one opponent and uses a standard 52-card bridge deck. Each card is assigned a numerical value. Cards numbered from 2 to 10 are assigned the values shown on the card. For example, a 7 of spades has a value of 7; a 3 of hearts has a value of 3. Face cards (kings, queens, and jacks) are each valued at 10, and an ace can be assigned a value of either 1 or 11, at the discretion of the player holding the card. At the outset of the game, two cards are dealt to the player and two cards to the dealer. Drawing an ace and any card with a point value of 10 is called *blackjack*. In most casinos, if the dealer draws blackjack, he or she automatically wins. What is the probability that the dealer will draw a blackjack?

3.14 What is the probability that you will be dealt a 5-card poker hand of four aces?

S E C T I O N 3.4

COMPOUND EVENTS

An event can often be viewed as a composition of two or more other events. Such events are called **compound events**; they can be formed (composed) in two ways.

DEFINITION 3.6
The **union** of two events A and B is the event that occurs if either A or B or both occur on a single performance of the experiment. We will denote the union of events A and B by the symbol $A \cup B$.

> **DEFINITION 3.7**
>
> The **intersection** of two events A and B is the event that occurs if both A and B occur on a single performance of the experiment. We will write $A \cap B$ for the intersection of events A and B.

EXAMPLE 3.16

Consider the die toss experiment. Define the following events:

A: {Toss an even number}
B: {Toss a number less than or equal to 3}

a. Describe $A \cup B$ for this experiment.
b. Describe $A \cap B$ for this experiment.
c. Calculate $P(A \cup B)$ and $P(A \cap B)$ assuming the die is fair.

SOLUTION

a. The union of A and B is the event that occurs if we observe either an even number, a number less than or equal to 3, or both on a single throw of the die. Consequently, the simple events in the event $A \cup B$ are those for which A occurs, B occurs, or both A and B occur. Testing the simple events in the entire sample space, we find that the collection of simple events in the union of A and B is

$$A \cup B = \{1, 2, 3, 4, 6\}$$

b. The intersection of A and B is the event that occurs if we observe both an even number and a number less than or equal to 3 on a single throw of the die. Testing the simple events to see which imply the occurrence of *both* events A and B, we see that the intersection contains only one simple event:

$$A \cap B = \{2\}$$

In other words, the intersection of A and B is the simple event Observe a 2.

c. Recalling that the probability of an event is the sum of the probabilities of the simple events of which the event is composed, we have

$$P(A \cup B) = P(1) + P(2) + P(3) + P(4) + P(6)$$

$$= \frac{1}{6} + \frac{1}{6} + \frac{1}{6} + \frac{1}{6} + \frac{1}{6} = \frac{5}{6}$$

and

$$P(A \cap B) = P(2) = \frac{1}{6}$$ ∎

Unions and intersections also can be defined for more than two events. For example, the event $A \cup B \cup C$ represents the union of three events, A, B, and C. This event, which includes the set of simple events in A, B, or C, will occur if any one or more of the events A, B, or C occurs. Similarly, the intersection A

∩ B ∩ C is the event that all three of the events A, B, and C occur. Therefore, A ∩ B ∩ C is the set of simple events that are in all three of the events A, B, and C.

EXAMPLE 3.17

Refer to Example 3.16 and define the event

C: {Toss a number greater than 1}

Find the simple events in

a. $A \cup B \cup C$ **b.** $A \cap B \cap C$

where

A: {Toss an even number}
B: {Toss a number less than or equal to 3}

SOLUTION

a. Event C contains the simple events corresponding to tossing a 2, 3, 4, 5, or 6, and event B contains the simple events 1, 2, and 3. Therefore, the event that either A, B, or C occurs contains all six simple events in S, i.e., those corresponding to tossing a 1, 2, 3, 4, 5, or 6.
b. You can see that you will observe all of the events, A, B, and C only if you observe a 2. Therefore, the intersection $A \cap B \cap C$ contains the single simple event Toss a 2.   ■

SECTION 3.5

COMPLEMENTARY EVENTS

A very useful concept in the calculation of event probabilities is the notion of complementary events.

DEFINITION 3.8

The **complement*** of an event A is the event that A does not occur, i.e., the event consisting of all simple events that are not in event A. We will denote the complement of A by A^c.

An event A is a collection of simple events, and the simple events included in A^c are those that are not in A. Figure 3.4 demonstrates this. You will note from the figure that all simple events in S are included in *either* A or A^c, and that *no* simple event is in both A and A^c. This leads us to conclude that the probabilities of an event and its complement must sum to 1.

FIGURE 3.4
Venn Diagram of
Complementary Events

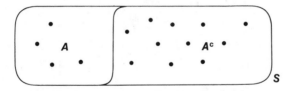

*Some texts use the symbol A' to denote the complement of an event A.

COMPLEMENTARY RELATIONSHIP

The sum of the probabilities of complementary events equals 1. That is,

$$P(A) + P(A^c) = 1$$

In many probability problems it will be easier to calculate the probability of the complement of the event of interest rather than the event itself. Then, since

$$P(A) + P(A^c) = 1$$

we can calculate $P(A)$ by using the relationship

$$P(A) = 1 - P(A^c)$$

EXAMPLE 3.18

Consider the experiment of tossing two fair coins. Calculate the probability of event

    A: {Observe at least one head}

by using the complementary relationship.

SOLUTION

We know that the event A: {Observe at least one head} consists of the simple events

    A: {HH, HT, TH}

The complement of A is defined as the event that occurs when A does not occur. Therefore,

    A^c: {Observe no heads} = {TT}

This complementary relationship is shown in Figure 3.5. Assuming the coins are balanced, we have

$$P(A^c) = P(TT) = \frac{1}{4}$$

and

$$P(A) = 1 - P(A^c) = 1 - \frac{1}{4} = \frac{3}{4}$$

FIGURE 3.5
Complementary Events in
the Toss of Two Coins

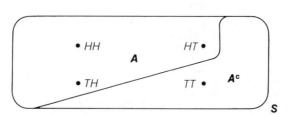

    ■

EXAMPLE 3.19

A fair coin is tossed ten times and the up face is recorded after each toss. Find the probability of the event

    A: {Observe at least one head}

SOLUTION

We will solve this problem by following the five steps for calculating probabilities of events (see Section 3.2).

Step 1 Define the experiment. The experiment is to record the results of the ten tosses of the coin.

Step 2 List the simple events. A simple event consists of a particular sequence of ten heads and tails. Thus, one simple event is

   HHTTTHTHTT

which denotes head on first toss, head on second toss, tail on third toss, etc. Others would be *HTHHHTTTTT* and *THHTHTHTTH*. There is obviously a very large number of simple events—too many to list. It can be shown (see Section 3.3) that there are $2^{10} = 1,024$ simple events for this experiment.

Step 3 Assign probabilities. Since the coin is fair, each sequence of heads and tails has the same chance of occurring and therefore all the simple events are equally likely. Then

$$P(\text{Each simple event}) = \frac{1}{1,024}$$

Step 4 Determine the simple events in event *A*. A simple event is in *A* if at least one *H* appears in the sequence of ten tosses. However, if we consider the complement of *A*, we find that

   A^c: {No heads are observed in ten tosses}

Thus, A^c contains only the simple event

   A^c: {*TTTTTTTTTT*}

and therefore

$$P(A^c) = \frac{1}{1,024}$$

Step 5 Since we know the probability of the complement of *A*, we use the relationship for complementary events:

$$P(A) = 1 - P(A^c) = 1 - \frac{1}{1,024} = \frac{1,023}{1,024} = .999$$

That is, we are virtually certain of observing at least one head in ten tosses of the coin.   ■

3.15 One game that is popular in many American casinos is roulette. Roulette is played by spinning a ball on a circular wheel that has been divided into 38 arcs of equal length; these bear the numbers 00, 0, 1, 2, . . . , 35, 36. The number on the arc at which the ball comes to rest is the outcome of one play of the game. The numbers are also colored in the following manner:

Red: 1 3 5 7 9 12 14 16 18
            19 21 23 25 27 30 32 34 36

Black: 2 4 6 8 10 11 13 15 17
          20 22 24 26 28 29 31 33 35

Green: 00 0

Players may place bets on the table in a variety of ways, including bets on odd, even, red, black, low (1–18), and high (19–36) outcomes. Consider the following events (00 and 0 are considered neither odd nor even):

A: {Outcome is an odd number}

B: {Outcome is a black number}

C: {Outcome is a high number}

List the simple events in
a. $A \cup B$ **b.** $A \cap C$ **c.** $B \cup C$ **d.** B^c **e.** $A \cap B \cap C$

3.16 Refer to Exercise 3.15. Find the probabilities of the events defined in parts **a–e** by summing the probabilities of the simple events in each.

3.17 An oil drilling venture involves the drilling of six wildcat oil wells in different parts of the country. Suppose that each drilling will produce either a dry well or an oil gusher. Assuming that the simple events for this experiment are equally likely, find the probability that at least one oil gusher will be discovered.

3.18 Enhanced protection against corrosion of steel sheet is a top priority of automakers. At Mazda Motor Corporation (Japan), there is a strong preference for thin, plated alloy coatings to improve protection against rust and adhesion. The accompanying table gives the breakdown of steel sheet usage in Mazda 626's exported to the United States. Suppose a single steel sheet is randomly selected from among those sheets used in the production of a Mazda 626 and we are interested in the type of steel sheet that is selected.
a. Define the experiment.
b. List the simple events for the experiment.
c. Assign probabilities to the simple events based on Mazda's steel sheet usage.
d. What is the probability that the steel sheet will be of the hot rolled, high strength type?
e. What is the probability that the steel sheet will be of the cold rolled type?
f. What is the probability that the steel sheet will not be plated?

TYPE OF STEEL SHEET	PERCENTAGE USED
Cold rolled	27
Cold rolled, high strength	12
Cold rolled, plated	30
Cold rolled, high strength, plated	15
Hot rolled	8
Hot rolled, high strength	5
Hot rolled, plated	3
TOTAL	100

Source: Chandler, H. E. "Materials Trends at Mazda Motor Corporation." *Metal Progress*, Vol. 129, No. 6, May 1986, p. 57 (Figure 3).

3.19 The game of craps is played with two dice. A player throws both dice, winning unconditionally if he produces a *natural* (the sum of the numbers showing on the two dice is 7 or 11), and losing unconditionally if he throws *craps* (a 2, 3, or 12).
   a. Find the probability that a player will throw a natural on the first toss of the dice.
   b. Find the probability that a player does not throw craps on the first toss of the dice.

| | | | | | | | | | | | |

S E C T I O N 3.6

CONDITIONAL PROBABILITY

The event probabilities we have discussed thus far give the relative frequencies of the occurrences of the events when the experiment is repeated a very large number of times. They are called **unconditional probabilities** because no special conditions are assumed other than those that define the experiment.

Sometimes we may wish to alter the probability of an event when we have additional knowledge that might affect its outcome. This revised probability is called the **conditional probability** of the event. For example, we have shown that the probability of observing an even number (event A) on a toss of a fair die is $\frac{1}{2}$. However, suppose you are given the information that on a particular throw of the die the result was a number less than or equal to 3 (event B). Would you still believe that the probability of observing an even number on that throw of the die is equal to $\frac{1}{2}$? If you reason that making the assumption that B has occurred reduces the sample space from six simple events to three simple events (namely, those contained in event B), the reduced sample space is as shown in Figure 3.6.

FIGURE 3.6
Reduced Sample Space for the Die Toss Experiment, Given That Event B Has Occurred

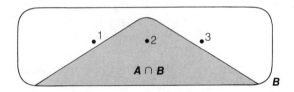

Since the only even number of the three numbers in the reduced sample space of event B is the number 2 and since the die is fair, we conclude that the probability that A occurs **given that B occurs** is one in three, or $\frac{1}{3}$. We will use the symbol $P(A \mid B)$ to represent the probability of event A given that event B occurs. For the die toss example, we write

$$P(A \mid B) = \frac{1}{3}$$

To get the probability of event A given that event B occurs, we proceed as follows: We divide the probability of the part of A that falls within the reduced sample space of event B, namely $P(A \cap B)$, by the total probability of the reduced sample space, namely $P(B)$. Thus, for the die toss example where event A: {Observe an even number} and event B: {Observe a number less than or equal to 3}, we find

$$P(A \mid B) = \frac{P(A \cap B)}{P(B)} = \frac{P(2)}{P(1) + P(2) + P(3)} = \frac{\frac{1}{6}}{\frac{3}{6}} = \frac{1}{3}$$

This formula for $P(A \mid B)$ is true in general:

FORMULA FOR CONDITIONAL PROBABILITY

To find the **conditional probability that event A occurs given that event B occurs**, divide the probability that *both A and B occur* by the probability that B occurs, that is,

$$P(A \mid B) = \frac{P(A \cup B)}{P(B)} \quad \text{where we assume that } P(B) \neq 0$$

EXAMPLE 3.20

Consider the following problem in process control. Suppose you are interested in the probability that a manufactured product (e.g., a small mechanical part) shipped to a buyer conforms to the buyer's specifications. Lots containing a large number of parts must pass inspection before they are accepted for shipment. [Assume that not all parts in a lot are inspected. For example, if the mean product characteristic (e.g., diameter) of a sample of parts selected from the lot falls within certain limits, the entire lot is accepted even though there may be one or more individual parts that fall outside specifications.] Let I represent the event that a lot passes inspection and let B represent the event that an individual part in a lot conforms to the buyer's specifications. Thus, $I \cap B$ is the simple event that the individual part is both shipped to the buyer (this happens when the lot containing the part passes inspection) and conforms to specifications, $I \cap B^c$ is the simple event that the individual part is shipped to the buyer but does not conform to specifications, etc. Assume the probabilities associated with the four simple events are as shown in the accompanying table:

SIMPLE EVENT	PROBABILITY
$I \cap B$	.80
$I \cap B^c$	.02
$I^c \cap B$	.15
$I^c \cap B^c$	.03

Find the probability that an individual part conforms to the buyer's specifications given that it is shipped to the buyer.

SOLUTION

If one part is selected from a lot of manufactured parts, what is the probability that the buyer will accept the part? In order to be accepted, the part must first be shipped to the buyer (i.e., the lot containing the part must pass inspection) *and* then the part must meet the buyer's specifications, so this *unconditional* probability is $P(I \cap B) = .80$.

In contrast, suppose you *know* that the selected part is from a lot which passes inspection. Now you are interested in the probability that the part conforms to specifications *given* that the part is shipped to the buyer, i.e., you want to determine the *conditional* probability $P(B \mid I)$. From the definition of conditional probability,

$$P(B \mid I) = \frac{P(I \cap B)}{P(I)}$$

where the event

I: {Part is shipped to the buyer}

contains the two simple events

$I \cap B$: {Part is shipped to buyer and conforms to specifications}

and

$I \cap B^c$: {Part is shipped to buyer but fails to meet specifications}

Recalling that the probability of an event is equal to the sum of the probabilities of its simple events, we obtain

$$P(I) = P(I \cap B) + P(I \cap B^c)$$
$$= .80 + .02 = .82$$

Then the conditional probability that a part conforms to specifications, given the part is shipped to the buyer, is

$$P(B \mid I) = \frac{P(I \cap B)}{P(I)} = \frac{.80}{.82} = .976$$

As we would expect, the probability that the part conforms to specifications, given that the part is shipped to the buyer, is higher than the unconditional probability that a part will be acceptable to the buyer. ∎

EXAMPLE 3.21

The investigation of consumer product complaints by the Federal Trade Commission (FTC) has generated much interest by manufacturers in the quality of their products. A manufacturer of food processors conducted an analysis of a large number of consumer complaints and found that they fell into the six categories shown in Table 3.5. If a consumer complaint is received, what is the probability that the cause of the complaint was product appearance given that the complaint originated during the guarantee period?

TABLE 3.5
Distribution of Product
Complaints

	REASON FOR COMPLAINT			TOTALS
	Electrical	Mechanical	Appearance	
During Guarantee Period	18%	13%	32%	63%
After Guarantee Period	12%	22%	3%	37%
TOTALS	30%	35%	35%	100%

SOLUTION

Let A represent the event that the cause of a particular complaint was product appearance and let B represent the event that the complaint occurred during the guarantee period. Checking Table 3.5, you can see that $(18 + 13 + 32)\% = 63\%$ of the complaints occurred during the guarantee time. Hence, $P(B) = .63$. The percentage of complaints that were caused by appearance and occurred during the guarantee time (the event $A \cap B$) is 32%. Therefore, $P(A \cap B) = .32$.

Using these probability values, we can calculate the conditional probability $P(A \mid B)$ that the cause of a complaint is appearance given that the complaint occurred during the guarantee time:

$$P(A \mid B) = \frac{P(A \cap B)}{P(B)} = \frac{.32}{.63} = .51$$

Consequently, you can see that slightly more than half the complaints that occurred during the guarantee time were due to scratches, dents, or other imperfections in the surface of the food processors.   ∎

EXERCISES 3.20–3.27

3.20 Refer to the game of roulette and the events described in Exercise 3.15. Find
   a. $P(A \mid B)$ **b.** $P(B \mid C)$ **c.** $P(C \mid A)$

3.21 Refer to the game of craps described in Exercise 3.19. A player casts the dice a single time.
   a. Given that the sum of the dice is odd, what is the probability that craps is thrown?
   b. Given that the player does not throw craps, what is the probability that the player throws a *double*—i.e., the same outcome on both dice?

3.22 The probability that a data-communications system will have high selectivity is .82, the probability that it will have high fidelity is .59, and the probability that it will have both is .33. Find the probability that a system with high fidelity will also have high selectivity.

3.23 A survey of users of word processors showed that 10% were dissatisfied with the word-processing system they are currently using. Half of those who were dissatisfied had purchased their systems from vendor A. It is also known that 20% of all those surveyed purchased their word-processing systems from vendor A. Given that a word processor was purchased from vendor A, what is the probability that the user is dissatisfied?

3.24 Along with the technological age comes the problem of workers being replaced by machines. A labor management organization wants to study the problem of workers displaced by automation within the industrial engineering field. Case reports for 100 workers whose loss of job is directly attributable to technological advances are selected within the industry. For each worker selected, it is determined whether he or she was given another job within the same company, found a job with another company in industrial engineering, found a job in a new field, or has been unemployed for longer than 6 months. In addition, the union status (union or non-union) of each worker is recorded, with the results shown in the table. A worker is to be selected from those surveyed.

	SAME COMPANY	NEW COMPANY (SAME FIELD)	NEW FIELD	UNEMPLOYED
Union	41	12	4	1
Non-union	16	9	11	6

a. If the selected worker found a job with a new company in the same field, what is the probability that the worker is a union member?

b. If the worker is not a union member, what is the probability that the worker has been unemployed for longer than 6 months?

3.25 Refer to the U.S. Army Corps of Engineers study on the DDT contamination of fish in the Tennessee River in Alabama (see Case Study 1.1). Part of the investigation focused on how far upstream the contaminated fish have migrated. (A fish is considered to be contaminated if its measured DDT concentration is greater than 5.0 parts per million.) Recall that Appendix III gives the DDT concentration, species, and capture location (in miles from the river's mouth) for each in a sample of 144 fish specimens. The accompanying table gives the number of contaminated fish found for each species–location combination. Suppose a contaminated fish is captured from the river.

		CAPTURE LOCATION		
		275–300	305–325	330–350
	Small-mouth buffalo	9	7	0
SPECIES	Large-mouth bass	0	0	1
	Channel catfish	31	23	6

a. Given that the fish is a channel catfish, what is the probability that it is captured 330–350 miles upstream?

b. Given that the fish is captured 275–300 miles upstream, what is the probability that it is a small-mouth buffalo?

OPTIONAL EXERCISES

3.26 Refer to Optional Exercise 3.13 and the game of blackjack. What is the probability that a player will win with blackjack?

3.27 The United States Nuclear Regulatory Commission assesses the safety risks associated with nuclear power plants. The commission has concluded that the probability of less than one latent cancer fatality (per year) due to core melt of a nuclear reactor is .00005. Suppose the probability that a core melt occurs during a given year is 1 in 100,000. Find the probability that at least one latent cancer fatality (per year) will occur as a result of a core melt of a nuclear reactor.

| | | | | | | | | | | | |

SECTION 3.7

PROBABILITIES OF UNIONS AND INTERSECTIONS

Since unions and intersections of events are themselves events, we can always calculate their probabilities by adding the probabilities of the simple events that compose them. However, when the probabilities of certain events are known, it is easier to use one or both of two rules to calculate the probability of unions and intersections. How and why these rules work will be illustrated by example.

EXAMPLE 3.22

A loaded (unbalanced) die is tossed and the up face is observed. The following two events are defined:

 A: {Observe an even number}

 B: {Observe a number less than 3}

Suppose that $P(A) = .4$, $P(B) = .2$, and $P(A \cap B) = .1$. Find $P(A \cup B)$. [*Note:* Assuming that we would know these probabilities in a practical situation is not very realistic, but the example will illustrate a point.]

SOLUTION

By studying the Venn diagram in Figure 3.7, we can obtain information that will help us find $P(A \cup B)$. We can see that

$$P(A \cup B) = P(1) + P(2) + P(4) + P(6)$$

Also, we know that

$$P(A) = P(2) + P(4) + P(6) = .4$$
$$P(B) = P(1) + P(2) = .2$$
$$P(A \cap B) = P(2) = .1$$

FIGURE 3.7
Venn Diagram for Die Toss

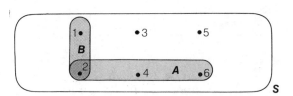

If we add the probabilities of the simple events that comprise events A and B, we find

$$P(A) + P(B) = \overbrace{P(2) + P(4) + P(6)}^{P(A)} + \overbrace{P(1) + P(2)}^{P(B)}$$
$$= \overbrace{P(1) + P(2) + P(4) + P(6)}^{P(A \cup B)} + \overbrace{P(2)}^{P(A \cap B)}$$

Thus, by subtraction, we have

$$P(A \cup B) = P(A) + P(B) - P(A \cap B)$$
$$= .4 + .2 - .1 = .5 \qquad \blacksquare$$

By studying the Venn diagram in Figure 3.8, you can see that the method used in Example 3.22 may be generalized to find the union of two events for any experiment. The probability of the union of two events, A and B, can always be obtained by summing $P(A)$ and $P(B)$ and subtracting $P(A \cap B)$. Note that we must subtract $P(A \cap B)$ because the simple event probabilities in $(A \cap B)$ have been included twice—once in $P(A)$ and once in $P(B)$.

FIGURE 3.8
Venn Diagram of Union

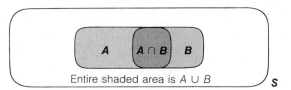

Entire shaded area is $A \cup B$

The formula for calculating the probability of the union of two events, often called the **additive rule of probability**, is given in the box.

ADDITIVE RULE OF PROBABILITY

The probability of the union of events A and B is the sum of the probabilities of events A and B minus the probability of the intersection of events A and B:

$$P(A \cup B) = P(A) + P(B) - P(A \cap B)$$

EXAMPLE 3.23

Records at an industrial plant show that 12% of all injured workers are admitted to a hospital for treatment, 16% are back on the job the next day, and 2% are both admitted to a hospital for treatment and back on the job the next day. If a worker is injured, what is the probability that the worker will be either admitted to a hospital for treatment, or back on the job the next day, or both?

SOLUTION

Consider the following events:

A: {An injured worker is admitted to the hospital for treatment}

B: {An injured worker returns to the job on the next day}

Then, from the information given in the statement of the example, we know that

$$P(A) = .12 \qquad P(B) = .16$$

and the probability of the event that an injured worker receives hospital treatment and returns to the job on the next day is

$$P(A \cap B) = .02$$

The event that an injured worker is admitted to the hospital, or returns to the job the next day, or both, is the union, $A \cup B$. The probability of $A \cup B$ is given by the additive rule of probability:

$$P(A \cup B) = P(A) + P(B) - P(A \cap B)$$
$$= .12 + .16 - .02 = .26$$

Thus, 26% of all injured workers either are admitted to the hospital, or return to the job the next day, or both.   ■

A very special relationship exists between events A and B when $A \cap B$ contains no simple events. In this case, we call the events A and B **mutually exclusive** events.

DEFINITION 3.9

Events A and B are **mutually exclusive** if $A \cap B$ contains no simple events.

Figure 3.9 shows a Venn diagram of two mutually exclusive events. The events A and B have no simple events in common, i.e., A and B cannot occur simultaneously, and $P(A \cap B) = 0$. Thus, we have the important relationship shown in the next box.

FIGURE 3.9
Venn Diagram of Mutually
Exclusive Events

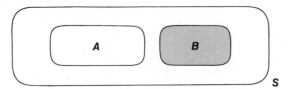

ADDITIVE RULE FOR MUTUALLY EXCLUSIVE EVENTS

If two events A and B are mutually exclusive, the probability of the union of A and B equals the sum of the probabilities of A and B:

$$P(A \cup B) = P(A) + P(B)$$

EXAMPLE 3.24

Consider the experiment of tossing two balanced coins. Find the probability of observing at least one head.

SOLUTION

Define the events

    A: {Observe at least one head}

    B: {Observe exactly one head}

    C: {Observe exactly two heads}

Note that $A = B \cup C$ and that $B \cap C$ contains no simple events (see Figure 3.10). Thus, B and C are mutually exclusive, so that

$$P(A) = P(B \cup C) = P(B) + P(C)$$
$$= \frac{1}{2} + \frac{1}{4} = \frac{3}{4}$$

FIGURE 3.10
Venn Diagram for Coin
Toss Experiment

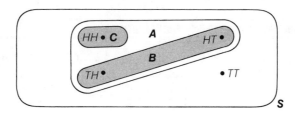

Although Example 3.24 is very simple, the concept of writing events with verbal descriptions that include the phrases "at least" or "at most" as unions of mutually exclusive events is a very useful one. This enables us to find the probability of the event by adding the probabilities of the mutually exclusive events.

The second rule of probability, which will help us find the probability of the intersection of two events, is illustrated by Example 3.25.

EXAMPLE 3.25

A data processor is interested in the event that a job will be processed immediately upon submission. This event is the intersection of the following two events:

    A: {The computer is functional}

    B: {The job will be processed immediately}

Based on available information, the data processor believes that the probability is .90 that the computer will be functional at any particular time and that the probability is .05 that the job will run immediately upon submission given that the computer is functional. That is,

    $P(A) = .90$ and $P(B \mid A) = .05$

Based on the information provided, what is the probability that a submitted job will be processed immediately? That is, find $P(A \cap B)$.

SOLUTION

As you will see, we have already developed a formula for finding the probability of an intersection of two events. Recall that the conditional probability of B given A is

$$P(B \mid A) = \frac{P(A \cap B)}{P(A)}$$

Multiplying both sides of this equation by $P(A)$, we obtain a formula for the probability of the intersection of events A and B. This is often called the **multiplicative rule of probability** and is given by

$$P(A \cap B) = P(A) \, P(B \mid A)$$

Thus,

$$P(A \cap B) = (.90)(.05) = .045$$

The probability that a submitted job will be processed immediately is .045. ∎

MULTIPLICATIVE RULE OF PROBABILITY

$$P(A \cap B) = P(A \mid B)P(B) = P(B \mid A)P(A)$$

EXAMPLE 3.26

Consider the experiment of tossing a fair coin twice and recording the up face on each toss. The following events are defined:

A: {First toss is a head}
B: {Second toss is a head}

Does *knowing* that event A has occurred affect the probability that B will occur?

SOLUTION

Intuitively the answer should be no, since what occurs on the first toss should in no way affect what occurs on the second toss. Let us check our intuition. Recall the sample space for this experiment:

1. Observe *HH*
2. Observe *HT*
3. Observe *TH*
4. Observe *TT*

Each of these simple events has a probability of $\frac{1}{4}$. Thus,

$$P(B) = P(HH) + P(TH) \quad \text{and} \quad P(A) = P(HH) + P(HT)$$
$$= \frac{1}{4} + \frac{1}{4} = \frac{1}{2} \qquad\qquad\qquad = \frac{1}{4} + \frac{1}{4} = \frac{1}{2}$$

Now, what is $P(B \mid A)$?

$$P(B \mid A) = \frac{P(A \cap B)}{P(A)} = \frac{P(HH)}{P(A)}$$

$$= \frac{\frac{1}{4}}{\frac{1}{2}} = \frac{1}{2}$$

We can now see that $P(B) = \frac{1}{2}$ and $P(B \mid A) = \frac{1}{2}$. Knowing that the first toss resulted in a head does not affect the probability that the second toss will be a head. The probability is $\frac{1}{2}$ whether or not we know the result of the first toss. When this occurs, we say that the two events A and B are **independent**. ∎

DEFINITION 3.10

Events A and B are **independent** if the occurrence of B does not alter the probability that A has occurred, i.e., events A and B are independent if

$P(A \mid B) = P(A)$

When events A and B are **independent** it will also be true that

$P(B \mid A) = P(B)$

Events that are not independent are said to be **dependent**.

EXAMPLE 3.27

Consider the experiment of tossing a fair die and define the following events:

A: {Observe an even number}

B: {Observe a number less than or equal to 4}

Are events A and B independent?

SOLUTION

The Venn diagram for this experiment is shown in Figure 3.11. We first calculate

$$P(A) = P(2) + P(4) + P(6) = \frac{1}{2}$$

$$P(B) = P(1) + P(2) + P(3) + P(4) = \frac{4}{6} = \frac{2}{3}$$

$$P(A \cap B) = P(2) + P(4) = \frac{2}{6} = \frac{1}{3}$$

FIGURE 3.11
Venn Diagram for
Example 3.27

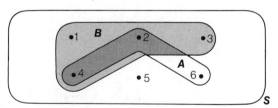

Now assuming B has occurred, the conditional probability of A given B is

$$P(A \mid B) = \frac{P(A \cap B)}{P(B)} = \frac{\frac{1}{3}}{\frac{2}{3}} = \frac{1}{2} = P(A)$$

Thus, assuming that event B occurs does not alter the probability of observing an even number—it remains $\frac{1}{2}$. Therefore, the events A and B are independent. Note that if we calculate the conditional probability of B given A, our conclusion is the same:

$$P(B \mid A) = \frac{P(A \cap B)}{P(A)} = \frac{\frac{1}{3}}{\frac{1}{2}} = \frac{2}{3} = P(B)$$ ∎

EXAMPLE 3.28

Refer to the consumer product complaint study in Example 3.21. The percentages of complaints of various types in the pre- and post-guarantee periods are shown in Table 3.5. Define the following events:

A: {Cause of complaint is product appearance}

B: {Complaint occurred during the guarantee term}

Are A and B independent events?

SOLUTION

Events A and B are independent if $P(A \mid B) = P(A)$. We calculated $P(A \mid B)$ in Example 3.21 to be .51, and from Table 3.5 we can see that

$$P(A) = .32 + .03 = .35$$

Therefore, $P(A \mid B)$ is not equal to $P(A)$, and A and B are not independent events.

∎

We will make three final points about independence. The first is that the property of independence, unlike the mutually exclusive property, cannot be shown on or gleaned from a Venn diagram and you cannot trust your intuition. In general, the only way to check for independence is by performing the calculations of the probabilities in the definition.

The second point concerns the relationship between the mutually exclusive and independence properties. Suppose that events A and B are mutually exclusive, as shown in Figure 3.9. Are these events independent or dependent? That is, does the assumption that B occurs alter the probability of the occurrence of A? It certainly does, because if we assume that B has occurred, it is impossible for A to have occurred simultaneously. Thus, **mutually exclusive events are dependent events**.

The third point is that the probability of the intersection of independent events is very easy to calculate. Referring to the formula for calculating the probability of an intersection, we find

$$P(A \cap B) = P(B)P(A \mid B)$$

Thus, since $P(A \mid B) = P(A)$ when A and B are independent, we have the useful rule stated in the box.

MULTIPLICATIVE RULE FOR INDEPENDENT EVENTS

If events A and B are independent, the probability of the intersection of A and B equals the product of the probabilities of A and B, i.e.,

$$P(A \cap B) = P(A)P(B)$$

In the die toss experiment, we showed in Example 3.27 that the events A: {Observe an even number} and B: {Observe a number less than or equal to 4} are independent if the die is fair. Thus,

$$P(A \cap B) = P(A)P(B) = \left(\frac{1}{2}\right)\left(\frac{2}{3}\right) = \frac{1}{3}$$

This agrees with the result

$$P(A \cap B) = P(2) + P(4) = \frac{2}{6} = \frac{1}{3}$$

that we obtained in the example.

EXAMPLE 3.29

In Example 3.3, a quality control engineer considered the problem of determining whether an assembly line is out of control. In the example, we discussed the problem of finding the probability that one, two, or, in general, k items arriving off the assembly line are defective. We are now ready to find the probability that both of two items arriving in succession off the line are defective. Suppose that the line is out of control and that 20% of the items being produced are defective.

a. If two items arrive in succession off the line, what is the probability that they are both defective?

b. If k items arrive in succession off the line, what is the probability that all are defective?

SOLUTION

a. Let D_1 be the event that item 1 is defective and let D_2 be a similar event for item 2. The event that *both* items will be defective is the intersection $D_1 \cap D_2$. Then, since it is not unreasonable to assume that the operating conditions of the items would be independent of one another, the probability that both will be defective is

$$P(D_1 \cap D_2) = P(D_1)P(D_2)$$
$$= (.2)(.2) = (.2)^2 = .04$$

b. Let D_i represent the event that the ith item arriving in succession off the line is defective. Then the event that all three of three items arriving in succession will be defective is the intersection of the event $D_1 \cap D_2$ (from part **a**) with the event D_3. Assuming independence of the events D_1, D_2, and D_3, we have

$$P(D_1 \cap D_2 \cap D_3) = P(D_1 \cap D_2)P(D_3)$$
$$= (.2)^2(.2) = (.2)^3 = .008$$

Noting the pattern, you can see that the probability that all k out of k arriving items are defective is the probability of $D_1 \cap D_2 \cap \cdots \cap D_k$, or

$$P(D_1 \cap D_2 \cap \cdots \cap D_k) = (.2)^k \quad \text{for } k = 1, 2, 3, \dots \qquad \blacksquare$$

EXERCISES 3.28–3.36

3.28 Reports have recently surfaced regarding the shortage of college faculty in the field of civil engineering. One proposed solution is to hire more practicing civil engineers as faculty members. However, this idea has generally been met with resistance in the academic world. In order to investigate this problem, 200 College of Engineering deans were asked to give their opinions regarding the main barrier to the hiring of practitioners to college faculty. In addition, the deans were asked whether their colleges stress "theory" or "applications" more heavily. The accompanying table shows the proportions of responses that fall into the respective categories. Suppose one college dean is to be selected from the 200 surveyed. Find the probability that:
a. The dean's college stresses applications.
b. The dean's college stresses theory and considers lack of a Ph.D. as the main barrier to hiring practitioners.
c. The dean's college stresses applications, or considers the salary demands of practitioners too high, or both.
d. The dean does not consider lack of teaching experience as a barrier to hiring practitioners.

	MAIN BARRIER TO HIRING PRACTITIONERS		
	Lack of Ph.D.	Salary Demands Too High	Lack of Teaching Experience
Stress Theory	.13	.10	.02
Stress Applications	.37	.21	.17

3.29 According to the National Highway Traffic Safety Administration, "as many as 9 out of 10 heavily used cars, such as those from the leasing companies, may have had their mileage rolled back when resold" (reported in the *Orlando Sentinel*, Apr. 13, 1984). Officials estimate that an altered odometer adds $750 to the price of a used car. Suppose you are considering buying a used car from an auto leasing company that has three cars available. Assume also that 90% of all used cars sold by the company have falsified odometer readings and that the three available cars represent a random sample of used cars sold by the company.
a. What is the probability that all three cars have falsified odometer readings?
b. What is the probability that none of the three cars has a falsified odometer reading?
c. Suppose a salesman claims that none of the three cars has a falsified odometer reading. What would you infer about the claim?

3.30 A two-component electronic system is connected in parallel so that it fails only if both of its components fail. The probability that the first component fails is .10. If the first component fails, the probability that the second component fails is .05. What is the probability that the two-component electronic system fails?

3.31 Managers of oil exploration portfolios make decisions on which prospects to pursue based, in part, on the level of risk associated with each venture. Kinchen (1986) examined the problem of risk analysis in oil exploration using the outcomes and associated probabilities for a single prospect, shown in the table.

OUTCOME (barrels)	PROBABILITY
0 (dry hole)	.60
50,000	.10
100,000	.15
500,000	.10
1,000,000	.05

Source: Kinchen, A. L. "Projected Outcomes of Exploration Programs Based on Current Program Status and the Impact of Prospects Under Consideration." *Journal of Petroleum Technology*, Vol. 38, No. 4, Apr. 1986, p. 462.

a. What is the probability that a single oil well prospect will result in no more than 100,000 barrels of oil?

b. What is the probability that a single oil well prospect will strike oil?

c. Kinchen also considered two identical oil well prospects. List the possible outcomes (i.e., simple events) if the two wells are drilled. Assume the outcomes listed in the table are the only ones possible for any one well. [*Hint:* One possible simple event is two dry holes.]

d. Use the information in the table to calculate the probabilities of the simple events in part c. (Assume that the individual outcomes of the two wells are independent of each other.)

e. Refer to part d. Find the probability that at least one of the two oil prospects strikes oil.

3.32 Refer to Exercise 3.19 and the game of craps. Consider the following events:

  A: {Player throws craps}

  B: {Player throws a natural}

  C: {Player throws 9, 10, or 11}

a. Which pairs of events, if any, are mutually exclusive?

b. Which pairs of events, if any, are independent?

3.33 The transport of neutral particles in an evacuated duct is an important aspect of nuclear fusion reactor design. In one experiment, particles entering through the duct ends streamed unimpeded until they collided with the inner duct wall. Upon colliding, they are either scattered (reflected) or absorbed by the wall (*Nuclear Science and Engineering*, May 1986). The reflection probability (i.e., the probability a particle is reflected off the wall) for one type of duct was found to be .16.

a. If two particles are released into the duct, find the probability that both will be reflected.

b. If five particles are released into the duct, find the probability that all five will be absorbed.

c. What assumption about the simple events in parts **a** and **b** is required to calculate the probabilities?

3.34 Experience has shown that a manufacturer of computer software produces, on the average, only 1 defective blank diskette in 100.

    a. Of the next three blank diskettes manufactured, what is the probability that all three will be nondefective?

    b. In general, if k blank diskettes are manufactured, what is the probability that at least one of the k will be defective?

3.35 Traditionally, geotechnical engineers have employed *working stress designs* (WSD) for designing structures safe from collapse. One study examined total safety of conventional WSD in three design areas: earthworks, earth-retaining structures and excavations, and foundations (*Canadian Geotechnical Journal*, Nov. 1985). The table gives the probability of failure in each of the design areas.

DESIGN AREA	PROBABILITY OF FAILURE
Earthworks	.01
Earth-retaining structures and excavations	.001
Foundations	.0001

    Consider a WSD comprised of the three design areas. (Assume that the failure of any one design area is independent of the failure of the others.)

    a. What is the probability of failure in either the earthworks or earth-retaining structures and excavations design area?

    b. What is the probability of failure in all three design areas?

OPTIONAL EXERCISE

3.36 In a recent article, Joseph Fiksel* writes on the problem of compensating victims of chronic diseases (such as cancer and birth defects) who are exposed to hazardous and toxic substances. The key to compensation, as far as the U.S. judicial system is concerned, is the *probability of causation* (i.e., the likelihood, for a person developing the disease, that the cause is due to exposure to the hazardous substance). Usually, the probability of causation must be greater than .50 in order for the court to award compensation. Fiksel gives examples on how to calculate the probability of causation for several different scenarios.

    a. "*Ordinary causation*," as defined by Fiksel, "describes a situation in which the presence of a single factor, such as asbestos insulation, is believed to cause an effect, such as mesothelioma." For this situation, define the following events:

        D: {Effect (disease) occurs}

        A: {Factor A present}

    Under ordinary causation, if *A* occurs, then *D* must occur. However, *D* can also occur when factor A is not present. The probability of causation for factor A, then,

*Fiksel, J. "Victim Compensation: Understanding the Problem of Indeterminate Causation." *Environmental Science and Technology*, May 1986. Copyright 1986 American Chemical Society. Reprinted with permission.

is the conditional probability $P(A \mid D)$. Show that the probability of causation for factor A is

$$P(A \mid D) = \frac{P_1 - P_0}{P_1}$$

where P_0 is the probability that the effect occurs when factor A is *not* present, and P_1 is the probability that the effect occurs when factor A is either present or not. [*Note:* To epidemiologists, P_1 is often called the *overall risk rate* for the disease and $P_1 - P_0$ is the *additional risk* attributable to the presence of factor A.] [*Hint:* The simple events for this experiment are $\{D^c \cap A^c, D \cap A, \text{ and } D \cap A^c\}$. Write P_0 and P_1 in terms of the simple events.]

b. Fiksel defines *simultaneous exclusive causation* as "a situation in which two or more causal factors are present but the resulting effect is caused by *one and only one* of these factors." Consider two factors, A and B, and let B be the event that factor B is present. In this situation, if either A or B occurs, then D must occur. However, both A and B cannot occur simultaneously (i.e., A and B are mutually exclusive). Assuming that D cannot occur when neither A nor B occurs, show that the probabilities of causations for factors A and B are, respectively,

$$P(A \mid D) = \frac{P_1}{P_1 + P_2} \quad \text{and} \quad P(B \mid D) = \frac{P_2}{P_1 + P_2}$$

where P_1 is the probability that the effect occurs when factor A is present and P_2 is the probability that the effect occurs when factor B is present. [*Hint:* The simple events for this experiment are $\{D^c \cap A^c \cap B^c, D \cap A \cap B^c, \text{ and } D \cap A^c \cap B\}$.]

c. *Simultaneous joint causation*, writes Fiksel, describes a more realistic situation "in which several factors can contribute in varying degrees to the occurrence of an effect. For example, a cigarette smoker who is exposed to radiation and chemical carcinogens in the workplace may develop a lung tumor. Whether the tumor was caused wholly by one factor or by a combination of factors is, at present, impossible to determine." For this case, consider two factors, A and B, which affect D independently. Also, assume that if either A, B, or both occur, then D must occur; and D cannot occur if neither A nor B occurs. Show that the probabilities of causation for factors A and B are, respectively,

$$P(A \mid D) = \frac{P_1}{P_1 + P_2 - P_1 P_2} \quad \text{and} \quad P(B \mid D) = \frac{P_2}{P_1 + P_2 - P_1 P_2}$$

where P_1 is the probability that the effect occurs when factor A is present and P_2 is the probability that the effect occurs when factor B is present. [*Hint:* The simple events for this experiment are $\{D^c \cap A^c \cap B^c, D \cap A \cap B^c, D \cap A^c \cap B, \text{ and } D \cap A \cap B\}$.]

SECTION 3.8

BAYES' RULE

An early attempt to employ probability in making inferences is the basis for a branch of statistical methodology known as **Bayesian statistical methods**. The logic employed by the English philosopher, the Reverend Thomas Bayes (1702–1761) is illustrated by Example 3.30.

EXAMPLE 3.30

A construction company employs three sales engineers. Engineers 1, 2, and 3 estimate the costs of 30%, 20%, and 50% respectively, of all jobs bid by the company. For $i = 1, 2, 3$, define A_i to be the event that a job is estimated by engineer i and define E to be the event that a serious error is made in estimating the cost. The following probabilities are known to describe the error rates of the engineers:

$$P(E \mid A_1) = .01$$
$$P(E \mid A_2) = .03$$
$$P(E \mid A_3) = .02$$

If a particular bid results in a serious error in estimating the job costs, which engineer was responsible? (Of course, we cannot answer this question with certainty, but statisticians who favor Bayesian logic would employ the following method for making this inference.)

SOLUTION

The occurrence of the event E is evidence upon which we wish to base an inference—namely, to infer which engineer committed the error. To solve this problem using Bayes' logic, we will calculate the conditional probabilities, $P(A_1 \mid E)$, $P(A_2 \mid E)$, and $P(A_3 \mid E)$. The engineer with the largest conditional probability is then selected as the one most likely to have estimated the cost of the job.

From the statement of the example, we have the following information:

$$P(A_1) = .30 \qquad P(E \mid A_1) = .01$$
$$P(A_2) = .20 \qquad P(E \mid A_2) = .03$$
$$P(A_3) = .50 \qquad P(E \mid A_3) = .02$$

Then

$$P(A_1 \cap E) = P(A_1)P(E \mid A_1) = (.30)(.01) = .003$$
$$P(A_2 \cap E) = P(A_2)P(E \mid A_2) = (.20)(.03) = .006$$
$$P(A_3 \cap E) = P(A_3)P(E \mid A_3) = (.50)(.02) = .010$$

The event E is the union of the three mutually exclusive events, $A_1 \cap E$, $A_2 \cap E$, and $A_3 \cap E$. Thus,

$$P(E) = P(A_1 \cap E) + P(A_2 \cap E) + P(A_3 \cap E)$$
$$= .003 + .006 + .010 = .019$$

We now apply the formula for conditional probability to obtain

$$P(A_1 \mid E) = \frac{P(A_1 \cap E)}{P(E)} = \frac{.003}{.019} = .158$$

$$P(A_2 \mid E) = \frac{P(A_2 \cap E)}{P(E)} = \frac{.006}{.019} = .316$$

$$P(A_3 \mid E) = \frac{P(A_3 \cap E)}{P(E)} = \frac{.010}{.019} = .526$$

Using Bayes' reasoning, we would choose engineer 3 as the one responsible for the error because the conditional probability of A_3, given the occurrence of the error (event E), is the largest of the three conditional probabilities. ∎

Bayes' method can be applied when sample information E may have been produced by any one of k mutually exclusive and exhaustive states of nature (or populations), $A_1, A_2, \ldots, A_k$. The formula for finding the appropriate conditional probabilities is given in the box.

BAYES' RULE

Given k mutually exclusive and exhaustive states of nature, $A_1, A_2, \ldots, A_k$, and an observed event E, then $P(A_i \mid E)$, for $i = 1, 2, \ldots, k$, is

$$P(A_i \mid E) = \frac{P(A_i \cap E)}{P(E)}$$

$$= \frac{P(A_i)P(E \mid A_i)}{P(A_1)P(E \mid A_1) + P(A_2)P(E \mid A_2) + \cdots + P(A_k)P(E \mid A_k)}$$

3.37 Refer to the table in Exercise 3.25 giving the number of contaminated fish found for each species–location combination. This information can be used to construct the accompanying table of probabilities. Each entry in the body of the table represents the (estimated) probability of observing a certain species of contaminated fish, *given* the capture location. For example, .225 in the upper left corner is the probability that the contaminated fish is a small-mouth buffalo, given that it was captured 275–300 miles upstream. The entries in the last row and last column represent unconditional probabilities. For example, .21 in the upper right corner is the probability that the contaminated fish is a small-mouth buffalo, and .52 in the lower left corner is the probability that the contaminated fish is captured 275–300 miles upstream.

		CAPTURE LOCATION			ALL LOCATIONS
		275–300	305–325	330–350	
	Small-mouth buffalo	.225	.23	0	.21
SPECIES	Large-mouth bass	0	0	.14	.01
	Channel catfish	.775	.77	.86	.78
ALL SPECIES		.52	.39	.09	

Given that a contaminated channel catfish is captured from the Tennessee River, find the probability that it was located:
a. 275–300 miles upstream.
b. 305–325 miles upstream.
c. 330–350 miles upstream.

3.38 A manufacturing operation utilizes two production lines to assemble electronic fuses. Both lines produce fuses at the same rate and generally produce 2.5% defective fuses. However, production line 1 recently suffered mechanical difficulty and produced 6.0% defectives during a 3-week period. This situation was not known until several lots of electronic fuses produced in this period were shipped to customers. If one of two fuses tested by a customer was found to be defective, what is the probability that the lot from which it came was produced on malfunctioning line 1? (Assume all the fuses in the lot were produced on the same line.)

3.39 The computing system at a large university is currently undergoing shutdown for repairs. Previous shutdowns have been due to either hardware failure, software failure, or power (electronic) failure. The system is forced to shut down 73% of the time when it experiences hardware problems, 12% of the time when it experiences software problems, and 88% of the time when it experiences electronic problems. Maintenance engineers have determined that the probabilities of hardware, software, and power problems are .01, .05, and .02, respectively. What is the probability that the current shutdown is due to hardware failure? Software failure? Power failure?

3.40 An important component of a personal computer (PC) is a microchip. The table gives the percentages of microchips that a certain PC manufacturer purchases from seven suppliers.

SUPPLIER	PERCENTAGE
S_1	.15
S_2	.05
S_3	.10
S_4	.20
S_5	.12
S_6	.20
S_7	.18

a. Suppose it is known that the proportions of defective microchips produced by the seven suppliers are .001, .0003, .0007, .006, .0002, .0002, and .001, respectively. If a single PC microchip failure is observed, which supplier is most likely responsible?

b. Suppose the seven suppliers produce defective microchips at the same rate, .0005. If a single PC microchip failure is observed, which supplier is most likely responsible?

PROBABILITY AND STATISTICS: AN EXAMPLE

We have introduced a number of new concepts in the preceding sections, and this makes the study of probability a particularly arduous task. It is therefore very important to establish clearly the connection between probability and statistics, which we will do in the remaining chapters. Although Bayes' rule demonstrates one way that probability can be used to make statistical inferences, traditional methods of statistical inference use probability in a slightly different way. In this section, we will present one brief example of this traditional approach to statistical inference so that you can begin to understand why some knowledge of probability is important in the study of statistics.

Suppose a firm that manufactures concrete studs is researching the hypothesis that its new chemically anchored studs achieve greater holding capacity and greater carrying load capacity than the more conventional, mechanically anchored studs. To test the hypothesis, three new chemical anchors are selected from a day's production and subjected to a durability test. Each of the three $\frac{1}{2}$-inch studs is drilled and set into a slab of 4,000 pounds per square inch stone aggregate concrete, and their tensile load capacities (in pounds) are recorded. It is known from many previous durability tests that the relative frequency distribution of tensile load capacities for mechanically anchored studs is mound-shaped, with a mean of 10,000 pounds and a standard deviation of 2,000 pounds. Suppose that all three of the chemically anchored studs tested have tensile strengths greater than 12,000 pounds. What can researchers for the firm conclude?

The relative frequency distribution of the tensile load capacities for mechanically anchored studs is shown in Figure 3.12. If the distribution is mound-shaped and approximately symmetric about the mean, we can conclude that approximately 16% of mechanical anchors will have tensile strengths over 12,000 pounds (see the Empirical Rule in Section 1.7). Now, define the events

A_1: {Stud 1 has tensile strength over 12,000 pounds}

A_2: {Stud 2 has tensile strength over 12,000 pounds}

A_3: {Stud 3 has tensile strength over 12,000 pounds}

We want to find $P(A_1 \cap A_2 \cap A_3)$, the probability that all three tested studs have tensile load capacities over 12,000 pounds.

FIGURE 3.12
Relative Frequency
Distribution for Tensile
Load Capacities

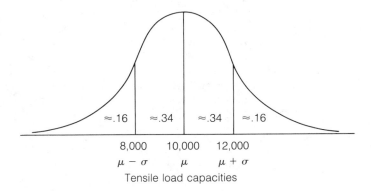

Since the studs are selected by chance from a large production, it may be plausible to assume that the events A_1, A_2, and A_3 are independent. That is,

$$P(A_2 \mid A_1) = P(A_2)$$

In words, knowing that the first stud has a tensile strength over 12,000 pounds does not affect the probability that the second stud has a tensile strength over 12,000 pounds. With the assumption of independence, we can calculate the probability of the intersection by multiplying the individual probabilities:

$$P(A_1 \cap A_2 \cap A_3) = P(A_1)P(A_2)P(A_3)$$
$$\approx (.16)(.16)(.16) = .004096$$

Thus, the probability that the firm's researchers will observe all three studs with tensile load capacity over 12,000 pounds is only about .004, *if the relative frequency distribution of tensile strengths for chemically anchored studs is no different from that for mechanically anchored studs.* If this event were to occur, the researchers might conclude that it lends credence to the theory that chemically anchored studs achieve greater carrying load capacity than mechanically anchored studs, *since it is so unlikely to occur if the distributions of tensile strength are the same.* Such a conclusion would be an application of the rare event approach to statistical inference, and you can see that the basic principles of probability play an important role.

│ │ │ │ │ │ │ │ │ │ │ │ │

EXERCISES 3.41–3.43

3.41 Refer to Exercise 3.34. Suppose that of the next four blank diskettes manufactured, at least one is defective. What would you infer about the claimed defective rate of .01? Explain.

3.42 Since 1960, parcels of land that may contain oil have been placed in a lottery with the winner receiving leasing rights (at $1 per acre per year) for a period of 10 years. United States citizens 21 years or older are eligible and may enter the lottery by paying a $10 filing fee to the Bureau of Land Management. For several months in 1980, however, the lottery was suspended in order to investigate a player who won three parcels of land in 1 month. The three lotteries had 1,836, 1,365, and 495 entries, respectively (*Orlando Sentinel Star*, Oct. 12, 1980). An Interior Department audit stated that "federal workers did a poor job of shaking the drum before the drawing." Based on your knowledge of probability and rare events, would you make the same inference as that made by the auditor? Explain.

3.43 In 1939, at the beginning of World War II, a group of British engineers and statisticians was formed in London to investigate the problem of the lethality of antiaircraft weapons.* One of the main goals of the research team was to assess the probability that a single shell would destroy (or cripple) the aircraft at which it was fired. Although a great deal of data existed at the time on ground-to-ground firing with artillery shells, little information was available on the accuracy of antiaircraft guns. Consequently, a series of trials was run in 1940 in which gun crews shot at free-flying (unpiloted) aircraft. When German aircraft began to bomb England later in that same year, however, the researchers found that the aiming errors of antiaircraft guns under battle stress were considerably greater than those estimated from trials. Let p be the probability that an antiaircraft shell strikes within a 30-foot radius of its target. Assume that under simulated conditions, $p = .45$.

   a. In an actual attack by a single German aircraft, suppose that three antiaircraft shells are fired and all three miss their target by more than 30 feet. Is it reasonable to conclude that in battle conditions p differs from .45?

   b. Answer part **a** assuming that you observe ten consecutive shots which all miss their target by more than 30 feet.

*Pearson, E. S. "Statistics and Probability Applied to Problems of Antiaircraft Fire in World War II." In *Statistics: A Guide to the Unknown*, 2nd ed. San Francisco: Holden-Day, 1978, pp. 474–482.

S E C T I O N 3.10

RANDOM SAMPLING

How a sample is selected from a population is of vital importance in statistical inference because the probability of an observed sample will be used to infer the characteristics of the sampled population. To illustrate, suppose you deal yourself 4 cards from a deck of 52 cards and all 4 cards are aces. Do you conclude that your deck is an ordinary bridge deck, containing only 4 aces, or do you conclude that the deck is stacked with more than 4 aces? It depends on how the cards were drawn. If the 4 aces are always placed at the top of a standard bridge deck, drawing 4 aces is not unusual—it is certain. On the other hand, if the cards are thoroughly mixed, drawing 4 aces in a sample of 4 cards is highly improbable. The point, of course, is that in order to use the observed sample of 4 cards to draw inferences about the population (the deck of 52 cards), you need to know how the sample was selected from the deck.

One of the simplest and most frequently employed sampling procedures produces what is known as a **random sample**.

DEFINITION 3.11

If *n* elements are selected from a population in such a way that every set of *n* elements in the population has an equal probability of being selected, the *n* elements are said to be a **random sample**.*

EXAMPLE 3.31

An experiment was conducted in which each of ten different antiscalants was added to an aliquot of brine. One of the ten brine solutions is to be selected, filtered, and the amount of silica determined. How would you select the brine solution so that the choice is random?

SOLUTION

If the choice is to be random, each brine solution must have the same probability of being drawn. That is, each solution should have a probability of $\frac{1}{10}$ of being selected. A method to achieve the objective of equal selection probabilities is to *thoroughly mix* the ten brine solutions and *blindly* pick one of the solutions. If this procedure were repeatedly used, each time replacing the selected solution, a particular solution should be chosen approximately $\frac{1}{10}$ of the time in a long series of draws. This method of sampling is known as **random sampling**. ∎

If a population is not too large and the elements can be numbered on slips of paper, poker chips, etc., you can physically mix the slips of paper or chips and remove *n* elements from the total. The numbers that appear on the chips selected would indicate the population elements to be included in the sample. Such a procedure will not guarantee a random sample, because it is often difficult to achieve a thorough mix, but it provides a reasonably good approximation to random sampling.

*Strictly speaking, this is a **simple random sample**. There are many different types of random samples. The simple random sample is the most common.

EXAMPLE 3.32

Suppose you want to randomly sample 5 (we will keep the number in the sample small to simplify our example) from a shipment of 100,000 bolts for quality control testing. Give a procedure for selecting this random sample.

SOLUTION

Since there are 100,000 bolts, it is not feasible to select a random sample by numbering slips of paper (or poker chips, etc.), mixing them, and choosing five. Instead, we will enlist the aid of Table 1 of Appendix II.

First, we number the bolts in the shipment from 1 to 100,000. Then, we turn to a page of Table 1, say the first page. (A partial reproduction of the first page of Table 1 is shown in Table 3.6.) Now, randomly select a starting number, say the random number appearing in the third row, second column. This number is 48360. Proceed down the second column to obtain the remaining four random numbers. The five selected random numbers are shaded in Table 3.6. Using the first five digits to represent the bolts from 1 to 99,999 and the number 00000 to represent bolt 100,000, you can see that the bolts numbered

48,360   93,093   39,975   6,907   72,905

should be included in your sample.

TABLE 3.6

Partial Reproduction of Table 1 of Appendix II

ROW \ COLUMN	1	2	3	4	5	6
1	10480	15011	01536	02011	81647	91646
2	22368	46573	25595	85393	30995	89198
3	24130	48360	22527	97265	76393	64809
4	42167	93093	06243	61680	07856	16376
5	37570	39975	81837	16656	06121	91782
6	77921	06907	11008	42751	27756	53498
7	99562	72905	56420	69994	98872	31016
8	96301	91977	05463	07972	18876	20922
9	89579	14342	63661	10281	17453	18103
10	85475	36857	53342	53988	53060	59533
11	28918	69578	88231	33276	70997	79936
12	63553	40961	48235	03427	49626	69445
13	09429	93969	52636	92737	88974	33488
14	10365	61129	87529	85689	48237	52267
15	07119	97336	71048	08178	77233	13916

Table 1 of Appendix II is just one example of a **table of random numbers**. Most samplers use such a table to obtain random samples. Random number tables are generated by computer in such a way that every number of the same length, i.e., containing the same number of digits, occurs with (approximately) equal probability. Further, the occurrence of any one number in a position is independent of any of the other numbers that appear in the table. To use a table of random numbers, number the N elements in the population from 1 to N. Then turn to Table 1 and select a starting number in the table. Proceeding from this number either across the rows or down the column, remove and record n numbers from the table. Use only the necessary number of digits in each random number to identify the element to be included in the sample.

3.44 Laboratory tests were conducted to compare the permeability of open-graded asphalt concrete with asphalt contents of 3% and 7% (*Journal of Testing and Evaluation*, July 1981). Eight batches of cement were prepared—four with a 3% asphalt mix and four with a 7% asphalt mix. Use Table 1 of Appendix II to randomly select the four batches that receive the 3% asphalt mix.

3.45 One of the most infamous examples of improper sampling was conducted in 1936 by the *Literary Digest* to determine the winner of the Landon–Roosevelt presidential election. The poll, which predicted Landon to be the winner, was conducted by sending ballots to a random sample of persons selected from among the names listed in the telephone directories of that year. In the actual election, Landon won in Maine and Vermont but lost in the remaining 46 states. The *Literary Digest's* erroneous forecast is believed to be the major reason for its eventual failure.

What was the cause of the *Digest's* erroneous forecast; i.e., why might the sampling procedure described above yield a sample of people whose opinions might be biased in favor of Landon?

3.46 Every 10 years the United States population census provides essential information about our nation and its people. The basic constitutional purpose of the census is to apportion the membership of the House of Representatives among the states. However, the census has many other important uses. For example, private business uses the census for plant location and marketing.

The 1980 census included questions on age, sex, race, marital status, family relationship, and income; this census was mailed to every household in the United States. In some cities, however, a series of questions was added for a 5% sample of the city's households. That is, each of a random sample of the city's households was mailed a census form that included additional questions. Suppose that a particular city contained 100,000 households and, of these, 5,000 were selected and mailed the longer census form.

a. If you worked for the Bureau of the Census and were assigned the task of selecting a random sample of 5,000 of the city's households, describe how you would proceed.

b. Suppose that one of the additional questions on the long form of the census concerned energy consumption. The city used this sample information to project the average energy consumption for the city's 100,000 households. Explain why it is important that the sample of 5,000 households be random.

c. Using the procedure you described in part a, randomly select a sample of 10 households from the 100,000 households in the city.

We have developed some of the basic tools of probability to enable us to assess the probabilities of various sample outcomes given a specific population structure. Although many of the examples we presented were of no practical importance, they accomplished their purpose if you now understand the concepts and definitions necessary for a basic understanding of probability.

In the next several chapters we will present probability models that can be used to solve practical problems. You will see that for most applications, we will need to make inferences about unknown aspects of these probability models, i.e., we will need to apply inferential statistics to the problem.

3.47 Researchers at the Upjohn Company have developed a new sustained-release tablet for a prescription drug. To determine the effectiveness of the tablet, the following experiment was conducted. Six tablets were randomly selected from each of 30 production lots. Each tablet was submersed in water and the percent dissolved was measured at 2, 4, 6, 8, 10, 12, 16, and 20 hours.*
  a. Find the total number of measurements (percent dissolved) recorded in the experiment.
  b. For each lot, the measurements at each time period are averaged. How many averages are obtained?

3.48 A state Department of Transportation (DOT) recently claimed that each of five bidders received equal consideration in the awarding of two road construction contracts and that, in fact, the two contract recipients were randomly selected from among the five bidders. Three of the bidders were large construction conglomerates and two were small specialty contractors. Suppose that both contracts were awarded to large construction conglomerates.
  a. What is the probability of this event occurring if, in fact, the DOT's claim is true?
  b. Is the probability computed in part **a** inconsistent with the DOT's claim that the selection was random?

3.49 The accompanying table summarizes the results of a recent report on the survival rates of U.S. corporations. The first column lists various ages (in years) of the firms and the second column gives the percentage of all U.S. corporations that survive to the specified age. Of those firms that survive to the specified age, the third column gives the percentages surviving at least 5 years beyond that age.

AGE	PERCENTAGE SURVIVING TO SPECIFIED AGE	PERCENTAGE SURVIVING AT LEAST 5 YEARS BEYOND SPECIFIED AGE
5	38 %	55%
10	21	65
15	14	70
20	10	73
25	7	76
50	2	83
75	1	86
100	.5	88

Source: Nystrom, P. C. and Starbuck, W. H. "To Avoid Organizational Crises, Unlearn." *Organizational Dynamics*, Spring 1984, Vol. 12, No. 4, p. 54. © 1984 American Management Association, New York. All rights reserved.

*Klassen, R. A. "The Application of Response Surface Methods to a Tablet Formulation Problem." Paper presented at Joint Statistical Meetings, American Statistical Association and Biometric Society, Aug. 1986, Chicago, Ill.

    a. Find the probability that a new corporation survives to age 15 years. (Assume that the survival rates remain unchanged.)

    b. Find the probability that a new corporation survives to at least age 15 years, given that it has survived to age 10 years.

    c. Find the probability that a new corporation does not survive to age 5 years (i.e., fails before 5 years).

3.50 Two newly designed data base management systems (DBMS), A and B, are being considered for marketing by a large computer software vendor. To determine whether DBMS users have a preference for one of the two systems, four of the vendor's customers are randomly selected and given the opportunity to evaluate the performances of each of the two systems. After sufficient testing, each user is asked to state which DBMS gave the better performance (measured in terms of CPU utilization, execution time, and disk access).

    a. Count the possible outcomes for this marketing experiment.

    b. If DBMS users actually have no preference for one system over the other (i.e., the performances of the two systems are identical), what is the probability that all four sampled users prefer system A?

    c. If all four customers express their preference for system A, can the software vendor infer that DBMS users in general have a preference for one of the two systems?

3.51 A traffic engineer conducted a study of the urban mass transportation habits of a city's workers. The study revealed the following: Fifteen percent of the city workers regularly drive their own car to work. Of those who do drive their own car to work, 80% would gladly switch to public mass transportation if it were available. Forty percent of the city workers live more than 3 miles from the center of the city.

    Suppose that one city worker is chosen at random. Define the events A, B, and C as follows:

    A: {The person regularly drives his or her own car to work}

    B: {The person would gladly switch to public mass transportation if it were available}

    C: {The person lives within 3 miles of the center of the city}

    a. Find $P(A)$.

    b. Find $P(B \mid A)$.

    c. Find $P(C)$.

    d. Explain whether the pairs of events, A and B, A and C, B and C, are mutually exclusive.

3.52 A study was conducted to examine the relationship between the cost structure and the mechanical properties of equiaxed grains in unidirectionally solidified ingots (*Metallurgical Transactions*, May 1986). Ingots composed of copper alloys were poured into one of three mold types (columnar, mixed, or equiaxed) with either a transverse or longitudinal orientation. From each ingot five tensile specimens were obtained at varying distances (10, 35, 60, 85, and 100 millimeters) from the ingot chill face and yield strength determined.

    a. How many strength measurements will be obtained if the experiment includes one ingot for each mold type–orientation combination?

    b. Suppose three of the ingots will be selected for further testing at the 100-mm distance. How many samples of three ingots can be selected from the total number of ingots in the experiment?

c. Use Table 1 of Appendix II to randomly select the three ingots for further testing.

d. Calculate the probability that the sample selected includes the three highest tensile strengths among all the ingots in the experiment.

e. Calculate the probability that the sample selected includes at least two of the three ingots with the highest tensile strengths.

3.53 Researchers at a major oil company have developed a new oil-drilling device. Six drilling sites are being considered as potential testing sites for the new device, each in a different state—Arizona, California, Louisiana, New Mexico, Oklahoma, and Texas. Due to the high cost of experimenting with this new device, only three testing sites will be used. These three sites will be chosen randomly from the six states.

a. What is the probability that both Arizona and Texas are selected for testing the new device?

b. What is the probability that California or Louisiana or both are selected for testing the new device?

c. Given that New Mexico is selected, what is the probability that Oklahoma is also selected?

3.54 Recently, the National Aeronautics and Space Administration (NASA) purchased a new solar-powered battery guaranteed to have a failure rate of only 1 in 20. A new system to be used in a space vehicle operates on one of these batteries. To increase the reliability of the system, NASA installed three batteries, each designed to operate if the preceding batteries in the chain fail. If the system is operated in a practical situation, what is the probability that all three batteries would fail?

3.55 A brewery utilizes two bottling machines, but they do not operate simultaneously. The second machine acts as a back-up system to the first machine, and operates only when the first breaks down during operating hours. The probability that the first machine breaks down during operating hours is .20. If in fact the first breaks down, then the second machine is turned on and has a probability of .30 of breaking down.

a. What is the probability that the brewery's bottling system is not working during operating hours?

b. The *reliability* of the bottling process is the probability that the system is working during operating hours. Find the reliability of the bottling process at the brewery.

3.56 The merging process from an acceleration lane to the through lane of a freeway constitutes an important aspect of traffic operation at interchanges. A study of parallel and tapered interchange ramps in Israel revealed the accompanying information on traffic lags (where a *lag* is defined as an interval of time between arrivals of major streams of vehicles) accepted and rejected by drivers in the merging lane.

TYPE OF INTERCHANGE LANE	TRAFFIC CONDITION ON FREEWAY	NUMBER OF MERGING DRIVERS ACCEPTING THE FIRST AVAILABLE LAG	NUMBER OF MERGING DRIVERS REJECTING THE FIRST AVAILABLE LAG
Tapered	Heavy traffic	16	115
	Little traffic	67	121
Parallel	Heavy traffic	40	139
	Little traffic	144	331

Source: Polus, A. and Livneh, M. "Vehicle Flow Characteristics on Acceleration Lanes." *Journal of Transportation Engineering*, Vol. III, No. 6, Nov. 1985, pp. 600–601 (Table 4).

a. What is the probability that a driver in a tapered merging lane with heavy traffic will accept the first available lag?

b. What is the probability that a driver in a parallel merging lane will reject the first available lag in traffic?

c. Given that a driver accepts the first available lag in little traffic, what is the probability that the driver is in a parallel merging lane?

3.57 The probability that a certain electronics component fails when first used is .10. If it does not fail immediately, the probability that it lasts for 1 year is .99. What is the probability that a new component will last 1 year?

3.58 There are three different managerial positions open in the computer lab of a large data-processing firm: data entry manager, data control manager, and computer operations manager. Suppose the firm will fill these positions from eight available candidates. (Assume that each of the eight candidates is qualified for all three managerial positions.) How many different ways can the firm fill the three positions?

3.59 In order to ensure delivery of its raw materials, a company has decided to establish a pattern of purchases with at least two potential suppliers. If five suppliers are available, how many choices (options) are available to the company?

3.60 A large research and development corporation is interested in providing an in-house continuing education program for its employees. It compiled the given table of percentages describing its 5,000 employees' current education level.

	HIGHEST DEGREE OBTAINED			
	High School Diploma	Bachelor's	Master's	Ph.D.
Males	5%	20%	12%	11%
Females	18%	15%	14%	5%

Suppose an employee is selected at random from the firm's 5,000 employees and the following events are defined:

A: {Employee chosen is a male}

B: {Employee chosen is a female}

C: {Highest degree obtained by the chosen employee is Ph.D.}

D: {Highest degree obtained by the chosen employee is master's}

E: {Highest degree obtained by the chosen employee is bachelor's}

F: {Highest degree obtained by the chosen employee is high school diploma}

Find:

a. $P(A \cup C)$ **b.** $P(B \cup F)$ **c.** $P(A \cap D)$ **d.** $P(E \cap B)$ **e.** $P(E \mid B)$

f. $P(A \mid D)$ **g.** Are A and D independent events?

h. Are E and B mutually exclusive events?

3.61 An assembler of computer terminals and modems uses parts from two sources. Company A supplies 80% of the parts and company B supplies the remaining 20% of the parts. From past experience, the assembler knows that 5% of the parts supplied by company A are defective and 3% of the parts supplied by company B are defective. An assembled modem selected at random is found to have a defective part. Which of the two companies is more likely to have supplied the defective part?

3.62 In an effort to assist the Occupational Safety and Health Administration in the development of federal safety standards, the Bureau of Labor Statistics conducted a survey of workers who suffer serious hand injuries on the job (*Engineering News-Record*, Mar. 3, 1983). The Bureau reported that, despite their relatively small numbers, carpenters account for 4% of all job-related hand injuries. The survey also indicated that 29% of the injured workers attributed their injuries to the pace at which they were working and 13% of those injured workers wearing hand protection claimed their gloves actually caused the accident. Assume that the events described above are independent.

    a. Find the probability that a worker with a job-related hand injury is a carpenter whose injury was caused by the pace at which he was working.

    b. Find the probability that a worker with a job-related hand injury does not attribute the cause of the accident to his protective hand gloves.

OPTIONAL SUPPLEMENTARY EXERCISES

3.63 Five construction companies each offer bids on three distinct Department of Transportation (DOT) contracts. A particular company will be awarded at most one DOT contract.

    a. How many different ways can the bids be awarded?

    b. Under the assumption that the simple events are equally likely, find the probability that company 2 is awarded a DOT contract.

    c. Suppose that companies 4 and 5 have submitted noncompetitive bids. If the contracts are awarded at random by the DOT, find the probability that both these companies receive contracts.

3.64 Refer to Exercise 3.19. In the two-dice game of craps, a player wins if he throws a *natural* (a 7 or 11) and loses if he throws *craps* (a 2, 3, or 12). However, if the sum of the two dice is 4, 5, 6, 8, 9, or 10 (each of these is known as a *point*), the player continues throwing the dice until the same outcome (point) is repeated (in which case the player wins), or the outcome 7 occurs (in which case the player loses). For example, if a player's first toss results in a 6, the player continues to toss the dice until a 6 or 7 occurs. If a 6 occurs first, the player wins. If a 7 occurs first, the player loses.

    a. What is the probability that a player throws a point on the first toss?

    b. If a player throws a point on the first toss, what is the probability the player wins the game on the next toss?

    c. If a player throws a point on the first toss, what is the probability the player loses the game on the next toss?

    d. Show that the probability a player wins the game (i.e., *makes a pass*) in two or fewer tosses is .327.

3.65 Consider 5-card poker hands dealt from a standard 52-card bridge deck. Two important events are:

    A: {You draw a flush}

    B: {You draw a straight}

    a. Find $P(A)$.   **b.** Find $P(B)$.

    c. The event that both A and B occur, i.e., $A \cap B$, is called a *straight flush*. Find $P(A \cap B)$.

[*Note:*   A *flush* consists of any 5 cards of the same suit. A *straight* consists of any 5 cards with values in sequence. In a straight, the cards may be of any suit and an ace may be considered as having a value of 1 or a value higher than a king.]

3.66 a. A professor asks his class to write a FORTRAN computer program that prints all three-letter sequences involving the five letters A, B, E, T, and O. How many different three-letter sequences will need to be printed?

b. Answer part **a** if the program is to be modified so that each three-letter sequence has at least one vowel.

3.67 The U.S. Command, Control, Communication and Intelligence (C^3I) System includes sensors (e.g., satellites and radars), communication links, and computer systems that allow gathering and processing of information that a missile attack on the continental United States may be on the way. The C^3I System contains two basic components: a warning system, designed to detect missile attacks on the United States, and a response system, designed to launch a counterattack. The response system cannot launch a missile unless a signal is detected by the warning system. However, if a warning signal is received, it is possible the response system will fail to launch a counterattack.

Paté-Cornell and Neu conducted a probabilistic analysis of the reliability of the C^3I System (*Risk Analysis*, Vol. 5, No. 2, 1985). In particular, the researchers were interested in the probabilities of an accidental strike by the C^3I System, i.e., the probability that the system launches a nuclear missile toward the U.S.S.R. based on a false alert.

Suppose the conditional probability that the response system will launch a missile toward the U.S.S.R., given a false alert from the warning system, is .90. Suppose also that the probability of false alert (i.e., the probability that the warning system sends a signal to the response system, given no Soviet attack is made) is .02. If the probability of a Soviet missile attack on the continental United States is .01, find the probability of an accidental strike by the C^3I System. [*Hint:*   For three events, *A, B,* and *C,* the formula

$$P(A \cap B \cap C) = P(C \mid A \cap B) \cdot P(B \mid A) \cdot P(A)$$

gives the probability of their intersection.]

REFERENCES

Feller, W. *An Introduction to Probability Theory and Its Applications*, 3rd ed. Vol. 1. New York: Wiley, 1968.

McClave, J. T. and Dietrich, F. H. II. *Statistics*, 3rd ed. San Francisco: Dellen Publishing Co., 1985.

Mendenhall, W., Scheaffer, R. L., and Wackerly, D. *Mathematical Statistics with Applications*, 3rd ed. Boston: Duxbury Press, 1986.

Parzen, E. *Modern Probability Theory and Its Applications*. New York: Wiley, 1960.

CHAPTER 4

OBJECTIVE

To explain what is meant by a discrete random variable, its probability distribution, and corresponding numerical descriptive measures; to present some useful discrete probability distributions and show how they can be used to solve practical problems.

CONTENTS

DISCRETE RANDOM VARIABLES

**DISCRETE RANDOM
VARIABLES**

As we noted in Chapter 1, the experimental events of greatest interest are often numerical, i.e., we conduct an experiment and observe the numerical value of some variable. If we repeat the experiment n times, we obtain a sample of quantitative data. To illustrate, suppose a manufactured product (e.g., a mechanical part) is sold in lots of 20 boxes of 12 items each. As a check on the quality of the product, a process control engineer randomly selects four from among the 240 items in a lot and checks to determine whether the items are defective. If more than one sampled item is found to be defective, the entire lot will be rejected.

The selection of four manufactured items from among 240 produces a sample space S that contains $\binom{240}{4}$ simple events, one corresponding to each possible combination of four items that might be selected from the lot. Although a description of a specific simple event would identify the four items acquired in a particular sample, the event of interest to the process control engineer is an observation on the variable y, the number of defective items among the four items that are tested. To each simple event in S, there corresponds one and only one value of the variable y. Therefore, a functional relation exists between the simple events in S and the values that y can assume. The event $y = 0$ is the collection of all simple events that contain no defective items. Similarly, the event $y = 1$ is the collection of all simple events in which one defective item is observed. Since the value that y can assume is a numerical event (i.e., an event defined by some number that varies in a random manner from one repetition of the experiment to another), it is called a **random variable**.

DEFINITION 4.1

A **random variable** is a numerical-valued function defined over a sample space.

The number y of defective items in a selection of four items from among 240 is an example of a **discrete random variable**, one that can assume a countable number of values. For our example, the random variable y may assume any of the five values, $y = 0, 1, 2, 3,$ or 4. As another example, the number y of jobs received by a computer center in a day is also a discrete random variable which could, theoretically, assume a value that is large beyond all bound. The possible values for this discrete random variable correspond to the nonnegative integers, $y = 0, 1, 2, 3, \ldots, \infty$, and the number of such values is countable.

DEFINITION 4.2

A **discrete random variable** y is one that can assume only a countable number of values.

Random variables observed in nature often possess similar characteristics and consequently can be classified according to type. In this chapter, we will study five different types of discrete random variables and will use the methods of Chapter 3 to derive the probabilities associated with their possible values. We will also begin to develop some intuitive ideas about how the probabilities of observed sample data can be used to make statistical inferences.

| | | | | | | | | | | |

SECTION 4.2

THE PROBABILITY DISTRIBUTION FOR A DISCRETE RANDOM VARIABLE

Since the values that a random variable y can assume are numerical events, we will want to calculate their probabilities. A table, formula, or graph that gives these probabilities is called the **probability distribution** for the random variable y. We will illustrate this concept using a simple coin tossing example.

EXAMPLE 4.1

A balanced coin is tossed twice and the number y of heads is observed. Find the probability distribution for y.

SOLUTION

Let H_i and T_i denote the observation of a head and a tail, respectively, on the ith toss, for $i = 1, 2$. The four simple events and the associated values of y are shown in Table 4.1.

TABLE 4.1

SIMPLE EVENT	DESCRIPTION	$P(E_i)$	NUMBER OF HEADS y
E_1	H_1H_2	$\frac{1}{4}$	2
E_2	H_1T_2	$\frac{1}{4}$	1
E_3	T_1H_2	$\frac{1}{4}$	1
E_4	T_1T_2	$\frac{1}{4}$	0

The event $y = 0$ is the collection of all simple events that yield a value of $y = 0$, namely, the single simple event E_4. Therefore, the probability that y assumes the value 0 is

$$P(y = 0) = p(0) = P(E_4) = \frac{1}{4}$$

The event $y = 1$ contains two simple events, E_2 and E_3. Therefore,

$$P(y = 1) = p(1) = P(E_2) + P(E_3) = \frac{1}{4} + \frac{1}{4} = \frac{1}{2}$$

Finally,

$$P(y = 2) = p(2) = P(E_1) = \frac{1}{4}$$

TABLE 4.2

Probability Distribution for
y, the Number of Heads in
Two Tosses of a Coin

y	$p(y)$
0	$\frac{1}{4}$
1	$\frac{1}{2}$
2	$\frac{1}{4}$
$\sum_y p(y) = 1$	

The probability distribution $p(y)$ is displayed in tabular form in Table 4.2 and as a probability histogram in Figure 4.1. As we will show in Section 4.5, this probability distribution can also be given by the formula

$$p(y) = \frac{\binom{2}{y}}{4}$$

where

$$p(0) = \frac{\binom{2}{0}}{4} = \frac{1}{4}$$

$$p(1) = \frac{\binom{2}{1}}{4} = \frac{2}{4} = \frac{1}{2}$$

$$p(2) = \frac{\binom{2}{2}}{4} = \frac{1}{4}$$

FIGURE 4.1

Probability Distribution for
y, the Number of Heads in
Two Tosses of a Coin

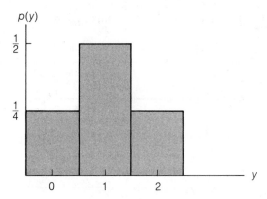

DEFINITION 4.3

The **probability distribution** for a discrete random variable y is a table, graph, or formula that gives the probability $p(y)$ associated with each possible value of y.

The probability distribution $p(y)$ for a discrete random variable must satisfy two properties. First, because $p(y)$ is a probability, it must assume a value in the interval $0 \leq p(y) \leq 1$. Second, the sum of the values of $p(y)$ over all values of y must equal 1. This is true because we assigned one and only one value of y to each simple event in S. It follows that the values that y can assume represent different sets of simple events and are, therefore, mutually exclusive events.

Summing $p(y)$ over all possible values of y is then equivalent to summing the probabilities of all simple events in S, and from Section 3.2, $P(S)$ is known to be equal to 1.

REQUIREMENTS FOR A DISCRETE PROBABILITY DISTRIBUTION

1. $0 \le p(y) \le 1$

2. $\sum_{\text{all } y} p(y) = 1$

To conclude this section, we will discuss the relationship between the probability distribution for a discrete random variable and the relative frequency distribution of data (discussed in Section 1.4). Suppose you were to toss two coins over and over again a very large number of times and record the number y of heads observed for each toss. A relative frequency distribution for the resulting collection of 0's, 1's, and 2's would be very similar to the probability distribution shown in Figure 4.1. In fact, if it were possible to repeat the experiment an infinitely large number of times, the two distributions would be almost identical. (We say "almost" identical because there are probably few, if any, perfectly balanced coins.) Thus, the probability distribution of Figure 4.1 provides a **model** for a conceptual population of values of y—the values of y that would be observed if the experiment were to be repeated an infinitely large number of times.

Commencing with Section 4.5, we will introduce a number of discrete random variables for many populations of data that occur in the physical, biological, social, and information sciences.

EXERCISES 4.1–4.4

4.1 The director of marketing for a small computer manufacturer believes the discrete probability distribution shown in the accompanying figure characterizes the number, y, of new computers the firm will lease next year.

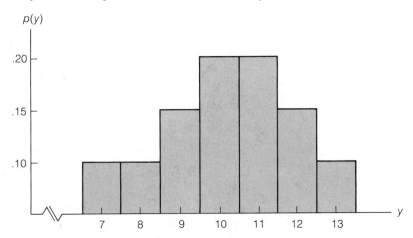

a. Is this a valid probability distribution? Explain.
b. Display the probability distribution in tabular form.
c. What is the probability that exactly 9 computers will be leased?
d. What is the probability that fewer than 12 computers will be leased?

4.2 Consider the segment of an electric circuit with three relays shown here. Current will flow from A to B if there is at least one closed path when the switch is thrown. Each of the three relays has an equally likely chance of remaining open or closed when the switch is thrown. Let y represent the number of relays that close when the switch is thrown.

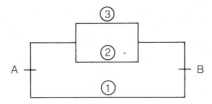

a. Find the probability distribution for y and display it in tabular form.
b. What is the probability that current will flow from A to B?

4.3 Refer to the *Metal Progress* (May 1986) study of steel sheet usage at Mazda Motor Corporation, introduced in Exercise 3.18. The table listing the seven steel types and percentages used in production is reproduced here. Suppose that three steel sheets are randomly selected from among those used in production of Mazda 626's. Find the probability distribution of y, the number of cold rolled sheets in the sample.

TYPE OF STEEL SHEET	PERCENTAGE USED
Cold rolled	27
Cold rolled, high strength	12
Cold rolled, plated	30
Cold rolled, high strength, plated	15
Hot rolled	8
Hot rolled, high strength	5
Hot rolled, plated	3
	100

Source: Chandler, H. E. "Materials Trends at Mazda Motor Corporation." *Metal Progress*, Vol. 129, No. 6, May 1986, p. 57 (Figure 3).

4.4 A quality control engineer samples five from a large lot of manufactured firing pins and checks for defects. Unknown to the inspector, three of the five sampled firing pins are defective. The engineer will test the five pins in a randomly selected order until a defective is observed (in which case the entire lot will be rejected). Let y be the number of firing pins the quality control engineer must test. Find the probability distribution of y.

SECTION 4.3

THE EXPECTED VALUE FOR A RANDOM VARIABLE *y* OR FOR A FUNCTION *g(y)* OF *y*

Since a probability distribution for a random variable *y* is a model for a population relative frequency distribution, we can describe it with numerical descriptive measures, such as its mean and standard deviation, and we can use Tchebysheff's theorem and the Empirical Rule to identify improbable values of *y*.

The **expected value** (or **mean**) of a random variable *y*, denoted by the symbol $E(y)$, is defined as follows:

DEFINITION 4.4

Let *y* be a discrete random variable with probability distribution $p(y)$. Then the **mean** or **expected value of y** is

$$\mu = E(y) = \sum_{\text{all } y} yp(y)$$

EXAMPLE 4.2

Refer to the two-coin tossing experiment of Example 4.1 and the probability distribution for the random variable *y*, shown in Figure 4.1. Demonstrate that the formula for $E(y)$ yields the mean of the probability distribution for the discrete random variable *y*.

SOLUTION

If we were to repeat the coin tossing experiment a large number of times—say, 400,000 times—we would expect to observe $y = 0$ heads approximately 100,000 times, $y = 1$ head approximately 200,000 times, and $y = 2$ heads approximately 100,000 times. If we calculate the mean value of these 400,000 values of *y*, we obtain

$$\mu \approx \frac{\sum y}{n} = \frac{100{,}000(0) + 200{,}000(1) + 100{,}000(2)}{400{,}000}$$

$$= 0\left(\frac{100{,}000}{400{,}000}\right) + 1\left(\frac{200{,}000}{400{,}000}\right) + 2\left(\frac{100{,}000}{400{,}000}\right)$$

$$= 0\left(\frac{1}{4}\right) + 1\left(\frac{1}{2}\right) + 2\left(\frac{1}{4}\right) = \sum_{\text{all } y} yp(y)$$ ∎

If *y* is a random variable, so also is any function $g(y)$ of *y*. The **expected value** of $g(y)$ is defined as follows:

DEFINITION 4.5

Let *y* be a discrete random variable with probability distribution $p(y)$ and let $g(y)$ be a function of *y*. Then the **mean** or **expected value of g(y)** is

$$E[g(y)] = \sum_{\text{all } y} g(y)p(y)$$

One of the most important functions of a discrete random variable *y* is its **variance**, i.e., the expected value of the squared deviation of *y* from its mean μ.

> **DEFINITION 4.6**
>
> Let y be a discrete random variable with probability distribution $p(y)$. Then the **variance of** y is
>
> $$\sigma^2 = E[(y - \mu)^2]$$
>
> The **standard deviation of** y is the positive square root of the variance of y:
>
> $$\sigma = \sqrt{\sigma^2}$$

EXAMPLE 4.3

Refer to the two-coin tossing experiment of Example 4.1 and the probability distribution for y, shown in Figure 4.1. Find the variance and standard deviation of y.

SOLUTION

In Example 4.2, we found that the mean value of y is $\mu = 1$. Then

$$\sigma^2 = E[(y - \mu)^2] = \sum_{y=0}^{2} (y - \mu)^2 p(y)$$

$$= (0 - 1)^2\left(\frac{1}{4}\right) + (1 - 1)^2\left(\frac{1}{2}\right) + (2 - 1)^2\left(\frac{1}{4}\right) = \frac{1}{2}$$

and

$$\sigma = \sqrt{\sigma^2} = \sqrt{\frac{1}{2}} = .707 \qquad\blacksquare$$

EXAMPLE 4.4

Refer to Example 4.3 and find the probability that y will fall in the interval $\mu \pm 2\sigma$.

SOLUTION

From Examples 4.2 and 4.3, we know that $\mu = 1$ and $\sigma = .707$. Then the interval $\mu \pm 2\sigma$ is $-.414$ to 2.414. Since y must assume one of only three values, $y = 0$, 1, and 2, all of which fall in the computed interval, the probability that y falls in the interval $\mu \pm 2\sigma$ is 1.0. Clearly, the Empirical Rule* (used in Chapter 1 to describe the variation for a finite set of data and the spread of its relative frequency histogram) provides an adequate description of the spread or variation in the probability distribution of Figure 4.1. $\qquad\blacksquare$

EXERCISES 4.5–4.11

4.5 Find the mean and variance of the probability distribution shown in Exercise 4.1.

4.6 Find the mean and variance of the probability distribution derived in Exercise 4.2.

---

*A variation of Tchebysheff's theorem, first presented in Chapter 1, states that the probability that a random variable y falls in the interval $\mu \pm k\sigma$, for $k \geq 1$, is at least equal to $(1 - 1/k^2)$. We omit the proof of this theorem.

4.7 Find the mean and variance of the probability distribution derived in Exercise 4.3.

4.8 Refer to Exercise 4.4. Suppose the cost of testing a single firing pin is $200.
a. What is the expected cost of inspecting the lot?
b. What is the variance?
c. Within what range would you expect the inspection cost to fall?

4.9 Refer to the oil exploration study discussed in Exercise 3.31. Kinchen (1986) gives an example in which a $50,000 exploration budget is allocated to a single prospect. The well can result in either a dry hole, 50,000 barrels (bbl), 100,000 bbl, 500,000 bbl, or 1,000,000 bbl, with probabilities and monetary outcomes shown in the table. Let y represent the monetary value of a single oil prospect. Find $E(y)$ and σ^2.

POSSIBLE RESULTS (bbl)	MONETARY OUTCOME ($)	PROBABILITY
Dry hole	−50,000	.60
50,000	−20,000	.10
100,000	30,000	.15
500,000	430,000	.10
1,000,000	950,000	.05

Source: Kinchen, A. L. "Projected Outcomes of Exploration Programs Based on Current Program Status and the Impact of Prospects Under Consideration." *Journal of Petroleum Technology*, Vol. 38, No. 4, April 1986, p. 462 (Table 1).

4.10 A company's marketing and accounting departments have determined that if the company markets its newly developed magnetron tube (the major component in microwave ovens), the contribution of the new tube to the firm's profit during the next 6 months is described by the probability distribution shown in the table. The company has decided it should market the new line of magnetron tubes if the expected contribution to profit for the next 6 months is over $10,000. Based on the probability distribution, will the company market the new tube?

PROFIT CONTRIBUTION	p(PROFIT CONTRIBUTION)
−$5,000[a]	.3
$10,000	.4
$30,000	.3

[a]A negative contribution is a loss.

TIME TO EVACUATE (nearest hour)	PROBABILITY
13	.04
14	.25
15	.40
16	.18
17	.10
18	.03

4.11 A panel of meteorological and civil engineers studying emergency evacuation plans for Florida's Gulf Coast in the event of a hurricane has estimated that it would take between 13 and 18 hours to evacuate people living in low-lying land with the probabilities shown in the table.
a. Calculate the mean and standard deviation of the probability distribution of the evacuation times.
b. Within what range would you expect the time to evacuate to fall?
c. Weather forecasters say they cannot accurately predict a hurricane landfall more than 14 hours in advance. If the Gulf Coast Civil Engineering Department waits

until the 14-hour warning before beginning evacuation, what is the probability that all residents of low-lying areas are evacuated safely (i.e., before the hurricane hits the Gulf Coast)?

| | | | | | | | | | | | |

SECTION 4.4

SOME USEFUL EXPECTATION THEOREMS

We now present three theorems that are especially useful in finding the expected value of a function of a random variable. We will leave the proofs of these theorems as optional exercises.

THEOREM 4.1

Let y be a discrete random variable with probability distribution $p(y)$ and let c be a constant. Then the expected value (or mean) of c is

$$E(c) = c$$

THEOREM 4.2

Let y be a discrete random variable with probability distribution $p(y)$ and let c be a constant. Then the expected value (or mean) of cy is

$$E(cy) = cE(y)$$

THEOREM 4.3

Let y be a discrete random variable with probability distribution $p(y)$, and let $g_1(y), g_2(y), \ldots, g_k(y)$ be functions of y. then

$$E[g_1(y) + g_2(y) + \cdots + g_k(y)] = E[g_1(y)] + E[g_2(y)] + \cdots E[g_k(y)]$$

Theorems 4.1–4.3 can be used to derive a simple formula for computing the variance of a random variable, as given by Theorem 4.4.

THEOREM 4.4

$$\sigma^2 = E(y^2) - \mu^2$$

Proof of Theorem 4.4   From Definition 4.6, we have the following expression for σ^2:

$$\sigma^2 = E[(y - \mu)^2] = E(y^2 - 2\mu y + \mu^2)$$

Applying Theorem 4.3 yields

$$\sigma^2 = E(y^2) + E(-2\mu y) + E(\mu^2)$$

We now apply Theorems 4.1 and 4.2 to obtain

$$\sigma^2 = E(y^2) - 2\mu E(y) + \mu^2 = E(y^2) - 2\mu(\mu) + \mu^2$$
$$= E(y^2) - 2\mu^2 + \mu^2$$
$$= E(y^2) - \mu^2$$

We will use Theorem 4.4 to derive the variances for some of the discrete random variables presented in the following sections. The method is demonstrated in Example 4.5.

EXAMPLE 4.5

Use Theorem 4.4 to find the variance for the random variable y of Example 4.1.

SOLUTION

In Example 4.3, we found the variance of y, the number of heads observed in the tossing of two coins, by finding $\sigma^2 = E[(y - \mu)^2]$ directly. Since this can be a tedious procedure, it is usually easier to find $E(y^2)$ and then use Theorem 4.4 to compute σ^2. For our example,

$$E(y^2) = \sum_{\text{all } y} y^2 p(y) = (0)^2\left(\frac{1}{4}\right) + (1)^2\left(\frac{1}{2}\right) + (2)^2\left(\frac{1}{4}\right) = 1.5$$

Substituting the value $\mu = 1$ (obtained in Example 4.2) into the statement of Theorem 4.4, we have

$$\sigma^2 = E(y^2) - \mu^2$$
$$= 1.5 - (1)^2 = .5$$

Note that this is the value of σ^2 that we obtained in Example 4.3.   ∎

In Sections 4.5–4.8, we will present some useful discrete probability distributions and will state without proof the mean, variance, and standard deviation for each. Some of these quantities will be derived in optional examples; other derivations will be left as optional exercises.

EXERCISES 4.12–4.16

4.12 Refer to Exercises 4.1 and 4.5. The manufacturer leases new computers at a cost of $15,000 per year. Find the mean and variance of the total amount the company will earn next year from leasing computers.

4.13 Refer to Exercise 4.4, where y is the number of firing pins tested in a sample of five selected from a large lot. Suppose the cost of inspecting a single pin is $300 if the pin is defective and $100 if not. Then the total cost C (in dollars) of the inspection is given by the equation $C = 200 + 100y$. Find the mean and variance of C.

OPTIONAL EXERCISES

4.14 Prove Theorem 4.1. [*Hint:* Use the fact that $\Sigma_{\text{all } y}\, p(y) = 1$.]

4.15 Prove Theorem 4.2. [*Hint:* The proof follows directly from Definition 4.5.]

4.16 Prove Theorem 4.3.

| | | | | | | | | | | |
S E C T I O N 4.5

**THE BINOMIAL
PROBABILITY
DISTRIBUTION**

Many real-life experiments are analogous to tossing an unbalanced coin a number n of times. Suppose that 80% of the jobs submitted to a data-processing center are of a statistical nature. Then selecting a random sample of ten submitted jobs would be analogous to tossing an unbalanced coin ten times, with the probability of tossing a head (drawing a statistical job) on a single trial equal to .80. Public opinion or consumer preference polls that elicit one of two responses—yes or no, approve or disapprove, etc.—are also analogous to the unbalanced coin tossing experiment if the number N in the population is large and if the sample size n is relatively small, say $.10N$ or less. All these experiments are particular examples of a **binomial experiment**. Such experiments and the resulting binomial random variables possess the characteristics stated in the box.

CHARACTERISTICS THAT DEFINE A BINOMIAL RANDOM VARIABLE

1. The experiment consists of n identical trials.
2. There are only two possible outcomes on each trial. We will denote one outcome by S (for Success) and the other by F (for Failure).
3. The probability of S remains the same from trial to trial. This probability will be denoted by p, and the probability of F will be denoted by q. Note that $p + q = 1$.
4. The trials are independent.
5. The binomial random variable y is the number of S's in n trials.

The binomial probability distribution, its mean, and its variance are shown in the following box.

**THE PROBABILITY DISTRIBUTION, MEAN, AND VARIANCE
FOR A BINOMIAL RANDOM VARIABLE**

The probability distribution for a binomial random variable is given by

$$p(y) = \binom{n}{y} p^y q^{n-y} \qquad (y = 0, 1, 2, \ldots, n)$$

where

p = Probability of a success on a single trial

$q = 1 - p$

n = Number of trials

y = Number of successes in n trials

$$\binom{n}{y} = \frac{n!}{y!(n-y)!}$$

The mean and variance of the binomial random variable are, respectively,

$$\mu = np \qquad \text{and} \qquad \sigma^2 = npq$$

The binomial probability distribution is derived as follows. A simple event for a binomial experiment consisting of n trials can be represented by the symbol

$$SFSFFFSSSF \ldots SFS$$

where the letter in the ith position, proceeding from left to right, denotes the outcome of the ith trial. Since we want to find the probability $p(y)$ of observing y successes in the n trials, we will need to sum the probabilities of all simple events that contain y successes (S's) and $(n - y)$ failures (F's). Such simple events would appear symbolically as

$$\overbrace{SSSS \ldots S}^{y} \overbrace{FF \ldots F}^{(n - y)}$$

or some different arrangement of these symbols.

Since the trials are independent, the probability of a *particular* simple event implying y successes is

$$P(\overbrace{SSS \ldots S}^{y} \overbrace{FF \ldots F}^{(n - y)}) = p^y q^{n-y}$$

The *number* of these equiprobable simple events is equal to the number of ways we can arrange the y S's and the $(n - y)$ F's in n positions corresponding to the n trials. This is equal to the number of ways of selecting y positions (trials) for the y S's from a total of n positions. This number, given by Theorem 3.4, is

$$\binom{n}{y} = \frac{n!}{y!(n - y)!}$$

We have determined the probability of each simple event that results in y successes, as well as the number of such events. We now sum the probabilities of these simple events to obtain

$$p(y) = \left(\begin{array}{c} \text{Number of simple events} \\ \text{implying } y \text{ successes} \end{array}\right) \left(\begin{array}{c} \text{Probability of one of these} \\ \text{equiprobable simple events} \end{array}\right)$$

or

$$p(y) = \binom{n}{y} p^y q^{n-y}$$

EXAMPLE 4.6

Tests for impurities commonly found in drinking water from private wells showed that 30% of all wells in a particular county have impurity A. If a random sample of five wells is selected from the large number of wells in the county, what is the probability that:

a. Exactly three will have impurity A?
b. At least three?
c. Fewer than three?

SOLUTION

The first step is to confirm that this experiment possesses the characteristics of a binomial experiment. The experiment consists of $n = 5$ trials, one corresponding to each randomly selected well. Each trial results in an S (the well contains impurity A) or an F (the well does not contain impurity A). Since the total number of wells in the county is large, the probability of drawing a single well and finding that it contains impurity A is equal to .3, and this probability will remain approximately the same (for all practical purposes) for each of the five selected wells. Further, since the sampling was random, we assume that the outcome on any one well is unaffected by the outcome of any other and that the trials are independent. Finally, we are interested in the number y of wells in the sample of $n = 5$ that contain impurity A. Therefore, the sampling procedure represents a binomial experiment with $n = 5$ and $p = .3$.

a. The probability of drawing exactly $y = 3$ wells containing impurity A is

$$p(y) = \binom{n}{y}p^y q^{n-y}$$

where $n = 5$, $p = .3$, and $y = 3$. Thus,

$$p(3) = \frac{5!}{3!2!}(.3)^3(.7)^2 = .13230$$

b. The probability of observing at least three wells containing impurity A is

$$P(y \geq 3) = p(3) + p(4) + p(5)$$

where

$$p(4) = \frac{5!}{4!1!}(.3)^4(.7)^1 = .02835$$

$$p(5) = \frac{5!}{5!0!}(.3)^5(.7)^0 = .00243$$

Since we found $p(3)$ in part **a**, we have

$$P(y \geq 3) = p(3) + p(4) + p(5)$$
$$= .13230 + .02835 + .00243 = .16308$$

c. Although $P(y < 3) = p(0) + p(1) + p(2)$, we can avoid calculating these probabilities by using the complementary relationship and the fact that $\sum_{y=0}^{n} p(y) = 1$. Therefore,

$$P(y < 3) = 1 - P(y \geq 3) = 1 - .16308 = .83692 \qquad \blacksquare$$

Tables that give partial sums of the form

$$\sum_{y=0}^{k} p(y)$$

for binomial probabilities are given in Table 2 of Appendix II, for $n = 5$, 10, 15, 20, and 25. For example, you will find that the partial sum given in the table for $n = 5$, in the row corresponding to $k = 2$ and the column corresponding to $p = .3$, is

$$\sum_{y=0}^{2} p(y) = p(0) + p(1) + p(2) = .837$$

This answer, correct to three decimal places, agrees with our answer to part c of Example 4.6.

EXAMPLE 4.7

Find the mean, variance, and standard deviation for a binomial random variable with $n = 20$ and $p = .6$. Construct the interval $\mu \pm 2\sigma$ and compute $P(\mu - 2\sigma < y < \mu + 2\sigma)$.

SOLUTION

Applying the formulas given previously, we have

$$\mu = np = 20(.6) = 12$$
$$\sigma^2 = npq = 20(.6)(.4) = 4.8$$
$$\sigma = \sqrt{4.8} = 2.19$$

The binomial probability distribution for $n = 20$ and $p = .6$ and the interval $\mu \pm 2\sigma$, or 7.62 to 16.38, are shown in Figure 4.2. The values of y that lie in

FIGURE 4.2
Binomial Probability
Distribution for y in
Example 4.7
$(n = 20, p = .6)$

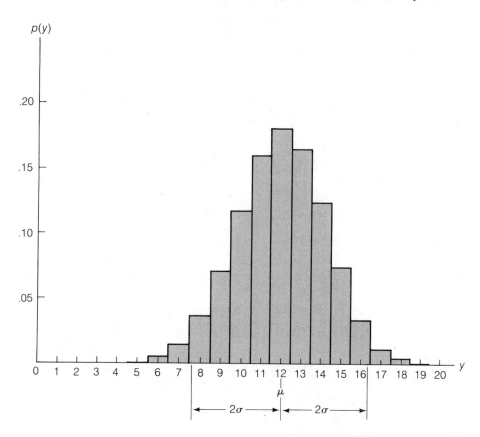

the interval $\mu \pm 2\sigma$ are 8, 9, ... , 16. Therefore,

$$P(\mu - 2\sigma < y < \mu + 2\sigma) = P(y = 8, 9, 10, \ldots, \text{ or } 16)$$
$$= \sum_{y=0}^{16} p(y) - \sum_{y=0}^{7} p(y)$$

We obtain the values of these partial sums from Table 2 of Appendix II:

$$P(\mu - 2\sigma < y < \mu + 2\sigma) = \sum_{y=0}^{16} p(y) - \sum_{y=0}^{7} p(y)$$
$$= .984 - .021 = .963$$

You can see that this result agrees with Tchebysheff's theorem and is close to the value of .95 specified by the Empirical Rule, discussed in Chapter 1. ∎

EXAMPLE 4.8 (OPTIONAL)   Derive the formula for the expected value for the binomial random variable.

SOLUTION   By Definition 4.4,

$$\mu = E(y) = \sum_{\text{all } y} yp(y) = \sum_{y=0}^{n} y \frac{n!}{y!(n-y)!} p^y q^{n-y}$$

The easiest way to sum these terms is to convert them into binomial probabilities and then use the fact that $\sum_{y=0}^{n} p(y) = 1$. Noting that the first term of the summation is equal to 0 (since $y = 0$), we have

$$\mu = \sum_{y=1}^{n} y \frac{n!}{[y(y-1) \cdots 3 \cdot 2 \cdot 1](n-y)!} p^y q^{n-y}$$

$$= \sum_{y=1}^{n} \frac{n!}{(y-1)!(n-y)!} p^y q^{n-y}$$

Because n and p are constants, we can use Theorem 4.2 to factor np out of the sum:

$$\mu = np \sum_{y=1}^{n} \frac{(n-1)!}{(y-1)!(n-y)!} p^{y-1} q^{n-y}$$

Let $z = (y - 1)$. Then when $y = 1$, $z = 0$ and when $y = n$, $z = (n - 1)$; thus,

$$\mu = np \sum_{y=1}^{n} \frac{(n-1)!}{(y-1)!(n-y)!} p^{y-1} q^{n-y}$$

$$= np \sum_{z=0}^{n-1} \frac{(n-1)!}{z![(n-1)-z]!} p^z q^{(n-1)-z}$$

The quantity inside the summation sign is $p(z)$, where z is a binomial random variable based on $(n - 1)$ trials. Therefore,

$$\sum_{z=0}^{n-1} p(z) = 1$$

and

$$\mu = np \sum_{z=0}^{n-1} p(z) = np(1) = np \qquad \blacksquare$$

| | | | | | | | | | | | | |

EXERCISES 4.17–4.29

4.17 Use the formula for the binomial probability distribution to find the probabilities for $n = 4$, $p = .5$, and $y = 0, 1, 2, 3,$ and 4.

4.18 Use the binomial probabilities given in Table 2 of Appendix II to find $p(y)$ for $n = 10$ and
a. $p = .1$ **b.** $p = .5$ **c.** $p = .9$
d. Construct graphs (similar to Figure 4.2) of the three probability distributions in parts **a–c**. Note the symmetry of the distribution for $p = .5$ and the skewness for $p = .1$ and $p = .9$.

4.19 Executives in the chemical industry claim that only 5% of all chemical plants in the United States discharge more than the EPA's suggested maximum amount of toxic waste into the air and water. Suppose that the EPA randomly samples 20 of the very large number of chemical plants for inspection. If in fact the executives' claim is true, what is the probability that the number y of plants in violation of the EPA's standard is:
a. Less than 1?   **b.** Less than or equal to 1?
c. Less than 2?   **d.** More than 1?
e. What would you infer about the executives' claim if the observed value of y is 3? Explain.

4.20 Refer to the neutral particle transport problem described in Exercise 3.33. Recall that particles released into an evacuated duct collide with the inner duct wall and are either scattered (reflected) with probability .16 or absorbed with probability .84 (*Nuclear Science and Engineering*, May 1986).
a. If four particles are released into the duct, what is the probability that all four will be absorbed by the inner duct wall? Exactly three of the four?
b. If 20 particles are released into the duct, what is the probability that at least 10 will be reflected by the inner duct wall? Exactly ten?

4.21 A particular system in a space vehicle must work properly in order for the spaceship to gain reentry into the earth's atmosphere. One component of the system operates successfully only 85% of the time. To increase the reliability of the system, four of the components will be installed in such a way that the system will operate successfully if at least one component is working successfully.
a. What is the probability that the system will fail? Assume the components operate independently.
b. If the system does in fact fail, what would you infer about the claimed 85% success rate of a single component?

4.22 A random sample of 25 computer programmers was selected to compare two full-screen-edit intelligent terminals, one manufactured by vendor A and the other by vendor B. After working with the two types of terminals, each programmer was asked to choose the terminal that he or she preferred. Let y be the number of sampled programmers who prefer the terminal manufactured by vendor A.

a. Is y a binomial random variable? Explain.

b. If the two terminals are, in fact, identical, then $p = .5$. Find $P(y \geq 19)$, assuming $p = .5$.

c. If you observe $y = 19$, what would you infer about the true value of p?

4.23 During the 1950's, a number of atomic weapons tests were conducted in the desert in Nevada. Since that time, estimates of radiation exposure to off-site populations, especially Utah, have been the subject of considerable scientific effort. The Surveillance, Epidemiology, and End Results (SEER) Registry collected data on incidence of thyroid cancer among Utah residents over the period 1973–1977. SEER found that the incidence rate of thyroid cancer among 50-year-old males is 3.89 per 100,000 population (*Health Physics*, Jan. 1986). This implies that the probability of a 50-year-old Utah male developing thyroid cancer is .0000389. In a random sample of 1,000 50-year-old Utah males, let y equal the number developing thyroid cancer.

a. Calculate the mean and variance of y.

b. Would you expect to observe at least one 50-year-old male with thyroid cancer among the 1,000? [*Hint:* Use Tchebysheff's theorem and the results of part **a.**]

4.24 An engineering development laboratory conducted an experiment to investigate the life characteristics of a new solar heating panel, designed to have a useful life of at least 5 years with probability $p = .95$. A random sample of 20 such solar panels was selected and the useful life of each was recorded.

a. What is the probability that exactly 18 will have a useful life of at least 5 years?

b. What is the probability that at most 10 will have a useful life of at least 5 years?

c. If only 10 of the 20 solar panels have a useful life of at least 5 years, what would you infer about the true value of p?

4.25 "A recent Louis Harris and Associates poll on public attitudes toward water pollution cleanup found overwhelming support for tougher clean water standards, even if it means more expensive goods and services or fewer jobs" (*Engineering News-Record*, Dec. 23, 1982). Forty-six percent of those surveyed believe that government regulations regarding pollution control are "not protective enough," while 60% felt that "Congress should make the Clean Water Act stricter." In response to a question about whether industry should be required to install best available technology (BAT) for pollution control, over 50% of those surveyed said they would rather endure factory shutdowns and lost jobs than waivers from BAT standards. Suppose ten people are randomly selected and asked to give their opinion regarding BAT pollution control. Assuming $p = .5$, find the probability that:

a. None would prefer factory shutdowns and lost jobs to waiving BAT standards.

b. At least five would prefer factory shutdowns and lost jobs to waiving BAT standards.

c. At least one would prefer factory shutdowns and lost jobs to waiving BAT standards.

OPTIONAL EXERCISES

4.26 For the binomial probability distribution $p(y)$, show that $\sum_{y=0}^{n} p(y) = 1$. [*Hint:* The binomial theorem, which pertains to the expansion of $(a + b)^n$, states that

$$(a + b)^n = \binom{n}{0}a^n + \binom{n}{1}a^{n-1}b + \binom{n}{2}a^{n-2}b^2 + \cdots + \binom{n}{n}b^n$$

Let $a = q$ and $b = p$.]

4.27 Show that, for a binomial random variable,

$$E[y(y-1)] = npq + \mu^2 - \mu$$

[*Hint:* Write the expected value as a sum, factor out $y(y-1)$, and then factor terms until each term in the sum is a binomial probability. Use the fact that $\Sigma_y\, p(y) = 1$ to sum the series.]

4.28 Use the results of Exercise 4.27 and the fact that

$$E[y(y-1)] = E(y^2 - y) = E(y^2) - E(y) = E(y^2) - \mu$$

to find $E(y^2)$ for a binomial random variable.

4.29 Use the results of Exercises 4.27 and 4.28, in conjunction with Theorem 4.4, to show that $\sigma^2 = npq$ for a binomial random variable.

SECTION 4.6

THE NEGATIVE BINOMIAL AND THE GEOMETRIC PROBABILITY DISTRIBUTIONS

Consider a series of trials identical to those described for the binomial experiment (Section 4.5) and let y equal the number of trials until the rth success is observed. The probability distribution for the random variable y is known as a **negative binomial distribution**. Its formula is given in the next box, together with the mean and variance for a negative binomial random variable.

THE PROBABILITY DISTRIBUTION, MEAN, AND VARIANCE FOR A NEGATIVE BINOMIAL RANDOM VARIABLE

The probability distribution for a negative binomial random variable is given by

$$p(y) = \binom{y-1}{y-r}p^r q^{y-r} \qquad (y = r, r+1, r+2, \ldots)$$

where

p = Probability of success on a single trial

$q = 1 - p$

y = Number of trials until the rth success is observed

The mean and variance of a negative binomial random variable are, respectively,

$$\mu = \frac{r}{p} \qquad \text{and} \qquad \sigma^2 = \frac{rq}{p^2}$$

The negative binomial probability distribution is often used for a discrete probability model for the length of time before some event occurs; for example, the length of time a customer must wait in line until receiving service, or the length of time until a piece of equipment fails. For this application, each unit of time represents a trial that can result in a success (S) or failure (F) and y represents the number of trials (time units) until the rth success is observed.

From the box, you can see that the negative binomial probability distribution is a function of two parameters, p and r. For the special case $r = 1$, the probability distribution of y is known as a **geometric probability distribution**.

THE PROBABILITY DISTRIBUTION, MEAN, AND VARIANCE
FOR A GEOMETRIC RANDOM VARIABLE

$$p(y) = pq^{y-1} \qquad (y = 1, 2, \ldots)$$

where y is the number of trials until the first success is observed

$$\mu = \frac{1}{p}$$

$$\sigma^2 = \frac{q}{p^2}$$

To derive the negative binomial probability distribution, note that every simple event that results in y trials until the rth success will contain $(y - r)$ F's and r S's, as depicted here:

$$\overbrace{\underbrace{F \quad F \quad S \quad F \quad F \ldots S \quad F}_{(y - r) \text{ F's and } (r - 1) \text{ S's}} \quad \overbrace{S}^{r\text{th } S}}$$

The number of different simple events that result in $(y - r)$ F's before the rth S is the number of ways that we can arrange the $(y - r)$ F's and $(r - 1)$ S's, namely,

$$\binom{(y - r) + (r - 1)}{y - r} = \binom{y - 1}{y - r}$$

Then, since the probability associated with each of these simple events is $p^r q^{y-r}$, we have

$$p(y) = \binom{y - 1}{y - r} p^r q^{y-r}$$

Examples 4.9 and 4.10 demonstrate the use of the negative binomial and the geometric probability distributions, respectively.

EXAMPLE 4.9

To attach the housing on a motor, a production line assembler must use an electrical hand tool to set and tighten four bolts. Suppose that the probability of setting and tightening a bolt in any 1-second time interval is $p = .8$. If the assembler fails in the first second, the probability of success during the second 1-second interval is .8, and so on.

a. Find the probability distribution of y, the length of time until a complete housing is attached.

b. Find $p(6)$.

c. Find the mean and variance of y.

SOLUTION

a. Since the housing contains $r = 4$ bolts, we will use the formula for the negative binomial probability distribution. Substituting $p = .8$ and $r = 4$ into the formula for $p(y)$, we obtain

$$p(y) = \binom{y-1}{y-r}p^r q^{y-r} = \binom{y-1}{y-4}(.8)^4(.2)^{y-4}$$

b. To find the probability that the complete assembly operation will require $y = 6$ seconds, we substitute $y = 6$ into the formula obtained in part **a** and find

$$p(y) = \binom{5}{2}(.8)^4(.2)^2 = (10)(.4096)(.04) = .16384$$

c. For this negative binomial distribution,

$$\mu = \frac{r}{p} = \frac{4}{.8} = 5 \text{ seconds}$$

and

$$\sigma^2 = \frac{rq}{p^2} = \frac{4(.2)}{(.8)^2} = 1.25$$ ■

EXAMPLE 4.10

A manufacturer uses electrical fuses in an electronic system. The fuses are purchased in large lots and tested sequentially until the first defective fuse is observed. Assume that the lot contains 10% defective fuses.

a. What is the probability that the first defective fuse will be one of the first five fuses tested?

b. Find the mean, variance, and standard deviation for y, the number of fuses tested until the first defective fuse is observed.

SOLUTION

a. The number y of fuses tested until the first defective fuse is observed is a geometric random variable with

$$p = .1 \quad \text{(probability that a single fuse is defective)}$$
$$q = 1 - p = .9$$

and

$$p(y) = pq^{y-1} \quad (y = 1, 2, \ldots)$$
$$= (.1)(.9)^{y-1}$$

The probability that the first defective fuse is one of the first five fuses tested is

$$P(y \le 5) = p(1) + p(2) + \cdots + p(5)$$
$$= (.1)(.9)^0 + (.1)(.9)^1 + \cdots + (.1)(.9)^4 = .41$$

b. The mean, variance, and standard deviation of this geometric random variable are

$$\mu = \frac{1}{p} = \frac{1}{.1} = 10$$

$$\sigma^2 = \frac{q}{p^2} = \frac{.9}{(.1)^2} = 90$$

$$\sigma = \sqrt{\sigma^2} = \sqrt{90} = 9.49$$

∎

EXERCISES 4.30–4.39

4.30 Suppose y is a negative binomial random variable. Calculate $p(y)$ for each of the following situations:
a. $p = .2$, $r = 2$, $y = 3$
b. $p = .5$, $r = 3$, $y = 5$
c. $p = .8$, $r = 3$, $y = 5$

4.31 a. Calculate $p(y)$ for the negative binomial probability distribution with $p = .6$ and $r = 3$, for $y = 6, 7, 8$, and 9.
b. Construct a probability histogram for $p(y)$.

4.32 Refer to the negative binomial probability distribution of Exercise 4.31.
a. Calculate μ and σ for the probability distribution.
b. Locate the points $\mu + 2\sigma$ and $\mu - 2\sigma$ on the y-axis of the graph in part **b** of Exercise 4.31. Find $P(\mu - 2\sigma \le y \le \mu + 2\sigma)$.

4.33 Let y be a geometric random variable with $p = .7$.
a. Calculate $p(y)$ for $y = 1, 2, \ldots, 5$.
b. Construct a probability histogram for $p(y)$.

4.34 Refer to Exercise 4.33.
a. Calculate μ and σ for the geometric probability distribution.
b. Locate the points $\mu + 2\sigma$ and $\mu - 2\sigma$ on the y-axis of the graph in part **b** of Exercise 4.33. Find $P(\mu - 2\sigma \le y \le \mu + 2\sigma)$.

4.35 Refer to Exercise 4.20. If neutral particles are released one at a time into the evacuated duct, find the probability that more than five particles will need to be released until we observe two particles reflected by the inner duct wall.

4.36 Assume that hitting oil at one drilling location is independent of success at another and that, in a particular region, the probability of success at an individual location is .3.
a. What is the probability that a driller will hit oil on or before the third drilling?
b. If y is the number of drillings until the first success occurs, find the mean and standard deviation of y.
c. Is it likely that y will exceed 10? Explain.

4.37 Refer to Exercise 4.36. Suppose the drilling company believes that a venture will be profitable if the number of wells drilled until the second success occurs is less than or equal to 7. Find the probability that the venture will be successful.

4.38 Refer to Exercise 4.23. Let y represent the number of 50-year-old Utah males examined until the first incidence of thyroid cancer is detected.

    a. Find $P(y = 1,000)$.
    b. Find the mean and variance of y.
    c. Is it likely that y will exceed 100,000? Explain.

OPTIONAL EXERCISE

4.39 Let y be a negative binomial random variable with parameters r and p. Then it can be shown that $w = y - r$ is also a negative binomial random variable, where w represents the number of failures before the rth success is observed. Use the facts that

$$E(y) = \frac{r}{p} \quad \text{and} \quad \sigma_y^2 = \frac{rq}{p^2}$$

to show that

$$E(w) = \frac{rq}{p} \quad \text{and} \quad \sigma_w^2 = \frac{rq}{p^2}$$

[*Hint:* Use Theorems 4.1, 4.2, and 4.3.]

| | | | | | | | | | | | |
S E C T I O N 4.7

**THE
HYPERGEOMETRIC
PROBABILITY
DISTRIBUTION**

When we are sampling from a finite population of Successes and Failures (such as a finite population of consumer preference responses or a finite collection of observations in a shipment containing nondefective and defective manufactured products), the assumptions for a binomial experiment are satisfied exactly only if the result of each trial is observed and then replaced in the population before the next observation is made. This method of sampling is called **sampling with replacement**. However, in practice, we usually **sample without replacement**, i.e., we randomly select n different elements from among the N elements in the population. As noted in Section 4.5, when N is large and n/N is small (say, less than .1), the probability of drawing an S remains approximately the same from one trial to another, the trials are (essentially) independent, and the probability distribution for the number of successes, y, is *approximately* a binomial probability distribution. However, when N is small or n/N is large, we would want to use the exact probability distribution for y. This distribution, known as a **hypergeometric probability distribution**, is the topic of this section. The defining characteristics and probability distribution for a hypergeometric random variable are stated in the boxes.

CHARACTERISTICS THAT DEFINE A HYPERGEOMETRIC RANDOM VARIABLE

1. The experiment consists of randomly drawing n elements without replacement from a set of N elements, r of which are S's (for Success) and $(N - r)$ of which are F's (for Failure).

2. The hypergeometric random variable y is the number of S's in the draw of n elements.

**THE PROBABILITY DISTRIBUTION, MEAN, AND VARIANCE
FOR A HYPERGEOMETRIC RANDOM VARIABLE**

The hypergeometric probability distribution is given by

$$p(y) = \frac{\binom{r}{y}\binom{N-r}{n-y}}{\binom{N}{n}} \qquad [y = \text{Maximum } [0, n - (N-r)], \ldots, \\ \text{Minimum } (r, n)]$$

where

N = Total number of elements

r = Number of S's in the N elements

n = Number of elements drawn

y = Number of S's drawn in the n elements

The mean and variance of a hypergeometric random variable are, respectively,

$$\mu = \frac{nr}{N} \qquad \sigma^2 = \frac{r(N-r)n(N-n)}{N^2(N-1)}$$

To derive the hypergeometric probability distribution, we first note that the total number of simple events in S is equal to the number of ways of selecting n elements from N, namely $\binom{N}{n}$. *A simple event implying y successes* will be a selection of n elements in which y are S's and $(n-y)$ are F's. Since there are r S's from which to choose, the number of different ways of selecting y of them is $a = \binom{r}{y}$. Similarly, the number of ways of selecting $(n-y)$ F's from among the total of $(N-r)$ is $b = \binom{N-r}{n-y}$. We now apply Theorem 3.1 to determine the number of ways of selecting y S's and $(n-y)$ F's—that is, the number of simple events implying y successes:

$$a \cdot b = \binom{r}{y}\binom{N-r}{n-y}$$

Finally, since the selection of any one set of n elements is as likely as any other, all the simple events are equiprobable and thus,

$$p(y) = \frac{\text{Number of simple events that imply } y \text{ successes}}{\text{Number of simple events}} = \frac{\binom{r}{y}\binom{N-r}{n-y}}{\binom{N}{n}}$$

EXAMPLE 4.11

An experiment is conducted to select a suitable catalyst for the commercial production of ethylenediamine (EDA), a product used in soaps. Suppose a chemical engineer randomly selects three catalysts for testing from among a group of ten catalysts, six of which have low acidity and four of which have high acidity.

a. Find the probability that no highly acidic catalyst is selected.
b. Find the probability that exactly one highly acidic catalyst is selected.

SOLUTION

Let y be the number of highly acidic catalysts selected. Then y is a hypergeometric random variable with $N = 10$, $n = 3$, $r = 4$, and

$$p(y) = \frac{\binom{4}{y}\binom{6}{3-y}}{\binom{10}{3}}$$

a. $p(0) = \dfrac{\binom{4}{0}\binom{6}{3}}{\binom{10}{3}} = \dfrac{(1)(20)}{120} = \dfrac{1}{6}$

b. $p(1) = \dfrac{\binom{4}{1}\binom{6}{2}}{\binom{10}{3}} = \dfrac{(4)(15)}{120} = \dfrac{1}{2}$ ■

EXAMPLE 4.12

Refer to Example 4.11.

a. Find μ, σ^2, and σ for the random variable y.
b. Find $P(\mu - 2\sigma < y < \mu + 2\sigma)$.

SOLUTION

a. Since y is a hypergeometric random variable with $N = 10$, $n = 3$, and $r = 4$, the mean and variance are

$$\mu = \frac{nr}{N} = \frac{(3)(4)}{10} = 1.2$$

$$\sigma^2 = \frac{r(N-r)n(N-n)}{N^2(N-1)} = \frac{4(10-4)3(10-3)}{(10)^2(10-1)}$$

$$= \frac{(4)(6)(3)(7)}{(100)(9)} = .56$$

The standard deviation is

$$\sigma = \sqrt{.56} = .75$$

b. The probability distribution and the interval $\mu \pm 2\sigma$, or $-.3$ to 2.7, are shown in Figure 4.3. The only possible value of y that falls outside the interval is $y = 3$. Therefore,

$$P(\mu - 2\sigma < y < \mu + 2\sigma) = 1 - p(3) = 1 - \frac{\binom{4}{3}\binom{6}{0}}{\binom{10}{3}}$$

$$= 1 - \frac{4}{120} = .967$$

FIGURE 4.3
Probability Distribution for
y in Example 4.12

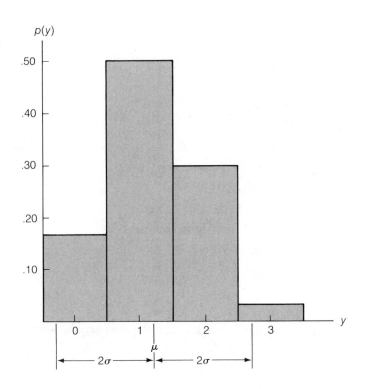

EXAMPLE 4.13   Refer to Example 4.11. Find the mean, μ, of the random variable y.

SOLUTION   By Definition 4.4,

$$\mu = E(y) = \sum_{\text{all } y} yp(y) = \sum_{y=0}^{3} y\frac{\binom{4}{y}\binom{6}{3-y}}{120}$$

Using the values of $p(y)$ calculated in Examples 4.11 and 4.12, and

$$p(2) = \frac{\binom{4}{2}\binom{6}{1}}{120} = \frac{(6)(6)}{120} = \frac{3}{10}$$

we obtain by substitution:

$$\mu = 0p(0) + 1p(1) + 2p(2) + 3p(3)$$
$$= 0 + 1\left(\frac{1}{2}\right) + 2\left(\frac{3}{10}\right) + 3\left(\frac{1}{30}\right) = 1.2$$

Note that this is the value we obtained in Example 4.12 by applying the formula given in the previous box. ∎

EXERCISES 4.40–4.45

4.40 Suppose that y is a hypergeometric random variable. Compute $p(y)$ for each of the following cases:
a. $N = 5$, $n = 3$, $r = 4$, $y = 1$ **b.** $N = 10$, $n = 5$, $r = 3$, $y = 3$
c. $N = 3$, $n = 2$, $r = 2$, $y = 2$ **d.** $N = 4$, $n = 2$, $r = 2$, $y = 0$

4.41 Suppose y is a hypergeometric random variable with $N = 12$, $n = 8$, and $r = 7$.
a. Display the probability distribution for y in tabular form.
b. Compute μ and σ for y.
c. Graph $p(y)$ and locate μ and the interval $\mu \pm 2\sigma$ on the graph.
d. What is the probability that y will fall within the interval $\mu \pm 2\sigma$?

4.42 Use the results of Exercise 4.41 to find the following probabilities:
a. $P(y = 1)$ **b.** $P(y = 4)$ **c.** $P(y \leq 4)$
d. $P(y \geq 5)$ **e.** $P(y < 3)$ **f.** $P(y \geq 8)$

4.43 An investigative task force established by the Environmental Protection Agency was scheduled to investigate 20 industrial firms to check for violations of pollution control regulations. However, budget cutbacks have drastically reduced the size of the task force and they will be able to investigate only three of the 20 firms. If it is known that five of the firms are actually operating in violation of regulations, find the probability that:
a. None of the three sampled firms will be found in violation of regulations.
b. All three firms investigated will be found in violation of regulations.
c. At least one of the three firms will be operating in violation of pollution control regulations.

4.44 If you are purchasing small lots of a manufactured product and it is very costly to test a single item, it may be desirable to test a sample of items from the lot rather than every item in the lot. Such a sampling plan would be based on a hypergeometric probability distribution. For example, suppose each lot contains ten items. You decide to sample four items per lot and reject the lot if you observe one or more defectives.
a. If the lot contains one defective item, what is the probability that you will accept the lot?

b. What is the probability that you will accept the lot if it contains four defective items?

OPTIONAL EXERCISE

4.45 Show that the mean of a hypergeometric random variable is $\mu = nr/N$. [*Hint:* Show that

$$\frac{y\binom{r}{y}\binom{N-r}{n-y}}{\binom{N}{n}} = \frac{\frac{nr}{N}\binom{r-1}{y-1}\binom{N-1-(r-1)}{n-1-(y-1)}}{\binom{N-1}{n-1}}$$

and then use the fact that

$$\frac{\binom{r-1}{y-1}\binom{N-1-(r-1)}{n-1-(y-1)}}{\binom{N-1}{n-1}}$$

is the hypergeometric probability distribution for $z = (y-1)$, where z is the number of S's in $(n-1)$ trials, with a total of $(r-1)$ S's in $(N-1)$ elements.]

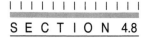

SECTION 4.8

THE POISSON PROBABILITY DISTRIBUTION

The **Poisson probability distribution**, named for the French mathematician S. D. Poisson (1781–1840), provides a model for the relative frequency of the number of "rare events" that occur in a unit of time, area, volume, etc. The number of new jobs submitted to a computer in any one minute, the number of fatal accidents per month in a manufacturing plant, and the number of visible defects in a diamond are variables whose relative frequency distributions can be approximated well by Poisson probability distributions. The characteristics of a Poisson random variable are listed in the box.

CHARACTERISTICS OF A POISSON RANDOM VARIABLE

1. The experiment consists of counting the number y of times a particular event occurs during a given unit of time, or in a given area or volume (or weight, distance, or any other unit of measurement).
2. The probability that an event occurs in a given unit of time, area, or volume is the same for all the units.
3. The number of events that occur in one unit of time, area, or volume is independent of the number that occur in other units.
4. The mean (or expected) number of events in each unit will be denoted by the Greek letter lambda, λ.

The formulas for the probability distribution, the mean, and the variance of a Poisson random variable are shown in the next box. You will note that the formula involves the quantity $e = 2.71828\ldots$, the base of natural logarithms. Values of e^{-y}, needed to compute values of $p(y)$, are given in Table 3 of Appendix II.

PROBABILITY DISTRIBUTION, MEAN, AND VARIANCE FOR A POISSON RANDOM VARIABLE

The probability distribution for a Poisson random variable is given by

$$p(y) = \frac{\lambda^y e^{-\lambda}}{y!} \qquad (y = 0, 1, 2, \ldots)$$

where

$\quad \lambda = $ Mean number of events during the given time period

$\quad e = 2.71828\ldots$

The mean and variance of a Poisson random variable are, respectively,

$$\mu = \lambda \qquad \text{and} \qquad \sigma^2 = \lambda$$

EXAMPLE 4.14

Suppose the number y of cracks per concrete specimen for a particular type of cement mix has approximately a Poisson probability distribution. Furthermore, assume that the average number of cracks per specimen is 2.5.

a. Find the mean and standard deviation of y, the number of cracks per concrete specimen.
b. Find the probability that a randomly selected concrete specimen has exactly five cracks.
c. Find the probability that a randomly selected concrete specimen has two or more cracks.

SOLUTION

a. The mean and variance of a Poisson random variable are both equal to λ. Thus, for this example

$$\mu = \lambda = 2.5 \qquad \sigma^2 = \lambda = 2.5$$

Then the standard deviation is

$$\sigma = \sqrt{2.5} = 1.58$$

b. We want the probability that a concrete specimen has exactly five cracks. The probability distribution for y is

$$p(y) = \frac{\lambda^y e^{-\lambda}}{y!}$$

Then, since $\lambda = 2.5$, $y = 5$, and $e^{-2.5} = .082085$ (from Table 3 of Appendix II),

$$p(5) = \frac{(2.5)^5 e^{-2.5}}{5!} = \frac{(2.5)^5(.082085)}{5 \cdot 4 \cdot 3 \cdot 2 \cdot 1} = .067$$

c. To find the probability that a concrete specimen has two or more cracks, we need to find

$$P(y \geq 2) = p(2) + p(3) + p(4) + \cdots = \sum_{y=2}^{\infty} p(y)$$

In order to find the probability of this event, we must consider the complementary event. Thus,

$$
\begin{aligned}
P(y \geq 2) &= 1 - P(y \leq 1) = 1 - [p(0) + p(1)] \\
&= 1 - \frac{(2.5)^0 e^{-2.5}}{0!} - \frac{(2.5)^1 e^{-2.5}}{1!} \\
&= 1 - \frac{1(.082085)}{1} - \frac{2.5(.082085)}{1} \\
&= 1 - .287 = .713
\end{aligned}
$$

According to our Poisson model, the probability that a concrete specimen has two or more cracks is .713.

The probability distribution for y is shown in Figure 4.4 for y values between 0 and 9. The mean $\mu = 2.5$ and the interval $\mu \pm 2\sigma$, or $-.7$ to 5.7, are indicated.

FIGURE 4.4

Poisson Probability Distribution for y in Example 4.14

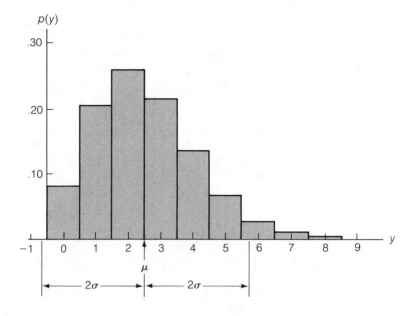

The Poisson probability distribution is related to and can be used to approximate a binomial probability distribution when n is large and $\mu = np$ is small, say, $np \leq 7$. The proof of this fact is beyond the scope of this text, but it can be found in Feller (1968).

EXAMPLE 4.15   Let y be a binomial random variable with $n = 25$ and $p = .1$.

a. Use Table 2 of Appendix II to determine the exact value of $P(y \leq 1)$.

b. Find the Poisson approximation to $P(y \leq 1)$. [*Note:* Although we would prefer to compare the Poisson approximation to binomial probabilities for larger values of n, we are restricted in this example by the limitations of Table 2.]

SOLUTION   **a.** From Table 2 of Appendix II, with $n = 25$ and $p = .1$, we have

$$P(y \leq 1) = \sum_{y=0}^{1} p(y) = .271$$

b. Since $n = 25$ and $p = .1$, we will approximate $p(y)$ using a Poisson probability distribution with mean

$$\lambda = np = (25)(.1) = 2.5$$

Then,

$$p(y) = \frac{(2.5)^y e^{-2.5}}{y!}$$

where, from Table 3 of Appendix II, $e^{-2.5} = .082085$. Substituting into the formula for $p(y)$ yields

$$p(0) \approx \frac{(2.5)^0 e^{-2.5}}{0!} = \frac{e^{-2.5}}{1} = .082$$

$$p(1) \approx \frac{(2.5)^1 e^{-2.5}}{1!} = (2.5)e^{-2.5} = .205$$

Therefore,

$$P(y \leq 1) = p(0) + p(1) \approx .082 + .205 = .287$$

This approximation, .287, to the exact value of $P(y \leq 1) = .271$, is reasonably good considering that the approximation procedure is usually applied to binomial probability distributions for which n is much larger than 25.   ■

EXAMPLE 4.16   Show that the expected value of a Poisson random variable is λ.

SOLUTION   By Definition 4.4, we have

$$E(y) = \sum_{\text{all } y} yp(y) = \sum_{y=0}^{\infty} y \frac{\lambda^y e^{-\lambda}}{y!}$$

The first term of this series will equal 0, because $y = 0$. Therefore,

$$E(y) = \sum_{y=0}^{\infty} \frac{y \lambda^y e^{-\lambda}}{y!} = \sum_{y=1}^{\infty} \frac{\lambda^y e^{-\lambda}}{(y-1)!} = \sum_{y=1}^{\infty} \frac{\lambda \cdot \lambda^{y-1} e^{-\lambda}}{(y-1)!}$$

Factoring the constant λ outside the summation and letting $z = (y - 1)$, we obtain

$$E(y) = \lambda \sum_{z=0}^{\infty} \frac{\lambda^z e^{-\lambda}}{z!} = \lambda \sum_{z=0}^{\infty} p(z)$$

where z is a Poisson random variable with mean λ. Hence,

$$E(y) = \lambda \sum_{z=0}^{\infty} p(z) = \lambda(1) = \lambda$$

∎

EXERCISES 4.46–4.57

4.46 Suppose y is a random variable for which a Poisson probability distribution provides a good characterization. Compute the following:
a. $P(y \leq 2)$, when $\lambda = 2$
b. $P(y = 1)$, when $\lambda = 5$
c. $P(y \geq 1)$, when $\lambda = 3$
d. $P(y = 0)$, when $\lambda = 9$

4.47 Suppose y is a random variable for which a Poisson probability distribution with $\lambda = 5.5$ provides a good characterization.
a. Graph $p(y)$ for $y = 0, 1, 2, \ldots, 9, 10$.
b. Find μ and σ for the random variable y, and locate μ and the interval $\mu \pm 2\sigma$ on the graph of part **a**.
c. What is the probability that y will fall within the interval $\mu \pm 2\sigma$?

4.48 Suppose y is a binomial random variable with $n = 100$ and $p = .02$. Use the Poisson approximation to the binomial distribution to find:
a. $P(y \geq 3)$ **b.** $P(y = 5)$ **c.** $P(y = 0)$

4.49 A can company reports that the number of breakdowns per 8-hour shift on its machine-operated assembly line follows a Poisson distribution, with a mean of 1.5.
a. What is the probability of exactly two breakdowns during the midnight shift?
b. What is the probability of fewer than two breakdowns during the afternoon shift?
c. What is the probability of no breakdowns during three consecutive 8-hour shifts? (Assume that the machine operates independently across shifts.)

4.50 A recent study of natural rock slope movements in the Canadian Rockies over the past 5,000 years revealed that the number of major rockslides per 100 square kilometers had an expected value of 1.57 (*Canadian Geotechnical Journal*, Nov. 1985).
a. Find the mean and standard deviation of y, the number of major rockslides per 100 square kilometers in the Canadian Rockies over a 5,000-year period.
b. What is the probability of observing three or more major rockslides per 100 square kilometers over a 5,000-year period?

4.51 The random variable y, the number of cars that arrive at an intersection during a specified period of time, often possesses (approximately) a Poisson probability distribution. When the mean arrival rate λ is known, the Poisson probability distribution can be used to aid a traffic engineer in the design of a traffic control system. Suppose you estimate that the mean number of arrivals per minute at the intersection is one car per minute.
a. What is the probability that in a given minute, the number of arrivals will equal three or more?

b. Can you assure the engineer that the number of arrivals will rarely exceed three per minute?

4.52 The number y of computer input errors per minute made by a particular computer programmer has a Poisson distribution with an average of .75 error per minute.
a. Find the mean and variance of y.
b. What is the probability that the programmer will make at least one error in a particular minute?
c. What is the probability that the programmer will make no errors in a particular minute?

4.53 A discharge (or response) rate of auditory nerve fibers (recorded as the number of spikes per 200 milliseconds of noise burst) is used to measure the effect of acoustic stimuli in the auditory nerve. An empirical study of auditory nerve fiber response rates in cats resulted in a mean of 15 spikes/ms (*Journal of the Acoustical Society of America*, Feb. 1986). Let y represent the auditory nerve fiber response rate for a randomly selected cat in the study.
a. If y is approximately a Poisson random variable, find the mean and standard deviation of y.
b. Assuming y is Poisson, what is the approximate probability that y exceeds 27 spikes/ms?
c. In the study, the variance of y was found to be "substantially smaller" than 15 spikes/ms. Is it reasonable to expect y to follow a Poisson process? How will this affect the probability computed in part **b**?

4.54 The Environmental Protection Agency (EPA) issues standards on air and water pollution that vitally affect the safety of consumers and the operations of industry. For example, the EPA states that manufacturers of vinyl chloride and similar compounds must limit the amount of these chemicals in plant air emissions to 10 parts per million. Suppose you represent one of the manufacturers and you know that the mean emission of vinyl chloride for your plant is 4 parts per million.
a. If the parts per million of vinyl chloride in air follows a Poisson probability distribution and y is the parts per million for a particular sample, what is the standard deviation of y for your plant?
b. If the mean parts per million for your plant is in fact equal to 4, is it likely that a sample would yield a value of y that would exceed the EPA limits? Explain.

4.55 A random sample of 300 people is selected and each is asked whether they would prefer to own a personal computer (PC) or a video cassette recorder (VCR). Assuming that 2% of the public would prefer a PC, show how you could approximate the probability that at most 30 people in the sample would prefer a PC.

OPTIONAL EXERCISES

4.56 Show that for a Poisson random variable y,

$$E(y^2) = \lambda^2 + \lambda$$

[*Hint:* First derive the result $E[y(y-1)] = \lambda^2$ from the fact that

$$E[y(y-1)] = \sum_{y=0}^{\infty} y(y-1)\frac{\lambda^y e^{-\lambda}}{y!} = \lambda^2 \sum_{y=2}^{\infty} \frac{\lambda^{y-2} e^{-\lambda}}{(y-2)!} = \lambda^2 \sum_{z=0}^{\infty} \frac{\lambda^z e^{-\lambda}}{z!}$$

Then apply the result $E[y(y-1)] = E(y^2) - E(y)$.]

4.57 Show that for a Poisson random variable y, $\sigma^2 = \lambda$. [*Hint:* Use the result of Exercise 4.56 and Theorem 4.4.]

<hr>

SECTION 4.9

MOMENTS AND MOMENT GENERATING FUNCTIONS (OPTIONAL)

The **moments** of a random variable can be used to completely describe its probability distribution.

DEFINITION 4.7

The **k**th **moment** of a random variable y, **taken about the origin,** is denoted by the symbol μ'_k and defined to be

$$\mu'_k = E(y^k) \qquad (k = 0, 1, 2, \ldots)$$

DEFINITION 4.8

The **k**th **moment** of a random variable y, **taken about its mean,** is denoted by the symbol μ_k and defined to be

$$\mu_k = E[(y - \mu)^k]$$

You have already encountered two important moments of random variables. The mean of a random variable is $\mu'_1 = \mu$ and the variance is $\mu_2 = \sigma^2$. Other moments about the origin or about the mean can be used to measure the lack of symmetry or the tendency of a distribution to possess a large peak near the center. In fact, if all of the moments of a discrete random variable exist, they completely define its probability distribution. This fact is often used to prove that two random variables possess the same probability distributions. For example, if two discrete random variables, x and y, possess moments about the origin, $\mu'_{1x}, \mu'_{2x}, \mu'_{3x}, \ldots$ and $\mu'_{1y}, \mu'_{2y}, \mu'_{3y}, \ldots$, respectively, and if all corresponding moments are equal, i.e., if $\mu'_{1x} = \mu'_{1y}$, $\mu'_{2x} = \mu'_{2y}$, etc., then the two discrete probability distributions, $p(x)$ and $p(y)$, are identical.

The moments of a discrete random variable can be found directly using Definition 4.7, but as Examples 4.8 and 4.16 indicate, summing the series needed to find $E(y)$, $E(y^2)$, etc., can be tedious. Sometimes the difficulty in finding the moments of a random variable can be alleviated by using the **moment generating function** of the random variable.

DEFINITION 4.9

The **moment generating function**, $m(t)$, of a discrete random variable y is defined to be

$$m(t) = E(e^{ty})$$

The moment generating function of a discrete random variable is simply a mathematical expression that condenses all the moments into a single formula. In order to extract specific moments from it, we first note that, by Definition 4.9,

$$E(e^{ty}) = \sum_{\text{all } y} e^{ty} p(y)$$

where

$$e^{ty} = 1 + ty + \frac{(ty)^2}{2!} + \frac{(ty)^3}{3!} + \frac{(ty)^4}{4!} + \cdots$$

Then, if μ_i' is finite for $i = 1, 2, 3, 4, \ldots,$

$$m(t) = E(e^{ty}) = \sum_{\text{all } y} e^{ty} p(y) = \sum_{\text{all } y} \left[1 + ty + \frac{(ty)^2}{2!} + \frac{(ty)^3}{3!} + \cdots \right] p(y)$$

$$= \sum_{\text{all } y} \left[p(y) + typ(y) + \frac{t^2}{2!} y^2 p(y) + \frac{t^3}{3!} y^3 p(y) + \cdots \right]$$

Now apply Theorems 4.2 and 4.3 to obtain

$$m(t) = \sum_{\text{all } y} p(y) + t \sum_{\text{all } y} yp(y) + \frac{t^2}{2!} \sum_{\text{all } y} y^2 p(y) + \cdots$$

But, by Definition 4.7, $\sum_{\text{all } y} y^k p(y) = \mu_k'$. Therefore,

$$m(t) = 1 + t\mu_1' + \frac{t^2}{2!}\mu_2' + \frac{t^3}{3!}\mu_3' + \cdots$$

This indicates that if we have the moment generating function of a random variable and can expand it into a power series in t, i.e.,

$$m(t) = 1 + a_1 t + a_2 t^2 + a_3 t^3 + \cdots$$

then it follows that the coefficient of t will be $\mu_1' = \mu$, the coefficient of t^2 will be $\mu_2'/2!$, and, in general, the coefficient of t^k will be $\mu_k'/k!$.

If we cannot easily expand $m(t)$ into a power series in t, we can find the moments of y by differentiating $m(t)$ with respect to t and then setting t equal to 0. Thus,

$$\frac{dm(t)}{dt} = \frac{d}{dt}\left(1 + t\mu_1' + \frac{t^2}{2!}\mu_2' + \frac{t^3}{3!}\mu_3' + \cdots \right)$$

$$= \left(0 + \mu_1' + \frac{2t}{2!}\mu_2' + \frac{3t^2}{3!}\mu_3' + \cdots \right)$$

Letting $t = 0$, we obtain

$$\left. \frac{dm(t)}{dt} \right]_{t=0} = (\mu_1' + 0 + 0 + \cdots) = \mu_1' = \mu$$

Taking the second derivative of $m(t)$ with respect to t yields

$$\frac{d^2m(t)}{dt^2} = \left(0 + \mu_2' + \frac{3!}{3!}t\mu_3' + \cdots\right)$$

Then, letting $t = 0$, we obtain

$$\frac{d^2m(t)}{dt^2}\bigg]_{t=0} = (\mu_2' + 0 + 0 + \cdots) = \mu_2'$$

Theorem 4.5 describes how to extract μ_k' from the moment generating function $m(t)$.

THEOREM 4.5

If $m(t)$ exists, then the kth moment about the origin is equal to

$$\mu_k' = \frac{d^km(t)}{dt^k}\bigg]_{t=0}$$

To illustrate the use of the moment generating function, consider the following examples.

EXAMPLE 4.17   Derive the moment generating function for a binomial random variable.

SOLUTION   The moment generating function is given by

$$m(t) = E(e^{ty}) = \sum_{y=0}^{n} e^{ty}p(y) = \sum_{y=0}^{n} e^{ty}\binom{n}{y}p^yq^{n-y} = \sum_{y=0}^{n} \binom{n}{y}(pe^t)^yq^{n-y}$$

We now recall the binomial theorem (see Optional Exercise 4.26):

$$(a + b)^n = \sum_{y=0}^{n} \binom{n}{y}a^yb^{n-y}$$

Letting $a = pe^t$ and $b = q$ yields the desired result:

$$m(t) = (pe^t + q)^n$$ ∎

EXAMPLE 4.18   Use Theorem 4.5 to derive $\mu_1' = \mu$ and μ_2' for the binomial random variable.

SOLUTION   From Theorem 4.5,

$$\mu_1' = \mu = \frac{dm(t)}{dt}\bigg]_{t=0} = n(pe^t + q)^{n-1}(pe^t)\bigg]_{t=0}$$

$$= n(pe^0 + q)^{n-1}(pe^0)$$

But $e^0 = 1$. Therefore,

$$\mu_1' = \mu = n(p + q)^{n-1}p = n(1)^{n-1}p = np$$

Similarly,

$$\mu_2' = \frac{d^2 m(t)}{dt^2}\bigg]_{t=0} = np\frac{d}{dt}[e^t(pe^t + q)^{n-1}]\bigg]_{t=0}$$

$$= np[e^t(n-1)(pe^t + q)^{n-2}pe^t + (pe^t + q)^{n-1}e^t]\bigg]_{t=0}$$

$$= np[(1)(n-1)(1)p + (1)(1)] = np[(n-1)p + 1]$$

$$= np(np - p + 1) = np(np + q) = n^2p^2 + npq \qquad \blacksquare$$

EXAMPLE 4.19 Use the results of Example 4.18, in conjunction with Theorem 4.4, to derive the variance of a binomial random variable.

SOLUTION By Theorem 4.4,

$$\sigma^2 = E(y^2) - \mu^2 = \mu_2' - (\mu_1')^2$$

Substituting the values of μ_2' and $\mu_1' = \mu$ from Example 4.18 yields

$$\sigma^2 = n^2p^2 + npq - (np)^2 = npq \qquad \blacksquare$$

As demonstrated in Examples 4.18 and 4.19, it is easier to use the moment generating function to find μ_1' and μ_2' for a binomial random variable than to find $\mu_1' = E(y)$ and $\mu_2' = E(y^2)$ separately. You have to sum only a single series to find $m(t)$. This is also the best method for finding μ_1' and μ_2' for many other random variables, but not for all.

The probability distributions, means, variances, and moment generating functions for some useful discrete random variables are summarized in Table 4.3.

TABLE 4.3 Some Useful Discrete Random Variables

RANDOM VARIABLE	$p(y)$	μ	σ^2	$m(t)$
Binomial	$p(y) = \binom{n}{y}p^y q^{n-y}$ where $q = 1 - p$ $y = 0, 1, \ldots, n$	np	npq	$(pe^t + q)^n$
Hypergeometric	$p(y) = \dfrac{\binom{r}{y}\binom{N-r}{n-y}}{\binom{N}{n}}$	$\dfrac{nr}{N}$	$\dfrac{r(N-r)n(N-n)}{N^2(N-1)}$	(Not given)
Poisson	$p(y) = \dfrac{\lambda^y e^{-\lambda}}{y!}$ $y = 0, 1, 2, \ldots$	λ	λ	$e^{\lambda(e^t - 1)}$
Geometric	$p(y) = p(1-p)^{y-1}$ $y = 1, 2, \ldots$	$\dfrac{1}{p}$	$\dfrac{1-p}{p^2}$	$\dfrac{pe^t}{1 - (1-p)e^t}$
Negative binomial	$p(y) = \binom{y-1}{r-1}p^r(1-p)^{y-r}$ $y = r, r+1, \ldots$	$\dfrac{r}{p}$	$\dfrac{r(1-p)}{p^2}$	$\left(\dfrac{pe^t}{1 - (1-p)e^t}\right)^r$

4.58 Derive the moment generating function of the Poisson random variable. [*Hint:* Write

$$m(t) = E(e^{ty}) = \sum_{y=0}^{\infty} e^{ty}\frac{\lambda^y e^{-\lambda}}{y!}$$

$$= e^{-\lambda} \sum_{y=0}^{\infty} \frac{(\lambda e^t)^y}{y!} = e^{-\lambda}e^{\lambda e^t} \sum_{y=0}^{\infty} \frac{(\lambda e^t)^y e^{-\lambda e^t}}{y!}$$

Then note that the quantity being summed is a Poisson probability with parameter λe^t.]

4.59 Use the result of Exercise 4.58 to derive the mean and variance of the Poisson distribution.

4.60 Use the moment generating function given in Table 4.3 to derive the mean and variance of a geometric random variable.

SECTION 4.10

SUMMARY

This chapter introduces the concepts of numerical events and discrete random variables. A **random variable** is a rule that assigns one and only one value of a variable y to each simple event in the sample space. A random variable is said to be **discrete** if it can assume only a countable number of values.

The **probability distribution** of a discrete random variable is a table, graph, or formula that gives the probability associated with each value of y. The **expected value** $E(y) = \mu$ is the mean of this probability distribution and $E[(y - \mu)^2] = \sigma^2$ is its **variance**.

Five random variables—the **binomial**, the **negative binomial**, the **geometric**, the **hypergeometric**, and the **Poisson**—were presented, along with their probability distributions. We noted the physical characteristics of experiments that generate these random variables and identified some practical sampling situations that fit, to a reasonable degree of approximation, these experimental conditions. We gave the mean and variance for each of the random variables, showed how μ and σ provide measures of the location and variation of the probability distributions, and, in some cases, derived these quantities. Finally, we showed how the probability distribution can be used to calculate probabilities and, thereby, to evaluate the likelihood of the occurrence of some numerical events.

4.61 The economic risks taken by engineering-related businesses can be classified as being either *pure risks* or *speculative risks*. A pure risk is faced when there is a chance of incurring an economic loss but no chance of gain. A speculative risk is faced when there is a chance of gain as well as a chance of loss. Risk is sometimes measured by computing the variance or standard deviation of the probability distribution that describes the potential gains or losses of the firm. This follows from the fact that the greater the variation in potential outcomes, the greater is the uncertainty faced by the firm. On the other hand, the smaller the variation, the more predictable are the firm's gains or losses. The two discrete probability distributions given in the table were developed from historical data. They describe the potential total physical damage losses next year to the computerized robots that operate at two different industrial engineering firms. Both firms have ten industrial robots, and both have the same expected loss next year.

FIRM A		FIRM B	
Loss Next Year	Probability	Loss Next Year	Probability
$ 0	.01	$ 0	.00
500	.01	200	.01
1,000	.01	700	.02
1,500	.02	1,200	.02
2,000	.35	1,700	.15
2,500	.30	2,200	.30
3,000	.25	2,700	.30
3,500	.02	3,200	.15
4,000	.01	3,700	.02
4,500	.01	4,200	.02
5,000	.01	4,700	.01

a. Verify that both firms have the same expected total physical damage loss.
b. Compute the standard deviation of both probability distributions and determine which firm faces the greater risk of physical damage to its industrial robots next year.
c. Was part b concerned with measuring speculative risk or pure risk? Explain.

4.62 Suppose you are a purchasing officer for a data-processing company. You have purchased 50,000 terminal switches and have been guaranteed by the supplier that the shipment will contain no more than .1% defectives. To check the shipment, you randomly sample 500 switches, test them, and find that four are defective.
a. Assuming the supplier's claim is true, compute μ and σ for the number of defectives in a sample of 500 switches.
b. If the supplier's claim is true, is it likely you would have found four defective switches in the sample?
c. Based on this sample, what inference would you make concerning the supplier's guarantee?

4.63 Lesser developed countries experiencing rapid population growth often face severe traffic control problems in their large cities. Traffic engineers have determined that elevated rail systems may provide a feasible solution to these traffic woes. Studies indicate that the number of maintenance related shutdowns of the elevated rail system in a particular country has a mean equal to 6.5 per month.
a. Find the probability that at least five shutdowns of the elevated rail system will occur next month in the country.
b. Find the probability that exactly four shutdowns will occur next month.

4.64 Two of the five mechanical engineers employed by the county sanitation department have experience in the design of steam turbine power plants. You have been instructed to choose randomly two of the five engineers to work on a project for a new power plant.

a. What is the probability that you will choose the two engineers with experience in the design of steam turbine power plants?

b. What is the probability that you will choose at least one of the engineers with such experience?

4.65 Refer to Exercise 4.9. Suppose a $100,000 exploration budget is divided equally between two identical and independent oil prospects, with probabilities and monetary outcomes as shown in the table given in Exercise 4.9.

a. Let x represent the sum of monetary values of the two oil prospects. Find the probability distribution for x.

b. Find $E(x)$ and σ^2. Compare these values to your results from part **a**.

c. What is the probability of doubling the $100,000 investment in the two oil prospects? Compare this to the probability of doubling the $50,000 investment in a single oil prospect in Exercise 4.9.

d. What is the probability of "gambler's ruin" (i.e., two dry holes) in the two oil prospects? Compare this to the probability of "gambler's ruin" in a single oil prospect in Exercise 4.9.

4.66 Refer to the *Mining Engineering* (Apr. 1986) study on underground coal mine safety, discussed in Exercise 1.16. Research revealed that "intermediate injuries," i.e., disabling injuries resulting from falls of roof and slides, haulage, machinery, and electrical and explosive accidents, constitute 41% of all disabling injuries and 98% of all fatal injuries in underground coal mines.

a. Find the probability that in a random sample of five disabling injuries, exactly three were intermediate injuries.

b. Find the probability that at least two of the five disabling injuries were intermediate injuries.

c. In a random sample of five fatal injuries, find the probability that at least two were intermediate injuries.

4.67 The manufacturer of a price-reading optical scanner claims that the probability it will misread the price of any product by misreading the "bar code" on a product's label is .001. At the time one of the scanners was installed in a supermarket, the store manager tested its performance. Let y be the number of trials (i.e., the number of prices read by the scanner) until the first misread price is observed.

a. Find the probability distribution for y. (Assume the trials represent independent events.)

b. If the manufacturer's claim is correct, what is the probability that the scanner will not misread a price until after the fifth price is read?

c. If in fact the third price is misread, what inference would you make about the manufacturer's claim? Explain.

4.68 An important function in any business is long-range planning. Additions to a firm's physical plant, for example, cannot be achieved overnight; their construction must be planned years in advance. Anticipating a substantial growth in residential customers over the next 25 years, an electric company is planning today for the power plant it will need 25 years hence. It obviously cannot be certain exactly how many residential customers it will need to serve in 25 years, but the company describes its projected need with the probability distribution shown in the accompanying table.

y = Number of Residential Customers	p(y)
10,000	.05
16,000	.20
23,000	.45
30,000	.20
36,000	.05
40,000	.05

a. Find the mean and standard deviation of the number of residential customers the electrical company will need to serve in 25 years.

b. Suppose that the cost of building a power plant is $60 per customer. Find the mean and standard deviation of the cost the electric company must incur to provide service for future customers.

4.69 In recent years, the use of the telephone as a data collection instrument for public opinion polls has been steadily increasing. However, one of the major factors bearing on the extent to which the telephone will become an acceptable data collection tool in the future is the *refusal rate*—i.e., the percentage of the eligible subjects actually contacted who refuse to take part in the poll. Suppose that past records indicate a refusal rate of 20% in a large city. A poll of 25 city residents is to be taken and y is the number of residents contacted by telephone who refuse to take part in the poll.

a. Find the mean and variance of y.

b. Find $P(y \leq 5)$.

c. Find $P(y > 10)$.

4.70 When radar was first introduced during World War II, it was very difficult for an operator manning the screen to distinguish a static interference blip from an actual enemy aircraft blip. Although the operator did not want to sound an alarm needlessly, failure to alert the defenses could have serious consequences. Records indicate that 60% of all observed blips represented enemy aircraft. Suppose that during a particular siege there were five blips spotted on the screen at different points in time and the radar operator alerted the defenses on each occasion. Assume that the events are independent and compute the probability of each of the following events:

a. Radar operator made the correct decision on all five occasions

b. Radar operator made the correct decision on at least three occasions

c. Radar operator was incorrect all five times (and therefore sounded five false alarms)

4.71 A study of vehicle flow characteristics on acceleration lanes (i.e., merging ramps) at a major freeway in Israel found that one out of every six vehicles uses less than one-third of the acceleration lane before merging into traffic (*Journal of Transportation Engineering*, Nov. 1985). Suppose we monitor the location of the merge for the next five vehicles that enter the acceleration lane.

a. What is the probability that none of the vehicles will use less than one-third of the acceleration lane?

b. What is the probability that exactly two of the vehicles will use less than one-third of the acceleration lane?

4.72 Refer to Exercise 4.71. Suppose that the number of vehicles using the acceleration lane per minute has a mean equal to 1.1.

a. What is the probability that more than two vehicles will use the acceleration lane in the next minute?

b. What is the probability that exactly three vehicles will use the acceleration lane in the next minute?

4.73 An advertisement for a data base management system (DBMS) claims that this particular DBMS is preferred over all others by 40% of all DBMS users.

a. What is the probability that fewer than 4 of 20 randomly selected DBMS users prefer the advertiser's system, if in fact the advertised claim is true?

b. What is the probability that between 5 and 15 (inclusive) of 20 randomly selected DBMS users prefer the advertiser's system, if the claim is true?

c. If only 3 of the 20 sampled DBMS users prefer the advertiser's brand, what inference would you make? Why?

4.74 Military intelligence is currently testing a new missile protection system that is composed of four independent radar screens. The probability that any one of the radar screens will detect an airplane is .9. If an enemy airplane enters unauthorized territory, what is the probability that:

a. All four of the radar screens will detect the airplane?

b. At least one of the radar screens will detect the airplane?

c. None of the screens will detect the enemy plane?

4.75 Computer technology has developed to the point where most industrial "robots" are programmed to operate through microprocessors. The probability that one such computerized robot breaks down during any one 8-hour shift is .2. Find the probability that the robot will operate for at most five shifts before breaking down twice.

4.76 "The sufficiently prolonged continuation of a low probability makes a given outcome inevitable," writes A. J. Coale in *Population and Development Review* (Sept. 1985). The "inevitable" event Coale is specifically referring to is a nuclear war. Experts agree that the probability of a nuclear war occurring in a given year is small, but not 0. According to Coale, then, "over hundreds of years this makes nuclear war virtually certain." Suppose the probability of a nuclear war occurring in any given year is only .01. What is the probability of a nuclear war occurring in the next 5 years? 10 years? 15 years? 20 years?

OPTIONAL SUPPLEMENTARY EXERCISES

4.77 Suppose the random variable y has a moment generating function given by

$$m(t) = \frac{1}{5}e^t + \frac{2}{5}e^{2t} + \frac{2}{5}e^{3t}$$

a. Find the mean of y.

b. Find the variance of y.

4.78 Let y be a geometric random variable with the probability distribution given in Table 4.3. Show that $E(y) = 1/p$. [*Hint:* Write

$$E(y) = p \sum_{y=1}^{\infty} yq^{y-1} \quad \text{where } q = 1 - p$$

and note that

$$\frac{dq^y}{dq} = yq^{y-1}$$

Thus,

$$E(y) = p \sum_{y=1}^{\infty} yq^{y-1} = p\frac{d}{dq}\left(\sum_{y=1}^{\infty} q^y\right)$$

Then use the fact that

$$\sum_{y=1}^{\infty} q^y = \frac{q}{1-q}$$

(The sum of this infinite series is given in most mathematical handbooks.)]

4.79 The probability generating function $P(t)$ for a discrete random variable y is defined to be

$$P(t) = E(t^y) = p_0 + p_1t + p_2t^2 + \cdots$$

where $p_i = P(y = i)$.
a. Find $P(t)$ for the Poisson distribution. [*Hint:* Write

$$E(t^y) = \sum_{y=0}^{\infty} \frac{(\lambda t)^y e^{-\lambda}}{y!} = e^{\lambda(t-1)} \sum_{y=0}^{\infty} \frac{(\lambda t)^y e^{-\lambda t}}{y!}$$

and note that the quantity being summed is a Poisson probability with mean λt.]
b. Use the facts that

$$E(y) = \frac{dP(t)}{dt}\bigg]_{t=1} \qquad \text{and} \qquad E[y(y-1)] = \frac{d^2P(t)}{dt^2}\bigg]_{t=1}$$

to derive the mean and variance of a Poisson random variable.

REFERENCES

Feller, W. *An Introduction to Probability Theory and Its Applications*, Vol. I, 3rd ed. New York: Wiley, 1968.

Hogg, R. V. and Craig, A. T. *Introduction to Mathematical Statistics*, 4th ed. New York: Macmillan, 1978.

Mendenhall, W., Scheaffer, R. L., and Wackerly, D. *Mathematical Statistics with Applications*, 3rd ed. Boston: Duxbury Press, 1986.

Mood, A. M., Graybill, F. A., and Boes, D. C. *Introduction to the Theory of Statistics*, 3rd ed. New York: McGraw-Hill, 1963.

Mosteller, F., Rourke, R. E. K., and Thomas, G. B. *Probability with Statistical Applications*, 2nd ed. Reading, Mass.: Addison-Wesley, 1970.

Parzen, E. *Modern Probability Theory and Its Applications*. New York: Wiley, 1964.

Parzen, E. *Stochastic Processes*. San Francisco: Holden-Day, 1962.

Standard Mathematical Tables, 17th ed. Cleveland: Chemical Rubber Company, 1969.

C H A P T E R 5

OBJECTIVE

To distinguish between continuous and discrete random variables and their respective probability distributions; to present some useful continuous probability distributions and show how they can be used to solve some practical problems.

CONTENTS

CONTINUOUS RANDOM VARIABLES

S E C T I O N 5.1

**CONTINUOUS
RANDOM VARIABLES**

Many random variables observed in real life are not discrete random variables because the number of values that they can assume is not countable. For example, the waiting time y (in minutes) for a submitted data-processing job to be completed could, in theory, assume any of the uncountably infinite number of values in the interval $0 < y < \infty$. The daily rainfall at some location, the strength (in pounds per square inch) of a steel bar, and the intensity of sunlight at a particular time of the day are other examples of random variables that can assume any one of the uncountably infinite number of points in one or more intervals on the real line. In contrast to discrete random variables, such variables are called **continuous random variables**.

The preceding discussion identifies the difference between discrete and continuous random variables, but it fails to point to a practical problem. It is impossible to assign a finite amount of probability to each of the uncountable number of points in a line interval in such a way that the sum of the probabilities is 1. Therefore, the distinction between discrete and continuous random variables is usually based on the difference in their **cumulative distribution functions**.

DEFINITION 5.1

The **cumulative distribution function $F(y_0)$** for a random variable y is equal to the probability

$$F(y_0) = P(y \leq y_0)$$

For a discrete random variable, the cumulative distribution function is the cumulative sum of $p(y)$, from the smallest value that y can assume, to a value of y_0. For example, from the cumulative sums in Table 2 of Appendix II, we obtain the following values of $F(y)$ for a binomial random variable with $n = 5$ and $p = .5$:

$$F(0) = P(y \leq 0) = \sum_{y=0}^{0} p(y) = p(0) = .031$$

$$F(1) = P(y \leq 1) = \sum_{y=0}^{1} p(y) = .188$$

$$F(2) = P(y \leq 2) = \sum_{y=0}^{2} p(y) = .500$$

$$F(3) = P(y \leq 3) = .812$$

$$F(4) = P(y \leq 4) = .969$$

$$F(5) = P(y \leq 5) = 1$$

A graph of $p(y)$ is shown in Figure 5.1. The value of $F(y_0)$ is equal to the sum of the areas of the probability rectangles from $y = 0$ to $y = y_0$. The probability $F(3)$ is shaded in the figure.

FIGURE 5.1

Probability Distribution for a Binomial Random Variable ($n = 5$, $p = .5$); Shaded Area Corresponds to $F(3)$

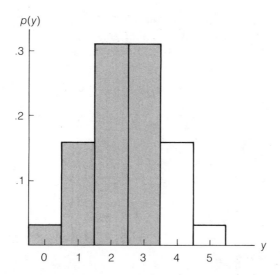

A graph of the cumulative distribution function for the binomial random variable with $n = 5$ and $p = .5$, shown in Figure 5.2, illustrates an important property of the cumulative distribution functions for all discrete random variables: *They are step functions.* For example, $F(y)$ is equal to .031 until, as y increases, it reaches $y = 1$. Then $F(y)$ jumps abruptly to $F(1) = .188$. The value of $F(y)$ then remains constant as y increases until y reaches $y = 2$. Then $F(y)$ rises abruptly to $F(2) = .500$. Thus, $F(y)$ is a discontinuous function which jumps upward at a countable number of points ($y = 0, 1, 2, 3$, and 4).

In contrast to the cumulative distribution function for a discrete random variable, the cumulative distribution function $F(y)$ for a continuous random variable is a **monotonically increasing** continuous function of y. This means that $F(y)$ is

FIGURE 5.2

Cumulative Distribution Function $F(y)$ for a Binomial Random Variable ($n = 5$, $p = .5$)

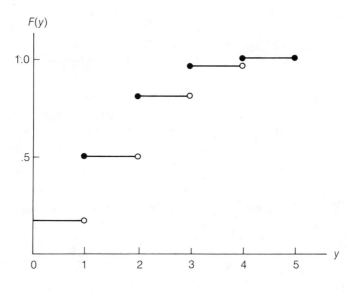

a continuous function such that if $y_a < y_b$, then $F(y_a) \leq F(y_b)$, i.e., as y increases, $F(y)$ never decreases. A graph of the cumulative distribution function for a continuous random variable might appear as shown in Figure 5.3.

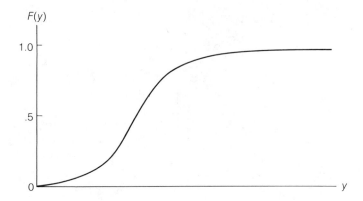

DEFINITION 5.2

A **continuous random variable** y is one that has the following three properties:

1. The cumulative distribution function, $F(y)$, is continuous.
2. y takes on an uncountably infinite number of values.
3. The probability that y equals any one particular value is 0.

S E C T I O N   5.2

**THE DENSITY
FUNCTION FOR A
CONTINUOUS
RANDOM VARIABLE**

In Chapter 1, we described a large set of data by means of a relative frequency distribution. If the data represent measurements on a continuous random variable and if the amount of data is very large, we can reduce the width of the class intervals until the distribution appears to be a smooth curve. A **probability density function** is a theoretical model for this distribution.

DEFINITION 5.3

If $F(y)$ is the cumulative distribution function for a continuous random variable y, then the **density function** $f(y)$ for y is

$$f(y) = \frac{dF(y)}{dy}$$

The density function for a continuous random variable y, the model for some real-life population of data, will usually be a smooth curve as shown in Figure 5.4. It follows from Definition 5.3 that

$$F(y) = \int_{-\infty}^{y} f(t)\, dt$$

FIGURE 5.4
Density Function $f(y)$ for a Continuous Random Variable

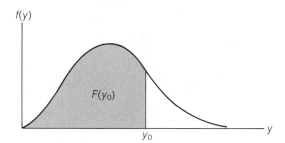

Thus, the cumulative area under the curve between $-\infty$ and a point y_0 is equal to $F(y_0)$.

The density function for a continuous random variable must always satisfy the two properties given in the box.

PROPERTIES OF A DENSITY FUNCTION

1. $f(y) \geq 0$

2. $\int_{-\infty}^{\infty} f(y) \, dy = F(\infty) = 1$

EXAMPLE 5.1

Let c be a constant and consider the density function

$$f(y) = \begin{cases} cy & \text{if } 0 \leq y \leq 1 \\ 0 & \text{elsewhere} \end{cases}$$

Find the value of c.

SOLUTION

Since $\int_{-\infty}^{\infty} f(y) \, dy$ must equal 1, we have

$$\int_{-\infty}^{\infty} f(y) \, dy = \int_{0}^{1} cy \, dy = c\frac{y^2}{2}\Big]_{0}^{1} = c\left(\frac{1}{2}\right) = 1$$

Solving for c yields $c = 2$, and thus, $f(y) = 2y$. A graph of $f(y)$ is shown in Figure 5.5.

FIGURE 5.5
Graph of the Density Function $f(y)$ for Example 5.1

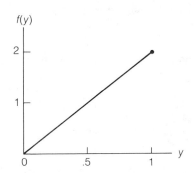

EXAMPLE 5.2

Refer to Example 5.1. Find the cumulative distribution function for the random variable y. Then find $F(.2)$ and $F(.7)$.

SOLUTION

By Definition 5.3, it follows that

$$F(y) = \int_{-\infty}^{y} f(t)\,dt = \int_{0}^{y} 2t\,dt = 2\left(\frac{t^2}{2}\right)\Big]_{0}^{y} = y^2$$

Then

$$F(.2) = P(y \le .2) = (.2)^2 = .04$$
$$F(.7) = P(y \le .7) = (.7)^2 = .49$$

The value of $F(y)$ when $y = .7$—i.e., $F(.7)$—is the shaded area in Figure 5.6.

FIGURE 5.6
Graph of the Density
Function $f(y)$ for Example
5.2; Shaded Area
Corresponds to $F(.7)$

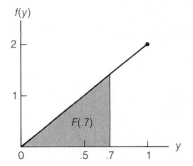

Many of the continuous random variables with applications in statistics have density functions whose integrals cannot be expressed in closed form. They can only be approximated by numerical methods. Tables of areas under several such density functions are presented in Appendix II and will be introduced as required.

EXERCISES 5.1–5.4

5.1 Let c be a constant and consider the density function

$$f(y) = \begin{cases} cy^2 & \text{if } 0 \le y \le 2 \\ 0 & \text{elsewhere} \end{cases}$$

　a. Find the value of c.
　b. Find the cumulative distribution function $F(y)$.
　c. Compute $F(1)$.
　d. Compute $F(.5)$.

5.2 Let c be a constant and consider the density function

$$f(y) = \begin{cases} c(2 - y) & \text{if } 0 \le y \le 1 \\ 0 & \text{elsewhere} \end{cases}$$

　a. Find the value of c.
　b. Find the cumulative distribution function $F(y)$.
　c. Compute $F(.4)$.

5.3 Let c be a constant and consider the density function

$$f(y) = \begin{cases} ce^{-y} & \text{if } y > 0 \\ 0 & \text{elsewhere} \end{cases}$$

a. Find the value of c.
b. Find the cumulative distribution function $F(y)$.
c. Compute $F(2.6)$.
d. Show that $F(0) = 0$ and $F(\infty) = 1$.

OPTIONAL EXERCISE

5.4 Continuous probability distributions provide theoretical models for the lifelength of a component (e.g., computer chip, light bulb, automobile, air conditioning unit, and so on). Often, it is important to know whether or not it is better to replace periodically an old component with a new component. For example, for certain types of light bulbs an old bulb that has been in use for a while tends to have a longer lifelength than a new bulb. Let y represent the lifelength of some component with cumulative distribution function $F(y)$. Then the "life" distribution $F(y)$ is considered **new better than used** (NBU) if

$$\bar{F}(x + y) \leq \bar{F}(x)\bar{F}(y) \quad \text{for all } x, y \geq 0$$

where $\bar{F}(y) = 1 - F(y)$ (*Microelectronics and Reliability*, Jan. 1986). Alternatively, a "life" distribution $F(y)$ is **new worse than used** (NWU) if

$$\bar{F}(x + y) \geq \bar{F}(x)\bar{F}(y) \quad \text{for all } x, y \geq 0$$

a. Consider the density function

$$f(y) = \begin{cases} y/2 & \text{if } 0 < y < 2 \\ 0 & \text{elsewhere} \end{cases}$$

Find the "life" distribution, $F(y)$.
b. Determine whether the "life" distribution $F(y)$ is NBU or NWU.

SECTION 5.3

EXPECTED VALUES FOR CONTINUOUS RANDOM VARIABLES

You will recall from your study of calculus that integration is a summation process. Thus, finding the integral

$$F(y_0) = \int_{-\infty}^{y_0} f(t) \, dt$$

for a continuous random variable is analogous to finding the sum

$$F(y_0) = \sum_{y \leq y_0} p(y)$$

for a discrete random variable. Then it is natural to employ the same definitions for the expected value of a continuous random variable y, for the expected value of a function $g(y)$, and for the variance of y that were given for a discrete random variable in Section 4.3. The only difference is that we will substitute the integration symbol for the summation symbol. It also can be shown (proof omitted) that the expectation theorems of Section 4.4 hold for discrete random variables. We now summarize these definitions and theorems and present some examples of their use.

DEFINITION 5.4

Let y be a continuous random variable with density function $f(y)$. Then the **expected value of y** is

$$E(y) = \int_{-\infty}^{\infty} yf(y)\, dy$$

DEFINITION 5.5

Let y be a continuous random variable with density function $f(y)$ and let $g(y)$ be any function of y. Then the expected value of $g(y)$ is

$$E[g(y)] = \int_{-\infty}^{\infty} g(y)f(y)\, dy$$

THEOREM 5.1

Let c be a constant and let y be a continuous random variable. Then

$$E(c) = c$$

THEOREM 5.2

Let c be a constant and let y be a continuous random variable. Then

$$E(cy) = cE(y)$$

THEOREM 5.3

Let y be a continuous random variable and let $g_1(y), g_2(y), \ldots, g_k(y)$ be k functions of y. Then

$$E[g_1(y) + g_2(y) + \cdots + g_k(y)] = E[g_1(y)] + E[g_2(y)] + \cdots + E[g_k(y)]$$

THEOREM 5.4

Let y be a continuous random variable with $E(y) = \mu$. Then

$$\sigma^2 = E[(y - \mu)^2] = E(y^2) - \mu^2$$

EXAMPLE 5.3

Refer to Example 5.1. Find the mean and standard deviation for the continuous random variable y.

SOLUTION

Recall that $f(y) = 2y$. Therefore,

$$E(y) = \int_{-\infty}^{\infty} yf(y)\ dy = \int_0^1 y(2y)\ dy = \int_0^1 2y^2\ dy = \frac{2y^3}{3}\Big]_0^1 = \frac{2}{3}$$

$$E(y^2) = \int_{-\infty}^{\infty} y^2 f(y)\ dy = \int_0^1 y^2(2y)\ dy = \int_0^1 2y^3\ dy = \frac{2y^4}{4}\Big]_0^1 = \frac{1}{2}$$

Then, by Theorem 5.4,

$$\sigma^2 = E(y^2) - \mu^2 = \frac{1}{2} - \left(\frac{2}{3}\right)^2 = .0556$$

and thus

$$\sigma = \sqrt{.0556} = .24$$

∎

EXAMPLE 5.4

Refer to Examples 5.1 and 5.3. The interval $\mu \pm 2\sigma$ is shown on the graph of $f(y)$ in Figure 5.7. Find $P(\mu - 2\sigma < y < \mu + 2\sigma)$.

SOLUTION

From Example 5.3, we have $\mu = \frac{2}{3} \approx .67$ and $\sigma = .24$. Therefore, $\mu - 2\sigma = .19$ and $\mu + 2\sigma = 1.15$. Since $P(y > 1) = 0$, we want to find the probability $P(y > .19)$, corresponding to the shaded area in Figure 5.7:

$$P(\mu - 2\sigma < y < \mu + 2\sigma) = P(y > .19) = \int_{.19}^1 f(y)\ dy$$

$$= \int_{.19}^1 2y\ dy = y^2\Big]_{.19}^1 = 1 - (.19)^2 = .96$$

FIGURE 5.7
Graph Showing the Interval $\mu \pm 2\sigma$ for $f(y) = 2y$

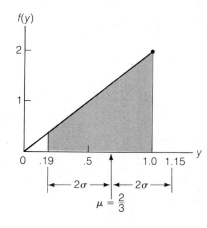

∎

In Chapter 1, we applied the Empirical Rule to mound-shaped relative frequency distributions of data. The Empirical Rule may also be applied to mound-shaped theoretical—i.e., probability—distributions. As examples in the preceding chapters demonstrate, the percentage (or proportion) of a data set in the interval $\mu \pm 2\sigma$ is usually very close to .95, the value specified by the Empirical Rule. This is certainly true for the probability distribution considered in Example 5.4.

EXAMPLE 5.5

Let y be a continuous random variable with probability density function

$$f(y) = \begin{cases} \dfrac{e^{-y/2}}{2} & \text{if } 0 \leq y < \infty \\ 0 & \text{elsewhere} \end{cases}$$

Find the mean, variance, and standard deviation of y. (This density function is known as the **exponential probability distribution**.)

SOLUTION

The mean of the random variable y is given by

$$\mu = E(y) = \int_{-\infty}^{\infty} yf(y) \, dy = \int_{0}^{\infty} \frac{ye^{-y/2}}{2} dy$$

In order to compute this definite integral, we use the following general formula, found in most mathematical handbooks:*

$$\int ye^{ay} \, dy = \frac{e^{ay}}{a^2}(ay - 1)$$

By substituting $a = -\frac{1}{2}$, we obtain

$$\mu = \frac{1}{2}(4) = 2$$

To find σ^2, we will first find $E(y^2)$ by making use of the general formula†

$$\int y^m e^{ay} \, dy = \frac{y^m e^{ay}}{a} - \frac{m}{a} \int y^{m-1} e^{ay} \, dy$$

Then with $a = -\frac{1}{2}$ and $m = 2$, we can write

$$E(y^2) = \int_{-\infty}^{\infty} y^2 f(y) \, dy = \int_{0}^{\infty} \frac{y^2 e^{-y/2}}{2} \, dy = \frac{1}{2}(16) = 8$$

---

*See, for example, *Standard Mathematical Tables* (1969). Otherwise, the result can be derived using integration by parts:

$$\int ye^{ay} \, dy = \frac{ye^{ay}}{a} - \int \frac{e^{ay}}{a} \, dy$$

†This result is also derived using integration by parts.

Thus, by Theorem 5.4,

$$\sigma^2 = E(y^2) - \mu^2 = 8 - (2)^2 = 4$$

and

$$\sigma = \sqrt{4} = 2$$ ∎

EXAMPLE 5.6

A graph of the density function of Example 5.5 is shown in Figure 5.8. Find $P(\mu - 2\sigma < y < \mu + 2\sigma)$.

FIGURE 5.8

Graph of the Density
Function of Example 5.5

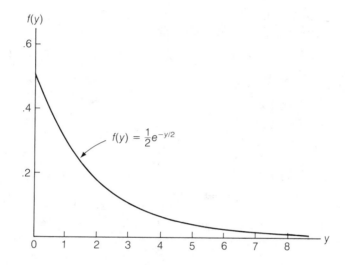

$$f(y) = \frac{1}{2}e^{-y/2}$$

SOLUTION

We showed in Example 5.5 that $\mu = 2$ and $\sigma = 2$. Therefore, $\mu - 2\sigma = 2 - 4 = -2$ and $\mu + 2\sigma = 6$. Since $f(y) = 0$ for $y < 0$,

$$P(\mu - 2\sigma < y < \mu + 2\sigma) = \int_0^6 f(y)\, dy = \int_0^6 \frac{e^{-y/2}}{2}\, dy$$

$$= -e^{-y/2}\Big]_0^6 = 1 - e^{-3}$$

$$= 1 - .049787 = .950213$$

The Empirical Rule of Chapter 1 would suggest that a good approximation to this probability is .95. You can see that for the negative exponential density function, the approximation is very close to the exact probability, .950213.   ∎

In many practical situations, we will know the variance (or standard deviation) of a random variable y and will want to find the standard deviation of cy, where c is a constant. For example, we might know the standard deviation of the weight y in ounces of a particular type of computer chip and want to find the standard deviation of the weight in grams. Since 1 ounce = 28.349527 grams, we would want to find the standard deviation of cy, where $c = 28.349527$. The variance of cy is given by Theorem 5.5.

THEOREM 5.5

Let y be a random variable* with mean μ and variance σ^2. Then the variance of cy is

$$\sigma^2_{cy} = c^2\sigma^2$$

Proof of Theorem 5.5   From Theorem 5.2, we know that $E(cy) = cE(y) = c\mu$. Using the definition of the variance of a random variable, we can write

$$\sigma^2_{cy} = E[(cy - c\mu)^2] = E\{[c(y - \mu)]^2\} = E[c^2(y - \mu)^2]$$

Then, by Theorem 5.2,

$$\sigma^2_{cy} = c^2E[(y - \mu)^2]$$

But, $E[(y - \mu)^2] = \sigma^2$. Therefore,

$$\sigma^2_{cy} = c^2\sigma^2$$

As an example of the application of this theorem, suppose that the variance of the weight y of a computer chip is 1.1 (ounces)^2. Then the variance of the weight of the chip in grams is equal to $(28.349527)^2(1.1) \approx 884.1$ (grams)^2. The standard deviation of the weight in grams is equal to $\sqrt{884.1} = 29.7$ grams.

EXERCISES 5.5–5.12

5.5 Find μ and σ^2 for the continuous random variable of Exercise 5.1. Then compute $P(\mu - 2\sigma < y < \mu + 2\sigma)$ and compare to the Empirical Rule.

5.6 Find μ and σ^2 for the continuous random variable of Exercise 5.2. Then compute $P(\mu - 2\sigma < y < \mu + 2\sigma)$ and compare to the Empirical Rule.

5.7 Find μ and σ^2 for the continuous random variable of Exercise 5.3. Then compute $P(\mu - 2\sigma < y < \mu + 2\sigma)$ and compare to the Empirical Rule.

5.8 The amount of time y (in minutes) that a commuter train is late is a continuous random variable with probability density

$$f(y) = \begin{cases} \dfrac{3}{500}(25 - y^2) & \text{if } -5 < y < 5 \\ 0 & \text{elsewhere} \end{cases}$$

[*Note:* A negative value of y means that the train is early.]
a. Find the mean and variance of the amount of time in minutes the train is late.
b. Find the mean and variance of the amount of time in hours the train is late.
c. Find the mean and variance of the amount of time in seconds the train is late.

*This theorem applies to discrete or continuous random variables.

OPTIONAL EXERCISES

5.9 Prove Theorem 5.1.

5.10 Prove Theorem 5.2.

5.11 Prove Theorem 5.3.

5.12 Prove Theorem 5.4.

| | | | | | | | | | | | | |

S E C T I O N 5.4

THE UNIFORM PROBABILITY DISTRIBUTION

Suppose you were to randomly select a number y represented by a point in the interval $a \leq y \leq b$. Then y is called a **uniform random variable** and its density function, mean, and variance are as shown in the box.

THE DENSITY FUNCTION, MEAN, AND VARIANCE FOR A UNIFORM RANDOM VARIABLE y

$$f(y) = \begin{cases} \dfrac{1}{b-a} & \text{if} \quad a \leq y \leq b \\ 0 & \text{elsewhere} \end{cases}$$

$$\mu = \frac{a+b}{2} \qquad \sigma^2 = \frac{(b-a)^2}{12}$$

EXAMPLE 5.7

Suppose the research department of a steel manufacturer believes that one of the company's rolling machines is producing sheets of steel of varying thickness. The thickness is a uniform random variable with values between 150 and 200 millimeters. Any sheets less than 160 millimeters thick must be scrapped, since they are unacceptable to buyers.

a. Calculate the mean and standard deviation of y, the thickness of the sheets produced by this machine. Then graph the probability distribution, and show the mean on the horizontal axis. Also show 1 and 2 standard deviation intervals around the mean.

b. Calculate the fraction of steel sheets produced by this machine that have to be scrapped.

SOLUTION

a. To calculate the mean and standard deviation for y, we substitute 150 and 200 millimeters for a and b, respectively, in the formulas. Thus,

$$\mu = \frac{a+b}{2} = \frac{150 + 200}{2} = 175 \text{ millimeters}$$

and

$$\sigma = \frac{b-a}{\sqrt{12}} = \frac{200 - 150}{\sqrt{12}} = \frac{50}{3.464} = 14.43 \text{ millimeters}$$

The uniform probability distribution is

$$f(y) = \frac{1}{b - a} = \frac{1}{200 - 150} = \frac{1}{50}$$

The graph of this function is shown in Figure 5.9. The mean and 1 and 2 standard deviation intervals around the mean are shown on the horizontal axis.

FIGURE 5.9

Frequency Function for y in Example 5.7

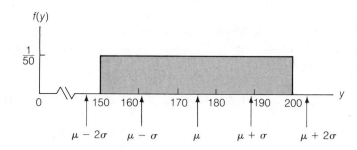

b. To find the fraction of steel sheets produced by the machine that have to be scrapped, we must find the probability that y, the thickness, is less than 160 millimeters. As indicated in Figure 5.10, we need to calculate the area under the frequency function $f(y)$ between the points $a = 150$ and $c = 160$. This is the area of a rectangle with base $160 - 150 = 10$ and height $\frac{1}{50}$. The fraction that has to be scrapped is then

$$P(y < 160) = (\text{Base})(\text{Height}) = (10)\left(\frac{1}{50}\right) = \frac{1}{5}$$

That is, 20% of all the sheets made by this machine must be scrapped.

FIGURE 5.10

The Probability That the Sheet Thickness, y, is Between 150 and 160 Millimeters

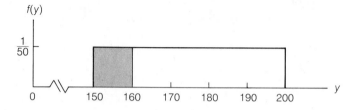

The random numbers in Table 1 of Appendix II were generated by a computer program that randomly selects values of y from a uniform distribution. (However, the random numbers are terminated at some specified decimal place.) One of the most important applications of the uniform distribution is described in Chapter 7. There, we will use it, along with a computer program that generates random numbers, to simulate the sampling of many other types of random variables.

5.13 A manufacturing company has developed a fuel efficient machine that combines pressure washing with steam cleaning. It is designed to deliver 7 gallons of cleaner per minute at 1,000 pounds per square inch for pressure washing. In fact, it delivers an amount at random anywhere between 6.5 and 7.5 gallons per minute. Assume that y, the amount of cleaner delivered, is a uniform random variable with probability density

$$f(y) = \begin{cases} 1 & \text{if } 6.5 \le y \le 7.5 \\ 0 & \text{elsewhere} \end{cases}$$

a. Find the mean and standard deviation of y. Then graph $f(y)$, showing the locations of the mean and 1 and 2 standard deviation intervals around the mean.
b. Find the probability that more than 7.2 gallons of cleaner are dispensed per minute.

5.14 The amount of time y between pauses on a full-screen-edit terminal (i.e., the time required for the terminal to process an edit command and make the corrections on the screen) is uniformly distributed between .5 and 2.25 seconds.
a. Find the mean and variance of y.
b. Locate the interval $\mu \pm 2\sigma$ on a graph of the probability distribution and compute $P(\mu - 2\sigma < y < \mu + 2\sigma)$. Compare your result with Tchebysheff's theorem.
c. What is the probability the terminal will process an edit command and make the appropriate corrections on the screen in less than 1 second?

5.15 The Department of Transportation (DOT) has determined that the winning (low) bid y (in dollars) on a road construction contract has a uniform distribution with probability density function

$$f(y) = \begin{cases} \dfrac{5}{8d} & \text{if } \dfrac{2d}{5} \le y \le 2d \\ 0 & \text{elsewhere} \end{cases}$$

where d is the DOT estimate of the cost of the job.
a. Find the mean and standard deviation of y. Then graph $f(y)$, showing the locations of the mean and 1 and 2 standard deviation intervals around the mean.
b. What fraction of the winning bids on road construction contracts are less than the DOT estimate?

OPTIONAL EXERCISES

5.16 Assume that the random variable y is uniformly distributed over the interval $a \le y \le b$. Verify the following:

a. $\mu = \dfrac{a + b}{2}$ and $\sigma^2 = \dfrac{(b - a)^2}{12}$

b. $F(y) = \begin{cases} \dfrac{y - a}{b - a} & \text{if } a \le y \le b \\ 0 & \text{if } y < a \\ 1 & \text{if } y > b \end{cases}$

5.17 Assume that y is uniformly distributed over the interval $0 \leq y \leq 1$. Show that, for $a \geq 0$, $b \geq 0$, and $(a + b) \leq 1$,

$$P(a < y < a + b) = b$$

5.18 Show that the uniform distribution is *new better than used* (NBU) over the interval (0, 1). (See Optional Exercise 5.4 for the definition of NBU.)

S E C T I O N 5.5

THE NORMAL PROBABILITY DISTRIBUTION

The **normal** (or **Gaussian**) **density function** was proposed by C. F. Gauss (1777–1855) as a model for the relative frequency distribution of *errors*, such as errors of measurement. Amazingly, this bell-shaped curve provides an adequate model for the relative frequency distributions of data collected from many different scientific areas and, as we will show in Chapter 7, it models the probability distributions of many statistics that we will use for making inferences.

The normal random variable possesses a density function characterized by two parameters. This density function, its mean, and its variance are shown in the box.

THE DENSITY FUNCTION, MEAN, AND VARIANCE FOR A NORMAL RANDOM VARIABLE y

$$f(y) = \frac{1}{\sigma\sqrt{2\pi}}e^{-(y-\mu)^2/2\sigma^2} \qquad -\infty < y < \infty$$

The parameters μ and σ^2 are the mean and variance, respectively, of the normal random variable y.

There is an infinite number of normal density functions—one for each combination of μ and σ. The mean μ measures the location of the distribution and the standard deviation σ measures its spread. Several different normal density functions are shown in Figure 5.11.

FIGURE 5.11
Several Normal Distributions, with Different Means and Standard Deviations

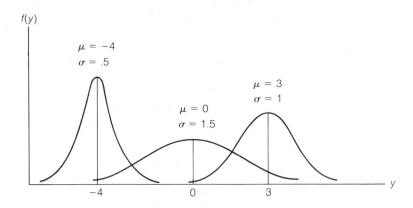

A closed-form expression cannot be obtained for the integral of the normal density function. Consequently, areas under the normal curve must be obtained by using approximation procedures. These areas are given in Table 4 of Appendix II for the **standard normal variable,**

$$z = \frac{y - \mu}{\sigma}$$

which has mean equal to 0 and variance equal to 1.* In practical terms, z is the distance between the value of a normal random variable y and its mean μ, measured in units of its standard deviation σ.

The entries in Table 4 of Appendix II are the areas under the normal curve between the mean, $z = 0$, and a value of z to the right of the mean (see Figure 5.12). To find the area under the normal curve between $z = 0$ and, say $z = 1.33$, move down the left column of Table 4 to the row corresponding to $z = 1.3$. Then move across the top of the table to the column marked .03. The entry at the intersection of this row and column gives the area $A = .4082$. Because the normal curve is symmetric about the mean, areas to the left of the mean are equal to the corresponding areas to the right of the mean. For example, the area A between the mean $z = 0$ and $z = -.68$ is equal to the area between $z = 0$ and $z = -.68$. This area will be found in Table 4 at the intersection of the row corresponding to 0.6 and the column corresponding to .08 as $A = .2517$.

FIGURE 5.12

Standard Normal Density Function Showing the Tabulated Areas Given in Table 4 of Appendix II

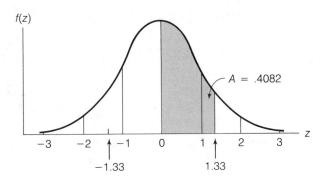

EXAMPLE 5.8

Suppose y is a normally distributed random variable with mean 10 and standard deviation 2.1.

a. Find $P(y \geq 11)$.
b. Find $P(7.6 \leq y \leq 12.2)$.

SOLUTION

a. The value $y = 11$ corresponds to a z value of

$$z = \frac{y - \mu}{\sigma} = \frac{11 - 10}{2.1} = .48$$

*The proof that the mean and variance of z are 0 and 1, respectively, is left as an optional exercise.

and thus, $P(y \geq 11) = P(z \geq .48)$. The area under the standard normal curve corresponding to this probability is shaded in Figure 5.13. Since the normal curve is symmetric about $z = 0$ and the total area beneath the curve is 1, the area to the right of $z = 0$ is equal to .5. Thus, the shaded area is equal to $(.5 - A)$, where A is the tabulated area corresponding to $z = .48$. The area A, given in Table 4 of Appendix II, is .1844. Therefore,

$$P(y \geq 11) = .5 - A = .5 - .1844 = .3156$$

FIGURE 5.13

Standard Normal Distribution for Example 5.8; Shaded Area Is $P(y \geq 11)$

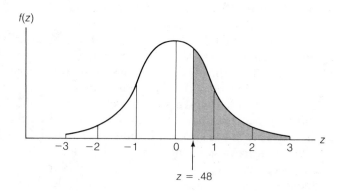

b. The values $y_1 = 7.6$ and $y_2 = 12.2$ correspond to the z values

$$z_1 = \frac{y_1 - \mu}{\sigma} = \frac{7.6 - 10}{2.1} = -1.14$$

$$z_2 = \frac{y_2 - \mu}{\sigma} = \frac{12.2 - 10}{2.1} = 1.05$$

The probability $P(7.6 \leq y \leq 12.2) = P(-1.14 \leq z \leq 1.05)$ is the shaded area shown in Figure 5.14. It is equal to the sum of A_1 and A_2, the areas corresponding to z_1 and z_2, respectively, where $A_1 = .3729$ and $A_2 = .3531$. Therefore,

$$P(7.6 \leq y \leq 12.2) = A_1 + A_2 = .3729 + .3531 = .7260$$

FIGURE 5.14

Standard Normal Distribution for Example 5.8

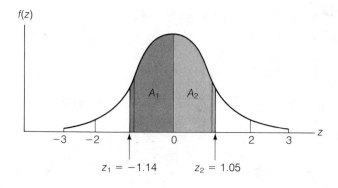

EXAMPLE 5.9

The U.S. Department of Agriculture (USDA) has recently patented a process that uses a bacterium for removing bitterness from citrus juices (*Chemical Engineering*, Feb. 3, 1986). In theory, almost all the bitterness could be removed by the process, but for practical purposes the USDA aims at 50% overall removal. Suppose a USDA spokesman claims that the percentage of bitterness removed from an 8-ounce glass of freshly squeezed citrus juice is normally distributed with mean 50.1 and standard deviation 10.4. To test this claim, the bitterness removal process is applied to a randomly selected 8-ounce glass of citrus juice. Find the probability that the process removes less than 33.7% of the bitterness.

SOLUTION

The value $y = 33.7$ corresponds to the value of the standard normal random variable:

$$z = \frac{y - \mu}{\sigma} = \frac{33.7 - 50.1}{10.4} = -1.58$$

Therefore, $P(y \leq 33.7) = P(z \leq -1.58)$, the shaded area in Figure 5.15, is equal to .5 minus the area A that corresponds to $z = 1.58$. Therefore, the probability that the process removes less than 33.7% of the bitterness is

$$P(y \leq 33.7) = .5 - .4429 = .0571$$

FIGURE 5.15
The Probability That Percentage of Bitterness Removed Is Less Than 33.7% in Example 5.9

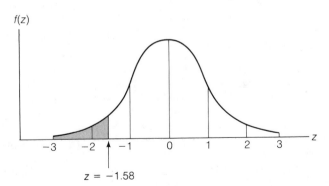

$z = -1.58$

EXAMPLE 5.10

Refer to Example 5.9. If the test on the single glass of citrus juice yielded a bitterness removal percentage of 33.7, would you tend to doubt the USDA spokesman's claim?

SOLUTION

Given the sample information, we have several choices. We could conclude that the spokesman's claim is true, i.e., that the mean percentage of bitterness removed for the new process is 50.1% and that we have just observed a *rare event*, one that would occur with a probability of only .0571. Or, we could conclude that the spokesman's claim for the mean percentage is too high, i.e., that the true mean is less than 50.1%. Or, perhaps the assumed value of σ or the assumption of normality may be in error. Given a choice, we think you will agree that there is reason to doubt the USDA spokesman's claim. ∎

| | | | | | | | | | | | | |

EXERCISES 5.19–5.26

5.19 Use Table 4 of Appendix II to find the following probabilities for a standard normal random variable:

a. $P(.5 < z < 1.5)$ **b.** $P(-1.75 < z < -.28)$
c. $P(-2.32 < z < .11)$ **d.** $P(z > .27)$
e. $P(z < -1.33)$ **f.** $P(z < 1.71)$

5.20 Find the value of the standard normal random variable z, call it z_0, such that:

a. $P(z > z_0) = .05$ **b.** $P(z > z_0) = .025$
c. $P(z > z_0) = .80$ **d.** $P(z < z_0) = .0013$
e. $P(z < z_0) = .97$ **f.** $P(z < z_0) = .5596$

5.21 State governments are tightening their antitrust laws and imposing stiffer penalties on highway construction bid riggers. If it is known that the bids received on a 110-mile highway construction project are normally distributed with a mean of $290 million and a standard deviation of $40 million, find the probability that a contractor selected at random would have a bid of less than $200 million.

5.22 Steel used for water pipelines is often coated on the inside with cement mortar to prevent corrosion. In a study of the mortar coatings of a pipeline used in a water transmission project in California (*Transportation Engineering Journal*, Nov. 1979), the mortar thickness was specified to be $\frac{7}{16}$ inch. A very large number of thickness measurements produced a mean equal to .635 inch and a standard deviation equal to .082 inch. If the thickness measurements were normally distributed, approximately what percentage were less than $\frac{7}{16}$ inch?

5.23 A standard fluorescent tube has a lifelength that is normally distributed with a mean of 7,000 hours and a standard deviation of 1,000 hours. A competitor has developed a compact fluorescent lighting system that will fit into incandescent sockets. It claims that the new compact tube has a normally distributed lifelength with a mean of 7,500 hours and a standard deviation of 1,200 hours.

a. Which fluorescent tube is more likely to have a lifelength greater than 9,000 hours?

b. Which tube is more likely to have a lifelength less than 5,000 hours?

5.24 The distribution of the demand (in number of units per unit time) for a product can often be approximated by a normal probability distribution. For example, a communication cable company has determined that the number of push-button terminal switches demanded daily has a normal distribution with mean 200 and standard deviation 50.

a. On what percentage of days will the demand be less than 90 switches?

b. On what percentage of days will the demand fall between 225 and 275 switches?

c. Based on cost considerations, the company has determined that its best strategy is to produce a sufficient number of switches so that it will fully supply demand on 94% of all days. How many terminal switches should the company produce per day?

5.25 Refer to the *Journal of the Acoustical Society of America* (Feb. 1986) study of auditory nerve response rates in cats, discussed in Exercise 4.53. A key question addressed by the research is whether rate changes (i.e., changes in number of spikes per burst of noise) produced by tones in the presence of background noise are large enough to detect reliably. That is, can the tone be detected reliably when background noise is present? In the theory of signal detection, the problem involves a comparison of two probability distributions. Let y represent the auditory nerve response rate (i.e., the number of spikes observed) under two conditions: when the stimulus is background noise only (N) and when the stimulus is a tone plus background noise (T). The probability distributions for y under the two conditions are represented by the density functions, $f_N(y)$ and $f_T(y)$, respectively, where we assume that the mean response rate under the background-noise-only condition is less than the mean response rate under the tone-plus-noise condition, i.e., $\mu_N < \mu_T$. In this situation, an observer sets a threshold C and decides that a tone is present if $y \geq C$ and decides that no tone is present if $y < C$. Assume that $f_N(y)$ and $f_T(y)$ are both normal density functions with means $\mu_N = 10.1$ spikes per burst and $\mu_T = 13.6$ spikes per burst, respectively, and equal variances $\sigma_N^2 = \sigma_T^2 = 2$.

a. For a threshold of $C = 11$ spikes per burst, find the probability of detecting the tone given that the tone is present. (This is known as the *detection probability*.)

b. For a threshold of $C = 11$ spikes per burst, find the probability of detecting the tone given that only background noise is present. (This is known as the *probability of false alarm*.)

c. Usually, it is desirable to maximize detection probability while minimizing false alarm probability. Can you find a value of C that will both increase the detection probability (part **a**) and decrease the probability of false alarm (part **b**)?

OPTIONAL EXERCISE

5.26 Let y be a normal random variable with mean μ and variance σ^2. Show that

$$z = \frac{y - \mu}{\sigma}$$

has mean 0 and variance 1. [*Hint:* Apply Theorems 5.2–5.4.]

| | | | | | | | | | | | |

S E C T I O N   5.6

GAMMA-TYPE PROBABILITY DISTRIBUTIONS

Many random variables, such as the length of the useful life of a computer, can assume only nonnegative values. The relative frequency distributions for data of this type can often be modeled by **gamma-type density functions**. The formulas for a gamma density function, its mean, and its variance are shown in the box on page 196.

The formula for the gamma density function contains two parameters, α and β. The parameter β, known as a **scale parameter**, reflects the size of the units in which y is measured. (It performs the same function as the parameter σ that appears in the formula for the normal density function.) The parameter α is known as a shape parameter. Changing its value changes the shape of the gamma distribution. This enables us to obtain density functions of many different shapes to model relative frequency distributions of experimental data. Computer graphs of the gamma density function for $\alpha = 1$, 3, and 5, with $\beta = 1$, are shown in Figure 5.16.

**THE PROBABILITY DENSITY FUNCTION, MEAN, AND
VARIANCE FOR A GAMMA-TYPE RANDOM VARIABLE**

The probability density function for a gamma-type random variable is given by

$$f(y) = \begin{cases} \dfrac{y^{\alpha-1}e^{-y/\beta}}{\beta^\alpha \Gamma(\alpha)} & \text{if } 0 \le y < \infty; \quad \alpha > 0; \quad \beta > 0 \\ 0 & \text{elsewhere} \end{cases}$$

where

$$\Gamma(\alpha) = \int_0^\infty y^{\alpha-1}e^{-y}\,dy$$

The mean and variance of a gamma-type random variable are, respectively,

$$\mu = \alpha\beta \qquad \sigma^2 = \alpha\beta^2$$

It can be shown (proof omitted) that $\Gamma(\alpha) = (\alpha - 1)\Gamma(\alpha - 1)$ and that $\Gamma(\alpha) = (\alpha - 1)!$ when α is a positive integer. Values of $\Gamma(\alpha)$ for $1.0 \le \alpha \le 2.0$ are presented in Table 5 of Appendix II.

FIGURE 5.16   Graphs of Gamma Density Functions for $\alpha = 1$, 3, and 5; $\beta = 1$

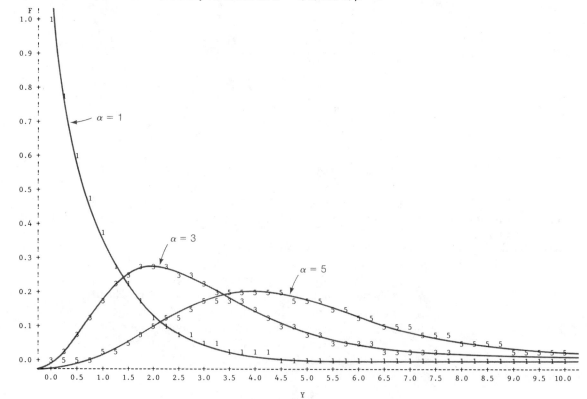

Except for the special case where α is an integer, we cannot obtain a closed-form expression for the integral of the gamma density function. Consequently, the cumulative distribution function for a gamma random variable, called an **incomplete gamma function**, must be obtained using approximation procedures with the aid of a computer. Values of this function are given in *Tables of the Incomplete Gamma Function* (1956).

A gamma-type random variable that plays an important role in statistics is the **chi-square random variable**. Chi-square values and corresponding areas under the chi-square density function are given in Table 7 of Appendix II. We will discuss the use of this table in Chapter 8.

When $\alpha = 1$, the gamma density function is known as an *exponential distribution*.* This important density function is employed as a model for the relative frequency distribution of the length of time between arrivals at a service counter (computer center, supermarket checkout counter, hospital clinic, etc.) when the probability of a customer arrival in any one unit of time is equal to the probability of arrival during any other. It is also used as a model for the length of life of industrial equipment or products when the probability that an "old" component will operate at least t additional time units, given it is now functioning, is the same as the probability that a "new" component will operate at least t time units. Equipment subject to periodic maintenance and parts replacement often exhibits this property of "never growing old."

THE DENSITY FUNCTION, MEAN, AND VARIANCE FOR A CHI-SQUARE RANDOM VARIABLE

A **chi-square (χ^2) random variable** is a gamma-type random variable with $\alpha = \nu/2$ and $\beta = 2$:

$$f(\chi^2) = c(\chi^2)^{(\nu/2)-1}e^{-\chi^2/2} \qquad (0 \le \chi^2 < \infty)$$

where

$$c = \frac{1}{2^{\nu/2}\Gamma\left(\frac{\nu}{2}\right)}$$

The mean and variance of a chi-square random variable are, respectively,

$$\mu = \nu \qquad \sigma^2 = 2\nu$$

The parameter ν is called the **number of degrees of freedom** for the chi-square distribution.

The exponential distribution is related to the Poisson probability distribution. In fact, it can be shown (proof omitted) that if the number of arrivals at a service

*The exponential distribution was encountered in Examples 5.5 and 5.6 of Section 5.3.

counter follows a Poisson probability distribution with the mean number of arrivals per unit time equal to $1/\beta$, then the density function for the length of time y between any pair of successive arrivals will be an exponential distribution with mean equal to β, i.e.,

$$f(y) = \frac{e^{-y/\beta}}{\beta} \qquad (0 \leq y < \infty)$$

THE DENSITY FUNCTION, MEAN, AND VARIANCE FOR AN EXPONENTIAL RANDOM VARIABLE

An **exponential distribution** is a gamma density function with $\alpha = 1$:

$$f(y) = \frac{e^{-y/\beta}}{\beta} \qquad (0 \leq y < \infty)$$

with mean and variance

$$\mu = \beta \qquad \sigma^2 = \beta^2$$

EXAMPLE 5.11

From past experience, a manufacturer knows that the relative frequency distribution of the length of time (in months) between major customer product complaints can be modeled by a gamma density function with $\alpha = 2$ and $\beta = 4$. Fifteen months after the manufacturer tightened its quality control requirements, the first complaint arrived. Does this suggest that the mean time between major customer complaints may have increased?

SOLUTION

We want to determine whether the observed value of $y = 15$ months, or some larger value of y, would be improbable if, in fact, $\alpha = 2$ and $\beta = 4$. We do not give a table of areas under the gamma density function in this text, but we can obtain some idea of the magnitude of $P(y \geq 15)$ by calculating the mean and standard deviation for the gamma density function when $\alpha = 2$ and $\beta = 4$. Thus,

$$\mu = \alpha\beta = (2)(4) = 8$$
$$\sigma^2 = \alpha\beta^2 = (2)(4)^2 = 32$$
$$\sigma = 5.7$$

Since $y = 15$ months lies barely more than 1 standard deviation beyond the mean ($\mu + \sigma = 8 + 5.7 = 13.7$ months), we would not regard 15 months as an unusually large value of y. Consequently, we would conclude that there is insufficient evidence to indicate that the company's new quality control program has been effective in increasing the mean time between complaints. We will present formal statistical procedures for answering this question in later chapters. ■

EXAMPLE 5.12 (OPTIONAL)

Show that the mean for a gamma-type random variable is equal to $\mu = \alpha\beta$.

SOLUTION   We first write

$$E(y) = \int_{-\infty}^{\infty} yf(y) \, dy = \int_{0}^{\infty} y \frac{y^{\alpha-1}e^{-y/\beta}}{\beta^{\alpha}\Gamma(\alpha)} \, dy = \int_{0}^{\infty} \frac{y^{(\alpha+1)-1}e^{-y/\beta}}{\beta^{\alpha}\Gamma(\alpha)} \, dy$$

Multiplying and dividing the integrand by $\alpha\beta$ and using the fact that $\Gamma(\alpha) = (\alpha - 1)\Gamma(\alpha - 1)$, we obtain

$$E(y) = \alpha\beta \int_{0}^{\infty} \frac{y^{(\alpha+1)-1}e^{-y/\beta}}{(\alpha\beta)\beta^{\alpha}\Gamma(\alpha)} \, dy = \alpha\beta \int_{0}^{\infty} \frac{y^{(\alpha+1)-1}e^{-y/\beta}}{\beta^{\alpha+1}\Gamma(\alpha + 1)} \, dy$$

The integrand is a gamma density function with parameters $(\alpha + 1)$ and β. Therefore, since the integral of any density function over $-\infty < y < \infty$ is equal to 1, we conclude

$$E(y) = \alpha\beta(1) = \alpha\beta$$ ∎

EXERCISES 5.27–5.38

5.27 Suppose a random variable y has a probability distribution given by

$$f(y) = \begin{cases} cy^2e^{-y/2} & \text{if } y > 0 \\ 0 & \text{elsewhere} \end{cases}$$

Find the value of c that makes $f(y)$ a density function.

5.28 Researchers have discovered that the maximum flood level (in millions of cubic feet per second) over a 4-year period for the Susquehanna River at Harrisburg, Pennsylvania, follows approximately a gamma distribution with $\alpha = 3$ and $\beta = .07$ (*Journal of Quality Technology*, Jan. 1986).
  a. Find the mean and variance of the maximum flood level over a 4-year period for the Susquehanna River.
  b. The researchers arrived at their conclusions about the maximum flood level distribution by observing maximum flood levels over 20 4-year periods, from 1890–1969. Suppose that over the 4-year period 1982–1985 the maximum flood level was observed to be .60 million cubic feet per second. Would you expect to observe a value this high from a gamma distribution with $\alpha = 3$ and $\beta = .07$? What can you infer about the maximum flood level distribution for the 4-year period 1982–1985?

5.29 The lifetime y (in hours) of the central processing unit of a certain type of microcomputer is an exponential random variable with parameter $\beta = 1{,}000$.
  a. Find the mean and variance of the lifetime of the central processing unit.
  b. What is the probability that a central processing unit will have a lifetime of at least 2,000 hours?
  c. What is the probability that a central processing unit will have a lifetime of at most 1,500 hours?

5.30 The length of time between breakdowns of an essential piece of equipment is an important factor in deciding on the amount of auxiliary equipment needed to assure continuous service. A machine room foreman believes the time between breakdowns of a particular electrical generator is best approximated by an exponential distribution with mean equal to 10 days.

a. What is the standard deviation of this negative exponential distribution?

b. Assuming that the foreman has correctly characterized the distribution for the time between breakdowns and that the generator broke down today, what is the probability that the generator will break down again within the next 14 days?

c. What is the probability that the generator will operate for more than 20 days without a breakdown?

5.31 In finding and correcting errors in a computer program (*debugging*) and determining the program's reliability, computer software experts have noted the importance of the distribution of the time until the next program error is found. Suppose that this random variable has a gamma distribution with parameter $\alpha = 1$. One computer programmer believes that the mean time between finding program errors is $\beta = 24$ days. Suppose that a programming error is found today.

a. Assuming that $\beta = 24$, find the probability that it will take at least 60 days to discover the next programming error.

b. If the next programming error takes at least 60 days to find, what would you infer about the programmer's claim that the mean time between the detection of programming errors is $\beta = 24$ days? Why?

5.32 The length of time y (in minutes) required to generate a human reaction to tear gas formula A has a gamma distribution with $\alpha = 2$ and $\beta = 2$. The distribution for formula B is also gamma, but with $\alpha = 1$ and $\beta = 4$.

a. Find the mean length of time required to generate a human reaction to tear gas formula A. Find the mean for formula B.

b. Find the variances for both distributions.

c. Which tear gas has a higher probability of generating a human reaction in less than 1 minute? [*Hint:* You may use the fact that

$$\int ye^{-y/2}\, dy = -2ye^{-y/2} + \int 2e^{-y/2}\, dy$$

This result is derived by integration by parts.]

5.33 Vardeman and Ray (*Technometrics*, May 1985) suggest that the number of industrial accidents can be modeled by an exponential distribution. Suppose the number of accidents per hour at an industrial plant is exponentially distributed with mean $\beta = .5$.

a. What is the probability that at least one accident will occur in a randomly selected hour at the industrial plant?

b. What is the probability that less than two accidents will occur in a randomly selected hour at the industrial plant?

OPTIONAL EXERCISES

5.34 Show that the variance of a gamma distribution with parameters α and β is $\alpha\beta^2$.

5.35 Let y have an exponential distribution with mean β. Show that $P(y > a) = e^{-a/\beta}$. [*Hint:* Find $F(a) = P(y \le a)$.]

5.36 Refer to the concepts of *new better than used* (NBU) and *new worse than used* (NWU) in Optional Exercise 5.4. Show that the exponential distribution satisfies both the NBU and NWU properties. (Such a "life" distribution is said to be *new same as used* or *memoryless*.)

5.37 Show that $\Gamma(\alpha) = (\alpha - 1)\Gamma(\alpha - 1)$.

5.38 We have stated that a chi-square random variable has a gamma-type density with $\alpha = \nu/2$ and $\beta = 2$. Find the mean and variance of a chi-square random variable.

SECTION 5.7

THE WEIBULL PROBABILITY DISTRIBUTION

In Section 5.6, we noted that the gamma density function can be used to model the distribution of the length of life (failure time) of manufactured components, equipment, etc. Another distribution used by engineers for the same purpose is known as the **Weibull distribution**.*

THE PROBABILITY DENSITY FUNCTION, MEAN, AND VARIANCE FOR A WEIBULL RANDOM VARIABLE

$$f(y) = \begin{cases} \dfrac{\alpha}{\beta}y^{\alpha-1}e^{-y^{\alpha}/\beta} & \text{if } 0 \le y < \infty; \quad \alpha > 0; \quad \beta > 0 \\ 0 & \text{elsewhere} \end{cases}$$

$$\mu = \beta^{1/\alpha}\Gamma\left(\frac{\alpha + 1}{\alpha}\right)$$

$$\sigma^2 = \beta^{2/\alpha}\left[\Gamma\left(\frac{\alpha + 2}{\alpha}\right) - \Gamma^2\left(\frac{\alpha + 1}{\alpha}\right)\right]$$

The Weibull density function contains two parameters, α and β. The **scale parameter**, β, reflects the size of the units in which the random variable y is measured. The parameter α is the shape parameter. By changing the value of the shape parameter α, we can generate a widely varying set of curves to model real-life failure time distributions. For the case $\alpha = 1$, we obtain the exponential distribution of Section 5.6. The graphs of Weibull density functions for different values of α and β are shown in Figure 5.17 (page 202).

In addition to providing a good model for the failure time distributions of many manufactured items, the Weibull distribution is easy to use. A closed-form expression for its cumulative distribution function exists and can be used to obtain areas under the Weibull curve. Example 5.13 will illustrate the procedure.

---

*See Weibull (1951).

FIGURE 5.17   Graphs of Weibull Density Functions

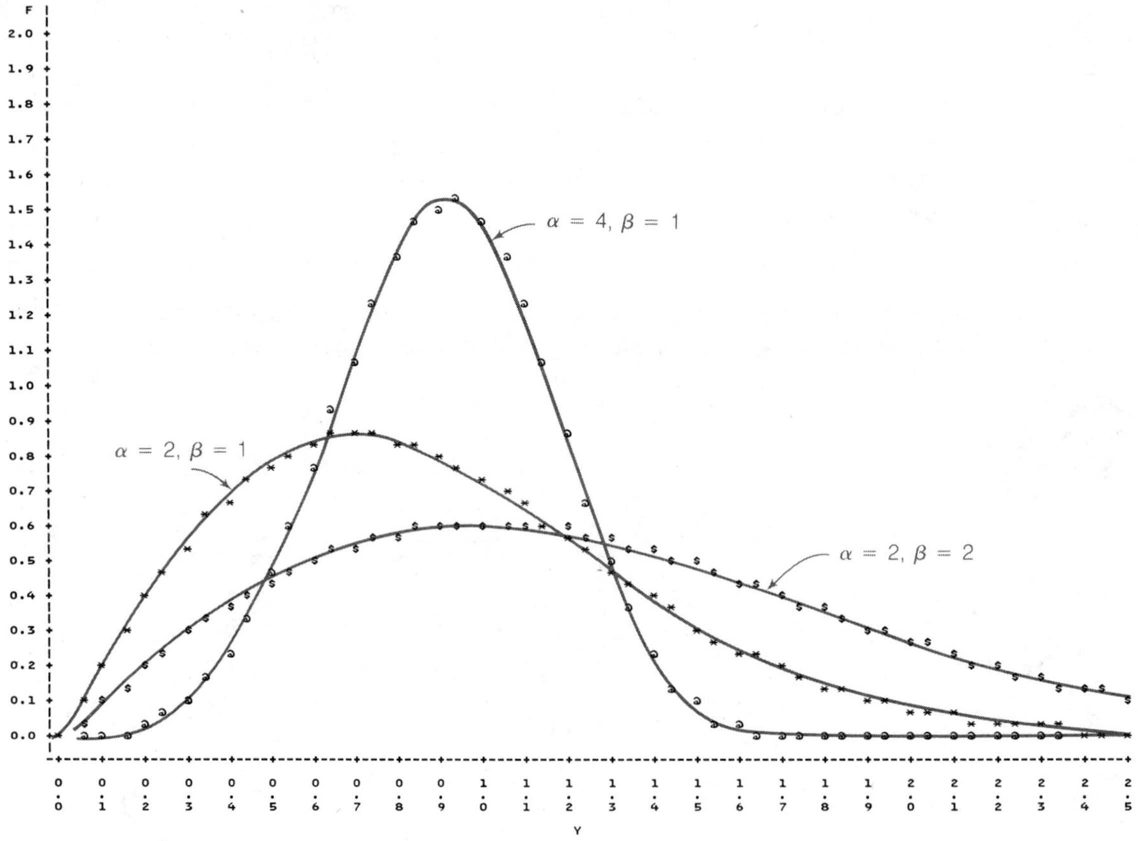

EXAMPLE 5.13

The length of life (in hours) of a drill bit used in a manufacturing operation has a Weibull distribution with $\alpha = 2$ and $\beta = 100$. Find the probability that a drill bit will fail before 8 hours of usage.

SOLUTION

The cumulative distribution function for a Weibull distribution is

$$F(y_0) = \int_0^{y_0} f(y)\, dy = \int_0^{y_0} \frac{\alpha}{\beta} y^{\alpha-1} e^{-y^\alpha/\beta}\, dy$$

By making the transformation $z = y^\alpha$, we have $dz = \alpha y^{\alpha-1}\, dy$ and the integral reduces to

$$F(y_0) = 1 - e^{-z/\beta} = 1 - e^{-y_0^\alpha/\beta}$$

To find the probability that y is less than 8 hours, we calculate

$$P(y < 8) = F(8) = 1 - e^{-(8)^\alpha/\beta}$$
$$= 1 - e^{-(8)^2/100} = 1 - e^{-.64}$$

Interpolating between $e^{-.60}$ and $e^{-.65}$ in Table 3 of Appendix II, or using a calculator with the e function, we find $e^{-.64} \approx .527$. Therefore, the probability that a drill bit will fail before 8 hours is

$$P(y < 8) = 1 - e^{-.64} = 1 - .527 = .473 \qquad \blacksquare$$

EXAMPLE 5.14   Refer to Example 5.13. Find the mean life of the drill bits.

SOLUTION   Substituting $\alpha = 2$ and $\beta = 100$ into the formula for the mean of a Weibull random variable yields

$$\mu = \beta^{1/\alpha} \Gamma\left(\frac{\alpha + 1}{\alpha}\right) = (100)^{1/2} \Gamma\left(\frac{2 + 1}{2}\right) = 10\Gamma(1.5)$$

From Table 5 of Appendix II, we find $\Gamma(1.5) = .88623$. Therefore, the mean life of the drill bits is

$$\mu = (10)\Gamma(1.5) = (10)(.88623) = 8.8623 \approx 8.86 \text{ hours} \qquad \blacksquare$$

EXERCISES 5.39–5.48

5.39 Refer to Example 5.13. Calculate the values of $f(y)$ for $y = 2, 5, 8, 11, 14,$ and 17. Plot the points $(y, f(y))$ and construct a graph of the failure time distribution of the drill bits.

5.40 Refer to Examples 5.13 and 5.14. Calculate the variance of the failure time distribution. Then find the probability that the length of life of a drill bit will fall within 2 hours of its mean.

5.41 Suppose the random variable y has a Weibull density function with $\alpha = 4$ and $\beta = 100$.
  a. Find $F(5)$.
  b. Find $P(y \geq 3)$.
  c. Find μ and σ.
  d. Find $P(\mu - 2\sigma \leq y \leq \mu + 2\sigma)$.

5.42 Based on extensive testing, a manufacturer of washing machines believes that the distribution of the time (in years) until a major repair is required has a Weibull distribution with $\alpha = 2$ and $\beta = 4$.
  a. If the manufacturer guarantees all machines against a major repair for 2 years, what proportion of all new washers will have to be repaired under the guarantee?
  b. Find the mean and standard deviation of the length of time until a major repair is required.
  c. Find $P(\mu - 2\sigma \leq y \leq \mu + 2\sigma)$.
  d. Is it likely that y will exceed 6 years?

5.43 *Wind models* are used in engineering design for wind energy analysis and design limit wind speeds. A widely used model of wind speed y (in miles per second) is the Weibull density function with parameters $\alpha = 1$ and $\beta = v/2$, where v (the characteristic speed) is the 63.21 percentile of the wind speed distribution (*Atmospheric*

Environment, Vol. 18, No. 10, 1984). At a particular site in Great Britain, the characteristic wind speed is known to be $v = 11.3$ miles per second. Use the Weibull wind model to find:

a. $E(y)$ and σ^2.

b. The probability that wind speed y is less than 6 miles per second.

c. The probability that wind speed y is greater than 10 miles per second.

5.44 The length of time (in months after maintenance) until failure of a bank's surveillance television equipment has a Weibull distribution with $\alpha = 2$ and $\beta = 60$. If the bank wants the probability of a breakdown before the next scheduled maintenance to be .05, how frequently should the equipment receive periodic maintenance?

5.45 Japanese electrical engineers have recently developed a sophisticated radar system called the moving target detector (MTD), designed to reject ground clutter, rain clutter, birds, and other interference (*IEE Proceedings*, Aug. 1984). The researchers show that the magnitude y of the Doppler frequency of a radar-received signal obeys a Weibull distribution with parameters $\alpha = 2$ and β.

a. Find $E(y)$.

b. Find σ^2.

c. Give an expression for the probability that the magnitude y of the Doppler frequency exceeds some constant C.

OPTIONAL EXERCISES

5.46 Show that for the Weibull distribution,

$$\mu = \beta^{1/\alpha}\Gamma\left(\frac{\alpha + 1}{\alpha}\right)$$

5.47 Show that for the Weibull distribution,

$$E(y^2) = \beta^{2/\alpha}\Gamma\left(\frac{\alpha + 2}{\alpha}\right)$$

Then use the relationship $\sigma^2 = E[(y - \mu)^2] = E(y^2) - \mu^2$ to show that

$$\sigma^2 = \beta^{2/\alpha}\left[\Gamma\left(\frac{\alpha + 2}{\alpha}\right) - \Gamma^2\left(\frac{\alpha + 1}{\alpha}\right)\right]$$

5.48 Show that the Weibull distribution with $\alpha = 2$ and $\beta > 0$ is *new better than used* (NBU). (See Optional Exercise 5.4 for the definition of NBU.)

SECTION 5.8

BETA-TYPE PROBABILITY DISTRIBUTIONS

Recall from Section 5.6 that the gamma density function provides a model for the relative frequency distribution of a random variable that possesses a fixed lower limit, but which can become infinitely large. In contrast, the **beta density function**, also characterized by two parameters, possesses finite lower and upper limits. We will give these limits as 0 and 1, but the density function, with modification, can be defined over any specified finite interval. The probability density function, the mean, and the variance for a beta-type random variable are shown in the box on page 206. Graphs of beta density functions for ($\alpha = 2$, $\beta = 4$), ($\alpha = 2$, $\beta = 2$), and ($\alpha = 3$, $\beta = 2$) are shown in Figure 5.18.

FIGURE 5.18 Graphs of Beta Density Functions

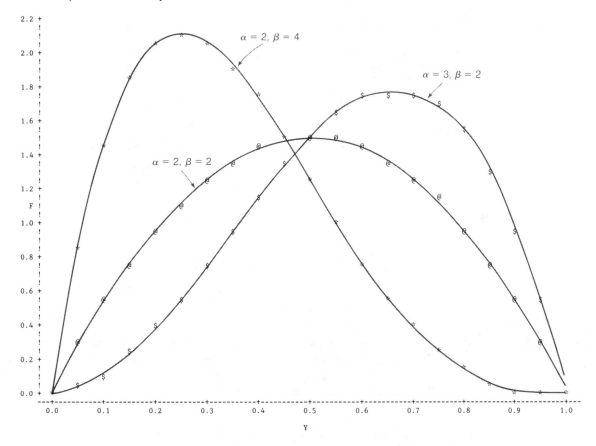

The cumulative distribution function $F(y)$ of a beta density function is called an **incomplete beta function.** Values of this function for various values of y, α, and β are given in *Tables of the Incomplete Beta Function* (1956). For the special case where α and β are integers, it can be shown that

$$F(p) = \int_0^p \frac{y^{\alpha-1}(1 - y)^{\beta-1}}{B(\alpha, \beta)}\, dy = \sum_{y=\alpha}^n p(y)$$

where $p(y)$ is a binomial probability distribution with parameters p and $n = (\alpha + \beta - 1)$. Recall that tables giving the cumulative sums of binomial probabilities are given in Table 2 of Appendix II, for $n = 5, 10, 15, 20$, and 25. More extensive tables of these probabilities are listed in the references at the end of the chapter.

THE PROBABILITY DENSITY FUNCTION, MEAN, AND VARIANCE FOR A BETA RANDOM VARIABLE

The probability density function for a beta-type random variable is given by

$$f(y) = \begin{cases} \dfrac{y^{\alpha-1}(1-y)^{\beta-1}}{B(\alpha, \beta)} & \text{if} \quad 0 \le y \le 1; \quad \alpha > 0; \quad \beta > 0 \\ 0 & \text{elsewhere} \end{cases}$$

where

$$B(\alpha, \beta) = \int_0^1 y^{\alpha-1}(1-y)^{\beta-1}\, dy = \frac{\Gamma(\alpha)\Gamma(\beta)}{\Gamma(\alpha + \beta)}$$

The mean and variance of a beta random variable are, respectively,

$$\mu = \frac{\alpha}{\alpha + \beta} \qquad \sigma^2 = \frac{\alpha\beta}{(\alpha + \beta)^2(\alpha + \beta + 1)}$$

[Recall that

$$\Gamma(a) = \int_0^\infty y^{a-1}e^{-y}\, dy$$

and $\Gamma(\alpha) = (\alpha - 1)!$ when α is a positive integer.]

EXAMPLE 5.15

Data collected over time on the utilization of a computer core (as a proportion of the total capacity) were found to possess a relative frequency distribution that could be approximated by a beta density function with $\alpha = 2$ and $\beta = 4$. Find the probability that the proportion of the core being used at any particular time will be less than .20.

SOLUTION

The probability that the proportion of the core being utilized will be less than $p = .2$ is

$$F(p) = \int_0^p \frac{y^{\alpha-1}(1-y)^{\beta-1}}{B(\alpha, \beta)}\, dy = \sum_{y=\alpha}^n p(y)$$

where $p(y)$ is a binomial probability distribution with $n = (\alpha + \beta - 1) = (2 + 4 - 1) = 5$ and $p = .2$. Therefore,

$$F(.2) = \sum_{y=2}^5 p(y) = 1 - \sum_{y=0}^1 p(y)$$

From Table 2 of Appendix II for $n = 5$ and $p = .2$, we find that

$$\sum_{y=0}^1 p(y) = .737$$

Therefore, the probability that the computer core will be less than 20% occupied at any particular time is

$$F(.2) = 1 - \sum_{y=0}^{1} p(y) = 1 - .737 = .263$$ ∎

EXERCISES 5.49–5.56

5.49 A continuous random variable y has a beta distribution with probability density

$$f(y) = \begin{cases} cy^5(1 - y)^2 & \text{if} \quad 0 \le y \le 1 \\ 0 & \text{elsewhere} \end{cases}$$

Find the value of c that will make $f(y)$ a density function.

5.50 The proportion y of a data-processing company's yearly hardware repair budget allocated to repair its tri-log color printer has an approximate beta distribution with parameters $\alpha = 2$ and $\beta = 9$.
 a. Find the mean and variance of y.
 b. Compute the probability that for any randomly selected year, at least 40% of the hardware repair budget is used to repair the color printer.
 c. What is the probability that at most 10% of the yearly repair budget is used for the color printer?

5.51 An investigation into pollution control expenditures of industrial firms found that the annual percentage of plant capacity shutdown attributable to environmental and safety regulation has an approximate beta distribution with $\alpha = 1$ and $\beta = 25$.
 a. Find the mean and variance of the annual percentage of plant capacity shutdown due to environmental and safety regulation.
 b. Find the probability that more than 1% of plant capacity shutdown is attributable to environmental and safety regulation.

5.52 An important property of certain products that are in powder or granular form is their particle size distribution. For example, refractory cements are adversely affected by too high a proportion of coarse granules, which can lead to weaknesses due to poor packing. G. H. Brown (*Journal of Quality Technology*, July 1985) showed that the beta distribution provides an adequate model for the percentage y of refractory cement granules in bulk form that are coarse. Suppose you are interested in controlling the proportion y of coarse refractory cement in a lot, where y has a beta distribution with parameters $\alpha = \beta = 2$.
 a. Find the mean and variance of y.
 b. If you will accept the lot only if less than 10% of refractory cement granules are coarse, find the probability of lot acceptance.

5.53 Suppose the proportion of small data-processing firms that make a profit during their first year of operation possesses a relative frequency distribution that can be approximated by the beta density with $\alpha = 5$ and $\beta = 6$.
 a. Find the probability that at most 60% of all small data-processing firms make a profit during their first year of operation.
 b. Find the probability that at least 80% of all small data-processing firms make a profit during their first year of operation.

OPTIONAL EXERCISES

5.54 Verify that the mean of a beta density with parameters α and β is given by $\mu = \alpha/(\alpha + \beta)$.

5.55 Show that if y has a beta density with $\alpha = 1$ and $\beta = 1$, then y is uniformly distributed over the interval $0 \le y \le 1$.

5.56 Show that the beta distribution with $\alpha = 2$ and $\beta = 1$ is *new better than used* (NBU). (See Optional Exercise 5.4 for the definition of NBU.)

| | | | | | | | | | | | |

S E C T I O N   5.9

MOMENTS AND MOMENT GENERATING FUNCTIONS (OPTIONAL)

The **moments** and **moment generating functions** for continuous random variables are defined in exactly the same way as for discrete random variables, except that the expectations involve integration.* The relevance and applicability of a moment generating function $m(t)$ are the same in the continuous case, as we now illustrate with two examples.

EXAMPLE 5.16

Find the moment generating function for a gamma-type random variable.

SOLUTION

The moment generating function is given by

$$m(t) = E(e^{ty}) = \int_0^\infty e^{ty} \frac{y^{\alpha-1} e^{-y/\beta}}{\beta^\alpha \Gamma(\alpha)}\, dy$$

$$= \int_0^\infty \frac{y^{\alpha-1} e^{-y(1/\beta-t)}}{\beta^\alpha \Gamma(\alpha)}\, dy = \int_0^\infty \frac{y^{\alpha-1} e^{-y/[\beta/(1-\beta t)]}}{\beta^\alpha \Gamma(\alpha)}\, dy$$

An examination of this integrand indicates that we can convert it into a gamma density function with parameters α and $\beta/(1 - \beta t)$, by factoring $1/\beta^\alpha$ out of the integral and multiplying and dividing by $[\beta/(1 - \beta t)]^\alpha$. Therefore,

$$m(t) = \frac{1}{\beta^\alpha} \left(\frac{\beta}{1 - \beta t} \right)^\alpha \int_0^\infty \frac{y^{\alpha-1} e^{-y/[\beta/(1-\beta t)]}}{\left(\dfrac{\beta}{1 - \beta t} \right)^\alpha \Gamma(\alpha)}\, dy$$

The integral of this gamma density function is equal to 1. Therefore,

$$m(t) = \frac{1}{(1 - \beta t)^\alpha}(1) = \frac{1}{(1 - \beta t)^\alpha} \qquad \blacksquare$$

EXAMPLE 5.17

Refer to Example 5.16. Use $m(t)$ to find μ_1' and μ_2'. Use the results to derive the mean and variance of a gamma-type random variable.

*Moments and moment generating functions for discrete random variables are discussed in Optional Section 4.9.

SOLUTION The first two moments about the origin, evaluated at $t = 0$, are

$$\mu_1' = \mu = \frac{dm(t)}{dt}\bigg]_{t=0} = \frac{-\alpha(-\beta)}{(1-\beta t)^{\alpha+1}}\bigg]_{t=0} = \alpha\beta$$

and

$$\mu_2' = \frac{d^2m(t)}{dt^2}\bigg]_{t=0}$$
$$= -\frac{\alpha\beta(\alpha+1)(-\beta)}{(1-\beta t)^{\alpha+2}}\bigg]_{t=0} = \alpha(\alpha+1)\beta^2$$

Then, applying Theorem 5.4, we obtain

$$\sigma^2 = E(y^2) - \mu^2 = \mu_2' - \mu^2$$
$$= \alpha(\alpha+1)\beta^2 - \alpha^2\beta^2 = \alpha\beta^2 \qquad \blacksquare$$

Some useful probability density functions, with their means, variances, and moment generating functions, are summarized in Table 5.1.

TABLE 5.1 Some Useful Continuous Random Variables

RANDOM VARIABLE	PROBABILITY DENSITY FUNCTION	MEAN	VARIANCE	MOMENT GENERATING FUNCTION
Uniform	$f(y) = \dfrac{1}{b-a}$ $\quad a \le y \le b$	$\dfrac{a+b}{2}$	$\dfrac{(b-a)^2}{12}$	$\dfrac{e^{tb} - e^{ta}}{t(b-a)}$
Normal	$f(y) = \dfrac{e^{-(y-\mu)^2/2\sigma^2}}{\sigma\sqrt{2\pi}}$ $\quad -\infty < y < \infty$	μ	σ^2	$e^{\mu t + (t^2\sigma^2/2)}$
Gamma	$f(y) = \dfrac{y^{\alpha-1}e^{-y/\beta}}{\beta^\alpha\Gamma(\alpha)}$ $\quad 0 \le y < \infty$	$\alpha\beta$	$\alpha\beta^2$	$(1-\beta t)^{-\alpha}$
Exponential	$f(y) = \dfrac{1}{\beta}e^{-y/\beta}$ $\quad 0 \le y < \infty$	β	β^2	$\dfrac{1}{(1-\beta t)}$
Weibull	$f(y) = \dfrac{\alpha}{\beta}y^{\alpha-1}e^{-y^\alpha/\beta}$ $\quad 0 \le y < \infty$	$\beta^{1/\alpha}\Gamma\left(\dfrac{\alpha+1}{\alpha}\right)$	$\beta^{2/\alpha}\left[\Gamma\left(\dfrac{\alpha+2}{\alpha}\right) - \Gamma^2\left(\dfrac{\alpha+1}{\alpha}\right)\right]$	$\beta^{t/\alpha}\Gamma(1+t/\alpha)$
Chi-square	$f(\chi^2) = \dfrac{(\chi^2)^{(\nu/2)-1}e^{-\chi^2/2}}{2^{\nu/2}\Gamma\left(\dfrac{\nu}{2}\right)}$ $\quad 0 \le \chi^2 < \infty$	ν	2ν	$(1-2t)^{-\nu/2}$
Beta	$f(y) = \dfrac{\Gamma(\alpha+\beta)}{\Gamma(\alpha)\Gamma(\beta)}y^{\alpha-1}(1-y)^{\beta-1}$ $\quad 0 \le y \le 1$	$\dfrac{\alpha}{\alpha+\beta}$	$\dfrac{\alpha\beta}{(\alpha+\beta)^2(\alpha+\beta+1)}$	Closed-form expression does not exist.

5.57 Use the moment generating function $m(t)$ of the normal density to find μ_1' and μ_2'. Then use these results to show that a normal random variable has mean μ and variance σ^2.

5.58 Verify that the moment generating function of a chi-square random variable with ν degrees of freedom is

$$m(t) = (1 - 2t)^{-\nu/2}$$

[*Hint:* Use the fact that a chi-square random variable has a gamma-type density function with $\alpha = \nu/2$ and $\beta = 2$.]

5.59 Verify that the moment generating function of a uniform random variable on the interval $a \le y \le b$ is

$$m(t) = \frac{e^{tb} - e^{ta}}{t(b - a)}$$

5.60 Consider a continuous random variable y with density

$$f(y) = \begin{cases} e^y & \text{if } y < 0 \\ 0 & \text{elsewhere} \end{cases}$$

a. Find the moment generating function $m(t)$ of y.
b. Use the result of part **a** to find the mean and variance of y.

SECTION 5.10

SUMMARY

Continuous random variables are defined to be those that possess continuous **cumulative distribution functions**. From a practical point of view, they are variables that can assume values corresponding to the infinitely large number of points contained in one or more intervals on the real line.

The relative frequency distribution for a population of data associated with a continuous random variable can be modeled using a **probability density function**. A density function $f(y)$, usually a continuous smooth curve, must satisfy the following properties:

$$f(y) \ge 0 \quad \text{and} \quad \int_{-\infty}^{\infty} f(y)\, dy = 1$$

The **expected value** (or **mean**) of a continuous random variable y or of any function $g(y)$ of y is defined in the same manner as for discrete random variables, except that integration is substituted for summation.

Although there are many different types of continuous random variables, most relative frequency distributions of data in the engineering and computer sciences can be modeled by one of the five probability density functions presented in this chapter: the **uniform distribution**, the **normal distribution**, the **gamma distribution**, the **Weibull distribution**, and the **beta distribution**. We gave some practical examples of the use of these density functions, and will subsequently show that they play a very important role in statistical inference.

5.61 The continuous random variable y has a probability distribution given by

$$f(y) = \begin{cases} cye^{-y^2} & \text{if } y > 0 \\ 0 & \text{elsewhere} \end{cases}$$

a. Find the value of c that makes $f(y)$ a probability density.
b. Find $F(y)$.
c. Compute $P(y > 2.5)$.

5.62 The problem of passenger congestion prompted a large international airport to install a monorail connecting its main terminal to the three concourses, A, B, and C. The engineers designed the monorail so that the amount of time a passenger at concourse B must wait for a monorail car has a uniform distribution ranging from 0 to 10 minutes.

a. Find the mean and variance of y, the time a passenger at concourse B must wait for the monorail. (Assume that the monorail travels sequentially from concourse A, to concourse B, to concourse C, back to concourse B, and then returns to concourse A. The route is then repeated.)
b. If it takes the monorail 1 minute to go from concourse to concourse, find the probability that a hurried passenger can reach concourse A less than 4 minutes after arriving at the monorail station at concourse B.

5.63 Many products are mass produced on automated assembly lines. The probability distribution of the length of time between the arrivals of successive manufactured components off the assembly line is often (approximately) exponential. Suppose the mean time between arrivals of magnetron tubes manufactured on an assembly line is 20 seconds.

a. What is the probability that a particular interarrival time (the time between the arrivals of two magnetron tubes) is less than 10 seconds?
b. What is the probability that the next four interarrival times are all less than 10 seconds?
c. What is the probability that an interarrival time will exceed 1 minute?

5.64 The metropolitan airport commission is considering the establishment of limitations on the extent of noise pollution around a local airport. At the present time the noise level per jet takeoff in one neighborhood near the airport is approximately normally distributed with a mean of 100 decibels and a standard deviation of 6 decibels.

a. What is the probability that a randomly selected jet will generate a noise level greater than 108 decibels in this neighborhood?
b. What is the probability that a randomly selected jet will generate a noise level of exactly 100 decibels?
c. Suppose a regulation is passed that requires jet noise in this neighborhood to be lower than 105 decibels 95% of the time. Assuming the standard deviation of the noise distribution remains the same, how much will the mean noise level have to be lowered to comply with the regulation?

5.65 The length of time required to assemble a photoelectric cell is normally distributed with a mean of 18.1 minutes and a standard deviation of 1.3 minutes. What is the probability that it will require more than 20 minutes to assemble a cell?

5.66 Suppose that the fraction of defective modems shipped by a data-communications vendor has an approximate beta distribution with $\alpha = 5$ and $\beta = 21$.
  a. Find the mean and variance of the fraction of defective modems per shipment.
  b. What is the probability that a randomly selected shipment will contain at least 30% defectives?
  c. What is the probability that a randomly selected shipment will contain no more than 5% defectives?

5.67 W. Nelson (*Journal of Quality Technology*, July 1985) suggests that the Weibull distribution usually provides a better representation for the lifelength of a product than the exponential distribution. Nelson used a Weibull distribution with $\alpha = 1.5$ and $\beta = 110$ to model the lifelength y of a roller bearing (in thousands of hours).
  a. Find the probability that a roller bearing of this type will have a service life of less than 12.2 thousand hours.
  b. Recall that a Weibull distribution with $\alpha = 1$ is an exponential distribution. Nelson claims that very few products have an exponential life distribution, although such a distribution is commonly applied. Calculate the probability from part **a** using the exponential distribution. Compare your answer to that obtained in part **a**.

5.68 Teleconferences, electronic mail, and word processors are among the tools that can reduce the length of business meetings. A recent survey indicated that the percentage reduction y in time spent by business professionals in meetings due to automated office equipment is approximately normally distributed with mean equal to 15% and standard deviation equal to 4%.
  a. What proportion of all business professionals with access to automated office equipment have reduced their time in meetings by more than 22%?
  b. What proportion of all business professionals with access to automated office equipment have reduced their time in meetings by 10% or less?

5.69 Suppose we are counting events that occur according to a Poisson distribution, such as the number of data-processing jobs submitted to a computer center. If it is known that exactly one such event has occurred in a given interval of time, say $(0, t)$, then the actual time of occurrence is uniformly distributed over this interval. Suppose that during a given 30-minute period, one data-processing job was submitted. Find the probability that the job was submitted during the last 5 minutes of the 30-minute period.

5.70 The shelf-life of a product is a random variable that is related to consumer acceptance and, ultimately, to sales and profit. Suppose the shelf-life of bread is best approximated by an exponential distribution with mean equal to 2 days. What fraction of the loaves stocked today would you expect to be saleable (i.e., not stale) 3 days from now?

5.71 The percentage y of impurities per batch in a certain chemical product is a beta random variable with probability density

$$f(y) = \begin{cases} 90y^8(1 - y) & \text{if } 0 \le y \le 1 \\ 0 & \text{elsewhere} \end{cases}$$

  a. What are the values of α and β?
  b. Compute the mean and variance of y.
  c. A batch with more than 80% impurities cannot be sold. What is the probability that a randomly selected batch cannot be sold because of excessive impurities?

5.72 The lifelength y (in years) of a memory chip in a mainframe computer is a Weibull random variable with probability density

$$f(y) = \begin{cases} \dfrac{1}{8} y e^{-y^2/16} & \text{if } 0 \le y < \infty \\ 0 & \text{elsewhere} \end{cases}$$

a. What are the values of α and β?

b. Compute the mean and variance of y.

c. Find the probability that a new memory chip will not fail before 6 years.

5.73 *System downtime* is defined as the fraction of time a computer system is inoperative due to hardware and/or software failure. Suppose that system downtime y (in hours) at a university computer center has a Weibull distribution with $\alpha = 2$ and $\beta = 2$. When the system is down for longer than 1 hour, all current working files are lost. If the system goes down while a user is accessing a working file, what is the probability that the file will be recovered?

5.74 Each year the top marlin fishermen from around the world compete in the Hawaiian International Bluefish Tournament. In 1984, one fisherman landed a 987-pound Pacific blue marlin—a world record until a check showed his fishing line was a few pounds over the 80-pound limit. The "80 pounds" refers to the strength of the fishing line—i.e., the weight the line is tested to hold outside water. Suppose the actual strength of manufactured "80-pound-test line" is normally distributed with a mean of 80 pounds and a standard deviation of .2 pound.

a. What is the probability that an "80-pound-test line" randomly selected from the production process will have a strength of at least 1 pound over the 80-pound limit?

b. Based on the probability computed in part **a**, is it likely that the fisherman actually used "80-pound-test line" to catch the 987-pound blue marlin? Explain.

OPTIONAL EXERCISES

5.75 In an effort to aid engineers seeking to predict the efficiency of a solar-powered device, Olseth and Skartveit (*Solar Energy*, Vol. 33, No. 6, 1984) developed a model for daily insolation y at sea-level locations within the temperate storm belt. In order to account for both "clear sky" and "overcast" days, the researchers constructed a probability density function for y (measured as a percentage) using a linear combination of two modified gamma distributions:

$$f(y) = wg(y, \lambda_1) + (1 - w)g(1 - y, \lambda_2), \quad 0 < y < 1$$

where

$$g(y) = \frac{(1 - y)e^{\lambda y}}{\displaystyle\int_0^1 (1 - y)e^{\lambda y}\, dy}$$

λ_1 = mean insolation of "clear sky" days

λ_2 = mean insolation of "cloudy" days

and w is a weighting constant, $0 \le w \le 1$. Show that

$$\int_0^1 f(y)\, dy = 1$$

5.76 Let $m_y(t)$ be the moment generating function of a continuous random variable y. If a and b are constants, show that:

a. $m_{y+a}(t) = E[e^{(y+a)t}] = e^{at}m_y(t)$

b. $m_{by}(t) = E[e^{(by)t}] = m_y(bt)$

c. $m_{[(y+a)/b]}(t) = E[e^{(y+a)t/b}] = e^{at/b}m_y\left(\dfrac{t}{b}\right)$

REFERENCES

Hogg, R. V. and Craig, A. T. *Introduction to Mathematical Statistics*, 4th ed. New York: Macmillan, 1978.

Mendenhall, W., Scheaffer, R. L., and Wackerly, D. *Mathematical Statistics with Applications*, 3rd ed. Boston: Duxbury Press, 1986.

Mood, A. M., Graybill, F. A., and Boes, D. *Introduction to the Theory of Statistics*, 3rd ed. New York: McGraw-Hill, 1974.

Parzen, E. *Modern Probability Theory and Its Applications*. New York: Wiley, 1964.

Pearson, K. *Tables of the Incomplete Beta Function*. New York: Cambridge University Press, 1956.

Pearson, K. *Tables of the Incomplete Gamma Function*. New York: Cambridge University Press, 1956.

Standard Mathematical Tables, 17th ed. Cleveland: Chemical Rubber Company, 1969.

Tables of the Binomial Probability Distribution. Department of Commerce, National Bureau of Standards, Applied Mathematics Series 6, 1950.

Weibull, W. "A Statistical Distribution Function of Wide Applicability." *Journal of Applied Mechanics*, 18 (1951), pp. 293–297.

OBJECTIVE

To introduce the concepts of a bivariate probability distribution, covariance, and independence; to show you how to find the expected value and variance of a linear function of random variables.

CONTENTS

BIVARIATE PROBABILITY DISTRIBUTIONS

**BIVARIATE
PROBABILITY
DISTRIBUTIONS FOR
DISCRETE RANDOM
VARIABLES**

In Chapter 3, we learned that the probability of the intersection of two events (i.e., the event that both A and B occur) is equal to

$$P(A \cap B) = P(A)P(B \mid A) = P(B)P(A \mid B)$$

If we assign two numbers to each point in the sample space—one corresponding to the value of a discrete random variable y_1, and the second to a discrete random variable y_2—then specific values of y_1 and y_2 represent two numerical events. The probability of the intersection of these two events is obtained by replacing the symbol A by y_1 and the symbol B by y_2:

$$P(A \cap B) = p(y_1, y_2) = p_1(y_1)p_2(y_2 \mid y_1) = p_2(y_2)p_1(y_1 \mid y_2)$$

[*Note:* To distinguish between the probability distributions, we will always use the subscript 1 when we refer to the probability distribution of y_1 and the subscript 2 when we refer to the probability distribution of y_2.]

A table, graph, or formula that gives the probability of the intersection (y_1, y_2) for all values of y_1 and y_2 is called the **joint probability distribution** of y_1 and y_2. The probability distribution $p_1(y_1)$ gives the probabilities of observing specific values of y_1 and, similarly, $p_2(y_2)$ gives the probabilities of the discrete random variable y_2. Thus, $p_1(y_1)$ and $p_2(y_2)$, called **marginal probability distributions** for y_1 and y_2, respectively, are the familiar unconditional probability distributions for discrete random variables encountered in Chapter 4.

Finally, the probability of the numerical event y_1, given that the event y_2 occurred, is the conditional probability of y_1 given y_2. A table, graph, or formula that gives these probabilities for all values of y_2 is called the **conditional probability distribution** for y_1 given y_2 and is denoted by the symbol $p_1(y_1 \mid y_2)$.

DEFINITION 6.1

The **joint probability distribution** $p(y_1, y_2)$ for two discrete random variables, y_1 and y_2, is a table, graph, or formula that gives the values of $p(y_1, y_2)$ for every combination of values of y_1 and y_2.

EXAMPLE 6.1

Consider the bivariate (two-variable) joint probability distribution shown in Table 6.1. The numbers in the cells are the values of $p(y_1, y_2)$ corresponding to pairs of values of the discrete random variables y_1 and y_2, for $y_1 = 1, 2, 3, 4$, and $y_2 = 0, 1, 2, 3$. Find the marginal probability distribution $p_1(y_1)$ for the discrete random variable y_1.

TABLE 6.1

		y_1			
		1	2	3	4
	0	0	.10	.20	.10
	1	.03	.07	.10	.05
y_2	2	.05	.10	.05	0
	3	0	.10	.05	0

SOLUTION

To find the marginal probability distribution for y_1, we need to find $P(y_1 = 1)$, $P(y_1 = 2)$, $P(y_1 = 3)$, and $P(y_1 = 4)$. Since $y_1 = 1$ can occur when $y_2 = 0, 1, 2$, or 3 occurs, then $P(y_1 = 1) = p_1(1)$ is calculated by summing the probabilities of four mutually exclusive events:

$$P(y_1 = 1) = p_1(1) = p(1, 0) + p(1, 1) + p(1, 2) + p(1, 3)$$

Substituting the values for $p(y_1, y_2)$ given in Table 6.1, we obtain

$$P(y_1 = 1) = p_1(1) = 0 + .03 + .05 + 0 = .08$$

Similarly,

$$P(y_1 = 2) = p_1(2) = p(2, 0) + p(2, 1) + p(2, 2) + p(2, 3)$$
$$= .10 + .07 + .10 + .10 = .37$$
$$P(y_1 = 3) = p_1(3) = p(3, 0) + p(3, 1) + p(3, 2) + p(3, 3)$$
$$= .20 + .10 + .05 + .05 = .40$$
$$P(y_1 = 4) = p_1(4) = p(4, 0) + p(4, 1) + p(4, 2) + p(4, 3)$$
$$= .10 + .05 + 0 + 0 = .15$$

The marginal probability distribution $p_1(y_1)$ is given in the following table:

y_1	1	2	3	4
$p_1(y_1)$	.08	.37	.40	.15

Note from the table that $\sum_{y_1=1}^{4} p_1(y_1) = 1$. ∎

Example 6.1 shows that the marginal probability distribution for a discrete random variable y_1 may be obtained by summing $p(y_1, y_2)$ over all values of y_2. It also shows that the sum of the joint probabilities $p(y_1, y_2)$ over all combinations of values of y_1 and y_2 is equal to 1.

DEFINITION 6.2

Let y_1 and y_2 be discrete random variables and let $p(y_1, y_2)$ be their joint probability distribution. Then the **marginal (unconditional) probability distributions** of y_1 and y_2 are, respectively,

$$p_1(y_1) = \sum_{y_2} p(y_1, y_2) \quad \text{and} \quad p_2(y_2) = \sum_{y_1} p(y_1, y_2)$$

[*Note:* We will use the symbol $\sum_{y_i}$ to denote summation over all values of y_i, for $i = 1, 2$.]

REQUIREMENTS FOR A DISCRETE JOINT PROBABILITY DISTRIBUTION

1. $p(y_1, y_2) \geq 0$ for all values of y_1 and y_2

2. $\sum_{y_2} \sum_{y_1} p(y_1, y_2) = 1$

EXAMPLE 6.2

Refer to Example 6.1. Find the conditional probability distribution of y_1 given $y_2 = 2$.

SOLUTION

There are four conditional probability distributions of y_1—one for each value of y_2. From Chapter 3, we know that

$$P(A \mid B) = \frac{P(A \cap B)}{P(B)}$$

If we let a value of y_1 correspond to the event A and a value of y_2 to the event B, then it follows that

$$p_1(y_1 \mid y_2) = \frac{p(y_1, y_2)}{p_2(y_2)}$$

or, when $y_2 = 2$,

$$p_1(y_1 \mid 2) = \frac{p(y_1, 2)}{p_2(2)}$$

Therefore,

$$p_1(1 \mid 2) = \frac{p(1, 2)}{p_2(2)}$$

From Table 6.1, we obtain $p(1, 2) = .05$ and $P(y_2 = 2) = p_2(2) = .2$. Therefore,

$$p_1(1 \mid 2) = \frac{p(1, 2)}{p_2(2)} = \frac{.05}{.20} = .25$$

Similarly,

$$p_1(2 \mid 2) = \frac{p(2, 2)}{p_2(2)} = \frac{.10}{.20} = .50$$

$$p_1(3 \mid 2) = \frac{p(3, 2)}{p_2(2)} = \frac{.05}{.20} = .25$$

$$p_1(4 \mid 2) = \frac{p(4, 2)}{p_2(2)} = \frac{0}{.20} = 0$$

Therefore, the conditional probability distribution of y_1, given that $y_2 = 2$, is as shown in the following table:

y_1	1	2	3	4
$p_1(y_1 \mid 2)$	.25	.50	.25	0

∎

Note from Example 6.2 that the sum of the conditional probabilities $p_1(y_1 \mid 2)$ over all values of y_1 is equal to 1. Thus, a conditional probability distribution satisfies the requirements that all probability distributions must satisfy:

$$p_1(y_1 \mid y_2) \geq 0 \quad \text{and} \quad \sum_{y_1} p_1(y_1 \mid y_2) = 1$$

Similarly,

$$p_2(y_2 \mid y_1) \geq 0 \quad \text{and} \quad \sum_{y_2} p_2(y_2 \mid y_1) = 1$$

DEFINITION 6.3

Let y_1 and y_2 be discrete random variables and let $p(y_1, y_2)$ be their joint probability distribution. Then the **conditional probability distributions** for y_1 and y_2 are defined as follows:

$$p_1(y_1 \mid y_2) = \frac{p(y_1, y_2)}{p_2(y_2)} \quad \text{and} \quad p_2(y_2 \mid y_1) = \frac{p(y_1, y_2)}{p_1(y_1)}$$

In the preceding discussion, we defined the bivariate joint marginal and conditional probability distributions for two discrete random variables, y_1 and y_2. The concepts can be extended to any number of discrete random variables. Thus, we could define a third random variable y_3 by assigning a value of y_3 to each point in the sample space. The joint probability distribution $p(y_1, y_2, y_3)$ would be a table, graph, or formula that gives the values of $p(y_1, y_2, y_3)$, the event that the intersection (y_1, y_2, y_3) occurs, for all combinations of values of y_1, y_2, and y_3. In general, the joint probability distribution for two or more discrete random variables is called a **multivariate probability distribution**. We will denote the discrete joint multivariate probability distribution for $y_1, y_2, \ldots, y_n$ by the symbol $p(y_1, y_2, \ldots, y_n)$ and will use it in definitions that follow. However, we will confine our discussion and examples to the less complex bivariate case.

EXERCISES 6.1–6.5

6.1 The joint probability distribution $p(y_1, y_2)$ for two discrete random variables, y_1 and y_2, is given in the accompanying table.

		y_1					
		0	1	2	3	4	5
y_2	0	0	.050	.025	0	.025	0
	1	.200	.050	0	.300	0	0
	2	.100	0	0	0	.100	.150

a. Find the marginal probability distribution $p_1(y_1)$ for y_1.
b. Find the marginal probability distribution $p_2(y_2)$ for y_2.
c. Find the conditional probability distribution $p_1(y_1 \mid y_2)$.
d. Find the conditional probability distribution $p_2(y_2 \mid y_1)$.

6.2 Consider the experiment of tossing a pair of dice. Let y_1 be the outcome (i.e., the number of dots appearing face up) on the first die and let y_2 be the outcome on the second die.

a. Find the joint probability distribution $p(y_1, y_2)$.
b. Find the marginal probability distributions $p_1(y_1)$ and $p_2(y_2)$.
c. Find the conditional probability distributions $p_1(y_1 \mid y_2)$ and $p_2(y_2 \mid y_1)$.
d. Compare the probability distributions of parts **b** and **c**. What phenomenon have you observed?

6.3 A special delivery truck travels from point A to point B and back over the same route each day. There are three traffic lights on this route. Let y_1 be the number of red lights the truck encounters on the way to delivery point B and let y_2 be the number of red lights the truck encounters on the way back to delivery point A. A traffic engineer has determined the joint probability distribution of y_1 and y_2 shown in the table.

		y_1			
		0	1	2	3
	0	.01	.02	.07	.01
	1	.03	.06	.10	.06
y_2	2	.05	.12	.15	.08
	3	.02	.09	.08	.05

a. Find the marginal probability distribution of y_2.
b. Given that the truck encounters $y_1 = 2$ red lights on the way to delivery point B, find the probability distribution of y_2.

OPTIONAL EXERCISES

6.4 From a group of three data-processing managers, two senior systems analysts, and two quality control engineers, three people are to be randomly selected to form a committee that will study the feasibility of adding computer graphics at a consulting firm. Let y_1 denote the number of data-processing managers and y_2 the number of senior systems analysts selected for the committee.

a. Find the joint probability distribution of y_1 and y_2.
b. Find the marginal distribution of y_1.

6.5 Let y_1 and y_2 be two discrete random variables with joint probability distribution $p(y_1, y_2)$. Define

$$F_1(a) = P(y_1 \le a) \quad \text{and} \quad F_1(a \mid y_2) = P(y_1 \le a \mid y_2)$$

Verify each of the following:

a. $F_1(a) = \sum_{y_1 \le a} \sum_{y_2} p(y_1, y_2)$

b. $F_1(a \mid y_2) = \dfrac{\sum_{y_1 \le a} p(y_1, y_2)}{p_2(y_2)}$

| | | | | | | | | | | | | |
SECTION 6.2

BIVARIATE PROBABILITY DISTRIBUTIONS FOR CONTINUOUS RANDOM VARIABLES

As we have noted in Chapters 4 and 5, definitions and theorems that apply to discrete random variables apply as well to continuous random variables. The only difference is that the probabilities for discrete random variables are summed, whereas those for continuous random variables are integrated. As we proceed through this chapter, we will define and develop concepts in the context of discrete random variables and will use them to justify equivalent definitions and theorems pertaining to continuous random variables.

DEFINITION 6.4

The **bivariate joint probability density function** $f(y_1, y_2)$ for two continuous random variables y_1 and y_2 is one that satisfies the following properties:

1. $f(y_1, y_2) \geq 0$ for all values of y_1 and y_2

2. $\int_{-\infty}^{\infty} \int_{-\infty}^{\infty} f(y_1, y_2) = 1$

DEFINITION 6.5

Let $f(y_1, y_2)$ be the joint density function for y_1 and y_2. Then the **marginal density functions** for y_1 and y_2 are

$$f_1(y_1) = \int_{-\infty}^{\infty} f(y_1, y_2)\, dy_2 \quad \text{and} \quad f_2(y_2) = \int_{-\infty}^{\infty} f(y_1, y_2)\, dy_1$$

DEFINITION 6.6

Let $f(y_1, y_2)$ be the joint density function for y_1 and y_2. Then the **conditional density functions** for y_1 and y_2 are

$$f_1(y_1 \mid y_2) = \frac{f(y_1, y_2)}{f_2(y_2)} \quad \text{and} \quad f_2(y_2 \mid y_1) = \frac{f(y_1, y_2)}{f_1(y_1)}$$

EXAMPLE 6.3

Suppose the joint density function for two continuous random variables, y_1 and y_2, is given by

$$f(y_1, y_2) = \begin{cases} cy_1 & \text{if } 0 \leq y_1 \leq 1; 0 \leq y_2 \leq 1 \\ 0 & \text{elsewhere} \end{cases}$$

Determine the value of the constant c.

SOLUTION

A graph of $f(y_1, y_2)$ traces a three-dimensional wedge-shaped figure over the unit square ($0 \leq y_1 \leq 1$ and $0 \leq y_2 \leq 1$) in the (y_1, y_2)-plane, as shown in

Figure 6.1. The value of c is chosen so that $f(y_1, y_2)$ satisfies the property

$$\int_{-\infty}^{\infty} \int_{-\infty}^{\infty} f(y_1, y_2) \, dy_1 dy_2 = 1$$

Performing this integration yields

$$\int_{-\infty}^{\infty} \int_{-\infty}^{\infty} f(y_1, y_2) \, dy_1 dy_2 = \int_{0}^{1} \int_{0}^{1} cy_1 \, dy_1 dy_2$$

$$= c \int_{0}^{1} \int_{0}^{1} y_1 \, dy_1 dy_2 = c \int_{0}^{1} \frac{y_1^2}{2} \Big]_{0}^{1} dy_2$$

$$= c \int_{0}^{1} \frac{1}{2} \, dy_2 = \frac{c}{2} y_2 \Big]_{0}^{1} = \frac{c}{2}$$

Setting this quantity equal to 1 and solving for c, we obtain

$$1 = \frac{c}{2} \quad \text{or} \quad c = 2$$

Therefore,

$$f(y_1, y_2) = 2y_1 \quad \text{for} \quad 0 \le y_1 \le 1 \text{ and } 0 \le y_2 \le 1$$

FIGURE 6.1
Graph of the Joint Density
Function for Example 6.3

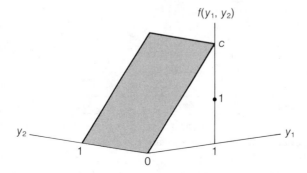

EXAMPLE 6.4

Refer to Example 6.3 and find the marginal density function for y_1. Show that

$$\int_{-\infty}^{\infty} f_1(y_1) = 1$$

SOLUTION

By Definition 6.5,

$$f_1(y_1) = \int_{-\infty}^{\infty} f(y_1, y_2) \, dy_2 = 2 \int_{0}^{1} y_1 \, dy_2 = 2y_1 y_2 \Big]_{y_2=0}^{y_2=1} = 2y_1$$

Thus,

$$\int_{-\infty}^{\infty} f_1(y_1) \, dy_1 = 2 \int_{0}^{1} y_1 \, dy_1 = 2\left(\frac{y_1^2}{2}\right) \Big]_{0}^{1} = 1$$

EXAMPLE 6.5

Refer to Example 6.3 and show that the marginal density function for y_2 is a uniform distribution.

SOLUTION

The marginal density function for y_2 is given by

$$f_2(y_2) = \int_{-\infty}^{\infty} f(y_1, y_2)\, dy_1 = 2 \int_0^1 y_1\, dy_1 = 2\frac{y_1^2}{2}\Big]_0^1 = 1$$

Therefore, $f_2(y_2)$ is a uniform distribution defined over the interval $0 \le y_2 \le 1$.

∎

EXAMPLE 6.6

Refer to Examples 6.3–6.5. Find the conditional density function for y_1 given y_2 and show that it satisfies the property

$$\int_{-\infty}^{\infty} f_1(y_1 \mid y_2)\, dy_1 = 1$$

SOLUTION

Using the marginal density function $f_2(y_2) = 1$ (obtained in Example 6.5) and Definition 6.6, we derive the conditional density function as follows:

$$f_1(y_1 \mid y_2) = \frac{f(y_1, y_2)}{f_2(y_2)} = \frac{2y_1}{1} = 2y_1$$

We now show that the integral of $f_1(y_1 \mid y_2)$ over all values of y_1 is equal to 1:

$$\int_0^1 f_1(y_1 \mid y_2)\, dy_1 = 2 \int_0^1 y_1\, dy_1 = 2\frac{y_1^2}{2}\Big]_0^1 = 1$$

∎

EXAMPLE 6.7

Suppose the joint density function for y_1 and y_2 is

$$f(y_1, y_2) = \begin{cases} cy_1 & \text{if } 0 \le y_1 \le y_2;\ 0 \le y_2 \le 1 \\ 0 & \text{elsewhere} \end{cases}$$

Find the value of c.

SOLUTION

Refer to Figure 6.1. If we pass a plane through the wedge, diagonally between the points $(0, 0)$ and $(1, 1)$, and perpendicular to the (y_1, y_2)-plane, then the slice lying along the y_2-axis will have a shape similar to that of the given density function (graphed in Figure 6.2). The value of c will be larger than the value found in Example 6.3 because the volume of the solid shown in Figure 6.2 must equal 1.

FIGURE 6.2
Graph of the Joint Density
Function for Example 6.7

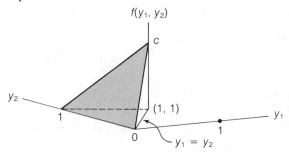

We find c by integrating $f(y_1, y_2)$ over the triangular region (shown in Figure 6.3) defined by $0 \leq y_1 \leq y_2$ and $0 \leq y_2 \leq 1$, setting this integral equal to 1, and solving for c:

$$\int_{-\infty}^{\infty}\int_{-\infty}^{\infty} f(y_1, y_2)\, dy_1 dy_2 = \int_0^1 \int_0^{y_2} cy_1\, dy_1 dy_2 = c \int_0^1 \frac{y_1^2}{2}\Big]_0^{y_2} dy_2$$

$$= c \int_0^1 \frac{y_2^2}{2}\, dy_2 = c\left(\frac{y_2^3}{6}\right)\Big]_0^1 = \frac{c}{6}$$

Setting this quantity equal to 1 and solving for c yields $c = 6$ and thus, $f(y_1, y_2) = 6y_1$ over the region of interest.

FIGURE 6.3

Region of Integration for Example 6.7

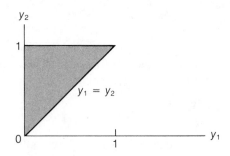

The joint density function for more than two random variables, say y_1, $y_2, \ldots, y_n$, is denoted by the symbol $f(y_1, y_2, \ldots, y_n)$. Marginal and conditional density functions are defined in a manner similar to that employed for the bivariate case, but we will not need them for the discussion that follows.

EXERCISES 6.6–6.10

6.6 Let y_1 and y_2 have the joint density

$$f(y_1, y_2) = \begin{cases} y_1 + cy_2 & \text{if } 1 \leq y_1 \leq 2; 0 \leq y_2 \leq 1 \\ 0 & \text{elsewhere} \end{cases}$$

where c is a constant.
a. Find the value of c that makes $f(y_1, y_2)$ a probability density function.
b. Find the marginal density for y_2 and show that

$$\int_{-\infty}^{\infty} f_2(y_2)\, dy_2 = 1$$

c. Find $f_1(y_1 \mid y_2)$, the conditional density for y_1 given y_2.

6.7 Let y_1 and y_2 have the joint density

$$f(y_1, y_2) = \begin{cases} cy_1 y_2 & \text{if } 0 \leq y_1 \leq 1; 0 \leq y_2 \leq 1 \\ 0 & \text{elsewhere} \end{cases}$$

a. Find the value of c that makes $f(y_1, y_2)$ a probability density function.
b. Find the marginal densities $f_1(y_1)$ and $f_2(y_2)$.
c. Find the conditional densities $f_1(y_1 \mid y_2)$ and $f_2(y_2 \mid y_1)$.

6.8 The joint density of y_1, the total time (in minutes) between a computer job's arrival in the job queue and its leaving the system after execution, and y_2, the time (in minutes) the job waits in the queue before being executed, is

$$f(y_1, y_2) = \begin{cases} ce^{-y_1^2} & \text{if } 0 \leq y_2 \leq y_1; 0 \leq y_1 \leq \infty \\ 0 & \text{elsewhere} \end{cases}$$

a. Find the value of c that makes $f(y_1, y_2)$ a probability density function.
b. Find the marginal density for y_1 and show that

$$\int_{-\infty}^{\infty} f_1(y_1)\, dy_1 = 1$$

c. Show that the conditional density for y_2 given y_1 is a uniform distribution over the interval $0 \leq y_2 \leq y_1$.

OPTIONAL EXERCISES

6.9 Let y_1 and y_2 be two continuous random variables with joint probability density $f(y_1, y_2)$. The joint distribution function $F(a, b)$ is defined as follows:

$$F(a, b) = P(y_1 \leq a, y_2 \leq b) = \int_{-\infty}^{a} \int_{-\infty}^{b} f(y_1, y_2)\, dy_2 dy_1$$

Verify each of the following:
a. $F(-\infty, -\infty) = F(-\infty, y_2) = F(y_1, -\infty) = 0$
b. $F(\infty, \infty) = 1$
c. If $a_2 \geq a_1$ and $b_2 \geq b_1$, then

$$F(a_2, b_2) - F(a_1, b_2) \geq F(a_2, b_1) - F(a_1, b_1)$$

6.10 Let y_1 and y_2 be two continuous random variables with joint probability density

$$f(y_1, y_2) = \begin{cases} ce^{-(y_1 + y_2)} & \text{if } 0 \leq y_1 < \infty; 0 \leq y_2 < \infty \\ 0 & \text{otherwise} \end{cases}$$

a. Find the value of c.
b. Find $f_1(y_1)$.
c. Find $f_2(y_2)$.
d. Find $f_1(y_1 \mid y_2)$.
e. Find $f_2(y_2 \mid y_1)$.
f. Find $P(y_1 \leq 1 \text{ and } y_2 \leq 1)$.

SECTION 6.3

THE EXPECTED VALUE OF FUNCTIONS OF TWO OR MORE RANDOM VARIABLES

The statistics that we will subsequently use for making inferences are computed from the data contained in a sample. The sample measurements can be viewed as observations on n random variables, $y_1, y_2, \ldots, y_n$, where y_1 represents the first measurement in the sample, y_2 represents the second measurement, etc. Since the sample statistics are functions of the random variables $y_1, y_2, \ldots, y_n$, they also will be random variables and will possess probability distributions. In order to describe these distributions, we will define the expected value (or mean) of functions of two or more random variables and present three expectation theorems that correspond to those given in Chapter 5. The definitions and theorems will be given in the bivariate context, but they can be written in general for any number of random variables by substituting corresponding multivariate functions and notation.

DEFINITION 6.7

Let $g(y_1, y_2)$ be a function of the random variables y_1 and y_2. Then the **expected value (mean)** of $g(y_1, y_2)$ is defined to be

$$E[g(y_1, y_2)] = \begin{cases} \displaystyle\sum_{y_2}\sum_{y_1} g(y_1, y_2)p(y_1, y_2) & \text{if } y_1 \text{ and } y_2 \text{ are discrete} \\[2ex] \displaystyle\int_{-\infty}^{\infty}\int_{-\infty}^{\infty} g(y_1, y_2)f(y_1, y_2)\, dy_1 dy_2 & \text{if } y_1 \text{ and } y_2 \text{ are continuous} \end{cases}$$

Suppose $g(y_1, y_2)$ is a function of only one of the random variables, say y_1. We will show that, in the discrete situation, the expected value of this function possesses the same meaning as in Chapter 5. Let $g(y_1, y_2)$ be a function of y_1 only, i.e., $g(y_1, y_2) = g(y_1)$. Then

$$E[g(y_1)] = \sum_{y_1}\sum_{y_2} g(y_1)p(y_1, y_2)$$

Summing first over y_2 (in which case, y_1 is regarded as a constant that can be factored outside the summation sign), we obtain

$$E[g(y_1)] = \sum_{y_1} g(y_1) \sum_{y_2} p(y_1, y_2)$$

But, by Definition 6.2, $\sum_{y_2} p(y_1, y_2)$ is the marginal probability distribution for y_1. Therefore,

$$E[g(y_1)] = \sum_{y_1} g(y_1)p_1(y_1)$$

You can verify that this is the same expression given for $E[g(y_1)]$ in Definition 4.5. An analogous result holds (proof omitted) if y_1 and y_2 are continuous random variables. Thus, if (μ_1, σ_1^2) and (μ_2, σ_2^2) denote the means and variances of y_1 and y_2, respectively, then the bivariate expectations for functions of either y_1 or y_2 will equal the corresponding expectations given in Chapter 5, i.e., $E(y_1) = \mu_1$, $E[(y_1 - \mu_1)^2] = \sigma_1^2$, etc.

It can be shown (proof omitted) that the three expectation theorems of Chapter 5 hold for bivariate and, in general, for multivariate probability distributions. We will use these theorems in Sections 6.5 and 6.7.

THEOREM 6.1

Let c be a constant. Then the expected value of c is

$$E(c) = c$$

THEOREM 6.2

Let c be a constant and let $g(y_1, y_2)$ be a function of the random variables y_1 and y_2. Then the expected value of $cg(y_1, y_2)$ is

$$E[cg(y_1, y_2)] = cE[g(y_1, y_2)]$$

THEOREM 6.3

Let $g_1(y_1, y_2), g_2(y_1, y_2), \ldots, g_k(y_1, y_2)$ be k functions of the random variables y_1 and y_2. Then the expected value of the sum of these functions is

$$E[g_1(y_1, y_2) + g_2(y_1, y_2) + \cdots + g_k(y_1, y_2)]$$
$$= E[g_1(y_1, y_2)] + E[g_2(y_1, y_2)] + \cdots + E[g_k(y_1, y_2)]$$

EXERCISES 6.11–6.17

6.11 Refer to Exercise 6.3.
  a. On the average, how many red lights should the truck expect to encounter on the way to delivery point B, i.e., what is $E(y_1)$?
  b. The total number of red lights encountered over the entire route—that is, going to point B and back to point A—is $(y_1 + y_2)$. Find $E(y_1 + y_2)$.

6.12 Refer to Exercise 6.7.
  a. Find $E(y_1 - y_2)$.
  b. Find $E(3y_2)$.

OPTIONAL EXERCISES

6.13 Refer to Exercise 6.6.
  a. Find $E(y_1)$.   **b.** Find $E(y_2)$.
  c. Find $E(y_1 + y_2)$.   **d.** Find $E(y_1 y_2)$.

6.14 Refer to Exercise 6.10.
  a. Find $E(y_1)$.   **b.** Find $E(y_2)$.
  c. Find $E(y_1 + y_2)$.   **d.** Find $E(y_1 y_2)$.

6.15 Let y_1 and y_2 be two continuous random variables with joint probability distribution $f(y_1, y_2)$. Consider the function $g(y_1)$. Show that

$$E[g(y_1)] = \int_{-\infty}^{\infty} g(y_1) f_1(y_1) \, dy_1$$

6.16 Prove Theorems 6.1–6.3 for discrete random variables y_1 and y_2.

6.17 Prove Theorems 6.1–6.3 for continuous random variables y_1 and y_2.

SECTION 6.4

INDEPENDENCE

In Chapter 3 we learned that two events A and B are said to be independent if $P(A \cap B) = P(A)P(B)$. Then, since the values assumed by two discrete random variables, y_1 and y_2, represent two numerical events, it follows that y_1 and y_2 are **independent** if $p(y_1, y_2) = p_1(y_1)p_2(y_2)$. Two continuous random variables are said to be independent if they satisfy a similar criterion.

DEFINITION 6.8

Let y_1 and y_2 be discrete random variables with joint probability distribution $p(y_1, y_2)$. Then y_1 and y_2 are said to be **independent** if and only if

$$p(y_1, y_2) = p_1(y_1)p_2(y_2) \qquad \text{for all pairs of values of } y_1 \text{ and } y_2$$

DEFINITION 6.9

Let y_1 and y_2 be continuous random variables with joint density function $f(y_1, y_2)$ and marginal density functions $f_1(y_1)$ and $f_2(y_2)$. Then y_1 and y_2 are said to be **independent** if and only if

$$f(y_1, y_2) = f_1(y_1)f_2(y_2) \qquad \text{for all pairs of values of } y_1 \text{ and } y_2$$

EXAMPLE 6.8

Refer to Example 6.3 and determine whether y_1 and y_2 are independent.

SOLUTION

From Examples 6.3–6.5, we have the following results:

$$f(y_1, y_2) = 2y_1 \qquad f_1(y_1) = 2y_1 \qquad f_2(y_2) = 1$$

Therefore,

$$f_1(y_1)f_2(y_2) = (2y_1)(1) = 2y_1 = f(y_1, y_2)$$

and, by Definition 6.9, y_1 and y_2 are independent random variables.   ■

EXAMPLE 6.9

Refer to Example 6.7 and determine whether y_1 and y_2 are independent.

SOLUTION

From Example 6.7, we determined that $f(y_1, y_2) = 6y_1$ when $0 \leq y_1 \leq y_2$ and $0 \leq y_2 \leq 1$. Therefore,

$$f_1(y_1) = \int_{-\infty}^{\infty} f(y_1, y_2)\, dy_2 = \int_{y_1}^{1} 6y_1\, dy_2 = 6y_1 y_2 \Big]_{y_1}^{1}$$
$$= 6y_1(1 - y_1) \qquad \text{where } 0 \leq y_1 \leq 1$$

Similarly,

$$f_2(y_2) = \int_{-\infty}^{\infty} f(y_1, y_2)\, dy_1 = \int_{0}^{y_2} 6y_1\, dy_1 = \frac{6y_1^2}{2} \Big]_{0}^{y_2}$$
$$= 3y_2^2 \qquad \text{where } 0 \leq y_2 \leq 1$$

You can see that $f_1(y_1)f_2(y_2) = 18y_1(1 - y_1)y_2^2$ is *not* equal to $f(y_1, y_2)$. Therefore, y_1 and y_2 are *not* independent random variables. ∎

Theorem 6.4 points to a useful consequence of independence.

THEOREM 6.4

If y_1 and y_2 are independent random variables, then

$$E(y_1y_2) = E(y_1)E(y_2)$$

Proof of Theorem 6.4   We will prove the theorem for the discrete case. The proof for the continuous case is identical, except that integration is substituted for summation. By the definition of expected value, we have

$$E(y_1y_2) = \sum_{y_2} \sum_{y_1} y_1y_2p(y_1, y_2)$$

But, since y_1 and y_2 are independent, we can write $p(y_1, y_2) = p_1(y_1)p_2(y_2)$. Therefore,

$$E(y_1y_2) = \sum_{y_2} \sum_{y_1} y_1y_2p_1(y_1)p_2(y_2)$$

If we sum first with respect to y_1, then we can treat y_2 and $p_2(y_2)$ as constants and apply Theorem 6.2 to factor them out of the sum as follows:

$$E(y_1y_2) = \sum_{y_2} y_2p_2(y_2) \sum_{y_1} y_1p_1(y_1)$$

But,

$$\sum_{y_1} y_1p_1(y_1) = E(y_1) \quad \text{and} \quad \sum_{y_2} y_2p_2(y_2) = E(y_2)$$

Therefore, $E(y_1y_2) = E(y_1)E(y_2)$.

EXERCISES 6.18–6.24

6.18 Refer to Exercise 6.1. Are y_1 and y_2 independent?

6.19 Refer to Exercise 6.2. Are y_1 and y_2 independent?

6.20 Refer to Exercise 6.6. Are y_1 and y_2 independent?

6.21 Refer to Exercise 6.7. Are y_1 and y_2 independent?

6.22 The lifelength y (in hundreds of hours) for fuses used in a televideo computer terminal has an exponential distribution with mean $\beta = 5$. Each terminal requires two such fuses—one acting as a backup that comes into use only when the first fuse fails.
a. If two such fuses have independent lifelengths y_1 and y_2, find the joint density $f(y_1, y_2)$.
b. The total effective lifelength of the two fuses is $(y_1 + y_2)$. Find the expected total effective lifelength of a pair of fuses in a televideo computer terminal.

OPTIONAL EXERCISES

6.23 Prove Theorem 6.4 for the continuous case.

6.24 Let y_1 and y_2 denote the lifetimes of two different types of components in an electronic system. The joint density of y_1 and y_2 is given by

$$f(y_1, y_2) = \begin{cases} \frac{1}{8} y_1 e^{-(y_1 + y_2)/2} & \text{if } y_1 > 0; y_2 > 0 \\ 0 & \text{elsewhere} \end{cases}$$

Show that y_1 and y_2 are independent. [*Hint:*   A theorem in multivariate probability theory states that y_1 and y_2 are independent if we can write

$$f(y_1, y_2) = g(y_1)h(y_2)$$

where $g(y_1)$ is a nonnegative function of y_1 only and $h(y_2)$ is a nonnegative function of y_2 only.]

SECTION 6.5

THE COVARIANCE OF TWO RANDOM VARIABLES

When we think of two variables y_1 and y_2 being related, we usually imagine a relationship in which y_2 increases as y_1 increases or y_2 decreases as y_1 increases. In other words, we tend to think in terms of **linear relationships**.

If y_1 and y_2 are random variables and we collect a sample of n pairs of values (y_1, y_2), it is unlikely that the plotted data points would fall exactly on a straight line. Instead, they would probably scatter about a straight line. If the points lie very close to a straight line, as in Figures 6.4a and 6.4b, we think of the linear relationship between y_1 and y_2 as being very strong. If they are widely scattered about a line, as in Figure 6.4c, we think of the relationship as weak. How can we measure the strength of the linear relationship between two random variables, y_1 and y_2?

FIGURE 6.4

Linear Relationships Between y_1 and y_2

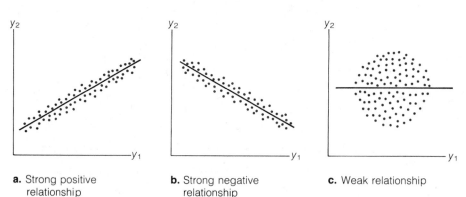

a. Strong positive relationship

b. Strong negative relationship

c. Weak relationship

One way to measure the strength of a linear relationship is to calculate the cross-product of the deviations $(y_1 - \mu_1)(y_2 - \mu_2)$ for each data point. These cross-products will be positive when the data points are in the upper right or lower left quadrant of Figure 6.5 and negative when the points are in the upper left or lower right quadrant. If all the points lie close to a line with positive slope,

FIGURE 6.5
Signs of the
Cross-Products
$(y_1 - \mu_1)(y_2 - \mu_2)$

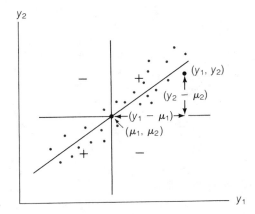

as in Figure 6.4a, almost all the cross-products $(y_1 - \mu_1)(y_2 - \mu_2)$ will be positive and their mean value will be relatively large and positive. Similarly, if all the points lie close to a line with a negative slope, as in Figure 6.4b, the mean value of $(y_1 - \mu_1)(y_2 - \mu_2)$ will be a relatively large negative number. However, if the linear relationship between y_1 and y_2 is relatively weak, as in Figure 6.4c, the points will fall in all four quadrants, some cross-products $(y_1 - \mu_1)(y_2 - \mu_2)$ will be positive, some will be negative, and their mean value will be relatively small—perhaps very close to 0. This leads to the following definition of a measure of the strength of the linear relationship between two random variables.

DEFINITION 6.10

The **covariance** of two random variables, y_1 and y_2, is defined to be

$$\text{Cov}(y_1, y_2) = E[(y_1 - \mu_1)(y_2 - \mu_2)]$$

THEOREM 6.5

$$\text{Cov}(y_1, y_2) = E(y_1 y_2) - \mu_1 \mu_2$$

Proof of Theorem 6.5   By Definition 6.10, we can write

$$\text{Cov}(y_1, y_2) = E[(y_1 - \mu_1)(y_2 - \mu_2)]$$
$$= E(y_1 y_2 - \mu_1 y_2 - \mu_2 y_1 + \mu_1 \mu_2)$$

Applying Theorems 6.1, 6.2, and 6.3 yields

$$\text{Cov}(y_1, y_2) = E(y_1 y_2) - \mu_1 E(y_2) - \mu_2 E(y_1) + \mu_1 \mu_2$$
$$= E(y_1 y_2) - \mu_1 \mu_2 - \mu_1 \mu_2 + \mu_1 \mu_2$$
$$= E(y_1 y_2) - \mu_1 \mu_2$$

EXAMPLE 6.10

Find the covariance of the random variables y_1 and y_2 of Example 6.3.

SOLUTION

The variables have joint density function $f(y_1, y_2) = 2y_1$ when $0 \le y_1 \le 1$ and $0 \le y_2 \le 1$. Then

$$E(y_1 y_2) = \int_0^1 \int_0^1 (y_1 y_2) 2 y_1 \, dy_1 \, dy_2$$

$$= \int_0^1 2 \left(\frac{y_1^3}{3} \right) \Big]_0^1 y_2 \, dy_2 = \frac{2}{3} \int_0^1 y_2 \, dy_2 = \frac{2}{3} \left(\frac{y_2^2}{2} \right) \Big]_0^1 = \frac{1}{3}$$

In Examples 6.4 and 6.5, we obtained the marginal density functions $f_1(y_1) = 2y_1$ and $f_2(y_2) = 1$. Therefore,

$$\mu_1 = E(y_1) = \int_0^1 y_1 f_1(y_1) \, dy_1 = \int_0^1 y_1 (2y_1) \, dy_1 = \frac{2y_1^3}{3} \Big]_0^1 = \frac{2}{3}$$

Furthermore, since y_2 is a uniform random variable defined over the interval $0 \le y_2 \le 1$ (see Example 6.5), it follows from Section 5.4 that $\mu_2 = \frac{1}{2}$. Then

$$\text{Cov}(y_1, y_2) = E(y_1 y_2) - \mu_1 \mu_2 = \frac{1}{3} - \left(\frac{2}{3} \right)\left(\frac{1}{2} \right) = 0 \qquad \blacksquare$$

Example 6.10 demonstrates an important result: If y_1 and y_2 are independent, then their covariance will equal 0. However, *the converse is not true*.

THEOREM 6.6

If two random variables y_1 and y_2 are independent, then

$$\text{Cov}(y_1, y_2) = 0$$

The proof of Theorem 6.6, which follows readily from Theorem 6.5, is left as an optional exercise.

SECTION 6.6

THE CORRELATION COEFFICIENT ρ

If the covariance between two random variables is positive, then y_2 tends to increase as y_1 increases. If the covariance is negative, then y_2 tends to decrease as y_1 increases. But what can we say about the numerical value of the covariance? We know that a covariance equal to 0 means that there is no linear relationship between y_1 and y_2, but when the covariance is nonzero, its absolute value will depend on the units of measurement of y_1 and y_2. To overcome this difficulty, we define a standardized version of the covariance known as the **coefficient of correlation**.

DEFINITION 6.11

The **coefficient of correlation** ρ for two random variables y_1 and y_2 is

$$\rho = \frac{\text{Cov}(y_1, y_2)}{\sigma_1 \sigma_2}$$

where σ_1 and σ_2 are the standard deviations of y_1 and y_2, respectively.

Since ρ is equal to the covariance divided by the product of two positive quantities, σ_1 and σ_2, it will have the same sign as the covariance but, in addition, it will (proof omitted) assume a value in the interval $-1 \leq \rho \leq 1$. Values of $\rho = -1$ and $\rho = 1$ imply perfect straight-line relationships between y_1 and y_2, the former with a negative slope and the latter with a positive slope. A value of $\rho = 0$ implies no linear relationship between y_1 and y_2.

PROPERTY OF THE CORRELATION COEFFICIENT

$$-1 \leq \rho \leq 1$$

EXERCISES 6.25–6.36

6.25 Find the covariance of the random variables y_1 and y_2 in Exercise 6.1.

6.26 Find the covariance of the random variables y_1 and y_2 in Exercise 6.2.

6.27 Find the covariance of the random variables y_1 and y_2 in Exercise 6.6.

6.28 Find the covariance of the random variables y_1 and y_2 in Exercise 6.7.

6.29 Refer to Exercise 6.3. Find the coefficient of correlation ρ for y_1 and y_2.

6.30 Refer to Exercise 6.6. Find the coefficient of correlation ρ for y_1 and y_2.

6.31 Refer to Exercise 6.7. Find the coefficient of correlation ρ for y_1 and y_2.

OPTIONAL EXERCISES

6.32 Commercial kerosene is stocked in a bulk tank at the beginning of each week. Because of limited supplies, the proportion y_1 of the capacity of the tank available for sale and the proportion y_2 of the capacity of the tank actually sold during the week are continuous random variables. Their joint distribution is given by

$$f(y_1, y_2) = \begin{cases} 4y_1^2 & \text{if } 0 \leq y_2 \leq y_1; \ 0 \leq y_1 \leq 1 \\ 0 & \text{elsewhere} \end{cases}$$

Find the covariance of y_1 and y_2.

6.33 Prove Theorem 6.6 for the discrete case.

6.34 Prove Theorem 6.6 for the continuous case.

		y_1	
	-1	0	$+1$
-1	$\frac{1}{12}$	$\frac{2}{12}$	$\frac{1}{12}$
$y_2 \quad 0$	$\frac{2}{12}$	0	$\frac{2}{12}$
$+1$	$\frac{1}{12}$	$\frac{2}{12}$	$\frac{1}{12}$

6.35 As an illustration of why the converse of Theorem 6.6 is not true, consider the joint distribution of two discrete random variables, y_1 and y_2, shown in the accompanying table. Show that $\text{Cov}(y_1, y_2) = 0$, but that y_1 and y_2 are dependent.

6.36 Find the covariance of y_1 and y_2 for the random variables of Exercise 6.10.

SECTION 6.7

THE EXPECTED VALUE AND VARIANCE OF LINEAR FUNCTIONS OF RANDOM VARIABLES

Many experiments are conducted in the engineering and computer sciences in order to develop a mathematical model explaining the relationship between two or more variables. The usual objective is to be able to predict the value of one of the variables, y, given specific values of the other variables. In the methodology most frequently employed for fitting a model to multivariable data sets—**regression analysis** (the topic of Chapters 10–12)—estimates of the model parameters are **linear functions** of the observed sample y values.

DEFINITION 6.12

Let $y_1, y_2, \ldots, y_n$ be random variables and let $a_1, a_2, \ldots, a_n$ be constants. Then ℓ is a **linear function** of $y_1, y_2, \ldots, y_n$ if

$$\ell = a_1 y_1 + a_2 y_2 + \cdots + a_n y_n$$

The expected value (mean) and variance of a linear function of $y_1, y_2, \ldots, y_n$ may be computed using the formulas presented in Theorem 6.7.

THEOREM 6.7

The Expected Value $E(\ell)$ and Variance $V(\ell)$* of a Linear Function of y_1, $y_2, \ldots, y_n$

Suppose the means and variances of $y_1, y_2, \ldots, y_n$ are (μ_1, σ_1^2), $(\mu_2, \sigma_2^2), \ldots, (\mu_n, \sigma_n^2)$, respectively. If $\ell = a_1 y_1 + a_2 y_2 + \cdots + a_n y_n$, then

$$E(\ell) = a_1 \mu_1 + a_2 \mu_2 + \cdots + a_n \mu_n$$

and

$$
\begin{aligned}
\sigma_\ell^2 = V(\ell) = {} & a_1^2 \sigma_1^2 + a_2^2 \sigma_2^2 + \cdots + a_n^2 \sigma_n^2 \\
& + 2a_1 a_2 \text{Cov}(y_1, y_2) + 2a_1 a_3 \text{Cov}(y_1, y_3) + 2a_1 a_n \text{Cov}(y_1, y_n) \\
& + 2a_2 a_3 \text{Cov}(y_2, y_3) + \cdots + 2a_2 a_n \text{Cov}(y_2, y_n) \\
& + \cdots + 2a_{n-1} a_n \text{Cov}(y_{n-1}, y_n)
\end{aligned}
$$

---

*In the preceding sections, we have used different subscripts on the symbol σ^2 to denote the variances of different random variables. This notation is cumbersome if the random variable is a function of several other random variables. Consequently, we will use the notation $\sigma_{()}^2$ or $V()$ interchangeably to denote a variance.

Proof of Theorem 6.7 By Theorem 6.3, we know

$$E(\ell) = E(a_1 y_1) + E(a_2 y_2) + \cdots + E(a_n y_n)$$

Then, by Theorem 6.2,

$$E(\ell) = a_1 E(y_1) + a_2 E(y_2) + \cdots + a_n E(y_n)$$
$$= a_1 \mu_1 + a_2 \mu_2 + \cdots + a_n \mu_n$$

Similarly,

$$
\begin{aligned}
V(\ell) &= E\{[\ell - E(\ell)]^2\} \\
&= E[(a_1 y_1 + a_2 y_2 + \cdots + a_n y_n - a_1 \mu_1 - a_2 \mu_2 - \cdots - a_n \mu_n)^2] \\
&= E\{[a_1(y_1 - \mu_1) + a_2(y_2 - \mu_2) + \cdots + a_n(y_n - \mu_n)]^2\} \\
&= E[a_1^2(y_1 - \mu_1)^2 + a_2^2(y_2 - \mu_2)^2 + \cdots + a_n^2(y_n - \mu_n)^2 \\
&\quad + 2a_1 a_2(y_1 - \mu_1)(y_2 - \mu_2) + 2a_1 a_3(y_1 - \mu_1)(y_3 - \mu_3) \\
&\quad + \cdots + 2a_{n-1}a_n(y_{n-1} - \mu_{n-1})(y_n - \mu_n)] \\
&= a_1^2 E[(y_1 - \mu_1)^2] + \cdots + a_n^2 E[(y_n - \mu_n)^2] \\
&\quad + 2a_1 a_2 E[(y_1 - \mu_1)(y_2 - \mu_2)] + 2a_1 a_3 E[(y_1 - \mu_1)(y_3 - \mu_3)] \\
&\quad + \cdots + 2a_{n-1}a_n E[(y_{n-1} - \mu_{n-1})(y_n - \mu_n)]
\end{aligned}
$$

By the definitions of variance and covariance, we have

$$E[(y_i - \mu_i)^2] = \sigma_i^2 \quad \text{and} \quad E[(y_i - \mu_i)(y_j - \mu_j)] = \text{Cov}(y_i, y_j)$$

Therefore,

$$
\begin{aligned}
V(\ell) &= a_1^2 \sigma_1^2 + a_2^2 \sigma_2^2 + \cdots + a_n^2 \sigma_n^2 + 2a_1 a_2 \text{Cov}(y_1, y_2) + 2a_1 a_3 \text{Cov}(y_1, y_3) \\
&\quad + \cdots + 2a_2 a_3 \text{Cov}(y_2, y_3) + \cdots + 2a_{n-1}a_n \text{Cov}(y_{n-1}, y_n)
\end{aligned}
$$

EXAMPLE 6.11

Suppose y_1, y_2, and y_3 are random variables with $(\mu_1 = 1, \sigma_1^2 = 2)$, $(\mu_2 = 3, \sigma_2^2 = 1)$, $(\mu_3 = 0, \sigma_3^2 = 4)$, $\text{Cov}(y_1, y_2) = -1$, $\text{Cov}(y_1, y_3) = 2$, and $\text{Cov}(y_2, y_3) = 1$. Find the mean and variance of

$$\ell = 2y_1 + y_2 - 3y_3$$

SOLUTION

The linear function

$$\ell = 2y_1 + y_2 - 3y_3$$

has coefficients $a_1 = 2$, $a_2 = 1$, and $a_3 = -3$. Then by Theorem 6.7,

$$
\begin{aligned}
E(\ell) &= a_1 \mu_1 + a_2 \mu_2 + a_3 \mu_3 \\
&= (2)(1) + (1)(3) + (-3)(0) = 5 \\
V(\ell) &= a_1^2 \sigma_1^2 + a_2^2 \sigma_2^2 + a_3^2 \sigma_3^2 \\
&\quad + 2a_1 a_2 \text{Cov}(y_1, y_2) + 2a_1 a_3 \text{Cov}(y_1, y_3) + 2a_2 a_3 \text{Cov}(y_2, y_3) \\
&= (2)^2(2) + (1)^2(1) + (-3)^2(4) \\
&\quad + 2(2)(1)(-1) + 2(2)(-3)(2) + 2(1)(-3)(1) \\
&= 11
\end{aligned}
$$

These results indicate that the probability distribution of ℓ is centered about $E(\ell) = \mu_\ell = 5$ and that its spread is measured by $\sigma_\ell = \sqrt{V(\ell)} = \sqrt{11} = 3.3$. If we were to randomly select values of y_1, y_2, and y_3, we would expect the value of ℓ to fall in the interval $\mu_\ell \pm 2\sigma_\ell$, or -1.6 to 11.6, according to the Empirical Rule. ∎

EXAMPLE 6.12

Let $y_1, y_2, \ldots, y_n$ be a sample of n independent observations selected from a population with mean μ and variance σ^2. Find the expected value and variance of the sample mean, $\bar{y}$.

SOLUTION

The sample measurements, $y_1, y_2, \ldots, y_n$, can be viewed as observations on n independent random variables, where y_1 corresponds to the first observation, y_2 to the second, etc. Therefore, the sample mean $\bar{y}$ will be a random variable with a probability distribution (or density function).

By writing

$$\bar{y} = \frac{\sum_{i=1}^{n} y_i}{n} = \frac{y_1}{n} + \frac{y_2}{n} + \cdots + \frac{y_n}{n}$$

we see that $\bar{y}$ is a linear function of $y_1, y_2, \ldots, y_n$, with $a_1 = \frac{1}{n}, a_2 = \frac{1}{n}, \ldots, a_n = \frac{1}{n}$. Since $y_1, y_2, \ldots, y_n$ are independent, it follows from Theorem 6.6 that the covariance of y_i and y_j, for all pairs with $i \neq j$, will equal 0. Therefore, we can apply Theorem 6.7 to obtain

$$E(\bar{y}) = \left(\frac{1}{n}\right)\mu + \left(\frac{1}{n}\right)\mu + \cdots + \left(\frac{1}{n}\right)\mu = \frac{n\mu}{n} = \mu$$

$$V(\bar{y}) = \left(\frac{1}{n}\right)^2 \sigma^2 + \left(\frac{1}{n}\right)^2 \sigma^2 + \cdots + \left(\frac{1}{n}\right)^2 \sigma^2 = \left(\frac{n}{n^2}\right)\sigma^2 = \frac{\sigma^2}{n}$$

∎

EXAMPLE 6.13

Suppose that the population of Example 6.12 has mean $\mu = 10$ and variance $\sigma^2 = 4$. Describe the probability distribution for a sample mean based on $n = 25$ observations.

SOLUTION

From Example 6.12, we know that the probability distribution of the sample mean will have mean and variance

$$E(\bar{y}) = \mu = 10 \quad \text{and} \quad \sigma_{\bar{y}}^2 = V(\bar{y}) = \frac{\sigma^2}{n} = \frac{4}{25}$$

and thus,

$$\sigma_{\bar{y}} = \sqrt{V(\bar{y})} = \sqrt{\frac{4}{25}} = \frac{2}{5} = .4$$

Therefore, the probability distribution of $\bar{y}$ will be centered about its mean, $\mu = 10$, and most of the distribution will fall in the interval $\mu \pm 2\sigma_{\bar{y}}$, or $10 \pm 2(.4)$, or 9.2 to 10.8. We will learn more about the properties of the probability distribution of $\bar{y}$ in Chapter 7. ∎

6.37 Suppose that y_1, y_2, and y_3 are random variables with ($\mu_1 = 0$, $\sigma_1^2 = 2$), ($\mu_2 = -1$, $\sigma_2^2 = 3$), ($\mu_3 = 5$, $\sigma_3^2 = 9$), $\text{Cov}(y_1, y_2) = 1$, $\text{Cov}(y_1, y_3) = 4$, and $\text{Cov}(y_2, y_3) = -2$. Find the mean and variance of

$$\ell = \frac{1}{2}y_1 - y_2 + 2y_3$$

6.38 Suppose that y_1, y_2, y_3, and y_4 are random variables with

$$E(y_1) = 2 \quad V(y_1) = 4 \quad \text{Cov}(y_1, y_2) = -1 \quad \text{Cov}(y_2, y_3) = 0$$
$$E(y_2) = 4 \quad V(y_2) = 8 \quad \text{Cov}(y_1, y_3) = 1 \quad \text{Cov}(y_2, y_4) = 2$$
$$E(y_3) = -1 \quad V(y_3) = 6 \quad \text{Cov}(y_1, y_4) = \frac{1}{2} \quad \text{Cov}(y_3, y_4) = 0$$
$$E(y_4) = 0 \quad V(y_4) = 1$$

Find the mean and variance of

$$\ell = -3y_1 + 2y_2 + 6y_3 - y_4$$

6.39 Refer to Exercise 6.2. Find the mean and variance of $(y_1 + y_2)$, the sum of the dots showing on the two dice.

6.40 Refer to Exercises 6.3 and 6.11. Find the variance of $(y_1 + y_2)$. Within what range would you expect $(y_1 + y_2)$ to fall?

6.41 Refer to Exercises 6.7 and 6.12. Find the variance of $(y_1 - y_2)$.

OPTIONAL EXERCISES

6.42 A particular manufacturing process yields a proportion p of defective items in each lot. The number y of defectives in a random sample of n items from the process follows a binomial distribution. Find the expected value and variance of $\hat{p} = y/n$, the fraction of defectives in the sample. [*Hint:* Write $\hat{p}$ as a linear function of a single random variable y, i.e., $\hat{p} = a_1 y$, where $a_1 = 1/n$.]

6.43 Let $y_1, y_2, \ldots, y_n$ be a sample of n independent observations selected from a gamma distribution with $\alpha = 1$ and $\beta = 2$. Show that the expected value and variance of the sample mean $\bar{y}$ are identical to the expected value and variance of a gamma distribution with parameters $\alpha = n$ and $\beta = 2$.

In Chapter 4, we defined a single random variable over a sample space. In this chapter, we extend this concept by defining two or more random variables, y_1, $y_2, \ldots, y_n$, over a sample space. The probability of the intersection of the discrete numerical events, $y_1, y_2, \ldots, y_n$, is represented by a **joint probability distribution** $p(y_1, y_2, \ldots, y_n)$. Analogously, we define a **joint density function**, $f(y_1, y_2, \ldots, y_n)$, that enables us to calculate probabilities associated with continuous random variables $y_1, y_2, \ldots, y_n$.

The final objective of this chapter was to enable us to obtain the mean and variance of a **linear function** of random variables. This will be particularly useful in subsequent chapters when the random variables represent independent sample observations from a single population.

6.44 The management of a bank must decide whether to install a commercial loan decision-support system (an on-line management information system) to aid its analysts in making commercial loan decisions. Past experience shows that y_1, the additional number (per year) of correct loan decisions—accepting good loan applications and rejecting those that would eventually be defaulted—attributable to the decision-support system, and y_2, the lifetime (in years) of the decision-support system, have the joint probability distribution shown in the table.

		y_1									
		0	10	20	30	40	50	60	70	80	90
	1	.001	.002	.002	.025	.040	.025	.005	.005	0	0
	2	.005	.005	.010	.075	.100	.075	.050	.030	.030	.025
y_2	3	0	0	0	.025	.050	.080	.050	.080	.040	.030
	4	0	.001	.002	.005	.010	.025	.010	.003	.001	.001
	5	0	.002	.005	.005	.020	.030	.015	0	0	0

a. Find the marginal probability distributions, $p_1(y_1)$ and $p_2(y_2)$.
b. Find the conditional probability distribution, $p_1(y_1 \mid y_2)$.
c. Given that the decision-support system is in its third year of operation, find the probability that at least 40 additional correct loan decisions will be made.
d. Find the expected lifetime of the decision-support system, i.e., find $E(y_2)$.
e. Are y_1 and y_2 correlated? Are y_1 and y_2 independent?
f. Each correct loan decision contributes approximately $25,000 to the bank's profit. Compute the mean and standard deviation of the additional profit attributable to the decision-support system. [*Hint:* Use the marginal distribution $p_1(y_1)$.]

6.45 Suppose that y_1 and y_2, the proportions of an 8-hour workday that two gas station attendants actually spend on performing their assigned duties, have joint probability density

$$f(y_1, y_2) = \begin{cases} y_1 + y_2 & \text{if } 0 \le y_1 \le 1; 0 \le y_2 \le 1 \\ 0 & \text{elsewhere} \end{cases}$$

a. Find the marginal probability distributions, $f_1(y_1)$ and $f_2(y_2)$.
b. Verify that

$$\int_{-\infty}^{\infty} f_1(y_1)\, dy_1 = 1 \quad \text{and} \quad \int_{-\infty}^{\infty} f_2(y_2)\, dy_2 = 1$$

c. Find the conditional probability distributions, $f_1(y_1 \mid y_2)$ and $f_2(y_2 \mid y_1)$.
d. Verify that

$$\int_{-\infty}^{\infty} f_1(y_1 \mid y_2)\, dy_1 = 1 \quad \text{and} \quad \int_{-\infty}^{\infty} f_2(y_2 \mid y_1)\, dy_2 = 1$$

e. Are y_1 and y_2 correlated? Are y_1 and y_2 independent?

f. The proportion d of "dead" time (i.e., time when no assigned duties are performed) for the two attendants is given by the relation $d = 1 - (y_1 + y_2)/2$. Find $E(d)$ and $V(d)$. Within what limits would you expect d to fall?

6.46 Concrete experiences a characteristic marked increase in "creep" when it is heated for the first time under load. An experiment was conducted to investigate the transient thermal strain behavior of concrete (*Magazine of Concrete Research*, Dec. 1985). Two variables thought to affect thermal strain are y_1, rate of heating (degrees Centigrade per minute), and y_2, level of load (percentage of initial strength). Concrete specimens are prepared and tested under various combinations of heating rate and load, and the thermal strain is determined for each. Suppose the joint probability distribution for y_1 and y_2 for those specimens that yielded acceptable results is given in the table. Suppose a concrete specimen is randomly selected from among those in the experiment that yielded acceptable thermal strain behavior.

		\multicolumn{5}{c}{y_1 (°C/minute)}
		.1	.2	.3	.4	.5
y_2	0	.17	.11	.07	.05	.05
	10	.10	.06	.05	.02	.01
	20	.09	.04	.03	.01	0
	30	.08	.04	.02	0	0

a. Find the probability that the concrete specimen was heated at a rate of .3°C/minute.

b. Given that the concrete specimen was heated at .3°C/minute, find the probability that the specimen had a load of 20%.

c. Are rate of heating y_1 and level of load y_2 correlated?

d. Are rate of heating y_1 and level of load y_2 independent?

OPTIONAL EXERCISES

6.47 Let y_1 and y_2 be two continuous random variables with joint density $f(y_1, y_2)$. Show that

$$f_2(y_2 \mid y_1)f_1(y_1) = f_1(y_1 \mid y_2)f_2(y_2)$$

6.48 Let y_1 and y_2 be uncorrelated random variables. Verify each of the following:
a. $V(y_1 + y_2) = V(y_1 - y_2)$
b. $\text{Cov}[(y_1 + y_2), (y_1 - y_2)] = V(y_1) - V(y_2)$

6.49 Suppose three continuous random variables have the joint distribution

$$f(y_1, y_2, y_3) = \begin{cases} c(y_1 + y_2)e^{-y_3} & \text{if } 0 \le y_1 \le 1; 0 \le y_2 \le 2; y_3 > 0 \\ 0 & \text{elsewhere} \end{cases}$$

a. Find the value of c that makes $f(y_1, y_2, y_3)$ a probability density.

b. Are the three variables independent? [*Hint:* If $f(y_1, y_2, y_3) = f_1(y_1)f_2(y_2)f_3(y_3)$, then y_1, y_2, and y_3 are independent.]

REFERENCES

Hoel, P. G. *Introduction to Mathematical Statistics*, 4th ed. New York: Wiley, 1971.

Hogg, R. V. and Craig, A. T. *Introduction to Mathematical Statistics*, 4th ed. New York: Macmillan, 1978.

Mendenhall, W., Scheaffer, R. L., and Wackerly, D. D. *Mathematical Statistics with Applications*, 3rd ed. Boston: Duxbury Press, 1986.

Mood, A. M., Graybill, F. A., and Boes, D. *Introduction to the Theory of Statistics*, 3rd ed. New York: McGraw-Hill, 1974.

OBJECTIVE

To present two methods for finding the probability distribution (sampling distribution) of a statistic; to identify the sampling distributions for some useful statistics.

CONTENTS

SAMPLING DISTRIBUTIONS

SECTION 7.1

SAMPLING DISTRIBUTIONS

As noted in Chapter 6, the n measurements in a sample can be viewed as observations on n random variables, $y_1, y_2, \ldots, y_n$. Consequently, the sample mean $\bar{y}$, the sample variance s^2, and other statistics are functions of random variables—functions that we will use in the following chapters to make inferences about population parameters. Thus, a primary reason for presenting the theory of probability and probability distributions in the preceding chapters was to enable us to find and evaluate the properties of the probability distribution of a statistic. This probability distribution is often called the **sampling distribution** of the statistic. As is the case for a single random variable, its mean is the expected value of the statistic. Its standard deviation is called the **standard error** of the statistic.

DEFINITION 7.1

The **sampling distribution** of a statistic is its probability distribution.

DEFINITION 7.2

The **standard error** of a statistic is the standard deviation of its sampling distribution.

The mathematical techniques for finding the sampling distribution of a statistic are difficult to apply and except for very simple examples, are beyond the scope of this text. We will introduce this topic in Section 7.2 in order to develop a procedure that will enable us to use a computer to generate random samples from theoretical populations of data. We will use this simulated sampling procedure to draw many samples of a specified size, calculate the value of a statistic for each sample, and then form a relative frequency histogram of these values. The resulting relative frequency histogram will be an approximation to the sampling distribution of the statistic.

Even if we are unable to find the exact mathematical form of the probability distribution of a statistic and are unable to approximate it using simulation, we can always find its mean and variance using the methods of Chapters 4–6. Then we can obtain an approximate description of the sampling distribution by applying Tchebysheff's theorem and the Empirical Rule.

SECTION 7.2

PROBABILITY DISTRIBUTIONS OF FUNCTIONS OF RANDOM VARIABLES

There are essentially three methods for finding the density function for a function of one or more random variables. Two of these—the moment generating function method and the transformation method—are beyond the scope of this text, but a discussion of them can be found in the references at the end of the chapter. The third method, which we will call the **cumulative distribution function method**, will be demonstrated with examples.

Suppose g is a function of one or more random variables. The cumulative distribution function method finds the density function for g by first finding the probability $P(g \leq g_0)$, which (dropping the subscript 0) is equal to $F(g)$. The density function $f(g)$ is then found by differentiating $F(g)$ with respect to g. We will demonstrate the method in Examples 7.1 and 7.2.

EXAMPLE 7.1

Suppose the random variable y has density function

$$f(y) = \begin{cases} \dfrac{e^{-y/\beta}}{\beta} & \text{if} \quad 0 \leq y < \infty \\ 0 & \text{elsewhere} \end{cases}$$

and let $g(y) = y^2$. Find the density function for g.

SOLUTION

A graph of $g = y^2$ is shown in Figure 7.1. We will denote the cumulative distribution functions of g and y as $G(g)$ and $F(y)$, respectively. We note from the figure that g will be less than g_0 whenever y is less than y_0; it follows that $P(g \leq g_0) = G(g_0) = F(y_0)$. Since $g = y^2$, we have $y_0 = \sqrt{g_0}$ and

$$F(y_0) = F(\sqrt{g_0}) = \int_{-\infty}^{\sqrt{g_0}} f(y)\, dy = \int_{0}^{\sqrt{g_0}} \frac{e^{-y/\beta}}{\beta}\, dy = -e^{-y/\beta}\Big]_0^{\sqrt{g_0}} = 1 - e^{-(\sqrt{g_0}/\beta)}$$

Therefore, the cumulative distribution function for g is

$$F(g) = 1 - e^{-(\sqrt{g}/\beta)}$$

and the density function for g is

$$\frac{dF(g)}{dg} = f(g) = \frac{g^{-1/2}e^{-(\sqrt{g}/\beta)}}{2\beta}$$

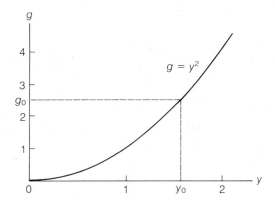

FIGURE 7.1
A Graph of $g = y^2$

EXAMPLE 7.2 (OPTIONAL)

If the random variables y_1 and y_2 possess a uniform joint density function over the unit square, then $f(y_1, y_2) = 1$ for $0 \leq y_1 \leq 1$ and $0 \leq y_2 \leq 1$. Find the density function for the sum $g = y_1 + y_2$.

SOLUTION

Each value of g corresponds to a series of points on the line $g_0 = y_1 = y_2$ (see Figure 7.2). Written in the slope–intercept form, $y_2 = g_0 - y_1$, this is the equation of a line with slope equal to -1 and y-intercept equal to g_0. The values of g that are less than or equal to g_0 are those corresponding to points (y_1, y_2) below the line $g_0 = y_1 + y_2$. (This area is shaded in Figure 7.2.) Then, for values of the y-intercept g_0, $0 \le g_0 \le 1$, the probability that g is less than or equal to g_0 is equal to the volume of a solid over the shaded area shown in the figure. We

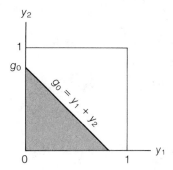

FIGURE 7.2
A Graph Showing the Region of Integration to Find $F(g_0)$, $0 \le g_0 \le 1$

could find this probability by multiple integration, but it is easier to obtain it with the aid of geometry. Each of the two equal sides of the triangle has length g_0. Therefore, the area of the shaded triangular region is $g_0^2/2$, the height of the solid over the region is $f(y_1, y_2) = 1$, and the volume is

$$P(g \le g_0) = F(g_0) = \frac{g_0^2}{2}$$

We now drop the subscript to obtain

$$F(g) = \frac{g^2}{2} \qquad (0 \le g \le 1)$$

The equation for $F(g)$ is different over the interval $1 \le g \le 2$. The probability $P(g \le g_0) = F(g_0)$ is the integral of $f(y_1, y_2) = 1$ over the shaded area shown in Figure 7.3. The integral can be found by subtracting from 1 the volume corresponding to the small triangular (nonshaded) area that lies above the line $g_0 = y_1 + y_2$. To find the length of one side of this triangle, we need to locate the point where the line $g_0 = y_1 + y_2$ intersects the line $y_2 = 1$. Substituting $y_2 = 1$ into the equation of the line, we find

$$g_0 = y_1 + 1 \qquad \text{or} \qquad y_1 = g_0 - 1$$

The point $(g_0 - 1, 1)$ is shown on Figure 7.3. The two equal sides of the triangle each have length $d = 1 - (g_0 - 1) = 2 - g_0$ and the area of the triangle lying above the line $g_0 = y_1 + y_2$ is then

$$\text{Area} = \frac{1}{2}(\text{Base})(\text{Height}) = \frac{1}{2}(2 - g_0)(2 - g_0) = \frac{(2 - g_0)^2}{2}$$

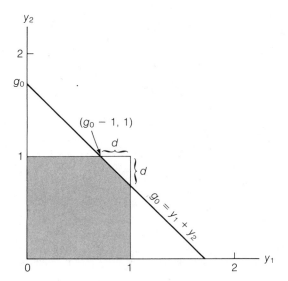

FIGURE 7.3

A Graph Showing the Region of Integration to Find $F(g_0)$, $1 \le g_0 \le 2$

Since the height of the solid constructed over the triangle is $f(y_1, y_2) = 1$, the probability that g lies above the line $g_0 = y_1 + y_2$ is $(2 - g_0)^2/2$. Subtracting this probability from 1, we find the probability that g lies below the line to be

$$F(g_0) = P(g \le g_0) = 1 - \frac{(2 - g_0)^2}{2}$$

We drop the subscript and simplify to obtain

$$F(g) = -1 + 2g - \frac{g^2}{2} \qquad (1 \le g \le 2)$$

The density function for the sum of the two random variables y_1 and y_2 is now obtained by differentiating $F(g)$:

$$f(g) = \frac{dF(g)}{dg} = \frac{d\left(\frac{g^2}{2}\right)}{dg} = g \qquad\qquad (0 \le g \le 1)$$

$$f(g) = \frac{dF(g)}{dg} = \frac{d\left(-1 + 2g - \frac{g^2}{2}\right)}{dg} = 2 - g \qquad (1 \le g \le 2)$$

Graphs of the cumulative distribution function and the density function for $g = y_1 + y_2$ are shown in Figures 7.4a and 7.4b (page 246), respectively. Note that the area under the density function over the interval $0 \le g \le 2$ is equal to 1.

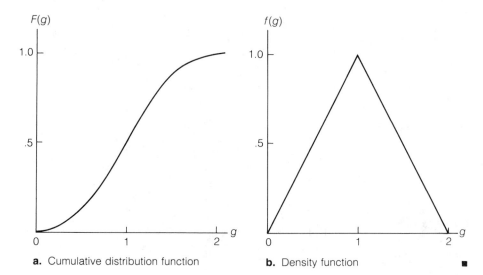

FIGURE 7.4
Graphs of the Cumulative
Distribution Function and
Density Function for
$g = y_1 + y_2$

a. Cumulative distribution function   **b.** Density function   ∎

One of the most useful functions of a single continuous random variable is the cumulative distribution function itself. We will show that if y is a continuous random variable with density function $f(y)$ and cumulative distribution function $F(y)$, then $g = F(y)$ has a uniform probability distribution over the interval $0 \le g \le 1$. Using a computer program for generating random numbers, we can generate a random sample of g values. For each value of g, we can solve for the corresponding value of y using the equation $g = F(y)$ and, thereby, obtain a random sample of y values from a population modeled by the density function $g(y)$. We will present this important transformation as a theorem, prove it, and then demonstrate its use with an example.

THEOREM 7.1

Let y be a continuous random variable with density function $f(y)$ and cumulative distribution $F(y)$. Then the density function of $g = F(y)$ will be a uniform distribution defined over the interval $0 \le g \le 1$, i.e., $f(g) = 1$.

Proof of Theorem 7.1   Figure 7.5 shows the graph of $g = F(y)$ for a continuous random variable y. You can see from the figure that there is a one-to-one correspondence between y values and g values, and that values of y corresponding to values of g in the interval $0 \le g \le g_0$ will be those in the interval $0 \le y \le y_0$. Therefore,

$$P(g \le g_0) = P(y \le y_0) = F(y_0)$$

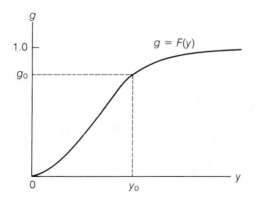

FIGURE 7.5
Cumulative Distribution
Function $F(y)$

But since $g = F(y)$, we have $F(y_0) = g_0$. Therefore, we can write

$$P(g \leq g_0) = F(y_0) = g_0$$

The cumulative distribution function for g is obtained by dropping the subscript:

$$F(g) = g$$

Finally, we differentiate to obtain the density function:

$$f(g) = \frac{dF(g)}{dg} = 1$$

EXAMPLE 7.3

Use Theorem 7.1 to generate a random sample of $n = 3$ observations from an exponential distribution with $\beta = 2$.

SOLUTION

The density function for the exponential distribution with $\beta = 2$ is

$$f(y) = \begin{cases} \dfrac{e^{-y/2}}{2} & \text{if} \quad 0 \leq y < \infty \\ 0 & \text{elsewhere} \end{cases}$$

and the cumulative distribution function is

$$F(y) = \int_{-\infty}^{y} f(t)\, dt = \int_{0}^{y} \frac{e^{-t/2}}{2}\, dt = -e^{-t/2} \Big]_{0}^{y} = 1 - e^{-y/2}$$

If we let $g = F(y) = 1 - e^{-y/2}$, then Theorem 7.1 tells us that g has a uniform density function over the interval $0 \leq g \leq 1$.

To draw a random number y from the exponential distribution, we first randomly draw a value of g from the uniform distribution. This can be done by drawing a random number from Table 1 of Appendix II. Suppose, for example, that we draw the random number 10480. This corresponds to the random selection of the value $g_1 = .10480$ from a uniform distribution over the interval

$0 \le g \le 1$. Substituting this value of g_1 into the formula for $g = F(y)$ and solving for y, we obtain

$$g_1 = F(y) = 1 - e^{-y_1/2}$$
$$.10480 = 1 - e^{-y_1/2}$$
$$e^{-y_1/2} = .8952$$
$$\frac{-y_1}{2} = -.111$$
$$y_1 = .222$$

If the next two random numbers selected are 22368 and 24130, then the corresponding values of the uniform random variable are $g_2 = .22368$ and $g_3 = .24130$. By substituting these values into the formula $g = 1 - e^{-y/2}$, you can verify that $y_2 = .506$ and $y_3 = .552$. Thus, $y_1 = .222$, $y_2 = .506$, and $y_3 = .552$ represent three randomly selected observations on an exponential random variable with mean equal to 2.   ■

EXERCISES 7.1–7.7

7.1 Consider the density function

$$f(y) = \begin{cases} 2y & \text{if } 0 \le y \le 1 \\ 0 & \text{elsewhere} \end{cases}$$

Find the density function of $g(y)$, where:
a. $g(y) = y^2$ **b.** $g(y) = 2y - 1$ **c.** $g(y) = 1/y$

7.2 Consider the density function

$$f(y) = \begin{cases} e^{-(y-3)} & \text{if } y > 3 \\ 0 & \text{elsewhere} \end{cases}$$

Find the density function of $g(y)$, where:
a. $g(y) = e^{-y}$ **b.** $g(y) = y - 3$ **c.** $g(y) = y/3$

7.3 The amount y of paper used per day by a line printer at a university computing center has an exponential distribution with mean equal to 5 boxes (i.e., $\beta = 5$). The daily cost of the paper is proportional to $g(y) = (3y + 2)$. Find the probability density function of the daily cost of paper used by the line printer.

7.4 An environmental engineer has determined that the amount y (in parts per million) of pollutant per water sample collected near the discharge tubes of an island power plant has probability density function

$$f(y) = \begin{cases} \dfrac{1}{10} & \text{if } 0 < y < 10 \\ 0 & \text{elsewhere} \end{cases}$$

A new cleaning device has been developed to help reduce the amount of pollution discharged into the ocean. It is believed that the amount $g(y)$ of pollutant discharged when the device is operating will be related to y by

$$g(y) = \begin{cases} \dfrac{y}{2} & \text{if } 0 < y < 5 \\[2mm] \dfrac{2y - 5}{2} & \text{if } 5 < y < 10 \end{cases}$$

Find the probability density function of $g(y)$.

7.5 Use Theorem 7.1 to draw a random sample of $n = 5$ observations from a beta distribution with $\alpha = 2$ and $\beta = 1$.

7.6 Use Theorem 7.1 to draw a random sample of $n = 5$ observations from a distribution with probability density function

$$f(y) = \begin{cases} e^y & \text{if } y < 0 \\ 0 & \text{elsewhere} \end{cases}$$

OPTIONAL EXERCISES

7.7 The total time y_1 (in minutes) from the time a computer job is submitted until its run is completed and the time y_2 the job waits in the job queue before being run have the joint density function

$$f(y_1, y_2) = \begin{cases} e^{-y_1} & \text{if } 0 \le y_2 \le y_1 < \infty \\ 0 & \text{elsewhere} \end{cases}$$

The CPU time for the job (i.e., the length of time the job is in control of the computer's central processing unit) is given by the difference $g = y_1 - y_2$. Find the density function of a job's CPU time. [*Hint:*   You may use the facts that

$$P(g \le g_0) = P(g \le g_0, y_1 > g_0) + P(g \le g_0, y_1 \le g_0)$$
$$= P(y_1 - g_0 \le y_2 \le y_1, g_0 < y_1 < \infty) + P(0 \le y_2 \le y_1, 0 \le y_1 \le g_0)$$
$$= \int_{g_0}^{\infty} \int_{y_1 - g_0}^{y_1} e^{-y_1} \, dy_2 dy_1 + \int_0^{g_0} \int_0^{y_1} e^{-y_1} \, dy_2 dy_1$$

and $\int ye^{-y} \, dy = -ye^{-y} + \int e^{-y} \, dy$ in determining the density function.]

S E C T I O N   7.3

**FINDING A SAMPLING
DISTRIBUTION BY
SIMULATION**

We have explained in Section 7.1 that a statistic g is a function of the n sample measurements, $y_1, y_2, \ldots, y_n$, and we have shown in Section 7.2 how we can use probability theory and mathematics to find its sampling distribution. However, the mathematical problem of finding $F(g)$ is often very difficult to solve. When such a situation occurs, we may be able to find an approximation to $f(g)$ by computer simulation.

To illustrate the procedure, we will approximate the sampling distribution for the sum $g = y_1 + y_2$ of a sample of $n = 2$ observations from a uniform distribution over the interval $0 \le y \le 1$. Recall that we found an exact expression for this sampling distribution in Optional Example 7.2. Thus, we will be able to compare our simulated sampling distribution with the exact form of the sampling distribution shown in Figure 7.4b.

To begin the simulation procedure, we used the computer to generate 10,000 pairs of random numbers, with each pair representing a sample (y_1, y_2) from the uniform distribution over the interval $0 \leq y \leq 1$. We then programmed the computer to calculate the sum $g = y_1 + y_2$ for each of the 10,000 pairs. A computer-generated relative frequency histogram for the 10,000 values of g is shown in Figure 7.6. By comparing Figures 7.4b and 7.6, you can see that the simulated sampling distribution provides a good approximation to the true probability distribution of the sum of a sample of $n = 2$ observations from a uniform distribution.

FIGURE 7.6 Simulated Sampling Distribution for the Sum of Two Observations from a Uniform Distribution

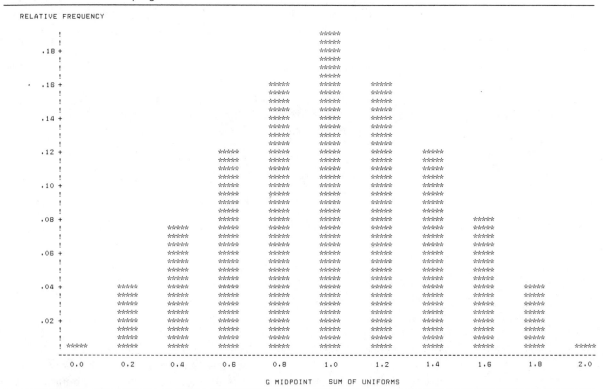

EXAMPLE 7.4   Simulate the sampling distribution of the sample mean

$$\bar{y} = \frac{y_1 + y_2 + y_3 + y_4 + y_5}{5}$$

for a sample of $n = 5$ observations drawn from each of the three probability distributions shown in Figure 7.7:

a. Uniform $(0 \le y \le 1)$
b. Normal $(\mu = 0, \sigma = 1)$
c. Exponential $(\beta = 1)$

Repeat the procedure for $n = 10, 15, 25,$ and 100.

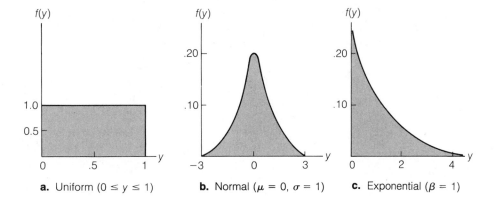

FIGURE 7.7
Three Probability
Distributions of
Example 7.4

a. Uniform $(0 \le y \le 1)$ **b.** Normal $(\mu = 0, \sigma = 1)$ **c.** Exponential $(\beta = 1)$

SOLUTION

We first obtained 1,000 computer-generated random samples of size $n = 5$ from each of the three probability distributions and programmed the computer to compute the mean

$$\bar{y} = \frac{y_1 + y_2 + y_3 + y_4 + y_5}{5}$$

for each sample. Relative frequency histograms for the 1,000 values of $\bar{y}$ obtained from each of the probability distributions are shown in Figure 7.8a (page 252). Note their shapes for this small value of n.

The relative frequency histograms of $\bar{y}$ based on samples of size $n = 10, 15, 25,$ and 100, also simulated by computer, are shown in Figures 7.8b, 7.8c, 7.8d, and 7.8e, respectively. Note that the values of $\bar{y}$ tend to cluster about the mean of the probability distribution from which the sample was taken ($\mu = \frac{1}{2}$ for the uniform, $\mu = 0$ for the normal, and $\mu = 1$ for the exponential). Furthermore, as n increases, there is less variation in the sampling distribution. You can also see from the figures that as the sample size increases, the shape of the sampling distribution of $\bar{y}$ tends toward the shape of the normal distribution (symmetric and mound-shaped), regardless of the shape of the probability distribution from which the sample was selected. ∎

In Section 7.4 we generalize the results of Example 7.4 in the form of a theorem.

FIGURE 7.8a   Relative Frequency Distribution of a Random Sample of 1,000 Sample Means (*n* = 5)

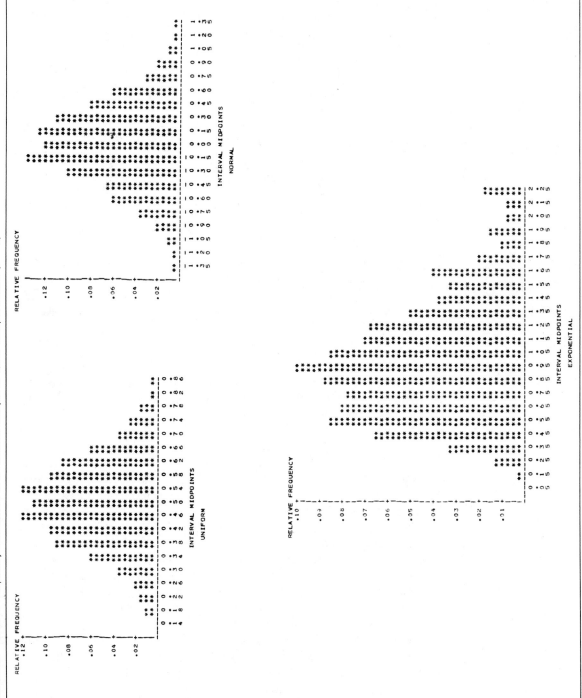

FIGURE 7.8b Relative Frequency Distribution of a Random Sample of 1,000 Sample Means ($n = 10$)

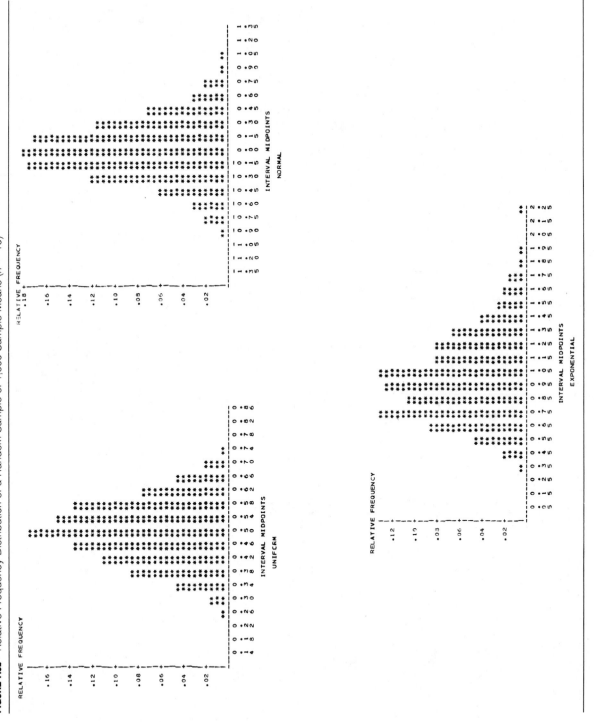

FIGURE 7.8c   Relative Frequency Distribution of a Random Sample of 1,000 Sample Means ($n = 15$)

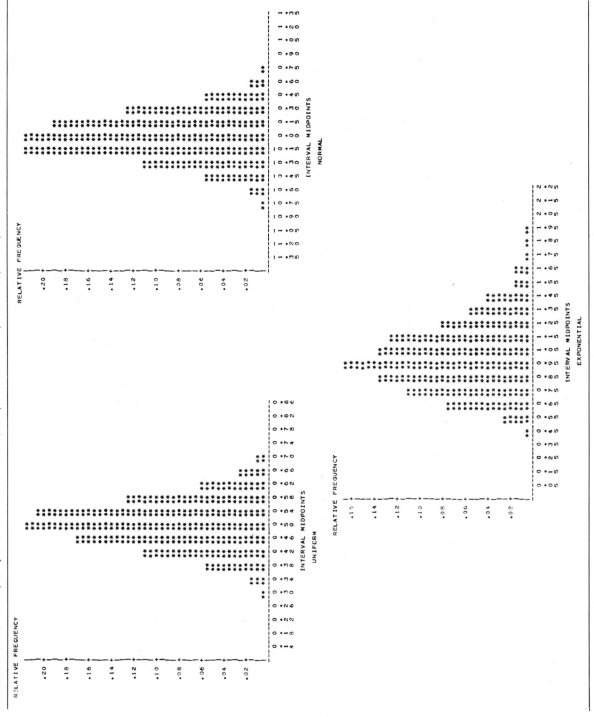

FIGURE 7.8d Relative Frequency Distribution of a Random Sample of 1,000 Sample Means ($n = 25$)

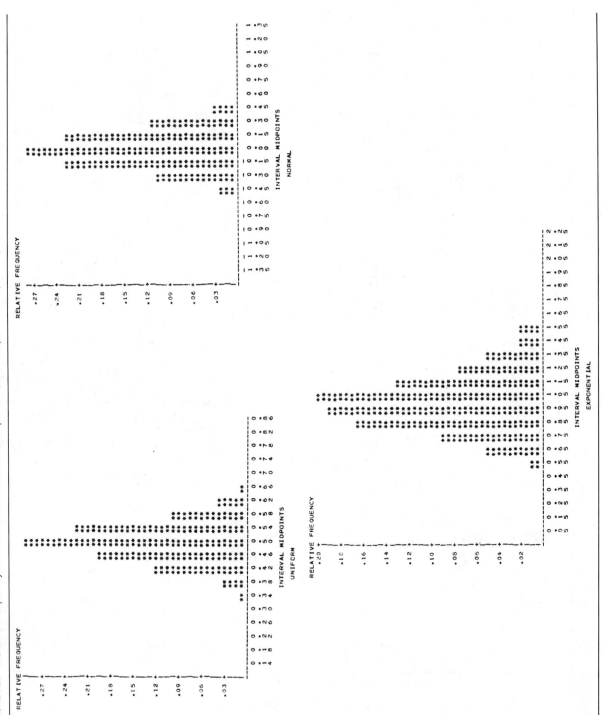

FIGURE 7.8e   Relative Frequency Distribution of a Random Sample of 1,000 Sample Means ($n = 100$)

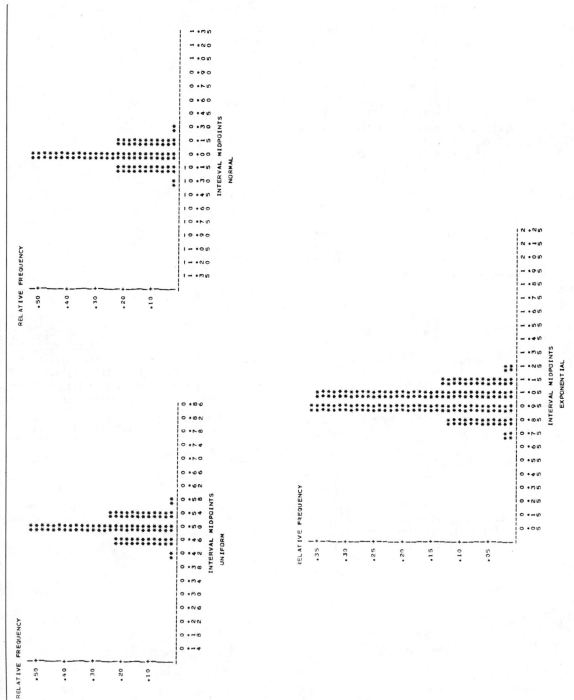

The simulation of the sampling distribution of the sample mean based on independent random samples from uniform, normal, and exponential distributions in Example 7.4 illustrated the ideas embodied in one of the most important theorems in statistics. The following version of the theorem applies to the sampling distribution of the sample mean.

THEOREM 7.2

THE CENTRAL LIMIT THEOREM If random samples of n observations, y_1, $y_2, \ldots, y_n$, are drawn from a population with finite mean μ and variance σ^2, then, when n is sufficiently large, the sampling distribution of the sample mean can be approximated by a normal density function with mean $\mu_{\bar{y}} = \mu$ and standard deviation $\sigma_{\bar{y}} = \sigma/\sqrt{n}$.

The significance of the central limit theorem is that we can use the normal distribution to approximate the sampling distribution of the sample mean $\bar{y}$ as long as the population possesses a finite mean and variance and the number n of measurements in the sample is sufficiently large. How large the sample size must be will depend on the nature of the sampled population. You can see from our simulated experiment in Example 7.4 that the sampling distribution of $\bar{y}$ tends to become very nearly normal for sample sizes as small as $n = 25$ for the uniform, normal, and exponential population distributions. When the population distribution is symmetric about its mean, the sampling distribution of $\bar{y}$ will be mound-shaped and nearly normal for sample sizes as small as $n = 10$. In addition, if the sampled population possesses a normal distribution, then the sampling distribution of $\bar{y}$ will be a normal density function, regardless of the sample size. (This may be seen in Figure 7.8.) In fact, it can be shown that the sampling distribution of *any linear function* of normally distributed random variables, even those that are correlated and have different means and variances, is a normal distribution. This important result is presented (without proof) in Theorem 7.3 (page 258) and illustrated in an example.

EXAMPLE 7.5

Suppose you select independent random samples from two normal populations, n_1 observations from population 1 and n_2 observations from population 2. If the means and variances for populations 1 and 2 are (μ_1, σ_1^2) and (μ_2, σ_2^2), respectively, and if $\bar{y}_1$ and $\bar{y}_2$ are the corresponding sample means, find the distribution of the difference $(\bar{y}_1 - \bar{y}_2)$.

SOLUTION

Since $\bar{y}_1$ and $\bar{y}_2$ are both linear functions of normally distributed random variables, they will be normally distributed by Theorem 7.3. The means and variances of the sample means (see Example 6.12) are

$$E(\bar{y}_i) = \mu_i \quad \text{and} \quad V(\bar{y}_i) = \frac{\sigma_i^2}{n_i} \quad (i = 1, 2)$$

THEOREM 7.3

Let $a_1, a_2, \ldots, a_n$ be constants and let $y_1, y_2, \ldots, y_n$ be n normally distributed random variables with $E(y_i) = \mu_i$, $V(y_i) = \sigma_i^2$, and $\text{Cov}(y_i, y_j) = \sigma_{ij}$ $(i = 1, 2, \ldots, n)$. Then the sampling distribution of

$$\ell = a_1y_1 + a_2y_2 + \cdots + a_ny_n$$

possesses a normal density function with mean and variance*

$$E(\ell) = \mu = a_1\mu_1 + a_2\mu_2 + \cdots + a_n\mu_n$$

and

$$\begin{aligned} V(\ell) = {}& a_1^2\sigma_1^2 + a_2^2\sigma_2^2 + \cdots + a_n^2\sigma_n^2 \\ & + 2a_1a_2\sigma_{12} + 2a_1a_3\sigma_{13} + \cdots + 2a_1a_n\sigma_{1n} \\ & + 2a_2a_3\sigma_{23} + \cdots + 2a_2a_n\sigma_{2n} \\ & + \cdots + 2a_{n-1}a_n\sigma_{n-1,n} \end{aligned}$$

Then, $\ell = \bar{y}_1 - \bar{y}_2$ is a linear function of two normally distributed random variables, $\bar{y}_1$ and $\bar{y}_2$. According to Theorem 7.3, ℓ will be normally distributed with

$$E(\ell) = \mu_\ell = E(\bar{y}_1) - E(\bar{y}_2) = \mu_1 - \mu_2$$
$$V(\ell) = \sigma_\ell^2 = (1)^2V(\bar{y}_1) + (-1)^2V(\bar{y}_2) + 2(1)(-1)\text{Cov}(\bar{y}_1, \bar{y}_2)$$

But, since the samples were independently selected, $\bar{y}_1$ and $\bar{y}_2$ are independent and $\text{Cov}(\bar{y}_1, \bar{y}_2) = 0$. Therefore,

$$V(\ell) = \frac{\sigma_1^2}{n_1} + \frac{\sigma_2^2}{n_2}$$

We have shown that $(\bar{y}_1 - \bar{y}_2)$ is a normally distributed random variable with mean $(\mu_1 - \mu_2)$ and variance $(\sigma_1^2/n_1 + \sigma_2^2/n_2)$. ∎

Typical applications of the central limit theorem, however, involve samples selected from nonnormal or unknown populations, as illustrated in Examples 7.6 and 7.7.

EXAMPLE 7.6

A manufacturer of aluminum foil claims that its 75-foot roll has a mean length of 75.05 feet per roll and a standard deviation of .12 foot. To check this claim, a consumer group plans to randomly sample 36 of the company's 75-foot rolls, measure the length of each, and compute the sample mean length.

a. Assuming the manufacturer's claim is true, describe the sampling distribution of the sample mean.

*The formulas for the mean and variance of a linear function of any random variables, $y_1, y_2, \ldots, y_n$, were given in Theorem 6.7.

b. Assuming the manufacturer's claim is true, what is the probability that the sample mean will be less than 75 feet?

c. Suppose the sample mean actually equals 74.97 feet. Can this evidence be used to refute the manufacturer's claim? Explain.

SOLUTION

a. According to the manufacturer, the distribution of the lengths of its 75-foot roll of aluminum foil has mean $\mu = 75.05$ feet and standard deviation $\sigma = .12$ foot. By Theorem 7.2 (the central limit theorem), the sampling distribution of the sample mean $\bar{y}$ is approximately normal with mean

$$\mu_{\bar{y}} = \mu = 75.05$$

and standard deviation

$$\sigma_{\bar{y}} = \frac{\sigma}{\sqrt{n}} = \frac{.12}{\sqrt{36}} = .02$$

b. We want to calculate $P(\bar{y} < 75)$. Since $\bar{y}$ has an approximate normal distribution, we have

$$P(\bar{y} < 75) = P\left(\frac{\bar{y} - \mu_{\bar{y}}}{\sigma_{\bar{y}}} < \frac{75 - \mu_{\bar{y}}}{\sigma_{\bar{y}}}\right)$$

$$\approx P\left(z < \frac{75 - 75.05}{.02}\right) = P(z < -2.5)$$

where z is a standard normal random variable. Using Table 4 of Appendix II, we obtain

$$P(z < -2.5) = 1 - .4938 = .0062$$

Therefore, $P(\bar{y} < 75) \approx .0062$.

c. Based on the probability from part b, a sample mean less than 75 feet is very unlikely to occur if the values of μ and σ given by the manufacturer are correct. Since we observed $\bar{y} = 74.97$, there is evidence to refute the manufacturer's claim. That is, either the true mean length μ is less than 75.05 feet or the standard deviation σ is greater than .12 foot. ∎

EXAMPLE 7.7

Consider a binomial experiment with n trials and probability of success p on each trial. The number y of successes divided by the number n of trials is called the **sample proportion of successes** and is denoted by the symbol $\hat{p} = y/n$. Explain why the random variable

$$z = \frac{\hat{p} - p}{\sqrt{\dfrac{pq}{n}}}$$

has approximately a standard normal distribution for large values of n.

SOLUTION

If we denote the outcome of the ith trial as y_i ($i = 1, 2, \ldots, n$), where

$$y_i = \begin{cases} 1 & \text{if outcome is a success} \\ 0 & \text{if outcome is a failure} \end{cases}$$

then the number y of successes in n trials is equal to the sum

$$\sum_{i=1}^{n} y_i$$

Therefore, $\hat{p} = y/n$ is a sample mean and, according to Theorem 7.2, $\hat{p}$ will be approximately normally distributed when the sample size n is large. To find the expected value and variance of $\hat{p}$, we can view $\hat{p}$ as a linear function of a single random variable y:

$$\hat{p} = \ell = a_1 y_1 = \left(\frac{1}{n}\right) y \qquad \text{where } a_1 = \frac{1}{n} \text{ and } y_1 = y$$

We now apply Theorem 6.7 to obtain $E(\ell)$ and $V(\ell)$:

$$E(\hat{p}) = \frac{1}{n} E(y) = \frac{1}{n}(np) = p$$

$$V(\hat{p}) = \left(\frac{1}{n}\right)^2 V(y) = \frac{1}{n^2}(npq) = \frac{pq}{n}$$

Therefore,

$$z = \frac{\hat{p} - p}{\sqrt{\dfrac{pq}{n}}}$$

is equal to the deviation between a normally distributed random variable $\hat{p}$ and its mean p, expressed in units of its standard deviation, $\sqrt{pq/n}$. This satisfies the definition of a standard normal random variable given in Section 5.5. ∎

The central limit theorem also applies to the sum of a sample of n measurements subject to the conditions stated in Theorem 7.2. The only difference is that the approximating normal distribution will have mean $n\mu$ and variance $n\sigma^2$.

MODIFICATION OF THEOREM 7.2: THE CENTRAL LIMIT THEOREM FOR SUMS

If random samples of n observations, $y_1, y_2, \ldots, y_n$, are drawn from a population with finite mean μ and variance σ^2, then, when n is sufficiently large, the sampling distribution of the sum

$$\sum_{i=1}^{n} y_i$$

can be approximated by a normal density function with mean $\mu_{\Sigma y_i} = n\mu$ and $\sigma^2_{\Sigma y_i} = n\sigma^2$.

In Section 7.5 we apply the central limit theorem for sums to show that the normal density function can be used to approximate the binomial probability distribution when the number n of trials is large.

7.8 Let $\bar{y}_{25}$ represent the mean of a random sample of size $n = 25$ from a probability distribution with unknown density $f(y)$, mean $\mu = 17$, and standard deviation $\sigma = 10$. Similarly, let $\bar{y}_{100}$ represent the mean of a random sample of size $n = 100$ selected from the same probability distribution.
a. Describe the sampling distributions of $\bar{y}_{25}$ and $\bar{y}_{100}$.
b. Which of the probabilities, $P(15 < \bar{y}_{25} < 19)$ or $P(15 < \bar{y}_{100} < 19)$, would you expect to be larger?
c. Calculate approximations to the two probabilities of part **b**.

7.9 General trace organic monitoring describes the process in which water engineers analyze water samples for various types of organic material (e.g., contaminants). One such contaminant, commonly found in treated surface water, is the pesticide trihalomethane (THM). General trace organic monitoring at the Bedford (England) water treatment works revealed a mean THM level of 51 μg/l and a standard deviation of 14 μg/l (*Journal of the Institution of Water Engineers and Scientists*, Feb. 1986). Assume that these figures represent the population mean μ and standard deviation σ, respectively. Suppose we collect 45 water samples (called water "profiles") at the Bedford plant and measure the THM level in each.
a. Describe the sampling distribution of $\bar{y}$, the mean THM level of the 45 water profiles.
b. Find the probability that $\bar{y}$ exceeds 52 μg/l.
c. Find the probability that $\bar{y}$ falls between 49.5 and 50.5 μg/l.

7.10 Many species of terrestrial frogs that hibernate at or near the ground surface can survive prolonged exposure to low winter temperatures. In freezing conditions, the frog's body temperature, called its *supercooling temperature*, remains relatively higher due to an accumulation of glycerol in its body fluids. Studies have shown that the supercooling temperature of terrestrial frogs frozen at $-6°C$ has a relative frequency distribution with a mean of $-2.18°C$ and a standard deviation of .32°C (*Science*, May 1983). Consider the mean supercooling temperature, $\bar{y}$, of a random sample of $n = 42$ terrestrial frogs frozen at $-6°C$.
a. Find the probability that $\bar{y}$ exceeds $-2.05°C$.
b. Find the probability that $\bar{y}$ falls between $-2.20°C$ and $-2.10°C$.

7.11 This year a large architectural and engineering consulting firm began a program of compensating its management personnel for sick days not used. The firm decided to pay each manager a bonus for every unused sick day. In past years, the number y of sick days used per manager per year had a probability distribution with mean $\mu = 9.2$ and variance $\sigma^2 = 3.24$. To determine whether the compensation program has effectively reduced the mean number of sick days used, the firm randomly sampled $n = 80$ managers and recorded y, the number of sick days used by each at year's end.
a. Assuming the compensation program was *not* effective in reducing the average number of sick days used, find the probability that $\bar{y}$, the mean number of sick days used by the sample of 80 managers, is less than 8.80 days, i.e., find $P(\bar{y} < 8.80)$.

b. If you observe $\bar{y} < 8.80$, what inference would you make about the effectiveness of the compensation program?

7.12 Dioxin, often described as the most toxic chemical known, is created as a by-product in the manufacture of herbicides such as Agent Orange. Scientists have found that .000005 gram (five-millionths of a gram) of dioxin—a dot barely visible to the human eye—is a lethal dose for experimental guinea pigs in more than half the animals tested, making dioxin 2,000 times more toxic than strychnine. Assume that the amount of dioxin required to kill a guinea pig has a relative frequency distribution with mean $\mu = .000005$ gram and standard deviation $\sigma = .000002$ gram. Consider an experiment in which the amount of dioxin required to kill each of $n = 50$ guinea pigs is measured, and the sample mean $\bar{y}$ is computed.
a. Calculate $\mu_{\bar{y}}$ and $\sigma_{\bar{y}}$.
b. Find the probability that the mean amount of dioxin required to kill the 50 guinea pigs is larger than .0000053 gram.

7.13 Engineers responsible for the design and maintenance of aircraft pavements traditionally use pavement-quality concrete or Marshall asphalt surfaces. A study was conducted at Luton Airport (United Kingdom) to assess the suitability of concrete blocks as a surface for aircraft pavements (*Proceedings of the Institute of Civil Engineers, Part I*, Apr. 1986). The original pavement-quality concrete of the western end of the runway was overlaid with 80-mm thick concrete blocks and a series of plate bearing tests was carried out to determine the load classification number (LCN)—a measure of breaking strength—of the surface. Let $\bar{y}$ represent the mean LCN of a sample of 25 concrete block sections on the western end of the runway.
a. Prior to resurfacing, the mean LCN of the original pavement-quality concrete of the western end of the runway was known to be $\mu = 60$. Assume the standard deviation was $\sigma = 10$. If the mean strength of the new concrete block surface is no different from that of the original surface, find the probability that $\bar{y}$, the sample mean LCN of the 25 concrete block sections, exceeds 65.
b. The plate bearing tests on the new concrete block surface resulted in $\bar{y} = 73$. Based on this result, what can you infer about the true mean LCN of the new surface?

7.14 A large freight elevator can transport a maximum of 10,000 pounds (5 tons). Suppose a load of cargo containing 45 boxes must be transported via the elevator. Experience has shown that the weight y of a box of this type of cargo follows a probability distribution with mean $\mu = 200$ pounds and standard deviation $\sigma = 55$ pounds. What is the probability that all 45 boxes can be loaded onto the freight elevator and transported simultaneously? [*Hint:* Find $P(\Sigma_{i=1}^{45} y_i \leq 10,000).$]

OPTIONAL EXERCISES

7.15 If y has a χ^2 distribution with n degrees of freedom (see Definition 5.6), then y could be represented by $y = \Sigma_{i=1}^{n} x_i$, where the x_i's are independent χ^2 distributions, each with 1 degree of freedom.
a. Show that $z = (y - n)/\sqrt{2n}$ has approximately a standard normal distribution for large values of n.
b. If y has a χ^2 distribution with 30 degrees of freedom, find the approximate probability that y falls within 2 standard deviations of its mean, i.e., find $P(\mu - 2\sigma < y < \mu + 2\sigma)$.

7.16 Let $\hat{p}_1$ be the sample proportion of successes in a binomial experiment with n_1 trials and let $\hat{p}_2$ be the sample proportion of successes in a binomial experiment with n_2 trials, conducted independently of the first. Let p_1 and p_2 be the corresponding population parameters. Show that

$$z = \frac{\hat{p}_1 - \hat{p}_2 - (p_1 - p_2)}{\sqrt{\dfrac{p_1 q_1}{n_1} + \dfrac{p_2 q_2}{n_2}}}$$

has approximately a standard normal distribution for large values of n_1 and n_2.

SECTION 7.5

NORMAL APPROXIMATION TO THE BINOMIAL DISTRIBUTION

Consider the binomial random variable y with parameters n and p. Recall that y has mean $\mu = np$ and variance $\sigma^2 = npq$. We showed in Example 7.7 that the number y of successes in n trials can be regarded as a sum consisting of n values of 0 and 1, with each 0 and 1 representing the outcome (failure or success, respectively) of a particular trial, i.e.,

$$y = \sum_{i=1}^{n} y_i \qquad \text{where } y_i = \begin{cases} 1 & \text{if success*} \\ 0 & \text{if failure} \end{cases}$$

Then, according to the central limit theorem for sums (modification of Theorem 7.2), the binomial probability distribution $p(y)$ should become more nearly normal as n becomes larger.

The normal approximation to a binomial probability distribution is reasonably good even for small samples—say, n as small as 10—when $p = .5$ and the distribution of y is therefore symmetric about its mean $\mu = np$. When p is near 0 (or 1), the binomial probability distribution will tend to be skewed to the right (or left), but this skewness will disappear as n becomes large. In general, the approximation will be good when n is large enough so that $\mu - 2\sigma = np - 2\sqrt{npq}$ and $\mu + 2\sigma = np + 2\sqrt{npq}$ both lie between 0 and n.

CONDITION REQUIRED TO APPLY A NORMAL APPROXIMATION TO A BINOMIAL PROBABILITY DISTRIBUTION

The approximation will be good if both $\mu - 2\sigma = np - 2\sqrt{npq}$ and $\mu + 2\sigma = np + 2\sqrt{npq}$ lie between 0 and n.

EXAMPLE 7.8

Let y be a binomial probability distribution with $n = 10$ and $p = .5$.

a. Graph $p(y)$ and superimpose on the graph a normal distribution with $\mu = np$ and $\sigma = \sqrt{npq}$.

b. Use Table 2 of Appendix II to find $P(y \le 4)$.

c. Use the normal approximation to the binomial probability distribution to find an approximation to $P(y \le 4)$.

---

*The random variable y_i is known as a **Bernoulli random variable** with probability distribution $p(y_i) = p y q^{1-y}$, $y = 0, 1$. The binomial probability distribution can be derived as the sum of n independent Bernoulli random variables.

SOLUTION

a. The graphs of $p(y)$ and a normal distribution with

$$\mu = np = (10)(.5) = 5$$

and

$$\sigma = \sqrt{npq} = \sqrt{(10)(.5)(.5)} = 1.58$$

are shown in Figure 7.9. Note that both $\mu - 2\sigma = 1.84$ and $\mu + 2\sigma = 8.16$ lie between 0 and $n = 10$. The normal density function with $\mu = 5$ and $\sigma = 1.58$ thus provides a good approximation to $p(y)$.

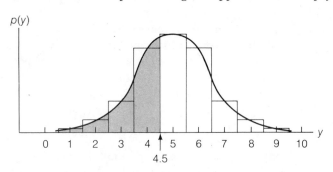

FIGURE 7.9

A Binomial Probability Distribution ($n = 10$, $p = .5$) and the Approximating Normal Distribution ($\mu = np = 5$ and $\sigma = \sqrt{npq} = 1.58$)

b. From Table 2 of Appendix II, we obtain

$$\sum_{y=0}^{4} p(y) = .377$$

c. By examining Figure 7.9, you can see that $P(y \leq 4)$ is the area under the normal curve to the left of $y = 4.5$. Note that the area to the left of $y = 4$ would *not* be appropriate because it would omit half the probability rectangle corresponding to $y = 4$. We need to add .5 to 4 before calculating the probability in order to correct for the fact that we are using a continuous probability distribution to approximate a discrete probability distribution. The value .5 is called the **continuity correction factor** for the normal approximation to the binomial probability (see box). The z value corresponding to the corrected value $y = 4.5$ is

$$z = \frac{y - \mu}{\sigma} = \frac{4.5 - 5}{1.58} = \frac{-.5}{1.58} = -.32$$

The area between $z = 0$ and $z = .32$, given in Table 4 of Appendix II, is $A = .1255$. Therefore,

$$P(y \leq 4) \approx .5 - A = .5 - .1255 = .3745$$

Thus, the normal approximation to $P(y \leq 4) = .377$ is quite good, although n is as small as 10. The sample size would have to be larger in order to apply the approximation if p were not equal to .5. ∎

CONTINUITY CORRECTION FOR THE NORMAL APPROXIMATION TO A BINOMIAL PROBABILITY

Let y be a binomial random variable with parameters n and p, and let z be a standard random variable. Then

$$P(y \leq a) \approx P\left(z < \frac{(a + .5) - np}{\sqrt{npq}}\right)$$

$$P(y \geq a) \approx P\left(z > \frac{(a - .5) - np}{\sqrt{npq}}\right)$$

$$P(a \leq y \leq b) \approx P\left(\frac{(a - .5) - np}{\sqrt{npq}} < z < \frac{(b + .5) - np}{\sqrt{npq}}\right)$$

EXERCISES 7.17–7.21

7.17 Let y be a binomial random variable with $n = 15$ and $p = .3$.
  a. Use Table 2 of Appendix II to find $P(y \leq 8)$.
  b. Use the normal approximation to the binomial probability distribution to find an approximation to $P(y \leq 8)$. Compare to your answer in part **a**.

7.18 According to the U.S. Department of Labor, women are vastly underrepresented in high-technology jobs related to computers. In the category of computer systems analyst—the fastest growing and highest paid computer-related job—women constitute only 30% of the work force (*Wall Street Journal*, Sept. 16, 1985).
  a. Use the normal approximation to the binomial to calculate the probability that more than half of a random sample of 20 computer systems analysts are women.
  b. Use the normal approximation to the binomial to calculate the probability that five or less of a random sample of 20 computer systems analysts are women.
  c. Use the binomial tables to calculate the exact probabilities in parts **a** and **b**. Does the normal distribution provide a good approximation to the binomial distribution?

7.19 The merging process from an acceleration lane to the through lane of a freeway constitutes an important aspect of traffic operation at interchanges. A study of parallel interchange ramps in Israel revealed that many drivers do not use the entire length of parallel lanes for acceleration, but seek as soon as possible an appropriate gap in the major stream of traffic for merging (*Transportation Engineering*, Nov. 1985). At one site (Yavneh), 54% of the drivers use less than half the lane length available before merging. Suppose we plan to monitor the merging patterns of a random sample of 330 drivers at the Yavneh site.
  a. What is the approximate probability that fewer than 100 of the drivers will use less than half the acceleration lane length before merging?
  b. What is the approximate probability that 200 or more of the drivers will use less than half the acceleration lane length before merging?

7.20 A problem with items that are mass produced is quality control. The production process must be monitored to ensure that the rate of defective items is kept at an acceptably low level. One method of dealing with this problem is **lot acceptance sampling**, in which a random sample of items produced is selected and each item in the sample is carefully tested. The entire lot of items is then accepted or rejected, based on the number of defectives observed in the sample. Suppose a manufacturer of pocket calculators randomly chooses 200 stamped circuits from a day's production and determines y, the number of defective circuits in the sample. If a sample defective rate of 6% or less is considered acceptable and, unknown to the manufacturer, 8% of the entire day's production of circuits is defective, find the approximate probability that the lot of stamped circuits will be rejected.

7.21 How well does a college engineering degree prepare you for the workplace? A 2-year nationwide survey of engineers and engineering managers in "specific high-demand" industries revealed that only 34% believe that their companies make good use of their learned skills (*Chemical Engineering*, Feb. 3, 1986). In a random sample of 50 engineers and engineering managers, consider the number y who believe that their employer makes good use of their college engineering background. Find the approximate probability that:

a. $y \leq 10$ **b.** $y \geq 25$ **c.** $20 \leq y \leq 30$

SECTION 7.6

SAMPLING DISTRIBUTIONS RELATED TO THE NORMAL DISTRIBUTION

The sampling distributions of statistics based on random samples of observations from a normal population often possess interesting mathematical properties. We will present several well-known results without proof. Proofs using the methodology of Chapter 6 can be found in the references at the end of this chapter.

THEOREM 7.4

If a random sample of n observations, $y_1, y_2, \ldots, y_n$, is selected from a normal distribution with mean μ and variance σ^2, then the sampling distribution of

$$\chi^2 = \frac{(n - 1)s^2}{\sigma^2}$$

has a chi-square density function (see Definition 5.6) with $\nu = (n - 1)$ degrees of freedom.

DEFINITION 7.3

Let χ_1^2 and χ_2^2 be chi-square random variables with ν_1 and ν_2 degrees of freedom, respectively. If χ_1^2 and χ_2^2 are independent, then

$$F = \frac{\chi_1^2/\nu_1}{\chi_2^2/\nu_2}$$

is said to have an **F distribution** with ν_1 numerator degrees of freedom and ν_2 denominator degrees of freedom.

DEFINITION 7.4

Let z be a standard normal random variable and χ^2 be a chi-square random variable with ν degrees of freedom. If z and χ^2 are independent, then

$$t = \frac{z}{\sqrt{\chi^2/\nu}}$$

is said to possess a **Student's t distribution** (or, simply t **distribution**) with ν degrees of freedom.

THEOREM 7.5

If χ_1^2 and χ_2^2 are independent chi-square random variables with ν_1 and ν_2 degrees of freedom, respectively, then the sum $(\chi_1^2 + \chi_2^2)$ has a chi-square distribution with $(\nu_1 + \nu_2)$ degrees of freedom.

The sampling distributions for the F and t statistics can be derived using the methods of Section 7.2. Both sampling distributions are related to the density function for a beta-type random variable (see Section 5.8) and it can be shown (proof omitted) that a t distribution is actually a special case of an F distribution with $\nu_1 = 1$ and ν_2 degrees of freedom. Neither of the cumulative distribution functions can be obtained in closed form. Consequently, we dispense with the equations of the density functions and present useful values of the statistics and corresponding areas in tabulated form in Appendix II. We will introduce these tables in Chapter 8.

Of what value are the χ^2, F, and t statistics? In Chapter 8 we will develop methods for estimating the values of population parameters. One way of making these estimates is to acquire a statistic that involves only the parameter that you want to estimate. When the parameter of interest is the variance of a normal population, the chi-square statistic satisfies this property because it is a function of the single parameter σ^2. Another such statistic is the subject of Example 7.9.

EXAMPLE 7.9

Suppose that $\bar{y}$ and s^2 are the mean and variance of a random sample of n observations from a normally distributed population with mean μ and variance σ^2. It can be shown (proof omitted) that $\bar{y}$ and s^2 are statistically independent when the sampled population has a normal distribution. Use this result to show that

$$t = \frac{\bar{y} - \mu}{s/\sqrt{n}}$$

possesses a t distribution with $\nu = (n - 1)$ degrees of freedom.*

---

*The result was first published in 1908 by W. S. Gosset, who wrote under the pen name of Student. Thereafter, this statistic became known as Student's t.

SOLUTION   We know from Theorem 7.3 that $\bar{y}$ is normally distributed with mean μ and variance σ^2/n. Therefore,

$$z = \frac{\bar{y} - \mu}{\sigma/\sqrt{n}}$$

is a standard normal random variable. We also know from Theorem 7.4 that

$$\chi^2 = \frac{(n-1)s^2}{\sigma^2}$$

is a χ^2 random variable with $\nu = (n-1)$ degrees of freedom. Then, using Definition 7.4 and the information that $\bar{y}$ and s^2 are independent, we conclude that

$$t = \frac{z}{\sqrt{\chi^2/\nu}} = \frac{\dfrac{\bar{y} - \mu}{\sigma/\sqrt{n}}}{\sqrt{\dfrac{(n-1)s^2}{\sigma^2}\Big/(n-1)}} = \frac{\bar{y} - \mu}{s/\sqrt{n}}$$

has a Student's t distribution with $\nu = (n-1)$ degrees of freedom. ∎

Theorem 7.4 and Example 7.9 identify the sampling distributions of two statistics that will play important roles in statistical inference. Others are presented without proof in Table 7.1. All are based on random sampling from normally distributed populations. The results contained in Table 7.1 will be needed in Chapter 8.

TABLE 7.1a. Sampling Distributions of Statistics Based on a Random Sample from a Single Normally Distributed Population with Mean μ and Variance σ^2

STATISTIC	SAMPLING DISTRIBUTION	ADDITIONAL ASSUMPTIONS	BASIS OF DERIVATION OF SAMPLING DISTRIBUTION
$\chi^2 = \dfrac{(n-1)s^2}{\sigma^2}$	Chi-square with $\nu = (n-1)$ degrees of freedom	None	Methods of Section 7.2
$t = \dfrac{\bar{y} - \mu}{s/\sqrt{n}}$	Student's t with $\nu = (n-1)$ degrees of freedom	None	Theorems 7.3–7.4 and Definition 7.4

b. Sampling Distributions of Statistics Based on Independent Random Samples of n_1 and n_2 Observations, Respectively, from Normally Distributed Populations with Parameters (μ_1, σ_1^2) and (μ_2, σ_2^2)

STATISTIC	SAMPLING DISTRIBUTION	ADDITIONAL ASSUMPTIONS	BASIS OF DERIVATION OF SAMPLING DISTRIBUTION
$\chi^2 = \dfrac{(n_1 + n_2 - 2)s^2}{\sigma^2}$ where $s^2 = \dfrac{(n_1 - 1)s_1^2 + (n_2 - 1)s_2^2}{n_1 + n_2 - 2}$	Chi-square with $\nu = (n_1 + n_2 - 2)$ degrees of freedom	$\sigma_1^2 = \sigma_2^2 = \sigma^2$	Theorems 7.4–7.5
$t = \dfrac{(\bar{y}_1 - \bar{y}_2) - (\mu_1 - \mu_2)}{s\sqrt{\dfrac{1}{n_1} + \dfrac{1}{n_2}}}$ where $s^2 = \dfrac{(n_1 - 1)s_1^2 + (n_2 - 1)s_2^2}{n_1 + n_2 - 2}$	Student's t with $\nu = (n_1 + n_2 - 2)$ degrees of freedom	$\sigma_1^2 = \sigma_2^2 = \sigma^2$	Theorems 7.3–7.4 and Definition 7.4
$F = \left(\dfrac{s_1^2}{s_2^2}\right)\left(\dfrac{\sigma_2^2}{\sigma_1^2}\right)$	F distribution with $\nu_1 = (n_1 - 1)$ numerator degrees of freedom and $\nu_2 = (n_2 - 1)$ denominator degrees of freedom	None	Theorem 7.4 and Definition 7.3

| | | | | | | | | | | | | |

EXERCISES 7.22–7.24   OPTIONAL EXERCISES

7.22 Let $y_1, y_2, \ldots, y_n$ be a random sample of n_1 observations from a normal distribution with mean μ_1 and variance σ_1^2. Let $x_1, x_2, \ldots, x_{n_2}$ be a random sample of n_2 observations from a normal distribution with mean μ_2 and variance σ_2^2. Assuming the samples were independently selected, show that

$$F = \left(\frac{s_1^2}{s_2^2}\right)\left(\frac{\sigma_2^2}{\sigma_1^2}\right)$$

has an F distribution with $\nu_1 = (n_1 - 1)$ numerator degrees of freedom and $\nu_2 = (n_2 - 1)$ denominator degrees of freedom.

7.23 Let s_1^2 and s_2^2 be the variances of independent random samples of sizes n_1 and n_2 selected from normally distributed populations with parameters (μ_1, σ^2) and (μ_2, σ^2), respectively. Thus, the populations have different means, but a common variance σ^2. To estimate the common variance, we can combine information from both samples and use the **pooled estimator**

$$s^2 = \frac{(n_1 - 1)s_1^2 + (n_2 - 1)s_2^2}{n_1 + n_2 - 2}$$

Use Theorems 7.4 and 7.5 to show that $(n_1 + n_2 - 2)s^2/\sigma^2$ has a chi-square distribution with $\nu = (n_1 + n_2 - 2)$ degrees of freedom.

7.24 Let $\bar{y}_1$ and $\bar{y}_2$ be the means of independent random samples of sizes n_1 and n_2 selected from normally distributed populations with parameters (μ_1, σ^2) and (μ_2, σ^2), respectively. If

$$s^2 = \frac{(n_1 - 1)s_1^2 + (n_2 - 1)s_2^2}{n_1 + n_2 - 2}$$

show that

$$t = \frac{(\bar{y}_1 - \bar{y}_2) - (\mu_1 - \mu_2)}{s\sqrt{\dfrac{1}{n_1} + \dfrac{1}{n_2}}}$$

has a Student's t distribution with $\nu = (n_1 + n_2 - 2)$ degrees of freedom.

SECTION 7.7

GENERATING RANDOM NUMBERS BY COMPUTER (OPTIONAL)

Most statistical computer software packages have algorithms for generating random numbers. In this section we demonstrate how to use the random number generators in the SAS System and show how the resulting random numbers can be used to simulate the sampling distribution of a particular statistic—e.g., the sample mean. Table 7.2 at the end of the section gives a brief description of the random number generators in SPSS^x, BMDP, and Minitab.

The random number generator for the SAS computer package is accessed with the key word RANUNI. The RANUNI function will generate uniform random numbers in the interval from 0 to 1. For example, the SAS statement

```
U = RANUNI(0);
```

generates a single number, called U, with a value between 0 and 1. SAS uses the **seed**, the integer value in parentheses following RANUNI, to generate the first of a series of random numbers. The programmer arbitrarily chooses this seed, which can be any numeric constant. By changing the seed, you change the set of random numbers generated.

Some sampling experiments and most simulation techniques involve the selection of a large number—say, 10,000—of random numbers. In SAS, DO–END statements (similar to DO loops in FORTRAN) are particularly useful for this purpose. Example 7.10 demonstrates the use of DO–END statements in random sampling.

EXAMPLE 7.10

In Example 3.32 (Section 3.10) we used a table of random numbers (Table 1 of Appendix II) to select a random sample of five bolts from a population of 100,000 bolts. Show how we can use the computer to accomplish the same objective.

SOLUTION

Consider the SAS statements shown in Program 7.1. The SAS commands create a data set called SELECT, consisting of five values of the variable NUMBER, where NUMBER is a uniform random number selected from the interval 0 to 1.

PROGRAM 7.1
SAS Program for
Generating Five Uniform
Random Numbers

Command
line

```
1   DATA SELECT;
2   DO N = 1 TO 5;
3   NUMBER = RANUNI(0);     Data entry instructions: Generate five random
4   OUTPUT;                 numbers on the interval 0 to 1
5   END;
6   PROC PRINT;             }Print the five random numbers
```

For example, the five values of NUMBER generated by SAS might be .01388672, .67221005, .74111320, .00961745, and .22596071. By dropping the decimal point and using the first five digits to represent the bolts, we see that the bolts numbered 1,388, 67,221, 74,111, 961, and 22,596 should be included in our sample. ∎

The SAS statements used to generate the sampling distribution for the sum of a sample of $n = 2$ observations from a uniform distribution (Section 7.3) are shown in Example 7.11.

EXAMPLE 7.11

Write the SAS program that generated Figure 7.6, the simulated sampling distribution for the sum $g = y_1 + y_2$, where (y_1, y_2) is a sample of size $n = 2$ from a uniform distribution over the interval $0 \leq y \leq 1$.

SOLUTION

Recall that we used the simulation procedure of Section 7.3 to generate 10,000 pairs of the form (y_1, y_2). The appropriate SAS statements are shown in Program 7.2.

PROGRAM 7.2
SAS Program for
Simulating the Sampling
Distribution of the Sum
$g = y_1 + y_2$

Command
line

```
1   DATA SIMULATE;
2   DO N = 1 TO 10000;
3   Y1 = RANUNI(0);               Data entry instructions: Generate 10,000 values of
4   Y2 = RANUNI(22713);           the sum g = y₁ + y₂
5   G = Y1 + Y2;
6   OUTPUT;
7   END;                          Statistical analysis
8   PROC CHART; VBAR G/TYPE=PERCENT;}  instructions: Construct a
                                       relative frequency histogram
```
∎

EXAMPLE 7.12

Refer to Example 7.3. Write a SAS program that will generate a random sample of $n = 50$ observations from an exponential distribution with $\beta = 2$.

SOLUTION

Recall from Example 7.3 that we must first randomly draw $n = 50$ values from the uniform distribution. Call these values $g_1, g_2, \ldots, g_{50}$. The 50 values of the exponential random variable, $y_1, y_2, \ldots, y_{50}$, are then obtained using the relation

$$g_i = 1 - e^{-y_i/2} \quad (i = 1, 2, \ldots, 50)$$

TABLE 7.2 Random Number Generators for Four Computer Packages: SAS, Minitab, BMDP, and SPSS[x]

PACKAGE	COMMAND LINE	STATEMENTS/COMMANDS/FUNCTIONS	DESCRIPTION

SAS	1
	2
	3
	4
	5
		`DATA UNIFORM;`
		`DO K = 1 TO N;`
		`Y = RANUNI(I);`
		`OUTPUT; END;`
		`PROC PRINT;`
			Generates N uniform random numbers on the interval 0 to 1

			[*Note:* The seed I can be any numeric value.]

	1
	2
	3
	4
	5
		`DATA NORMAL;`
		`DO K = 1 TO N;`
		`Y = M + S*RANNOR(I);`
		`OUTPUT; END;`
		`PROC PRINT;`
			Generates N random numbers from a normal distribution with μ = M and σ = S.

Minitab	1 2	`URANDOM N VALUES, IN COLUMN C` `PRINT VALUES IN COLUMN C`	Generates N uniform random numbers on the interval 0 to 1
	1 2	`NRANDOM N VALUES, MEAN M, ST DEV S, IN COLUMN C` `PRINT VALUES IN COLUMN C`	[*Note:* The data are stored in column C (e.g., C1 or C2) of the worksheet.]
			Generates N random numbers from a normal distribution with μ = M and σ = S

BMDP	1
	2
	3
	4
	5
	6
	7
		`BMDP 1D`
		`/ PROBLEM TITLE IS 'UNIFORM RANDOM NUMBERS'.`
		`/ INPUT VARIABLES = 0, CASE = N,`
		`/ VARIABLE NAMES = Y, ADD = 1,`
		`/ TRANSFORM Y = RNDU(I).`
		`/ PRINT DATA.`
		`/ END`
			Generates N uniform random numbers on the interval 0 to 1

			[*Note:* The seed I must be a large positive odd integer.]

	1
	2
	3
	4
	5
	6
	7
		`BMDP 1D`
		`/ PROBLEM TITLE IS 'NORMAL RANDOM NUMBERS'.`
		`/ INPUT VARIABLES = 0, CASE = N,`
		`/ VARIABLE NAMES = Y, ADD = 1,`
		`/ TRANSFORM Y = M + S*RNDG(I).`
		`/ PRINT DATA.`
		`/ END`
			Generates N random numbers from a normal distribution with μ = M and σ = S

SPSS[x]	1
	2
	3
	4
	5
	
	.
	.
	.
	.
		`DATA LIST    FREE / X`
		`COMPUTE      Y=UNIFORM(1)`
		`PRINT / X Y`
		`BEGIN DATA`
		`1 2 3 4 5 6 . . .`
		
		`. . . N`
		
		`END DATA`
			Generates N uniform random numbers on the interval 0 to 1

	1
	2
	3
	4
	5
	
	.
	.
	.
	.
		`DATA LIST    FREE / X`
		`COMPUTE      Y=M + NORMAL(S)`
		`PRINT / X Y`
		`BEGIN DATA`
		`1 2 3 4 5 6 . . .`
		
		`. . . N`
		
		`END DATA`
			Generates N random numbers from a normal distribution with μ = M and σ = S

or

$$y_i = -2 \ln(1 - g) \qquad (i = 1, 2, \ldots, 50)$$

The SAS statements required to accomplish this are shown in Program 7.3. The data set named EXP contains 50 values of G and 50 values of Y, where G is a uniform random variable and Y is an exponential random variable with parameter $\beta = 2$.

PROGRAM 7.3
SAS Program for Generating a Random Sample of 50 Observations from an Exponential Distribution

```
Command
line
  1    DATA EXP;
  2    DO N = 1 TO 50;
  3    G = RANUNI(55203);          Data entry instructions:
  4    Y = (-2)*LOG(1-G);          Generate 50 values from
  5    OUTPUT;                     the exponential distribution
  6    END;
  7    PROC PRINT;}  Print the results
```

■

We have used the distribution function approach of Section 7.2 to generate random numbers from the exponential distribution. However, this method is more difficult to apply if we want to generate random numbers from a normal distribution, because the distribution function for the normal density cannot be written in closed form. Fortunately, most computer packages have an algorithm for randomly generating normal observations. In SAS, the RANNOR function generates normal random observations with mean 0 and variance 1. For example, the statement

```
Y = RANNOR(721881);
```

generates a normal random variable with $\mu = 0$ and $\sigma^2 = 1$, while the statement

```
Y = 16 + 2*RANNOR(9998803);
```

generates a normal random variable with $\mu = 16$ and $\sigma^2 = (2)^2 = 4$. To select a random sample from the normal distribution, use the RANNOR random number generator within DO–END statements as illustrated in Examples 7.11 and 7.12.

Random number generators for the Minitab, BMDP, and SPSS^x computer packages are similar to those of SAS. A brief summary of the use of these four generators is given in Table 7.2. Consult the references at the end of this chapter for more details.

EXERCISES 7.25–7.27

7.25 Refer to Exercise 7.5. Use the computer to generate a random sample of $n = 100$ observations from a beta distribution with $\alpha = 2$ and $\beta = 1$.

7.26 Use the computer to simulate the sampling distribution of the sample mean

$$\bar{y} = \frac{y_1 + y_2 + \cdots + y_{50}}{50}$$

where $y_1, y_2, \ldots, y_{50}$ are uniform random variables on the interval 0 to 1.

7.27 Refer to Exercise 7.6. Use the computer to generate a random sample of $n = 100$ observations from a distribution with probability density

$$f(y) = \begin{cases} e^y & \text{if } y < 0 \\ 0 & \text{elsewhere} \end{cases}$$

Repeat the procedure 1,000 times and compute the sample mean $\bar{y}$ for each of the 1,000 samples of size $n = 100$. Then generate (by computer) a relative frequency histogram for the 1,000 sample means. Does your result agree with the theoretical sampling distribution described by the central limit theorem?

SECTION 7.8

SUMMARY

In the following chapters, we will use sample statistics to make inferences about population parameters; the properties of these statistics will be determined by their probability distributions. The probability distribution of a statistic is called its **sampling distribution**.

A **simulation procedure** may be used to approximate the sampling distribution for a statistic. Random samples of a fixed size are drawn from a known population of data. The value of some statistic—say, the sample mean $\bar{y}$—is computed for each sample. The relative frequency distribution of the values of the statistic, generated by repeated sampling, approximates the probability distribution of the statistic.

Evidence of the major role that the normal distribution plays in statistical inference is given by the **central limit theorem**, Theorem 7.3, and the related χ^2, F, **and** t **distributions**. The central limit theorem explains why many statistics, especially those based on large samples, possess sampling distributions that can be approximated by a normal density function. Theorem 7.3, which states that linear functions of normally distributed random variables will be normally distributed, provides further explanation for the common occurrence of normally distributed sampling distributions. The χ^2, t, and F statistics are approximated when sampling from normally distributed populations. You will encounter them frequently in the statistical methodology to be developed in the following chapters.

SUPPLEMENTARY EXERCISES 7.28–7.46

7.28 Consider the density function

$$f(y) = \begin{cases} 3y^2 & \text{if } 0 \le y \le 1 \\ 0 & \text{elsewhere} \end{cases}$$

Find the density function of $g(y)$, where:
a. $g(y) = \sqrt{y}$ **b.** $g(y) = 3 - y$ **c.** $g(y) = -\ln(y)$

7.29 A supplier of home heating oil has a 250-gallon tank that is filled at the beginning of each week. Since the weekly demand for the oil increases steadily up to 100 gallons and then levels off between 100 and 250 gallons, the probability distribution of the weekly demand y (in hundreds of gallons) can be represented by

$$f(y) = \begin{cases} \dfrac{y}{2} & \text{if } 0 \le y \le 1 \\ \dfrac{1}{2} & \text{if } 1 \le y \le 2.5 \\ 0 & \text{elsewhere} \end{cases}$$

If the supplier's profit is given by $g(y) = 10y - 2$, find the probability density function of g.

7.30 To determine whether a metal lathe producing machine bearings is properly adjusted, a random sample of 36 bearings is collected and the diameter of each is measured. If the standard deviation of the diameter of the machine bearings measured over a long period of time is .001 inch, what is the probability that the mean diameter $\bar{y}$ of the sample of 36 bearings will lie within .0001 inch of the population mean diameter of the bearings?

7.31 Refer to Exercise 7.30. Suppose the mean diameter of the bearings produced by the machine is supposed to be .5 inch. The company decides to use the sample mean (from Exercise 7.30) to decide whether the process is in control—i.e., whether it is producing bearings with a mean diameter of .5 inch. The machine will be considered out of control if the mean of the sample of $n = 36$ diameters is less than .4994 inch or larger than .5006 inch. If the true mean diameter of the bearings produced by the machine is .501 inch, what is the probability that the test will fail to imply that the process is out of control?

7.32 The distribution of the number of characters printed per second by a particular kind of line printer at a computer terminal has the following parameters: $\mu = 45$ characters per second, $\sigma = 2$ characters per second.
   a. Describe the sampling distribution of the mean number of characters printed per second for random samples of 1-minute intervals.
   b. Find the approximate probability that the sample mean for a random sample of 60 seconds will be between 44.5 and 45.3 characters per second.
   c. Find the approximate probability that the sample mean for a random sample of 60 seconds will be less than 44 characters per second.

7.33 The determination of the percent canopy closure of a forest is essential for wildlife habitat assessment, watershed runoff estimation, erosion control, and other forest management activities. One way in which geoscientists estimate percent forest canopy closure is through the use of a satellite sensor called the Landsat Thematic Mapper. A study of the percent canopy closure in the San Juan National Forest (Colorado) was conducted by examining Thematic Mapper Simulator (TMS) data collected by aircraft at various forest sites (*IEEE Transactions on Geoscience and Remote Sensing*, Jan. 1986). The mean and standard deviation of the readings obtained from TMS Channel 5 were found to be 121.74 and 27.52, respectively.
   a. Let $\bar{y}$ be the mean TMS reading for a sample of 32 forest sites. Assuming the figures given above are population values, describe the sampling distribution of $\bar{y}$.
   b. Use the sampling distribution of part **a** to find the probability that $\bar{y}$ falls between 118 and 130.

7.34 Use Theorem 7.1 to draw a random sample of $n = 5$ observations from a population with probability density function given by

$$f(y) = \begin{cases} 2(y - 1) & \text{if } 1 \le y < 2 \\ 0 & \text{elsewhere} \end{cases}$$

7.35 Use Theorem 7.1 to draw a random sample of $n = 5$ observations from a population with probability density function given by

$$f(y) = \begin{cases} 2ye^{-y^2} & \text{if } 0 < y < \infty \\ 0 & \text{elsewhere} \end{cases}$$

7.36 Shear block tests on epoxy-repaired timber indicate that the probability distribution of the bond strengths of parallel grain, mill lumber specimens has a mean of 1,312 pounds per square inch (psi) and a standard deviation of 422 psi (*Journal of Structural Engineering*, Feb. 1986). Suppose a sample of 100 epoxy-repaired timber specimens is randomly selected and the bond strength of each is determined.
   a. Describe the sampling distribution of $\bar{y}$, the mean bond strength of the sample of 100 epoxy-repaired timber specimens.
   b. Compute $P(\bar{y} \ge 1,418)$.
   c. If the actual sample mean is computed to be $\bar{y} = 1,418$, what would you infer about the shear block test results?

7.37 The temperature at which an indoor fire-prevention sprinkler system is triggered depends on such factors as the size of the room and the humidity. The manufacturer of one such sprinkler system claims that the distribution of the temperature at which its sprinklers are triggered has an average of 125°F and a standard deviation of 1.5°F. Before installing these sprinklers in a warehouse that stores highly flammable materials, a ceramic engineer wants to check the validity of the manufacturer's claim. The engineer plans to subject ten randomly selected sprinklers to a series of temperature-controlled experiments. The sprinkler system will *not* be installed if the sprinklers tested fail to activate at an average temperature within .5° of the claimed mean of 125°. What is the probability that the engineer will not install the sprinkler system even if the true mean activation temperature is actually 125°? (Assume that the standard deviation reported by the manufacturer is correct.)

7.38 From past experience, a computer programmer knows that the cost of running a certain type of job has a relative frequency distribution with a mean of $1.20 and a standard deviation of $.44. Suppose the programmer needs to run 20 more jobs of this type, but has only $25.00 remaining in the computer account. What is the approximate probability that the programmer will be able to run all 20 jobs without depleting the computer account?

7.39 Refer to the problem of transporting neutral particles in a nuclear fusion reactor, described in Exercise 3.33. Recall that particles released into a certain type of evacuated duct collide with the inner duct wall and are either scattered (reflected) with probability .16 or absorbed with probability .84 (*Nuclear Science and Engineering*, May 1986). Suppose 2,000 neutral particles are released into an unknown type of evacuated duct in a nuclear fusion reactor. Of these, 280 are reflected. What is the approximate probability that as few as 280 (i.e., 280 or fewer) of the 2,000 neutral particles would be reflected off the inner duct wall if the reflection probability of the evacuated duct is .16?

7.40 The manufacturer of a new instant-picture camera claims that its product has "the world's fastest-developing color film by far." Extensive laboratory testing has shown that the relative frequency distribution for the time it takes the new instant camera to begin to reveal the image after shooting has a mean of 9.8 seconds and a standard deviation of .55 second. Suppose 50 of these cameras are randomly selected from the production line and tested. The time until the image is first revealed, y, is recorded for each.

    a. Describe the sampling distribution of $\bar{y}$, the mean time it takes the sample of 50 cameras to begin to reveal the image.

    b. Find the probability that the mean time until the image is first revealed for the 50 sampled cameras is greater than 9.70 seconds.

    c. If the mean and standard deviation of the population relative frequency distribution for the times until the cameras begin to reveal the image are correct, would you expect to observe a value of $\bar{y}$ less than 9.55 seconds? Explain.

7.41 According to a recent survey, 74% of the petroleum engineers who are currently employed are located in Texas, Oklahoma, Louisiana, or California. What is the approximate probability that at least 352 of 500 randomly selected employed petroleum engineers work in one of these four states?

OPTIONAL SUPPLEMENTARY EXERCISES

7.42 The waiting time y until delivery of a new component for a data-processing unit is uniformly distributed over the interval from 1 to 5 days. The cost c (in hundreds of dollars) of this delay to the purchaser is given by $c = (2y^2 + 3)$. Find the probability that the cost of delay is at least \$2,000, i.e., compute $P(c \geq 20)$.

7.43 Let y_1 and y_2 be a sample of $n = 2$ observations from a gamma random variable with parameters $\alpha = 1$ and arbitrary β, and corresponding density function

$$f(y_i) = \begin{cases} \dfrac{1}{\beta}e^{-y_i/\beta} & \text{if } y_i > 0 \quad (i = 1, 2) \\ 0 & \text{elsewhere} \end{cases}$$

Show that the sum $g = (y_1 + y_2)$ is also a gamma random variable with parameters $\alpha = 2$ and β. [*Hint:* You may use the result

$$P(g \leq g_0) = P(0 < y_2 \leq g - y_1, 0 \leq y_1 < g) = \int_0^g \int_0^{g-y_1} f(y_1, y_2)\, dy_2\, dy_1$$

Then use the fact that

$$f(y_1, y_2) = f(y_1)f(y_2)$$

since y_1 and y_2 are independent.]

7.44 Let y have an exponential density with mean β. Show that $g(y) = 2y/\beta$ has a χ^2 density with $\nu = 2$ degrees of freedom.

7.45 The lifetime y of an electronic component of a home minicomputer has a *Rayleigh density*, given by

$$f(y) = \begin{cases} \left(\dfrac{2y}{\beta}\right)e^{-y^2/\beta} & \text{if } y > 0 \\ 0 & \text{elsewhere} \end{cases}$$

Find the probability density function for $g(y) = y^2$ and identify the type of density function. [*Hint:* You may use the result

$$\int \frac{2y}{\beta} e^{-y^2/\beta} \, dy = -e^{-y^2/\beta}$$

in determining the density function for $g(y)$.]

7.46 Let y_1 and y_2 be a random sample of $n = 2$ observations from a normal distribution with mean μ and variance σ^2.

a. Show that

$$z = \frac{y_1 - y_2}{\sqrt{2}\sigma}$$

has a standard normal distribution.

b. Given the result in part **a**, show that z^2 possesses a χ^2 distribution with 1 degree of freedom. [*Hint:* First show that $s^2 = (y_1 - y_2)^2/2$; then apply Theorem 7.4.]

REFERENCES

BMDPC: User's Guide to BMDP on the IBM PC. Los Angeles: BMDP Statistical Software, 1986.

BMDP User's Digest, 2nd ed. MaryAnn Hill, ed. Los Angeles: BMDP Statistical Software, 1982.

Dixon, W. J., Brown, M. B., Engelman, L., Frane, J. W., Hill, M. A., Jennrich, R. I., and Toporek, J. D. *BMDP Statistical Software*, 1985 ed. Berkeley: University of California Press.

Freedman, D., Pisani, R., and Purves, R. *Statistics.* New York: W. W. Norton and Co., 1978.

Hogg, R. V. and Craig, A. T. *Introduction to Mathematical Statistics*, 4th ed. New York: Macmillan, 1978.

McClave, J. T. and Dietrich, F. H. II. *Statistics*, 3rd ed. San Francisco: Dellen, 1985.

Mendenhall, W., Scheaffer, R. L., and Wackerly, D. D. *Mathematical Statistics with Applications*, 3rd ed. Boston: Duxbury, 1986.

Mendenhall, W. *Introduction to Probability and Statistics*, 7th ed. Boston: Duxbury, 1987.

Norusis, M. J. *SPSS/PC+: SPSS for the IBM PC/XT/AT*, 1986 ed. SPSS, Inc., Suite 3000, 444 N. Michigan Ave., Chicago, Ill. 60611.

Norusis, M. J. *The SPSS Guide to Data Analysis*, 1986 ed. SPSS, Inc., Suite 3000, 444 N. Michigan Ave., Chicago, Ill. 60611.

Ryan, T. A., Joiner, B. L., and Ryan, B. F. *Minitab Reference Manual.* Minitab Project, University Park, Pa., 1985.

Ryan T. A., Joiner, B. L., and Ryan, B. F. *Minitab Student Handbook*, 2nd ed., Boston: Duxbury, 1985.

SAS Procedures Guide for Personal Computers, Version 6, 1986. SAS Institute, Inc., Box 8000, Cary, N.C. 27511.

SAS User's Guide: Statistics, Version 5, 1985. A. A. Ray, ed. SAS Institute, Inc., Box 8000, Cary, N.C. 27511.

Snedecor, G. W. and Cochran, W. G. *Statistical Methods*, 7th ed. Ames, Iowa: Iowa State University Press, 1980.

SPSS^X User's Guide. 1983 ed. SPSS, Inc., Suite 3000, 444 N. Michigan Ave., Chicago, Ill. 60611.

C H A P T E R 8

OBJECTIVE

To explain the basic concepts of statistical estimation; to present some estimators and to illustrate their use in practical sampling situations involving one or two samples.

CONTENTS

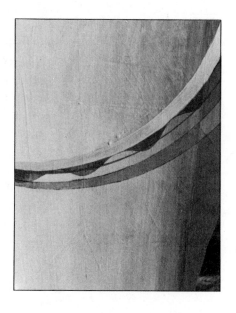

ESTIMATION

ESTIMATORS

An inference about a population parameter can be made in either of two ways—we can estimate its value or we can make a decision about it. To illustrate, we can estimate the mean number μ of jobs submitted per hour to a data-processing center or we might want to decide whether the mean μ exceeds some value—say, 60. The method for making a decision about one or more population parameters, called a **statistical test of a hypothesis**, is the topic of Chapter 9. This chapter will be concerned with **estimation**.

Suppose we want to estimate some population parameter, which we denote by θ. For example, θ could be a population mean μ, a population variance σ^2, or the probability $F(a)$ that an observation selected from the population is less than or equal to the value a. A **point estimator**, designated by the symbol $\hat{\theta}$ (i.e., we place a "hat" over the symbol of a parameter to denote its estimator), is a rule or formula that tells us how to use the observations in a sample to compute a single number (a point) that serves as an **estimate** of the value of θ. For example, the mean $\bar{y}$ of a random sample of n observations, $y_1, y_2, \ldots, y_n$, selected from a population is a point estimator of the population mean μ—i.e., $\hat{\mu} = \bar{y}$. Similarly, the sample variance s^2 is a point estimator of σ^2—i.e., $\hat{\sigma}^2 = s^2$.

DEFINITION 8.1

A **point estimator** is a rule or formula that tells us how to calculate an estimate based on the measurements contained in a sample. The number that results from the calculation is called a **point estimate**.

Another way to estimate the value of a population parameter θ is to use an interval estimator. An **interval estimator** is a rule, usually expressed as a formula, for calculating two points from the sample data. The objective is to form an interval which we think contains θ. For example, if we estimate the mean number μ of jobs submitted to a data-processing center to be between 40 and 60 jobs per hour, then the interval 40 to 60 is an interval estimate of μ.

DEFINITION 8.2

An **interval estimator** is a formula that tells us how to use sample data to calculate an interval that estimates a population parameter.

In this chapter, we will identify desirable properties of point and interval estimators, explain how to compare two or more estimators for a single parameter, and show how to measure how good a single estimate actually is. In addition, we will present methods for finding both point and interval estimators, give the formulas for some useful estimators, and show how they can be used in practical situations.

Since a point estimator is calculated from a sample, it possesses a sampling distribution. The sampling distribution of a point estimator completely describes its properties. For example, according to the central limit theorem, the sampling distribution for a sample mean will be approximately normally distributed for large sample sizes, say $n = 30$ or more, with mean μ and standard error $\sigma/\sqrt{n}$ (see Figure 8.1). The figure shows that a sample mean $\bar{y}$ is equally likely to fall above or below μ and that the probability is approximately .95 that it will not deviate from μ by more than $2\sigma_{\bar{y}} = 2\sigma/\sqrt{n}$. These characteristics identify the two most desirable properties of estimators.

FIGURE 8.1
Sampling Distribution of a
Sample Mean for Large
Samples

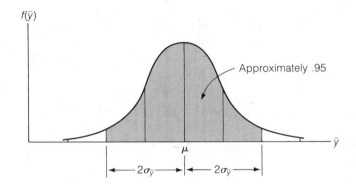

In general, we would like the sampling distribution of an estimator to be centered over the parameter being estimated. If the mean of the sampling distribution of an estimator $\hat{\theta}$ is equal to the estimated parameter θ, then the estimator is said to be **unbiased**. If not, the estimator is said to be **biased**. The sample mean is an unbiased estimator of the population mean μ. Sampling distributions for unbiased and biased estimators are shown in Figures 8.2a and 8.2b, respectively.

FIGURE 8.2
Sampling Distributions for
Unbiased and Biased
Estimators of θ

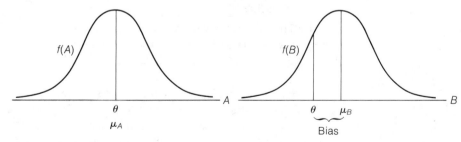

a. Estimator A is unbiased.   **b.** Estimator B is biased.

DEFINITION 8.3

An estimator $\hat{\theta}$ of a parameter θ is **unbiased** if $E(\hat{\theta}) = \theta$. If $E(\hat{\theta}) \neq \theta$, the estimator is said to be **biased**.

DEFINITION 8.4

The **bias B** of an estimator $\hat{\theta}$ is equal to the difference between the mean $E(\hat{\theta})$ of the sampling distribution of $\hat{\theta}$ and θ, i.e.,

$$B = E(\hat{\theta}) - \theta$$

In addition to unbiasedness, we would like the sampling distribution of an estimator to have **minimum variance**, i.e., we want the spread of the sampling distribution to be as small as possible so that estimates will tend to fall close to θ.

Figure 8.3 portrays the sampling distributions of two unbiased estimators, A and B, with A having smaller variance than B. An unbiased estimator that has the minimum variance among all unbiased estimators is called the **minimum variance unbiased estimator (MVUE)**.

FIGURE 8.3

Sampling Distributions for Two Unbiased Estimators of θ with Different Variances

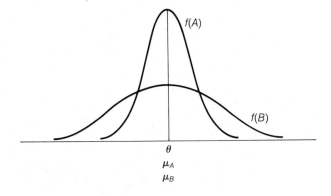

DEFINITION 8.5

The **minimum variance unbiased estimator (MVUE)** of a parameter θ is the estimator $\hat{\theta}$ that has the smallest variance of all unbiased estimators.

Sometimes we cannot achieve both unbiasedness and minimum variance in the same estimator. For example, Figure 8.4 shows a biased estimator A with slight bias, but with a smaller variance than the MVUE B. In such a case, we prefer the estimator that minimizes the **mean squared error**, the mean of the squared deviations between $\hat{\theta}$ and θ:

Mean squared error for $\hat{\theta}$: $E[(\hat{\theta} - \theta)^2]$

It can be shown (proof omitted) that

$$E[(\hat{\theta} - \theta)^2] = V(\hat{\theta}) + B^2$$

FIGURE 8.4
Sampling Distributions of
Biased Estimator A and
MVUE B

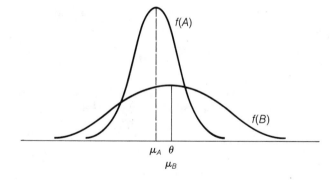

Therefore, if $\hat{\theta}$ is unbiased, i.e., if $B = 0$, then the mean squared error is equal to $V(\hat{\theta})$. Furthermore, when $B = 0$, the estimator $\hat{\theta}$ that yields the smallest mean squared error is also the MVUE for θ.

EXAMPLE 8.1

Let $y_1, y_2, \ldots, y_n$ be a random sample of n observations from a normal distribution with mean μ and variance σ^2. Show that the sample variance

$$s^2 = \frac{\sum_{i=1}^{n} (y_i - \bar{y})^2}{n - 1}$$

is an unbiased estimator of the population variance σ^2. Find the standard deviation of the sampling distribution of s^2.

SOLUTION

From Theorem 7.4, we know that

$$\frac{(n - 1)s^2}{\sigma^2} = \chi^2$$

where χ^2 is a chi-square random variable with $\nu = (n - 1)$ degrees of freedom. Rearranging terms yields

$$s^2 = \frac{\sigma^2}{(n - 1)}\chi^2$$

from which it follows that

$$E(s^2) = E\left[\frac{\sigma^2}{(n - 1)}\chi^2\right]$$

Applying Theorem 5.2, we obtain

$$E(s^2) = \frac{\sigma^2}{(n - 1)}E(\chi^2)$$

We know from Section 5.6 that $E(\chi^2) = \nu$ and $V(\chi^2) = 2\nu$; thus

$$E(s^2) = \frac{\sigma^2}{(n-1)}(\nu) = \frac{\sigma^2}{(n-1)}(n-1) = \sigma^2$$

Therefore, by Definition 8.3, we conclude that s^2 is an unbiased estimator of σ^2.

To find the variance of σ^2, we write s^2 as a linear function of a single random variable, χ^2, and apply Theorem 6.7. Thus,

$$V(s^2) = V\left[\frac{\sigma^2}{(n-1)}\chi^2\right] = \frac{\sigma^4}{(n-1)^2}V(\chi^2)$$

$$= \frac{\sigma^4}{(n-1)^2}(2\nu) = \frac{\sigma^4(2)(n-1)}{(n-1)^2} = \frac{2\sigma^4}{n-1}$$

Therefore, the standard deviation of the sampling distribution of s^2 is

$$\sigma_{s^2} = \sigma^2\sqrt{\frac{2}{n-1}}$$

The sampling distribution of s^2 is shown in Figure 8.5. We would expect most estimates of σ^2 to lie within $2\sigma_{s^2}$ or $2\sigma^2\sqrt{2/(n-1)}$ of σ^2. You will notice that the standard error σ_{s^2} is a function of σ^2, the parameter that we want to estimate. In a real-life sampling situation, we would not know the value of σ^2. We could, however, approximate its value with the estimate s^2 when the sample size n is large.

FIGURE 8.5
Sampling Distribution
of s^2

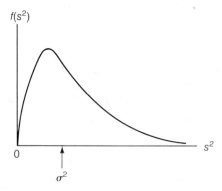

EXAMPLE 8.2   In Example 8.1, we showed that the sample variance s^2 is an unbiased estimator of the population variance σ^2 when the sampled population is normally distributed. Show that the sample variance s^2, based on a random sample of n observations, is an unbiased estimator of σ^2, *regardless of the nature of the sampled population.*

SOLUTION

By the definition of sample variance, we have

$$s^2 = \sum_{i=1}^{n} \frac{(y_i - \bar{y})^2}{n - 1}$$

Adding and subtracting the mean μ yields

$$s^2 = \frac{1}{n - 1} \sum_{i=1}^{n} [(y_i - \mu) - (\bar{y} - \mu)]^2$$

$$= \frac{1}{(n - 1)} \sum_{i=1}^{n} [(y_i - \mu)^2 - 2(y_i - \mu)(\bar{y} - \mu) + (\bar{y} - \mu)^2]$$

$$= \frac{1}{n - 1} \left[\sum_{i=1}^{n} (y_i - \mu)^2 - 2(\bar{y} - \mu) \sum_{i=1}^{n} (y_i - \mu) + n(\bar{y} - \mu)^2 \right]$$

Now we note that

$$-2(\bar{y} - \mu) \sum_{i=1}^{n} (y_i - \mu) = -2n(\bar{y} - \mu) \left(\frac{\sum_{i=1}^{n} y_i - n\mu}{n} \right) = -2n(\bar{y} - \mu)^2$$

Therefore,

$$s^2 = \frac{1}{n - 1} \left[\sum_{i=1}^{n} (y_i - \mu)^2 - n(\bar{y} - \mu)^2 \right]$$

Since each y value, $y_1, y_2, \ldots, y_n$, was randomly selected from a population with mean μ and variance σ^2, it follows that

$$E[(y_i - \mu)^2] = \sigma^2 \qquad (i = 1, 2, \ldots, n)$$

and

$$E[(\bar{y} - \mu)^2] = \sigma_{\bar{y}}^2 = \frac{\sigma^2}{n}$$

Taking the expected value of s^2, we obtain

$$E(s^2) = E\left\{ \frac{1}{n - 1} \left[\sum_{i=1}^{n} (y_i - \mu)^2 - n(\bar{y} - \mu)^2 \right] \right\}$$

$$= \frac{1}{n - 1} \left\{ E\left[\sum_{i=1}^{n} (y_i - \mu)^2 \right] - E[n(\bar{y} - \mu)^2] \right\}$$

$$= \frac{1}{n - 1} \left\{ \sum_{i=1}^{n} [E(y_i - \mu)^2] - nE(\bar{y} - \mu)^2] \right\}$$

$$= \frac{1}{n - 1} \left[\sum_{i=1}^{n} \sigma^2 - n\left(\frac{\sigma^2}{n}\right) \right] = \frac{1}{n - 1}(n\sigma^2 - \sigma^2)$$

$$= \left(\frac{n - 1}{n - 1}\right)\sigma^2 = \sigma^2$$

This shows that, regardless of the nature of the sampled population, s^2 is an unbiased estimator of σ^2. ∎

| | | | | | | | | | | | |

EXERCISES 8.1–8.6

OPTIONAL EXERCISES

8.1 Let y_1, y_2, y_3 be a random sample from an exponential distribution with mean θ, i.e., $E(y_i) = \theta$, $i = 1, 2, 3$. Consider three estimators of θ:

$$\hat{\theta}_1 = \bar{y} \qquad \hat{\theta}_2 = y_1 \qquad \hat{\theta}_3 = \frac{y_1 + y_2}{2}$$

a. Show that all three estimators are unbiased.
b. Which of the estimators has the smallest variance? [*Hint:* Recall that, for an exponential distribution, $V(y_i) = \theta^2$.]

8.2 Let $y_1, y_2, y_3, \ldots, y_n$ be a random sample from a Poisson distribution with mean λ, i.e., $E(y_i) = \lambda$, $i = 1, 2, \ldots, n$. Consider four estimators of λ:

$$\hat{\lambda}_1 = \bar{y} \qquad\qquad \hat{\lambda}_2 = n(y_1 + y_2) + \cdots + y_n)$$

$$\hat{\lambda}_3 = \frac{y_1 + y_2}{2} \qquad \hat{\lambda}_4 = \frac{y_1}{n}$$

a. Which of the four estimators are unbiased?
b. Of the unbiased estimators, which has the smallest variance? [*Hint:* Recall that, for a Poisson distribution, $V(y_i) = \lambda$.]

8.3 Suppose y has a binomial distribution with parameters n and p.
a. Show that $\hat{p} = y/n$ is an unbiased estimator of p.
b. Find the variance of $\hat{p}$.

8.4 Let $y_1, y_2, \ldots, y_n$ be a random sample from a gamma distribution with parameters $\alpha = 2$ and β unknown.
a. Show that $\bar{y}$ is a biased estimator of β. Compute the bias.
b. Show that $\hat{\beta} = \bar{y}/2$ is an unbiased estimator of β.
c. Find the variance of $\hat{\beta} = \bar{y}/2$. [*Hint:* Recall that, for a gamma distribution, $E(y_i) = 2\beta$ and $V(y_i) = 2\beta^2$.]

8.5 Show that $E[(\hat{\theta} - \theta)^2] = V(\hat{\theta}) + B^2$, where the bias $B = E(\hat{\theta}) - \theta$. [*Hint:* Write $(\hat{\theta} - \theta) = [\hat{\theta} - E(\hat{\theta})] + [E(\hat{\theta}) - \theta]$.]

8.6 Let y_1 be a sample of size 1 from a uniform distribution over the interval from 2 to θ.
a. Show that y_1 is a biased estimator of θ and compute the bias.
b. Show that $2(y_1 - 1)$ is an unbiased estimator of θ.
c. Find the variance of $2(y_1 - 1)$.

| | | | | | | | | | | | |

SECTION 8.3

FINDING POINT ESTIMATORS: METHODS OF ESTIMATION

There are a number of different methods for finding point estimators of parameters. Two classical methods, the **method of moments** and the **method of maximum likelihood**, are the main topics of this section. These techniques produce the estimators of the population parameters encountered in Sections 8.5–8.11. A discussion of other methods for finding point estimators is beyond the scope of this text; we give a brief description of these other methods and refer you to the references given at the end of this chapter.

METHOD OF MOMENTS The method of estimation that we have employed thus far is to use sample numerical descriptive measures to estimate their population parameters. For example, we used the sample mean $\bar{y}$ to estimate the population mean μ. From Definition 4.7, we know that the parameter $E(y) = \mu$ is the first moment about the origin or, as it is sometimes called, the **first population moment**. Similarly, we define the **first sample moment** as

$$\bar{y} = \frac{\sum_{i=1}^{n} y_i}{n}$$

The general technique of using sample moments to estimate their corresponding population moments is called the **method of moments**.

DEFINITION 8.6

Let $y_1, y_2, \ldots, y_n$ represent a random sample of size n from some probability distribution (discrete or continuous). The **kth population moment** and **kth sample moment** are defined as follows:

kth population moment: $E(y^k)$

kth sample moment: $\bar{y} = \dfrac{\sum_{i=1}^{n} y_i^k}{n}$

For the case $k = 1$, the first population moment is $E(y) = \mu$ and the first sample moment is $\bar{y}$.

DEFINITION 8.7

Let $y_1, y_2, \ldots, y_n$ represent a random sample of size n from a probability distribution (discrete or continuous) with parameters $\theta_1, \theta_2, \ldots, \theta_m$. Then the **moment estimators**, $\hat{\theta}_1, \hat{\theta}_2, \ldots, \hat{\theta}_m$, are obtained by equating the first m sample moments to the corresponding first m population moments:

$$E(y) = \frac{1}{n} \sum y_i$$

$$E(y^2) = \frac{1}{n} \sum y_i^2$$

$$\vdots$$

$$E(y^m) = \frac{1}{n} \sum y_i^m$$

and solving for $\theta_1, \theta_2, \ldots, \theta_m$. (Note that the first m population moments will be functions of $\theta_1, \theta_2, \ldots, \theta_m$.)

For the special case $m = 1$, the moment estimator of θ is some function of the sample mean $\bar{y}$.

EXAMPLE 8.3

The response rate y of auditory nerve fibers in cats has an approximate Poisson distribution with unknown mean λ (*Journal of the Acoustical Society of America*, Feb. 1986). Suppose the auditory nerve fiber response rate (recorded as number of spikes per 200 milliseconds of noise burst) was measured in each of a random sample of ten cats. The data are given below:

$$15.1 \quad 14.6 \quad 12.0 \quad 19.2 \quad 16.1 \quad 15.5 \quad 11.3 \quad 18.7 \quad 17.1 \quad 17.2$$

Calculate a point estimate for the mean response rate λ using the method of moments.

SOLUTION

We have only one parameter, λ, to estimate; therefore, the moment estimator is found by setting the first population moment, $E(y)$, equal to the first sample moment, $\bar{y}$. For the Poisson distribution, $E(y) = \lambda$; hence, the moment estimator is

$$\hat{\lambda} = \bar{y}$$

For this example,

$$\bar{y} = \frac{15.1 + 14.6 + \cdots + 17.2}{10} = 15.68$$

Thus, our estimate of the mean auditory nerve fiber response rate λ is 15.68 spikes per 200 milliseconds of noise burst. ∎

EXAMPLE 8.4 (OPTIONAL)

The time y until failure from fatigue cracks for underground cable possesses an approximate gamma probability distribution with parameters α and β (*IEEE Transactions on Energy Conversion*, Mar. 1986). Let $y_1, y_2, \ldots, y_n$ be a random sample of n observations on the random variable y. Find the moment estimators of α and β.

SOLUTION

Since we must estimate two parameters, α and β, the method of moments requires that we set the first two population moments equal to their corresponding sample moments. From Section 5.6, we know that for the gamma distribution

$$\mu = E(y) = \alpha\beta$$
$$\sigma^2 = \alpha\beta^2$$

Also, from Theorem 4.4, $\sigma^2 = E(y^2) - \mu^2$. Thus, $E(y^2) = \sigma^2 + \mu^2$. Then for the gamma distribution, the first two population moments are

$$E(y) = \alpha\beta$$
$$E(y^2) = \sigma^2 + \mu^2 = \alpha\beta^2 + (\alpha\beta)^2$$

Setting these equal to their respective sample moments, we have

$$\hat{\alpha}\hat{\beta} = \bar{y}$$
$$\hat{\alpha}\hat{\beta}^2 + (\hat{\alpha}\hat{\beta})^2 = \frac{\sum y_i^2}{n}$$

Substituting $\bar{y}$ for $\hat{\alpha}\hat{\beta}$ in the second equation, we obtain

$$\bar{y}\hat{\beta} + (\bar{y})^2 = \frac{\sum y_i^2}{n}$$

or,

$$\bar{y}\hat{\beta} = \frac{\sum y_i^2}{n} - (\bar{y})^2$$

$$= \frac{\sum y_i^2 - n(\bar{y})^2}{n} = \frac{\sum y_i^2 - \frac{\left(\sum y_i\right)^2}{n}}{n}$$

$$= s^2$$

Our two equations are now reduced to

$$\hat{\alpha}\hat{\beta} = \bar{y}$$

$$\bar{y}\hat{\beta} = s^2$$

Solving these equations simultaneously, we obtain the moment estimators

$$\hat{\beta} = \frac{s^2}{\bar{y}} \quad \text{and} \quad \hat{\alpha} = \frac{\bar{y}^2}{s^2} = \left(\frac{\bar{y}}{s}\right)^2 \qquad \blacksquare$$

METHOD OF MAXIMUM LIKELIHOOD The method of maximum likelihood and an exposition of the properties of maximum likelihood estimators are the results of work by Sir Ronald A. Fisher (1890–1962). Fisher's logic can be seen by considering the following example: If we randomly select a sample of n observations, y_1, $y_2, \ldots, y_n$, of a discrete random variable y and if the probability distribution $p(y)$ is a function of a single parameter θ, then the probability of observing these n independent values of y is

$$p(y_1, y_2, \ldots, y_n) = p(y_1)p(y_2) \cdots p(y_n)$$

Fisher called this joint probability of the sample values, $y_1, y_2, \ldots, y_n$, the **likelihood L** of the sample, and suggested that one should choose as an estimate of θ the value of θ that maximizes L. If the likelihood L of the sample is a function of two parameters, say θ_1 and θ_2, then the maximum likelihood estimates of θ_1 and θ_2 are the values that maximize L. The notion is easily extended to the situation in which L is a function of more than two parameters.

DEFINITION 8.8

a. The **likelihood L** of a sample of n observations, $y_1, y_2, \ldots, y_n$, is the joint probability function $p(y_1, y_2, \ldots, y_n)$ when $y_1, y_2, \ldots, y_n$ are discrete random variables.

b. The **likelihood L** of a sample of n observations, $y_1, y_2, \ldots, y_n$, is the joint density function $f(y_1, y_2, \ldots, y_n)$ when $y_1, y_2, \ldots, y_n$ are continuous random variables.

Theorem 8.1 follows directly from the definition of independence and Definitions 6.8 and 6.9.

THEOREM 8.1

a. Let $y_1, y_2, \ldots, y_n$ represent a random sample of n observations on a random variable y. Then $L = p(y_1)p(y_2) \cdots p(y_n)$ when y is a discrete random variable with probability distribution $p(y)$.

b. Let $y_1, y_2, \ldots, y_n$ represent a random sample of n observations on a random variable y. Then $L = f(y_1)f(y_2) \cdots f(y_n)$ when y is a continuous random variable with density function $f(y)$.

DEFINITION 8.9

Let L be the likelihood of a sample, where L is a function of the parameters $\theta_1, \theta_2, \ldots, \theta_k$. Then the **maximum likelihood estimators** of $\theta_1, \theta_2, \ldots, \theta_k$ are the values of $\theta_1, \theta_2, \ldots, \theta_k$ that maximize L.

To simplify our explanation of how to find a maximum likelihood estimator, we will assume that L is a function of a single parameter θ. Then, from differential calculus, we know that the value of θ that maximizes (or minimizes) L is the value for which $\dfrac{dL}{d\theta} = 0$. Obtaining this solution, which always yields a maximum (proof omitted), can be difficult because L is usually the product of a number of quantities involving θ. Differentiating a sum is easier than differentiating a product, so we attempt to maximize the logarithm of L rather than L itself. Since the logarithm of L is a monotonically increasing function of L, L will be maximized by the same value of θ that maximizes its logarithm. We illustrate the procedure in Examples 8.5 and 8.6.

EXAMPLE 8.5

Let $y_1, y_2, \ldots, y_n$ be a random sample of n observations on a random variable y with the exponential density function

$$f(y) = \begin{cases} \dfrac{e^{-y/\beta}}{\beta} & \text{if } 0 \le y < \infty \\ 0 & \text{elsewhere} \end{cases}$$

Determine the maximum likelihood estimator of β.

SOLUTION

Since $y_1, y_2, \ldots, y_n$ are independent random variables, we have

$$L = f(y_1)f(y_2) \cdots f(y_n)$$
$$= \left(\frac{e^{-y_1/\beta}}{\beta}\right)\left(\frac{e^{-y_2/\beta}}{\beta}\right) \cdots \left(\frac{e^{-y_n/\beta}}{\beta}\right)$$
$$= \frac{e^{-\sum_{i=1}^{n} y_i/\beta}}{\beta^n}$$

Taking the natural logarithm of L yields

$$\ln(L) = \ln(e^{-\Sigma_{i=1}^{n} y_i/\beta}) - n \ln(\beta) = -\frac{\sum\limits_{i=1}^{n} y_i}{\beta} - n \ln(\beta)$$

Then

$$\frac{d \ln(L)}{d\beta} = \frac{\sum\limits_{i=1}^{n} y_i}{\beta^2} - \frac{n}{\beta}$$

Setting this derivative equal to 0 and solving for $\hat{\beta}$, we obtain

$$\frac{\sum\limits_{i=1}^{n} y_i}{\hat{\beta}^2} - \frac{n}{\hat{\beta}} = 0 \quad \text{or} \quad n\hat{\beta} = \sum\limits_{i=1}^{n} y_i$$

This yields

$$\hat{\beta} = \frac{\sum\limits_{i=1}^{n} y_i}{n} = \bar{y}$$

Therefore, the maximum likelihood estimator (MLE) of β is the sample mean $\bar{y}$, i.e., $\hat{\beta} = \bar{y}$. ∎

EXAMPLE 8.6 (OPTIONAL)　Let $y_1, y_2, \ldots, y_n$ be a random sample of n observations on the random variable y, where $f(y)$ is a normal density function with mean μ and variance σ^2. Find the maximum likelihood estimators of μ and σ^2.

SOLUTION　Since $y_1, y_2, \ldots, y_n$ are independent random variables, it follows that

$$L = f(y_1)f(y_2) \cdots f(y_n)$$
$$= \left(\frac{e^{-(y_1-\mu)^2/(2\sigma^2)}}{\sigma\sqrt{2\pi}}\right)\left(\frac{e^{-(y_2-\mu)^2/(2\sigma^2)}}{\sigma\sqrt{2\pi}}\right) \cdots \left(\frac{e^{-(y_n-\mu)^2/(2\sigma^2)}}{\sigma\sqrt{2\pi}}\right)$$
$$= \frac{e^{-\Sigma_{i=1}^{n}(y_i-\mu)^2/(2\sigma^2)}}{\sigma^n(2\pi)^{n/2}}$$

and

$$\ln(L) = -\frac{\sum\limits_{i=1}^{n}(y_i - \mu)^2}{2\sigma^2} - \frac{n}{2}\ln(\sigma^2) - \frac{n}{2}\ln(2\pi)$$

Taking derivatives of $\ln(L)$ with respect to μ and σ and setting them equal to 0 yields

$$\frac{d \ln(L)}{d\mu} = \frac{\sum\limits_{i=1}^{n} 2(y_i - \hat{\mu})}{2\hat{\sigma}^2} - 0 - 0 = 0$$

and

$$\frac{d\,\ln(L)}{d\sigma^2} = \frac{\sum_{i=1}^{n}(y_i - \hat{\mu})^2}{2\hat{\sigma}^4} - \frac{n}{2}\left(\frac{1}{\hat{\sigma}^2}\right) - 0 = 0$$

The values of μ and σ^2 that maximize L [and hence $\ln(L)$] will be the simultaneous solution of these two equations. The first equation reduces to

$$\sum_{i=1}^{n}(y_i - \hat{\mu}) = 0 \quad \text{or} \quad \sum_{i=1}^{n} y_i - n\hat{\mu} = 0$$

and it follows that

$$n\hat{\mu} = \sum_{i=1}^{n} y_i \quad \text{and} \quad \hat{\mu} = \bar{y}$$

Substituting $\hat{\mu} = \bar{y}$ into the second equation and multiplying by $2\hat{\sigma}^2$, we obtain

$$\frac{\sum_{i=1}^{n}(y_i - \bar{y})^2}{\hat{\sigma}^2} = n \quad \text{or} \quad \hat{\sigma}^2 = \frac{\sum_{i=1}^{n}(y_i - \bar{y})^2}{n}$$

Therefore, the maximum likelihood estimators of μ and σ^2 are

$$\hat{\mu} = \bar{y} \quad \text{and} \quad \hat{\sigma}^2 = \frac{\sum_{i=1}^{n}(y_i - \bar{y})^2}{n}$$

Note that the maximum likelihood estimator of σ^2 is equal to the sum of squares of deviations $\sum_{i=1}^{n}(y_i - \bar{y})^2$ divided by n, while the sample variance s^2 uses a divisor of $(n-1)$. We showed in Examples 8.1 and 8.2 that s^2 is an unbiased estimator of σ^2. Therefore, the maximum likelihood estimator

$$\hat{\sigma}^2 = \frac{\sum_{i=1}^{n}(y_i - \bar{y})^2}{n} = \frac{(n-1)}{n}s^2$$

is a biased estimator of σ^2.   ∎

Fisher showed that maximum likelihood estimators possess some very unusual properties. As the sample size n becomes larger and larger, the sampling distribution of a maximum likelihood estimator $\hat{\theta}$ tends to become more and more nearly normal, with mean equal to θ and a variance that is equal to or less than the variance of *any other* estimator. Although these properties of maximum likelihood estimators pertain only to estimates based on large samples, they tend to provide support for the maximum likelihood method of estimation. The properties of maximum likelihood estimators based on small samples can be acquired by using the methods of Chapters 4, 5, and 6 to derive their sampling distributions or, at the very least, to acquire their means and variances.

There are several other techniques available for finding point estimators in addition to the method of moments and the method of maximum likelihood. One of these, the **method of least squares**, finds the estimate of θ that minimizes the mean square error (MSE),

$$\text{MSE} = E(\hat{\theta} - \theta)^2$$

The method of least squares—a very important estimation technique—is the topic of Chapters 10–12. Some of the other methods are briefly described below; consult the references at the end of this chapter if you want to learn more about their use.

JACKKNIFE ESTIMATORS Tukey (1958) developed a "leave-one-out-at-a-time" approach to estimation that is gaining increasing acceptance among practitioners, called the **jackknife**.* Let $y_1, y_2, \ldots, y_n$ be a sample of size n from a population with parameter θ. An estimate $\hat{\theta}_{(i)}$ is obtained by omitting the ith observation (i.e., y_i) and computing the estimate based on the remaining $(n-1)$ observations. This calculation is performed for each observation in the data set, and the procedure results in n estimates of θ: $\hat{\theta}_{(1)}, \hat{\theta}_{(2)}, \ldots, \hat{\theta}_{(n)}$. The **jackknife estimator** of θ is then some suitably chosen linear combination (e.g., a weighted average) of the n estimates. Application of the jackknife is suggested for situations where we are likely to have biased samples or find it difficult to assess the variability of the more traditional estimators.

ROBUST ESTIMATORS Many of the estimators discussed in Sections 8.5–8.11 are based on the assumption that the sampled population is approximately normal. When the distribution of the sampled population deviates greatly from normality, such estimators do not have desirable properties (e.g., unbiasedness and minimum variance). An estimator that performs well for a very wide range of probability distributions is called a **robust estimator**. For example, a robust estimate of the population mean μ, called the **M-estimator**, compares favorably to the sample mean $\bar{y}$ when the sampled population is normal and is considerably better than $\bar{y}$ when the population is heavy-tailed. See Mosteller and Tukey (1977) and DeVore (1982) for a good practical discussion of robust estimation techniques.

BAYES ESTIMATORS The classical approach to estimation is based on the concept that the unknown parameter θ is a constant. All the information available to us about θ is contained in the random sample $y_1, y_2, \ldots, y_n$ selected from the relevant population. In contrast, the **Bayesian** approach to estimation regards θ as a random variable with some known (**prior**) probability distribution $g(\theta)$. The sample information is used to modify the prior distribution on θ to obtain the **posterior** distribution, $f(\theta \mid y_1, y_2, \ldots, y_n)$. The **Bayes estimator** of θ is then the mean of the posterior probability distribution [see Mendenhall, Scheaffer, and Wackerly (1981)].

*The procedure derives its name from the Boy Scout jackknife, which serves as a handy tool in a variety of situations when specialized techniques may not be available.

OPTIONAL EXERCISES

8.7 Let $y_1, y_2, \ldots, y_n$ be a random sample of n observations from a Poisson distribution with probability function

$$p(y) = \frac{e^{-\lambda}\lambda^y}{y!} \quad (y = 0, 1, 2, \ldots)$$

a. Find the maximum likelihood estimator of λ.
b. Is the maximum likelihood estimator unbiased?

8.8 A binomial experiment consisting of n trials resulted in observations $y_1, y_2, \ldots, y_n$, where

$$y_i = \begin{cases} 1 & \text{if the } i\text{th trial was a success} \\ 0 & \text{if not} \end{cases}$$

and $P(y_i = 1) = p$, $P(y_i = 0) = 1 - p$. Let $y = \sum_{i=1}^{n} y_i$ be the number of successes in n trials.
a. Find the moment estimator of p.
b. Is the moment estimator unbiased?
c. Find the maximum likelihood estimator of p. [*Hint:* $L = \binom{n}{y}p^y(1 - p)^{n-y}$.]
d. Is the maximum likelihood estimator unbiased?

8.9 Let $y_1, y_2, \ldots, y_n$ be a random sample of n observations from an exponential distribution with density

$$f(y) = \begin{cases} \frac{1}{\beta}e^{-y/\beta} & \text{if } y > 0 \\ 0 & \text{if otherwise} \end{cases}$$

a. Find the moment estimator of β.
b. Is the moment estimator unbiased?
c. Find $V(\hat{\beta})$.

8.10 Let $y_1, y_2, \ldots, y_n$ be a random sample of n observations on a random variable y, where $f(y)$ is a gamma density function with $\alpha = 2$ and unknown β:

$$f(y) = \begin{cases} \frac{ye^{-y/\beta}}{\beta^2} & \text{if } y > 0 \\ 0 & \text{otherwise} \end{cases}$$

a. Find the maximum likelihood estimator of β.
b. Find $E(\hat{\beta})$ and $V(\hat{\beta})$.

8.11 Refer to Exercise 8.10.
a. Find the moment estimator of β.
b. Find $E(\hat{\beta})$ and $V(\hat{\beta})$.

8.12 Let $y_1, y_2, \ldots, y_n$ be a random sample of n observations from a normal distribution with mean 0 and unknown variance σ^2. Find the maximum likelihood estimator of σ^2.

S E C T I O N   8.4

**FINDING INTERVAL
ESTIMATORS: THE
PIVOTAL METHOD**

In Section 8.1, we defined an interval estimator as a rule that tells how to use the sample observations to calculate two numbers that define an interval that will enclose the estimated parameter with a high probability. The resulting random interval (random, because the sample observations used to calculate the endpoints of the interval are random variables) is called a **confidence interval** and the probability that it contains the estimated parameter is called its **confidence coefficient**. If a confidence interval has a confidence coefficient equal to .95, we call it a 95% confidence interval. If the confidence coefficient is .99, the interval is said to be a 99% confidence interval, etc. A more practical interpretation of the confidence coefficient for a confidence interval is given later in this section.

DEFINITION 8.10

The **confidence coefficient** for a confidence interval is equal to the probability that the random interval will contain the estimated parameter.

One way to find a confidence interval for a parameter θ is to acquire a **pivotal statistic**, a statistic that is a function of the sample values and the single parameter θ. Because many statistics are approximately normally distributed when the sample size n is large (central limit theorem), we can construct confidence intervals for their expected values using the standard normal random variable z as a pivotal statistic.

To illustrate, let $\hat{\theta}$ be a statistic with a sampling distribution that is approximately normally distributed for large samples with mean $E(\hat{\theta}) = \theta$ and standard error $\sigma_{\hat{\theta}}$. Then,

$$z = \frac{\hat{\theta} - \theta}{\sigma_{\hat{\theta}}}$$

is a standard normal random variable. Since z is also a function of only the sample statistic $\hat{\theta}$ and the parameter θ, we will use it as a pivotal statistic. To derive a confidence interval for θ, we first make a probability statement about the pivotal statistic. To do this, we locate values $z_{\alpha/2}$ and $-z_{\alpha/2}$ that place a probability of $\alpha/2$ in each tail of the z distribution (see Figure 8.6, page 296), i.e., $P(z > z_{\alpha/2}) = \alpha/2$. It can be seen from Figure 8.6 that

$$P(-z_{\alpha/2} \le z \le z_{\alpha/2}) = 1 - \alpha$$

Substituting the expression for z into the probability statement and using some simple algebraic operations on the inequality, we obtain

$$
\begin{aligned}
P(-z_{\alpha/2} \le z \le z_{\alpha/2}) &= P\left(-z_{\alpha/2} \le \frac{\hat{\theta} - \theta}{\sigma_{\hat{\theta}}} \le z_{\alpha/2}\right) \\
&= P(-z_{\alpha/2}\sigma_{\hat{\theta}} \le \hat{\theta} - \theta \le z_{\alpha/2}\sigma_{\hat{\theta}}) \\
&= P(-\hat{\theta} - z_{\alpha/2}\sigma_{\hat{\theta}} \le -\theta \le -\hat{\theta} + z_{\alpha/2}\sigma_{\hat{\theta}}) \\
&= P(\hat{\theta} - z_{\alpha/2}\sigma_{\hat{\theta}} \le \theta \le \hat{\theta} + z_{\alpha/2}\sigma_{\hat{\theta}}) = 1 - \alpha
\end{aligned}
$$

FIGURE 8.6

Locating $z_{\alpha/2}$

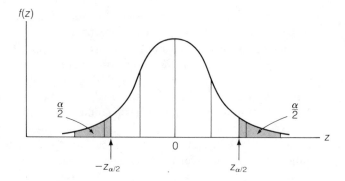

Therefore, the probability that the interval formed by

$$\text{LCL} = \hat{\theta} - z_{\alpha/2}\sigma_{\hat{\theta}} \qquad \text{to} \qquad \text{UCL} = \hat{\theta} + z_{\alpha/2}\sigma_{\hat{\theta}}$$

will enclose θ is equal to $(1 - \alpha)$. The quantities LCL and UCL are called the **lower** and **upper confidence limits**, respectively, for the confidence interval. The confidence coefficient for the interval will be $(1 - \alpha)$.

The derivation of a large-sample $(1 - \alpha)100\%$ confidence interval for θ is summarized in Theorem 8.2.

THEOREM 8.2

Let $\hat{\theta}$ be normally distributed for large samples with $E(\hat{\theta}) = \theta$ and standard error $\sigma_{\hat{\theta}}$. Then a $(1 - \alpha)100\%$ confidence interval for θ is

$$\hat{\theta} - z_{\alpha/2}\sigma_{\hat{\theta}} \qquad \text{to} \qquad \hat{\theta} + z_{\alpha/2}\sigma_{\hat{\theta}}$$

The large-sample confidence interval can also be acquired intuitively by examining Figure 8.7. The z value corresponding to an area $A = .475$—i.e., the z value that places area $\alpha = .025$ in the upper tail of the z distribution—is (see Table 4 of Appendix II) $z_{.025} = 1.96$. Therefore, the probability that $\hat{\theta}$ will lie within $1.96\sigma_{\hat{\theta}}$ of θ is .95. You can see from Figure 8.7 that whenever $\hat{\theta}$ falls within the interval $\theta \pm 1.96\sigma_{\hat{\theta}}$, then the interval $\hat{\theta} \pm 1.96\sigma_{\hat{\theta}}$ will enclose θ. Therefore, $\hat{\theta} \pm 1.96\sigma_{\hat{\theta}}$ yields a 95% confidence interval for θ.

FIGURE 8.7

The Sampling Distribution
of θ for Large Samples

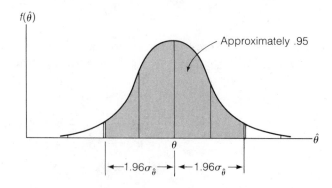

We may encounter one slight difficulty when we attempt to apply this confidence interval in practice. It is often the case that $\sigma_{\hat{\theta}}$ is a function of the parameter θ that we are attempting to estimate. However, when the sample size n is large (which we have assumed throughout the derivation), we can substitute the estimate $\hat{\theta}$ for the parameter θ to obtain an approximate value for $\sigma_{\hat{\theta}}$.

In Example 8.7 we will use a pivotal statistic to find a confidence interval for μ when the sample size is small ($n < 30$).

EXAMPLE 8.7

Let $\bar{y}$ and s^2 be the sample mean and variance based on a random sample of n observations ($n < 30$) from a normal distribution with mean μ and variance σ^2. Find a 95% confidence interval for μ.

SOLUTION

A pivotal statistic for μ can be constructed using the t statistic of Chapter 7. By Definition 7.4,

$$t = \frac{z}{\sqrt{\chi^2/\nu}}$$

where z and χ^2 are independent random variables and χ^2 is based on ν degrees of freedom. We know that $\bar{y}$ is normally distributed and that

$$z = \frac{\bar{y} - \mu}{\sigma/\sqrt{n}}$$

is a standard normal random variable. From Theorem 7.4, it follows that

$$\frac{(n-1)s^2}{\sigma^2} = \chi^2$$

is a chi-square random variable with $\nu = (n-1)$ degrees of freedom. We state (without proof) that $\bar{y}$ and s^2 are independent when they are based on a random sample selected from a normal distribution. Therefore, z and χ^2 will be independent random variables. Substituting the expressions for z and χ^2 into the formula for t, we obtain

$$t = \frac{z}{\sqrt{\chi^2/\nu}} = \frac{\dfrac{\bar{y} - \mu}{\sigma/\sqrt{n}}}{\sqrt{\dfrac{(n-1)s^2}{\sigma^2} \Big/ (n-1)}} = \frac{\bar{y} - \mu}{s/\sqrt{n}}$$

Note that the pivotal statistic is a function only of μ and the sample statistics $\bar{y}$ and s^2.

The next step in finding a confidence interval for μ is to make a probability statement about the pivotal statistic t. We will select two values of t, call them $t_{\alpha/2}$ and $-t_{\alpha/2}$, that correspond to probabilities of $\alpha/2$ in the upper and lower tails, respectively, of the t distribution (see Figure 8.8, page 298). From Figure 8.8, it can be seen that

$$P(-t_{\alpha/2} \leq t \leq t_{\alpha/2}) = 1 - \alpha$$

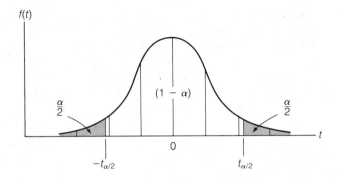

Substituting the expression for t into the probability statement, we obtain

$$P(-t_{\alpha/2} \le t \le t_{\alpha/2}) = P\left(-t_{\alpha/2} \le \frac{\bar{y} - \mu}{s/\sqrt{n}} \le t_{\alpha/2}\right) = 1 - \alpha$$

Multiplying the inequality within the brackets by $s/\sqrt{n}$, we obtain

$$P\left[-t_{\alpha/2}\left(\frac{s}{\sqrt{n}}\right) \le \bar{y} - \mu \le t_{\alpha/2}\left(\frac{s}{\sqrt{n}}\right)\right] = 1 - \alpha$$

Subtracting $\bar{y}$ from each part of the inequality yields

$$P\left[-\bar{y} - t_{\alpha/2}\left(\frac{s}{\sqrt{n}}\right) \le -\mu \le \bar{y} + t_{\alpha/2}\left(\frac{s}{\sqrt{n}}\right)\right] = 1 - \alpha$$

Finally, we multiply each term of the inequality by (-1), thereby reversing the inequality signs. The result is

$$P\left[\bar{y} - t_{\alpha/2}\left(\frac{s}{\sqrt{n}}\right) \le \mu \le \bar{y} + t_{\alpha/2}\left(\frac{s}{\sqrt{n}}\right)\right] = 1 - \alpha$$

Therefore, a $(1 - \alpha)100\%$ confidence interval for μ when n is small is

$$\bar{y} - t_{\alpha/2}\left(\frac{s}{\sqrt{n}}\right) \qquad \text{to} \qquad \bar{y} + t_{\alpha/2}\left(\frac{s}{\sqrt{n}}\right)$$ ■

We now apply the confidence interval derived in Example 8.7 to a practical situation.

EXAMPLE 8.8

Chemical plants must be regulated to prevent the poisoning of fish in nearby rivers or streams. One of the measurements made on fish to evaluate the potential toxicity of chemicals is the length reached by adults. If a river or stream is inhabited by an abundance of adult fish with lengths less than the average adult length of their species, we have strong evidence that the river is being chemically contaminated. A chemical plant, under investigation for chlorine poisoning of a stream, has hired a biologist to estimate the mean length of fathead minnows (the main inhabitants of the stream) exposed to 20 micrograms of chlorine per liter of water. The biologist captures 20 newborn fathead minnows from the stream and rears them in aquaria with this chlorine concentration. The length of each (in millimeters) is measured after a 10-week maturation period, with the following results:

$$\bar{y} = 27.5 \qquad s = 2.6$$

Construct a 95% confidence interval for the true mean length of fathead minnows reared in chlorine-contaminated water. Assume that the lengths of the fathead minnows are approximately normal.

SOLUTION

Recall that the sampling distribution of the t statistic depends on its degrees of freedom, ν. The tabulated values t_a, such that $P(t \geq t_a) = a$, are given in Table 6 of Appendix II, for values of ν from 1 to 29, as well as the value of t_a when ν becomes infinitely large. An abbreviated version of this table is shown in Table 8.1. For example, suppose a t statistic is based on $\nu = 4$ degrees of freedom (df) and we want to find the value t_a that places probability $a = .025$ in the upper tail of the t distribution. The appropriate value, shaded in Table 8.1, is $t_{.025} = 2.776$.

TABLE 8.1
An Abbreviated Version of Table 6 of Appendix II

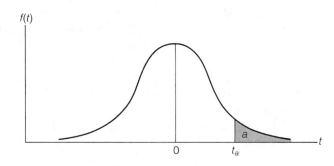

DEGREES OF FREEDOM	$t_{.100}$	$t_{.050}$	$t_{.025}$	$t_{.010}$	$t_{.005}$
1	3.078	6.314	12.706	31.821	63.657
2	1.886	2.920	4.303	6.965	9.925
3	1.638	2.353	3.182	4.541	5.841
4	1.533	2.132	2.776	3.747	4.604
5	1.476	2.015	2.571	3.365	4.032
6	1.440	1.943	2.447	3.143	3.707
7	1.415	1.895	2.365	2.998	3.499
8	1.397	1.860	2.306	2.896	3.355
9	1.383	1.833	2.262	2.821	3.250
10	1.372	1.812	2.228	2.764	3.169
11	1.363	1.796	2.201	2.718	3.106
12	1.356	1.782	2.179	2.681	3.055
13	1.350	1.771	2.160	2.650	3.012
14	1.345	1.761	2.145	2.624	2.977
15	1.341	1.753	2.131	2.602	2.947

For our example, $n = 20$ and t will possess $(n - 1) = 19$ degrees of freedom. Since we want to find a $95\% = (1 - \alpha)100\%$ confidence interval for the mean length μ of fathead minnows, $\alpha = .05$ and we need to find the value $t_{.025}$ corresponding to $a = .025$ and 19 degrees of freedom. This value is given in Table 6 of Appendix II as $t_{\alpha/2} = t_{.025} = 2.093$. Then the confidence interval is

$$\bar{y} \pm t_{.025}\left(\frac{s}{\sqrt{n}}\right) = 27.5 \pm 2.093\left(\frac{2.60}{\sqrt{20}}\right)$$

$$= 27.5 \pm 1.22 \quad \text{or} \quad (26.28, 28.72)$$

Since the confidence coefficient is .95, we say that we are 95% confident that the interval from 26.28 to 28.72 millimeters contains the true mean length, μ, of fathead minnows reared in chlorine-contaminated water. ∎

To demonstrate the interpretation of a confidence interval, we programmed a computer to draw 1,000 samples of size $n = 10$ from a normal distribution with mean $\mu = 10$ and variance $\sigma^2 = 1$. A 95% confidence interval for μ (derived in Example 8.7) was computed for each of the 1,000 samples. These are shown in Table 8.2. Only the 50 intervals that are starred (*) fail to enclose the mean $\mu = 10$. The proportion that enclose μ, .95, is exactly equal to the confidence coefficient. This explains why we are reasonably confident that the interval calculated in Example 8.8 (26.28, 28.72), encloses the true value of μ. *If we were to employ our interval estimator on repeated occasions, 95% of the intervals constructed would contain μ.*

Confidence intervals for population parameters other than the population mean can be derived using the pivotal method outlined in this section. The estimators and pivotal statistics for many of these parameters are well known. In Sections 8.5–8.11 we give the confidence interval formulas for several population parameters that are commonly encountered in practice.

EXERCISES 8.13–8.22

8.13 Use Table 6 of Appendix II to determine the values of $t_{\alpha/2}$ that would be used in the construction of a confidence interval for a population mean for each of the following combinations of confidence coefficient and sample size:
a. Confidence coefficient .99, $n = 18$
b. Confidence coefficient .95, $n = 10$
c. Confidence coefficient .90, $n = 15$

8.14 It can be shown (proof omitted) that as the sample size n increases, the t distribution tends to normality and the value t_a, such that $P(t > t_a) = a$, approaches the value z_a, such that $P(z > z_a) = a$. Use Table 6 of Appendix II to verify that as the sample size n gets infinitely large, $t_{.05} = z_{.05}$, $t_{.025} = z_{.025}$, and $t_{.01} = z_{.01}$.

TABLE 8.2 One Thousand 95% Confidence Intervals for the Mean of a Normal Distribution ($\mu = 10$, $\sigma^2 = 1$).
[*Note:* Starred (*) intervals fail to include $\mu = \dagger 0$.]

SAMPLE	LCL	UCL	SAMPLE	LCL	UCL	SAMPLE	LCL	UCL	SAMPLE	LCL	UCL	SAMPLE	LCL	UCL
1	9.574	11.183	2	9.370	11.137	3	9.425	10.403	4	9.356	10.777	5	9.956	11.455
6	8.807	10.589	7	9.188	10.588	8	9.798	11.390	9	10.057	11.561*	10	9.793	11.009
11	9.180	10.848	12	9.735	10.399	13	9.518	10.785	14	9.872	11.003	15	9.028	10.907
16	9.414	11.107	17	9.603	10.816	18	9.469	10.896	19	8.758	9.889*	20	9.617	11.285
21	8.926	10.389	22	9.710	10.512	23	8.847	10.667	24	9.148	10.675	25	9.722	11.017
26	9.575	11.221	27	8.820	10.664	28	9.222	11.015	29	9.525	10.717	30	9.036	10.802
31	8.758	10.151	32	9.043	10.650	33	9.819	10.958	34	9.085	10.610	35	8.847	10.692
36	9.590	10.551	37	9.826	11.063	38	9.506	10.615	39	9.322	10.401	40	8.892	10.239
41	9.519	10.812	42	9.023	10.640	43	9.560	10.651	44	8.885	10.327	45	9.901	11.353
46	9.348	10.324	47	9.188	10.766	48	9.173	10.563	49	9.039	10.567	50	9.234	10.864
51	8.799	10.503	52	9.870	11.382	53	8.706	10.830	54	9.690	11.002	55	9.340	10.740
56	9.244	10.666	57	9.910	11.487	58	8.690	10.111	59	9.149	10.661	60	9.320	10.032
61	9.052	11.085	62	9.593	10.985	63	9.103	10.768	64	9.429	10.023	65	9.275	10.638
66	9.139	11.033	67	9.621	10.916	68	9.464	10.571	69	9.717	10.938	70	8.895	10.638
71	9.496	11.260	72	9.124	10.437	73	9.416	10.718	74	8.516	10.296	75	8.991	10.290
76	9.225	10.554	77	8.641	10.451	78	9.598	11.359	79	9.443	11.217	80	9.384	11.055
81	9.088	10.592	82	9.777	11.142	83	9.160	10.683	84	8.969	10.122	85	9.275	10.372
86	9.676	10.772	87	9.075	10.507	88	9.425	10.059	89	9.382	10.761	90	9.646	10.648
91	9.414	10.410	92	9.661	10.909	93	9.240	10.679	94	9.138	10.469	95	9.497	10.824
96	8.871	10.627	97	9.402	10.923	98	9.241	10.695	99	8.832	10.790	100	9.546	10.472
101	9.604	10.705	102	9.315	10.519	103	9.306	10.572	104	9.129	10.750	105	9.566	10.658
106	9.490	10.558	107	9.052	10.387	108	9.464	10.969	109	9.899	10.919	110	9.045	10.284
111	9.112	10.341	112	9.593	10.836	113	8.726	10.515	114	9.511	10.916	115	9.535	10.938
116	9.526	10.690	117	8.848	10.395	118	8.765	10.126	119	9.151	10.696	120	8.764	10.382
121	9.598	10.578	122	8.988	10.286	123	9.435	10.890	124	8.833	10.335	125	9.644	10.744
126	9.589	10.533	127	9.284	10.174	128	9.110	10.522	129	9.502	10.728	130	9.144	11.044
131	9.344	10.851	132	9.915	11.372	133	9.252	10.399	134	9.833	11.188	135	9.268	10.219
136	9.681	10.804	137	9.082	10.719	138	9.374	10.198	139	9.303	10.781	140	9.046	10.329
141	9.191	10.640	142	9.777	10.812	143	8.622	10.513	144	9.175	10.931	145	9.227	10.863
146	9.167	10.455	147	9.323	11.067	148	9.148	10.282	149	9.169	10.360	150	9.635	10.989
151	9.354	10.900	152	8.950	10.546	153	8.911	9.986*	154	9.039	10.502	155	9.765	10.749
156	9.584	10.838	157	8.566	9.779*	158	9.114	10.508	159	9.312	10.971	160	9.262	10.390
161	9.518	10.579	162	9.621	11.306	163	9.183	10.142	164	8.686	10.564	165	9.249	10.030
166	9.129	11.088	167	9.740	10.844	168	9.201	10.795	169	8.680	10.580	170	9.442	10.740
171	9.382	10.690	172	9.542	11.062	173	9.201	10.730	174	9.485	11.251	175	8.640	10.463
176	9.385	10.516	177	9.241	10.585	178	9.495	10.810	179	9.859	11.136	180	9.356	10.975
181	8.769	10.727	182	8.884	10.759	183	9.011	10.829	184	9.401	10.731	185	8.637	10.468
186	9.157	10.439	187	9.925	11.065	188	9.427	11.199	189	9.550	11.040	190	9.729	10.887
191	9.014	11.275	192	9.277	10.485	193	9.390	10.611	194	9.358	10.997	195	9.229	10.976
196	9.153	10.465	197	8.475	10.120	198	9.596	10.469	199	9.485	11.083	200	9.238	10.721
201	8.930	10.040	202	9.121	10.960	203	9.596	10.440	204	9.595	11.125	205	9.136	10.383
206	9.380	10.707	207	9.045	10.714	208	8.894	10.830	209	9.029	10.821	210	9.320	10.355
211	8.958	9.899*	212	9.283	10.236	213	9.280	10.681	214	8.606	10.343	215	9.585	11.183
216	9.312	10.209	217	9.511	10.891	218	9.733	10.805	219	9.037	10.317	220	9.777	10.910
221	9.302	10.266	222	9.349	10.646	223	9.252	10.943	224	9.682	11.676	225	8.773	10.697
226	9.316	10.725	227	9.756	10.685	228	9.544	10.478	229	9.523	10.876	230	9.444	10.866
231	9.138	10.986	232	9.534	11.093	233	9.455	10.602	234	8.962	10.305	235	9.374	10.571
236	9.830	10.732	237	9.778	11.217	238	9.597	11.095	239	8.985	10.318	240	8.915	10.549
241	9.020	10.123	242	9.810	10.923	243	8.951	10.315	244	10.451	11.345*	245	9.170	10.339
246			247	9.296	10.288	248	9.510	10.359	249	9.031	10.354	250	9.428	11.241

TABLE 8.2 (continued)

SAMPLE	(LCL , UCL)	SAMPLE	(LCL , UCL)	SAMPLE	(LCL , UCL)	SAMPLE	(LCL , UCL)	SAMPLE	(LCL , UCL)
251	(9.445 , 10.761)	252	(9.697 , 10.937)	253	(9.494 , 11.246)	254	(9.200 , 10.861)	255	(8.904 , 10.378)
256	(9.129 , 10.713)	257	(9.383 , 10.142)	258	(9.879 , 10.792)	259	(8.852 , 10.912)	260	(9.725 , 10.517)
261	(9.626 , 11.264)	262	(8.700 , 10.547)	263	(8.911 , 10.488)	264	(9.289 , 10.694)	265	(9.229 , 10.756)
266	(9.188 , 10.753)	267	(9.396 , 11.143)	268	(9.225 , 11.391)	269	(9.005 , 10.153)	270	(9.196 , 10.505)
271	(9.208 , 10.526)	272	(8.902 , 10.389)	273	(8.742 , 11.019)	274	(9.069 , 10.410)	275	(9.501 , 10.632)
276	(9.324 , 10.635)	277	(9.488 , 11.056)	278	(9.277 , 10.547)	279	(9.408 , 10.679)	280	(9.329 , 10.839)
281	(8.636 , 9.820)*	282	(9.646 , 10.639)	283	(9.403 , 10.742)	284	(9.216 , 10.454)	285	(8.598 , 9.849)*
286	(9.266 , 11.348)	287	(9.208 , 10.449)	288	(9.113 , 10.901)	289	(8.934 , 10.334)	290	(9.306 , 10.454)
291	(9.573 , 11.202)	292	(9.063 , 10.685)	293	(10.229 , 11.040)*	294	(9.254 , 11.018)	295	(9.137 , 10.709)
296	(8.815 , 10.211)	297	(9.007 , 10.592)	298	(8.787 , 10.315)	299	(9.260 , 10.962)	300	(9.319 , 10.876)
301	(8.995 , 10.614)	302	(9.104 , 10.095)	303	(9.306 , 10.346)	304	(9.239 , 10.968)	305	(9.113 , 9.934)*
306	(9.390 , 10.522)	307	(9.639 , 10.848)	308	(9.209 , 10.601)	309	(8.866 , 11.200)	310	(9.597 , 11.396)
311	(8.391 , 10.287)	312	(8.964 , 10.791)	313	(9.645 , 10.769)	314	(9.373 , 10.817)	315	(9.420 , 10.751)
316	(9.312 , 11.081)	317	(8.901 , 10.141)	318	(9.730 , 10.525)	319	(9.636 , 11.284)	320	(9.291 , 10.781)
321	(9.723 , 10.775)	322	(9.249 , 10.688)	323	(9.113 , 10.160)	324	(9.109 , 10.434)	325	(9.329 , 11.007)
326	(8.959 , 10.226)	327	(9.664 , 10.516)	328	(9.856 , 11.101)	329	(9.345 , 10.956)	330	(8.908 , 10.843)
331	(9.693 , 11.499)	332	(9.423 , 11.238)	333	(8.978 , 10.768)	334	(9.534 , 11.050)	335	(9.576 , 11.082)
336	(9.371 , 10.638)	337	(8.950 , 11.271)	338	(9.276 , 10.557)	339	(9.310 , 10.619)	340	(9.073 , 11.076)
341	(9.281 , 10.795)	342	(9.744 , 10.505)	343	(9.542 , 10.813)	344	(8.913 , 10.316)	345	(9.414 , 11.246)
346	(9.661 , 10.722)	347	(8.724 , 10.361)	348	(9.224 , 11.179)	349	(9.354 , 10.569)	350	(9.318 , 10.665)
351	(9.174 , 10.895)	352	(9.615 , 11.003)	353	(9.121 , 10.696)	354	(9.517 , 10.884)	355	(8.730 , 10.328)
356	(8.942 , 10.684)	357	(8.890 , 10.399)	358	(9.640 , 11.146)	359	(9.349 , 11.253)	360	(9.522 , 10.452)
361	(9.136 , 10.555)	362	(8.923 , 10.764)	363	(9.737 , 11.513)	364	(9.159 , 10.257)	365	(9.736 , 11.236)
366	(9.156 , 10.699)	367	(9.515 , 11.037)	368	(9.175 , 10.724)	369	(9.475 , 10.408)	370	(9.108 , 10.933)
371	(9.135 , 10.924)	372	(9.144 , 10.631)	373	(8.854 , 10.319)	374	(8.680 , 10.513)	375	(9.529 , 10.391)
376	(9.886 , 11.506)	377	(9.633 , 10.961)	378	(9.252 , 11.411)	379	(9.218 , 10.469)	380	(9.648 , 10.653)
381	(9.193 , 10.900)	382	(9.331 , 10.868)	383	(9.072 , 10.634)	384	(8.974 , 10.078)	385	(9.170 , 10.652)
386	(9.063 , 10.490)	387	(9.047 , 10.484)	388	(9.202 , 10.194)	389	(9.475 , 10.655)	390	(9.553 , 10.720)
391	(9.396 , 11.115)	392	(9.236 , 10.644)	393	(8.785 , 10.183)	394	(9.145 , 10.765)	395	(9.340 , 11.055)
396	(9.123 , 10.534)	397	(9.828 , 11.058)	398	(9.386 , 10.229)	399	(9.234 , 10.500)	400	(9.072 , 10.405)
401	(9.680 , 10.752)	402	(9.487 , 10.849)	403	(9.539 , 11.137)	404	(9.795 , 11.293)	405	(9.566 , 10.792)
406	(8.983 , 10.842)	407	(9.410 , 10.964)	408	(9.892 , 10.949)	409	(9.097 , 11.117)	410	(9.229 , 11.201)
411	(9.451 , 10.924)	412	(9.530 , 10.756)	413	(9.328 , 11.021)	414	(9.512 , 10.590)	415	(9.027 , 10.805)
416	(8.982 , 10.591)	417	(9.059 , 10.856)	418	(8.971 , 10.620)	419	(9.236 , 10.456)	420	(9.085 , 10.768)
421	(8.900 , 10.358)	422	(9.604 , 11.044)	423	(10.265 , 11.443)*	424	(9.101 , 10.972)	425	(9.229 , 10.903)
426	(9.092 , 10.530)	427	(8.971 , 10.457)	428	(9.116 , 11.071)*	429	(9.579 , 11.107)	430	(9.066 , 10.596)
431	(8.892 , 10.710)	432	(9.684 , 11.258)	433	(8.919 , 10.350)	434	(9.226 , 11.093)	435	(9.012 , 10.969)
436	(8.582 , 10.107)	437	(9.106 , 9.925)*	438	(8.820 , 10.324)	439	(9.031 , 10.282)	440	(9.206 , 10.572)
441	(9.473 , 10.449)	442	(9.075 , 10.210)	443	(9.500 , 11.252)	444	(9.513 , 10.446)	445	(8.878 , 10.616)
446	(8.818 , 10.675)	447	(9.399 , 11.045)	448	(8.961 , 10.221)	449	(9.866 , 10.829)	450	(9.655 , 10.463)
451	(9.835 , 10.905)	452	(9.397 , 10.600)	453	(9.073 , 10.202)	454	(9.006 , 10.141)	455	(9.322 , 10.270)
456	(9.596 , 10.959)	457	(9.479 , 11.106)	458	(9.978 , 11.042)	459	(9.637 , 10.857)	460	(8.759 , 10.644)
461	(9.574 , 11.400)	462	(9.186 , 10.611)	463	(9.646 , 11.112)	464	(9.587 , 10.604)	465	(9.907 , 10.885)
466	(9.211 , 10.779)	467	(9.169 , 10.604)	468	(9.094 , 11.083)	469	(8.755 , 10.473)	470	(9.661 , 11.289)
471	(9.798 , 11.143)	472	(9.358 , 11.244)	473	(9.485 , 10.234)	474	(9.044 , 10.299)	475	(9.358 , 11.257)
476	(9.325 , 10.594)	477	(8.857 , 10.668)	478	(9.487 , 10.684)	479	(8.896 , 10.333)	480	(9.382 , 10.697)
481	(9.298 , 10.465)	482	(9.594 , 10.939)	483	(9.297 , 10.780)	484	(9.242 , 10.483)	485	(9.074 , 9.904)*
486	(9.284 , 10.777)	487	(9.125 , 10.438)	488	(9.081 , 10.442)	489	(8.961 , 10.458)	490	(9.420 , 10.687)
491	(9.284 , 10.471)	492	(9.462 , 10.845)	493	(9.274 , 10.676)	494	(9.527 , 10.925)	495	(9.796 , 11.205)
496	(9.080 , 10.854)	497	(9.399 , 10.606)	498	(8.593 , 10.554)	499	(9.527 , 10.925)	500	(9.340 , 10.413)

TABLE 8.2 (continued)

SAMPLE	LCL	UCL	SAMPLE	LCL	UCL	SAMPLE	LCL	UCL	SAMPLE	LCL	UCL	SAMPLE	LCL	UCL
501	8.907	10.531	502	9.243	10.401	503	9.270	10.787	504	9.331	11.183	505	9.200	10.602
506	9.422	10.835	507	9.279	10.687	508	8.884	10.683	509	9.190	10.255	510	8.733	10.404
511	9.441	10.889	512	9.070	10.392	513	8.946	9.772*	514	8.799	10.236	515	9.070	10.830
516	8.984	10.365	517	9.183	10.387	518	9.236	11.276	519	9.826	11.197	520	8.906	10.672
521	9.755	10.996	522	9.400	10.880	523	9.374	11.072	524	9.288	11.211	525	9.412	11.000*
526	9.028	10.300	527	8.647	10.285	528	9.190	10.014	529	9.238	10.465	530	10.003	11.190*
531	9.644	10.974	532	9.679	10.738	533	8.559	10.836	534	9.895	11.131	535	9.653	11.007
536	9.769	11.144	537	8.837	10.136	538	9.939	11.341	539	9.553	10.853	540	9.351	10.552
541	9.532	11.320	542	9.262	10.728	543	8.864	10.042	544	9.052	10.482	545	9.551	10.610
546	9.564	11.060	547	9.699	10.912	548	8.915	9.854*	549	8.801	10.648	550	9.111	9.913*
551	9.811	11.558	552	8.593	10.128	553	8.612	11.456	554	9.555	10.986	555	9.567	10.666
556	9.399	10.979	557	9.168	11.195	558	9.270	11.128	559	9.197	10.554	560	8.985	10.492
561	9.067	10.622	562	8.843	10.484	563	9.346	10.862	564	8.692	10.475	565	9.413	10.583
566	9.416	10.678	567	9.451	11.099	568	9.339	11.306	569	8.933	10.698	570	9.212	10.368
571	8.894	10.438	572	9.161	10.964	573	9.841	10.734	574	8.990	10.541	575	8.530	10.038
576	9.687	11.005	577	9.131	10.759	578	9.167	10.523	579	9.301	10.507	580	9.379	10.872
581	9.298	10.917	582	8.407	10.030	583	9.080	10.305	584	9.043	10.509	585	9.636	11.032
586	9.562	10.527	587	9.224	10.394	588	9.439	10.278	589	9.320	10.482	590	9.363	11.047
591	9.136	10.521	592	9.059	10.320	593	8.686	10.710	594	9.280	10.267	595	9.251	10.964
596	8.693	10.114	597	8.712	11.075	598	9.340	11.485	599	8.244	9.684*	600	9.583	10.992
601	9.232	10.346	602	9.014	10.458	603	9.861	10.548	604	9.139	11.097	605	9.060	10.269
606	9.712	11.648	607	8.963	10.055	608	8.991	10.898	609	9.540	10.769	610	9.822	11.243
611	9.338	10.357	612	8.632	10.201	613	9.371	10.527	614	9.155	10.582	615	8.806	10.919
616	9.182	10.488	617	9.403	10.755	618	9.199	10.590	619	9.016	10.844	620	9.321	11.077
621	9.475	10.651	622	9.481	10.701	623	9.661	10.332	624	9.358	10.812	625	9.046	10.679
626	9.948	10.907	627	8.649	9.996*	628	9.201	10.885	629	9.195	10.908	630	9.460	10.435
631	9.222	10.772	632	9.757	10.880	633	9.926	10.879	634	9.027	10.425	635	8.436	10.011
636	9.160	10.474	637	9.723	11.075	638	8.597	10.962	639	10.024	10.931*	640	8.475	10.397
641	8.710	10.702	642	10.038	11.678*	643	9.706	11.277*	644	9.028	11.275	645	9.395	10.414
646	9.283	10.641	647	8.628	10.107	648	9.456	10.965	649	9.999	11.336	650	8.587	10.063
651	9.616	11.090	652	9.403	10.537	653	10.263	10.597	654	9.325	10.717	655	9.795	10.737
656	9.669	11.778	657	9.739	10.636	658	9.285	11.385*	659	9.210	10.552	660	9.384	10.962
661	9.041	10.347	662	9.380	10.846	663	9.950	10.174	664	9.602	10.584	665	9.092	10.439
666	9.475	10.844	667	9.192	10.844	668	10.134	11.064	669	8.523	10.431	670	9.657	11.222
671	8.710	10.470	672	8.854	10.039	673	8.833	10.898	674	9.500	10.956	675	9.546	10.782
676	9.115	10.545	677	9.005	10.434	678	9.783	9.846*	679	9.384	10.647	680	9.783	11.586
681	9.160	10.040	682	8.822	10.238	683	9.374	11.275	684	8.895	10.274	685	8.986	10.854
686	8.665	10.523	687	8.630	10.270	688	9.914	10.562	689	8.787	10.323	690	9.483	10.850
691	9.369	10.797	692	9.271	10.776	693	8.715	10.482	694	8.764	10.481	695	8.934	10.053
696	9.280	10.143	697	8.354	9.985*	698	9.599	10.578*	699	8.488	10.224	700	9.278	10.213*
701	9.247	10.552	702	9.043	10.327	703	9.578	11.239*	704	8.815	10.387	705	8.786	9.901*
706	9.029	10.654	707	9.731	10.545	708	9.143	10.619	709	9.254	10.501	710	9.045	10.718
711	9.552	10.664	712	10.039	11.511*	713	9.670	10.914	714	9.491	10.669	715	9.589	10.869
716	8.900	10.986	717	9.557	10.872	718	8.845	10.155	719	9.316	11.041	720	9.420	10.645
721	9.839	10.896	722	9.264	10.440	723	10.020	10.306	724	9.235	11.200	725	9.194	11.026
726	9.497	10.654	727	9.212	10.904	728	9.328	10.393	729	9.563	10.672	730	9.646	11.070
731	9.082	10.194	732	9.171	10.781	733	9.016	10.914	734	8.944	10.604	735	9.028	10.664
736	9.074	9.924*	737	9.080	10.604	738	8.641	10.155	739	10.052	11.660*	740	9.806	10.831
741	9.051	10.307	742	9.180	10.632	743	9.181	10.306	744	9.751	10.867	745	9.162	10.615
746	9.088	10.658	747	9.184	9.982*	748	8.697	10.393	749	9.292	10.612	750	9.331	10.615

TABLE 8.2 (continued)

SAMPLE	LCL	UCL	SAMPLE	LCL	UCL	SAMPLE	LCL	UCL	SAMPLE	LCL	UCL	SAMPLE	LCL	UCL
751	9.138	10.361	752	9.604	11.201	753	8.921	10.326	754	8.943	10.219	755	9.222	10.216
756	9.530	10.981	757	9.248	10.720	758	9.646	10.700	759	8.895	10.036	760	9.618	10.742
761	9.290	10.929	762	9.504	10.942	763	9.053	10.474	764	9.754	10.946	765	9.198	10.351
766	9.146	10.468	767	9.180	10.399	768	9.177	10.305	769	9.130	10.580	770	9.960	11.238
771	8.694	10.742	772	9.463	10.594	773	9.348	11.102	774	9.224	10.726	775	9.622	11.160
776	9.082	10.291	777	9.352	10.366	778	9.604	11.415	779	8.366	9.595*	780	9.278	10.527
781	10.024	11.043*	782	9.247	10.508	783	10.053	11.078*	784	8.640	10.792	785	8.498	10.529
786	9.486	11.021	787	9.215	10.090	788	9.647	11.227	789	8.559	10.444	790	9.859	11.008
791	9.867	10.967	792	9.095	10.364	793	8.815	10.275	794	8.648	10.216	795	9.388	10.640
796	8.862	10.274	797	9.218	10.439	798	9.299	10.668	799	9.015	10.139	800	9.263	10.718
801	9.502	11.150	802	9.598	11.290	803	9.843	11.204	804	9.377	10.387	805	9.655	10.713
806	8.571	9.804*	807	9.369	10.523	808	8.432	10.584	809	9.305	10.629	810	9.612	10.451
811	9.253	9.991*	812	9.060	10.301	813	9.323	11.395	814	9.261	10.791	815	9.197	10.769
816	9.425	10.722	817	9.166	10.566	818	9.511	10.630	819	9.185	10.674	820	9.439	10.561
821	9.795	11.330	822	9.491	11.104	823	9.133	10.491	824	9.459	10.787	825	9.383	10.426
826	9.276	10.493	827	9.528	10.964*	828	8.961	10.897	829	8.814	10.037	830	9.321	9.934*
831	9.430	10.786	832	10.506	11.206*	833	9.033	10.450	834	9.641	11.223	835		
836	9.046	10.512	837	9.281	10.414	838	8.707	10.181	839	9.870	11.157	840		
841	9.058	10.378	842	9.480	11.349	843	8.897	10.717	844	9.611	10.216	845		
846	9.350	10.886	847	9.411	10.844	848	8.984	10.566	849	8.968	10.537	850	8.722	10.380
851	9.054	10.647	852	8.873	9.791*	853	10.021	11.515*	854	9.554	11.099	855	8.524	10.378
856	8.781	10.739	857	9.385	10.910	858	8.945	10.416	859	9.183	10.624	860	9.462	10.607
861	9.099	10.434	862	9.331	10.806	863	9.771	10.995	864	9.327	10.731	865	8.963	10.438
866	9.259	11.270	867	9.211	10.519	868	9.821	11.420	869	9.335	10.513	870	9.078	10.210
871	10.080	10.769*	872	9.375	10.590	873	8.535	9.890*	874	9.414	10.751	875	8.877	9.994*
876	9.587	10.795	877	9.121	10.960	878	9.486	10.822	879	10.293	11.456*	880	8.812	10.421
881	9.058	10.909	882	8.990	10.079	883	9.580	11.051	884	9.185	10.505	885	9.367	10.074
886	9.301	10.096	887	9.194	10.273	888	9.278	11.004	889	9.658	10.170	890	9.234	10.570
891	8.630	10.978	892	9.842	11.724	893	9.504	10.998	894	9.287	10.866	895	9.518	10.493
896	9.986	10.907	897	9.758	11.048	898	9.687	10.993	899	9.381	10.822	900	9.784	10.718
901	9.114	10.575	902	8.869	10.508	903	9.363	10.595	904	9.252	10.618	905	9.780	10.687
906	9.147	10.241	907	9.448	10.569	908	9.330	10.693	909	9.096	10.499	910	9.368	11.079
911	9.047	10.283	912	9.036	10.381	913	9.655	11.262	914	9.400	9.964*	915	9.159	10.773
916	9.456	10.747	917	9.445	10.250	918	9.270	10.158	919	9.419	10.101	920	9.155	10.204
921	9.736	11.113	922	9.445	10.763	923	9.423	10.674	924	8.777	10.774	925	9.106	10.613
926	9.087	10.368	927	9.079	10.049	928	9.245	10.969	929	9.096	10.402	930		
931	9.603	10.961	932	9.511	11.157	933	9.650	10.768	934	9.149	10.002	935	10.015	11.540*
936	9.676	10.788	937	9.700	11.167	938	9.615	11.085	939	9.555	10.694	940	9.382	10.570
941	8.498	9.897*	942	9.216	10.406	943	9.140	10.459	944	9.543	10.540	945	8.824	10.638
946	9.523	10.824	947	9.147	10.940	948	9.068	10.536	949	9.119	10.172	950	8.709	10.920
951	9.850	11.410	952	9.729	10.705	953	9.067	10.090	954	9.599	11.064	955	9.753	10.594
956	9.501	10.523	957	9.598	10.543	958	9.220	10.626	959	8.391	9.950*	960	9.629	10.871
961	9.105	10.574	962	9.504	10.752	963	9.137	10.475	964	9.303	10.910	965	9.563	10.745
966	9.161	10.453	967	9.487	10.788	968	9.531	11.014	969	8.920	10.599	970	9.058	10.440
971	9.409	10.760	972	8.981	10.301	973	9.097	10.186	974	8.674	10.776	975	9.010	10.959
976	8.714	10.521	977	9.176	10.770	978	9.263	10.555	979	8.700	10.244	980	9.334	10.496
981	9.577	10.873	982	9.383	11.605	983	9.462	10.826	984	9.367	10.726	985	8.657	10.698
986	9.436	10.970	987	9.532	10.247	988	9.309	10.876	989	9.536	10.799	990	9.827	10.044
991	8.834	9.807*	992	8.672	10.769	993	8.974	10.373	994	9.169	10.891	995	8.704	
996	9.713	10.932	997	9.169	10.769	998	9.595	10.769	999	9.648	10.762	1000	9.029	10.684

8.15 Let y be the number of successes in a binomial experiment with n trials and probability of success p. Assuming that n is large, use the sample proportion of successes $\hat{p} = y/n$ to form a confidence interval for p with confidence coefficient $(1 - \alpha)$. [*Hint:* Start with the pivotal statistic

$$z = \frac{\hat{p} - p}{\sqrt{\dfrac{\hat{p}\hat{q}}{n}}}$$

and use the fact (proof omitted) that for large n, z is approximately a standard normal random variable.]

8.16 Let $y_1, y_2, \ldots, y_n$ be a random sample from a Poisson distribution with mean λ. Suppose we use $\bar{y}$ as an estimator of λ. Derive a $(1 - \alpha)100\%$ confidence interval for λ. [*Hint:* Start with the pivotal statistic

$$z = \frac{\bar{y} - \lambda}{\sqrt{\lambda/n}}$$

and show that for large samples, z is approximately a standard normal random variable. Then substitute $\bar{y}$ for λ in the denominator (why can you do this?) and follow the pivotal method of Example 8.7.]

8.17 Let $y_1, y_2, \ldots, y_n$ be a random sample of n observations from an exponential distribution with mean β. Derive a large-sample confidence interval for β. [*Hint:* Start with the pivotal statistic

$$z = \frac{\bar{y} - \beta}{\beta/\sqrt{n}}$$

and show that for large samples, z is approximately a standard normal random variable. Then substitute $\bar{y}$ for β in the denominator (why can you do this?) and follow the pivotal method of Example 8.7.]

OPTIONAL EXERCISES

8.18 Let $\bar{y}_1$ and s_1^2 be the sample mean and sample variance, respectively, of n_1 observations randomly selected from a population with mean μ_1 and variance σ_1^2. Similarly, define $\bar{y}_2$ and s_2^2 for an independent random sample of n_2 observations from a population with mean μ_2 and σ_2^2. Derive a large-sample confidence interval for $(\mu_1 - \mu_2)$. [*Hint:* Start with the pivotal statistic

$$z = \frac{(\bar{y}_1 - \bar{y}_2) - (\mu_1 - \mu_2)}{\sqrt{\dfrac{\sigma_1^2}{n_1} + \dfrac{\sigma_2^2}{n_2}}}$$

and show that for large samples, z is approximately a standard normal random variable. Substitute s_1^2 for σ_1^2 and s_2^2 for σ_2^2 (why can you do this?) and follow the pivotal method of Example 8.7.]

8.19 Let $(\bar{y}_1, s_1^2)$ and $(\bar{y}_2, s_2^2)$ be the means and variances of two independent random samples of sizes n_1 and n_2, respectively, selected from normal populations with different means, μ_1 and μ_2, but with a common variance, σ^2.

a. Show that $E(\bar{y}_1 - \bar{y}_2) = \mu_1 - \mu_2$.

b. Show that

$$V(\bar{y}_1 - \bar{y}_2) = \sigma^2\left(\frac{1}{n_1} + \frac{1}{n_2}\right)$$

c. Explain why

$$z = \frac{(\bar{y}_1 - \bar{y}_2) - (\mu_1 - \mu_2)}{\sigma\sqrt{\dfrac{1}{n_1} + \dfrac{1}{n_2}}}$$

is a standard normal random variable.

8.20 Refer to Exercise 8.19. According to Theorem 7.4,

$$\chi_1^2 = \frac{(n_1 - 1)s_1^2}{\sigma^2} \quad \text{and} \quad \chi_2^2 = \frac{(n_2 - 1)s_2^2}{\sigma^2}$$

are independent chi-square random variables with $(n_1 - 1)$ and $(n_2 - 1)$ df, respectively. Show that

$$\chi^2 = \frac{(n_1 - 1)s_1^2 + (n_2 - 1)s_2^2}{\sigma^2}$$

is a chi-square random variable with $(n_1 + n_2 - 2)$ df.

8.21 Refer to Exercises 8.19 and 8.20. The pooled estimator of the common variance σ^2 is given by

$$s^2 = \frac{(n_1 - 1)s_1^2 + (n_2 - 1)s_2^2}{n_1 + n_2 - 2}$$

Show that

$$t = \frac{(\bar{y}_1 - \bar{y}_2) - (\mu_1 - \mu_2)}{s\sqrt{\dfrac{1}{n_1} + \dfrac{1}{n_2}}}$$

has a Student's t distribution with $(n_1 + n_2 - 2)$ df. [*Hint:* Recall that $t = z/\sqrt{\chi^2/\nu}$ has a Student's t distribution with ν df and use the results of Exercises 8.19c and 8.20.]

8.22 Use the pivotal statistic t given in Exercise 8.21 to derive a $(1 - \alpha)100\%$ small-sample confidence interval for $(\mu_1 - \mu_2)$.

From our discussions in Section 8.3, we already know that a useful point estimate of the population mean μ is $\bar{y}$, the sample mean. According to the central limit theorem (Theorem 7.2) we also know that for sufficiently large n (say, $n \geq 30$), the sampling distribution of the sample mean $\bar{y}$ is approximately normal with $E(\bar{y}) = \mu$ and $V(\bar{y}) = \sigma^2/n$. The fact that $E(\bar{y}) = \mu$ implies that $\bar{y}$ is an unbiased estimator of μ. Furthermore, it can be shown (proof omitted) that $\bar{y}$ has the smallest variance among all unbiased estimators of μ. Hence, $\bar{y}$ is the MVUE for μ. Therefore, it is not surprising that $\bar{y}$ is considered the best estimator of μ.

Since $\bar{y}$ is approximately normal for large n, we can apply Theorem 8.2 to construct a large-sample $(1 - \alpha)100\%$ confidence interval for μ. Substituting $\hat{\theta} = \bar{y}$ and $\sigma_{\hat{\theta}} = \sigma/\sqrt{n}$ into the confidence interval formula given in Theorem 8.2, we obtain the formula given in the box.

LARGE-SAMPLE $(1 - \alpha)100\%$ CONFIDENCE INTERVAL FOR A POPULATION MEAN, μ

$$\bar{y} \pm z_{\alpha/2}\sigma_{\bar{y}} = \bar{y} \pm z_{\alpha/2}\left(\frac{\sigma}{\sqrt{n}}\right)$$

where $z_{\alpha/2}$ is the z value that locates an area of $\alpha/2$ to its right, σ is the standard deviation of the population from which the sample was selected, n is the sample size, and $\bar{y}$ is the value of the sample mean.

[*Note:* When the value of σ is unknown (as will usually be the case), the sample standard deviation s may be used to approximate σ in the formula for the confidence interval. The approximation is generally quite satisfactory when $n \geq 30$.]

Assumptions: None (since the central limit theorem guarantees that $\bar{y}$ is approximately normal regardless of the distribution of the sampled population)

EXAMPLE 8.9

A public utilities company is considering increasing the price of electricity during peak-load periods of the day (9:00 A.M. to 4:00 P.M.) and reducing the price during off-peak periods. The company believes that this revised pricing structure will encourage customers to conserve energy during the period when electrical consumption is the highest, and that it will eventually lead to an overall reduction in monthly consumption. To investigate the effectiveness of the plan, the company randomly selected $n = 45$ customers and recorded the total electrical consumption (in kilowatt-hours) of each during a month in which the revised pricing structure was in effect. The following summary statistics were calculated:

$$\bar{y} = 2{,}003 \qquad s = 388$$

Construct a 90% confidence interval for μ, the true mean monthly electrical consumption of customers under the revised pricing policy. Interpret the interval.

SOLUTION

For a confidence coefficient of $1 - \alpha = .90$, we have $\alpha = .10$ and $\alpha/2 = .05$; therefore, a 90% confidence interval for μ is given by

$$\bar{y} \pm z_{\alpha/2}\left(\frac{\sigma}{\sqrt{n}}\right) = \bar{y} \pm z_{.05}\left(\frac{\sigma}{\sqrt{n}}\right)$$

$$\approx \bar{y} \pm z_{.05}\left(\frac{s}{\sqrt{n}}\right)$$

$$= 2{,}003 \pm z_{.05}\left(\frac{388}{\sqrt{45}}\right)$$

where $z_{.05}$ is the z value corresponding to an upper-tail area of .05. From Table 4 of Appendix II, $z_{.05} = 1.645$. Then the desired interval is

$$2{,}003 \pm z_{.05}\left(\frac{388}{\sqrt{45}}\right) = 2{,}003 \pm 1.645\left(\frac{388}{\sqrt{45}}\right)$$

$$= 2{,}003 \pm 95.15$$

or 1,907.85 to 2,098.15 kilowatt-hours per month. We are 90% confident that the interval (1,907.85, 2,098.15) encloses the true value of μ. If the population mean monthly consumption of the utility's customers *under the regular pricing policy* had been, say, 2,200 kilowatt-hours, then we would infer that the revised pricing policy has effectively reduced monthly electrical usage, since all values within the 90% confidence interval fall below 2,200. ∎

Sometimes, time or cost limitations may restrict the number of sample observations that may be obtained for estimating μ. In the case of small samples ($n < 30$), the following two problems arise:

1. Since the central limit theorem applies only to large samples, we are not able to assume that the sampling distribution of $\bar{y}$ is approximately normal. Therefore, we cannot apply Theorem 8.2. For small samples, the sampling distribution of $\bar{y}$ depends on the particular form of the relative frequency distribution of the population being sampled.
2. The sample standard deviation s may not be a satisfactory approximation to the population standard deviation σ if the sample size is small.

Fortunately, we may proceed with estimation techniques based on small samples if we can assume that the population from which the sample is selected has an approximate normal distribution. If this assumption is valid, then we can use the procedure of Example 8.7 to construct a confidence interval for μ. The general form of a small-sample confidence interval for μ, based on the Student's t distribution, is as shown in the next box.

SMALL-SAMPLE ($1 - \alpha$)100% CONFIDENCE INTERVAL FOR THE POPULATION MEAN, μ

$$\bar{y} \pm t_{\alpha/2}\left(\frac{s}{\sqrt{n}}\right)$$

where the distribution of t is based on $(n - 1)$ degrees of freedom.

Assumption: The population from which the sample is selected has an approximate normal distribution.

EXAMPLE 8.10

Suppose a regional computer center wants to evaluate the performance of its disk memory system. One measure of performance is the average time between failures of its disk drive. To estimate this value, the center recorded the time between failures for a random sample of 20 disk drive failures. The following sample statistics were computed:

$$\bar{y} = 1{,}762 \text{ hours} \qquad s = 215 \text{ hours}$$

Estimate the true mean time between failures with a 99% confidence interval.

SOLUTION

For a confidence coefficient of $1 - \alpha = .99$, we have $\alpha = .01$ and $\alpha/2 = .005$. Since the sample size is small ($n = 20$), our estimation technique requires the assumption that the time y between disk drive failures of the disk memory system has an approximately normal distribution (i.e., the sample of 20 disk drive failure times is selected from a normal population).

Applying the formula for a small-sample confidence interval for μ, we obtain

$$\bar{y} \pm t_{\alpha/2}\left(\frac{s}{\sqrt{n}}\right) = \bar{y} \pm t_{.005}\left(\frac{s}{\sqrt{n}}\right)$$

$$= 1{,}762 \pm t_{.005}\left(\frac{215}{\sqrt{20}}\right)$$

where $t_{.005}$ is the value corresponding to an upper-tail area of .005 in the Student's t distribution based on $(n - 1) = 19$ degrees of freedom. From Table 6 of Appendix II, the required t value is $t_{.005} = 2.861$. Substitution of this value yields

$$1{,}762 \pm t_{.005}\left(\frac{215}{\sqrt{20}}\right) = 1{,}762 \pm (2.861)\left(\frac{215}{\sqrt{20}}\right)$$

$$= 1{,}762 \pm 137.54$$

or, 1,624.46 to 1,899.54 hours. Thus, if the distribution of the time between failures of the disk drive system is approximately normal, then we can be 99% confident that the interval (1,624.46, 1,899.54) encloses μ, the true mean time between failures. ∎

Before concluding this section, we will comment on the assumption that the sampled population is normally distributed. In the real world, we rarely know whether a sampled population has an exact normal distribution. However, empirical studies indicate that moderate departures from this assumption do not seriously affect the confidence coefficients for small-sample confidence intervals. For example, if the population of times between failure for the disk drive system of Example 8.10 has a distribution that is mound-shaped but nonnormal, it is likely that the actual confidence coefficient for the 99% confidence interval will be close to .99—at least close enough to be of practical use. As a consequence, the small-sample confidence interval given in the box is frequently used by experimenters when estimating the population mean of a nonnormal distribution as long as the distribution is mound-shaped and only moderately skewed. For populations that depart greatly from normality, however, other estimation techniques, such as robust estimation or the jackknife, are recommended.

8.23 Molded-rubber expansion joints used in heating and air-conditioning systems are manufactured to withstand high pressure. A rubber company has purchased new machinery to produce joints with 5-inch internal diameters. For a random sample of 50 molded-rubber expansion joints, the mean internal diameter is $\bar{y} = 4.985$ inches and the standard deviation is $s = .03$ inch. Estimate the mean internal diameter of all expansion joints produced by the new machinery with a 95% confidence interval.

8.24 The Geothermal Loop Experimental Facility, located in the Salton Sea in southern California, is a U.S. Department of Energy operation for studying the feasibility of generating electricity from the hot, highly saline brines of the Salton Sea. Operating experience has shown that the saline brines leave silica scale deposits on metallic plant piping, causing excessive plant outages. Jacobsen et al. (*Journal of Testing and Evaluation*, Vol. 9, No. 2, Mar. 1981, pp. 82–92) have found that scaling can be reduced somewhat by adding chemical solutions to the brine. In one screening experiment, each of five antiscalants was added to an aliquot of brine, and the solutions were filtered. A silica determination (parts per million of silicon dioxide) was made on each filtered sample after a holding time of 24 hours, with the following results:

229   255   280   203   229

a. Compute $\bar{y}$ and s for the sample.
b. Estimate the mean amount of silicon dioxide present in the five antiscalant solutions. Use a 90% confidence interval.
c. What assumption is required for the interval estimate of part **b** to be valid?

8.25 A *construction cost index*, a measure of the average change in construction costs over time, is interpreted as the percentage increase in construction costs relative to a base year. One such index, with 1913 as the base year, was computed in February 1986 for a sample of 20 cities in the United States. The results yielded an average construction cost index of $\bar{y} = 4,001.46$ and a standard deviation of $s = 605.26$. Estimate μ, the mean construction cost index of all cities in the United States for February 1986, using a 99% confidence interval.

8.26 What do college recruiters think are the most important topics to be covered in a job interview? To answer this and other questions, Taylor and Sniezek elicited the opinions of recruiters interviewing at a small midwestern college and a large midwestern university (*Journal of Occupational Psychology*, 1984). Recruiters were asked to rate on a 105-point scale the importance of each in a list of 25 interview topics [where 0 = least important (can often be omitted without hurting the interview), 52.5 = average importance (can sometimes be omitted without hurting the interview), and 105 = most important (can never be omitted without hurting the interview)]. The topic concerning "applicant's skill in communicating ideas to others" received the highest ratings of the $n = 58$ college recruiters who returned the questionnaire. The sample mean rating and sample standard deviation for this topic were $\bar{y} = 84.84$ and $s = 15.67$, respectively.
a. Give a point estimate for the true mean rating of the importance of "applicant's skill in communicating ideas to others" by all college recruiters.
b. Use the sample information to construct a 95% confidence interval for the true mean rating.
c. What is the confidence coefficient for the interval of part **b**? Interpret this value.

8.27 An evaluation of trace metal chemistry and cycling in an acidic Adirondack lake was reported in *Environmental Science & Technology* (Dec. 1985). Twenty-four (24) water samples were collected from Darts Lake, New York, and analyzed for concentration of both lead and aluminum particulates.
 a. The lead concentration measurements had a mean of 9.9 nmol/l and a standard deviation of 8.4 nmol/l. Calculate a 99% confidence interval for the true mean lead concentration in water samples collected from Darts Lake.
 b. The aluminum concentration measurements had a mean of 6.7 nmol/l and a standard deviation of 10.8 nmol/l. Calculate a 99% confidence interval for the true mean aluminum concentration in water samples collected from Darts Lake.
 c. What assumptions are necessary for the intervals of parts **a** and **b** to be valid?

8.28 Studies have shown that more than 12 million tin-coated steel cans are removed from the municipal waste streams of our cities and recycled each day. Suppose it is desired to estimate the mean number of tin cans recovered from mixed refuse per year in American cities. A random sample of eight American cities yielded the following summary statistics on number of tin cans (in millions) recovered per city last year:

$$\bar{y} = 105.7 \qquad s = 9.3$$

 a. Construct a 95% confidence interval for the true mean number of tin cans removed annually from mixed refuse for recycling in American cities.
 b. What assumption is required for the interval estimate of part **a** to be valid?

SECTION 8.6

ESTIMATION OF THE DIFFERENCE BETWEEN TWO POPULATION MEANS: INDEPENDENT SAMPLES

In Section 8.5, we learned how to estimate the parameter μ from a single population. We now proceed to a technique for using the information in two samples to estimate the difference between two population means, $(\mu_1 - \mu_2)$, when the samples are collected independently. For example, we may want to compare the mean starting salaries for bachelor's degree graduates of the University of Florida in the Colleges of Engineering and Education, or the mean operating costs of automobiles with rotary engines and standard engines, or the mean failure times of two electronic components. The technique to be presented is a straightforward extension of that used for estimation of a single population mean.

Suppose we select independent random samples of sizes n_1 and n_2 from populations with means μ_1 and μ_2, respectively. Intuitively, we want to use the difference between the sample means, $(\bar{y}_1 - \bar{y}_2)$, to estimate $(\mu_1 - \mu_2)$. In Example 7.5, we showed that

$$E(\bar{y}_1 - \bar{y}_2) = \mu_1 - \mu_2$$
$$V(\bar{y}_1 - \bar{y}_2) = \frac{\sigma_1^2}{n_1} + \frac{\sigma_2^2}{n_2}$$

You can see that $(\bar{y}_1 - \bar{y}_2)$ is an unbiased estimator for $(\mu_1 - \mu_2)$. Further, it can be shown (proof omitted) that $V(\bar{y}_1 - \bar{y}_2)$ is smallest among all unbiased estimators, i.e., $(\bar{y}_1 - \bar{y}_2)$ is the MVUE for $(\mu_1 - \mu_2)$.

According to the central limit theorem, $(\bar{y}_1 - \bar{y}_2)$ will also be approximately normal for large n_1 and n_2 regardless of the distributions of the sampled populations. Thus, we can apply Theorem 8.2 to construct a large-sample confidence interval for $(\mu_1 - \mu_2)$. The procedure for forming a large-sample confidence interval for $(\mu_1 - \mu_2)$ appears in the accompanying box.

LARGE-SAMPLE $(1 - \alpha)100\%$ CONFIDENCE INTERVAL FOR $(\mu_1 - \mu_2)$

$$(\bar{y}_1 - \bar{y}_2) \pm z_{\alpha/2}\sigma_{(\bar{y}_1-\bar{y}_2)} = (\bar{y}_1 - \bar{y}_2) \pm z_{\alpha/2}\sqrt{\frac{\sigma_1^2}{n_1} + \frac{\sigma_2^2}{n_2}}$$

$$\approx (\bar{y}_1 - \bar{y}_2) \pm z_{\alpha/2}\sqrt{\frac{s_1^2}{n_1} + \frac{s_2^2}{n_2}}$$

[*Note:* We have used the sample variances s_1^2 and s_2^2 as approximations to the corresponding population parameters.]

Assumptions:
1. The two random samples are selected in an independent manner from the target populations. That is, the choice of elements in one sample does not affect, and is not affected by, the choice of elements in the other sample.
2. The sample sizes n_1 and n_2 are sufficiently large in order for the central limit theorem to apply. (We recommend $n_1 \geq 30$ and $n_2 \geq 30$.)

EXAMPLE 8.11

It is desired to estimate the difference between the mean starting salaries for all bachelor's degree graduates of the University of Florida in the Colleges of Engineering and Education during the past year. The following information is available.*

1. A random sample of 40 starting salaries for College of Engineering graduates produced a sample mean of $22,653 and a standard deviation of $4,172.
2. A random sample of 30 starting salaries for College of Education graduates produced a sample mean of $12,291 and a standard deviation of $3,864.

Construct a 95% confidence interval for the difference between mean starting salaries for graduates of the two colleges. Interpret the interval.

SOLUTION

We will let the subscript 1 refer to the College of Engineering and the subscript 2 to the College of Education. We will also define the following notation:

μ_1 = Population mean starting salary of all bachelor's degree graduates of the College of Engineering

μ_2 = Population mean starting salary of all bachelor's degree graduates of the College of Education

*The data for this example were provided by Mr. Maurice Mayberry, Director, Career Resource Center, University of Florida.

Similarly, let $\bar{y}_1$ and $\bar{y}_2$ denote the respective sample means; s_1 and s_2, the respective sample standard deviations; and n_1 and n_2, the respective sample sizes. The given information may be summarized as in Table 8.3.

TABLE 8.3
Summary of Information for Example 8.11

	COLLEGE OF ENGINEERING	COLLEGE OF EDUCATION
Sample size	$n_1 = 40$	$n_2 = 30$
Sample mean	$\bar{y}_1 = \$22,653$	$\bar{y}_2 = \$12,291$
Sample standard deviation	$s_1 = \$4,172$	$s_2 = \$3,864$

The general form of a 95% confidence interval for $(\mu_1 - \mu_2)$, based on large, independent samples from the target populations, is given by

$$(\bar{y}_1 - \bar{y}_2) \pm z_{.025}\sqrt{\frac{\sigma_1^2}{n_1} + \frac{\sigma_2^2}{n_2}}$$

Recall that $z_{.025} = 1.96$ and use the information in Table 8.3 to make the following substitutions to obtain the desired confidence interval:

$$(22,653 - 12,291) \pm 1.96\sqrt{\frac{\sigma_1^2}{40} + \frac{\sigma_2^2}{30}}$$

$$\approx (22,653 - 12,291) \pm 1.96\sqrt{\frac{(4,172)^2}{40} + \frac{(3,864)^2}{30}}$$

$$= 10,362 \pm 1,893$$

or ($8,469, $12,255).

The use of this method of estimation produces confidence intervals that will enclose $(\mu_1 - \mu_2)$, the difference between population means, 95% of the time. Hence, we can be reasonably confident that the mean starting salary of College of Engineering bachelor's degree graduates during the past year was between $8,469 and $12,255 higher than the mean starting salary of College of Education bachelor's degree graduates. ∎

A confidence interval for $(\mu_1 - \mu_2)$, based on small samples from each population, is derived using Student's t distribution. As was the case when estimating a single population mean from information in a small sample, we must make specific assumptions about the relative frequency distributions of the two populations, as indicated in the box on page 314. These assumptions are required if either $n_1 < 30$ or $n_2 < 30$.

SMALL-SAMPLE $(1 - \alpha)100\%$ CONFIDENCE INTERVAL FOR $(\mu_1 - \mu_2)$

$$(\bar{y}_1 - \bar{y}_2) \pm t_{\alpha/2} \sqrt{s_p^2\left(\frac{1}{n_1} + \frac{1}{n_2}\right)}$$

where

$$s_p^2 = \frac{(n_1 - 1)s_1^2 + (n_2 - 1)s_2^2}{n_1 + n_2 - 2}$$

and the value of $t_{\alpha/2}$ is based on $(n_1 + n_2 - 2)$ degrees of freedom.

Assumptions: 1. Both of the populations from which the samples are selected have relative frequency distributions that are approximately normal.
2. The variance σ_1^2 and σ_2^2 of the two populations are equal.
3. The random samples are selected in an independent manner from the two populations.

Note that this procedure requires that the samples be selected from two normal populations which have equal variances (i.e., $\sigma_1^2 = \sigma_2^2 = \sigma^2$). Since we are assuming the variances are equal, we construct an estimate of σ^2 based on the information contained in *both* samples. This **pooled estimate** is denoted by s_p^2 and is computed as shown in the box.

EXAMPLE 8.12

Woelfl et al. (1981) conducted laboratory tests to investigate the stability and permeability of open-graded asphalt concrete. In one part of the experiment, four concrete specimens were prepared for asphalt contents of 3% and 7% by total weight of mix. The water permeability of each concrete specimen was determined by flowing de-aired water across the specimen and measuring the amount of water loss. The permeability measurements (recorded in inches per hour) for the eight concrete specimens are shown in Table 8.4. Find a 95% confidence interval for the difference between the mean permeabilities of concrete made with asphalt contents of 3% and 7%. Interpret the interval.

TABLE 8.4
Permeability
Measurements for 3% and
7% Asphalt Concrete,
Example 8.12

ASPHALT CONTENT	3%	1,189	840	1,020	980
	7%	853	900	733	785

Source: Woelfl, G., Wei, I., Faulstich, C., and Litwack, H. "Laboratory Testing of Asphalt Concrete for Porous Pavements." *Journal of Testing and Evaluation*, Vol. 9, No. 4, July 1981, pp. 175–181. Copyright American Society for Testing and Materials.

SOLUTION

First, we calculate the means and variances of the two samples. For the 3% asphalt,

$$\bar{y}_1 = \frac{\sum y_i}{n} = \frac{1,189 + 840 + 1,020 + 980}{4} = \frac{4,029}{4} = 1,007.25$$

$$s_1^2 = \frac{\sum y_i^2 - \dfrac{\left(\sum y_i\right)^2}{n}}{n - 1}$$

$$= \frac{(1,189)^2 + (840)^2 + (1,020)^2 + (980)^2 - \dfrac{(4,029)^2}{4}}{3} = 20,636.92$$

For the 7% asphalt,

$$\bar{y}_2 = \frac{\sum y_i}{n} = \frac{853 + 900 + 733 + 785}{4} = \frac{3,271}{4} = 817.75$$

$$s_2^2 = \frac{\sum y_i^2 - \dfrac{\left(\sum y_i\right)^2}{n}}{n - 1}$$

$$= \frac{(853)^2 + (900)^2 + (733)^2 + (785)^2 - \dfrac{(3,271)^2}{4}}{3} = 5,420.92$$

Since both samples are small ($n_1 = n_2 = 4$), the procedure requires the assumption that the two samples of permeability measurements are independently and randomly selected from normal populations with equal variances. The 95% small-sample confidence interval is

$$(\bar{y}_1 - \bar{y}_2) \pm t_{.025} \sqrt{s_P^2 \left(\frac{1}{n_1} + \frac{1}{n_2}\right)}$$

$$= (1,007.25 - 817.75) \pm t_{.025} \sqrt{s_P^2 \left(\frac{1}{4} + \frac{1}{4}\right)}$$

where $t_{.025} = 2.447$ is obtained from the t distribution (Table 6 of Appendix II) based on $n_1 + n_2 - 2 = 4 + 4 - 2 = 6$ degrees of freedom, and

$$s_P^2 = \frac{(n_1 - 1)s_1^2 + (n_2 - 1)s_2^2}{n_1 + n_2 - 2} = \frac{3(20,636.92) + 3(5,420.92)}{6}$$

$$= 13,028.92$$

is the pooled sample variance. Substitution yields the interval

$$(1,007.25 - 817.75) \pm 2.447 \sqrt{13,028.92\left(\frac{1}{4} + \frac{1}{4}\right)}$$

$$= 189.5 \pm 197.50$$

or, -8.00 to 387.00. Thus, we are 95% confident that the interval $(-8, 387)$ encloses the true difference between the mean permeabilities of the two types of concrete. Since the interval includes 0, we are unable to conclude that the two means differ.   ∎

As with the one-sample case, the assumptions required for estimating $(\mu_1 - \mu_2)$ with small samples do not have to be satisfied exactly for the interval estimate to be useful in practice. Slight departures from these assumptions do not seriously affect the level of confidence in the procedure. For example, when the variances σ_1^2 and σ_2^2 of the sampled populations are unequal, researchers have found that the small-sample confidence interval for $(\mu_1 - \mu_2)$ given in the box will still yield valid results in practice as long as the two populations are normal and the sample sizes are equal, i.e., $n_1 = n_2$. In the case where $\sigma_1^2 \neq \sigma_2^2$ and $n_1 \neq n_2$, an approximate confidence interval for $(\mu_1 - \mu_2)$ can be constructed by modifying the degrees of freedom associated with the t distribution.

We conclude this section by giving the approximate small-sample confidence intervals for $(\mu_1 - \mu_2)$ for two situations when the assumption of equal variances is violated: $n_1 = n_2$ and $n_1 \neq n_2$.

APPROXIMATE SMALL-SAMPLE $(1 - \alpha)$100% CONFIDENCE INTERVAL FOR $(\mu_1 - \mu_2)$ WHEN $\sigma_1^2 \neq \sigma_2^2$

$$n_1 = n_2 = n: \quad (\bar{y}_1 - \bar{y}_2) \pm t_{\alpha/2} \sqrt{\frac{s_1^2}{n} + \frac{s_2^2}{n}}$$

where the distribution of t depends on $\nu = n_1 + n_2 - 2 = 2(n - 1)$ degrees of freedom.

$$n_1 \neq n_2: \quad (\bar{y}_1 - \bar{y}_2) \pm t_{\alpha/2} \sqrt{\frac{s_1^2}{n_1} + \frac{s_2^2}{n_2}}$$

where the distribution of t has degrees of freedom equal to

$$\nu = \frac{(s_1^2/n_1 + s_2^2/n_2)^2}{\dfrac{(s_1^2/n_1)^2}{n_1 - 1} + \dfrac{(s_2^2/n_2)^2}{n_2 - 1}}$$

[*Note:* In the case of $n_1 \neq n_2$, the value of ν will not generally be an integer. Round ν down to the nearest integer in order to use the t table (Table 6 of Appendix II).]

Assumptions: 1. Both of the populations from which the samples are selected have relative frequency distributions that are approximately normal.
2. The random samples are selected in an independent manner from the two populations.

In the next section we consider confidence intervals for $(\mu_1 - \mu_2)$ for dependent samples.

EXERCISES 8.29–8.34

8.29 Agricultural experts in Israel have developed a new method of irrigation, called *fertigation*, in which fertilizer is added to water and the mixture is dripped periodically onto the roots of the plants. Very little water—a precious commodity in Israel—is wasted, and the nutrients go directly where they are needed. In order to test this new process, 100 acres were randomly selected and their historical yields were recorded. The fertigation process was then applied to the new crop and the new yields were recorded. The accompanying table summarizes the results.

	BEFORE FERTIGATION	AFTER FERTIGATION
Sample size	100	100
Mean yield	40%	75%
Standard deviation	8%	6%

a. Estimate the difference between the true mean yields before and after fertigation. Use a 90% confidence interval.
b. Interpret the confidence interval of part **a**.

8.30 Suppose a regional computer center wants to evaluate the performance of its disk memory system. One measure of performance is the average time between failures of a disk drive. Since the computer center operates two disk drives, it wants to estimate $(\mu_1 - \mu_2)$, the difference in mean time between failures of the two disk drives. Independent random samples of $n_1 = 10$ and $n_2 = 15$ failures produced the following statistics:

DISK DRIVE 1	DISK DRIVE 2
$\bar{y}_1 = 92$ hours	$\bar{y}_2 = 108$ hours
$s_1 = 16$ hours	$s_2 = 12$ hours

Estimate $(\mu_1 - \mu_2)$ with a 95% confidence interval. Which of the two disk drives appears to give better performance?

8.31 The methodology for conducting a stress analysis of newly designed timber structures is well known. However, few data are available on the actual or allowable stress for repairing damaged structures. Consequently, design engineers often propose a repair scheme (e.g., gluing) without any knowledge of its structural effectiveness. To partially fill this void, a stress analysis was conducted on epoxy-repaired truss joints (*Journal of Structural Engineering*, Feb. 1986). Tests were conducted on epoxy-bonded truss joints made of various species of wood in order to determine actual glue-line shear stress (recorded in pounds per square inch). Summary information for independent

	SOUTHERN PINE	PONDEROSA PINE
Sample size	100	47
Mean shear stress (psi)	1,312	1,352
Standard deviation	422	271

Source: Avent, R. R. "Design Criteria for Epoxy Repair of Timber Structures." *Journal of Structural Engineering*, Vol. 112, No. 2, Feb. 1986, p. 232.

random samples of Southern Pine and Ponderosa Pine truss joints is given in the accompanying table. Estimate the difference between the mean shear strengths of epoxy-repaired truss joints for the two species of wood with a 90% confidence interval.

8.32 To investigate the possible link between fluoride content of drinking water and cancer, Yiamouyiannis and Burk (1977) recorded cancer death rates (number of deaths per 100,000 population) from 1952–1969 in 20 selected U.S. cities—the ten largest fluoridated cities and the ten largest cities not fluoridated by 1969. Maritz and Jarrett (*Applied Statistics*, Feb. 1983) used the data collected by Yiamouyiannis and Burk to calculate for each city the annual rate of increase in cancer death rate over this 18-year period for each of four age groups: under 25, 25–44, 45–64, and 65 or older. The data for the 45–64 age group are reproduced in the accompanying table.

FLUORIDATED		NONFLUORIDATED	
City	Annual Increase in Cancer Death Rate	City	Annual Increase in Cancer Death Rate
Chicago	1.0640	Los Angeles	.8875
Philadelphia	1.4118	Boston	1.7358
Baltimore	2.1115	New Orleans	1.0165
Cleveland	1.9401	Seattle	.4923
Washington	3.8772	Cincinnati	4.0155
Milwaukee	−.4561	Atlanta	−1.1744
St. Louis	4.8359	Kansas City	2.8132
San Francisco	1.8875	Columbus	1.7451
Pittsburgh	4.4964	Newark	−.5676
Buffalo	1.4045	Portland	2.4471

Source: Maritz, J. S. and Jarrett, R. G. "The Use of Statistics to Examine the Association Between Fluoride in Drinking Water and Cancer Death Rates." *Applied Statistics*, Vol. 32, No. 2, 1983, pp. 97–101.

a. Construct a 95% confidence interval for the difference between the mean annual increases in cancer death rates for fluoridated and nonfluoridated cities.

b. Interpret the interval obtained in part a.

c. What assumptions are necessary for the validity of the interval estimation procedure and any inferences derived from it? Do you think these assumptions are satisfied?

8.33 Suppose you want to estimate the difference in annual operating costs for automobiles with rotary engines and those with standard engines. You find 8 owners of cars with rotary engines and 12 owners of cars with standard engines who have purchased their cars within the last 2 years and are willing to participate in the experiment. Each of the 20 owners keeps accurate records of the amount spent on operating his or her car (including gasoline, oil, repairs, etc.) for a 12-month period. All costs are recorded on a per 1,000-mile basis to adjust for differences in mileage driven during the 12-month period. The results are summarized in the table.

ROTARY	STANDARD
$n_1 = 8$	$n_2 = 12$
$\bar{y}_1 = \$56.96$	$\bar{y}_2 = \$52.73$
$s_1 = \$4.85$	$s_2 = \$6.35$

a. Estimate the true difference $(\mu_1 - \mu_2)$ in the mean operating costs per 1,000 miles between cars with rotary and cars with standard engines, using a 90% confidence interval. Assume $\sigma_1^2 = \sigma_2^2$.

b. Repeat part **a**, but assume $\sigma_1^2 \neq \sigma_2^2$.

8.34 *Sintering*, one of the most important techniques of materials science, is used to convert powdered material into a porous solid body. The following two measures characterize the final product:

V_V = Percentage of total volume of final product that is solid

$$= \left(\frac{\text{Solid volume}}{\text{Porous volume + Solid volume}}\right) \cdot 100$$

S_V = Solid-pore interface area per unit volume of the product

When $V_V = 100\%$, the product is completely solid—i.e., it contains no pores. Both V_V and S_V are estimated by a microscopic examination of polished cross sections of sintered material. The accompanying table gives the mean and standard deviation of the values of S_V (in squared centimeters per cubic centimeter) and V_V (percentage) for $n = 100$ specimens of sintered nickel for two different sintering times.

TIME	S_V		V_V	
	$\bar{y}$	s	$\bar{y}$	s
10 minutes	736.0	181.9	96.73	2.1
150 minutes	299.5	161.0	97.82	1.5

Data and experimental information provided by Guoquan Liu while visiting at the University of Florida in 1983.

a. Find a 95% confidence interval for the mean change in S_V between sintering times of 10 minutes and 150 minutes. What inference would you make concerning the difference in mean sintering times?

b. Repeat part **a** for V_V.

SECTION 8.7

ESTIMATION OF THE DIFFERENCE BETWEEN TWO POPULATION MEANS: MATCHED PAIRS

The large- and small-sample procedures for estimating the difference between two population means presented in Section 8.6 were based on the assumption that the samples were randomly and independently selected from the target populations. Sometimes we can obtain more information about the difference between population means, $(\mu_1 - \mu_2)$, by selecting **paired observations**.

For example, suppose you want to compare two methods for drying concrete using samples of five cement mixes with each method. One method of sampling would be to randomly select ten mixes from among all available mixes and then randomly assign five to drying method 1 and five to drying method 2. The strength measurements (in psi) obtained after conducting a series of strength tests would represent independent random samples of strengths attained by concrete specimens dried by the two different methods. The difference between the mean strength measurements, $(\mu_1 - \mu_2)$, could be estimated using the confidence interval procedure described in Section 8.6.

A better method of sampling would be to match the concrete specimens in pairs according to type of mix. From each mix pair, one specimen would be randomly selected to be dried by method 1; the other specimen would be assigned to be dried by method 2, as shown in Table 8.5. Then the differences between **matched pairs** of strength measurements should provide a clearer picture of the difference in strengths for the two drying methods because the matching would tend to cancel the effects of the factors that formed the basis of the matching (i.e., the effects of the different cement mixes).

TABLE 8.5

Set-up of the Matched-Pairs Design for Comparing Two Methods of Drying Concrete

TYPE OF MIX	METHOD 1	METHOD 2
A	Specimen 2	Specimen 1
B	Specimen 2	Specimen 1
C	Specimen 1	Specimen 2
D	Specimen 2	Specimen 1
E	Specimen 1	Specimen 2

In a matched-pairs experiment, the symbol μ_d is commonly used to denote the mean difference between matched pairs of measurements, where $\mu_d = (\mu_1 - \mu_2)$. Once the differences in the sample are calculated, a confidence interval for μ_d is identical to the confidence interval for the mean of a single population given in Section 8.5.

The procedure for estimating the difference between two population means based on matched-pairs data when the sample is small (as will usually be the case in practice) is given in the box.

SMALL-SAMPLE $(1 - \alpha)100\%$ CONFIDENCE INTERVAL FOR $\mu_d = (\mu_1 - \mu_2)$: MATCHED PAIRS

Let $d_1, d_2, \ldots, d_n$ represent the differences between the pairwise observations in a random sample of n matched pairs. Then the small-sample confidence interval for $\mu_d = (\mu_1 - \mu_2)$ is

$$\bar{d} \pm t_{\alpha/2}\left(\frac{s_d}{\sqrt{n}}\right)$$

where $\bar{d}$ is the mean of the n sample differences, s_d is their standard deviation, and $t_{\alpha/2}$ is based on $(n - 1)$ degrees of freedom.

Assumptions: 1. The sample paired observations are randomly selected from the target population of paired observations.
2. The population of paired differences is normally distributed.

EXAMPLE 8.13

One desirable characteristic of water pipes is that the quality of water they deliver be equal to or near the quality of water entering the system at the water treatment plant. A type of ductile iron pipe has provided an excellent water delivery system for the St. Louis County Water Company. The chlorine levels of water emerging from the South water treatment plant and at the Fire Station (Fenton Zone 13) were measured over a 12-month period, with the results shown in Table 8.6. Find a 95% confidence interval for the mean difference in monthly chlorine content between the two locations.

TABLE 8.6 Chlorine Content Data for Example 8.13

							MONTH						
		Jan.	Feb.	Mar.	Apr.	May	June	July	Aug.	Sept.	Oct.	Nov.	Dec.
LOCATION	South Plant	2.0	2.0	2.1	1.9	1.7	1.8	1.7	1.9	2.0	2.0	2.1	2.1
	Fire Station	2.2	2.2	2.1	2.0	1.9	1.9	1.8	1.7	1.9	1.9	1.8	2.0
DIFFERENCE		−.2	−.2	0	−.1	−.2	−.1	−.1	.2	.1	.1	.3	.1

Source: "St. Louis County Standardizes Pipe and Procedures for Reliability." Staff Report, Water and Sewage Works, Dec. 1980.

SOLUTION

Since the chlorine levels at the two plants were recorded over the same 12 months, the data are collected as matched pairs. We want to estimate $\mu_d = (\mu_1 - \mu_2)$, where

μ_1 = Mean monthly chlorine level at the South Plant

μ_2 = Mean monthly chlorine level at the Fire Station

The differences between pairs of monthly chlorine levels are computed as

d = (South Plant level) − (Fire Station level)

and are shown in the last row of Table 8.6. The mean, variance, and standard deviation of the 12 differences are

$$\bar{d} = \frac{\sum d}{n} = \frac{-.1}{12} = -.0083$$

$$s_d^2 = \frac{\sum d^2 - \frac{\left(\sum d\right)^2}{n}}{n - 1} = \frac{.31 - \frac{(-.1)^2}{12}}{11} = \frac{.31 - .00083}{11} = .0281$$

$$s_d = \sqrt{s_d^2} = \sqrt{.0281} = .1676$$

The value of $t_{.025}$, based on $(n - 1) = (12 - 1) = 11$ degrees of freedom, is given in Table 6 of Appendix II as $t_{.025} = 2.201$. Substituting these values into the formula for the confidence interval, we obtain

$$\bar{d} \pm t_{.025}\left(\frac{s_d}{\sqrt{n}}\right)$$

$$= -.0083 \pm 2.201\left(\frac{.1676}{\sqrt{12}}\right)$$

$$= -.0083 \pm .1065$$

or (−.1148, .0982). We estimate with 95% confidence that the difference between the mean monthly chlorine levels of water at the two St. Louis locations falls within the interval from −.1148 to .0982. Since 0 is within the interval, there is insufficient evidence to conclude there is a difference between the two means.

∎

Before leaving this discussion of the analysis of matched-pair observations, we want to stress that the pairing of the experimental units (the objects upon which the measurements are taken) must be performed *before* the data are collected. Recall that the objective is to compare two methods of "treating" the experimental units. By using the matched pairs of units that have similar characteristics, we are able to cancel out the effects of the variables used to match the pairs.

EXERCISES 8.35–8.38

8.35 A federal traffic safety researcher was hired to ascertain the effect of wearing safety devices (shoulder harnesses, seat belts) on reaction times to peripheral stimuli. To investigate this question, he randomly selected 15 subjects from the students enrolled in a driver education program. Each subject performed a simulated driving task that allowed reaction times to be recorded under two conditions, wearing a safety device (restrained condition) and no safety device (unrestrained condition). Thus, each subject received two reaction-time scores, one for the restrained condition and one for the unrestrained condition. The data (in hundredths of a second) are shown in the accompanying table.

DRIVER	1	2	3	4	5	6	7	8	9	10	11	12	13	14	15
Restrained	36.7	37.5	39.3	44.0	38.4	43.1	36.2	40.6	34.9	31.7	37.5	42.8	32.6	36.8	38.0
Unrestrained	36.1	35.8	38.4	41.7	38.3	42.6	33.6	40.9	32.5	30.7	37.4	40.2	33.1	33.6	37.5

a. Construct a 98% confidence interval for the difference between mean reaction-time scores for the restrained and unrestrained drivers.

b. What assumptions are necessary for the validity of the interval estimation procedure of part a?

c. Based on the interval of part a, what would you infer about the mean reaction times for the driving conditions?

8.36 Wall and Peterson (1986) developed a heat transfer model for predicting winter heat loss in wastewater treatment clarifiers. Part of their analysis involved a comparison of clear-sky solar irradiation for horizontal surfaces at different sites in the midwest. The day-long solar irradiation levels (in BTU/sq. ft.) at two midwestern locations of different latitudes (St. Joseph, Missouri, and Iowa Great Lakes) were recorded on each of seven clear-sky winter days. The data are given in the table. Find a 95% confidence interval for the mean difference between the day-long clear-sky solar irradiation levels at the two sites.

DATE	ST. JOSEPH, MO	IOWA GREAT LAKES
December 21	782	593
January 6	965	672
January 21	948	750
February 6	1,181	988
February 21	1,414	1,226
March 7	1,633	1,462
March 21	1,852	1,698

Source: Wall, D. J. and Peterson, G. "Model for Winter Heat Loss in Uncovered Clarifiers." *Journal of Environmental Engineering*, Vol. 112, No. 1, Feb. 1986, p. 128.

8.37 Many households are installing fuel-saving devices on shower heads to cut down on the use of hot water. Ten households were randomly selected to participate in a study to investigate the effectiveness of this equipment. Measurements (in gallons per minute) of water usage were recorded before and after the installation of the fuel saver at each household. The results are provided below.

HOUSEHOLD	BEFORE	AFTER
1	5.21	4.66
2	4.33	2.02
3	2.09	2.17
4	5.72	4.98
5	7.31	4.23
6	3.96	1.55
7	4.88	5.00
8	5.60	2.75
9	7.35	3.68
10	10.95	7.71

a. Construct a 99% confidence interval for the difference between the mean water usage before installation of the fuel-saving device and the mean usage after installation. Interpret the interval.

b. Is there evidence that the fuel-saving device decreases water usage? Explain.

8.38 Medical researchers believe that exposure to dust from cotton bract induces respiratory disease in susceptible field workers. An experiment was conducted to determine the effect of air-dried green cotton bract extract (GBE) on the cells of non-dust exposed mill workers (*Environmental Research*, Feb. 1986). Blood samples taken on eight workers were incubated with varying concentrations of GBE. After a short period of time, the cyclic AMP level (a measure of cell activity expressed in picomoles per million cells) of each blood sample was measured. The data for two GBE concentrations, 0 mg/ml (salt buffer, control solution) and .2 mg/ml, are reproduced in the table on page 324. [Note that one blood sample was taken from each worker, with one aliquot exposed to the salt buffer solution and the other to the GBE.]

WORKER	GBE CONCENTRATION (mg/ml)	
	0	.2
A	8.8	4.4
B	13.0	5.7
C	9.2	4.4
D	6.5	4.1
F	9.1	4.4
H	17.0	7.9

Source: Butcher, B. T., Reed, M. A., and O'Neil, C. E. "Biochemical and Immuno-logic Characterization of Cotton Bract Extract and Its Effect on *in Vitro* Cyclic AMP Production." *Environmental Research*, Vol. 39, No. 1, Feb. 1986, p. 119.

a. Find a 95% confidence interval for the mean difference between the cyclic AMP levels of blood samples exposed to the two concentrations of GBE.

b. Based on the interval obtained in part **a**, is there evidence that exposure to GBE blocks cell activity?

SECTION 8.8

ESTIMATION OF A POPULATION PROPORTION

We will now consider the method for estimating the binomial proportion p of successes—that is, the proportion of elements in a population that have a certain characteristic. For example, a quality control inspector may be interested in the proportion of defective items produced on an assembly line; or a supplier of heating oil may be interested in the proportion of homes in its service area that are heated by natural gas.

A logical candidate for a point estimate of the population proportion p is the sample proportion $\hat{p} = y/n$, where y is the number of observations in a sample of size n that have the characteristic of interest (i.e., y is the number of "successes"). In Example 7.7 we showed that for large n, $\hat{p}$ is approximately normal with mean

$$E(\hat{p}) = p$$

and variance

$$V(\hat{p}) = \frac{pq}{n}$$

Therefore, $\hat{p}$ is an unbiased estimator of p and (proof omitted) has the smallest variance among all unbiased estimators; that is, $\hat{p}$ is the MVUE for p. Since $\hat{p}$ is approximately normal, we can use it as a pivotal statistic and apply Theorem 8.2 to derive the formula for a large-sample confidence interval for p shown in the box.

LARGE-SAMPLE $(1 - \alpha)100\%$ CONFIDENCE INTERVAL FOR A POPULATION PROPORTION, p

$$\hat{p} \pm z_{\alpha/2}\sigma_{\hat{p}} \approx \hat{p} \pm z_{\alpha/2}\sqrt{\frac{\hat{p}\hat{q}}{n}}$$

where $\hat{p}$ is the sample proportion of observations with the characteristic of interest, and $\hat{q} = 1 - \hat{p}$.

[*Note:* The interval is approximate since we must substitute the sample $\hat{p}$ and $\hat{q}$ for the corresponding population values for $\sigma_{\hat{p}}$.]

Assumption: The sample size n is sufficiently large so that the approximation is valid. As a rule of thumb, the condition of a "sufficiently large" sample size will be satisfied if the interval $\hat{p} \pm 2\sigma_{\hat{p}}$ does not contain 0 or 1.

Note that we must substitute $\hat{p}$ and $\hat{q}$ into the formula for $\sigma_{\hat{p}} = \sqrt{pq/n}$ in order to construct the interval. This approximation will be valid as long as the sample size n is sufficiently large. Many researchers adopt the rule of thumb that n is "sufficiently large" if the interval $\hat{p} \pm 2\sqrt{\hat{p}\hat{q}/n}$ does not contain 0 or 1.

EXAMPLE 8.14

Stainless steels are frequently used in chemical plants to handle corrosive fluids. However, these steels are especially susceptible to stress corrosion cracking in certain environments. In a sample of 295 steel alloy failures that occurred in oil refineries and petrochemical plants in Japan over the last 10 years, 118 were caused by stress corrosion cracking and corrosion fatigue (Yamamoto and Kagawa, *Materials Performance*, June 1981). Construct a 95% confidence interval for the true proportion of alloy failures caused by stress corrosion cracking.

SOLUTION

The sample proportion of alloy failures caused by corrosion is

$$\hat{p} = \frac{\text{Number of alloy failures in sample caused by corrosion}}{\text{Number of alloy failures in sample}}$$

$$= \frac{118}{295} = .4$$

Thus, $\hat{q} = 1 - .4 = .6$. The approximate 95% confidence interval is then

$$\hat{p} \pm z_{.025}\sqrt{\frac{\hat{p}\hat{q}}{n}} = .4 \pm 1.96\sqrt{\frac{(.4)(.6)}{295}} = .4 \pm .056$$

or (.344, .456). [Note that the approximation is valid since the interval does not include 0 or 1.]

We are 95% confident that the interval from .344 to .456 encloses the true proportion of alloy failures that were caused by corrosion. If we repeatedly selected random samples of $n = 295$ alloy failures and constructed a 95% confidence interval based on each sample, then we would expect 95% of the confidence intervals constructed to contain p.   ∎

It should be noted that small-sample procedures are available for the estimation of a population proportion p. The details are not included in our discussion, however, because most surveys in actual practice use samples that are large enough to employ the procedure of this section.

EXERGISES 8.39–8.44

8.39 A survey of 50 data-processing firms was conducted to estimate the proportion of all data-processing firms that have implemented new software aids to improve their productivity. Of the 50 sampled firms, 12 indicated that they have implemented new software aids. Find a 95% confidence interval for p, the true proportion of all data-processing firms that have implemented new software aids to improve productivity.

8.40 According to a recent report by the U.S. Surgeon General, electrical engineers have the lowest smoking rate among all workers surveyed (*IEEE Spectrum*, April 1986). Only 16% of the male electrical engineers in the sample smoke cigarettes regularly. Assuming the sample contained $n = 50$ male electrical engineers, construct a 90% confidence interval for the proportion of all male electrical engineers who smoke regularly.

8.41 According to a study conducted by the California Division of Labor Research and Statistics (*Engineering News Record*, Mar. 10, 1983), roofing is one of the most hazardous occupations. Of 2,514 worker injuries that caused absences for a full workday or shift after the injury, 23% were attributable to falls from high elevations on level surfaces, 21% to falling hand tools or other materials, 19% to overexertion, and 20% to burns or scalds. Assume that the 2,514 injuries can be regarded as a random sample from the population of all roofing injuries in California.
  a. Construct a 95% confidence interval for the proportion of all injuries that are due to falls.
  b. Construct a 95% confidence interval for the proportion of all injuries that result in burns or scalds.

8.42 As part of a cooperative research agreement between the United States and Japan, a full-scale reinforced concrete building was designed and tested under simulated earthquake loading conditions in Japan (*Journal of Structural Engineering*, Jan. 1986). For one part of the study, several U.S. design engineers were asked to evaluate the new design. Of the 48 engineers surveyed, 36 believed the shear wall of the structure to be too lightly reinforced. Find a 95% confidence interval for the true proportion of U.S. design engineers who consider the shear wall of the building too lightly reinforced.

8.43 A large oil company recently won a multimillion dollar damage claim against the supplier of a 50-mile stretch of defective undersea pipe. The line was laid as part of a twin pipeline project for oil and gas. To support its claim, the oil company instructed a team of divers to examine 100 randomly selected 1-foot sectors of the undersea pipeline. The divers reported that 18 of the sampled pipeline sectors had minor cracks. The oil company claimed that, although these cracks did not pose an immediate threat, they did present a potential site for corrosion. Estimate p, the true proportion of the 50-mile stretch of pipeline that is defective, using a 99% confidence interval.

8.44 The pesticide Temik is used for controlling insects that feed on potatoes, oranges, and other crops. According to federal standards, drinking water wells with levels of Temik above 1 part per billion are considered contaminated. The accompanying table lists the results of tests for Temik contamination conducted in five states over the past few years. For each state, construct a 95% confidence interval for the true proportion of wells contaminated with Temik.

STATE	NUMBER OF WELLS TESTED	NUMBER OF CONTAMINATED WELLS
New York	10,500	2,750
Wisconsin	700	105
Maine	124	82
Florida	825	4
Virginia	76	17

SECTION 8.9

ESTIMATION OF THE DIFFERENCE BETWEEN TWO POPULATION PROPORTIONS

This section extends the method of Section 8.8 to the case in which we want to estimate the difference between two binomial proportions. For example, we may be interested in comparing the proportion p_1 of defective items produced by machine 1 to the proportion p_2 of defective items produced by machine 2.

Let y_1 and y_2 represent the numbers of successes in two independent binomial experiments with samples of size n_1 and n_2, respectively. To estimate the difference $(p_1 - p_2)$, where p_1 and p_2 are binomial parameters—i.e., the probabilities of success in the two independent binomial experiments—consider the proportion of successes in each of the samples:

$$\hat{p}_1 = \frac{y_1}{n_1} \quad \text{and} \quad \hat{p}_2 = \frac{y_2}{n_2}$$

Intuitively, we would expect $(\hat{p}_1 - \hat{p}_2)$ to provide a reasonable estimate of $(p_1 - p_2)$. Since $(\hat{p}_1 - \hat{p}_2)$ is a linear function of the binomial random variables y_1 and y_2, where $E(y_i) = n_i p_i$ and $V(y_i) = n_i p_i q_i$, we have

$$E(\hat{p}_1 - \hat{p}_2) = E(\hat{p}_1) - E(\hat{p}_2) = E\left(\frac{y_1}{n_1}\right) - E\left(\frac{y_2}{n_2}\right)$$

$$= \frac{1}{n_1}E(y_1) - \frac{1}{n_2}E(y_2) = \frac{1}{n_1}(n_1 p_1) - \frac{1}{n_2}(n_2 p_2)$$

$$= p_1 - p_2$$

and

$$V(\hat{p}_1 - \hat{p}_2) = V(\hat{p}_1) + V(\hat{p}_2) - 2\,\text{Cov}(\hat{p}_1, \hat{p}_2)$$

$$= V\left(\frac{y_1}{n_1}\right) + V\left(\frac{y_2}{n_2}\right) - 0 \qquad \text{since } y_1 \text{ and } y_2 \text{ are independent}$$

$$= \frac{1}{n_1^2}V(y_1) + \frac{1}{n_2^2}V(y_2)$$

$$= \frac{1}{n_1^2}(n_1 p_1 q_1) + \frac{1}{n_2^2}(n_2 p_2 q_2)$$

$$= \frac{p_1 q_1}{n_1} + \frac{p_2 q_2}{n_2}$$

Thus, $(\hat{p}_1 - \hat{p}_2)$ is an unbiased estimator of $(p_1 - p_2)$ and in addition, it has minimum variance (proof omitted).

The central limit theorem also guarantees that, for sufficiently large sample sizes n_1 and n_2, the sampling distribution of $(\hat{p}_1 - \hat{p}_2)$ will be approximately normal. It follows (Theorem 8.2) that a large-sample confidence interval for $(p_1 - p_2)$ may be obtained as shown in the box.

LARGE-SAMPLE $(1 - \alpha)100\%$ CONFIDENCE INTERVAL FOR $(p_1 - p_2)$

$$(\hat{p}_1 - \hat{p}_2) \pm z_{\alpha/2}\sigma_{(\hat{p}_1 - \hat{p}_2)} \approx (\hat{p}_1 - \hat{p}_2) \pm z_{\alpha/2}\sqrt{\frac{\hat{p}_1 \hat{q}_1}{n_1} + \frac{\hat{p}_2 \hat{q}_2}{n_2}}$$

where $\hat{p}_1$ and $\hat{p}_2$ are the sample proportions of observations with the characteristic of interest.

[*Note:* We have followed the usual procedure of substituting the sample values $\hat{p}_1$, $\hat{q}_1$, $\hat{p}_2$, and $\hat{q}_2$ for the corresponding population values required for $\sigma_{(\hat{p}_1 - \hat{p}_2)}$.

Assumption: The samples are sufficiently large that the approximation is valid. As a general rule of thumb, we will require that the intervals

$$\hat{p}_1 \pm 2\sqrt{\frac{\hat{p}_1 \hat{q}_1}{n_1}} \qquad \text{and} \qquad \hat{p}_2 \pm 2\sqrt{\frac{\hat{p}_2 \hat{q}_2}{n_2}}$$

do not contain 0 or 1.

Note that we must substitute the values of $\hat{p}_1$ and $\hat{p}_2$ for p_1 and p_2, respectively, to obtain an estimate of $\sigma_{(\hat{p}_1 - \hat{p}_2)}$. As in the one-sample case, this approximation is reasonably accurate when both n_1 and n_2 are sufficiently large, i.e., if the intervals

$$\hat{p}_1 \pm 2\sqrt{\frac{\hat{p}_1 \hat{q}_1}{n_1}} \qquad \text{and} \qquad \hat{p}_2 \pm 2\sqrt{\frac{\hat{p}_2 \hat{q}_2}{n_2}}$$

do not contain 0 or 1.

EXAMPLE 8.15

A traffic engineer conducted a study of vehicular speeds on a segment of street that had the posted speed limit changed several times. When the posted speed limit on the street was 30 miles per hour, the engineer monitored the speeds of 100 randomly selected vehicles traversing the street and observed 49 violations of the speed limit. After the speed limit was raised to 35 miles per hour, the engineer again monitored the speeds of 100 randomly selected vehicles and observed 19 vehicles in violation of the speed limit. Find a 99% confidence interval for $(p_1 - p_2)$, where p_1 is the true proportion of vehicles that (under similar driving conditions) exceed the lower speed limit (30 miles per hour) and p_2 is the true proportion of vehicles that (under similar driving conditions) exceed the higher speed limit (35 miles per hour). Interpret the interval.

SOLUTION

In this example,

$$\hat{p}_1 = \frac{49}{100} = .49 \quad \text{and} \quad \hat{p}_2 = \frac{19}{100} = .19$$

Note that the intervals

$$\hat{p}_1 \pm 2\sqrt{\frac{\hat{p}_1\hat{q}_1}{n_1}} = .49 \pm 2\sqrt{\frac{(.49)(.51)}{100}}$$

$$= .49 \pm .10$$

$$\hat{p}_2 \pm 2\sqrt{\frac{\hat{p}_2\hat{q}_2}{n_2}} = .19 \pm 2\sqrt{\frac{(.19)(.81)}{100}}$$

$$= .19 \pm .08$$

do not contain 0 or 1. Thus, we can apply the approximation for a large-sample confidence interval for $(p_1 - p_2)$.

For a confidence interval of $(1 - \alpha) = .99$, we have $\alpha = .01$ and $z_{\alpha/2} = z_{.005} = 2.58$ (from Table 4 of Appendix II). Substitution into the confidence interval formula yields:

$$(\hat{p}_1 - \hat{p}_2) \pm z_{\alpha/2}\sqrt{\frac{\hat{p}_1\hat{q}_1}{n_1} + \frac{\hat{p}_2\hat{q}_2}{n_2}}$$

$$= (.49 - .19) \pm 2.58\sqrt{\frac{(.49)(.51)}{100} + \frac{(.19)(.81)}{100}}$$

$$= .30 \pm .164$$

Our interpretation is that the true difference, $(p_1 - p_2)$, falls between .136 and .464 with 99% confidence. Since the lower bound on our estimate is positive (.136), we are fairly confident that the proportion of all vehicles in violation of the lower speed limit (30 miles per hour) exceeds the corresponding proportion in violation of the higher speed limit (35 miles per hour) by at least .136. ∎

Small-sample estimation procedures for $(p_1 - p_2)$ will not be discussed here for the reasons outlined at the end of Section 8.8.

8.45 Some companies have their own marketing research departments, while others contract the services of a marketing research firm. The marketing research department of a large manufacturer of word processors was charged with the responsibility of determining user preferences regarding its newly developed voice and touch data-entry terminal system (system A) relative to the industry leader (system B). A random sample of 205 users of word processors was selected and asked to state their preferences between the two systems. The results indicated that 119 prefer system A and 86 prefer system B. Find a 90% confidence interval for the true percentage of users of word processors who prefer system A over system B. Interpret the interval.

8.46 The nuclear mishap on Three Mile Island near Harrisburg, Pennsylvania, on March 28, 1979, forced many local residents to evacuate their homes—some temporarily, others permanently. In order to assess the impact of the accident on the area population, a questionnaire was designed and mailed to a sample of 150 households within 2 weeks after the accident occurred. Residents were asked how they felt both before and after the accident about having some of their electricity generated from nuclear power. The summary results are provided in the table.

| | ATTITUDE TOWARD NUCLEAR POWER | | | TOTALS |
	Favor	Oppose	Indifferent	
Before accident	62	35	53	150
After accident	52	72	26	150

Source: Brown, A. et al. *Final Report on a Survey of Three Mile Island Area Residents*. Department of Geography, Michigan State University, Aug. 1979.

  a. Construct a 99% confidence interval for the difference in the true proportions of Three Mile Island residents who favor nuclear power before and after the accident.
  b. Construct a 99% confidence interval for the difference in the true proportions of Three Mile Island residents who oppose nuclear power before and after the accident.

8.47 The Egyptian National Scientific and Technical Information Network (ENSTINET) operates an online database search service of existing U.S. databases. A database "search" occurs when a specific request is executed by ENSTINET during a single session. In situations when the search produces irrelevant or no output, the search is "rerun." According to *Information Processing & Management* (Vol. 22, No. 3, 1986), ENSTINET performed 342 database searches in 1982, of which 40 were rerun. In 1985, 83 of 2,117 searches required reruns. Assuming that the two samples of database searches are independent and random, construct a 95% confidence interval for the difference between the proportions of database search reruns performed by ENSTINET in 1982 and 1985. Interpret the interval.

8.48 One of American industry's most fundamental problems—the stagnation in productivity—has economic experts seeking methods of "reindustrializing" the United States. One possible answer to the productivity stagnation may be industrial robots (*Time*, Dec. 8, 1980). An industrial robot has a control and memory system, often in the form of a minicomputer, which enables it to be programmed to carry out a number of work routines faster and more efficiently than a human. Since the Japanese now operate most of the robots in the world, it is decided to estimate $(p_1 - p_2)$, the difference between the proportions of U.S. and Japanese firms that currently employ at least one industrial robot. Suppose random samples of U.S. and Japanese firms are selected, and the numbers of firms employing at least one industrial robot are recorded, with the sample sizes and results summarized in the accompanying table. Estimate the true difference in proportions $(p_1 - p_2)$, using a 95% confidence interval. Which country has the higher proportion of firms that employ industrial robots?

	U.S.	JAPAN
Number of firms sampled	$n_1 = 75$	$n_2 = 50$
Number of sampled firms that employ at least one industrial robot	16	22

8.49 A recent telephone survey was conducted to evaluate the job market for graduating computer science and data-processing majors. One question of interest is whether graduates of metropolitan New York colleges have an edge on graduates of colleges in other areas of the country because of the concentration of large user and corporate headquarters in New York City. Eighty New York City graduates and 65 graduates of colleges in other parts of the country were polled concerning their job status at the time of graduation. Fifty-seven New York graduates indicated that they had secured employment with a computer-oriented firm, while 43 of the graduates from other areas indicated they had found a job in their field. Establish a 90% confidence interval for the true difference in the proportions of graduates from New York City and other areas who had secured employment with a computer-oriented firm at the time of graduation.

| | | | | | | | | | | | |

S E C T I O N   8.10

ESTIMATION OF A POPULATION VARIANCE

In the previous sections we considered interval estimates for population means and proportions. In this section we discuss confidence intervals for a population variance σ^2, and, in Section 8.11, confidence intervals for the ratio of two variances, σ_1^2/σ_2^2. Unlike means and proportions, the pivotal statistics for variances do not possess a normal (z) distribution or a t distribution. In addition, certain assumptions are required regardless of the sample size.

Let $y_1, y_2, \ldots, y_n$ be a random sample from a normal distribution with mean μ and variance σ^2. From Theorem 7.4, we know that

$$\chi^2 = \frac{(n-1)s^2}{\sigma^2}$$

possesses a chi-square distribution with $(n - 1)$ degrees of freedom. Confidence intervals for σ^2 are based on the pivotal statistic, χ^2.

Upper-tail areas of the chi-square distribution have been tabulated and are given in Table 7 of Appendix II, a portion of which is reproduced in Table 8.7 (page 322). The table gives the values of χ^2, denoted χ_a^2, that locate an area a in the upper tail of the chi-square distribution, i.e., $P(\chi^2 > \chi_a^2) = a$. Thus, for

$n = 10$ and an upper-tail area of $a = .05$, we have $n - 1 = 9$ degrees of freedom and $\chi_a^2 = \chi_{.05}^2 = 16.9190$ (shaded in Table 8.7).

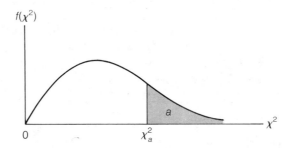

DEGREES OF FREEDOM	$\chi_{.100}^2$	$\chi_{.050}^2$	$\chi_{.025}^2$	$\chi_{.010}^2$	$\chi_{.005}^2$
1	2.70554	3.84146	5.02389	6.63490	7.87944
2	4.60517	5.99147	7.37776	9.21034	10.5966
3	6.25139	7.81473	9.34840	11.3449	12.8381
4	7.77944	9.48773	11.1433	13.2767	14.8602
5	9.23635	11.0705	12.8325	15.0863	16.7496
6	10.6446	12.5916	14.4494	16.8119	18.5476
7	12.0170	14.0671	16.0128	18.4753	20.2777
8	13.3616	15.5073	17.5346	20.0902	21.9550
9	14.6837	16.9190	19.0228	21.6660	23.5893
10	15.9871	18.3070	20.4831	23.2093	25.1882
11	17.2750	19.6751	21.9200	24.7250	26.7569
12	18.5494	21.0261	23.3367	26.2170	28.2995
13	19.8119	22.3621	24.7356	27.6883	29.8194
14	21.0642	23.6848	26.1190	29.1413	31.3193
15	22.3072	24.9958	27.4884	30.5779	32.8013
16	23.5418	26.2962	28.8454	31.9999	34.2672
17	24.7690	27.5871	30.1910	33.4087	35.7185
18	25.9894	28.8693	31.5264	34.8053	37.1564
19	27.2036	30.1435	32.8523	36.1908	38.5822

Unlike the z and t distributions, the chi-square distribution is not symmetric about 0. In order to find values of χ^2 that locate an area a in the lower tail of the distribution, we must find χ_{1-a}^2, where $P(\chi^2 > \chi_{1-a}^2) = 1 - a$. For example, the value of χ^2 that places an area $a = .05$ in the lower tail of the distribution when df $= 9$ is $\chi_{1-a}^2 = \chi_{.95}^2 = 3.32511$ (see Table 7 of Appendix II). We use this fact to write a probability statement for the pivotal statistic χ^2:

$$P(\chi_{1-\alpha/2}^2 \le \chi^2 \le \chi_{\alpha/2}^2) = 1 - \alpha$$

where $\chi_{\alpha/2}^2$ and $\chi_{(1-\alpha/2)}^2$ are tabulated values of χ^2 that place a probability of $\alpha/2$ in each tail of the chi-square distribution (see Figure 8.9).

FIGURE 8.9
The Location of $\chi^2_{(1-\alpha/2)}$
and $\chi^2_{\alpha/2}$ for a Chi-Square
Distribution

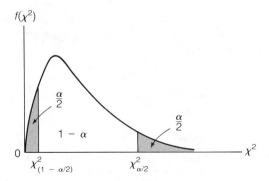

Substituting $[(n-1)s^2]/\sigma^2$ for χ^2 in the probability statement and performing some simple algebraic manipulations, we obtain

$$P\left(\chi^2_{(1-\alpha/2)} \leq \frac{(n-1)s^2}{\sigma^2} \leq \chi^2_{\alpha/2}\right)$$

$$= P\left(\frac{\chi^2_{(1-\alpha/2)}}{(n-1)s^2} \leq \frac{1}{\sigma^2} \leq \frac{\chi^2_{\alpha/2}}{(n-1)s^2}\right)$$

$$= P\left(\frac{(n-1)s^2}{\chi^2_{\alpha/2}} \leq \sigma^2 \leq \frac{(n-1)s^2}{\chi^2_{(1-\alpha/2)}}\right) = 1-\alpha$$

Thus, a $(1-\alpha)100\%$ confidence interval for σ^2 is

$$\frac{(n-1)s^2}{\chi^2_{\alpha/2}} \leq \sigma^2 \leq \frac{(n-1)s^2}{\chi^2_{(1-\alpha/2)}}$$

A $(1-\alpha)100\%$ CONFIDENCE INTERVAL FOR A POPULATION VARIANCE, σ^2

$$\frac{(n-1)s^2}{\chi^2_{\alpha/2}} \leq \sigma^2 \leq \frac{(n-1)s^2}{\chi^2_{(1-\alpha/2)}}$$

where $\sigma^2_{\alpha/2}$ and $\chi^2_{(1-\alpha/2)}$ are values of χ^2 that locate an area of $\alpha/2$ to the right and $\alpha/2$ to the left, respectively, of a chi-square distribution based on $(n-1)$ degrees of freedom.

Assumption: The population from which the sample is selected has an approximate normal distribution.

Note that the assumption of normality is required regardless of whether the sample size n is large or small.

EXAMPLE 8.16

A quality control supervisor in a cannery knows that the exact amount each can contains will vary, since there are certain uncontrollable factors that affect the amount of fill. The mean fill per can is important, but equally important is the variation σ^2 of the amount of fill. If σ^2 is large, some cans will contain too little and others too much. In order to estimate the variation of fill at the cannery, the supervisor randomly selects ten cans and weighs the contents of each. The following results are obtained:

$$\bar{y} = 7.98 \text{ ounces} \qquad s = .04 \text{ ounce}$$

Constuct a 90% confidence interval for the true variation in fill of cans at the cannery.

SOLUTION

The supervisor wishes to estimate σ^2, the population variance of the amount of fill. A $(1 - \alpha)100\%$ confidence interval for σ^2 is

$$\frac{(n - 1)s^2}{\chi^2_{\alpha/2}} \leq \sigma^2 \leq \frac{(n - 1)s^2}{\chi^2_{(1-\alpha/2)}}$$

For the confidence interval to be valid, we must assume that the sample of observations (amounts of fill) is selected from a normal population. For this example, $(1 - \alpha) = .90$ and $\alpha/2 = .10/2 = .05$. Therefore, we need the tabulated values $\chi^2_{.05}$ and $\chi^2_{.95}$ for $(n - 1) = 9$ df. These values are

$$\chi^2_{.05} = 16.9190 \qquad \text{and} \qquad \chi^2_{.95} = 3.32511$$

Substituting into the formula, we obtain

$$\frac{(10 - 1)(.04)^2}{16.9190} \leq \sigma^2 \leq \frac{(10 - 1)(.04)^2}{3.32511}$$
$$.000851 \leq \sigma^2 \leq .004331$$

We are 90% confident that the true variance in amount of fill of cans at the cannery falls between .000851 and .004331. The quality control supervisor could use this interval to check whether the variation of fill at the cannery is too large and in violation of government regulatory specifications. ∎

EXERCISES 8.50–8.54

8.50 For each of the following combinations of a and degrees of freedom (df), find the value of chi-square, χ^2_a, that places an area a in the upper tail of the chi-square distribution:
a. $a = .05$, df = 7   b. $a = .10$, df = 16
c. $a = .01$, df = 10   d. $a = .025$, df = 8
e. $a = .005$, df = 5

8.51 A machine used to fill beer cans must operate so that the amount of beer actually dispensed varies very little. If too much beer is released, the cans will overflow, causing waste. If too little beer is released, the cans will not contain enough beer, causing complaints from customers. A random sample of the fills for 20 cans yielded a standard deviation of .07 ounce. Estimate the true variance of the fills using a 95% confidence interval.

8.52 An interlaboratory study was conducted to determine the variation in the measured level of polychlorinated biphenyls (PCBs) in environmentally contaminated sediments (*Analytical Chemistry*, Nov. 1985). Samples of sediment from New Bedford Harbor (Massachusetts) known to be contaminated with PCBs were collected and aliquot solutions prepared. For one part of the study, the PCB concentration in each of a random sample of five aliquots was determined by a single laboratory using the Webb–McCall procedure. The analysis yielded a mean PCB concentration of 56 mg/kg and a standard deviation of .45 mg/kg. Find a 90% confidence interval for the variance in the Webb–McCall-measured PCB levels of contaminated sediment.

8.53 An experiment was conducted to investigate the precision of measurements of a saturated solution of iodine after an extended period of continuous stirring. The data shown in the table represent $n = 10$ iodine concentration measurements on the same solution. The population variance σ^2 measures the variability—i.e., the precision—of a measurement. Use the data to find a 95% confidence interval for σ^2.

RUN	CONCENTRATION	RUN	CONCENTRATION
1	5.507	6	5.527
2	5.506	7	5.504
3	5.500	8	5.490
4	5.497	9	5.500
5	5.506	10	5.497

8.54 Geologists analyze fluid inclusions in rock in order to infer the compositions of fluids present when the rocks crystallized. A new technique, called laser Raman microprobe (LRM) spectroscopy, has been developed for this purpose. An experiment was conducted to estimate the precision of the LRM technique (*Applied Spectroscopy*, Feb. 1986). A chip of natural Brazilian quartz with several artificially produced fluid inclusions was subjected to LRM spectroscopy. The amount of liquid carbon dioxide (CO_2) present in the inclusion was recorded for the same inclusion on four different days. The data (in mole percentage) are given below:

86.6   84.6   85.5   85.9

a. Obtain an estimate of the precision of the LRM technique by constructing a 99% confidence interval for the variation in the CO_2 concentration measurements.

b. What assumption is required for the interval estimate to be valid?

S E C T I O N 8.11

ESTIMATION OF THE RATIO OF TWO POPULATION VARIANCES

The common statistical procedure for comparing two population variances, σ_1^2 and σ_2^2, makes an inference about the ratio σ_1^2/σ_2^2. This is because the sampling distribution of the estimator of σ_1^2/σ_2^2 is well known when the samples are randomly and independently selected from two normal populations. Under these assumptions, a confidence interval for σ_1^2/σ_2^2 is based on the pivotal statistic

$$F = \frac{\chi_1^2/\nu_1}{\chi_2^2/\nu_2}$$

where χ_1^2 and χ_2^2 are chi-square random variables with $\nu_1 = (n_1 - 1)$ and $\nu_2 = (n_2 - 1)$ degrees of freedom, respectively. Substituting $(n - 1)s^2/\sigma^2$ for χ^2 (see Theorem 7.4) we may write

$$F = \frac{\chi_1^2/\nu_1}{\chi_2^2/\nu_2} = \frac{\dfrac{(n_1 - 1)s_1^2}{\sigma_1^2} \Big/ (n_1 - 1)}{\dfrac{(n_2 - 1)s_2^2}{\sigma_2^2} \Big/ (n_2 - 1)}$$

$$= \frac{s_1^2/\sigma_1^2}{s_2^2/\sigma_2^2} = \left(\frac{s_1^2}{s_2^2}\right)\left(\frac{\sigma_2^2}{\sigma_1^2}\right)$$

From Definition 7.3, we know that F has an F distribution with $\nu_1 = (n_1 - 1)$ numerator degrees of freedom and $\nu_2 = (n_2 - 1)$ denominator degrees of freedom. An F distribution can be symmetric about its mean, skewed to the left, or skewed to the right; its exact shape depends on the degrees of freedom associated with s_1^2 and s_2^2, i.e., $(n_1 - 1)$ and $(n_2 - 1)$.

In order to establish lower and upper confidence limits for σ_1^2/σ_2^2, we need to be able to find tabulated F values corresponding to the tail areas of the distribution. The *upper-tail* F values can be found in Tables 8, 9, 10, and 11 of Appendix II for $a = .10, .05, .025$, and $.01$, respectively. Table 9 of Appendix II is partially reproduced in Table 8.8. The columns of Tables 8–11 of Appendix II correspond to various degrees of freedom for the numerator sample variance, s_1^2, in the pivotal statistic, while the rows correspond to the degrees of freedom for the denominator sample variance, s_2^2. For example, with numerator degrees of freedom $\nu_1 = 7$ and denominator degrees of freedom $\nu_2 = 9$, we have $F_{.05} = 3.29$ (shaded in Table 8.8). Thus, $a = .05$ is the tail area to the right of 3.29 in the F distribution with 7 numerator df and 9 denominator df, i.e., $P(F > F_{.05}) = .05$.

Lower-tail values of the F distribution are not given in Tables 8–11 of Appendix II. However, it can be shown (proof omitted) that

$$F_{1-a(\nu_1,\nu_2)} = \frac{1}{F_{a(\nu_2,\nu_1)}}$$

where $F_{1-a(\nu_1,\nu_2)}$ is the F value that cuts off an area a in the lower tail of an F distribution based on ν_1 numerator and ν_2 denominator degrees of freedom, and $F_{a(\nu_2,\nu_1)}$ is the F value that cuts off an area a in the *upper* tail of an F distribution based on ν_2 numerator and ν_1 denominator degrees of freedom. For example, suppose we want to find the value that locates an area $a = .05$ in the *lower* tail of an F distribution with $\nu_1 = 7$ and $\nu_2 = 9$. That is, we want to find $F_{1-a(\nu_1,\nu_2)} = F_{.95(7,9)}$. First, we find the upper-tail values, $F_{.05(9,7)} = 3.68$, from Table 8.8. (Note that we must switch the numerator and denominator degrees of freedom to obtain this value.) Then, we calculate

$$F_{.95(7,9)} = \frac{1}{F_{.05(9,7)}} = \frac{1}{3.68} = .272$$

TABLE 8.8 Abbreviated Version of Table 9 of Appendix II: Tabulated Values of the F Distribution, $\alpha = .05$

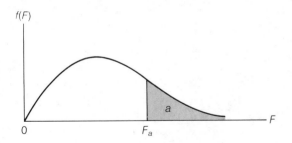

ν_2 \ ν_1	\multicolumn{9}{c}{NUMERATOR DEGREES OF FREEDOM}
	1	2	3	4	5	6	7	8	9
1	161.4	199.5	215.7	224.6	230.2	234.0	236.8	238.9	240.5
2	18.51	19.00	19.16	19.25	19.30	19.33	19.35	19.37	19.38
3	10.13	9.55	9.28	9.12	9.01	8.94	8.89	8.85	8.81
4	7.71	6.94	6.59	6.39	6.26	6.16	6.09	6.04	6.00
5	6.61	5.79	5.41	5.19	5.05	4.95	4.88	4.82	4.77
6	5.99	5.14	4.76	4.53	4.39	4.28	4.21	4.15	4.10
7	5.59	4.74	4.35	4.12	3.97	3.87	3.79	3.73	3.68
8	5.32	4.46	4.07	3.84	3.69	3.58	3.50	3.44	3.39
9	5.12	4.26	3.86	3.63	3.48	3.37	3.29	3.23	3.18
10	4.96	4.10	3.71	3.48	3.33	3.22	3.14	3.07	3.02
11	4.84	3.98	3.59	3.36	3.20	3.09	3.01	2.95	2.90
12	4.75	3.89	3.49	3.26	3.11	3.00	2.91	2.85	2.80
13	4.67	3.81	3.41	3.18	3.03	2.92	2.83	2.77	2.71
14	4.60	3.74	3.34	3.11	2.96	2.85	2.76	2.70	2.65

(DENOMINATOR DEGREES OF FREEDOM)

Using the notation established above, we can write a probability statement for the pivotal statistic F (see Figure 8.10):

$$P(F_{1-\alpha/2(\nu_1,\nu_2)} \leq F \leq F_{\alpha/2(\nu_1,\nu_2)}) = 1 - \alpha$$

FIGURE 8.10
F Distribution with
$\nu_1 = (n_1 - 1)$ and
$\nu_2 = (n_2 - 1)$

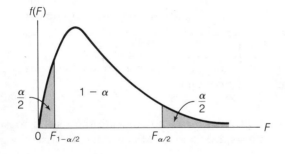

Letting $F_L = F_{1-\alpha/2}$ and $F_U = F_{\alpha/2}$, and substituting $(s_1^2/s_2^2)(\sigma_2^2/\sigma_1^2)$ for F, we obtain:

$$P(F_L \leq F \leq F_U) = P\left[F_L \leq \left(\frac{s_1^2}{s_2^2}\right)\left(\frac{\sigma_2^2}{\sigma_1^2}\right) \leq F_U\right]$$

$$= P\left(\frac{s_2^2}{s_1^2}F_L \leq \frac{\sigma_2^2}{\sigma_1^2} \leq \frac{s_2^2}{s_1^2}F_U\right)$$

$$= P\left(\frac{s_1^2}{s_2^2} \cdot \frac{1}{F_U} \leq \frac{\sigma_1^2}{\sigma_2^2} \leq \frac{s_1^2}{s_2^2} \cdot \frac{1}{F_L}\right) = 1 - \alpha$$

or,

$$P\left(\frac{s_1^2}{s_2^2} \cdot \frac{1}{F_{\alpha/2(\nu_1,\nu_2)}} \leq \frac{\sigma_1^2}{\sigma_2^2} \leq \frac{s_1^2}{s_2^2} \cdot \frac{1}{F_{1-\alpha/2(\nu_1,\nu_2)}}\right) = 1 - \alpha$$

Replacing $F_{1-\alpha/2(\nu_1,\nu_2)}$ with $1/F_{\alpha/2(\nu_2,\nu_1)}$, we obtain the final form of the confidence interval:

$$P\left(\frac{s_1^2}{s_2^2} \cdot \frac{1}{F_{\alpha/2(\nu_1,\nu_2)}} \leq \frac{\sigma_1^2}{\sigma_2^2} \leq \frac{s_1^2}{s_2^2} \cdot F_{\alpha/2(\nu_2,\nu_1)}\right) = 1 - \alpha$$

A $(1 - \alpha)100\%$ CONFIDENCE INTERVAL FOR THE RATIO OF TWO POPULATION VARIANCES, σ_1^2/σ_2^2

$$\frac{s_1^2}{s_2^2} \cdot \frac{1}{F_{\alpha/2(\nu_1,\nu_2)}} \leq \frac{\sigma_1^2}{\sigma_2^2} \leq \frac{s_1^2}{s_2^2}F_{\alpha/2(\nu_2,\nu_1)}$$

where $F_{\alpha/2(\nu_1,\nu_2)}$ is the value of F that locates an area $\alpha/2$ in the upper tail of the F distribution with $\nu_1 = (n_1 - 1)$ numerator and $\nu_2 = (n_2 - 1)$ denominator degrees of freedom, and $F_{\alpha/2(\nu_2,\nu_1)}$ is the value of F that locates an area $\alpha/2$ in the upper tail of the F distribution with $\nu_2 = (n_2 - 1)$ numerator and $\nu_1 = (n_1 - 1)$ denominator degrees of freedom.

Assumptions: 1. Both of the populations from which the samples are selected have relative frequency distributions that are approximately normal.
2. The random samples are selected in an independent manner from the two populations.

As in the one-sample case, normal populations must be assumed regardless of the sizes of the two samples.

EXAMPLE 8.17   Suppose an investor wants to compare the risks associated with two different computer stocks, IBM and COMSAT, where the risk of a given stock is measured by the variation in daily price changes. The investor obtains independent random samples of 21 daily price changes for IBM and 25 daily price changes for COMSAT. The sample results are summarized in the accompanying table. Compare the risks associated with the two stocks by forming a 95% confidence interval for the ratio of the true population variances, σ_1^2/σ_2^2.

IBM	COMSAT
$n_1 = 21$	$n_2 = 25$
$\bar{y}_1 = .585$	$\bar{y}_2 = .572$
$s_1 = .023$	$s_2 = .014$

SOLUTION

First, we must assume that the distributions of the daily price changes for the two computer stocks are both approximately normal. Since we want a 95% confidence interval, the value of $\alpha/2$ is .025, and we need to find $F_{.025(\nu_1,\nu_2)}$ and $F_{.025(\nu_2,\nu_1)}$. The sample sizes are $n_1 = 21$ and $n_2 = 25$; thus, $F_{.025(\nu_1,\nu_2)}$ is based on $\nu_1 = (n_1 - 1) = 20$ numerator df and $\nu_2 = (n_2 - 1) = 24$ denominator df. Consulting Table 10 of Appendix II, we obtain $F_{.025(20,24)} = 2.33$. In contrast, $F_{.025(\nu_2,\nu_1)}$ is based on $\nu_2 = (n_2 - 1) = 24$ numerator df and $\nu_1 = (n_1 - 1) = 20$ denominator df; hence (from Table 10 of Appendix II), $F_{.025(24,20)} = 2.41$. Substituting the values for s_1, s_2, $F_{.025(\nu_1,\nu_2)}$, and $F_{.025(\nu_2,\nu_1)}$ into the confidence interval formula, we have

$$\frac{(.023)^2}{(.014)^2}\left(\frac{1}{2.33}\right) \leq \frac{\sigma_1^2}{\sigma_2^2} \leq \frac{(.023)^2}{(.014)^2}(2.41)$$

$$1.16 \leq \frac{\sigma_1^2}{\sigma_2^2} \leq 6.50$$

We estimate with 95% confidence that the ratio σ_1^2/σ_2^2 of the true population variances will fall between 1.16 and 6.50. Since all the values within the interval $(1.16, 6.50)$ are greater than 1.0, we can be fairly confident that the risk associated with IBM stock, as measured by σ_1^2, is greater than the risk associated with COMSAT stock, as measured by σ_2^2. ∎

EXERCISES 8.55–8.60

8.55 Find $F_{.05}$ for an F distribution with:
  a. Numerator df = 7,   denominator df = 25
  b. Numerator df = 10,   denominator df = 8
  c. Numerator df = 30,   denominator df = 60
  d. Numerator df = 15,   denominator df = 4

8.56 Find F_a for an F distribution with 15 numerator df and 12 denominator df for the following values of a:
  a. $a = .025$ **b.** $a = .05$ **c.** $a = .10$

8.57 Refer to the data given in Exercise 8.31. Construct a 90% confidence interval for the ratio of the shear stress variances of epoxy-repaired truss joints for the two species of wood. Based on this interval, is there evidence to indicate that the two shear stress variances differ? Explain.

8.58 A firm has been experimenting with two different physical arrangements of its assembly line. It has been determined that both arrangements yield approximately the same average number of finished units per day. To obtain an arrangement that produces greater process control, you suggest that the arrangement with the smaller variance in the number of finished units produced per day be permanently adopted. Two independent random samples yield the results shown in the table.

ASSEMBLY LINE 1	ASSEMBLY LINE 2
$n_1 = 21$ days	$n_2 = 21$ days
$s_1^2 = 1,432$	$s_2^2 = 3,761$

  a. Construct a 90% confidence interval for the ratio of the variances of the number of finished units for the two assembly line arrangements.
  b. What assumptions must be satisfied to ensure the validity of the interval estimate of part **a**?
  c. Based on the interval of part **a**, which of the two arrangements would you recommend?

8.59 Refer to the cancer death rate increases for fluoridated and nonfluoridated cities given in Exercise 8.32. Find a 95% confidence interval for the ratio of the variances of the cancer death rate increases for the two groups of cities. Based on the interval, does it appear that the assumption of equal variances required to conduct the analysis of Exercise 8.32 is satisfied?

8.60 Refer to the PCB study described in Exercise 8.52. Recall that level of PCB was measured in each of a sample of five aliquots using the Webb–McCall procedure. Another sample of five aliquots of sediment was measured for PCBs using a diffferent procedure, called the Aroclor Standard comparison. Summary statistics on PCB concentration for the two samples are given in the table.

	WEBB–McCALL	AROCLOR STANDARD
Sample size	5	5
Mean PCB concentration (mg/kg)	56	60
Standard deviation	.45	.89

Source: Alford-Stevens, A. L., Budde, W. L., and Bellar, T. A. "Interlaboratory Study on Determination of Polychlorinated Biphenyls in Environmentally Contaminated Sediments." *Analytical Chemistry*, Vol. 57, No. 13, Nov. 1985, p. 2454. Reprinted with permission from *Analytical Chemistry*. Copyright 1985 American Chemical Society.

  a. Construct a 90% confidence interval for the ratio of the variances in the PCB levels measured by the two techniques.
  b. What assumptions are required for the interval estimate to be valid?

SECTION 8.12

CHOOSING THE
SAMPLE SIZE

One of the first problems encountered when applying statistics in a practical situation is to decide on the number of measurements to include in the sample(s). The solution to this problem depends on the answers to the following questions: Approximately how wide do you want your confidence interval to be? What confidence coefficient do you require?

You have probably noticed that the half-widths of many of the confidence intervals presented in Sections 8.5–8.11 are functions of the sample size and the estimated standard error of the point estimator involved. For example, the half-width H of the small-sample confidence interval for μ is

$$H = t_{\alpha/2}\left(\frac{s}{\sqrt{n}}\right)$$

where $t_{\alpha/2}$ depends on the sample size n and s is a statistic computed from the sample data. Since we will not know s before selecting the sample and we have no control over its value, the easiest way to decrease the width of the confidence interval is to increase the sample size n. Generally speaking, the larger the sample size, the more information you will acquire and the smaller will be the width of the confidence interval. We illustrate the procedure for selecting the sample size in Examples 8.18 and 8.19.

EXAMPLE 8.18

As part of a Department of Energy (DOE) survey, American families will be randomly selected and questioned about the amount of money they spent last year on home heating oil or gas. Of particular interest to the DOE is the average amount μ spent last year on heating fuel. If the DOE wants the estimate of μ to be correct to within \$10 with a confidence coefficient of .95, how many families should be included in the sample?

SOLUTION

The DOE wants to obtain an interval estimate of μ, with confidence coefficient equal to $(1 - \alpha) = .95$ and half-width of the interval equal to 10. The half-width of a large-sample confidence interval for μ is

$$H = z_{\alpha/2}\sigma_{\bar{y}} = z_{\alpha/2}\left(\frac{\sigma}{\sqrt{n}}\right)$$

In this example we have $H = 10$ and $z_{\alpha/2} = z_{.025} = 1.96$. In order to solve the equation for n, we need to know σ. But, as will usually be the case in practice, σ is unknown. Suppose, however, that the DOE knows from past records that the yearly amounts spent on heating fuel have a range of approximately \$520. Then we could approximate σ by letting the range equal to 4σ.* Thus,

$$4\sigma \approx 520 \quad \text{or} \quad \sigma \approx 130$$

---

*From the Empirical Rule, we expect about 95% of the observations to fall between $\mu - 2\sigma$ and $\mu + 2\sigma$. Thus,

$$\text{Range} \approx (\mu + 2\sigma) - (\mu - 2\sigma) = 4\sigma$$

Solving for n, we have

$$H = z_{\alpha/2}\left(\frac{\sigma}{\sqrt{n}}\right) \quad \text{or} \quad 10 = 1.96\left(\frac{130}{\sqrt{n}}\right)$$

or

$$n = \frac{(1.96)^2(130)^2}{(10)^2} \approx 650$$

Consequently, the DOE will need to elicit responses from 650 American families in order to estimate the mean amount spent on home heating fuel last year to within $10 with 95% confidence. Since this would require an extensive and costly survey, the DOE might decide to allow a larger half-width (say, $H = 15$ or $H = 20$) in order to reduce the sample size, or the DOE might decrease the desired confidence coefficient. The important point is that the experimenter can obtain an idea of the sampling effort necessary to achieve a specified precision in the final estimate by determining the approximate sample size *before* the experiment is begun.   ∎

EXAMPLE 8.19

A production supervisor suspects a difference exists between the proportions p_1 and p_2 of defective items produced by two different machines. Experience has shown that the proportion defective for each of the two machines is in the neighborhood of .03. If the supervisor wants to estimate the difference in the proportions correct to within .005 with probability .95, how many items must be randomly sampled from the production of each machine? (Assume that you want $n_1 = n_2 = n$.)

SOLUTION

Since we want to estimate $(p_1 - p_2)$ with a 95% confidence interval, we will use $z_{\alpha/2} = z_{.025} = 1.96$. In order for the estimate to be correct to within .005, the half-width of the confidence interval must equal .005. Then, letting $p_1 = p_2 = .03$ and $n_1 = n_2 = n$, we find the required sample size per machine by solving the following equation for n:

$$H = z_{\alpha/2}\sigma_{(\hat{p}_1 - \hat{p}_2)} \quad \text{or} \quad H = z_{\alpha/2}\sqrt{\frac{p_1 q_1}{n_1} + \frac{p_2 q_2}{n_2}}$$

$$.005 = 1.96\sqrt{\frac{(.03)(.97)}{n} + \frac{(.03)(.97)}{n}}$$

$$.005 = 1.96\sqrt{\frac{2(.03)(.97)}{n}}$$

$$n = \frac{(1.96)^2(2)(.03)(.97)}{(.005)^2} \approx 8,944$$

You can see that this may be a tedious sampling procedure. If the supervisor insists on estimating $(p_1 - p_2)$ correct to within .005 with probability equal to .95, approximately 9,000 items will have to be inspected for each machine.   ∎

You can see from the calculations in Example 8.19 that $\sigma_{(\hat{p}_1-\hat{p}_2)}$ (and hence the solution, $n_1 = n_2 = n$) depends on the actual (but unknown) values of p_1 and p_2. In fact, the required sample size $n_1 = n_2 = n$ is largest when $p_1 = p_2 = .5$. Therefore, if you have no prior information on the approximate values of p_1 and p_2, use $p_1 = p_2 = .5$ in the formula for $\sigma_{(\hat{p}_1-\hat{p}_2)}$. If p_1 and p_2 are in fact close to .5, then the resulting values of n_1 and n_2 will be correct. If p_1 and p_2 differ substantially from .5, then your solutions for n_1 and n_2 will be larger than needed. Consequently, using $p_1 = p_2 = .5$ when solving for n_1 and n_2 is a conservative procedure because the sample sizes n_1 and n_2 will be at least as large as (and probably larger than) needed.

The formulas for calculating the sample size(s) required for estimating the parameters μ, $(\mu_1 - \mu_2)$, p, and $(p_1 - p_2)$ are summarized in the boxes. Sample size calculations for variances require more sophisticated techniques and are beyond the scope of this text.

CHOOSING THE SAMPLE SIZE FOR ESTIMATING A POPULATION MEAN μ TO WITHIN H UNITS WITH PROBABILITY $(1 - \alpha)$

$$n = \left(\frac{z_{\alpha/2}\sigma}{H}\right)^2$$

[*Note:* The population standard deviation σ will usually have to be approximated.]

CHOOSING THE SAMPLE SIZES FOR ESTIMATING THE DIFFERENCE $(\mu_1 - \mu_2)$ BETWEEN A PAIR OF POPULATION MEANS CORRECT TO WITHIN H UNITS WITH PROBABILITY $(1 - \alpha)$

$$n_1 = n_2 = \left(\frac{z_{\alpha/2}}{H}\right)^2(\sigma_1^2 + \sigma_2^2)$$

where n_1 and n_2 are the numbers of observations sampled from each of the two populations, and σ_1^2 and σ_2^2 are the variances of the two populations.

CHOOSING THE SAMPLE SIZE FOR ESTIMATING A POPULATION PROPORTION p TO WITHIN H UNITS WITH PROBABILITY $(1 - \alpha)$

$$n = \left(\frac{z_{\alpha/2}}{H}\right)^2 pq$$

where p is the value of the population proportion that you are attempting to estimate and $q = 1 - p$.

[*Note:* This technique requires previous estimates of p and q. If none are available, use $p = q = .5$ for a conservative choice of n.]

CHOOSING THE SAMPLE SIZES FOR ESTIMATING THE DIFFERENCE $(p_1 - p_2)$ BETWEEN TWO POPULATION PROPORTIONS TO WITHIN H UNITS WITH PROBABILITY $(1 - \alpha)$

$$n_1 = n_2 = \left(\frac{z_{\alpha/2}}{H}\right)^2 (p_1 q_1 + p_2 q_2)$$

where p_1 and p_2 are the proportions for populations 1 and 2, respectively, and n_1 and n_2 are the numbers of observations to be sampled from each population.

EXERCISES 8.61–8.65

8.61 The federal government requires states to certify that they are enforcing the 55 miles per hour speed limit and that motorists are driving at that speed. A state is in jeopardy of losing millions of dollars in federal road funds if more than 60% of its vehicles on 55 miles per hour highways are exceeding the speed limit. The state highway patrol conducts 70 radar surveys each year at a total of 50 sites to estimate the proportion p of vehicles exceeding 55 miles per hour. Each sample survey involves at least 400 vehicles.

a. How large a sample should be selected at a particular site to estimate p to within 3% with 90% confidence? Last year approximately 60% of all vehicles exceeded 55 miles per hour.

b. The highway patrol also estimates μ, the average speed of vehicles on state highways. Accordingly, it wants to know whether the sample size determined in part **a** is large enough to also estimate μ to within .25 mile per hour with 90% confidence. Assume that the standard deviation of vehicle speeds is approximately 2 miles per hour. How large a sample should be taken at a particular site to estimate μ with the desired reliability?

8.62 A consumer protection agency wants to compare the work of two electrical contractors in order to evaluate their safety records. The agency plans to inspect residences in which each of these contractors has done the wiring to estimate the difference in the proportions of residences that are electrically deficient. Suppose the proportions of deficient work are expected to be about .10 for both contractors. How many homes should be inspected to estimate the difference in proportions to within .05 with 90% confidence?

8.63 A large steel corporation conducted an experiment to compare the average iron contents of two consignments of lumpy iron ore. In accordance with industrial standards, n increments of iron ore were randomly selected from each consignment and measured for iron content. From previous experiments it is known that iron contents vary over a range of roughly 3%. How large should n be if the steel company wants to estimate the difference in mean iron contents of the two consignments correct to within .05% with 95% confidence? [*Hint:* To obtain an approximate value for σ_1 and σ_2, set $\sigma_1 = \sigma_2 = \sigma$ and set Range $= 4\sigma$. Then $3 \approx 4\sigma$ and $\sigma \approx \frac{3}{4}$]

8.64 *Materials requirements planning (MRP) systems* are computerized planning and control systems for manufacturing operations. Since their introduction in the mid-1960's, MRP systems have been used to manage raw materials and work-in-process inventories while improving customer service. Suppose you want to estimate the proportion p of manufacturing firms that use MRP systems. Approximately how large a sample would be required to estimate p to within .02 with a confidence coefficient of .95? (Use a conservative estimate of $p \approx .5$ in your calculations.)

OPTIONAL EXERCISE

8.65 When determining the sample size required to estimate p, show that the sample size n is largest when $p = .5$.

SECTION 8.13

SUMMARY

Estimation is a procedure for inferring the value(s) of one (or more) population parameters. An **estimator**, a rule that tells how to calculate a particular **estimate** of a parameter based on information contained in a sample, can be one of two types. A **point estimator** uses the sample data to calculate a single number that serves as an estimate of a population parameter. An **interval estimator** uses the sample data to calculate two numbers that define an interval that is intended to enclose the estimated parameter with some predetermined probability.

Point and interval estimators can be acquired intuitively; it seems reasonable to use sample statistics to estimate the corresponding population parameters (the **method of moments**). In addition, point estimators can be acquired using the **method of maximum likelihood** (Section 8.3) or the **method of least squares** (Chapter 10); interval estimators can be constructed using **pivotal statistics** and the procedure illustrated in Section 8.4.

It is possible to acquire many different estimators for the same parameter. For example, we could estimate a population mean using the sample mean $\bar{y}$, the sample median, or, as a third possibility, the average of the largest and smallest measurements in a sample. A comparison of the sampling distributions of different estimators will enable us to choose the best of a group. In general, we prefer point estimators that are **unbiased** and possess **minimum variance**, i.e., **minimum variance unbiased estimators** (MVUE). For a given **confidence coefficient**, we prefer interval estimators with a mean interval width that is small and subject to a small amount of variation.

We presented a number of point and interval estimators and demonstrated how they can be applied in practical situations. (These results are summarized in Tables 8.9a and 8.9b, pages 346–347). By reviewing the examples, you can see that estimation as a method of inference attempts to answer the question, "What is the value of the parameter θ?" We will approach inference-making from a different point of view in Chapter 9.

TABLE 8.9a Summary of Estimation Procedures: One-Sample Case

PARAMETER θ	ESTIMATOR $\hat{\theta}$	$E(\hat{\theta})$	$\sigma_{\hat{\theta}}$	APPROXIMATION TO $\sigma_{\hat{\theta}}$	$(1-\alpha)100\%$ CONFIDENCE INTERVAL	SAMPLE SIZE	ADDITIONAL ASSUMPTIONS
Mean μ	$\bar{y}$	μ	$\dfrac{\sigma}{\sqrt{n}}$	$\dfrac{s}{\sqrt{n}}$	$\bar{y} \pm z_{\alpha/2}\left(\dfrac{s}{\sqrt{n}}\right)$	$n \geq 30$	None
					$\bar{y} \pm t_{\alpha/2}\left(\dfrac{s}{\sqrt{n}}\right)$ where $t_{\alpha/2}$ is based on $(n-1)$ df	$n < 30$	Normal population
Binomial proportion p	$\hat{p} = \dfrac{y}{n}$	p	$\sqrt{\dfrac{pq}{n}}$	$\sqrt{\dfrac{\hat{p}\hat{q}}{n}}$	$\hat{p} \pm z_{\alpha/2}\sqrt{\dfrac{\hat{p}\hat{q}}{n}}$	n large enough so that the interval $\hat{p} \pm 2\sqrt{\dfrac{\hat{p}\hat{q}}{n}}$ does not contain 0 or 1	None
Variance σ^2	s^2	σ^2	Not needed	Not needed	$\dfrac{(n-1)s^2}{\chi^2_{\alpha/2}} \leq \sigma^2 \leq \dfrac{(n-1)s^2}{\chi^2_{(1-\alpha/2)}}$ where $\chi^2_{\alpha/2}$ and $\chi^2_{(1-\alpha/2)}$ are the tabulated values of χ^2, given in Table 7 of Appendix II, that locate $\alpha/2$ in each tail of the chi-square distribution with $(n-1)$ df, i.e., $P(\chi^2 \geq \chi^2_{\alpha/2}) = \alpha/2$ and $P(\chi^2 \geq \chi^2_{(1-\alpha/2)}) = 1 - \alpha/2$	All n	Normal population

TABLE 8.9b Summary of Estimation Procedures: Two-Sample Case

PARAMETER θ	ESTIMATOR $\hat{\theta}$	$E(\hat{\theta})$	$\sigma_{\hat{\theta}}$	APPROXIMATION TO $\sigma_{\hat{\theta}}$	$(1-\alpha)100\%$ CONFIDENCE INTERVAL	SAMPLE SIZES	ADDITIONAL ASSUMPTIONS
$(\mu_1 - \mu_2)$ Difference between population means: Independent samples	$(\bar{y}_1 - \bar{y}_2)$ Difference between sample means	$(\mu_1 - \mu_2)$	$\sqrt{\dfrac{\sigma_1^2}{n_1} + \dfrac{\sigma_2^2}{n_2}}$ $\sqrt{\sigma^2\left(\dfrac{1}{n_1} + \dfrac{1}{n_2}\right)}$	$\sqrt{\dfrac{s_1^2}{n_1} + \dfrac{s_2^2}{n_2}}$ $\sqrt{s_p^2\left(\dfrac{1}{n_1} + \dfrac{1}{n_2}\right)}$ where $s_p^2 = \dfrac{(n_1 - 1)s_1^2 + (n_2 - 1)s_2^2}{n_1 + n_2 - 2}$	$(\bar{y}_1 - \bar{y}_2) \pm z_{\alpha/2}\sqrt{\dfrac{s_1^2}{n_1} + \dfrac{s_2^2}{n_2}}$ $(\bar{y}_1 - \bar{y}_2) \pm t_{\alpha/2}\sqrt{s_p^2\left(\dfrac{1}{n_1} + \dfrac{1}{n_2}\right)}$ where $t_{\alpha/2}$ is based on $(n_1 + n_2 - 2)$ df	$n_1 \geq 30,\ n_2 \geq 30$ Either $n_1 < 30$ or $n_2 < 30$ or both	None Both populations normal with equal variances $(\sigma_1^2 = \sigma_2^2)$
$\mu_d = (\mu_1 - \mu_2)$ Difference between population means: Matched pairs	$\bar{d} = \Sigma d_i/n$ Mean of sample differences	μ_d	$\dfrac{\sigma_d}{\sqrt{n_d}}$	$\dfrac{s_d}{\sqrt{n_d}}$ where s_d is the standard deviation of the sample of differences	$\bar{d} \pm t_{\alpha/2}\left(\dfrac{s_d}{\sqrt{n_d}}\right)$ where $t_{\alpha/2}$ is based on $(n_d - 1)$ df	All n_d (If $n_d \geq 30$, then $t_{\alpha/2}$ will be approximately equal to $z_{\alpha/2}$.)	Population of differences d_i is normal
$(p_1 - p_2)$ Difference between two binomial parameters	$(\hat{p}_1 - \hat{p}_2)$ Difference between the sample proportions $\hat{p}_1 = y_1/n_1$ and $\hat{p}_2 = y_2/n_2$	$(p_1 - p_2)$	$\sqrt{\dfrac{p_1 q_1}{n_1} + \dfrac{p_2 q_2}{n_2}}$	$\sqrt{\dfrac{\hat{p}_1 \hat{q}_1}{n_1} + \dfrac{\hat{p}_2 \hat{q}_2}{n_2}}$	$(\hat{p}_1 - \hat{p}_2) \pm z_{\alpha/2}\sqrt{\dfrac{\hat{p}_1 \hat{q}_1}{n_1} + \dfrac{\hat{p}_2 \hat{q}_2}{n_2}}$	n_1 and n_2 large enough so that the intervals $\hat{p}_1 \pm 2\sqrt{\dfrac{\hat{p}_1 \hat{q}_1}{n_1}}$ and $\hat{p}_2 \pm 2\sqrt{\dfrac{\hat{p}_2 \hat{q}_2}{n_2}}$ do not contain 0 or 1	Independent samples
σ_1^2/σ_2^2 Ratio of population variances	s_1^2/s_2^2 Ratio of sample variances	Not needed	Not needed	Not needed	$\left(\dfrac{s_1^2}{s_2^2}\right)\dfrac{1}{F_{\alpha/2(\nu_1, \nu_2)}} \leq \dfrac{\sigma_1^2}{\sigma_2^2} \leq \left(\dfrac{s_1^2}{s_2^2}\right)F_{\alpha/2(\nu_2, \nu_1)}$ where $F_{\alpha/2(\nu_1, \nu_2)}$ and $F_{\alpha/2(\nu_2, \nu_1)}$ are the tabulated values of F (Tables 8, 9, 10, and 11 of Appendix II) that place an area equal to $\alpha/2$ in the upper tail of the F distribution, where $F_{\alpha/2(\nu_1, \nu_2)}$ is based on $\nu_1 = (n_1 - 1)$ numerator and $\nu_2 = (n_2 - 1)$ denominator degrees of freedom, and $F_{\alpha/2(\nu_2, \nu_1)}$ is based on $\nu_2 = (n_2 - 1)$ numerator and $\nu_1 = (n_1 - 1)$ denominator degrees of freedom	All n_1 and n_2	Independent samples from two normal populations

8.66 The National Aeronautics and Space Administration (NASA) is continually testing the components of its spacecraft. Suppose NASA desires to estimate the mean lifetime of a particular mechanical component used in the space shuttle *Columbia*. Due to the prohibitive cost, only ten components can be tested under simulated space conditions. The lifetimes (in hours) of the components were recorded with the following results:

$$\bar{y} = 1{,}173.6 \qquad s = 36.3$$

Estimate the mean lifetime of the mechanical components with a 95% confidence interval. Assume that the population of lifetimes of the mechanical component has an approximately normal relative frequency distribution.

8.67 When new instruments are developed to perform chemical analyses of products (food, medicine, etc.), they are usually evaluated with respect to two criteria: accuracy and precision. *Accuracy* refers to the ability of the instrument to identify correctly the nature and amounts of a product's components. *Precision* refers to the consistency with which the instrument will identify the components of the same material. Thus, a large variability in the identification of a single sample of a product indicates a lack of precision. Suppose a pharmaceutical firm is considering two brands of an instrument designed to identify the components of certain drugs. As part of a comparison of precision, ten test-tube samples of a well-mixed batch of a drug are selected and then five are analyzed by instrument A and five by instrument B. The data shown in the table are the percentages of the primary component of the drug given by the instruments.

INSTRUMENT A	INSTRUMENT B
43	46
48	49
37	43
52	41
45	48

a. Construct a 90% confidence interval that will enable the firm to compare the precision of the two instruments.
b. Based on the interval estimate of part **a**, what would you infer about the precision of the two instruments?
c. What assumptions must be satisfied to ensure the validity of any inferences derived from the interval estimate?

8.68 The February 10, 1983, edition of the *Engineering News Record* reported:

Union mechanical contractors concerned about the financial condition of the pension plans they pay into may not have as much to worry about as they thought, according to a survey of the plans conducted by the United Association of Plumbers and Pipefitters. The survey was ordered in response to MCAA concerns that unfunded vested benefits of the union's pension plans might total as much as $800 million. Although the union's report did not disclose the total unfunded liability of the 210 plumber, pipefitter and sprinkler fitter pension plans examined, it said that 130 plans (61.9%) currently have no unfunded vested liability . . . .

Use a 95% confidence interval to estimate the true proportion of pension plans with no unfunded vested liability.

8.69 A regional data center serving a major university and surrounding city wants to estimate the variability in the number of computer jobs it processes per day. A random sample of 18 days during the past 4 months was selected and the number of jobs processed each day was recorded. The following sample statistics were computed:

$$\bar{y} = 1,266 \qquad s = 175$$

Estimate the true variability in the number of jobs processed per day using a 99% confidence interval. What assumptions are necessary for the validity of the estimation technique?

8.70 Refer to the LRM spectroscopy experiment described in Exercise 8.54. The amount of liquid CO_2 present in each of two different fluid inclusions (named FREO and FRITZ) was measured on each of four randomly selected days. The data are reproduced in the table. Construct a 95% confidence interval for the mean difference between the CO_2 concentration readings (in mole percentage) of the two fluid inclusions.

DAY	INCLUSION FREO	INCLUSION FRITZ
1	86.6	83.8
2	84.6	85.3
3	85.5	84.6
4	85.9	83.4

Source: Wopenka, B. and Pasteris, J. D. "Limitations to Quantitative Analysis of Fluid Inclusions in Geological Samples by Laser Raman Microprobe Spectroscopy." *Applied Spectroscopy*, Vol. 40, No. 2, Feb. 1986, p. 149.

8.71 Some power plants are located near rivers or oceans so that the available water can be used for cooling the condensers. As part of an environmental impact study, suppose a power company wants to estimate the difference in mean water temperature between the discharge of its plant and the offshore waters. How many sample measurements must be taken at each site in order to estimate the true difference between means to within .2°C with 95% confidence? Assume the range in readings will be about 4°C at each site and the same number of readings will be taken at each site.

8.72 A survey was conducted to compare the average salaries of data-processing managers employed in the financial sectors of two areas of the country—the South and the East Central regions. Independent random samples of 250 data-processing managers working in the South and 600 working in the East Central area were selected and asked to reveal their annual salaries. The results are summarized in the accompanying table.

SOUTH	EAST CENTRAL
$\bar{y}_1 = \$32,300$	$\bar{y}_2 = \$30,800$
$s_1 = \$4,700$	$s_2 = \$3,200$

a. Construct a 99% confidence interval for the difference in the true mean salaries of data-processing managers working in the two areas.

b. According to the interval of part **a**, in which region are data-processing managers paid the higher average salary?

8.73 A study was conducted to compare the attitudes of American and Soviet college students on nuclear war (*Gainesville Sun*, Jan. 30, 1985). One hundred Harvard University and 100 Moscow State University students took part in the survey. One question asked whether the students believe nuclear weapons are an effective deterrent against war. Sixty-three of the Harvard students responded affirmatively compared to only 3 of the Moscow State students.

a. Calculate a 95% confidence interval for the difference between the proportions of Harvard and Moscow State students who believe that nuclear weapons are an effective deterrent against war. Interpret the interval.

b. How could the width of the interval obtained in part **a** be reduced?

c. Although Harvard students were recruited randomly for the study, there is speculation that the Soviet students were selected much more carefully. How could the nonrandom Soviet sample bias the results of part **a**?

8.74 A process was designed for producing stainless steel hydraulic valves with a nickel content of 13% by weight. The manufacturer randomly samples 16 stainless steel hydraulic valves and finds their mean nickel content is 12.4% and the standard deviation is 1.23%. Find a 95% confidence interval for the variability of nickel content by weight of stainless steel hydraulic valves. What assumptions must you make in order to form the interval?

8.75 Two alloys, A and B, are used in the manufacture of steel bars. Suppose a steel producer wants to compare the two alloys on the basis of average load capacity, where the load capacity of a steel bar is defined as the maximum load (weight in tons) it can support without breaking. Steel bars containing alloy A and steel bars containing alloy B were randomly selected and tested for load capacity. The results are summarized in the accompanying table.

ALLOY A	ALLOY B
$n_1 = 11$	$n_2 = 17$
$\bar{y}_1 = 43.7$	$\bar{y}_2 = 48.5$
$s_1^2 = 24.4$	$s_2^2 = 19.9$

a. Find a 99% confidence interval for the difference between the true average load capacities for the two alloys.

b. For the interval of part **a** to be valid, what assumptions must be satisfied?

c. Interpret the interval of part **a**. Can you conclude that the average load capacities for the two alloys are different?

d. How many steel bars of each type should be sampled in order to estimate the true difference in average load capacities to within 2 tons with 99% confidence? (Assume $n_1 = n_2 = n$.)

8.76 Refer to Exercise 8.75. Suppose the steel producer also wants to compare the variabilities of the load capacities for the two alloys. Estimate the ratio of the variances of the load capacities using a 95% confidence interval.

8.77 A small commercial builder contracts skilled electricians to do all his electrical work and pays them $16 an hour. However, he is contemplating offering the electricians a set fee of $2,400 per job contracted. To determine if this plan is economically feasible, the builder needs to estimate the average length of time it takes an electrician to complete a job. A random sample of 250 electricians yields a mean contract length of 14.1 days and a standard deviation of 2.3 days per contract.

   a. Estimate the mean time required for an electrician to complete a contract. Use a 99% confidence interval.

   b. Use the result of part **a** to obtain a 99% confidence interval for the mean cost to the builder per electrician per contract. Interpret the result. (Assume the electricians work, on the average, 10 hours per day.)

8.78 A Pentagon statistician is evaluating a prototype bomber to see if it can strike on target more often than the existing bomber can. Two independent samples of size 50 each are obtained with the results shown in the table.

	PROTOTYPE BOMBER	EXISTING BOMBER
Number of bomber runs	$n_1 = 50$	$n_2 = 50$
Number of times the target was hit	42	31

   a. Estimate the difference between the proportions of runs in which the target is hit for the prototype bomber and the existing bomber, using a 90% confidence interval.

   b. Does it appear that the new prototype bomber can strike on target more often than the existing bomber can? Explain.

8.79 A recent development in do-it-yourself home improvement is a special type of foam insulation that may be installed to make a home more energy efficient. However, it has been suggested that, in some instances, the foam (which usually contains a derivative of formaldehyde) produces a vapor that may be carcinogenic. The National Cancer Institute wants to estimate the proportion of recently insulated homes in a particular state that have this type of foam insulation. In order to obtain this estimate, a random sample of 100 homes in which insulation had recently been added was taken and the type of insulation was noted. The results show that 24 of the 100 homes had foam insulation.

   a. Construct a 95% confidence interval for the true proportion of recently insulated homes in this state that have the foam insulation.

   b. Interpret the interval in terms of the problem.

   c. How would the width of the confidence interval in part **a** change if the confidence coefficient were increased from .95 to .99?

8.80 Will premium gasoline provide an increase in the mileage per gallon obtained by your automobile in comparison with the mileage for standard gasoline? Is the higher price of premium gasoline worth the increase (assuming an increase exists)? To assist in answering these questions, a government agency randomly selected 20 automobiles from its fleet and conducted a matched-pairs experiment. Each automobile was operated until it consumed 100 gallons of gasoline, but two test runs were performed. On one run regular gasoline was used; on the other, premium was used. At the conclusion of the experiment, the miles per gallon were calculated for each automobile on each of the two runs and the differences between the regular and premium mileages calculated. The mean and standard deviation of the 20 sample differences were $\bar{d} = -2.7$, $s_d = 1.3$. Estimate the difference in mean gasoline mileage between regular and premium gasolines using a 95% confidence interval. Interpret the interval estimate.

OPTIONAL SUPPLEMENTARY EXERCISES

8.81 Let $y_1, y_2, \ldots, y_n$ denote a random sample from a uniform distribution with probability density

$$f(y) = \begin{cases} 1 & \text{if } \theta \le y \le \theta + 1 \\ 0 & \text{elsewhere} \end{cases}$$

a. Show that $\bar{y}$ is a biased estimator of θ, and compute the bias.
b. Find $V(\bar{y})$.
c. What function of $\bar{y}$ is an unbiased estimator of θ?

8.82 Let $\bar{y}_1$ be the mean of a random sample of n_1 observations from a Poisson distribution with mean λ_1 and let $\bar{y}_2$ be the mean of a random sample of n_2 observations from a Poisson distribution with mean λ_2. Assume the samples are independent.
a. Show that $(\bar{y}_1 - \bar{y}_2)$ is an unbiased estimator of $(\lambda_1 - \lambda_2)$.
b. Find $V(\bar{y}_1 - \bar{y}_2)$. How could you estimate this variance?
c. Construct a large-sample $(1 - \alpha)100\%$ confidence interval for $(\lambda_1 - \lambda_2)$. [*Hint:* Consider

$$z = \frac{(\bar{y}_1 - \bar{y}_2) - (\lambda_1 - \lambda_2)}{\sqrt{\dfrac{\bar{y}_1}{n_1} + \dfrac{\bar{y}_2}{n_2}}}$$

as a pivotal statistic.]

8.83 Suppose y is a random sample of size $n = 1$ from a gamma distribution with parameters $\alpha = 1$ and arbitrary β.
a. Show that $2y/\beta$ has a gamma distribution with parameters $\alpha = 1$ and $\beta = 2$. [*Hint:* Use the distribution function approach of Section 7.2.]
b. Use the result of part a to show that $2y/\beta$ has a chi-square distribution with 2 degrees of freedom. [*Hint:* The result follows directly from Section 5.6.]
c. Derive a 95% confidence interval for β using $2y/\beta$ as a pivotal statistic.

8.84 Suppose y is a random sample of size $n = 1$ from a normal distribution with mean 0 and unknown variance σ^2.
a. Show that y^2/σ^2 has a chi-square distribution with 1 degree of freedom. [*Hint:* The result follows directly from Theorem 7.4.]
b. Derive a 95% confidence interval for σ^2 using y^2/σ^2 as a pivotal statistic.

8.85 A confidence interval for θ is said to be *unbiased* if the expected value of the interval midpoint is equal to θ.
a. Show that the small-sample confidence interval for μ,

$$\bar{y} - t_{\alpha/2}\left(\frac{s}{\sqrt{n}}\right) \le \mu \le \bar{y} + t_{\alpha/2}\left(\frac{s}{\sqrt{n}}\right)$$

is unbiased.

b. Show that the confidence interval for σ^2,

$$\frac{(n-1)s^2}{\chi^2_{\alpha/2}} \le \sigma^2 \le \frac{(n-1)s^2}{\chi^2_{(1-\alpha/2)}}$$

is biased.

8.86 Suppose y is a single observation from a normal distribution with mean μ and variance 1. Use y to find a 95% confidence interval for μ. [*Hint:* Start with the pivotal statistic $z = (y - \mu)$. Since z is a standard normal random variable,

$$P(-z_{.025} \le y - \mu \le z_{.025}) = .95$$

Follow the method of Example 8.7.]

8.87 Suppose y is a single observation from a uniform distribution defined on the interval from 0 to θ. Find a 95% confidence limit LCL for θ such that $P(\text{LCL} < \theta < \infty) = .95$. [*Hint:* Start with the pivotal statistic y/θ and show (using the method of Chapter 7) that y/θ is uniformly distributed on the interval from 0 to 1. Then observe that

$$P\left(0 < \frac{y}{\theta} < .95\right) = \int_0^{.95} (1)\, dy = .95$$

and proceed to obtain LCL.]

REFERENCES

Devore, J. L. *Probability and Statistics for Engineering and the Sciences.* Monterey, California: Brooks/Cole, 1982. Chapter 6.

Freedman, D., Pisani, R., and Purves, R. *Statistics.* New York: W. W. Norton and Co., 1978.

Hoel, P. G. *Introduction to Mathematical Statistics*, 4th ed. New York: Wiley, 1971.

Hogg, R. V. and Craig, A. T. *Introduction to Mathematical Statistics*, 4th ed. New York: Macmillan, 1978.

McClave, J. T. and Dietrich, F. H. II. *Statistics*, 3rd ed. San Francisco: Dellen, 1985.

Mendenhall, W. *Introduction to Probability and Statistics*, 6th ed. Boston: Duxbury, 1983.

Mendenhall, W., Scheaffer, R. L., and Wackerly, D. D. *Mathematical Statistics with Applications*, 2nd ed. Boston: Duxbury, 1981.

Mood, A. M., Graybill, F. A., and Boes, D. *Introduction to the Theory of Statistics*, 3rd ed. New York: McGraw-Hill, 1974.

Mosteller, F. and Tukey, J. W. *Data Analysis and Regression.* Reading, Massachusetts: Addison-Wesley, 1977. Chapters 8 and 10.

Snedecor, G. W. and Cochran, W. G. *Statistical Methods*, 7th ed. Ames, Iowa: Iowa State University Press, 1980.

Tukey, J. W. "Bias and Confidence in Not-Quite Large Samples." *Annals of Mathematical Statistics*, Vol. 29, 1958, p. 614.

C H A P T E R 9

OBJECTIVE

To introduce the basic concepts of a statistical test of a hypothesis; to present statistical tests for several common population parameters and to illustrate their use in practical sampling situations.

CONTENTS

TESTS OF HYPOTHESES

STATISTICAL TESTS OF HYPOTHESES

As stated in Chapter 8, there are two ways to make inferences about population parameters. We can estimate their values (the subject of Chapter 8) or we can make decisions about them. Making decisions about population parameters—**testing hypotheses** about their values—is the topic of this chapter.

The need to decide whether a particular claim or statement is correct is often the motivation for a statistical test of a hypothesis. For example, suppose an investigator for the Environmental Protection Agency (EPA) wants to test a chemical company's claim that the mean level μ of a certain type of pollutant released into the atmosphere is less than 3 parts per million. If 3 parts per million is the upper limit allowed by the EPA, the investigator would want to use sample data (daily pollution measurements) to decide whether the company is violating the law, i.e., to decide whether $\mu > 3$. Or, suppose that a manufacturer purchases terminal fuses in lots of 10,000 and that the supplier of the fuses guarantees that no more than 1% of the fuses in any given lot are defective. Since the manufacturer cannot test each of the 10,000 fuses in a lot, he must decide whether to accept or reject a lot based on an examination of a sample of fuses selected from the lot. If the number y of defective fuses in a sample of, say $n = 100$, is large, he will reject the lot and send it back to the supplier. Thus, he wants to decide whether the proportion p of defectives in the lot exceeds .01, based on information contained in a sample.

We now return to the EPA example to introduce the concepts involved in a test of a hypothesis. We will use a method analogous to proof by contradiction. The theory that we want to support, called the **alternative** (or **research**) **hypothesis**, is that $\mu > 3$, where μ is the true mean level of pollution, in parts per million. The alternative hypothesis is denoted by the symbol H_a. The theory contradictory to the alternative hypothesis, that μ is at most equal to 3, say $\mu = 3$, is called the **null hypothesis** and is denoted by the symbol H_0. Thus, we hope to show support for the alternative hypothesis, $H_a: \mu > 3$, by obtaining sample evidence indicating that the null hypothesis, $H_0: \mu = 3$, is false.

The decision to reject or not to reject the null hypothesis is based on a statistic, called a **test statistic**, computed from sample data. For example, suppose we plan to base our decision on a sample of $n = 30$ daily pollution readings. If the sample mean $\bar{y}$ of the 30 pollution measurements is much larger than 3, we would tend to reject the null hypothesis and conclude that $\mu > 3$. However, if $\bar{y}$ is smaller than 3, say $\bar{y} = 2.8$ parts per million, we would not have sufficient evidence to refute the null hypothesis. Thus, the sample mean $\bar{y}$ serves as a test statistic. The values that $\bar{y}$ can assume will be divided into two sets. Those larger than some specified value, say $\bar{y} \geq 3.1$, will imply rejection of the null hypothesis and acceptance of the alternative hypothesis. This set of values of the test statistic is known as the **rejection region** for the test. A test of the null hypothesis, $H_0: \mu = 3$, against the alternative hypothesis, $H_a: \mu > 3$, employing the sample mean $\bar{y}$ as a test statistic and $\bar{y} \geq 3.1$ as a rejection region, represents one particular test that possesses specific properties. If we change the rejection region to $\bar{y} \geq 3.2$, we obtain a different test with different properties.

The preceding discussion indicates that a statistical test consists of the four elements summarized in the box.

ELEMENTS OF A STATISTICAL TEST

1. **Null hypothesis**, H_0, about one or more population parameters
2. **Alternative hypothesis**, H_a, that we will accept if we decide to reject the null hypothesis
3. **Test statistic**, computed from sample data
4. **Rejection region**, indicating the values of the test statistic that will imply rejection of the null hypothesis

---

In Section 9.2 we will show how to evaluate the reliability of a statistical test, how to compare one test with another, and how to evaluate the reliability of a particular test decision. We will apply the results to several practical examples.

| | | | | | | | | | | | |

S E C T I O N 9.2

EVALUATING THE PROPERTIES OF A STATISTICAL TEST

TABLE 9.1
Conclusions and Consequences for Testing a Hypothesis

Since a statistical test can result in one of only two outcomes—rejecting or not rejecting the null hypothesis—the test conclusion is subject to only two types of error, as illustrated in Table 9.1.

		TRUE STATE OF NATURE	
		H_0 true (H_a false)	H_0 false (H_a true)
DECISION	Reject H_0	Type I error	Correct decision
	Do not reject H_0	Correct decision	Type II error

You might reject the null hypothesis if, in fact, it is true. Or you might decide not to reject the null hypothesis if, in fact, it is false. The probabilities of making these two types of errors measure the risks of making incorrect decisions when we perform a test of hypothesis and, consequently, provide measures of the goodness of this inferential decision-making procedure.

DEFINITION 9.1

Rejecting the null hypothesis if it is true is a **Type I error**. The probability of making a Type I error is denoted by the symbol α.

DEFINITION 9.2

Failing to reject the null hypothesis if it is false is a **Type II error**. The probability of making a Type II error is denoted by the symbol β.

EXAMPLE 9.1

A manufacturer of minicomputers believes that it can sell a particular software package to more than 20% of the buyers of its computers. Ten prospective purchasers of the computer were randomly selected and questioned about their interest in the software package. Of these, four indicated that they planned to buy the package. Does this sample provide sufficient evidence to indicate that more than 20% of the computer purchasers will buy the software package?

SOLUTION

Let p be the true proportion of all prospective computer buyers who will purchase the software package. Since we want to show that $p > .2$, we choose H_a: $p > .2$ for the alternative hypothesis and H_0: $p = .2$ for the null hypothesis. We will use the binomial random variable y, the number of prospective purchasers in the sample who plan to buy the software, as the test statistic and will reject H_0: $p = .2$ if y is large. A graph of $p(y)$ for $n = 10$ and $p = .2$ is shown in Figure 9.1.

FIGURE 9.1

Graph of $p(y)$ for $n = 10$ and $p = .2$, i.e., if the Null Hypothesis Is True

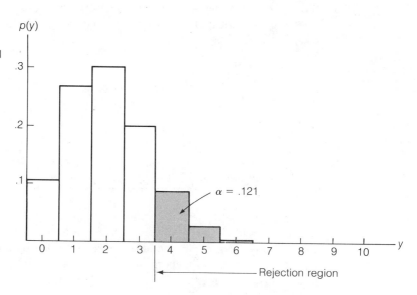

Large values of y will support the alternative hypothesis, H_a: $p > .2$, but what values of y should we include in the rejection region? Suppose that we select values of $y \geq 4$ as the rejection region. Then the elements of the test are:

H_0: $p = .2$

H_a: $p > .2$

Test statistic: y

Rejection region: $y \geq 4$

To conduct the test, we note that the observed value of y, $y = 4$, falls in the rejection region. Thus, for this test procedure, we reject the null hypothesis, H_0: $p = .2$, and conclude that the manufacturer is correct, i.e., $p > .2$. ∎

EXAMPLE 9.2

What is the probability that the statistical test procedure of Example 9.1 would lead us to an incorrect decision if, in fact, the null hypothesis is true?

SOLUTION

We will calculate the probability α that the test procedure would lead us to make a Type I error, i.e., to reject H_0 if, in fact, H_0 is true. This is the probability that y falls in the rejection region if in fact $p = .2$:

$$\alpha = P(y \geq 4 \text{ if in fact } p = .2) = 1 - \sum_{y=0}^{3} p(y)$$

The partial sum $\sum_{y=0}^{3} p(y)$ for a binomial random variable with $n = 10$ and $p = .2$ is given in Table 2 of Appendix II as .879. Therefore,

$$\alpha = 1 - \sum_{y=0}^{3} p(y) = 1 - .879 = .121$$

The probability that the test procedure would lead us to conclude that $p > .2$, if in fact it is not, is .121. This probability corresponds to the area of the shaded region in Figure 9.1. ∎

In Example 9.1, we computed the probability α of committing a Type I error. The probability β of making a Type II error, i.e., failing to detect a value of p greater than .2, depends on the value of p. For example, if $p = .20001$, it will be very difficult to detect this small deviation from the null hypothesized value of $p = .2$. In contrast, if $p = 1.0$, then *every* prospective purchaser of the minicomputer will want to buy the software package, and in such a case it will be very evident from the sample information that $p > .2$. We will illustrate the procedure for calculating β in Example 9.3.

EXAMPLE 9.3

Refer to Example 9.2 and suppose that p is actually equal to .60. What is the probability β that the test procedure will fail to reject H_0: $p = .2$ if, in fact, $p = .6$?

SOLUTION

The binomial probability distribution $p(y)$ for $n = 10$ and $p = .6$ is shown in Figure 9.2 (page 360). The probability that we will fail to reject H_0 is equal to the probability that $y = 0, 1, 2,$ or 3, i.e., the probability that y does not fall in the rejection region. This probability, β, corresponds to the shaded area under the probability histogram in the figure. Therefore,

$$\beta = P(y \leq 3 \text{ if in fact } p = .6) = \sum_{y=0}^{3} p(y) \quad \text{for } n = 10 \text{ and } p = .6$$

FIGURE 9.2
Graph of $p(y)$ for $n = 10$ and $p = .6$, i.e., if the Alternative Hypothesis Is True

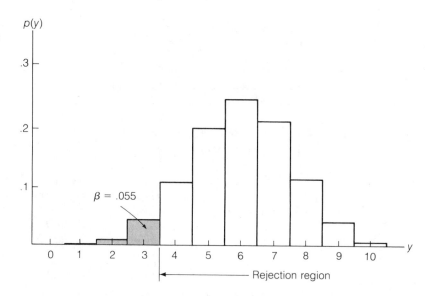

This partial sum, given in Table 2 of Appendix II for a binomial random variable with $n = 10$ and $p = .6$, is .055. Therefore, the probability that we will fail to reject H_0: $p = .2$ if p is as large as .6 is $\beta = .055$. ∎

Another important property of a statistical test is its ability to detect departures from the null hypothesis when they exist. This is measured by the probability of rejecting H_0 when, in fact, H_0 is false. Note that this probability is simply $(1 - \beta)$:

$$P(\text{Reject } H_0 \text{ when } H_0 \text{ is false}) = 1 - P(\text{Accept } H_0 \text{ when } H_0 \text{ is false})$$
$$= 1 - P(\text{Type II error})$$
$$= 1 - \beta$$

The probability $(1 - \beta)$ is called the **power of the test**. The higher the power, the greater the probability of detecting departures from H_0 when they exist.

DEFINITION 9.3

The **power** of a statistical test, $(1 - \beta)$, is the probability of rejecting the null hypothesis H_0 when, in fact, H_0 is false.

EXAMPLE 9.4

Refer to the test of hypothesis in Example 9.1. Find the power of the test if in fact $p = .3$.

SOLUTION

From Definition 9.3, the power of the test is the probability $(1 - \beta)$. The probability of making a Type II error, i.e., failing to reject H_0: $p = .2$, if in fact $p = .3$, will be larger than the value of β calculated in Example 9.3 because $p = .3$ is much closer to the hypothesized value of $p = .2$. Thus,

$$\beta = P(y \le 3 \text{ if in fact } p = .3) = \sum_{y=0}^{3} p(y) \quad \text{for } n = 10 \text{ and } p = .3$$

The value of this partial sum, given in Table 2 of Appendix II for a binomial random variable with $n = 10$ and $p = .3$, is .650. Therefore, the probability that we will fail to reject H_0: $p = .2$ if in fact $p = .3$ is $\beta = .650$ and the power of the test is $(1 - \beta) = (1 - .650) = .350$. You can see that the closer the actual value of p is to the hypothesized null value, the more unlikely it is that we will reject H_0: $p = .2$. ∎

The preceding examples indicate how we can calculate α and β for a simple statistical test and thereby measure the risks of making Type I and Type II errors. These probabilities describe the properties of this inferential decision-making procedure and enable us to compare one test with another. For two tests, each with a rejection region selected so that α is equal to some specified value, say .10, we would select the test that, for a specified alternative, has the smaller risk of making a Type II error, i.e., one that has the smaller value of β. This is equivalent to choosing the test with the higher power.

We will present a number of statistical tests in the following sections. In each case, the probability α of making a Type I error is known, i.e., α is selected by the experimenter and the rejection region is determined accordingly. In contrast, the value of β for a specific alternative is often difficult to calculate. This explains why we attempt to show that H_a is true by showing that the data do not support H_0. We hope that the sample evidence will support the alternative (or research) hypothesis. If it does, we will be concerned only about making a Type I error, i.e., rejecting H_0 if it is true. The probability α of committing such an error will be known.

EXERCISES 9.1–9.6

9.1 Define α and β for a statistical test of hypothesis.

9.2 Explain why each of the following statements is incorrect:
a. The probability that the null hypothesis is correct is equal to α.
b. If the null hypothesis is rejected, then the test proves that the alternative hypothesis is correct.
c. In all statistical tests of hypothesis, $\alpha + \beta = 1$.

9.3 Pascal is a high-level programming language used frequently in minicomputers and microprocessors. An experiment was conducted to investigate the proportion of Pascal variables that are *array* variables (in contrast to *scalar* variables, which are less efficient in terms of execution time). Twenty variables are randomly selected from a set of Pascal programs and y, the number of array variables, is recorded. Suppose we want to test the hypothesis that Pascal is a more efficient language than Algol, in which 20% of the variables are array variables. That is, we will test H_0: $p = .20$ against H_a: $p > .20$, where p is the probability of observing an array variable on each trial. (Assume that the 20 trials are independent.)
a. Find α for the rejection region $y \ge 8$.
b. Find α for the rejection region $y \ge 5$.

c. Find β for the rejection region $y \geq 8$ if $p = .5$. [*Note:*   Past experience has shown that approximately half the variables in most Pascal programs are array variables.]

d. Find β for the rejection region $y \geq 5$ if $p = .5$.

e. Which of the rejection regions, $y \geq 8$ or $y \geq 5$, is more desirable if you want to minimize the probability of a Type I error? Type II error?

9.4 Refer to Exercise 9.3.

a. Find the rejection region of the form $y \geq a$ so that α is approximately equal to .01.

b. For the rejection region determined in part **a**, find the power of the test, if in fact $p = .4$.

c. For the rejection region determined in part **a**, find the power of the test, if in fact $p = .7$.

9.5 A manufacturer of power meters, which are used to regulate energy thresholds of a data-communications system, claims that when its production process is operating correctly, only 10% of the power meters will be defective. A vendor has just received a shipment of 25 power meters from the manufacturer. Suppose the vendor wants to test H_0: $p = .10$ against H_a: $p > .10$, where p is the true proportion of power meters that are defective. Use $y \geq 6$ as the rejection region.

a. Determine the value of α for this test procedure.

b. Find β if in fact $p = .2$. What is the power of the test for this value of p?

c. Find β if in fact $p = .4$. What is the power of the test for this value of p?

OPTIONAL EXERCISE

9.6 Show that for a fixed sample size n, α increases as β decreases, and vice versa.

SECTION 9.3

FINDING STATISTICAL TESTS: AN EXAMPLE OF A LARGE-SAMPLE TEST

In order to find a statistical test about one or more population parameters, we must (1) find a suitable test statistic and (2) specify a rejection region. One method for finding a reasonable test statistic for testing a hypothesis was proposed by R. A. Fisher. For example, suppose we want to test a hypothesis about the sole parameter θ of a probability function $p(y)$ or density function $f(y)$, and let L represent the likelihood of the sample. Then to test the null hypothesis, H_0: $\theta = \theta_0$, Fisher's **likelihood ratio test statistic** is

$$\lambda = \frac{\text{Likelihood assuming } \theta = \theta_0}{\text{Likelihood assuming } \theta = \hat{\theta}} = \frac{L(\theta_0)}{L(\hat{\theta})}$$

where $\hat{\theta}$ is the maximum likelihood estimator of θ. Fisher reasoned that if θ differs from θ_0, then the value of the likelihood L when $\theta = \hat{\theta}$ will be larger than when $\theta = \theta_0$. Thus, the rejection region for the test contains values of λ that are small—say, smaller than some value λ_R.

If you are interested in learning more about Fisher's likelihood ratio test, consult the references at the end of this chapter. Fortunately, most of the statistics that we would choose intuitively for test statistics are functions of the corresponding likelihood ratio statistic λ. These are the pivotal statistics used to construct confidence intervals in Chapter 8.

Recall that most of the pivotal statistics in Chapter 8 have approximately normal sampling distributions for large samples. This fact allows us to easily derive a large-sample statistical test of hypothesis. To illustrate, suppose that we want to test a hypothesis, H_0: $\theta = \theta_0$, about a parameter θ and that the estimator $\hat{\theta}$ possesses a normal sampling distribution with mean θ and standard deviation $\sigma_{\hat{\theta}}$. We will further assume that $\sigma_{\hat{\theta}}$ is known or that we can obtain a good approximation for it when the sample size(s) is (are) large. It can be shown (proof omitted) that the likelihood ratio test statistic λ reduces to the standard normal variable z:

$$z = \frac{\hat{\theta} - \theta_0}{\sigma_{\hat{\theta}}}$$

The location of the rejection region for this test can be deduced by examining the formula for the test statistic z. The farther $\hat{\theta}$ departs from θ_0, i.e., the larger the absolute value of the deviation $|\hat{\theta} - \theta_0|$, the greater will be the weight of evidence to indicate that θ is not equal to θ_0. If we want to detect values of θ larger than θ_0, i.e., H_a: $\theta > \theta_0$, we locate the rejection region in the upper tail of the sampling distribution of the standard normal z test statistic (see Figure 9.3a, page 364). If we want to detect only values of θ less than θ_0, i.e., H_a: $\theta < \theta_0$, we locate the rejection region in the lower tail of the z distribution (see Figure 9.3b). These two tests are called **one-tailed statistical tests** because the entire rejection region is located in only one tail of the z distribution. However, if we want to detect *either* a value of θ larger than θ_0 or a value smaller than θ_0, i.e., H_a: $\theta \neq \theta_0$, we locate the rejection region in both the upper and the lower tails of the z distribution (see Figure 9.3c). This is called a **two-tailed statistical test**.

The large-sample statistical test that we have described is summarized in the box. Many of the population parameters and test statistics discussed in Sections 9.5–9.11 satisfy the assumptions of this test. We will illustrate the use of the test with a practical example on the population mean μ.

A LARGE-SAMPLE TEST BASED ON THE STANDARD NORMAL z TEST STATISTIC

ONE-TAILED TEST	TWO-TAILED TEST
H_0: $\theta = \theta_0$	H_0: $\theta = \theta_0$
H_a: $\theta > \theta_0$	H_a: $\theta \neq \theta_0$
(or H_a: $\theta < \theta_0$)	
Test statistic: $z = \dfrac{\hat{\theta} - \theta_0}{\sigma_{\hat{\theta}}}$	Test statistic: $z = \dfrac{\hat{\theta} - \theta_0}{\sigma_{\hat{\theta}}}$
Rejection region: $z > z_\alpha$	Rejection region: $z < -z_{\alpha/2}$
(or $z < -z_\alpha$)	or $z > z_{\alpha/2}$
where $P(z > z_\alpha) = \alpha$	where $P(z > z_{\alpha/2}) = \alpha/2$

FIGURE 9.3 Rejection Regions for One- and Two-Tailed Tests

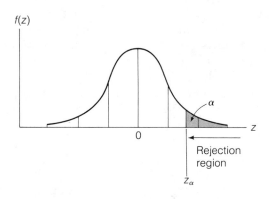

a. One-tailed test;
H_a: $\theta > \theta_0$

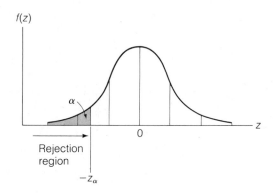

b. One-tailed test;
H_a: $\theta < \theta_0$

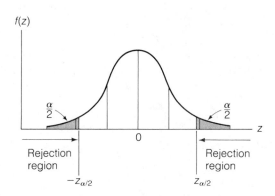

c. Two-tailed test;
H_a: $\theta \neq \theta_0$

EXAMPLE 9.5 The Department of Highway Improvements, responsible for repairing a 25-mile stretch of interstate highway, wants to design a surface that will be structurally efficient. One important consideration is the volume of heavy freight traffic on the interstate. State weigh stations report that the average number of heavy-duty trailers traveling on a 25-mile segment of the interstate is 72 per hour. However, the section of highway to be repaired is located in an urban area and the department engineers believe that the volume of heavy freight traffic for this particular sector is greater than the average reported for the entire interstate. In order to validate this theory, the department monitors the highway for 50 1-hour periods randomly selected throughout the month. Suppose the sample mean and standard deviation of the heavy freight traffic for the 50 sampled hours are

$$\bar{y} = 74.1 \qquad s = 13.3$$

Do the data support the department's theory? Use $\alpha = .10$.

SOLUTION

For this example, the parameter of interest is μ, the average number of heavy-duty trailers traveling on the 25-mile stretch of interstate highway. Recall that the sample mean $\bar{y}$ is used to estimate μ, and that for large n, $\bar{y}$ has an approximately normal sampling distribution. Thus, we can apply the large-sample test outlined in the box.

The elements of the test are

H_0: $\mu = 72$

H_a: $\mu > 72$

Test statistic: $z = \dfrac{\bar{y} - 72}{\sigma_{\bar{y}}} = \dfrac{\bar{y} - 72}{\sigma/\sqrt{n}} \approx \dfrac{\bar{y} - 72}{s/\sqrt{n}}$

Rejection region: $z > 1.28$

We now substitute the sample statistics into the test statistic to obtain

$$z \approx \frac{74.1 - 72}{13.3/\sqrt{50}} = 1.12$$

Thus, although the average number of heavy freight trucks per hour in the sample exceeds the state's average by more than 2, the z value of 1.12 does not fall in the rejection region (see Figure 9.4). Therefore, this sample does not provide sufficient evidence at $\alpha = .10$ to support the Department of Highway Improvements' theory. ∎

What is the risk of making an incorrect decision in Example 9.5? If we reject the null hypothesis then we know that the probability of making a Type I error (rejecting H_0 if it is true) is $\alpha = .10$. However, we failed to reject the null

FIGURE 9.4
Location of the Test
Statistic for Example 9.5

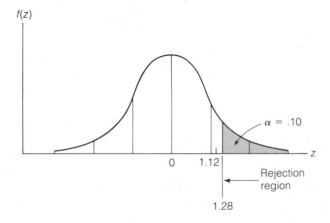

hypotheses in Example 9.5 and, consequently, we must be concerned about the possibility of making a Type II error (failing to reject H_0 if in fact it is false). We will evaluate the risk of making a Type II error in Example 9.6.

EXAMPLE 9.6

Refer to Example 9.5. If the mean number μ of heavy freight trucks traveling a particular 25-mile stretch of interstate highway is in fact 78 per hour, what is the probability that the test procedure of Example 9.5 would fail to detect it? That is, what is the probability β that we would fail to reject H_0: $\mu = 72$ if μ is actually equal to 78?

SOLUTION

In order to calculate β for the large-sample z test, we need to specify the rejection region in terms of the point estimator $\hat{\theta}$, where, for this example, $\hat{\theta} = \bar{y}$. From Figure 9.4, you can see that the rejection region consists of values of $z \geq 1.28$. To determine the value of $\bar{y}$ corresponding to $z = 1.28$, we substitute into the equation

$$z = \frac{\bar{y} - \mu_0}{\sigma/\sqrt{n}} \approx \frac{\bar{y} - \mu_0}{s/\sqrt{n}} \quad \text{or} \quad 1.28 = \frac{\bar{y} - 72}{13.3/\sqrt{50}}$$

Solving for $\bar{y}$, we obtain $\bar{y} = 74.41$. Therefore, the rejection region for the test is $z \geq 1.28$ or, equivalently, $\bar{y} \geq 74.41$.

FIGURE 9.5
The Probability β of Making a Type II Error if $\mu = 78$ in Example 9.6

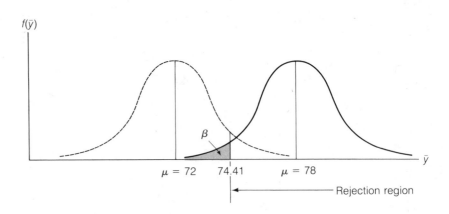

The dotted curve in Figure 9.5 is the sampling distribution for $\bar{y}$ if H_0: $\mu = 72$ is true. This curve was used to locate the rejection region for $\bar{y}$ (and, equivalently, z), i.e., values of $\bar{y}$ contradictory to H_0: $\mu = 72$. The solid curve is the sampling distribution for $\bar{y}$ if $\mu = 78$. Since we want to find β if H_0 is in fact false and $\mu = 78$, we want to find the probability that $\bar{y}$ does not fall in the rejection region if $\mu = 78$. This probability corresponds to the shaded area under the solid curve for values of $\bar{y} < 74.41$. To find this area under the normal curve, we need to find the area A corresponding to

$$z = \frac{\bar{y} - 78}{\sigma/\sqrt{n}} \approx \frac{74.41 - 78}{13.3/\sqrt{50}} = -1.91$$

The value of A, given in Table 4 of Appendix II, is .4719. Then from Figure 9.5, it can be seen that

$$\beta = .5 - A = .5 - .4719 = .0281$$

Therefore, the probability of failing to reject H_0: $\mu = 72$ if μ is, in fact, as large as $\mu = 78$, is only .0281. ■

Example 9.6 illustrates that it is not too difficult to calculate β for various alternatives for the large-sample z test. However, it may be extremely difficult to calculate β for other tests. Although sophisticated techniques are available for evaluating the risk of making a Type II error when the exact value of β is unavailable or is difficult to calculate, they are beyond the scope of this text. Consult the references at the end of this chapter if you are interested in learning about these methods.

EXERCISES 9.7–9.9

OPTIONAL EXERCISES

9.7 Suppose $y_1, y_2, \ldots, y_n$ is a random sample from a normal distribution with unknown mean μ and variance $\sigma^2 = 1$, i.e.,

$$f(y) = \frac{1}{\sqrt{2\pi}} e^{-(y-\mu)^2/2}$$

Show that the likelihood L of the sample is

$$L(\mu) = \left(\frac{1}{\sqrt{2\pi}}\right)^n e^{-\sum_{i=1}^n (y_i - \mu)^2/2}$$

9.8 Refer to Optional Exercise 9.7. Suppose we want to test H_0: $\mu = 0$ against the alternative H_a: $\mu > 0$. Since the estimator of μ is $\hat{\mu} = \bar{y}$, the likelihood ratio test statistic is

$$\lambda = \frac{L(\mu_0)}{L(\hat{\mu})} = \frac{L(0)}{L(\bar{y})}$$

Show that

$$\lambda = e^{-n(\bar{y})^2/2}$$

[*Hint:* Use the fact that $\sum_{i=1}^n (y_i - \bar{y})^2 = \sum_{i=1}^n y_i^2 - n\bar{y}^2$.]

9.9 Refer to Optional Exericises 9.7 and 9.8. Show that the rejection region $\lambda \leq \lambda_\alpha$ is equivalent to the rejection region $\bar{y} \geq \bar{y}_\alpha$, where $P(\lambda \leq \lambda_\alpha) = \alpha$ and $P(\bar{y} \geq \bar{y}_\alpha) = \alpha$. [*Hint:* Use the fact that $e^{-a^2} \to 0$ as $a \to \infty$.]

SECTION 9.4

CHOOSING THE NULL AND ALTERNATIVE HYPOTHESES

Now that you have conducted a large-sample statistical test of hypothesis and have seen how to calculate the value of β, the probability of failing to reject H_0: $\theta = \theta_0$ if θ is in fact equal to some alternative value, $\theta = \theta_a$, the logic for choosing the null and alternative hypotheses may make more sense to you. The theory that we want to support (or detect if true) is chosen as the alternative hypothesis because, if the data support H_a (i.e., if we reject H_0), we immediately know the value of α, the probability of incorrectly rejecting H_0 if it is true. In contrast, if we choose the null hypothesis as the theory that we want to support, and if the data support this theory, i.e., the test leads to nonrejection of H_0, then we would have to investigate the values of β for some specific alternatives. Clearly, we want to avoid this tedious and sometimes extremely difficult task. This explains why we always construct a test so that the theory that we want to support (or detect) is H_a.

Another issue that arises in a practical situation is whether to conduct a one- or a two-tailed test. The decision depends on what you want to detect. For example, suppose you operate a chemical plant that produces a variable amount y of product per day and that if μ, the mean value of y, is less than 100 tons per day, you will eventually be bankrupt. If μ exceeds 100 tons per day, you are financially safe. In order to determine whether your process is leading to financial disaster, you will want to detect whether $\mu < 100$ tons, and you will conduct a one-tailed test of H_0: $\mu = 100$ versus H_a: $\mu < 100$. If you were to conduct a two-tailed test for this situation, you would reduce your chance of detecting values of μ less than 100 tons, i.e., you would increase the values of β for alternative values of $\mu < 100$ tons.

As a different example, suppose you have designed a new drug so that its mean potency is some specific level, say 10%. As the mean potency tends to exceed 10%, you lose money. If it is less than 10% by some specified amount, the drug becomes ineffective as a pharmaceutical (and you lose money). To conduct a test of the mean potency μ for this situation, you would want to detect values of μ either larger than or smaller than $\mu = 10$. Consequently, you would select H_a: $\mu \neq 10$ and conduct a two-tailed statistical test. These examples demonstrate that a statistical test is an attempt to detect departures from H_0; the key to the test is to define the specific alternatives that you want to detect.

In Sections 9.5–9.11 we will present applications of the hypothesis-testing logic developed in this chapter. The cases to be considered are those for which we developed estimation procedures in Chapter 8. Since the theory and reasoning involved are based on the developments of Chapter 8 and Sections 9.1–9.4, we will present only a summary of the hypothesis-testing procedure for one-tailed and two-tailed tests in each situation.

HYPOTHESIS TESTS ABOUT A POPULATION MEAN

In Example 9.5 we developed a large-sample test for a population mean based on the standard normal z statistic. The elements of this test are summarized in the box.

LARGE-SAMPLE ($n \geq 30$) TEST OF HYPOTHESIS ABOUT A POPULATION MEAN μ

ONE-TAILED TEST	TWO-TAILED TEST

ONE-TAILED TEST

H_0: $\mu = \mu_0$

H_a: $\mu > \mu_0$
  (or H_a: $\mu < \mu_0$)

Test statistic:

$$z = \frac{\bar{y} - \mu_0}{\sigma_{\bar{y}}} \approx \frac{\bar{y} - \mu_0}{s/\sqrt{n}}$$

Rejection region:

$$z > z_\alpha \quad (\text{or } z < -z_\alpha)$$

TWO-TAILED TEST

H_0: $\mu = \mu_0$

H_a: $\mu \neq \mu_0$

Test statistic:

$$z = \frac{\bar{y} - \mu_0}{\sigma_{\bar{y}}} \approx \frac{\bar{y} - \mu_0}{s/\sqrt{n}}$$

Rejection region:

$$z < -z_{\alpha/2} \text{ or } z > z_{\alpha/2}$$

where z_α is the z value such that $P(z > z_\alpha) = \alpha$; and $z_{\alpha/2}$ is the z value such that $P(z > z_{\alpha/2}) = \alpha/2$. [*Note:* μ_0 is our symbol for the particular numerical value specified for μ in the null hypothesis.]

Assumptions: None (since the central limit theorem guarantees that $\bar{y}$ is approximately normal regardless of the distribution of the sampled population)

EXAMPLE 9.7

Humerus bones from the same species of animal tend to have approximately the same length-to-width ratios. When fossils of humerus bones are discovered, archeological engineers can often determine the species of animal by examining the length-to-width ratios of the bones. It is known that species A has a mean ratio of 8.5. Suppose 41 fossils of humerus bones were unearthed at an archeological site in East Africa, which species A is believed to have inhabited. (Assume that the unearthed bones are all from the same unknown species.) The length-to-width ratios of the bones were measured and are summarized as follows:

$$\bar{y} = 9.25 \qquad s = 1.16$$

We wish to test the hypothesis that μ, the population mean ratio of all bones of this particular species, is equal to 8.5 against the alternative that it is different from 8.5, i.e., we wish to test whether the unearthed bones are from species A. Suppose we also want a very small chance of rejecting H_0, if in fact μ is equal to 8.5. That is, it is important that we avoid making a Type I error. The hypothesis-testing procedure that we have developed gives us the advantage of being able to choose any significance level that we desire. Since the significance level, α, is also the probability of a Type I error, we will choose α to be very small. In general, researchers who consider a Type I error to have very serious practical consequences should perform the test at a very low α value—say, $\alpha = .01$. Other researchers may be willing to tolerate an α value as high as .10 if a Type I error is not deemed a serious error to make in practice.

Test whether μ, the population mean ratio, is different from 8.5, using a significance level of $\alpha = .01$.

SOLUTION

We formulate the following hypotheses:

$H_0: \quad \mu = 8.5$

$H_a: \quad \mu \neq 8.5$

The sample size exceeds 30; thus, we may proceed with the large-sample test about μ.

At significance level $\alpha = .01$, we will reject the null hypothesis for this two-tailed test if

$$z < -z_{\alpha/2} = -z_{.005} \quad \text{or if} \quad z > z_{\alpha/2} = z_{.005}$$

i.e., if $z < -2.58$ or if $z > 2.58$. This rejection region is shown in Figure 9.6.

FIGURE 9.6

Rejection Region for Example 9.7

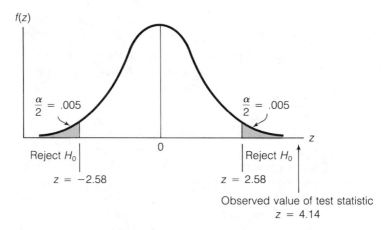

The value of the test statistic is computed as follows:

$$z \approx \frac{\bar{y} - \mu_0}{s/\sqrt{n}} = \frac{9.25 - 8.5}{1.16/\sqrt{41}} = 4.14$$

Since this value lies within the rejection region (see Figure 9.6), we reject H_0 and conclude that the mean length-to-width ratio of all humerus bones of this particular species is significantly different from 8.5. If the null hypothesis is in fact true (i.e., if $\mu = 8.5$), then the probability that we have incorrectly rejected it is equal to $\alpha = .01$. ∎

The *practical* implications of the result obtained in Example 9.7 remain to be studied further. Perhaps the animal discovered at the archeological site is of some species other than A. Alternatively, the unearthed humerus bones may have larger than normal length-to-width ratios due to unusual feeding habits of species A.

It is not always the case that a statistically significant result implies a practically significant result. The researcher must retain his or her objectivity and judge the practical significance using, among other criteria, his or her knowledge of the subject matter and the phenomenon under investigation.

A small-sample statistical test for making inferences about a population mean is (like its associated confidence interval of Section 8.5) based on the assumption that the sample data are independent observations on a normally distributed random variable. The test statistic is the pivotal t statistic given in Section 8.5. The elements of the statistical test are listed in the accompanying box. As we suggested in Chapter 8, the small-sample test will possess the properties specified in the box even if the sampled population is moderately nonnormal.

SMALL-SAMPLE ($n < 30$) TEST OF HYPOTHESIS ABOUT A POPULATION MEAN μ

ONE-TAILED TEST	TWO-TAILED TEST

ONE-TAILED TEST

H_0: $\mu = \mu_0$

H_a: $\mu > \mu_0$
  (or H_a: $\mu < \mu_0$)

Test statistic:

$$t = \frac{\bar{y} - \mu_0}{s/\sqrt{n}}$$

Rejection region:

  $t > t_\alpha$ (or $t < -t_\alpha$)

TWO-TAILED TEST

H_0: $\mu = \mu_0$

H_a: $\mu \neq \mu_0$

Test statistic:

$$t = \frac{\bar{y} - \mu_0}{s/\sqrt{n}}$$

Rejection region:

  $t < t_{\alpha/2}$ or $t > t_{\alpha/2}$

where the distribution of t is based on $(n - 1)$ degrees of freedom; t_α is the t value such that $P(t > t_\alpha) = \alpha$; and $t_{\alpha/2}$ is the t value such that $P(t > t_{\alpha/2}) = \alpha/2$.

Assumption: The relative frequency distribution of the population from which the sample was selected is approximately normal.

EXAMPLE 9.8

A major car manufacturer wants to test a new engine to determine whether it meets new air pollution standards. The mean emission, μ, of all engines of this type must be less than 20 parts per million of carbon. Ten engines are manufactured for testing purposes, and the mean and standard deviation of the emissions for this sample of engines are determined to be

$$\bar{y} = 17.1 \text{ parts per million} \qquad s = 3.0 \text{ parts per million}$$

Do the data supply sufficient evidence to allow the manufacturer to conclude that this type of engine meets the pollution standard? Assume that the manufacturer is willing to risk a Type I error with probability $\alpha = .01$.

SOLUTION

The manufacturer wants to establish the research hypothesis that the mean emission level, μ, for all engines of this type is less than 20 parts per million. The elements of this small-sample one-tailed test are

H_0: $\mu = 20$

H_a: $\mu < 20$

Test statistic: $t = \dfrac{\bar{y} - 20}{s/\sqrt{n}}$

Assumption: The relative frequency distribution of the population of emission levels for all engines of this type is approximately normal.

Rejection region: For $\alpha = .01$ and df $= (n - 1) = 9$, reject H_0 if
$$t < -t_{.01} = -2.821 \quad \text{(see Figure 9.7)}$$

FIGURE 9.7
Rejection Region for
Example 9.8

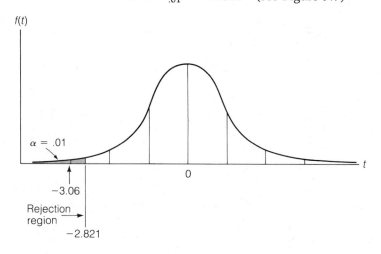

We now calculate the test statistic:

$$t = \frac{\bar{y} - 20}{s/\sqrt{n}} = \frac{17.1 - 20}{3.0/\sqrt{10}} = -3.06$$

Since the calculated t falls in the rejection region, the manufacturer concludes that $\mu < 20$ parts per million and the new engine type meets the pollution standard. Are you satisfied with the reliability associated with this inference? The probability is only $\alpha = .01$ that the test would support the research hypothesis if in fact it were false.   ∎

EXERCISES 9.10–9.17

9.10 A machine is set to produce bolts with a mean length of 1 inch. Bolts that are too long or too short do not meet the customer's specifications and must be rejected. To avoid producing too many rejects, the bolts produced by the machine are sampled from time to time and tested as a check to determine whether the machine is still operating properly, i.e., producing bolts with a mean length of 1 inch. Suppose 50 bolts have been sampled, and $\bar{y} = 1.02$ inches and $s = .04$ inch. Does the sample evidence indicate that the machine is producing bolts with a mean length not equal to 1 inch; i.e., is the production process out of control? Test using $\alpha = .01$.

9.11 Results of the second National Health and Nutrition Examination Survey indicate that the mean blood lead concentration of individuals between the ages of 6 months and 74 years is 14 μg/dl (*Analytical Chemistry*, Feb. 1986). However, the blood lead concentration in black children under the age of 5 years was found to be significantly higher than this figure. Suppose that in a random sample of 200 black children below the age of 5 years, the mean blood lead concentration is 21 μg/dl and the standard deviation is 10 μg/dl. Is there sufficient evidence to indicate that the true mean blood lead concentration in young black children is greater than 14 μg/dl? Test using $\alpha = .01$.

9.12 The EPA sets a limit of 5 parts per million on PCB (a dangerous substance) in water. A major manufacturing firm producing PCB for electrical insulation discharges small amounts from the plant. The company management, attempting to control the amount of PCB in its discharge, has given instructions to halt production if the mean amount of PCB in the effluent exceeds 3 parts per million. A random sampling of 50 water specimens produced the following statistics:

$$\bar{y} = 3.1 \text{ parts per million} \qquad s = .5 \text{ part per million}$$

a. Do these statistics provide sufficient evidence to halt the production process? Use $\alpha = .01$.

b. If you were the plant manager, would you want to use a large or a small value for α for the test in part **a**? Explain.

9.13 The building specifications in a certain city require that the sewer pipe used in residential areas have a mean breaking strength of more than 2,500 pounds per lineal foot. A manufacturer who would like to supply the city with sewer pipe has submitted a bid and provided the following additional information: An independent contractor randomly selected seven sections of the manufacturer's pipe and tested each for breaking strength. The results (pounds per lineal foot) are shown below:

2,610   2,750   2,420   2,510   2,540   2,490   2,680

Is there sufficient evidence to conclude that the manufacturer's sewer pipe meets the required specifications? Use a significance level of $\alpha = .10$.

9.14 Scientists have labelled benzene, a chemical solvent commonly used to synthesize plastics, as a possible cancer-causing agent. Studies have shown that people who work with benzene more than 5 years have 20 times the incidence of leukemia than the general population. As a result, the federal government has lowered the maximum allowable level of benzene in the workplace from 10 parts per million (ppm) to 1 ppm (reported in *Florida Times-Union*, Apr. 2, 1984). Suppose a steel manufacturing plant, which exposes its workers to benzene daily, is under investigation by the Occupational Health and Safety Commission. Twenty air samples, collected over a period of 1 month and examined for benzene content, yielded the following summary statistics:

$$\bar{y} = 2.1 \text{ ppm} \qquad s = 1.7 \text{ ppm}$$

a. Is the steel manufacturing plant in violation of the new government standards? Test the hypothesis that the mean level of benzene at the steel manufacturing plant is greater than 1 ppm, using $\alpha = .05$.

b. What assumption is required for the hypothesis test to be valid?

OPTIONAL EXERCISES

9.15 Refer to Examples 9.5 and 9.6. Find the value of β for $\mu_a = 74$.

9.16 Refer to Exercise 9.10.
  a. Find the value of β for $\mu_a = 1.015$.
  b. Find the power of the test for $\mu_a = 1.045$.

9.17 Refer to Optional Exercises 9.7–9.9. Show that the rejection region for the likelihood ratio test is given by $z > z_\alpha$, where $P(z > z_\alpha) = \alpha$. [*Hint:* Under the assumption that H_0: $\mu = 0$ is true, show that $\sqrt{n}(\bar{y})$ is a standard normal random variable.]

S E C T I O N 9.6

HYPOTHESIS TESTS ABOUT THE DIFFERENCE BETWEEN TWO POPULATION MEANS: INDEPENDENT SAMPLES

Consider independent random samples from two populations with means μ_1 and μ_2, respectively. When the sample sizes are large (i.e., $n_1 \geq 30$ and $n_2 \geq 30$), a test of hypothesis for the difference between the population means ($\mu_1 - \mu_2$) is based on the pivotal z statistic given in Section 8.6. A summary of the large-sample test is provided in the box.

LARGE-SAMPLE TEST OF HYPOTHESIS ABOUT ($\mu_1 - \mu_2$)

ONE-TAILED TEST	TWO-TAILED TEST
H_0: $(\mu_1 - \mu_2) = D_0$	H_0: $(\mu_1 - \mu_2) = D_0$
H_a: $(\mu_1 - \mu_2) > D_0$	H_a: $(\mu_1 - \mu_2) \neq D_0$
[or H_a: $(\mu_1 - \mu_2) < D_0$]	

Test statistic:

$$z = \frac{(\bar{y}_1 - \bar{y}_2) - D_0}{\sigma_{(\bar{y}_1 - \bar{y}_2)}}$$

$$\approx \frac{(\bar{y}_1 - \bar{y}_2) - D_0}{\sqrt{\dfrac{s_1^2}{n_1} + \dfrac{s_2^2}{n_2}}}$$

Test statistic:

$$z = \frac{(\bar{y}_1 - \bar{y}_2) - D_0}{\sigma_{(\bar{y}_1 - \bar{y}_2)}}$$

$$\approx \frac{(\bar{y}_1 - \bar{y}_2) - D_0}{\sqrt{\dfrac{s_1^2}{n_1} + \dfrac{s_2^2}{n_2}}}$$

Rejection region: *Rejection region:*

$z > z_\alpha$ (or $z < -z_\alpha$) $z < -z_{\alpha/2}$ or $z > z_{\alpha/2}$

[*Note:* D_0 is our symbol for the particular numerical value specified for $(\mu_1 - \mu_2)$ in the null hypothesis. In many practical applications, we wish to hypothesize that there is no difference between the population means; in such cases, $D_0 = 0$.]

Assumptions: 1. The sample sizes n_1 and n_2 are sufficiently large—say, $n_1 \geq 30$ and $n_2 \geq 30$.
  2. The two samples are selected randomly and independently from the target populations.

EXAMPLE 9.9

To reduce costs, a bakery has implemented a new leavening process for preparing commercial bread loaves. Loaves of bread were randomly sampled and analyzed for calorie content both before and after implementation of the new process. A summary of the results of the two samples is shown in the table. Do these samples provide sufficient evidence to conclude that the mean number of calories per loaf has decreased since the new leavening process was implemented? Test using $\alpha = .05$.

NEW PROCESS	OLD PROCESS
$n_1 = 50$	$n_2 = 30$
$\bar{y}_1 = 1{,}255$ calories	$\bar{y}_2 = 1{,}330$ calories
$s_1 = 215$ calories	$s_2 = 238$ calories

SOLUTION

We can best answer this question by performing a test of a hypothesis. Defining μ_1 as the mean calorie content per loaf manufactured by the new process and μ_2 as the mean calorie content per loaf manufactured by the old process, we will attempt to support the research (alternative) hypothesis that $\mu_2 > \mu_1$ [i.e., that $(\mu_1 - \mu_2) < 0$]. Thus, we will test the null hypothesis that $(\mu_1 - \mu_2) = 0$, rejecting this hypothesis if $(\bar{y}_1 - \bar{y}_2)$ equals a large negative value. The elements of the test are as follows:

H_0: $(\mu_1 - \mu_2) = 0$ (i.e., $D_0 = 0$)

H_a: $(\mu_1 - \mu_2) < 0$ (i.e., $\mu_1 < \mu_2$)

Test statistic: $z = \dfrac{(\bar{y}_1 - \bar{y}_2) - D_0}{\sigma_{(\bar{y}_1 - \bar{y}_2)}} = \dfrac{(\bar{y}_1 - \bar{y}_2) - 0}{\sigma_{(\bar{y}_1 - \bar{y}_2)}}$

(since both n_1 and n_2 are greater than or equal to 30)

Rejection region: $z < -z_\alpha = -1.645$ (see Figure 9.8, page 376)

Assumptions: The two samples of bread loaves are independently selected.

We now calculate

$$z = \frac{(\bar{y}_1 - \bar{y}_2) - 0}{\sigma_{(\bar{y}_1 - \bar{y}_2)}} = \frac{(1{,}255 - 1{,}330)}{\sqrt{\dfrac{\sigma_1^2}{n_1} + \dfrac{\sigma_2^2}{n_2}}}$$

$$\approx \frac{-75}{\sqrt{\dfrac{s_1^2}{n_1} + \dfrac{s_2^2}{n_2}}} = \frac{-75}{\sqrt{\dfrac{(215)^2}{50} + \dfrac{(238)^2}{30}}} = \frac{-75}{53.03} = -1.41$$

As you can see in Figure 9.8, the calculated z value does not fall in the rejection region. The samples do not provide sufficient evidence, with $\alpha = .05$, to conclude that the new process yields a loaf with fewer mean calories. ∎

When the sample sizes n_1 and n_2 are inadequate to permit use of the large-sample procedure of Example 9.9, modifications may be made to perform a small-sample test of hypothesis about the difference between two population means.

FIGURE 9.8

Rejection Region for
Example 9.9

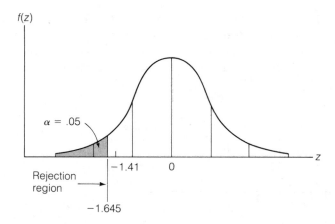

The test procedure is based on assumptions that are, again, more restrictive than in the large-sample case. The elements of the hypothesis test and the assumptions required are listed in the box.

SMALL-SAMPLE TEST OF HYPOTHESIS ABOUT $(\mu_1 - \mu_2)$

ONE-TAILED TEST

H_0: $(\mu_1 - \mu_2) = D_0$

H_a: $(\mu_1 - \mu_2) > D_0$

    [or H_a: $(\mu_1 - \mu_2) < D_0$]

Test statistic:

$$t = \frac{(\bar{y}_1 - \bar{y}_2) - D_0}{\sqrt{s_p^2\left(\dfrac{1}{n_1} + \dfrac{1}{n_2}\right)}}$$

Rejection region:

$t > t_\alpha$ (or $t < -t_\alpha$)

TWO-TAILED TEST

H_0: $(\mu_1 - \mu_2) = D_0$

H_a: $(\mu_1 - \mu_2) \neq D_0$

Test statistic:

$$t = \frac{(\bar{y}_1 - \bar{y}_2) - D_0}{\sqrt{s_p^2\left(\dfrac{1}{n_1} + \dfrac{1}{n_2}\right)}}$$

Rejection region:

$t < -t_{\alpha/2}$ or $t > t_{\alpha/2}$

where

$$s_p^2 = \frac{(n_1 - 1)s_1^2 + (n_2 - 1)s_2^2}{n_1 + n_2 - 2}$$

and the distribution of t is based on $(n_1 + n_2 - 2)$ degrees of freedom.

Assumptions: 1. The populations from which the samples are selected both have approximately normal relative frequency distributions.

               2. The variances of the two populations are equal, i.e., $\sigma_1^2 = \sigma_2^2$.

               3. The random samples are selected in an independent manner from the two populations.

EXAMPLE 9.10

Computer response time is defined as the length of time a user has to wait for the computer to access information on the disk. Suppose a data center wants to compare the average response times of its two computer disk drives. If μ_1 is the mean response time of disk 1 and μ_2 is the mean response time of disk 2, we want to detect a difference between μ_1 and μ_2—if such a difference exists. Therefore, we want to test the null hypothesis

$$H_0: \quad (\mu_1 - \mu_2) = 0$$

against the alternative hypothesis

$$H_a: \quad (\mu_1 - \mu_2) \neq 0 \quad (\text{i.e., } \mu_1 > \mu_2 \text{ or } \mu_1 < \mu_2)$$

Independent random samples of 13 response times for disk 1 and 15 response times for disk 2 were selected. A summary of the data (recorded in milliseconds) is given in the table. Is there sufficient evidence to indicate a difference between the mean response times of the two disk drives? Test using $\alpha = .05$.

DISK 1	DISK 2
$n_1 = 13$	$n_2 = 15$
$\bar{y}_1 = 68$	$\bar{y}_2 = 53$
$s_1 = 18$	$s_2 = 16$

SOLUTION

We first calculate

$$s_p^2 = \frac{(n_1 - 1)s_1^2 + (n_2 - 1)s_2^2}{n_1 + n_2 - 2}$$

$$= \frac{(13 - 1)(18)^2 + (15 - 1)(16)^2}{13 + 15 - 2}$$

$$= \frac{7{,}472}{26} = 287.38$$

Then, if we can assume that the distributions of the response times for the two disk drives are both approximately normal with equal variances, the test statistic is

$$t = \frac{(\bar{y}_1 - \bar{y}_2) - D_0}{\sqrt{s_p^2\left(\frac{1}{n_1} + \frac{1}{n_2}\right)}} = \frac{(68 - 53) - 0}{\sqrt{287.38\left(\frac{1}{13} + \frac{1}{15}\right)}}$$

$$= \frac{15}{6.42} = 2.34$$

Since the observed value of t ($t = 2.34$) falls in the rejection region (see Figure 9.9, page 378), the samples provide sufficient evidence to indicate that the mean response times differ for the two disk drives. Or, we say that the test results are statistically significant at the $\alpha = .05$ level of significance. Because the rejection was in the positive or upper tail of the t distribution, it appears that the mean response time for disk drive 1 exceeds that for disk drive 2.

FIGURE 9.9
Rejection Region for a
Two-Tailed t Test:
$\alpha = .05$, df $= 26$

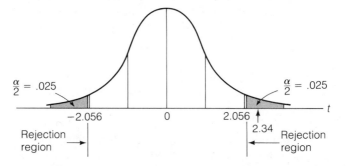

Recall from Section 8.6 that valid small-sample inferences about $(\mu_1 - \mu_2)$ can still be made when the assumption of equal variances is violated. We conclude this section by giving the modifications required to obtain approximate small-sample tests about $(\mu_1 - \mu_2)$ when $\sigma_1^2 \neq \sigma_2^2$ for the two cases described in Section 8.6: $n_1 = n_2$ and $n_1 \neq n_2$.

MODIFICATIONS TO SMALL-SAMPLE TESTS ABOUT $(\mu_1 - \mu_2)$ WHEN $\sigma_1^2 \neq \sigma_2^2$

$n_1 = n_2 = n$

Test statistic:

$$t = \frac{(\bar{y}_1 - \bar{y}_2) - D_0}{\sqrt{\dfrac{s_1^2}{n_1} + \dfrac{s_2^2}{n_2}}} = \frac{(\bar{y}_1 - \bar{y}_2) - D_0}{\sqrt{\dfrac{1}{n}(s_1^2 + s_2^2)}}$$

Degrees of freedom: $\nu = n_1 + n_2 - 2 = 2(n - 1)$

$n_1 \neq n_2$

Test statistic:

$$t = \frac{(\bar{y}_1 - \bar{y}_2) - D_0}{\sqrt{\dfrac{s_1^2}{n_1} + \dfrac{s_2^2}{n_2}}}$$

Degrees of freedom: $\nu = \dfrac{(s_1^2/n_1 + s_2^2/n_2)^2}{\left[\dfrac{(s_1^2/n_1)^2}{n_1 - 1} + \dfrac{(s_2^2/n_2)^2}{n_2 - 1}\right]}$

Note: The value of ν will generally not be an integer. Round down to the nearest integer in order to use the t table (Table 6 of Appendix II).

EXERCISES 9.18–9.25

9.18 The percentage of body fat can be a good indicator of an individual's energy metabolic status and general health. In a recent study of the percentage of body fat of college students in India, two groups of healthy male students, from urban and rural colleges in eastern India, were independently and randomly selected. The percentage of body fat in each was measured, with the results summarized in the table. Does the sample information provide sufficient evidence to conclude that the mean percentage of body fat in healthy male college students residing in urban areas of India differs from the corresponding mean for students residing in rural areas? Use a significance level of $\alpha = .05$.

URBAN STUDENTS	RURAL STUDENTS
$n_1 = 193$	$n_2 = 188$
$\bar{y}_1 = 12.07$	$\bar{y}_2 = 11.04$
$s_1 = 3.04$	$s_2 = 2.63$

Source: Bandyopadhyay, B. and Chattopadhyay, H. "Body Fat in Urban and Rural Male College Students of Eastern India." *American Journal of Physical Anthropology*, Jan. 1981, 54, pp. 119–122.

9.19 The Metro Atlanta Rapid Transit Authority (MARTA) has recently implemented a 6-week bus driver training program designed to reduce the tardiness of buses on regularly scheduled routes. In order to gauge the effectiveness of the training program, a study was conducted before and after the program was instituted.

MARTA authorities were stationed at 30 randomly selected bus stops (involving 30 different buses) in the metro-Atlanta area before the program was implemented to observe tardiness (in minutes) of the scheduled bus arrivals. Similar data were collected at 35 randomly selected bus stops (involving 35 different buses) after the 6-week training program. (The tardiness of a bus which arrived early or on time would be recorded as 0 minutes.) The results are summarized in the accompanying table. MARTA authorities are interested in determining if the mean tardiness of bus arrivals at metro-Atlanta bus stops has decreased significantly since the implementation of the training program. Perform a test of hypothesis for MARTA. Use a significance level of $\alpha = .02$.

BEFORE TRAINING PROGRAM	AFTER TRAINING PROGRAM
$n_1 = 30$	$n_2 = 35$
$\bar{y}_1 = 5.25$ minutes	$\bar{y}_2 = 2.37$ minutes
$s_1 = 1.88$ minutes	$s_2 = 1.45$ minutes

9.20 An industrial plant wants to determine which of two types of fuel—gas or electric—will produce more useful energy at the lower cost. One measure of economical energy production, called the *plant investment per delivered quad*, is calculated by taking the amount of money (in dollars) invested in the particular utility by the plant, and dividing by the delivered amount of energy (in quadrillion British thermal units). The smaller this ratio, the less an industrial plant pays for its delivered energy.

Random samples of 11 plants using electrical utilities and 16 plants using gas utilities were taken, and the plant investment/quad was calculated for each. The results are summarized in the table. Do these data provide sufficient evidence at the $\alpha = .05$ level of significance to indicate a difference in the average investment/quad between the plants using gas and those using electrical utilities? What assumptions are required for the procedure you used to be valid?

	ELECTRIC	GAS
Sample size	11	16
Mean investment/quad (billions)	$44.5	$34.5
Variance	76.4	63.8

9.21 Refer to Exercise 9.20. Conduct the test if it is known that the plant investment/quad variances differ for gas and electric utilities.

9.22 The lining of steel water pipelines with cement mortar, a process introduced in France approximately 140 years ago, provides protection against pipeline corrosion. In order to resist cracking, the emplaced mortar must satisfy certain thickness and strength specifications. Researchers recently conducted a study of the mortar lining of pipeline installed for the San Bernardino Valley Municipal Water District. To determine whether the mortar satisfied strength specifications, specimens of mortar were collected and cured—some for a period of 7 days, others for a period of 28 days. Then each specimen was subjected to compressive and tensile strength tests. A summary of the strength data contained in four samples, each consisting of strength measurements (in pounds per square inch) on 50 mortar specimens, is given in the accompanying table.

	COMPRESSIVE STRENGTH Cure Time		TENSILE STRENGTH Cure Time	
	7 days	28 days	7 days	28 days
Sample mean	8,477	10,404	621	737
Sample standard deviation	820	928	48	55

Source: Aroni, S. and Fletcher, G. "Observations on Mortar Lining of Steel Pipelines." *Transportation Engineering Journal*, Nov. 1979.

a. Is there sufficient evidence to indicate that the mean compressive strength of mortar specimens cured for 28 days is greater than the corresponding mean for specimens cured for 7 days? Test using $\alpha = .05$.

b. Is there sufficient evidence to indicate the mean tensile strength of mortar specimens cured for 28 days is greater than the corresponding mean for specimens cured for 7 days? Test using $\alpha = .05$.

9.23 A field experiment was conducted to ascertain the impact of desert granivores (seed-eaters) on the density and distribution of seeds in the soil. Since some desert rodents are known to hoard seeds in surface caches, the study was specifically designed to determine if these caches eventually produce more seedlings, on the average, than an adjacent control area. Forty small areas excavated by rodents were located and covered with plastic cages to prevent rodents from reusing the caches. A caged control area was set up adjacent to each of the caged caches. The numbers of seedlings germinating from the caches and from the control areas were then observed. A summary of the data is provided in the table. Is there sufficient evidence (at $\alpha = .05$) to indicate that the average number of seedlings germinating from the seed caches of desert rodents is significantly higher than the corresponding average for the control areas?

CACHES	CONTROL AREAS
$n_1 = 40$	$n_2 = 40$
$\bar{y}_1 = 5.3$	$\bar{y}_2 = 2.7$
$s_1 = 1.3$	$s_2 = .7$

Source: Reichman, O. J. "Desert Granivore Foraging and Its Impact on Seed Densities and Distributions." *Ecology*, Dec. 1979, *60*, pp. 1085–1092. Copyright 1979, the Ecological Society of America. Reprinted by permission.

9.24 A major oil company has developed a new gasoline additive that is designed to increase average gas mileage in subcompact cars. Before marketing the new additive, the company conducts the following experiment. Fifty subcompact cars are randomly selected and divided into two groups of 25 cars each. The gasoline additive is dispensed into the tanks of the cars in one group but not the other. The miles per gallon obtained by each car in the study is then recorded. The data are summarized in the accompanying table. Is there sufficient evidence at $\alpha = .10$ for the oil company to claim that the average gas mileage obtained by subcompact cars with the additive is greater than the average gas mileage obtained by subcompact cars without the additive? What assumptions are required for the procedure you used to be valid?

WITHOUT ADDITIVE	WITH ADDITIVE
$n_1 = 25$	$n_2 = 25$
$\bar{y}_1 = 28.4$ miles per gallon	$\bar{y}_2 = 32.1$ miles per gallon
$s_1 = 9.5$ miles per gallon	$s_2 = 12.7$ miles per gallon

9.25 Refer to Exercise 9.24. Conduct the test if it is known that the gas mileage variances differ for cars with and without the additive.

| | | | | | | | | | | |

S E C T I O N 9.7

HYPOTHESIS TEST ABOUT THE DIFFERENCE BETWEEN TWO POPULATION MEANS: MATCHED PAIRS

We explained in Section 8.7 that it may be possible to acquire more information on the difference between two population means by using data collected in matched pairs instead of independent samples. Consider, for example, an experiment to investigate the effectiveness of cloud seeding in the artificial production of rainfall. Two farming areas with similar past meteorological records were selected for the experiment. One is seeded regularly throughout the year, while the other is left unseeded. The monthly precipitation (in inches) at the farms will be recorded for 6 randomly selected months. The resulting data, matched on months, can be used to test a hypothesis about the difference between the mean monthly precipitation in the seeded and unseeded farm areas. The appropriate procedure is summarized in the box.

SMALL-SAMPLE TEST OF HYPOTHESIS ABOUT $(\mu_1 - \mu_2)$: MATCHED PAIRS

ONE-TAILED TEST

H_0: $(\mu_1 - \mu_2) = D_0$

H_a: $(\mu_1 - \mu_2) > D_0$
  [or H_a: $(\mu_1 - \mu_2) < D_0$]

Test statistic:

$$t = \frac{\bar{d} - D_0}{s_d/\sqrt{n}}$$

Rejection region:

$t > t_\alpha$ (or $t < -t_\alpha$)

TWO-TAILED TEST

H_0: $(\mu_1 - \mu_2) = D_0$

H_a: $(\mu_1 - \mu_2) \neq D_0$

Test statistic:

$$t = \frac{\bar{d} - D_0}{s_d/\sqrt{n}}$$

Rejection region:

$t < -t_{\alpha/2}$ or $t > t_{\alpha/2}$

[*Note:* D_0 is our symbol for the particular numerical value specified for $(\mu_1 - \mu_2)$ in the null hypothesis. In many practical applications, we want to hypothesize that there is no difference between the population means; in such cases, $D_0 = 0$.]

Assumptions: 1. The relative frequency distribution of the population of differences is approximately normal.
2. The paired differences are randomly selected from the population of differences.

EXAMPLE 9.11

Consider the cloud seeding experiment to compare monthly precipitation at the two farm areas. Do the data given in Table 9.2 provide sufficient evidence to

TABLE 9.2
Monthly Precipitation Data (in Inches) for Example 9.11

FARM AREA	1	2	3	4	5	6
Seeded	1.75	2.12	1.53	1.10	1.70	2.42
Unseeded	1.62	1.83	1.40	.75	1.71	2.33
d	.13	.29	.13	.35	−.01	.09

indicate that the mean monthly precipitation at the seeded farm area exceeds the corresponding mean for the unseeded farm area? Test using $\alpha = .05$.

SOLUTION

The first step in conducting the test is to calculate the difference d in monthly precipitation at the two farm areas for each month. These differences (where the observation for the unseeded farm area is subtracted from the observation for the seeded area within each pair) are shown in the last row of Table 9.2.

Next, we calculate the mean $\bar{d}$ and standard deviation s_d for this sample of $n = 6$ differences. The mean is computed as

$$\bar{d} = \frac{\Sigma d}{n} = \frac{.13 + .29 + .13 + .35 - .01 + .09}{6} = \frac{.98}{6} = .163$$

To find the variance, we first compute

$$\Sigma d^2 = (.13)^2 + (.29)^2 + (.13)^2 + (.35)^2 + (-.01)^2 + (.09)^2 = .2486$$

Substituting into the formula given in Section 1.7, we have

$$s_d^2 = \frac{\Sigma(d - \bar{d})^2}{n - 1} = \frac{\Sigma d^2 - \dfrac{\left(\Sigma d\right)^2}{n}}{n - 1} = \frac{.2486 - \dfrac{(.98)^2}{6}}{6 - 1} = .0177$$

and

$$s_d = \sqrt{s_d^2} = \sqrt{.0177} = .133$$

Let μ_1 and μ_2 represent the mean monthly precipitation values for the seeded and unseeded farm areas, respectively. Since we want to be able to detect H_a: $(\mu_1 - \mu_2) > 0$, we will conduct a one-tailed test of the null hypothesis H_0: $(\mu_1 - \mu_2) = 0$ (i.e., $D_0 = 0$). The test statistic will have a t distribution based on $(n - 1) = (6 - 1) = 5$ degrees of freedom. We will reject the null hypothesis if

$$t > t_{.05} = 2.015 \quad \text{(see Figure 9.10)}$$

FIGURE 9.10
Rejection Region for
Example 9.11

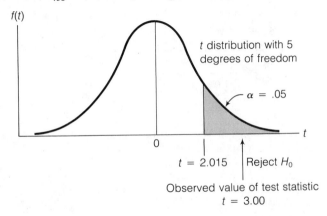

Substituting the values $\bar{d} = .163$ and $s_d = .133$ into the formula for the test statistic, we have

$$t = \frac{\bar{d} - D_0}{s_d/\sqrt{n}} = \frac{.163 - 0}{.133/\sqrt{6}} = 3.00$$

Since this value of the test statistic exceeds the critical value $t_{.05} = 2.015$, there is sufficient evidence (at $\alpha = .05$) to indicate that the mean monthly precipitation at the seeded farm area exceeds the mean for the unseeded farm area.   ∎

In this experiment of Example 9.11, why did we collect the data in matched pairs rather than use independent random samples of months, with some assigned to only the seeded area and others to only the unseeded area? The answer is that we expected some months to have more rain than others. To cancel out this variation from month to month, the experiment was designed so that precipitation at both farm areas would be recorded during the same months. Then both farm areas would be subjected to the same weather pattern in a given month. By comparing precipitation *within* each month, we were able to obtain more information on the difference in mean monthly precipitation than we could have obtained by independent random sampling.

| | | | | | | | | | | | | |

EXERCISES 9.26–9.29

9.26 Tetrachlorodibenzo-p-dioxin (TCDD) is a highly toxic substance found in industrial wastes. A study was conducted to determine the amount of TCDD present in the tissues of bullfrogs inhabiting the Rocky Branch Creek in central Arkansas, an area known to be contaminated by TCDD (*Chemosphere*, Feb. 1986). The level of TCDD (in parts per trillion) was measured in several specific tissues of four female bull frogs and the ratio of TCDD in the tissue to TCDD in the leg muscle of the frog was recorded for each. The relative ratios of contaminant for two tissues, the liver and the ovaries, are given for each of the four frogs in the accompanying table. According to the researchers, "the data set suggests that the [mean] relative level of TCDD in the ovaries of female frogs is higher than the [mean] level in the liver of the frogs." Test this claim using $\alpha = .05$.

FROG	A	B	C	D
Liver	11.0	14.6	14.3	12.2
Ovaries	34.2	41.2	32.5	26.2

Source: Korfmacher, W. A., Hansen, E. B., and Rowland, K. L. "Tissue Distribution of 2,3,7,8-TCDD in Bullfrogs Obtained from a 2,3,7,8-TCDD-Contaminated Area." *Chemosphere*, Vol. 15, No. 2, Feb. 1986, p. 125. Reprinted with permission. Copyright 1986, Pergamon Press, Ltd.

9.27 For the perception of speech, profoundly deaf persons rely mainly on speechreading, i.e., they perceive spoken language by observing the articulatory movements, facial expressions, and gestures of the speaker. Can speech perception be improved by supplementing the speechreader with auditorily presented information about the prosody of the speech signal? To investigate this phenomenon, ten normal-hearing subjects participated in an experiment in which they were asked to verbally reproduce sentences spoken but not heard on a video monitor (*Journal of the Acoustical Society of America*, Feb. 1986). The sentences were presented to the subjects under each of two conditions: (1) speechreading with information about the frequency and amplitude of the speech signal (denoted S + F + A), and (2) speechreading only (denoted S). For each of the ten subjects, the difference between the percentage of correctly reproduced syllables under condition S + F + A and under condition S was calculated. The mean and standard deviation of the differences are provided below:

$$\bar{d} = 20.4 \qquad s_d = 17.44$$

Test the hypothesis that the mean percentage of correct syllables under condition S + F + A exceeds the corresponding mean under condition S. Use $\alpha = .05$.

9.28 Two drugs used for the treatment of glaucoma were tested for effectiveness on seven diseased dogs. Drug A was administered to one eye (chosen randomly) of each dog and drug B to the other eye. Pressure measurements taken 1 hour later on both eyeballs of each dog are shown in the accompanying table. (The smaller the measurement, the more effective the drug.)

DOG	PRESSURE MEASUREMENT	
	Drug A	Drug B
1	.22	.18
2	.16	.14
3	.20	.21
4	.19	.12
5	.13	.13
6	.10	.06
7	.15	.12

a. Is there sufficient evidence (at $\alpha = .05$) to indicate a difference in mean pressure readings for the two drugs?

b. What assumptions are necessary for the validity of the hypothesis-testing technique?

9.29 One of the keys to occupational therapy is patient motivation. A study was conducted to determine whether *purposeful activity* (defined as tasks that are goal-directed) provides intrinsic motivation to exercise performance (*Journal of Occupational Therapy*, Mar. 1984). Twenty-six females were recruited to take part in the study. Each female subject was instructed to perform two similar exercises, jumping rope (the purposeful activity) and jumping without a rope (the nonpurposeful activity), until their perceived exertion level reached 17 on the RPE scale (i.e., until they had worked their bodies "very hard"). The length of time (in minutes) that each subject jumped was then recorded for each of the two exercises and the difference d_i (computed by subtracting the length of jumping time without rope from the length of jumping time with rope) was calculated. A summary of the 26 differences is provided below:

$$\bar{d} = 41.84 \text{ seconds} \qquad s_d = 110.28 \text{ seconds}$$

One theory held by occupational therapists is that those performing a purposeful activity are more motivated, and hence, tend to fatigue less easily. Test the hypothesis that the mean exercise time for the purposeful activity (jumping with a rope) exceeds the mean exercise time for the nonpurposeful activity (jumping without a rope). Use $\alpha = .05$.

SECTION 9.8

HYPOTHESIS TEST ABOUT A POPULATION PROPORTION

In Section 9.2 we gave several examples of a statistical test of hypothesis for a population proportion p. When the sample size is large, the sample proportion of successes $\hat{p}$ is approximately normal and the general formulas for conducting a large-sample z test (given in Section 9.3) can be applied.

The procedure for testing a hypothesis about a population proportion, p, based on a large sample from the target population is described in the box. (Recall that p represents the probability of success in a binomial experiment.) In order for the procedure to be valid, the sample size must be sufficiently large to guarantee approximate normality of the sampling distribution of the sample proportion, $\hat{p}$. As with confidence intervals, a general rule of thumb for determining whether n is "sufficiently large" is that the interval $\hat{p} \pm 2\sqrt{\hat{p}\hat{q}/n}$ does not contain 0 or 1.

LARGE-SAMPLE TEST OF HYPOTHESIS ABOUT A POPULATION PROPORTION

ONE-TAILED TEST	TWO-TAILED TEST
H_0: $p = p_0$	H_0: $p = p_0$
H_a: $p > p_0$	H_a: $p \neq p_0$
(or H_a: $p < p_0$)	
Test statistic:	*Test statistic:*
$z = \dfrac{\hat{p} - p_0}{\sqrt{p_0 q_0/n}}$	$z = \dfrac{\hat{p} - p_0}{\sqrt{p_0 q_0/n}}$
Rejection region:	*Rejection region:*
$z > z_\alpha$ (or $z < -z_\alpha$)	$z < -z_{\alpha/2}$ or $z > z_{\alpha/2}$
where $q_0 = 1 - p_0$	where $q_0 = 1 - p_0$

Assumption: The interval $\hat{p} \pm 2\sqrt{\hat{p}\hat{q}/n}$ does not contain 0 or 1.

EXAMPLE 9.12

Controversy surrounds the use of weathering steel in the construction of highway bridges. Critics have recently cited serious corrosive problems with weathering steel and are currently urging states to prohibit its use in bridge construction. On the other hand, the steel corporations claim that these charges are exaggerated and report that 95% of all weathering steel bridges in operation show "good" performance, with no major corrosive damage. To test this claim, a team of engineers and steel industry experts evaluated 60 randomly selected weathering steel bridges and found 54 of them showing "good" performance. Is there evidence, at $\alpha = .05$, that the true proportion of weathering steel highway bridges that show "good" performance is less than .95, the figure quoted by the steel corporations?

SOLUTION

The parameter of interest is a population proportion, p. We want to test

$$H_0: \quad p = .95$$
$$H_a: \quad p < .95$$

where p is the true proportion of all weathering steel highway bridges that show "good" performance.

At significance level $\alpha = .05$, the null hypothesis will be rejected if

$$z < -z_{.05}$$

that is, H_0 will be rejected if

$$z < -1.645 \quad \text{(see Figure 9.11)}$$

The sample proportion of bridges that show "good" performance is

$$\hat{p} = \frac{54}{60} = .90$$

FIGURE 9.11
Rejection Region for
Example 9.12

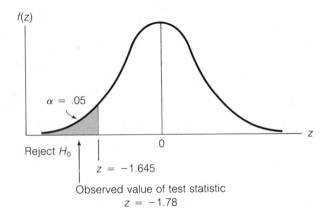

$f(z)$

$\alpha = .05$

Reject H_0

$z = -1.645$

Observed value of test statistic
$z = -1.78$

Thus, the test statistic has the value

$$z = \frac{\hat{p} - p_0}{\sqrt{p_0 q_0 / n}} = \frac{.90 - .95}{\sqrt{(.95)(.05)/60}} = -1.78$$

The null hypothesis can be rejected (at $\alpha = .05$), since the computed value of z falls within the rejection region. There is sufficient evidence to support the hypothesis that the proportion of weathering steel highway bridges that show "good" performance is less than .95.

[Note that the interval

$$\hat{p} \pm 2\sqrt{\hat{p}\hat{q}/n} = .90 \pm 2\sqrt{\frac{(.90)(.10)}{60}}$$

$$= .90 \pm .077$$

does not contain 0 or 1. Thus, the sample size is large enough to guarantee the validity of the hypothesis test.]   ∎

Although small-sample procedures are available for testing hypotheses about a population proportion, the details are omitted from our discussion. It is our experience that they are of limited utility, since most surveys of binomial populations (for example, opinion polls) performed in the real world use samples that are large enough to employ the techniques of this section.

EXERCISES 9.30–9.34

9.30 Distortions that occur on a computer graphics terminal screen are often due to data being lost in the communications linkage process between the terminal and the computer. A manufacturer of a new data-communications error controller claims that the chance of losing data with the controller in operation is only .01. To test this claim, the communications link between a graphics terminal and computer is monitored with the error controller in operation. Of a random sample of 200 on-screen graphic items, six were distorted because of data errors in the communications link. Does the sample evidence refute the manufacturer's claim? Use $\alpha = .05$.

9.31 As part of the evaluation for an environmental impact statement of proposed hydroelectric design on the Stikine River in British Columbia, researchers conducted preliminary investigations of the effects of human-induced disturbances on the behavior of the resident mountain goat population (*Environmental Management*, Mar. 1983). Goat responses to exploration activities, including close-flying helicopters, fixed-wing aircraft, human bipedal movement, and loud blasts from geological drilling activities, were recorded for $n = 804$ goats. The researchers observed that 265 goats displayed a severe flight response to local rock or plant cover. Test the hypothesis that over 30% of the resident mountain goats will show a severe response to human-induced disturbances. Use $\alpha = .05$.

9.32 Computers are now being used in the job interviewing process to elicit detailed information from applicants. The *Personnel Journal* (Aug. 1984) reports on a job interview program at Greentree Systems that runs on a personal computer. Job applicants sit at a computer terminal and respond to multiple-choice questions appearing on the screen. The computer profiles the answers and then suggests certain "probe" questions that can be used during the "live" personal interview. Interviewers at Greentree appreciate the computer assistance, but what about job applicants? A spokesman for Greentree claims that 85% of the job applicants prefer the computer multiple-choice questions to a face-to-face interview. Suppose that in a random sample of 200 job applicants interviewed by a computer, 162 state that they prefer the computer interview to a "live" interview. Is this sufficient evidence to dispute the spokesman's claim? Test using $\alpha = .01$.

9.33 Architects and engineers, faced with public-sector (i.e., government) cuts, are turning to private-sector clients to fill an increasing share of their workloads. According to some researchers, the decrease in popularity of public-sector work among small, medium, and large architecture–engineering (A–E) firms has been dramatic. Two years ago, one-third of all A–E firms reported they relied on public sector projects for most (if not all) of their work. In a recent survey of 60 A–E firms, ten indicated that they depended so heavily on government contracts. Do the sample data provide sufficient evidence to conclude that the percentage of A–E firms that rely heavily on public-sector clients has declined during the past 2 years? Use $\alpha = .05$.

9.34 The strength of a pesticide dosage is often measured by the proportion of pests the dosage will kill. A particular dosage of rat poison is fed to 250 rats. Of these rats, 215 died due to the poison. Test the hypothesis that the true proportion of rats that will succumb to the dosage is larger than .85. Use a significance level of $\alpha = .10$.

S E C T I O N   9.9

HYPOTHESIS TEST ABOUT THE DIFFERENCE BETWEEN TWO POPULATION PROPORTIONS

The method for performing a large-sample test of hypothesis about $(p_1 - p_2)$, the difference between two binomial proportions, is outlined in the accompanying box.

When testing the null hypothesis that $(p_1 - p_2)$ equals some specified difference—say, D_0—we make a distinction between the case $D_0 = 0$ and the case $D_0 \neq 0$. For the special case $D_0 = 0$, i.e., when we are testing $H_0: (p_1 - p_2) = 0$ or, equivalently, $H_0: p_1 = p_2$, the best estimate of $p_1 = p_2 = p$ is found by dividing the total number of successes in the combined samples by the total number of observations in the two samples. That is, if y_1 is the number of successes in sample 1 and y_2 is the number of successes in sample 2, then

$$\hat{p} = \frac{y_1 + y_2}{n_1 + n_2}$$

In this case, the best estimate of the standard deviation of the sampling distribution of $(\hat{p}_1 - \hat{p}_2)$ is found by substituting $\hat{p}$ for both p_1 and p_2:

$$\sigma_{(\hat{p}_1 - \hat{p}_2)} = \sqrt{\frac{p_1 q_1}{n_1} + \frac{p_2 q_2}{n_2}} \approx \sqrt{\frac{\hat{p}\hat{q}}{n_1} + \frac{\hat{p}\hat{q}}{n_2}} = \sqrt{\hat{p}\hat{q}\left(\frac{1}{n_1} + \frac{1}{n_2}\right)}$$

LARGE-SAMPLE TEST OF HYPOTHESIS ABOUT $(p_1 - p_2)$

ONE-TAILED TEST

H_0: $(p_1 - p_2) = D_0$

H_a: $(p_1 - p_2) > D_0$
     [or H_a: $(p_1 - p_2) < D_0$]

Test statistic:

$$z = \frac{(\hat{p}_1 - \hat{p}_2) - D_0}{\sigma_{(\hat{p}_1 - \hat{p}_2)}}$$

Rejection region:

$z > z_\alpha$ (or $z < -z_\alpha$)

where

TWO-TAILED TEST

H_0: $(p_1 - p_2) = D_0$

H_a: $(p_1 - p_2) \neq D_0$

Test statistic:

$$z = \frac{(\hat{p}_1 - \hat{p}_2) - D_0}{\sigma_{(\hat{p}_1 - \hat{p}_2)}}$$

Rejection region:

$z < -z_{\alpha/2}$ or $z > z_{\alpha/2}$

$$\sigma_{(\hat{p}_1 - \hat{p}_2)} = \sqrt{\frac{p_1 q_1}{n_1} + \frac{p_2 q_2}{n_2}}$$

When $D_0 \neq 0$, calculate $\sigma_{(\hat{p}_1 - \hat{p}_2)}$ using $\hat{p}_1$ and $\hat{p}_2$:

$$\sigma_{(\hat{p}_1 - \hat{p}_2)} \approx \sqrt{\frac{\hat{p}_1 \hat{q}_1}{n_1} + \frac{\hat{p}_2 \hat{q}_2}{n_2}}$$

where $\hat{q}_1 = 1 - \hat{p}_1$ and $\hat{q}_2 = 1 - \hat{p}_2$.

For the special case where $D_0 = 0$, calculate

$$\sigma_{(\hat{p}_1 - \hat{p}_2)} \approx \sqrt{\hat{p}\hat{q}\left(\frac{1}{n_1} + \frac{1}{n_2}\right)}$$

where the total number of successes in the combined samples is $(y_1 + y_2)$ and

$$\hat{p}_1 = \hat{p}_2 = \hat{p} = \frac{y_1 + y_2}{n_1 + n_2}$$

Assumption: The intervals

$$\hat{p}_1 \pm 2\sqrt{\frac{\hat{p}_1 \hat{q}_1}{n_1}} \quad \text{and} \quad \hat{p}_2 \pm 2\sqrt{\frac{\hat{p}_2 \hat{q}_2}{n_2}}$$

do not contain 0 or 1.

For all cases in which $D_0 \neq 0$ [for example, when testing H_0: $(p_1 - p_2) = .2$], we use $\hat{p}_1$ and $\hat{p}_2$ in the formula for $\sigma_{(\hat{p}_1 - \hat{p}_2)}$. However, in most practical situations, we will want to test for a difference between proportions—that is, we will want to test H_0: $(p_1 - p_2) = 0$.

The sample sizes n_1 and n_2 must be sufficiently large to ensure that the sampling distributions of $\hat{p}_1$ and $\hat{p}_2$, and hence of the difference $(\hat{p}_1 - \hat{p}_2)$, are approximately normal. The rule of thumb used to determine if the sample sizes are "sufficiently large" is the same as that given in Section 8.9, namely that the intervals

$$\hat{p}_1 \pm 2\sqrt{\frac{\hat{p}_1\hat{q}_1}{n_1}} \quad \text{and} \quad \hat{p}_2 \pm 2\sqrt{\frac{\hat{p}_2\hat{q}_2}{n_2}}$$

do not contain 0 or 1. [*Note:* If the sample sizes are not sufficiently large, p_1 and p_2 can be compared using a technique to be discussed in Chapter 15.]

EXAMPLE 9.13

Recently there have been intensive campaigns encouraging people to save energy by car-pooling to work. Some cities have created an incentive for car-pooling by designating certain highway traffic lanes as "car-pool only" (i.e., only cars with two or more passengers can use these lanes). In order to evaluate the effectiveness of this plan, toll booth personnel in one city monitored 2,000 randomly selected cars in 1984 (before the car-pool only lanes were established) and 1,500 cars in 1986 (after the car-pool only lanes were established). The results of the study are shown in the table, where y_1 and y_2 represent the numbers of cars with two or more passengers (i.e., car-pool riders) in the 1986 and 1984 samples, respectively. Do the data indicate that the fraction of cars with car-pool riders has increased over this period? Use $\alpha = .05$.

1986	1984
$n_1 = 1,500$	$n_2 = 2,000$
$y_1 = 576$	$y_2 = 652$

SOLUTION

If we define p_1 and p_2 as the true proportions of cars with car-pool riders in 1986 and 1984, the elements of our test are:

H_0: $(p_1 - p_2) = 0$

H_a: $(p_1 - p_2) > 0$

(The test is one-tailed since we are interested only in determining whether the proportion of cars with car-pool riders has increased.)

Test statistic: $z = \dfrac{(\hat{p}_1 - \hat{p}_2) - 0}{\sigma_{(\hat{p}_1 - \hat{p}_2)}}$

Rejection region: $\alpha = .05$

$z > z_\alpha = z_{.05} = 1.645$ (see Figure 9.12, page 392)

We now calculate the sample proportions of cars with car-pool riders:

$$\hat{p}_1 = \frac{576}{1,500} = .384 \qquad \hat{p}_2 = \frac{652}{2,000} = .326$$

Note that neither of the intervals

$$\hat{p}_1 \pm 2\sqrt{\frac{\hat{p}_1\hat{q}_1}{n_1}} = .384 \pm 2\sqrt{\frac{(.384)(.616)}{1,500}} = .384 \pm .025$$

FIGURE 9.12
Rejection Region for
Example 9.13

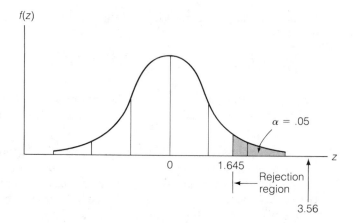

$$\hat{p}_2 \pm 2\sqrt{\frac{\hat{p}_2 \hat{q}_2}{n_2}} = .326 \pm 2\sqrt{\frac{(.326)(.674)}{2,000}} = .326 \pm .021$$

includes 0 or 1. Therefore, the samples are sufficiently large to guarantee the validity of the test.

The test statistic is

$$z = \frac{(\hat{p}_1 - \hat{p}_2) - 0}{\sigma_{(\hat{p}_1 - \hat{p}_2)}} \approx \frac{(\hat{p}_1 - \hat{p}_2)}{\sqrt{\hat{p}\hat{q}\left(\frac{1}{n_1} + \frac{1}{n_2}\right)}}$$

where

$$\hat{p} = \frac{y_1 + y_2}{n_1 + n_2} = \frac{576 + 652}{1,500 + 2,000} = .351$$

Thus,

$$z = \frac{.384 - .326}{\sqrt{(.351)(.649)\left(\frac{1}{5,000} + \frac{1}{2,000}\right)}} = \frac{.058}{.0164} = 3.56$$

There is sufficient evidence at $\alpha = .05$ to conclude that the proportion of all cars with car-pool riders has increased over the 1984–1986 period. We could place a confidence interval on $(p_1 - p_2)$ if we were interested in estimating the extent of the increase.   ∎

EXERCISES 9.35–9.39

9.35 Every 5 years the Mechanics Division of ASEE conducts a nationwide survey on undergraduate mechanics education at colleges and universities. In 1985, 66 of the 100 colleges surveyed covered fluid statics in their undergraduate engineering program, compared to 43% in the 1975 survey (*Engineering Education*, Apr. 1986).

Assuming that 100 colleges were also surveyed in 1975, conduct a test to determine whether the percentage of colleges covering fluid statics increased from 1975 to 1985. Use $\alpha = .01$.

9.36 Researchers at Mount Sinai Medical Center in New York believe that "Lou Gehrig's disease" (a neurological disorder that slowly paralyzes and kills its victims, one of the first of whom was New York Yankees' baseball slugger Lou Gehrig in 1941), may be linked to household pets, especially small dogs. A five-member medical team has found that 72% of the afflicted patients studied had small household dogs at least 20 years before contracting Lou Gehrig's disease. In contrast, only 33% of a healthy control group had pet dogs early in life. (Assume that 100 afflicted patients and 100 healthy controls were studied.) Is there sufficient evidence to indicate that the percentage of all patients with Lou Gehrig's disease who had pet dogs early in life is larger than the corresponding percentage for nonafflicted people? Test using $\alpha = .01$.

9.37 The nuclear mishap on Three Mile Island near Harrisburg, Pennsylvania, on March 28, 1979, forced many local residents to evacuate their homes—some temporarily, others permanently. In order to assess the impact of the accident on the area population, a questionnaire was designed and mailed to a sample of 150 households within 2 weeks after the accident occurred. Residents were asked how they felt both before and after the accident about having some of their electricity generated from nuclear power. The summary results are provided in the table. Is there sufficient evidence to indicate a difference in the proportions of Three Mile Island residents who favor nuclear power before and after the accident? Use $\alpha = .01$.

	ATTITUDE TOWARD NUCLEAR POWER			TOTALS
	Favor	Oppose	Indifferent	
Before accident	62	35	53	150
After accident	52	72	26	150

Source: Brown, A. et al. *Final Report on a Survey of Three Mile Island Area Residents*. Department of Geography, Michigan State University, Aug. 1979.

9.38 A study was conducted to determine the impact of a multifunction workstation (MFWS) on the way managers work (*Datamation*, Feb. 15, 1986). Two groups of managers at a St. Louis-based defense agency took part in the survey: a test group consisting of 12 managers who currently use MFWS software and a control group of 25 non-MFWS users. One question on the survey concerned the information sources of the managers. In the test group (MFWS users), 4 of the 12 managers reported that their major source of information is the computer, while 2 of the 25 in the control group (non-MFWS users) rely on the computer as their major source of information. Is there evidence of a difference between the proportions of MFWS users and non-MFWS users who rely on the computer as their major information source? Test using $\alpha = .10$.

9.39 Many "solar" homes waste the sun's valuable energy. To be efficient, a solar house must be specially designed to trap solar radiation and use it effectively. Basically, energy-efficient solar heating systems can be categorized into two groups, *passive* solar heating systems and *active* solar heating systems. In a passive solar heating system, the house itself is a solar energy collector, while in an active solar heating system, elaborate mechanical equipment is used to convert the sun's rays into heat.

Suppose we want to determine whether there is a difference between the proportions of passive solar and active solar heating systems that require less than 200 gallons of oil per year in fuel consumption. Independent random samples of 50 passive and 50 active solar-heated homes are selected and the numbers that required less than 200 gallons of oil last year are noted, with the results given in the table. Is there evidence of a difference between the proportions of passive and active solar-heated homes that required less than 200 gallons of oil in fuel consumption last year? Test at a level of significance of $\alpha = .02$.

	PASSIVE SOLAR	ACTIVE SOLAR
Number of homes	50	50
Number that required less than 200 gallons of oil last year	37	46

SECTION 9.10

HYPOTHESIS TEST ABOUT A POPULATION VARIANCE

Recall from Section 8.10 that the pivotal statistic for estimating a population variance σ^2 does not possess a normal (z) distribution for large samples. Therefore, we cannot apply the large-sample procedure outlined in Section 9.3 when testing hypotheses about σ^2.

When the sample is selected from a normal population, however, the pivotal statistic possesses a chi-square (χ^2) distribution and the test can be conducted as outlined in the box. Note that the assumption of normality is required regardless of whether the sample size n is large or small.

TEST OF HYPOTHESIS ABOUT A POPULATION VARIANCE σ^2

ONE-TAILED TEST

H_0: $\sigma^2 = \sigma_0^2$

H_a: $\sigma^2 > \sigma_0^2$

    [or H_a: $\sigma^2 < \sigma_0^2$]

Test statistic:

$$\chi^2 = \frac{(n-1)s^2}{\sigma_0^2}$$

Rejection region:

$$\chi^2 > \chi_\alpha^2 \quad (\text{or } \chi^2 < \chi_{1-\alpha}^2)$$

TWO-TAILED TEST

H_0: $\sigma^2 = \sigma_0^2$

H_a: $\sigma^2 \neq \sigma_0^2$

Test statistic:

$$\chi^2 = \frac{(n-1)s^2}{\sigma_0^2}$$

Rejection region:

$$\chi^2 < \chi_{1-\alpha/2}^2 \text{ or } \chi^2 > \chi_{\alpha/2}^2$$

where χ_α^2 and $\chi_{1-\alpha}^2$ are values of χ^2 that locate an area of α to the right and α to the left, respectively, of a chi-square distribution based on $(n-1)$ degrees of freedom.

[*Note:* σ_0^2 is our symbol for the particular numerical value specified for σ^2 in the null hypothesis.]

Assumption: The population from which the random sample is selected has an approximate normal distribution.

EXAMPLE 9.14

Refer to Example 8.16, concerning the variability of the amount of fill at a cannery. Suppose regulatory agencies specify that the standard deviation of the amount of fill should be less than .1 ounce. The quality control supervisor sampled $n = 10$ cans and calculated $s = .04$. Does this value of s provide sufficient evidence to indicate that the standard deviation σ of the fill measurements is less than .1 ounce?

SOLUTION

Since the null and alternative hypotheses must be stated in terms of σ^2 (rather than σ), we will want to test the null hypothesis that $\sigma^2 = .01$ against the alternative that $\sigma^2 < .01$. Therefore, the elements of the test are

$$H_0: \quad \sigma^2 = .01$$
$$H_a: \quad \sigma^2 < .01$$

Test statistic: $\quad \chi^2 = \dfrac{(n-1)s^2}{\sigma_0^2}$

Rejection region: The smaller the value of s^2 we observe, the stronger the evidence in favor of H_a. Thus, we reject H_0 for "small values" of the test statistic. With $\alpha = .05$ and 9 df, the χ^2 value for rejection is found in Table 7 of Appendix II and pictured in Figure 9.13. We will reject H_0 if $\chi^2 < 3.32511$.

Remember that the area given in Table 7 of Appendix II is the area to the *right* of the numerical value in the table. Thus, to determine the lower-tail value that has $\alpha = .05$ to its *left*, we use the $\chi^2_{.95}$ column in Table 7.
Since

$$\chi^2 = \frac{(n-1)s^2}{\sigma_0^2} = \frac{9(.04)^2}{.01} = 1.44$$

is less than 3.32511, the supervisor can conclude that the variance of the population of all amounts of fill is less than .01 ($\sigma < 0.1$) with 95% confidence. As usual, the confidence is in the procedure used—the χ^2 test. If this procedure is

FIGURE 9.13
Rejection Region for
Example 9.14

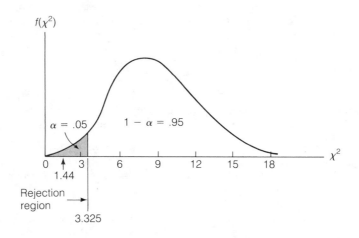

repeatedly used, it will incorrectly reject H_0 only 5% of the time. Thus, the quality control supervisor is confident in the decision that the cannery is operating within the desired limits of variability.   ∎

9.40 Polychlorinated biphenyls (PCBs), used in the manufacture of large electrical transformers and capacitors, are extremely hazardous contaminants when released into the environment. The Environmental Protection Agency (EPA) is experimenting with a new device for measuring PCB concentration in fish. In order to check the precision of the new instrument, seven PCB readings were taken on the same fish sample. The data are recorded here (in parts per million):

$$6.2 \quad 5.8 \quad 5.7 \quad 6.3 \quad 5.9 \quad 5.8 \quad 6.0$$

Suppose the EPA requires an instrument that yields PCB readings with a variance of less than .1. Does the new instrument meet the EPA's specifications? Test at $\alpha = .05$.

9.41 The most common method of disinfecting water for potable use is free residual chlorination. Recently, preammoniation (i.e., the addition of ammonia to the water prior to applying free chlorine) has received considerable attention as an alternative treatment. In one study, 44 water specimens treated with preammoniation were found to have a mean effluent turbidity of 1.8 and a standard deviation of .16 (*American Water Works Journal*, Jan. 1986). Is there sufficient evidence to indicate that the variance of the effluent turbidity in water specimens disinfected by the preammoniation method exceeds .0016? (The value .0016 represents the known effluent turbidity variance of water specimens treated with free chlorine.) Test using $\alpha = .01$.

9.42 In any canning process, a manufacturer will lose money if the cans contain either significantly more or significantly less than is claimed on the label. Accordingly, canners pay close attention to the amount of their product being dispensed by the can-filling machines. Consider a company that produces a fast-drying rubber cement in 32-ounce aluminum cans. A quality control inspector is interested in testing whether the variance of the amount of rubber cement dispensed into the cans is more than .3. If so, the dispensing machine is in need of adjustment. Since inspection of the canning process requires that the dispensing machines be shut down, and shutdowns for any lengthy period of time cost the company thousands of dollars in lost revenue, the inspector is able to obtain a random sample of only ten cans for testing. After measuring the weights of their contents, the inspector computes the following summary statistics:

$$\bar{x} = 31.55 \text{ ounces} \qquad s = .48 \text{ ounce}$$

a. Does the sample evidence indicate that the dispensing machines are in need of adjustment? Test at significance level $\alpha = .05$.

b. What assumption is necessary for the hypothesis test of part **a** to be valid?

9.43 A new gun-like apparatus has been devised to replace the needle in administering vaccines. The apparatus, which is connected to a large supply of the vaccine, can be set to inject different amounts of the serum, but the variance in the amount of serum injected in a given person must not be greater than .06 to ensure proper inoculation. A random sample of 25 injections resulted in a variance of .135. Do the data provide sufficient evidence to indicate the gun is not working properly? Use $\alpha = .10$.

HYPOTHESIS TEST ABOUT THE RATIO OF TWO POPULATION VARIANCES

As in the one-sample case, the pivotal statistic for comparing two population variances, σ_1^2 and σ_2^2, has a nonnormal sampling distribution. Recall from Section 8.11 that the ratio of the sample variances s_1^2/s_2^2 possesses, under certain conditions, an F distribution.

The elements of the hypothesis test for the ratio of two population variances, σ_1^2/σ_2^2, are given in the box.

TEST OF HYPOTHESIS FOR THE RATIO OF TWO POPULATION VARIANCES σ_1^2/σ_2^2

ONE-TAILED TEST

H_0: $\dfrac{\sigma_1^2}{\sigma_2^2} = 1$ (i.e., $\sigma_1^2 = \sigma_2^2$)

H_a: $\dfrac{\sigma_1^2}{\sigma_2^2} > 1$ (i.e., $\sigma_1^2 > \sigma_2^2$)

$\left[\text{or, } H_a: \dfrac{\sigma_1^2}{\sigma_2^2} < 1 \quad (\text{i.e., } \sigma_1^2 < \sigma_2^2)\right]$

Test statistic:

$F = \dfrac{s_1^2}{s_2^2} \left[\text{or, } F = \dfrac{s_2^2}{s_1^2}\right]$

TWO-TAILED TEST

H_0: $\dfrac{\sigma_1^2}{\sigma_2^2} = 1$ (i.e., $\sigma_1^2 = \sigma_2^2$)

H_a: $\dfrac{\sigma_1^2}{\sigma_2^2} \neq 1$ (i.e., $\sigma_1^2 \neq \sigma_2^2$)

Test statistic:

$F = \dfrac{\text{Larger sample variance}}{\text{Smaller sample variance}}$

$= \begin{cases} \dfrac{s_1^2}{s_2^2} & \text{when } s_1^2 > s_2^2 \\ \dfrac{s_2^2}{s_1^2} & \text{when } s_2^2 > s_1^2 \end{cases}$

Rejection region:

$F > F_\alpha$

Rejection region:

$F > F_{\alpha/2}$

where F_α and $F_{\alpha/2}$ are values that locate area α and $\alpha/2$, respectively, in the upper tail of the F-distribution with ν_1 = numerator degrees of freedom (i.e., the df for the sample variance in the numerator) and ν_2 = denominator degrees of freedom (i.e., the df for the sample variance in the denominator).

Assumptions: 1. Both of the populations from which the samples are selected have relative frequency distributions that are approximately normal.
2. The random samples are selected in an independent manner from the two populations.

EXAMPLE 9.15

Heavy doses of ethylene oxide (ETO) in rabbits have been shown to alter significantly the DNA structure of cells. Although it is a known mutagen and suspected carcinogen, ETO is used quite frequently in sterilizing hospital supplies. A study was conducted to investigate the effect of ETO on hospital personnel involved with the sterilization process. Thirty-one subjects were randomly selected and assigned to one of two tasks. Eighteen subjects were assigned the task of opening the sterilization package that contains ETO (task 1). The remaining 13 subjects were assigned the task of opening and unloading the sterilizer gun filled with ETO (task 2). After the tasks were performed, researchers measured the amount of ETO (in milligrams) present in the bloodstream of each subject. A summary of the results appears in the table. Do the data provide sufficient evidence to indicate a difference in the variability of the ETO levels in subjects assigned to the two tasks? Test using $\alpha = .10$.

	TASK 1	TASK 2
Sample size	18	13
Mean	5.90	5.60
Standard deviation	1.93	3.10

SOLUTION

Let

σ_1^2 = Population variance of ETO levels in subjects assigned task 1

σ_2^2 = Population variance of ETO levels in subjects assigned task 2

In order for this test to yield valid results, we must assume that both samples of ETO levels come from normal populations and that the samples are independent.

The hypotheses of interest are then

$$H_0: \quad \frac{\sigma_1^2}{\sigma_2^2} = 1 \quad (\sigma_1^2 = \sigma_2^2)$$

$$H_a: \quad \frac{\sigma_1^2}{\sigma_2^2} \neq 1 \quad (\sigma_1^2 \neq \sigma_2^2)$$

The nature of the F tables given in Appendix II affects the form of the test statistic. To form the rejection region for a two-tailed F test we want to make certain that the upper tail is used, because only the upper-tail values of F are shown in Tables 8, 9, 10, and 11 of Appendix II. To accomplish this, **we will always place the larger sample variance in the numerator of the F test statistic.** This has the effect of doubling the tabulated value for α, since we double the probability that the F ratio will fall in the upper tail by always placing the larger sample variance in the numerator. That is, we make the test two-tailed by putting the larger variance in the numerator rather than establishing rejection regions in both tails.

Thus, for our example, we have a denominator s_1^2 with df $= n_1 - 1 = 17$ and a numerator s_2^2 with df $= n_2 - 1 = 12$. Therefore, the test statistic will be

$$F = \frac{\text{Larger sample variance}}{\text{Smaller sample variance}} = \frac{s_2^2}{s_1^2}$$

and we will reject $H_0: \sigma_1^2 = \sigma_2^2$ for $\alpha = .10$ when the calculated value of F exceeds the tabulated value:

$$F_{\alpha/2} = F_{.05} = 2.38$$

We can now calculate the value of the test statistic and complete the analysis:

$$F = \frac{s_2^2}{s_1^2} = \frac{(3.10)^2}{(1.93)^2} = \frac{9.61}{3.72} = 2.58$$

When we compare this to the rejection region shown in Figure 9.14, we see that $F = 2.58$ falls in the rejection region. Therefore, the data provide sufficient evidence to indicate that the population variances differ. It appears that hospital personnel involved with opening the sterilization package (task 1) have less variable ETO levels than those involved with opening and unloading the sterilizer gun (task 2).

FIGURE 9.14

Rejection Region for Example 9.15

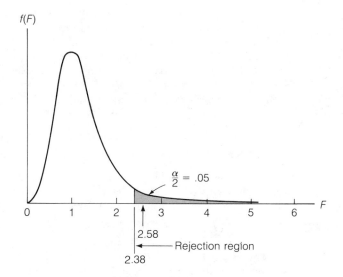

What would you have concluded in Example 9.15 if the value of F calculated from the samples had not fallen in the rejection region? Would you conclude that the null hypothesis of equal variances is true? No, because then you risk the possibility of a Type II error (failing to reject H_0 if H_a is true) without knowing the value of β, the probability of failing to reject H_0: $\sigma_1^2 = \sigma_2^2$ if in fact it is false. Since we will not consider the calculation of β for specific alternatives, when the F statistic does not fall in the rejection region, we simply conclude that insufficient sample evidence exists to refute the null hypothesis that $\sigma_1^2 = \sigma_2^2$.

Example 9.15 illustrates the technique for calculating the test statistic and rejection region for a two-tailed test in order to avoid the problem of locating an F-value in the lower tail of the F distribution. In a one-tailed test this is much easier to accomplish since we can control how we specify the ratio of the population variances in H_0 and H_a. That is, we can always make a one-tailed test an *upper-tailed* test. For example, if we want to test whether σ_1^2 is greater than σ_2^2, then we write the alternative hypothesis as

$$H_a: \quad \frac{\sigma_1^2}{\sigma_2^2} > 1 \quad (\text{i.e., } \sigma_1^2 > \sigma_2^2)$$

and the appropriate test statistic is $F = s_1^2/s_2^2$. Conversely, if we want to test whether σ_1^2 is less than σ_2^2 (i.e, whether σ_2^2 is greater than σ_1^2), we write

$$H_a: \quad \frac{\sigma_2^2}{\sigma_1^2} > 1 \quad (\text{i.e., } \sigma_2^2 > \sigma_1^2)$$

and the corresponding test statistic is $F = s_2^2/s_1^2$.

EXERCISES 9.44–9.49

9.44 An experiment was conducted to study the effect of reinforced flanges on the torsional capacity of reinforced concrete T-beams (*Journal of the American Concrete Institute*, Jan.–Feb. 1986). Several different types of T-beams were used in the experiment, each type having a different flange width. The beams were tested under combined torsion and bending until failure (cracking). One variable of interest is the cracking torsion moment at the top of the flange of the T-beam. Cracking torsion moments for eight beams with 70-cm slab widths and eight beams with 100-cm slab widths are recorded below:

70-cm slab width: 6.00, 7.20, 10.20, 13.20, 11.40, 13.60, 9.20, 11.20
100-cm slab width: 6.80, 9.20, 8.80, 13.20, 11.20, 14.90, 10.20, 11.80

a. Is there evidence of a difference in the variation in the cracking torsion moments of the two types of T-beams? Use $\alpha = .10$.
b. What assumptions are required for the test to be valid?

9.45 A computer programmer wants to compare the variability in the cost of running jobs using time-sharing options (TSO) and terminal control programming (TCP) from a computer terminal. Independent random samples of 20 TSO jobs and 15 TCP jobs were selected and their costs recorded. The results are summarized in the accompanying table. Is there evidence of a difference in the variability of costs of TSO and TCP jobs? Use $\alpha = .05$.

TSO	TCP
$n_1 = 20$	$n_2 = 15$
$\bar{y}_1 = \$1.75$	$\bar{y}_2 = \$2.13$
$s_1 = \$1.02$	$s_2 = \$.89$

9.46 Refer to the speechreading study introduced in Exercise 9.27. A second experiment was conducted to compare the variability in the sentence perception of normal-hearing individuals with no prior experience in speechreading to those with experience in speechreading. The sample consisted of 24 inexperienced and 12 experienced subjects. All subjects were asked to verbally reproduce sentences under several conditions, one of which was speechreading supplemented with sound-pressure information. A summary of the results (percentage of correct syllables) for the two groups is given in the table. Conduct a test to determine whether the variance in the percentage of correctly reproduced syllables differs between the two groups of speechreaders. Test using $\alpha = .10$.

INEXPERIENCED SPEECHREADERS	EXPERIENCED SPEECHREADERS
$n_1 = 24$	$n_2 = 12$
$\bar{y}_1 = 87.1$	$\bar{y}_2 = 86.1$
$s_1 = 8.7$	$s_2 = 12.4$

Source: Breeuwer, M. and Plomp, R. "Speechreading Supplemented with Auditorily Presented Speech Parameters." *Journal of the Acoustical Society of America*, Vol. 79, No. 2, Feb. 1986, p. 487.

9.47 Refer to the general trace organic monitoring study discussed in Exercise 7.9. The total organic carbon (TOC) level was measured in water samples collected at two sewage treatment sites in England. The accompanying table gives the summary information on the TOC levels (measured in mg/l) found in the rivers adjacent to the two sewage facilities. Since the river at the Foxcote sewage treatment works was subject to periodic spillovers, not far upstream of the plant's intake, it is believed that the TOC levels found at Foxcote will have greater variation than the levels at Bedford. Does the sample information support this hypothesis? Test at $\alpha = .05$.

BEDFORD	FOXCOTE
$n_1 = 61$	$n_2 = 52$
$\bar{y}_1 = 5.35$	$\bar{y}_2 = 4.27$
$s_1 = .96$	$s_2 = 1.27$

Source: Pinchin, M. J. "A Study of the Trace Organics Profiles of Raw and Potable Water Systems." *Journal of the Institute of Water Engineers & Scientists*, Vol. 40, No. 1, Feb. 1986, p. 87.

OPTIONAL EXERCISES

9.48 Suppose we want to test $H_0: \sigma_1^2 = \sigma_2^2$ versus $H_a: \sigma_1^2 \neq \sigma_2^2$. Show that the rejection region given by

$$\frac{s_1^2}{s_2^2} > F_{\alpha/2} \quad \text{or} \quad \frac{s_1^2}{s_2^2} < F_{(1-\alpha/2)}$$

where F depends on $\nu_1 = (n_1 - 1)$ df and $\nu_2 = (n_2 - 1)$ df, is equivalent to the rejection region given by

$$\frac{s_1^2}{s_2^2} > F_{\alpha/2} \quad \text{where } F \text{ depends on } \nu_1 \text{ numerator df and } \nu_2 \text{ denominator df}$$

$$\text{or} \quad \frac{s_2^2}{s_1^2} > F^*_{\alpha/2} \quad \text{where } F^* \text{ depends on } \nu_2 \text{ numerator df and } \nu_1 \text{ denominator df}$$

[*Hint:* Use the fact (proof omitted) that

$$F_{(1-\alpha/2)} = \frac{1}{F^*_{\alpha/2}}$$

where F depends on ν_1 numerator df and ν_2 denominator df and F^* depends on ν_2 numerator df and ν_1 denominator df.]

9.49 Use the results of Optional Exercise 9.48 to show that

$$P\left(\frac{\text{Larger sample variance}}{\text{Smaller sample variance}} > F_{\alpha/2}\right) = \alpha$$

where F depends on numerator df = [(Sample size for numerator sample variance) − 1] and denominator df = [(Sample size for denominator sample variance) − 1]. [*Hint:* First write

$$P\left(\frac{\text{Larger sample variance}}{\text{Smaller sample variance}} > F_{\alpha/2}\right) = P\left(\frac{s_1^2}{s_2^2} > F_{\alpha/2} \quad \text{or} \quad \frac{s_2^2}{s_1^2} > F_{\alpha/2}\right)$$

Then use the fact that $P(F > F_{\alpha/2}) = \alpha/2$.]

SECTION 9.12

THE OBSERVED SIGNIFICANCE LEVEL FOR A TEST

According to the statistical test procedures described in the preceding sections, the rejection region and the corresponding value of α are selected prior to conducting the test and the conclusion is stated in terms of rejecting or not rejecting the null hypothesis. A second method of presenting the result of a statistical test is one that reports the extent to which the test statistic disagrees with the null hypothesis and leaves the reader the task of deciding whether to reject the null hypothesis. This measure of disagreement is called the **observed significance level** (or **p-value**) for the test.* It is also sometimes referred to as the **attained significance level** for the test.

DEFINITION 9.4

The **observed significance level**, or **p-value**, for a specific statistical test is the probability (assuming H_0 is true) of observing a value of the test statistic that is at least as contradictory to the null hypothesis, and supportive of the alternative hypothesis, as the one computed from the sample data.

EXAMPLE 9.16

SOLUTION

Find the observed significance level for the statistical test of Example 9.5.

In Example 9.5, we tested a hypothesis about the mean μ of the number of heavy freight trucks per hour using a particular 25-mile stretch of interstate highway. Since we wanted to detect values of μ larger than $\mu_0 = 72$, we conducted a one-tailed test, rejecting H_0 for large values of $\bar{y}$, or equivalently, large values of z. The observed value of z, computed from the sample of $n = 50$ randomly selected 1-hour periods, was $z = 1.12$. Since any value of z larger than $z = 1.12$ would be even more contradictory to H_0, the observed significance level for the test is

$$p\text{-value} = P(z \geq 1.12)$$

*The term *p-value* or *probability value* was coined by users of statistical methods. The p in the expression *p-value* should not be confused with the binomial parameter p.

FIGURE 9.15

Finding the p-Value for an Upper-Tailed Test When $z = 1.12$

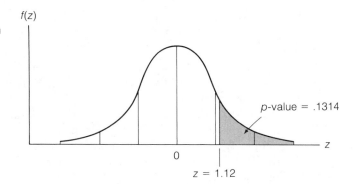

This value corresponds to the shaded area in the upper tail of the z distribution shown in Figure 9.15. The area A corresponding to $z = 1.12$, given in Table 4 of Appendix II, is .3686. Therefore, the observed significance level is

$$p\text{-value} = P(z \geq 1.12) = .5 - A = .5 - .3686 = .1314$$

This result indicates that the probability of observing a z value at least as contradictory to H_0 as the one observed in this test (if H_0 is in fact true) is .1314. We would therefore fail to reject H_0 for any preselected value of α less than .1314. ∎

EXAMPLE 9.17

Suppose that the test of Example 9.5 had been a two-tailed test, i.e., suppose that the alternative of interest had been H_a: $\mu \neq 72$. Find the observed significance level for the test.

SOLUTION

If the test were two-tailed, either very large or very small values of z would be contradictory to the null hypothesis H_0: $\mu = 72$. Consequently, values of $z \geq 1.12$ or $z \leq -1.12$ would be more contradictory to H_0 than the observed value of $z = 1.12$. Therefore, the observed significance level for the test (shaded in Figure 9.16) is

$$p\text{-value} = P(z \geq 1.12) + P(z \leq -1.12)$$
$$= 2(.1314) = .2628$$

FIGURE 9.16

Finding the p-Value for a Two-Tailed Test When $z = 1.12$

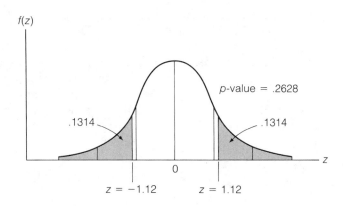

∎

EXAMPLE 9.18

SOLUTION

Find the observed significance level for the test described in Example 9.8.

The test of Example 9.8 was a lower-tailed test of H_0: $\mu = 20$ versus H_a: $\mu < 20$. Since the value of t computed from the sample data was $t = -3.06$, the observed significance level (or p-value) for the test is equal to the probability that t would assume a value less than or equal to -3.06, if in fact H_0 were true. This is equal to the area in the lower tail of the t distribution (shaded in Figure 9.17). To find this area, i.e., the p-value for the test, we consult the t table (Table 6 of Appendix II). Unlike the table of areas under the normal curve, Table 6 gives only the t values corresponding to the areas .100, .050, .025, .010, and .005. Therefore, we can only approximate the p-value for the test. Since the observed t value was based on 9 degrees of freedom, we use the df = 9 row in Table 6 and move across the row until we reach the t values that are closest to the observed $t = -3.06$. [*Note:* We ignore the minus sign.] The t values corresponding to p-values of .010 and .005 are 2.821 and 3.250, respectively. Since the observed t value falls between $t_{.010}$ and $t_{.005}$, the p-value for the test lies between .005 and .010. We could interpolate to more accurately locate the p-value for the test, but it is easier and adequate for our purposes to choose the larger area as the p-value and report it as .010. Thus, we would reject the null hypothesis, H_0: $\mu = 20$ parts per million, for any value of α larger than .01.

FIGURE 9.17
The Observed
Significance Level for the
Test of Example 9.18

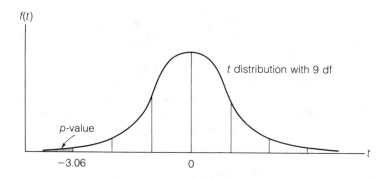

■

When publishing the results of a statistical test of hypothesis in journals, case studies, reports, etc., many researchers make use of p-values. Instead of selecting α a priori and then conducting a test as outlined in this chapter, the researcher may compute and report the value of the appropriate test statistic and its associated p-value. It is left to the reader of the report to judge the significance of the result, i.e., the reader must determine whether to reject the null hypothesis in favor of the alternative hypothesis, based on the reported p-value. Usually, the null hypothesis will be rejected if the observed significance level is *less* than the fixed significance level α chosen by the reader. There are two inherent advantages of reporting test results in this manner: (1) Readers are permitted to select the maximum value of α that they would be willing to tolerate if they actually carried out a standard test of hypothesis in the manner outlined in this chapter, and (2)

It is an easy way to present the results of test calculations performed by a computer. Some statistical computer programs perform the calculations for a test, give the observed value of the test statistic, and leave it to the reader to formulate a conclusion. Others give the observed significance level for the test, a procedure that makes it easy for the user to decide whether to reject the null hypothesis.

EXERCISES 9.50–9.53	**9.50** For a large-sample test of H_0: $\theta = \theta_0$ versus H_a: $\theta > \theta_0$, compute the p-value associated with each of the following test statistic values:

9.50 For a large-sample test of H_0: $\theta = \theta_0$ versus H_a: $\theta > \theta_0$, compute the p-value associated with each of the following test statistic values:
   a. $z = 1.96$ **b.** $z = 1.645$ **c.** $z = 2.67$ **d.** $z = 1.25$

9.51 For a large-sample test of H_0: $\theta = \theta_0$ versus H_a: $\theta \neq \theta_0$, compute the p-value associated with each of the following test statistic values:
   a. $z = -1.01$ **b.** $z = -2.37$ **c.** $z = 4.66$ **d.** $z = 1.45$

9.52 Compute and interpret the p-value for the test of Example 9.9.

9.53 Compute and interpret the p-values for the tests conducted in the following exercises:
   a. Exercise 9.13   **b.** Exercise 9.18   **c.** Exercise 9.20
   d. Exercise 9.28   **e.** Exercise 9.30   **f.** Exercise 9.39

SECTION 9.13

TESTS OF HYPOTHESES ABOUT MEANS USING THE COMPUTER (OPTIONAL)

In this section we present the computer commands for conducting tests of hypotheses concerning population means. Most statistical program packages can perform small-sample t tests about a single population mean, the difference between two population means for independent samples, and the difference between the population means for matched pairs. For each case, we discuss the SAS commands in detail, and then list the commands (if available) for SPSS^X, Minitab, and BMDP in table form. When the sample size(s) is (are) large, these tests still apply since the t distribution closely approximates the z distribution for large n.

STUDENT'S t TEST: SINGLE MEAN

Consider the problem of testing the hypothesis that the mean yearly salary of College of Engineering deans at southern universities exceeds \$50,000, i.e.,

H_0: $\mu = 50,000$

H_a: $\mu > 50,000$

In order to perform the test, a random sample of $n = 6$ salaries is selected. These salaries are given below:

   \$46,000   \$39,500   \$46,500   \$54,000   \$62,000   \$54,700

Since the sample size is small, the appropriate procedure to use is a Student's t test.

The commands for conducting this small-sample t test using the SAS System are given in Program 9.1.

PROGRAM 9.1

SAS Commands for Conducting the Student's t Test of H_0: $\mu = 50{,}000$ vs. H_a: $\mu > 50{,}000$

Command line

```
1   DATA DEANS;
2   INPUT SALARY @@;
3   NSALARY=SALARY-50000;
4   CARDS;
5   46000 39500 46500 54000 62000 54700
6   PROC MEANS T PRT;
7   VAR NSALARY;
```

Data entry instructions: Create the data set containing the sample of 6 salaries

Statistical analysis instructions: Conduct Student's t test

Command lines 1–5 The data set DEANS (arbitrarily named) includes six values of the variable SALARY and of a transformed variable called NSALARY, where NSALARY is computed by subtracting the hypothesized mean (50,000) from each value of SALARY. The reason for this transformation will be made clear in the following paragraph.

Command lines 6–7 The PROC MEANS statement (line 6) commands SAS to conduct a one-sample t test on the values of the variable specified in the VAR statement (line 7). SAS will test the null hypothesis

H_0: Mean of variable specified in VAR statement $= 0$

In this example, we specify the transformed variable NSALARY; thus, SAS conducts a test of

H_0: $\mu_{\text{NSALARY}} = 0$

Since $\mu_{\text{NSALARY}} = \mu_{\text{SALARY}} - 50{,}000$, the test is equivalent to

H_0: $\mu_{\text{SALARY}} = 50{,}000$

the desired test.

The output from the MEANS procedure is shown in Figure 9.18. Note that the value of the t statistic is reported ($t = .14$) as well as the observed significance level (p-value) for a two-tailed test (.8957). The reported p-value for a one-tailed test is p-value/2. For example, if we wished to test H_0: $\mu = 50{,}000$ against the one-tailed alternative H_a: $\mu > 50{,}000$, the observed significance level of the test would be p-value/2 = .8957/2 = .4478. In either case (one-tailed or two-tailed), there is insufficient evidence to reject the null hypothesis.

FIGURE 9.18

Output from the SAS Program 9.1

```
STATISTICAL ANALYSIS SYSTEM

VARIABLE      T        PR>|T|
NSALARY      0.14      0.8957
```

The appropriate commands for conducting a one-sample Student's t test using Minitab and BMDP are listed in Table 9.3. Their respective outputs are very similar to the SAS output in Figure 9.18. For example, each package will report the calculated value of the t statistic and the observed significance level of the test. A one-sample Student's t test in SPSS^X is not available.

TABLE 9.3   General Commands for Conducting a One-Sample Student's t Test of H_0: $\mu = \mu_0$ in SAS, BMDP, and Minitab

PACKAGE	COMMAND LINE	STATEMENTS/COMMANDS/FUNCTIONS	DESCRIPTION
a. SAS	1	`DATA HYPTEST;`	Y = Variable of interest (e.g., salary)
	2	`INPUT Y @@;`	
	3	`NEWY=Y-XXXXX;`	
	4	`CARDS;`	Select the hypothesized value of μ (μ_0) and substitute for XXXXX.
		input data values	
	5	`PROC MEANS T PRT;`	NEWY = Transformed variable $(Y - \mu_0)$ that is to be tested.
	6	`VAR NEWY;`	
b. BMDP	1	`BMDP 3D`	Y = Variable of interest (e.g., salary)
	2	`/ PROBLEM TITLE IS 'ONE SAMPLE T TEST'.`	
	3	`/ INPUT VARIABLES=1.`	
	4	`        FORMAT=STREAM.`	Select the hypothesized value of μ (μ_0) and substitute for XXXXX.
	5	`/ VARIABLE NAMES=Y.`	
	6	`        ADD=NEW.`	
	7	`/ TRANSFORM NEWY=Y-XXXXX;`	NEWY = Transformed variable $(Y - \mu_0)$ that is to be tested.
	8	`/ TEST VARIABLES=NEWY.`	
	9	`/ END`	
		input data values	
	10	`/END`	
c. Minitab	1	`SET DATA IN C1`	Select hypothesized value of μ (μ_0) and substitute the value for XXXXX.
		input data values	
	2	`TTEST OF MU=XXXXX, ALT.=0, ON DATA IN C1`	ALT.=0 implies a two-tailed test is to be performed (i.e., H_a: $\mu \neq \mu_0$).
			For a lower-tailed test H_a: $\mu < \mu_0$, use ALT.=-1.
			For an upper-tailed test H_a: $\mu > \mu_0$, use ALT.=1.

STUDENT'S t TEST: DIFFERENCE BETWEEN TWO MEANS, INDEPENDENT SAMPLES

Recall that a test of the equality of two population means H_0: $\mu_1 = \mu_2$ using small samples relies on the assumption that the variances of the two populations are equal (see Section 9.6). The t statistic is then computed using a pooled estimate of the variance. Three of the four statistical software packages discussed in this text—SPSS^X, Minitab, and BMDP—have specific commands for a "pooled" two-sample t test.

To conduct a two-sample t test in SAS, you must use the analysis of variance (ANOVA) procedure for an independent sampling design to be discussed in Chapter 13. The two-sample t test commands for SPSS^X, Minitab, and BMDP are listed in Table 9.4 (page 408). (Refer to Chapter 13 for the appropriate SAS statements.) The output for each of the computer packages includes the value of the t statistic and the two-tailed observed significance level (p-value) of the test.

TABLE 9.4 General Commands for Conducting a Two-Sample t Test of $H_0: (\mu_1 - \mu_2) = 0$ Based on Independent Samples; Minitab, BMDP, and SPSS^x

PACKAGE	COMMAND LINE	STATEMENTS/COMMANDS/FUNCTIONS	DESCRIPTION
a. Minitab	1	`SET DATA FOR FIRST SAMPLE IN C1`	ALT.=0 implies a two-tailed test is to be performed (i.e., $H_a: (\mu_1 - \mu_2) \neq 0$).
		input data values	
	2	`SET DATA FOR SECOND SAMPLE IN C2`	For a lower-tailed test H_a: $(\mu_1 - \mu_2) < 0$, use ALT.=−1.
		input data values	For an upper-tailed test H_a: $(\mu_1 - \mu_2) > 0$, use ALT.=1.
	3	`POOLED T, ALT.=0, ON DATA IN C1,C2`	
b. BMDP	1	`BMDP 3D`	Y = Variable of interest (e.g., salary)
	2	`/ PROBLEM TITLE IS 'TWO-SAMPLE T TEST'.`	
	3	`/ INPUT VARIABLES ARE 2,`	SAMPLE = Grouping variable name
	4	`         FORMAT IS FREE.`	= 1, if observation from first sample
	5	`/ VARIABLE NAMES ARE Y, SAMPLE.`	2, if observation from second sample
	6	`         GROUPING IS SAMPLE.`	
	7	`/ GROUP CODES(SAMPLE) ARE 1,2.`	
	8	`         NAMES(SAMPLE) ARE FIRST, SECOND.`	
	9	`/ END`	
		input data values	
	10	`/END`	
c. SPSS^x	1	`DATA LIST   FREE/Y, SAMPLE`	Y = Variable of interest (e.g., salary)
	2	`T-TEST     GROUPS=SAMPLE/VARIABLES=Y`	
	3	`BEGIN DATA`	SAMPLE = Grouping variable name
	4	input data values	= 1, if observation from first sample
		`END DATA`	2, if observation from second sample

STUDENT'S t TEST: DIFFERENCE BETWEEN POPULATION MEANS, MATCHED PAIRS

A test of the equality of two population means $H_0: (\mu_1 - \mu_2) = 0$ using a matched-pairs experiment requires that you first calculate the difference, d_i, between the measurements associated with each member of the matched pair (see Section 9.7). The t statistic is then computed using the mean and standard deviation of the sample differences. Since the mean of the differences, μ_d, is equivalent to $(\mu_1 - \mu_2)$, the analysis for a matched-pairs experiment is identical to the test for a single mean, $H_0: \mu_d = 0$.

The commands for conducting a test of hypothesis using a matched-pairs experiment in SAS, Minitab, and BMDP are almost identical to the commands for a Student's t test for a single mean given in Table 9.3. The only difference is that you will need to specify additional commands to calculate the differences, d_i, between the measurements in each pair. SPSS^x is the only one of the four packages that calculates the differences internally. The SAS, Minitab, BMDP, and

SPSS^X commands for conducting a Student's t test using matched pairs are listed in Table 9.5. The output for each of the computer packages includes the value of the t statistic and the two-tailed significance level (p-value) of the test.

TABLE 9.5 General Commands for Conducting a Student's t Test Using Matched Pairs: H_0: $\mu_d = 0$ (i.e., $\mu_1 - \mu_2 = 0$); SAS, BMDP, Minitab, SPSS^X

PACKAGE	COMMAND LINE	STATEMENTS/COMMANDS/FUNCTIONS	DESCRIPTION
a. SAS	1 2 3 4 5 6	DATA HYPTEST; INPUT Y1 Y2; D=Y1-Y2; CARDS; input data values PROC MEANS T PRT; VAR D;	Y1 and Y2 are the measure-ments for each member of the matched pair. D = Difference between the measurements in each pair

b. BMDP	1
	2
	3
	4
	5
	6
	7
	8
	9
	
	10
		BMDP 3D
		/ PROBLEM TITLE IS 'MATCHED PAIRS'.
		/ INPUT VARIABLES = 2.
		    FORMAT=FREE.
		/ VARIABLE NAMES=Y1,Y2,D.
		    ADD=NEW.
		/ TRANSFORM D=Y1-Y2.
		/ TEST VARIABLES=D.
		/ END
		
		    input data values
		
		/END
			Y1 and Y2 are the measurements for each member of the matched pair.
			
			D = Difference between the measurements in each pair

c. Minitab	1 2 3	SET DATA IN C1 C2 input data values SUBTRACT C2 FROM C1 PUT IN C3 TTEST OF MU=0, ALT.=0, ON DATA IN C3	Data in columns C1 and C2 correspond to the two measurements for each matched pair. Data in column C3 are the differences between the two measurements. ALT.=0 implies a two-tailed test, H_a: $\mu_d \neq 0$. For a lower-tailed test H_a: $\mu_d < 0$, use ALT.=−1. For an upper-tailed test H_a: $\mu_d > 0$, use ALT.=1.
d. SPSS^X	1 2 3 4	DATA LIST FREE/Y1,Y2 T-TEST PAIRS=Y1 Y2 BEGIN DATA input data values END DATA	Y1 and Y2 are the measure-ments for each member of the matched pairs.

SUMMARY

This chapter presents the basic concepts of a statistical **test of a hypothesis** about one or more population parameters. Tests of hypotheses are used when the ultimate practical objective of an inference is to reach a decision about the value(s) of the parameter(s). We can evaluate the goodness of the inference in terms of α and β, the probabilities of making incorrect decisions.

The close relationship between estimation and hypothesis testing is apparent when we compare the statistics employed for these two purposes. The statistics used to construct confidence intervals for parameters in Chapter 8 were then used to test hypotheses about the same parameters in Chapter 9. These tests are summarized in Tables 9.6(a) and 9.6(b).

In the following chapters, we will present some very useful methodology for analyzing multivariable experiments. As you will subsequently learn, the confidence intervals and tests that we will employ are based on an assumption of normality. Thus, the statistics that we will use to construct confidence intervals and test hypotheses possess sampling distributions that are the familiar t, χ^2, and F distributions of Chapters 7, 8, and 9.

TABLE 9.6(a) Summary of Hypothesis Tests: One-Sample Case

PARAMETER (θ)	NULL HYPOTHESIS (H_0)	POINT ESTIMATOR ($\hat{\theta}$)	TEST STATISTIC	SAMPLE SIZE	ADDITIONAL ASSUMPTIONS
μ	$\mu = \mu_0$	$\bar{y}$	$z = \dfrac{\bar{y} - \mu_0}{\sigma/\sqrt{n}} \approx \dfrac{\bar{y} - \mu_0}{s/\sqrt{n}}$	$n \geq 30$	None
			$t = \dfrac{\bar{y} - \mu_0}{s/\sqrt{n}}$ where t is based on $\nu = (n-1)$ degrees of freedom	$n < 30$	Normal population
p	$p = p_0$	$\hat{p} = \dfrac{y}{n}$	$z = \dfrac{\hat{p} - p_0}{\sqrt{\dfrac{p_0 q_0}{n}}}$	n large enough so that the interval $\hat{p} \pm 2\sqrt{\dfrac{\hat{p}\hat{q}}{n}}$ is between 0 and 1	None
σ^2	$\sigma^2 = \sigma_0^2$	s^2	$\chi^2 = \dfrac{(n-1)s^2}{\sigma_0^2}$ where χ^2 has a chi-square distribution with $\nu = (n-1)$ degrees of freedom	All n	Normal population

TABLE 9.6(b) Summary of Hypothesis Tests: Two-Sample Case

PARAMETER (θ)	NULL HYPOTHESIS (H_0)	POINT ESTIMATOR ($\hat{\theta}$)	TEST STATISTIC	SAMPLE SIZE	ADDITIONAL ASSUMPTIONS
$(\mu_1 - \mu_2)$ Independent samples	$(\mu_1 - \mu_2) = D_0$ (If we want to detect a difference between μ_1 and μ_2, then $D_0 = 0$.)	$(\bar{y}_1 - \bar{y}_2)$	$z = \dfrac{(\bar{y}_1 - \bar{y}_2) - D_0}{\sqrt{\dfrac{\sigma_1^2}{n_1} + \dfrac{\sigma_2^2}{n_2}}}$ $\approx \dfrac{(\bar{y}_1 - \bar{y}_2) - D_0}{\sqrt{\dfrac{s_1^2}{n_1} + \dfrac{s_2^2}{n_2}}}$	$n_1 \geq 30,\ n_2 \geq 30$	None
			$t = \dfrac{(\bar{y}_1 - \bar{y}_2) - D_0}{\sqrt{s_p^2\left(\dfrac{1}{n_1} + \dfrac{1}{n_2}\right)}}$ where t is based on $\nu = n_1 + n_2 - 2$ degrees of freedom and $s_p^2 = \dfrac{(n_1 - 1)s_1^2 + (n_2 - 1)s_2^2}{n_1 + n_2 - 2}$	Either $n_1 < 30$ or $n_2 < 30$ or both	Both populations normal with equal variances ($\sigma_1^2 = \sigma_2^2$) (For situations in which $\sigma_1^2 \neq \sigma_2^2$, see the modifications listed in the box on page 378).
$\mu_d = (\mu_1 - \mu_2)$ Matched pairs	$\mu_d = D_0$ (If we want to detect a difference between μ_1 and μ_2, then $D_0 = 0$.)	$\bar{d} = \sum_{i=1}^{n} d_i / n$ Mean of sample differences	$t = \dfrac{\bar{d} - D_0}{s_d / \sqrt{n_d}}$ where t is based on $\nu = (n_d - 1)$ degrees of freedom	All n_d (If $n_d \geq 30$, then the standard normal (z) test may be used.)	Population of differences d_i is normal
$(p_1 - p_2)$	$(p_1 - p_2) = D_0$ (If we want to detect a difference between p_1 and p_2, then $D_0 = 0$.)	$(\hat{p}_1 - \hat{p}_2)$	For $D_0 = 0$: $z = \dfrac{(\hat{p}_1 - \hat{p}_2)}{\sqrt{\hat{p}\hat{q}\left(\dfrac{1}{n_1} + \dfrac{1}{n_2}\right)}}$ where $\hat{p} = \dfrac{y_1 + y_2}{n_1 + n_2}$ For $D_0 \neq 0$: $z = \dfrac{(\hat{p}_1 - \hat{p}_2) - D_0}{\sqrt{\dfrac{\hat{p}_1\hat{q}_1}{n_1} + \dfrac{\hat{p}_2\hat{q}_2}{n_2}}}$	n_1 and n_2 large enough so that the intervals $\hat{p}_1 \pm 2\sqrt{\dfrac{\hat{p}_1\hat{q}_1}{n_1}}$ and $\hat{p}_2 \pm 2\sqrt{\dfrac{\hat{p}_2\hat{q}_2}{n_2}}$ do not contain 0 or 1	Independent samples
$\dfrac{\sigma_1^2}{\sigma_2^2}$	$\dfrac{\sigma_1^2}{\sigma_2^2} = 1$ (i.e., $\sigma_1^2 = \sigma_2^2$)	$\dfrac{s_1^2}{s_2^2}$	For $H_a: \sigma_1^2 > \sigma_2^2$: $F = \dfrac{s_1^2}{s_2^2}$ For $H_a: \sigma_2^2 > \sigma_1^2$: $F = \dfrac{s_2^2}{s_1^2}$ For $H_a: \sigma_1^2 \neq \sigma_2^2$: $F = \dfrac{\text{Larger } s^2}{\text{Smaller } s^2}$ where the distribution of F is based on $\nu_1 =$ numerator degrees of freedom and $\nu_2 =$ denominator degrees of freedom	All n_1 and n_2	Independent random samples from normal populations

9.54 Suppose you want to determine whether users of data processors have a preference between word processors A and B. If users have no preference for either of the two word processors (i.e., if the two systems are identical), then the probability p that a user prefers system A is $p = .5$. Let y be the number of users in a sample of ten who prefer system A, and suppose you want to test H_0: $p = .5$ against H_a: $p \neq .5$. One possible test procedure is to reject H_0 if $y \leq 1$ or $y \geq 8$.
a. Find α for this test.
b. Find β if $p = .4$. What is the power of the test?
c. Find β if $p = .8$. What is the power of the test?

In Exercises 9.55–9.59, formulate the appropriate null and alternative hypotheses.

9.55 A manufacturer of fishing line wants to show that the mean breaking strength of a competitor's 22-pound line is really less than 22 pounds.

9.56 A craps player who has experienced a long run of bad luck at the craps table wants to test whether the casino dice are "loaded," i.e., whether the proportion of "sevens" occurring in many tosses of the two dice is different from $\frac{1}{6}$ (if the dice are fair, the probability of tossing a "seven" is $\frac{1}{6}$).

9.57 Each year, *Computerworld* magazine reports the Datapro ratings of all computer software vendors. Vendors are rated on a scale from 1 to 4 (1 = poor, 4 = excellent) in such areas as reliability, efficiency, ease of installation, and ease of use by a random sample of software users. A software vendor wants to determine whether its product has a higher mean Datapro rating than a rival vendor's product.

9.58 Testing the thousands of new drug compounds for the few that might be effective is known in the pharmaceutical industry as *drug screening*. Abandoning a drug that is in fact useful is termed a *false negative*, while scheduling a drug for more extensive (and expensive) testing if it is in fact useless is termed a *false positive*. Suppose that a drug compound has been developed as a possible cure for cancer in mice and that the drug will be considered for further testing if more than 10% of all cancerous mice treated with the drug are cured. The goal of the researchers is to determine whether the drug effects a cure.

9.59 When it is operating correctly, a metal lathe produces machine bearings with a mean diameter of $\frac{1}{2}$ inch. Otherwise, the process is out of control. A quality control inspector wants to check whether the process is out of control.

9.60 The ion balance of our atmosphere has a significant effect on human health. A high concentration of positive ions in a room can induce fatigue, stress, and respiratory problems in the room's occupants. However, research has shown that introduction of additional negative ions into the room's atmosphere (through a negative ion generator), in combination with constant ventilation, restores the natural balance of ions that is conducive to human health. One experiment was conducted as follows. One hundred employees of a large factory were randomly selected and divided into two groups of 50 each. Both groups were told that they would be working in an atmosphere with an ion balance controlled through negative ion generators. However, unknown to the employees, the generators were switched on only in the experimental group's work area. At the end of the day, the number of employees reporting migraine, nausea, fatigue, faintness, or some other physical discomfort was recorded for each group. The results are summarized in the accompanying table.

	EXPERIMENTAL GROUP (ION GENERATORS ON)	CONTROL GROUP (ION GENERATORS OFF)
Number in sample	$n_1 = 50$	$n_2 = 50$
Number in sample who experience some type of physical discomfort	3	12

a. Perform a test of hypothesis to determine if the proportion of employees in the experimental group who experience some type of physical discomfort at the end of the day is significantly less than the corresponding proportion for the control group. Use a significance level of $\alpha = .03$.

b. Compute the p-value for this test.

9.61 The quality control department of a paper company measures the brightness (a measure of reflectance) of finished paper on a periodic basis throughout the day. Two instruments that are available to measure the paper specimens are subject to error, but they can be adjusted so that the mean readings for a control paper specimen are the same for both instruments. Suppose you are concerned about the precision of the two instruments—namely, that instrument 2 is less precise than instrument 1. To check this theory, five measurements of a single paper sample are made on both instruments. The data are shown in the table. Do the data provide sufficient evidence to indicate that instrument 2 is less precise than instrument 1? Test using $\alpha = .05$.

INSTRUMENT 1	INSTRUMENT 2
29	26
28	34
30	30
28	32
30	28

9.62 The testing department of a tire and rubber company schedules truck and passenger tires for durability tests. Currently, tires are scheduled twice weekly on flexible processors (machines that can handle either truck or passenger tires) using the shortest processing time (SPT) approach. Under SPT, the tire with the shortest processing time is scheduled first. Company researchers have developed a new scheduling rule which they believe will reduce the average flow time (i.e., the average completion time of a test) and lead to a reduction in the average tardiness of a scheduled test. In order to compare the two scheduling rules, 64 tires were randomly selected and divided into two groups of equal size. One set of tires was scheduled using SPT, the other using the proposed rule. A summary of the flow times and tardiness (in hours) of the tire tests is provided in the table on page 414.

	FLOW TIME		TARDINESS	
	Mean	Variance	Mean	Variance
SPT	158.28	8,532.80	5.26	452.09
Proposed rule	117.07	5,208.53	4.52	319.41

a. Is there sufficient evidence at $\alpha = .05$ to conclude that the average flow time is less under the proposed scheduling rule than under the SPT approach?

b. Is there sufficient evidence at $\alpha = .05$ to conclude that the proposed scheduling rule will lead to a reduction in the average tardiness of tire tests?

9.63 Refer to the reinforced concrete T-beam cracking experiment described in 9.44. The experimental results were compared to the theoretical results obtained using the failure surface method of predicting ultimate load capacity. The actual and theoretical ultimate torsion moments for six T-beams with 40-cm slab widths are given in the table. Conduct a test to determine whether the experimental mean ultimate torsion moment differs from the theoretical mean ultimate torsion moment. Use $\alpha = .05$.

T-BEAM	1	2	3	4	5	6
Experimental result	4.70	5.20	5.40	5.40	4.30	4.80
Theoretical result	4.63	4.65	5.60	5.60	3.62	3.62

Source: Zararis, P. D. and Penelis, G. Jr. "Reinforced Concrete T-Beams in Torsion and Bending." *Journal of the American Concrete Institute*, Vol. 83, No. 1, Jan.–Feb. 1986, p. 153.

9.64 A problem that occurs with certain types of mining is that some byproducts tend to be mildly radioactive and these products sometimes get into our freshwater supply. The EPA has issued regulations concerning a limit on the amount of radioactivity in supplies of drinking water. Particularly, the maximum level for naturally occurring radiation is 5 picocuries per liter of water. A random sample of 24 water specimens from a city's water supply produced the sample statistics $\bar{y} = 4.61$ picocuries per liter and $s = .87$ picocurie per liter.

a. Do these data provide sufficient evidence to indicate that the mean level of radiation is safe (below the maximum level set by the EPA)? Test using $\alpha = .01$.

b. Why should you want to use a small value of α for the test in part a?

c. Calculate the value of β for the test if $\mu_a = 4.5$ picocuries per liter of water.

d. Calculate and interpret the p-value for the test.

9.65 Usually, when trees grown in greenhouses are replanted in their natural habitat, there is only a 50% survival rate. However, a recent General Telephone and Electronics (GTE) advertisement claimed that trees grown in a particular environment ideal for plant growth have a 95% survival rate when replanted. These trees are grown inside a mountain in Idaho where the air temperature, carbon dioxide content, and humidity are all constant, and there are no major disease or insect problems. A key growth ingredient—light—is supplied by specially made GTE Sylvania Super-Metalarc lamps. These lights help the young trees develop a more fibrous root system that aids in the transplantation. Suppose that we want to challenge GTE's claim, i.e., we want to test whether the true proportion of all trees grown inside the Idaho mountain that survive when replanted in their natural habitat is less than .95. We randomly sample 50 of the trees grown in the controlled environment, replant the trees in their natural habitat, and observe that 46 of the trees survive. Perform the test at a level of significance of $\alpha = .01$.

9.66 A *parallel processor*, or *paracomputer*, consists of autonomous processing elements (PEs) sharing a central memory. Researchers at New York University have recently designed such a paracomputer, called the NYU Ultracomputer. To assess the impact of network delay on overall ultracomputer performance, the researchers simulated central memory access time for sample instructions from a parallel version of a NASA weather program. Two sets of access times were simulated—one set processed with 16 processing elements, the other set with 48 processing elements. With 16 PEs, the average central memory access time was 8.94 seconds, while with 48 PEs the average central memory access time was 8.83 seconds. Assume that $n = 1,000$ instructions were simulated for each of the two programs, with standard deviations equal to 3.10 and 3.50, respectively. This information was not provided in the researchers' report. Is there sufficient evidence to indicate a difference between the average central memory access times of instructions processed with 16 and 48 PEs? Test using $\alpha = .05$.

9.67 It is essential in the manufacture of machinery to utilize parts that conform to specifications. In the past, diameters of the ball bearings produced by a certain manufacturer had a variance of .00156. To cut costs, the manufacturer instituted a less expensive production method. The variance of the diameters of 100 randomly sampled bearings produced by the new process was .00211. Do the data provide sufficient evidence to indicate that diameters of ball bearings produced by the new process are more variable than those produced by the old process? Test at $\alpha = .05$.

9.68 The Federal Trade Commission (FTC) conducts periodic tests on the tar and nicotine content and carbon monoxide emission of each brand of cigarette sold in the United States. Suppose an investigator for the FTC suspects there is a significant difference between the mean amounts of carbon monoxide in smoke emitted from Marlboro and Kool brand cigarettes. To test his theory, he selects independent random samples of cigarettes of each brand, determines the amount (in milligrams) of carbon monoxide in the smoke emitted from each, and summarizes his results as shown in the accompanying table. Test the investigator's belief, using significance level $\alpha = .01$.

MARLBORO	KOOL
$n_1 = 30$	$n_2 = 40$
$\bar{y}_1 = 16.4$	$\bar{y}_2 = 15.5$
$s_1 = 1.2$	$s_2 = 1.1$

9.69 The use of computer equipment in business is growing at a phenomenal rate. A recent study revealed that 184 of 616 working adults now regularly use a personal computer, microcomputer, computer terminal, or word processor on the job (*Journal of Advertising Research*, Apr./May 1984). Is this sufficient evidence to indicate that the proportion of all working adults who regularly use computer equipment on the job exceeds 25%? Test using $\alpha = .05$.

9.70 The means and standard deviations shown in the table summarize information on the strengths (modules of rupture at ground line, in pounds per square inch) for two types of wooden poles used by the utility industry. Do the data provide sufficient evidence to indicate a difference in the mean strengths of wooden poles made from coastal Douglas fir and southern pine? Test using $\alpha = .01$.

SPECIES	SAMPLE SIZE	SAMPLE MEAN	SAMPLE STANDARD DEVIATION
Coastal Douglas fir	118	8,380	644.62
Southern pine	147	8,870	611.72

Source: Goodman, J. R., Vanderbilt, M. D., and Criswell, M. E. "Reliability-Based Design of Wood Transmission Line Structures," *Journal of Structural Engineering*, Vol. 109, No. 3, 1983, pp. 690–704.

9.71 The accompanying table provides data on the theoretical (calculated) and experimental values of the vapor pressures for dibenzothiophene, a heterocycloaromatic compound similar to those found in coal tar. If the theoretical model for computer vapor pressure is a good model of reality, the true mean difference between the experimental and calculated values of vapor pressure for a given temperature will equal 0.

TEMPERATURE (°C)	VAPOR PRESSURE (TORR)		TEMPERATURE (°C)	VAPOR PRESSURE (TORR)	
	Experimental	Calculated		Experimental	Calculated
100.60	.282	.276	116.69	.669	.695
101.36	.314	.307	119.38	.834	.805
104.60	.335	.350	121.08	.890	.882
106.44	.404	.390	123.61	1.01	1.01
108.70	.422	.444	124.90	1.07	1.08
110.96	.513	.505	127.74	1.26	1.25
112.62	.554	.554	130.24	1.42	1.43
115.21	.642	.640	131.75	1.55	1.54

Source: Edwards, D. R. and Prausnitz, J. M. "Vapor Pressures of Some Sulphur-Containing, Coal-Related Compounds." *Journal of Chemical and Engineering Data*, Vol. 26, 1981, pp. 121–124. Copyright 1981 American Chemical Society. Reprinted with permission.

a. Do the data provide sufficient evidence to indicate that the mean difference differs from 0? Test using $\alpha = .05$.

b. Calculate and interpret the *p*-value for the test.

9.72 Data-processing firms that are seeking to replace their minicomputer or microprocessor tend to be loyal to their present vendor, i.e., it is very likely they will purchase a new minicomputer from the vendor that sold them the one they are currently using. Thus, one objective of minicomputer vendors is to identify those firms where a vendor change is being considered. Suppose it is decided to compare the proportions of small businesses and large businesses in the replacement market that are considering a change in vendor. Independent random samples of 50 small businesses and 60 large businesses, all in the replacement market, were selected and the number considering a change in vendor was recorded for each group. The results are given in the table.

Is there evidence that the proportions of firms considering a change in vendor differ for the two groups? Test at $\alpha = .10$.

	SMALL BUSINESSES	LARGE BUSINESSES
Number sampled	50	60
Number considering a change in vendor	6	11

9.73 Two relatively new energy-saving concepts in home building are solar-powered homes and earth-sheltered homes. An individual is drawing up plans for a new home and wants to compare expected annual heating costs for the two types of innovation. Independent random samples of solar-powered homes (which receive 50% of their energy from the sun) and earth-sheltered homes yielded the accompanying summary data on annual heating costs. (You may assume the homes were comparable with respect to size, climatic conditions, etc.) Is there evidence (at $\alpha = .05$) that the variability in the annual costs of heating earth-sheltered homes is significantly less than the variability in the annual costs of heating 50% solar-powered homes?

SOLAR-POWERED	EARTH-SHELTERED
$n_1 = 12$	$n_2 = 6$
$\bar{y}_1 = \$285$	$\bar{y}_2 = \$234$
$s_1 = \$55$	$s_2 = \$26$

9.74 Recently, a new miniaturized component for use in microcomputers has been developed. These components greatly increase the speed of the microcomputers, but may not last long enough to be practical. It is known that the mean lifetime of the larger component used in a microcomputer manufactured by a successful firm is 500 hours. However, a random sample of 18 newly built miniaturized components yielded the following summary statistics on lifetime:

$$\bar{y} = 482 \text{ hours} \qquad s = 59 \text{ hours}$$

Is there evidence to indicate that the mean lifetime of the new miniaturized component is significantly less than the average of 500 hours for the larger components? Test using $\alpha = .01$. What assumption is required for the hypothesis test to be valid?

9.75 Heat stress in dairy cows can have a dramatic negative effect on milk production. High temperatures tend to reduce a cow's food intake, which in turn reduces milk yield. Researchers in the IFAS Dairy Research Unit and the Department of Agricultural Engineering at the University of Florida have developed design criteria for the construction of shade structures which they believe will help alleviate heat stress for dairy cows. In one experiment, 31 Holstein cows in the last trimester of pregnancy were divided into two groups. Sixteen cows were given access to a shade structure and the remaining 15 cows were denied shade. Researchers recorded the 100-day milk yield (in pounds) of each cow after calving. The mean milk yields of the two groups are shown in the table (page 418). Is there sufficient evidence to indicate a difference between the mean milk yields of cows given access to shade and cows denied shade? Use $\alpha = .10$. (Assume the standard deviations of milk yields are equal to 40 pounds for both groups.)

	SHADE	NO SHADE
Sample size	16	15
Mean	367.4	330.8

Source: "Minimizing Heat Stress for Dairy Cows." *Florida Agricultural Research 83*, Vol. 2, No. 1, Winter 1983, pp. 10–13.

REFERENCES

BMDP User's Digest, 2nd ed. MaryAnn Hill, ed. Los Angeles: BMDP Statistical Software, 1982.

BMDPC: User's Guide to BMDP on the IBM PC. BMDP Statistical Software, Inc., Los Angeles, CA 90025.

Dixon, W. J., Brown, M. B., Engelman, L., Frane, J. W., Hill, M. A., Jennrich, R. I., and Toporek, J. D. *BMDP Statistical Software*, 1985 ed. Berkeley: University of California Press.

Freedman, D., Pisani, R., and Purves, R. *Statistics*. New York: W. W. Norton and Co., 1978.

Hoel, P. G. *Introduction to Mathematical Statistics*, 4th ed. New York: Wiley, 1971.

Hogg, R. V. and Craig, A. T. *Introduction to Mathematical Statistics*, 4th ed. New York: Macmillan, 1978.

McClave, J. T. and Dietrich, F. H. II. *Statistics*, 3rd ed. San Francisco: Dellen, 1985.

Mendenhall, W. *Introduction to Probability and Statistics*, 7th ed. Boston: Duxbury, 1987.

Mendenhall, W., Scheaffer, R. L., and Wackerly, D. D. *Mathematical Statistics with Applications*, 2nd ed. Boston: Duxbury, 1981.

Mood, A. M., Graybill, F. A., and Boes, D. *Introduction to the Theory of Statistics*, 3rd ed. New York: McGraw-Hill, 1974.

Norusis, M. J. *SPSS/PC+: SPSS for the IBM PC/XT/AT*, 1986 ed. SPSS, Inc., Suite 3000, 444 N. Michigan Avenue, Chicago, Ill. 60611.

Norusis, M. J. *The SPSS Guide to Data Analysis*, 1986 ed. SPSS, Inc., Suite 3000, 444 N. Michigan Avenue, Chicago, Ill. 60611.

Ryan, T. A., Joiner, B. L., and Ryan, B. F. *Minitab Reference Manual*. Minitab Project, University Park, Pa., 1985.

Ryan, T. A., Joiner, B. L., and Ryan, B. F. *Minitab Student Handbook*, 2nd ed. Boston: Duxbury, 1985.

SAS Procedures Guide for Personal Computers, Version 6 ed., 1986. SAS Institute, Inc., Box 8000, Cary, N.C. 27511.

SAS User's Guide: Basics, Version 5 ed., 1985. SAS Institute, Inc., Box 8000, Cary, N.C. 27511.

SAS User's Guide: Statistics, Version 5 ed., 1985. SAS Institute, Inc., Box 8000, Cary, N.C. 27511.

Snedecor, G. W. and Cochran, W. G. *Statistical Methods*, 7th ed. Ames, Iowa: Iowa State University Press, 1980.

SPSS^X User's Guide, 1983 ed. SPSS, Inc., Suite 3000, 444 N. Michigan Ave., Chicago, Ill. 60611.

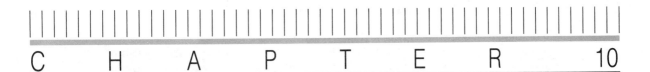

C H A P T E R 10

OBJECTIVE

To present the basic concepts of regression analysis based on a simple linear relation between a response y and a single predictor variable x.

CONTENTS

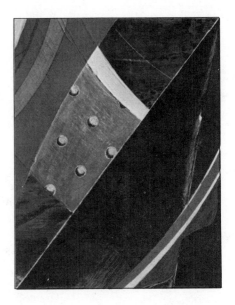

SIMPLE LINEAR REGRESSION ANALYSIS

INTRODUCTION

One of the most important applications of statistics involves estimating the mean value of a response variable y or predicting some future value of y based on knowledge of a set of related independent* variables, $x_1, x_2, \ldots, x_k$.

For example, the manager of a data-processing center might want to relate the waiting time y (the **dependent variable**) between the time a job is submitted to a computer and the time it is completed to such variables as the number and sizes of the jobs already awaiting execution and the size of the job being submitted (the **independent variables**). The objective would be to develop a **prediction equation** (or **model**) that expresses y as a function of the independent variables. This would enable the manager to predict y for specific values of the independent variables and, ultimately, to use knowledge derived from a study of the prediction equation to institute policies to control the waiting time.

As another example, an engineer might want to relate the rate of malfunction y of a mechanical assembler to such variables as its speed of operation and the assembler operator. The objective would be to develop a prediction equation relating the dependent variable y to the independent variables and to use the prediction equation to predict the value of the rate of malfunction y for various combinations of speed of operation and operator.

The models used to relate a dependent variable y to the independent variables $x_1, x_2, \ldots, x_k$, are called **regression models** or **linear statistical models** because they express the mean value of y for given values of $x_1, x_2, \ldots, x_k$, as a linear function of a set of unknown parameters. These parameters are estimated from sample data using a process to be explained in Section 10.3.

The concepts of a regression analysis are introduced in this chapter using a very simple regression model—one that relates y to a single independent variable x. We will learn how to fit this model to a set of data using the **method of least squares** and will examine in detail the different types of inferences that can result from a regression analysis. In Chapters 11–12, we will apply this knowledge to help us understand the theoretical and practical implications of a **multiple regression analysis**—the problem of relating y to two or more independent variables.

The preceding chapters provide a foundation for a study of applied statistics. Although the derivations of the sampling distributions for many of the statistics that we will subsequently encounter are mathematically beyond the scope of this text, the knowledge that you have acquired will help you to understand how they are derived and, in many instances, will enable you to find the means and variances of these sampling distributions.

Your theoretical knowledge will be useful for another reason. The theory of statistics, like the theories of physics, engineering, economics, etc., is only a model for reality. It exactly explains reality only when the assumptions of the methodology are exactly satisfied. Since this situation rarely occurs, the application of statistics (or physics, engineering, economics, etc.) to the solution of real-world problems is an art. Thus, in order to apply theory to the real world, you must know the extent to which deviations from assumptions will affect the

---

*When the word *independent* is applied to the variances $x_1, x_2, \ldots, x_k$, we mean independent in an algebraic rather than probabilistic sense.

resulting statistical inferences and you must be able to adapt the model and methodology to the conditions of a practical problem. A basic understanding of the theory underlying the methodology will help you to do this.

| | | | | | | | | | | | | |

SECTION 10.2

A SIMPLE LINEAR REGRESSION MODEL: ASSUMPTIONS

In order to simplify our discussion, we will postulate the following fictitious situation and data set. Suppose that the developer of a new insulation material wants to determine the amount of compression that would be produced on a 2-inch thick specimen of the material when subjected to different amounts of pressure. Five experimental pieces of the material were tested under different pressures. The values of x (in units of 10 pounds per square inch) and the resulting amounts of compression y (in units of .1 inch) are given in Table 10.1. A plot of the data, called a **scattergram**, is shown in Figure 10.1.

TABLE 10.1

Compression Versus Pressure for an Insulation Material

SPECIMEN	PRESSURE x	COMPRESSION y
1	1	1
2	2	1
3	3	2
4	4	2
5	5	4

FIGURE 10.1

Scattergram for the Data in Table 10.1

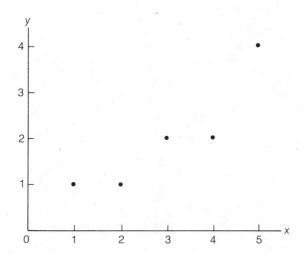

Suppose we believe that the value of y tends to increase in a linear manner as x increases. Then we could select a model relating y to x by drawing a line through the points in Figure 10.1. Such a **deterministic model**—one that does not allow for errors of prediction—might be adequate if all of the points of Figure 10.1 fell on the fitted line. However, you can see that this idealistic situation will not occur for the data of Table 10.1. No matter how you draw a line through the points of Figure 10.1, at least some of the points will deviate substantially from the fitted line.

The solution to the preceding problem is to construct a **probabilistic model** relating y to x—one that acknowledges the random variation of the data points about a line. One type of probabilistic model, a **simple linear regression model**, makes the assumption that the mean value of y for a given value of x graphs as a straight line and that points deviate about this **line of means** by a random (positive or negative) amount equal to ε, i.e.,

$$ y = \underbrace{\beta_0 + \beta_1 x}_{\substack{\text{Mean value of } y \\ \text{for a given } x}} + \underbrace{\varepsilon}_{\substack{\text{Random} \\ \text{error}}} $$

where β_0 and β_1 are unknown parameters of the deterministic (nonrandom) portion of the model. If we assume that the points deviate above and below the line of means, with some deviations positive, some negative, and with $E(\varepsilon) = 0$, then the mean value of y is

$$ E(y) = E(\beta_0 + \beta_1 x + \varepsilon) = \beta_0 + \beta_1 x + E(\varepsilon) = \beta_0 + \beta_1 x $$

Therefore, the mean value of y for a given value of x, represented by the symbol $E(y)$,* graphs as a straight line with y-intercept equal to β_0 and slope equal to β_1. A graph of the hypothetical line of means, $E(y) = \beta_0 + \beta_1 x$, is shown in Figure 10.2.

FIGURE 10.2

A Graph of the Data Points of Table 10.1 and the Hypothetical Line of Means, $E(y) = \beta_0 + \beta_1 x$

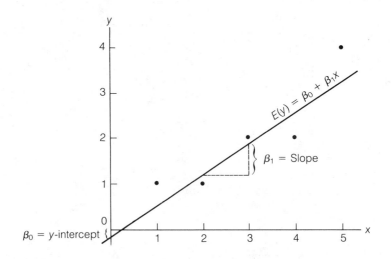

*The mean value of y for a given value of x should be denoted by the symbol $E(y \mid x)$. However, this notation becomes cumbersome when the model contains more than one independent variable. Consequently, we will abbreviate the notation and represent $E(y \mid x)$ by the symbol $E(y)$.

A SIMPLE LINEAR REGRESSION MODEL

$$y = \beta_0 + \beta_1 x + \varepsilon$$

where

$y = $ **Dependent variable** (variable to be modeled— sometimes called the **response variable**)

$x = $ **Independent variable** (variable used as a **predictor** of y)

ε (epsilon) $= $ Random error component

β_0 (beta-zero) $= $ y-intercept of the line, i.e., point at which the line intercepts or cuts through the y-axis (see Figure 10.2)

β_1 (beta-one) $= $ Slope of the line, i.e., amount of increase (or decrease) in the deterministic component of y for every 1-unit increase in x (see Figure 10.2)

In order to fit a simple linear regression model to a set of data, we must find estimators for the unknown parameters, β_0 and β_1, of the line of means, $E(y) = \beta_0 + \beta_1 x$. Since the sampling distributions of these estimators will depend on the probability distribution of the random error ε, we must first make specific assumptions about its properties. These assumptions, summarized below, are basic to every statistical regression analysis.

ASSUMPTION 1   The mean of the probability distribution of ε is 0. That is, the average of the errors over an infinitely long series of experiments is 0 for each setting of the independent variable x. This assumption implies that the mean value of y, $E(y)$, for a given value of x is $E(y) = \beta_0 + \beta_1 x$.

ASSUMPTION 2   The variance of the probability distribution of ε is constant for all settings of the independent variable x. For our straight-line model, this assumption means that the variance of ε is equal to a constant, say σ^2, for all values of x.

ASSUMPTION 3   The probability distribution of ε is normal.

ASSUMPTION 4   The errors associated with any two different observations are independent. That is, the error associated with one value of y has no effect on the errors associated with other y values.

The implications of the first three assumptions can be seen in Figure 10.3 (page 424), which shows distributions of errors for three particular values of x, namely x_1, x_2, and x_3. Note that the relative frequency distributions of the errors are normal, with a mean of 0, and a constant variance σ^2 (all the distributions shown have the same amount of spread or variability). The straight line shown in Figure 10.3 is the mean value of y for a given value of x, $E(y) = \beta_0 + \beta_1 x$.

FIGURE 10.3

The Probability
Distribution of ε

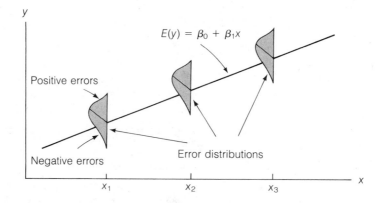

Various techniques exist for checking the validity of these assumptions, and there are remedies to be applied when they appear to be invalid. In actual practice, the assumptions need not hold exactly in order for least squares estimators and test statistics (to be described subsequently) to possess the measures of reliability that we would expect from a regression analysis. The assumptions will be satisfied adequately for many practical applications.

SECTION 10.3

ESTIMATING β_0 AND β_1: THE METHOD OF LEAST SQUARES

In order to choose the "best-fitting" line for a set of data, we must estimate the unknown parameters, β_0 and β_1, of the simple linear regression model. These estimators could be found using the method of maximum likelihood (Section 8.3), but the easiest method—and one that is intuitively appealing—is the **method of least squares**. When the assumptions of Section 10.2 are satisfied, then the maximum likelihood and the least squares estimators of β_0 and β_1 are identical.

The reasoning behind the method of least squares can be seen by examining Figure 10.4, which shows a line drawn on the scattergram of the data points of Table 10.1. The vertical line segments represent **deviations** of the points from the line. You can see by shifting a ruler around the graph that it is possible to find many lines for which the sum of deviations (or **errors**) is equal to 0, but it can be shown that there is one and only one line for which the **sum of squares of the deviations** is a minimum. The sum of squares of the deviations is called the **sum of squares for error** and is denoted by the symbol SSE. The line is called the **least squares line**, the **regression line**, or **least squares prediction equation**.

To find the least squares line for a set of data, assume that we have a sample of n data points which can be identified by corresponding values of x and y, say (x_1, y_1), (x_2, y_2), . . . , (x_n, y_n). For example, the $n = 5$ data points shown in Table 10.1 are (1, 1), (2, 1), (3, 2), (4, 2), and (5, 4). The straight-line model for the response y in terms of x is

$$y = \beta_0 + \beta_1 x + \varepsilon$$

FIGURE 10.4

Graph Showing the
Deviations of the Points
About a Line

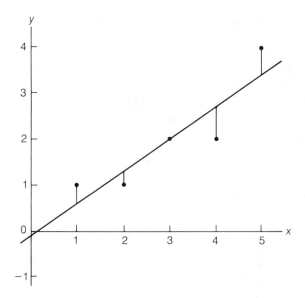

The line of means is $E(y) = \beta_0 + \beta_1 x$ and the fitted line, which we hope to find, is represented as $\hat{y} = \hat{\beta}_0 + \hat{\beta}_1 x$. Thus, $\hat{y}$ is an estimator of the mean value of y, $E(y)$, and a predictor of some future value of y; and $\hat{\beta}_0$ and $\hat{\beta}_1$ are estimators of β_0 and β_1, respectively.

For a given data point, say the point (x_i, y_i), the observed value of y is y_i and the predicted value of y would be obtained by substituting x_i into the prediction equation:

$$\hat{y}_i = \hat{\beta}_0 + \hat{\beta}_1 x_i$$

And the deviation of the ith value of y from its predicted value is

$$(y_i - \hat{y}_i) = [y_i - (\hat{\beta}_0 + \hat{\beta}_1 x_i)]$$

Then the sum of squares of the deviations of the y values about their predicted values for all of the n data points is

$$\text{SSE} = \sum_{i=1}^{n} [y_i - (\hat{\beta}_0 + \hat{\beta}_1 x_i)]^2$$

The quantities $\hat{\beta}_0$ and $\hat{\beta}_1$ that make the SSE a minimum are called the **least squares estimates** of the population parameters β_0 and β_1, and the prediction equation $\hat{y} = \hat{\beta}_0 + \hat{\beta}_1 x$ is called the **least squares line**.

DEFINITION 10.1

The **least squares line** is one that has a smaller SSE than any other straight-line model.

The values of $\hat{\beta}_0$ and $\hat{\beta}_1$ that minimize

$$\text{SSE} = \sum_{i=1}^{n} [y_i - (\hat{\beta}_0 + \hat{\beta}_1 x_i)]^2$$

are obtained by setting the two partial derivatives, $\partial \text{SSE}/\partial \hat{\beta}_0$ and $\partial \text{SSE}/\partial \hat{\beta}_1$, equal to 0 and solving the resulting simultaneous linear system of **least squares equations**. To illustrate, we first compute the partial derivatives:

$$\frac{\partial \text{SSE}}{\partial \hat{\beta}_0} = \sum_{i=1}^{n} 2[y_i - (\hat{\beta}_0 + \hat{\beta}_1 x_i)](-1)$$

$$\frac{\partial \text{SSE}}{\partial \hat{\beta}_1} = \sum_{i=1}^{n} 2[y_i - (\hat{\beta}_0 + \hat{\beta}_1 x_i)](-x_i)$$

Setting these partial derivatives equal to 0 and simplifying, we obtain the least squares equations:

$$\sum_{i=1}^{n} y_i - \sum_{i=1}^{n} \hat{\beta}_0 - \hat{\beta}_1 \sum_{i=1}^{n} x_i = \sum_{i=1}^{n} y_i - n\hat{\beta}_0 - \hat{\beta}_1 \sum_{i=1}^{n} x_i = 0$$

$$\sum_{i=1}^{n} x_i y_i - \hat{\beta}_0 \sum_{i=1}^{n} x_i - \hat{\beta}_1 \sum_{i=1}^{n} x_i^2 = 0$$

or

$$n\hat{\beta}_0 + \hat{\beta}_1 \sum_{i=1}^{n} x_i = \sum_{i=1}^{n} y_i$$

$$\hat{\beta}_0 \sum_{i=1}^{n} x_i + \hat{\beta}_1 \sum_{i=1}^{n} x_i^2 = \sum_{i=1}^{n} x_i y_i$$

Solving this pair of simultaneous linear equations for β_0 and β_1, we obtain (proof omitted) the formulas shown in the box.

FORMULAS FOR THE LEAST SQUARES ESTIMATES

Slope: $\quad \hat{\beta}_1 = \dfrac{\text{SS}_{xy}}{\text{SS}_{xx}}$

y-intercept: $\quad \hat{\beta}_0 = \bar{y} - \hat{\beta}_1 \bar{x}$

where

$$\text{SS}_{xy} = \sum_{i=1}^{n} (x_i - \bar{x})(y_i - \bar{y}) = \sum_{i=1}^{n} x_i y_i - \frac{\left(\sum_{i=1}^{n} x_i\right)\left(\sum_{i=1}^{n} y_i\right)}{n}$$

$$\text{SS}_{xx} = \sum_{i=1}^{n} (x_i - \bar{x})^2 = \sum_{i=1}^{n} x_i^2 - \frac{\left(\sum_{i=1}^{n} x_i\right)^2}{n}$$

n = Sample size

EXAMPLE 10.1

Calculate the least squares estimates of β_0 and β_1 for the data of Table 10.1.

SOLUTION

Preliminary computations for finding the least squares line for the insulation compression data are contained in Table 10.2. We can now calculate*

$$SS_{xy} = \sum x_i y_i - \frac{\left(\sum x_i\right)\left(\sum y_i\right)}{5} = 37 - \frac{(15)(10)}{5}$$

$$= 37 - 30 = 7$$

TABLE 10.2
Preliminary Computations for the Insulation Compression Example

x_i	y_i	x_i^2	$x_i y_i$
1	1	1	1
2	1	4	2
3	2	9	6
4	2	16	8
5	4	25	20
Totals $\sum x_i = 15$	$\sum y_i = 10$	$\sum x_i^2 = 55$	$\sum x_i y_i = 37$

$$SS_{xx} = \sum x_i^2 - \frac{\left(\sum x_i\right)^2}{5} = 55 - \frac{(15)^2}{5}$$

$$= 55 - 45 = 10$$

Then, the slope of the least squares line is

$$\hat{\beta}_1 = \frac{SS_{xy}}{SS_{xx}} = \frac{7}{10} = .7$$

and the y-intercept is

$$\hat{\beta}_0 = \bar{y} - \hat{\beta}_1 \bar{x} = \frac{\sum y_i}{5} - \hat{\beta}_1 \frac{\left(\sum x_i\right)}{5}$$

$$= \frac{10}{5} - (.7)\frac{(15)}{5}$$

$$= 2 - (.7)(3) = 2 - 2.1 = -.1$$

The least squares line is thus

$$\hat{y} = \hat{\beta}_0 + \hat{\beta}_1 x = -.1 + .7x$$

*Since summations will be used extensively from this point on, we will omit the limits on Σ when the summation includes all the measurements in the sample, i.e., when the symbol is $\Sigma_{i=1}^n$, we will write Σ.

The graph of this line is shown in Figure 10.5.

FIGURE 10.5
The Line $\hat{y} = -.1 + .7x$
Fit to the Data

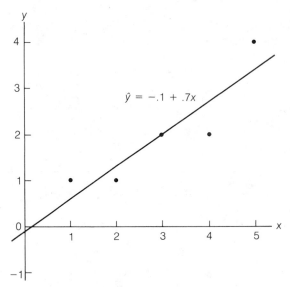

The observed and predicted values of y, the deviations of the y values about their predicted values, and the squares of these deviations for the data of Table 10.1 are shown in Table 10.3. Note that the sum of squares of the deviations, SSE, is 1.10. This is smaller than the value of SSE that would be obtained by fitting any other possible straight line to the data.

TABLE 10.3

Comparing Observed and Predicted Values for the Least Squares Model

x	y	$\hat{y} = -.1 + .7x$	$(y - \hat{y})$	$(y - \hat{y})^2$
1	1	.6	$(1 - .6) = .4$	.16
2	1	1.3	$(1 - 1.3) = -.3$	.09
3	2	2.0	$(2 - 2.0) = 0$	.00
4	2	2.7	$(2 - 2.7) = -.7$	.49
5	4	3.4	$(4 - 3.4) = .6$	.36
			Sum of errors = 0	SSE = 1.10

To summarize, we have defined the best-fitting straight line to be the one that satisfies the least squares criterion; that is, the sum of the squared errors will be smaller than for any other straight-line model. This line is called the **least squares line**, and its equation is called the **least squares prediction equation**.

EXERCISES 10.1–10.13

10.1 In each case graph the line that passes through the points.

    a. (0, 2) and (2, 6)   **b.** (0, 4) and (2, 6)

    c. (0, −2) and (−1, −6)   **d.** (0, −4) and (3, −7)

10.2 The equation for a straight line (deterministic) is

$$y = \beta_0 + \beta_1 x$$

If the line passes through the point $(0, 1)$, then $x = 0$, $y = 1$ must satisfy the equation. That is,

$$1 = \beta_0 + \beta_1(0)$$

Similarly, if the line passes through the point $(2, 3)$, then $x = 2$, $y = 3$ must satisfy the equation:

$$3 = \beta_0 + \beta_1(2)$$

Use these two equations to solve for β_0 and β_1, and find the equation of the line that passes through the points $(0, 1)$ and $(2, 3)$.

10.3 Find the equations of the lines passing through the points in Exercise 10.1.

10.4 Plot the following lines:
a. $y = 3 + 2x$ **b.** $y = 1 + x$ **c.** $y = -2 + 3x$
d. $y = 5x$ **e.** $y = 4 - 2x$

10.5 Give the slope and y-intercept for each of the lines defined in Exercise 10.4.

10.6 Use the method of least squares to fit a straight line to these six data points:

x	1	2	3	4	5	6
y	1	2	2	3	5	5

a. What are the least squares estimates of β_0 and β_1?
b. Plot the data points and graph the least squares line on the scattergram.

10.7 Use the method of least squares to fit a straight line to these five data points:

x	-2	-1	0	1	2
y	4	3	3	1	-1

a. What are the least squares estimates of β_0 and β_1?
b. Plot the data points and graph the least squares line on the scattergram.

10.8 It has been a characteristic of the computer industry that hardware prices decline as technology advances. Suppose a manufacturer of personal computers is interested in describing the list price trend of its personal computer system over the last 5 years. The data shown in the table were supplied by the manufacturer.

YEAR x	LIST PRICE y
1982	$6,100
1983	4,600
1984	3,800
1985	2,500
1986	1,900

a. Find the least squares line describing the list price trend over the 5-year period.
b. Plot the data points and graph the least squares line as a check on your calculations.

10.9 The Federal Communications Commission (FCC) specifies that radiated electromagnetic emissions from digital devices are to be measured in an open-field test site. In order to verify test-site acceptability, the site attenuation (i.e., the transmission loss from the input of one half-wave dipole to the output of another when both dipoles are positioned over the ground plane) must be evaluated. A study conducted at a test site in Fort Collins, Colorado yielded the accompanying data on site attenuation (in decibels) and transmission frequency (in megahertz) for dipoles at a distance of 3 meters.

TRANSMISSION FREQUENCY x, MHz	SITE ATTENUATION y, dBL
50	11.5
100	15.8
200	18.2
300	22.6
400	26.2
500	27.1
600	29.5
700	30.7
800	31.3
900	32.6
1,000	34.9

Source: Bennett, W. S. "An Error Analysis of the FCC Site-Attenuation Approximation." *IEEE Transactions on Electromagnetic Compatibility,* Vol. EMC-27, No. 3, Aug. 1985, p. 113 (Table IV). © 1985 IEEE.

a. Construct a scattergram for the data. Does it appear that x and y are linearly related?

b. Find the least squares line relating site attenuation y to transmission frequency x.

c. Plot the least squares line on your scattergram as a check on your calculations.

10.10 A state attorney general is investigating the possibility of collusive bidding among the state's road construction contractors. One aspect of the investigation involves a comparison of the number of contractors bidding on a job and the resultant winning (lowest) bid price. The sample data listed in the accompanying table were supplied by the state's Department of Transportation.

JOB	NUMBER OF BIDDERS x	WINNING BID PRICE y, millions of dollars
1	5	4.0
2	2	6.5
3	8	3.5
4	3	7.1
5	2	7.3
6	3	5.6
7	11	2.1
8	6	3.2

a. Construct a scattergram for the data.
b. Assuming the relationship between the variables is best described by a straight line, use the method of least squares to estimate the y-intercept and slope of the line.
c. Plot the least squares line on your scattergram.
d. According to your least squares line, what is the expected low bid price on a road construction job on which four contractors bid? [*Note:* A measure of the reliability of this prediction will be discussed in Section 10.9.]

10.11 A new computer software query package has been designed to achieve more efficient access and maintenance of large-scale data sets. Efficiency is measured in terms of the number of disk I/O's (called *storage blocks*) required to access and maintain the data set; the smaller the number of blocks that are read, the faster the operation takes place. To evaluate the performance of the new software system, the number of disk I/O's required to access a large-scale data set was recorded for each of a sample of 15 data sets of various sizes (where size is measured as the number of records in the data set). The results are shown in the table.

DATA SET	NUMBER OF RECORDS x, thousands	NUMBER OF DISK I/O's y, thousands
1	350	36
2	200	20
3	450	45
4	50	5
5	400	40
6	150	18
7	350	38
8	300	32
9	150	21
10	500	54
11	100	11
12	400	43
13	200	19
14	50	7
15	250	26

a. Find the least squares line relating y to x.
b. Plot the data and graph the least squares line as a check on your calculations.
c. Use your least squares line to estimate the number of disk I/O's required to access and maintain a data set with 220,000 observations. [*Note:* A measure of the reliability of this prediction will be discussed in Section 10.9.]

OPTIONAL EXERCISES

10.12 Consider the pair of simultaneous linear equations:

$$n\hat{\beta}_0 + \hat{\beta}_1 \sum x_i = \sum y_i$$
$$\hat{\beta}_0 \sum x_i + \hat{\beta}_1 \sum x_i^2 = \sum x_i y_i$$

Derive the formulas for the least squares estimates, $\hat{\beta}_0$ and $\hat{\beta}_1$.

10.13 The maximum likelihood estimator of the mean μ of a normal distribution is the sample mean $\bar{y}$. Consider the model $E(y) = \mu$. Show that the least squares estimator of μ is also $\bar{y}$. [*Hint:* Minimize $SSE = \Sigma(y_i - \hat{\mu})^2$ with respect to $\hat{\mu}$.]

S E C T I O N   10.4

PROPERTIES OF THE LEAST SQUARES ESTIMATORS

An examination of the formulas for the least squares estimators reveals that they are linear functions of the observed y values, $y_1, y_2, \ldots, y_n$. Since we have assumed (Section 10.2) that the random errors associated with these y values, $\varepsilon_1, \varepsilon_2, \ldots, \varepsilon_n$, are independent, normally distributed random variables with mean 0 and variance σ^2, it follows that the y values will be normally distributed with mean $E(y) = \beta_0 + \beta_1 x$ and variance σ^2 and that $\hat{\beta}_0$ and $\hat{\beta}_1$ will possess sampling distributions that are normally distributed (Theorem 7.3).

The mean and the variance of the sampling distribution of $\hat{\beta}_1$ are given in Section 10.6. We will illustrate how they are acquired in Example 10.2.

EXAMPLE 10.2

Find the mean and variance of the sampling distribution of $\hat{\beta}_1$.

SOLUTION

The quantity SS_{xx} that appears in the formula for $\hat{\beta}_1$ involves only the x values, which are assumed to be known—i.e., nonrandom. Therefore, SS_{xx} can be treated as a constant when we find the expected value of $\hat{\beta}_1$. In contrast, SS_{xy} is a function of the random variables, $y_1, y_2, \ldots, y_n$. Thus,

$$SS_{xy} = \sum (x_i - \bar{x})(y_i - \bar{y}) = \sum [(x_i - \bar{x})(y_i) - (x_i - \bar{x})\bar{y}]$$
$$= \sum (x_i - \bar{x})y_i - \bar{y} \sum (x_i - \bar{x})$$

But

$$\sum (x_i - \bar{x}) = \sum x_i - n\bar{x} = \sum x_i - \sum x_i = 0$$

Therefore, $SS_{xy} = \Sigma(x_i - \bar{x})y_i$. Substituting this quantity into the formula for $\hat{\beta}_1$, we obtain

$$\hat{\beta}_1 = \frac{SS_{xy}}{SS_{xx}} = \frac{1}{SS_{xx}} \sum (x_i - \bar{x})y_i$$

$$= \frac{(x_1 - \bar{x})}{SS_{xx}}y_1 + \frac{(x_2 - \bar{x})}{SS_{xx}}y_2 + \cdots + \frac{(x_n - \bar{x})}{SS_{xx}}y_n$$

This shows that $\hat{\beta}_1$ is a linear function of the normally distributed random variables, $y_1, y_2, \ldots, y_n$. The coefficients, $a_1, a_2, \ldots, a_n$, of the random variables in the linear function are

$$a_1 = \frac{(x_1 - \bar{x})}{SS_{xx}} \qquad a_2 = \frac{(x_2 - \bar{x})}{SS_{xx}} \qquad \cdots \qquad a_n = \frac{(x_n - \bar{x})}{SS_{xx}}$$

The final step in finding the mean $E(\hat{\beta}_1)$ and the variance $V(\hat{\beta}_1)$ of the sampling distribution of $\hat{\beta}_1$ is to apply Theorem 6.7, which gives the rule for finding the mean and the variance of a linear function of random variables. Thus,

$$E(\hat{\beta}_1) = E\left[\frac{(x_1 - \bar{x})}{SS_{xx}}y_1 + \frac{(x_2 - \bar{x})}{SS_{xx}}y_2 + \cdots + \frac{(x_n - \bar{x})}{SS_{xx}}y_n\right]$$

where $y_1, y_2, \ldots, y_n$ are obtained by substituting the appropriate values of x into the formula for the linear model, i.e.,

$$y_1 = \beta_0 + \beta_1 x_1 + \varepsilon_1 \quad \text{and} \quad E(y_1) = \beta_0 + \beta_1 x_1$$
$$y_2 = \beta_0 + \beta_1 x_2 + \varepsilon_2 \quad \text{and} \quad E(y_2) = \beta_0 + \beta_1 x_2$$
$$\vdots \qquad\qquad\qquad\qquad \vdots$$
$$y_n = \beta_0 + \beta_1 x_n + \varepsilon_n \quad \text{and} \quad E(y_n) = \beta_0 + \beta_1 x_n$$

Therefore,

$$E(\hat{\beta}_1) = \frac{(x_1 - \bar{x})}{SS_{xx}}E(y_1) + \frac{(x_2 - \bar{x})}{SS_{xx}}E(y_2) + \cdots + \frac{(x_n - \bar{x})}{SS_{xx}}E(y_n)$$

$$= \frac{(x_1 - \bar{x})}{SS_{xx}}(\beta_0 + \beta_1 x_1) + \frac{(x_2 - \bar{x})}{SS_{xx}}(\beta_0 + \beta_1 x_2) + \cdots + \frac{(x_n - \bar{x})}{SS_{xx}}(\beta_0 + \beta_1 x_n)$$

$$= \frac{\beta_0}{SS_{xx}}\sum (x_i - \bar{x}) + \frac{\beta_1}{SS_{xx}}\sum (x_i - \bar{x})x_i$$

But,

$$SS_{xx} = \sum (x_i - \bar{x})^2 = \sum [(x_i - \bar{x})x_i - \bar{x}(x_i - \bar{x})]$$
$$= \sum (x_i - \bar{x})x_i - \bar{x}\sum (x_i - \bar{x})$$

Since we have already shown that $\Sigma(x_i - \bar{x}) = 0$, we have $SS_{xx} = \Sigma(x_i - \bar{x})x_i$ and therefore,

$$E(\hat{\beta}_1) = 0 + \frac{\beta_1}{SS_{xx}}(SS_{xx}) = \beta_1$$

This shows that $\hat{\beta}_1$ is an unbiased estimator of β_1.

Applying the formula given in Theorem 6.7 for finding the variance of a linear function of random variables, and remembering that the covariance between any pair of y values will equal 0 because all pairs of y values are assumed to be independent, we have

$$V(\hat{\beta}_1) = \frac{(x_1 - \bar{x})^2}{(SS_{xx})^2}V(y_1) + \frac{(x_2 - \bar{x})^2}{(SS_{xx})^2}V(y_2) + \cdots + \frac{(x_n - \bar{x})^2}{(SS_{xx})^2}V(y_n)$$

According to the assumptions made in Section 10.2, $V(y_1) = V(y_2) = \cdots = V(y_n) = \sigma^2$. Therefore,

$$V(\hat{\beta}_1) = \frac{(x_1 - \bar{x})^2}{(SS_{xx})^2}\sigma^2 + \frac{(x_2 - \bar{x})^2}{(SS_{xx})^2}\sigma^2 + \cdots + \frac{(x_n - \bar{x})^2}{(SS_{xx})^2}\sigma^2$$

$$= \frac{\sigma^2 \sum (x_i - \bar{x})^2}{(SS_{xx})^2} = \sigma^2 \frac{SS_{xx}}{(SS_{xx})^2} = \frac{\sigma^2}{SS_{xx}}$$

and

$$\sigma_{\hat{\beta}_1} = \frac{\sigma}{\sqrt{SS_{xx}}}$$ ∎

We will use the results of Example 10.2 in Section 10.6 to test hypotheses about and to construct a confidence interval for the slope β_1 of a regression line. The practical implications of these inferences will also be explained.

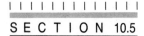

**EXERCISES
10.14–10.16**

OPTIONAL EXERCISES

10.14 We showed in Example 10.2 that $\hat{\beta}_1$, the least squares estimator of the slope β_1, is an unbiased estimator of β_1, i.e., $E(\hat{\beta}_1) = \beta_1$. Use the result to show that $E(\hat{\beta}_0) = \beta_0$.

10.15 Show that

$$\hat{\beta}_0 = \bar{y} - \hat{\beta}_1 \bar{x} = \sum \left[\frac{1}{n} - \frac{\bar{x}(x_i - \bar{x})}{SS_{xx}} \right] y_i$$

[*Hint:* Note that

$$\hat{\beta}_1 = \frac{SS_{xy}}{SS_{xx}} = \frac{\sum (x_i - \bar{x})(y_i - \bar{y})}{SS_{xx}}$$

$$= \frac{\sum (x_i - \bar{x}) y_i}{SS_{xx}} - \frac{\bar{y} \sum (x_i - \bar{x})}{SS_{xx}}$$

$$= \frac{\sum (x_i - \bar{x}) y_i}{SS_{xx}}$$

since $\sum (x_i - \bar{x}) = 0$.]

10.16 In Optional Exercise 10.15, you showed that $\hat{\beta}_0$ could be written as a linear function of independent random variables. Use Theorem 6.7 to show that

$$V(\hat{\beta}_0) = \frac{\sigma^2}{n} \left(\frac{\sum x_i^2}{SS_{xx}} \right)$$

SECTION 10.5

AN ESTIMATOR OF σ^2

In most practical situations, the variance σ^2 of the random error ε will be unknown and must be estimated from the sample data. Since σ^2 measures the variation of the y values about the line $E(y) = \beta_0 + \beta_1 x$, it seems intuitively reasonable to estimate σ^2 by dividing SSE by an appropriate number. Theorem 10.1, an extension of Theorem 7.4, will be useful in obtaining an unbiased estimator.

THEOREM 10.1

Let $s^2 = \text{SSE}/(n - 2)$. Then, when the assumptions of Section 10.2 are satisfied, the statistic

$$\chi^2 = \frac{\text{SSE}}{\sigma^2} = \frac{(n - 2)s^2}{\sigma^2}$$

possesses a chi-square distribution with $\nu = (n - 2)$ degrees of freedom.

From Theorem 10.1, it follows that

$$s^2 = \frac{\chi^2 \sigma^2}{n - 2}$$

Then

$$E(s^2) = \frac{\sigma^2}{n - 2} E(\chi^2)$$

where $E(\chi^2) = \nu = (n - 2)$. Therefore,

$$E(s^2) = \frac{\sigma^2}{n - 2}(n - 2) = \sigma^2$$

and we conclude that s^2 is an unbiased estimator of σ^2.

The procedure used in Table 10.3 to calculate SSE can lead to large rounding errors. The formula for s^2 and an appropriate method for calculating SSE are shown in the accompanying box. We will illustrate the calculation of s^2 with Example 10.3.

ESTIMATION OF σ^2

$$s^2 = \frac{\text{SSE}}{\text{Degrees of freedom for error}} = \frac{\text{SSE}}{n - 2}$$

where

$$\text{SSE} = \sum (y_i - \hat{y}_i)^2 = \text{SS}_{yy} - \hat{\beta}_1 \text{SS}_{xy}$$

$$\text{SS}_{yy} = \sum (y_i - \bar{y})^2 = \sum y_i^2 - \frac{\left(\sum y_i\right)^2}{n}$$

Warning: When performing these calculations, you may be tempted to round the calculated values of SS_{yy}, $\hat{\beta}_1$, and SS_{xy}. Be certain to carry at least six significant figures for each of these quantities to avoid substantial errors in the calculation of SSE.

EXAMPLE 10.3   Estimate σ^2 for the data of Table 10.1.

SOLUTION   In the insulation compression example, we previously calculated SSE = 1.10 for the least squares line $\hat{y} = -.1 + .7x$. Recalling that there were $n = 5$ data points, we have $n - 2 = 5 - 2 = 3$ df for estimating σ^2. Thus,

$$s^2 = \frac{\text{SSE}}{n - 2} = \frac{1.10}{3} = .367$$

is the estimated variance, and

$$s = \sqrt{.367} = .61$$

is the estimated standard deviation of ε.   ∎

You may be able to obtain an intuitive feeling for s by recalling the interpretation given to a standard deviation in Chapter 1 and remembering that the least squares line estimates the mean value of y for a given value of x. Since s measures the spread of the distribution of y values about the least squares line, we should not be surprised to find that most of the observations lie within $2s$ or $2(.61) = 1.22$ of the least squares line. For this simple example (only five data points), all five data points fall within $2s$ of the least squares line. In Section 10.9, we will use s to evaluate the error of prediction when the least squares line is used to predict a value of y to be observed for a given value of x.

INTERPRETATION OF s, THE ESTIMATED STANDARD DEVIATION OF ε

We expect most of the observed y values to lie within $2s$ of their respective least squares predicted values, $\hat{y}$.

EXERCISES 10.17–10.24

10.17 Suppose you fit a least squares line to nine data points and calculate SSE = .219. Find s^2, the estimator of σ^2, the variance of the random error term ε.

10.18 Calculate SSE and s^2 for the least squares lines plotted in:
   a. Exercise 10.6   **b.** Exercise 10.7   **c.** Exercise 10.8
   d. Exercise 10.9   **e.** Exercise 10.10   **f.** Exercise 10.11

10.19 As part of a study on the rate of combustion of artificial graphite in humid air flow, an experiment was conducted to investigate oxygen diffusivity through a water vapor mixture. Sample mixtures of nitrogen and oxygen were prepared with .017 mole fraction of water at nine different temperatures, and the oxygen diffusivity was measured for each. The data are reproduced in the table.

TEMPERATURE	OXYGEN DIFFUSIVITY	TEMPERATURE	OXYGEN DIFFUSIVITY
x	y	x	y
1,000	1.69	1,500	3.39
1,100	1.99	1,600	3.79
1,200	2.31	1,700	4.21
1,300	2.65	1,800	4.64
1,400	3.01		

Source: Matsui, K., Tsuji, H., and Makino, A. "The Effects of Water Vapor Concentration on the Rate of Combustion of an Artificial Graphite in Humid Air Flow." *Combustion and Flame*, Vol. 50, 1983, pp. 107–118. Copyright 1983 by The Combustion Institute. Reprinted by permission of Elsevier Science Publishing Co., Inc.

a. Fit a simple linear model relating mean oxygen diffusivity, $E(y)$, to the temperature, x.
b. Plot the data points on a scattergram.
c. Compute SSE and s^2.

10.20 An engineer conducted a study to determine whether there is a linear relationship between the breaking strength, y, of wooden beams and the specific gravity, x, of the wood. Ten randomly selected beams of the same cross-sectional dimensions were stressed until they broke. The breaking strength and the specific gravity of the wood are shown in the table for each of the ten beams.

BEAM	BREAKING STRENGTH	SPECIFIC GRAVITY
	y	x
1	11.14	.499
2	12.74	.558
3	13.13	.604
4	11.51	.441
5	12.38	.550
6	12.60	.528
7	11.13	.418
8	11.70	.480
9	11.02	.406
10	11.41	.467

a. Find the least squares prediction equation relating the breaking strength y of a wooden beam to the beam's specific gravity x.
b. Plot the data and graph the least squares line.
c. Predict the breaking strength of a beam if its specific gravity is .590. [*Note:* A measure of reliability for this prediction will be discussed in Section 10.9.]
d. Calculate SSE and s^2.

10.21 Calculate SSE for the data of Table 10.1, using the formula

$$\text{SSE} = \text{SS}_{yy} - \hat{\beta}_1 \text{SS}_{xy}$$

and verify that it equals the value (1.10) computed in Table 10.3.

10.22 The thermogravimetric balance (TG) is a new technique developed to evaluate the thermal behavior of chemical compounds. Abou El Naga and Salem (1986) compared the TG technique to the standard method of evaluating the thermooxidation stability of base oils and their additive blends (for example, transformer oils, turbine oils and transmission oils). For each of a sample of ten base oils, the amount y of oxidative compounds formed at the oxidation point was determined using the TG technique and the total percentage x of oxidation products determined by the standard method. The results of the experiment are shown in the accompanying table.

BASE OIL	TG TECHNIQUE: AMOUNT OF OXIDATIVE COMPOUNDS y, % weight	STANDARD METHOD: TOTAL OXIDATION PRODUCTS x, %
1	25.4	2.3
2	27.11	2.5
3	28.0	2.65
4	17.9	1.3
5	18.9	1.45
6	22.9	1.9
7	30.8	3.3
8	18.6	1.4
9	24.4	2.1
10	29.8	2.9

Source: Abou El Naga, H. H. and Salem, A. E. M. "Base Oils Thermooxidation." *Lubrication Engineering*, Vol. 24, No. 4, Apr. 1986, p. 213. Reprinted by permission of the American Society of Lubrication Engineers. All rights reserved.

a. Fit a simple linear model relating amount y of oxidative compounds determined by the TG technique to the total percentage x of oxidation products determined by the standard method.
b. Plot the data points and least squares line on a scattergram.
c. Compute SSE and s^2.

OPTIONAL EXERCISES

10.23 Show that $V(s^2) = 2\sigma^4/(n-2)$. [*Hint:*   The result follows from Theorem 10.1 and the fact that $V(\chi^2) = 2\nu$.]

10.24 Verify that SSE $= \Sigma(y_i - \hat{y}_i)^2 = SS_{yy} - \hat{\beta}_1 SS_{xy}$.

SECTION 10.6

ASSESSING THE UTILITY OF THE MODEL: MAKING INFERENCES ABOUT THE SLOPE β_1

Refer again to the data of Table 10.1 and suppose that the compression of the insulation material is *completely unrelated* to the pressure. What could be said about the values of β_0 and β_1 in the hypothesized probabilistic model

$$y = \beta_0 + \beta_1 x + \varepsilon$$

if x contributes no information for the prediction of y? The implication is that the mean of y, i.e., the deterministic part of the model $E(y) = \beta_0 + \beta_1 x$, does not change as x changes. Regardless of the value of x, you always predict the same value of y. In the straight-line model, this means that the true slope, β_1, is equal to 0. Therefore, to test the null hypothesis that x contributes no information for the prediction of y against the alternative hypothesis that these variables are linearly related with a slope differing from 0, we test

H_0: $\beta_1 = 0$

H_a: $\beta_1 \neq 0$

If the data support the alternative hypothesis, we will conclude that x does contribute information for the prediction of y using the straight-line model [although the true relationship between $E(y)$ and x could be more complex than a straight line]. Thus, to some extent, this is a test of the utility of the hypothesized model.

The appropriate test statistic is found by considering the sampling distribution of $\hat{\beta}_1$, the least squares estimator of the slope β_1. The sampling distribution of this statistic (discussed in Section 10.4) is described in the box.

SAMPLING DISTRIBUTION OF $\hat{\beta}_1$

If we make the four assumptions about ε (see Section 10.2), then the sampling distribution of $\hat{\beta}_1$, the least squares estimator of slope, will be a normal distribution with mean β_1 (the true slope) and standard deviation

$$\sigma_{\hat{\beta}_1} = \frac{\sigma}{\sqrt{SS_{xx}}} \quad \text{(see Figure 10.6)}$$

FIGURE 10.6
Sampling Distribution of $\hat{\beta}_1$

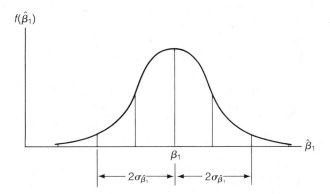

Since σ will usually be unknown, the appropriate test statistic will generally be a Student's t statistic formed as follows:

$$t = \frac{\hat{\beta}_1 - \text{Hypothesized value of } \beta_1}{s_{\hat{\beta}_1}} \quad \text{where } s_{\hat{\beta}_1} = \frac{s}{\sqrt{SS_{xx}}}$$

$$= \frac{\hat{\beta}_1 - 0}{s/\sqrt{SS_{xx}}}$$

Note that we have substituted the estimator s for σ, and then formed $s_{\hat{\beta}_1}$ by dividing s by $\sqrt{SS_{xx}}$. The number of degrees of freedom associated with this t statistic is the same as the number of degrees of freedom associated with s. Recall that this will be $(n-2)$ df when the hypothesized model is a straight line (see Section 10.5).

The setup of our test of the utility of the model is summarized in the box.

A TEST OF MODEL UTILITY

ONE-TAILED TEST	**TWO-TAILED TEST**
H_0: $\beta_1 = 0$	H_0: $\beta_1 = 0$
H_a: $\beta_1 < 0$	H_a: $\beta_1 \neq 0$
(or H_a: $\beta_1 > 0$)	

Test statistic: $t = \dfrac{\hat{\beta}_1}{s_{\hat{\beta}_1}} = \dfrac{\hat{\beta}_1}{s/\sqrt{SS_{xx}}}$ Test statistic: $t = \dfrac{\hat{\beta}_1}{s_{\hat{\beta}_1}} = \dfrac{\hat{\beta}_1}{s/\sqrt{SS_{xx}}}$

Rejection region: $t < -t_\alpha$ Rejection region:
 (or $t > t_\alpha$) $t < -t_{\alpha/2}$ or $t > t_{\alpha/2}$

where t_α is based on $(n-2)$ df   where $t_{\alpha/2}$ is based on $(n-2)$ df

The values of t_a such that $P(t \geq t_a) = a$ are given in Table 6 of Appendix II.

Assumptions: The four assumptions about ε listed in Section 10.2.

EXAMPLE 10.4

Refer to Examples 10.1 and 10.3 and test the hypothesis that $\beta_1 = 0$.

SOLUTION

For the insulation compression example, we will choose $\alpha = .05$ and, since $n = 5$, df $= (n-2) = 5 - 2 = 3$. Then the rejection region for the two-tailed test is

$$t < -t_{.025} \quad \text{or} \quad t > t_{.025}$$

where $t_{.025}$, given in Table 6 of Appendix II, is $t_{.025} = 3.182$. We previously calculated $\hat{\beta}_1 = .7$, $s = 61$, and $SS_{xx} = 10$. Thus.

$$t = \frac{\hat{\beta}_1}{s/\sqrt{SS_{xx}}} = \frac{.7}{.61/\sqrt{10}} = \frac{.7}{.19} = 3.7$$

Since this calculated t value falls in the upper-tail rejection region (see Figure 10.7), we reject the null hypothesis and conclude that the slope β_1 is not 0. The sample evidence indicates that x contributes information for the prediction of y using a linear model for the relationship between compression and pressure.

FIGURE 10.7

Rejection Region and
Calculated t Value for
Testing Whether the
Slope $\beta_1 = 0$

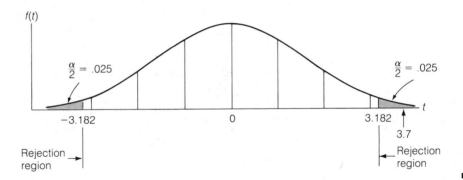

Another way to make inferences about the slope β_1 is to estimate it using a confidence interval. This interval is formed as shown in the box.

A $(1 - \alpha)100\%$ CONFIDENCE INTERVAL FOR THE SLOPE β_1

$$\hat{\beta}_1 \pm t_{\alpha/2}s_{\hat{\beta}_1} \qquad \text{where} \quad s_{\hat{\beta}_1} = \frac{s}{\sqrt{SS_{xx}}}$$

and $t_{\alpha/2}$ is based on $(n - 2)$ df

EXAMPLE 10.5

Find a 95% confidence interval for β_1 in Example 10.1.

SOLUTION

For the insulation compression example, a 95% confidence interval for the slope β_1 is

$$\hat{\beta}_1 \pm t_{.025}s_{\hat{\beta}_1} = .7 \pm 3.182\left(\frac{s}{\sqrt{SS_{xx}}}\right)$$

$$= .7 \pm 3.182\left(\frac{.61}{\sqrt{10}}\right) = .7 \pm .61$$

Thus, we estimate that the interval from .09 to 1.31 includes the slope parameter β_1.

Since all the values in this interval are positive, it appears that β_1 is positive and that the mean of y, $E(y)$, increases as x increases. However, the rather large width of the confidence interval reflects the small number of data points (and, consequently, a lack of information) in the experiment. We would expect a narrower interval if the sample size were increased.   ■

Before concluding this section, we call your attention to the similarity between the t statistic for testing hypotheses about β_1 and the t statistic for testing hypotheses about the means of normal populations in Chapter 9. Also note the similarity of the corresponding confidence intervals. In each case, the general form of the test statistic is

$$t = \frac{\text{Parameter estimator} - \text{Its hypothesized mean}}{\text{Estimated standard error of the estimator}}$$

and the general form of the confidence interval is

Point estimator $\pm t_{\alpha/2}$(Estimated standard error of the estimator)

In the Optional Exercises of this section, we outline the procedure for acquiring the t statistic for testing hypotheses about and constructing confidence intervals for β_1.

**EXERCISES
10.25–10.35**

10.25 Do the data provide sufficient evidence to indicate that β_1 differs from 0 for the least squares analyses in the following exercises? Use $\alpha = .05$.
  a. Exercise 10.6   **b.** Exercise 10.7
  c. Exercise 10.8   **d.** Exercise 10.9
  e. Exercise 10.10   **f.** Exercise 10.11

10.26 Do the data in Exercise 10.19 provide sufficient evidence to indicate that oxygen diffusivity, y, tends to increase as the temperature, x, increases (i.e., that $\beta_1 > 0$)? Test using $\alpha = .05$.

10.27 Do the data in Exercise 10.20 support the theory that breaking strength increases as the specific gravity of a beam increases? Test using $\alpha = .05$.

10.28 Do the data in Exercise 10.22 provide sufficient evidence to indicate that the amount of oxidative compounds measured using the TG technique is positively linearly related to the total percentage of oxidation products determined by the standard method? Test using $\alpha = .01$.

10.29 An electronics firm specializing in videogames is interested in determining whether a linear relationship exists between the amount it spends on television advertising and total sales. The data listed in the table are available.

MONTH	SALES thousands of dollars	TELEVISION ADVERTISING EXPENDITURES thousands of dollars
January	50	.5
February	90	.9
March	30	.4
April	90	.7
May	91	1.1
June	95	.75
July	95	.8

a. Find the least squares line for the given data. Plot the data on a scattergram and graph the line as a check on your calculations.

b. Letting $\alpha = .05$, test the null hypothesis that $\beta_1 = 0$. What alternative hypothesis would you select for this test? Draw appropriate conclusions concerning the adequacy of a linear model to describe the relationship between sales and television expenditures.

c. Construct a 90% confidence interval for the slope parameter in the hypothesized linear model.

d. Interpret the confidence interval and explain what it tells you about the relationship between sales and television advertising expenditure.

10.30 A nuclear engineer has been assigned the task of developing a model to predict peak power load at a nuclear power plant. Initially, the engineer will model peak power load as a function of the high temperature for the day, based on the theory that higher temperatures result in higher peak power loads. The high temperature and peak power load were observed for a random sample of six days and are listed in the table. Do the data provide sufficient evidence to indicate that peak power load increases as daily high temperature increases?

DAY	HIGH TEMPERATURE x, °F	PEAK POWER LOAD y, megawatts
1	92	207
2	84	139
3	95	211
4	102	273
5	88	156
6	97	244

10.31 As part of a computer system performance evaluation, a system manager is interested in predicting the response time for computer terminals. *Terminal response time* is defined as the length of time (in seconds) it takes the computer to respond to a command sent from a computer terminal by pressing one of the terminal's program function keys. Although many variables influence terminal response time, the system manager will model the response time as a function of the number of simultaneous users (i.e., the number of users who are accessing the computer's central processing unit at the same time the command was sent). The manager has collected the sample data given in the accompanying table.

NUMBER OF SIMULTANEOUS USERS x	TERMINAL RESPONSE TIME y, seconds
1	.22
2	.59
3	1.01
4	1.36
5	1.42

a. Find the least squares line relating y to x.

b. Construct a 95% confidence interval to estimate the mean increase in terminal response time for each additional simultaneous user.

c. Interpret the confidence interval in part **b**. What does it tell you about the relationship between terminal response time and the number of simultaneous users?

10.32 An experiment was conducted to study the stress corrosion cracking of Type 304 stainless steel in a simulated boiling water reactor environment (*Transactions of the ASME*, Jan. 1986). Six specimens of stainless steel were annealed and sensitized in 289°C water with dissolved oxygen and sulfate under various stress intensity factors (i.e., loads). The table gives the maximum load and resulting crack growth rate (in meters per second) for the six specimens.

MAXIMUM LOAD x, MPa · m$^{1/2}$	30.0	35.6	41.5	50.2	55.5	61.1
CRACK GROWTH RATE y, m/s × 10^{10}	1.0	2.2	3.9	5.8	5.0	14.0

Source: Park, J. Y., Ruther, W. E., Kassner, T. F., and Shack, W. J. "Stress Corrosion Crack Growth Rates in Type 304 Stainless Steel in Simulated BWR Environments." *Transactions of the American Society of Mechanical Engineers*, Vol. 108, No. 1, Jan. 1986, p. 23 (Table 4).

a. Is there sufficient evidence to indicate that cracking growth rate increases linearly with maximum load? Test using $\alpha = .10$.

b. Estimate the mean increase in cracking growth rate for every 1-unit increase in maximum load using a 90% confidence interval.

OPTIONAL EXERCISES

10.33 Explain why

$$z = \frac{\hat{\beta}_1 - \beta_1}{\sigma_{\hat{\beta}_1}} = \frac{\hat{\beta}_1 - \beta_1}{\sigma/\sqrt{SS_{xx}}}$$

is normally distributed with mean 0 and variance 1 when the four assumptions of Section 10.2 are satisfied.

10.34 It can be shown (proof omitted) that the least squares estimates, $\hat{\beta}_0$ and $\hat{\beta}_1$, are independent (in a probabilistic sense) of s^2. Use this fact, in conjunction with Theorem 10.1 and the result of Optional Exercise 10.33, to show that

$$t = \frac{\hat{\beta}_1 - \beta_1}{s/\sqrt{SS_{xx}}}$$

has a Student's t distribution with $\nu = (n - 2)$ df.

10.35 Use the t statistic in Optional Exercise 10.34 as a pivotal statistic to derive a $(1 - \alpha)100\%$ confidence interval for β_1.

The claim is often made that the crime rate and the unemployment rate are "highly correlated." Another popular belief is that the Gross National Product (GNP) and the rate of inflation are "correlated." Some people even believe that the Dow Jones Industrial Average and the lengths of fashionable skirts are "correlated." Here, the term *correlation* implies a relationship between two variables.

The **Pearson product moment correlation coefficient** r, defined in the box, provides a quantitative measure of the strength of the linear relationship between x and y, just as does the least squares slope $\hat{\beta}_1$. However, unlike the slope, the correlation coefficient r is *scaleless*. The value of r is always between -1 and $+1$, no matter what the units of x and y are.

DEFINITION 10.2

The **Pearson product moment coefficient of correlation** r is a measure of the strength of the linear relationship between two variables x and y. It is computed (for a sample of n measurements on x and y) as follows:

$$r = \frac{SS_{xy}}{\sqrt{SS_{xx}SS_{yy}}}$$

Since both r and $\hat{\beta}_1$ provide information about the utility of the model, it is not surprising that there is a similarity in their computational formulas. Particularly, note that SS_{xy} appears in the numerators of both expressions and, since both denominators are always positive, r and $\hat{\beta}_1$ will always be of the same sign (either both positive or both negative). A value of r near or equal to 0 implies little or no linear relationship between y and x. In contrast, the closer r is to 1 or -1, the stronger the linear relationship between y and x. And, if $r = 1$ or $r = -1$, all the points fall exactly on the least squares line. Positive values of r imply that y increases as x increases; negative values imply that y decreases as x increases. See Figure 10.8 on page 446.

EXAMPLE 10.6

A computer hardware firm wants to know the correlation between the size of its sales force and its yearly sales revenue. The records for the past 10 years are examined, and the results listed in Table 10.4 are obtained. Calculate the coefficient of correlation r for the data.

TABLE 10.4

YEAR	NUMBER OF SALESPEOPLE x	SALES y, million dollars	YEAR	NUMBER OF SALESPEOPLE x	SALES y, million dollars
1977	15	1.35	1982	29	2.93
1978	18	1.63	1983	30	3.41
1979	24	2.33	1984	32	3.26
1980	22	2.41	1985	35	3.63
1981	25	2.63	1986	38	4.15

FIGURE 10.8

Values of r and Their Implications

a. Positive r: y increases as x increases

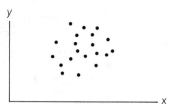

b. r near 0; little or no linear relationship between y and x

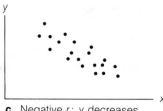

c. Negative r: y decreases as x increases

d. $r = 1$: a perfect positive relationship between y and x

e. $r = -1$: a perfect negative relationship between y and x

f. r near 0: little or no linear relationship between y and x

SOLUTION

We need to calculate SS_{xy}, SS_{xx}, and SS_{yy}:

$$SS_{xy} = \sum x_i y_i - \frac{\left(\sum x_i\right)\left(\sum y_i\right)}{10} = 800.62 - \frac{(268)(27.73)}{10} = 57.456$$

$$SS_{xx} = \sum x_i^2 - \frac{\left(\sum x_i\right)^2}{10} = 7,668 - \frac{(268)^2}{10} = 485.6$$

$$SS_{yy} = \sum y_i^2 - \frac{\left(\sum y_i\right)^2}{10} = 83.8733 - \frac{(27.73)^2}{10} = 6.97801$$

Then, the coefficient of correlation is

$$r = \frac{SS_{xy}}{\sqrt{SS_{xx}SS_{yy}}} = \frac{57.456}{\sqrt{(485.6)(6.97801)}} = \frac{57.456}{58.211} = .99$$

Thus, the size of the sales force and sales revenue are very highly correlated—at least over the past 10 years. The implication is that a strong positive linear

FIGURE 10.9
Scattergram for
Example 10.6

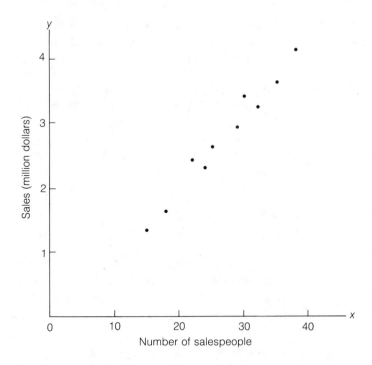

relationship exists between these variables (see Figure 10.9). We must be careful, however, not to jump to any unwarranted conclusions. For instance, the firm may be tempted to conclude that the best thing it can do to increase sales is to hire a large number of new salespeople. The implication of such a conclusion is that there is a **causal** relationship between the two variables. However, **high correlation does not imply causality**. Many other factors, such as the firm's growing expertise, economic inflation (1986 dollars are worth less than 1977 dollars), and a wider line of products and services, may contribute. ∎

WARNING

High correlation does not imply causality. If a large positive or negative value of the sample correlation coefficient r is observed, it is incorrect to conclude that a change in x causes a change in y. The only valid conclusion is that a linear trend *may* exist between x and y.

Keep in mind that the correlation coefficient r measures the correlation between x values and y values in the sample, and that a similar linear coefficient of correlation exists for the population from which the data points were selected. The **population correlation coefficient** is denoted by the symbol ρ (rho). As you might expect, ρ is estimated by the corresponding sample statistic, r. Or, rather than estimating ρ, we might want to test the hypothesis $H_0: \rho = 0$ against $H_a:$ $\rho \neq 0$, i.e., test the hypothesis that x contributes no information for the prediction of y using the straight-line model against the alternative that the two variables are at least linearly related. However, we have already performed this identical test in Section 10.6 when we tested $H_0: \beta_1 = 0$ against $H_a: \beta_1 \neq 0$. It can be shown that the null hypothesis $H_0: \rho = 0$ is equivalent to the hypothesis $H_0: \beta_1 = 0$. When we tested the null hypothesis $H_0: \beta_1 = 0$ in connection with the insulation compression example, the data led to a rejection of the hypothesis for $\alpha = .05$. This implies that the null hypothesis of a zero linear correlation between the two variables (pressure and compression) can also be rejected at $\alpha = .05$. The only real difference between the least squares slope $\hat{\beta}_1$ and the coefficient of correlation r is the measurement scale. Therefore, the information they provide about the utility of the least squares model is to some extent redundant. Furthermore, the slope β_1 gives us additional information on the amount increase (or decrease) in y for every 1-unit increase in x. For this reason, we recommend using the slope to make inferences about the existence of a positive or negative linear relationship between two variables. For those who prefer to test for a linear relationship between two variables using the coefficient of correlation r, we outline the procedure in the box.

TEST OF HYPOTHESIS FOR LINEAR CORRELATION

ONE-TAILED TEST	**TWO-TAILED TEST**
$H_0: \quad \rho = 0$	$H_0: \quad \rho = 0$
$H_a: \quad \rho > 0$	$H_a: \quad \rho \neq 0$
$\quad$ (or $H_a: \quad \rho < 0$)	
Test statistic: $\quad r$	*Test statistic:* $\quad r$
Rejection region: $\quad r > r_\alpha$	*Rejection region:*
$\quad$ (or $r < -r_\alpha$)	$\quad r > r_{\alpha/2}$ or $r < -r_{\alpha/2}$

where the distribution of the sample correlation coefficient r depends on the sample size n, and r_α and $r_{\alpha/2}$ are the critical values obtained from Table 12 of Appendix II, such that

$$P(r > r_\alpha) = \alpha \quad \text{and} \quad P(r > r_{\alpha/2}) = \alpha/2$$

SECTION 10.8

THE COEFFICIENT OF DETERMINATION

Another way to measure the contribution of x in predicting y is to consider how much the errors of prediction of y can be reduced by using the information provided by x.

To illustrate, suppose a sample of data has the scattergram shown in Figure 10.10a. If we assume that x contributes no information for the prediction of y,

FIGURE 10.10 A Comparison of the Sum of Squares of Deviations for Two Models

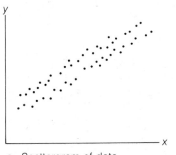

a. Scattergram of data

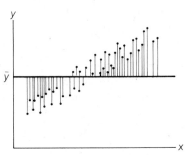

b. Assumption: x contributes no information for predicting y; $\hat{y} = \bar{y}$

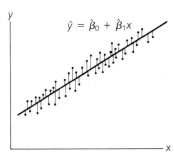

c. Assumption: x contributes information for predicting y; $\hat{y} = \hat{\beta}_0 + \hat{\beta}_1 x$

the best prediction for a value of y is the sample mean $\bar{y}$, which graphs as the horizontal line shown in Figure 10.10b. The vertical line segments in Figure 10.10b are the deviations of the points about the mean $\bar{y}$. Note that the sum of squares of deviations for the model $\hat{y} = \bar{y}$ is $SS_{yy} = \Sigma (y_i - \bar{y})^2$.

Now suppose that you fit a least squares line to the same set of data and locate the deviations of the points about the line as shown in Figure 10.10c. Compare the deviations about the prediction lines in parts b and c in Figure 10.10. You can see that:

1. If x contributes little or no information for the prediction of y, then the sums of squares of deviations for the two lines,

$$SS_{yy} = \Sigma (y_i - \bar{y})^2 \quad \text{and} \quad SSE = \Sigma (y_i - \hat{y}_i)^2$$

will be nearly equal.

2. If x does contribute information for the prediction of y, then SSE will be smaller than SS_{yy}. In fact, if all the points fall on the least squares line, then SSE = 0.

A convenient way of measuring how well the least squares equation $\hat{y} = \hat{\beta}_0 + \hat{\beta}_1 x$ performs as a predictor of y is to compute the reduction in the sum of squares of deviations that can be attributed to x, expressed as a proportion of SS_{yy}. This quantity, called the **coefficient of determination**, is

$$\frac{SS_{yy} - SSE}{SS_{yy}}$$

In simple linear regression it can be shown that this quantity is equal to the square of the simple linear coefficient of correlation r.

DEFINITION 10.3

The **coefficient of determination** is

$$r^2 = \frac{SS_{yy} - SSE}{SS_{yy}} = 1 - \frac{SSE}{SS_{yy}}$$

It represents the proportion of the sum of squares of deviations of the y values about their predicted values that can be attributed to a linear relation between y and x. (In simple linear regression, it may also be computed as the square of the coefficient of correlation r.)

Note that r^2 is always between 0 and 1, because r is between -1 and $+1$. Thus, $r^2 = .60$ means that the sum of squares of deviations of the y values about their predicted values has been reduced 60% by the use of $\hat{y}$, instead of $\bar{y}$, to predict y.

EXAMPLE 10.7

Calculate the coefficient of determination for the insulation compression example. The data are repeated in Table 10.5.

TABLE 10.5

PRESSURE x, 10 pounds per square inch	COMPRESSION y, .1 inch
1	1
2	1
3	2
4	2
5	4

SOLUTION

We first calculate

$$SS_{yy} = \sum y_i^2 - \frac{\left(\sum y_i\right)^2}{5} = 26 - \frac{(10)^2}{5} = 26 - 20 = 6$$

From previous calculations, we have

$$SSE = \sum (y_i - \hat{y}_i)^2 = 1.10$$

Then, the coefficient of determination is given by

$$r^2 = \frac{SS_{yy} - SSE}{SS_{yy}} = \frac{6.0 - 1.1}{6.0} = \frac{4.9}{6.0} = .82$$

So we know that by using the pressure x to predict compression y with the least squares line $\hat{y} = -.1 + .7x$, the total sum of squares of deviations of the five sample y values about their predicted values has been reduced 82% by the use of the linear predictor $\hat{y}$.   ∎

CASE STUDY 10.1

As evidenced by the cost overruns of public building projects, the initial estimate of the ultimate cost of a structure is often rather poor. These estimates usually rely on a precise definition of the proposed building in terms of working drawings and specifications. However, cost estimators do not take random error into account, so that no measure of reliability is possible for the deterministic estimates. Crandall and Cedercreutz (1976) propose the use of a probabilistic model to make cost estimates. They use regression models to relate cost to independent variables such as volume, amount of glass, floor area, etc. Crandall and Cedercreutz's rationale for choosing this approach is that "one of the principal merits of the least squares regression model, for the purpose of preliminary cost estimating, is the method of dealing with anticipated error." They go on to point out that when random error is anticipated, "statistical methods, such as regression analysis, attack the problem head on."

Crandall and Cedercreutz initially focused on the cost of mechanical work (heating, ventilating, and plumbing), since this part of the total cost is generally difficult to predict. Conventional cost estimates rely heavily on the amount of ductwork and piping used in construction, but this information is not precisely known until too late to be of use to the cost estimator. One of several models discussed was a simple linear model relating mechanical cost to floor area. Based on the data associated with 26 factory and warehouse buildings, the least squares prediction equation given in Figure 10.11 was found. It was concluded that floor area and mechanical cost are linearly related, since the t statistic (for testing H_0: $\beta_1 = 0$) was found to equal 3.61, which is significant with an α as small as .002. Thus, floor area should be useful when predicting the mechanical cost of a factory

FIGURE 10.11
Simple Linear Model Relating Cost to Floor Area

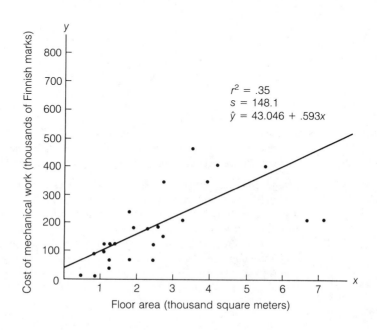

$r^2 = .35$
$s = 148.1$
$\hat{y} = 43.046 + .593x$

or warehouse. In addition, the regression model enables the reliability of the predicted cost to be assessed.

The value of the coefficient of determination r^2 was found to be .35. This tells us that only 35% of the variation among mechanical costs is accounted for by the differences in floor areas. Since there is only one independent variable in the model, this relatively small value of r^2 should not be too surprising. If other variables related to mechanical cost were included in the model, they would probably account for a significant portion of the remaining 65% of the variation in mechanical cost not explained by floor area. In the next chapter we discuss this important aspect of relating a response to more than one independent variable.

| | | | | | | | | | | | | |

**EXERCISES
10.36–10.46**

10.36 Find the correlation coefficient and the coefficient of determination for the sample data of Exercise 10.20 and interpret your results.

10.37 Refer to the oxygen diffusivity experiment described in Exercise 10.19. The data for the nine sample mixtures of nitrogen and oxygen are reproduced in the accompanying table.

TEMPERATURE x	OXYGEN DIFFUSIVITY y
1,000	1.69
1,100	1.99
1,200	2.31
1,300	2.65
1,400	3.01
1,500	3.39
1,600	3.79
1,700	4.21
1,800	4.64

a. Calculate r and r^2. Interpret their values.

b. Conduct a test to determine whether temperature and oxygen diffusivity are positively correlated. Use $\alpha = .05$. Do your results agree with Exercise 10.26?

10.38 The electroencephalogram (EEG) is a device used to measure brain waves. Neurologists have found that the peak EEG frequency in normal children increases with age. In one study, 287 normal children ranging from 2 to 16 years were instructed to hold a 65-gram weight in the palm of their outstretched hand for a brief but unspecified time. The peak EEG frequency (measured in hertz) was then recorded for each child. The data were then grouped according to age of the children, and the average peak frequency was calculated for each age group. The data appear in the accompanying table.

AGE x, years	AVERAGE PEAK EEG FREQUENCY y, hertz	AGE x, years	AVERAGE PEAK EEG FREQUENCY y, hertz
2	5.33	10	7.28
3	5.75	11	7.06
4	5.80	12	7.60
5	5.60	13	7.45
6	6.00	14	8.23
7	5.78	15	8.50
8	5.90	16	9.38
9	6.23		

Source: Tryon, W. W. "Developmental Equations for Postural
Tremor." *Science*, Vol. 215, No. 2, pp. 300–301, 1982. Copyright 1982 by the AAAS.

a. Construct a scattergram for the data. After examining the scattergram, do you
think that x and y are correlated? If correlation is present, is it positive or negative?
b. Find the correlation coefficient r and interpret its value.
c. Do the data provide sufficient evidence to indicate that x and y are linearly
correlated? Test using $\alpha = .05$.

10.39 Data on monthly sales (y), price per unit during the month (x_1), and amount spent
on advertising (x_2), for home computers are shown in the table for a 5-month period.
Based on this sample, which variable—price or advertising expenditure—appears
to provide more information about sales? Explain.

MONTH	TOTAL MONTHLY SALES y, thousands of dollars	PRICE PER UNIT x_1, thousands of dollars	AMOUNT SPENT ON ALL FORMS OF ADVERTISING x_2, hundreds of dollars
June	40	3.0	6.0
July	50	2.2	5.0
August	55	2.1	8.0
September	30	3.3	7.5
October	45	2.5	5.5

10.40 In 1984, federal government outlays for elementary, secondary, and vocational education were cut, yet Scholastic Aptitude Test (SAT) scores increased. Has such a
relationship existed in the past? According to *Fortune* (Oct. 29, 1984), "for the
decade ending in 1984 . . . federal spending on education is strongly and negatively
correlated with both verbal and math SAT scores." The correlation coefficient
between verbal scores and federal spending on education for the past $n = 10$ years
is $r = -.92$, while the correlation between math scores and federal spending is
$r = -.71$.

a. Do the data support *Fortune*'s claim that verbal SAT scores and federal spending
on education are "strongly and negatively correlated"? Test using $\alpha = .05$.

b. Do the data support *Fortune*'s claim that math SAT scores and federal spending on education are "strongly and negatively correlated"? Test using $\alpha = .05$.

c. Calculate the coefficient of determination for a straight-line model relating verbal SAT scores to federal spending. Interpret this value.

d. Calculate the coefficient of determination for a straight-line model relating math SAT scores to federal spending. Interpret this value.

10.41 In an analysis of the costs of construction, labor and material costs are two basic components. Changes in the component costs will lead, of course, to changes in total construction costs.

MONTH	CONSTRUCTION COST y	INDEX OF ALL CONSTRUCTION MATERIALS x
January	193.2	180.0
February	193.1	181.7
March	193.6	184.1
April	195.1	185.3
May	195.6	185.7
June	198.1	185.9
July	200.9	187.7
August	202.7	189.6

a. Use the data in the table to find a measure of the importance of the materials component. Do this by determining the proportion of the sum of squares of deviations of the y values about their mean that can be explained by a linear relationship between the construction cost index and the material cost index.

b. Do the data provide sufficient evidence to indicate a nonzero correlation between y and x? Use $\alpha = .05$.

10.42 A major portion of the effort expended in developing commercial computer software is associated with program testing. A study was undertaken to assess the potential usefulness of various product- and process-related variables in identifying error-prone software (*IEEE Transactions on Software Engineering*, Apr. 1985). A straight-line model relating the number y of module defects to the number x of unique operands in the module was fit to the data collected for a sample of software modules. The coefficient of determination for this analysis was $r^2 = .74$.

a. Interpret the value of r^2.

b. Based on this value, would you infer that the straight-line model is a useful predictor of number y of module defects? Explain.

10.43 Use the method of least squares and the sample data in the table to model the relationship between the number of items produced by a particular manufacturing process and the total variable cost involved in production. Find the coefficient of determination and explain its significance in the context of this problem.

TOTAL OUTPUT y	TOTAL VARIABLE COST x, dollars
10	10
15	12
20	20
20	21
25	22
30	20
30	19

OPTIONAL EXERCISES

10.44 Verify that

$$\hat{\beta}_1 = r \sqrt{\frac{SS_{yy}}{SS_{xx}}} \quad \text{and} \quad SSE = SS_{yy}(1 - r^2)$$

10.45 Use the result of Optional Exercise 10.44 to show that

$$\frac{\hat{\beta}_1}{s/\sqrt{SS_{xx}}} = \frac{r\sqrt{n - 2}}{\sqrt{1 - r^2}}$$

10.46 A test of H_0: $\rho = 0$ can also be conducted using the test statistic

$$t = \frac{r\sqrt{n - 2}}{\sqrt{1 - r^2}}$$

where t has a Student's t distribution with $\nu = (n - 2)$ df. Derive this result using the result of Optional Exercise 10.45 and the fact that testing H_0: $\rho = 0$ is equivalent to testing H_0: $\beta_1 = 0$.

| | | | | | | | | | | | | |

S E C T I O N   10.9

USING THE MODEL FOR ESTIMATION AND PREDICTION

If we are satisfied that a useful model has been found to describe the relationship between the compression of the insulation material and compressive pressure, we are ready to accomplish the original objectives for building the model: using it to estimate or to predict the amount of compression for a particular level of compressive pressure.

The most common uses of a probabilistic model can be divided into two categories. **The first is the use of the model for estimating the mean value of y, $E(y)$, for a specific value of x.** For our example, we may want to estimate the mean amount of compression for all specimens of insulation subjected to a compressive pressure of 40 ($x = 4$) pounds per square inch. **The second use of the model entails predicting a particular y value for a given x.** That is, if we decide to install the insulation in a particular piece of equipment in which we think it will be subjected to a pressure of 40 pounds per square inch, we will want to predict the insulation compression for this particular specimen of insulation material.

In the case of estimating a mean value of y, we are attempting to estimate the mean result of a very large number of experiments at the given x value. In the second case, we are trying to predict the outcome of a single experiment at the given x value. In which of these model uses do you expect to have more success, i.e., which value—the mean or individual value of y—can we estimate (or predict) with greater accuracy?

Before answering this question, we first consider the problem of choosing an estimator (or predictor) of the mean (or individual) y value. We will use the least squares model

$$\hat{y} = \hat{\beta}_0 + \hat{\beta}_1 x$$

both to estimate the mean value of y and to predict a particular value of y for a given value of x. For our example, we found

$$\hat{y} = -.1 + .7x$$

so that the estimated mean compression of all specimens of insulation when x = 4 (compressive pressure of 40 pounds per square inch) is

$$\hat{y} = -.1 + .7(4) = 2.7$$

or .27 inch (the units of y are tenths of an inch). The identical value is used to predict the y value when x = 4. That is, both the estimated mean value and the predicted value of y equal $\hat{y} = 2.7$ when x = 4, as shown in Figure 10.12.

FIGURE 10.12

Estimated Mean Value and Predicted Individual Value of Compression y for x = 4

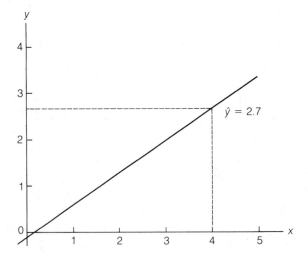

The difference in these two model uses lies in the relative accuracy of the estimate and the prediction. These accuracies are best measured by the repeated sampling errors of the least squares line when it is used as an estimator and as a predictor, respectively. These errors are given in the box.

**SAMPLING ERRORS FOR THE ESTIMATOR OF THE MEAN OF *y*
AND THE PREDICTOR FOR AN INDIVIDUAL *y***

1. The standard deviation of the sampling distribution of the estimator $\hat{y}$ of the mean value of y at a particular value of x, say x_p, is

$$\sigma_{\hat{y}} = \sigma\sqrt{\frac{1}{n} + \frac{(x_p - \bar{x})^2}{SS_{xx}}}$$

where σ is the standard deviation of the random error ε.

2. The standard deviation of the prediction error for the predictor $\hat{y}$ of an individual y value for $x = x_p$ is

$$\sigma_{(y-\hat{y})} = \sigma\sqrt{1 + \frac{1}{n} + \frac{(x_p - \bar{x})^2}{SS_{xx}}}$$

where σ is the standard deviation of the random error ε.

The true value of σ will rarely be known. Thus, we estimate σ by s and calculate the estimation and prediction intervals as shown in the following boxes.

A $(1 - \alpha)$100% CONFIDENCE INTERVAL FOR THE MEAN VALUE OF *y* FOR $x = x_p$

$$\hat{y} \pm t_{\alpha/2}(\text{Estimated standard deviation of } \hat{y})$$

or

$$\hat{y} \pm t_{\alpha/2}s\sqrt{\frac{1}{n} + \frac{(x_p - \bar{x})^2}{SS_{xx}}}$$

where $t_{\alpha/2}$ is based on $(n - 2)$ df

A $(1 - \alpha)$100% PREDICTION INTERVAL FOR AN INDIVIDUAL *y* FOR $x = x_p$

$$\hat{y} \pm t_{\alpha/2}[\text{Estimated standard deviation of } (y - \hat{y})]$$

or

$$\hat{y} \pm t_{\alpha/2}s\sqrt{1 + \frac{1}{n} + \frac{(x_p - \bar{x})^2}{SS_{xx}}}$$

where $t_{\alpha/2}$ is based on $(n - 2)$ df

EXAMPLE 10.8

Find a 95% confidence interval for the mean insulation compression when the pressure is 40 pounds per square inch.

SOLUTION

For a compressive pressure of 40 pounds per square inch, $x_p = 4$ and, since $n = 5$, df $= n - 2 = 3$. Then the confidence interval for the mean value of y is

$$\hat{y} \pm t_{\alpha/2} s \sqrt{\frac{1}{n} + \frac{(x_p - \bar{x})^2}{SS_{xx}}}$$

or

$$\hat{y} \pm t_{.025} s \sqrt{\frac{1}{5} + \frac{(4 - \bar{x})^2}{SS_{xx}}}$$

Recall that $\hat{y} = 2.7$, $s = .61$, $\bar{x} = 3$, and $SS_{xx} = 10$. From Table 6 of Appendix II, $t_{.025} = 3.182$. Thus, we have

$$2.7 \pm (3.182)(.61) \sqrt{\frac{1}{5} + \frac{(4 - 3)^2}{10}} = 2.7 \pm (3.182)(.61)(.55)$$

$$= 2.7 \pm 1.1$$

We estimate that the interval from .16 inch to .38 inch encloses the mean amount of compression when the insulation is subjected to a compressive pressure of 40 pounds per square inch. Note that we used a small amount of data for purposes of illustration in fitting the least squares line and that the width of the interval could be decreased by using a larger number of data points.   ∎

EXAMPLE 10.9

Predict the amount of compression for an individual piece of insulation material subjected to a compressive pressure of 40 pounds per square inch. Use a 95% prediction interval.

SOLUTION

To predict the compression for a particular piece of insulation material for which $x_p = 4$, we calculate the 95% prediction interval as

$$\hat{y} \pm t_{\alpha/2} s \sqrt{1 + \frac{1}{n} + \frac{(x_p - \bar{x})^2}{SS_{xx}}} = 2.7 \pm (3.182)(.61) \sqrt{1 + \frac{1}{5} + \frac{(4 - 3)^2}{10}}$$

$$= 2.7 \pm (3.182)(.61)(1.14) = 2.7 \pm 2.2$$

Therefore, we predict that the compression for the piece of insulation material will fall in the interval from .05 inch to .49 inch. As in the case for the confidence interval for the mean value of y, the prediction interval for y is quite large. This is because we have chosen a simple example (only five data points) to fit the least squares line. The width of the prediction interval could be reduced by using a larger number of data points.   ∎

A comparison of the confidence interval for the mean value of y and the prediction interval for some future value of y for a compressive pressure of 40 pounds per square inch ($x = 4$) is illustrated in Figure 10.13. It is important to note that the prediction interval for an individual value of y will always be wider than the confidence interval for a mean value of y. You can see this by examining the formulas for the two intervals and you can see it in Figure 10.13.

FIGURE 10.13

A 95% Confidence
Interval for Mean
Compression and a
Prediction Interval for
Compression When $x = 4$

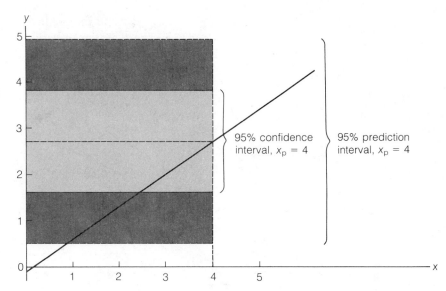

The error in estimating the mean value of y, $E(y)$, for a given value of x, say x_p, is the distance between the least squares line and the true line of means, $E(y) = \beta_0 + \beta_1 x$. This error, $[\hat{y} - E(y)]$, is shown in Figure 10.14. In contrast, the error $(y_p - \hat{y})$ in predicting some future value of y is the sum of the two errors—the error of estimating the mean of y, $E(y)$, shown in Figure 10.14, plus the random error that is a component of the value of y to be predicted (see Figure 10.15, page 460). Consequently, the error of predicting a particular value of y will usually be larger than the error of estimating the mean value of y for a particular value of x. Note from their formulas that both the error of estimation and the error of prediction take their smallest values when $x_p = \bar{x}$. The farther x lies from $\bar{x}$, the larger will be the errors of estimation and prediction. You can see why this is true by noting the deviations for different values of x between

FIGURE 10.14

Error of Estimating the
Mean Value of y for a
Given Value of x

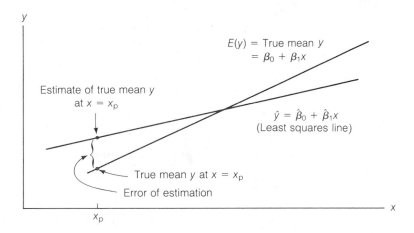

the line of means $E(y) = \beta_0 + \beta_1 x$ and the predicted line of means $\hat{y} = \hat{\beta}_0 + \hat{\beta}_1 x$ shown in Figure 10.15. The deviation is larger at the extremities of the interval where the largest and smallest values of x in the data set occur.

FIGURE 10.15

Error of Predicting a
Future Value of y for a
Given Value of x

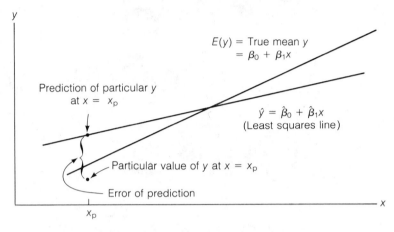

To conclude this section, we will find the variance of the value of $\hat{y}$ when $x = x_p$. This variance plays an important role in developing the confidence interval for $E(y)$ when $x = x_p$ and the prediction interval for a particular value of y when $x = x_p$.

EXAMPLE 10.10

Find the variance of $\hat{y}$ when $x = x_p$.

SOLUTION

When $x = x_p$, we have $\hat{y} = \hat{\beta}_0 + \hat{\beta}_1 x_p$. Adding and subtracting $\hat{\beta}_1 \bar{x}$ yields

$$\hat{y} = \hat{\beta}_0 + \hat{\beta}_1 \bar{x} + \hat{\beta}_1 x_p - \hat{\beta}_1 \bar{x} = \hat{\beta}_0 + \hat{\beta}_1 \bar{x} + \hat{\beta}_1 (x_p - \bar{x})$$

where $\hat{\beta}_0 = \bar{y} - \hat{\beta}_1 \bar{x}$. Substituting this value of $\hat{\beta}_0$ into the expression for $\hat{y}$, we obtain

$$\hat{y} = \bar{y} - \hat{\beta}_1 \bar{x} + \hat{\beta}_1 \bar{x} + \hat{\beta}_1 (x_p - \bar{x}) = \bar{y} + \hat{\beta}(x_p - \bar{x})$$

The next step is to express $\hat{y}$ as a linear function of the random y values, $y_1, y_2, \ldots, y_n$, so that we can obtain $V(\hat{y})$ as the variance of a linear function of independent random variables. We now write

$$\hat{y} = \bar{y} + \hat{\beta}_1 (x_p - \bar{x}) = \sum \frac{y_i}{n} + \frac{(x_p - \bar{x})}{SS_{xx}} \sum (x_i - \bar{x}) y_i$$

$$= \sum \frac{y_i}{n} + \sum \frac{(x_p - \bar{x})(x_i - \bar{x})}{SS_{xx}} y_i$$

We can now express $\hat{y}$ as a single summation:

$$\hat{y} = \sum \left[\frac{1}{n} + \frac{(x_p - \bar{x})(x_i - \bar{x})}{SS_{x\dot{x}}} \right] y_i$$

i.e., $\hat{y}$ is a linear function of the independent random variables, $y_1, y_2, \ldots, y_n$, where the coefficient of y_i is

$$\left[\frac{1}{n} + \frac{(x_p - \bar{x})(x_i - \bar{x})}{SS_{xx}}\right]$$

Then, by Theorem 6.7,

$$V(\hat{y}) = \sum \left[\frac{1}{n} + \frac{(x_p - \bar{x})(x_i - \bar{x})}{SS_{xx}}\right]^2 V(y_i)$$

where $V(y_i) = \sigma^2$, $i = 1, 2, \ldots, n$. Therefore,

$$V(\hat{y}) = \sum \left[\frac{1}{n^2} + \frac{2}{n}\frac{(x_p - \bar{x})(x_i - \bar{x})}{SS_{xx}} + \frac{(x_p - \bar{x})^2(x_i - \bar{x})^2}{(SS_{xx})^2}\right]\sigma^2$$

$$= \left[\frac{n}{n^2} + \frac{2}{n}\frac{(x_p - \bar{x})}{SS_{xx}}\sum (x_i - \bar{x}) + \frac{(x_p - \bar{x})^2}{(SS_{xx})^2}\sum (x_i - \bar{x})^2\right]\sigma^2$$

$$= \left[\frac{1}{n} + \frac{(x_p - \bar{x})^2}{(SS_{xx})^2}SS_{xx}\right]\sigma^2 \quad \text{since } \sum (x_i - \bar{x}) = 0$$

$$= \sigma^2\left[\frac{1}{n} + \frac{(x_p - \bar{x})^2}{SS_{xx}}\right]$$

You can see that this agrees with the formula for $V(\hat{y})$ given earlier in this section. ∎

EXERCISES 10.47–10.57

10.47 Refer to Exercise 10.10. Find a 95% confidence interval for the mean winning bid price when four contractors bid on a road construction project.

10.48 Refer to Exercise 10.29. Television advertising for videogames next month will be $750.
a. Find a 90% prediction interval for videogame sales next month.
b. Find a 90% confidence interval for the mean monthly videogame sales when television advertising is $750 per month.
c. Compare and comment on the sizes of the intervals in parts a and b.
d. Could you reduce the size of either or both intervals by increasing your sample size? Explain.

10.49 Explain why for a particular x value, the prediction interval for an individual y value will always be wider than the confidence interval for the mean value of y.

10.50 Explain why the confidence interval for the mean value of y for a particular x value, say x_p, gets wider the farther x_p is from $\bar{x}$. What are the implications of this phenomenon for estimation and prediction?

10.51 Refer to Exercise 10.30. Find a 90% confidence interval for the mean peak power load during all days with a high temperature of 90°F.

10.52 In planning for an initial orientation meeting with new computer science majors, the chairman of the Computer and Information Sciences Department wants to emphasize the importance of doing well in the major courses in order to get better-paying jobs after graduation. To support this point, the chairman plans to show that there is a strong positive correlation between starting salaries for recent computer science graduates and their grade-point averages in the major courses. Records for seven of last year's computer science graduates are selected at random and are given in the table on page 462.

GRADE-POINT AVERAGE IN MAJOR COURSES x	STARTING SALARY y, thousands of dollars
2.58	15.8
3.27	23.8
3.85	25.5
3.50	25.2
3.33	24.5
2.89	22.6
2.23	17.6

a. Find the least squares prediction equation.
b. Plot the data and graph the line as a check on your calculations.
c. Find a 95% prediction interval for a graduate whose grade-point average is 3.2.
d. What is the mean starting salary for graduates with grade-point averages equal to 3.0? Use a 95% confidence interval.

10.53 An automated system for marking large numbers of student computer programs, called AUTOMARK, has been used successfully at McMaster University in Ontario, Canada. AUTOMARK takes into account both program correctness and program style when marking student assignments. To evaluate the effectiveness of the automated system, AUTOMARK was used to grade the FORTRAN77 assignments of a class of 33 students. These grades were then compared to the grades assigned by the instructor. The results are shown in the accompanying table.

AUTOMARK GRADE x	INSTRUCTOR GRADE y	AUTOMARK GRADE x	INSTRUCTOR GRADE y
12.2	10	17.8	17
10.6	11	18.0	17
15.1	12	18.2	17
16.2	12	18.4	17
16.6	12	18.6	17
16.6	13	19.0	17
17.2	14	19.3	17
17.6	14	19.5	17
18.2	14	19.7	17
16.5	15	18.6	18
17.2	15	19.0	18
18.2	15	19.2	18
15.1	16	19.4	18
17.2	16	19.6	18
17.5	16	20.1	18
18.6	16	19.2	19
18.8	16		

Source: Redish, K. A. and Smyth, W. F. "Program Style Analysis: A Natural Byproduct of Program Compilation." *Communications of the Association for Computing Machinery*, Vol. 29, No. 2, Feb. 1986, p. 132 (Figure 4). Copyright 1986, Association for Computing Machinery, Inc.

a. Find the least squares prediction equation for the straight-line model relating instructor grade y to AUTOMARK grade x.
b. Is there sufficient evidence to indicate that the model is useful for predicting y? Test using $\alpha = .10$.
c. Calculate a 95% prediction interval for the instructor-assigned grade of a FORTRAN77 assignment that received an AUTOMARK score of 17.5. Interpret the interval.

OPTIONAL EXERCISES

10.54 Suppose you want to predict some future value of y when $x = x_p$ using the prediction equation $\hat{y} = \hat{\beta}_0 + \hat{\beta}_1 x$. The error of prediction will be the difference between the actual value of y_p and the predicted value $\hat{y}$, i.e.,

Error of prediction $= y_p - \hat{y}$

a. Explain why the error of prediction will be normally distributed.
b. Find the expected value and the variance of the error of prediction.

10.55 Explain why

$$z = \frac{\text{Error of prediction}}{\text{Standard devation of the error}}$$

$$= \frac{y_p - \hat{y}}{\sigma_{(y_p - \hat{y})}} = \frac{y_p - \hat{y}}{\sigma\sqrt{1 + \dfrac{1}{n} + \dfrac{(x_p - \bar{x})^2}{SS_{xx}}}}$$

is a standard normal random variable.

10.56 Show that

$$t = \frac{\text{Error of prediction}}{\text{Estimated standard deviation of the error}}$$

$$= \frac{y_p - \hat{y}}{s\sqrt{1 + \dfrac{1}{n} + \dfrac{(x_p - \bar{x})^2}{SS_{xx}}}}$$

has a Student's t distribution with $\nu = (n - 2)$ df.

10.57 Use the t statistic in Optional Exercise 10.56 as a pivotal statistic to derive a $(1 - \alpha)100\%$ prediction interval for y_p.

| | | | | | | | | | | |

S E C T I O N 10.10

SIMPLE LINEAR REGRESSION: AN EXAMPLE

In the previous sections we have presented the basic elements necessary to fit and use a straight-line regression model. In this section we will assemble these elements by applying them in an example.

Suppose a fire insurance company wants to relate the amount of fire damage in major residential fires to the distance between the residence and the nearest fire station. The study is to be conducted in a large suburb of a major city; a sample of 15 recent fires in this suburb is selected. The amount of damage y and the distance x between the fire and the nearest fire station are recorded for each fire. The results are given in Table 10.6 (page 464).

TABLE 10.6
Fire Damage Data

DISTANCE FROM FIRE STATION x, miles	FIRE DAMAGE y, thousands of dollars
3.4	26.2
1.8	17.8
4.6	31.3
2.3	23.1
3.1	27.5
5.5	36.0
.7	14.1
3.0	22.3
2.6	19.6
4.3	31.3
2.1	24.0
1.1	17.3
6.1	43.2
4.8	36.4
3.8	26.1

STEP 1   First, we hypothesize a model to relate fire damage y to the distance x from the nearest fire station. We will hypothesize a straight-line probabilistic model:

$$y = \beta_0 + \beta_1 x + \varepsilon$$

STEP 2   Next, we use the data to estimate the unknown parameters in the deterministic component of the hypothesized model. We make some preliminary calculations:

$$SS_{xx} = \sum x_i^2 - \frac{\left(\sum x_i\right)^2}{15} = 196.16 - \frac{(49.2)^2}{15}$$

$$= 196.160 - 161.376 = 34.784$$

$$SS_{yy} = \sum y_i^2 - \frac{\left(\sum y_i\right)^2}{15} = 11{,}376.48 - \frac{(396.2)^2}{15}$$

$$= 11{,}376.480 - 10{,}464.963 = 911.517$$

$$SS_{xy} = \sum x_i y_i - \frac{\left(\sum x_i\right)\left(\sum y_i\right)}{15} = 1{,}470.65 - \frac{(49.2)(396.2)}{15}$$

$$= 1{,}470.650 - 1{,}299.536 = 171.114$$

Then the least squares estimates of the slope β_1 and intercept β_0 are

$$\hat{\beta}_1 = \frac{SS_{xy}}{SS_{xx}} = \frac{171.114}{34.784} = 4.919$$

$$\hat{\beta}_0 = \bar{y} - \hat{\beta}_1 \bar{x} = \frac{396.2}{15} - 4.919\left(\frac{49.2}{15}\right)$$

$$= 26.413 - (4.919)(3.28) = 26.413 - 16.134$$

$$= 10.279$$

And the least squares equation is

$$\hat{y} = 10.279 + 4.919x$$

This prediction equation is graphed in Figure 10.16, along with a plot of the data points.

FIGURE 10.16

Least Squares Model for
the Fire Damage Data

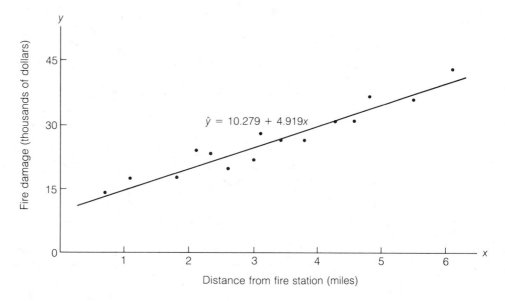

STEP 3   Now, we specify the probability distribution of the random error component ε. The assumptions about the distribution will be identical to those listed in Section 10.2. Although we know that these assumptions are not completely satisfied (they rarely are for any practical problem), we are willing to assume they are approximately satisfied for this example. We have to estimate the variance σ^2 of ε, so we calculate

$$SSE = \sum (y_i - \hat{y}_i)^2 = SS_{yy} - \hat{\beta}_1 SS_{xy}$$

where the last expression represents a shortcut formula for SSE. Thus,

$$SSE = 911.517 - (4.919)(171.114) = 911.517 - 841.709766 = 69.807234^*$$

To estimate σ^2, we divide SSE by the degrees of freedom available for error, $(n - 2)$. Thus,

$$s^2 = \frac{SSE}{n-2} = \frac{69.807234}{15-2} = 5.3698 \quad \text{and} \quad s = \sqrt{5.3698} = 2.32$$

The value of s implies that most of the observed fire damage values (y's) will fall within $2s = 4.64$ thousand dollars of their respective predicted values.

*We have used the exact (not rounded) values for SS_{yy}, $\hat{\beta}_1$, and SS_{xy} to calculate SSE for this example. In problems where rounding is necessary, at least six significant figures should be carried for these quantities. Otherwise, the calculated value of SSE may be substantially in error.

STEP 4   We can now check the utility of the hypothesized model, that is, whether x really contributes information for the prediction of y using the straight-line model. First test the null hypothesis that the slope β_1 is 0, i.e., that there is no linear relationship between fire damage and the distance from the nearest fire station. We test:

$$H_0: \quad \beta_1 = 0$$
$$H_a: \quad \beta_1 \neq 0$$

Test statistic: $t = \dfrac{\hat{\beta}_1 - 0}{s_{\hat{\beta}_1}} = \dfrac{\hat{\beta}_1}{s/\sqrt{SS_{xx}}}$

Assumptions: Those made about ε in Section 10.2.

For $\alpha = .05$, we will reject H_0 if

$$t > t_{\alpha/2} \quad \text{or} \quad t < - t_{\alpha/2}$$

where for $n = 15$, df $= (n - 2) = (15 - 2) = 13$ and $t_{.025} = 2.160$. We then calculate the t statistic:

$$t = \frac{\hat{\beta}_1}{s_{\hat{\beta}_1}} = \frac{\hat{\beta}_1}{s/\sqrt{SS_{xx}}} = \frac{4.919}{2.32/\sqrt{34.784}} = \frac{4.919}{.393} = 12.5$$

This large t value leaves little doubt that distance between the fire and the fire station contributes information for the prediction of fire damage. In fact, it appears that fire damage increases as the distance increases.

We gain additional information about the relationship by forming a confidence interval for the slope β_1. A 95% confidence interval is

$$\hat{\beta}_1 \pm t_{.025} s_{\hat{\beta}_1} = 4.919 \pm (2.160)(.393)$$
$$= 4.919 \pm .849 = (4.070, 5.768)$$

We estimate that the interval from \$4,070 to \$5,768 encloses the mean increase (β_1) in fire damage per additional mile distance from the fire station.

Another measure of the utility of the model is the coefficient of correlation r:

$$r = \frac{SS_{xy}}{\sqrt{SS_{xx}SS_{yy}}} = \frac{171.114}{\sqrt{(34.784)(911.517)}} = \frac{171.114}{178.062} = .96$$

The high correlation confirms our conclusion that β_1 differs from 0; it appears that fire damage and distance from the fire station are highly correlated.

The coefficient of determination is

$$r^2 = (.96)^2 = .92$$

which implies that 92% of the sum of squares of deviations of the y values about $\bar{y}$ is explained by the distance x between the fire and the fire station. All signs point to a strong linear relationship between x and y.

STEP 5   We are now prepared to use the least squares model. Suppose the insurance company wants to predict the fire damage if a major residential fire were to occur 3.5 miles from the nearest fire station, i.e., $x_p = 3.5$. The predicted value is

$$\hat{y} = \hat{\beta}_0 + \hat{\beta}_1 x_p = 10.279 + (4.919)(3.5)$$
$$= 10.279 + 17.216 = 27.5$$

(we round to the nearest tenth to be consistent with the units of the original data in Table 10.6). If we want a 95% prediction interval, we calculate

$$\hat{y} \pm t_{.025} s \sqrt{1 + \frac{1}{n} + \frac{(x_p - \bar{x})^2}{SS_{xx}}} = 27.5 \pm (2.16)(2.32) \sqrt{1 + \frac{1}{15} + \frac{(3.5 - 3.28)^2}{34.784}}$$
$$= 27.5 \pm (2.16)(2.32)\sqrt{1.0681}$$
$$= 27.5 \pm 5.2 = (22.3, 32.7)$$

The model yields a 95% prediction interval for fire damage in a major residential fire 3.5 miles from the nearest station of $22,300 to $32,700.

Caution: We would not use this prediction model to make predictions for homes less than .7 mile or more than 6.1 miles from the nearest fire station. A look at the data in Table 10.6 reveals that all the x values fall between .7 and 6.1. It is dangerous to use the model to make predictions outside the region in which the sample data fall. A straight line might not provide a good model for the relationship between the mean value of y and the value of x when stretched over a wider range of x values.

WARNING

Using the least squares prediction equation to estimate the mean value of y or to predict a particular value of y for values of x that fall outside the range of the values of x contained in your sample data may lead to errors of estimation or prediction that are much larger than expected. Although the least squares model may provide a very good fit to the data over the range of x values contained in the sample, it could give a poor representation of the true model for values of x outside this region.

| | | | | | | | | | | | |

SECTION 10.11

A COMPUTER PRINTOUT FOR SIMPLE LINEAR REGRESSION

The computations required for a simple linear regression analysis, even if performed with the aid of a pocket or desk calculator, can become tedious and cumbersome, especially if the number of measurements is large. Many institutions have installed statistical computer program packages which fit a straight-line regression model by the method of least squares. These computer packages enable the user to greatly decrease the burden of calculation. In this section, we locate, discuss, and interpret the elements of a simple linear regression on a computer printout from one of the more popular statistical computer program packages, called the SAS® System.* At your institution, other computer packages (such as BMDP, Minitab, and SPSS^X) may be installed. These packages also have simple linear regression procedures available and their respective printouts are similar to that of the SAS System. (Computer printouts for BMDP, Minitab, and SPSS^X are shown and discussed in Chapter 11.) Regardless of the package(s) to which you have access, you should be able to understand the regression results after reading the discussion that follows.

The SAS printout for the simple linear regression analysis of the data relating fire damage to distance from fire station in Table 10.6 is shown in Figure 10.17.

First, the estimates of the y-intercept and the slope are found about halfway down the printout on the left-hand side, under the column labeled Parameter Estimate and in the rows labeled INTERCEP and X, respectively. The values are $\hat{\beta}_0 = 10.277929$ and $\hat{\beta}_1 = 4.919331$. When rounded to three decimal places, these quantities agree with our calculations in Section 10.10.

Next, we find the measures of variability: SSE, s^2, and s. They are shaded in the upper portion of the printout. SSE is found under the column heading Sum of Squares and in the row labeled Error: SSE = 69.75098. The estimate of the error variance σ^2 is under the column heading Mean Square and in the row labeled Error: $s^2 = 5.36546$. The estimate of the standard deviation σ is directly to the right of the heading Root MSE: $s = 2.31635$. Again, all values (after rounding) agree with our corresponding hand-computed quantities.

The coefficient of determination is shown (shaded) under the heading R-Square in the upper portion of the printout: $r^2 = .9235$. Again, to two decimal places, this agrees with our hand-calculated value. The coefficient of correlation r is not given on the printout.

The t statistic for testing $H_0: \beta_1 = 0$ versus $H_a: \beta_1 \neq 0$ is given (shaded) in the center of the page under the column heading T for H0: Parameter = 0 in the row corresponding to X. The value $t = 12.525$ agrees with our computed value when we used the formula

$$t = \frac{\hat{\beta}_1}{s_{\hat{\beta}_1}} = \frac{\hat{\beta}_1}{s/\sqrt{SS_{xx}}}$$

To determine which hypothesis this test statistic supports, we can establish a rejection region using the t table (Table 6 of Appendix II), just as we did in Section 10.10. However, the printout makes this unnecessary, because the observed significance level, or p-value, is shown (shaded) immediately to the right of the t statistic, under the column heading PROB > |T|. Remember that if the observed significance level is less than the α value you select, then the test statistic supports the alternative hypothesis at that level. For example, if we select $\alpha = .05$ in this example, the observed significance level of .0001 given on the printout indicates that we should reject H_0. We can conclude that there is sufficient evidence at $\alpha = .05$ to infer that a linear relationship between fire damage and distance from the station is useful for predicting damage.

*SAS is the registered trademark of SAS Institute, Inc., Cary, North Carolina, U.S.A.

FIGURE 10.17 SAS Printout for the Fire Damage Linear Regression

---

FIRE DAMAGE EXAMPLE
STRAIGHT—LINE MODEL WITH PREDICTION INTERVALS

Model: MODEL1
Dep Variable: Y

Analysis of Variance

Source	DF	Sum of Squares	Mean Square	F Value	Prob>F
Model	1	841.76636	841.76636	156.886	0.0001
Error	13	69.75098	5.36546		
C Total	14	911.51733			

Root MSE	2.31635	R-Square	0.9235	
Dep Mean	26.41333	Adj R-Sq	0.9176	
C.V.	8.76961			

Parameter Estimates

Variable	DF	Parameter Estimate	Standard Error	T for H0: Parameter=0	Prob > |T|
INTERCEP	1	10.277929	1.42027781	7.237	0.0001
X	1	4.919331	0.39274775	12.525	0.0001

Obs	X	Y	Predict Value	Residual	Lower95% Predict	Upper95% Predict
1	3.4	26.2000	27.0037	-0.8037	21.8344	32.1729
2	1.8	17.8000	19.1327	-1.3327	13.8141	24.4514
3	4.6	31.3000	32.9068	-1.6068	27.6186	38.1951
4	2.3	23.1000	21.5924	1.5076	16.3577	26.8271
5	3.1	27.5000	25.5279	1.9721	20.3573	30.6984
6	5.5	36.0000	37.3342	-1.3342	31.8334	42.8351
7	0.7	14.1000	13.7215	0.3785	8.1087	19.3342
8	3	22.3000	25.0359	-2.7359	19.8622	30.2097
9	2.6	19.6000	23.0682	-3.4682	17.8678	28.2686
10	4.3	31.3000	31.4311	-0.1311	26.1908	36.6713
11	2.1	24.0000	20.6085	3.3915	15.3442	25.8729
12	1.1	17.3000	15.6892	1.6108	10.1999	21.1785
13	6.1	43.2000	40.2858	2.9142	34.5906	45.9811
14	4.8	36.4000	33.8907	2.5093	28.5640	39.2175
15	3.8	26.1000	28.9714	-2.8714	23.7843	34.1585
16 *	3.5	.	27.4956	.	22.3239	32.6672

Sum of Residuals   -3.73035E-14
Sum of Squared Residuals   69.7510
Predicted Resid SS (Press)   93.2117

If you wish to conduct a one-tailed test, the observed significance level is half that given on the printout (assuming that the sign of the test statistic agrees with the alternative hypothesis). Thus, if we were testing H_0: $\beta_1 = 0$ versus H_a: $\beta_1 > 0$ in this example, the observed significance level would be $\frac{1}{2}(.0001) = .00005$.

Predicted y values and the corresponding prediction intervals are given in the lower portion of the SAS printout. To find the 95% prediction interval for the fire damage y when the distance from the fire station is $x = 3.5$ miles, first locate the value 3.5 in the column labeled X (the last value in the column). The prediction is given in the center column labeled Predict Value in the row corresponding to 3.5: $\hat{y} = 27.4956$. The lower and upper confidence bounds are given in the columns headed Lower and Upper 95% Predict, respectively: Lower = 22.3239 and Upper = 32.6672. Again, all values agree after rounding with our calculations in Section 10.10.

In Chapter 12 we will discuss the interpretation of those portions of the SAS printout which were not mentioned here. However, the important elements of a simple linear regression analysis have been located, and you should be able to use this discussion as a guide to interpreting simple linear regression computer printouts. Details on how to perform a simple linear regression analysis on the computer using any one of the four program packages, SAS, SPSS^X, Minitab, or BMDP, are provided in Chapter 11.

SECTION 10.12

SUMMARY

We have introduced an extremely useful tool in this chapter—the **method of least squares** for fitting a prediction equation to a set of data. This procedure, along with associated statistical tests and estimations, is called a **regression analysis**. In five steps we showed how to use sample data to build a model relating a dependent variable y to a single independent variable x.

STEPS TO FOLLOW IN A SIMPLE LINEAR REGRESSION ANALYSIS

1. The first step is to hypothesize a **probabilistic model**. In this chapter, we confined our attention to the **straight-line model**, $y = \beta_0 + \beta_1 x + \varepsilon$.

2. The second step is to use the method of least squares to estimate the unknown parameters in the **deterministic component**, $\beta_0 + \beta_1 x$. The least squares estimates yield a model $\hat{y} = \hat{\beta}_0 + \hat{\beta}_1 x$ with a **sum of squared errors (SSE)** that is smaller than the SSE for any other straight-line model.

3. The third step is to specify the probability distribution of the **random error component** ε.

4. The fourth step is to assess the utility of the hypothesized model. Included here are making inferences about the **slope** β_1, calculating the **coefficient of correlation** r, and calculating the **coefficient of determination** r^2.

5. Finally, if we are satisfied with the model, we are prepared to use it. We used the model to **estimate the mean y value**, $E(y)$, for a given x value and to **predict an individual y value** for a specific value of x.

The following chapter will develop more fully the concepts introduced in this chapter.

SUPPLEMENTARY
EXERCISES
10.58–10.67

10.58 At temperatures approaching absolute zero (273 degrees below zero Celsius), helium exhibits traits that defy many laws of conventional physics. An experiment has been conducted with helium in solid form at various temperatures near absolute zero. The solid helium is placed in a dilution refrigerator along with a solid impure substance, and the proportion (by weight) of the impurity passing through the solid helium is recorded. (This phenomenon of solids passing directly through solids is known as *quantum tunnelling*.) The data are given in the table.

PROPORTION OF IMPURITY PASSING THROUGH HELIUM y	TEMPERATURE x, °C
.315	−262
.202	−265
.204	−256
.620	−267
.715	−270
.935	−272
.957	−272
.906	−272
.985	−273
.987	−273

a. Construct a scattergram of the data.
b. Find the least squares line for the data and plot it on your scattergram.
c. Define β_1 in the context of this problem.
d. Test the hypothesis (at $\alpha = .05$) that temperature contributes no information for the prediction of the proportion of impurity passing through helium when a linear model is used. Draw the appropriate conclusions.
e. Find a 90% confidence interval for β_1. Interpret your results.
f. Find the coefficient of correlation for the given data.
g. Find the coefficient of determination for the linear model you constructed in part b. Interpret your result.
h. Find a 99% prediction interval for the proportion of impurity passing through helium when the temperature is set at −270°C.
i. Estimate the mean proportion of impurity passing through helium when the temperature is set at −270°C. Use a 99% confidence interval.

10.59 The operating characteristics of zinc bromine batteries are partially dependent upon the vapor pressure of the bromine complex used in the battery. The accompanying data (page 472) provide information on the relationship between vapor pressure of a bromine complex (tetramethylammonium bromide) as a function of its temperature.

TEMPERATURE x, °C	VAPOR PRESSURE y, mm Hg	TEMPERATURE x, °C	VAPOR PRESSURE y, mm Hg
0	0.7	25.5	28.7
5.1	4.3	32.0	35.6
10.4	11.0	39.5	49.0
14.7	17.4	49.7	67.0

Source: Bajpal, S. N. "Vapor Pressures of Bromine-Quaternary Ammonium Salt Complexes for Zinc Bromine Battery Applications." *Journal of Chemical and Engineering Data*, Vol. 26, 1981, pp. 2–4. Copyright 1981, American Chemical Society. Reprinted with permission.

a. Construct a scattergram for the data.
b. Find the least squares fit of the model $y = \beta_0 + \beta_1 x + \varepsilon$.
c. Compute r^2 and interpret its value.
d. Compute s and interpret its value.
e. Is there evidence that temperature is a useful linear predictor of vapor pressure? Test using $\alpha = .01$.
f. Estimate the mean increase in vapor pressure y for an increase of 1°C in bromine temperature x. Use a 99% confidence interval.
g. Find a 99% prediction interval for the vapor pressure of the bromine complex used in a battery when the bromine temperature is 30°C.

10.60 The data in the table give the market share for a new user-friendly computer as a function of television advertising expenditures.

MONTH	MARKET SHARE y, %	TELEVISION ADVERTISING EXPENDITURE x, thousands of dollars
January	15	23
March	17	25
May	13	21
July	14	24
September	16	26

a. Use the methods of this chapter to find the least squares line relating market share to television advertising expenditure. Plot the data and graph the least squares line as a check on your calculations.
b. Do the data provide sufficient evidence to indicate that x contributes information for the prediction of y? Test using $\alpha = .10$.
c. Find a 95% confidence interval for β_1 and interpret your result.
d. Find a 90% confidence interval for the expected market share when $25,000 is spent on television advertising.
e. Find a 95% prediction interval for the market share that will be obtained when $23,000 is spent in television advertising.

10.61 In order to satisfy the Public Service Commission's energy conservation requirements, an electric utility company must develop a reliable model for projecting the number of residential electric customers in its service area. The first step is to study the effect of changing population on electric customers. The information shown in the table was obtained for the service area from 1977 to 1986.

YEAR	POPULATION IN SERVICE AREA x, hundreds	RESIDENTIAL ELECTRIC CUSTOMERS IN SERVICE AREA y
1977	262	14,041
1978	319	16,953
1979	361	18,984
1980	381	19,870
1981	405	20,953
1982	439	22,538
1983	472	23,985
1984	508	25,641
1985	547	27,365
1986	592	29,967

a. Find the least squares line relating residential electric customers to population in service area.
b. Plot the data and graph the line as a check on your calculations.
c. Does x contribute information for the prediction of y?
d. Calculate r and r^2 and interpret their values.
e. Find a 90% confidence interval for the mean number of electric customers when the service area population is 72,900.
f. Suppose the 1990 population projection was published to be 72,900. Find a 90% prediction interval for the projected number of residential electric customers.
g. Refer to your answers to parts e and f. Discuss the problems associated with using the least squares line to forecast the number of residential electric customers in 1990.

10.62 The accompanying table shows a portion of the experimental data obtained in a study of the radial tension strength of concrete pipe. The concrete pipe used for the experiment had an inside diameter of 84 inches and a wall thickness of approximately 8.75 inches. In addition, it was reinforced with cold drawn wire. The response y is the load (in pounds per foot) until the first crack in a pipe specimen was observed. The independent variable x is the age of the specimen (in days) at the time of the test.

LOAD lb/ft	AGE days	LOAD lb/ft	AGE days
11,450	20	10,540	25
10,420	20	9,470	31
11,142	20	9,190	31
10,840	25	9,540	31
11,170	25		

Source: Heger, F. J. and McGrath, T. J. "Radial Tension Strength of Pipe and Other Curved Flexural Members." *Journal of the American Concrete Institute*, Vol. 80, No. 1, 1983, pp. 33–39.

a. Do the data provide sufficient evidence to indicate that the load required before the first crack is observed decreases linearly with the age of the pipe? Test using $\alpha = .05$.

b. Estimate the mean load required before the first crack is observed for 28-day-old pipe specimens. Use a 95% confidence interval.

c. Comment on the width of the interval you constructed in part **b**.

10.63 To investigate the relationship between yield of potatoes, y, and level of fertilizer application, x, an experimenter divided a field into eight plots of equal size and applied differing amounts of fertilizer to each. The yield of potatoes (in pounds) and the fertilizer application (in pounds) were recorded for each plot. The data are shown in the table.

x	1	1.5	2	2.5	3	3.5	4	4.5
y	25	31	27	28	36	35	32	34

a. Is there sufficient evidence to indicate that yield y and level of fertilizer application x are positively correlated? Test at $\alpha = .01$.

b. Calculate r^2 and interpret its value.

c. Construct a 99% confidence interval for the mean yield of potatoes from plots to which 3.75 pounds of fertilizer have been applied.

10.64 "In the analysis of urban transportation systems it is important to be able to estimate expected travel time between locations." Cook and Russell (*Transportation Research*, June 1980) collected data in the city of Tulsa on the urban travel times and distances between locations for two types of vehicles—large hoist compactor trucks and passenger cars. A simple linear regression analysis was conducted for both sets of data (y = urban travel time in minutes, x = distance between locations in miles) with the results shown in the accompanying table.

PASSENGER CARS	TRUCKS
$\hat{y} = 2.50 + 1.93x$	$\hat{y} = 1.85 + 3.86x$
$r^2 = .676$; p-value $< .05$	$r^2 = .758$; p-value $< .01$

Source: Cook, T. M. and Russell, R. A. "Estimating Urban Travel Times: A Comparative Study." *Transportation Research*, 14A, June 1980, pp. 173–175. Copyright 1980, Pergamon Press, Ltd. Reprinted with permission.

a. Is there sufficient evidence to indicate that distance between locations is linearly related to urban travel time for passenger cars? Test at $\alpha = .05$.

b. Is there sufficient evidence to indicate that distance between locations is linearly related to urban travel time for trucks? Test at $\alpha = .01$.

c. Interpret the value of r^2 for the two prediction equations.

d. Estimate the mean urban travel time for all passenger cars traveling a distance of 3 miles on Tulsa's highways.

e. Predict the urban travel time for a particular truck traveling a distance of 5 miles on Tulsa's highways.

f. Explain how we could attach a measure of reliability to the inferences derived in parts **d** and **e**.

10.65 The Environmental Protection Agency establishes industrial and occupational standards for the ambient air quality of total suspended particulates. The high-volume air sampler—the standard device used for sampling total suspended particulates—collects suspended particulates on large filters. The name *high-volume* is derived from the fact that the air sampler has a high sampling flow rate (measured in standard cubic meters per minute). Because of this high flow rate, large quantities of particles are collected over a 24-hour sampling period. However, the flow rate will vary depending on the pressure drop (in inches of water) across the filter medium. An experiment was conducted to investigate the relationship between flow rate and pressure drop. Eight sampling environments in which the high-volume air sampler was implemented yielded the measurements on average flow rate (y) and pressure drop across filter (x) listed in the accompanying table.

FLOW RATE y	PRESSURE DROP x
.92	10
1.25	15
.60	8
1.13	12
1.56	18
1.10	13
.65	9
1.33	15

a. Use the data to develop a simple linear model for predicting the average flow rate of the high-volume air sampler based on the pressure drop of the filter.
b. Is the model of part **a** useful for predicting flow rate? (Use $\alpha = .05$.)
c. Using a 95% prediction interval, predict the flow rate in a sampling environment in which the pressure drop across the filter is 11 inches of water.

10.66 Passive and active solar energy systems are becoming viable options to home builders as installation and operating costs decrease. Laminated solar modules utilize high-quality, single crystal silicon solar cells, connected electrically in series, to deliver a specified power output. Research was conducted to investigate the relationship between the solar cell temperature (°C) rise above ambient and the amount of insulation (megawatts per square centimeter). Data collected for six solar cells sampled under identical experimental conditions are recorded in the table.

TEMPERATURE RISE ABOVE AMBIENT y	INSULATION x
9	25
25	70
20	50
12	30
15	45
22	60

a. Fit a least squares line to the data.
b. Plot the data and graph the line as a check on your calculations.
c. Calculate r and r^2. Interpret their values.
d. Is the model useful for predicting temperature rise above ambient? (Use $\alpha = .01$).
e. Estimate the mean temperature rise above ambient for solar cells with insulation of 35 megawatts per square centimeter. Use a 99% confidence interval.

10.67 A study was conducted to examine the inhibiting properties of the sodium salts of phosphoric acid on the corrosion of iron. The data shown in the table provide a measure of corrosion of Armco iron in tap water containing various concentrations of NaPO$_4$ inhibitor.

CONCENTRATION OF NaPO$_4$ x, parts per million	MEASURE OF CORROSION RATE y	CONCENTRATION OF NaPO$_4$ x, parts per million	MEASURE OF CORROSION RATE y
2.50	7.68	26.20	.93
5.03	6.95	33.00	.72
7.60	6.30	40.00	.68
11.60	5.75	50.00	.65
13.00	5.01	55.00	.56
19.60	1.43		

Source: Andrzejaczek, B. J. "Mechanism of Action of Acidified Sodium Phosphate Solution as a Corrosion Inhibitor of Iron in Tap Water." *British Corrosion Journal*, Vol. 14, No. 3, 1979, pp. 176–178.

a. Construct a scattergram for the data.
b. Suppose we regard the 11 data points as a sample. Fit the linear model $y = \beta_0 + \beta_1 x_1 + \varepsilon$ to the data.
c. Does the model of part **b** provide an adequate fit? Test using $\alpha = .05$.
d. Construct a 95% confidence interval for the mean corrosion rate of iron in tap water in which the concentration of NaPO$_4$ is 20 parts per million.

REFERENCES

Chou, Ya-lun. *Statistical Analysis with Business and Economic Applications*, 2nd ed. New York: Holt, Rinehart, and Winston, 1975.

Crandall, J. S. and Cedercreutz, M. "Preliminary Cost Estimates for Mechanical Work." *Building Systems Design*, 73, Oct.–Nov. 1976, pp. 35–51.

Draper, N. and Smith, H. *Applied Regression Analysis*, 2nd ed. New York: Wiley, 1981.

Miller, R. B. and Wichern, D. W. *Intermediate Business Statistics: Analysis of Variance, Regression, and Time Series*. New York: Holt, Rinehart, and Winston, 1977.

Neter, J., Wasserman, W., and Kutner, M. H. *Applied Linear Statistical Models*. Homewood, Ill.: Richard D. Irwin, 1985.

OBJECTIVE

To extend the methods of Chapter 10 to develop a procedure for predicting a response y based on the values of two or more independent variables; to illustrate the types of practical inferences that can be drawn from this type of analysis.

CONTENTS

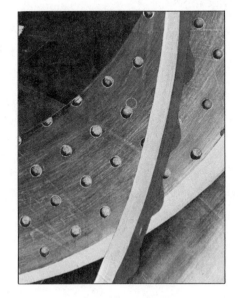

MULTIPLE REGRESSION ANALYSIS

| | | | | | | | | | | | |

**LINEAR MODELS FOR
A MULTIPLE
REGRESSION
ANALYSIS**

The models for a **multiple regression analysis** are similar to the simple linear regression model except that they contain more terms. For example, suppose we think that the mean time $E(y)$ required to perform a data-processing job increases as the computer utilization increases and that the relationship is curvilinear. Instead of using the **straight-line model**, $E(y) = \beta_0 + \beta_1 x_1$, to model the relationship, we might use the **quadratic model**

$$E(y) = \beta_0 + \beta_1 x_1 + \beta_2 x_1^2$$

where x_1 is a variable that measures computer utilization. A quadratic model, often referred to as a **second-order linear model**, graphs as a parabola and allows for some curvature in the relationship (see Figure 11.1), in contrast to a straight-line or **first-order linear model**.

FIGURE 11.1

Graph of a Second-Order
Linear Model

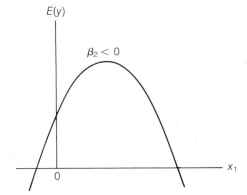

If, in addition, we think that the mean time required to process a job is also related to the size x_2 of the job, we could include x_2 in the model. A graph of $E(y)$ as a function of x_1 and x_2 traces a surface (called a **response surface**) over the (x_1, x_2)-plane. For example, the first-order model

$$E(y) = \beta_0 + \beta_1 x_1 + \beta_2 x_2$$

traces a planar surface over the (x_1, x_2)-plane (see Figure 11.2).

For our example (and for most real-life applications), we would expect curvature in the response surface and would use a second-order linear model

$$E(y) = \beta_0 + \beta_1 x_1 + \beta_2 x_2 + \beta_3 x_1 x_2 + \beta_4 x_1^2 + \beta_5 x_2^2$$

to model the relationship. A graph of a typical second-order response surface is shown in Figure 11.3.

All the models that we have written so far are called **linear models** because $E(y)$ is a linear function of the unknown parameters, $\beta_0, \beta_1, \beta_2, \ldots$. The model

$$E(y) = \beta_0 e^{-\beta_1 x}$$

is *not* a linear model because $E(y)$ is not a *linear* function of the unknown model parameters, β_0 and β_1.

FIGURE 11.2
Computer-Generated
Graph of a First-Order
Model

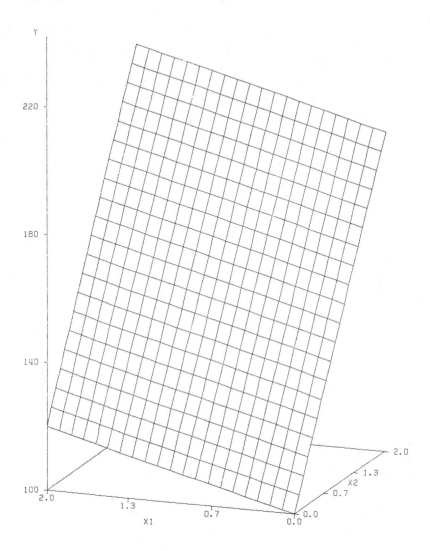

FIGURE 11.3
Graph of a Second-Order
Response Surface

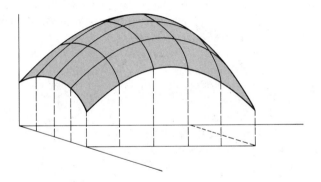

The independent variables that measure computer utilization (x_1) and size of job (x_2) are both quantitative variables—i.e., they measure the amount or quantity of something. We can also enter qualitative (nonquantitative) independent variables into the model. For example, suppose we expect the mean length of time required to process a job to be related to the day of the week that the job is processed. The qualitative independent variable, day of the week, can be entered into the model using **dummy** (or **index**) **variables**. Neglecting other terms (those corresponding to the quantitative independent variables) in the model, we would write

$$E(y) = \beta_0 + \beta_1 x_1 + \beta_2 x_2 + \beta_3 x_3 + \beta_4 x_4 + \beta_5 x_5 + \beta_6 x_6$$

where

$$x_1 = \begin{cases} 1 & \text{if the observation is made on Sunday} \\ 0 & \text{if not} \end{cases}$$

$$x_2 = \begin{cases} 1 & \text{if the observation is made on Monday} \\ 0 & \text{if not} \end{cases}$$

$$\vdots$$

$$x_6 = \begin{cases} 1 & \text{if the observation is made on Friday} \\ 0 & \text{if not} \end{cases}$$

The dummy variables enter the appropriate β parameter (which may be positive or negative), depending on the day of the week. Thus, on Sunday, $x_1 = 1$, $x_2 = x_3 = \cdots = x_6 = 0$ and the mean value of y is

$$E(y) = \beta_0 + \beta_1(1) = \beta_0 + \beta_1$$

Similarly, the mean value of y for Friday is

$$E(y) = \beta_0 + \beta_6$$

All of the dummy variables are assigned a value equal to 0 when an observation is made on Saturday. The mean value of y for Saturday is

$$E(y) = \beta_0$$

In general, qualitative variables are entered into a model using dummy variables—one fewer than the number of levels that the qualitative variable may assume. (We will have more to say about dummy variables in Chapter 12.)

Selecting an appropriate regression model for a particular situation is very important. If you try to fit a straight line (simple linear regression model) through a set of data points that plot as a curve, you will obtain a poor fit to the data. No statistical methods will compensate for poor model selection. We will defer further discussion of model construction until Chapter 12. In this chapter, we will assume that you have selected a reasonable model for your situation and will concentrate on the procedure for fitting the model to a set of data and on the associated methods of statistical inference.

Once a linear model has been chosen to relate $E(y)$ to a set of independent variables, the steps of a multiple regression analysis parallel those of a simple regression analysis. The only differences are that the mathematical theory is beyond the scope of this text and the computations are considerably more complex. In the following sections we will summarize the assumptions underlying a multiple regression analysis, present the methods for estimating and testing hypotheses about the model parameters, and show how to find a confidence interval for $E(y)$ or a prediction interval for y for specific values of the independent variables. Since most multiple regression analyses are performed on a computer, we will demonstrate how to interpret the output produced by several widely used multiple regression computer software packages.

SECTION 11.2

MODEL ASSUMPTIONS

After we have selected the deterministic portion of a regression model—i.e., a model for $E(y)$—we add a component ε to compensate for random error.

$$y = E(y) + \varepsilon$$

This component must obey the assumptions of the simple linear regression model—namely, that it is normally distributed with mean 0 and variance equal to σ^2. Further, we assume that the random errors associated with any pair of y values are independent.

In order to be able to present formulas for the parameter estimates, we need to write $E(y)$ in a standard form. Thus, we will let

$$E(y) = \beta_0 + \beta_1 x_1 + \beta_2 x_2 + \cdots + \beta_k x_k$$

be a model containing β_0 and k terms involving the predictor variables. The x values that appear in the model are those of Section 11.1. For example, x_2 could be x_1^2, x_3 could be $\sin(x_1)$, etc. The essential points are that the quantities x_1, $x_2, \ldots, x_k$ can be measured without error when a value of y is observed and that they do not involve any unknown parameters.

The linear regression model and associated assumptions are summarized in the box.

ASSUMPTIONS FOR A MULTIPLE REGRESSION ANALYSIS

1. $y = E(y) + \varepsilon$
  $= \beta_0 + \beta_1 x_1 + \beta_2 x_2 + \cdots + \beta_k x_k + \varepsilon$
2. For any given set of values of $x_1, x_2, \ldots, x_k$, ε has a normal probability distribution with mean equal to 0 and variance equal to σ^2.
3. The random errors are independent (in a probabilistic sense).

**FITTING THE MODEL:
THE METHOD OF
LEAST SQUARES**

The method of fitting a multiple regression model is identical to that of fitting the first-order (straight-line) model. Thus, we will use the method of least squares and choose estimates of $\beta_0, \beta_1 \ldots, \beta_k$ that minimize

$$\text{SSE} = \sum (y_i - \hat{y}_i)^2 = \sum [y_i - (\hat{\beta}_0 + \hat{\beta}_1 x_{1i} + \hat{\beta}_2 x_{2i} + \cdots + \hat{\beta}_k x_{ki})]^2$$

As in the case of the straight-line model, the sample estimates $(\hat{\beta}_0, \hat{\beta}_1, \ldots, \hat{\beta}_k)$ that minimize SSE will be obtained as solutions to the system of simultaneous linear equations

$$\frac{\partial \text{SSE}}{\partial \hat{\beta}_0} = 0 \qquad \frac{\partial \text{SSE}}{\partial \hat{\beta}_1} = 0 \qquad \cdots \qquad \frac{\partial \text{SSE}}{\partial \hat{\beta}_k} = 0$$

To illustrate the nature of this system, we will examine the first equation. Taking the partial derivative of SSE with respect to $\hat{\beta}_0$, we obtain

$$\frac{\partial \text{SSE}}{\partial \hat{\beta}_0} = 2 \sum [y_i - (\hat{\beta}_0 + \hat{\beta}_1 x_{1i} + \cdots + \hat{\beta}_k x_{ki})] \, (-1)$$

Setting $\frac{\partial \text{SSE}}{\partial \hat{\beta}_0}$ equal to 0 yields

$$\sum y_i - \left(n\hat{\beta}_0 + \sum x_{1i}\hat{\beta}_1 + \sum x_{2i}\hat{\beta}_2 + \cdots + \sum x_{ki}\hat{\beta}_k \right) = 0$$

or

$$n\hat{\beta}_0 + \left(\sum x_{1i} \right) \hat{\beta}_1 + \left(\sum x_{2i} \right) \hat{\beta}_2 + \cdots + \left(\sum x_{ki} \right) \hat{\beta}_k = \sum y_i$$

As in the case of simple linear regression, this is a linear equation in $\hat{\beta}_0, \hat{\beta}_1, \ldots, \hat{\beta}_k$. The k remaining least squares equations, all linear equations in $\hat{\beta}_0, \hat{\beta}_1, \ldots, \hat{\beta}_k$, are

$$\left(\sum x_{1i} \right) \hat{\beta}_0 + \left(\sum x_{1i} \right)^2 \hat{\beta}_1 + \left(\sum x_{1i} x_{2i} \right) \hat{\beta}_2 + \cdots + \left(\sum x_{1i} x_{ki} \right) \hat{\beta}_k = \sum x_{1i} y_i$$

$$\left(\sum x_{2i} \right) \hat{\beta}_0 + \left(\sum x_{1i} x_{2i} \right) \hat{\beta}_1 + \left(\sum x_{2i} \right)^2 \hat{\beta}_2 + \cdots + \left(\sum x_{1i} x_{ki} \right) \hat{\beta}_k = \sum x_{2i} y_i$$

$$\vdots$$

$$\left(\sum x_{ki} \right) \hat{\beta}_0 + \left(\sum x_{1i} x_{ki} \right) \hat{\beta}_1 + \qquad \cdots \qquad + \left(\sum x_{ki} \right)^2 \hat{\beta}_k = \sum x_{ki} y_i$$

As you can see, writing the $(k + 1)$ least squares linear equations is a task; solving them simultaneously by hand is even more difficult. An easy way to express the equations and to solve them is to use matrix algebra, but the inevitable computations are best performed on a computer.

In the following sections, we use matrix algebra to give formulas for the least squares estimates, SSE, test statistics, confidence intervals, and prediction intervals. Their use will be illustrated with simple numerical examples. (You may want to review the concepts in Appendix I, Matrix Algebra, before reading the remainder of this chapter.) To conclude, we will examine the multiple regression computer printouts for several popular statistical software packages and will present some examples of multiple regression computer analyses.

|||||||||||||
SECTION 11.4

THE LEAST SQUARES EQUATIONS AND THEIR SOLUTION

In order to apply matrix algebra to a regression analysis, we must place the data in matrices in a particular pattern. We will suppose that the linear model is

$$y = \beta_0 + \beta_1 x_1 + \beta_2 x_2 + \cdots + \beta_k x_k + \varepsilon$$

where $x_1, x_2, \ldots$ could actually represent the squares, cubes, cross products, or other functions of predictor variables, and ε is a random error. We will assume that we have collected n data points, i.e., n values of y and corresponding values of $x_1, x_2, \ldots, x_k$, and that these are denoted as shown in Table 11.1. Then the two data matrices Y and X are as shown in the box.

TABLE 11.1

DATA POINT	y value	x_1	x_2	$\cdots$	x_k
1	y_1	x_{11}	x_{21}	$\cdots$	x_{k1}
2	y_2	x_{12}	x_{22}	$\cdots$	x_{k2}
$\vdots$	$\vdots$	$\vdots$	$\vdots$		$\vdots$
n	y_n	x_{1n}	x_{2n}	$\cdots$	x_{kn}

THE DATA MATRICES Y AND X AND THE $\hat{\beta}$ MATRIX

$$Y = \begin{bmatrix} y_1 \\ y_2 \\ y_3 \\ \vdots \\ y_n \end{bmatrix} \quad X = \begin{bmatrix} 1 & x_{11} & x_{21} & \cdots & x_{k1} \\ 1 & x_{12} & x_{22} & \cdots & x_{k2} \\ 1 & x_{13} & x_{23} & \cdots & x_{k3} \\ \vdots & \vdots & \vdots & & \vdots \\ 1 & x_{1n} & x_{2n} & \cdots & x_{kn} \end{bmatrix} \quad \hat{\beta} = \begin{bmatrix} \hat{\beta}_0 \\ \hat{\beta}_1 \\ \hat{\beta}_2 \\ \vdots \\ \hat{\beta}_k \end{bmatrix}$$

Notice that the first column in the X matrix is a column of 1's. Thus, we are inserting a value of x, namely x_0, as the coefficient of β_0, where x_0 is a variable always equal to 1. Therefore, there is one column in the X matrix for each β parameter. Also, remember that a particular data point is identified by specific rows of the Y and X matrices. For example, the y value y_3 for data point 3 is in the third row of the Y matrix, and the corresponding values of $x_1, x_2, \ldots, x_k$ appear in the third row of the X matrix.

The $\hat{\beta}$ matrix shown in the box contains the least squares estimates (which we are attempting to obtain) of the coefficients $\beta_0, \beta_1, \ldots, \beta_k$ of the linear model

$$y = \beta_0 + \beta_1 x_1 + \beta_2 x_2 + \cdots + \beta_k x_k + \varepsilon$$

Using the Y and X data matrices, their transposes, and the $\hat{\boldsymbol{\beta}}$ matrix, we can write the least squares equations as:

LEAST SQUARES MATRIX EQUATION

$$(X'X)\hat{\boldsymbol{\beta}} = X'Y$$

Thus, $(X'X)$ is the coefficient matrix of the least squares estimates $\hat{\beta}_0$, $\hat{\beta}_1$, $\ldots$, $\hat{\beta}_k$, and $X'Y$ gives the matrix of constants that appear on the right-hand side of the equality signs.

The solution, which follows from Appendix I.3,* is

LEAST SQUARES SOLUTION

$$\hat{\boldsymbol{\beta}} = (X'X)^{-1}X'Y$$

Thus, to solve the least squares matrix equation, the computer calculates $(X'X)$, $(X'X)^{-1}$, $X'Y$, and, finally, the product $(X'X)^{-1}X'Y$. We will illustrate this process using the data for the insulation compression example from Section 10.2.

EXAMPLE 11.1

Find the least squares line for the insulation compression data repeated in Table 11.2.

TABLE 11.2

Compression Versus Pressure for an Insulation Material

SPECIMEN	PRESSURE x	COMPRESSION y
1	1	1
2	2	1
3	3	2
4	4	2
5	5	4

SOLUTION

The model is

$$y = \beta_0 + \beta_1 x_1 + \varepsilon$$

and the Y, X, and $\hat{\boldsymbol{\beta}}$ matrices are

$$Y = \begin{bmatrix} 1 \\ 1 \\ 2 \\ 2 \\ 4 \end{bmatrix} \quad X = \begin{matrix} \;x_0\;\;x_1 \\ \begin{bmatrix} 1 & 1 \\ 1 & 2 \\ 1 & 3 \\ 1 & 4 \\ 1 & 5 \end{bmatrix} \end{matrix} \quad \hat{\boldsymbol{\beta}} = \begin{bmatrix} \hat{\beta}_0 \\ \hat{\beta}_1 \end{bmatrix}$$

*In the notation of Appendix I.3, $A = X'X$, $V = \hat{\boldsymbol{\beta}}$, and $G = X'Y$. Then the solution to the equation $AV = G$ is $V = A^{-1}G$.

Then,

$$X'X = \begin{bmatrix} 1 & 1 & 1 & 1 & 1 \\ 1 & 2 & 3 & 4 & 5 \end{bmatrix} \begin{bmatrix} 1 & 1 \\ 1 & 2 \\ 1 & 3 \\ 1 & 4 \\ 1 & 5 \end{bmatrix} = \begin{bmatrix} 5 & 15 \\ 15 & 55 \end{bmatrix}$$

$$X'Y = \begin{bmatrix} 1 & 1 & 1 & 1 & 1 \\ 1 & 2 & 3 & 4 & 5 \end{bmatrix} \begin{bmatrix} 1 \\ 1 \\ 2 \\ 2 \\ 4 \end{bmatrix} = \begin{bmatrix} 10 \\ 37 \end{bmatrix}$$

The last matrix that we need is $(X'X)^{-1}$. This matrix can be found by using a packaged computer program or by using the method of Appendix I.4. Thus, you would find

$$(X'X)^{-1} = \begin{bmatrix} 1.1 & -.3 \\ -.3 & .1 \end{bmatrix}$$

Then the solution to the least squares equation is

$$\hat{\boldsymbol{\beta}} = (X'X)^{-1}X'Y = \begin{bmatrix} 1.1 & -.3 \\ -.3 & .1 \end{bmatrix} \begin{bmatrix} 10 \\ 37 \end{bmatrix} = \begin{bmatrix} -.1 \\ .7 \end{bmatrix}$$

Thus, $\hat{\beta}_0 = -.1$, $\hat{\beta}_1 = .7$, and the prediction equation is

$$\hat{y} = -.1 + .7x$$

You can verify that this is the same answer as obtained in Section 10.3. ∎

EXAMPLE 11.2

An electrical utility company wants to predict the monthly power usage of a home as a function of the size x of the home based on the model

$$y = \beta_0 + \beta_1 x + \beta_2 x^2 + \varepsilon$$

Find the least squares estimates of β_0, β_1, and β_2. The data are shown in Table 11.3.

TABLE 11.3
Data for Power
Usage Study

SIZE OF HOME x, square feet	MONTHLY USAGE y, kilowatt-hours
1,290	1,182
1,350	1,172
1,470	1,264
1,600	1,493
1,710	1,571
1,840	1,711
1,980	1,804
2,230	1,840
2,400	1,956
2,930	1,954

SOLUTION

The Y, X, and $\hat{\beta}$ matrices are shown below:

$$
Y = \begin{bmatrix} 1,182 \\ 1,172 \\ 1,264 \\ 1,493 \\ 1,571 \\ 1,711 \\ 1,804 \\ 1,840 \\ 1,956 \\ 1,954 \end{bmatrix}
\qquad
X = \begin{array}{ccc} x_0 & x & x^2 \\ \begin{bmatrix} 1 & 1,290 & 1,664,100 \\ 1 & 1,350 & 1,822,500 \\ 1 & 1,470 & 2,160,900 \\ 1 & 1,600 & 2,560,000 \\ 1 & 1,710 & 2,924,100 \\ 1 & 1,840 & 3,385,600 \\ 1 & 1,980 & 3,920,400 \\ 1 & 2,230 & 4,972,900 \\ 1 & 2,400 & 5,760,000 \\ 1 & 2,930 & 8,584,900 \end{bmatrix} \end{array}
$$

Then:

$$
X'X = \begin{bmatrix} 10 & 18,800 & 37,755,400 \\ 18,800 & 37,755,400 & 8,093.9 \times 10^7 \\ 37,755,400 & 8,093.9 \times 10^7 & 1.843 \times 10^{14} \end{bmatrix}
$$

$$
X'Y = \begin{bmatrix} 15,947 \\ 31,283,250 \\ 6.53069 \times 10^{10} \end{bmatrix}
$$

And, using a standard computer package, we obtain

$$
(X'X)^{-1} = \begin{bmatrix} 26.9156 & -.027027 & 6.3554 \times 10^{-6} \\ -.027027 & 2.75914 \times 10^{-5} & -6.5804 \times 10^{-9} \\ 6.3554 \times 10^{-6} & -6.5804 \times 10^{-9} & 1.5934 \times 10^{-12} \end{bmatrix}
$$

Finally, performing the multiplication, we obtain

$$
\begin{aligned}
\hat{\beta} &= (X'X)^{-1}X'Y \\
&= \begin{bmatrix} 26.9156 & -.027027 & 6.3554 \times 10^{-6} \\ -.027027 & 2.75914 \times 10^{-5} & -6.5804 \times 10^{-9} \\ 6.3554 \times 10^{-6} & -6.5804 \times 10^{-9} & 1.5934 \times 10^{-12} \end{bmatrix} \begin{bmatrix} 15.947 \\ 31,283,250 \\ 6.53069 \times 10^{10} \end{bmatrix} \\
&= \begin{bmatrix} -1,216.14389 \\ 2.39893 \\ -.00045 \end{bmatrix}
\end{aligned}
$$

Thus, $\hat{\beta}_0 = -1,216.14389$, $\hat{\beta}_1 = 2.39893$, $\hat{\beta}_2 = -.00045$, and the prediction equation is

$$
\hat{y} = -1,216.14389 + 2.39893x - .00045x^2
$$

■

11.1 Use the method of least squares to fit a straight line to the five data points shown in the table.

x	-2	-1	0	1	2
y	4	3	3	1	-1

a. Construct Y and X matrices for the data.
b. Find $X'X$ and $X'Y$.
c. Find the least squares estimates $\hat{\beta} = (X'X)^{-1}X'Y$. [*Note:* See Theorem I.1 in Appendix I for information on finding $(X'X)^{-1}$.]
d. Give the prediction equation.

11.2 Use the method of least squares to fit the model $E(y) = \beta_0 + \beta_1 x$ to the six data points given in the accompanying table.

x	1	2	3	4	5	6
y	1	2	2	3	5	5

a. Construct Y and X matrices for the data.
b. Find $X'X$ and $X'Y$.
c. Verify that

$$(X'X)^{-1} = \begin{bmatrix} \frac{13}{15} & -\frac{7}{35} \\ -\frac{7}{35} & \frac{2}{35} \end{bmatrix}$$

d. Find the $\hat{\beta}$ matrix.
e. Give the prediction equation.

11.3 An experiment was conducted in which two y observations were collected for each of five values of x, as listed in the table. Use the method of least squares to fit the second-order model, $E(y) = \beta_0 + \beta_1 x + \beta_2 x^2$, to the ten data points.

x	-2		-1		0		1		2	
y	1.1	1.3	2.0	2.1	2.7	2.8	3.4	3.6	4.1	4.0

a. Give the dimensions of the Y and X matrices.
b. Verify that

$$(X'X)^{-1} = \begin{bmatrix} \frac{17}{70} & 0 & -\frac{1}{14} \\ 0 & \frac{1}{20} & 0 \\ -\frac{1}{14} & 0 & \frac{1}{28} \end{bmatrix}$$

c. Find the $\hat{\beta}$ matrix and the least squares prediction equation.
d. Plot the data points and graph the prediction equation.

11.4 A study was conducted to examine the relationship between the cast structure and the mechanical properties of equiaxed grains in unidirectionally solidified ingots (*Metallurgical Transactions*, May 1986). Ingots composed of aluminum, copper, and titanium alloys (the equiaxed structure) were poured into molds and cooled by water until solidification. From each ingot, five tensile specimens were obtained at varying distances from the ingot chill face and the grain size (i.e., number of grains) was determined for each specimen. The data for one such ingot are given in the accompanying table. Suppose we are interested in fitting the quadratic model

$$y = \beta_0 + \beta_1 x + \beta_2 x^2 + \varepsilon$$

where

y = Grain size
x = Distance from chill face

DISTANCE FROM CHILL FACE x, cm	GRAIN SIZE y, mm
1.0	.24
3.5	.38
6.0	.44
8.5	.61
11.0	.75

Source: Kato, H. and Cahoon, J. R. "Tensile Properties of Directionally Solidified Al-4 Wt Pct Cu Alloys with Columnar and Equiaxed Grains." *Metallurgical Transactions*, Vol. 17A, No. 5, May 1986, p. 830 (Table II).

a. Construct Y and X matrices for the data.
b. Find $X'X$ and $X'Y$.
c. Use the technique outlined in Appendix I.4 to find $(X'X)^{-1}$.
d. Find the $\hat{\beta}$ matrix and the least squares prediction equation.
e. Plot the data points and graph the prediction equation.

| | | | | | | | | | | | |
SECTION 11.5

PROPERTIES OF THE LEAST SQUARES ESTIMATORS

In Section 11.4 we noted that the $\hat{\beta}$ matrix is equal to the product of the $(X'X)^{-1}X'$ matrix and the Y matrix:

$$\hat{\beta} = [(X'X)^{-1}X']Y$$

Since elements in the $\hat{\beta}$ matrix (i.e., the estimators $\hat{\beta}_0, \hat{\beta}_1, \ldots, \hat{\beta}_k$) are obtained by multiplying the rows of $(X'X)^{-1}X'$ by the column matrix Y, it follows that $\hat{\beta}_0$ will equal the product of the first row of $(X'X)^{-1}X'$ and the Y matrix and, in general, $\hat{\beta}_i$ will equal the product of the $(i + 1)$st row of $(X'X)^{-1}X'$ and Y. Therefore, for $i = 0, 1, 2, \ldots, k$, $\hat{\beta}_i$ is a linear function of n normally distributed random variables, $y_1, y_2, \ldots, y_n$, and, by Theorem 7.3, $\hat{\beta}_i$ possesses a normal sampling distribution.

Derivation of the means and variances of the sampling distributions of $\hat{\beta}_0$, $\hat{\beta}_1, \ldots, \hat{\beta}_k$ is beyond the scope of this text. However, it can be shown that the least squares estimators provide unbiased estimates of $\beta_0, \beta_1, \ldots, \beta_k$, i.e.,

$$E(\hat{\beta}_i) = \beta_i \quad \text{for } i = 0, 1, 2, \ldots, k$$

The standard errors and covariances of the estimators $\hat{\beta}_0, \hat{\beta}_1, \ldots, \hat{\beta}_k$, are determined by the elements of the $(X'X)^{-1}$ matrix. Thus, if we denote the $(X'X)^{-1}$ matrix as

$$(X'X)^{-1} = \begin{bmatrix} c_{00} & c_{01} & \cdots & c_{0k} \\ c_{10} & c_{11} & \cdots & c_{1k} \\ c_{20} & c_{21} & c_{22} & \cdots & c_{2k} \\ \vdots & \vdots & \vdots & & \vdots \\ c_{k0} & \cdot & \cdot & \cdot & c_{kk} \end{bmatrix}$$

then it can be shown (proof omitted) that the standard errors of the sampling distributions of $\hat{\beta}_0, \hat{\beta}_1, \ldots, \hat{\beta}_k$ are

$$\sigma_{\hat{\beta}_0} = \sigma\sqrt{c_{00}}$$
$$\sigma_{\hat{\beta}_1} = \sigma\sqrt{c_{11}}$$
$$\sigma_{\hat{\beta}_2} = \sigma\sqrt{c_{22}}$$
$$\vdots$$
$$\sigma_{\hat{\beta}_k} = \sigma\sqrt{c_{kk}}$$

where σ is the standard deviation of the random error ε. In other words, the diagonal elements of $(X'X)^{-1}$ gives the values of $c_{00}, c_{11}, \ldots, c_{kk}$ that are required for finding the standard errors of the estimators $\hat{\beta}_0, \hat{\beta}_1, \ldots, \hat{\beta}_k$.

The properties of the sampling distributions of the least squares estimators are summarized in the box.

THEOREM 11.1

PROPERTIES OF THE SAMPLING DISTRIBUTION OF $\hat{\beta}_i$ ($i = 0, 1, 2, \ldots, k$) The sampling distribution of $\hat{\beta}_i$ ($i = 0, 1, 2, \ldots, k$) is normal with

$$E(\hat{\beta}_i) = \beta_i \qquad V(\hat{\beta}_i) = c_{ii}\sigma^2 \qquad \sigma_{\hat{\beta}_i} = \sigma\sqrt{c_{ii}}$$

The off-diagonal elements of the $(X'X)^{-1}$ matrix determine the covariances of $\hat{\beta}_0, \hat{\beta}_1, \ldots, \hat{\beta}_k$. Thus, it can be shown that the covariance of two parameter estimators, say $\hat{\beta}_i$ and $\hat{\beta}_j$ (where $i \neq j$), is equal to

$$\text{Cov}(\hat{\beta}_i, \hat{\beta}_j) = c_{ij}\sigma^2 = c_{ji}\sigma^2$$

For example, $\text{Cov}(\hat{\beta}_0, \hat{\beta}_2) = c_{02}\sigma^2 = c_{20}\sigma^2$ and $\text{Cov}(\hat{\beta}_2, \hat{\beta}_3) = c_{23}\sigma^2 = c_{32}\sigma^2$. These covariances are necessary in order to determine the variance of the prediction equation

$$\hat{y} = \hat{\beta}_0 + \hat{\beta}_1 x_1 + \hat{\beta}_2 x_2 + \cdots + \hat{\beta}_k x_k$$

or of any other linear function of $\hat{\beta}_0, \hat{\beta}_1, \ldots, \hat{\beta}_k$. They will also play a role in finding a confidence interval for $E(y)$ and a prediction interval for y in Sections 11.9 and 11.10.

SECTION 11.6

ESTIMATING σ^2, THE VARIANCE OF ε

You will recall that the variances of the estimators of all of the β parameters and of $\hat{y}$ will depend on the value of σ^2, the variance of the random error ε that appears in the linear model. Since σ^2 will rarely be known in advance, we must use the sample data to estimate its value.

ESTIMATOR OF σ^2, THE VARIANCE OF ε IN A MULTIPLE REGRESSION MODEL

$$s^2 = \frac{\text{SSE}}{n - \text{Number of } \beta \text{ parameters in model}}$$

where

$$\text{SSE} = Y'Y - \hat{\beta}'X'Y$$

We will demonstrate the use of these formulas with the insulation compression data of Example 11.1.

EXAMPLE 11.3

Find SSE for the insulation compression data of Example 11.1.

SOLUTION

From Example 11.1 we have

$$\hat{\beta} = \begin{bmatrix} -.1 \\ .7 \end{bmatrix} \quad \text{and} \quad X'Y = \begin{bmatrix} 10 \\ 37 \end{bmatrix}$$

Then,

$$Y'Y = \begin{bmatrix} 1 & 1 & 2 & 2 & 4 \end{bmatrix} \begin{bmatrix} 1 \\ 1 \\ 2 \\ 2 \\ 4 \end{bmatrix} = 26 \quad \text{and} \quad \hat{\beta}'X'Y = \begin{bmatrix} -.1 & .7 \end{bmatrix} \begin{bmatrix} 10 \\ 37 \end{bmatrix} = 24.9$$

So

$$\text{SSE} = Y'Y - \hat{\beta}'X'Y = 26 - 24.9 = 1.1$$

(Note that this is the same answer as was obtained in Section 10.3.)
   Finally,

$$s^2 = \frac{\text{SSE}}{n - \text{Number of } \beta \text{ parameters in model}} = \frac{1.1}{5 - 2} = .367$$

This estimate is needed to construct a confidence interval for β_1, to test a hypothesis concerning its value, or to construct a confidence interval for the mean compression $E(y)$ for a given compressive pressure x.   ∎

You will not be surprised to learn that the sampling distribution of s^2 is related to the chi-square distribution. In fact, Theorems 7.4 and 10.1 are special cases of Theorem 11.2 (proof omitted).

THEOREM 11.2

Consider the linear model

$$y = \beta_0 + \beta_1 x_1 + \beta_2 x_2 + \cdots + \beta_k x_k + \varepsilon$$

which contains $(k + 1)$ unknown β parameters that must be estimated. If the assumptions of Section 11.2 are satisfied, then the statistic

$$\chi^2 = \frac{SSE}{\sigma^2} = \frac{[n - (k + 1)]s^2}{\sigma^2}$$

has a chi-square distribution with $\nu = [n - (k + 1)]$ degrees of freedom.

Using Theorem 11.2 we can show that s^2 is an unbiased estimator of σ^2:

$$E(s^2) = E\left\{\frac{\chi^2 \sigma^2}{[n - (k + 1)]}\right\} = \frac{\sigma^2}{[n - (k + 1)]}E(\chi^2)$$

where $E(\chi^2) = \nu = [n - (k + 1)]$. Therefore,

$$E(s^2) = \left(\frac{\sigma^2}{[n - (k + 1)]}\right)[n - (k + 1)] = \sigma^2$$

and we conclude that s^2 is an unbiased estimator of σ^2.

SECTION 11.7

CONFIDENCE INTERVALS AND TESTS OF HYPOTHESES FOR $\beta_0, \beta_1, \ldots, \beta_k$

A $(1 - \alpha)100\%$ confidence interval for a model parameter β_i $(i = 0, 1, 2, \ldots, k)$ can be constructed (see the Optional Exercises of this section) using the pivotal method and the t statistic

$$t = \frac{\hat{\beta}_i - \beta_i}{s\sqrt{c_{ii}}}$$

The quantity $s\sqrt{c_{ii}}$ is the estimated standard error of $\hat{\beta}_i$ and is obtained by replacing σ by s in the formula for the standard error. The resulting confidence interval for β_i takes the same form as the small-sample confidence interval for a population mean given in Section 8.7.

A $(1 - \alpha)100\%$ CONFIDENCE INTERVAL FOR β_i

$\hat{\beta}_i \pm t_{\alpha/2}(\text{Estimated standard error of } \hat{\beta}_i)$

or

$\hat{\beta}_i \pm t_{\alpha/2}s\sqrt{c_{ii}}$

where $t_{\alpha/2}$ is based on the number of degrees of freedom associated with s.

Similarly, the test statistic for testing the null hypothesis $H_0: \beta_i = 0$ is

$$t = \frac{\hat{\beta}_i}{\text{Estimated standard error of } \hat{\beta}_i} = \frac{\hat{\beta}_i}{s\sqrt{c_{ii}}}$$

The test is summarized in the box.

TEST OF AN INDIVIDUAL PARAMETER COEFFICIENT IN THE MULTIPLE REGRESSION MODEL

$y = \beta_0 + \beta_1 x_1 + \beta_2 x_2 + \cdots + \beta_k x_k + \varepsilon$

ONE-TAILED TEST

$H_0: \quad \beta_i = 0$
$H_a: \quad \beta_i > 0$
$\quad (\text{or } \beta_i < 0)$

Test statistic:* $\quad t = \dfrac{\hat{\beta}_i}{s_{\hat{\beta}_i}} = \dfrac{\hat{\beta}_i}{s\sqrt{c_{ii}}}$

Rejection region: $\quad t > t_\alpha$
$\quad\quad\quad (\text{or } t < -t_\alpha)$

TWO-TAILED TEST

$H_0: \quad \beta_i = 0$
$H_a: \quad \beta_i \neq 0$

Test statistic:* $\quad t = \dfrac{\hat{\beta}_i}{s_{\hat{\beta}_i}} = \dfrac{\hat{\beta}_i}{s\sqrt{c_{ii}}}$

Rejection region:
$\quad\quad t > t_{\alpha/2} \text{ or } t < -t_{\alpha/2}$

where

$n = $ Number of observations
$k = $ Number of independent variables in the model

and $t_{\alpha/2}$ is based on $[n - (k + 1]$ df

Assumptions: See Section 11.2 for the assumptions about the probability distribution of the random error component ε.

EXAMPLE 11.4

Refer to Example 11.1 and find the estimated standard error for the sampling distribution of $\hat{\beta}_1$, the estimator of the slope of the line β_1. Then give a 95% confidence interval for β_1.

*To test the null hypothesis that a parameter β_i equals some value other than zero, say $H_0: \beta_i = \beta_{i0}$, use the test statistic $t = (\hat{\beta}_i - \beta_{i0})/s_{\hat{\beta}_i}$. All other aspects of the test will be described in the box.

SOLUTION

The $(X'X)^{-1}$ matrix for the least squares solution of Example 11.1 was

$$(X'X)^{-1} = \begin{bmatrix} 1.1 & -3 \\ -.3 & .1 \end{bmatrix}$$

Therefore, $c_{00} = 1.1$, $c_{11} = .1$, and the estimated standard error for $\hat{\beta}_1$ is

$$s_{\hat{\beta}_1} = s\sqrt{c_{11}} = \sqrt{.367}(\sqrt{.1}) = .192$$

The value for s, $\sqrt{.367}$, was obtained from Example 11.3.

A 95% confidence interval for β_1 is

$$\hat{\beta}_1 \pm t_{\alpha/2} s\sqrt{c_{11}}$$

$$.7 \pm (3.182)(.192) = (.09, 1.31)$$

The t value, $t_{.025}$, is based on $(n - 2) = 3$ df. Observe that this is the same confidence interval as the one obtained in Example 10.5 (Section 10.6). ■

The SAS printout for a multiple regression analysis is identical to the printout for a simple linear regression analysis, except that it contains estimates, test statistic values, etc. for all parameters of the multiple regression model. In Example 11.5, we will present the computer printout for the multiple regression analysis for the power usage data of Example 11.2, and will compare some of our computed values with those shown on the printout.

EXAMPLE 11.5

Refer to Example 11.2 and the least squares solution for fitting power usage y to the size of a home x using the model

$$y = \beta_0 + \beta_1 x + \beta_2 x^2 + \varepsilon$$

a. Compute the estimated standard error for $\hat{\beta}_1$ and compare this result with the SAS regression analysis printout shown in Figure 11.4.

b. Compute the value of the test statistic for testing H_0: $\beta_2 = 0$. Compare this with the value given in the computer printout shown in Figure 11.4.

c. Test H_0: $\beta_2 = 0$ against H_a: $\beta_2 \neq 0$. State your conclusions.

FIGURE 11.4 SAS Computer Printout for the Power Usage Data of Examples 11.2 and 11.5

DEPENDENT VARIABLE: Y

SOURCE	DF	SUM OF SQUARES	MEAN SQUARE	F VALUE	PR > F	R-SQUARE	C.V.
MODEL	2	831069.54637065	415534.77318533	189.71	0.0001	0.981885	2.9348
ERROR	7	15332.55362935	2190.36480419		ROOT MSE		Y MEAN
CORRECTED TOTAL	9	846402.10000000			46.80133336		1594.70000000

SOURCE	DF	TYPE I SS	F VALUE	PR > F	DF	TYPE III SS	F VALUE	PR > F
X	1	703957.18342042	321.39	0.0001	1	208574.93309206	95.22	0.0001
X*X	1	127112.36295023	58.03	0.0001	1	127112.36295023	58.03	0.0001

PARAMETER	ESTIMATE	T FOR H0: PARAMETER=0	PR > :T:	STD ERROR OF ESTIMATE
INTERCEPT	-1216.14388700	-5.01	0.0016	242.80636850
X	2.39893018	9.76	0.0001	0.24583560
X*X	-0.00045004	-7.62	0.0001	0.00005908

SOLUTION

The values of $\hat{\beta}_0$, $\hat{\beta}_1$, and $\hat{\beta}_2$ given in the ESTIMATE column of the computer printout shown in Figure 11.4 are the same as those obtained in Example 11.2. Therefore, the fitted model is

$$\hat{y} = -1{,}216.14389 + 2.39893x - .00045x^2$$

The $(X'X)^{-1}$ matrix, obtained in Example 11.2, is

$$(X'X)^{-1} = \begin{bmatrix} 26.9156 & -.027027 & 6.3554 \times 10^{-6} \\ -.027027 & 2.75914 \times 10^{-5} & -6.5804 \times 10^{-9} \\ 6.3554 \times 10^{-6} & -6.5804 \times 10^{-9} & 1.5934 \times 10^{-12} \end{bmatrix}$$

From $(X'X)^{-1}$ we see that

$$c_{00} = 26.9156$$
$$c_{11} = 2.75914 \times 10^{-5}$$
$$c_{22} = 1.5934 \times 10^{-12}$$

and from the printout, $s^2 = 2{,}190.3648$ (shaded under MEAN SQUARE for ERROR), and $s = 46.801$ (shaded under ROOT MSE).

a. The estimated standard error of $\hat{\beta}_1$ is

$$s_{\hat{\beta}_1} = s\sqrt{c_{11}}$$
$$= (46.801)\sqrt{2.75914 \times 10^{-5}} = .24583$$

Notice that this agrees with the value of $s_{\hat{\beta}_1}$ shaded in the computer printout (Figure 11.4) under the column labeled STD ERROR OF ESTIMATE in the row labeled X.

b. The value of the test statistic for testing H_0: $\beta_2 = 0$ is

$$t = \frac{\hat{\beta}_2}{s\sqrt{c_{22}}} = \frac{-.00045}{(46.801)\sqrt{1.5934 \times 10^{-12}}} = -7.62$$

Notice that this value of the t statistic agrees with the value given in the column headed T FOR H0: PARAMETER $= 0$ shown in the printout (Figure 11.4).

c. To test H_0: $\beta_2 = 0$ against the alternative hypothesis H_a: $\beta_2 \neq 0$, we will conduct a two-tailed t test and reject H_0 if $t > t_{\alpha/2}$ or $t < -t_{\alpha/2}$. For our example, t has $[n - (k + 1)] = 10 - 3 = 7$ degrees of freedom. (This number of degrees of freedom is shaded in the DF column and ERROR row of the computer printout shown in Figure 11.4.) Therefore, for $\alpha = .05$, we will reject H_0 (see Table 6 of Appendix II) if $t > 2.365$ or $t < -2.365$. Since the observed value of $t = -7.62$ is less than -2.365, there is evidence to indicate that $\beta_2 \neq 0$, i.e., that x^2 contributes information for the prediction of y. The parameter β_2 produces curvature in a graph of the second-order model. Therefore, the practical implication of the test conclusion is that there is evidence to indicate curvature in the model.   ■

The SAS printout shown in Figure 11.4 (Example 11.5) also gives the two-tailed observed significance level (i.e., p-value) for each t test. These values appear under the column headed PR>|T|. The shaded observed significance level .0001 corresponds to the quadratic term, and this implies that we would reject H_0: $\beta_2 = 0$ in favor of H_a: $\beta_2 \neq 0$ at any α level larger than .0001. [For one-sided alternatives (e.g., H_a: $\beta_2 < 0$), the p-value is half that given on the printout, i.e., p-value $= \frac{1}{2}$ (PR>|T|).]

| | | | | | | | | | | | | |

EXERCISES 11.5–11.13

11.5 Do the data given in Exercise 11.1 provide sufficient evidence to indicate that x contributes information for the prediction of y? Test H_0: $\beta_1 = 0$ against H_a: $\beta_1 \neq 0$ using $\alpha = .05$.

11.6 Find a 90% confidence interval for the slope of the line of Exercise 11.5.

11.7 Do the data given in Exercise 11.3 provide sufficient evidence to indicate curvature in the model for $E(y)$? Test H_0: $\beta_2 = 0$ against H_a: $\beta_2 \neq 0$ using $\alpha = .10$.

11.8 Refer to the *Metallurgical Transactions* study described in Exercise 11.4. Conduct a test to determine whether downward curvature exists in the relationship between grain size (y) and distance from chill face (x). Use $\alpha = .05$.

11.9 Laboratory tests were conducted to determine the effect of asphalt content on the stability and permeability of open-graded asphalt concrete (*Journal of Testing and Evaluation*, July 1981). Four concrete specimens were prepared for each of the following asphalt contents (percentage by total weight of mix): 3, 4, 5, 6, 7, and 8. The water permeability of each concrete specimen was determined by flowing de-aired water across the specimen and measuring the amount of water loss. The permeability measurements (in inches per hour) for the 24 concrete specimens are shown in the table.

ASPHALT CONTENT x, %	PERMEABILITY y, in/hr	ASPHALT CONTENT x, %	PERMEABILITY y, in/hr
3	1,189	6	707
3	840	6	927
3	1,020	6	1,067
3	980	6	822
4	1,440	7	853
4	1,227	7	900
4	1,022	7	733
4	1,293	7	585
5	1,227	8	395
5	1,180	8	270
5	980	8	310
5	1,210	8	208

Source: Woelfl, G., Wei, I., Faulstich, C., and Litwack, H. "Laboratory Testing of Asphalt Concrete for Porous Pavements." *Journal of Testing and Evaluation*, Vol. 9, No. 4, July 1981, pp. 175–181. Copyright American Society for Testing and Materials.

a. Plot the points in a scattergram.

b. Use the methods of Chapter 10 to fit the first-order model

$$y = \beta_0 + \beta_1 x + \varepsilon$$

What do you conclude about the utility of the model?

c. The SAS printout for the quadratic model

$$y = \beta_0 + \beta_1 x + \beta_2 x^2 + \varepsilon$$

is shown here. Is there sufficient evidence to indicate that the quadratic term should be included in the model? Test using $\alpha = .05$.

SAS Computer Printout for Exercise 11.9

DEPENDENT VARIABLE: Y

SOURCE	DF	SUM OF SQUARES	MEAN SQUARE	F VALUE	PR > F	R-SQUARE	C.V.
MODEL	2	2203970.74940483	1101985.37470241	52.85	0.0001	0.834253	16.2057
ERROR	21	437878.20892851	20851.34328231		ROOT MSE		Y MEAN
CORRECTED TOTAL	23	2641848.95833334			144.39994211		891.04166667

SOURCE	DF	TYPE I SS	F VALUE	PR > F	DF	TYPE III SS	F VALUE	PR > F
X	1	1580554.88928571	75.80	0.0001	1	380956.70871787	18.27	0.0003
X*X	1	623415.86011912	29.90	0.0001	1	623415.86011912	29.90	0.0001

PARAMETER	ESTIMATE	T FOR H0: PARAMETER=0	PR > :T:	STD ERROR OF ESTIMATE
INTERCEPT	-48.55535714	-0.14	0.8871	337.93244741
X	560.46339286	4.27	0.0003	131.12226955
X*X	-64.61160714	-5.47	0.0001	11.81649402

11.10 An experiment was conducted to investigate the effect of temperature (T) and pressure (P) on the yield y of a chemical. Each of the two factors, temperature and pressure, was held constant at two levels—T at $50°$ and $70°$, P at 10 pounds per square inch and 20 pounds per square inch—and the yield of each of the four combinations was measured. In order to simplify the calculations, the two factors were coded to produce two independent variables, x_1 and x_2:

$$x_1 = \begin{cases} 1 & \text{if } T = 70 \\ -1 & \text{if } T = 50 \end{cases} \quad \text{and} \quad x_2 = \begin{cases} 1 & \text{if } P = 20 \\ -1 & \text{if } P = 10 \end{cases}$$

The results are shown in the accompanying table.

		x_2	
		-1	1
x_1	-1	24.5	28.4
	1	22.1	16.7

a. Fit the linear model

$$E(y) = \beta_0 + \beta_1 x_1 + \beta_2 x_2$$

to the data.

b. Do the data provide sufficient evidence to indicate that temperature contributes information for the prediction of yield of chemical? Use $\alpha = .10$.

c. Do the data provide sufficient evidence to indicate that pressure contributes information for the prediction of yield of chemical? Use $\alpha = .10$.

d. Find a 90% confidence interval for β_1.

e. Find a 90% confidence interval for β_2.

11.11 The *Canadian Geotechnical Journal* (Aug. 1985) reported on a study to investigate the reliability of the use of fragmented Queenston Shale, a compaction shale, as a rockfill construction material. In particular, the researchers wanted to estimate the stress–strain relationship of the fragmented material. Based on a graph shown in the paper, the accompanying data were reproduced on deviatoric stress and axial strain for wet shale specimens.

DEVIATORIC STRESS y, kPa	AXIAL STRAIN x, %
500	1.0
2,000	2.8
2,750	4.3
3,500	6.0
4,375	7.5
4,875	9.0
5,250	10.5
6,000	13.5
6,625	16.7
7,000	19.8
7,125	23.0
7,000	26.0
7,125	27.5

Source: Caswell, R. H. and Trak, B. "Some Geotechnical Characteristics of Fragmented Queenston Shale." *Canadian Geotechnical Journal*, Vol. 22, No. 3, Aug. 1985, pp. 403–408.

a. Plot the data on a scattergram. What type of relationship appears to exist?
b. The quadratic model $E(y) = \beta_0 + \beta_1 x + \beta_2 x^2$ was fit to the data, with the results shown in the accompanying SAS printout. Test the hypothesis that deviatoric stress y increases with axial strain x at a decreasing rate. Use $\alpha = .05$.

SAS Printout for Exercise 11.11

```
                        ANALYSIS OF VARIANCE

                     SUM OF          MEAN
     SOURCE    DF    SQUARES         SQUARE       F VALUE    PROB>F

     MODEL      2   57287428.97    28643714.49    802.791    0.0001
     ERROR     10     356801.80    35680.17987
     C TOTAL   12   57644230.77

              ROOT MSE      188.892       R-SQUARE     0.9938
              DEP MEAN     4932.692       ADJ R-SQ     0.9926
              C.V.         3.829389

                        PARAMETER ESTIMATES

                     PARAMETER       STANDARD      T FOR H0:
   VARIABLE   DF     ESTIMATE          ERROR     PARAMETER=0    PROB > |T|

   INTERCEP    1     248.63581      147.80771        1.682        0.1235
   X           1     619.76311       25.80667772     24.016       0.0001
   XX          1     -13.75231582     0.87153259    -15.779       0.0001
```

c. Give the observed significance level for the test of part **b** and interpret its value.

d. Locate the estimate of σ on the printout and interpret its value.

OPTIONAL EXERCISES

11.12 If the assumptions of Section 11.2 are satisfied, it can be shown that s^2 is independent of $\hat{\beta}_i$, the least squares estimator of β_i. Use this fact, along with Theorems 11.1 and 11.2 to show that

$$t = \frac{\hat{\beta}_i - \beta_i}{s\sqrt{c_{ii}}}$$

has a t distribution with $[n - (k + 1)]$ degrees of freedom.

11.13 Use the t statistic given in Optional Exercise 11.12, in conjunction with the pivotal method, to derive the formula (given in Section 11.7) for a $(1 - \alpha)100\%$ confidence interval for β_i.

| | | | | | | | | | | | | |

S E C T I O N   11.8

CHECKING THE UTILITY OF A MODEL: R^2 AND THE ANALYSIS OF VARIANCE F TEST

Conducting t tests on each β parameter in a model is not a good way to determine whether a model is contributing information for the prediction of y. If we were to conduct a series of t tests to determine whether the independent variables are contributing to the predictive relationship, it is very likely that we would make one or more errors in deciding which terms to retain in the model and which to exclude. For example, suppose that all the β parameters (except β_0) are in fact equal to 0. Although the probability of concluding that any single β parameter differs from 0 is only α, the probability of rejecting *at least one* true null hypothesis in a set of t tests is much higher. You can see why this is true by considering the following analogy: The probability of observing a head on a single toss of a coin is only .5, but the probability of observing *at least one* head in five tosses of a coin is .97. Thus, in multiple regression models for which a large number of independent variables are being considered, conducting a series of t tests may include a large number of insignificant variables and exclude some useful ones. If we want to test the utility of a multiple regression model, we will need a global test (one that encompasses all the β parameters). We would also like to find some statistical quantity that measures how well the model fits the data.

We begin with the easier problem—finding a measure of how well a linear model fits a set of data. For this we use the multiple regression equivalent of r^2, the coefficient of determination for the straight-line model (Chapter 10). Thus, we define the **sample multiple coefficient of determination R^2** as

$$R^2 = 1 - \frac{\sum(y_i - \hat{y}_i)^2}{\sum(y_i - \bar{y})^2} = 1 - \frac{\text{SSE}}{\text{SS}_{yy}}$$

where $\hat{y}_i$ is the predicted value of y_i for the model. Just as for the simple linear model, R^2 is a sample statistic that represents the fraction of the sample variation of the y values (measured by SS_{yy}) that is attributable to the regression model.

Thus, $R^2 = 0$ implies a complete lack of fit of the model to the data, and $R^2 = 1$ implies a perfect fit, with the model passing through every data point. In general, the larger the value of R^2, the better the model fits the data.

DEFINITION 11.1

The **multiple coefficient of determination**, R^2, is defined as

$$R^2 = 1 - \frac{SSE}{SS_{yy}}$$

where $SSE = \Sigma (y_i - \hat{y}_i)^2$, $SS_{yy} = \Sigma (y_i - \bar{y})^2$, and $\hat{y}_i$ is the predicted value of y_i for the multiple regression model.

To illustrate, the value $R^2 = .982$ for the electrical usage example is indicated in Figure 11.5. This very high value of R^2 implies that 98.2% of the sample variation is attributable to, or explained by, the independent variable (home size) x. Thus, R^2 is a sample statistic that tells how well the model fits the data, and thereby represents a measure of the utility of the entire model.

FIGURE 11.5 SAS Printout for Electrical Usage Example

DEPENDENT VARIABLE: Y

SOURCE	DF	SUM OF SQUARES	MEAN SQUARE	F VALUE	PR > F	R-SQUARE	C.V.
MODEL	2	831069.54637065	415534.77318533	189.71	0.0001	0.981885	2.9348
ERROR	7	15332.55362935	2190.36480419		ROOT MSE		Y MEAN
CORRECTED TOTAL	9	846402.10000000			46.80133336		1594.70000000

SOURCE	DF	TYPE I SS	F VALUE	PR > F	DF	TYPE III SS	F VALUE	PR > F
X	1	703957.18342042	321.39	0.0001	1	208574.93309206	95.22	0.0001
X*X	1	127112.36295023	58.03	0.0001	1	127112.36295023	58.03	0.0001

PARAMETER	ESTIMATE	T FOR H0: PARAMETER=0	PR > :T:	STD ERROR OF ESTIMATE
INTERCEPT	-1216.14388700	-5.01	0.0016	242.80636850
X	2.39893018	9.76	0.0001	0.24583560
X*X	-0.00045004	-7.62	0.0001	0.00005908

The fact that R^2 is a sample statistic implies that it can be used to make inferences about the utility of the entire model for predicting y values for specific settings of the independent variables. In particular, for the electrical usage data, the test

H_0: $\beta_1 = \beta_2 = 0$

H_a: At least one of the parameters β_1 and β_2 is nonzero

would formally test the utility of the overall model. The test statistic used to test this null hypothesis is

Test statistic: $F = \dfrac{\text{Mean square for model}}{\text{Mean square for error}}$

$$= \frac{SS(Model)/k}{SSE/[n - (k + 1)]}$$

where n is the number of data points, k is the number of parameters in the model (not including β_0), and SS(Model) = SS(Total) − SSE. When H_0 is true, this F test statistic will have an F probability distribution with k df in the numerator and $[n - (k + 1)]$ df in the denominator. The upper-tail values of the F distribution are given in Tables 8, 9, 10, and 11 of Appendix II.

It can be shown (proof omitted) that an equivalent form of this test statistic is

$$F = \frac{R^2/k}{(1 - R^2)/[n - (k + 1)]}$$

Therefore, the F test statistic becomes large as the coefficient of determination R^2 becomes large. To determine how large F must be before we can conclude at a given value of α that the model is useful for predicting y, we set up the rejection region as follows:

Rejection region: $F > F_\alpha$ where $\nu_1 = k$ df, $\nu_2 = n - (k + 1)$ df

This test procedure is summarized in the box.

THE ANALYSIS OF VARIANCE F TEST: TESTING THE OVERALL UTILITY OF THE MODEL $E(y) = \beta_0 + \beta_1 x_1 + \beta_2 x_2 + \cdots + \beta_k x_k$

H_0: $\beta_1 = \beta_2 = \cdots = \beta_k = 0$

H_a: At least one of the parameters, $\beta_1, \beta_2, \ldots, \beta_k$, differs from 0

Test statistic: $F = \dfrac{R^2/k}{(1 - R^2)/[n - (k + 1)]}$

$\qquad\qquad = \dfrac{\text{Mean square for model}}{\text{Mean square for error}} = \dfrac{\text{SS(Model)}/k}{\text{SSE}/[n - (k + 1)]}$

Rejection region: $F > F_\alpha$, where $\nu_1 = k$ and $\nu_2 = [n - (k + 1)]$

Assumptions: See Section 11.2 for the assumptions about the probability distribution of the random error component ε.

EXAMPLE 11.6

Refer to the electrical usage example (Example 11.5) and test to determine whether the model contributes information for the prediction of y.

SOLUTION

For the electrical usage example, $n = 10$, $k = 2$, and $n - (k + 1) = 7$. At $\alpha = .05$, we will reject H_0: $\beta_1 = \beta_2 = 0$ if

$\qquad F > F_{.05}$ where $\nu_1 = 2$ and $\nu_2 = 7$ (the degrees of freedom are shaded on the printout in Figure 11.5)

or

$\qquad F > 4.74$ (see Figure 11.6)

FIGURE 11.6

Rejection Region for the
F statistic with $\nu_1 = 2$,
$\nu_2 = 7$, and $\alpha = .05$

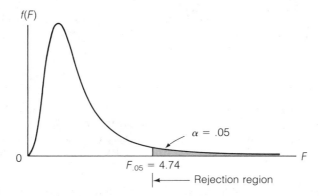

From the computer printout (Figure 11.5) we find that the computed F (shaded in the upper right region under F VALUE) is 189.71. Since this value greatly exceeds the tabulated value of 4.74, we conclude that at least one of the model coefficients β_1 and β_2 is nonzero. Therefore, this F test indicates that the second-order model $y = \beta_0 + \beta_1 x + \beta_2 x^2 + \varepsilon$ is useful for predicting electrical usage.

We could arrive at the same decision by checking the observed significance level (p-value) of the F test, given as PR>F in the SAS printout. This value (shaded in Figure 11.5) indicates that we will reject H_0 for any α greater than .0001.

To summarize the discussion in this section, the value of R^2 is an indicator of how well the prediction equation fits the data. More importantly, it can be used (in the F statistic) to determine whether the data provide sufficient evidence to indicate that the model contributes information for the prediction of y. Intuitive evaluations of the contribution of the model based on the computed value of R^2 must be examined with care. The value of R^2 will increase as more and more variables are added to the model. Consequently, you could force R^2 to take a value very close to 1 even though the model contributes no information for the prediction of y. In fact, R^2 will equal 1 when the number of terms in the model equals the number of data points. Therefore, **you should not rely solely on the value of R^2 to tell you whether the model is useful for predicting y. Use the analysis of variance F test.**

**EXERCISES
11.14–11.21**

11.14 Most companies institute rigorous safety programs in order to assure employee safety. Suppose that 60 reports of accidents over the last year at a company are randomly selected, and that the number of hours the employee had worked before the accident occurred, x, and the amount of time the employee lost from work, y, are recorded. A quadratic model is proposed to investigate a fatigue hypothesis that more serious accidents occur near the end of a day than near the beginning. Thus the proposed model

$$E(y) = \beta_0 + \beta_1 x + \beta_2 x^2$$

was fitted to the data, and part of the computer printout appears on page 502.

Portion of the Computer
Printout for Exercise 11.14

SOURCE	DF	SUM OF SQUARES	MEAN SQUARE	F VALUE
MODEL	2	112.110	56.055	1.28
ERROR	57	2496.201	43.793	R-SQUARE
TOTAL	59	2608.311		.0430

a. Do the data support the fatigue hypothesis? Use $\alpha = .05$ to test whether the proposed model is useful in predicting the lost work time, y.

b. Does the result of the test in part a necessarily mean that no fatigue factor exists? Explain. [*Hint:*   The true model of the relationship between y and x may include higher-order or other terms.]

11.15 Refer to the *IEEE Transactions on Software Engineering* (April 1985) study on identifying error-prone software, Exercise 10.42. A multiple regression analysis was conducted to identify the computer module-related variables (called metrics) useful for predicting the number y of discovered module defects. For a certain product written in PL/S language, the following model was fit to data collected for $n = 253$ modules:

$$E(y) = \beta_0 + \beta_1 x_1 + \beta_2 x_2$$

where

x_1 = Number of unique operands in the module

x_2 = Number of conditional statements, loops, and Boolean operators in the module

The multiple coefficient of determination for the model was $R^2 = .78$. Is there sufficient evidence to indicate that the model is useful for predicting the number y of defects in modules of the software product? Test using $\alpha = .05$.

11.16 Suppose you fit the model

$$y = \beta_0 + \beta_1 x_1 + \beta_2 x_2 + \beta_3 x_1 x_2 + \beta_4 x_1^2 + \beta_5 x_2^2 + \varepsilon$$

to $n = 30$ data points and obtain

$$SSE = .37 \quad \text{and} \quad R^2 = .89$$

a. Do the values of SSE and R^2 suggest that the model provides a good fit to the data? Explain.

b. Is the model of any use in predicting y? Test the null hypothesis that

$$E(y) = \beta_0$$

that is,

$$H_0: \quad \beta_1 = \beta_2 = \cdots = \beta_5 = 0$$

against the alternative hypothesis

$$H_a: \quad \text{At least one of the parameters } \beta_1, \beta_2, \ldots, \beta_5 \text{ is nonzero}$$

Use $\alpha = .05$.

11.17 A study was conducted at Union Carbide to identify the optimal catalyst preparation conditions in the conversion of monoethanolamine (MEA) to ethylenediamine

(EDA), a substance used commercially in soaps.* For each of ten preselected catalysts, the following experimental variables were measured:

y = Rate of conversion of MEA to EDA

x_1 = Atom ratio of metal used in the experiment

x_2 = Reduction temperature

$x_3 = \begin{cases} 1 & \text{if high acidity support used} \\ 0 & \text{if low acidity support used} \end{cases}$

The data for the $n = 10$ experiments were used to fit the multiple regression model $E(y) = \beta_0 + \beta_1 x_1 + \beta_2 x_2 + \beta_3 x_3$. The results are summarized below:

$$\hat{y} = 40.2 - .808x_1 - 6.38x_2 - 4.45x_3 \qquad R^2 = .899$$

$$s_{\hat{\beta}_1} = .231 \qquad s_{\hat{\beta}_2} = 1.93 \qquad s_{\hat{\beta}_3} = .99$$

a. Is there sufficient evidence to indicate that the model is useful for predicting rate of conversion y? Test using $\alpha = .01$.

b. Conduct a test to determine whether atom ratio x_1 is a useful predictor of rate of conversion y. Use $\alpha = .05$.

c. Construct a 95% confidence interval for β_2. Interpret the interval.

11.18 Suppose you fit the model

$$y = \beta_0 + \beta_1 x_1 + \beta_2 x_2 + \beta_3 x_3 + \varepsilon$$

to $n = 20$ data points and obtain

$$\sum (y_i - \hat{y}_i)^2 = 225 \quad \text{and} \quad \sum (y_i - \bar{y})^2 = 305$$

a. Find R^2. Does the value of R^2 suggest that the model provides a good fit to the data? Explain.

b. Test the null hypothesis

$$H_0: \quad \beta_1 = \beta_2 = \beta_3 = 0$$

against the alternative hypothesis

$$H_a: \quad \text{At least one of the } \beta \text{ parameters is nonzero}$$

Use $\alpha = .05$.

11.19 Refer to Exercise 10.64. In an attempt to improve upon the ability of the model to predict urban travel times, Cook and Russell (*Transportation Research*, June 1980) added a second independent variable—weighted average speed limit between the two urban locations. The proposed model takes the form

$$y = \beta_0 + \beta_1 x_1 + \beta_2 x_2 + \varepsilon$$

where

y = Urban travel time (minutes)

x_1 = Distance between locations (miles)

x_2 = Weighted speed limit between locations (miles per hour)

*Source: Hansen, J. L. and Best, D. C. "How to Pick a Winner." Paper presented at Joint Statistical Meetings, American Statistical Association and Biometric Society, August 1986, Chicago, Illinois.

This model was fitted to the data pertaining to passenger cars and to the data pertaining to trucks, with the results shown in the accompanying table.

PASSENGER CARS	TRUCKS
$\hat{y} = 5.46 + 2.15x_1 - .09x_2$	$\hat{y} = 4.84 + 3.92x_1 - .09x_2$
$R^2 = .687; n = 567$	$R^2 = .771; n = 918$

Source: Cook, T. M. and Russell, R. A. "Estimating Urban Travel Times: A Comparative Study." *Transportation Research, 14A,* June 1980, pp. 173–175. Copyright 1980, Pergamon Press, Ltd. Reprinted with permission.

a. Is the model useful for predicting the urban travel times of passenger cars? Use $\alpha = .05$.
b. Is the model useful for predicting the urban travel times of trucks? Use $\alpha = .05$.

11.20 The Florida Department of Transportation (DOT) wants to develop a model relating bid price for a road construction project to length of the road to be built or repaired and number of bidders. Since the DOT believes that the bid price increases linearly with road length and number of bidders, the following model is hypothesized:

$$y = \beta_0 + \beta_1 x_1 + \beta_2 x_2 + \varepsilon$$

where

y = Bid price (thousands of dollars)
x_1 = Length of road (miles)
x_2 = Number of bidders

Data collected on bid price, road length, and number of bidders for 32 randomly selected construction projects were used to fit the model, and a portion of the SAS printout is shown here.

SAS Printout for Exercise 11.20

SOURCE	DF	SUM OF SQUARES	MEAN SQUARE	F VALUE	PR > F
MODEL	2	4277159.70740504	2138579.85170252	120.65	0.0001
ERROR	29	514034.51534496	17725.32811534		ROOT MSE
CORRECTED TOTAL	31	4791194.21875000	R-SQUARE		133.13650181
			0.892713		

PARAMETER	ESTIMATE	T FOR H0: PARAMETER = 0	PR > ¦T¦	STD ERROR OF ESTIMATE
INTERCEPT	-1336.72205214	-7.71	0.0001	173.35612607
X1	12.73619884	14.11	0.0001	0.90238049
X2	85.81513260	9.86	0.0001	8.70575681

a. Is the model useful for estimating mean bid price? Use $\alpha = .01$.
b. Test the hypothesis that the mean bid price increases as the number of bidders increases for road construction projects of the same length. Use $\alpha = .01$.

11.21 Because the coefficient of determination R^2 always increases when a new independent variable is added to the model, it is tempting to include many variables in a model to force R^2 to be near 1. However, doing so reduces the degrees of freedom available for estimating σ^2, which adversely affects our ability to make reliable inferences. As an example, suppose you want to predict the CPU time of a computer job using eighteen predictor variables (such as size of job, time of submission, and estimated lines of print). You fit the model

$$y = \beta_0 + \beta_1 x_1 + \beta_2 x_2 + \cdots + \beta_{17} x_{17} + \beta_{18} x_{18} + \varepsilon$$

where y = CPU time and $x_1, x_2, \ldots, x_{18}$ are the predictor variables. Using the relevant information on $n = 20$ jobs to fit the model, you obtain $R^2 = .95$. Test to determine whether this value of R^2 is large enough for you to infer that this model is useful—i.e., that at least one term in the model is important for predicting CPU time. Use $\alpha = .05$.

SECTION 11.9

A CONFIDENCE INTERVAL FOR A LINEAR FUNCTION OF THE β PARAMETERS; A CONFIDENCE INTERVAL FOR $E(y)$

Suppose we were to postulate that the mean value of the productivity y of a company is related to the size of the company, x, and that the relationship could be modeled by the expression

$$E(y) = \beta_0 + \beta_1 x + \beta_2 x^2$$

A graph of $E(y)$ might appear as shown in Figure 11.7.

We might have several reasons for collecting data on the productivity and size of a set of n companies and finding the least squares prediction equation,

$$\hat{y} = \hat{\beta}_0 + \hat{\beta}_1 x + \hat{\beta}_2 x^2$$

For example, we might want to estimate the mean productivity for a company of a given size (say, $x = 2$). That is, we might want to estimate

$$E(y) = \beta_0 + \beta_1 x + \beta_2 x^2$$
$$= \beta_0 + 2\beta_1 + 4\beta_2 \quad \text{where } x = 2$$

FIGURE 11.7
Graph of Mean
Productivity $E(y)$

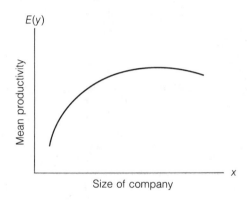

Size of company

Or we might want to estimate the marginal increase in productivity, the slope of a tangent to the curve, when $x = 2$ (see Figure 11.8). The marginal productivity for y when $x = 2$ is the rate of change of $E(y)$ with respect to x, evaluated at $x = 2$.* The marginal productivity for a value of x, denoted by the symbol $dE(y)/dx$, can be shown (proof omitted) to be

$$\frac{dE(y)}{dx} = \beta_1 + 2\beta_2 x$$

Therefore, the marginal productivity at $x = 2$ is

$$\frac{dE(y)}{dx} = \beta_1 + 2\beta_2(2) = \beta_1 + 4\beta_2$$

FIGURE 11.8
Marginal Productivity

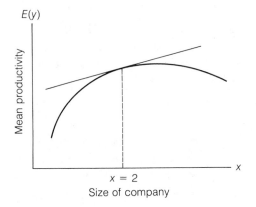

Note that for $x = 2$, both $E(y)$ and the marginal productivity are *linear* functions of the unknown parameters β_0, β_1, β_2 in the model. The problem we pose in this section is that of finding confidence intervals for linear functions of β parameters or testing hypotheses concerning their values. The information necessary to solve this problem is rarely given in a standard multiple regression analysis computer printout, but we can find these confidence intervals or values of the appropriate test statistics from knowledge of $(X'X)^{-1}$.

We will suppose that we have a model,

$$y = \beta_0 + \beta_1 x_1 + \cdots + \beta_k x_k + \varepsilon$$

and that we are interested in making an inference about a linear function of the β parameters, say

$$a_0\beta_0 + a_1\beta_1 + \cdots + a_k\beta_k$$

---

*The marginal productivity for y given x is the first derivative of $E(y) = \beta_0 + \beta_1 x + \beta_2 x^2$ with respect to x.

where $a_0, a_1, \ldots, a_k$ are known constants. Further, we will use the corresponding linear function of least squares estimates,

$$\ell = a_0\hat{\beta}_0 + a_1\hat{\beta}_1 + \cdots + a_k\hat{\beta}_k$$

as our best estimate of $a_0\beta_0 + a_1\beta_1 + \cdots + a_k\beta_k$.

We recall from Section 11.5 that the least squares estimators, $\hat{\beta}_0$, $\hat{\beta}_1$, $\hat{\beta}_2$, $\ldots$, $\hat{\beta}_k$, are normally distributed with

$$E(\hat{\beta}_i) = \beta_i$$
$$V(\hat{\beta}_i) = c_{ii}\sigma^2 \quad (i = 0, 1, 2, \ldots, k)$$

and covariances

$$\text{Cov}(\hat{\beta}_i, \hat{\beta}_j) = c_{ij}\sigma^2 \quad (i \neq j)$$

It then follows by Theorem 7.2 that

$$\ell = a_0\hat{\beta}_0 + a_1\hat{\beta}_1 + \cdots + a_k\hat{\beta}_k$$

is normally distributed with mean, variance, and standard deviation as given by Theorem 11.3.

THEOREM 11.3

PROPERTIES OF THE SAMPLING DISTRIBUTION OF $\ell = a_0\hat{\beta}_0 + a_1\hat{\beta}_1 + \cdots + a_k\hat{\beta}_k$

The sampling distribution of ℓ is normal with

$$E(\ell) = a_0\beta_0 + a_1\beta_1 + \cdots + a_k\beta_k$$
$$V(\ell) = [a'(X'X)^{-1}a]\sigma^2$$
$$\sigma_\ell = \sqrt{V(\ell)} = \sigma\sqrt{a'(X'X)^{-1}a}$$

where σ is the standard deviation of ε, $(X'X)^{-1}$ is the inverse matrix obtained in fitting the least squares model, and

$$a = \begin{bmatrix} a_0 \\ a_1 \\ a_2 \\ \vdots \\ a_k \end{bmatrix}$$

Theorem 11.3 indicates that ℓ is an unbiased estimator of

$$E(\ell) = a_0\beta_0 + a_1\beta_1 + \cdots + a_k\beta_k$$

and that its sampling distribution would appear as shown in Figure 11.9 (page 508).

FIGURE 11.9
Sampling Distribution for ℓ

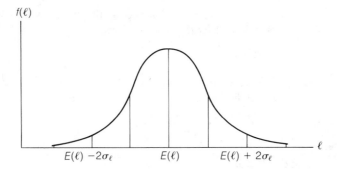

It can be shown that a $(1 - \alpha)100\%$ confidence interval for $E(\ell)$ is as shown in the box.

A $(1 - \alpha)100\%$ CONFIDENCE INTERVAL FOR $E(\ell)$

$$\ell \pm t_{\alpha/2}s\sqrt{a'(X'X)^{-1}a}$$

where

$$E(\ell) = a_0\beta_0 + a_1\beta_1 + \cdots + a_k\beta_k$$

$$\ell = a_0\hat{\beta}_0 + a_1\hat{\beta}_1 + \cdots + a_k\hat{\beta}_k \qquad a = \begin{bmatrix} a_0 \\ a_1 \\ a_2 \\ \vdots \\ a_k \end{bmatrix}$$

s and $(X'X)^{-1}$ are obtained from the least squares procedure, and $t_{\alpha/2}$ is based on the number of degrees of freedom associated with s.

The linear function of the β parameters that is most often the focus of our attention is

$$E(y) = \beta_0 + \beta_1 x_1 + \cdots + \beta_k x_k$$

That is, we want to find a confidence interval for $E(y)$ for specific values of x_1, $x_2, \ldots, x_k$. For this special case, $\ell = \hat{y}$ and the a matrix is

$$a = \begin{bmatrix} 1 \\ x_1 \\ x_2 \\ \vdots \\ x_k \end{bmatrix}$$

where the symbols $x_1, x_2, \ldots, x_k$ in the a matrix indicate the specific numerical values assumed by these variables. Thus, the procedure for forming a confidence interval for $E(y)$ is as shown in the box.

A $(1 - \alpha)$100% CONFIDENCE INTERVAL FOR $E(y)$

$$\ell \pm t_{\alpha/2} s \sqrt{a'(X'X)^{-1}a}$$

where

$$E(y) = \beta_0 + \beta_1 x_1 + \cdots + \beta_k x_k$$

$$\ell = \hat{y} = \hat{\beta}_0 + \hat{\beta}_1 x_1 + \cdots + \hat{\beta}_k x_k \qquad a = \begin{bmatrix} 1 \\ x_1 \\ x_2 \\ \vdots \\ x_k \end{bmatrix}$$

s and $(X'X)^{-1}$ are obtained from the least squares analysis, and $t_{\alpha/2}$ is based on the number of degrees of freedom associated with s, namely $[n - (k + 1)]$.

EXAMPLE 11.7

Refer to the data of Example 11.1 for insulation compression y and compressive pressure x. Find a 95% confidence interval for the mean compression $E(y)$ when the pressure is $x = 4$.

SOLUTION

The confidence interval for $E(y)$ for a given value of x is

$$\hat{y} \pm t_{\alpha/2} s \sqrt{a'(X'X)^{-1}a}$$

Consequently, we need to find and substitute the values of $a'(X'X)^{-1}a$, $t_{\alpha/2}$, and $\hat{y}$ into this formula. Since we want to estimate

$$\begin{aligned} E(y) &= \beta_0 + \beta_1 x \\ &= \beta_0 + \beta_1(4) \quad \text{when } x = 4 \\ &= \beta_0 + 4\beta_1 \end{aligned}$$

it follows that the coefficients of β_0 and β_1 are $a_0 = 1$ and $a_1 = 4$, and thus,

$$a = \begin{bmatrix} 1 \\ 4 \end{bmatrix}$$

From Examples 11.1 and 11.3, we have $\hat{y} = -.1 + .7x$, $s^2 = .367$, $s = .61$, and

$$(X'X)^{-1} = \begin{bmatrix} 1.1 & -.3 \\ -.3 & .1 \end{bmatrix}$$

Then,

$$a'(X'X)^{-1}a = \begin{bmatrix} 1 & 4 \end{bmatrix} \begin{bmatrix} 1.1 & -.3 \\ -.3 & .1 \end{bmatrix} \begin{bmatrix} 1 \\ 4 \end{bmatrix}$$

We first calculate

$$a'(X'X)^{-1} = [1 \quad 4] \begin{bmatrix} 1.1 & -.3 \\ -.3 & .1 \end{bmatrix} = [-.1 \quad .1]$$

Then,

$$a'(X'X)^{-1}a = [-.1 \quad .1] \begin{bmatrix} 1 \\ 4 \end{bmatrix} = .3$$

The t value, $t_{.025}$, based on 3 df is 3.182. So, a 95% confidence interval for the mean compression of the insulation material when subjected to a pressure of 4 (that is, 40 pounds per square inch) is

$$\hat{y} \pm t_{\alpha/2}s\sqrt{a'(X'X)^{-1}a}$$

Since $\hat{y} = -.1 + .7x = -.1 + (.7)(4) = 2.7$, the 95% confidence interval for $E(y)$ when $x = 4$ is

$$2.7 \pm (3.182)(.61)\sqrt{.3}$$
$$2.7 \pm 1.1$$

or 1.6 to 3.8 inches. Notice that this is exactly the same result as obtained in Example 10.8. ∎

EXAMPLE 11.8

An engineer recorded a measure of productivity y and the size x for each of 100 companies producing cement. A regression model,

$$y = \beta_0 + \beta_1 x + \beta_2 x^2 + \varepsilon$$

fit to the $n = 100$ data points produced the following results:

$$\hat{y} = 2.6 + .7x - .2x^2$$

where x is coded to take values in the interval $-2 < x < 2$, and

$$(X'X)^{-1} = \begin{bmatrix} .0025 & .0005 & -.0070 \\ .0005 & .0055 & 0 \\ -.0070 & 0 & .0050 \end{bmatrix} \qquad s = .14$$

Find a 95% confidence interval for the marginal increase in productivity given that the coded size of a plant is $x = 1.5$.

SOLUTION

The mean value of y for a given value of x is

$$E(y) = \beta_0 + \beta_1 x + \beta_2 x^2$$

Therefore, the marginal increase in y for $x = 1.5$ is

$$\frac{dE(y)}{dx} = \beta_1 + 2\beta_2 x$$
$$= \beta_1 + 2(1.5)\beta_2$$

Or,

$$E(\ell) = \beta_1 + 3\beta_2 \quad \text{when } x = 1.5$$

Note from the prediction equation, $\hat{y} = 2.6 + .7x - .2x^2$, that $\hat{\beta}_1 = .7$ and $\hat{\beta}_2 = -.2$. Therefore,

$$\ell = \hat{\beta}_1 + 3\hat{\beta}_2 = .7 + 3(-.2) = .1$$

and

$$a = \begin{bmatrix} a_0 \\ a_1 \\ a_2 \end{bmatrix} = \begin{bmatrix} 0 \\ 1 \\ 3 \end{bmatrix}$$

We next calculate

$$a'(X'X)^{-1}a = \begin{bmatrix} 0 & 1 & 3 \end{bmatrix} \begin{bmatrix} .0025 & .0005 & -.0070 \\ .0005 & .0055 & 0 \\ -.0070 & 0 & .0050 \end{bmatrix} \begin{bmatrix} 0 \\ 1 \\ 3 \end{bmatrix} = .0505$$

Then, since s will be based on $n - (k + 1) = 100 - 3 = 97$ df, $t_{.025} \approx 1.96$, and a 95% confidence interval for the marginal increase in productivity when $x = 1.5$ is

$$\ell \pm t_{.025}s\sqrt{a'(X'X)^{-1}a}$$

or

$$.1 \pm (1.96)(.14)\sqrt{.0505}$$

$$.1 \pm .062$$

Thus, the marginal increase in productivity, the slope of the tangent to the curve

$$E(y) = \beta_0 + \beta_1 x + \beta_2 x^2$$

is estimated to lie in the interval $.1 \pm .062$ at $x = 1.5$. A graph of $\hat{y} = 2.6 + .7x - .2x^2$ is shown in Figure 11.10.

FIGURE 11.10
A Graph of
$\hat{y} = 2.6 + .7x - .2x^2$

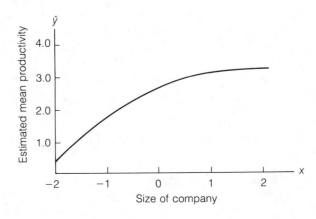

Size of company

**A PREDICTION
INTERVAL FOR SOME
VALUE OF *y* TO BE
OBSERVED IN THE
FUTURE**

We have indicated that two of the most important applications of the least squares predictor $\hat{y}$ are estimating the mean value of y (the topic of the preceding section) and predicting a new value of y, yet unobserved, for specific values of x_1, x_2, $\ldots$, x_k. The difference between these two inferential problems (when each would be pertinent) was explained in Chapter 10, but we will give another example to make certain that the distinction is clear at this point.

Suppose you are the manager of a manufacturing plant and that y, the daily profit, is a function of various process variables x_1, x_2, $\ldots$, x_k. Suppose you want to know how much money you would make *in the long run* if the x's are set at specific values. For this case, you would be interested in finding a confidence interval for the mean profit per day, $E(y)$. In contrast, suppose you planned to operate the plant for only one more day! Then you would be interested in predicting the value of y, the profit associated with tomorrow's production.

We have indicated that the error of prediction is always larger than the error of estimating $E(y)$. You can see this by comparing the formula for the prediction interval (shown in the box) with the formula for the confidence interval for $E(y)$ that was given in Section 11.9.

A $(1 - \alpha)100\%$ PREDICTION INTERVAL FOR y

$$\hat{y} \pm t_{\alpha/2}s\sqrt{1 + a'(X'X)^{-1}a}$$

where

$$\hat{y} = \hat{\beta}_0 + \hat{\beta}_1 x_1 + \cdots + \hat{\beta}_k x_k$$

s and $(X'X)^{-1}$ are obtained from the least squares analysis,

$$a = \begin{bmatrix} 1 \\ x_1 \\ x_2 \\ \cdot \\ \cdot \\ \cdot \\ x_k \end{bmatrix}$$

contains the numerical values of x_1, x_2, $\ldots$, x_k and $t_{\alpha/2}$ is based on the number of degrees of freedom associated with s, namely $[n - (k + 1)]$.

EXAMPLE 11.9

Refer to the insulation compression example (Example 11.7). Find a 95% prediction interval for the compression of a particular piece of insulation when it is to be subjected to a pressure of 40 pounds per square inch ($x = 4$).

SOLUTION

The 95% prediction interval for the compression of this particular piece of insulation is

$$\hat{y} \pm t_{\alpha/2}s\sqrt{1 + a'(X'X)^{-1}a}$$

From Example 11.7, when $x = 4$, $\hat{y} = -.1 + .7x = -.1 + (.7)(4) = 2.7$, $s = .61$, $t_{.025} = 3.182$, and $a'(X'X)^{-1}a = .3$. Then the 95% prediction interval for *y* is

$$2.7 \pm (3.182)(.61)\sqrt{1 + .3}$$
$$2.7 \pm 2.2 \quad \text{or} \quad .05 \text{ to } .49 \text{ inch} \qquad\blacksquare$$

EXERCISES
11.22–11.34

11.22 Refer to Exercise 11.1. Find a 90% confidence interval for $E(y)$ when $x = 1$. Interpret the interval.

11.23 Refer to Exercise 11.1. Suppose you plan to observe *y* for $x = 1$. Find a 90% prediction interval for that value of *y*. Interpret the interval.

11.24 Refer to Exercise 11.2. Find a 90% confidence interval for $E(y)$ when $x = 2$. Interpret the interval.

11.25 Refer to Exercise 11.2. Find a 90% confidence interval for a value of *y* to be observed in the future when $x = 2$. Interpret the interval.

11.26 Refer to Exercise 11.3. Find a 90% confidence interval for the mean value of *y* when $x = 1$. Interpret the interval.

11.27 Refer to Exercise 11.3. Find a 90% prediction interval for a value of *y* to be observed in the future when $x = 1$.

11.28 Refer to Exercise 11.10. Find a 95% confidence interval for the mean yield $E(y)$ when the temperature is set at $50°$ ($x_1 = -1$) and pressure is set at 20 pounds per square inch ($x_2 = 1$). Interpret the interval.

11.29 Refer to Exercise 11.10. Find a 95% prediction interval for a yield *y* to be observed in the future when the temperature is set at $50°$ ($x_1 = -1$) and pressure is set at 20 pounds per square inch ($x_2 = 1$). Interpret the interval.

OPTIONAL EXERCISES

11.30 Since $\hat{\beta}_0, \hat{\beta}_1, \ldots, \hat{\beta}_k$ are independent of s^2, it follows that

$$\ell = a_0\hat{\beta}_0 + a_1\hat{\beta}_1 + \cdots + a_k\hat{\beta}_k$$

is independent of s^2. Use this fact and Theorems 11.2 and 11.3 to show that

$$t = \frac{\ell - E(\ell)}{s\sqrt{a'(X'X)^{-1}a}}$$

has a Student's *t* distribution with $[n - (k + 1)]$ degrees of freedom.

11.31 Let $\ell = \hat{y} = \hat{\beta}_0 + \hat{\beta}_1x_1 + \hat{\beta}_2x_2 + \cdots + \hat{\beta}_kx_k$. Use the *t* statistic of Optional Exercise 11.30, in conjunction with the pivotal method, to derive the formula (given in Section 11.9) for a $(1 - \alpha)100\%$ confidence interval for $E(y)$.

11.32 Let $\hat{y} = \hat{\beta}_0 + \hat{\beta}_1 x_1 + \hat{\beta}_2 x_2 + \cdots + \hat{\beta}_k x_k$ be the least squares prediction equation and let y be some observation to be obtained in the future.

a. Explain why $(\hat{y} - y)$ is normally distributed.

b. Show that

$$E(\hat{y} - y) = 0$$

and

$$V(\hat{y} - y) = [1 + a'(X'X)^{-1}a]\sigma^2$$

11.33 Show that

$$t = \frac{\hat{y} - y}{s\sqrt{1 + a'(X'X)^{-1}a}}$$

has a Student's t distribution with $[n - (k + 1)]$ degrees of freedom.

11.34 Use the result of Optional Exercise 11.33 and the pivotal method to derive the formula (given in Section 11.10) for a $(1 - \alpha)100\%$ prediction interval for y.

SECTION 11.11

COMPUTER PRINTOUTS FOR A MULTIPLE REGRESSION ANALYSIS

The formulas for a multiple regression analysis presented in the previous sections can become quite cumbersome (sometimes physically impossible) to use for large data sets and/or complex models. Consequently, in practice most researchers resort to the use of computers. We have highlighted the key elements of a multiple regression analysis as they appear in the SAS printout. However, there are a multitude of statistical software packages capable of performing multiple regression; some of the most popular, in addition to the SAS System, are BMDP, Minitab, and SPSS^X.

The multiple regression computer programs for these packages may differ in what they are programmed to do, how they do it, and the appearance of their printouts, but all of them print the basic outputs needed for a regression analysis. For example, some will compute confidence intervals for $E(y)$ and prediction intervals for y; others will not. However, all test the null hypotheses that the individual β parameters equal 0 using Student's t tests and all give the least squares estimates, the values of SSE, s^2, and other pertinent information.

To illustrate, the Minitab, SAS, BMDP, and SPSS^X regression analysis computer printouts for Example 11.2 are shown in Figure 11.11. In that example, we fit the quadratic model

$$y = \beta_0 + \beta_1 x + \beta_2 x^2 + \varepsilon$$

to $n = 10$ data points, where

$y =$ Monthly usage (kilowatt-hours)
$x =$ Size of home (square feet)

The data are reproduced for convenience in Table 11.4.

TABLE 11.4
Data for Power
Usage Study

SIZE OF HOME x, square feet	MONTHLY USAGE y, kilowatt-hours
1,290	1,182
1,350	1,172
1,470	1,264
1,600	1,493
1,710	1,571
1,840	1,711
1,980	1,804
2,230	1,840
2,400	1,956
2,930	1,954

FIGURE 11.11   Computer printouts for Example 11.2
a. Minitab Regression Printout

```
THE REGRESSION EQUATION IS
Y = -1216. + 2.40 X1 -0.0005 X2
                                        ST. DEV.    T-RATIO =
          COLUMN       COEFFICIENT      OF COEF.    COEF/S.D.
          --            -1216.1          242.8       -5.00
X1    C1               2.3989           0.2458        9.75
X2    C3            -0.00045004      0.00005907      -7.61

THE ST. DEV. OF Y ABOUT REGRESSION LINE IS
S = 46.80
WITH (  10- 3) = 7 DEGREES OF FREEDOM

R-SQUARED = 98.2 PERCENT
R-SQUARED = 97.7 PERCENT, ADJUSTED FOR D.F.

ANALYSIS OF VARIANCE
  DUE TO      DF            SS        MS=SS/DF
REGRESSION    2          831069       415534
RESIDUAL      7           15332        2190
TOTAL         9          846401

FURTHER ANALYSIS OF VARIANCE
SS EXPLAINED BY EACH VARIABLE WHEN ENTERED IN THE ORDER GIVEN
  DUE TO      DF            SS
REGRESSION    2          831069
C1            1          703956
C3            1          127112

          X1             Y      PRED. Y    ST.DEV.
ROW       C1             C2      VALUE      PRED. Y    RESIDUAL    ST.RES.
 10      2930         1954.0    1949.1      44.7         4.8       0.35 X

X DENOTES AN OBS. WHOSE X VALUE GIVES IT LARGE INFLUENCE.

DURBIN-WATSON STATISTIC = 2.08
```

b. SAS Regression Printout

DEPENDENT VARIABLE: Y

SOURCE	DF	SUM OF SQUARES	MEAN SQUARE	F VALUE	PR > F	R-SQUARE	C.V.
MODEL	2	831069.54637065	415534.77318533	189.71	0.0001	0.981885	2.9348
ERROR	7	15332.55362935	2190.36480419		ROOT MSE		Y MEAN
CORRECTED TOTAL	9	846402.10000000			46.80133336		1594.70000000

SOURCE	DF	TYPE I SS	F VALUE	PR > F	DF	TYPE III SS	F VALUE	PR > F
X	1	703957.18342042	321.39	0.0001	1	208574.93309206	95.22	0.0001
X*X	1	127112.36295023	58.03	0.0001	1	127112.36295023	58.03	0.0001

PARAMETER	ESTIMATE	T FOR HO: PARAMETER=0	PR > :T:	STD ERROR OF ESTIMATE
INTERCEPT	-1216.14388700	-5.01	0.0016	242.80636850
X	2.39893018	9.76	0.0001	0.24583560
X*X	-0.00045004	-7.62	0.0001	0.00005908

c. BMDP Regression Printout

```
DEPENDENT VARIABLE. . . . . . . . . . . .     2 Y
 TOLERANCE . . . . . . . . . . . . . . .   0.0010
ALL DATA CONSIDERED AS A SINGLE GROUP

MULTIPLE R              0.9909      STD. ERROR OF EST.      46.8154
MULTIPLE R-SQUARE       0.9819
```

ANALYSIS OF VARIANCE

	SUM OF SQUARES	DF	MEAN SQUARE	F RATIO	P(TAIL)
REGRESSION	831060.5625	2	415530.2500	189.594	0.0000
RESIDUAL	15341.7656	7	2191.6807		

VARIABLE		COEFFICIENT	STD. ERROR	STD. REG COEFF	T	P(2 TAIL)	TOLERANCE
INTERCEPT		-1216.04077					
X1	1	2.39883	0.24590	4.049	9.755	0.0000	0.01503
X2	3	-0.45002E-03	0.59093E-04	-3.161	-7.615	0.0001	0.01503

d. SPSS^X Regression Printout

```
VARIABLE LIST NUMBER 1    LISTWISE DELETION OF MISSING DATA

EQUATION NUMBER 1    DEPENDENT VARIABLE..  Y

BEGINNING BLOCK NUMBER  1.  METHOD:  ENTER     X1     X2

VARIABLE(S) ENTERED ON STEP NUMBER  1..    X2
                                    2..    X1
```

MULTIPLE R	.99090	ANALYSIS OF VARIANCE	DF	SUM OF SQUARES	MEAN SQUARE
R SQUARE	.98189	REGRESSION	2	831069.54637	415534.77319
ADJUSTED R SQUARE	.97671	RESIDUAL	7	15332.55363	2190.36480
STANDARD ERROR	46.80133				

```
                        F =    189.71030      SIGNIF F =  .0000
```

------------------ VARIABLES IN THE EQUATION ------------------

VARIABLE	B	SE B	BETA	T	SIG T
X2	-4.50040E-04	5.9077E-05	-3.16102	-7.618	.0001
X1	2.39893	.24584	4.04915	9.758	.0000
(CONSTANT)	-1216.14389	242.80637		-5.009	.0016

INDEPENDENT VARIABLES Notice that the Minitab printout gives the prediction equation at the top of the printout. The independent variables, shown in the prediction equation and listed on the left side of the printout are x_1 and x_2. Thus, Minitab treats the product x^2 as a second independent variable, x_2, which must be computed before the fitting begins. For this reason, the Minitab prediction equation will always appear on the printout as first-order even though some of the independent variables shown in the prediction equation may actually be the squares or cross products of other independent variables. The inclusion of the squares or cross products of independent variables is treated in the same manner in the BMDP and SPSS^X programs shown in Figure 11.11. The SAS program is the only one of these four that can be instructed to include these terms automatically, and they appear in the printout with an asterisk ($*$) that indicates multiplication. Thus, in the SAS printout, x^2 is printed as X$*$X.

PARAMETER ESTIMATES The estimates of the regression coefficients appear opposite the identifying variable in the Minitab and BMDP columns titled COEFFICIENT, in the SAS column titled ESTIMATE, and in the SPSS^X column titled B. Compare the estimates given in these four columns. Note that the Minitab printout gives the estimates with a much lesser degree of accuracy (fewer decimal places) than the SAS, BMDP, and SPSS^X printouts. (Ignore the columns titled BETA in the SPSS^X printout and STD. REG COEFF in the BMDP printout. These are standardized estimates and will not be discussed in this text.)

STANDARD ERRORS OF ESTIMATES The estimated standard errors of the estimates are given in the Minitab column titled ST. DEV. OF COEF., in the SAS column titled STD ERROR OF ESTIMATE, in the BMDP column titled STD. ERROR, and in the SPSS^X column titled SE B.

***t* TESTS FOR INDIVIDUAL β PARAMETERS** The values of the test statistics for testing H_0: $\beta_i = 0$ (for $i = 1, 2$) are shown in the Minitab column titled T-RATIO = COEF/S.D., in the SAS column titled T FOR H0: PARAMETER=0, and in the BMDP and SPSS^X columns titled T. Note that the computed t values shown in the Minitab, SPSS^X, SAS, and BMDP columns are identical (except for the number of decimal places) and that Minitab does not give the observed significance level of the test. Consequently, to draw conclusions from the Minitab printout, you must compare the computed values of t with the critical values given in a t table (Table 6 of Appendix II). In contrast, the SAS printout gives the observed significance level for each t test in the column titled PR$>|T|$, BMDP gives this value in the column titled P(2TAIL), and SPSS^X in the column titled SIG T. Note that these observed significance levels have been computed assuming that the tests are two-tailed. The observed significance level for a one-tailed test would equal half the value shown.

SSE AND s^2 The Minitab printout gives the value SSE = 15332 under the ANALYSIS OF VARIANCE column headed SS and in the row identified as RESIDUAL. The value of $s^2 = 2{,}190$ is shown in the same row under the column headed MS = SS/DF, and the degrees of freedom DF appears in the same row as 7. The corresponding values are shown at the top of the SAS printout in the row labeled ERROR and in the columns designated SUM OF SQUARES, MEAN SQUARE, and DF, respectively. These quantities appear with similar headings in the SPSS^X and BMDP printouts.

COEFFICIENT OF DETERMINATION R^2 The value of R^2, as defined in Section 11.8, is given in the Minitab printout as 98.2 PERCENT (we defined this quantity as a ratio where $0 \leq R^2 \leq 1$). It is given in the top right corner of the SAS printout as 0.981885 (R-SQUARE), at the top of the BMDP printout as .9819 (MULTIPLE R-SQUARE), and in the left column of the SPSS^X printout as .98189 (R SQUARE). (Ignore the quantities shown in the Minitab as R^2 ADJUSTED FOR D.F. and in the SPSS^X printout as ADJUSTED R SQUARE. These quantities are adjusted for the degrees of freedom associated with the total sum of squares and SSE and are not used or discussed in this text.)

F TEST FOR OVERALL MODEL UTILITY　　The F statistic for testing the utility of the overall model—that is, for testing the null hypothesis that all model parameters (except β_0) equal 0—is shown under the title F VALUE as 189.71 in the top center of the SAS printout. In addition, the SAS printout gives the observed significance level of this F test under PR>F. Similarly, the F value and corresponding observed significance level are given in the BMDP printout under F RATIO and P(TAIL), respectively, and in the SPSS^X printout under the ANALYSIS OF VARIANCE table. The F statistic for testing the utility of the overall model is not given in the Minitab printout. If you are using Minitab and want to obtain the value of this statistic, you must compute it using one of the formulas given in the box in Section 11.8:

$$F = \frac{R^2/k}{(1 - R^2)/[n - (k + 1)]}$$

or

$$F = \frac{\text{Mean square for regression}}{\text{Mean square for error (or residual)}}$$
$$= \frac{\text{Mean square for regression}}{s^2}$$

These quantities are given in the Minitab printout under the column marked MS = SS/DF. Thus,

$$F = \frac{415{,}534}{2{,}190} = 189.7$$

This value agrees with the values given in the SAS, BMDP, and SPSS^X printouts.

We will not comment on the merits or drawbacks of the various packages because you will have to use the software available at your computer center and become familiar with its output. Most of the computer printouts are similar and it is relatively easy to learn how to read one output after you have become familiar with another. We have used different packages in the examples to help you with this. The commands necessary to call the multiple regression programs for the four packages are discussed in Section 11.14.*

---

*The BMDP multiple regression package analyzes data differently from the other three packages. BMDP takes a **stepwise** approach: If a certain variable does not meet a specified criterion, it will be excluded from the model. Thus, users of BMDP should be aware of the possibility that not all independent variables may appear on the printout.

| | | | | | | | | | | | |

S E C T I O N 11.12

SOME PITFALLS: ESTIMABILITY, MULTICOLLINEARITY, AND EXTRAPOLATION

There are several problems you should be aware of when constructing a prediction model for some response y. A few of the most important will be discussed in this section.

PROBLEM 1: PARAMETER ESTIMABILITY   Suppose you want to fit a model relating the strength y of a new type of plastic fitting to molding temperature x. We propose the first-order model

$$E(y) = \beta_0 + \beta_1 x$$

Now, suppose we mold a sample of three plastic fittings, each at a temperature of 300°F. The data are graphed in Figure 11.12. You can see the problem: The parameters of the line cannot be estimated when all the data are concentrated at a single x value. Recall that it takes two points (x values) to fit a straight line. Thus, the parameters are not estimable when only one x value is observed.

FIGURE 11.12

Plastic Strength and Molding Temperature Data

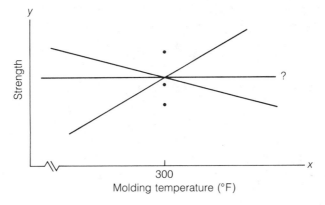

A similar problem would occur if we attempted to fit the second-order model

$$E(y) = \beta_0 + \beta_1 x + \beta_2 x^2$$

to a set of data for which only one or two different x values were observed (see Figure 11.13, page 520). At least three different x values must be observed before a second-order model can be fit to a set of data (that is, before all three parameters are estimable). In general, the number of levels of x must be at least one more than the order of the polynomial in x that you want to fit. Remember, also, that the sample size n must be sufficiently large to allow degrees of freedom for estimating σ^2.

REQUIREMENTS FOR FITTING A pTH-ORDER POLYNOMIAL REGRESSION MODEL
$$E(y) = \beta_0 + \beta_1 x + \beta_2 x^2 + \cdots + \beta_p x^p$$ **1.** The number of levels of x must be greater than or equal to $(p + 1)$. **2.** The sample size n must be greater than $(p + 1)$ in order to allow sufficient degrees of freedom for estimating σ^2.

FIGURE 11.13

Only Two *x* Values
Observed—Second-Order
Model Is Not Estimable

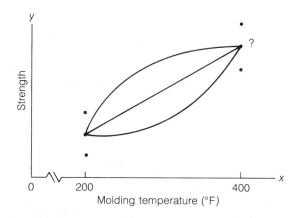

Since many variables observed in nature cannot be controlled by the researcher, the independent variables will almost always be observed at a sufficient number of levels to permit estimation of the model parameters. However, when the computer program you use suddenly refuses to fit a model, the problem is probably inestimable parameters.

PROBLEM 2: PARAMETER INTERPRETATION   Given that the parameters of the model are estimable, it is important to interpret the parameter estimates correctly. A typical misconception is that $\hat{\beta}_i$ *always* measures the effect of x_i on $E(y)$, *independent* of the other x variables in the model. This may be true for some models, but is not true in general. Generally, the interpretation of an individual β parameter becomes increasingly more difficult as the model becomes more complex. In Chapter 12, we give the β interpretations for a number of different multiple regression models.

Another misconception about parameter estimates is that a statistically significant $\hat{\beta}_i$ value establishes a *cause-and-effect* relationship between $E(y)$ and x_i. That is, if $\hat{\beta}_i$ is found to be significantly greater than 0, then some practitioners would infer that an increase in x_i *causes* an increase in the mean response, $E(y)$. However, we warned in Section 10.7 about the dangers of inferring a causal relationship between two variables. There may be many other independent variables (some of which we may have included in our model, some of which we may have omitted) that affect the mean response. Unless we can control the values of these other variables, we are uncertain about what is actually causing the observed increase in y. In Chapter 13, we introduce the notion of **designed experiments**, where the values of the independent variables are set in advance before the value of y is observed. Only with such an experiment can a cause-and-effect relationship be established.

PROBLEM 3: MULTICOLLINEARITY   Often, two or more of the independent variables used in the model for $E(y)$ will contribute redundant information. That is, the independent variables will be correlated with each other. For example, suppose we want to construct a model to predict the gasoline mileage rating, y, of a truck as a function of its load, x_1, and the horsepower, x_2, of its engine. In general, you would expect heavier loads to require greater horsepower and to result in lower mileage ratings. Thus, although both x_1 and x_2 contribute information for the prediction of mileage rating, some of the information is overlapping, because x_1 and x_2 are correlated. When the independent variables are correlated, we say that **multicollinearity** exists. In practice, it is not uncommon to observe correlations among the independent variables. However, a few problems arise when serious multicollinearity is present in the regression analysis.

First, high correlations among the independent variables increase the likelihood of rounding errors in the calculations of the β estimates, standard errors, and so forth.* Second, the regression results may be confusing and misleading.

To illustrate, if the gasoline mileage rating model

$$E(y) = \beta_0 + \beta_1 x_1 + \beta_2 x_2$$

were fit to a set of data, we might find that the t values for both $\hat{\beta}_1$ and $\hat{\beta}_2$ (the least squares estimates) are nonsignificant. However, the F test for H_0: $\beta_1 = \beta_2 = 0$ would probably be highly significant. The tests may seem to be contradictory, but really they are not. The t tests indicate that the contribution of one variable, say $x_1 = $ load, is not significant after the effect of $x_2 = $ horsepower has been discounted (because x_2 is also in the model). The significant F test, on the other hand, tells us that at least one of the two variables is making a contribution to the prediction of y (i.e., either β_1, β_2, or both differ from 0). In fact, both are probably contributing, but the contribution of one overlaps with that of the other.

Multicollinearity can also have an effect on the signs of the parameter estimates. More specifically, a value of $\hat{\beta}_i$ may have the opposite sign from what is expected. For example, we expect the signs of both of the parameter estimates for the gasoline mileage rating model to be negative, yet the regression analysis for the model might yield the estimates $\hat{\beta}_1 = .2$ and $\hat{\beta}_2 = -.7$. The positive value of $\hat{\beta}_1$ seems to contradict our expectation that heavy loads will result in lower mileage ratings. However, it is dangerous to interpret a β coefficient when the independent variables are correlated. Because the variables contribute redundant information, the effect of load (x_1) on mileage rating is measured only partially by $\hat{\beta}_1$. Also, we warned in the discussion of Problem 2 that we cannot establish a cause-and-effect relationship between y and the predictor variables based on **observational data** (data for which the values of the independent variables are uncontrolled). By attempting to interpret the value $\hat{\beta}_1$, we are really trying to establish a cause-and-effect relationship between y and x_1 (by suggesting that a heavy load x_1 will *cause* a lower mileage rating y).

How can you avoid the problems of multicollinearity in regression analysis? One way is to conduct a designed experiment so that the levels of the x variables are uncorrelated. Unfortunately, time and cost constraints may prevent you from collecting data in this manner. For these and other reasons, most data collected in scientific studies are observational. Since observational data frequently consist

---

*The result is due to the fact that, in the presence of severe multicollinearity, the computer has difficulty inverting the $(X'X)$ matrix.

of correlated independent variables, you will need to recognize when multicollinearity is present and, if necessary, make modifications in the regression analysis.

Several methods are available for detecting multicollinearity in regression. A simple technique is to calculate the coefficient of correlation r between each pair of independent variables in the model and use the procedure outlined in Section 10.7 to test for evidence of positive or negative correlation. If one or more of the r values is statistically different from 0, the variables in question are correlated and a severe multicollinearity problem may exist.* Other indications of the presence of multicollinearity include those mentioned in the beginning of this section—namely, nonsignificant t tests for the individual β parameters when the F test for overall model adequacy is significant, and parameter estimates with opposite signs from what is expected. More formal methods for detecting multicollinearity exist, but are beyond the scope of this text. (Consult the references given at the end of the chapter if you want to learn more about these methods.)

The methods for detecting multicollinearity are summarized in the accompanying box. We illustrate the use of these statistics in Example 11.10.

DETECTING MULTICOLLINEARITY IN THE REGRESSION MODEL

$$E(y) = \beta_0 + \beta_1 x_1 + \beta_2 x_2 + \cdots + \beta_k x_k$$

The following are indicators of multicollinearity:

1. Significant correlations between pairs of independent variables in the model
2. Nonsignificant t tests for the individual β parameters when the F test for overall model adequacy $H_0: \beta_1 = \beta_2 = \cdots = \beta_k = 0$ is significant
3. Opposite signs (from what is expected) in the estimated parameters

---

EXAMPLE 11.10

The Federal Trade Commission annually ranks varieties of domestic cigarettes according to their tar, nicotine, and carbon monoxide contents. The U.S. surgeon general considers each of these three substances hazardous to a smoker's health. Past studies have shown that increases in the tar and nicotine contents of a cigarette are accompanied by an increase in the carbon monoxide emitted from the cigarette smoke. Table 11.5 contains tar, nicotine, and carbon monoxide contents (in milligrams) and weight (in grams) for a sample of 25 (filter) brands tested in 1983. Suppose we want to model carbon monoxide content, y, as a function of tar content, x_1, nicotine content, x_2, and weight, x_3, using the model

$$E(y) = \beta_0 + \beta_1 x_1 + \beta_2 x_2 + \beta_3 x_3$$

---

*Remember that r measures only the pairwise correlation between x values. Three variables, x_1, x_2, and x_3, may be highly correlated as a group, but may not exhibit large pairwise correlations. Thus, multicollinearity may be present even when all pairwise correlations are not significantly different from 0.

The model is fit to the 25 data points in Table 11.5 and a portion of the resulting SAS printout is shown in Figure 11.14 (page 524). Examine the printout. Do you detect any signs of multicollinearity?

TABLE 11.5

FTC Cigarette Data for Example 11.10

BRAND	TAR x_1, milligrams	NICOTINE x_2, milligrams	WEIGHT x_3, grams	CARBON MONOXIDE y, milligrams
Alpine	14.1	.86	.9853	13.6
Benson & Hedges	16.0	1.06	1.0938	16.6
Bull Durham	29.8	2.03	1.1650	23.5
Camel Lights	8.0	.67	.9280	10.2
Carlton	4.1	.40	.9462	5.4
Chesterfield	15.0	1.04	.8885	15.0
Golden Lights	8.8	.76	1.0267	9.0
Kent	12.4	.95	.9225	12.3
Kool	16.6	1.12	.9372	16.3
L&M	14.9	1.02	.8858	15.4
Lark Lights	13.7	1.01	.9643	13.0
Marlboro	15.1	.90	.9316	14.4
Merit	7.8	.57	.9705	10.0
Multifilter	11.4	.78	1.1240	10.2
Newport Lights	9.0	.74	.8517	9.5
Now	1.0	.13	.7851	1.5
Old Gold	17.0	1.26	.9186	18.5
Pall Mall Light	12.8	1.08	1.0395	12.6
Raleigh	15.8	.96	.9573	17.5
Salem Ultra	4.5	.42	.9106	4.9
Tareyton	14.5	1.01	1.0070	15.9
True	7.3	.61	.9806	8.5
Viceroy Rich Lights	8.6	.69	.9693	10.6
Virginia Slims	15.2	1.02	.9496	13.9
Winston Lights	12.0	.82	1.1184	14.9

Source: Federal Trade Commission (1983).

SOLUTION

First, notice that a test of H_0: $\beta_1 = \beta_2 = \beta_3 = 0$ is highly significant. The F value (shaded on the printout) is very large ($F = 78.984$) and the observed significance level of the test (also shaded) is small (p-value = .0001). Therefore, we can reject H_0 for any α greater than .0001 and conclude that at least one of the parameters, β_1, β_2, and β_3, is nonzero. The t tests for two of the three individual β's, however, are nonsignificant. (The p-values for these tests are shaded on the printout.) Unless tar is the only one of the three variables useful for predicting carbon monoxide content, these results are the first indication of a potential multicollinearity problem.

A second clue to the presence of multicollinearity is the negative value for $\hat{\beta}_2$ (shaded on the printout), $\hat{\beta}_2 = -2.63$. From past studies, we expect carbon monoxide content (y) to increase when nicotine content (x_2) increases—that is, we expect a *positive* relationship between y and x_2, not a negative one.

FIGURE 11.14   Portion of the SAS Printout for Example 11.10

```
DEP VARIABLE: CO
                                        ANALYSIS OF VARIANCE

                              SUM OF        MEAN
                  SOURCE  DF  SQUARES       SQUARE      F VALUE    PROB>F

                  MODEL    3  495.25781    165.08594    78.984     0.0001
                  ERROR   21   43.89258562   2.09012312
                  C TOTAL 24  539.15040

                  ROOT MSE      1.445726    R-SQUARE     0.9186
                  DEP MEAN     12.528       ADJ R-SQ     0.9070
                  C.V.         11.53996

                                        PARAMETER ESTIMATES

                    PARAMETER      STANDARD     T FOR H0:
       VARIABLE  DF  ESTIMATE      ERROR        PARAMETER=0    PROB > |T|

       INTERCEP  1   3.20219002    3.46175473    0.925         0.3655
       TAR       1   0.96257386    0.24224436    3.974         0.0007
       NICOTINE  1  -2.63166111    3.90055745   -0.675         0.5072
       WEIGHT    1  -0.13048185    3.88534182   -0.034         0.9735
```

All signs indicate that a serious multicollinearity problem exists. To confirm our suspicions, we calculated the coefficient of correlation r for each of the three pairs of independent variables in the model. These values are given in Table 11.6. You can see that tar content (x_1) and nicotine content (x_2) appear to be highly correlated ($r = .977$), while weight (x_3) appears to be moderately correlated with both tar content ($r = .491$) and nicotine content ($r = .500$). In fact, all three sample correlations exceed the critical value, $r_{.025} = .423$, for a two-tailed test of $H_0: \rho = 0$ conducted at $\alpha = .05$ with $n - 2 = 23$ df.

TABLE 11.6

Correlation Coefficients for the Three Pairs of Independent Variables in Example 11.10

PAIR	r
x_1, x_2	.977
x_1, x_3	.491
x_2, x_3	.500

Once you have detected that a multicollinearity problem exists, there are several alternative measures available for solving the problem. The appropriate measure to take depends on the severity of the multicollinearity and the ultimate goal of the regression analysis.

Some researchers, when confronted with highly correlated independent variables, choose to include only one of the correlated variables in the final model. One way of deciding which variable to include is by using **stepwise regression**, a topic discussed in Chapter 12. Generally, only one (or a small number) of a set of multicollinear independent variables will be included in the regression model by the stepwise regression procedure. This procedure tests the parameter associated with each variable in the presence of all the variables already in the model. For example, in fitting the gasoline mileage rating model introduced earlier, if at one step the variable representing truck load is included as a significant variable in the prediction of the mileage rating, the variable representing horsepower will probably never be added in a future step. Thus, if a set of independent variables is thought to be multicollinear, some screening by stepwise regression may be helpful.

If you are interested in using the model for estimation and prediction, you may decide not to drop any of the independent variables from the model. In the presence of multicollinearity, we have seen that it is dangerous to interpret the individual β's for the purpose of establishing cause and effect. However, confidence intervals for $E(y)$ and prediction intervals for y generally remain unaffected **as long as the values of the independent variables used to predict y follow the same pattern of multicollinearity exhibited in the sample data.** That is, you must take strict care to ensure that the values of the x variables fall within the experimental region. (We will discuss this problem in further detail in Problem 4.) Alternatively, if your goal is to establish a cause-and-effect relationship between y and the independent variables, you will need to conduct a designed experiment to break up the pattern of multicollinearity.

When fitting a polynomial regression model [for example, the second-order model $E(y) = \beta_0 + \beta_1 x + \beta_2 x^2$], the independent variables $x_1 = x$ and $x_2 = x^2$ will often be correlated. If the correlation is high, the computer solution may result in extreme rounding errors. For this model, the solution is not to drop one of the independent variables but to transform the x variable in such a way that the correlation between the coded x and x^2 values is substantially reduced. Coding the independent quantitative variables in polynomial regression models is a topic discussed in Chapter 12.

SOLUTIONS TO SOME PROBLEMS CREATED BY MULTICOLLINEARITY

1. Drop one or more of the correlated independent variables from the final model. A screening procedure such as stepwise regression (see Chapter 12) is helpful in determining which variables to drop.
2. If you decide to keep all the independent variables in the model:
   a. Avoid making inferences about the individual β parameters (such as establishing a cause-and-effect relationship between y and the predictor variables).
   b. Restrict inferences about $E(y)$ and future y values to values of the independent variables that fall within the experimental region (see Problem 4).
3. If your ultimate objective is to establish a cause-and-effect relationship between y and the predictor variables, use a designed experiment (see Chapter 13).
4. To reduce rounding errors in polynomial regression models, code the independent variables so that first-, second-, and higher-order terms for a particular x variable are not highly correlated (see Chapter 12).

PROBLEM 4: PREDICTION OUTSIDE THE EXPERIMENTAL REGION   The fitted regression model enables us to construct a confidence interval for $E(y)$ and a prediction interval for y for values of the independent variable only within the region of experimentation, i.e., within the range of values of the independent variables used in the experiment. For example, suppose that you conduct experiments on the mean strength of the plastic fittings (see Figure 11.13) at several different temperatures in the interval 200°F to 400°F. The regression model that you fit to the data is valid for estimating $E(y)$ or for predicting values of y for values of x in the range 200°F $\leq x \leq$ 400°F. However, if you attempt to extrapolate beyond the experimental region, you risk the possibility that the fitted model is no longer a good approximation to the mean strength of the plastic (see Figure 11.15). For example, the plastic may become too brittle when formed at 500°F and possess no strength at all. Estimating and predicting outside of the experimental region are sometimes necessary. If you do so, keep in mind the possibility of a large extrapolation error.

FIGURE 11.15

Using a Regression Model Outside the Experimental Region

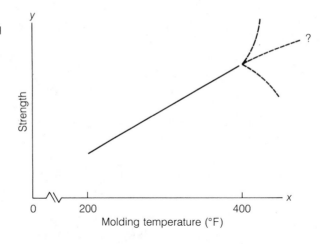

PROBLEM 5: CORRELATED ERRORS   Another problem associated with using a regression model to predict an economic variable y or some quality measure y of an industrial product, based on independent variables $x_1, x_2, \ldots, x_k$, arises from the fact that the data are obtained at equidistant points in time. These observations may be correlated, which in turn often causes the prediction errors of the regression model to be correlated. Thus, the assumption of independent errors is violated, and the model tests and prediction intervals are no longer valid. The solution to this problem is to construct a **time series model** that accounts for the correlated errors. Texts that discuss the analysis of time series are included in the references at the end of this chapter.

| | | | | | | | | | | | | |

**EXERCISES
11.35–11.41**

11.35 Why is it dangerous to predict y for values of the independent variables that fall outside the experimental region?

11.36 Discuss the problems that result when multicollinearity is present in a regression analysis.

11.37 How can you detect multicollinearity?

11.38 What remedial measures are available when multicollinearity is detected?

11.39 The management of an engineering consultant firm is considering the possibility of setting up its own market research department rather than continuing to use the services of a market research firm. Management wants to know what salary should be paid to a market researcher, based on years of experience. An independent consultant has proposed the quadratic model

$$E(y) = \beta_0 + \beta_1 x + \beta_2 x^2$$

where

y = Annual salary (thousands of dollars)

x = Years of experience

In order to fit the model, the consultant randomly sampled three market researchers at other firms and recorded the information given in the accompanying table. Give your opinion regarding the adequacy of the proposed model.

	y	x
Researcher 1	40	2
Researcher 2	25	1
Researcher 3	42	3

11.40 A bioengineer wants to model the amount (y) of carbohydrate solubilized during steam processing of peat as a function of temperature (x_1), exposure time (x_2), and pH value (x_3). Data collected for each of 15 peat samples were used to fit the model

$$E(y) = \beta_0 + \beta_1 x_1 + \beta_2 x_2 + \beta_3 x_3$$

A summary of the regression results follows:

$$\hat{y} = -3,000 + 3.2x_1 - .4x_2 - 1.1x_3 \qquad R^2 = .93$$

$$s_{\hat{\beta}_1} = 2.4 \qquad s_{\hat{\beta}_2} = .6 \qquad s_{\hat{\beta}_3} = .8$$

$$r_{12} = .92 \qquad r_{13} = .87 \qquad r_{23} = .81$$

Based on these results, the bioengineer concludes that none of the three independent variables, x_1, x_2, and x_3, is a useful predictor of carbohydrate amount, y. Do you agree with this statement? Explain.

11.41 Refer to the FTC cigarette data of Example 11.10. The data are reproduced on page 528 for convenience.

BRAND	TAR x_1, milligrams	NICOTINE x_2, milligrams	WEIGHT x_3, grams	CARBON MONOXIDE y, milligrams
Alpine	14.1	.86	.9853	13.6
Benson & Hedges	16.0	1.06	1.0938	16.6
Bull Durham	29.8	2.03	1.1650	23.5
Camel Lights	8.0	.67	.9280	10.2
Carlton	4.1	.40	.9462	5.4
Chesterfield	15.0	1.04	.8885	15.0
Golden Lights	8.8	.76	1.0267	9.0
Kent	12.4	.95	.9225	12.3
Kool	16.6	1.12	.9372	16.3
L&M	14.9	1.02	.8858	15.4
Lark Lights	13.7	1.01	.9643	13.0
Marlboro	15.1	.90	.9316	14.4
Merit	7.8	.57	.9705	10.0
Multifilter	11.4	.78	1.1240	10.2
Newport Lights	9.0	.74	.8517	9.5
Now	1.0	.13	.7851	1.5
Old Gold	17.0	1.26	.9186	18.5
Pall Mall Light	12.8	1.08	1.0395	12.6
Raleigh	15.8	.96	.9573	17.5
Salem Ultra	4.5	.42	.9106	4.9
Tareyton	14.5	1.01	1.0070	15.9
True	7.3	.61	.9806	8.5
Viceroy Rich Lights	8.6	.69	.9693	10.6
Virginia Slims	15.2	1.02	.9496	13.9
Winston Lights	12.0	.82	1.1184	14.9

Source: Federal Trade Commission (1983).

a. Fit the model $E(y) = \beta_0 + \beta_1 x_1$ to the data. Is there evidence that tar content (x_1) is useful for predicting carbon monoxide content (y)?

b. Fit the model $E(y) = \beta_0 + \beta_2 x_2$ to the data. Is there evidence that nicotine content (x_2) is useful for predicting carbon monoxide content (y)?

c. Fit the model $E(y) = \beta_0 + \beta_3 x_3$ to the data. Is there evidence that weight (x_3) is useful for predicting carbon monoxide content (y)?

d. Compare the signs of $\hat{\beta}_1$, $\hat{\beta}_2$, and $\hat{\beta}_3$ in the models of parts **a**, **b**, and **c**, respectively, to the signs of the $\hat{\beta}$'s in the multiple regression model fit in Example 11.10. The fact that the $\hat{\beta}$'s change dramatically when the independent variables are removed from the model is another indication of a serious multicollinearity problem.

SECTION 11.13

RESIDUAL ANALYSIS

An analysis of **residuals**, the differences ($y - \hat{y}$) between the y values and their corresponding predicted values, often provides information that can lead to modifications and improvements in a regression model.

One method for analyzing the residuals in a regression analysis is to plot the value of each residual versus the corresponding value of the independent variable, x. (If the model contains more than one independent variable, a plot would be

> **DEFINITION 11.2**
>
> A regression **residual** is defined as the difference between an observed y value and its corresponding predicted value:
>
> $$\text{Residual} = (y - \hat{y})$$

constructed for each of the independent variables.) A second method is to plot each residual versus its predicted value. The following example and discussion illustrate the use of these two types of plots.

EXAMPLE 11.11

Fit a first-order linear model to the data shown in Table 11.7. Then calculate the residuals, plot them versus x, and analyze the plot.

TABLE 11.7
Data for Example 11.11

x	y
0	1
1	4
2	6
3	8
4	9
5	10
6	10
7	8

SOLUTION

The least squares equation for the data can be obtained by using the formulas of Section 10.3 by performing a standard regression analysis on a computer. The resulting prediction equation is

$$\hat{y} = 3.167 + 1.095x$$

(For the sake of brevity, we omit the calculations needed to find $\hat{\beta}_0$ and $\hat{\beta}_1$.) Substituting each value of x into this prediction equation, we can calculate $\hat{y}$ and the corresponding residual, $y - \hat{y}$. The value of x, the predicted value of $\hat{y}$, and the residual $(y - \hat{y})$ are shown in Table 11.8 for each of the data points.

TABLE 11.8
Calculation of the Residuals for the Simple Linear Regression Analysis of Example 11.11

x	y	$\hat{y}$	$(y - \hat{y})$
0	1	3.167	−2.167
1	4	4.262	−.262
2	6	5.357	.643
3	8	6.452	1.548
4	9	7.548	1.452
5	10	8.643	1.357
6	10	9.738	.262
7	8	10.833	−2.833

One way to analyze the residuals from Table 11.8 is to plot them versus the independent variable x. This plot is shown in Figure 11.16a. Instead of varying in a random pattern as x increases, the values of the residuals cycle from negative to positive to negative. This cyclical behavior is present because we have fit a first-order (straight-line) linear model to data for which a second-order model is appropriate. A plot of the data points on a graph of $\hat{y}$ versus x is shown in Figure 11.16b. The residuals are represented by the vertical bars between the data points and the fitted line. Those below the $\hat{y}$ line are negative; those above it are positive. Figure 11.16 shows why fitting the wrong model to a set of data can produce patterns in the residuals when they are plotted versus an independent variable. For this simple example, the nonrandom (in this case, cyclical) behavior of the residuals can be eliminated by fitting a second-order model to the data. In general, certain patterns in the values of the residuals may suggest a need to modify the deterministic portion of the regression model, but the exact change that is needed may not always be obvious.   ∎

FIGURE 11.16

Analysis of Residuals for Example 11.11

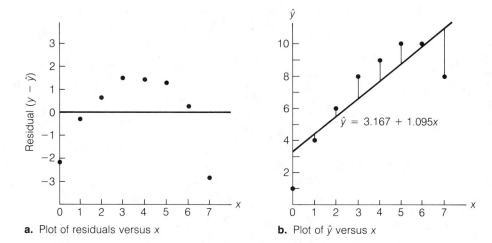

a. Plot of residuals versus x

b. Plot of $\hat{y}$ versus x

A plot of the residuals may sometimes reveal the fact that the variance of y is not stable (as required by the assumptions stated in Section 11.2. For example, a plot of the residuals versus an independent variable x may display a pattern as shown in Figure 11.17. In the figure, the range in values of the residuals increases

FIGURE 11.17

Residual Plot Showing Changes in the Variance of y

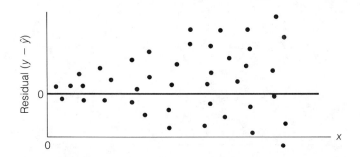

as x increases, thus indicating that the variance of the response variable y becomes larger as x increases in value.

Residual plots of the type shown in Figure 11.17 are not uncommon because the variance of y often depends on the mean value of y. Variables that represent counts per unit of area, volume, time, etc. (i.e., Poisson random variables) are cases in point. For a Poisson random variable, the variance of y is equal to $E(y) = \mu$, i.e., $\sigma_y^2 = \mu$.

Since $\hat{y}$ is an estimator of $E(y)$, a plot of the residuals versus $\hat{y}$ may indicate how the range of the residuals (and hence, σ_y) varies as $E(y)$ increases. If the plot assumes the pattern shown in Figure 11.18a and if you think it is possible

FIGURE 11.18

Plots of the Residuals Versus $\hat{y}$ for Poisson and Binomial Response Variables

a. Poisson

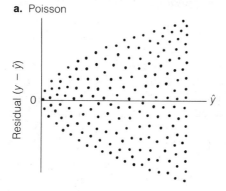

b. Binomial

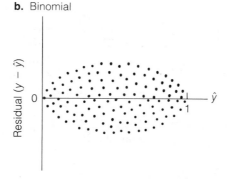

c. Multiplicative

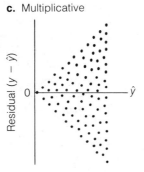

that y is approximately a Poisson random variable, you may be able to stabilize the variance of the response by fitting $\sqrt{y}$ (instead of y) to the independent variables. Similarly, if y is a percentage or proportion, $\hat{p} = y/n$, we would expect $\sigma_{\hat{p}} = \sqrt{p(1 - p)/n}$ to be small when p is near 0 or 1 and to reach a maximum when p is equal to .5. A plot of the residuals versus $\hat{y}$ for this type of data would appear as shown in Figure 11.18b. To stabilize the variance of this type of data, fit $y^* = \sin^{-1} \sqrt{y}$, where y is expressed in radians.

A third situation that requires a variance-stabilizing transformation occurs when the response variable y follows a **multiplicative model**. Unlike the **additive** models discussed so far, in this model the dependent variable is written as the *product* of its mean and the random error component:

$$y = [E(y)] \cdot \varepsilon$$

The variance of this response will grow proportionally to the square of the mean, i.e., $\text{Var}(y) = [E(y)]^2\sigma^2$, where σ^2 is the variance of ε. Data subject to multiplicative errors produce a pattern of residuals about $\hat{y}$ like that shown in Figure 11.18c. The appropriate transformation for this type of data is $y^* = \log(y)$.

The three variance-stabilizing transformations we have discussed are summarized in Table 11.9.

TABLE 11.9

Transformations to Stabilize the Variance of a Response

RESIDUAL PLOT	TYPE OF DATA	CHARACTERISTICS	TRANSFORMATION
As shown in Figure 11.18a	Poisson	Counts per unit of time, distance, volume, etc.	$y^* = \sqrt{y}$
As shown in Figure 11.18b	Binomial	Proportions, percentages, or numbers of successes for a fixed number n of trials	$y^* = \sin^{-1} \sqrt{y}$ where y is proportion
As shown in Figure 11.18c	Multiplicative	Business and economic data	$y^* = \log(y)$

In addition, residual plots can also be used to detect **outliers**, values of y that appear to be in disagreement with the model. Since almost all values of y should lie within 3σ of $E(y)$, the mean values of y, we would expect most of them to lie within $3s$ of $\hat{y}$. If a residual is larger than $3s$ (in absolute value), we consider it an outlier and seek background information which might explain the reason for its large value.

DEFINITION 11.3

A residual that is larger than $3s$ (in absolute value) is considered to be an **outlier**.

To detect outliers we can construct horizontal lines located a distance of $3s$ above and below 0 (see Figure 11.19) on a residual plot. Any residual falling outside the band formed by these lines would be considered an outlier. We would then initiate an investigation to seek the cause of the departure of such observations from expected behavior.

FIGURE 11.19

3s Lines Used to
Locate Outliers

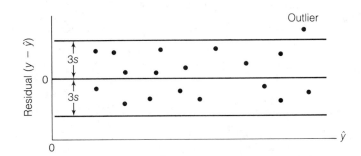

Although some analysts advocate elimination of outliers, regardless of whether cause can be assigned, others encourage the correction of only those outliers that can be traced to specific causes. The best philosophy is probably a compromise between these extremes. For example, before deciding the fate of an outlier you may want to determine how much influence it has on the regression analysis. When an accurate outlier (i.e., an outlier that is not due to recording or measurement error) is found to have a dramatic effect on the regression analysis, it may be the model and not the outlier that is suspect. Omission of important independent variables or higher-order terms could be the reason why the model is not predicting well for the outlying observation. Several sophisticated numerical techniques are available for identifying outlying influential observations. Consult the references given at the end of this chapter for a discussion of these techniques.

**EXERCISES
11.42–11.46**

11.42 A first-order model is fit to the data shown in the table, with the following results:

$$\hat{y} = 2.588 + .541x \qquad s = .356$$

x	-2	-2	-1	-1	0	0	1	1	2	2	3	3
y	1.1	1.3	2.0	2.1	2.7	2.8	3.4	3.6	4.0	3.9	3.8	3.6

a. Calculate the residuals for the model.
b. Plot the residuals versus x. Do you detect any trends? If so, what does the pattern suggest about the model?
c. Plot the residuals versus $\hat{y}$. Identify any outliers on the plot.

d. Refer to the residual plot constructed in part **c**. Do you detect any trends? If so, what does the pattern suggest about the model?

11.43 The accompanying table gives the engine size and miles per gallon ratings for 11 gasoline-fueled subcompact and compact cars. (The engine sizes are in total cubic inches of cylinder volume.) In an attempt to predict gas mileage from the engine size of subcompact and compact cars, the first-order model

$$y = \beta_0 + \beta_1 x + \varepsilon$$

is fit to the data. The resulting least squares model is $\hat{y} = 49.333 - .1802x$.

CAR	CYLINDER VOLUME x	MILES PER GALLON y
VW Rabbit	97	42
Mazda GLC	91	35
Plymouth Horizon	105	30
BMW	196	16
SAAB	121	19
Honda Accord	107	27
Dodge Colt	86	37
Renault	100	26
Toyota Celica	144	25
Datsun 200SX	119	28
Honda Civic	81	33

a. Calculate the regression residuals for this model.
b. Plot the residuals against cylinder volume, x.
c. Do you detect any distinctive patterns or trends in this plot?
d. What does your answer to part **c** suggest about model adequacy or the usual assumptions made about the random error term?

11.44 Underinflated or overinflated tires can increase tire wear and decrease gas mileage. A new tire was tested for wear at different pressures with the results shown in the table. Suppose you are interested in modeling the relationship between y and x.

PRESSURE x, pounds per square inch	MILEAGE y, thousands
30	29
31	32
32	36
33	38
34	37
35	33
36	26

a. Fit the straight-line model $y = \beta_0 + \beta_1 x + \varepsilon$ to the data.
b. Calculate the residuals for the model.

c. Plot the residuals versus x. Do you detect any trends? If so, what does the pattern suggest about the model?

d. Fit the quadratic model $y = \beta_0 + \beta_1 x + \beta_2 x^2 + \varepsilon$ to the data using an available multiple regression computer program package. Has the addition of the quadratic term improved model adequacy?

11.45 Breakdowns of machines that produce steel cans are very costly. The more break-downs, the fewer cans produced, and the smaller the company's profits. To help anticipate profit loss, the owners of a can company would like to find a model that will predict the number of breakdowns on the assembly line. The model proposed by the company's statisticians is the following:

$$y = \beta_0 + \beta_1 x_1 + \beta_2 x_2 + \beta_3 x_3 + \beta_4 x_4 + \varepsilon$$

where y is the number of breakdowns per 8-hour shift,

$$x_1 = \begin{cases} 1 & \text{if afternoon shift} \\ 0 & \text{otherwise} \end{cases} \qquad x_2 = \begin{cases} 1 & \text{if midnight shift} \\ 0 & \text{otherwise} \end{cases}$$

x_3 is the temperature of the plant (°F), and x_4 is the number of inexperienced personnel working on the assembly line. After the model is fit using the least squares procedure, the residuals are plotted. The trend in the plot reveals that the response variable y may be of the Poisson type, and hence that a square-root transformation is necessary in order to achieve a variance that is approximately the same for all settings of the independent variables. The regression analysis for the transformed model

$$y^* = \sqrt{y} = \beta_0 + \beta_1 x_1 + \beta_2 x_2 + \beta_3 x_3 + \beta_4 x_4 + \varepsilon$$

produces the prediction equation

$$\hat{y}^* = 1.3 + .008 x_1 - .13 x_2 + .0025 x_3 + .26 x_4$$

a. Use the equation to predict the number of breakdowns during the midnight shift if the temperature of the plant at that time is 87°F and if there is only one inexperienced worker on the assembly line.

b. A 95% prediction interval for y^* when $x_1 = 0$, $x_2 = 0$, $x_3 = 90°F$, and $x_4 = 2$ is $(1.965, 2.125)$. For those same values of the independent variables, find a 95% prediction interval for y, the number of breakdowns per 8-hour shift.

c. A 95% confidence interval for $E(y^*)$ when $x_1 = 0$, $x_2 = 0$, $x_3 = 90°F$, and $x_4 = 2$ is $(1.987, 2.107)$. Using only the information given in this problem, is it possible to find a 95% confidence interval for $E(y)$? Explain.

11.46 The manager of a retail appliance store wants to model the proportion of appliance owners who decide to purchase a service contract for a specific major appliance. Since the manager believes that the proportion y decreases with age x of the appliance (in years), he will fit the first-order model

$$E(y) = \beta_0 + \beta_1 x$$

A sample of 50 purchasers of new appliances are contacted about the possibility of purchasing a service contract. Fifty owners of 1-year-old machines, and 50 owners each of 2-, 3-, and 4-year-old machines are also contacted. One year later, another survey is conducted in a similar manner. The proportion y of owners deciding to purchase the service policy is shown in the table on page 536.

AGE OF APPLIANCE x, years	PROPORTION BUYING SERVICE CONTRACT y
0	.94
0	.96
1	.7
1	.76
2	.6
2	.4
3	.24
3	.3
4	.12
4	.1

a. Fit the first-order model to the data.

b. Calculate the residuals and construct a residual plot versus $\hat{y}$.

c. What does the plot from part **b** suggest about the variance of y?

d. Explain how you could stabilize the variances.

e. Refit the model using the appropriate variance-stabilizing transformation. Plot the residuals for the transformed model and compare to the plot obtained in part b. Does the assumption of homoscedasticity appear to be satisfied?

| | | | | | | | | | | | | |

SECTION 11.14

RUNNING A MULTIPLE REGRESSION ANALYSIS ON THE COMPUTER (OPTIONAL)

If you want to perform a multiple regression analysis on the computer, you will need to become familiar with the multiple regression procedure commands of the statistical software package available at your institution. In this section we show you the commands that produced the SAS regression printout shown in Figure 11.11b. Then we provide a summary of the general regression commands of the four statistical program packages—SAS, Minitab, SPSS^X, and BMDP.

Refer to the electrical usage data given in Table 11.3. In Example 11.2, we used the SAS multiple regression package to fit the model

$$E(y) = \beta_0 + \beta_1 x + \beta_2 x^2$$

where

y = Monthly usage

x = Size of home

The program statements that produced the printout shown in Figure 11.11b are given in Program 11.1.

Commands 1–13 comprise the data entry instructions of the SAS program. A data set named USAGE is created; it includes ten observations on two variables, Y and X. Once the data set has been created, only two additional SAS statements are necessary to fit a multiple regression model.

The PROC GLM statement (command 14) calls the SAS General Linear Models procedure. The GLM procedure uses the principle of least squares to fit multiple regression (general linear) models. The only other required statement is the

PROGRAM 11.1

SAS Statements That
Produced the Printout of
Figure 11.11b

Command line		
1	DATA USAGE;	Data entry instructions:
2	INPUT X Y;	Create the electrical
3	CARDS;	usage data set
4	1290 1182	
5	1350 1172	
.	. .	Input data values: One observation (home) per line
.	. .	
.	. .	
13	2930 1954	
14	PROC GLM;	Statistical analysis instructions:
15	MODEL Y = X X*X;	Fit a multiple regression model to the data

MODEL statement (command 15). The MODEL statement specifies the dependent variable to the left of the equals sign and the list of the independent variables to the right. Notice that the second-order term in the model is entered as X*X. Whenever you want to include squared terms or other higher-order terms in your model, use an asterisk (*) to indicate multiplication. For example, x_1^2 is typed as X1*X1, $x_1 x_2$ is typed X1*X2, and $x_1 x_2 x_3$ is typed X1*X2*X3.

The independent variable (size of the home) in the electrical usage model is quantitative. If you want to include a qualitative variable in the model, you must first create dummy variables for the qualitative variable using a 0–1 system of coding.* (We will discuss the 0–1 system of coding for qualitative variables in more detail in Chapter 12.) This is most easily accomplished by reading in the values of the dummy variables directly from the data cards. For example, suppose x_3 is a qualitative variable for the type of insulation used in the home, where

$$x_3 = \begin{cases} 1 & \text{if insulation A} \\ 0 & \text{if insulation B} \end{cases}$$

To include x_3 on the data set, keypunch the INPUT statement as

INPUT X1 X2 X3 Y;

and then enter 0's and 1's for x_3 in the appropriate columns of the data cards. The variable x_3 can then be entered in the MODEL statement as a single independent variable or as part of a higher-order term.

The multiple regression commands for other statistical computer packages differ slightly from the SAS commands. We present a summary of only the minimal statements *necessary* for fitting a multiple regression model with these packages in Programs 11.2b–d (page 538). Some of these packages have options that enable the user to generate much more than the output shown in Figure 11.11 [including, for example, confidence intervals for $E(y)$, prediction intervals for y, and residual plots]. We recommend that you consult the reference manuals for the particular

---

*SAS will automatically code the qualitative variables of a data set if a CLASSES statement is included after the PROC GLM card. For more details on the use of the CLASSES card, consult the references given at the end of this chapter.

PROGRAM 11.2

Multiple Regression
Programs for the Model
$E(y) = \beta_0 + \beta_1 x_1 + \beta_2 x_1^2$

package you want to use before running a multiple regression analysis. A list of these computer manuals is given in the references at the end of this chapter.

a. SAS

Command
line

```
1    DATA NEW;      ⎫ Data entry instructions
2    INPUT Y X1;    ⎭
3    CARDS;
     [Input data values]
4    PROC GLM;
5    MODEL Y = X1 X1*X1/P CLI;  ⎫ Regression analysis, prediction intervals,
                                 ⎭ and residuals
```

b. SPSS^x

Command
line

```
1    DATA LIST FREE/Y X1 ⎫ Data entry instructions
2    COMPUTE X2 = X1*X1  ⎭
3    REGRESSION VARIABLES = Y,X1,X2/          ⎫ Regression
4             CRITERIA = TOLERANCE (.00001)/  ⎪ analysis, predicted
5             DEPENDENT = Y/ENTER X1,X2/      ⎬ values, and
6             CASEWISE= ALL/                  ⎪ residuals
7    BEGIN DATA                               ⎭
     [Input data values]
8    END DATA
```

c. Minitab

Command
line

```
1    READ Y IN C1, X1 IN C2           ⎫
     [Input data values]              ⎬ Data entry instructions
2    MULTIPLY C2 BY C2, PUT IN C3     ⎭
3    REGRESS C1 ON 2 PREDICTORS, C2 C3  ⎫ Regression analysis, pre-
                                         ⎭ dicted values, and residuals
```

d. BMDP*

Command
line

```
1    BMDP 1R
2    / PROBLEM TITLE IS 'REGRESSION'.   ⎫
3    / INPUT VARIABLES = 2,             ⎪
4           FORMAT IS FREE.             ⎬ Data entry instructions
5    / VARIABLE NAMES = Y,X1            ⎪
6           ADD= 1.                     ⎭
7    / TRANSFORM X2 = X1*X1.
8    / REGRESS DEPENDENT = Y.           ⎫ Regression analysis,
9           INDEPENDENT = X1,X2.        ⎬ predicted values,
10          TOL = .001.                 ⎭ and residuals
11   / PRINT DATA.
12   / END
     [Input data values]
13   /END
```

*If the observed significance level corresponding to an independent variable is too high, the BMDP multiple regression routine may not include the variable in the model. Consult the BMDP references for information on how you can be sure to include all the independent variables in the model.

We have discussed some of the methodology of **multiple regression analysis**, a technique for modeling a dependent variable y as a function of several independent variables $x_1, x_2, \ldots, x_k$. The steps we follow in constructing and using multiple regression models are much the same as those for the simple straight-line models:

1. The form of the probabilistic model is hypothesized.
2. The model coefficients are estimated using least squares.
3. The probability distribution of ε is specified and σ^2 is estimated.
4. The utility of the model is checked using the analysis of variance F test and the multiple coefficient of determination R^2. The t tests on individual β parameters aid in deciding the final form of the model.
5. An analysis of residuals is conducted to determine if the data comply with the assumptions in step 3.
6. If the model is deemed useful and the assumptions are satisfied, it may be used to make estimates and to predict values of y to be observed in the future.

We have covered steps 2–6 in this chapter, assuming that the model was specified. Chapter 12 is devoted to step 1—model construction.

11.47 Use the method of least squares to fit a straight line to the six data points listed in the table.

x	-5	-3	-1	1	3	5
y	1.1	1.9	3.0	3.8	5.1	6.0

a. Construct Y and X matrices for the data.
b. Find $X'X$ and $X'Y$.
c. Find the least squares estimates,

$$\hat{\beta} = (X'X)^{-1}X'Y$$

[*Note:* See Theorem I.1 in Appendix I for information on finding $(X'X)^{-1}$.]
d. Give the prediction equation.
e. Find SSE and s^2.
f. Does the model contribute information for the prediction of y? Test $H_0: \beta_1 = 0$. Use $\alpha = .05$.
g. Find r^2 and interpret its value.
h. Find a 90% confidence interval for $E(y)$ when $x = .5$. Interpret the interval.

11.48 An experiment was conducted to investigate the effect of extrusion pressure P and temperature at extrusion T on the strength y of a new type of plastic. Two plastic specimens were prepared for each of five combinations of pressure and temperature. The specimens were then tested in random order, and the breaking strength for each specimen was recorded. The independent variables were coded as follows to simplify computations:

$$x_1 = \frac{P - 200}{10} \qquad x_2 = \frac{T - 400}{25}$$

The $n = 10$ data points are listed in the table.

y	x_1	x_2
5.2; 5.0	-2	2
.3; $-.1$	-1	-1
-1.2; -1.1	0	-2
2.2; 2.0	1	-1
6.2; 6.1	2	2

a. Give the Y and X matrices needed to fit the model $y = \beta_0 + \beta_1 x_1 + \beta_2 x_2 + \varepsilon$.
b. Find the least squares prediction equation.
c. Find SSE and s^2.
d. Does the model contribute information for the prediction of y? Test using $\alpha = .05$.
e. Find R^2 and interpret its value.
f. Test the null hypothesis that $\beta_1 = 0$. Use $\alpha = .05$. What is the practical implication of the test?
g. Find a 90% confidence interval for the mean strength of the plastic for $x_1 = -2$ and $x_2 = 2$.
h. Suppose a single specimen of the plastic is to be installed in the engine mount of a Douglas DC-10 aircraft. Find a 90% prediction interval for the strength of this specimen if $x_1 = -2$ and $x_2 = 2$.

11.49 J. Vuorinen carried out a series of experiments to gather information on the coefficient of permeability of concrete (*Magazine of Concrete Research*, Sept. 1985). In one experiment, the outflow of water from the pores of a concrete specimen after it had been under saturating water pressure for a period of time was recorded for different combinations of concrete permeability and porosity. The resulting water quantities after different lapses of time for one permeability–porosity combination are given in the table.

TIME t, seconds	WATER OUTFLOW w, grams per cylinder
201	3.88
325	4.93
525	6.42
775	7.80
975	8.72
1,200	9.60

Source: Vuorinen, J. "Applications of Diffusion Theory to Permeability Tests on Concrete, Part II: Pressure-Saturation Test on Concrete and Coefficient of Permeability." *Magazine of Concrete Research*, Vol. 37, No. 132, Sept. 1985, p. 156 (Table II.1).

a. According to Vuorinen, "the quantity of water discharged is approximately in linear relationship with the square root of time" for most of the permeability–

porosity combinations. Using the formulas given in Section 10.3, fit the following model to the data in the table:

$$E(w) = \alpha_0 + \alpha_1\sqrt{t}$$

b. Is there sufficient evidence to indicate that quantity of water outflow and the square root of time are linearly related? Test using $\alpha = .10$.

11.50 Refer to Exercise 11.49. Vuorinen fit the water outflow-time model for each of nine permeability–porosity combinations and used the results to develop a model for the coefficient of permeability of concrete, y. Specifically, he fit the model*

$$E(y) = \beta_0 + \beta_1 x_1 + \beta_2 x_2$$

where

$x_1 =$ Porosity of the cement

$x_2 =$ Estimated slope coefficient ($\hat{\alpha}_1$) of the corresponding water outflow-time regression line

The data are reproduced here and the SAS printout for the analysis is shown on page 542.

COEFFICIENT OF PERMEABILITY y, (meters per second) $\times 10^{-11}$	POROSITY x_1	ESTIMATED WATER OUTFLOW-TIME SLOPE COEFFICIENT x_2
1.00	.050	.903
1.00	.035	.722
1.00	.025	.590
.10	.050	.345
.10	.035	.282
.10	.025	.233
.01	.050	.103
.01	.035	.091
.01	.025	.078

Source: Vuorinen, J. "Applications of Diffusion Theory to Permeability Tests on Concrete, Part II: Pressure-Saturation Test on Concrete and Coefficient of Permeability." *Magazine of Concrete Research*, Vol. 37, No. 132, Sept. 1985, p. 156 (Table II.1).

a. Give the least squares prediction equation.
b. Conduct a test of overall model utility. Interpret the p-value of the test.
c. Is there evidence that concrete porosity x_1 is a useful predictor of coefficient of permeability y? Test using $\alpha = .05$.
d. Is there evidence that the estimated water outflow-time slope is a useful predictor of coefficient of permeability y? Test using $\alpha = .05$.
e. Locate R^2 on the printout and interpret its value.
f. Locate the estimate of σ on the printout and interpret its value.

*In actuality, Vuorinen fit the logarithmic model
$$\log(y) = \beta_0 + \beta_1\log(x_1) + \beta_2\log(x_2) + \varepsilon$$

542   11   MULTIPLE REGRESSION ANALYSIS

SAS Printout for Exercise 11.50

ANALYSIS OF VARIANCE

SOURCE	DF	SUM OF SQUARES	MEAN SQUARE	F VALUE	PROB>F
MODEL	2	1.65931863	0.82965932	35.843	0.0005
ERROR	6	0.13888137	0.02314689		
C TOTAL	8	1.79820000			

ROOT MSE	0.152141	R-SQUARE	0.9228	
DEP MEAN	0.37	ADJ R-SQ	0.8970	
C.V.	41.1192			

PARAMETER ESTIMATES

VARIABLE	DF	PARAMETER ESTIMATE	STANDARD ERROR	T FOR H0: PARAMETER=0	PROB > |T|
INTERCEP	1	0.13202076	0.19005130	0.695	0.5133
X1	1	-9.30712231	5.05702529	-1.840	0.1153
X2	1	1.55756304	0.18396157	8.467	0.0001

11.51 Refer to Exercise 11.2. Suppose we obtained two replications of the experiment—i.e., two values of y were observed for each of the six values of x. The data are shown in the accompanying table.

x	1		2		3		4		5		6	
y	1.1	.5	1.8	2.0	2.0	2.9	3.8	3.4	4.1	5.0	5.0	5.8

a. Suppose (as in Exercise 11.2) you want to fit the model $E(y) = \beta_0 + \beta_1 x$. Construct Y and X matrices for the data. [*Hint:* Remember, the Y matrix must be of dimensions 12×1.]
b. Find $X'X$ and $X'Y$.
c. Compare the $X'X$ matrix for two replications of the experiment with the $X'X$ matrix obtained for a single replication (part **b** of Exercise 11.2). What is the relationship between the elements in the two matrices?
d. Observe the $(X'X)^{-1}$ matrix for a single replication (see part **c** of Exercise 11.2). Verify that the $(X'X)^{-1}$ matrix for two replications contains elements that are equal to $\frac{1}{2}$ of the values of the corresponding elements in the $(X'X)^{-1}$ matrix for a single replication of the experiment. [*Hint:* Show that the product of the $(X'X)^{-1}$ matrix (for two replications) and the $X'X$ matrix from part **c** equals the identity matrix **I**.]
e. Find the prediction equation.
f. Find SSE and s^2.
g. Do the data provide sufficient information to indicate that x contributes information for the prediction of y? Test using $\alpha = .05$.
h. Find r^2 and interpret its value.

11.52 Refer to Exercise 11.51.
a. Find a 90% confidence interval for $E(y)$ when $x = 4.5$. Interpret the interval.

b. Suppose we want to predict the value of y if, in the future, $x = 4.5$. Find a 90% prediction interval for y and interpret the interval.

11.53 Suppose you fit the model

$$y = \beta_0 + \beta_1 x_1 + \beta_2 x_1^2 + \beta_3 x_2 + \beta_4 x_3 + \varepsilon$$

to $n = 25$ data points and find that

$$\hat{\beta}_0 = 1.26 \qquad \hat{\beta}_1 = -2.43 \qquad \hat{\beta}_2 = .05 \qquad \hat{\beta}_3 = .62 \qquad \hat{\beta}_4 = 1.81$$

$$\text{SSE} = .41 \qquad R^2 = .83$$

$$s_{\hat{\beta}_1} = 1.21 \qquad s_{\hat{\beta}_2} = .16 \qquad s_{\hat{\beta}_3} = .26 \qquad s_{\hat{\beta}_4} = 1.49$$

a. Is there sufficient evidence to conclude that at least one of the parameters, β_1, β_2, β_3, or β_4, is nonzero? Test using $\alpha = .05$.
b. Test H_0: $\beta_1 = 0$ against H_a: $\beta_1 < 0$. Use $\alpha = .05$.
c. Test H_0: $\beta_2 = 0$ against H_a: $\beta_2 > 0$. Use $\alpha = .05$.
d. Test H_0: $\beta_3 = 0$ against H_a: $\beta_3 \neq 0$. Use $\alpha = .05$.

11.54 A first-order model is fit to the data shown in the table, with the following results:

$$\hat{y} = -3.179 + 2.491x \qquad s = 4.154$$

x	2	4	7	10	12	15	18	20	21	25
y	5	10	12	22	25	27	39	50	47	65

a. Calculate the residuals for the model.
b. Plot the residuals versus x. Do you detect any trends? If so, what does the pattern suggest about the model?
c. Plot the residuals versus $\hat{y}$. Identify any outliers on the plot.
d. Refer to the residual plot constructed in part c. Do you detect any trends? If so, what does the pattern suggest about the model?

11.55 One of the best methods of improving the common resistance of chromium deposits on steel is to apply a uniformly porous layer. An experiment was conducted to investigate the relationship between porosity (number of pores per square centimeter) and chromium deposit thickness (μm). The data are shown here and the SAS computer printout for fitting a second-order model to the data is on page 544.

x	.1	.15	.25	.42	.53
y	50	30	20	13	4.5

SAS Computer Printout for Exercise 11.55

DEPENDENT VARIABLE: Y

SOURCE	DF	SUM OF SQUARES	MEAN SQUARE	F VALUE	PR > F	R-SQUARE	C.V.
MODEL	2	1135.47927571	567.73963786	12.27	0.0753	0.924657	28.9425
ERROR	2	92.52072429	46.26036214		ROOT MSE		Y MEAN
CORRECTED TOTAL	4	1228.00000000			6.80149705		23.50000000

SOURCE	DF	TYPE I SS	F VALUE	PR > F	DF	TYPE III SS	F VALUE	PR > F
X	1	1043.95220030	22.57	0.0416	1	220.89380767	4.78	0.1605
X*X	1	91.52707541	1.98	0.2948	1	91.52707541	1.98	0.2948

PARAMETER	ESTIMATE	T FOR H0: PARAMETER=0	PR > :T:	STD ERROR OF ESTIMATE
INTERCEPT	66.95119596	4.78	0.0411	14.00261798
X	-243.52342917	-2.19	0.1605	111.44317949
X*X	245.97681060	1.41	0.2948	174.87345376

OBSERVATION	X	OBSERVED VALUE	PREDICTED VALUE	RESIDUAL	LOWER 95% CL FOR MEAN	UPPER 95% CL FOR MEAN
1	0.1	50.00000000	45.05862115	4.94137885	20.39913466	69.71810764
2	0.15	30.00000000	35.95715982	-5.95715982	18.67580288	53.23851677
3	0.25	20.00000000	21.44300933	-1.44300933	-0.39198771	43.27976637
4	0.42	13.00000000	8.06166510	4.93833490	-12.45990659	28.58323679
5	0.53	4.50000000	6.97866460	-2.47866460	-20.67289424	34.63022344
6 *	0.3	.	16.03208016	.	-7.98479009	40.04895042

* OBSERVATION WAS NOT USED IN THIS ANALYSIS

a. Write a second-order model relating mean porosity $E(y)$ to chromium deposit thickness x.

b. Give the prediction equation for y.

c. Find SSE, s^2, and R^2.

d. A 95% confidence interval for $E(y)$ when the thickness x is .3 (μm) is shown on the printout in the row corresponding to observation 6. Interpret this confidence interval.

e. Why is the confidence interval for $E(y)$ in part **d** so wide?

11.56 Suppose you used Minitab to fit the model

$$y = \beta_0 + \beta_1 x_1 + \beta_2 x_2 + \varepsilon$$

to $n = 15$ data points and you obtained the computer printout shown here.

Minitab Computer Printout for Exercise 11.56

```
THE REGRESSION EQUATION IS
Y = 90.1 - 1.84 X1 + .285 X2

                                            ST. DEV.    T-RATIO=
                 COLUMN     COEFFICIENT     OF COEF.    COEF/S.D.
                   --             90.1         23.1        3.90
      X1           C2           -1.836         .367       -5.01
      X2           C3            .285          .231        1.24
THE ST. DEV. OF Y ABOUT REGRESSION LINE IS
S =        10.7
WITH(   15-3) = 12 DEGREES OF FREEDOM

R-SQUARED = 91.6 PERCENT
R-SQUARED = 90.2 PERCENT,ADJUSTED FOR D.F.

ANALYSIS OF VARIANCE

DUE TO          DF      SS      MS=SS/DF
REGRESSION       2    14801.     7400.
RESIDUAL        12     1364.      114.
TOTAL           14    16165.
```

a. What is the least squares prediction equation?

b. Find R^2 and interpret its value.

c. Is there sufficient evidence to indicate that the model is useful for predicting y? Conduct an F test using $\alpha = .05$.

d. Test $H_0: \beta_1 = 0$ against $H_a: \beta_1 \neq 0$. Use $\alpha = .05$.

11.57 Poly (perfluoropropyleneoxide), i.e., PPFPO, is a viscous liquid used extensively in the electronics industry as a lubricant. In a study reported in *Applied Spectroscopy* (Jan. 1986), the infrared reflectance spectra properties of PPFPO were examined. The optical density (y) for the prominent infrared absorption of PPFPO was recorded for different experimental settings of band frequency (x_1) and film thickness (x_2) in a Perkin–Elmer Model 621 infrared spectrometer. The results are given in the accompanying table.

OPTICAL DENSITY y	BAND FREQUENCY x_1, cm^{-1}	FILM THICKNESS x_2, milligrams
.231	740	1.1
.107	740	.62
.053	740	.31
.129	805	1.1
.069	805	.62
.030	805	.31
1.005	980	1.1
.559	980	.62
.321	980	.31
2.948	1,235	1.1
1.633	1,235	.62
.934	1,235	.31

Source: Pacansky, J., England, C. D., and Waltman, R. "Infrared Spectroscopic Studies of Poly (Perfluoropropyleneoxide) on Gold Substrates: A Classical Dispersion Analysis for the Refractive Index." *Applied Spectroscopy*, Vol. 40, No. 1, Jan. 1986, p. 9 (Table I).

a. If you have access to a multiple regression program package, fit the model

$$E(y) = \beta_0 + \beta_1 x_1 + \beta_2 x_2$$

b. Is the model useful for predicting optical density y?

c. Use the program package to find a 95% prediction interval for the optical density of PPFPO with band frequency $x_1 = 1,000$ cm^{-1} and film thickness $x_2 = .50$ mg.

11.58 Before accepting a job, a computer at a major university estimates the cost of running the job in order to see if the user's account contains enough money to cover the cost. As part of the job submission, the user must specify estimated values for two variables—central processing unit (CPU) time and number of lines printed. While the CPU time required and the lines printed do not account for the complete cost of the run, it is thought that knowledge of their values should allow a good prediction of job cost. The following model is proposed to explain the relationship of lines printed and CPU time to job cost:*

$$E(y) = \beta_0 + \beta_1 x_1 + \beta_2 x_2 + \beta_3 x_1 x_2$$

where

y = Job cost

x_1 = Lines printed

x_2 = CPU time (tenths of a second)

Records from 20 previous runs were used to fit this model. A portion of the SAS printout is shown here.

SAS Printout for Exercise 11.58

SOURCE	DF	SUM OF SQUARES	MEAN SQUARE	F VALUE	PR > F
MODEL	3	43.25090461	14.41696820	84.96	0.0001
ERROR	16	2.71515039	0.16969690	R-SQUARE	ROOT MSE
CORRECTED TOTAL	19	45.96605500		0.940931	0.41194283

PARAMETER	ESTIMATE	T FOR H0: PARAMETER = 0	PR > ¦T¦	STD ERROR OF ESTIMATE
INTERCEPT	0.04564705	0.22	0.8313	0.21082636
X1	0.00078505	5.80	0.0001	0.00013537
X2	0.23737262	7.50	0.0001	0.03163301
X1*X2	-0.00003809	-2.99	0.0086	0.00001273

X1	X2	PREDICTED VALUE	LOWER 95% CL FOR MEAN	UPPER 95% CL FOR MEAN
2000	42	8.38574865	7.32284845	9.44864885

a. Identify the least squares model that was fitted to the data.
b. What are the values of SSE and s^2 (estimate of σ^2) for the data?
c. What do we mean by the statement: This value of SSE (see part b) is minimum?

11.59 Refer to Exercise 11.58 and the portion of the SAS printout shown.
a. Is there evidence that the overall model is useful for predicting job cost? Test at $\alpha = .05$.
b. What assumptions are necessary for the validity of the test conducted in part a?

11.60 Refer to Exercise 11.58 and the portion of the SAS printout shown. Use a 95% confidence interval to estimate the mean cost of computer jobs that print 2,000 lines and require 4.2 seconds of CPU time.

11.61 A naval base is considering modifying or adding to its fleet of 48 standard aircraft. The final decision regarding the type and number of aircraft to be added depends on a comparison of cost versus effectiveness of the modified fleet. Consequently, the naval base would like to model the projected percentage increase y in fleet effectiveness by the end of the decade as a function of the cost x of modifying the fleet. A first proposal is the quadratic model

$$E(y) = \beta_0 + \beta_1 x + \beta_2 x^2$$

*We will discuss this model, called an **interaction model**, in detail in Chapter 12.

The data provided in the accompanying table were collected on ten naval bases of a similar size that recently expanded their fleets. The data were used to fit the model, and the SAS printout of the multiple regression analysis is also reproduced.

PERCENTAGE IMPROVEMENT AT END OF DECADE y	COST OF MODIFYING FLEET x, millions of dollars
18	125
32	160
9	80
37	162
6	110
3	90
30	140
10	85
25	150
2	50

SAS Printout for Exercise 11.61

```
DEPENDENT VARIABLE: Y
SOURCE            DF      SUM OF SQUARES    MEAN SQUARE     F VALUE      PR > F       R-SQUARE        C.V.
MODEL             2       1368.77500634     684.38750317    33.08        0.0003       0.904318        26.4450
ERROR             7       144.82499366      20.68928481                  ROOT MSE                     Y MEAN
CORRECTED TOTAL   9       1513.60000000                                  4.54854755                   17.20000000

SOURCE            DF      TYPE I SS         F VALUE   PR > F      DF       TYPE III SS     F VALUE    PR > F
X                 1       1274.82024333     61.62     0.0001      1        20.79693794     1.01       0.3494
X*X               1       93.95476301       4.54      0.0706      1        93.95476301     4.54       0.0706

                                            T FOR H0:        PR > :T:       STD ERROR OF
PARAMETER         ESTIMATE                  PARAMETER=0                     ESTIMATE
INTERCEPT         10.65903604               0.73             0.4876         14.55009061
X                 -0.28160568               -1.00            0.3494         0.28087588
X*X               0.00267194                2.13             0.0706         0.00125383
```

a. Interpret the value of R^2 on the printout.
b. Find the value of s and interpret its value.
c. Perform a test of overall model adequacy. Use $\alpha = .05$.
d. Is there sufficient evidence to conclude that the percentage improvement y increases more quickly for more costly fleet modifications than for less costly fleet modifications? Test with $\alpha = .05$.

11.62 Refer to Exercise 11.14. Suppose the company persists in using the quadratic model, despite its apparent lack of utility. The fitted model is

$$\hat{y} = 12.3 + .25x - .0033x^2$$

where $\hat{y}$ is the predicted time lost (days) and x is the number of hours worked prior to an accident.

a. Use the model to estimate the mean number of days missed by all employees who have an accident after 6 hours of work.
b. Suppose the 95% confidence interval for the estimated mean in part a is determined to be (1.35, 26.01). What is the interpretation of this interval? Does this interval reconfirm your conclusion about this model in Exercise 11.14?

11.63 A large manufacturing firm wants to determine whether a relationship exists between the number of work-hours an employee misses per year and the employee's annual wages. A sample of 15 employees produced the data in the accompanying table. A first-order model was fit to the data with the results:

$$\hat{y} = 222.64 - 9.60x \qquad r^2 = .073$$

EMPLOYEE	WORK-HOURS MISSED y	ANNUAL WAGES x, thousands of dollars
1	49	12.8
2	36	14.5
3	127	8.3
4	91	10.2
5	72	10.0
6	34	11.5
7	155	8.8
8	11	17.2
9	191	7.8
10	6	15.8
11	63	10.8
12	79	9.7
13	543	12.1
14	57	21.2
15	82	10.9

a. Interpret the value of r^2.

b. Calculate and plot the regression residuals. What do you notice?

c. After searching through its employees' files, the firm has found that employee #13 had been fired but that his name had not been removed from the active employee payroll. This explains the large accumulation of work-hours missed (543) by that employee. In view of this fact, what is your recommendation concerning this outlier?

d. Refit the model to the data, excluding the outlier, and find the least squares line. Calculate r^2 and comment on model adequacy.

11.64 Plastics made under different environmental conditions are known to have differing strengths. A scientist would like to know which combination of temperature and pressure yields a plastic with a high breaking strength. A small preliminary experiment was run at two pressure levels and two temperature levels. The following model was proposed:

$$E(y) = \beta_0 + \beta_1 x_1 + \beta_2 x_2$$

where

y = Breaking strength (pounds)

x_1 = Temperature (°F)

x_2 = Pressure (pounds per square inch)

A sample of $n = 16$ observations yielded

$$\hat{y} = 226.8 + 4.9x_1 + 1.2x_2$$

with

$$s_{\hat{\beta}_1} = 1.11 \qquad s_{\hat{\beta}_2} = .27$$

Do the data indicate that pressure is an important predictor of breaking strength? Test using $\alpha = .05$.

11.65 Poplar trees are known by ecologists and forestry experts for their rapid growth rate and cross-breeding capacity. In fact, because of this high growth potential, clones of certain hybrid poplars are being considered as potential sources of fiber and fuel. Associated with this rapid growth rate is the poplar's need for an abundance of water—the transpiration rate of poplars (on the basis of leaf area) may be twice as high as that of other deciduous trees.

In a recent investigation of water relations in poplars (*Ecology*, Feb. 1981), two plants from each of two poplar clones (identified by the numbers 5263 and 5271) and three plants from each of two other poplar clones (identified by the numbers 5331 and 5319) were selected from a group of cuttings planted at the Hugo Sauer Nursery near Rhinelander, Wisconsin. (The four clones were chosen because of their differing growth rates.) Since clones of poplar trees may differ in root penetration, and thus in their capacity to extract water from the soil, the researchers decided to examine the relationship between the soil water potential of the clones and their transpiration rates. For a period of approximately 2 weeks, the soil water potential y (in megapascals) and transpirational flux density x (in micrograms per centimeter squared per second) were measured each day for individually selected leaves of the plants. The data for each of the four clones were to be analyzed separately. The researchers fit the quadratic model

$$y = \beta_0 + \beta_1 x + \beta_2 x^2 + \varepsilon$$

to the four data sets using multiple regression. The results are summarized in the table.

CLONE	LEAST SQUARES PREDICTION EQUATION	$s_{\hat{\beta}_1}$	$s_{\hat{\beta}_2}$	R^2	n
5263	$\hat{y} = -1.47 - 2.53x + .12x^2$	.14	.01	.62	418
5319	$\hat{y} = -1.48 - 1.90x + .08x^2$	.14	.01	.59	417
5331	$\hat{y} = -1.11 - 2.43x + .14x^2$	.19	.02	.53	315
5271	$\hat{y} = -1.67 - 1.89x + .07x^2$	.10	.01	.68	315

Source: Pallardy, S. G. and Kozlowski, T. T. "Water Relations of *Populus* Clones." *Ecology*, Feb. 1981, 62, pp. 159–169. Copyright 1981, the Ecological Society of America.

For each of the four clones:
a. Interpret the value of R^2.
b. Test the hypothesis that the overall model is useful for predicting the soil water potential of leaves from the poplar clone. Use $\alpha = .05$.

c. Is there evidence of curvature in the response model relating soil water potential y to transpirational flux density x? Use $\alpha = .05$.

d. List any assumptions required for the validity of the tests conducted in parts **b** and **c**.

11.66 *Sintering*, one of the most important techniques of materials science, is used to convert a powdered material into a porous solid body. The following two measures characterize the final product:

V_v = Percentage of total volume of final product that is solid

$$= \left(\frac{\text{Solid volume}}{\text{Porous volume} + \text{Solid volume}}\right) \cdot 100$$

S_v = Solid-pore interface area per unit volume of the product

When $V_v = 100\%$, the product is completely solid—i.e., it contains no pores. Both V_v and S_v are estimated by a microscopic examination of polished cross sections of sintered material. Generally, the longer a powdered material is sintered, the more solid will be the product. Thus, we would expect S_v to decrease and V_v to increase as the sintering time is increased. The accompanying table gives the mean and standard deviation of the values of S_v (in squared centimeters per cubic centimeter) and V_v (percentage) for 100 specimens of sintered nickel for six different sintering times.*

SAMPLE	TIME	S_v		V_v	
	minutes	Mean	Standard Deviation	Mean	Standard Deviation
1	1.0	1,076.5	295.0	95.83	1.2
2	10.0	736.0	181.9	96.73	2.1
3	28.5	509.4	154.7	97.38	2.1
4	150.0	299.5	161.0	97.82	1.5
5	450.0	165.0	110.4	99.03	1.3
6	1,000.0	72.9	76.6	99.49	1.1

a. Plot the sample means of the S_v measurements versus sintering time. Hypothesize a linear model relating mean S_v to sintering time x.

b. Plot the sample means of the S_v measurements versus V_v. Hypothesize a linear model relating mean V_v to sintering time x.

c. Suppose you were to fit a linear model relating $E(S_v)$ to sintering time x. Explain why the data may violate the assumptions of Section 11.2.

11.67 Ignoring the possible violation of assumptions (see part c of Exercise 11.66), we will fit a second-order model to the $n = 6$ sample means, i.e., we will fit

$$E(S_v) = \beta_0 + \beta_1 x + \beta_2 x^2$$

where x is the sintering time. The SAS printout for the regression analysis is shown here.

*Data and experimental information provided by Guoquan Liu while visiting at the University of Florida in 1983.

SAS Computer Printout for Exercise 11.67

DEPENDENT VARIABLE: S

SOURCE	DF	SUM OF SQUARES	MEAN SQUARE	F VALUE	PR > F	R-SQUARE	C.V.
MODEL	2	548985.24236433	274492.62118216	4.82	0.1155	0.762822	50.0538
ERROR	3	170691.61263567	56897.20421189		ROOT MSE		S MEAN
CORRECTED TOTAL	5	719676.85500000			238.53134849		476.55000000

SOURCE	DF	TYPE I SS	F VALUE	PR > F	DF	TYPE III SS	F VALUE	PR > F
TIME	1	412210.33355679	7.24	0.0743	1	268935.31141092	4.73	0.1180
TIME*TIME	1	136774.90880754	2.40	0.2188	1	136774.90880754	2.40	0.2188

PARAMETER	ESTIMATE	T FOR H0: PARAMETER=0	PR > :T:	STD ERROR OF ESTIMATE
INTERCEPT	779.37835004	5.59	0.0113	139.41448552
TIME	-2.35199961	-2.17	0.1180	1.08182944
TIME*TIME	0.00166336	1.55	0.2188	0.00107282

OBSERVATION	TIME	OBSERVED VALUE	PREDICTED VALUE	RESIDUAL	LOWER 95% CL FOR MEAN	UPPER 95% CL FOR MEAN
1	1	1076.50000000	777.02801379	299.47198621	335.39009700	1218.66593057
2	10	736.00000000	756.02468977	-20.02468977	331.88327670	1180.16610284
3	28.5	509.40000000	713.69742406	-204.29742406	319.89587834	1107.49896977
4	150	299.50000000	464.00397516	-164.50397516	74.95319626	853.05475406
5	450	165.00000000	57.80862727	107.19137273	-630.46367508	746.08092962
6	1000	72.90000000	90.73726996	-17.83726996	-666.70440223	848.17894215

a. Graph the prediction equation and plot the data points.
b. Is the model adequate for predicting S_v? Test using $\alpha = .05$.
c. Confidence intervals for $E(S_v)$ are shown on the printout for each sintering time. Explain why some of the confidence intervals will be conservative and others will be nonconservative. Explain which confidence intervals might be expected to fall in each of the two categories.

11.68 Refer to Exercise 11.66. A second-order model relating V_v to sintering time x is given by

$$E(V_v) = \beta_0 + \beta_1 x + \beta_2 x^2$$

Refer to the accompanying SAS computer printout for the regression analysis.

SAS Computer Printout for Exercise 11.68

DEPENDENT VARIABLE: V

SOURCE	DF	SUM OF SQUARES	MEAN SQUARE	F VALUE	PR > F	R-SQUARE	C.V.
MODEL	2	8.58051304	4.29025652	13.61	0.0313	0.900696	0.5747
ERROR	3	0.94602029	0.31534010		ROOT MSE		V MEAN
CORRECTED TOTAL	5	9.52653333			0.56155151		97.71333333

SOURCE	DF	TYPE I SS	F VALUE	PR > F	DF	TYPE III SS	F VALUE	PR > F
TIME	1	7.18359279	22.78	0.0175	1	3.29208494	10.44	0.0482
TIME*TIME	1	1.39692025	4.43	0.1260	1	1.39692025	4.43	0.1260

PARAMETER	ESTIMATE	T FOR H0: PARAMETER=0	PR > :T:	STD ERROR OF ESTIMATE
INTERCEPT	96.55086769	294.17	0.0001	0.32821017
TIME	0.00822903	3.23	0.0482	0.00254685
TIME*TIME	-5.3157979E-06	-2.10	0.1260	0.00000253

OBSERVATION	TIME	OBSERVED VALUE	PREDICTED VALUE	RESIDUAL	LOWER 95% CL FOR MEAN	UPPER 95% CL FOR MEAN
1	1	95.83000000	96.55909140	-0.72909140	95.51938555	97.59879726
2	10	96.73000000	96.63262644	0.09737356	95.63411094	97.63114195
3	28.5	97.38000000	96.78107739	0.59892261	95.85398813	97.70816666
4	150	97.82000000	97.66561731	0.15438269	96.74971232	98.58152230
5	450	99.03000000	99.17748383	-0.14748383	97.55715027	100.79781740
6	1000	99.49000000	99.46410362	0.02589638	97.68093123	101.24727601

    a. Graph the prediction equation and plot the data points.

    b. Is there sufficient evidence to indicate that the quadratic term should be included in the model? Test using $\alpha = .05$.

    c. Confidence intervals for $E(V_y)$ are shown on the printout. Find and interpret the confidence interval for $E(V_y)$ at sintering time 150 minutes.

11.69 Refer to Exercise 11.66. The unstable values of the standard deviations for S_y shown in the table indicate a strong possibility that the standard regression assumption of equal variances is violated for the second-order model of Exercise 11.67. We can satisfy this assumption by transforming the response to a new response that has a constant variance. For this exercise, consider the log transform* $S_v^* = \log(S_v)$ and fit the model

$$E(S_v^*) = \beta_0 + \beta_1 x$$

Refer to the SAS printout for the regression analysis of the log transform model shown here.

SAS Computer Printout for Exercise 11.69

DEPENDENT VARIABLE: LS

SOURCE	DF	SUM OF SQUARES	MEAN SQUARE	F VALUE	PR > F	R-SQUARE	C.V.
MODEL	1	4.38661445	4.38661445	28.72	0.0058	0.877753	6.7163
ERROR	4	0.61093382	0.15273346		ROOT MSE		LS MEAN
CORRECTED TOTAL	5	4.99754827			0.39081128		5.81884709

SOURCE	DF	TYPE I SS	F VALUE	PR > F	DF	TYPE III SS	F VALUE	PR > F
TIME	1	4.38661445	28.72	0.0058	1	4.38661445	28.72	0.0058

PARAMETER	ESTIMATE	T FOR H0: PARAMETER=0	PR > :T:	STD ERROR OF ESTIMATE
INTERCEPT	6.46771675	32.29	0.0001	0.20028753
TIME	-0.00237464	-5.36	0.0058	0.00044310

OBSERVATION	TIME	OBSERVED VALUE	PREDICTED VALUE	RESIDUAL	LOWER 95% CL INDIVIDUAL	UPPER 95% CL INDIVIDUAL
1	1	6.98147032	6.46534211	0.51612820	5.24643507	7.68424916
2	10	6.60129012	6.44397038	0.15725974	5.22805905	7.65988170
3	28.5	6.23323356	6.40003959	-0.16680602	5.18999122	7.61008795
4	150	5.70211442	6.11152114	-0.40940673	4.92976479	7.29327750
5	450	5.10594547	5.39912993	-0.29318446	4.20714050	6.59111937
6	1000	4.28908864	4.09307938	0.19600926	2.61900161	5.56715714

    a. Graph the prediction equation and plot the data points.

    b. Is the model adequate for predicting $\log(S_v)$? Test using $\alpha = .05$.

    c. Prediction intervals for the transformed response $\log(S_v)$ are shown on the printout. The predicted value of S_v is the antilog,

$$\hat{S}_v = e^{\widehat{\log(S_v)}}$$

To obtain prediction intervals for S_v, take the antilogs of the endpoints of the intervals.[†] Find a 95% prediction interval for S_v when the sintering time is 150 minutes.

---

*To see the stabilizing effect of the log transform, use your calculator to take the logs of the standard deviations for S_v shown in the table. Note that the transformed values appear to be much less variable.

[†]Unfortunately, you cannot take antilogs to find the confidence interval for the mean response $E(y)$. This is because the mean value of $\log(y)$ is not equal to the logarithm of the mean of y.

REFERENCES

Barnett, V. and Lewis, T. *Outliers in Statistical Data.* New York: Wiley, 1978.

Belsley, D. A., Kuh, E., and Welsch, R. E. *Regression Diagnostics: Identifying Influential Data and Sources of Collinearity.* New York: Wiley, 1980.

BMDPC: User's Guide to BMDP on the IBM PC. BMDP Statistical Software, Inc. Los Angeles, Calif. 90025.

Bowerman, B. L. and O'Connell, T. *Forecasting and Time Series.* North Scituate, Mass.: Duxbury Press, 1979.

Box, G. E. P. and Jenkins, G. M. *Time Series Analysis, Forecasting and Control.* San Francisco: Holden-Day, Inc., 1970.

Chou, Ya-Iun. *Statistical Analysis with Business and Economic Applications,* 2nd ed. New York: Holt, Rinehart, and Winston, 1975.

Dixon, W. J., Brown, M. B., Engelman, L., Frane, J. W., Hill, M. A., Jennrich, R. I., and Toporek, J. D. *BMDP Statistical Software,* 1985 ed. Berkeley: University of California Press.

Draper, N. R. and Smith, H. *Applied Regression Analysis,* 2nd ed. New York: Wiley, 1981.

Fuller, W. *Introduction to Statistical Time Series.* New York: Wiley, 1976.

Graybill, F. A. *Theory and Application of the Linear Model.* North Scituate, Mass.: Duxbury, 1976.

Hamburg, M. *Statistical Analysis for Decision-Making,* 2nd ed. New York: Harcourt Brace Jovanovich, 1977.

Kleinbaum, D. and Kupper, L. *Applied Regression Analysis and Other Multivariable Methods.* North Scituate, Mass.: Duxbury, 1978.

Mendenhall, W. *Introduction to Linear Models and the Design and Analysis of Experiments.* Belmont, Ca.: Wadsworth, 1968.

Mendenhall, W. and Sincich, T. *A Second Course in Business Statistics: Regression Analysis,* 2nd ed. San Francisco: Dellen, 1986.

Miller, R. B. and Wichern, D. W. *Intermediate Business Statistics: Analysis of Variance, Regression, and Time Series.* New York: Holt, Rinehart, and Winston, 1977.

Neter, J., Wasserman, W., and Kutner, M. H. *Applied Linear Statistical Models.* Homewood, Ill.: Richard D. Irwin, 1985.

Norusis, M. J. *The SPSS Guide to Data Analysis,* 1986 ed. SPSS, Inc., Suite 3000, 444 N. Michigan Avenue, Chicago, Ill. 60611.

Norusis, M. J. *SPSS/PC+: SPSS for the IBM PC/XT/AT,* 1986 ed. SPSS, Inc., Suite 3000, 444 N. Michigan Avenue, Chicago, Ill. 60611.

Ryan, T. A., Joiner, B. L., and Ryan, B. F. *Minitab Reference Manual.* University Park, Pa.: Minitab Project, 1985.

Ryan, T. A., Joiner, B. L., and Ryan, B. F. *Minitab Student Handbook,* 2nd ed. Boston: Duxbury Press, 1985.

SAS Procedures Guide for Personal Computers, Version 6 ed., 1986. SAS Institute, Inc., Box 8000, Cary, N.C. 27511.

SAS Statistics Guide for Personal Computers, Version 6 ed., 1986. SAS Institute, Inc., Box 8000, Cary, N.C. 27511.

SAS User's Guide: Statistics, Version 5 ed. (1985). SAS Institute, Inc., Box 8000, Cary, N.C. 27511.

SPSS^X User's Guide, 1983 ed. SPSS, Inc., Suite 3000, 444 N. Michigan Ave., Chicago, Ill. 60611.

Weisberg, S. *Applied Linear Regression.* New York: Wiley, 1980.

Winkler, R. L. and Hays, W. L. *Statistics: Probability, Inference, and Decision,* 2nd ed. New York: Holt, Rinehart, and Winston, 1975.

Younger, M. S. *A First Course in Linear Regression,* 2nd ed. Boston: Duxbury, 1985.

OBJECTIVE

To show you why the choice of the deterministic portion of a linear model is crucial to the acquisition of a good prediction equation; to present some basic concepts and procedures for constructing good linear models.

CONTENTS

INTRODUCTION TO MODEL BUILDING

INTRODUCTION

We have indicated in Chapters 10 and 11 that the first step in the construction of a regression model is to hypothesize the form of the deterministic portion of the probabilistic model. This **model building**, or model construction, stage is the key to the success (or failure) of the regression analysis. If the hypothesized model does not reflect, at least approximately, the true nature of the relationship between the mean response $E(y)$ and the independent variables $x_1, x_2, \ldots, x_k$, the modeling effort will usually be unrewarded.

By *model building*, we mean writing a model that will provide a good fit to a set of data and that will give good estimates of the mean value of y and good predictions of future values of y for given values of the independent variables. To illustrate, suppose you want to relate the demand y for a given brand of personal computer to advertising expenditure x, and (unknown to you) the second-order model

$$E(y) = \beta_0 + \beta_1 x + \beta_2 x^2$$

would permit you to predict y with a very small error of prediction (see Figure 12.1a). Unfortunately, you have erroneously chosen the first-order model

$$E(y) = \beta_0 + \beta_1 x$$

to explain the relationship between y and x (see Figure 12.1b).

FIGURE 12.1

Two Models for Relating Demand y to Advertising Expenditure x

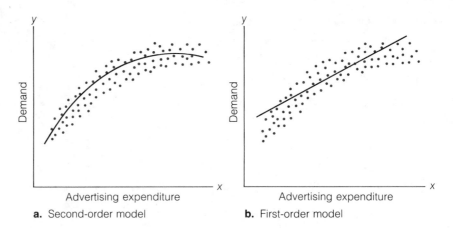

a. Second-order model   **b.** First-order model

The consequence of choosing the wrong model is clearly demonstrated by comparing Figures 12.1a and 12.1b. The errors of prediction for the second-order model are relatively small in comparison to those for the first-order model. The lesson to be learned from this simple example is clear. Choosing a good set of independent (predictor) variables $x_1, x_2, \ldots, x_k$ will not guarantee a good prediction equation. In addition to selecting independent variables that contain information about y, you must specify an equation relating y to $x_1, x_2, \ldots, x_k$ that will provide a good fit to your data.

In the following sections, we will present some useful models for relating a response y to one or more predictor variables.

| | | | | | | | | | | | |

SECTION 12.2

THE TWO TYPES OF INDEPENDENT VARIABLES: QUANTITATIVE AND QUALITATIVE

Two types of variables may arise in experimental situations: **quantitative** and **qualitative**. For the types of regression analyses considered in this text the dependent variable will always be quantitative, but the independent variables may be either quantitative or qualitative. As you will see, the way an independent variable enters the model depends on its type.

DEFINITION 12.1

A **quantitative** independent variable is one that assumes numerical values corresponding to the points on a line. An independent variable that is not quantitative is called **qualitative**.

The waiting time before a computer begins to process data, the number of defects in a product, and the kilowatt-hours of electricity used per day are all examples of quantitative independent variables. On the other hand, suppose three different styles of packaging, A, B, and C, are used by a manufacturer. This independent variable is qualitative, since it is not measured on a numerical scale. Certainly, the style of packaging is an independent variable that may affect the sales of a product, and we would want to include it in a model describing the product's sales, y.

DEFINITION 12.2

The different intensity settings of an independent variable are called its **levels**.

For a quantitative independent variable, the levels correspond to the numerical values it assumes. For example, if the number of defects in a product ranges from 0 to 3, the independent variable has four levels: 0, 1, 2, and 3.

The levels of a qualitative variable are not numerical. They can be defined only by describing them. For example, the independent variable for the style of packaging was observed at three levels: A, B, and C.

EXAMPLE 12.1

Suppose we want to predict the salary of a corporate executive at a high-technology firm as a function of the following four independent variables:

a. Number of years of experience
b. Sex of the employee
c. Firm's net asset value
d. Rank of the employee

For each of these independent variables, give its type and describe the levels you would expect to observe.

SOLUTION

a. The independent variable for the number of years of experience is quantitative, since its values are numerical. We would expect to observe levels ranging from 0 to 40 (approximately) years.

b. The independent variable for sex is qualitative, since its levels can be described only by the nonnumerical labels "female" and "male."

c. The independent variable for the firm's net asset value is quantitative, with a large number of possible levels corresponding to the range of dollar values representing various firms' net asset values.

d. Suppose the independent variable for the rank of the employee is observed at three levels: supervisor, assistant vice president, and vice president. Since we cannot assign a realistic numerical measure of relative importance to each position, rank is a qualitative independent variable.   ■

Quantitative independent variables are treated differently from qualitative variables in regression modeling. In the next section, we will begin our discussion of how quantitative variables are used in the modeling effort.

| | | | | | | | | | | | | |

EXERCISES 12.1–12.3

12.1 Companies keep personnel files that contain important information on each employee's background. The data in these files could be used to predict employee performance ratings. Identify the independent variables listed below as qualitative or quantitative. For qualitative variables, suggest several levels that might be observed. For quantitative variables, give a range of values (levels) for which the variable might be observed.
  a. Age
  b. Years of experience with the company
  c. Highest educational degree
  d. Job classification
  e. Marital status
  f. Religious preference
  g. Salary
  h. Sex

12.2 An experiment was conducted to investigate the sheet flow rate of a land waste treatment plant. Classify each of the following independent variables as quantitative or qualitative and describe the levels the variables might assume.
  a. Amount of rainfall
  b. Method of treatment
  c. Irrigation rate
  d. Slope of grass mat
  e. Type of sod

12.3 Consider the following variables related to running a computer job. Classify each variable as quantitative or qualitative and describe the levels each variable might assume.
  a. CPU time
  b. Software system

    c. Lines of output

    d. Job cost

    e. Date of submission

SECTION 12.3

MODELS WITH A SINGLE QUANTITATIVE INDEPENDENT VARIABLE

The most common linear models relating y to a single quantitative independent variable x are those derived from a polynomial expression of the type shown in the box. Specific models, obtained by assigning particular values to p, are listed below.

FORMULA FOR A pTH-ORDER POLYNOMIAL WITH ONE QUANTITATIVE INDEPENDENT VARIABLE

$$E(y) = \beta_0 + \beta_1 x + \beta_2 x^2 + \beta_3 x^3 + \cdots + \beta_p x^p$$

where p is an integer and $\beta_0, \beta_1, \ldots, \beta_p$ are unknown parameters that must be estimated.

---

1. FIRST-ORDER MODEL

$$E(y) = \beta_0 + \beta_1 x$$

Interpretation of model parameters

    β_0: y-intercept; the value of $E(y)$ when $x = 0$

    β_1: Slope of the line; the change in $E(y)$ for a 1-unit increase in x

General comments   The first-order model is used when you expect the rate of change in y per unit change in x to remain fairly stable over the range of values of x for which you wish to predict y (see Figure 12.2). Most relationships between $E(y)$ and x are curvilinear, but the curvature over the range of values of x for which you wish to predict y may be very slight. When this occurs, a first-order (straight-line) model should provide a good fit to your data.

FIGURE 12.2

Graph of a First-Order Model

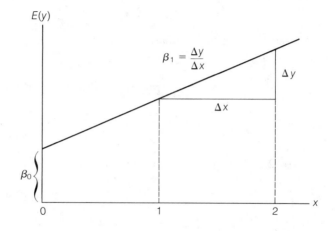

2. SECOND-ORDER MODEL

$$E(y) = \beta_0 + \beta_1 x + \beta_2 x^2$$

Interpretation of model parameters

β_0: y-intercept; the value of $E(y)$ when $x = 0$

β_1: Shift parameter; changing the value of β_1 shifts the parabola to the right or left (increasing the value of β_1 causes the parabola to shift to the right)

β_2: Rate of curvature

General comments A second-order model traces a parabola, one that opens either downward ($\beta_2 < 0$) or upward ($\beta_2 > 0$), as shown in Figure 12.3. Since most relationships will possess some curvature, a second-order model will often be a good choice to relate y to x.

FIGURE 12.3
The Graphs of Two
Second-Order Models

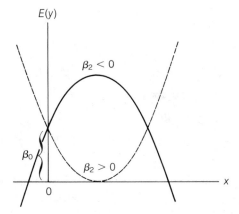

3. THIRD-ORDER MODEL

$$E(y) = \beta_0 + \beta_1 x + \beta_2 x^2 + \beta_3 x^3$$

Interpretation of model parameters

β_0: y-intercept; the value of $E(y)$ when $x = 0$

β_3: The magnitude of β_3 controls the rate of reversal of curvature for the curve

General comments Reversals in curvature are not common, but such relationships can be modeled by third- and higher-order polynomials. As can be seen in Figure 12.3, a second-order model contains no reversals in curvature. The slope continues to either increase or decrease as x increases and produces either a trough or a peak. A third-order model (see Figure 12.4) contains one reversal

FIGURE 12.4

The Graphs of Two
Third-Order Models

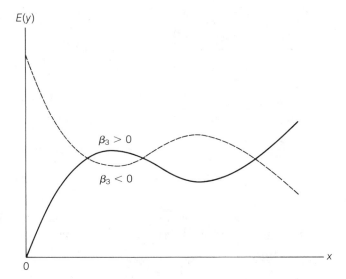

$E(y)$

$\beta_3 > 0$

$\beta_3 < 0$

0

x

in curvature and produces one peak and one trough. In general, the graph of a pth-order polynomial will contain a total of $(p - 1)$ peaks and troughs.

Most functional relationships in nature seem to be smooth (except for random error), that is, they are not subject to rapid and irregular reversals in direction. Consequently, the second-order polynomial model is perhaps the most useful of those described above. To develop a better understanding of how this model is used, consider the following example.

EXAMPLE 12.2

Power companies have to be able to predict the peak power load at their various stations in order to operate effectively. The peak power load is the maximum amount of power that must be generated each day to meet demand.

Suppose a power company located in the southern part of the United States decides to model daily peak power load, y, as a function of the daily high temperature, x, and the model is to be constructed for the summer months when demand is greatest. Although we would expect the peak power load to increase as the high temperature increases, the *rate* of increase in $E(y)$ might also increase as x increases. That is, a 1-unit increase in high temperature from 100°F to 101°F might result in a larger increase in power demand than would a 1-unit increase from 80°F to 81°F. Therefore, we postulate the second-order model

$$E(y) = \beta_0 + \beta_1 x + \beta_2 x^2$$

and we expect β_2 to be positive.

A random sample of 25 summer days is selected, and the data are shown in Table 12.1 (page 562). Fit a second-order model using these data, and test the hypothesis that the power load increases at an increasing *rate* with temperature— i.e., that $\beta_2 > 0$.

TABLE 12.1
Power Load Data

TEMPERATURE (°F)	PEAK LOAD (megawatts)	TEMPERATURE (°F)	PEAK LOAD (megawatts)	TEMPERATURE (°F)	PEAK LOAD (megawatts)
94	136.0	106	178.2	76	100.9
96	131.7	67	101.6	68	96.3
95	140.7	71	92.5	92	135.1
108	189.3	100	151.9	100	143.6
67	96.5	79	106.2	85	111.4
88	116.4	97	153.2	89	116.5
89	118.5	98	150.1	74	103.9
84	113.4	87	114.7	86	105.1
90	132.0				

SOLUTION

The SAS printout shown in Figure 12.5 gives the least squares fit of the second-order model using the data in Table 12.1. The prediction equation is

$$\hat{y} = 385.048 - 8.293x + .05982x^2$$

A plot of this equation and the observed values is given in Figure 12.6.

We now test to determine whether the sample value, $\hat{\beta}_2 = .05982$, is large enough to conclude *in general* that the power load increases at an increasing rate with temperature:

$H_0: \quad \beta_2 = 0$

$H_a: \quad \beta_2 > 0$

Test statistic: $\quad t = \dfrac{\hat{\beta}_2}{s_{\hat{\beta}_2}}$

For $\alpha = .05$, $n = 25$, and $k = 2$, we will reject H_0 if $t > t_{.05}$, where $t_{.05} = 1.717$ (from Table 6 of Appendix II) has $[n - (k + 1)] = 22$ degrees of freedom. From Figure 12.5 the calculated value of t is 7.93. Since this value exceeds $t_{.05} = 1.717$, we reject H_0 at $\alpha = .05$ and conclude that the mean power load increases at an increasing rate with temperature.

FIGURE 12.5   Portion of the SAS Printout for the Second-Order Model of Example 12.2

SOURCE	DF	SUM OF SQUARES	MEAN SQUARE	F VALUE	PR > F
MODEL	2	15011.77199776	7505.88599888	259.69	0.0001
ERROR	22	635.87840224	28.90356374		ROOT MSE
				R-SQUARE	
CORRECTED TOTAL	24	15647.65040000		0.959363	5.37620347

PARAMETER	ESTIMATE	T FOR H0: PARAMETER = 0	PR > ¦T¦	STD ERROR OF ESTIMATE
INTERCEPT	385.04809323	6.98	0.0001	55.17243578
TEMP	-8.29252680	-6.38	0.0001	1.29904502
TEMP*TEMP	0.05982337	7.93	0.0001	0.00754855

FIGURE 12.6
Plot of the Observations
and the Second-Order
Least Squares Fit

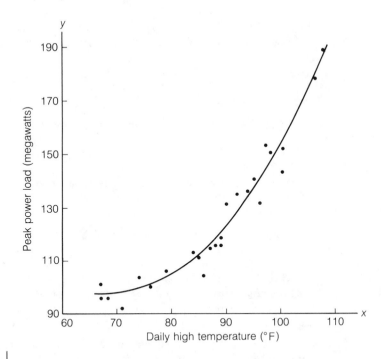

Daily high temperature (°F)

Peak power load (megawatts)

■

EXERCISES 12.4–12.12

12.4 Graph the following polynomials and identify the order of each on your graph:
  a. $E(y) = 2 + 3x$ **b.** $E(y) = 2 + 3x^2$
  c. $E(y) = 1 + 2x + 2x^2 + x^3$ **d.** $E(y) = 2x + 2x^2 + x^3$
  e. $E(y) = 2 - 3x^2$ **f.** $E(y) = -2 + 3x$

12.5 The accompanying graphs depict pth-order polynomials for one independent variable.

i. $E(y)$

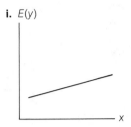

ii. $E(y)$

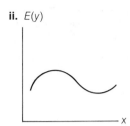

iii. $E(y)$

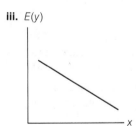

iv. $E(y)$

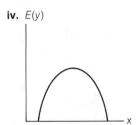

a. For each graph, identify the order of the polynomial.

b. Using the parameters β_0, β_1, β_2, etc., write an appropriate model relating $E(y)$ to x for each graph.

c. The signs (+ or −) of many of the parameters in the models of part **b** can be determined by examining the graphs. Give the signs of those parameters that can be determined.

12.6 Consider the following polynomial model:

$$E(y) = 5 - 3x + x^2$$

a. Give the order of this polynomial.

b. Sketch the curve corresponding to the equation for $E(y)$.

c. How would the graph change if the coefficient of x^2 were negative rather than positive?

12.7 Consider the following polynomial model:

$$E(y) = 2 - 4x$$

a. Give the order of this polynomial.

b. Sketch the curve corresponding to the equation for $E(y)$.

c. How would the graph change if the coefficient of x were positive instead of negative?

12.8 The amount of pressure used to produce a certain plastic is thought to be related to the strength of the plastic. Researchers believe that, as pressure is increased, the strength of the plastic increases until, at some point, increases in pressure will have a detrimental effect on strength. Write a model to relate the strength, y, of the plastic to pressure, x, that would reflect the above beliefs. Sketch the model.

12.9 Underinflated or overinflated tires can increase tire wear. A new tire was tested for wear at different pressures with the results shown in the table.

PRESSURE x, pounds per square inch	MILEAGE y, thousands
30	29
31	32
32	36
33	38
34	37
35	33
36	26

a. Plot the data on a scattergram.

b. If you were given only the information for $x = 30, 31, 32, 33$, what kind of model would you suggest? For $x = 33, 34, 35, 36$? For all the data?

12.10 An experiment was conducted to relate the turnaround time of a computer job (i.e., the length of time between the submission of a job and the time at which the job's output is printed) to the number of seconds of CPU time specified on the job card. It is suspected that turnaround time will increase more slowly for smaller jobs (jobs with few CPU seconds) than for larger jobs (jobs with many CPU seconds). Write an appropriate model relating mean turnaround time to CPU time.

12.11 Air pollution regulations for power plants are often written so that the maximum amount of pollution that can be emitted increases as the plant's output increases. Assuming this is true, write a model relating the maximum amount of pollution permitted (in parts per million) to a plant's output (in megawatts).

12.12 An engineer has proposed the following model to describe the relationship between the number of acceptable items produced per day (output) and the number of work-hours expended per day (input) in a particular production process:

$$y = \beta_0 + \beta_1 x + \beta_2 x^2 + \varepsilon$$

where

y = Number of acceptable items produced per day

x = Number of work-hours per day

A portion of the Minitab computer printout that results from fitting this model to a sample of 25 weeks of production data is shown here. Test the hypothesis that as amount of input increases, the amount of output also increases but at a decreasing rate. Do the data provide sufficient evidence to indicate that the *rate* of increase in output per unit increase of input decreases as the input increases? Test using $\alpha = .05$.

Minitab Computer Printout for Exercise 12.12

```
THE REGRESSION EQUATION IS
Y = -6.17 + 2.04 X1 - .0323 X2

                                     ST. DEV.    T-RATIO =
         COLUMN    COEFFICIENT       OF COEF.    COEF/S.D.
         -              -6.173          1.666       -3.71
X1       C2              2.036           .185       11.02
X2       C3            -.03231          .00489       -6.60


THE ST. DEV. OF Y ABOUT REGRESSION LINE IS
S =     1.243
WITH (25 - 3) = 22 DEGREES OF FREEDOM


R-SQUARED = 95.5 PERCENT
R-SQUARED = 95.1 PERCENT, ADJUSTED FOR D.F.


ANALYSIS OF VARIANCE

DUE TO         DF          SS       MS=SS/DF
REGRESSION      2     718.168      359.084
RESIDUAL       22      33.992        1.545
TOTAL          24     752.160
```

MODELS WITH TWO QUANTITATIVE INDEPENDENT VARIABLES

Like models with a single quantitative independent variable, models with two quantitative independent variables are classified as first-order, second-order, and so forth. Since we rarely encounter third- or higher-order relationships in practice, we focus our discussion on first- and second-order models.

1. FIRST-ORDER MODEL

$$E(y) = \beta_0 + \beta_1 x_1 + \beta_2 x_2$$

Interpretation of model parameters:

β_0: y-intercept; the value of $E(y)$ when $x_1 = x_2 = 0$

β_1:   Change in $E(y)$ for a 1-unit increase in x_1, when x_2 is held fixed

β_2:   Change in $E(y)$ for a 1-unit increase in x_2, when x_1 is held fixed

FIGURE 12.7
Computer-Generated Graph of a First-Order Model

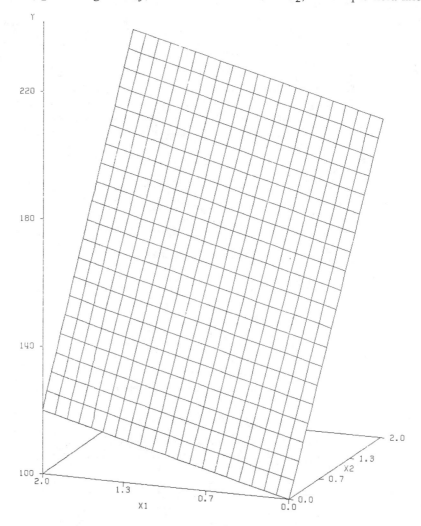

General comments The graph in Figure 12.7 traces a **response surface** (in contrast to the **response curve** that is used to relate $E(y)$ to a *single* quantitative variable). In particular, a first-order model relating $E(y)$ to two independent quantitative variables, x_1 and x_2, graphs as a plane in a three-dimensional space. The plane traces the value of $E(y)$ for every combination of values (x_1, x_2) that correspond to points in the (x_1, x_2)-plane. Most response surfaces in the real world are well behaved (smooth) and they have curvature. Consequently, a first-order model is appropriate only if the response surface is fairly flat over the (x_1, x_2)-region that is of interest to you.

The assumption that a first-order model will adequately characterize the relationship between $E(y)$ and the variables x_1 and x_2 is equivalent to assuming that x_1 and x_2 do not "interact"; that is, you assume that the effect on $E(y)$ of a change in x_1 (for a fixed value of x_2) is the same regardless of the value of x_2 (and vice versa). Thus, "no interaction" is equivalent to saying that the effect of changes in one variable (say, x_1) on $E(y)$ is *independent* of the value of the second variable (say, x_2). For example, if we assign values to x_2 in a first-order model, the graph of $E(y)$ as a function of x_1 would produce parallel lines as shown in Figure 12.8. These lines, called **contour lines**, show the contours of the surface when it is sliced by three planes, each of which is parallel to the $[E(y), x_1]$-plane, at distances $x_2 = 1, 2,$ and 3 from the origin.

DEFINITION 12.3

Two variables x_1 and x_2 are said to **interact** if the change in $E(y)$ for a 1-unit change in x_1 (when x_2 is held fixed) is dependent on the value of x_2.

FIGURE 12.8

A Graph Indicating No Interaction Between x_1 and x_2

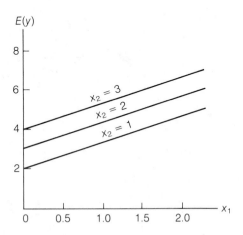

2. AN INTERACTION MODEL (SECOND-ORDER)

$$E(y) = \beta_0 + \beta_1 x_1 + \beta_2 x_2 + \beta_3 x_1 x_2$$

Interpretation of model parameters

β_0: y-intercept; the value of $E(y)$ when $x_1 = x_2 = 0$

β_1 and β_2: Changing β_1 and β_2 causes the surface to shift along the x_1 and x_2 axes

β_3: Controls the rate of twist in the ruled surface (see Figure 12.9)

When one independent variable is held fixed, the model produces straight lines with the following slopes:

$\beta_1 + \beta_3 x_2$: Change in $E(y)$ for a 1-unit increase in x_1, when x_2 is held fixed

$\beta_2 + \beta_3 x_1$: Change in $E(y)$ for a 1-unit increase in x_2, when x_1 is held fixed

FIGURE 12.9

Computer-Generated Graph for an Interaction Model (Second-Order)

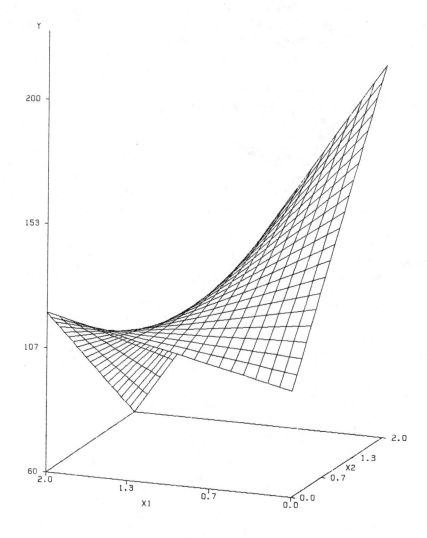

General comments This model is said to be second-order because the order of the highest-order (x_1x_2) term in x_1 and x_2 is 2; i.e., the sum of the exponents of x_1 and x_2 equals 2. This interaction model traces a ruled surface in a three-dimensional space (see Figure 12.9). You could produce such a surface by placing a pencil perpendicular to a line and moving it along the line, while rotating it around the line. The resulting surface would appear as a twisted plane. A graph of $E(y)$ as a function of x_1 for given values of x_2 (say, $x_2 = 1$, 2, and 3) produces nonparallel contour lines (see Figure 12.10), thus indicating that the change in $E(y)$ for a given change in x_1 is dependent on the value of x_2 and, therefore, that x_1 and x_2 interact. Interaction is an extremely important concept because it is easy to get in the habit of fitting first-order models and individually examining the relationships between $E(y)$ and each of a set of independent variables, x_1, $x_2, \ldots, x_k$. Such a procedure is meaningless when interaction exists (which is, at least to some extent, almost always the case), and it can lead to gross errors in interpretation. For example, suppose that the relationship between $E(y)$ and x_1 and x_2 is as shown in Figure 12.10 and that you have observed y for each of the $n = 9$ combinations of values of x_1 and x_2, ($x_1 = 0$, 1, 2, and $x_2 = 1$, 2, 3). If you fit a first-order model in x_1 and x_2 to the data, the fitted plane would be (except for random error) approximately parallel to the (x_1, x_2)-plane, thus suggesting that x_1 and x_2 contribute very little information about $E(y)$. That this is not the case is clearly indicated by Figure 12.10. Fitting a first-order model to the data would not allow for the twist in the true surface and would therefore give a false impression of the relationship between $E(y)$ and x_1 and x_2. The procedure for detecting interaction between two independent variables can be seen by examining the model. The interaction model differs from the noninteraction first-order model only in the inclusion of the $\beta_3x_1x_2$ term:

Interaction model: $\quad E(y) = \beta_0 + \beta_1x_1 + \beta_2x_2 + \beta_3x_1x_2$

First-order model: $\quad E(y) = \beta_0 + \beta_1x_1 + \beta_2x_2$

FIGURE 12.10
A Graph Indicating
Interaction Between
x_1 and x_2

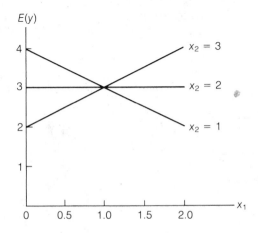

Therefore, to test for the presence of interaction, we test

H_0: $\beta_3 = 0$ (no interaction)

against the alternative hypothesis

H_a: $\beta_3 \neq 0$ (interaction)

using the familiar Student's t test of Section 11.7.

3. A COMPLETE SECOND-ORDER MODEL

$$E(y) = \beta_0 + \beta_1 x_1 + \beta_2 x_2 + \beta_3 x_1 x_2 + \beta_4 x_1^2 + \beta_5 x_2^2$$

Interpretation of model parameters

β_0: y-intercept; the value of $E(y)$ when $x_1 = x_2 = 0$

β_1 and β_2:   Changing β_1 and β_2 causes the surface to shift along the x_1 and the x_2 axes

β_3:   The value of β_3 controls the rotation of the surface

β_4 and β_5:   Signs and values of these parameters control the type of surface and the rates of curvature

The following three types of surfaces may be produced by a second-order model:

β_4 and β_5 positive: A paraboloid that opens upward (Figure 12.11a)

β_4 and β_5 negative: A paraboloid that opens downward (Figure 12.11b)

β_4 and β_5 differ in sign: A saddle-shaped surface (Figure 12.11c)

General comments   A complete second-order model is the three-dimensional equivalent of a second-order model in a single quantitative variable. Instead of tracing parabolas, it traces paraboloids and saddle surfaces. Since you fit only a portion of the complete surface to your data, a complete second-order model

FIGURE 12.11
Graphs of Three
Second-Order Surfaces

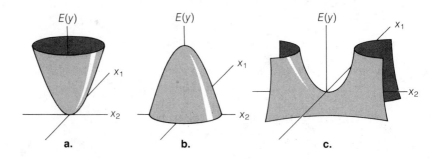

a.   b.   c.

provides a very large variety of gently curving surfaces. It is a good choice for a model if you expect curvature in the response surface relating $E(y)$ to x_1 and x_2.

12.13 **a.** Write a first-order model relating $E(y)$ to two quantitative independent variables, x_1 and x_2.

b. Modify the model you constructed in part **a** to include an interaction term.

c. Modify the model you constructed in part **b** to make it a complete second-order model.

12.14 Suppose the true relationship between $E(y)$ and the quantitative independent variables x_1 and x_2 is described by the following first-order model:

$$E(y) = 4 - x_1 + 2x_2$$

a. Describe the corresponding response surface.

b. Plot the contour lines of the response surface for $x_1 = 2, 3, 4$, where $0 \le x_2 \le 5$.

c. Plot the contour lines of the response surface for $x_2 = 2, 3, 4$, where $0 \le x_1 \le 5$.

d. Use the contour lines you plotted in parts **b** and **c** to explain how changes in the settings of x_1 and x_2 affect $E(y)$.

e. Use your graph from part **b** to determine how much $E(y)$ changes when x_1 is changed from 4 to 2 and x_2 is simultaneously changed from 1 to 2.

12.15 Suppose the true relationship between $E(y)$ and the quantitative independent variables x_1 and x_2 is

$$E(y) = 4 - x_1 + 2x_2 + x_1 x_2$$

Answer the questions posed in Exercise 12.14. Explain the effect of the interaction term on the mean response $E(y)$.

12.16 An energy conservationist wants to develop a model that will estimate the mean annual gasoline consumption in the United States (in millions of barrels), y, as a function of two independent variables:

$x_1 = $ Number of cars (millions) in use during year

$x_2 = $ Number of trucks (millions) in use during year

a. Identify the independent variables as quantitative or qualitative.

b. Write the first-order model for $E(y)$.

c. Write the complete second-order model for $E(y)$.

d. With respect to the model of part **c**, specify the null and alternative hypotheses you would employ in testing for the presence of interaction between x_1 and x_2.

12.17 The dissolved oxygen content, y, in rivers and streams is related to the amount, x_1, of nitrogen compounds per liter of water and the temperature, x_2, of the water. Write the complete second-order model relating $E(y)$ to x_1 and x_2.

12.18 An exploration seismologist wants to develop a model that will allow him to estimate the average signal-to-noise ratio of an earthquake's seismic wave, y, as a function of two independent variables:

x_1 = Frequency (cycles per second)

x_2 = Amplitude of the wavelet

a. Identify the independent variables as quantitative or qualitative.
b. Write the first-order model for $E(y)$.
c. Write a model for $E(y)$ that contains all first-order and interaction terms. Sketch typical response curves showing $E(y)$, the mean signal-to-noise ratio, versus x_2, the amplitude of the wavelet, for different values of x_1 (assume that x_1 and x_2 interact).
d. Write the complete second-order model for $E(y)$.

12.19 Researchers at the Upjohn Company utilized multiple regression analysis in the development of a sustained-release tablet.* One of the objectives of the research was to develop a model relating the dissolution y of a tablet (i.e., the percentage of the tablet dissolved over a specified period of time) to the following independent variables:

x_1 = Excipient level (i.e., amount of nondrug ingredient in the tablet)

x_2 = Process variable (e.g., machine setting under which tablet is processed)

a. Write the complete second-order model for $E(y)$.
b. Write a model that hypothesizes straight-line relationships between $E(y)$, x_1, and x_2. Assume that x_1 and x_2 do not interact.
c. Repeat part b, but add interaction to the model.
d. For the model in part c, what is the slope of the linear relationship between $E(y)$ and x_1 for fixed x_2?
e. For the model in part c, what is the slope of the linear relationship between $E(y)$ and x_2 for fixed x_1?

12.20 Refer to Exercise 12.10. Consider a second independent variable—number of users in the computer system. Write a model relating mean turnaround time, $E(y)$, to CPU time x_1 and number of users x_2 assuming:
a. A complete second-order model is appropriate.
b. A first-order model with interaction is appropriate.

SECTION 12.5

CODING QUANTITATIVE INDEPENDENT VARIABLES (OPTIONAL)

In fitting higher-order polynomial regression models (e.g., second- or third-order models), it is often a good practice to code the quantitative independent variables. For example, suppose one of the independent variables in a regression analysis is temperature, T, and T is observed at three levels: 50°, 100°, and 150°. We can code (or transform) the temperature measurements using the formula

*Source: Klassen, R. A. "The Application of Response Surface Methods to a Tablet Formulation Problem." Paper presented at Joint Statistical Meetings, American Statistical Association and Biometric Society, August 1986, Chicago, Ill.

$$x = \frac{T - 100}{50}$$

Then the coded levels $x = -1$, 0, and 1 correspond to the original levels 50°, 100°, and 150°.

In a general sense, **coding** means transforming a set of independent variables (qualitative or quantitative) into a new set of independent variables. For example, if we observe two independent variables,

$T = $ Temperature

$P = $ Pressure

then we can transform the two independent variables, T and P, into two new coded variables, x_1 and x_2, where x_1 and x_2 are related to T and P by two functional equations,

$$x_1 = f_1(T, P) \qquad x_2 = f_2(T, P)$$

The functions f_1 and f_2, which are frequently expressed as equations, establish a one-to-one correspondence between combinations of levels of T and P with combinations of the coded values of x_1 and x_2.

Since qualitative independent variables are not numerical, it is necessary to code their values in order to fit the regression model. However, you might ask why we would bother to code the quantitative independent variables. There are two related reasons for coding quantitative variables. At first glance, it would appear that a computer would be oblivious to the values assumed by the independent variables in a regression analysis, but this is not the case. Recall from Section 11.3 that the computer must calculate the $(X'X)^{-1}$ matrix in order to obtain the least squares estimates of the model parameters. Considerable rounding error may occur during the inversion process if the numbers in the $(X'X)$ matrix vary greatly in absolute value. This can produce sizable errors in the computed values of the least squares estimates, $\hat{\beta}_0$, $\hat{\beta}_1$, $\hat{\beta}_2$, ... Coding makes it computationally easier for the computer to invert the matrix, thus leading to more accurate estimates.

A second reason for coding quantitative variables pertains to the problem of multicollinearity discussed in Section 11.12. When polynomial regression models (e.g., second-order models) are fit, the problem of multicollinearity is unavoidable, especially when higher-order terms are fit. For example, consider the quadratic model

$$E(y) = \beta_0 + \beta_1 x + \beta_2 x^2$$

If the range of the values of x is narrow, then the two variables, $x_1 = x$ and $x_2 = x^2$, will generally be highly correlated. As we pointed out in Section 11.12, the likelihood of rounding errors in the regression coefficients is increased in the presence of multicollinearity.

The best way to cope with the rounding error problem is to:

1. Code the quantitative variable so that the new coded origin is in the center of the coded values. For example, by coding temperature, T, as

$$x = \frac{T - 100}{50}$$

we obtain coded values $-1, 0, 1$. This places the coded origin, 0, in the middle of the range of coded values (-1 to 1).

2. Code the quantitative variable so that the range of the coded values is approximately the same for all coded variables. You need not hold exactly to this requirement. The range of values for one independent variable could be double or triple the range of another without causing any difficulty, but it would not be desirable to have a sizable disparity in the ranges, say a ratio of 100 to 1.

When the data are observational (the values assumed by the independent variables are uncontrolled), the coding procedure described in the next box satisfies, reasonably well, these two requirements. The coded variable u is similar to the standardized normal z statistic of Section 5.5. Thus, the u value is the deviation (the distance) between an x value and the mean of the x values, $\bar{x}$, expressed in units of s_x.* Since we know that most (approximately 95%) measurements in a set will lie within 2 standard deviations of their mean, it follows that most of the coded u values will lie in the interval -2 to $+2$.

CODING PROCEDURE FOR OBSERVATIONAL DATA

Let

x = Uncoded quantitative independent variable

u = Coded quantitative independent variable

Then if x takes values $x_1, x_2, \ldots, x_n$ for the n data points in the regression analysis, let

$$u_i = \frac{x_i - \bar{x}}{s_x}$$

where s_x is the standard deviation of the x values, i.e.,

$$s_x = \sqrt{\frac{\sum_{i=1}^{n} (x_i - \bar{x})^2}{n - 1}}$$

---

*The divisor of the deviation, $x - \bar{x}$, need not equal s_x exactly. Any number approximately equal to s_x would suffice.

If you apply this coding to each quantitative variable, the range of values for each will be approximately -2 to $+2$. The variation in the absolute values of the elements of the coefficient matrix will be moderate, and rounding errors generated in finding the inverse of the matrix will be reduced. Additionally, the correlation between x and x^2 will be reduced.

EXAMPLE 12.3

Exercise 10.41 gives observational data on the index of building construction costs per month as a function of the index of the cost of construction materials (other components of construction costs would be labor, the cost of money, and so forth). The data are shown in Table 12.2.

TABLE 12.2
Index of Building
Construction Costs

MONTH	CONSTRUCTION COST[a] y	INDEX OF ALL CONSTRUCTION MATERIALS[b] x
January	193.2	180.0
February	193.1	181.7
March	193.6	184.1
April	195.1	185.3
May	195.6	185.7
June	198.1	185.9
July	200.9	187.7
August	202.7	189.6

[a]Source: United States Department of Commerce, Bureau of the Census.
[b]Source: United States Department of Labor, Bureau of Labor Statistics. Tables were given in Tables E-1 (p. 43) and E-2 (p. 44), respectively, in *Construction Review*, United States Department of Commerce, Oct. 1976, 22 (8).

a. Give the equation relating the coded variable u to the index of construction materials x using the coding system for observational data.
b. Calculate the coded values, u, for the eight x values.
c. Find the sum of the $n = 8$ values for u.

SOLUTION

a. We first find $\bar{x}$ and s_x:

$$\bar{x} = \frac{\sum\limits_{i=1}^{n} x_i}{n} = \frac{1,480.0}{8} = 185.0$$

$$\sum_{i=1}^{n}(x_i - \bar{x})^2 = \sum_{i=1}^{n} x_i^2 - \frac{\left(\sum\limits_{i=1}^{n} x_i\right)^2}{n} = 273,866.54 - \frac{(1,480.0)^2}{8} = 66.54$$

$$s_x = \sqrt{\frac{\sum_{i=1}^{n} (x_i - \bar{x})^2}{n - 1}} = \sqrt{\frac{66.54}{7}} = 3.08$$

Then the equation relating u and x is

$$u = \frac{x - 185.0}{3.08}$$

b. When $x = 180.0$,

$$u = \frac{x - 185.0}{3.08} = \frac{180.0 - 185.0}{3.08} = -1.62$$

Similarly, when $x = 181.7$,

$$u = \frac{x - 185.0}{3.08} = \frac{181.7 - 185.0}{3.08} = -1.07$$

Table 12.3 gives the coded values for all $n = 8$ observations. [*Note:* You can see that all the $n = 8$ values for u lie in the interval from -2 to $+2$.]

TABLE 12.3

Coded Values of x, Example 12.3

INDEX x	CODED VALUES u
180.0	−1.62
181.7	−1.07
184.1	−.29
185.3	.10
185.7	.23
185.9	.29
187.7	.88
189.6	1.49

c. If you ignore rounding error, the sum of the $n = 8$ values for u will equal 0. This is because the sum of the deviations of a set of measurements about their mean is always equal to 0. ∎

To illustrate the advantage of coding, consider fitting the second-order model

$$E(y) = \beta_0 + \beta_1 x + \beta_2 x^2$$

to the data of Example 12.3. It can be shown that the coefficient of correlation between the two variables, x and x^2, is $r = .999$. However, the coefficient of correlation between the corresponding coded values, u and u^2, is only $r = -.203$. Thus, we can avoid potential rounding error caused by multicollinearity by fitting, instead, the model

$$E(y) = \beta_0^* + \beta_1^* u + \beta_2^* u^2$$

Other methods of coding have been developed to reduce rounding errors and multicollinearity. One of the more complex coding systems involves fitting **orthogonal polynomials**. An orthogonal system of coding guarantees that the coded independent variables will be uncorrelated. For a discussion of orthogonal polynomials, consult the references given at the end of this chapter.

EXERCISES
12.21–12.24

12.21 As part of the first-year evaluation for new salespeople, a large computer software firm projects the second-year sales for each salesperson based on his or her sales for the first year. Data for $n = 8$ salespeople are shown in the table

FIRST-YEAR SALES x, thousands of dollars	SECOND-YEAR SALES y, thousands of dollars
75.2	99.3
91.7	125.7
100.3	136.1
64.2	108.6
81.8	102.0
110.2	153.7
77.3	108.8
80.1	105.4

a. Give the equation relating the coded variable u to the first-year sales, x, using the coding system for observational data.
b. Calculate the coded values, u.
c. Calculate the coefficient of correlation r between the variables x and x^2.
d. Calculate the coefficient of correlation r between the variables u and u^2. Compare this value to the value computed in part **c**.
e. If you have access to a statistical computer package, fit the model

$$E(y) = \beta_0 + \beta_1 u + \beta_2 u^2$$

12.22 Suppose you want to use the coding system for observational data to fit a second-order model to the tire pressure–automobile mileage data of Exercise 12.9, which are repeated in the table.

PRESSURE x, pounds per square inch	MILEAGE y, thousands
30	29
31	32
32	36
33	38
34	37
35	33
36	26

   a. Give the equation relating the coded variable u to pressure, x, using the coding system for observational data.

   b. Calculate the coded values, u.

   c. Calculate the coefficient of correlation r between the variables x and x^2.

   d. Calculate the coefficient of correlation r between the variables u and u^2. Compare this value to the value computed in part **c**.

   e. If you have access to a statistical computer package, fit the model

$$E(y) = \beta_0 + \beta_1 u + \beta_2 u^2$$

12.23 Refer to the *Journal of Testing and Evaluation* study on permeability of open-graded asphalt, described in Exercise 11.9. The data for the analysis are repeated in the accompanying table.

ASPHALT CONTENT $x,\%$	PERMEABILITY y, in/hr	ASPHALT CONTENT $x,\%$	PERMEABILITY y, in/hr
3	1,189	6	707
3	840	6	927
3	1,020	6	1,067
3	980	6	822
4	1,440	7	853
4	1,227	7	900
4	1,022	7	733
4	1,293	7	585
5	1,227	8	395
5	1,180	8	270
5	980	8	310
5	1,210	8	208

Source: Woelfl, G., Wei, I., Faulstich, C., and Litwack, H. "Laboratory Testing of Asphalt Concrete for Porous Pavements." *Journal of Testing and Evaluation*, Vol. 9, No. 4, July 1981, pp. 175–181.

   a. Give the equation relating the coded variable u to asphalt content, x, using the coding system for observational data.

   b. Calculate the coded values, u.

   c. Calculate the coefficient of correlation r between the variables x and x^2.

   d. Calculate the coefficient of correlation r between the variables u and u^2. Compare this value to the value computed in part **c**.

   e. If you have access to a statistical computer package, fit the model

$$E(y) = \beta_0 + \beta_1 u + \beta_2 u^2$$

12.24 Refer to the *Applied Spectroscopy* study on the infrared reflectance spectra properties of poly (perfluoropropyleneoxide), described in Exercise 11.57. The data for the analysis are repeated in the table.

OPTICAL DENSITY y	BAND FREQUENCY x_1, cm^{-1}	FILM THICKNESS x_2, milligrams
.231	740	1.1
.107	740	.62
.053	740	.31
.129	805	1.1
.069	805	.62
.030	805	.31
1.005	980	1.1
.559	980	.62
.321	980	.31
2.948	1235	1.1
1.633	1235	.62
.934	1235	.31

Source: Pacansky, J., England, C. D., and Waltman, R. "Infrared Spectroscopic Studies of Poly (Perfluoropropylene-oxide) on Gold Substrates: A Classical Dispersion Analysis for the Refractive Index." *Applied Spectroscopy*, Vol. 40, No. 1, Jan. 1986, p. 9 (Table I).

Suppose you want to fit the complete second-order model

$$E(y) = \beta_0 + \beta_1 x_1 + \beta_2 x_2 + \beta_3 x_1 x_2 + \beta_4 x_1^2 + \beta_5 x_2^2$$

using the coding system given in this section.

a. Give the coded values u_1 and u_2 for x_1 and x_2, respectively.
b. Compare the coefficient of correlation between x_1 and x_1^2 with the coefficient of correlation between u_1 and u_1^2.
c. Compare the coefficient of correlation between x_2 and x_2^2 with the coefficient of correlation between u_2 and u_2^2.
d. Give the prediction equation.

SECTION 12.6

MODEL BUILDING: TESTING PORTIONS OF A MODEL

The presentation of models with one and with two quantitative independent variables raises a very general question. Do certain terms in the model contribute more information than others for the prediction of y?

To illustrate, suppose you have collected data on a response, y, and two quantitative independent variables, x_1 and x_2, and you are considering the use of either a first-order or a second-order model to relate $E(y)$ to x_1 and x_2. Will the second-order model provide better predictions of y than the first-order model? To answer this question, examine the two models, and note that the second-order model contains all terms contained in the first-order model plus three additional terms—those involving β_3, β_4, and β_5:

First-order model: $E(y) = \beta_0 + \beta_1 x_1 + \beta_2 x_2$

$$\text{Second-order model:} \quad E(y) = \beta_0 + \beta_1 x_1 + \beta_2 x_2 + \overbrace{\beta_3 x_1 x_2 + \beta_4 x_1^2 + \beta_5 x_2^2}^{\text{Second-order terms}}$$

Therefore, asking whether the second-order (or *complete*) model contributes more information for the prediction of y than the first-order (or *reduced*) model is equivalent to asking whether at least one of the parameters, β_3, β_4, or β_5, differs from 0—i.e., whether the terms involving β_3, β_4, and β_5 should be retained in the model. Therefore, to test whether the second-order terms should be included in the model, we test the null hypothesis

$$H_0: \quad \beta_3 = \beta_4 = \beta_5 = 0$$

(i.e., the second-order terms do not contribute information for the prediction of y) against the alternative hypothesis

$$H_a: \quad \text{At least one of the parameters, } \beta_3, \beta_4, \text{ or } \beta_5, \text{ differs from 0}$$

(i.e., at least one of the second-order terms contributes information for the prediction of y).

The procedure for conducting this test is intuitive: First, we use the method of least squares to fit the first-order model and calculate the corresponding sum of squares for error, SSE_1 (the sum of squares of the deviations between observed and predicted y values). Next, we fit the second-order model and calculate its sum of squares for error, SSE_2. Then, we compare SSE_1 to SSE_2 by calculating the difference $SSE_1 - SSE_2$. If the second-order terms contribute to the model, then SSE_2 should be much smaller than SSE_1, and the difference $SSE_1 - SSE_2$ will be large. The larger the difference, the greater the weight of evidence that the second-order model provides better predictions of y than does the first-order model.

The sum of squares for error will always decrease when new terms are added to the model. The question is whether this decrease is large enough to conclude that it is due to more than just an increase in the number of model terms and to chance. To test the null hypothesis that the parameters of the second-order terms, β_3, β_4, and β_5 simultaneously equal 0, we use an F statistic calculated as follows:

$$F = \frac{\text{Drop in SSE/Number of } \beta \text{ parameters being tested}}{s^2 \text{ for the second-order model}}$$

$$= \frac{(SSE_1 - SSE_2)/3}{SSE_2/[n - (5 + 1)]}$$

When the assumptions listed in Section 11.2 about the error term ε are satisfied and the β parameters for the second-order terms are all 0 (i.e., H_0 is true), this F statistic has an F distribution with $\nu_1 = 3$ and $\nu_2 = n - 6$ degrees of freedom. Note that ν_1 is the number of β parameters being tested and ν_2 is the number of degrees of freedom associated with s^2 in the second-order model.

If the second-order terms *do* contribute to the model (i.e., H_a is true), we expect the F statistic to be large. Thus, we use a one-tailed test and reject H_0 if F exceeds some critical value, F_α. A summary of the steps used in testing the null hypothesis that a set of model parameters are all equal to 0 is shown in the box.

**F TEST FOR TESTING THE NULL HYPOTHESIS:
EACH OF A SET OF β-PARAMETERS EQUALS 0**

Reduced model: $E(y) = \beta_0 + \beta_1 x_1 + \cdots + \beta_g x_g$

Complete model:

$$E(y) = \beta_0 + \beta_1 x_1 + \cdots + \beta_g x_g + \beta_{g+1} x_{g+1} + \cdots + \beta_k x_k$$

H_0: $\beta_{g+1} = \beta_{g+2} = \cdots = \beta_k = 0$

H_a: At least one of the β parameters under test is nonzero

Test statistic: $F = \dfrac{(\text{SSE}_1 - \text{SSE}_2)/(k - g)}{\text{SSE}_2/[n - (k + 1)]}$

where

   SSE_1 = Sum of squared errors for the reduced model

   SSE_2 = Sum of squared errors for the complete model

   $k - g$ = Number of β parameters specified in H_0

   $k + 1$ = Number of β parameters in the complete model

      n = Sample size

Rejection region: $F > F_\alpha$

where

   $\nu_1 = k - g$ = Degrees of freedom for the numerator

   $\nu_2 = n - (k + 1)$ = Degrees of freedom for the denominator

---

EXAMPLE 12.4

Many companies manufacture products (e.g., steel, paint, gasoline) that are at least partially chemically produced. In many instances, the quality of the finished product is a function of the temperature and pressure at which the chemical reactions take place. Suppose you wanted to model the quality, y, of a product as a function of the temperature, x_1, and the pressure, x_2, at which it is produced. Four inspectors independently assign a quality score between 0 and 100 to each product, and then the quality, y, is calculated by averaging the four scores. An experiment is conducted by varying temperature between 80 and 100°F and pressure between 50 and 60 pounds per square inch. The resulting data are given in Table 12.4 (page 582).

TABLE 12.4
Temperature, Pressure, and Quality of the Finished Product

x_1 (°F)	x_2 (pounds per square inch)	y	x_1 (°F)	x_2 (pounds per square inch)	y	x_1 (°F)	x_2 (pounds per square inch)	y
80	50	50.8	90	50	63.4	100	50	46.6
80	50	50.7	90	50	61.6	100	50	49.1
80	50	49.4	90	50	63.4	100	50	46.4
80	55	93.7	90	55	93.8	100	55	69.8
80	55	90.9	90	55	92.1	100	55	72.5
80	55	90.9	90	55	97.4	100	55	73.2
80	60	74.5	90	60	70.9	100	60	38.7
80	60	73.0	90	60	68.8	100	60	42.5
80	60	71.2	90	60	71.3	100	60	41.4

a. Fit a second-order model to the data.
b. Sketch the response surface.
c. Do the data provide sufficient evidence to indicate that the second-order terms contribute information for the prediction of y?

SOLUTION

a. The complete second-order model is

$$E(y) = \beta_0 + \beta_1 x_1 + \beta_2 x_2 + \beta_3 x_1 x_2 + \beta_4 x_1^2 + \beta_5 x_2^2$$

The data in Table 12.4 were used to fit this model, and a portion of the SAS output is shown in Figure 12.12.

FIGURE 12.12 Portion of the SAS Printout for Example 12.4

SOURCE	DF	SUM OF SQUARES	MEAN SQUARE	F VALUE	PR > F
MODEL	5	8402.26453714	1680.45290743	596.32	0.0001
ERROR	21	59.17842582	2.81802028		ROOT MSE
CORRECTED TOTAL	26	8461.44296296		R-SQUARE	1.67869601
				0.993006	

PARAMETER	ESTIMATE	T FOR H0: PARAMETER = 0	PR > |T|	STD ERROR OF ESTIMATE
INTERCEPT	-5127.89907417	-46.49	0.0001	110.29601483
X1	31.09638889	23.13	0.0001	1.34441322
X2	139.74722222	44.50	0.0001	3.14005411
X1*X2	-0.14550000	-15.01	0.0001	0.00969196
X1*X1	-0.13338889	-19.46	0.0001	0.00685325
X2*X2	-1.14422222	-41.74	0.0001	0.02741299

The least squares prediction equation is

$$\hat{y} = -5{,}127.90 + 31.10x_1 + 139.75x_2 - .146x_1x_2 - .133x_1^2 - 1.14x_2^2$$

FIGURE 12.13

Plot of Second-Order
Least Squares Model
for Example 12.4

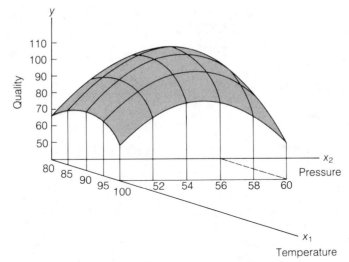

b. A three-dimensional graph of this prediction model is shown in Figure 12.13. Note that the mean quality seems to be greatest for temperatures of about 85–90°F and for pressures of about 55–57 pounds per square inch.* Further experimentation in these ranges might lead to a more precise determination of the optimal temperature–pressure combination.

c. To determine whether the data provide sufficient evidence to indicate that the second-order terms contribute information for the prediction of y, we test

$$H_0: \quad \beta_3 = \beta_4 = \beta_5 = 0$$

against the alternative hypothesis,

$$H_a: \quad \text{At least one of the parameters, } \beta_3, \beta_4, \text{ or } \beta_5, \text{ differs from } 0$$

The first step in conducting the test is to drop the second-order terms out of the complete (second-order) model and fit the reduced model

$$E(y) = \beta_0 + \beta_1 x_1 + \beta_2 x_2$$

to the data. The SAS computer printout for this procedure is shown in Figure 12.14 (page 584). You can see that the sums of squares for error, given in Figures 12.12 and 12.14 for the complete and reduced models, respectively, are

$$SSE_2 = 59.17842582$$

$$SSE_1 = 6{,}671.50851852$$

and that s^2 for the complete model is

$$s_2^2 = 2.81802028$$

---

*We can estimate the values of temperature and pressure that maximize quality in the least squares model by solving $\partial \hat{y} / \partial x_1 = 0$ and $\partial \hat{y} / \partial x_2 = 0$ for x_1 and x_2. These estimated optimal values are $x_1 = 86.25°F$ and $x_2 = 55.58$ pounds per square inch.

FIGURE 12.14 SAS Computer Printout for the Reduced (First-Order) Model in Example 12.4

```
DEPENDENT VARIABLE: Y

SOURCE                    DF    SUM OF SQUARES      MEAN SQUARE     F VALUE
MODEL                      2    1789.93444444      894.96722222       3.22
ERROR                     24    6671.50851852      277.97952160     PR > F
CORRECTED TOTAL           26    8461.44296296                       0.0577

R-SQUARE             C.V.          ROOT MSE           Y MEAN
0.211540           24.8984      16.67271788        66.96296296

                            T FOR H0:     PR > !T!     STD ERROR OF
PARAMETER      ESTIMATE     PARAMETER=0                  ESTIMATE

INTERCEPT    106.08518519        1.90      0.0700       55.94500427
X1            -0.91611111       -2.33      0.0285        0.39297973
X2             0.78777778        1.00      0.3262        0.78595946
```

Recall that $n = 27$, $k = 5$, and $g = 2$. Therefore, the calculated value of the F statistic, based on $\nu_1 = k - g = 3$ and $\nu_2 = n - (k + 1) = 21$ degrees of freedom is

$$F = \frac{(\text{SSE}_1 - \text{SSE}_2)/(k - g)}{\text{SSE}_2/[n - (k + 1)]} = \frac{(\text{SSE}_1 - \text{SSE}_2)/(k - g)}{s_2^2}$$

where $\nu_1 = k - g$ is equal to the number of parameters involved in H_0 and s_2^2 is the value of s^2 for the complete model. Therefore,

$$F = \frac{(6{,}671.50851852 - 59.17842582)/3}{2.81802028} = 782.1$$

The final step in the test is to compare this computed value of F with the tabulated value based on $\nu_1 = 3$ and $\nu_2 = 21$ degrees of freedom. If we choose $\alpha = .05$, then $F_{.05} = 3.07$. Since the computed value of F falls in the rejection region (see Figure 12.15)—i.e., it exceeds $F_{.05} = 3.07$—we reject H_0 and conclude that at least one of the second-order terms contributes information

FIGURE 12.15

Rejection Region for the F Test H_0: $\beta_3 = \beta_4 = \beta_5 = 0$

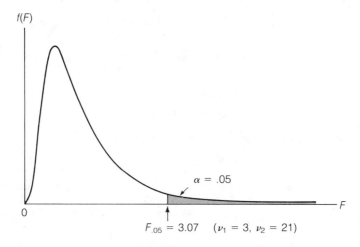

$f(F)$

$\alpha = .05$

0

F

$F_{.05} = 3.07$ $(\nu_1 = 3, \nu_2 = 21)$

for the prediction of y. In other words, the data support the contention that the curvature we see in the response surface is not due simply to random variation in the data. The second-order model appears to provide better predictions of y than does a first-order model.   ∎

Example 12.4 demonstrates the motivation for testing a hypothesis that each one of a set of β parameters equals 0 and it also demonstrates the procedure. Other applications of this test will appear in the following sections.

12.25 Suppose you fit the regression model

$$y = \beta_0 + \beta_1 x_1 + \beta_2 x_2 + \beta_3 x_1 x_2 + \beta_4 x_1^2 + \beta_5 x_2^2 + \varepsilon$$

to $n = 30$ data points and you wish to test

$$H_0: \quad \beta_3 = \beta_4 = \beta_5 = 0$$

a. State the alternative hypothesis, H_a.
b. Explain in detail how you would find the quantities necessary to compute the F statistic for this test of hypothesis.
c. What are the numerator and denominator degrees of freedom associated with the F statistic?

12.26 Suppose you fit the complete and reduced models for the hypothesis test described in Exercise 12.25 and obtain $SSE_1 = 246.1$ and $SSE_2 = 215.2$. Conduct the hypothesis test and interpret the results of your test. Test using $\alpha = .05$.

12.27 A large data-processing firm rates the performance of each member of its technical staff once a year. Each person is rated on a scale of 0 to 100 by his or her immediate supervisor, and this merit rating is used to determine the size of the person's pay raise for the coming year. The firm's personnel department is interested in developing a regression model to help them forecast the merit rating that an applicant for a technical position will receive after being employed 3 years. The data-processing firm proposes to use the following model to forecast the merit ratings of applicants who have just completed their graduate studies and who have no prior related job experience:

$$E(y) = \beta_0 + \beta_1 x_1 + \beta_2 x_2 + \beta_3 x_1 x_2 + \beta_4 x_1^2 + \beta_5 x_2^2$$

where

$y =$ Applicant's merit rating after 3 years

$x_1 =$ Applicant's grade-point average (GPA) in graduate school

$x_2 =$ Applicant's total score (verbal plus quantitative) on the Graduate Record Examination (GRE)

A random sample of $n = 40$ employees who have been on the technical staff of the data-processing firm more than 3 years was selected. Each employee's merit rating after 3 years, graduate school GPA, and total score on the GRE were recorded. The proposed model was fit to these data with the aid of a computer. A portion of the resulting computer printout is shown at the top of page 586.

SOURCE	DF	SUM OF SQUARES	MEAN SQUARE
MODEL	5	4911.56	982.31
ERROR	34	1830.44	53.84
TOTAL	39	6742.00	R-SQUARE
			0.73

The reduced model $E(y) = \beta_0 + \beta_1 x_1 + \beta_2 x_2$ was also fit to the same data. The resulting computer printout is partially reproduced below:

SOURCE	DF	SUM OF SQUARES	MEAN SQUARE
MODEL	2	3544.84	1772.42
ERROR	37	3197.16	86.41
TOTAL	39	6742.00	R-SQUARE
			0.53

a. Identify the appropriate null and alternative hypotheses to test whether the complete (second-order) model contributes information for the prediction of y.

b. Conduct the test of hypothesis given in part **a**. Test using $\alpha = .05$. Interpret the results in the context of this problem.

c. Identify the appropriate null and alternative hypotheses to test whether the complete model contributes more information than the reduced (first-order) model for the prediction of y.

d. Conduct the test of hypothesis given in part **c**. Test using $\alpha = .05$. Interpret the results in the context of this problem.

e. Which model, if either, would you use to predict y? Explain.

12.28 Refer to Exercise 12.16, in which an energy conservationist wants to develop a regression model to forecast annual gasoline consumption in the United States. The complete and reduced models for the test that you described in part **d** of Exercise 12.16 were fit to $n = 25$ data points. The resulting values for SSE_1 and SSE_2 were 1,065.9 and 400.6, respectively.

a. Conduct the test to determine whether the data present sufficient evidence to indicate interaction between x_1 and x_2. Test using $\alpha = .05$.

b. Which model seems better for forecasting annual gasoline consumption? Why?

12.29 Refer to Exercise 12.18, in which an exploration seismologist wants to develop a regression model for estimating the mean signal-to-noise ratio of seismic waves from earthquakes. The model under consideration is a complete second-order model:

$$E(y) = \beta_0 + \beta_1 x_1 + \beta_2 x_2 + \beta_3 x_1 x_2 + \beta_4 x_1^2 + \beta_5 x_2^2$$

where

y = Signal-to-noise ratio

x_1 = Frequency of wavelet

x_2 = Amplitude of wavelet

Following is a portion of the computer printout that results from fitting this model to $n = 12$ data points:

SOURCE	DF	SUM OF SQUARES	MEAN SQUARE
MODEL	5	38638.97	7727.79
ERROR	6	159.94	26.66
TOTAL	11	38798.91	R-SQUARE
			0.996

The reduced first-order model

$$E(y) = \beta_0 + \beta_1 x_1 + \beta_2 x_2$$

was also fit to the same data and the resulting computer printout is partially reproduced below:

SOURCE	DF	SUM OF SQUARES	MEAN SQUARE
MODEL	2	36704.5	18352.2
ERROR	9	2094.4	232.7
TOTAL	11	38798.9	R-SQUARE
			0.946

Is there sufficient evidence to conclude that a second-order model contributes more information for the prediction of y than does a first-order model? Test using $\alpha = .05$.

Suppose we want to write a model for the mean profit, $E(y)$, per sales dollar of a construction company as a function of the sales engineer who estimates and bids on a job (for the purpose of explanation, we will ignore other independent variables that might affect the response.) Further, suppose there are three sales engineers: Jones, Smith, and Adams. Then Sales engineer is a single qualitative variable with three levels corresponding to Jones, Smith, and Adams. Note that with a qualitative independent variable, we cannot attach a quantitative meaning to a given level. All we can do is describe it.

To simplify our notation, let μ_A be the mean profit per sales dollar for Jones, and let μ_B and μ_C be the corresponding mean profits for Smith and Adams. Our objective is to write a single prediction equation that will give the mean value of y for the three sales engineers. This can be done as follows:

$$E(y) = \beta_0 + \beta_1 x_1 + \beta_2 x_2$$

$$x_1 = \begin{cases} 1 & \text{if Smith is the sales engineer} \\ 0 & \text{if Smith is not the sales engineer} \end{cases}$$

$$x_2 = \begin{cases} 1 & \text{if Adams is the sales engineer} \\ 0 & \text{if Adams is not the sales engineer} \end{cases}$$

The variables x_1 and x_2 are not meaningful independent variables as for the case of the models with quantitative independent variables. Instead, they are **dummy (indicator) variables** that make the model function. To see how they work, let $x_1 = 0$ and $x_2 = 0$. This condition will apply when we are seeking the mean response for Jones (neither Smith nor Adams is the sales engineer; hence, it must be Jones). Then the mean value of y when Jones is the sales engineer is

$$\mu_A = E(y) = \beta_0 + \beta_1(0) + \beta_2(0) = \beta_0$$

This tells us that the mean profit per sales dollar for Jones is β_0. Or, it means that $\beta_0 = \mu_A$.

Now suppose we want to represent the mean response, $E(y)$, when Smith is the sales engineer. Checking the dummy variable definitions, we see that we should let $x_1 = 1$ and $x_2 = 0$:

$$\mu_B = E(y) = \beta_0 + \beta_1 x_1 + \beta_2 x_2 = \beta_0 + \beta_1(1) + \beta_2(0) = \beta_0 + \beta_1$$

or, since $\beta_0 = \mu_A$,

$$\mu_B = \mu_A + \beta_1$$

Then it follows that the interpretation of β_1 is

$$\beta_1 = \mu_B - \mu_A$$

which is the difference in the mean profit per sales dollar for Jones and Smith.

Finally, if we want the mean value of y when Adams is the sales engineer, we set $x_1 = 0$ and $x_2 = 1$:

$$\mu_C = E(y) = \beta_0 + \beta_1(0) + \beta_2(1) = \beta_0 + \beta_2$$

or, since $\beta_0 = \mu_A$,

$$\mu_C = \mu_A + \beta_2$$

Then it follows that the interpretation of β_2 is

$$\beta_2 = \mu_C - \mu_A$$

Note that we were able to describe *three levels* of the qualitative variable with only *two dummy variables*. This is because the mean of the base level (Jones, in this case) is accounted for by the intercept β_0.

Since Sales engineer is a qualitative variable, we will use a bar graph to show the value of mean profit $E(y)$ for the three levels of Sales engineer (see Figure 12.16). In particular, note that the height of the bar, $E(y)$, for each level of Sales

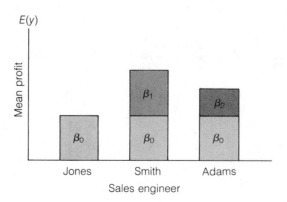

FIGURE 12.16

Bar Chart Comparing $E(y)$ for Three Sales Engineers

engineer is equal to the sum of the model parameters shown in the preceding equations. You can see that the height of the bar corresponding to Jones is β_0; i.e., $E(y) = \beta_0$. Similarly, the heights of the bars corresponding to Smith and Adams are $E(y) = \beta_0 + \beta_1$ and $E(y) = \beta_0 + \beta_2$, respectively.*

Now, carefully examine the model with a single qualitative independent variable at three levels, because we will use exactly the same pattern for any number of levels. Also, the interpretation of the parameters will always be the same.

One level is selected as the base level (we used Jones as level A). Then, for the one–zero system of coding† for the dummy variables,

$$\mu_A = \beta_0$$

The coding for all dummy variables is as follows: To represent the mean value of y for a particular level, let that dummy variable equal 1; otherwise, the dummy variable is set equal to 0. Using this system of coding, we have

$$\mu_B = \beta_0 + \beta_1$$
$$\mu_C = \beta_0 + \beta_2$$
$$\vdots$$

Because $\mu_A = \beta_0$, any other model parameter will represent the difference in means for that level and the base level:

$$\beta_1 = \mu_B - \mu_A$$
$$\beta_2 = \mu_C - \mu_A$$
$$\vdots$$

*Either β_1 or β_2, or both, could be negative. If, for example, β_1 were negative, the height of the bar corresponding to Smith would be *reduced* (rather than increased) from the height of the bar for Jones by the amount β_1. Figure 12.16 is constructed assuming that β_1 and β_2 are positive quantities.

†We do not have to use a one–zero system of coding for the dummy variables. Any two-value system will work, but the interpretation given to the model parameters will depend on the code. Using the one–zero system makes the model parameters easy to interpret.

The general procedure is given in the accompanying box.

PROCEDURE FOR WRITING A MODEL WITH ONE QUALITATIVE INDEPENDENT VARIABLE AT k LEVELS

$$E(y) = \beta_0 + \beta_1 x_1 + \beta_2 x_2 + \cdots + \beta_{k-1} x_{k-1}$$

where x_i is the dummy variable for level i and

$$x_i = \begin{cases} 1 & \text{if } E(y) \text{ is the mean for level } i \\ 0 & \text{otherwise} \end{cases}$$

Then, for this system of coding

$$\mu_A = \beta_0$$
$$\mu_B = \beta_0 + \beta_1$$
$$\mu_C = \beta_0 + \beta_2$$
$$\mu_D = \beta_0 + \beta_3$$
$$\vdots$$

Also, note that

$$\beta_1 = \mu_B - \mu_A$$
$$\beta_2 = \mu_C - \mu_A$$
$$\beta_3 = \mu_D - \mu_A$$
$$\vdots$$

EXAMPLE 12.5

A large consulting firm markets a computerized system for monitoring road construction bids to various state departments of transportation. Since the high cost of maintaining the system is partially absorbed by the firm, the firm wants to compare the mean annual maintenance costs accrued by system users in three different states: Kansas, Kentucky, and Texas. A sample of ten users is selected from each state installation and the maintenance cost accrued by each is recorded, as shown in Table 12.5. Do the data provide sufficient evidence to indicate that the mean annual maintenance costs accrued by system users differ for the three state installations?

SOLUTION

The model relating $E(y)$ to the single qualitative variable, State installation, is

$$E(y) = \beta_0 + \beta_1 x_1 + \beta_2 x_2$$

where

$$x_1 = \begin{cases} 1 & \text{if Kentucky} \\ 0 & \text{if not} \end{cases}$$

$$x_2 = \begin{cases} 1 & \text{if Texas} \\ 0 & \text{if not} \end{cases}$$

TABLE 12.5
Annual Maintenance
Costs

| | STATE INSTALLATION | | |
	1: Kansas	2: Kentucky	3: Texas
	$ 198	$ 563	$ 385
	126	314	693
	443	483	266
	570	144	586
	286	585	178
	184	377	773
	105	264	308
	216	185	430
	465	330	644
	203	354	515
Totals	$2,796	$3,599	$4,778

and

$$\beta_1 = \mu_2 - \mu_1$$
$$\beta_2 = \mu_3 - \mu_1$$

where μ_1, μ_2, and μ_3 are the mean responses for Kansas, Kentucky, and Texas, respectively. Testing the null hypothesis that the means for the three states are equal, i.e., $\mu_1 = \mu_2 = \mu_3$, is equivalent to testing

$$H_0: \quad \beta_1 = \beta_2 = 0$$

because if $\beta_1 = \mu_2 - \mu_1 = 0$ and $\beta_2 = \mu_3 - \mu_1 = 0$, then μ_1, μ_2, and μ_3 must be equal. The alternative hypothesis is

$$H_a: \quad \text{At least one of the parameters, } \beta_1 \text{ or } \beta_2, \text{ differs from 0}$$

There are two ways to conduct this test. We can fit the complete model shown above and the reduced model (deleting the terms involving β_1 and β_2),

$$E(y) = \beta_0$$

and conduct the F test described in the preceding section (we leave this as an exercise for you). Or, we can use the F test of the complete model (Section 11.8), which tests the null hypothesis that all parameters in the model, with the exception of β_0, equal 0. Either way you conduct the test, you will obtain the same computed value of F, the value shown on the SAS printout for a test of the complete model. The SAS printout for fitting the complete model,

$$E(y) = \beta_0 + \beta_1 x_1 + \beta_2 x_2$$

is shown in Figure 12.17 (page 592), and the value of the F statistic for testing the complete model, $F = 3.48$, is shaded. We will want to compare this value with the tabulated value of F based on $\nu_1 = 2$ and $\nu_2 = 27$ degrees of freedom. If we choose $\alpha = .05$, we will reject $H_0: \beta_1 = \beta_2 = 0$ if the computed value of F exceeds $F_{.05} = 3.35$. Since the computed value of F, $F = 3.48$, exceeds

FIGURE 12.17   SAS Computer Printout for Example 12.5

```
DEPENDENT VARIABLE: Y
SOURCE                  DF      SUM OF SQUARES         MEAN SQUARE       F VALUE      PR > F       R-SQUARE          C.V.
MODEL                    2      198772.46666667      99386.23333333        3.48       0.0452       0.205038       45.3632
ERROR                   27      770670.90000000      28543.36666667                   ROOT MSE                     Y MEAN
CORRECTED TOTAL         29      969443.36666667                                      168.94782232             372.43333333

SOURCE          DF           TYPE I SS       F VALUE     PR > F          DF         TYPE III SS      F VALUE     PR > F
X1               1         2356.26666667        0.08     0.7761           1       32240.45000000        1.13     0.2973
X2               1       196416.20000000        6.88     0.0141           1      196416.20000000        6.88     0.0141

                                       T FOR H0:      PR > :T:        STD ERROR OF
PARAMETER         ESTIMATE          PARAMETER=0                        ESTIMATE
INTERCEPT      279.60000000                5.23        0.0001          53.42599243
X1              80.30000000                1.06        0.2973          75.55576307
X2             198.20000000                2.62        0.0141          75.55576307
```

$F_{.05} = 3.35$, we reject H_0 and conclude that at least one of the parameters, β_1 or β_2, differs from 0. Or, equivalently, we conclude that the data provide sufficient evidence to indicate that the mean user maintenance cost does vary among the three state installations.   ∎

We make two additional comments about Example 12.5. A regression analysis is not the easiest way to analyze these data (unless you have ready access to a computer and a good regression program). A simpler procedure for calculating the value of the F statistic, known as an **analysis of variance**, is described in Chapter 13. If you choose to analyze the data by fitting complete and reduced models (Section 12.6), you will find that the least squares estimate of β_0 in the reduced model,

$$E(y) = \beta_0$$

is $\bar{y}$, the mean of all $n = 30$ observations, and that the sum of squares for error for the reduced model is

$$\text{SSE}_1 = \sum(y_i - \hat{y}_i)^2 = \sum(y_i - \bar{y})^2 = 969{,}443.367$$

This value is shown in the SAS printout in Figure 12.17 as the SUM OF SQUARES corresponding to CORRECTED TOTAL. We leave the remaining steps, calculating the drop in SSE and the resulting F statistic, to you. The value you obtain should be exactly the same as the value of F shown in the SAS printout in Figure 12.17.

**EXERCISES
12.30–12.34**

12.30   The following model was used to relate $E(y)$ to a single qualitative variable with four levels:

$$E(y) = \beta_0 + \beta_1 x_1 + \beta_2 x_2 + \beta_3 x_3$$

where

$$x_1 = \begin{cases} 1 & \text{if level 2} \\ 0 & \text{if not} \end{cases} \qquad x_2 = \begin{cases} 1 & \text{if level 3} \\ 0 & \text{if not} \end{cases} \qquad x_3 = \begin{cases} 1 & \text{if level 4} \\ 0 & \text{if not} \end{cases}$$

This model was fit to $n = 30$ data points and the following result was obtained:

$$\hat{y} = 10.2 - 4x_1 + 12x_2 + 2x_3$$

Use the least squares prediction equation to find the estimate of $E(y)$ for each level of the qualitative independent variable.

12.31 Refer to Exercise 12.30. Specify the null and alternative hypotheses you would employ to test whether $E(y)$ is the same for all four levels of the independent variable.

12.32 An electrical engineer wants to compare the mean lifelengths (in hours) of five different brands of magnetron tubes. Data are gathered on ten magnetron tubes selected at random from each of the five brands. Write a model that will give the mean lifelength for the five brands and interpret all the β parameters used in the model.

12.33 Due to the hot, humid weather conditions in Florida, the growth rates of beef cattle and the milk production of dairy cows typically decline during the summer. However, agricultural and environmental engineers have found that a well-designed shade structure can significantly increase the milk production of dairy cows. In one experiment, 30 cows were selected and divided into three groups of ten cows each. Group 1 cows were provided with a man-made shade structure, group 2 cows with tree shade, and group 3 cows with no shade. Of interest was the mean milk production (in gallons) of the cows in each group.
  a. Identify the independent variables in the experiment.
  b. Write a model relating the mean milk production, $E(y)$ to the independent variables. Identify and code all dummy variables.
  c. Interpret the β parameters of the model.

12.34 TexaSoft, a manufacturer of video arcade and home computer games, wants to model weekly sales, y, as a function of game product. The firm markets four games: Trilogy, Set the Hostages Free, Queen of Hearts, and Squirm.
  a. Write a model relating mean weekly sales, $E(y)$, to game product.
  b. Interpret the β parameters of the model.
  c. In terms of the β parameters, what are the mean weekly sales for the Queen of Hearts video game?

COMPARING THE SLOPES OF TWO OR MORE LINES

Suppose you want to relate the mean monthly sales, $E(y)$, of a company to monthly advertising expenditure, x, for three different advertising media—say, newspaper, radio, and television—and you wish to use first-order (straight-line) models to model the responses for all three media. Graphs of these three relationships might appear as shown in Figure 12.18 (page 594).

Since the lines in Figure 12.18 are hypothetical, a number of practical questions arise. Is one advertising medium as effective as any other; that is, do the three mean sales lines differ for the three advertising media? Do the increases in mean sales per dollar increase in advertising differ for the three advertising media; that is, do the slopes of the three lines differ? Note that each of the two practical questions has been rephrased into a question about the parameters that define the three lines of Figure 12.18. To answer them, we must write a single linear statistical model that will characterize the three lines of Figure 12.18. Then the practical questions can be answered by testing hypotheses about the model parameters.

FIGURE 12.18

Graphs of the Relationship Between Mean Sales, $E(y)$, and Advertising Expenditure, x

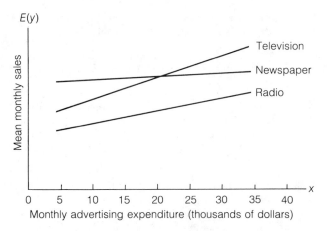

In the example above, the response (monthly sales) is a function of *two* independent variables, one quantitative (advertising expenditure, x) and one qualitative (type of medium). We will examine the different models that can be constructed relating $E(y)$ to these two independent variables.

1. The straight-line relationship between mean sales, $E(y)$, and advertising expenditure is the same for all three media; that is, a single line will describe the relationship between $E(y)$ and advertising expenditure, x_1, for all the media (see Figure 12.19).

$$E(y) = \beta_0 + \beta_1 x_1$$
$$x_1 = \text{Advertising expenditure}$$

FIGURE 12.19

The Relationship Between $E(y)$ and x_1 Is the Same for All Media

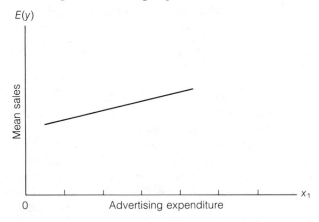

2. The straight lines relating mean sales, $E(y)$, to advertising expenditure, x_1, differ from one medium to another, but the rate of increase in mean sales per unit increase in dollar advertising expenditure, x_1, is the same for all media. That is, the lines are parallel but possess different y-intercepts (see Figure 12.20).

$$E(y) = \beta_0 + \beta_1 x_1 + \beta_2 x_2 + \beta_3 x_3$$

$$x_1 = \text{Advertising expenditure}$$

$$x_2 = \begin{cases} 1 & \text{if radio medium} \\ 0 & \text{if not} \end{cases}$$

$$x_3 = \begin{cases} 1 & \text{if television medium} \\ 0 & \text{if not} \end{cases}$$

FIGURE 12.20

Parallel Response Lines for the Three Media

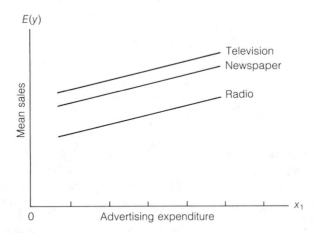

Notice that this model is essentially a combination of a first-order model with a single quantitative variable and the model with a single qualitative variable:

First-order model with a single
quantitative variable: $\qquad E(y) = \beta_0 + \boxed{\beta_1 x_1}$

Model with a single
qualitative variable
at three levels: $\qquad E(y) = \beta_0 + \boxed{\beta_2 x_2 + \beta_3 x_3}$

where x_1, x_2, and x_3 are defined as above. The model described implies no interaction between the two independent variables, advertising expenditure x_1 and the qualitative variable, type of advertising medium. The change in $E(y)$ for a 1-unit change in x_1 is identical (i.e., the slopes of the lines are equal) for all three advertising media. The terms corresponding to each of the independent variables are called **main effect terms** because they imply no interaction.

3. The straight lines relating mean sales, $E(y)$, to advertising expenditure, x_1, differ for the three advertising media; that is, the intercepts and slopes differ for the three lines (see Figure 12.21, page 596). As you will see, this interaction model is obtained by adding interaction terms (those involving the cross product terms, one each from each of the two independent variables):

$$E(y) = \beta_0 + \overbrace{\beta_1 x_1}^{\substack{\text{Main effect,} \\ \text{advertising} \\ \text{expenditure}}} + \overbrace{\beta_2 x_2 + \beta_3 x_3}^{\substack{\text{Main effect,} \\ \text{type of medium}}} + \overbrace{\beta_4 x_1 x_2 + \beta_5 x_1 x_3}^{\text{Interaction}}$$

FIGURE 12.21

Different Response Lines
for the Three Media

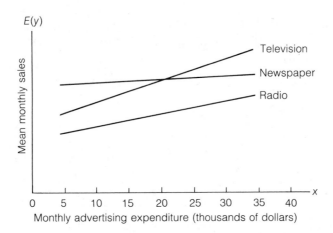

Note that each of the preceding models is obtained by adding terms to model 1, the single first-order model used to model the responses for all three media. Model 2 is obtained by adding the main effect terms for the qualitative variable, type of medium; and model 3 is obtained by adding the interaction terms to model 2.

Will a single line (Figure 12.19) characterize the responses for all three media, or do the three response lines differ as shown in Figure 12.21? A test of the null hypothesis that a single first-order model adequately describes the relationship between $E(y)$ and advertising expenditure x_1 for all three media is a test of the null hypothesis that the parameters of model 3, β_2, β_3, β_4, and β_5, equal 0; i.e.,

$$H_0: \quad \beta_2 = \beta_3 = \beta_4 = \beta_5 = 0$$

This hypothesis can be tested by fitting the complete model (model 3) and the reduced model (model 1) and conducting an F test, as described in Section 12.6.

Suppose we assume that the response lines for the three media will differ but wonder whether the data present sufficient evidence to indicate differences in the slopes of the lines. To test the null hypothesis that model 2 adequately describes the relationship between $E(y)$ and advertising expenditure x_1, we want to test

$$H_0: \quad \beta_4 = \beta_5 = 0$$

that is, that the two independent variables, advertising expenditure x_1 and the qualitative variable, type of medium, do not interact. This test can be conducted by fitting the complete model (model 3) and the reduced model (model 2), calculating the drop in the sum of squares for error, and conducting an F test.

EXAMPLE 12.6

Substitute the appropriate values of the dummy variables in model 3 to obtain the equations of the three response lines in Figure 12.21.

SOLUTION

The complete model that characterizes the three lines in Figure 12.21 is

$$E(y) = \beta_0 + \beta_1 x_1 + \beta_2 x_2 + \beta_3 x_3 + \beta_4 x_1 x_2 + \beta_5 x_1 x_3$$

where

$x_1 = $ Advertising expenditure

$$x_2 = \begin{cases} 1 & \text{if radio medium} \\ 0 & \text{if not} \end{cases}$$

$$x_3 = \begin{cases} 1 & \text{if television medium} \\ 0 & \text{if not} \end{cases}$$

Examining the coding, you can see that $x_2 = x_3 = 0$ when the advertising medium is newspaper. Substituting these values into the expression for $E(y)$, we obtain the newspaper medium line as:

Newspaper medium line

$$E(y) = \beta_0 + \beta_1 x_1 + \beta_2(0) + \beta_3(0) + \beta_4 x_1(0) + \beta_5 x_1(0)$$
$$= \beta_0 + \beta_1 x_1$$

Similarly, we substitute the appropriate values of x_2 and x_3 into the expression for $E(y)$ to obtain:

Radio medium line

$$E(y) = \beta_0 + \beta_1 x_1 + \beta_2(1) + \beta_3(0) + \beta_4 x_1(1) + \beta_5 x_1(0)$$

$$= \underbrace{(\beta_0 + \beta_2)}_{y\text{-intercept}} + \underbrace{(\beta_1 + \beta_4)}_{\text{Slope}} x_1$$

Television medium line

$$E(y) = \beta_0 + \beta_1 x_1 + \beta_2(0) + \beta_3(1) + \beta_4 x_1(0) + \beta_5 x_1(1)$$

$$= \underbrace{(\beta_0 + \beta_3)}_{y\text{-intercept}} + \underbrace{(\beta_1 + \beta_5)}_{\text{Slope}} x_1$$

If you were to fit model 3, obtain estimates of $\beta_0, \beta_1, \ldots, \beta_5$, and substitute them into the equations for the three media lines shown above, you would obtain exactly the same prediction equations as you would obtain if you fit three separate straight lines, one to each of the three sets of media data. You may ask why we would not fit the three lines separately. Why fit a model (model 3) that combines all three lines into the same equation? The answer is that you need to use this procedure if you want to use statistical tests to compare the three media lines. We need to be able to express a practical question about the lines in terms of a hypothesis that each of a set of parameters in the model equals 0. You could not do this if you perform three separate regression analyses and fit a line to each set of media data.   ■

EXAMPLE 12.7

An industrial psychologist conducted an experiment to investigate the relationship between worker productivity and a measure of salary incentive for two manufacturing plants, one, A, with union representation and the other, B, with nonunion representation. The productivity, y, per worker was measured by recording the number of acceptable machined castings that a worker could produce in a 4-week, 40 hour-per-week period. The incentive was the amount, x_1, of bonus (in cents per casting) paid for all castings produced in excess of 1,000 per worker for the 4-week period. Nine workers were selected from each plant and three from each group of nine were assigned to receive a 20¢ bonus per casting, three a 30¢ bonus, and three a 40¢ bonus per casting. The productivity data for the 18 workers, three for each plant type and incentive combination, are shown in Table 12.6.

TABLE 12.6
Productivity Data for
Example 12.7

TYPE OF PLANT	INCENTIVE		
	20¢/casting	30¢/casting	40¢/casting
Union	1,435 1,512 1,491	1,583 1,529 1,610	1,601 1,574 1,636
Nonunion	1,575 1,512 1,488	1,635 1,589 1,661	1,645 1,616 1,689

a. Plot the data points and graph the prediction equations for the two productivity lines. Assume that the relationship between mean productivity and incentive is first-order.

b. Do the data provide sufficient evidence to indicate a difference in worker response to incentives between the two plants?

SOLUTION

If we assume that a first-order model* is adequate to detect a change in mean productivity, $E(y)$, as a function of incentive, x_1, then the model that produces two productivity lines, one for each plant, is

$$E(y) = \beta_0 + \beta_1 x_1 + \beta_2 x_2 + \beta_3 x_1 x_2$$

where

x_1 = Incentive

$x_2 = \begin{cases} 1 & \text{if nonunion plant} \\ 0 & \text{if union plant} \end{cases}$

a. The SAS computer printout for the regression analysis is shown in Figure 12.22. The prediction equation is obtained by reading the parameter estimates from the printout:

$$\hat{y} = 1,365.833 + 6.217x_1 + 47.778x_2 + .033x_1 x_2$$

The prediction equation for the union plant can be obtained by substituting $x_2 = 0$ into the general prediction equation. Then

*Although the model contains a term involving $x_1 x_2$, it is first-order (graphs as a straight line) in the quantitative variable x_1. The variable x_2 is a dummy variable that introduces or deletes terms in the model. The order of a term is determined only by the quantitative variables that appear in the term.

FIGURE 12.22 SAS Computer Printout for the Complete Model of Example 12.7

```
DEPENDENT VARIABLE: Y
SOURCE                DF    SUM OF SQUARES      MEAN SQUARE    F VALUE
MODEL                  3     57332.38888889   19110.79629630    11.46
ERROR                 14     23349.22222223    1667.80158730    PR > F
CORRECTED TOTAL       17     80681.61111112                     0.0005

R-SQUARE              C.V.          ROOT MSE          Y MEAN
0.710600            2.5901       40.83872656    1576.72222222

                                  T FOR H0:      PR > !T!    STD ERROR OF
PARAMETER           ESTIMATE     PARAMETER=0                    ESTIMATE

INTERCEPT        1365.83333333        26.35       0.0001     51.83641257
X1                  6.21666667         3.73       0.0022      1.66723403
X2                 47.77777778         0.65       0.5251     73.30775769
X1*X2               0.03333333         0.01       0.9889      2.35782498
```

$$\hat{y} = \hat{\beta}_0 + \hat{\beta}_1 x_1 + \hat{\beta}_2(0) + \hat{\beta}_3 x_1(0)$$
$$= \hat{\beta}_0 + \hat{\beta}_1 x_1$$
$$= 1{,}365.833 + 6.217 x_1$$

Similarly, the prediction equation for the nonunion plant is obtained by substituting $x_2 = 1$ into the general prediction equation. Then,

$$\hat{y} = \hat{\beta}_0 + \hat{\beta}_1 x_1 + \hat{\beta}_2 x_2 + \hat{\beta}_3 x_1 x_2$$
$$= \hat{\beta}_0 + \hat{\beta}_1 x_1 + \hat{\beta}_2(1) + \hat{\beta}_3 x_1(1)$$

$$= \overbrace{(\hat{\beta}_0 + \hat{\beta}_2)}^{y\text{-intercept}} + \overbrace{(\hat{\beta}_1 + \hat{\beta}_3)}^{\text{Slope}} x_1$$
$$= (1{,}365.833 + 47.778) + (6.217 + .033) x_1$$
$$= 1{,}413.611 + 6.250 x_1$$

The graphs of these prediction equations are shown in Figure 12.23.

FIGURE 12.23
Graphs of the Prediction
Equations for the Two
Productivity Lines of
Example 12.7

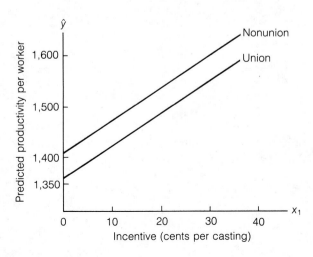

b. To determine whether the data provide sufficient evidence to indicate a difference in worker responses to incentives for the two plants, we test the null hypothesis that a *single* line characterizes the relationship between productivity per worker and the amount of incentive, x_1, against the alternative hypothesis that we need two separate lines to characterize the relationship—one for each plant. If there is no difference in mean response $E(y)$ to x_1 between the two plants, then we do not need type of plant in the model; i.e., we do not need the terms involving x_2. Therefore, we want to test

$$H_0: \quad \beta_2 = \beta_3 = 0$$

against the alternative hypothesis

$$H_a: \quad \text{At least one of the two parameters, } \beta_2 \text{ or } \beta_3, \text{ differs from } 0$$

The SAS computer printout for fitting the reduced model,

$$E(y) = \beta_0 + \beta_1 x_1$$

to the data is shown in Figure 12.24. Reading SSE_2 and SSE_1 from Figures 12.22 and 12.24, respectively, we obtain

Complete model: $SSE_2 = 23{,}349.22$

Reduced model: $SSE_1 = 34{,}056.28$

Drop in SSE: $SSE_1 - SSE_2 = 10{,}707.06$

FIGURE 12.24   SAS Computer Printout for the Reduced Model of Example 12.7

DEPENDENT VARIABLE: Y				
SOURCE	DF	SUM OF SQUARES	MEAN SQUARE	F VALUE
MODEL	1	46625.33333333	46625.33333333	21.91
ERROR	16	34056.27777778	2128.51736111	PR > F
CORRECTED TOTAL	17	80681.61111112		0.0003

R-SQUARE	C.V.	ROOT MSE	Y MEAN
0.577893	2.9261	46.13585765	1576.72222222

PARAMETER	ESTIMATE	T FOR H0: PARAMETER=0	PR > |T|	STD ERROR OF ESTIMATE
INTERCEPT	1389.72222222	33.56	0.0001	41.40819949
X1	6.23333333	4.68	0.0003	1.33182749

The value of s^2 for the complete model is obtained from Figure 12.22:

$$s^2 = 1{,}667.80$$

Substituting these values, along with $k = 3$ and $g = 1$, into the formula for the F statistic yields

$$F = \frac{(SSE_1 - SSE_2)/(k - g)}{s^2}$$

$$= \frac{(10{,}707.06)/2}{1{,}667.80} = 3.21$$

The numerator degrees of freedom (the number of parameters involved in H_0) is 2 and the denominator degrees of freedom (the number associated with s^2 in the complete model) is 14. If we choose $\alpha = .05$, the tabulated value of $F_{.05}$, given in Table 9 of Appendix II is 3.74. Since the computed value, $F = 3.21$, is less than the tabulated value, $F_{.05} = 3.74$, there is insufficient evidence (at $\alpha = .05$) to indicate a difference in worker response to incentives between the two plants. Therefore, there is no evidence to indicate that two different lines, one for each plant, are needed to describe the relationship between the mean productivity per worker, $E(y)$, and the amount of incentive, x_1. ∎

EXAMPLE 12.8

Refer to Example 12.7 and explain how you would determine whether the data provide sufficient evidence to indicate that the incentive x_1 affects mean productivity.

SOLUTION

If incentive did not affect mean productivity, we would not need terms involving x_1 in the model. Therefore, we would test the null hypothesis

$$H_0: \quad \beta_1 = \beta_3 = 0$$

against the alternative hypothesis

$$H_a: \quad \text{At least one of the parameters, } \beta_1 \text{ or } \beta_3, \text{ differs from 0}$$

We would fit the reduced model,

$$E(y) = \beta_0 + \beta_2 x_2$$

to the data and find SSE_1. The values of SSE_2 and s^2 for the complete model would be the same as those used in Example 12.7. Finally, you would calculate the value of the F statistic and compare it with a tabulated value of F based on $\nu_1 = 2$ and $\nu_2 = 14$ degrees of freedom. If the test leads to rejection of H_0, you have evidence to indicate that the increase in mean productivity that appears to be present in the graphs in Figure 12.23 is not due to chance variation. ∎

**EXERCISES
12.35–12.46**

12.35 Write a first-order model that relates $E(y)$ to one quantitative independent variable.

12.36 Add the main effect terms for one qualitative variable at three levels to the model of Exercise 12.35.

12.37 Add terms to the model of Exercise 12.36 to allow for interaction between the quantitative and qualitative independent variables.

12.38 Under what circumstances will the response lines of the model in Exercise 12.37 be parallel?

12.39 Under what circumstances will the model of Exercise 12.37 have only one response line?

12.40 In recent years, many companies have converted to the metric system of measurement. In an attempt to quantify some of the characteristics of companies that have converted to metric production, B. D. Phillips, H. A. G. Lakhani, and S. L. George analyzed data collected on 350 small manufacturers for a recently completed U.S. Metric Board study (*Technological Forecasting and Social Change*, Apr. 1984). One of the research objectives was to investigate the relationship between the percentage y of metric work performed by a company, age x_1 of the company (in years), and cost x_2 of metric conversion, where

$$x_2 = \begin{cases} 1 & \text{if cost over \$10,000} \\ 0 & \text{if not} \end{cases}$$

A first-order linear model was fit to the $n = 350$ data points with the following results:*

$$\hat{y} = 70.9770 - .2167x_1 - 13.2768x_2$$
$$R^2 = .0576 \qquad t \text{ (for } H_0\text{: } \beta_1 = 0) = -1.56 \qquad t \text{ (for } H_0\text{: } \beta_2 = 0) = -2.71$$

a. Sketch the least squares relationship between $\hat{y}$ and x_1 for the two levels of metric conversion cost.
b. Is there evidence that the model is useful for predicting percentage y of metric work performed? Test using $\alpha = .01$.
c. Is there evidence that percentage y of metric work performed decreases as age x_1 of the company increases, for companies with the same cost x_2 of conversion? Test using $\alpha = .05$.
d. Write an interaction model for percentage y of metric work performed.
e. How will interaction between age x_1 and cost x_2, if determined to be significant, affect the graphs constructed in part **a**?

12.41 A company is studying three different safety programs, A, B, and C, in an attempt to reduce the number of work-hours lost due to accidents. Each program is to be tried at three of the company's nine factories, and the plan is to monitor the lost work-hours, y, for a 1-year period beginning 6 months after the new safety program is instituted.
a. Write a main effects model relating $E(y)$ to the lost work-hours, x_1, the year before the plan is instituted and to the type of program that is instituted.
b. In terms of the model parameters from part **a**, what hypothesis would you test to determine whether the mean work-hours lost differ for the three safety programs?

12.42 Refer to Exercise 12.41. After the three safety programs have been in effect for 18 months, the complete main effects model is fit to the $n = 9$ data points. Using safety program A as the base level, the following results are obtained:

$$\hat{y} = -2.1 + .88x_1 - 150x_2 + 35x_3 \qquad SSE = 1,527.27$$

Then the reduced model $E(y) = \beta_0 + \beta_1 x_1$ is fit, with the result

$$\hat{y} = 15.3 + .84x_1 \qquad SSE = 3,113.14$$

Test to determine whether the mean work-hours lost differ for the three programs. Use $\alpha = .05$.

*Reprinted by permission of the publisher. Copyright 1984 by Elsevier Science Publishing Co., Inc.

12.43 A construction company is experimenting with three different cement mixes—dry, damp, and wet—for laying concrete. Since the compressive strength of a concrete slab varies as a function of hardening time and cement mix, the following main effects model is proposed:

$$E(y) = \beta_0 + \beta_1 x_1 + \beta_2 x_2 + \beta_3 x_3$$

where

y = Compressive strength (thousands of pounds per square inch)

x_1 = Hardening time of cement mix (days)

$x_2 = \begin{cases} 1 & \text{if damp cement} \\ 0 & \text{if not} \end{cases}$ $x_3 = \begin{cases} 1 & \text{if wet cement} \\ 0 & \text{if not} \end{cases}$

Dry cement is the base level.

a. What hypothesis would you test to determine whether mean compressive strength differs for the three cement mixes?

b. Using data collected for a sample of 50 batches of concrete, the main effects model is fit, with the result SSE = 140.5. Then the reduced model $E(y) = \beta_0 + \beta_1 x_1$ is fit to the same data, with the result SSE = 183.2. Test the hypothesis you formulated in part **a**. Use $\alpha = .05$

c. Explain how you would test the hypothesis that the slope of the linear relationship between mean compressive strength $E(y)$ and hardening time x_1 varies according to type of cement mix.

12.44 The Florida Citrus Commission is interested in evaluating the performance of two orange juice extractors, brand A and brand B. It is believed that the size of the fruit used in the test may influence the juice yield (amount of juice per pound of oranges) obtained by the extractors. The commission wants to find a regression model relating the mean yield, $E(y)$, to the type of orange juice extractor (brand A or brand B) and the size of orange (diameter), x_1.

a. Identify the independent variables as qualitative or quantitative.

b. Write a model that describes the relationship between $E(y)$ and size of orange as two parallel lines, one for each brand of extractor.

c. Modify the model of part **b** to permit the slopes of the two lines to differ.

d. Sketch typical response lines for the model of part **b**. Do the same for the model of part **c**. Label your graphs carefully.

e. Specify the null and alternative hypotheses you would employ to determine whether the model of part **c** provides more information for predicting yield than does the model of part **b**.

f. Explain how you would obtain the quantities necessary to compute the F statistic that would be employed in testing the hypotheses you described in part **e**.

12.45 Since glass is not prone to radiation damage, encapsulation of waste in glass is considered to be one of the most promising solutions to the problem of low-level nuclear waste in the environment. However, glass undergoes chemical changes when exposed to extreme environmental conditions and certain of its constituents leach into the surroundings. In addition, these chemical reactions may possibly weaken the glass. These concerns led to a study undertaken jointly by the Department of

Materials Science and Engineering at the University of Florida and the U.S. Department of Energy to assess the utility of glass as a waste encapsulant material.* Corrosive chemical solutions (called corrosion baths) were prepared and applied directly to glass samples containing one of three types of waste (TDS-3A, FE, and AL) and the chemical reactions observed over time. A few of the key variables measured were:

y = Amount of silicon (in parts per million) found in solution at end of experiment. (This is both a measure of the degree of breakdown in the glass and a proxy for the amount of radioactive species released into the environment.)

x_1 = Temperature (°C) of the corrosion bath

$$x_2 = \begin{cases} 1 & \text{if waste type TDS-3A} \\ 0 & \text{if not} \end{cases} \qquad x_3 = \begin{cases} 1 & \text{if waste type FE} \\ 0 & \text{if not} \end{cases}$$

Waste type AL is the base level. Suppose we want to model amount y of silicon as a function of temperature (x_1) and type of waste (x_2, x_3).

a. Write a model that proposes parallel straight-line relationships between amount of silicon and temperature, one line for each of the three waste types.

b. Add terms for the interaction between temperature and waste type to the model of part **a**.

c. Refer to the model of part **b**. For each waste type, give the slope of the line relating amount of silicon to temperature.

d. Explain how you could test for the presence of temperature–waste type interaction.

12.46 Due to the dramatic decline in the cost of computer hardware, it is becoming economically feasible to build computers with thousands of processors. However, the scheduling of computer jobs on these advanced computers can be a difficult task. Parallel scheduling algorithms have been designed to solve this problem. A *parallel algorithm* is a set of scheduling instructions designed to minimize the number of tardy jobs in the system and to minimize the mean finish time of the entire job stream. Suppose three different scheduling algorithms (A, B, and C) have been proposed for minimizing the mean finish time of n jobs in a system with a large number of processors.

a. Write a main effects model with interaction to relate the mean finish time, $E(y)$, to the number of jobs (x_1) and scheduling algorithm (A, B, or C).

b. The model of part **a** was fit to data collected on 12 simulated systems (four systems for each of the three algorithms) with the results shown in the accompanying printout. Test whether the model is useful in predicting mean finish time. Use $\alpha = .05$.

Computer Printout for Exercise 12.46

SOURCE	DF	SUM OF SQUARES	MEAN SQUARE	F	PR > F	R-SQUARE
MODEL	5	87.473	17.495	4.90	0.0394	0.803120
ERROR	6	21.443	3.574			
CORRECTED TOTAL	11	108.916				

*The background information for this exercise was provided by Dr. David Clark, Department of Materials Science and Engineering, University of Florida.

c. Write the main effects model (with no interaction) relating mean finish time to number of jobs and scheduling algorithm.

d. The main effects (reduced) model was fit to the data and produced SSE = 38.289. Does this provide sufficient evidence at the $\alpha = .05$ level of significance to indicate that the interaction terms should be kept in the model?

S E C T I O N   12.9

COMPARING TWO OR MORE RESPONSE CURVES

Suppose we think that the relationship between mean monthly sales, $E(y)$, and advertising expenditure, x_1 (Section 12.8), is second-order. The scenario for writing the models for this situation is as follows:

1. The mean sales curves are identical for all three advertising media; that is, a single second-order curve will suffice to describe the relationship between $E(y)$ and x_1 for all the media (see Figure 12.25).

$$E(y) = \beta_0 + \beta_1 x_1 + \beta_2 x_1^2$$

$x_1 = $ Advertising expenditure

FIGURE 12.25
The Relationship Between $E(y)$ and x_1 Is the Same for All Media

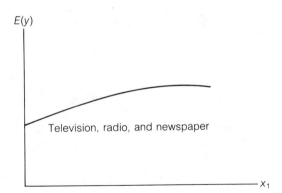

Television, radio, and newspaper

2. The response curves have the same shapes but different y-intercepts (see Figure 12.26).

FIGURE 12.26
The Response Curves Have the Same Shapes but Different y-intercepts

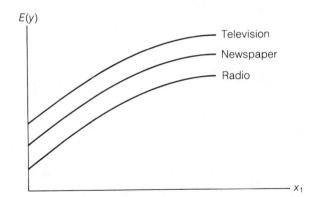

Television
Newspaper
Radio

$$E(y) = \beta_0 + \beta_1 x_1 + \beta_2 x_1^2 + \beta_3 x_2 + \beta_4 x_3$$

$$x_1 = \text{Advertising expenditure}$$

$$x_2 = \begin{cases} 1 & \text{if radio medium} \\ 0 & \text{if not} \end{cases}$$

$$x_3 = \begin{cases} 1 & \text{if television medium} \\ 0 & \text{if not} \end{cases}$$

3. The response curves for the three advertising media are different (i.e., advertising expenditure and type of medium interact), as shown in Figure 12.27.

$$E(y) = \beta_0 + \beta_1 x_1 + \beta_2 x_1^2 + \beta_3 x_2 + \beta_4 x_3$$
$$+ \beta_5 x_1 x_2 + \beta_6 x_1 x_3 + \beta_7 x_1^2 x_2 + \beta_8 x_1^2 x_3$$

FIGURE 12.27
The Response Curves for
the Three Media Differ

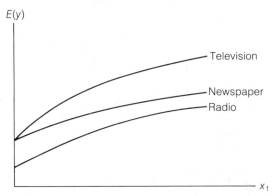

EXAMPLE 12.9

Give the equation of the complete second-order model for the radio advertising medium.

SOLUTION

Model 3 characterizes the relationship between $E(y)$ and x_1 for the radio advertising medium (see the coding) when $x_2 = 1$ and $x_3 = 0$. Substituting these values into model 3, we obtain

$$\begin{aligned} E(y) &= \beta_0 + \beta_1 x_1 + \beta_2 x_1^2 + \beta_3 x_2 + \beta_4 x_3 + \beta_5 x_1 x_2 + \beta_6 x_1 x_3 + \beta_7 x_1^2 x_2 + \beta_8 x_1^2 x_3 \\ &= \beta_0 + \beta_1 x_1 + \beta_2 x_1^2 + \beta_3(1) + \beta_4(0) + \beta_5 x_1(1) + \beta_6 x_1(0) + \beta_7 x_1^2(1) + \beta_8 x_1^2(0) \\ &= \underbrace{(\beta_0 + \beta_3)}_{y\text{-intercept}} + \underbrace{(\beta_1 + \beta_5)}_{\text{shift}} x_1 + \underbrace{(\beta_2 + \beta_7)}_{\text{rate of curvature}} x_1^2 \end{aligned}$$ ∎

EXAMPLE 12.10

What null hypothesis about the parameters of model 3 would you test if you want to determine whether the second-order curves for the three media are identical?

SOLUTION

If the curves were identical, we would not need the independent variable Type of medium in the model; that is, we would delete all terms involving x_2 and x_3. This would produce model 1,

$$E(y) = \beta_0 + \beta_1 x_1 + \beta_2 x_1^2$$

and the null hypothesis would be

$$H_0: \quad \beta_3 = \beta_4 = \beta_5 = \beta_6 = \beta_7 = \beta_8 = 0$$ ∎

EXAMPLE 12.11

Suppose we assume that the response curves for the three media differ but we want to know whether the second-order terms contribute information for the prediction of y. Or, equivalently, will a second-order model give better predictions than a first-order model?

SOLUTION

The only difference between model 3 and a first-order model are those terms involving x_1^2. Therefore, the null hypothesis, "the second-order terms contribute no information for the prediction of y," is equivalent to

$$H_0: \quad \beta_2 = \beta_7 = \beta_8 = 0$$ ∎

Examples 12.10 and 12.11 identify two tests that answer practical questions concerning a collection of second-order models. Other comparisons among the curves can be made by testing appropriate sets of model parameters (see the exercises).

The models described in the preceding sections provide only an introduction to statistical modeling. Models can be constructed to relate $E(y)$ to any number of quantitative and/or qualitative independent variables. You can compare response curves and surfaces for different levels of a qualitative variable or for different combinations of levels of two or more qualitative independent variables.

**EXERCISES
12.47–12.57**

12.47 Write a complete second-order model that relates $E(y)$ to one quantitative independent variable.

12.48 Add the main effect terms for one qualitative variable at three levels to the model of Exercise 12.47.

12.49 Add terms to the model of Exercise 12.48 to allow for interaction between the quantitative and qualitative independent variables.

12.50 Under what circumstances will the response curves of the model in Exercise 12.49 possess the same shape but have different y-intercepts?

12.51 Under what circumstances will the response curves of the model in Exercise 12.49 be parallel lines?

12.52 Under what circumstances will the response curves of the model in Exercise 12.49 be identical?

12.53 An equal-rights group has charged that women are being discriminated against in terms of salary structure in a state university system. It is thought that a complete second-order model will be adequate to describe the relationship between salary and years of experience for both men and women. A sample is to be taken from the records for faculty members (all of equal rank) within the system and the following model is to be fit.*

*In practice, we would include other variables in the model. We include only two here to simplify the exercise.

$$E(y) = \beta_0 + \beta_1 x_1 + \beta_2 x_1^2 + \beta_3 x_2 + \beta_4 x_1 x_2 + \beta_5 x_1^2 x_2$$

where

y = Annual salary (in thousands of dollars)

x_1 = Experience (years)

$x_2 = \begin{cases} 1 & \text{if female} \\ 0 & \text{if male} \end{cases}$

a. What hypothesis would you test to determine whether the *rate* of increase of mean salary with experience is different for males and females?

b. What hypothesis would you test to determine whether there are differences in mean salaries that are attributable to sex?

12.54 Refer to Exercise 12.53 and the model that was proposed to describe the relationship between salary and years of experience for both men and women. The following is a portion of the computer printout that results from fitting this model to a sample of 200 faculty members in the university system:

SOURCE	DF	SUM OF SQUARES	MEAN SQUARE
MODEL	5	2351.70	470.34
ERROR	194	783.90	4.04
TOTAL	199	3135.60	R-SQUARE
			0.750

The reduced model $E(y) = \beta_0 + \beta_1 x_1 + \beta_2 x_1^2$ is fit to the same data and the resulting computer printout is partially reproduced below:

SOURCE	DF	SUM OF SQUARES	MEAN SQUARE
MODEL	2	2340.37	1170.185
ERROR	197	795.23	4.04
TOTAL	199	3135.60	R-SQUARE
			0.746

Is there sufficient evidence to support the claim that the mean salary of faculty members is dependent on sex? Use $\alpha = .05$.

12.55 Refer to Exercise 12.43, in which a model relating mean compressive strength of concrete to hardening time and type of cement mix was proposed.

a. Write a second-order model that allows different response curves for the three types of cement mixes.

b. Explain how you would test the hypothesis that the three response curves have the same shape, but different y-intercepts.

12.56 Refer to Exercise 12.46 in which we considered models relating the mean finish time of a set of computer jobs to the number of jobs and scheduling algorithm.

a. Write the complete second-order model. Sketch the response curves for this model.

b. What hypothesis would you test to determine whether the response curves differ for the three scheduling algorithms?

c. The complete second-order model of part **a** was fit to the 12 data points, with the result SSE = 16.225. The reduced model

$$E(y) = \beta_0 + \beta_1 x_1 + \beta_2 x_1^2$$

was also fit, with the result SSE = 27.109. Conduct the test of part **b**. Use $\alpha = .05$.

12.57 An operations manager is interested in modeling $E(y)$, the expected length of time per month (in hours) that a machine will be shut down for repairs as a function of the type of machine (001 or 002) and the age of the machine (in years). The manager has proposed the following model:

$$E(y) = \beta_0 + \beta_1 x_1 + \beta_2 x_1^2 + \beta_3 x_2$$

where

x_1 = Age of machine

$x_2 = \begin{cases} 1 & \text{if machine type 001} \\ 0 & \text{if machine type 002} \end{cases}$

Data were obtained on $n = 20$ machine breakdowns and were used to estimate the parameters of the above model. A portion of the regression analysis computer printout is shown below:

SOURCE	DF	SUM OF SQUARES	MEAN SQUARE
MODEL	3	2396.364	798.788
ERROR	16	128.586	8.037
TOTAL	19	2524.950	R-SQUARE
			0.949

The reduced model $E(y) = \beta_0 + \beta_1 x_1 + \beta_2 x_2$ was fit to the same data. The regression analysis computer printout is partially reproduced below:

SOURCE	DF	SUM OF SQUARES	MEAN SQUARE
MODEL	2	2342.42	1171.21
ERROR	17	182.53	10.74
TOTAL	19	2524.95	R-SQUARE
			0.928

Is there sufficient evidence to conclude that the second-order (x_1^2) term in the model proposed by the operations manager is necessary? Test using $\alpha = .05$.

MODEL BUILDING: STEPWISE REGRESSION

In building a model to describe a response variable y, we must choose the important terms to be included in the model. The list of potentially important independent variables, with their associated main effect and interaction terms, may be extremely large. Therefore, we need some objective method of screening out those that are not important. The screening procedure that we present in this chapter is known as a **stepwise regression analysis**.

The most commonly used stepwise regression procedure, available in most popular computer packages, works as follows: The user first identifies the response, y, and the set of potentially important independent variables, x_1, x_2, . . . , x_k, where k will generally be large. (Note that this set of variables could represent both first- and higher-order terms, as well as any interaction terms that might be important information contributors.) The response and independent variables are then entered into the computer, and the stepwise procedure begins.

STEP 1   The computer fits all possible one-variable models of the form

$$E(y) = \beta_0 + \beta_1 x_i$$

to the data. For each model, the test of the null hypothesis

$$H_0: \quad \beta_1 = 0$$

against the alternative hypothesis

$$H_a: \quad \beta_1 \neq 0$$

is conducted using the t (or the equivalent F) test for a single β parameter. The independent variable that produces the largest (absolute) t value is declared the best one-variable predictor of y. Call this independent variable x_1.

STEP 2   The stepwise program now begins to search through the remaining $(k - 1)$ independent variables for the best two-variable model of the form

$$E(y) = \beta_0 + \beta_1 x_1 + \beta_2 x_i$$

This is done by fitting all two-variable models containing x_1 and each of the other $(k - 1)$ options for the second variable x_i. The t values for the test H_0: $\beta_2 = 0$ are computed for each of the $(k - 1)$ models (corresponding to the remaining independent variables x_i, $i = 2, 3, . . . , k$), and the variable having the largest t is retained. Call this variable x_2.

At this point, some computer packages diverge in methodology. The better packages now go back and check the t value of $\hat{\beta}_1$ after $\hat{\beta}_2 x_2$ has been added to the model. If the t value has become nonsignificant at some specified α level (say $\alpha = .10$), the variable x_1 is removed and a search is made for the independent variable with a β parameter that will yield the most significant t value in the presence of $\hat{\beta}_2 x_2$. Other packages do not recheck $\hat{\beta}_1$, but proceed directly to step 3.

The best-fitting plane may yield a different value for $\hat{\beta}_1$ than that obtained in step 1, because $\hat{\beta}_1$ and $\hat{\beta}_2$ may be correlated. Thus, both the value of $\hat{\beta}_1$ and its significance usually change from step 1 to step 2. For this reason, the computer packages that recheck the t values at each step are preferred.

STEP 3   The stepwise procedure now checks for a third independent variable to include in the model with x_1 and x_2. That is, we seek the best model of the form

$$E(y) = \beta_0 + \beta_1 x_1 + \beta_2 x_2 + \beta_3 x_i$$

To do this, we fit all the $(k - 2)$ models using x_1, x_2, and each of the $(k - 2)$ remaining variables, x_i, as a possible x_3. The criterion is again to include the independent variable with the largest t value. Call this best third variable x_3.

The better programs now recheck the t values corresponding to the x_1 and x_2 coefficients, replacing the variables that have t values that have become nonsignificant. This procedure is continued until no further independent variables can be found that yield significant t values (at the specified α level) in the presence of the variables already in the model.

The result of the stepwise procedure is a model containing only those terms with t values that are significant at the specified α level. Thus, in most practical situations, only several of the large number of independent variables will remain. However, it is very important *not* to jump to the conclusion that all the independent variables important for predicting y have been identified or that the unimportant independent variables have been eliminated. Remember, the stepwise procedure is using only *sample estimates* of the true model coefficients (β's) to select the important variables. An extremely large number of single β parameter t tests have been conducted, and the probability is very high that one or more errors have been made in including or excluding variables. That is, we have very probably included some unimportant independent variables in the model (Type I errors) and eliminated some important ones (Type II errors).

There is a second reason why we might not have arrived at a good model. When we choose the variables to be included in the stepwise regression, we may often omit high-order terms (to keep the number of variables manageable). Consequently, we may have initially omitted several important terms from the model. Thus, we should recognize stepwise regression for what it is: an objective screening procedure.

Now, we will consider interactions and quadratic terms (for quantitative variables) among variables screened by the stepwise procedure. It would be best to develop this response surface model with a second set of data independent of that used for the screening, so the results of the stepwise procedure can be partially verified with new data. However, this is not always possible, because in many practical modeling situations only a small amount of data is available.

Remember, do not be deceived by the impressive looking t values that result from the stepwise procedure—it has retained only the independent variables with the largest t values. Also, if you have used a main effects model for your stepwise procedure, remember that it may be greatly improved by the addition of interaction and quadratic terms.

EXAMPLE 12.12

Suppose we want to use multiple regression to model executive salary as a function of experience, education, sex, and other factors. A preliminary step in the construction of the model is to determine the most important independent variables. Ten independent variables to be considered are listed in Table 12.7. Since it would be very difficult to perform a regression analysis on a complete second-order model using ten independent variables, we need to eliminate those variables (or terms) that do not contribute much information for the prediction of salary. We will use the salary data for a sample of 100 executives to decide which of the ten variables should be included in the construction of the final model for executive salaries. The computer printouts that we show in the solution are based on this data set.

TABLE 12.7

Independent Variables in Example 12.12

INDEPENDENT VARIABLE	DESCRIPTION
x_1	Experience (years)—quantitative
x_2	Education (years)—quantitative
x_3	Sex (1 if male, 0 if female)—qualitative
x_4	Number of employees supervised—quantitative
x_5	Corporate assets (millions of dollars)—quantitative
x_6	Board member (1 if yes, 0 if no)—qualitative
x_7	Age (years)—quantitative
x_8	Company profits (past 12 months, millions of dollars)—quantitative
x_9	Has international responsibility (1 if yes, 0 if no)—qualitative
x_{10}	Company's total sales (past 12 months, millions of dollars)—quantitative

SOLUTION

We will use stepwise regression with the main effects of the ten independent variables to identify the most important variables. The dependent variable y is the natural logarithm of the executive salaries. The SAS stepwise regression printout is shown in Figure 12.28. SAS uses an F statistic, rather than a t statistic, in the stepwise procedure. It can be shown (proof omitted) that the square of a Student's t statistic with ν df is equal to an F statistic with 1 df in the numerator and ν df in the denominator. Thus, $t_{\alpha/2}^2 = F_\alpha$, where t is based on ν df and F has 1 numerator df and ν denominator df.

Note that the first variable included in the model is x_4, Number of employees supervised. At the second step, x_5, Corporate assets, enters the model. At the sixth step, x_6, a dummy variable for the qualitative variable Board member or not, is brought into the model. However, because the observed significance level (.2295) of the F statistic for x_6 is greater than the preassigned $\alpha = .10$, x_6 is removed from the model. Thus, at step 7 the procedure indicates that the five-variable model including x_1, x_2, x_3, x_4, and x_5 is best. That is, none of the other independent variables can meet the $\alpha = .10$ criterion for admission to the model. Therefore, in our final modeling effort, we will develop a model using these independent variables. ∎

FIGURE 12.28 SAS Stepwise Regression Computer Printout for Example 12.12

STEP 1
VARIABLE X4 ENTERED   R-SQUARE = 0.42071677

	DF	SUM OF SQUARES	MEAN SQUARE	F	PROB>F
REGRESSION	1	11.46864285	11.46864285	71.17	0.0001
ERROR	98	15.79112802	0.16113396		
TOTAL	99	27.25977087			

	B VALUE	STD ERROR	F	PROB>F
INTERCEPT	10.20077500			
X4 (EMPLOYEES SUPERVISED)	0.00057284	0.00006790	71.17	0.0001

STEP 2
VARIABLE X5 ENTERED   R-SQUARE = 0.78299675

	DF	SUM OF SQUARES	MEAN SQUARE	F	PROB>F
REGRESSION	2	21.34431198	10.67215599	175.00	0.0001
ERROR	97	5.91545889	0.06098411		
TOTAL	99	27.25977087			

	B VALUE	STD ERROR	F	PROB>F
INTERCEPT	9.87702903			
X4 (EMPLOYEES SUPERVISED)	0.00058353	0.00004178	195.06	0.0001
X5 (ASSETS)	0.00183730	0.00014438	161.94	0.0001

STEP 3
VARIABLE X1 ENTERED   R-SQUARE = 0.89667614

	DF	SUM OF SQUARES	MEAN SQUARE	F	PROB>F
REGRESSION	3	24.44318616	8.14772872	277.71	0.0001
ERROR	96	2.81658471	0.02933942		
TOTAL	99	27.25977087			

	B VALUE	STD ERROR	F	PROB>F
INTERCEPT	9.66449288			
X1 (EXPERIENCE)	0.01870784	0.00182032	105.62	0.0001
X4 (EMPLOYEES SUPERVISED)	0.00055251	0.00002914	359.59	0.0001
X5 (ASSETS)	0.00191195	0.00010041	362.60	0.0001

(continued)

FIGURE 12.28 (continued)

STEP 4
VARIABLE X3 ENTERED   R-SQUARE = 0.94815717

	DF	SUM OF SQUARES	MEAN SQUARE	F	PROB>F
REGRESSION	4	25.84654710	6.46163678	434.37	0.0001
ERROR	95	1.41322377	0.01487604		
TOTAL	99	27.25977087			

	B VALUE	STD ERROR	F	PROB>F
INTERCEPT	9.400077349			
X1 (EXPERIENCE)	0.02074868	0.00131310	249.68	0.0001
X3 (SEX)	0.30011726	0.03089939	94.34	0.0001
X4 (EMPLOYEES SUPERVISED)	0.00055288	0.00002075	710.15	0.0001
X5 (ASSETS)	0.00190876	0.00007150	712.74	0.0001

STEP 5
VARIABLE X2 ENTERED   R-SQUARE = 0.96039323

	DF	SUM OF SQUARES	MEAN SQUARE	F	PROB>F
REGRESSION	5	26.18009940	5.23601988	455.87	0.0001
ERROR	94	1.07967147	0.01148587		
TOTAL	99	27.25977087			

	B VALUE	STD ERROR	F	PROB>F
INTERCEPT	8.85387930			
X1 (EXPERIENCE)	0.02141724	0.00116047	340.61	0.0001
X2 (EDUCATION)	0.03315807	0.00615303	29.04	0.0001
X3 (SEX)	0.31927842	0.02738298	135.95	0.0001
X4 (EMPLOYEES SUPERVISED)	0.00056061	0.00001829	939.84	0.0001
X5 (ASSETS)	0.00193684	0.00006304	943.98	0.0001

(continued)

FIGURE 12.28 (continued)

STEP 6
VARIABLE X6 ENTERED   R-SQUARE = 0.96100666

	DF	SUM OF SQUARES	MEAN SQUARE	F	PROB>F
REGRESSION	6	26.19682148	4.36613691	382.00	0.0001
ERROR	93	1.06294939	0.01142956		
TOTAL	99	27.25977087			

	B VALUE	STD ERROR	F	PROB>F
INTERCEPT	8.87509152			
X1 (EXPERIENCE)	0.02133460	0.00115963	338.48	0.0001
X2 (EDUCATION)	0.03272195	0.00614851	28.32	0.0001
X3 (SEX)	0.31093801	0.02817264	121.81	0.0001
X4 (EMPLOYEES SUPERVISED)	0.00055820	0.00001835	925.32	0.0001
X5 (ASSETS)	0.00193764	0.00006289	949.31	0.0001
X6 (BOARD)	0.03866226	0.03196369	1.46	0.2295

STEP 7
VARIABLE X6 REMOVED   R-SQUARE = 0.96039323

	DF	SUM OF SQUARES	MEAN SQUARE	F	PROB>F
REGRESSION	5	26.18009940	5.23601988	455.87	0.0001
ERROR	94	1.07967147	0.01148587		
TOTAL	99	27.25977087			

	B VALUE	STD ERROR	F	PROB>F
INTERCEPT	8.85387930			
X1 (EXPERIENCE)	0.02141724	0.00116047	340.61	0.0001
X2 (EDUCATION)	0.03315807	0.00615303	29.04	0.0001
X3 (SEX)	0.31927842	0.02738298	135.95	0.0001
X4 (EMPLOYEES SUPERVISED)	0.00056061	0.00001829	939.84	0.0001
X5 (ASSETS)	0.00193684	0.00006304	943.98	0.0001

12.58 There are six independent variables, x_1, x_2, x_3, x_4, x_5, and x_6, that might be useful in predicting a response y. A total of $n = 50$ observations are available, and it is decided to employ stepwise regression to help in selecting the independent variables that appear to be useful. The computer fits all possible one-variable models of the form

$$E(y) = \beta_0 + \beta_1 x_i$$

where x_i is the ith independent variable, $i = 1, 2, \ldots, 6$. The information in the accompanying table is provided from the computer printout.

INDEPENDENT VARIABLE	$\hat{\beta}_1$	$s_{\hat{\beta}_1}$
x_1	1.6	.42
x_2	−.9	.01
x_3	3.4	1.14
x_4	2.5	2.06
x_5	−4.4	.73
x_6	.3	.35

a. Which independent variable is declared the best one-variable predictor of y? Explain.
b. Would this variable be included in the model at this stage? Explain.
c. Describe the next phase that a stepwise procedure would execute.

12.59 Many power plants dump hot waste water into surrounding rivers, streams, and oceans, an action that may have an adverse effect on the marine life in the dumping areas. A marine biologist was hired by the EPA to determine whether the hot water runoff from a particular power plant located near a large gulf is having an adverse effect on the marine life in the area. In the initial phase of the study, the biologist's goal is to acquire a prediction equation for the number of marine animals located at certain predesignated areas, or stations, in the gulf. Based on past experience, the biologist considered the following environmental factors as predictors for the number of animals at a particular station:

x_1 = Temperature of water (TEMP)
x_2 = Salinity of water (SAL)
x_3 = Dissolved oxygen content of water (DO)
x_4 = Turbidity index, a measure of the turbidity of the water (TI)
x_5 = Depth of the water at the station (ST_DEPTH)
x_6 = Total weight of sea grasses in sampled area (TGRSWT)

As a preliminary step in the construction of this model, the biologist used a stepwise regression procedure to identify the most important of these six variables. A total of 716 samples were taken at different stations in the gulf, producing the accompanying SAS printout. (The response measured was y, the log of the number of marine animals found in the sampled area.)

SAS Printout for Exercise 12.59

STEP 1
VARIABLE ST_DEPTH ENTERED   R-SQUARE = 0.12227337

	DF	SUM OF SQUARES	MEAN SQUARE	F	PROB>F
REGRESSION	1	57.44041114	57.44041114	99.47	0.0001
ERROR	714	412.32998120	0.57749297		
TOTAL	715	469.77039233			

	B VALUE	STD ERROR	F	PROB>F
INTERCEPT	8.38559344			
ST_DEPTH	-0.43678519	0.04379580	99.47	0.0001

STEP 2
VARIABLE TGRSWT ENTERED   R-SQUARE = 0.18211026

	DF	SUM OF SQUARES	MEAN SQUARE	F	PROB>F
REGRESSION	2	85.55000871	42.77500435	79.38	0.0001
ERROR	713	384.22038363	0.53887852		
TOTAL	715	469.77039233			

	B VALUE	STD ERROR	F	PROB>F
INTERCEPT	8.07681529			
ST_DEPTH	-0.35355301	0.04384775	65.02	0.0001
TGRSWT	0.00271332	0.00037568	52.16	0.0001

STEP 3
VARIABLE TI ENTERED   R-SQUARE = 0.18700047

	DF	SUM OF SQUARES	MEAN SQUARE	F	PROB>F
REGRESSION	3	87.84728446	29.28242815	54.59	0.0001
ERROR	712	381.92310787	0.53640886		
TOTAL	715	469.77039233			

	B VALUE	STD ERROR	F	PROB>F
INTERCEPT	7.38863937			
TI	0.65773503	0.31782817	4.28	0.0389
ST_DEPTH	-0.31451227	0.04764144	43.58	0.0001
TGRSWT	0.00261166	0.00037802	47.73	0.0001

(continued)

SAS Printout for Exercise 12.59 (continued)

STEP 4
VARIABLE DO ENTERED   R-SQUARE = 0.18892367

	DF	SUM OF SQUARES	MEAN SQUARE	F	PROB>F
REGRESSION	4	88.75074870	22.18768717	41.40	0.0001
ERROR	711	381.01964364	0.53589261		
TOTAL	715	469.77039233			

	B VALUE	STD ERROR	F	PROB>F
INTERCEPT	7.22576380			
DO	0.01769145	0.01362532	1.69	0.1946
TI	0.67347023	0.31790625	4.49	0.0345
ST_DEPTH	-0.30417372	0.04827962	39.69	0.0001
TGRSWT	0.00266958	0.00038047	49.23	0.0001

STEP 5
VARIABLE DO REMOVED   R-SQUARE = 0.18700047

	DF	SUM OF SQUARES	MEAN SQUARE	F	PROB>F
REGRESSION	3	87.84728446	29.28242815	54.59	0.0001
ERROR	712	381.92310787	0.53640886		
TOTAL	715	469.77039233			

	B VALUE	STD ERROR	F	PROB>F
INTERCEPT	7.38863937			
TI	0.65773503	0.31782817	4.28	0.0389
ST_DEPTH	-0.31451227	0.04764144	43.58	0.0001
TGRSWT	0.00261166	0.00037802	47.73	0.0001

a. According to the SAS printout, which of the six independent variables should be used in the model? (Use $\alpha = .10$.)

b. Are we able to assume that the marine biologist has identified all the important independent variables for the prediction of y? Why?

c. Using the variables identified in part **a**, write the first-order model with interaction that may be used to predict y.

d. How would the marine biologist determine whether the model specified in part **c** was better than the first-order model?

e. Note the small value of R^2. What action might the biologist take to improve the model?

S E C T I O N   12.11

STEPWISE REGRESSION ON THE COMPUTER (OPTIONAL)

Stepwise regression routines are available in each of the four statistical software packages discussed in this text. Although their methodologies and outputs may differ slightly, each provides a statistical screening procedure for eliminating unimportant variables from a large set of potential predictor variables. In this section we present the commands necessary to call the stepwise regression routines of the SAS, Minitab, SPSS^X, and BMDP packages.

EXAMPLE 12.13

Suppose we want to write the SAS program that produced the stepwise regression printout shown in Figure 12.28. Unlike the multiple regression (GLM) procedure, the SAS stepwise regression routine will not accept any variables that are non-numerical. Thus, the programmer has a choice of either entering numerical values for qualitative variables on the input data lines (for example, 1 for male and 0 for female) or reading the qualitative variable as a character variable (for example, M for male and F for female) and then creating the corresponding dummy variables within a DATA step. Although the first alternative is probably easier to implement, the second method is more convenient because it allows us to use the original set of input data values (with character values for qualitative variables) to fit a multiple regression model with a CLASSES statement (see Section 11.14). Since running a stepwise regression is only a first step in the model building process, we illustrate the second method in this example.

SOLUTION

The SAS program that generated the printout of Figure 12.28 is shown in Program 12.1.

PROGRAM 12.1
SAS Program That Generated the Printout of Figure 12.28

Command line

```
 1     DATA EXECS;
 2     INPUT Y X1 X2 SEX $ X4 X5 BOARD $ X7 X8 INT $ X10;
 3     IF SEX = 'M' THEN X3 = 1; ELSE X3 = 0;
 4     IF BOARD = 'Y' THEN X6 = 1; ELSE X6 = 0;
 5     IF INT = 'Y' THEN X9 = 1; ELSE X9 = 0;
 6     CARDS;
 7     50000 10 6 M 19 12.7 N 46 2.2 Y 5.1   ⎫ Input data
 8     41000 8 4 F 5 3.0 Y 37 .7 N 1.2        ⎬ values: One
 •                                            ⎪ observation
 •                                            ⎨ (executive)
 •                                            ⎪ per line
106    68000 15 4 M 121 40.1 Y 55 9.6 Y 6.3  ⎭
107    PROC STEPWISE;
108    MODEL Y = X1-X10/STEPWISE;
```

Command lines 1–6 represent the data entry instructions. Note that our data set, named EXECS, includes observations on a single dependent variable (Y) and ten independent variables. Since we want to read the qualitative independent variables SEX, BOARD, and INT as character (or nonnumeric) variables, a dollar sign follows the name of each on the INPUT card (line 2). This allows us to enter letters—M or F for SEX, N (no) or Y (yes) for both BOARD and INT— in the appropriate columns of the input data lines (lines 7–106).

The dummy variables for SEX, BOARD, and INT are created using IF-THEN-ELSE statements in lines 3–5. (Remember, the commands used to create new variables on a data set must follow the INPUT statement, but precede the CARDS statement.) Since each qualitative variable is at two levels, we need only a single dummy variable for each. Using a 0–1 system of coding, we have

$$X3 = \begin{cases} 1 & \text{if SEX = 'M'} \\ 0 & \text{if SEX = 'F'} \end{cases} \quad X6 = \begin{cases} 1 & \text{if BOARD = 'Y'} \\ 0 & \text{if BOARD = 'N'} \end{cases} \quad X9 = \begin{cases} 1 & \text{if INT = 'Y'} \\ 0 & \text{if INT = 'N'} \end{cases}$$

The last two lines of the program, lines 107–108, represent the statistical analysis instructions. The key word STEPWISE in the PROC statement (line 107) calls the SAS stepwise regression routine. The MODEL statement (line 108) identifies the dependent and independent variables to be used in the stepwise regression. Note that the names of the dummy variables for the qualitative variables, rather than the names of the original character variables, are entered here. The key word STEPWISE must also appear in the MODEL statement following the slash (/).* ∎

The stepwise regression commands for Minitab, SPSS^X, and BMDP are given in Programs 12.2–12.4. These packages produce output very similar to the SAS printout shown in Figure 12.28. However, these packages are not as flexible as SAS in terms of data input; in particular, the programmer must code numerical values for qualitative variables on the data cards.

PROGRAM 12.2

SPSS^X Stepwise Regression Program for the Executive Salary Example

Command line	
1	DATA LIST FREE/Y, X1 TO X10
2	REGRESSION VARIABLES = Y, X1 TO X10/
3	DEPENDENT = Y/STEPWISE/
4	BEGIN DATA
5	50000 10 6 1 19 12.7 0 46 2.2 1 5.1
6	41000 8 4 0 5 3.0 1 37 .7 0 1.2
⋮	
104	68000 15 4 1 121 40.1 1 55 9.6 1 6.3
105	END DATA

Input data values: One observation (executive) per line (lines 5, 6 ... 104)

*This is a user option. STEPWISE is just one method of variable selection available in SAS. Other methods are FORWARD and BACKWARD. Consult the *SAS User's Guide* (1985) for more information on these methodologies and the differences in their outputs.

PROGRAM 12.3

Minitab Stepwise
Regression Program for
the Executive Salary
Example

Command
line

```
    1    READ Y IN C1, INDEPENDENTS IN C2-C11
    2    50000 10 6 1 19 12.7 0 46 2.2 1 5.1     ⎫ Input data
    3    41000 8 4 0 5 3.0 1 37 .7 0 1.2         ⎬ values: One
    .                                            ⎪ observation
    .                                            ⎨ (executive)
    .                                            ⎭ per line
  101    68000 15 4 1 121 40.1 1 55 9.6 1 6.3
  102    STEPWISE REGRESS C1 ON PREDICTORS C2-C11
```

PROGRAM 12.4

BMDP Stepwise
Regression Program for
the Executive Salary
Example

Command
line

```
    1    BMDP 2R
    2    / PROBLEM TITLE IS 'STEPWISE REGRESSION'.
    3    / INPUT VARIABLES = 11.
    4            FORMAT IS FREE.
    5    / VARIABLE NAMES = Y,X1,X2,X3,X4,X5,X6,X7,X8,X9,X10.
    6    / REGRESS DEPENDENT = Y.
    7    / END
    8    50000 10 6 1 19 12.7 0 46 2.2 1 5.1     ⎫
    9    41000 8 4 0 5 3.0 1 37 .7 0 1.2         ⎬ Input values:
    .                                            ⎪ One observation
    .                                            ⎨ (executive) per line
    .                                            ⎪
  107    68000 15 4 1 121 40.1 1 55 9.6 1 6.3    ⎭
  108    /END
```

| | | | | | | | | | | |

S E C T I O N 12.12

SUMMARY

Although this chapter provides only an introduction to the very important topic of **model building**, it will enable you to construct many interesting and useful models for engineering and information science phenomena. You can build on this foundation and, with experience, develop competence in this fascinating area of statistics. Successful model building requires a delicate blend of knowledge of the process being modeled, geometry, and formal statistical testing.

The first step in model building is to **identify the response variable y and the set of independent variables**. Each independent variable is then classified as either **quantitative** or **qualitative**, and **dummy variables** are defined to represent the qualitative independent variables. If the total number of independent variables is large, you may want to use **stepwise regression** to screen out those that do not seem important for the prediction of y.

When the number of independent variables is manageable, the model builder is ready to begin a systematic effort. At least **second-order models**, those containing **two-way interactions and quadratic terms** in the quantitative variables, should be considered. Remember that a model with no interaction terms implies that each of the independent variables affects the response independently of the other independent variables. Quadratic terms add curvature to the contour curves when $E(y)$ is plotted as a function of the independent variable. The F test for testing a set of β parameters aids in deciding the final form of the prediction model.

Many problems can arise in regression modeling, and the intermediate steps are often tedious and frustrating. However, the end result of a careful and determined modeling effort is very rewarding—you will have a better understanding of the process generating the dependent variable y and a predictive model for y.

SUPPLEMENTARY EXERCISES
12.60–12.73

12.60 Suppose you fit the regression model

$$E(y) = \beta_0 + \beta_1 x_1 + \beta_2 x_2 + \beta_3 x_2^2 + \beta_4 x_1 x_2 + \beta_5 x_1 x_2^2$$

to $n = 35$ data points and want to test the null hypothesis

$$H_0: \quad \beta_4 = \beta_5 = 0$$

a. State the alternative hypothesis.
b. Explain in detail how to compute the F statistic needed to test the null hypothesis.
c. What are the numerator and denominator degrees of freedom associated with the F statistic in part **b**?
d. Give the rejection region for the test if $\alpha = .05$.

12.61 To model the relationship between y, a dependent variable, and x, an independent variable, a researcher has taken one measurement on y at each of three different x values. Drawing on his mathematical expertise, the researcher realizes that he can fit the second-order polynomial model

$$E(y) = \beta_0 + \beta_1 x + \beta_2 x^2$$

and it will pass exactly through all three points, yielding SSE = 0. The researcher, delighted with the "excellent" fit of the model, eagerly sets out to use it to make inferences. What problems will he encounter in attempting to make inferences?

12.62 Due to the increase in gasoline prices, many service stations are offering self-service gasoline at reduced prices. Suppose an oil company wants to model the mean monthly gasoline sales, $E(y)$, of its affiliated stations as a function of the type of service they offer: self-service, full-service, or both.
a. How many dummy variables will be needed to describe the qualitative independent variable Type of service?
b. Write the main effects model relating $E(y)$ to the type of service. Describe the coding of the dummy variables.

12.63 Many companies must accurately estimate their costs before a job is begun in order to acquire a contract and make a profit. For example, a heating and plumbing contractor may base cost estimates for new homes on the total area of the house and whether central air conditioning is to be installed.
a. Write a main effects model relating the mean cost of material and labor, $E(y)$, to the area and central air conditioning variables.
b. Write a complete second-order model for the mean cost as a function of the same two variables.
c. What hypothesis would you test to determine whether the second-order terms are useful for predicting mean cost?

d. Explain how you would compute the F statistic needed to test the hypothesis of part **c**.

12.64 Refer to Exercise 12.63. The contractor samples 25 recent jobs and fits both the complete second-order model (part **b**) and the reduced main effects model (part **a**), so that a test can be conducted to determine whether the additional complexity of the second-order model is necessary. The resulting SSE and R^2 values are shown in the table.

	SSE	R^2
Main effects	8.548	.950
Second-order	6.133	.964

a. Is there sufficient evidence to conclude that the second-order terms are important for predicting the mean cost? Use $\alpha = .05$.

b. Suppose the contractor decides to use the main effects model to predict costs. Use the global F test (Section 11.8) to determine whether the main effects model is useful for predicting costs.

12.65 One factor that must be considered in developing a shipping system that is beneficial to both the customer and the seller is time of delivery. A manufacturer of farm equipment can ship its products by either rail or truck. Quadratic models are thought to be adequate in relating time of delivery to distance traveled for both modes of transportation. Consequently, it has been suggested that the following model be fit:

$$E(y) = \beta_0 + \beta_1 x_1 + \beta_2 x_1^2 + \beta_3 x_2 + \beta_4 x_1 x_2 + \beta_5 x_1^2 x_2$$

where

$y = $ Shipping time

$x_1 = $ Distance to be shipped

$x_2 = \begin{cases} 1 & \text{if rail} \\ 0 & \text{if truck} \end{cases}$

a. What hypothesis would you test to determine whether the data indicate that the quadratic distance terms are useful in the model, i.e., whether curvature is present in the relationship between mean delivery time and distance?

b. What hypothesis would you test to determine whether there is a difference between mean delivery times by rail and truck?

12.66 Refer to Exercise 12.65. Suppose the model is fit to a total of 50 observations on delivery time. The sum of squared errors is SSE $= 226.12$. Then, the reduced model

$$E(y) = \beta_0 + \beta_1 x_1 + \beta_2 x_1^2$$

is fit to the same data, and SSE $= 259.34$. Test to determine whether the data indicate that the mean delivery time differs for rail and truck deliveries. Use $\alpha = .05$.

12.67 A 40-year-old masonry duplex structure has recently undergone a passive solar retrofit with features including insulated exterior walls, heat distribution systems, storm sashes, and air-lock entries. In order to gauge the effectiveness of the improvements, architectural engineers monitored the winter energy usage of the structure for 2 years prior to the retrofit and for 2 years after the retrofit. The engineers want to use the data to fit a regression model relating monthly energy usage y (therms per billing day) to weather intensity x_1 (ddh/bd) and x_2, where

$$x_2 = \begin{cases} 1 & \text{if prior to retrofit} \\ 0 & \text{if after retrofit} \end{cases}$$

a. Write the complete second-order model for $E(y)$.
b. Graph the contour lines for the model of part a.
c. Hypothesize a first-order model that allows for a constant difference between the mean monthly usage prior to and after the retrofit at different levels of weather intensity.
d. Graph the contour lines for the model of part c.

12.68 One of the most promising methods for extracting crude oil employs a carbon dioxide (CO_2) flooding technique. CO_2, when flooded into oil pockets, enhances oil recovery by displacing the crude oil. In a microscopic investigation of the CO_2 flooding process, flow tubes were dipped into sample oil pockets containing a known amount of oil. The oil pockets were flooded with CO_2 and the percentage of oil displaced was recorded. The experiment was conducted at three different flow pressures and three different dipping angles. The displacement test data are recorded in the accompanying table.

PRESSURE x_1, pounds per square inch	DIPPING ANGLE x_2, degrees	OIL RECOVERY y, percentage
1,000	0	60.58
1,000	15	72.72
1,000	30	79.99
1,500	0	66.83
1,500	15	80.78
1,500	30	89.78
2,000	0	69.18
2,000	15	80.31
2,000	30	91.99

Source: Wang, G. C. "Microscopic Investigation of CO_2 Flooding Process." *Journal of Petroleum Technology*, Vol. 34, No. 8, Aug. 1982, pp. 1789–1797.

a. Write the complete second-order model relating percentage oil recovery y to pressure x_1 and dipping angle x_2.
b. Plot the sample data on a scattergram, with percentage oil recovery y on the vertical axis and pressure x_1 on the horizontal axis. Connect the points corresponding to the same value of dipping angle x_2. Based on the scattergram, do you believe a complete second-order model is appropriate?

c. The SAS printout for the interaction model

$$y = \beta_0 + \beta_1 x_1 + \beta_2 x_2 + \beta_3 x_1 x_2 + \varepsilon$$

is provided here. Give the prediction equation for this model.

SAS Computer Printout for Exercise 12.68

DEPENDENT VARIABLE: Y

SOURCE	DF	SUM OF SQUARES	MEAN SQUARE	F VALUE	PR > F	R-SQUARE	C.V.
MODEL	3	843.19083333	281.06361111	44.67	0.0005	0.964031	3.2616
ERROR	5	31.45996667	6.29199333		ROOT MSE		Y MEAN
CORRECTED TOTAL	8	874.65080000			2.50838461		76.90666667

SOURCE	DF	TYPE I SS	F VALUE	PR > F	DF	TYPE III SS	F VALUE	PR > F
X1	1	132.44601667	21.05	0.0059	1	35.54320667	5.65	0.0634
X2	1	707.85481667	112.50	0.0001	1	28.58640115	4.54	0.0862
X1*X2	1	2.89000000	0.46	0.5280	1	2.89000000	0.46	0.5280

PARAMETER	ESTIMATE	T FOR H0: PARAMETER=0	PR > :T:	STD ERROR OF ESTIMATE
INTERCEPT	54.50000000	10.83	0.0001	5.03415841
X1	0.00769667	2.38	0.0634	0.00323831
X2	0.55411111	2.13	0.0862	0.25996282
X1*X2	0.00011333	0.68	0.5280	0.00016723

d. Construct a plot similar to the scattergram of part b, but use the predicted values from the interaction model on the vertical axis. Compare the two plots. Do you believe the interaction model will provide an adequate fit?

e. Check model adequacy using a statistical test with $\alpha = .05$.

f. Is there evidence of interaction between pressure x_1 and dipping angle x_2? Test using $\alpha = .05$.

12.69 Researchers recently conducted an analysis of bus travel demand in Albuquerque, New Mexico, a city selected because of its unique multicentered "Sun Tran" public transit system. One aspect of the study involved the development of a multiple regression model for predicting y, the home-origin trip rate (that is, the number of home-origin trips per 1,000 residents) of a Sun Tran subzone urban area. The following five independent variables, all designed to measure transit level of service (travel time), were entered into the model as main effects:

x_1 = Composite in-vehicle travel time to reach major destination (minutes)

x_2 = Composite transit wait time (minutes)

x_3 = Composite number of transfers required to reach major destination

x_4 = Number of transit routes serving the Sun Tran zone

$x_5 = \begin{cases} 1 & \text{if Sun Tran zone at end of major regional transportation corridor} \\ 0 & \text{if not} \end{cases}$

Data collected from a survey of the city's bus passengers in each of 298 Sun Tran planning analysis zones were used to fit the model

$$E(y) = \beta_0 + \beta_1 x_1 + \beta_2 x_2 + \beta_3 x_3 + \beta_4 x_4 + \beta_5 x_5$$

with the results shown in the accompanying printout (page 626).

SOURCE	DF	SS	MS	F
MODEL	5	32774	6555	87.45
ERROR	292	21886	75	
TOTAL	297	54660		

VARIABLE	PARAMETER ESTIMATE	STD ERROR OF ESTIMATE
INTERCEPT	22.0189	--
X1	-0.1807	0.0389
X2	-0.2498	0.1207
X3	-4.6910	1.7020
X4	3.6745	0.4027
X5	22.5201	3.5959

Source: Adapted from Nelson, D. and O'Neil, K. "Analyzing Demand for Grid System Transit." *Transportation Quarterly*, Vol. 37, No. 1, Jan. 1983, pp. 41–56. Reprinted with permission of the ENO Foundation for Transportation, Inc.

a. Write the least squares prediction equation.
b. Compute and interpret the value of R^2.
c. Compute and interpret the value of s.
d. Is the model useful for predicting home-origin trip rate y? Test using $\alpha = .05$.
e. Is there evidence that home-origin trip rate y decreases as in-vehicle travel time x_1 increases and the remaining independent variables are held constant? Test using $\alpha = .05$.
f. Construct a 95% confidence interval for β_4. Interpret the interval.

OPTIONAL SUPPLEMENTARY EXERCISES

[*Note:*   Starred (*) exercises require the use of a computer.]

12.70 The data in the table were obtained from an experiment designed to investigate the relationship between the yield, y, of potatoes and the levels of three minerals in the soil, x_1, x_2, and x_3. [*Note:*   The mineral levels have been coded by subtracting an appropriate constant from each of the x values.]

y	x_1	x_2	x_3	y	x_1	x_2	x_3
16.40	-1	-1	-1	2.75	1	1	1
13.51	1	-1	-1	14.33	-1.682	0	0
14.41	-1	1	-1	5.44	1.682	0	0
9.38	1	1	-1	19.80	0	-1.682	0
10.77	-1	-1	1	20.00	0	1.682	0
11.78	1	-1	1	9.37	0	0	-1.682
4.11	-1	1	1	10.03	0	0	1.682

*a. Fit a second-order polynomial,

$$y = \beta_0 + \beta_1 x_1 + \beta_2 x_2 + \beta_3 x_3 + \beta_4 x_1 x_2 + \beta_5 x_1 x_3$$
$$+ \beta_6 x_2 x_3 + \beta_7 x_1^2 + \beta_8 x_2^2 + \beta_9 x_3^2 + \varepsilon$$

by least squares.

*b. Test the hypothesis $H_0: \beta_4 = \beta_5 = \cdots = \beta_9 = 0$. That is, test whether the data provide sufficient evidence to indicate that a second-order model contributes more information for the prediction of yield than a first-order model. Test using $\alpha = .05$.

12.71 Five varieties of peas are currently being tested in Ohio to determine which is best suited for production in that state. A field was divided into 20 plots, with each variety of peas being planted in four plots. The yields in bushels of peas produced from each plot are shown in the accompanying table.

VARIETY OF PEAS				
A	B	C	D	E
26.2	29.2	29.1	21.3	20.1
24.3	28.1	30.8	22.4	19.3
21.8	27.3	33.9	24.3	19.9
28.1	31.2	32.8	21.8	22.1

a. Write a model for the data to reflect the yield in bushels as a function of pea variety and interpret all parameters in the model.

*b. Fit the proposed model to the data.

c. Do the data provide sufficient evidence to indicate the model is useful for predicting harvest yield? Use $\alpha = .05$.

12.72 Refer to Exercise 12.57. The data used to fit the operations manager's complete and reduced models are displayed in the table.

DOWN TIME hours per month	MACHINE AGE x_1, years	MACHINE TYPE	x_2
10	1.0	001	1
20	2.0	001	1
30	2.7	001	1
40	4.1	001	1
9	1.2	001	1
25	2.5	001	1
19	1.9	001	1
41	5.0	001	1
22	2.1	001	1
12	1.1	001	1
10	2.0	002	0
20	4.0	002	0
30	5.0	002	0
44	8.0	002	0
9	2.4	002	0
25	5.1	002	0
20	3.5	002	0
42	7.0	002	0
20	4.0	002	0
13	2.1	002	0

a. Use these data to test the null hypothesis that $\beta_1 = \beta_2 = 0$. Test using $\alpha = .10$.
b. Interpret the results of the test in the context of the problem.

*12.73 A firm has developed a new type of light bulb and is interested in evaluating its performance in order to decide whether to market the bulb. It is known that the level of light output of the bulb depends on the cleanliness of its surface area and the length of time the bulb has been in operation. The data in the accompanying table have been obtained. Use these data and the procedures you learned in this chapter to build a regression model that relates drop in light output to bulb surface cleanliness and length of operation.

DROP IN LIGHT OUTPUT percent original output	BULB SURFACE C = Clean, D = Dirty	LENGTH OF OPERATION hours
0	C	0
16	C	400
22	C	800
27	C	1,200
32	C	1,600
36	C	2,000
38	C	2,400
0	D	0
4	D	400
6	D	800
8	D	1,200
9	D	1,600
11	D	2,000
12	D	2,400

REFERENCES

BMDPC: User's Guide to BMDP on the IBM PC. BMDP Statistical Software, Inc. Los Angeles, Calif. 90025.
BMDP User's Digest, 2nd ed. MaryAnn Hill, ed. Los Angeles: BMDP Statistical Software, 1982.
Daniel, C. and Wood, F. *Fitting Equations to Data*, 2nd ed. New York: Wiley, 1980.
Dixon, W. J., Brown, M. B., Engelman, L., Frane, J. W., Hill, M. A., Jennrich, R. I., and Toporek, J. D. *BMDP Statistical Software*, 1985 ed. Berkeley: University of California Press.
Draper, N. and Smith, H. *Applied Regression Analysis*, 2nd ed. New York: Wiley, 1981.
Graybill, F. A. *Theory and Application of the Linear Model*. North Scituate, Mass.: Duxbury, 1976.
Mendenhall, W. *Introduction to Linear Models and the Design and Analysis of Experiments*. Belmont, Ca.: Wadsworth, 1968.
Mendenhall, W. and Sincich, T. *A Second Course in Business Statistics: Regression Analysis*, 2nd ed. San Francisco: Dellen Publishing Co., 1986.
Norusis, M. J. *The SPSS Guide to Data Analysis*, 1986 ed. SPSS, Inc., Suite 3000, 444 N. Michigan Avenue, Chicago, Ill. 60611.
Norusis, M. J. *SPSS/PC+: SPSS for the IBM PC/XT/AT*, 1986 ed. SPSS, Inc., Suite 3000, 444 N. Michigan Avenue, Chicago, Ill. 60611.

Ryan, T. A., Joiner, B. L., and Ryan, B. F. *Minitab Reference Manual*. University Park, Pa.: Minitab Project, 1985.

Ryan, T. A., Joiner, B. L., and Ryan, B. F. *Minitab Student Handbook*, 2nd ed. Boston: Duxbury Press, 1985.

SAS Procedures Guide for Personal Computers, Version 6 ed., 1986. SAS Institute, Inc., Box 8000, Cary, N.C. 27511.

SAS Statistics Guide for Personal Computers, Version 6 ed., 1986. SAS Institute, Inc., Box 8000, Cary, N.C. 27511.

SAS User's Guide: Statistics, Version 5 ed. (1985). A. A. Ray, ed. SAS Institute, Inc., Box 8000, Cary, N.C. 27511.

SPSS^X User's Guide, 1983 ed. SPSS, Inc., Suite 3000, 444 N. Michigan Ave., Chicago, Ill. 60611.

OBJECTIVE

To present a method for analyzing multivariable designed experiments; to identify its overlapping features with and its relation to regression analysis.

CONTENTS

THE ANALYSIS OF VARIANCE FOR DESIGNED EXPERIMENTS

| | | | | | | | | | | | |

SECTION 13.1

INTRODUCTION

In the preceding sections we learned how to analyze multivariable sample data using a multiple regression analysis. When the data have been obtained according to certain specified sampling procedures, they are easy to analyze and also may contain more information pertinent to the population parameters than could be obtained using simple random sampling. The procedure for selecting sample data is called the **design of an experiment**. When the experiment is properly designed, certain types of multivariable data can be analyzed using either a multiple regression analysis or an **analysis of variance**. The objective of this chapter is to introduce some aspects of experimental design and the analysis of the data from such experiments using an analysis of variance. In Section 13.8, we illustrate the relationship between this method of analysis and multiple regression.

| | | | | | | | | | | | |

SECTION 13.2

EXPERIMENTAL DESIGN

The study of experimental design originated in England and, in its early years, was associated solely with agricultural experimentation. The need for experimental design in agriculture was very clear: It takes a full year to obtain a single observation on the yield of a new variety of wheat. Consequently, the need to save time and money led to a study of ways to obtain more information using smaller samples. Similar motivations led to its subsequent acceptance and wide use in all fields of scientific experimentation. Despite this fact, the terminology associated with experimental design clearly indicates its early association with the biological sciences.

Independent variables that may be related to a response variable y are called **factors**. The value—that is, the intensity setting—assumed by a factor in an experiment is called a **level**. For example, suppose that an experiment is conducted to measure the hardness y of a new type of plastic as a function of two factors, the pressure and temperature at the time of molding. If the hardness of the plastic is measured at pressures 200, 300, and 400 pounds per square inch (psi) and at temperatures 200 and 300 degrees Fahrenheit (°F), then pressure is at three levels and temperature is at two levels. The combinations of levels of the factors for which y will be observed are called **treatments**. For example, if the hardness of the new plastic is measured for each of the six pressure–temperature combinations, (200 psi, 200°F), (300 psi, 200°F), (400 psi, 200°F), (200 psi, 300°F), (300 psi, 300°F), and (400 psi, 300°F), then the experiment would involve six treatments. The term *treatments* is used to describe the factor–level combinations to be included in an experiment because many experiments involve treating or doing something to alter the nature of the **experimental unit**, the object upon which a measurement is made. Thus, we might view the six pressure–temperature combinations as treatments of the experimental units of plastic used in the hardness experiment.

DEFINITION 13.1

The independent variables that are related to a response variable y are called **factors**.

DEFINITION 13.2

The intensity setting of a factor is called a **level**.

DEFINITION 13.3

A **treatment** is a particular combination of levels of the factors involved in an experiment.

The design of an experiment involves the following four steps:

1. Select the factors to be included in the experiment and identify the parameters that are the object of the study. Usually, the target parameters are the population means associated with the factor–level combinations (i.e., treatments).
2. Decide how much information you want to acquire—that is, decide upon the magnitude of the standard error(s) that you desire.
3. Choose the treatments (the factor–level combinations) to be included in the experiment and determine the number of observations to be made for each treatment.
4. Decide how the treatments will be assigned to the experimental units.

Steps 3 and 4 control the quantity of information in an experiment. We shall explain how this is done in Section 13.3.

SECTION 13.3

CONTROLLING THE INFORMATION IN AN EXPERIMENT

The problem of acquiring good experimental data is analogous to the problem faced by a communications engineer. The receipt of any signal, verbal or otherwise, depends on the volume of the signal and the amount of the background noise. The greater the volume of the signal, the greater will be the amount of information transmitted to the receiver. Conversely, the amount of information transmitted is reduced when the background noise is great. These intuitive thoughts about the factors that affect the information in an experiment are supported by the following fact: The standard errors of most estimators are proportional to σ (a measure of data variation or noise) and inversely proportional to the sample size (a measure of the volume of the signal).

Step 3 in the design of an experiment (see Section 13.2), the selection of the treatments to be included in an experiment and the specification of the sample sizes, determines the volume of nature's signal. You must select the treatments so that the observed values of y provide information on the parameters of interest. Then the larger the treatment sample sizes, the greater will be the quantity of information in the experiment. Is it possible to observe y and obtain no information on a parameter of interest? The answer is "yes." To illustrate, suppose that you attempt to fit a first-order model

$$E(y) = \beta_0 + \beta_1 x$$

to a set of $n = 10$ data points, all of which were observed for a single value of

FIGURE 13.1

Data Set with $n = 10$ Responses, All at $x = 5$

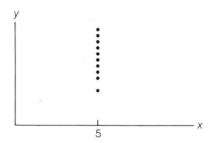

x, say $x = 5$. The data points might appear as shown in Figure 13.1. Clearly, there is no possibility of fitting a line to these data points. The only way to obtain information on β_0 and β_1 is to observe y for *different* values of x. Consequently, the $n = 10$ data points in this example contain absolutely no information on the parameters β_0 and β_1.

Step 4 in the design of an experiment provides an opportunity to reduce the noise (or experimental error) in an experiment. Known sources of data variation can be reduced or eliminated by **blocking**—that is, observing all treatments within relatively homogeneous **blocks** of experimental material. For example, suppose we want to compare the length of time required to assemble a digital watch using three different methods of assembly (treatments). One method for conducting the experiment would be to randomly assign five assemblers to each method of assembly and record the assembly time for each of the 15 assemblers included in the experiment. Using this method, we introduce an unwanted source of variation, namely, the portion contributed by the differences in dexterity of the 15 assemblers.

A second method of conducting the experiment allows us to block out the variation from assembler to assembler. We could use only five assemblers and require each to assemble three watches—one for each of the three methods of assembly. If a particular assembler is by nature slow or fast, then this characteristic will be reflected in the measurements obtained for all three assembly methods. In comparing the assembly times for two different assembly methods, we hope to cancel out the contribution of the worker to each of the measurements.

SECTION 13.4

THE LOGIC BEHIND AN ANALYSIS OF VARIANCE

Once the data for a designed experiment have been collected, we will want to use the sample information to make inferences about the population means associated with the various treatments. The method used to compare the treatment means is known as **analysis of variance**, or **ANOVA**. The concept behind an analysis of variance can be explained using the following simple example.

Consider an experiment with a single factor at two levels (that is, two treatments). Suppose we want to decide whether the two treatment means differ based on the means of two independent random samples, each containing $n_1 = n_2 = 5$ measurements, and that the y-values appear as in Figure 13.2. Note that the five circles on the left are plots of the y-values for sample 1 and the five solid

FIGURE 13.2
Plots of Data for
Two Samples

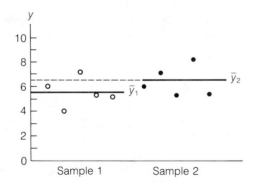

dots on the right are plots of the y-values for sample 2. Also, observe the horizontal lines that pass through the means for the two samples, $\bar{y}_1$ and $\bar{y}_2$. Do you think the plots provide sufficient evidence to indicate a difference between the corresponding population means?

If you are uncertain whether the population means differ for the data in Figure 13.2, examine the situation for two different samples in Figure 13.3(a). We think that you will agree that for these data, it appears that the population means differ. Examine a third case in Figure 13.3(b). For these data, it appears that there is little or no difference between the population means.

FIGURE 13.3
Plots of Data for
Two Cases

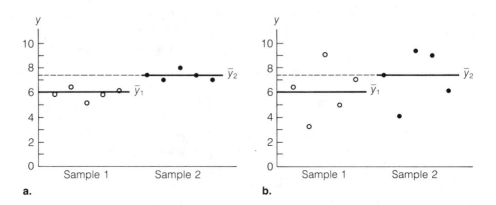

a.   b.

What elements of Figures 13.2 and 13.3 did we intuitively use to decide whether the data indicate a difference between the population means? The answer to the question is that we visually compared the distance (the variation) *between* the sample means to the variation *within* the y-values for each of the two samples. Since the difference between the sample means in Figure 13.3(a) is large relative to the within-sample variation, we inferred that the population means differ. Conversely, in Figure 13.3(b), the variation between the sample means is small relative to the within-sample variation and therefore there is little evidence to infer that the means are significantly different.

The variation within samples is measured by the pooled s^2 that we computed for the independent random samples t test of Section 9.6, namely

Within-sample variation: $s^2 = \dfrac{\displaystyle\sum_{i=1}^{n_1} (y_{i1} - \bar{y}_1)^2 + \sum_{i=1}^{n_2} (y_{i2} - \bar{y}_2)^2}{n_1 + n_2 - 2}$

$$= \dfrac{\text{SSE}}{n_1 + n_2 - 2}$$

where y_{i1} is the ith observation in sample 1 and y_{i2} is the ith observation in sample 2. The quantity in the numerator of s^2 is often denoted **SSE**, the **sum of squared errors**. As with regression analysis, SSE measures unexplained variability. But in this case, it measures variability *unexplained* by the differences between the sample means.

A measure of the between-sample variation is given by the weighted sum of squares of deviations of the individual sample means about the mean for all 10 observations, $\bar{y}$, divided by the number of samples minus 1, i.e.,

Between-sample variation: $\dfrac{n_1(\bar{y}_1 - \bar{y})^2 + n_2(\bar{y}_2 - \bar{y})^2}{2 - 1} = \dfrac{\text{SST}}{1}$

The quantity in the numerator is often denoted **SST**, the **sum of squares for treatments**, since it measures the variability *explained* by the differences between the sample means of the two treatments.

For this experimental design, SSE and SST sum to a known total, namely

$$\text{SS(Total)} = \sum (y_i - \bar{y})^2$$

Also, the ratio

$$F = \dfrac{\text{Between-sample variation}}{\text{Within-sample variation}}$$

$$= \dfrac{\text{SST}/1}{\text{SSE}/(n_1 + n_2 - 2)}$$

has an F distribution with $\nu_1 = 1$ and $\nu_2 = n_1 + n_2 - 2$ degrees of freedom (df) and therefore can be used to test the null hypothesis of no difference between the treatment means. The additivity property of the sums of squares led early researchers to view this analysis as a **partitioning** of $\text{SS(Total)} = \Sigma(y_i - \bar{y})^2$ into sources corresponding to the factors included in the experiment and to SSE. The simple formulas for computing the sums of squares, the additivity property, and the form of the test statistic made it natural for this procedure to be called an **analysis of variance**. We demonstrate the analysis of variance procedures for three special types of experimental designs in Sections 13.5, 13.6, and 13.7.

| | | | | | | | | | | |

S E C T I O N   13.5

THE ANALYSIS OF VARIANCE FOR A COMPLETELY RANDOMIZED DESIGN

A **completely randomized design** is one that involves a comparison of the means for a number, say p, of treatments, based on independent random samples of n_1, $n_2, \ldots, n_p$ observations, drawn from populations associated with treatments 1, 2, $\ldots$, p, respectively. Thus, we want to make inferences about p population means where μ_i is the mean of the population of measurements associated with treatment i, for $i = 1, 2, \ldots, p$. The null hypothesis to be tested is that the p population means are equal, i.e., H_0: $\mu_1 = \mu_2 = \cdots = \mu_p$, and the alternative hypothesis we wish to detect is that at least two of the treatment means differ.

DEFINITION 13.4

A **completely randomized design** to compare p treatment means is one in which independent random samples are drawn from each of the p populations.

FIGURE 13.4
The Partitioning of SS(Total) for a Completely Randomized Design

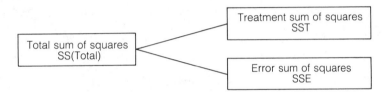

An analysis of variance provides an easy way to analyze the data from a completely randomized design. This analysis partitions SS(Total) into two components, SSE and SST (see Figure 13.4). Recall that the quantity SST denotes the sum of squares for treatments and measures the variation explained by the differences between the treatment means. The sum of squares for error, SSE, is a measure of the unexplained variability, obtained by calculating a pooled measure of the variability *within* the p samples. If the treatment means truly differ, then SSE should be substantially smaller than SST. We compare the two sources of variability by forming an F statistic:

$$F = \frac{\text{SST}/(p-1)}{\text{SSE}/(n-p)} = \frac{\text{MST}}{\text{MSE}}$$

where n is the total number of measurements. The numerator of the F statistic, MST = SST/$(p-1)$, denotes **mean square for treatments** and is based on $(p-1)$ degrees of freedom—one for each of the p treatments minus 1 for the estimation of the overall mean. The denominator of the F statistic, MSE = SSE/$(n-p)$, denotes **mean square for error** and is based on $(n-p)$ degrees of freedom—one for each of the n measurements minus 1 for each of the p treatment means being estimated. Thus, F is based on $\nu_1 = (p-1)$ and $\nu_2 = (n-p)$ degrees of freedom. If the computed value of F exceeds the upper critical value, F_α, we reject H_0 and conclude that at least two of the treatment means differ.

The notation used in an analysis of variance of a completely randomized design is given in Table 13.1. The computing formulas and the analysis of variance F test are given in the accompanying boxes. In this section we will illustrate the analysis of a completely randomized design using the analysis of variance formulas shown here. The regression approach to conducting an analysis of variance is illustrated in optional Section 13.9.

COMPUTING FORMULAS FOR THE ANALYSIS OF VARIANCE
FOR A COMPLETELY RANDOMIZED DESIGN

$$\text{Sum of all } n \text{ measurements} = \sum_{i=1}^{n} y_i$$

$$\text{Mean of all } n \text{ measurements} = \bar{y}$$

$$\text{Sum of squares of all } n \text{ measurements} = \sum_{i=1}^{n} y_i^2$$

$$\text{CM} = \text{Correction for mean}$$

$$= \frac{(\text{Total of all observations})^2}{\text{Total number of observations}} = \frac{\left(\sum_{i=1}^{n} y_i\right)^2}{n}$$

$$\text{SS(Total)} = \text{Total sum of squares}$$

$$= (\text{Sum of squares of all observations}) - \text{CM}$$

$$= \sum_{i=1}^{n} y_i^2 - \text{CM}$$

$$\text{SST} = \text{Sum of squares for treatments}$$

$$= \left(\begin{array}{c}\text{Sum of squares of treatment totals with} \\ \text{each square divided by the number of} \\ \text{observations for that treatment}\end{array}\right) - \text{CM}$$

$$= \frac{T_1^2}{n_1} + \frac{T_2^2}{n_2} + \cdots + \frac{T_p^2}{n_p} - \text{CM}$$

$$\text{SSE} = \text{Sum of squares for error}$$

$$= \text{SS(Total)} - \text{SST}$$

$$\text{MST} = \text{Mean square for treatments}$$

$$= \frac{\text{SST}}{p - 1}$$

$$\text{MSE} = \text{Mean square for error}$$

$$= \frac{\text{SSE}}{n - p}$$

TABLE 13.1

Summary of Notation for a Completely Randomized Design

	POPULATIONS (TREATMENTS)				
	1	2	3	$\cdots$	p
Mean	μ_1	μ_2	μ_3 $\cdots$		μ_p
Variance	σ_1^2	σ_2^2	σ_3^2 $\cdots$		σ_p^2

	INDEPENDENT RANDOM SAMPLES				
	1	2	3	$\cdots$	p
Sample size	n_1	n_2	n_3	$\cdots$	n_p
Sample totals	T_1	T_2	T_3	$\cdots$	T_p
Sample means, $\dfrac{T_i}{n_i}$	$\bar{T}_1$	$\bar{T}_2$	$\bar{T}_3$	$\cdots$	$\bar{T}_p$

Total number of measurements $= n = n_1 + n_2 + n_3 + \cdots + n_p$

ANALYSIS OF VARIANCE F TEST FOR A COMPLETELY RANDOMIZED DESIGN WITH p TREATMENTS

H_0: $\mu_1 = \mu_2 = \cdots = \mu_p$ (i.e., there is no difference in the treatment means)

H_a: At least two of the treatment means differ

Test statistic: $F = \dfrac{\text{MST}}{\text{MSE}} = \dfrac{\text{MST}}{s^2}$

Rejection region: $F > F_\alpha$, where F is based on $\nu_1 = (p - 1)$ and $\nu_2 = (n - p)$ degrees of freedom

---

EXAMPLE 13.1

An experiment was conducted to compare the wearing qualities of three types of paint when subjected to the abrasive action of a slowly rotating cloth-surfaced wheel. Ten paint specimens were tested for each paint type, and the number of hours until visible abrasion was apparent was recorded for each specimen. The data are shown in Table 13.2. Is there sufficient evidence to indicate a difference

TABLE 13.2

Wear Data For Three Types of Paint

PAINT TYPE		
1	2	3
148	513	335
76	264	643
393	433	216
520	94	536
236	535	128
134	327	723
55	214	258
166	135	380
415	280	594
153	304	465

in the mean time until abrasion is visibly evident for the three paint types? Conduct an analysis of variance test using $\alpha = .05$.

SOLUTION

The experiment involves a single factor, paint type, which is at three levels. Thus, we have a completely randomized design with $p = 3$ treatments. Let μ_1, μ_2, and μ_3 represent the mean abrasion times for paint types 1, 2, and 3, respectively. Then we want to test

$$H_0: \quad \mu_1 = \mu_2 = \mu_3$$

against

$$H_a: \quad \text{At least two of the three means differ}$$

From Table 13.2, the totals for the three samples are computed to be $T_1 = 2{,}296$, $T_2 = 3{,}099$, and $T_3 = 4{,}278$. We also have

$$\sum_{i=1}^{n} y_i = T_1 + T_2 + T_3 = 9{,}673$$

$$\sum_{i=1}^{n} y_i^2 = (148)^2 + (76)^2 + \cdots + (465)^2 = 4{,}088{,}341$$

Then, following the order of calculations listed earlier, we find

$$\text{CM} = \frac{\left(\sum_{i=1}^{n} y_i\right)^2}{n} = \frac{(9{,}673)^2}{30} = 3{,}118{,}897.633$$

$$\text{SS(Total)} = \sum_{i=1}^{n} y_i^2 - \text{CM}$$

$$= 4{,}088{,}341 - 3{,}118{,}897.633$$

$$= 969{,}443.367$$

$$\text{SST} = \frac{T_1^2}{n_1} + \frac{T_2^2}{n_2} + \frac{T_3^2}{n_3} - \text{CM}$$

$$= \frac{(2{,}296)^2}{10} + \frac{(3{,}099)^2}{10} + \frac{(4{,}278)^2}{10} - 3{,}118{,}897.633$$

$$= 3{,}317{,}670.1 - 3{,}118{,}897.633 = 198{,}772.467$$

$$\text{SSE} = \text{SS(Total)} - \text{SST} = 969{,}443.367 - 198{,}772.467$$

$$= 770{,}670.9$$

$$\text{MST} = \frac{\text{SST}}{p-1} = \frac{198{,}772.467}{2} = 99{,}386.234$$

$$\text{MSE} = \frac{\text{SSE}}{n-p} = \frac{770{,}670.9}{27} = 28{,}543.367$$

Then the value of the test statistic is

$$F = \frac{\text{MST}}{\text{MSE}} = \frac{99,386.234}{28,543.367} = 3.48$$

and the rejection region for the test is $F > F_{.05}$ (see Figure 13.5). The value of $F_{.05}$, based on $\nu_1 = p - 1 = 3 - 1 = 2$ and $\nu_2 = n - p = 30 - 3 = 27$ degrees of freedom, is 3.35. Since the computed value of F exceeds the critical value, there is evidence to indicate a difference in the mean time to visible abrasion for the three paint types.   ∎

FIGURE 13.5
Rejection Region for
Example 13.1;
Numerator df = 2,
Denominator df = 27,
$\alpha = .05$

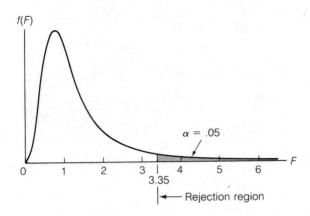

The results of an analysis of variance (ANOVA) are often summarized in tabular form. The general form of an ANOVA table for a completely randomized design is shown in Table 13.3. The column head "Source" refers to the source of variation, and for each source, "df" refers to the degrees of freedom, "SS" to the sum of squares, "MS" to the mean square, and "F" to the F statistic comparing the treatment mean square to the error mean square. Table 13.4 is the ANOVA summary table corresponding to the analysis of variance data for Example 13.1.

TABLE 13.3
ANOVA Summary Table
for a Completely
Randomized Design

SOURCE	df	SS	MS	F
Treatments	$p - 1$	SST	MST	MST/MSE
Error	$n - p$	SSE	MSE	
Total	$n - 1$	SS(Total)		

TABLE 13.4
ANOVA Summary Table
for Example 13.1

SOURCE	df	SS	MS	F
Paint types	2	198,772.467	99,386.234	3.48
Error	27	770,670.900	28,543.367	
Total	29	969,443.367		

Because the completely randomized design involves the selection of independent random samples, we can find a confidence interval for a single treatment mean or for the difference between two treatment means using the methods of Chapter 8. The estimate of σ^2 will be based on the pooled sum of squares within all p samples; that is,

$$\text{MSE} = s^2 = \frac{\text{SSE}}{n - p}$$

This is the same quantity that is used as the denominator for the analysis of variance F test. The formulas for the relevant confidence intervals of Chapter 8 are reproduced in the box.

CONFIDENCE INTERVALS FOR MEANS: COMPLETELY RANDOMIZED DESIGN

Single treatment mean (say, treatment i):

$$\bar{T}_i \pm t_{\alpha/2}\left(\frac{s}{\sqrt{n_i}}\right)$$

Difference between two treatment means (say, treatments i and j):

$$(\bar{T}_i - \bar{T}_j) \pm t_{\alpha/2}s\sqrt{\frac{1}{n_i} + \frac{1}{n_j}}$$

where $s = \sqrt{\text{MSE}}$ and $t_{\alpha/2}$ is the tabulated value of t (Table 6 of Appendix II) that locates $\alpha/2$ in the upper tail of the t distribution and has $(n - p)$ df (the degrees of freedom associated with error in the ANOVA).

EXAMPLE 13.2

Refer to Example 13.1 and find a 95% confidence interval for the mean time to abrasion for paint type 1.

SOLUTION

From Table 13.4, we have

$\text{MSE} = 28,543.367$

Then,

$$s = \sqrt{\text{MSE}} = \sqrt{28,543.367} = 168.9$$

The sample mean time to abrasion for paint type 1 is

$$\bar{T}_1 = \frac{T_1}{n_1} = \frac{2,296}{10} = 229.6 \text{ hours}$$

The tabulated value of $t_{.025}$ for 27 df (the same as for the MSE) is 2.052. So, a 95% confidence interval for μ_1, the mean time (in hours) to abrasion for paint type 1, is

$$\bar{T}_1 + t_{\alpha/2}\left(\frac{s}{\sqrt{n_1}}\right) = 229.6 \pm 2.052\left(\frac{168.9}{\sqrt{10}}\right)$$

or (120.0, 339.2).

Note that this confidence interval is quite wide—probably too wide to be of any practical value. The reason why the interval is so wide can be seen in the large amount of variation in wear for the specimens for each type of paint. For example, the time to abrasion for specimens of paint type 1 varies from 55 to 520 hours. The more variable the data, the larger will be the value of s that appears in the confidence interval and the wider will be the confidence interval. Consequently, if you want to obtain a more accurate estimate of treatment means with a narrower confidence interval, you will have to select larger samples for each paint type. ∎

EXAMPLE 13.3

Refer again to Example 13.1. Find a 95% confidence interval for the difference in mean time to abrasion between paint types 1 and 3.

SOLUTION

The mean time to abrasion for the sample of paint type 3 is

$$\bar{T}_3 = \frac{T_3}{n_3} = \frac{4,278}{10} = 427.8$$

and, from Example 13.2, $\bar{T}_1 = 229.6$. The tabulated t value, $t_{.025}$, is the same as for Example 13.2, namely 2.052. Then, the 95% confidence interval for $(\mu_3 - \mu_1)$, the difference in mean time to abrasion between paint types 1 and 3, is

$$(\bar{T}_3 - \bar{T}_1) \pm t_{.025}s\sqrt{\frac{1}{n_3} + \frac{1}{n_1}} = (427.8 - 229.6) \pm (2.052)(168.9)\sqrt{\frac{1}{10} + \frac{1}{10}}$$

or (43.2, 353.2).

As for the confidence interval for a single mean, the confidence interval for the difference $(\mu_3 - \mu_1)$ is very wide. This is due to the excessive within-sample variation. To obtain a narrower confidence interval, the sample sizes for the three paint types must be increased. However, the fact that the interval contains only positive numbers means we can conclude, with 95% confidence, that the mean time to abrasion for paint type 3 exceeds the mean time to abrasion for paint type 1. ∎

EXERCISES 13.1–13.11

13.1 Independent random samples were selected from three normally distributed populations with common (but unknown) variance, σ^2. The data are shown in the table (page 644).

SAMPLE 1	SAMPLE 2	SAMPLE 3
3.1	5.4	1.1
4.3	3.6	.2
1.2	4.0	3.0
	2.9	

a. Compute the appropriate sums of squares and mean squares and fill in the appropriate entries in the analysis of variance table:

SOURCE	df	SS	MS	F
Treatments				
Error				
Total				

b. Test the hypothesis that the population means are equal (i.e., $\mu_1 = \mu_2 = \mu_3$) against the alternative hypothesis that at least one mean is different from the other two. Test using $\alpha = .05$.

c. Find a 90% confidence interval for $(\mu_2 - \mu_3)$. Interpret the interval.

d. What would happen to the width of the confidence interval in part c if you quadrupled the number of observations in the two samples?

e. Find a 95% confidence interval for μ_2.

f. Approximately how many observations would be required if you wanted to be able to estimate a population mean correct to within .4 with probability equal to .95?

13.2 A partially completed ANOVA table for a completely randomized design is shown here:

SOURCE	df	SS	MS	F
Treatments	4	24.7		
Error				
Total	34	62.4		

a. Complete the ANOVA table.

b. How many treatments are involved in the experiment?

c. Do the data provide sufficient evidence to indicate a difference among the population means? Test using $\alpha = .10$.

d. Find the approximate observed significance level for the test in part c, and interpret it.

e. Suppose that $\bar{T}_1 = 3.7$ and $\bar{T}_2 = 4.1$. Do the data provide sufficient evidence to indicate a difference between μ_1 and μ_2? Assume that there are seven observations for each treatment. Test using $\alpha = .10$.

f. Refer to part e. Find a 90% confidence interval for $(\mu_1 - \mu_2)$.

g. Refer to part e. Find a 90% confidence interval for μ_1.

13.3 Vanadium (V) is a recently recognized essential trace element. An experiment was conducted to compare the concentrations of V in biological materials using isotope dilution mass spectrometry (*Analytical Chemistry*, Nov. 1985).

The accompanying table gives the quantities of V (measured in nanograms per gram) in dried samples of oyster tissue, citrus leaves, bovine liver, and human serum.

OYSTER TISSUE	CITRUS LEAVES	BOVINE LIVER	HUMAN SERUM
2.35	2.32	.39	.10
1.30	3.07	.54	.17
.34	4.09	.30	.14
			.16
			.16

Source: Fassett, J. D. and Kingston, H. M. "Determination of Nanogram Quantities of Vanadium in Biological Material by Isotope Dilution Thermal Ionization Mass Spectrometry with Ion Counting Detection." *Analytical Chemistry*, Vol. 57, No. 13, Nov. 1985, p. 2475 (Table II). Copyright 1985 American Chemical Society. Reprinted with permission.

a. Construct an ANOVA table for the data.
b. Is there sufficient evidence (at $\alpha = .05$) to indicate that the mean V concentrations differ among the four biological materials?
c. Estimate the mean V concentration in human serum with a 95% confidence interval.
d. Estimate the difference between the mean V concentrations in oyster tissue and citrus leaves with a 95% confidence interval.

13.4 As oil drilling costs rise at unprecedented rates, the task of measuring drilling performance becomes essential to a successful oil company. One method of lowering drilling costs is to increase drilling speed. Researchers at Cities Service Co. have developed a drill bit, called the PD-1, which they believe penetrates rock at a faster rate than any other bit on the market. It is decided to compare the speed of the PD-1 with the two fastest drill bits known, the IADC 1-2-6 and the IADC 5-1-7, at 12 drilling locations in Texas. Four drilling sites were randomly assigned to each bit, and the rate of penetration (RoP) in feet per hour (fph) was recorded after drilling 3,000 feet at each site. The data are given in the table. On the basis of this information, can Cities Service Co. conclude that the mean RoP differs for at least two of the three drill bits? Test at the $\alpha = .05$ level of significance.

PD-1	IADC 1-2-6	IADC 5-1-7
35.2	25.8	14.7
30.1	29.7	28.9
37.6	26.6	23.3
34.3	30.1	16.2

13.5 An excessive amount of ozone in the air is indicative of air pollution. Six air samples were collected from each of four locations in the industrial Midwest and measured for their content of ozone. The amounts of ozone (in parts per million) are shown in the accompanying table.
a. Perform an analysis of variance and construct an analysis of variance table for the data.
b. Do the data provide sufficient evidence to indicate differences in the mean ozone content among the four locations? Use $\alpha = .05$.
c. Find a 95% confidence interval for the mean ozone content at location 1.

LOCATION			
1	2	3	4
.08	.15	.13	.05
.10	.09	.10	.11
.09	.11	.15	.07
.07	.10	.09	.09
.09	.08	.09	.11
.06	.13	.17	.08

d. Find a 95% confidence interval for the difference in the mean ozone content between locations 1 and 3.

e. Suppose you want to estimate the difference in mean ozone content between two locations correct to within .01 part per million with probability approximately equal to .95. How many air samples would be required at each location? [*Hint:* Refer to Section 8.12.]

CM	CS	TE	TI
4	6	5	8
7	9	5	4
5	5	7	8
6	7	8	10
8	6	7	3

Source: Data are simulated values based on the group means reported in *Human Factors*, Feb. 1984. Copyright 1984 by the Human Factors Society, Inc. and reproduced by permission.

13.6 The display consoles of modern computer-based systems use many abbreviated words in order to accommodate the large volume of information to be displayed. Therefore, operators must learn to decode each abbreviation quickly and accurately. An experiment was conducted to determine the optimal method for abbreviating any specific set of words on the sonar consoles used at the Naval Submarine Medical Research Laboratory in Groton, Connecticut (*Human Factors*, Feb. 1984). Of the 20 Navy and civilian personnel who took part in the study, five were highly familiar with the sonar system. The 15 subjects unfamiliar with the system were randomly divided into three groups of five. Thus, the study consisted of a total of four groups (one experienced and three inexperienced groups), with five subjects per group. The experienced group and one inexperienced group (denoted TE and TI, respectively) were assigned to learn the simple method of abbreviation. One of the remaining inexperienced groups was assigned the conventional single abbreviation method (denoted CS), while the other was assigned the conventional multiple abbreviation method (denoted CM). Each subject was then given a list of 75 abbreviations to learn, one at a time, through the display console of a minicomputer. The number of trials until the subject accurately decoded at least 90% of the words on the list was recorded. Do the data provide sufficient evidence to indicate differences among the mean numbers of trials required for the four groups? Test using $\alpha = .05$.

		RIVER		
1	2	3	4	5
2	4	12	7	13
3	6	9	5	9
1	3	11	5	15
5	5	8	9	10
	7			11
				7

13.7 Polychlorinated biphenyls (PCBs), used in the manufacture of large electrical transformers and capacitors, are extremely hazardous contaminants when released into the environment. Samples of fish were taken from each of five rivers and analyzed for PCB concentration (in parts per million). The data are shown in the table. (Some of the analyses were unproductive, so the sample sizes vary from river to river.)

a. Perform an analysis of variance and construct an analysis of variance table.

b. Do the data provide sufficient evidence to indicate differences in the mean PCB concentration in fish for the five rivers? Test using $\alpha = .05$.

c. Find a 95% confidence interval for the difference in the mean PCB concentration in fish between rivers 1 and 2.

d. Suppose you want to estimate the difference in the mean PCB concentration in fish between rivers 1 and 2 correct to within .5 part per million with probability approximately equal to .95. How many fish would have to be included in each sample? (Assume that the sample sizes are to be equal.)

OPTIONAL EXERCISES

13.8 It can be shown that

$$\text{SST} = \sum_{i=1}^{p} n_i(\bar{T}_i - \bar{y})^2 = \sum_{i=1}^{p} \frac{T_i^2}{n_i} - \text{CM}$$

Verify this identity for $p = 3$ treatments.

13.9 The small-sample estimation and test procedures for comparing two population means discussed in Chapters 8 and 9 were based on independent random sampling—that is, a completely randomized design. For both estimation and test procedures, we used a pooled estimate of σ^2, namely

$$s^2 = \frac{(n_1 - 1)s_1^2 + (n_2 - 1)s_2^2}{n_1 + n_2 - 2}$$

$$= \frac{\sum_{i=1}^{n_1} (y_{i1} - \bar{y}_1)^2 + \sum_{i=1}^{n_2} (y_{i2} - \bar{y}_2)^2}{n_1 + n_2 - 2}$$

[*Note:* In the notation of this chapter, $\bar{y}_1 = \bar{T}_1$, $\bar{y}_2 = \bar{T}_2$, and, in general, $\bar{y}_i = \bar{T}_i$.] The numerator in this expression is SSE. In general, regardless of the number of treatments, it can be shown (proof omitted) that

$$SSE = (n_1 - 1)s_1^2 + (n_2 - 1)s_2^2 + \cdots + (n_p - 1)s_p^2$$

$$= \sum_{i=1}^{n_1} (y_{i1} - \bar{y}_1)^2 + \sum_{i=1}^{n_2} (y_{i2} - \bar{y}_2)^2 + \cdots + \sum_{i=1}^{n_p} (y_{ip} - \bar{y}_p)^2$$

Refer to Exercise 13.4. Calculate the sum of squares of deviations of the y values in each sample about their respective sample means. Then calculate SSE using the above formula and verify that it is the same value obtained in Exercise 13.4.

13.10 The means and standard deviations listed in the table provide information on the strengths (modules of rupture at ground line) for five types of wooden poles used by the utility industry. The data used to obtain the means and standard deviations (recorded in pounds per square inch) were collected according to a completely randomized design (independent random samples).

SPECIES	SAMPLE SIZE	SAMPLE MEAN	SAMPLE STANDARD DEVIATION
Northern white cedar	28	3,660	203.33
Western red cedar	387	5,550	298.39
Pacific silver fir	103	5,420	313.29
Coastal Douglas fir	118	8,380	644.62
Southern pine	147	8,870	611.72

Source: Goodman, J. R., Vanderbilt, M. D., and Criswell, M. E. "Reliability-Based Design of Wood Transmission Line Structures." *Journal of Structural Engineering*, Vol. 109, No. 3, 1983, pp. 690–704.

a. Find the totals for each sample and the total for all 783 strength measurements. Then compute CM and SST.

b. Since we do not know the value of $\sum_{i=1}^{n} y_i^2$, calculate SSE using the pooled method:

$$SSE = \sum_{i=1}^{n_1} (y_{i1} - \bar{y}_1)^2 + \sum_{i=1}^{n_2} (y_{i2} - \bar{y}_2)^2 + \cdots + \sum_{i=1}^{n_5} (y_{i5} - \bar{y}_5)^2$$

$$= (n_1 - 1)s_1^2 + (n_2 - 1)s_2^2 + \cdots + (n_5 - 1)s_5^2$$

c. Find SS(Total).

d. Construct an analysis of variance table for the data.

e. Do the data provide sufficient evidence to indicate differences in the mean strengths among the five types of poles? Test using $\alpha = .05$.

f. Find a 90% confidence interval for the mean strength of Northern white cedar poles.

g. Find a 90% confidence interval for the difference in mean strengths between Northern white cedar and Southern pine poles.

13.11 Unlike most other commonly used engineering materials, concrete experiences a characteristic marked increase in "creep" when it is heated for the first time under load. To investigate this phenomenon, a study of the thermal strain behavior of concrete was conducted (*Magazine of Concrete Research*, Dec. 1985). Concrete specimens were prepared and a constant load applied to each. The test specimens were then heated to a specified temperature at a rate of 1°C per minute, with the specimens randomly assigned to one of five temperature settings (100°, 200°, 300°, 400°, and 500°C). For each specimen, the difference between the free (unloaded) thermal strain and load-induced thermal strain, called the *total thermal strain*, was calculated. The sample size, mean, and standard deviation of the total thermal strain values for each temperature setting are given in the table. [*Note:* Thermal strain is recorded in units $\times 10^6$.]

TEMPERATURE	NUMBER OF SPECIMENS	MEAN	STANDARD DEVIATION
100	16	52	55
200	16	112	108
300	16	143	127
400	16	186	136
500	14	257	178

Source: Khoury, G. A., Grainger, B. N., and Sullivan, P. J. E. "Strain of Concrete During First Heating to 600°C Under Load." *Magazine of Concrete Research*, Vol. 37, No. 133, Dec. 1985, p. 198 (Table 2).

a. Use the technique of Exercise 13.10 to construct an ANOVA summary table for the data.

b. Is there sufficient evidence to indicate that heating temperature affects the mean total thermal strain of concrete? Test using $\alpha = .01$.

| | | | | | | | | | | | | |

S E C T I O N 13.6

THE ANALYSIS OF VARIANCE FOR A RANDOMIZED BLOCK DESIGN

A **randomized block design** often provides more information per observation than the amount contained in a completely randomized design. For example, suppose you want to compare the length of time required to process a bank's daily receipts using three different computer programs, A, B, and C. A completely randomized design could be achieved by selecting 15 days and randomly assigning the receipts for 5 days to be processed using each of the programs.

A better way to conduct the experiment—one that contains more information on the mean processing times—would be to utilize the receipts for only 5 days and process the data for each day using each of the three programs. This *randomized block* procedure acknowledges the fact that the length of time required to process a day's receipts varies substantially from day to day depending on the level of the day's business, the complexity of the transactions, and so on. By comparing the processing time for each day, we eliminate day-to-day variation from the comparison.

The randomized block design that we have just described is shown diagrammatically in Figure 13.6. The figure shows that there are five jobs. Each job can be viewed as a **block** of three experimental units—runs on the computer—one corresponding to the use of each of the programs, A, B, and C. The blocks are said to be **randomized** because the treatments (computer programs) are randomly assigned to the experimental units within a block. For our example, the programs employed to process a day's receipts would be run in a random order to avoid bias introduced by other unknown and unmeasured variables that may affect the processing time.

FIGURE 13.6

Diagram for a Randomized Block Design Containing $b = 5$ Blocks and $p = 3$ Treatments

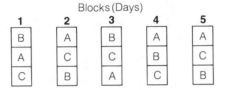

In general, a randomized block design to compare p treatments will contain b blocks, with each block containing p experimental units. Each treatment appears once in every block with the p treatments randomly assigned to the experimental units within each block.

DEFINITION 13.5

A **randomized block design** to compare p treatments involves b blocks, each containing p experimental units. The p treatments are randomly assigned to the units within each block, one unit per treatment.

The partitioning of SS(Total) for the randomized block design is most easily seen by examining Figure 13.7. Note that SS(Total) is now partitioned into three parts:

$$SS(Total) = SSB + SST + SSE$$

FIGURE 13.7

Partitioning of the Total Sum of Squares for the Randomized Block Design

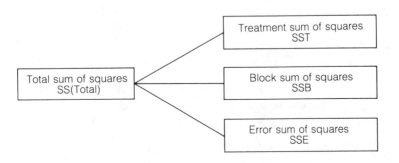

The notation that we will employ in the analysis of variance formulas is shown in Table 13.5. The formulas for calculating SST and SSB take the same pattern as the formula for calculating SST for the completely randomized design. Once these quantities have been calculated, we find SSE by subtraction:

$$SSE = SS(Total) - SST - SSB$$

The formulas for calculating the pertinent sums of squares and mean squares are given in the next box.

TABLE 13.5

Notation for the Analysis of Variance for a Randomized Block Design

	TREATMENT 1	TREATMENT 2	$\cdots$	TREATMENT p
Treatment Totals	T_1	T_2	$\cdots$	T_p
Treatment Means	$\bar{T}_1$	$\bar{T}_2$	$\cdots$	$\bar{T}_p$
	BLOCK 1	BLOCK 2	$\cdots$	BLOCK b
Block Totals	B_1	B_2	$\cdots$	B_b
Block Means	$\bar{B}_1$	$\bar{B}_2$	$\cdots$	$\bar{B}_b$

b = Number of measurements in a single treatment total
p = Number of measurements in a single block total
n = Total number of measurements in the complete experiment = $b \times p$

We are interested in using the randomized block design to test the same null and alternative hypotheses we tested using the completely randomized design:

H_0: $\mu_1 = \mu_2 = \cdots = \mu_p$

H_a: At least two treatment means differ

The test statistic, also identical to that used for the completely randomized design, is

$$F = \frac{MST}{MSE}$$

Since the sum of squares for blocks, SSB, measures the variation explained by the differences among the block means, we can also test

H_0:   The b block means are equal

H_a: At least two block means differ

using the test statistic

$$F = \frac{MSB}{MSE}$$

Although a test of a hypothesis concerning a difference among treatment means is our primary objective, the former test enables us to determine whether there is evidence of a difference among block means—that is, whether blocking is

**COMPUTING FORMULAS FOR THE ANALYSIS OF VARIANCE
FOR A RANDOMIZED BLOCK DESIGN**

$$\sum_{i=1}^{n} y_i = \text{Sum of all } n \text{ measurements}$$

$$\sum_{i=1}^{n} y_i^2 = \text{Sum of squares of all } n \text{ measurements}$$

$$CM = \text{Correction for mean}$$

$$= \frac{(\text{Total of all measurements})^2}{\text{Total number of measurements}} = \frac{\left(\sum_{i=1}^{n} y_i\right)^2}{n}$$

$$SS(\text{Total}) = \text{Total sum of squares}$$

$$= \text{Sum of squares of all measurements} - CM$$

$$= \sum_{i=1}^{n} y_i^2 - CM$$

$$SST = \text{Sum of squares for treatments}$$

$$= \left(\begin{array}{c}\text{Sum of squares of treatment totals with} \\ \text{each square divided by } b, \text{ the number of} \\ \text{measurements for that treatment}\end{array}\right) - CM$$

$$= \frac{T_1^2}{b} + \frac{T_2^2}{b} + \cdots + \frac{T_p^2}{b} - CM$$

$$SSB = \text{Sum of squares for blocks}$$

$$= \left(\begin{array}{c}\text{Sum of squares for block totals with} \\ \text{each square divided by } p, \text{ the number} \\ \text{of measurements in that block}\end{array}\right) - CM$$

$$= \frac{B_1^2}{p} + \frac{B_2^2}{p} + \cdots + \frac{B_b^2}{p} - CM$$

$$SSE = \text{Sum of squares for error}$$

$$= SS(\text{Total}) - SST - SSB$$

$$MST = \text{Mean square for treatments}$$

$$= \frac{SST}{p - 1}$$

$$MSB = \text{Mean square for blocks}$$

$$= \frac{SSB}{b - 1}$$

$$MSE = \text{Mean square for error}$$

$$= \frac{SSE}{n - p - b + 1}$$

really effective. If there are no differences among block means, then blocking will not reduce the variability in the experiment and, consequently, will be ineffective. *In fact, if there are no differences among block means, you will lose information by blocking because blocking reduces the number of degrees of freedom associated with s^2.*

The F tests for comparing block and treatment means are shown in the box. The use of the analysis of variance computational formulas and the tests are demonstrated in Examples 13.4 and 13.5.

ANALYSIS OF VARIANCE F TESTS FOR A RANDOMIZED BLOCK DESIGN

TEST FOR COMPARING TREATMENT MEANS

H_0: $\mu_1 = \mu_2 = \cdots = \mu_p$ (i.e., no differences among the p treatment means)

H_a: At least two of the treatment means differ

Test statistic: $F = \dfrac{\text{MST}}{\text{MSE}} = \dfrac{\text{MST}}{s^2}$

Rejection region: $F > F_\alpha$, where F is based on $\nu_1 = (p - 1)$ and $\nu_2 = (n - p - b + 1)$ degrees of freedom

TEST FOR COMPARING BLOCK MEANS

H_0:   There are no differences among the b block means

H_a: At least two of the block means differ

Test statistic: $F = \dfrac{\text{MSB}}{\text{MSE}} = \dfrac{\text{MSB}}{s^2}$

Rejection region: $F > F_\alpha$, where F is based on $\nu_1 = (b - 1)$ and $\nu_2 = (n - p - b + 1)$ degrees of freedom

---

EXAMPLE 13.4

Prior to submitting a bid for a construction job, companies prepare a detailed analysis of the estimated labor and materials costs required to complete the job. This estimate will depend on the estimator who performs the analysis. An estimate, if too high, will reduce the chance of acceptance of a company's bid price and, if too low, will reduce the profit or even cause the company to lose money on the job. A company that employs three job cost estimators wanted to compare the mean level of the estimators' estimates. This was done by having each estimator estimate the cost of the same four jobs. The data (in hundreds of thousands of dollars) are shown in Table 13.6. Perform an analysis of variance on the data and test to determine whether there is sufficient evidence to indicate differences among treatment and block means. Test using $\alpha = .05$.

SOLUTION

The data for this experiment were collected according to a randomized block design because we would expect estimates of the same job to be more nearly alike than estimates between jobs.

TABLE 13.6

Data for the Randomized Block Design of Example 13.4

		JOB 1	2	3	4	TOTALS	MEANS
ESTIMATOR	1	4.6	6.2	5.0	6.6	22.4	5.60
	2	4.9	6.3	5.4	6.8	23.4	5.85
	3	4.4	5.9	5.4	6.3	22.0	5.50
TOTALS		13.9	18.4	15.8	19.7	67.8	
MEANS		4.63	6.13	5.27	6.57		

The following calculations are required for the analysis of variance:

$$\sum_{i=1}^{n} y_i = 4.6 + 4.9 + \cdots + 6.3 = 67.8$$

$$\sum_{i=1}^{n} y_i^2 = (4.6)^2 + (4.9)^2 + \cdots + (6.3)^2 = 390.28$$

$$\text{CM} = \frac{\left(\sum_{i=1}^{n} y_i\right)^2}{n} = \frac{(67.8)^2}{12} = 383.07$$

$$\text{SS(Total)} = \sum_{i=1}^{n} y_i^2 - \text{CM} = 7.21$$

$$\text{SST} = \frac{\sum_{i=1}^{p} T_i^2}{b} - \text{CM}$$

$$= \frac{(22.4)^2 + (23.4)^2 + (22.0)^2}{4} - 383.07$$

$$= .26$$

$$\text{SSB} = \frac{\sum_{i=1}^{b} B_i^2}{p} - \text{CM} = \frac{(13.9)^2 + (18.4)^2 + (15.8)^2 + (19.7)^2}{3} - 383.07$$

$$= 6.763333$$

$$\text{SSE} = \text{SS(Total)} - \text{SST} - \text{SSB}$$

$$= 7.21 - .26 - 6.763333 = .186667$$

$$\text{MST} = \frac{\text{SST}}{p - 1} = \frac{.26}{2} = .13$$

$$\text{MSB} = \frac{\text{SSB}}{b - 1} = \frac{6.763333}{3} = 2.254$$

$$\text{MSE} = s^2 = \frac{\text{SSE}}{n - b - p + 1} = \frac{.186667}{12 - 4 - 3 + 1} = \frac{.186667}{6} = .031111$$

The sources of variation and their respective degrees of freedom, sums of squares, and mean squares for a randomized block design are summarized in the analysis of variance table shown in Table 13.7. The analysis of variance table for this example is shown in Table 13.8. Note that the degrees of freedom for the three sources of variation, Treatments, Blocks, and Error, sum to the degrees of freedom for SS(Total). Similarly, the sums of squares for the sources will always possess a sum equal to SS(Total).

TABLE 13.7

Analysis of Variance Table for a Randomized Block Design

SOURCE	df	SS	MS	F
Treatments	$p - 1$	SST	MST	MST/MSE
Blocks	$b - 1$	SSB	MSB	MSB/MSE
Error	$n - b - p + 1$	SSE	MSE	
Total	$n - 1$	SS(Total)		

TABLE 13.8

Analysis of Variance Table for Example 13.4

SOURCE	df	SS	MS	F
Treatments (Estimators)	2	.260	.130	4.19
Blocks (Jobs)	3	6.763	2.254	72.71
Error	6	.187	.031	
Total	11	7.210		

The values of the F statistics needed to test hypotheses about differences among the treatment and block means are shown in the last column of the analysis of variance table. Since the F statistic for testing treatment means is based on $\nu_1 = (p - 1) = 2$ and $\nu_2 = (n - b - p + 1) = 6$ degrees of freedom, we will reject the null hypothesis

H_0:   There are no differences among treatment means

if $F > F_{.05} = 5.14$. Since the computed value of F, 4.19, is less than $F_{.05}$, there is insufficient evidence, at the $\alpha = .05$ level of significance, to indicate differences among the mean level of estimates for the three estimators. The observed significance level for the test (obtained from a computer printout) is .07.

The F statistic for testing

H_0:   There are no differences among block means

is based on $\nu_1 = (b - 1) = 3$ and $\nu_2 = (n - b - p + 1) = 6$ degrees of freedom, and we will reject H_0 if $F > F_{.05} = 4.76$. Since the computed value of F, 72.71, exceeds $F_{.05}$, there is sufficient evidence to indicate differences among the mean estimates for the four different jobs.   ∎

Caution: The result of the test for the equality of block means must be interpreted with care, especially when the calculated value of the F test statistic does not fall in the rejection region. This does not necessarily imply that the block means are the same, i.e., that blocking is unimportant. Reaching this conclusion would be equivalent to accepting the null hypothesis, a practice we have carefully avoided due to the unknown probability of committing a Type II error (that is, of accepting H_0 when H_a is true). In other words, even when a test for block differences is inconclusive, we may still want to use the randomized block design in similar future experiments. If the experimenter believes that the experimental units are more homogeneous within blocks than among blocks, he or she should use the randomized block design regardless of whether the test comparing the block means shows them to be different.

Confidence intervals for the difference between a pair of treatment means are shown in the accompanying box.

CONFIDENCE INTERVALS FOR THE DIFFERENCE, $(\mu_i - \mu_j)$, BETWEEN A PAIR OF TREATMENT OR BLOCK MEANS: RANDOMIZED BLOCK DESIGN

TREATMENT MEANS

$$(\bar{T}_i - \bar{T}_j) \pm t_{\alpha/2}s\sqrt{\frac{2}{b}}$$

BLOCK MEANS

$$(\bar{B}_i - \bar{B}_j) \pm t_{\alpha/2}s\sqrt{\frac{2}{p}}$$

where

b = Number of blocks

p = Number of treatments

and $t_{\alpha/2}$ is based on $(n - b - p + 1)$ degrees of freedom

---

EXAMPLE 13.5

Refer to Example 13.4. Find a 90% confidence interval for the difference between the mean level of estimates for estimators 1 and 2.

SOLUTION

From Example 13.4, we know that $b = 4$, $\bar{T}_1 = 5.60$, $\bar{T}_2 = 5.85$, and $s^2 = $ MSE $= .031111$. The degrees of freedom associated with s^2 (and, therefore, with $t_{\alpha/2}$) is 6. Therefore, $s = \sqrt{s^2} = \sqrt{.031111} = .176$ and $t_{\alpha/2} = t_{.05} = 1.943$. Substituting these values into the formula for the confidence interval for $(\mu_1 - \mu_2)$, we obtain

$$(\bar{T}_1 - \bar{T}_2) \pm t_{\alpha/2}s\sqrt{\frac{2}{b}}$$

$$(5.60 - 5.85) \pm 1.943(.176)\sqrt{\frac{2}{4}}$$

$$-.250 \pm .241$$

or, $-.491$ to $-.009$. Since each unit represents $100,000, we estimate the difference between the mean level of job estimates for estimators 1 and 2 to be enclosed by the interval, $-$49,100$ to $-$900$. [*Note:* At first glance, this result may appear to contradict the result of the F test for comparing treatment means. However, the observed significance level of the F test (.07) implies that significant differences exist between the means at $\alpha = .10$, which is consistent with the fact that 0 is not within the 90% confidence interval.]   ∎

**EXERCISES
13.12–13.19**

13.12 A randomized block design was conducted to compare the mean responses for three treatments, A, B, and C, in four blocks. The data are shown in the table.

TREATMENT	BLOCK			
	1	2	3	4
A	3	6	1	2
B	5	7	4	6
C	2	3	2	2

a. Compute the appropriate sums of squares and mean squares and fill in the entries in the analysis of variance table shown below:

SOURCE	df	SS	MS	F
Treatments				
Blocks				
Error				
Total				

b. Do the data provide sufficient evidence to indicate a difference among treatment means? Test using $\alpha = .05$.
c. Do the data provide sufficient evidence to indicate that blocking was effective in reducing the experimental error? Test using $\alpha = .05$.
d. Find a 90% confidence interval for $(\mu_A - \mu_B)$.
e. What assumptions must the data satisfy to make the F tests in parts **b** and **c** valid?

13.13 The analysis of variance for a randomized block design produced the ANOVA table entries shown in the accompanying table.

SOURCE	df	SS	MS	F
Treatments	3	27.1		
Blocks	5		14.90	
Error		33.4		
Total				

a. Complete the ANOVA table.
b. Do the data provide sufficient evidence to indicate a difference among the treatment means? Test using $\alpha = .01$.
c. Do the data provide sufficient evidence to indicate that blocking was a useful design strategy to employ for this experiment? Explain.
d. If the sample means for treatments A and B are $\bar{T}_A = 9.7$ and $\bar{T}_B = 12.1$, respectively, find a 90% confidence interval for $(\mu_A - \mu_B)$. Interpret the interval.

13.14 An evaluation of diffusion bonding of zircaloy components is performed. The main objective is to determine which of three elements—nickel, iron, or copper—is the best bonding agent. A series of zircaloy components are bonded using each of the possible bonding agents. Since there is a great deal of variation in components machined from different ingots, a randomized block design is used, blocking on the ingots. A pair of components from each ingot are bonded together using each of the three agents, and the pressure (in units of 1,000 pounds per square inch) required to separate the bonded components is measured. The data are shown in the accompanying table.

INGOT	BONDING AGENT		
	Nickel	Iron	Copper
1	67.0	71.9	72.2
2	67.5	68.8	66.4
3	76.0	82.6	74.5
4	72.7	78.1	67.3
5	73.1	74.2	73.2
6	65.8	70.8	68.7
7	75.6	84.9	69.0

a. Is there evidence of a difference in pressure required to separate the components among the three bonding agents? Use $\alpha = .05$.
b. Form a 95% confidence interval to estimate the difference in mean pressure between nickel and iron. Interpret this interval.

13.15 The Perth (Australia) Metropolitan Water Authority recently completed construction of land pipeline for transporting domestic wastewaters from a primary treatment plant. During construction, the cement mortar lining of the pipeline was tested for cracking to determine whether autogenous healing will seal the cracks. Otherwise, expensive epoxy filling repairs would be necessary (*Proceedings of the Institute of Civil Engineers*, Apr. 1986). After cracks were observed in the pipeline, it was kept full of water for a period of 14 weeks. At each of 12 crack locations, crack widths were measured (in millimeters) after the 2nd, 6th, and 14th weeks of the wet period, as shown in the accompanying table.

CRACK LOCATION	CRACK WIDTH AFTER WETTING			
	0 Weeks	2 Weeks	6 Weeks	14 Weeks
1	.50	.20	.10	.10
2	.40	.20	.10	.10
3	.60	.30	.15	.10
4	.80	.40	.10	.10
5	.80	.30	.05	.05
6	1.00	.40	.05	.05
7	.90	.25	.05	.05
8	1.00	.30	.05	.10
9	.70	.25	.10	.10
10	.60	.25	.10	.05
11	.30	.15	.10	.05
12	.30	.14	.05	.05

Source: Cox, B. G. and Kelsall, K. J. "Construction of Cape Peron Ocean Outlet Perth, Western Australia." *Proceedings of the Institute of Civil Engineers*, Part 1, Vol. 80, Apr. 1986, p. 479 (Table 1).

a. Conduct a test to determine whether the mean crack widths differ for the four time periods. Test using $\alpha = .05$.

b. Construct a 95% confidence interval for the difference between the initial mean crack width (0 weeks) and the mean crack width after wetting for 14 weeks. Interpret the interval.

13.16 A highway paving firm employs three scratch cost engineers. Usually, only one engineer works on each contract, but it is advantageous to the company if the engineers are consistent enough so that it does not matter which of the three is assigned to estimate the cost of a particular contract. To check on the consistency of the engineers, several contracts are selected and all three engineers are asked to make cost estimates. The estimates (in thousands of dollars) for each contract by each engineer are given in the table.

CONTRACT	ENGINEER		
	A	B	C
1	27.3	26.5	28.2
2	66.7	67.3	65.9
3	104.8	102.1	100.8
4	87.6	85.6	86.5
5	54.5	55.6	55.9
6	58.7	59.2	60.1

a. Do these estimates provide sufficient evidence that the mean cost estimates for at least two of the engineers differ? Use $\alpha = .05$.

b. Find the approximate observed significance level for the test in part **a**, and interpret its value.

c. Present the complete ANOVA summary table for this experiment.

d. Use a 90% confidence interval to estimate the difference between the mean responses given by engineers B and C.

13.17 An experiment was conducted to compare two different methods of sampling to determine the iron content of iron ore. The first method, *manual sampling*, involved stopping the ore conveyor belt and removing a 1-meter length of ore from the belt. The second method, *mechanical sampling*, involved collecting the sample of ore (an increment) from the stream of ore falling from the end of the belt. The samples were matched by making certain that the ore for the mechanical sample came from approximately the same increment on the conveyor belt as the ore obtained for the manual sample. The data shown in the table are measurements on the iron ore content (percentage of increment) for 16 increments selected from a shipment of Chilean iron ore. (Only a portion of the experimental data is given here.)

INCREMENT NUMBER	SAMPLING METHOD		INCREMENT NUMBER	SAMPLING METHOD	
	Mechanical	Manual		Mechanical	Manual
1	62.66	63.92	9	61.75	62.03
2	62.87	63.64	10	63.15	62.17
3	63.22	63.64	11	63.08	64.34
4	63.01	63.27	12	63.22	62.30
5	62.10	62.94	13	63.22	63.50
6	63.43	64.61	14	63.08	62.73
7	63.22	62.87	15	62.87	61.89
8	63.57	64.20	16	61.68	62.10

Source: Sato, T., Ito, K., Chujo, S., and Takahashi, U. "Examples of Experiments on Systematic Sampling of Iron Ore." *Reports of Statistical Application Research*, Union of Japanese Scientists and Engineers, Vol. 18, No. 1, 1971.

a. What type of experimental design was employed for this experiment?

b. Perform an analysis of variance for the data and display the results in an analysis of variance table.

c. Let μ_1 be the mean of the population of measurements obtained by the mechanical sampling method and let μ_2 be the corresponding mean for the manual sampling method. Do the data provide sufficient evidence to indicate that μ_1 differs from μ_2? Test using $\alpha = .05$.

d. Find a 95% confidence interval for $\mu_1 - \mu_2$.

e. Compute the difference in the percentage iron measurements for each increment of ore. Then conduct the test of part **b** using a Student's t test for a matched-pairs experiment. [See Section 9.7.]

f. Show that the value of F computed in part **c** is equal to the square of the value of t computed in part **e**. Explain why the tests in parts **c** and **e** are equivalent.

g. Find a 95% confidence interval for $\mu_1 - \mu_2$ using the methods of Section 8.7. Verify that this gives the same confidence interval as that in part **d**.

13.18 A simulation study was conducted to investigate the machine performance of several new algorithms for functions in the FORTRAN computer program library (*IBM Journal of Research and Development*, Mar. 1986). The accompanying table gives the time per call (in microseconds) for several randomly selected scalar functions (averaged over 10,000 random arguments) on each of three different IBM System/370 machines.

FUNCTION	IBM 4331	IBM 4361	IBM 4341
EDUM	9.90	3.07	4.88
ACOS CIRC(O,PI)	179.62	33.28	33.23
SIN LINEAR(−PI,PI)	105.72	24.13	27.08
EXP LINEAR(−16,16)	254.82	39.14	37.46
D2DUM	13.47	4.63	5.72

Source: Agarwal, R. C. et al. "New Scalar and Vector Elementary Functions for the IBM System/370." *IBM Journal of Research and Development*, Vol. 30, No. 2, Mar. 1986, p. 139 (Table 4). Copyright 1986 by International Business Machines Corportion; reprinted with permission.

a. Treating the functions as blocks, construct an ANOVA summary table for this randomized block experiment.
b. Is there sufficient evidence to indicate that the mean function call times differ for the three IBM System/370 machines? Test using $\alpha = .10$.
c. Conduct a test to determine if blocking on functions was effective in removing an extraneous source of variation. Use $\alpha = .10$.

13.19 A power plant, which uses water from the surrounding bay for cooling its condensers, is required by the EPA to determine whether discharging its heated water into the bay has a detrimental effect on the plant life in the water. The EPA requests that the power plant make its investigation at three strategically chosen locations, called *stations*. Stations 1 and 2 are located near the plant's discharge tubes, while station 3 is located farther out in the bay. During one randomly selected day in each of four months, a diver descends to each of the stations, randomly samples a square meter area of the bottom, and counts the number of blades of different types of grasses present. The results for one important grass type are listed in the table.

MONTH	STATION		
	1	2	3
May	28	31	53
June	25	22	61
July	37	30	56
August	20	26	48

a. Is there sufficient evidence to indicate that the mean number of blades found per square meter per month differs for at least two of the three stations? Use $\alpha = .05$.

b. Is there sufficient evidence to indicate that the mean number of blades found per square meter differs among the 4 months? Use $\alpha = .05$.

c. Place a 90% confidence interval on the difference in means between stations 1 and 3.

S E C T I O N   13.7

THE ANALYSIS OF VARIANCE FOR A TWO-WAY CLASSIFICATION OF DATA: FACTORIAL EXPERIMENTS

A randomized block design is often called a **two-way classification of data** because it has the following characteristics:

1. It involves two independent variables—one factor and one direction of blocking.

2. Each level of one independent variable occurs with every level of the other independent variable.

A two-way classification of data always permits the display of the data in a two-way table, one containing r rows and c columns. For example, the data for the randomized block design of Example 13.4 are displayed in a two-way table containing $r = 3$ rows and $c = 4$ columns in Table 13.6. Each of the $rc = (3)(4) = 12$ cells of the table contains one observation.

The treatment selection for a two-factor experiment may also yield a two-way classification of data. For example, suppose you want to relate the mean number of defects on a finished item—say, a new desk top—to two factors, type of nozzle for the varnish spray gun and length of spraying time. Suppose further that you want to investigate the mean number of defects per desk for three types (three levels) of nozzles and for two lengths (two levels) of spraying time. If we choose the treatments for the experiment to include all combinations of the three levels of nozzle type with the two levels of spraying time, we will obtain a two-way classification of data. This selection of treatments is called a **complete 3 × 2 factorial experiment**.

DEFINITION 13.6

A **factorial experiment** is a method for selecting the treatments (that is, the factor–level combinations) to be included in an experiment. A complete factorial experiment is one in which observations are made for every combination of the factor levels.

If we were to include a third factor, say, paint type, at three levels, then a complete factorial experiment would include all $3 \times 2 \times 3 = 18$ combinations of nozzle type, spraying time, and paint type. The resulting collection of data would be called a **three-way classification of data**.

Factorial experiments are useful methods for selecting treatments because they permit us to make inferences about **factor interactions**. In this section we will learn how to perform an analysis of variance for a two-way classification of data. You will see that the computational procedure is the same for both the randomized block design and for the two-factor factorial experiment because each involves a two-way classification of data. You will also learn why the analysis of variance F tests differ and how these tests can be used to interpret the results of the experiment.

Suppose a two-way classification represents a two-factor factorial experiment with factor A at a levels and factor B at b levels. Further, assume that the ab treatments of the factorial experiment are replicated r times so that there are r observations for each of the ab treatment combinations (i.e., there are r observations in each of the ab cells of the two-way table). Then the total number of observations is $n = abr$ and the total sum of squares, SS(Total), can be partitioned into four parts, SS(A), SS(B), SS(AB), and SSE (see Figure 13.8). The first two sums of squares, SS(A) and SS(B), are called **main effect sums of squares** to distinguish them from the **interaction sum of squares**, SS(AB).

FIGURE 13.8

Partitioning of the Total Sum of Squares for a Two-Factor Factorial Experiment

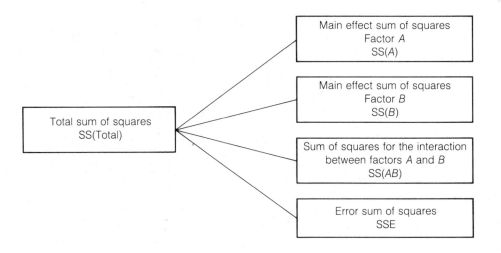

When the number of observations per cell for a two-way factorial experiment is the same for every cell (r observations per cell), the sums of squares and the degrees of freedom for the analysis of variance are additive:

$$SS(\text{Total}) = SS(A) + SS(B) + SS(AB) + SSE$$

and

$$abr - 1 = (a - 1) + (b - 1) + (a - 1)(b - 1) + ab(r - 1)$$

and the analysis of variance table would appear as shown in Table 13.9. Note that for a factorial experiment, the number r of observations per factor–level combination, must always be 2 or more (i.e., $r \geq 2$). Otherwise, you will not have any degrees of freedom for SSE.

TABLE 13.9 ANOVA Table for a Two-Way Classification of Data with r Observations per Cell: Neither Variable Is a Direction of Blocking

SOURCE	df	SS	MS	F
Main effects A	$(a - 1)$	SS(A)	$MS(A) = SS(A)/(a - 1)$	MS(A)/MSE
Main effects B	$(b - 1)$	SS(B)	$MS(B) = SS(B)/(b - 1)$	MS(B)/MSE
AB interaction	$(a - 1)(b - 1)$	SS(AB)	$MS(AB) = SS(AB)/[(a - 1)(b - 1)]$	MS(AB)/MSE
Error	$ab(r - 1)$	SSE	$MSE = SSE/[ab(r - 1)]$	
Total	$abr - 1$	SS(Total)		

If either A or B represents a direction of blocking, then the AB interaction terms are deleted from the ANOVA table and the degrees of freedom and sums of squares from lines 3 and 4 of Table 13.9 are combined to form a source of error variation. This is because the block–treatment interaction always represents experimental error.* The resulting analysis of variance would appear as shown in Table 13.10. Note that both the degrees of freedom and the sums of squares sum to those associated with the total sum of squares of deviations, SS(Total).

TABLE 13.10

ANOVA Table for a Two-Way Classification of Data with r Observations per Cell When B Is a Direction of Blocking

SOURCE	df	SS	MS	F
A	$a - 1$	SS(A)	$SS(A)/(a - 1)$	MS(A)/MSE
Blocks	$b - 1$	SS(B)	$SS(B)/(b - 1)$	MS(B)/MSE
Error	$abr - a - b + 1$	SSE	$SSE/(abr - a - b + 1)$	
Total	$abr - 1$	SS(Total)		

The notation used in the formulas for the respective sums of squares and the formulas for the sums of squares are given in the boxes on pages 664–665.

We have already explained (Section 13.6) how to conduct the F tests for determining whether a difference exists among treatment means (or block means) if the data are obtained from a randomized block design and the analysis of variance is as shown in Table 13.10. The confidence intervals for estimating the difference between two treatment means are also given in Section 13.6.

---

*The failure of the difference between a pair of treatments to remain the same from block to block is experimental error. If we think of a randomized block design in the context of a factorial experiment with treatments as one factor and blocks as another, then the failure of the difference between two treatments to remain the same from block to block is, by Definition 13.7, block–treatment interaction. In other words, in a randomized block design, block–treatment interaction and experimental error are synonymous.

NOTATION FOR THE ANALYSIS OF VARIANCE FOR A TWO-WAY CLASSIFICATION OF DATA

a = Number of levels of independent variable 1

b = Number of levels of independent variable 2

r = Number of measurements for each pair of levels of independent variables 1 and 2

A_i = Total of all measurements of independent variable 1 at level i ($i = 1, 2, \ldots, a$)

$\bar{A}_i$ = Mean of all measurements of independent variable 1 at level i ($i = 1, 2, \ldots, a$)

$\quad = \dfrac{A_i}{br}$

B_j = Total of all measurements of independent variable 2 at level j ($j = 1, 2, \ldots, b$)

$\bar{B}_j$ = Mean of all measurements of independent variable 2 at level j ($j = 1, 2, \ldots, b$)

$\quad = \dfrac{B_j}{ar}$

AB_{ij} = Total of all measurements at the ith level of independent variable 1 and at the jth level of independent variable 2 ($i = 1, 2, \ldots, a$; $j = 1, 2, \ldots, b$)

n = Total number of measurements = abr

If the data are obtained from a two-factor factorial experiment, the analysis of variance will appear as shown in Table 13.9. To test a hypothesis for any one of the three sources of variation (those corresponding to main effects of factor A, those corresponding to main effects of factor B, or those corresponding to the interaction between the two factors) you proceed in exactly the same manner as was done in earlier analyses, i.e., you divide the appropriate mean square by MSE and use this F ratio as a test statistic.

To estimate the mean for a single cell of the two-way table or the difference between the means for two cells (i.e., two different combinations of levels of the two factors), use the formulas in the boxes on pages 666–667.

Before we work through a numerical example of an analysis of variance for a factorial experiment, we need to understand the practical significance of the tests for factor interaction and factor main effects. We illustrate these concepts in Example 13.6.

COMPUTING FORMULAS FOR THE ANALYSIS OF VARIANCE FOR A TWO-WAY CLASSIFICATION OF DATA

CM = Correction for the mean

$$= \frac{(\text{Total of all } n \text{ measurements})^2}{n}$$

$$= \frac{\left(\sum_{i=1}^{n} y_i\right)^2}{n}$$

$SS(\text{Total})$ = Total sum of squares

= Sum of squares of all n measurements $-$ CM

$$= \sum_{i=1}^{n} y_i^2 - CM$$

$SS(A)$ = Sum of squares for main effects, independent variable 1

$$= \left(\begin{array}{c}\text{Sum of squares of the totals } A_1, A_2, \ldots, A_a \\ \text{divided by the number of measurements} \\ \text{in a single total, namely } br\end{array}\right) - CM$$

$$= \frac{\sum_{i=1}^{a} A_i^2}{br} - CM$$

$SS(B)$ = Sum of squares for main effects, independent variable 2

$$= \left(\begin{array}{c}\text{Sum of squares of the totals } B_1, B_2, \ldots, B_b \\ \text{divided by the number of measurements} \\ \text{in a single total, namely } ar\end{array}\right) - CM$$

$$= \frac{\sum_{i=1}^{b} B_i^2}{ar} - CM$$

$SS(AB)$ = Sum of squares for AB interaction

$$= \left(\begin{array}{c}\text{Sum of squares of the cell totals} \\ AB_{11}, AB_{12}, \ldots, AB_{ab} \text{ divided by} \\ \text{the number of measurements} \\ \text{in a single total, namely } r\end{array}\right) - SS(A) - SS(B) - CM$$

$$= \frac{\sum_{j=1}^{b}\sum_{i=1}^{a} AB_{ij}^2}{r} - SS(A) - SS(B) - CM$$

ANALYSIS OF VARIANCE F TESTS FOR A TWO-FACTOR FACTORIAL EXPERIMENT

TEST FOR FACTOR INTERACTION

H_0: No interaction between factors A and B

H_a: Factors A and B interact

Test statistic: $F = \dfrac{\text{MS}(AB)}{\text{MSE}} = \dfrac{\text{MS}(AB)}{s^2}$

Rejection region: $F > F_\alpha$, where F is based on $\nu_1 = (a - 1)(b - 1)$ and $\nu_2 = ab(r - 1)\text{df}$

TEST FOR MAIN EFFECTS FOR FACTOR A

H_0: There are no differences among the means for main effect A

H_a: At least two of the main effect A means differ

Test statistic: $F = \dfrac{\text{MS}(A)}{\text{MSE}} = \dfrac{\text{MS}(A)}{s^2}$

Rejection region: $F > F_\alpha$, where F is based on $\nu_1 = (a - 1)$ and $\nu_2 = ab(r - 1)\text{df}$

TEST FOR MAIN EFFECTS FOR FACTOR B

H_0: There are no differences among the means for main effect B

H_a: At least two of the main effect B means differ

Test statistic: $F = \dfrac{\text{MS}(B)}{\text{MSE}} = \dfrac{\text{MS}(B)}{s^2}$

Rejection region: $F > F_\alpha$, where F is based on $\nu_1 = (b - 1)$ and $\nu_2 = ab(r - 1)\text{df}$

$(1 - \alpha)100\%$ CONFIDENCE INTERVAL FOR THE MEAN OF A SINGLE CELL OF THE TWO-WAY TABLE: FACTORIAL EXPERIMENT

$$\bar{y}_{ij} \pm t_{\alpha/2}\frac{s}{\sqrt{r}}$$

where $\bar{y}_{ij}$ is the cell mean for the cell in the ith row, jth column,

$r = $ Number of measurements per cell

$s = \sqrt{\text{MSE}}$

and $t_{\alpha/2}$ is based on $ab(r - 1)$ df.

**$(1 - \alpha)100\%$ CONFIDENCE INTERVAL FOR THE DIFFERENCE
IN A PAIR OF CELL MEANS: FACTORIAL EXPERIMENT**

Let

$\bar{y}_1$ = Sample mean of the r measurements in the first cell

$\bar{y}_2$ = Sample mean of the r measurements in the second cell

Then, the $(1 - \alpha)100\%$ confidence interval for the difference between the cell means is

$$(\bar{y}_1 - \bar{y}_2) \pm t_{\alpha/2}s\sqrt{\frac{2}{r}}$$

where $s = \sqrt{\text{MSE}}$ and $t_{\alpha/2}$ is based on $ab(r - 1)$ df.

EXAMPLE 13.6

A company that stamps gaskets out of sheets of rubber, plastic, and other materials, wants to compare the mean number of gaskets produced per hour for two different types of stamping machines. Practically, the manufacturer wants to determine whether one machine is more productive than the other and, even more important, whether one machine is more productive in producing rubber gaskets while the other is more productive in producing plastic gaskets. To answer these questions, the manufacturer decides to conduct a 2×3 factorial experiment using three types of gasket materials, B_1, B_2, and B_3, with each of the two types of stamping machines, A_1 and A_2. Each machine is operated for three 1-hour time periods for each of the gasket materials, with the 18 1-hour time periods assigned to the six machine–material combinations in random order. (The purpose of the randomization is to eliminate the possibility that uncontrolled environmental factors might bias the results.) Suppose we have calculated and plotted the six treatment means. Two hypothetical plots of the six means are shown in Figures 13.9(a) and 13.9(b). The three means for stamping machine A_1 are connected by solid line segments and the corresponding three means for machine

FIGURE 13.9

Hypothetical Plot of the Means for the Six Machine–Material Combinations

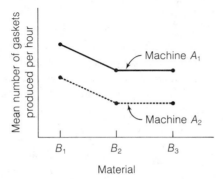

a. No interaction

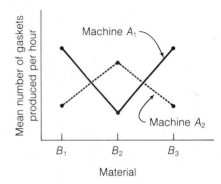

b. Interaction

A_2 by dashed line segments. What do these plots imply about the productivity of the two stamping machines?

SOLUTION

Figure 13.9(a) suggests that machine A_1 produces a larger number of gaskets per hour, regardless of the gasket material, and is therefore superior to machine A_2. On the average, machine A_1 stamps more cork (B_1) gaskets per hour than rubber or plastic, but the *difference* in the mean numbers of gaskets produced by the two machines remains approximately the same, regardless of the gasket material. Thus, the difference in the mean number of gaskets produced by the two machines is *independent* of the gasket material used in the stamping process.

In contrast to Figure 13.9(a), Figure 13.9(b) shows the productivity of machine A_1 to be greater than for machine A_2 when the gasket material is cork (B_1) or plastic (B_3). But the means are reversed for rubber (B_2) gasket material. For this material, machine A_2 produces, on the average, more gaskets per hour than machine A_1. Thus, Figure 13.9(b) illustrates a situation where the mean value of the response variable *depends* on the combination of the factor levels. When this situation occurs, we say that the factors *interact*. Thus, one of the most important objectives of a factorial experiment is to detect factor interaction if it exists.   ∎

DEFINITION 13.7

In a factorial experiment, when the difference in the mean levels of factor A depends on the different levels of factor B, we say that the factors A and B **interact**. If the difference is independent of the levels of B, then there is **no interaction** between A and B.

Tests for main effects are relevant only when no interaction exists between factors. Generally, the test for interaction is performed first. If there is evidence of factor interaction, then we will not perform the tests on the main effects. Rather, we will want to focus attention on the individual cell (treatment) means, perhaps locating one that is the largest or the smallest.

EXAMPLE 13.7

A manufacturer, whose daily supply of raw materials is variable and limited, can use the material to produce two different products in various proportions. The profit per unit of raw material obtained by producing each of the two products depends on the length of a product's manufacturing run and hence on the amount of raw material assigned to it. Other factors, such as worker productivity and machine breakdown, affect the profit per unit as well, but their net effect on profit is random and uncontrollable. The manufacturer has conducted an experiment to investigate the effect of the level of supply of raw materials, S, and the ratio of its assignment, R, to the two product manufacturing lines on the profit y per unit of raw material. The ultimate goal would be to be able to choose the best ratio R to match each day's supply of raw materials, S. The levels of supply of the raw material chosen for the experiment were 15, 18, and 21 tons; the levels of the ratio of allocation to the two product lines were $\frac{1}{2}$, 1, and 2. The response was the profit (in dollars) per unit of raw material supply obtained from a single day's production. Three replications of a complete 3×3 factorial experiment were conducted in a random sequence (i.e., a completely randomized design). The data for the 27 days are shown in Table 13.11.

TABLE 13.11

Data for Example 13.7

		RAW MATERIAL SUPPLY, TONS (S)		
		15	18	21
RATIO OF RAW MATERIAL ALLOCATION (R)	$\frac{1}{2}$	23, 20, 21	22, 19, 20	19, 18, 21
	1	22, 20, 19	24, 25, 22	20, 19, 22
	2	18, 18, 16	21, 23, 20	20, 22, 24

a. Calculate the appropriate sums of squares and construct an ANOVA table.
b. Do the data present sufficient evidence to indicate an interaction between supply S and ratio R?
c. Find a 95% confidence interval to estimate the mean profit per unit of raw materials when $S = 18$ tons and the ratio of allocation is $R = 1$.
d. Find a 95% confidence interval to estimate the difference in mean profit per unit of raw materials between $S = 18$, $R = \frac{1}{2}$, and $S = 18$, $R = 1$.

SOLUTION

a. The sums of squares for the ANOVA table are calculated as follows:

$$CM = \frac{(\text{Total of all } n \text{ measurements})^2}{n} = \frac{(558)^2}{27} = 11{,}532$$

$$SS(\text{Total}) = \sum_{i=1}^{n} y_i^2 - CM = 11{,}650 - 11{,}532 = 118$$

The next step is to construct a table showing the totals of y-values for each combination of levels of supply S and ratio R and then the totals for each level of S and for each level of R. These totals, computed from the raw data table, are shown in Table 13.12 (page 670). Then,

$$SS(\text{Supply}) = \frac{\sum_{i=1}^{3} S_i^2}{9} - CM = \frac{(177)^2 + (196)^2 + (185)^2}{9} - CM$$

$$= \frac{103{,}970}{9} - 11{,}532 = 20.22$$

$$SS(\text{Ratio}) = \frac{\sum_{j=1}^{3} R_j^2}{9} - CM = \frac{(183)^2 + (193)^2 + (182)^2}{9} - CM$$

$$= \frac{103{,}862}{9} - 11{,}532 = 8.22$$

$$SS(SR) = \sum_{j=1}^{3}\sum_{i=1}^{3} \frac{SR_{ij}^2}{3} - SS(\text{Supply}) - SS(\text{Ratio}) - CM$$

$$= \frac{(64)^2 + (61)^2 + (58)^2 + \cdots + (66)^2}{3} - 20.22 - 8.22 - 11{,}532$$

$$= \frac{34{,}820}{3} - 20.22 - 8.22 - 11{,}532 = 46.22^*$$

$$SSE = SS(\text{Total}) - SS(\text{Supply}) - SS(\text{Ratio}) - SS(SR)$$

$$= 118.00 - 20.22 - 8.22 - 46.22 = 43.33$$

The ANOVA table is given in Table 13.13.

TABLE 13.12
Totals for the Data
in Table 13.11

		RAW MATERIAL SUPPLY, TONS (S)			
		15	18	21	
RATIO OF RAW MATERIAL ALLOCATION (R)	$\frac{1}{2}$	64	61	58	$R_1 = 183$
	1	61	71	61	$R_2 = 193$
	2	52	64	66	$R_3 = 182$
		$S_1 = 177$	$S_2 = 196$	$S_3 = 185$	Total = 558

TABLE 13.13
ANOVA Table for
Example 13.7

SOURCE	df	SS	MS
Supply	2	20.22	10.11
Ratio	2	8.22	4.11
Supply–Ratio interaction	4	46.22	11.56
Error	18	43.33	2.41
Total	26	118.00	

b. To test the null hypothesis that supply and ratio do not interact, we use the test statistic

$$F = \frac{MS(SR)}{s^2} = \frac{11.56}{2.41} = 4.80$$

The degrees of freedom associated with MS(SR) and MSE are 4 and 18, respectively (given in Table 13.13). Therefore, we reject H_0 if $F > F_{.05}$, where $\nu_1 = 4$, $\nu_2 = 18$, and $F_{.05} = 2.93$. Since the computed value of F (4.80) exceeds $F_{.05}$, we reject H_0 and conclude that supply and ratio interact. The presence

*Using full accuracy in all sums of squares.

of interaction tells you that the mean profit depends on the particular combination of levels of supply S and ratio R. Consequently, there is little point in checking to see whether the means differ for the three levels of supply or whether they differ for the three levels of ratio (i.e., we will not perform the tests for main effects). For example, the supply level that gave the highest mean profit (over all levels of R) might not be the same supply–ratio level combination that produces the largest mean profit per unit of raw material.

c. A 95% confidence interval for the mean $E(y)$ when supply $S = 18$ and ratio $R = 1$ is

$$\bar{y}_{18,1} \pm t_{.025}\left(\frac{s}{\sqrt{r}}\right)$$

where $\bar{y}_{18,1}$ is the mean of the $r = 3$ values of y obtained for $S = 18$, $R = 1$ and $t_{.025} = 2.101$ is based on 18 df. Substituting, we obtain

$$\frac{71}{3} \pm 2.101 \left(\frac{1.55}{\sqrt{3}}\right)$$

$$23.67 \pm 1.88$$

Therefore, our interval estimate for the mean profit per unit of raw material where $S = 18$ and $R = 1$ is \$21.79 to \$25.55.

d. A 95% confidence interval for the difference in mean profit per unit of raw material for two different combinations of levels of S and R is

$$(\bar{y}_1 - \bar{y}_2) \pm t_{.025}s \sqrt{\frac{2}{r}}$$

where $\bar{y}_1$ and $\bar{y}_2$ represent the means of the $r = 3$ replications for the factor–level combinations $S = 18$, $R = \frac{1}{2}$ and $S = 18$, $R = 1$, respectively. Substituting, we obtain

$$\left(\frac{61}{3} - \frac{71}{3}\right) \pm (2.101)(1.55) \sqrt{\frac{2}{3}}$$

$$-3.33 \pm 2.66$$

Therefore, the interval estimate for the difference in mean profit per unit of raw material for the two factor–level combinations is $(-\$5.99, -\$.67)$. The negative values indicate that we estimate the mean for $S = 18$, $R = \frac{1}{2}$ to be less than the mean for $S = 18$, $R = 1$ by between \$.67 and \$5.99. ∎

The techniques illustrated in the solution of Example 13.7 would answer most of the practical questions you might have about profit per unit of Supply of raw materials if the two independent variables affecting the response were qualitative. But since both independent variables are quantitative, we can obtain much more information about their effect on response by performing a regression analysis. For example, the analysis of variance in Example 13.7 enables us to estimate the mean profit per unit of Supply for *only* the nine combinations of Supply–Ratio levels used in the factorial experiment. It will *not* permit us to estimate the mean response for some other combination of levels of the independent variables not included among the nine. For example, the prediction equation obtained from a regression analysis would enable you to estimate the mean profit per unit of Supply when $S = 17$, $R = 1$. You could not obtain this estimate from the analysis of variance in Example 13.7.

The prediction equation (i.e., the estimated response surface) found by regression analysis also contributes other information not provided by an analysis of variance. For example, we might want to estimate the rate of change in the mean profit, $E(y)$, for unit changes in S, R, or both for specific values of S and R. We will illustrate these applications of regression analysis to an analysis of variance in Section 13.8.

EXERCISES
13.20–13.26

13.20 The analysis of variance for a 3×2 factorial experiment, with four observations per treatment, produced the ANOVA table entries shown here.

SOURCE	df	SS	MS	F
A		100		
B	1			
AB	2		2.5	
Error			2.0	
Total		700		

a. Complete the ANOVA table.
b. Test for interaction between factor A and factor B. Use $\alpha = .05$.
c. Test for differences in main effect means for factor A. Use $\alpha = .05$.
d. Test for differences in main effect means for factor B. Use $\alpha = .05$.

13.21 The data for a 3×4 factorial experiment, with two observations per treatment, are shown in the accompanying table.

		FACTOR A			
		1	2	3	4
	1	5	7	6	5
		4	9	5	7
FACTOR B	2	6	10	5	9
		4	9	8	7
	3	8	7	5	6
		10	6	8	5

a. Perform an analysis of variance for the data and construct an analysis of variance table.
b. Do the data provide sufficient evidence of interaction between factor A and factor B? Test using $\alpha = .05$.
c. Find a 90% confidence interval for the mean value of y for the factor–level combination A_1, B_1.
d. Find a 90% confidence interval for the difference in the mean levels for factor–level combinations A_1, B_1 and A_2, B_1.

13.22 A study was conducted to investigate the effect of two factors on the mean level of sulfur content in coal: the laboratory conducting the analysis (seven levels) and the method of analysis (two levels). The 28 coal specimens used for the experiment, all from the same source, were randomly assigned, two to each of the 7×2 combinations of laboratory and method of analysis. The data are shown in the table.

		LABORATORY						
		L_1	L_2	L_3	L_4	L_5	L_6	L_7
METHOD OF ANALYSIS	A_1	.107	.127	.115	.108	.097	.114	.155
		.105	.122	.112	.108	.096	.119	.145
	A_2	.105	.127	.109	.117	.110	.116	.164
		.103	.124	.111	.115	.097	.122	.160

Source: Taguchi, G. "Signal Noise Ratio and its Application for Testing Material." *Reports of Statistical Application Research,* Union of Japanese Scientists and Engineers. Vol. 18, No. 4, 1971, pp. 21–33.

a. Perform an analysis of variance for the experiment and display the results in an analysis of variance table.
b. What is the practical significance of an interaction between method of analysis and laboratory?
c. Do the data provide sufficient evidence to indicate interaction between method of analysis and laboratory? Test using $\alpha = .05$.
d. Do the data provide sufficient evidence to indicate differences in the mean level of sulfur content readings from one laboratory to another? Test using $\alpha = .05$.
e. Do the data provide sufficient evidence to indicate a difference in the mean level of sulfur content between the two methods of analysis? Test using $\alpha = .05$.
f. Find a 95% confidence interval for the difference in the mean level of sulfur content for samples of coal analyzed in laboratories 1 and 2 using method of analysis A_1.

13.23 Refer to Example 13.6 and the factorial experiment designed to measure the effect of two factors, gasket material and stamping machine, on productivity of a manufacturing process. The data for the 2×3 factorial experiment, number of gaskets produced per hour (in thousands), are shown in the accompanying table (page 674).

		GASKET MATERIAL			TOTAL
		Cork, B_1	Rubber, B_2	Plastic, B_3	
STAMPING MACHINE	A_1	4.31 4.27 4.40	3.36 3.42 3.48	4.01 3.94 3.89	35.08
	A_2	3.94 3.81 3.99	3.91 3.80 3.85	3.48 3.53 3.42	33.73
TOTAL		24.72	21.82	22.27	68.81

a. Construct an ANOVA summary table.
b. Is there evidence of interaction between gasket material and stamping machine? Test using $\alpha = .01$.
c. Explain the practical significance of the result obtained in part b.
d. Based on the result from part c, would you recommend that tests for main effects be conducted? Explain.
e. Find a 95% confidence interval for the difference in the mean number of gaskets produced by machines A_1 and A_2, when stamping cork (B_1) gaskets. Interpret the interval.

13.24 An experiment was conducted to determine the effect of sintering time (two levels) on the compressive strength of two different metals. Five test specimens were sintered for each metal at each of the two sintering times. The data (in thousands of pounds per square inch) are shown in the accompanying table.

		SINTERING TIME					
		100 minutes			200 minutes		
METAL	1	17.1 15.2	16.5 16.7	14.9	19.4 17.2	18.9 20.7	20.1
	2	12.3 11.6	13.8 12.1	10.8	15.6 16.1	17.2 18.3	16.7

a. Perform an analysis of variance for the data and construct an analysis of variance table.
b. What is the practical significance of an interaction between sintering time and metal type?
c. Do the data provide sufficient evidence to indicate an interaction between sintering time and metal type? Test using $\alpha = .05$.
d. Find a 90% confidence interval for the mean compressive strength of metal 1 for a sintering time of 100 minutes.
e. Find a 90% confidence interval for the difference between the mean compressive strengths of metal 1 for the two different sintering times.

13.25 The chemical element antimony is sometimes added to tin–lead solder to replace the more expensive tin and to reduce the cost of soldering. A factorial experiment was conducted to determine how antimony affects the strength of the tin–lead solder joints (*Journal of Materials Science*, May 1986). Tin–lead solder specimens were prepared using one of four possible cooling methods (water-quenched, WQ; oil-quenched, OQ; air-blown, AB; and furnace-cooled, FC) and with one of four possible amounts of antimony (0%, 3%, 5%, and 10%) added to the composition. Three solder joints were randomly assigned to each of the 4 × 4 = 16 treatments and the shear strength of each measured. The experimental results are given in the table.

AMOUNT OF ANTIMONY (% weight)	COOLING METHOD	SHEAR STRENGTH (MPa)
0	WQ	17.6, 19.5, 18.3
0	OQ	20.0, 24.3, 21.9
0	AB	18.3, 19.8, 22.9
0	FC	19.4, 19.8, 20.3
3	WQ	18.6, 19.5, 19.0
3	OQ	20.0, 20.9, 20.4
3	AB	21.7, 22.9, 22.1
3	FC	19.0, 20.9, 19.9
5	WQ	22.3, 19.5, 20.5
5	OQ	20.9, 22.9, 20.6
5	AB	22.9, 19.7, 21.6
5	FC	19.6, 16.4, 20.5
10	WQ	15.2, 17.1, 16.6
10	OQ	16.4, 19.0, 18.1
10	AB	15.8, 17.3, 17.1
10	FC	16.4, 17.6, 17.6

Source: Tomlinson, W. J. and Cooper, G. A. "Fracture Mechanism of Brass/Sn-Pb-Sb solder Joints and the Effect of Production Variables on the Joint Strength." *Journal of Materials Science*, Vol. 21, No. 5, May 1986, p. 1731 (Table II). Copyright 1986 Chapman and Hall.

a. Construct an ANOVA summary table for the experiment.
b. Conduct a test to determine whether the two factors, amount of antimony and cooling method, interact. Use $\alpha = .01$.
c. Interpret the result obtained in part **b**.
d. If appropriate, conduct the tests for main effects. Use $\alpha = .01$.
e. Find a 99% confidence interval for the mean shear strength of tin–lead solder joints composed of 5% antimony and cooled with the air-blown method.
f. Find a 99% confidence interval for the difference between the mean shear strengths of 5% antimony–tin–lead solder joints cooled under two methods, water-quenched and oil-quenched.

OPTIONAL EXERCISE

13.26 How do women compare with men in their ability to perform laborious tasks that require strength? Some information on this question is provided in a study, by M. D. Phillips and R. L. Pepper, of the firefighting ability of men and women ("Shipboard

Fire-Fighting Performance of Females and Males," *Human Factors, 24,* 1982). Phillips and Pepper conducted a 2×2 factorial experiment to investigate the effect of the factor sex (male or female) and the factor weight (light or heavy) on the length of time required for a person to perform a particular firefighting task. Eight persons were selected for each of the $2 \times 2 = 4$ sex–weight categories of the 2×2 factorial experiment, and the length of time needed to complete the task was recorded for each of the 32 persons. The means and standard deviations of the four samples are shown in the table.

	LIGHT WEIGHT		HEAVY WEIGHT	
	Mean	Standard Deviation	Mean	Standard Deviation
Female	18.30	6.81	14.50	2.93
Male	13.00	5.04	12.25	5.70

Source: *Human Factors,* 1982, Vol. 24. Copyright 1982 by the Human Factors Society, Inc. and reproduced by permission.

a. Calculate the total of the $n = 8$ time measurements for each of the four categories of the 2×2 factorial experiment.
b. Calculate CM.
c. Use the results of parts **a** and **b** to calculate the sums of squares for sex, weight, and for the sex–weight interaction.
d. Calculate each sample variance. Then calculate the sums of squares of deviations *within* each sample for each of the four samples.
e. Calculate SSE. [*Hint:*   SSE is the pooled sum of squares of the deviations calculated in part **d**.]
f. Now that you know SS(Sex), SS(Weight), SS(Sex–Weight), and SSE, find SS(Total).
g. Summarize the calculations in an analysis of variance table.
h. Explain the practical significance of the presence (or absence) of sex–weight interaction. Do the data provide evidence of an interaction between sex and weight?
i. Construct a 95% confidence interval for the difference in mean time to complete the task between light men and light women. Interpret the interval.
j. Construct a 95% confidence interval for the difference in mean time to complete the task between heavy men and heavy women. Interpret the interval.

| | | | | | | | | | | | |

SECTION 13.8

THE RELATIONSHIP BETWEEN ANALYSIS OF VARIANCE AND REGRESSION

The preceding sections illustrated the analysis of variance approach to analyzing data collected from designed experiments. The analysis of variance sums of squares are easy to compute with the aid of a pocket or desk calculator. By forming a ratio of mean squares, we are able to test the hypothesis that a set of population means (treatment means, block means, or factor main effect means) are equal.

The same analysis can also be conducted using a multiple regression analysis. Each experimental design is associated with a linear model for the response y, called the **complete model**. The analysis of variance F test for testing a set of means is equivalent to a partial F test in regression in which the complete model is fit and compared to a **reduced model**. The difference between the SSEs for the two models, called the **drop in SSE**, is equal to the sum of squares for the appropriate source of variation (e.g., treatments) that appears in the numerator of the F statistic. Before you can apply regression analysis in an analysis of variance, you need to learn the appropriate complete and reduced models to fit for each type of experimental design.

Consider, first, the completely randomized design. Recall (from Section 13.5) that this design is one which involves a comparison of the means for p treatments based on independent random samples. Since we want to make inferences about the p population means, $\mu_1, \mu_2, \ldots, \mu_p$, the appropriate linear model for the response y is

$$E(y) = \beta_0 + \beta_1 x_1 + \beta_2 x_2 + \cdots + \beta_{p-1} x_{p-1}$$

where

$$x_1 = \begin{cases} 1 & \text{if treatment 2} \\ 0 & \text{if not} \end{cases} \qquad x_2 = \begin{cases} 1 & \text{if treatment 3} \\ 0 & \text{if not} \end{cases} \quad \cdots \quad x_{p-1} = \begin{cases} 1 & \text{if treatment } p \\ 0 & \text{if not} \end{cases}$$

and treatment 1 is the base level. Recall that this 0–1 system of coding implies that

$$\beta_0 = \mu_1$$
$$\beta_1 = \mu_2 - \mu_1$$
$$\beta_2 = \mu_3 - \mu_1$$
$$\vdots \qquad \vdots$$
$$\beta_{p-1} = \mu_p - \mu_1$$

The null hypothesis that the p population means are equal, i.e.,

$$H_0: \quad \mu_1 = \mu_2 = \cdots = \mu_p$$

is equivalent to the null hypothesis that all the treatment differences equal 0, i.e.,

$$H_0: \quad \beta_1 = \beta_2 = \cdots = \beta_{p-1} = 0$$

To test this hypothesis using regression, we compare the sum of squares for error, SSE_1, for the reduced model

$$E(y) = \beta_0$$

to the sum of squares for error, SSE_2, for the complete model

$$E(y) = \beta_0 + \beta_1 x_1 + \beta_2 x_2 + \cdots + \beta_{p-1} x_{p-1}$$

using the F statistic

$$F = \frac{(SSE_1 - SSE_2)/\text{Number of } \beta \text{ parameters in } H_0}{SSE_2/[n - (\text{Number of } \beta \text{ parameters in the complete model})]}$$

$$= \frac{(SSE_1 - SSE_2)/(p - 1)}{SSE_2/(n - p)}$$

The quantity, $SSE_1 - SSE_2$, in the numerator of the F statistic is equal to the sum of squares for treatments, SST, in an analysis of variance. Also, the sum of squares for error for the complete model, SSE_2, is equal to SSE in an analysis of variance. Thus,

$$F = \frac{(SSE_1 - SSE_2)/(p - 1)}{SSE_2/(n - p)} = \frac{SST/(p - 1)}{SSE/(n - p)} = \frac{MST}{MSE}$$

and, therefore, the analysis of variance and regression test statistics are identical.

We illustrate the regression approach to conducting an analysis of variance for a completely randomized design in the following example.

EXAMPLE 13.8

Refer to Example 13.1, where we compared the mean times until abrasion for three paint types. The experiment is a completely randomized design with three treatments: paint type 1, paint type 2, and paint type 3. Analyze the data shown in Table 13.2 using a regression analysis.

SOLUTION

The appropriate linear model for $p = 3$ treatments is

Complete model: $y = \beta_0 + \beta_1 x_1 + \beta_2 x_2 + \varepsilon$

where

$$x_1 = \begin{cases} 1 & \text{if paint type 2} \\ 0 & \text{if not} \end{cases}$$

$$x_2 = \begin{cases} 1 & \text{if paint type 3} \\ 0 & \text{if not} \end{cases}$$

The Minitab regression analysis for the complete model is shown in Figure 13.10. Note that the SSE shown in the printout, $SSE_2 = 770{,}669$, agrees (except for rounding) with the value of SSE calculated in Example 13.1.

The reduced model, obtained by omitting the two treatment parameters, is

Reduced model: $y = \beta_0 + \varepsilon$

Since we know that the least squares estimate of β_0 for this model is $\bar{y}$, it follows that the sum of squares for error for the reduced model is

$$SSE_1 = SS(\text{Total}) = \sum_{i=1}^{n} (y_i - \bar{y})^2 = \sum_{i=1}^{n} y_i^2 - \frac{\left(\sum_{i=1}^{n} y_i\right)^2}{n}$$

FIGURE 13.10
Minitab Computer Printout for the Completely Randomized Design, Example 13.8

```
THE REGRESSION EQUATION IS
Y =   230. + 80.3 X1 + 198. X2

                                  ST. DEV.   T-RATIO =
     COLUMN  COEFFICIENT   OF COEF.   COEF/S.D.
      --           229.6       53.4        4.30
X1   C1            80.3       75.6        1.06
X2   C2           198.2       75.6        2.62

THE ST. DEV. OF Y ABOUT REGRESSION LINE IS
S =       169.
WITH (  30- 3) =   27 DEGREES OF FREEDOM

R-SQUARED = 20.5 PERCENT
R-SQUARED = 14.6 PERCENT, ADJUSTED FOR D.F.

ANALYSIS OF VARIANCE

 DUE TO       DF      SS   MS=SS/DF
REGRESSION     2  198778.     99389.
RESIDUAL      27  770669.     28543.
TOTAL         29  969447.
```

This quantity is shown in Figure 13.10 as

$$\text{SS(Total)} = 969{,}447$$

Then the drop in the sum of squares for error that is attributable to a difference in treatment means is

$$\begin{aligned} \text{SST} &= \text{SSE}_1 - \text{SSE}_2 \\ &= 969{,}447 - 770{,}669 \\ &= 198{,}778 \end{aligned}$$

To test the null hypothesis H_0: $\beta_1 = \beta_2 = 0$ or, equivalently, H_0: $\mu_1 = \mu_2 = \mu_3$, we form the test statistic

$$\begin{aligned} F &= \frac{(\text{SSE}_1 - \text{SSE}_2)/(p-1)}{\text{SSE}_2/(n-p)} \\ &= \frac{198{,}778/2}{770{,}669/27} = 3.48 \end{aligned}$$

This value is identical to the analysis of variance F statistic computed in Example 13.1. For $\nu_1 = p - 1 = 3 - 1 = 2$ and $\nu_2 = n - p = 30 - 3 = 27$ df, $F_{.05} = 3.35$. Since the computed value of F, 3.48, exceeds the critical value, there is sufficient evidence (at $\alpha = .05$) to indicate that the means for the three paint types differ. ■

The regression approach to analyzing data from a completely randomized design is summarized in the box (page 680).

ANOVA *F* TEST FOR A COMPLETELY RANDOMIZED DESIGN WITH *p* TREATMENTS: REGRESSION APPROACH

H_0: $\beta_1 = \beta_2 = \cdots = \beta_{p-1} = 0$ (i.e., H_0: $\mu_1 = \mu_2 = \cdots = \mu_p$)

H_a: At least one of the β parameters listed in H_0 differs from 0
(i.e., H_a: At least two means differ)

Complete model: $E(y) = \beta_0 + \beta_1 x_1 + \beta_2 x_2 + \cdots + \beta_{p-1} x_{p-1}$

where

$$x_1 = \begin{cases} 1 & \text{if treatment 2} \\ 0 & \text{if not} \end{cases} \qquad x_2 = \begin{cases} 1 & \text{if treatment 3} \\ 0 & \text{if not} \end{cases}$$

$$\ldots, \quad x_{p-1} = \begin{cases} 1 & \text{if treatment } p \\ 0 & \text{if not} \end{cases}$$

Reduced model: $E(y) = \beta_0$

Test statistic: $F = \dfrac{(\text{SSE}_1 - \text{SSE}_2)/(p-1)}{\text{SSE}_2/(n-p)} = \dfrac{\text{MST}}{\text{MSE}}$

where

SSE_1 = SSE for reduced model

SSE_2 = SSE for complete model

Rejection region: $F > F_\alpha$, where the distribution of F is based on
$\nu_1 = p - 1$ and $\nu_2 = n - p$ degrees of freedom.

EXAMPLE 13.9

To further relate regression analysis and analysis of variance, refer to Examples 13.1 and 13.8. Note that

$$\beta_1 = \mu_2 - \mu_1$$

Therefore, since the analysis of variance estimate of $(\mu_2 - \mu_1)$ is $(\bar{T}_2 - \bar{T}_1)$, it follows that $\hat{\beta}_1$ must equal $(\bar{T}_2 - \bar{T}_1)$. Further, the estimated standard error of

$$\hat{\beta}_1 = \bar{T}_2 - \bar{T}_1$$

is by ANOVA formulas

$$s\sqrt{\frac{1}{n_1} + \frac{1}{n_2}}$$

Find $\hat{\beta}_1$ and its standard error in the regression printout for Example 13.8 (Figure 13.10), and verify that, in fact,

$$\hat{\beta}_1 = \bar{T}_2 - \bar{T}_1 \quad \text{and} \quad s_{\hat{\beta}_1} = s\sqrt{\frac{1}{n_1} + \frac{1}{n_2}}$$

SOLUTION

From Table 13.2, $\bar{T}_2 - \bar{T} = 309.9 - 229.6 = 80.30$. This is exactly the same as the value given for $\hat{\beta}_1$ in the regression printout (Figure 13.10). Similarly, in Example 13.1, we found $s = \sqrt{MSE} = 168.9$. Therefore,

$$s\sqrt{\frac{1}{n_1} + \frac{1}{n_2}} = 168.9\sqrt{\frac{1}{10} + \frac{1}{10}} = 75.53$$

Consulting the regression printout (Figure 13.10), you will see that the standard error of $\hat{\beta}_1$ is 75.6. (Again, the different values are due to rounding errors.)

The point of this example is that the completely randomized design may be analyzed using either a regression analysis or an analysis of variance. The conclusions will be identical. ∎

The regression approach to analyzing randomized block designs and factorial experiments is similar to that shown above. For each null hypothesis, complete and reduced models are fit. The drop in SSE, $SSE_1 - SSE_2$, always represents the analysis of variance SS for the appropriate source being tested. The ANOVA F tests for randomized block designs and factorial experiments using regression analysis are shown in the respective boxes (pages 682–683).

EXAMPLE 13.10

Refer to the Supply–Ratio 3×3 factorial experiment of Example 13.7.

a. Write the complete model for the experiment.
b. What hypothesis would you test to determine whether Supply and Ratio interact?

SOLUTION

a. Both factors, Supply and Ratio, are quantitative. According to the box on page 683, when the factors in a factorial experiment are quantitative, the main effects are represented by terms such as x, x^2, x^3, etc. Since each factor has three levels, we require two main effects, x and x^2, for each factor. (In general, the number of main effect terms will be one less than the number of levels for a factor.) Consequently, the complete factorial model for this 3×3 factorial experiment is

$$y = \beta_0 + \overbrace{\beta_1 x_1 + \beta_2 x_1^2}^{\text{Supply main effects}} + \overbrace{\beta_3 x_2 + \beta_4 x_2^2}^{\text{Ratio main effects}}$$
$$+ \underbrace{\beta_5 x_1 x_2 + \beta_6 x_1 x_2^2 + \beta_7 x_1^2 x_2 + \beta_8 x_1^2 x_2^2}_{\text{Supply–Ratio interaction}} + \varepsilon$$

where

$x_1 = $ Supply of raw material (in tons)

$x_2 = $ Ratio of allocation

Note that the interaction terms for the model are constructed by taking the products of the various main effect terms, one from each factor. For example,

---

ANOVA F TEST FOR A RANDOMIZED BLOCK DESIGN WITH p TREATMENTS AND b BLOCKS: REGRESSION APPROACH

TESTS FOR COMPARING TREATMENT MEANS

H_0: $\beta_1 = \beta_2 = \cdots = \beta_{p-1} = 0$
    (i.e., H_0:   The p treatment means are equal)

H_a: At least one of the β parameters listed in H_0 differs from 0
    (i.e., H_a: At least two treatment means differ)

Complete model:

$$E(y) = \beta_0 + \underbrace{\beta_1 x_1 + \cdots + \beta_{p-1} x_{p-1}}_{(p-1)\text{ treatment terms}} + \underbrace{\beta_p x_p + \cdots + \beta_{p+b-2} x_{p+b-2}}_{(b-1)\text{ block terms}}$$

where

$$x_1 = \begin{cases} 1 & \text{if treatment 2} \\ 0 & \text{if not} \end{cases} \quad \cdots \quad x_{p-1} = \begin{cases} 1 & \text{if treatment } p \\ 0 & \text{if not} \end{cases}$$

$$x_p = \begin{cases} 1 & \text{if block 2} \\ 0 & \text{if not} \end{cases} \quad \cdots \quad x_{p+b-2} = \begin{cases} 1 & \text{if block } b \\ 0 & \text{if not} \end{cases}$$

Reduced model: $E(y) = \beta_0 + \beta_p x_p + \cdots + \beta_{p+b-2} x_{p+b-2}$

Test statistic: $F = \dfrac{(SSE_1 - SSE_2)/(p-1)}{SSE_2/(n-p-b+1)} = \dfrac{MST}{s^2}$

where

SSE_1 = SSE for reduced model

SSE_2 = SSE for complete model

Rejection region: $F > F_\alpha$, where F is based on $\nu_1 = (p-1)$ and $\nu_2 = (n-p-b+1)$ degrees of freedom.

TESTS FOR COMPARING BLOCK MEANS

H_0: $\beta_p = \beta_{p+1} = \cdots = \beta_{p+b-2} = 0$
    (i.e., H_0:   The b block means are equal)

H_a: At least one of the β parameters listed in H_0 differs from 0
    (i.e., H_a: At least two block means differ)

Complete model: (See above)

Reduced model: $E(y) = \beta_0 + \beta_1 x_1 + \beta_2 x_2 + \cdots + \beta_{p-1} x_{p-1}$

Test statistic: $F = \dfrac{(SSE_1 - SSE_2)/(b-1)}{SSE_2/(n-p-b+1)} = \dfrac{MSB}{s^2}$

where

SSE_1 = SSE for reduced model

SSE_2 = SSE for complete model

Rejection region: $F > F_\alpha$, where F is based on $\nu_1 = (b-1)$ and $\nu_2 = (n-p-b+1)$ degrees of freedom.

ANOVA *F* TEST FOR INTERACTION IN A TWO-FACTOR FACTORIAL EXPERIMENT WITH FACTOR *A* AT *a* LEVELS AND FACTOR *B* AT *b* LEVELS: REGRESSION APPROACH

H_0: $\beta_{a+b-1} = \beta_{a+b} = \cdots = \beta_{ab-1} = 0$
(i.e., H_0: No interaction between factors A and B)

H_a: At least one of the β parameters listed in H_0 differs from 0
(i.e., H_a: Factors A and B interact)

Complete model:

$$E(y) = \overbrace{\beta_0 + \beta_1 x_1 + \cdots + \beta_{a-1} x_{a-1}}^{\text{Main effect } A \text{ terms}} + \overbrace{\beta_a x_a + \cdots + \beta_{a+b-2} x_{a+b-2}}^{\text{Main effect } B \text{ terms}}$$

$$+ \overbrace{\beta_{a+b-1} x_1 x_a + \beta_{a+b} x_1 x_{a+1} + \cdots \beta_{ab-1} x_{a-1} x_{a+b-2}}^{AB \text{ interaction terms}}$$

where*

$$x_1 = \begin{cases} 1 & \text{if level 2 of factor } A \\ 0 & \text{if not} \end{cases} \cdots x_{a-1} = \begin{cases} 1 & \text{if level } a \text{ of factor } A \\ 0 & \text{if not} \end{cases}$$

$$x_a = \begin{cases} 1 & \text{if level 2 of factor } B \\ 0 & \text{if not} \end{cases} \cdots x_{a+b-2} = \begin{cases} 1 & \text{if level } b \text{ of factor } B \\ 0 & \text{if not} \end{cases}$$

Reduced model:

$$E(y) = \overbrace{\beta_0 + \beta_1 x_1 + \cdots + \beta_{a-1} x_{a-1}}^{\text{Main effect } A \text{ terms}} + \overbrace{\beta_a x_a + \cdots + \beta_{a+b-2} x_{a+b-2}}^{\text{Main effect } B \text{ terms}}$$

Test statistic: $F = \dfrac{(\text{SSE}_1 - \text{SSE}_2)/[(a-1)(b-1)]}{\text{SSE}_2/[ab(r-1)]}$

where

$\text{SSE}_1 = $ SSE for reduced model

$\text{SSE}_2 = $ SSE for complete model

$r = $ Number of replications

Rejection region: $F > F_\alpha$, where F is based on $\nu_1 = (a-1)(b-1)$ and $\nu_2 = ab(r-1)$ df.

*Note: The independent variables, $x_1, x_2, \ldots, x_{a+b-2}$, are defined for an experiment in which both factors represent *qualitative* variables. When a factor is *quantitative*, you may choose to represent the main effects with quantitative terms such as x, x^2, x^3, and so forth.

we included terms involving the products of x_1 with x_2 and x_2^2. The remaining interaction terms were formed by multiplying x_1^2 with x_2 and x_2^2.

b. To test the null hypothesis that Supply and Ratio do not interact, we must test the null hypothesis that the interaction terms are not needed in the linear model of part **a**, i.e.,

$$H_0: \quad \beta_5 = \beta_6 = \beta_7 = \beta_8 = 0$$

This requires that we fit the reduced model

$$y = \beta_0 + \beta_1 x_1 + \beta_2 x_1^2 + \beta_3 x_2 + \beta_4 x_2^2$$

and perform the partial F test outlined in Section 12.6. The test statistic is

$$F = \frac{(SSE_1 - SSE_2)/4}{s^2}$$

where

$$SSE_1 = \text{SSE for reduced model}$$
$$SSE_2 = \text{SSE for complete model}$$
$$s^2 = \text{MSE for complete model}$$

■

EXAMPLE 13.11

Use the SAS computer software package to fit a second-order model to the data of the factorial experiment of Example 13.7.

SOLUTION

An analysis of variance for a complete two-variable factorial experiment *always* yields results, including an SSE, that would be obtained by fitting a complete factorial model of the type indicated in part **a** of Example 13.10. However, as indicated in Chapter 12, we would not expect the third- and fourth-order terms to contribute much to the model and would recommend a second-order model to fit the response surface of Example 13.7. The second-order model in two independent variables is

$$y = \beta_0 + \beta_1 x_1 + \beta_2 x_1^2 + \beta_3 x_2 + \beta_4 x_2^2 + \beta_5 x_1 x_2 + \varepsilon$$

where

$$x_1 = \text{Supply of raw material (in tons)}$$
$$x_2 = \text{Ratio of allocation}$$

The SAS regression analysis printout obtained by fitting this model to the data of Example 13.7 is shown in Figure 13.11. The estimates $\hat{\beta}_0 = -27.815$, $\hat{\beta}_1 = 5.944, \ldots, \hat{\beta}_5 = -2.296$ can be obtained from the printout. Hence, the second-order prediction equation relating Supply of raw materials, S, and Ratio of allocation, R, to profit per unit of raw material supply, y, is

$$\hat{y} = -27.815 + 5.944x_1 - 7.762x_2 + .746x_1x_2 - .185x_1^2 - 2.296x_2^2$$

FIGURE 13.11 SAS Computer Printout for Fitting a Second-Order Model to the Data of Example 13.11

DEPENDENT VARIABLE: PROFIT		PROFIT (IN CENTS)		
SOURCE	DF	SUM OF SQUARES	MEAN SQUARE	F VALUE
MODEL	5	63.50793651	12.70158730	4.89
ERROR	21	54.49206349	2.59486017	PR > F
CORRECTED TOTAL	26	118.00000000		0.0040

R-SQUARE	C.V.	ROOT MSE	PROFIT MEAN
0.538203	7.7945	1.61085697	20.66666667

PARAMETER	ESTIMATE	T FOR H0: PARAMETER=0	PR > ¦T¦	STD ERROR OF ESTIMATE
INTERCEPT	-27.81481481	-1.17	0.2557	23.80152168
S	5.94444444	2.25	0.0354	2.64418353
R	-7.76190476	-1.54	0.1389	5.04522969
S*R	0.74603175	3.68	0.0014	0.20294890
S*S	-0.18518519	-2.53	0.0193	0.07306996
R*R	-2.29629630	-1.71	0.1012	1.33939441

Note that the value of SSE given in Figure 13.11 differs from that obtained in the analysis of variance of Example 13.7. This is because we omitted the third- and fourth-order terms from the linear model when performing the regression analysis so that now SSE is based on a larger number of degrees of freedom, namely $[n - (k + 1)] = 27 - 6 = 21$ df. ∎

EXAMPLE 13.12

Do the data provide sufficient information to indicate that the complete factorial model given in Example 13.10 contributes more information for the prediction of y than the second-order model of Example 13.11?

SOLUTION

If the response to the question is "yes," then at least one of the parameters, β_6, β_7, or β_8, of the complete factorial model differs from 0 (i.e., they are needed in the model). Consequently, the null hypothesis, "The complete factorial model contributes no more information about y than that contributed by a second-order model" is equivalent to the hypothesis

$$H_0: \quad \beta_6 = \beta_7 = \beta_8 = 0$$

You will recall (from Section 12.6) that to test this hypothesis, we need to determine the drop in the sum of squares for error between the reduced model (i.e., the second-order model) and the complete model (i.e., the complete factorial model). From part a of Example 13.7, SSE for the complete model is $SSE_2 = 43.33$. From Figure 13.11, the reduced (second-order) model gives $SSE_1 = 54.492$. Therefore, the drop in SSE associated with β_6, β_7, and β_8, based on 3 df, is

$$Drop = SSE_1 - SSE_2 = 54.492 - 43.33 = 11.16$$

The F statistic needed for the test is

$$F = \frac{MS(\text{Drop})}{MSE(\text{Complete model})} = \frac{11.16/3}{2.41} = 1.54$$

[*Note:* MSE for the complete factorial model is given in Table 13.13.] Since MS(Drop) is based on 3 df and MSE(Complete model) is based on 18 df, we compare the computed value of F with $F_{.05} = 3.16$, where $F_{.05}$ is the tabulated value of F for 3 and 18 df. Since the computed value of F (1.54) is less than the critical value, $F_{.05} = 3.16$, we cannot reject the null hypothesis that $\beta_6 = \beta_7 = \beta_8 = 0$. That is, there is insufficient evidence to indicate that the third- and fourth-order terms associated with β_6, β_7, and β_8 contribute information for the prediction of y. ∎

EXAMPLE 13.13

Use the second-order model of Example 13.11 and find a 95% confidence interval for the mean profit per unit supply of raw material when $S = 17$, $R = 1$.

SOLUTION

[*Note:* To obtain this solution, you will need to use the methods of Chapter 11 and, particularly, you will need to use a computer program that prints the elements contained in the $(X'X)^{-1}$ matrix.] From Section 11.9, the 95% confidence interval for $E(y)$ is

$$\hat{y} \pm t_{\alpha/2}s\sqrt{a'(X'X)^{-1}a}$$

where $\hat{y}$ is obtained by substituting $S = x_1 = 17$ and $R = x_2 = 1$ into the prediction equation

$$\hat{y} = -27.815 + 5.944(17) - 7.762(1) + .746(17)(1) - .185(17)^2 - 2.296(1)^2$$
$$= 22.39$$

The value of s^2, based on 21 df, is given in the computer printout (Figure 13.11) as 2.595. Therefore, $s = 1.611$ and $t_{.025} = 2.080$. The **a** matrix contains the coefficients of $\hat{\beta}_0, \hat{\beta}_1, \ldots, \hat{\beta}_5$ used in the prediction equation to obtain $\hat{y}$. Therefore, when $x_1 = 17$ and $x_2 = 1$,

$$\mathbf{a} = \begin{bmatrix} 1 \\ x_1 \\ x_2 \\ x_1x_2 \\ x_1^2 \\ x_2^2 \end{bmatrix} = \begin{bmatrix} 1 \\ 17 \\ 1 \\ 17 \\ 289 \\ 1 \end{bmatrix}$$

The SAS computer printout of $(X'X)^{-1}$ is shown in Figure 13.12. Using this matrix and the **a** matrix shown above, we obtain the matrix product that appears in the formulas for the confidence interval:

$$a'(X'X)^{-1}a = .169165$$

Substituting this quantity, s, and $t_{\alpha/2}$ into the formula for the confidence interval, we obtain

$$\hat{y} \pm t_{\alpha/2}s\sqrt{a'(X'X)^{-1}a}$$

$$22.39 \pm (2.080)(1.611)(.411)$$

$$22.39 \pm 1.38$$

FIGURE 13.12 The $(X'X)^{-1}$ Matrix for Example 13.13

DEPENDENT VARIABLE: PROFIT

	INTERCEPT	S	R	S*R	S*S	R*R
INTERCEPT	218.32098765	-24.05555556	-8.33333333	0.33333333	0.65432099	0.86419753
S	-24.05555556	2.69444444	0.33333333	-0.01851852	-0.07407407	0.00000000
R	-8.33333333	0.33333333	9.80952381	-0.28571429	0.00000000	-1.77777778
S*R	0.33333333	-0.01851852	-0.28571429	0.01587302	-0.00000000	-0.00000000
S*S	0.65432099	-0.07407407	0.00000000	-0.00000000	0.00205761	-0.00000000
R*R	0.86419753	0.00000000	-1.77777778	-0.00000000	-0.00000000	0.69135802

Thus, we estimate (with confidence coefficient equal to .95) the mean profit per unit of supply to lie in the interval \$21.01 to \$23.77 when $S = 17$ tons and $R = 1$. Beyond this immediate result, you will note that this example illustrates the power and versatility of a regression analysis. That is, there is no way to obtain this estimate from the analysis of variance in Example 13.7. The computations necessary to obtain the confidence interval may appear to be tedious, but the matrix multiplications can easily be programmed for a computer. ∎

We conclude this section by pointing out another very powerful advantage of the regression approach to an analysis of variance. The ANOVA formulas for a factorial experiment and many other experimental designs can be used only when the sample sizes are equal. However, the regression approach applies to both equal and unequal sample sizes for the various factor–level combinations. A complete list of the advantages and disadvantages of the regression approach is provided in Section 13.14.

EXERCISES 13.27–13.33

13.27 Refer to Exercise 13.3.
  a. Give the complete and reduced models appropriate for conducting the ANOVA.
  b. If you have access to a statistical software computer package, fit the models specified in part **a** and compute the value of the F statistic. The value you obtain should agree with the value calculated in Exercise 13.3.

13.28 Refer to Exercise 13.15.
   a. Give the complete and reduced models appropriate for conducting the ANOVA.
   b. If you have access to a statistical software computer package, fit the models specified in part **a** and compute the value of the F statistic. The value you obtain should agree with the value calculated in Exercise 13.15.

13.29 Refer to Exercise 13.22.
   a. Give the complete and reduced models appropriate for conducting the ANOVA.
   b. If you have access to a statistical software computer package, fit the models specified in part **a** and compute the value of the F statistic. The value you obtain should agree with the value calculated in Exercise 13.22.

13.30 Refer to Exercise 13.24. An appropriate linear model for the experiment is

$$E(y) = \beta_0 + \beta_1 x_1 + \beta_2 x_2 + \beta_3 x_1 x_2$$

where

x_1 = Sintering time (in units of 100 minutes)

$$x_2 = \begin{cases} 1 & \text{if metal 1} \\ 0 & \text{if metal 2} \end{cases}$$

Fitting this model to the data using the SAS multiple regression package, we obtain the SAS printout shown here.

SAS Computer Printout for Exercise 13.30

DEPENDENT VARIABLE: Y

SOURCE	DF	SUM OF SQUARES	MEAN SQUARE	F VALUE	PR > F	R-SQUARE	C.V.
MODEL	3	131.41200000	43.80400000	34.77	0.0001	0.867017	6.9887
ERROR	16	20.15600000	1.25975000		ROOT MSE		Y MEAN
CORRECTED TOTAL	19	151.56800000			1.12238585		16.06000000

SOURCE	DF	TYPE I SS	F VALUE	PR > F	DF	TYPE III SS	F VALUE	PR > F
X1	1	76.83200000	60.99	0.0001	1	54.28900000	43.10	0.0001
X2	1	51.84200000	41.15	0.0001	1	14.79680000	11.75	0.0035
X1*X2	1	2.73800000	2.17	0.1598	1	2.73800000	2.17	0.1598

PARAMETER	ESTIMATE	T FOR H0: PARAMETER=0	PR > :T:	STD ERROR OF ESTIMATE
INTERCEPT	7.46000000	6.65	0.0001	1.12238585
X1	0.04660000	6.56	0.0001	0.00709859
X2	5.44000000	3.43	0.0035	1.58729329
X1*X2	-0.01480000	-1.47	0.1598	0.01003892

   a. Find the least squares prediction equation.
   b. Give the prediction equation relating the mean compressive strength to sintering time for metal 1.
   c. Give the prediction equation relating the mean compressive strength to sintering time for metal 2.
   d. Graph the prediction lines in parts **b** and **c**.
   e. What does the parameter β_3 measure?
   f. Test H_0: $\beta_3 = 0$ against the alternative hypothesis H_a: $\beta_3 \neq 0$. Use $\alpha = .05$.
   g. How does the test of part **f** relate to the test in part **c** of Exercise 13.24?

13.31 As part of a study on the rate of combustion of artificial graphite in humid air flow, researchers conducted an experiment to investigate oxygen diffusivity through a water vapor mixture. A 3×9 factorial experiment was conducted with mole fraction of water (H_2O) at three levels and temperature of the nitrogen–water mixture at nine levels. The data are shown in the accompanying table.

TEMPERATURE	MOLE FRACTION OF H_2O		
(K)	.0022	.017	.08
1,000	1.68	1.69	1.72
1,100	1.98	1.99	2.02
1,200	2.30	2.31	2.35
1,300	2.64	2.65	2.70
1,400	3.00	3.01	3.06
1,500	3.38	3.39	3.45
1,600	3.78	3.79	3.85
1,700	4.19	4.21	4.27
1,800	4.63	4.64	4.71

Source: Matsui, K., Tsuji, H., and Makino, A. "The Effects of Water Vapor Concentration on the Rate of Combustion of an Artificial Graphite in Humid Air Flow." *Combustion and Flame,* *50,* 1983, pp. 107–118. Copyright 1983 by The Combustion Institute. Reprinted by permission of Elsevier Science Publishing Co., Inc.

a. Explain why the traditional analysis of variance (using the formulas of Section 13.7) is inappropriate for the analysis of these data.

b. Write a second-order model relating mean oxygen diffusivity, $E(y)$, to temperature x_1 (in hundreds) and mole fraction x_2 (in thousandths).

c. The SAS computer printout for the regression analysis is shown on page 690. Find SSE, s^2, and SS(Total).

d. Find R^2 and verify that

$$R^2 = 1 - \frac{SSE}{SS(Total)}$$

Interpret the value of R^2

e. Suppose that temperature and mole fraction of H_2O do not interact. What does this imply about the relationship between $E(y)$ and x_1 and x_2?

f. Do the data provide sufficient information to indicate that temperature and mole fraction of H_2O interact? Test using $\alpha = .05$.

g. Give the least squares prediction equation for $E(y)$.

h. Substitute into the prediction equation to predict the mean diffusivity when the temperature of the process is 1,300°K and the mole fraction of water is .017.

i. The printout shows 95% confidence intervals for each of the 3×9 factor–level combinations. Find the 95% confidence interval for mean diffusivity when the temperature of the process is 1,300°K and the mole fraction of water is .017.

SAS Computer Printout for Exercise 13.31

DEPENDENT VARIABLE: Y

SOURCE	DF	SUM OF SQUARES	MEAN SQUARE	F VALUE	PR > F	R-SQUARE	C.V.
MODEL	5	24.85819086	4.97163817	99999.99	0.0000	0.999994	0.0865
ERROR	21	0.00014989	0.00000714		ROOT MSE		Y MEAN
CORRECTED TOTAL	26	24.85834074			0.00267159		3.08851852

SOURCE	DF	TYPE I SS	F VALUE	PR > F	DF	TYPE III SS	F VALUE	PR > F
X1	1	24.75312500	99999.99	0.0000	1	0.01164256	1631.21	0.0001
X2	1	0.01907267	2672.22	0.0001	1	0.00003123	4.38	0.0488
X1*X2	1	0.00109942	154.04	0.0001	1	0.00109942	154.04	0.0001
X1*X1	1	0.08489236	11894.04	0.0001	1	0.08489236	11894.04	0.0001
X2*X2	1	0.00000141	0.20	0.6617	1	0.00000141	0.20	0.6617

PARAMETER	ESTIMATE	T FOR H0: PARAMETER=0	PR > :T:	STD ERROR OF ESTIMATE
INTERCEPT	-0.28015040	-16.36	0.0001	0.01712523
X1	0.00100027	40.39	0.0001	0.00002477
X2	-0.28554904	-2.09	0.0488	0.13650623
X1*X2	-0.00073265	12.41	0.0001	0.00005903
X1*X1	9.5851371E-07	109.06	0.0001	0.00000001
X2*X2	0.55141443	0.44	0.6617	1.24232481

OBSERVATION	OBSERVED VALUE	PREDICTED VALUE	RESIDUAL	LOWER 95% CL FOR MEAN	UPPER 95% CL FOR MEAN
1	1.68000000	1.67961820	0.00038180	1.67624465	1.68299175
2	1.69000000	1.68639201	0.00360799	1.68327658	1.68950745
3	1.72000000	1.71792910	0.00207090	1.71413512	1.72172325
4	1.98000000	1.98109412	-0.00109412	1.97855499	1.98363325
5	1.99000000	1.98895226	0.00104774	1.98660597	1.99129854
6	2.02000000	2.02510514	-0.00510514	2.02225161	2.02795866
7	2.30000000	2.30174032	-0.00174032	2.29952051	2.30396012
8	2.31000000	2.31068278	-0.00068278	2.30855939	2.31280617
9	2.35000000	2.35145136	-0.00145136	2.34906787	2.35383486
10	2.64000000	2.64155679	-0.00155679	2.63936191	2.64375166
11	2.65000000	2.65158357	-0.00158357	2.64941268	2.65375447
12	2.70000000	2.69696707	0.00303213	2.69472049	2.69920524
13	3.00000000	3.00054353	-0.00054353	2.99832668	3.00276030
14	3.01000000	3.01165464	-0.00165464	3.00943780	3.01387149
15	3.06000000	3.06165464	-0.00165464	3.05943780	3.06387149
16	3.38000000	3.37870055	0.00129945	3.37650568	3.38089542
17	3.39000000	3.39089598	-0.00089598	3.38872509	3.39306688
18	3.45000000	3.44551169	0.00448831	3.44327432	3.44774907
19	3.78000000	3.77602784	0.00397216	3.77380804	3.77824764
20	3.79000000	3.78930760	0.00069240	3.78718421	3.79143099
21	3.85000000	3.84853902	0.00146098	3.84615552	3.85092251
22	4.19000000	4.19252541	-0.00252541	4.18998627	4.19506454
23	4.21000000	4.20688949	0.00311051	4.20454321	4.20923578
24	4.27000000	4.27073661	-0.00073661	4.26788309	4.27359014
25	4.63000000	4.62819325	0.00180675	4.62481970	4.63156680
26	4.64000000	4.64364166	-0.00364166	4.64052622	4.64675709
27	4.71000000	4.71210449	-0.00210449	4.70831042	4.71589855

SUM OF RESIDUALS	0.00000000
SUM OF SQUARED RESIDUALS	0.00014989
SUM OF SQUARED RESIDUALS - ERROR SS	-0.00000000
PRESS STATISTIC	0.00026687
FIRST ORDER AUTOCORRELATION	0.08830532
DURBIN-WATSON D	1.79286843

13.32 Suppose you plan to investigate the effect of hourly pay rate and length of workday on some measure y of worker productivity. Both pay rate and length of workday will be set at three levels and y will be observed for all combinations of these factors. Thus, a 3×3 factorial experiment will be employed.

a. Identify the factors and state whether they are quantitative or qualitative.

b. Identify the treatments to be employed in the experiment.

c. Write a complete factorial model for the experiment. [*Hint:* When the factors are quantitative, main effect terms include x and x^2 terms for each factor.]

d. What is the order of the model specified in part c?

e. Suppose you want to fit a second-order model to the data. Give the appropriate model.

f. If you have only one observation for each combination and you fit a complete factorial model to the data, how many degrees of freedom will be available for estimating σ^2?

g. Refer to part **f.** If you fit a second-order model to the data, how many degrees of freedom will be available for estimating σ^2?

h. Suppose you replicated the experiment and hence obtained two observations for each treatment. How many degrees of freedom would be available for estimating σ^2 if you fit: (1) the complete factorial model? (2) a second-order model?

i. Which model—complete factorial or second-order—do you think would be more appropriate for this experiment?

j. Explain how to conduct a test for interaction between pay rate and length of workday using the complete factorial model.

k. Explain how to conduct a test for interaction between pay rate and length of workday using the second-order model.

13.33 A $2 \times 2 \times 2 \times 2 = 2^4$ factorial experiment was conducted to investigate the effect of four factors on the light output, y, of flashbulbs. Two observations were taken for each of the factorial treatments. The factors are: amount of foil contained in a bulb (100 and 120 milligrams); speed of sealing machine (1.2 and 1.3 revolutions per minute); shift (day or night); machine operator (A or B). The data table for the two replications of the 2^4 factorial experiment is on page 692, and the SAS computer printout for the regression analysis is shown below.

SAS Printout for Exercise 13.33

DEPENDENT VARIABLE: UNITS

SOURCE	DF	SUM OF SQUARES	MEAN SQUARE	F VALUE 40.78
MODEL	15	745.46875000	49.69791667	PR > F
ERROR	16	19.50000000	1.21875000	0.0001
CORRECTED TOTAL	31	764.96875000		

R-SQUARE	C.V.	ROOT MSE	UNITS MEAN
0.974509	11.0053	1.10397011	10.03125000

PARAMETER	ESTIMATE	T FOR H0: PARAMETER=0	PR > ITI	STD ERROR OF ESTIMATE
INTERCEPT	10.03125000	51.40	0.0001	0.19515619
SHIFT	0.21875000	1.12	0.2789	0.19515619
OPERATOR	-0.15625000	-0.80	0.4351	0.19515619
SPEED	-0.34375000	-1.76	0.0973	0.19515619
FOIL	4.71875000	24.18	0.0001	0.19515619
SHIFT*OPERATOR	0.78125000	4.00	0.0010	0.19515619
SHIFT*SPEED	-0.15625000	-0.80	0.4351	0.19515619
SHIFT*FOIL	0.15625000	0.80	0.4351	0.19515619
OPERATOR*SPEED	-0.03125000	-0.16	0.8748	0.19515619
OPERATOR*FOIL	0.28125000	1.44	0.1688	0.19515619
SPEED*FOIL	0.09375000	0.48	0.6375	0.19515619
SHIFT*OPERATOR*SPEED	0.15625000	0.80	0.4351	0.19515619
SHIFT*OPERATOR*FOIL	-0.03125000	-0.16	0.8748	0.19515619
SHIFT*SPEED*FOIL	-0.21875000	-1.12	0.2789	0.19515619
OPERATOR*SPEED*FOIL	-0.09375000	-0.48	0.6375	0.19515619
SHIFT*OPER*SPEE*FOIL	0.09375000	0.48	0.6375	0.19515619

		AMOUNT OF FOIL			
		100 milligrams		120 milligrams	
		SPEED OF MACHINE			
		1.2 rpm	1.3 rpm	1.2 rpm	1.3 rpm
DAY SHIFT	Operator B	6; 5	5; 4	16; 14	13; 14
	Operator A	7; 5	6; 5	16; 17	16; 15
NIGHT SHIFT	Operator B	8; 6	7; 5	15; 14	17; 14
	Operator A	5; 4	4; 3	15; 13	13; 14

To simplify computations, we let

$$x_1 = \frac{\text{Amount of foil} - 110}{10} \qquad x_2 = \frac{\text{Speed of machine} - 1.25}{.05}$$

so that x_1 and x_2 will take values -1 and $+1$. Also,

$$x_3 = \begin{cases} -1 & \text{if night shift} \\ 1 & \text{if day shift} \end{cases} \qquad x_4 = \begin{cases} -1 & \text{if machine operator } B \\ 1 & \text{if machine operator } A \end{cases}$$

a. Do the data provide sufficient evidence to indicate that any of the factors contribute information for the prediction of y? Give the results of a statistical test to support your answer.
b. Identify the factors that appear to affect the amount of light y in the flashbulbs.
c. Give the complete factorial model for y. [*Hint:* For a factorial experiment with four factors, the complete model includes main effects for each factor, two-way cross product terms, three-way cross product terms, and four-way cross product terms.]
d. How many degrees of freedom will be available for estimating σ^2?

SECTION 13.9

THE ANALYSIS OF VARIANCE FOR A *k*-WAY CLASSIFICATION OF DATA (OPTIONAL)

We explained in Section 13.7 that a k-way classification of data can arise when we run all combinations of the levels of k independent variables (blocks or factors). For example, a replicated (repeated) k-factor factorial experiment would result in a k-way classification of data if we assigned the treatments to the experimental units according to a completely randomized design. Or, a k-way classification of data would result if we randomly assigned the treatments of a $(k - 1)$-factor factorial experiment to the experimental units of a randomized block design. For example, if we assigned the $2 \times 3 = 6$ treatments of a 2×3 factorial experiment to blocks containing six experimental units each, the data would be arranged in a three-way classification, i.e., according to the two factors and the blocks.

Because of the complicated nature of the formulas required for calculating the sums of squares for main effects and interactions for an analysis of variance, the easiest way to perform an analysis of variance for a k-way classification of data is to use one of the popular computer program packages. All calculate the necessary sums of squares, mean squares, F values, etc., and present the results in an analysis of variance table. Consequently, we will learn how to interpret the computer output for an analysis of variance. If you are interested in the computational formulas, you will find them in the references.

EXAMPLE 13.14

Indicate the sources of variation and their associated degrees of freedom for a $2 \times 3 \times 3$ factorial experiment with $r = 3$ experimental units randomly assigned to each treatment.

SOLUTION

Denote the three factors as A, B, and C. Then the linear model for the experiment will contain one parameter corresponding to main effects for A, two each for B and C, $(1)(2) = 2$ each for the AB and AC interactions, $(2)(2) = 4$ for the BC interaction, and $(1)(2)(2) = 4$ for the three-way ABC interaction. Three-way interaction terms measure the failure of two-way interaction effects to remain the same from one level to another level of the third factor. The sources of variation and the respective degrees of freedom corresponding to these sets of parameters are shown in Table 13.14.

TABLE 13.14

Analysis of Variance Table for Example 13.14

SOURCE	df
Main effect A	1
Main effect B	2
Main effect C	2
AB interaction	2
AC interaction	2
BC interaction	4
ABC interaction	4
Error	36
Total	53

The degrees of freedom for SS(Total) will always equal $(n - 1)$—that is, n minus 1 degree of freedom for β_0. Since the degrees of freedom for all sources must sum to the degrees of freedom for SS(Total), it follows that the degrees of freedom for Error will equal the degrees of freedom for SS(Total), minus the sum of the degrees of freedom for main effects and interactions, i.e., $(n - 1) - 17$. Our experiment will contain three observations for each of the $2 \times 3 \times 3 = 18$ treatments; therefore, $n = (18)(3) = 54$, and the degrees of freedom for Error will equal $53 - 17 = 36$.

If data for this experiment were analyzed on a computer, the computer printout would show the analysis of variance table that we have constructed and would include the associated mean squares, values of the F test statistics, and their observed significance levels. Each F statistic would represent the ratio of the source mean square to MSE $= s^2$. ∎

EXAMPLE 13.15

A transistor manufacturer conducted an experiment to investigate the effects of three factors on productivity (measured in thousands of dollars of items produced) per 40-hour week. The factors were:

1. Length of work week (two levels):   five consecutive 8-hour days or four consecutive 10-hour days
2. Shift (two levels):   day or evening shift
3. Number of coffee breaks (three levels):   0, 1, or 2

The experiment was conducted over a 24-week period with the $2 \times 2 \times 3 = 12$ treatments assigned in a random manner to the 24 weeks. The data for this completely randomized design are shown in Table 13.15. Perform an analysis of variance for the data.

TABLE 13.15

Data for Example 13.15

		DAY SHIFT			NIGHT SHIFT		
		COFFEE BREAKS			COFFEE BREAKS		
		0	1	2	0	1	2
LENGTH OF WORK WEEK	4 days	94	105	96	90	102	103
		97	106	91	89	97	98
	5 days	96	100	82	81	90	94
		92	103	88	84	92	96

SOLUTION

The computer commands for the SAS analysis of variance for a k-way classification of data are given in optional Section 13.13. Using this computer package, we obtain the computer printout shown in Figure 13.13.

FIGURE 13.13   SAS Computer Printout for an Analysis of Variance for a Three-Way Classification of Data (Example 13.15)

```
DEPENDENT VARIABLE: DOLLARS
SOURCE            DF      SUM OF SQUARES        MEAN SQUARE      F VALUE      PR > F      R-SQUARE           C.V.
MODEL             11     1009.83333333         91.80303030      13.43        0.0001    4 0.924897          2.7686
ERROR             12   1   82.00000000          6.83333333                   ROOT MSE             DOLLARS MEAN
CORRECTED TOTAL   23     1091.83333333                                     3  2.61406452            94.41666667

SOURCE            DF       ANOVA SS        F VALUE    PR > F
SHIFT              1      48.16666667       7.05      0.0210
DAYS               1     204.16666667      29.88      0.0001
BREAKS             2     334.08333333      24.45      0.0001
SHIFT*DAYS         1    2  8.16666667       1.20      0.2958
SHIFT*BREAKS       2     385.58333333      28.21      0.0001
DAYS*BREAKS        2       8.08333333       0.59      0.5689
SHIFT*DAYS*BREAKS  2      21.58333333       1.58      0.2461
```

Pertinent sections of the computer printout are boxed and numbered, as follows:

1. The value of SS(Total), shown in the CORRECTED TOTAL row of box 1, is 1091.83333333. The degrees of freedom associated with this quantity is equal to $(n - 1) = (24 - 1) = 23$. Box 1 gives the partitioning (the analysis of variance) of this quantity into two sources of variation. The first source, MODEL, corresponds to the 11 parameters (all except β_0) in the model. The second source is ERROR. The degrees of freedom, sums of squares, and mean

squares for these quantities are shown in their respective columns. For example, MSE = 6.83333333. The F statistic for testing

$$H_0: \quad \beta_1 = \beta_2 = \cdots = \beta_{11} = 0$$

is based on $\nu_1 = 11$ and $\nu_2 = 12$ degrees of freedom and is shown on the printout as $F = 13.43$. The observed significance level, shown under PR > F, is 0.0001. This small observed significance level presents ample evidence to indicate that at least one of the three independent variables—shifts, number of days in a working week, or number of coffee breaks per day—contributes information for the prediction of mean productivity.

2. To determine which sets of parameters are actually contributing information for the prediction of y, we examine the breakdown (box 2) of SS(Model) into components corresponding to the sets of parameters for main effects SHIFTS, DAYS, and BREAKS, and two-way interactions, SHIFT*DAYS, SHIFT*BREAKS, and DAYS*BREAKS. The last MODEL source of variation corresponds to the set of all three-way SHIFT*DAYS*BREAKS parameters. Note that the degrees of freedom for these sources sum to 11, the number of degrees of freedom for MODEL. Similarly, the sum of the component sums of squares is equal to SS(MODEL). Box 2 does not give the mean squares associated with the sources, but it does give the F values associated with testing hypotheses concerning the set of parameters associated with each source. It also gives the observed significance levels of these tests. You can see that there is ample evidence to indicate the presence of a SHIFT*BREAKS interaction. The F tests associated with all three main effect parameter sets are also statistically significant at the $\alpha = .05$ level of significance. The practical implication of these results is that there is evidence to indicate that all three independent variables, shift, number of work days per week, and number of coffee breaks per day, contribute information for the prediction of productivity. The presence of a SHIFT*BREAKS interaction means that the effect of the number of breaks on productivity is not the same from shift to shift. Thus, the specific number of coffee breaks that might achieve maximum productivity on one shift might be different from the number of breaks that will achieve maximum productivity on the other shift.

3. Box 3 gives the value of $s = \sqrt{\text{MSE}} = 2.61406452$. This value would be used to construct a confidence interval to compare the difference between two of the 12 treatment means. The confidence interval for the difference between a pair of means, $(\mu_i - \mu_j)$, would be

$$(\bar{y}_i - \bar{y}_j) \pm t_{\alpha/2} s \sqrt{\frac{2}{r}}$$

where r is the number of replications of the factorial experiment within a completely randomized design. There were $r = 2$ observations for each of the 12 treatments (factor–level combinations) in this example.

4. Box 4 gives the value of R^2, a measure of how well the model fits the experimental data. It is of value primarily when the number of degrees of freedom

for error is large—say, at least 5 or 6. The larger the number of degrees of freedom for error, the greater will be its practical importance. The value of R^2 for this analysis, 0.924897, indicates that the model provides a fairly good fit to the data. It also suggests that the model could be improved by adding new predictor variables or, possibly, by including higher-order terms in the variables originally included in the model.   ∎

EXAMPLE 13.16

In a manufacturing process, a plastic rod is produced by heating a granular plastic to a molten state and then extruding it under pressure through a nozzle. An experiment was conducted to investigate the effect of two factors, extrusion temperature (°F) and pressure (pounds per square inch), on the rate of extrusion (inches per second) of the molded rod. A complete 2×2 factorial experiment (that is, with each factor at two levels) was conducted. Three batches of granular plastic were used for the experiment, with each batch (viewed as a block) divided into four equal parts. The four portions of granular plastic for a given batch were randomly assigned to the four treatments and this was repeated for each of the three batches, resulting in a 2×2 factorial experiment laid out in three blocks. The data are shown in Table 13.16. Perform an analysis of variance for these data.

TABLE 13.16
Data for Example 13.16

		BATCH (BLOCK)					
		1		2		3	
		PRESSURE		PRESSURE		PRESSURE	
		40	60	40	60	40	60
TEMPERATURE	200°	1.35	1.74	1.31	1.67	1.40	1.86
	300°	2.48	3.63	2.29	3.30	2.14	3.27

SOLUTION

This experiment consists of a three-way classification of the data corresponding to batches (blocks), pressure, and temperature. The analysis of variance for this 2×2 factorial experiment (four treatments) laid out in a randomized block design (three blocks) yields the sources and degrees of freedom shown in Table 13.17.

TABLE 13.17
Table of Sources and Degrees of Freedom for Example 13.16

SOURCE	df
Pressure (P)	1
Temperature (T)	1
Blocks	2
Pressure–temperature interaction	1
Error	6
Total	11

The linear model for the experiment is

$$E(y) = \beta_0 + \overbrace{\beta_1 x_1}^{\substack{\text{Main} \\ \text{effect} \\ P}} + \overbrace{\beta_2 x_2}^{\substack{\text{Main} \\ \text{effect} \\ T}} + \overbrace{\beta_3 x_1 x_2}^{\substack{PT \\ \text{inter-} \\ \text{action}}} + \overbrace{\beta_4 x_3 + \beta_5 x_4}^{\substack{\text{Block} \\ \text{terms}}}$$

where

$x_1 = \text{Pressure}$ $\qquad x_2 = \text{Temperature}$

$x_3 = \begin{cases} 1 & \text{if block 2} \\ 0 & \text{otherwise} \end{cases}$ $\qquad x_4 = \begin{cases} 1 & \text{if block 3} \\ 0 & \text{otherwise} \end{cases}$

The computer printout for the analysis of variance using the SAS computer software package is shown in Figure 13.14.

FIGURE 13.14 SAS Computer Printout for an Analysis of Variance for a 2 × 2 Factorial Experiment in a Randomized Block Design

DEPENDENT VARIABLE: RATE

SOURCE	DF	SUM OF SQUARES	MEAN SQUARE	F VALUE	PR > F	R-SQUARE	C.V.
MODEL	5	7.14938333	1.42987667	83.23	0.0001	0.985786	5.9489
ERROR	6	0.10308333	0.01718056		ROOT MSE		RATE MEAN
CORRECTED TOTAL	11	7.25246667			0.13107462		2.20333333

SOURCE	DF	ANOVA SS	F VALUE	PR > F
PRESSURE	1	1.68750000	98.22	0.0001
TEMP	1	5.04403333	293.59	0.0001
TEMP*PRESSURE	1	0.36053333	20.98	0.0038
BATCH	2	0.05731667	1.67	0.2654

The format of the SAS computer printout was explained in Example 13.15. You can see from the printout that the *F* test for the model was highly significant (observed significance level is .0001). Thus, there is ample evidence to indicate differences among the block means, or the treatment means, or both. Proceeding to the breakdown of the model sources, you can see that the values of the *F* statistics for pressure, temperature, and the temperature–pressure interaction are all highly significant (that is, their observed significance levels are very small). Therefore, all of the terms ($\beta_1 x_1$, $\beta_2 x_2$, and $\beta_3 x_1 x_2$) contribute information for the prediction of *y*.

The treatments in the experiment were assigned according to a randomized block design. Thus, we expected the extrusion of the plastic to vary from batch to batch. Because the *F* test for testing differences among block means was not statistically significant (the observed significance level, PR > F, was as large as .2654), there is insufficient evidence to indicate a difference in the mean extrusion of the plastic from batch to batch. Blocking does not appear to have increased the amount of information in the experiment. ∎

Examples 13.15 and 13.16 show you how to interpret the results of the analysis of variance of a *k*-way classification of data. We will summarize the assumptions underlying these analyses in Section 13.11 and then show you how to call and use some available statistical computer software packages in optional Section 13.13.

| | | | | | | | | | | | | | |

**EXERCISES
13.34–13.41**

13.34 In increasingly severe oil well environments, oil producers are interested in high-strength nickel alloys that are corrosion-resistant. Since nickel alloys are especially susceptible to hydrogen embrittlement, an experiment was conducted to compare the yield strengths of nickel alloy tensile specimens cathodically charged in a 4% sulfuric acid solution saturated with carbon disulfide, a hydrogen recombination poison. Two alloys were combined: inconel alloy (75% nickel composition) and incoloy (30% nickel composition). The alloys were tested under two material conditions (cold rolled and cold drawn), each at three different charging times (0, 25, and 50 days). Thus, a $2 \times 2 \times 3$ factorial experiment was conducted, with alloy type at two levels, material condition at two levels, and charging time at three levels. Two hydrogen-charged tensile specimens were prepared for each of the $2 \times 2 \times 3 = 12$ factor–level combinations. Their yield strengths (kilograms per square inch) are recorded in the accompanying table.

		ALLOY TYPE			
		INCONEL		INCOLOY	
		Cold rolled	Cold drawn	Cold rolled	Cold drawn
CHARGING TIME	0 days	53.4 52.6	47.1 49.3	50.6 49.9	30.9 31.4
	25 days	55.2 55.7	50.8 51.4	51.6 53.2	31.7 33.3
	50 days	51.0 50.5	45.2 44.0	50.5 50.2	29.7 28.1

a. The SAS analysis of variance printout for the data is shown here. Is there evidence of any interactions among the three factors? Test using $\alpha = .05$. [*Note:* This means that you must test all the interaction parameters. The drop in SSE appropriate for the test would be the sum of all interaction sums of squares.]

SAS Computer Printout for Exercise 13.34

DEPENDENT VARIABLE: YIELD

SOURCE	DF	SUM OF SQUARES	MEAN SQUARE	F VALUE	PR > F	R-SQUARE	C.V.
MODEL	11	1931.73458333	175.61223485	258.73	0.0001	0.995801	1.8019
ERROR	12	8.14500000	0.67875000		ROOT MSE		YIELD MEAN
CORRECTED TOTAL	23	1939.87958333			0.82386285		45.72083333

SOURCE	DF	ANOVA SS	F VALUE	PR > F
CHRGTIME	2	71.04083333	52.33	0.0001
MATERIAL	1	956.34375000	1408.98	0.0001
CHRGTIME*MATERIAL	2	4.17250000	3.07	0.0836
TYPE	1	552.00041667	813.26	0.0001
CHRGTIME*TYPE	2	7.98583333	5.88	0.0166
MATERIAL*TYPE	1	339.75375000	500.56	0.0001
CHRGTIM*MATERIA*TYPE	2	0.43750000	0.32	0.7306

b. Now examine the *F* tests shown on the printout for the individual interactions. Which, if any, of the interactions are statistically significant at the .05 level of significance?

13.35 Refer to Exercise 13.34. Since charging time is a quantitative factor, we could plot the strength y versus charging time x_1 for each of the four combinations of alloy type and material condition. This suggests that a prediction equation relating mean strength $E(y)$ to charging time x_1 may be useful. Consider the model

$$E(y) = \beta_0 + \beta_1 x_1 + \beta_2 x_1^2 + \beta_3 x_2 + \beta_4 x_3 + \beta_5 x_2 x_3$$
$$+ \beta_6 x_1 x_2 + \beta_7 x_1 x_3 + \beta_8 x_1 x_2 x_3$$
$$+ \beta_9 x_1^2 x_2 + \beta_{10} x_1^2 x_3 + \beta_{11} x_1^2 x_2 x_3$$

where

x_1 = Charging time

$$x_2 = \begin{cases} 1 & \text{if inconel alloy} \\ 0 & \text{if incoloy alloy} \end{cases} \qquad x_3 = \begin{cases} 1 & \text{if cold rolled} \\ 0 & \text{if cold drawn} \end{cases}$$

a. Using the model shown above, give the relationship between mean strength $E(y)$ and charging time x_1 for cold drawn incoloy alloy.

b. Using the model shown above, give the relationship between mean strength $E(y)$ and charging time x_1 for cold drawn inconel alloy.

c. Using the model shown above, give the relationship between mean strength $E(y)$ and charging time x_1 for cold rolled inconel alloy.

13.36 Refer to Exercises 13.34 and 13.35. The SAS multiple regression analysis for fitting the model to the data is shown here.

SAS Computer Printout for Exercise 13.36

DEPENDENT VARIABLE: Y

SOURCE	DF	SUM OF SQUARES	MEAN SQUARE	F VALUE	PR > F	R-SQUARE	C.V.
MODEL	11	1931.73458333	175.61223485	258.73	0.0001	0.995801	1.8019
ERROR	12	8.14500000	0.67875000		ROOT MSE		Y MEAN
CORRECTED TOTAL	23	1939.87958333			0.82386285		45.72083333

SOURCE	DF	TYPE I SS	F VALUE	PR > F	DF	TYPE III SS	F VALUE	PR > F
X1	1	16.00000000	23.57	0.0004	1	4.50173077	6.63	0.0243
X1*X1	1	55.04083333	81.09	0.0001	1	8.16750000	12.03	0.0046
X2	1	552.00041667	813.26	0.0001	1	290.70250000	428.29	0.0001
X3	1	956.34375000	1408.98	0.0001	1	364.81000000	537.47	0.0001
X2*X3	1	339.75375000	500.56	0.0001	1	102.24500000	150.64	0.0001
X1*X2	1	3.42250000	5.04	0.0444	1	2.19240385	3.23	0.0975
X1*X3	1	3.42250000	5.04	0.0444	1	0.02778846	0.04	0.8430
X1*X2*X3	1	0.25000000	0.37	0.5552	1	0.30769231	0.45	0.5135
X1*X1*X2	1	4.56333333	6.72	0.0235	1	3.30041667	4.86	0.0477
X1*X1*X3	1	0.75000000	1.10	0.3139	1	0.09375000	0.14	0.7166
X1*X1*X2*X3	1	0.18750000	0.28	0.6087	1	0.18750000	0.28	0.6087

PARAMETER	ESTIMATE	T FOR H0: PARAMETER=0	PR > |T|	STD ERROR OF ESTIMATE
INTERCEPT	31.15000000	53.47	0.0001	0.58255901
X1	0.15300000	2.58	0.0243	0.05940960
X1*X1	-0.00396000	-3.47	0.0046	0.00114158
X2	17.05000000	20.70	0.0001	0.82386285
X3	19.10000000	23.18	0.0001	0.82386285
X2*X3	-14.30000000	-12.27	0.0001	1.16511802
X1*X2	0.15100000	1.80	0.0975	0.08401786
X1*X3	0.01700000	0.20	0.8430	0.08401786
X1*X2*X3	-0.08000000	-0.67	0.5135	0.11801919
X1*X1*X2	-0.00356000	-2.21	0.0477	0.00161443
X1*X1*X3	0.00060000	0.37	0.7166	0.00161443
X1*X1*X2*X3	0.00120000	0.53	0.6087	0.00228316

a. Find the prediction equation.

b. Find the prediction equations for each of the four combinations of alloy type and material condition.

c. Plot the data points for each of the four combinations of alloy type and material condition. Graph the respective prediction equations.

13.37 Refer to Exercises 13.34–13.36. If the relationship between mean strength $E(y)$ and charging time x_1 is the same for all four combinations of alloy type and material condition, the appropriate model for $E(y)$ is

$$E(y) = \beta_0 + \beta_1 x_1 + \beta_2 x_1^2$$

Use the accompanying SAS computer printout for fitting this model to the data, together with the information in the printout of Exercise 13.36 to decide whether the data provide sufficient evidence to indicate differences among the second-order models relating $E(y)$ to x_1 for the four categories of alloy type and material condition. Test using $\alpha = .05$.

SAS Computer Printout for Exercise 13.37

```
DEPENDENT VARIABLE: Y
SOURCE                      DF      SUM OF SQUARES        MEAN SQUARE       F VALUE        PR > F       R-SQUARE              C.V.
MODEL                        2        71.04083333        35.52041667          0.40        0.6759       0.036621          20.6330
ERROR                       21      1868.83875000        88.99232143                      ROOT MSE                      Y MEAN
CORRECTED TOTAL             23      1939.87958333                                         9.43357416                45.72083333

SOURCE                      DF         TYPE I SS       F VALUE     PR > F      DF         TYPE III SS     F VALUE     PR > F
X1                           1        16.00000000         0.18     0.6759       1        36.22230769        0.41     0.5304
X1*X1                        1        55.04083333         0.62     0.4404       1        55.04083333        0.62     0.4404

                                           T FOR H0:                          STD ERROR OF
PARAMETER                ESTIMATE      PARAMETER=0      PR > :T:                ESTIMATE
INTERCEPT            45.65000000            13.69       0.0001              3.33527213
X1                    0.21700000             0.64       0.5304              0.34013235
X1*X1                -0.00514000            -0.79       0.4404              0.00653577
```

13.38 An experiment was conducted to investigate the effects of three factors—paper stock, bleaching compound, and coating type—on the whiteness of fine bond paper. Three paper stocks (factor A), four types of bleaches (factor B), and two types of coatings (factor C) were used for the experiment. Six paper specimens were prepared for each of the $3 \times 4 \times 2$ stock–bleach–coating combinations and a measure of whiteness was recorded.

a. Construct an analysis of variance table showing the sources of variation and the respective degrees of freedom.

b. Suppose MSE = .14, MS(AB) = .39, and the mean square for all interactions combined is .73. Do the data provide sufficient evidence to indicate any interactions among the three factors? Test using $\alpha = .05$.

c. Do the data present sufficient evidence to indicate an AB interaction? Test using $\alpha = .05$. From a practical point of view, what is the significance of an AB interaction?

d. Suppose SS(A) = 2.35, SS(B) = 2.71, and SS(C) = .72. Find SS(Total). Then find R^2 and interpret its value.

13.39 A study was conducted to evaluate the use of computer-assisted instruction (CAI) in teaching an introductory FORTRAN programming course (GE 102) in the College of Engineering at Oregon State University (*Engineering Education*, Feb. 1986). One of the objectives was to investigate the effect of four factors on a student's final exam score (y) in the course. The factors and their respective levels are given below:

Group (3 levels):

Control group (student receives no CAI)

Guided CAI (student receives CAI, but proceeds at same pace as normal class)

Self-paced CAI (student receives CAI and proceeds at his or her own pace)

Math background (4 levels):
  High school algebra
  Trigonometry
  Differential calculus
  Integral calculus

Computer background (3 levels):
  None (little or no exposure)
  Some (one programming language)
  Extensive (more than one programming language)

Grade in prerequisite course, GE 101 (3 levels):
  A, B, or C

a. How many treatments are associated with this four-way factorial experiment?
b. Write the complete factorial model for this experiment. [*Hint:* Use dummy variables to represent the factors.]
c. What hypothesis would you test to determine whether interaction exists among the four factors?

13.40 A 2 × 2 factorial experiment was conducted for each of 3 weeks to determine the effect of two factors, temperature and pressure, on the yield of a chemical. Temperature was set at 300° and 500°. The pressure maintained in the reactor was set at 100 and 200 pounds per square inch. Four days were randomly selected within each week and the four factor–level combinations were randomly assigned to them.
a. What type of design was used for this experiment?
b. Construct an analysis of variance table showing all sources and their respective degrees of freedom.

13.41 Refer to Exercise 13.40. The yield data for the 2 × 2 factorial experiment, laid out in three blocks of time, are shown in the accompanying table. The SAS computer printout for the analysis of variance is also shown.

		WEEK 1		WEEK 2		WEEK 3	
		TEMPERATURE		TEMPERATURE		TEMPERATURE	
		300	500	300	500	300	500
PRESSURE	100	64	73	65	72	62	70
	200	69	81	71	85	67	83

SAS Computer Printout for Exercise 13.41

```
DEPENDENT VARIABLE: YIELD
SOURCE            DF      SUM OF SQUARES      MEAN SQUARE      F VALUE      PR > F       R-SQUARE         C.V.
MODEL             5        613.50000000      122.70000000       72.41       0.0001      0.983699        1.8121
ERROR             6         10.16666667        1.69444444                   ROOT MSE                 YIELD MEAN
CORRECTED TOTAL  11        623.66666667                                     1.30170828               71.83333333

SOURCE            DF          ANOVA SS      F VALUE      PR > F
PRESSURE          1        209.33333333      122.95       0.0001
TEMP              1        363.00000000      214.23       0.0001
PRESSURE*TEMP     1         27.00000000       15.93       0.0072
WEEK              2         15.16666667        4.48       0.0646
```

a. Why does the analysis of variance table not include sources for the interaction of weeks with temperature and pressure?

b. Do the data provide sufficient evidence to indicate an interaction between temperature and pressure? Give the p-value for the test. What is the practical significance of this result?

c. Was blocking in time useful in increasing the amount of information in the experiment? That is, do the data provide sufficient evidence to indicate differences among the block means? Give the p-value for the test.

SECTION 13.10

A PROCEDURE FOR MAKING MULTIPLE COMPARISONS

Many practical experiments are conducted to determine the largest (or the smallest) mean in a set. For example, suppose that a chemist has developed five chemical solutions for removing a corrosive substance from a metal fitting. The chemist would then want to determine the solution that will remove the greatest amount of the corrosive substance from the fitting in a single application. Similarly, a production engineer might want to determine which among six machines or which among three foremen achieves the highest mean productivity per hour. A mechanical engineer might want to choose one engine, from among five, that is most efficient, and so on.

Choosing the treatment with the largest mean from among five treatments might appear to be a simple matter. We could make, for example, $n_1 = n_2 = \cdots = n_5 = 10$ observations on each treatment, obtain the sample means, $\bar{y}_1, \bar{y}_2, \ldots, \bar{y}_5$, and compare them using Student's t tests to determine whether differences exist among the pairs of means. However, there is a problem associated with this procedure: *A Student's t test, with its corresponding value of α, is valid only when the two treatments to be compared are selected prior to experimentation.* After you have looked at the data, you cannot use a Student's t statistic to compare the treatments for the largest and smallest sample means because they will always be farther apart, on the average, than any pair of treatments selected at random. Furthermore, if you conduct a series of t tests, each with a chance α of indicating a difference between a pair of means if in fact no difference exists, then the risk of making *at least one* Type I error in a series of t tests will be larger than the value of α specified for a single t test.

There are a number of procedures for comparing and ranking a group of treatment means. The one that we present, known as **Tukey's method for multiple comparisons**, utilizes the Studentized range

$$q = \frac{\bar{y}_{max} - \bar{y}_{min}}{s/\sqrt{n}}$$

(where $\bar{y}_{max}$ and $\bar{y}_{min}$ are the largest and smallest sample means, respectively) to determine whether the difference in any pair of sample means implies a difference in the corresponding treatment means. The logic behind this multiple comparisons procedure is that if we determine a critical value for the difference between the largest and smallest sample means, $|\bar{y}_{max} - \bar{y}_{min}|$, one that implies a difference in their respective treatment means, then any other pair of sample means that differ by as much as or more than this critical value would also imply a difference in the corresponding treatment means. Tukey's procedure selects this critical distance, ω, so that the probability of making one or more Type I errors (concluding that a difference exists between a pair of treatment means if, in fact, they are identical) is α. Therefore, the risk of making a Type I error applies to the whole procedure, i.e., to the comparisons of all pairs of means, rather than to a single comparison.

Tukey's procedure is based on the assumption that the p sample means are based on independent random samples, each containing an equal number n_t of observations. Then if $s = \sqrt{\text{MSE}}$ is the computed standard deviation for the analysis, the distance ω is

$$\omega = q(p, \nu)\frac{s}{\sqrt{n_t}}$$

The tabulated statistic $q_\alpha(p, \nu)$ is the critical value of the Studentized range, the value that locates α in the upper tail of the q distribution. This critical value depends on α, the number of treatment means involved in the comparison, and ν, the number of degrees of freedom associated with MSE. Values of $q(p, \nu)$ for $\alpha = .05$ and $\alpha = .01$ are given in Tables 13 and 14, respectively, of Appendix II.

TUKEY'S MULTIPLE COMPARISONS PROCEDURE

1. Calculate

$$\omega = q_\alpha(p, \nu)\frac{s}{\sqrt{n_t}}$$

where

p = Number of sample means

$s = \sqrt{\text{MSE}}$

ν = Number of degrees of freedom associated with MSE

n_t = Number of observations in each of the p samples

$q_\alpha(p, \nu)$ = Critical value of the Studentized range (Tables 13 and 14 of Appendix II)

2. Rank the p sample means. Any pair of sample means differing by more than ω implies a difference in the corresponding population means.

Assumptions: Samples were randomly and independently selected from normal populations with means $\mu_1, \mu_2, \ldots, \mu_p$, and common variance σ^2.

EXAMPLE 13.17

Refer to the $2 \times 2 \times 3$ factorial experiment described in Example 13.15. Compare the six population means for each shift using Tukey's multiple comparisons procedure. Use $\alpha = .05$.

SOLUTION

The sample means for the 12 factor–level combinations are shown in Table 13.18. Since the sample means in the table represent measures of productivity in the manufacture of transistors, we would want to find the length of work week and number of coffee breaks best suited for each shift. Consequently, we will rank the six means for each shift.

TABLE 13.18
Sample Means for the $p = 12$ Treatments of Example 13.15

		DAY SHIFT			NIGHT SHIFT		
		COFFEE BREAKS			COFFEE BREAKS		
		0	1	2	0	1	2
LENGTH OF	4 days	95.5	105.5	93.5	89.5	99.5	100.5
WORK WEEK	5 days	94.0	101.5	85.0	82.5	91.0	95.0

The first step in the ranking procedure is to calculate ω for $p = 6$ (since we are ranking only six treatment means for each shift), $n_t = 2$, $\alpha = .05$, and $s = 2.61$ (shown on the computer printout of Figure 13.13). Since MSE is based on $\nu = 12$ degrees of freedom, we have

$$q_{.05}(6, 12) = 4.75$$

and

$$\omega = q_{.05}(6, 12)\left(\frac{s}{\sqrt{n_t}}\right) = (4.75)\left(\frac{2.61}{\sqrt{2}}\right) = 8.77$$

Therefore, population means corresponding to pairs of sample means that differ by more than $\omega = 8.77$ will be judged to be different. The sample means for the day and night shifts are ranked as follows:

Day shift: 85.0, 93.5, 94.0, 95.5, $\overline{101.5, 105.5}$

Night shift: 82.5, 89.5, 91.0, $\overline{95.0, 99.5, 100.5}$

Using $\omega = 8.77$ as a yardstick to determine differences between pairs of treatments for the day shift, we see that there is no evidence to indicate a difference between any pairs of treatments that are adjacent in the ranking. There is evidence to indicate that the population mean of the treatment corresponding to a 4-day work week and one coffee break per day, with a sample mean equal to 105.5, is different from the treatments with sample means equal to 95.5 or less. There is no evidence to indicate a difference in the treatments corresponding to the two largest sample means, 101.5 and 105.5 (indicated by the overbar). Further experimentation would be required to determine whether the observed difference in these two sample means really implies a difference in the corresponding population means.

A similar comparison of the sample means for the night shift indicates that a 4-day week with two coffee breaks per workday produced the highest mean (100.5) productivity. There is evidence to indicate that this treatment mean differs from those with sample means equal to 91.0 or less. However, there is no evidence to indicate differences among the treatment means corresponding to the three largest sample means. As in the case of the day shift, further experimentation would be required to determine the best combination of work week and coffee breaks for this shift. ∎

EXERCISES 13.42–13.46

13.42 Refer to Exercise 13.4. Use Tukey's multiple comparisons procedure to compare the mean rates of penetration for the three types of drill bits. Identify the means that appear to differ. Use $\alpha = .05$.

13.43 Refer to Exercise 13.15. Use Tukey's multiple comparisons procedure to compare the mean crack widths for the four wetting periods. Identify the means that appear to differ. Use $\alpha = .05$.

13.44 Refer to Exercise 13.23. Use Tukey's multiple comparisons procedure to compare the productivity means for the $2 \times 3 = 6$ machine–material combinations. Identify the means that appear to differ. Use $\alpha = .05$.

13.45 Refer to Exercise 13.7. Use Tukey's multiple comparisons procedure to compare the mean PCB levels in fish for the five rivers. Identify the means that appear to differ. Use $\alpha = .05$. [Note: For this exercise, the sample sizes for the five treatments (rivers) are unequal. Conservatively, use the smallest sample size in your calculation of ω.]

13.46 Refer to Exercise 13.25. Use Tukey's multiple comparisons procedure to compare the mean shear strengths for the four antimony amounts. Identify the means that appear to differ. Use $\alpha = .01$.

SECTION 13.11

ASSUMPTIONS

For all the experiments and designs discussed in this chapter, the probabilistic model for the response variable y is the familiar regression linear model of Chapter 12. Therefore, the assumptions underlying the analyses are the same as those required for a regression analysis (see Section 11.2). The assumptions for each experimental design, stated in the terminology of an analysis of variance, are summarized in the boxes below and on page 706.

ASSUMPTIONS FOR A TEST TO COMPARE k POPULATION MEANS: COMPLETELY RANDOMIZED DESIGN

1. All k population probability distributions are normal.
2. The k population variances are equal.
3. The samples from each population are random and independent.

ASSUMPTIONS FOR A TEST TO COMPARE k POPULATION MEANS (TREATMENTS): RANDOMIZED BLOCK DESIGN

1. The population probability distribution of the difference between any pair of treatment observations within a block is approximately normal.
2. The variance of this difference is constant and the same for all pairs of observations.
3. The treatments are randomly assigned to the experimental units within each block.

ASSUMPTIONS FOR A TWO-WAY ANOVA: FACTORIAL EXPERIMENT

1. The population probability distribution of the observations for any factor–level combination is approximately normal.
2. The variance of the probability distribution is constant and the same for all factor–level combinations.
3. The treatments (factor–level combinations) are randomly assigned to the experimental units.
4. The observations for each factor–level combination represent independent random samples.

In most scientific applications the assumptions will not be satisfied exactly. However, these analysis of variance procedures are flexible in the sense that slight departures from the assumptions will not significantly affect the analysis or the validity of the resulting inferences.

We conclude this section by noting that the regression model

$$y = E(y) + \varepsilon = \beta_0 + \beta_1 x_1 + \beta_2 x_2 + \cdots + \beta_k x_k + \varepsilon$$

is often called a **fixed-effect model** because it contains only a single random component. All other components of the model are nonrandom or, in the language of analysis of variance, **fixed**. An analysis of variance based on models containing more than one random component—that is, **mixed models**—will be presented in Chapter 14.

SECTION 13.12

COMPUTER PRINTOUTS FOR AN ANALYSIS OF VARIANCE

The calculations in an analysis of variance can become very tedious when the number of observations is large or when the observations contain a large number of digits. If you have access to a computer, this difficulty can be easily circumvented by using the regression approach, i.e., by fitting complete and reduced models (see Section 13.8). However, even the regression approach requires you to perform some calculations (for the partial F tests) and you will need to fit at least two models to carry out the analysis. For these reasons, you may want to take advantage of the analysis of variance routines included in most statistical computer software packages.

All four packages discussed in this text have procedures for conducting an analysis of variance. The SAS, SPSS^X, BMDP, and Minitab packages are capable of analyzing experimental designs ranging from simple one-way classifications of data (completely randomized designs) to the more sophisticated factorial experiments.* In this section, we will examine the respective analysis of variance printouts for a factorial experiment. Since the printouts for completely randomized designs and randomized block designs appear similar to that of the factorial experiment, you should be able to understand the ANOVA results for any of these three experimental procedures.

Refer to the 3×3 factorial experiment of Example 13.7, where we investigated the effect of two factors, level of supply of raw materials (S) and ratio of its assignment (R) to the product manufacturing lines, on the profit (y) per unit of raw material. The SAS, SPSS^X, BMDP, and Minitab printouts for the data of Table 13.11 are shown in Figures 13.15(a)–(d), respectively (page 708).

You can see that all four printouts report the results in the form of an ANOVA summary table, giving the source of variation, degrees of freedom, and sum of squares. Note that SAS, SPSS^X, and BMDP [Figures 13.15(a)–(c), respectively] report the F-values and corresponding observed significance levels (i.e., p-values). Minitab is the only one of the four packages that does not give the value of the test statistic for the analysis of variance F test. Consequently, you will have to compute the appropriate F statistic by hand and find the corresponding rejection region if you use Minitab.

The SPSS^X and BMDP printouts [Figures 13.15(b)–(c), respectively] include a source of variation that does not appear on either the SAS [Figure 13.15(a)] or Minitab [Figure 13.15(d)] printouts. This source, titled CONSTANT in the SPSS^X printout and MEAN in the BMDP printout, represents the term β_0 in the complete model for the factorial experiment. Since the source of variation due to β_0 is associated with 1 degree of freedom, the sum of the entries in the DF column for both SPSS^X and BMDP will equal the sample size n (in this example, $n = 12$), rather than the traditional degrees of freedom ($n - 1$). Note also that the source of variation due to error is titled WITHIN CELLS in the SPSS^X printout [Figure 13.15(b)].

The SAS package is the only one of the four packages that reports the results of a test for the overall adequacy of the complete factorial model. Note that in the upper portion of the SAS printout [Figure 13.15(a)], SS(Total) is partitioned into two sources of variation, MODEL and ERROR. The source designated as MODEL represents the sum of the main effects for supply S, main effects for ratio R, and supply–ratio interaction effects. Therefore, the F-value given in the upper right of the printout is the value of the test statistic for testing overall model adequacy, i.e.,

H_0: All β coefficients (excluding β_0) in the complete factorial model equal 0

For instructions on how to access the ANOVA routines of SAS, SPSS^X, BMDP, and Minitab, consult Optional Section 13.13.

---

*Currently, Minitab is limited to two-way classifications of data (i.e., two-factor factorial experiments); SAS, SPSS^X, and BMDP are capable of analyzing the more general k-way classifications of data, where k is any integer.

FIGURE 13.15 Printouts for the 3 × 3 Factorial Experiment on the Data in Table 13.11

(a) SAS Printout

SOURCE	DF	SUM OF SQUARES	MEAN SQUARE	F VALUE	PR > F	R-SQUARE	C.V.
MODEL	8	74.66666667	9.33333333	3.88	0.0081	0.632768	7.5077
ERROR	18	43.33333333	2.40740741		ROOT MSE		Y MEAN
CORRECTED TOTAL	26	118.00000000			1.55158223		20.66666667

SOURCE	DF	ANOVA SS	F VALUE	PR > F
RATIO	2	8.22222222	1.71	0.2094
SUPPLY	2	20.22222222	4.20	0.0318
RATIO*SUPPLY	4	46.22222222	4.80	0.0082

(b) SPSS^x Printout

SOURCE OF VARIATION	SUM OF SQUARES	DF	MEAN SQUARE	F	SIG. OF F
WITHIN CELLS	43.33333	18	2.40741		
CONSTANT	11532.00000	1	11532.00000	4790.21538	.000
SUPPLY	20.22222	2	10.11111	4.20000	.032
RATIO	8.22222	2	4.11111	1.70769	.209
SUPPLY BY RATIO	46.22222	4	11.55556	4.80000	.008

(c) BMDP Printout

SOURCE	SUM OF SQUARES	DEGREES OF FREEDOM	MEAN SQUARE	F	TAIL PROB.
MEAN	11532.00000	1	11532.00000	4790.21	0.0000
SUPPLY	20.22222	2	10.11111	4.20	0.0318
RATIO	8.22222	2	4.11111	1.71	0.2094
SR	46.22222	4	11.55556	4.80	0.0082
1 ERROR	43.33333	18	2.40741		

(d) Minitab Printout

ANALYSIS OF VARIANCE

DUE TO	DF	SS	MS=SS/DF
SUPPLY	2	20.22	10.11
RATIO	2	8.22	4.11
SUPPLY*RATIO	4	46.22	11.56
ERROR	18	43.33	2.41
TOTAL	26	118.00	

S E C T I O N 13.13

COMPUTER PROGRAMS FOR THE ANALYSIS OF VARIANCE (OPTIONAL)

All four computer software packages discussed in this text possess analysis of variance routines. The SAS, SPSS^X, and BMDP packages are capable of analyzing experimental designs ranging from simple one-way classifications of data (completely randomized designs) to the more sophisticated k-way classifications (factorial experiments). Minitab, however, is limited to one-way and two-way classifications of data.

As we noted in the previous sections, each experimental design possesses an underlying probabilistic model for the response variable y. These models can become quite complex, especially for large factorial experiments. Fortunately, the four software packages we discuss do not require that the exact model be specified. Rather, only the sources of variation for the analysis of variance need be given in the appropriate command cards. For example, for a 2×3 factorial with temperature at two levels and pressure at three levels, we need to specify three sources of variation: (1) Temperature (main effect); (2) Pressure (main effect); and (3) Temperature–pressure interaction. The computer package will automatically fit the complete factorial model that corresponds to these sources of variation and produce an ANOVA summary table.

EXAMPLE 13.18

Using one of the statistical program packages, write a computer program that will produce an ANOVA summary table for the $2 \times 2 \times 3$ replicated factorial experiment of Example 13.15.

SOLUTION

Since we want to conduct an analysis of variance for a three-way factorial experiment, we must use either SAS, SPSS^X, or BMDP. (Minitab can analyze, at most, a two-way factorial.) The commands for the SAS, SPSS^X, and BMDP packages are given in Programs 13.1–13.3, respectively.

PROGRAM 13.1 SAS ANOVA Commands for a $2 \times 2 \times 3$ Factorial Experiment

Command line

```
1    DATA ITEMS;                                        Data entry instructions:
2    INPUT SHIFT $ BREAKS DAYS DOLLARS @@;              Read the data of
3    CARDS;                                             Table 13.15
4    DAY 0 4 94 DAY 0 4 97
5    DAY 0 5 96 DAY 0 5 92
6    DAY 1 4 105 DAY 1 4 106
7    DAY 1 5 100 DAY 1 5 103
8    DAY 2 4 96 DAY 2 4 91
9    DAY 2 5 82 DAY 2 5 88          Input data values: Two
10   NIGHT 0 4 90 NIGHT 0 4 89      observations (weeks) per line
11   NIGHT 0 5 81 NIGHT 0 5 84
12   NIGHT 1 4 102 NIGHT 1 4 97
13   NIGHT 1 5 90 NIGHT 1 5 92
14   NIGHT 2 4 103 NIGHT 2 4 98
15   NIGHT 2 5 94 NIGHT 2 5 96
16   PROC ANOVA;                                        Statistical analysis
17   CLASSES SHIFT BREAKS DAYS;                         instructions: Call the
18   MODEL DOLLARS = SHIFT BREAKS DAYS SHIFT*BREAKS     ANOVA procedure
19         SHIFT*DAYS BREAKS*DAYS SHIFT*BREAKS*DAYS;
```

PROGRAM 13.2   SPSS^X ANOVA Commands for a 2 × 2 × 3 Factorial Experiment

Command
line

```
 1    DATA LIST FREE/SHIFT,BREAKS,DAYS,DOLLARS}   Data entry instruction:
                                                 Read the data of Table 13.15
 2    MANOVA     DOLLARS BY SHIFT(1,2) BREAKS(0,2) DAYS(4,5)/    Statistical analysis
 3               DESIGN = SHIFT, BREAKS, DAYS, SHIFT BY BREAKS,  instructions: Call
 4                        SHIFT BY DAYS, BREAKS BY DAYS,         the analysis of
 5                        SHIFT BY BREAKS BY DAYS/               variance procedure
 6    BEGIN DATA
 7    1 0 4 94 1 0 4 97
 8    1 0 5 96 1 0 5 92
 9    1 1 4 105 1 1 4 106
10    1 1 5 100 1 1 5 103
11    1 2 4 96 1 2 4 91
12    1 2 5 82 1 2 5 88      Input data values: Two
13    2 0 4 90 2 0 4 89      observations (weeks) per line
14    2 0 5 81 2 0 5 84
15    2 1 4 102 2 1 4 97
16    2 1 5 90 2 1 5 92
17    2 2 4 103 2 2 4 98
18    2 2 5 94 2 2 5 96
19    END DATA
```

PROGRAM 13.3   BMDP ANOVA Commands for a 2 × 2 × 3 Factorial Experiment

Command
line

```
 1    BMDP 2V}   Statistical analysis instruction:
                 Call the analysis of variance procedure
 2    / PROBLEM     TITLE IS 'FACTORIAL EXAMPLE'.   Data entry instructions:
 3    / INPUT       VARIABLES=4.                    Read the data
 4                  FORMAT=STREAM.                  of Table 13.15
 5    / VARIABLE    NAMES=SHIFT,BREAKS,DAYS,DOLLARS.
 6    / DESIGN      DEPENDENT=DOLLARS.              Statistical analysis instructions:
 7                  GROUP=SHIFT,BREAKS,DAYS.        Specify ANOVA design
 8                  INCLUDE=1,2,3,12,13,23,123.
 9    / END
10    1 0 4 94 1 0 4 97
11    1 0 5 96 1 0 5 92
12    1 1 4 105 1 1 4 106
13    1 1 5 100 1 1 5 103
14    1 2 4 96 1 2 4 91
15    1 2 5 82 1 2 5 88      Input data values: Two
16    2 0 4 90 2 0 4 89      observations (weeks) per line
17    2 0 5 81 2 0 5 84
18    2 1 4 102 2 1 4 97
19    2 1 5 90 2 1 5 92
20    2 2 4 103 2 2 4 98
21    2 2 5 94 2 2 5 96
22    / END
```

The SAS PROC command that calls the ANOVA procedure is given in line 16 of Program 13.1. This line is followed by a CLASSES statement and a MODEL statement. The CLASSES statement (line 17) identifies the independent variables (factors) for the experiment. *All* independent variables (quantitative and qualitative) should be included on this line. (This is in contrast to the SAS GLM procedure, in which only qualitative variables are specified in the CLASSES statement.) The dependent variable and sources of variation for the ANOVA are specified in the MODEL statement (lines 18–19). The dependent variable (DOLLARS) appears to the left of the equals sign; the sources of variation appear to the right. Since we want to fit a complete factorial model, all two-way and three-way interactions are specified, in addition to the main effect sources. The factors that comprise an interaction are separated by an asterisk (for example, SHIFT*BREAKS).

The SPSS^X commands that call the ANOVA procedure are given in lines 2–5 of Program 13.2. The key word MANOVA on the first analysis line (line 2) commands SPSS^X to begin processing the data for an analysis of variance. The dependent and independent variables are listed after the MANOVA command on line 2. The dependent variable (DOLLARS) is followed by the key word BY and the independent variables (or factors). The range of the coded values of the factors on the data lines *must* be specified in parentheses after each factor. For example, the values of SHIFT are coded as 1 for Day and 2 for Night. Thus, the expression SHIFT(1,2) appears in the MANOVA statement. Lines 3–5 identify the sources of variation for the analysis of variance. Following the expression DESIGN =, we indicate the various sources of variation in the model, with different sources separated by a comma. Interactions are indicated with the key word BY (for example, SHIFT BY BREAKS).

The BMDP program identification statement (line 1 of Program 13.3) accesses the BMDP analysis of variance (2V) procedure. The particular type of ANOVA design is specified in the DESIGN paragraph (lines 6–8), which usually includes three sentences: DEPENDENT, GROUP, and INCLUDE. The DEPENDENT sentence (line 6) identifies the dependent variable (DOLLARS), and the GROUP sentence (line 7) identifies the independent variables or factors (SHIFT, BREAKS, DAYS) in the factorial experiment. Each of the factors is assigned a number corresponding to the order in which they appear in the GROUP sentence. For example, SHIFT is assigned the number 1, BREAKS the number 2, and DAYS the number 3. The particular combinations of the factors that make up the sources of variation are specified in the INCLUDE sentence (line 8) using these GROUP numbers. Thus, a 1 appearing in the INCLUDE sentence commands BMDP to include the main effect for the first GROUP variable (SHIFT). Similarly, a 2 represents the main effect for the second GROUP variable (BREAKS) and a 3 represents the main effect for the third GROUP variable (DAYS). Interactions are specified with combinations of the GROUP numbers. For example, 13 represents the two-way interaction between SHIFT(1) and DAYS(3), and 123 the three-way interaction between SHIFT(1), BREAKS(2), and DAYS(3).

The ANOVA computer printouts for the three program packages are shown in Figure 13.16. You can see that all three packages produce similar ANOVA summary tables. Note that the SPSS^X and BMDP printouts [Figures 13.16(b)–(c), respectively] include a source of variation that does not appear on the SAS printout [Figure 13.16(a)]. This source, titled CONSTANT in the SPSS^X program and MEAN in the BMDP program, represents the term β_0 in the factorial model. Since the source of variation due to β_0 is associated with 1 degree of freedom, the sum of the entries in the DF column for both SPSS^X and BMDP will equal the sample size n (in this example, $n = 24$), rather than the traditional total degrees of freedom of $(n - 1)$. Note also that the source of variation due to error is titled WITHIN CELLS in the SPSS^X program.

FIGURE 13.16   ANOVA Computer Printouts for a $2 \times 2 \times 3$ Factorial Experiment

(a) SAS

DEPENDENT VARIABLE: DOLLARS

SOURCE	DF	SUM OF SQUARES	MEAN SQUARE	F VALUE	PR > F	R-SQUARE	C.V.
MODEL	11	1009.83333333	91.80303030	13.43	0.0001	0.924897	2.7686
ERROR	12	82.00000000	6.83333333		ROOT MSE		DOLLARS MEAN
CORRECTED TOTAL	23	1091.83333333			2.61406452		94.41666667

SOURCE	DF	ANOVA SS	F VALUE	PR > F
SHIFT	1	48.16666667	7.05	0.0210
DAYS	1	204.16666667	29.88	0.0001
BREAKS	2	334.08333333	24.45	0.0001
SHIFT*DAYS	1	8.16666667	1.20	0.2958
SHIFT*BREAKS	2	385.58333333	28.21	0.0001
DAYS*BREAKS	2	8.08333333	0.59	0.5689
SHIFT*DAYS*BREAKS	2	21.58333333	1.58	0.2461

(b) SPSS^X

TESTS OF SIGNIFICANCE FOR DOLLARS USING SEQUENTIAL SUMS OF SQUARES

SOURCE OF VARIATION	SUM OF SQUARES	DF	MEAN SQUARE	F	SIG. OF F
WITHIN CELLS	82.00000	12	6.83333		
CONSTANT	213948.16667	1	213948.16667	31309.48780	.000
SHIFT	48.16667	1	48.16667	7.04878	.021
BREAKS	334.08333	2	167.04167	24.44512	.000
DAYS	204.16667	1	204.16667	29.87805	.000
SHIFT BY BREAKS	385.58333	2	192.79167	28.21341	.000
SHIFT BY DAYS	8.16667	1	8.16667	1.19512	.296
BREAKS BY DAYS	8.08333	2	4.04167	.59146	.569
SHIFT BY BREAKS BY DAYS	21.58333	2	10.79167	1.57927	.246

(c) BMDP

ANALYSIS OF VARIANCE FOR 1-ST
DEPENDENT VARIABLE - DOLLARS

	SOURCE	SUM OF SQUARES	DEGREES OF FREEDOM	MEAN SQUARE	F	TAIL PROB.
	MEAN	213948.16667	1	213948.16667	31309.48	0.0000
	SHIFT	48.16667	1	48.16667	7.05	0.0210
	BREAKS	334.08333	2	167.04167	24.45	0.0001
	DAYS	204.16667	1	204.16667	29.88	0.0001
	SB	385.58333	2	192.79167	28.21	0.0000
	SD	8.16667	1	8.16667	1.20	0.2958
	BD	8.08333	2	4.04167	0.59	0.5689
	SBD	21.58333	2	10.79167	1.58	0.2461
1	ERROR	82.00000	12	6.83333		

For each of the three computer packages, SAS, SPSS^X, and BMDP, we have described how to access the analysis of variance procedure and identify the sources of variation in the linear model. Since we wanted to fit a full factorial model with three factors, we included sources of variation for all main effects and all possible two-way and three-way interactions. However, each package includes a default option which will automatically fit a full factorial model. The SPSS^X and BMDP packages fit a full factorial model *unless* another design is specified. Thus, we could have omitted the DESIGN statement (lines 3–5) in the SPSS^X program (Program 13.2) and the INCLUDE sentence (line 8) in the BMDP program (Program 13.3). This is in contrast to SAS, which *always* requires a MODEL statement. However, the MODEL statement can be simplified when a full factorial model is desired. The MODEL statement (lines 18–19) of the SAS program (Program 13.1) can be written as follows:

```
MODEL DOLLARS = SHIFT | BREAKS | DAYS;
```

When the factors (independent variables) are separated by vertical bars ($|$), SAS will automatically fit a full factorial model.

These default options provide a convenient way to command the computer to fit a full factorial model without identifying the sources of variation due to two-way, three-way, . . . , and k-way interactions. However, before you use the default option, be sure that the full factorial model is the model that you want to fit. For example, in a factorial experiment laid out in a randomized block design, you do not want to include treatment–block interaction in the linear model. Thus, do not use the default option of your computer package's analysis of variance routine. Instead, you must specify the sources of variation in the manner outlined in Programs 13.1–13.3, making sure that you omit any block–treatment interaction terms.

As we mentioned earlier, Minitab cannot perform an ANOVA for a k-way classification of data, where $k > 2$. However, Minitab's one-way and two-way ANOVA routines are comparable to those of SAS, SPSS^X, and BMDP. For the benefit of those who have access only to the Minitab statistical software package, we give the Minitab commands for a two-way ANOVA in Program 13.4.

PROGRAM 13.4

Minitab ANOVA Commands for a Two-Factor Factorial Experiment

Command line	
1	`READ FACTOR1 IN C1, FACTOR2 IN C2, DEPENDENT IN C3` [Input data values]
2	`TWOWAY AOV ON C3, FACTORS IN C1,C2`

■

| | | | | | | | | | | | | |

SECTION 13.14

SUMMARY

An **analysis of variance** partitions the total sum of squares, SS(Total), into SSE and the drops in the sums of squares for error associated with sets of main effect and interaction parameters.

An analysis of variance possesses advantages and disadvantages in comparison to a regression analysis. *The major advantage of an analysis of variance is that it is easy to perform on a pocket or desk calculator. The disadvantages are its restrictions and limitations.* They are as follows:

1. *The set of analysis of variance formulas appropriate for a particular experimental design applies only to that design.* If the data collected are observational (i.e., the independent variables are uncontrolled), an analysis of variance is inappropriate. No deviations from the design are permitted. Consequently, the method is of value only for special types of designed experiments.
2. *In contrast to a regression analysis, the analysis of variance formulas change from one design to another.* (There is a pattern, but the pattern is usually not apparent to a beginner.) Thus, at first contact, the topic appears to be obscure and disorganized.
3. *An analysis of variance does not give you a prediction equation.* This is a great handicap when one (or more) of the independent variables is quantitative.
4. *Although a linear model is always implied in an analysis of variance, it is rarely presented or discussed when analyses of data have been performed.* Consequently, the thrust of an analysis of variance is often counter (although it need not be) to the notion of modeling, which is the modern quantitative way of analyzing business phenomena.

Since we have listed more disadvantages than advantages to an analysis of variance, it would suggest that we think that the methods are superfluous to a regression analysis. But this is not the case. You can often obtain the sums of squares associated with sets of model parameters using an analysis of variance in less time than it takes to enter the data for a regression analysis into a computer. And, if the experiment has been properly designed and *if all the independent variables are qualitative*, you can obtain all the information from the analysis of variance that you would obtain from a regression analysis and do it much faster. Most important, an analysis of variance is essential when y is affected by more than one source of random variation. Then the experimental design permits us to separate these sources and to estimate the variances of their respective random components. A discussion of these models is not included in this text.

Perhaps the most important point for you to note in this chapter is the following: *If your data can be modeled using a linear model that contains a single random error component (which is the model used throughout this text), then a regression analysis can do everything that an analysis of variance can do and it can do more!* But you will probably need a computer to do it. Regression analysis is simple, uses the same formulas for all analyses, is programmed for most electronic computers, and can be used to analyze data obtained from both designed and undesigned experiments. The repeated fitting of regression models for designed experiments is very inexpensive on large electronic computers. Consequently, a beginner may be advised to stick to regression analyses for analyzing the relationship between a set of independent variables and a response y if an electronic computer is available to perform the computations.

Another important application of an analysis of variance is in analyzing data for which some of the sources of variation can be attributed to random error. We will demonstrate this application in Chapter 14.

| | | | | | | | | | | | | | |

SUPPLEMENTARY EXERCISES 13.47–13.64

13.47 In the United States, two basic types of management attitudes prevail: Theory X bosses believe that workers are basically lazy and untrustworthy, and Theory Y managers hold that employees are hard-working, dependable individuals. Japanese firms take a third approach: Theory Z companies emphasize long-range planning, consensus decision-making, and strong, mutual worker–employer loyalty. Suppose we want to compare the hourly wage rates of workers at Theory X-, Y-, and Z-style corporations. Independent random samples of six engineering firms of each managerial philosophy were selected, and the starting hourly wage rates for laborers at each were recorded, as shown in the accompanying table.

MANAGERIAL ATTITUDE		
Theory X	Theory Y	Theory Z
5.20	6.25	5.50
5.20	6.80	5.75
6.10	6.87	4.60
6.00	7.10	5.36
5.75	6.30	5.85
5.60	6.35	5.90

a. Is there evidence of a difference among the mean starting hourly wages of engineers at Theory X-, Y-, and Z-style firms? Test at $\alpha = .05$.

b. Use Tukey's multiple comparisons procedure to compare the mean hourly starting wages for the three theories. Identify the means that appear to differ. Use $\alpha = .05$.

13.48 Due to increased energy shortages and costs, utility companies are stressing ways in which home and apartment utility bills can be cut. One utility company reached an agreement with the owner of a new apartment complex to conduct a test of energy saving plans for apartments. The tests were to be conducted before the apartments were rented. Four apartments were chosen that were identical in size, amount of shade, and direction faced. Four plans were to be tested, one on each apartment. The thermostat was set at 75°F in each apartment and the monthly utility bill was recorded for each of the 3 summer months. The results are listed in the table on page 716.

		TREATMENT			
		1	2	3	4
MONTH	June	$74.44	$68.75	$71.34	$65.47
	July	86.96	73.47	83.62	72.33
	August	82.00	71.23	79.98	70.87

Treatment 1: No insulation; no awnings

Treatment 2: Insulation in walls and ceilings; no awnings

Treatment 3: No insulation; awnings for windows

Treatment 4: Insulation in walls and ceilings; awnings for windows

a. Is there evidence of a difference among the mean monthly utility bills for the four treatments? Use $\alpha = .01$.

b. Is there evidence that blocking is important, i.e., that the mean bills differ for at least two of the months? Use $\alpha = .05$.

c. To determine whether awnings on the windows help reduce costs, place a 95% confidence interval on the difference between the means of treatments 2 and 4.

13.49 To compare the preferences of technicians for three brands of calculators, each technician was required to perform an identical series of calculations on each of the three calculators, A, B, and C. To avoid the possibility of fatigue, a suitable time period separated each set of calculations and the calculators were used in random order by each technician. A preference rating, based on a 0–100 scale, was recorded for each machine–technician combination. These data are shown in the accompanying table.

		CALCULATOR BRAND		
		A	B	C
TECHNICIAN	1	85	90	95
	2	70	70	75
	3	65	60	80

a. Do the data provide sufficient evidence to indicate a difference in technician preference among the three brands? Use $\alpha = .05$.

b. Why did the experimenter have each technician test all three calculators? Why not randomly assign three different technicians to each calculator?

13.50 The steam explosion of peat renders fermentable carbohydrates that have a number of potentially important industrial uses. A study of the steam explosion process was initiated to determine the optimum conditions for the release of fermentable carbohydrate (*Biotechnology and Bioengineering*, Feb. 1986). Triplicate samples of peat were treated for .5, 1.0, 2.0, 3.0, and 5.0 minutes at 170°, 200°, and 215°C, in the steam explosion process. Thus, the experiment consists of two factors—temperature at three levels and treatment time at five levels. The accompanying table gives the percentage of carbohydrate solubilized for each of the $3 \times 5 = 15$ peat samples.

TEMPERATURE (°C)	TIME (minutes)	CARBOHYDRATE SOLUBILIZED (%)
170	.5	1.3
170	1.0	1.8
170	2.0	3.2
170	3.0	4.9
170	5.0	11.7
200	.5	9.2
200	1.0	17.3
200	2.0	18.1
200	3.0	18.1
200	5.0	18.8
215	.5	12.4
215	1.0	20.4
215	2.0	17.3
215	3.0	16.0
215	5.0	15.3

Source: Forsberg, C. W. et al. "The Release of Fermentable Carbohydrate from Peat by Steam Explosion and Its Use in the Microbial Production of Solvents." *Biotechnology and Bioengineering*, Vol. 28, No. 2, Feb. 1986, p. 179 (Table I). Copyright 1986.

a. What type of experimental design was employed?

b. Explain why the traditional analysis of variance formulas are inappropriate for the analysis of these data.

c. Write a second-order model relating mean amount of carbohydrate solubilized, $E(y)$, to temperature (x_1) and time (x_2).

d. Explain how you could test the hypothesis that the two factors, temperature (x_1) and time (x_2), interact.

e. If you have access to a statistical software computer package, fit the model and perform the test for interaction.

13.51 To reduce the time spent in transferring materials from one location to another, three methods have been devised. With no previous information available on the effectiveness of these three approaches, a study is performed. Each approach is tried several times, and the amount of time to completion (in hours) is recorded in the table.

a. What type of experimental design was used?

b. Is there evidence that the mean time to completion of the task differs for at least two of the three methods? Use $\alpha = .01$.

c. Form a 95% confidence interval for the mean time to completion for method B.

13.52 The concentration of a catalyst used in producing grouted sand is thought to affect its strength. An experiment designed to investigate the effects of three different concentrations of the catalyst utilized five test specimens of grout per concentration.

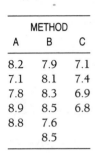

METHOD		
A	B	C
8.2	7.9	7.1
7.1	8.1	7.4
7.8	8.3	6.9
8.9	8.5	6.8
8.8	7.6	
	8.5	

The strength of a grouted sand was determined by placing the test specimen in a press and applying pressure until the specimen broke. The pressures required to break the specimens, expressed in pounds per square inch, are shown in the table.

CONCENTRATION OF CATALYST		
35%	40%	45%
5.9	6.8	9.9
8.1	7.9	9.0
5.6	8.4	8.6
6.3	9.3	7.9
7.7	8.2	8.7

a. Do the data provide sufficient evidence to indicate a difference in mean strength of the grouted sand among the three concentrations of catalyst? Test using $\alpha = .05$.

b. Find a 95% confidence interval for the difference in mean strength for specimens produced with a 35% concentration of catalyst versus those containing a 45% concentration of catalyst.

13.53 An experiment was conducted to compare the compressive strengths of two types of plastic. Eight specimens of plastic were tested for each type and the compressive strengths (in thousands of pounds per square inch) were recorded, with the results listed in the accompanying table.

PLASTIC TYPE	1	14.1	14.3	13.8	14.2	14.0	14.5	13.9	13.7
	2	14.2	13.9	13.8	14.3	14.1	13.4	13.8	14.0

a. Perform an analysis of variance for the data.

b. Do the data provide sufficient evidence to indicate a difference between the mean compressive strengths of the two plastics? Use an analysis of variance F test with $\alpha = .05$.

c. Find a 95% confidence interval for the difference between the mean compressive strengths of the two plastics.

d. How many plastic specimens of each type would have to be tested if you wanted to estimate the difference in mean strength correct to within .1 with probability approximately equal to .95? [*Hint:*   See Section 8.12.]

13.54 Refer to Exercise 13.53.

a. Construct a Student's t test statistic to test $H_0: \mu_1 = \mu_2$.

b. Show that the square of the computed value of t from part **a** is equal to the value of F computed in part **b** of Exercise 13.53.

c. Suppose that theoretical considerations make it impossible for the compressive strength of plastic type 2 to exceed the compressive strength for plastic type 1. What would you select as the alternative hypothesis for the test?

d. Suppose that you conducted the test in part **c** using $\alpha = .05$. Is this test equivalent to the F test in part **b** of Exercise 13.53?

13.55 The percentage of water removed from paper as it passes through a dryer depends on the temperature of the dryer and the speed of the paper passing through it.

A laboratory experiment was conducted to investigate the relationship between dryer temperature T at three levels and exposure time E (which is related to speed). A 3×3 factorial experiment was conducted with temperature at 100°, 120°, and 140°F and for exposure time T at 10, 20, and 30 seconds. Four paper specimens were prepared for each condition. The data (percentages of water removed) are shown in the accompanying table.

		TEMPERATURE (T)					
		100		120		140	
	10	24	26	33	33	45	49
		21	25	36	32	44	45
EXPOSURE TIME (E)	20	39	34	51	50	67	64
		37	40	47	52	68	65
	30	58	55	75	71	89	87
		56	53	70	73	86	83

a. Perform an analysis of variance for the data and construct an anlysis of variance table.

b. Do the data provide sufficient evidence to indicate that temperature and time interact? Test using $\alpha = .05$. What is the practical significance of this test?

c. Find a 90% confidence interval for the mean percentage of water removed when the temperature is 120°F and the exposure time is 20 seconds.

d. Find a 90% confidence interval for the difference in mean times for ($T = 120$, $E = 20$) and ($T = 120$, $E = 30$).

13.56 Refer to Exercise 13.55. Because both factors are quantitative, it would be useful to fit a linear model to the data and acquire a prediction equation that can be used to predict the percentage of water removed for various combinations of temperature and drying time. The SAS multiple regression printout for fitting the second-order model

$$E(y) = \beta_0 + \beta_1 x_1 + \beta_2 x_2 + \beta_3 x_1 x_2 + \beta_4 x_1^2 + \beta_5 x_2^2$$

to the data is shown here and on page 720.

SAS Computer Printout for Exercise 13.56

DEPENDENT VARIABLE: PCTWATER

SOURCE	DF	SUM OF SQUARES	MEAN SQUARE	F VALUE	PR > F	R-SQUARE	C.V.
MODEL	5	12658.11111111	2531.62222222	501.22	0.0001	0.988171	4.2967
ERROR	30	151.52777778	5.05092593			ROOT MSE	PCTWATER MEAN
CORRECTED TOTAL	35	12809.63888889				2.24742651	52.30555556

SOURCE	DF	TYPE I SS	F VALUE	PR > F	DF	TYPE III SS	F VALUE	PR > F
DRYTIME	1	8177.04166667	1618.92	0.0001	1	0.81917476	0.16	0.6900
TEMP	1	4374.00000000	865.98	0.0001	1	0.21867882	0.04	0.8366
DRYTIME*TEMP	1	81.00000000	16.04	0.0004	1	81.00000000	16.04	0.0004
DRYTIME*DRYTIME	1	23.34722222	4.62	0.0397	1	23.34722222	4.62	0.0397
TEMP*TEMP	1	2.72222222	0.54	0.4686	1	2.72222222	0.54	0.4686

PARAMETER	ESTIMATE	T FOR H0: PARAMETER=0	PR > :T:	STD ERROR OF ESTIMATE
INTERCEPT	-12.30555556	-0.42	0.6758	29.14201560
DRYTIME	-0.18750000	-0.40	0.6900	0.46558448
TEMP	0.10000000	0.21	0.8366	0.48059820
DRYTIME*TEMP	0.01125000	4.00	0.0004	0.00280928
DRYTIME*DRYTIME	0.01708333	2.15	0.0397	0.00794585
TEMP*TEMP	0.00145833	0.73	0.4686	0.00198646

(continued)

OBSERVATION	DRYTIME	TEMP	OBSERVED VALUE	PREDICTED VALUE	RESIDUAL	LOWER 95% CL FOR MEAN	UPPER 95% CL FOR MEAN
1	10	100	24.00000000	23.36111111	0.63888889	21.30136540	25.42085682
2	10	100	26.00000000	23.36111111	2.63888889	21.30136540	25.42085682
3	10	100	21.00000000	23.36111111	-2.36111111	21.30136540	25.42085682
4	10	100	25.00000000	23.36111111	1.63888889	21.30136540	25.42085682
5	10	120	33.00000000	34.02777778	-1.02777778	32.31725207	35.73830349
6	10	120	33.00000000	34.02777778	-1.02777778	32.31725207	35.73830349
7	10	120	36.00000000	34.02777778	1.97222222	32.31725207	35.73830349
8	10	120	32.00000000	34.02777778	-2.02777778	32.31725207	35.73830349
9	10	140	45.00000000	45.86111111	-0.86111111	43.80136540	47.92085682
10	10	140	49.00000000	45.86111111	3.13888889	43.80136540	47.92085682
11	10	140	44.00000000	45.86111111	-1.86111111	43.80136540	47.92085682
12	10	140	45.00000000	45.86111111	-0.86111111	43.80136540	47.92085682
13	20	100	39.00000000	37.86111111	1.13888889	36.15058540	39.57163682
14	20	100	34.00000000	37.86111111	-3.86111111	36.15058540	39.57163682
15	20	100	37.00000000	37.86111111	-0.86111111	36.15058540	39.57163682
16	20	100	40.00000000	37.86111111	2.13888889	36.15058540	39.57163682
17	20	120	51.00000000	50.77777778	0.22222222	49.06725207	52.48830349
18	20	120	50.00000000	50.77777778	-0.77777778	49.06725207	52.48830349
19	20	120	47.00000000	50.77777778	-3.77777778	49.06725207	52.48830349
20	20	120	52.00000000	50.77777778	1.22222222	49.06725207	52.48830349
21	20	140	67.00000000	64.86111111	2.13888889	63.15058540	66.57163682
22	20	140	64.00000000	64.86111111	-0.86111111	63.15058540	66.57163682
23	20	140	68.00000000	64.86111111	3.13888889	63.15058540	66.57163682
24	20	140	65.00000000	64.86111111	0.13888889	63.15058540	66.57163682
25	30	100	58.00000000	55.77777778	2.22222222	53.71803207	57.83752349
26	30	100	55.00000000	55.77777778	-0.77777778	53.71803207	57.83752349
27	30	100	56.00000000	55.77777778	0.22222222	53.71803207	57.83752349
28	30	120	53.00000000	55.77777778	-2.77777778	53.71803207	57.83752349
29	30	120	75.00000000	70.94444444	4.05555556	69.23391874	72.65497015
30	30	120	71.00000000	70.94444444	0.05555556	69.23391874	72.65497015
31	30	120	70.00000000	70.94444444	-0.94444444	69.23391874	72.65497015
32	30	120	73.00000000	70.94444444	2.05555556	69.23391874	72.65497015
33	30	140	89.00000000	87.27777778	1.72222222	85.21803207	89.33752349
34	30	140	87.00000000	87.27777778	-0.27777778	85.21803207	89.33752349
35	30	140	86.00000000	87.27777778	-1.27777778	85.21803207	89.33752349
36	30	140	83.00000000	87.27777778	-4.27777778	85.21803207	89.33752349

a. Find the prediction equation.

b. Estimate the mean percentage of water removed when $T = 120$ and $E = 20$. Why does this value differ from the sample mean of the four observations obtained for this factor–level combination?

c. Find SSE on the computer printout. Why does this differ from the value of SSE found in part **a** of Exercise 13.55?

d. Examine the computer printout and find the 95% confidence interval for the mean percentage of water removed when $T = 140$ and $E = 30$.

13.57 A chemist has run an experiment to study the effect of four treatments on the glass transition temperature (in degrees Kelvin) of a particular polymer compound. Raw material used to make this polymer is bought in small batches. The material is thought to be fairly uniform within a batch but variable between batches. Therefore, each treatment was run on samples from each batch with the results shown in the table.

		TREATMENT			
		1	2	3	4
	1	576	584	562	543
BATCH	2	515	563	522	536
	3	562	555	550	530

a. Do the data provide sufficient evidence to indicate a difference in mean temperature among the four treatments? Use $\alpha = .05$.

b. Is there sufficient evidence to indicate a difference in mean temperature among the three batches? Use $\alpha = .05$.

c. If the experiment were to be conducted again in the future, would you recommend any changes in the design of the experiment?

13.58 A company conducted an experiment to determine the effects of three types of incentive pay plans on worker productivity for both union and non-union workers. The company used plants in adjacent towns; one was unionized and the other was not. One-third of the production workers in each plant were assigned to each incentive plan. Then six workers were randomly selected from each group and their productivity (in number of items produced) was measured for a 1-week period. The six productivity measures for the 2×3 factor combinations are listed in the accompanying table.

| | | INCENTIVE PLAN | | |
		A	B	C
UNION AFFILIATION	Union	337 328 362 319 305 344	346 373 351 338 355 365	317 341 335 329 310 315
	Non-union	359 346 345 396 381 373	371 377 352 401 399 378	350 336 349 351 374 340

a. Perform an analysis of variance for the data and construct an analysis of variance table.

b. If an interaction between union affiliation and incentive plan is present, what implication would it have on the selection of an incentive plan for a particular company plant?

c. Do the data present sufficient evidence to indicate an interaction between union affiliation and incentive plan? Test using $\alpha = .05$.

d. Find a 90% confidence interval for the mean productivity for a unionized worker on incentive plan B.

e. Find a 90% confidence interval for the difference in mean productivity between union and non-union workers on incentive plan B.

13.59 *Acid rain* is considered by some environmentalists to be the nation's most serious environmental problem. It is formed by the combination of water vapor in clouds with nitrogen oxide and sulfuric dioxide emissions from the burning of coal, oil, and natural gas. The acidity of rain in central and northern Florida consistently ranges from 4.5 to 5 on the pH scale, a decidedly acid condition. To determine the effects of acid rain on the acidity of soils in a natural ecosystem, engineers at the University of Florida's Institute of Food and Agricultural Sciences irrigated experimental plots near Gainesville, Florida, with acid rain at two pH levels, 3.7 and 4.5. The acidity of the soil was then measured at three different depths, 0–15, 15–30, and 30–46 centimeters. Tests were conducted during three different time periods in 1981. The resulting soil pH values are shown in the table.

| | | APRIL 3, 1981 | | JUNE 16, 1981 | | JUNE 30, 1981 | |
| | | ACID RAIN pH | | ACID RAIN pH | | ACID RAIN pH | |
		3.7	4.5	3.7	4.5	3.7	4.5
SOIL DEPTH	0–15 cm	5.33	5.33	5.47	5.47	5.20	5.13
	15–30 cm	5.27	5.03	5.50	5.53	5.33	5.20
	30–46 cm	5.37	5.40	5.80	5.60	5.33	5.17

Source: "Acid Rain Linked to Growth of Coal-Fired Power." *Florida Agricultural Research 83*, Vol. 2, No. 1, Winter 1983.

Suppose we treat the experiment as a 2×3 factorial laid out in three blocks, where the factors are acid rain at two pH levels and soil depth at three levels, and the blocks are the three time periods. The SAS printout for the analysis of variance is provided here.

SAS Computer Printout for Exercise 13.59

```
DEPENDENT VARIABLE: SOILPH
SOURCE              DF       SUM OF SQUARES      MEAN SQUARE     F VALUE     PR > F      R-SQUARE        C.V.
MODEL                7          0.48685556        0.06955079       6.99      0.0034      0.830276      1.8616
ERROR               10          0.09952222        0.00995222                 ROOT MSE              SOILPH MEAN
CORRECTED TOTAL     17          0.58637778                                   0.09976083            5.35888889

SOURCE              DF          ANOVA SS        F VALUE       PR > F
DEPTH                2          0.06714444        3.37        0.0759
PH                   1          0.03042222        3.06        0.1110
DEPTH*PH             2          0.00781111        0.39        0.6854
DATE                 2          0.38147778       19.17        0.0004
```

a. Is there evidence of an interaction between pH level of acid rain and soil depth? Test using $\alpha = .05$.

b. Conduct a test to determine whether blocking over time was effective in removing an extraneous source of variation. Use $\alpha = .05$.

c. Construct a 95% confidence interval for the difference between the mean pH of soil irrigated by pH 3.7 acid rain and pH 4.5 acid rain, at a depth of 0–15 centimeters.

13.60 The data shown in the table are the results of an experiment conducted to investigate the effect of three factors on the percentage of ash in coal.

SAMPLE		A_1 Mojiri			A_2 Michel			A_3 Kairan			A_4 Met. Coke		
REPLICATION		X_1	X_2	X_3	X_1	X_2	X_3	X_1	X_2	X_3	X_1	X_2	X_3
B_1	C_1	7.30	7.35	7.42	10.69	10.58	10.72	12.20	12.27	12.23	9.99	10.02	9.95
	C_2	6.84	6.07	6.91	10.26	10.35	10.42	11.85	11.85	12.05	9.45	9.86	9.78
	C_3	7.05	6.49	7.24	10.61	10.08	10.31	12.34	11.74	11.44	9.76	9.79	9.77
	C_4	6.75	5.62	7.24	10.66	10.61	10.01	12.22	11.68	12.09	9.92	10.17	10.50
B_2	C_1	7.56	7.44	7.51	10.86	10.88	10.90	12.47	12.42	12.44	9.87	9.81	9.79
	C_2	7.10	7.37	7.32	10.45	10.62	10.87	12.47	12.28	12.04	9.46	9.60	9.62
	C_3	7.41	7.60	7.49	10.85	10.89	10.61	12.33	12.35	12.40	9.97	9.77	9.76
	C_4	7.29	7.62	7.43	10.68	11.58	10.60	12.04	12.21	12.51	9.76	10.10	9.61
B_3	C_1	7.51	7.64	7.58	10.30	10.68	10.73	12.42	12.41	12.39	9.97	10.02	10.01
	C_2	7.36	7.50	7.21	10.33	10.50	10.64	12.05	12.30	12.20	9.78	10.02	9.91
	C_3	7.56	7.55	7.47	10.73	10.75	10.84	12.44	12.30	12.26	9.88	9.90	10.06
	C_4	7.71	7.67	7.76	10.92	10.80	10.79	12.11	12.02	12.26	9.77	9.74	9.69
B_4	C_1	7.45	7.49	7.47	10.85	10.89	10.85	12.23	12.30	12.17	10.06	10.07	10.11
	C_2	7.15	7.68	7.18	10.37	10.79	10.71	11.52	12.17	11.82	9.71	9.86	9.78
	C_3	7.60	7.55	6.61	10.82	10.82	10.88	12.40	11.99	12.17	10.13	9.93	10.01
	C_4	8.06	7.05	7.57	11.26	10.56	10.31	11.96	11.87	12.06	10.01	9.98	9.84

Source: Fujimori, T. and Ishikawa, K. "Sampling Error on Taking Analysis-Sample of Coal After the Last Stage of a Reduction Process." *Reports of Statistical Application Research*, Union of Japanese Scientists and Engineers, Vol. 19, No. 4, 1972, pp. 22–32.

The three factors, each at four levels, were:

   Type of coal (factor A): Mojiri, Michel, Kairan, and Metallurgical Coke

   Maximum particle size (factor B): 246, 147, 74, and 48 microns

   Weight of selected coal specimen (factor C): 1 gram, 100 milligrams, 20 milligrams, and 5 milligrams

Three specimens were prepared for each of the $4 \times 4 \times 4 = 64$ factor–level combinations, yielding three replications of a complete $4 \times 4 \times 4$ factorial experiment.

a. Set up an analysis of variance table showing the sources and degrees of freedom for each.

b. Refer to the SAS computer printout for the analysis of variance shown here. Do the data provide evidence of any interactions among the factors? Test using $\alpha = .05$.

SAS Computer Printout for Exercise 13.60

DEPENDENT VARIABLE: ASH

SOURCE	DF	SUM OF SQUARES	MEAN SQUARE	F VALUE	PR > F	R-SQUARE	C.V.
MODEL	63	604.14238281	9.58956163	167.10	0.0001	0.987987	2.3946
ERROR	128	7.34586667	0.05738958		ROOT MSE		ASH MEAN
CORRECTED TOTAL	191	611.48824948			0.23956123		10.00411458

SOURCE	DF	ANOVA SS	F VALUE	PR > F
A	3	594.41387656	3452.51	0.0001
B	3	2.66601823	15.48	0.0001
A*B	9	2.89819635	5.61	0.0001
C	3	1.81288073	10.53	0.0001
A*C	9	0.25345052	0.49	0.8791
B*C	9	0.47214219	0.91	0.5159
A*B*C	27	1.62581823	1.05	0.4106

c. Does the mean level of coal ash obtained in the analysis depend on the weight of the coal specimen? Test using $\alpha = .05$.

d. Find 95% confidence intervals for the difference in the mean ash content between Mojiri and Michel coal at each of the four levels of maximum particle size.

OPTIONAL EXERCISES

13.61 In Exercise 9.22, we presented data on the compressive and tensile strength of mortar used to line steel water pipelines. In particular, we noted that mortar strength is expected to increase as the curing time of the mortar increases from 7 to 28 days. The compressive and tensile strength means and standard deviations, each based on the testing of samples of $n = 50$ specimens, are shown in the accompanying table.

	COMPRESSIVE STRENGTH		TENSILE STRENGTH	
	CURING TIME		CURING TIME	
	7 days	28 days	7 days	28 days
Sample Mean	$\bar{y}_1 = 8,477$	$\bar{y}_2 = 10,404$	$\bar{y}_1 = 621$	$\bar{y}_2 = 737$
Sample Standard Deviation	$s_1 = 820$	$s_2 = 928$	$s_1 = 48$	$s_2 = 55$
Sample Size	50	50	50	50

Source: Aroni, S. and Fletcher, G. "Observations on Mortar Lining of Steel Pipelines." *Transportation Engineering Journal,* Nov. 1979.

a. Refer to the compressive strength data and regard the two curing times as treatments. Find the total for all $n = 100$ observations. Then find CM and calculate SST.

b. Find SSE.

c. Find SS(Total).

d. Construct an analysis of variance table for the results of parts **a–c**.

13.62 Refer to Exercise 13.61. Suppose the researchers want to estimate the mean compressive strength of the mortar mix using a simple linear regression model to relate mean compressive strength $E(y)$ to curing time x over the time interval from 7 to 28 days.

a. Explain why the least squares line will pass through the points $(7, \bar{y}_1)$ and $(28, \bar{y}_2)$.

b. Find the least squares line.

c. Use the prediction equation of part **b** and the value of SSE found in Exercise 13.61 to find a 95% confidence interval for the mean compressive strength at $x = 20$ days.

d. Find r^2 and interpret its value.

13.63 Refer to Exercise 13.61.

a. Refer to the tensile strength data and regard the two curing times as treatments. Find the total for all $n = 100$ observations. Then find CM and calculate SST.

b. Find SSE.

13.64 Refer to Exercise 13.61. Suppose we use a simple linear regression model to relate mean tensile strength $E(y)$ to curing time x over the time interval from 7 to 28 days.

a. Explain why the least squares line will pass through the points $(7, \bar{y}_1)$ and $(28, \bar{y}_2)$.

b. Find the least squares line.

c. Use the prediction equation of part **b** and the value of SSE found in Exercise 13.63 to find a 95% confidence interval for the mean compressive strength at $x = 20$ days.

REFERENCES

Box, G., Hunter, W., and Hunter, S. *Statistics for Experimenters*. New York: John Wiley, 1978.

BMDP User's Digest, 2nd ed. MaryAnn Hill, ed. Los Angeles: BMDP Statistical Software, 1982.

BMDPC: User's Guide to BMDP on the IBM PC. BMDP Statistical Software, Inc., Los Angeles, Calif. 90025.

Cochran, W. G. and Cox, G. M. *Experimental Designs*, 2nd ed. New York: Wiley, 1957.

Davies, O. L. *Statistical Methods in Research and Production*, 3rd ed. London: Oliver and Boyd, 1958.

Davies, O. L. *The Design and Analysis of Industrial Experiments*, 2nd ed. New York: Hafner Publishing Co., 1956.

Devore, J. L. *Probability and Statistics for Engineering and the Sciences*. Monterey, Calif.: Brooks/Cole, 1982.

Dixon, W. J., Brown, M. B., Engelman, L., Frane, J. W., Hill, M. A., Jennrich, R. I., and Toporek, J. D. *BMDP Statistical Software*, 1985 ed. Berkeley: University of California Press.

Dunn, O. J. and Clark, V. *Applied Statistics: Analysis of Variance and Regression*. New York: John Wiley, 1974.

Johnson, N. and Leone, F. *Statistics and Experimental Design in Engineering and the Physical Sciences*, Vol. II, 2nd ed. New York: John Wiley, 1977.

Neter, J., Wasserman, W., and Kutner, M. H. *Applied Linear Statistical Models*, 2nd ed. Homewood, Ill.: Richard D. Irwin, 1985.

Norusis, M. J. *SPSS/PC+: SPSS for the IBM PC/XT/AT*, 1986 ed. SPSS, Inc., Suite 3000, 444 N. Michigan Avenue, Chicago, Ill. 60611.

Norusis, M. J. *The SPSS Guide to Data Analysis*, 1986 ed. SPSS, Inc., Suite 3000, 444 N. Michigan Avenue, Chicago, Ill. 60611.

Ott, L. *An Introduction to Statistical Methods and Data Analysis*. North Scituate: Duxbury Press, 1978.

Ryan, T. A., Joiner, B. L., and Ryan, B. F. *Minitab Reference Manual*. Minitab Project, University Park, Pa., 1985.

Ryan, T. A., Joiner, B. L., and Ryan, B. F. *Minitab Student Handbook*, 2nd ed. Boston: Duxbury, 1985.

SAS Procedures Guide for Personal Computers, Version 6 ed., 1986. SAS Institute, Inc., Box 8000, Cary, N. C. 27511.

SAS Statistics Guide for Personal Computers, Version 6 ed., 1986. SAS Institute, Inc., Box 8000, Cary, N. C. 27511.

SAS User's Guide: Basics, Version 5 ed., 1985. SAS Institute, Inc., Box 8000, Cary, N. C. 27511.

SAS User's Guide: Statistics, Version 5 ed., 1985. SAS Institute, Inc., Box 8000, Cary, N. C. 27511.

Scheffé, H. *The Analysis of Variance*. New York: Wiley, 1959.

SPSS^X User's Guide, 1983 ed. SPSS, Inc., Suite 3000, 444 N. Michigan Ave., Chicago, Ill. 60611.

Walpole, R. and Myers, R. *Probability and Statistics for Engineers and Scientists*, 3rd ed. New York: Macmillan, 1985.

OBJECTIVE

To present a useful application of an analysis of variance—partitioning the sum of squares for error into different sources of random variation; to show how nested sampling can be used to reduce the cost of sampling when we want to estimate a population mean.

CONTENTS

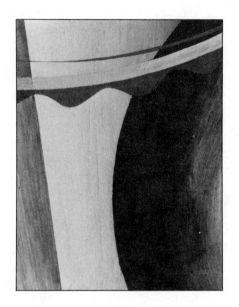

THE ANALYSIS OF VARIANCE FOR NESTED SAMPLING DESIGNS

INTRODUCTION

The random error ε in a fixed-effect model is intended to represent the contribution of many variables (most of them unknown) that affect the response variable y. We hope that the net effect of these variables on the response will assume the properties described in the assumptions listed in Section 11.2. Sometimes the random sources of variation that enter into the sum of squares for error can be partitioned into two or more sources. The following example illustrates this situation.

Suppose a pharmaceutical manufacturer wants to estimate the mean potency of a batch of an antibiotic. The potency reading produced by a piece of equipment will vary from observation to observation due to at least two sources of random error. Antibiotic that is being produced in a vat is not a homogeneous substance; the potency varies slightly from one location in the batch to another. In addition, the potency reading produced in the measurement process will vary from observation to observation due to equipment error. Thus, repeated measurements on the same specimen vary from one reading to another.

One way to separate and to estimate the magnitudes of these two sources of variation is to perform the sampling in two stages. First, we randomly select n_1 specimens from the batch. Then we measure the potency of each specimen n_2 times. Because n_2 second-stage sampling units are obtained from each first-stage or **primary unit** (see Figure 14.1), the sampling procedure is called a **nested sampling design**. It is also referred to as **subsampling**—that is, sampling within a sample.

FIGURE 14.1

Diagrammatic Representation of a Two-Stage Nested Sampling Design

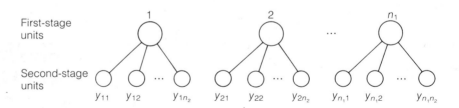

DEFINITION 14.1

A **two-stage nested sampling design** involves the random selection of n_1 first-stage (primary) units from a population. Subsamples of n_2 second-stage units are then randomly selected from within each primary unit.

Nested sampling can be expanded to any number of stages. For example, suppose that after the equipment reacts to a specimen's potency, an operator must reset a gauge before taking an individual reading. Thus, repeated readings of the equipment's reaction to a specimen will vary from one observation to another due to the operator's recalibration process. The magnitude of this third source of sampling error can be evaluated using a three-stage sampling design. In addition to the two stages previously described, for each measurement produced by the equipment's reaction to a specimen, the operator would be required to recalibrate and read the meter n_3 times. This three-stage nested sampling experiment is shown diagrammatically in Figure 14.2.

FIGURE 14.2

Diagrammatic Representation of a Three-Stage Nested Sampling Design

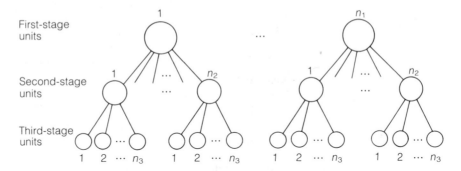

In Section 14.2, we will present a probabilistic model for a response obtained from a two-stage nested sampling design. Then we will show how to conduct an analysis of variance for the design and how to use the data to estimate and test hypotheses about the model parameters. We will present the analysis of variance and associated results for a three-stage nested sampling design in Section 14.3. In Section 14.4, we will demonstrate how to estimate the population mean μ and will show how the choice of the stage sample sizes, n_1 and n_2, affects the variance of the estimator.

| | | | | | | | | | | | | |
S E C T I O N   14.2

THE ANALYSIS OF VARIANCE FOR A TWO-STAGE NESTED SAMPLING DESIGN

Consider a two-stage nested sampling design consisting of n_2 second-stage units for each of n_1 first-stage units. Since each second-stage unit will yield one observation, the experiment will yield $n = n_1 n_2$ values of the response variable y.

We will let y_{ij} denote the observation on the jth second-stage unit ($j = 1, 2, \ldots, n_2$) within the ith first-stage unit ($i = 1, 2, \ldots, n_1$). The probabilistic model that we will use to describe this response is shown in the box. From a practical point of view, this model implies that y is equal to a constant, μ, plus

---

THE PROBABILISTIC MODEL FOR A TWO-STAGE NESTED SAMPLING DESIGN

$$y_{ij} = \mu + \alpha_i + \varepsilon_{ij} \qquad (i = 1, 2, \ldots, n_1; \quad j = 1, 2, \ldots, n_2)$$

where α_i and ε_{ij} are independent normally distributed random variables with

$$E(\alpha_i) = E(\varepsilon_{ij}) = 0$$
$$V(\alpha_i) = \sigma_\alpha^2$$
$$V(\varepsilon_{ij}) = \sigma^2$$

In addition, every pair of values, α_i and α_j ($i \neq j$), are independent. Similarly, pairs of values of ε are independent.

two random components, α_i and ε_{ij}. The response associated with every second-stage unit within the same first-stage unit i will read higher or lower than μ by the same random amount, α_i. The response y_{ij} associated with each second-stage unit will also be larger or smaller than $(\mu + \alpha_i)$ by an amount ε_{ij}. This random error will vary from one second-stage unit to another.

Because y_{ij} is equal to a constant (μ) plus the sum of two normally distributed random variables, it follows that y_{ij} is a normally distributed random variable with mean and variance

$$E(y_{ij}) = \mu + E(\alpha_i) + E(\varepsilon_{ij}) = \mu + 0 + 0 = \mu$$
$$V(y_{ij}) = V(\alpha_i) + V(\varepsilon_{ij}) + 2\text{Cov}(\varepsilon_{ij}, \alpha_i) = \sigma_\alpha^2 + \sigma^2 + 0$$
$$= \sigma_\alpha^2 + \sigma^2$$

Note that although all the random components of the model are independent of one another, the y values within the same first-stage unit will be correlated. To illustrate, the correlation between two observations from the ith first-stage unit is

$$\begin{aligned}
\text{Cov}(y_{ij}, y_{ik}) &= E\{[y_{ij} - E(y_{ij})][y_{ik} - E(y_{ik})]\} \\
&= E[(\mu + \alpha_i + \varepsilon_{ij} - \mu)(\mu + \alpha_i + \varepsilon_{ik} - \mu)] \\
&= E[(\alpha_i + \varepsilon_{ij})(\alpha_i + \varepsilon_{ik})] \\
&= E(\alpha_i^2 + \alpha_i\varepsilon_{ij} + \alpha_i\varepsilon_{ik} + \varepsilon_{ij}\varepsilon_{ik}) \\
&= E(\alpha_i^2) + E(\alpha_i\varepsilon_{ij}) + E(\alpha_i\varepsilon_{ik}) + E(\varepsilon_{ij}\varepsilon_{ik})
\end{aligned}$$

The last three expectations, which are covariances, equal 0 because the random components of the model are assumed to be independent. Then, since $E(\alpha_i) = 0$, it follows that $E(\alpha_i^2) = \sigma_\alpha^2$ and the covariance between two y values in the same first-stage unit is

$$\text{Cov}(y_{ij}, y_{ik}) = \sigma_\alpha^2$$

The analysis of variance for a nested sampling design partitions SS(Total) into two parts (see Table 14.1), one measuring the variability *between* the first-stage means and the second measuring the variability of the y values *within* the individual first-stage units.

The objectives of the analysis of variance are to obtain estimates of σ_α^2 and σ^2 and to determine whether $\sigma_\alpha^2 > 0$—that is, whether the variation among the first-stage (A) units exceeds the variation of the y values within first-stage units. The expected values of MS(A) and MS(B within A) are shown in the E(MS) column of Table 14.1. Unbiased estimates of σ_α^2 and σ^2 can be obtained from these mean squares. In addition, it can be shown (proof omitted) that when $\sigma_\alpha^2 = 0$,

$$F = \frac{\text{MS}(A)}{\text{MS}(B \text{ in } A)}$$

is an F statistic with $\nu_1 = n_1 - 1$ and $\nu_2 = n_1(n_2 - 1)$ degrees of freedom. The test of $H_0: \sigma_\alpha^2 = 0$ against $H_a: \sigma_\alpha^2 > 0$ is conducted in the same manner as the

TABLE 14.1   An Analysis of Variance Table for a Two-Stage Nested Sampling Design

SOURCE	df	SS	MS	E(MS)	F
First stage: A	$n_1 - 1$	SS(A)	MS(A)	$\sigma^2 + n_2\sigma_\alpha^2$	MS(A)/MS(B in A)
Second stage: B within A	$n_1(n_2 - 1)$	SS(B in A)	MS(B in A)	σ^2	
Total	$n_1 n_2 - 1$	SS(Total)			

F tests of Chapter 13. We will have more to say about the theory underlying this test in Section 14.5.

The notation for an analysis of variance for a two-stage nested sampling design, the formulas for computing the mean squares, and the F test are shown in the accompanying boxes. When you examine the formulas for calculating the sums of squares, note their similarity to the corresponding formulas for the analysis of variance of a replicated two-factor factorial experiment. If the first-stage units are viewed as one direction of classification, then the main effect sum of squares for this direction is SS(A). Then SS(B in A) can be calculated as SS(B in A) = SS(Total) − SS(A).

NOTATION FOR THE ANALYSIS OF VARIANCE OF A TWO-STAGE NESTED SAMPLING DESIGN

y_{ij} = Observation on the jth second-stage unit within the ith first-stage unit

n_1 = Number of first-stage units

n_2 = Number of second-stage units

$n = n_1 n_2$ = Total number of observations

A_i = Total of all observations in the ith first-stage unit

$\bar{A}_i$ = Mean of the n_2 observations in the ith first-stage unit

$\sum_{i=1}^{n_1} \sum_{j=1}^{n_2} y_{ij}$ = Total of all n observations

$\bar{y}$ = Mean of all n observations

ANALYSIS OF VARIANCE F TEST FOR A TWO-STAGE NESTED SAMPLING DESIGN

H_0: $\sigma_\alpha^2 = 0$

H_a: $\sigma_\alpha^2 > 0$

Test statistic: $F = \dfrac{\text{MS}(A)}{\text{MS}(B \text{ in } A)}$

Rejection region: $F > F_\alpha$
where F_α is the tabulated value for an F statistic with $\nu_1 = n_1 - 1$ and $\nu_2 = n_1(n_2 - 1)$ degrees of freedom.

CALCULATION FORMULAS FOR A TWO-STAGED NESTED SAMPLING DESIGN

$$CM = \text{Correction for the mean}$$

$$= \frac{(\text{Total of all observations})^2}{n} = \frac{\left(\sum_{i=1}^{n_1} \sum_{j=1}^{n_2} y_{ij}\right)^2}{n}$$

$$SS(\text{Total}) = \sum_{i=1}^{n_1} \sum_{j=1}^{n_2} (y_{ij} - \bar{y})^2$$

$$= (\text{Sum of squares of all observations}) - CM$$

$$= \sum_{i=1}^{n_1} \sum_{j=1}^{n_2} y_{ij}^2 - CM$$

$$SS(A) = n_2 \sum_{i=1}^{n_1} (\bar{A}_i - \bar{y})^2 = \sum_{i=1}^{n_1} \frac{A_i^2}{n_2} - CM$$

$$SS(B \text{ in } A) = SS(\text{Total}) - SS(A)$$

$$MS(A) = \frac{SS(A)}{n_1 - 1}$$

$$MS(B \text{ in } A) = \frac{SS(B \text{ in } A)}{n_1(n_2 - 1)}$$

EXAMPLE 14.1

The compressive strength of concrete depends on the proportion of water mixed with the cement, the mixing time, the thoroughness of the mixing process, and so on. Even though these variables are presumed fixed at values that will produce maximum compressive strength, they vary slightly from batch to batch and the compressive strength of the concrete varies accordingly. A state highway department conducted an experiment to compare the strength variation between batches to the strength variation of concrete specimens prepared within the same batch. Five concrete specimens were prepared for each of six batches. The compressive strength measurements (in thousands of pounds per square inch) are shown in Table 14.2. Perform an analysis of variance on the data and test $H_0: \sigma_\alpha^2 = 0$ against $H_a: \sigma_\alpha^2 > 0$, i.e., whether the batch-to-batch variation exceeds the within-batch variation.

TABLE 14.2

Compressive Strength Measurements for Concrete in Example 14.1

	BATCH					
	1	2	3	4	5	6
	5.01	4.74	4.99	5.64	5.07	5.90
	4.61	4.41	4.55	5.02	4.93	5.27
	5.22	4.98	4.87	4.89	4.81	5.65
	4.93	4.26	4.19	5.51	5.19	4.96
	5.37	4.80	4.77	5.17	5.48	5.39
Totals	25.14	23.19	23.37	26.23	25.48	27.17

SOLUTION

We perform the following preliminary calculations:

$$\sum_{i=1}^{n_1} \sum_{j=1}^{n_2} y_{ij} = \text{Total of all } n \text{ observations} = 150.58$$

$$\text{CM} = \frac{(\text{Total of all } n \text{ observations})^2}{n} = \frac{(150.58)^2}{30} = 755.81121333$$

$$\text{SS(Total)} = \sum_{i=1}^{n_1} \sum_{j=1}^{n_2} y_{ij}^2 - \text{CM} = 760.5572 - 755.81121333$$

$$= 4.74598667$$

$$\text{SS(A)} = \sum_{i=1}^{n_1} \frac{A_i^2}{n_2} - \text{CM}$$

$$= \frac{(25.14)^2 + (23.19)^2 + \cdots + (27.17)^2}{5} - 755.81121333$$

$$= 2.46974667$$

$$\text{SS(B in A)} = \text{SS(Total)} - \text{SS(A)}$$

$$= 4.74598667 - 2.46974667$$

$$= 2.27624000$$

$$\text{MS(A)} = \frac{\text{SS(A)}}{n_1 - 1} = \frac{2.46974667}{6 - 1} = .49394933$$

$$\text{MS(B in A)} = \frac{\text{SS(B in A)}}{n_1(n_2 - 1)} = \frac{2.27624000}{6(5 - 1)} = .09484333$$

The sums of squares, mean squares, and their respective degrees of freedom are shown in Table 14.3. The computed value of F for testing $H_0: \sigma_\alpha^2 = 0$ against $H_a: \sigma_\alpha^2 > 0$ is shown in the last column of Table 14.3. The tabulated value of F_α for $\alpha = .05$ with $\nu_1 = 5$ and $\nu_2 = 24$ degrees of freedom (given in Table 9 of Appendix II) is $F_{.05} = 2.62$. Since the computed value of F exceeds this value, there is evidence to indicate that $\sigma_\alpha^2 > 0$, i.e., that the variation between batches exceeds the variation within batches.

TABLE 14.3
Analysis of Variance Table for Example 14.1

SOURCE	df	SS	MS	F
A(Batch)	5	2.46974667	.49394933	5.21
B in A(Within batch)	24	2.27624000	.09484333	
Total	29			

The SAS computer printout for the analysis of variance of Example 14.1 is shown in Figure 14.3 (page 734). As in the case of the fixed-effect model, the breakdown of SS(Total) is presented in two tables. The first table (box 1) shows the partitioning of SS(Total) into two sources of variation, MODEL and ERROR.

FIGURE 14.3 SAS Computer Printout for the Analysis of Variance of Example 14.1

DEPENDENT VARIABLE: Y

	SOURCE	DF	SUM OF SQUARES	MEAN SQUARE	F VALUE	PR > F	R-SQUARE	C.V.
1	MODEL	5	2.46974667	0.49394933	5.21	0.0022	0.520386	6.1356
	ERROR	24	2.27624000	0.09484333		ROOT MSE		Y MEAN
	CORRECTED TOTAL	29	4.74598667			0.30796645		5.01933333

	SOURCE	DF	ANOVA SS	F VALUE	PR > F
2	A	5	2.46974667	5.21	0.0022

The portion corresponding to ERROR is always associated with the variation in the units of the last stage of a nested sampling design. Thus, for a two-stage design, ERROR corresponds to B in A. The source designated as MODEL represents the variation for any fixed-effect sources and the variation associated with all other stages. Since there is only one other stage (A) for this two-stage design, MODEL corresponds to the variation within first-stage sampling units—that is, A. The second table shown in Figure 14.3 (box 2) gives the breakdown of the MODEL source sum of squares into sums of squares corresponding to the stages not included in ERROR. Since we have only one other stage (A), it follows that the sum of squares for MODEL and the sum of squares for A are identical.

Example 14.1 illustrates the procedures for performing an analysis of variance for a two-stage nested sampling design and for testing to determine whether the batch-to-batch variance component, σ_α^2, is greater than 0. In Section 14.4 we will use the information contained in the analysis of variance to construct a $(1 - \alpha)100\%$ confidence interval for the population mean μ.

EXERCISES 14.1–14.6

14.1 A two-stage sampling design involving $n_1 = 4$ first-stage units and $n_2 = 3$ second-stage units per first-stage unit produced the data listed in the accompanying table.

FIRST-STAGE UNIT			
1	2	3	4
5	4	8	3
7	3	10	1
6	4	7	1

a. Perform an analysis of variance for the data and construct an analysis of variance table to display the results.
b. Estimate σ_α^2 and σ^2.
c. Do the data provide sufficient evidence to indicate that the variation between first-stage units exceeds the variation within second-stage units? Test using $\alpha = .05$.

14.2 The analysis of variance for a two-stage nested sampling design is summarized in the table.

SOURCE	df	SS
A	5	35.7
B in A	24	46.4

a. How many first-stage observations were included in the sample?
b. How many second-stage units were selected per first-stage unit?
c. Give the total number of observations obtained in the sample.

d. Write the probabilistic model for this sampling design.

e. Find estimates of σ_α^2 and σ^2.

f. Do the data provide sufficient evidence to indicate that $\sigma_\alpha^2 > 0$? Test using $\alpha = .05$.

14.3 Large highwall failures at a strip mine in Queensland, Australia, occur by the sliding of soft, black bands of clay, called black clay planes, near the base of the highwall. A study was conducted to determine whether the chemical and mineralogical properties of the black clay planes are similar to mudstone (*Engineering Geology*, Oct. 1985). Black clay and mudstone specimens were randomly selected at each of three randomly selected sites within the siltstone faces in the ramp area of the mine. The densities of the specimens (in kilograms per cubic meter) are recorded in the table.

SITE 1	SITE 2	SITE 3
2.06	2.09	2.07
1.84	2.03	2.04
2.47	2.01	1.90
2.12	2.04	2.00
2.00	2.41	2.64

Source: Seedsman, R. W. and Emerson, W. W. "The Formation of Planes of Weakness in the Highwall at Goonyella Mine, Queensland, Australia." *Engineering Geology*, Vol. 22, No. 2, Oct. 1985, p. 164 (Table I).

a. Construct an analysis of variance table for this two-stage nested sampling design.

b. Conduct a test to determine whether the variation in black clay and mudstone specimen densities between sites exceeds the variation within sites. Use $\alpha = .10$.

14.4 A two-stage nested sampling design was used to collect data to estimate the mean porosity of paper emerging from a paper machine. Ten patches of paper were randomly selected from the end of the paper roll and four porosity readings were made on each. The data are shown in the table.

PAPER PATCH	POROSITY READINGS			
1	974	978	976	975
2	981	985	978	986
3	1,014	1,012	1,018	1,010
4	990	996	989	988
5	1,012	1,009	1,011	1,012
6	978	980	974	982
7	988	979	986	983
8	1,004	1,001	1,008	1,008
9	989	984	982	983
10	999	1,002	998	1,003

a. Perform an analysis of variance for the data and construct an analysis of variance table to display the results.

b. Estimate σ_α^2 and σ^2.

c. Do the data provide sufficient evidence to indicate that the variation in porosity between patches exceeds the variation of porosity within patches? Test using $\alpha = .05$.

14.5 An experiment was conducted to monitor the resistivity of silicon monocrystals. The original data were collected according to a two-stage nested sampling design in which random samples of eight crystals were selected from among 30 lots. The measured resistivity of the crystals is recorded in the accompanying table for five of these lots.

LOT	MEASURED VALUES OF RESISTIVITY							
1	2.8	2.7	2.3	2.6	2.7	2.3	2.7	2.7
2	3.0	3.0	2.8	2.4	3.0	3.2	2.9	2.4
3	2.4	2.3	2.4	2.9	2.4	2.4	2.3	2.3
4	3.1	2.9	3.0	3.0	2.6	3.0	2.9	3.0
5	3.1	3.3	2.9	2.5	2.5	3.1	2.5	3.0

Source: Hoshide, M. "Optimization of Lot Size for Quality Assurance of Silicon Wafers." *Reports of Statistical Application Research*, Union of Japanese Scientists and Engineers, Vol. 19, No. 1, 1972, pp. 8–21.

a. Let σ_B^2 and σ_W^2 represent the components of between- and within-lot variances, respectively, of the resistivity readings. Estimate σ_B^2 and σ_W^2.

b. Do the data provide sufficient evidence to indicate that the variation in resistivity between lots exceeds the variation within lots? Test using $\alpha = .05$.

14.6 An automobile manufacturer conducted a study of the number of observed defects per new automobile delivered to its dealerships. Twenty-five days were randomly selected and ten automobiles were randomly selected from each day's production. After the cars arrived at the dealerships, the number of defects observed during preparation for sale and during the first 1,000 miles of use after purchase was recorded for each automobile. The data for 3 production days are shown in the table.

PRODUCTION DAYS		
1	2	3
14	13	7
17	18	11
12	10	12
8	15	10
15	19	6
12	13	9
10	16	7
13	12	11
9	10	6
7	14	13

a. Perform an analysis of variance for the data and construct an analysis of variance table to display the results.

b. Do the data provide sufficient evidence to indicate that the variation between days in the number of defects per new automobile exceeds the variation within days? Test using $\alpha = .05$.

SECTION 14.3

THE ANALYSIS OF VARIANCE FOR A THREE-STAGE NESTED SAMPLING DESIGN

We now assume that we have a three-stage sampling design containing n_1 first-stage units, n_2 second-stage units per first-stage unit, and n_3 third-stage units per second-stage unit. The total number of observations for this experiment is then $n = n_1 n_2 n_3$. The probabilistic model for a response obtained from a three-stage nested sampling design contains three random components, which represent the variation between first-, second-, and third-stage sampling units. We will let y_{ijk} denote the response on the kth third-stage unit within the jth second-stage and the ith first-stage units. The model for y_{ijk} is shown in the box.

THE PROBABILISTIC MODEL FOR A THREE-STAGE NESTED SAMPLING DESIGN

$$y_{ijk} = \mu + \alpha_i + \gamma_{ij} + \varepsilon_{ijk}$$

where α_i, γ_{ij}, and ε_{ijk} are independent, normally distributed random variables with

$$E(\alpha_i) = E(\gamma_{ij}) = E(\varepsilon_{ijk}) = 0$$
$$V(\alpha_i) = \sigma_\alpha^2$$
$$V(\gamma_{ij}) = \sigma_\gamma^2$$
$$V(\varepsilon_{ijk}) = \sigma^2$$

In addition, every pair of values, α_i and α_j $(i \neq j)$, are independent. Similarly, pairs of values of γ and ε are also independent.

The analysis of variance for a three-stage nested sampling design is an extension of the two-stage analysis. Before giving the computational formulas, we will examine the analysis of variance table shown in Table 14.4.

TABLE 14.4   Analysis of Variance Table for a Three-Stage Nested Sampling Design

SOURCE	df	SS	MS	E(MS)	F
First stage (A)	$n_1 - 1$	SS(A)	$\dfrac{SS(A)}{n_1 - 1}$	$\sigma^2 + n_3\sigma_\gamma^2 + n_2 n_3 \sigma_\alpha^2$	$\dfrac{MS(A)}{MS(B \text{ in } A)}$
Second stage $(B$ within $A)$	$n_1(n_2 - 1)$	SS$(B$ in $A)$	$\dfrac{SS(B \text{ in } A)}{n_1(n_2 - 1)}$	$\sigma^2 + n_3\sigma_\gamma^2$	$\dfrac{MS(B \text{ in } A)}{MS(C \text{ in } B)}$
Third stage $(C$ within $B)$	$n_1 n_2(n_3 - 1)$	SS$(C$ in $B)$	$\dfrac{SS(C \text{ in } B)}{n_1 n_2(n_3 - 1)}$	σ^2	
Total	$n_1 n_2 n_3 - 1$	SS(Total)			

NOTATION FOR THE ANALYSIS OF VARIANCE OF A THREE-STAGE NESTED SAMPLING DESIGN

y_{ijk} = Observation on the kth third-stage unit within the jth second-stage and the ith first-stage unit

n_1 = Number of first-stage units

n_2 = Number of second-stage units

n_3 = Number of third-stage units

$n = n_1 n_2 n_3$ = Total number of observations

A_i = Total of all observations in the ith first-stage unit

$\bar{A}_i$ = Mean of all observations in the ith first-stage unit

B_{ij} = Total of all observations in the jth second-stage unit within ith first-stage unit

$\bar{B}_{ij}$ = Mean of all observations in the jth second-stage unit within ith first-stage unit

$\sum_{i=1}^{n_1}\sum_{j=1}^{n_2}\sum_{k=1}^{n_3} y_{ijk}$ = Total of all n observations

$\bar{y}$ = Mean of all n observations

ANALYSIS OF VARIANCE F TESTS FOR A THREE-STAGE NESTED SAMPLING DESIGN

A TEST FOR FIRST-STAGE VARIATION

H_0: $\sigma_\alpha^2 = 0$

H_a: $\sigma_\alpha^2 > 0$

Test statistic: $F = \dfrac{MS(A)}{MS(B \text{ in } A)}$

Rejection region: $F > F_\alpha$
where F_α is the tabulated value for an F statistic that possesses $\nu_1 = n_1 - 1$ and $\nu_2 = n_1(n_2 - 1)$ degrees of freedom.

A TEST FOR SECOND-STAGE VARIATION

H_0: $\sigma_\gamma^2 = 0$

H_a: $\sigma_\gamma^2 > 0$

Test statistic: $F = \dfrac{MS(B \text{ in } A)}{MS(C \text{ in } B)}$

Rejection region: $F > F_\alpha$
where F_α is the tabulated value for an F statistic that possesses $\nu_1 = n_1(n_2 - 1)$ and $\nu_2 = n_1 n_2(n_3 - 1)$ degrees of freedom.

CALCULATION FORMULAS FOR A THREE-STAGE NESTED SAMPLING DESIGN

$$\text{CM} = \text{Correction for the mean}$$

$$= \frac{(\text{Total of all observations})^2}{n} = \frac{\left(\sum\limits_{i=1}^{n_1} \sum\limits_{j=1}^{n_2} \sum\limits_{k=1}^{n_3} y_{ijk}\right)^2}{n}$$

$$\text{SS(Total)} = \sum_{i=1}^{n_1} \sum_{j=1}^{n_2} \sum_{k=1}^{n_3} (y_{ijk} - \bar{y})^2$$

$$= (\text{Sum of squares of all observations}) - \text{CM}$$

$$= \sum_{i=1}^{n_1} \sum_{j=1}^{n_2} \sum_{k=1}^{n_3} y_{ijk}^2 - \text{CM}$$

$$\text{SS}(A) = n_2 n_3 \sum_{i=1}^{n_1} (\bar{A}_i - \bar{y})^2 = \sum_{i=1}^{n_1} \frac{A_i^2}{n_2 n_3} - \text{CM}$$

$$\text{SS}(B \text{ in } A) = n_3 \sum_{j=1}^{n_2} (\bar{B}_{1j} - \bar{A}_1)^2 + n_3 \sum_{j=1}^{n_2} (\bar{B}_{2j} - \bar{A}_2)^2$$

$$+ \cdots + n_3 \sum_{j=1}^{n_2} (\bar{B}_{n_1 j} - \bar{A}_{n_1})^2$$

$$= \sum_{i=1}^{n_1} \sum_{j=1}^{n_2} \frac{B_{ij}^2}{n_3} - \sum_{i=1}^{n_1} \frac{A_i^2}{n_2 n_3}$$

Note: Whenever totals are squared and summed, the divisor is equal to the number of observations in a *single* total. Thus, there are n_3 observations in a second-stage total and $n_2 n_3$ in a first-stage total.

$$\text{SS}(C \text{ in } B) = \text{SS(Total)} - \text{SS}(A) - \text{SS}(B \text{ in } A)$$

$$\text{MS}(A) = \frac{\text{SS}(A)}{n_1 - 1}$$

$$\text{MS}(B \text{ in } A) = \frac{\text{SS}(B \text{ in } A)}{n_1(n_2 - 1)}$$

$$\text{MS}(C \text{ in } B) = \frac{\text{SS}(C \text{ in } B)}{n_1 n_2(n_3 - 1)}$$

When certain assumptions are made concerning σ_α^2 and σ_γ^2, the ratios of mean squares are F statistics with degrees of freedom, ν_1 and ν_2, corresponding to the numerator and denominator mean squares, respectively. For example, if $\sigma_\gamma^2 = 0$, then $E[\text{MS}(B \text{ in } A)] = E[\text{MS}(C \text{ in } B)]$ and

$$F = \frac{\text{MS}(B \text{ in } A)}{\text{MS}(C \text{ in } B)}$$

has an F distribution with $\nu_1 = n_1(n_2 - 1)$ and $\nu_2 = n_1n_2(n_3 - 1)$ degrees of freedom. This statistic is used to test $H_0: \sigma_\gamma^2 = 0$ against $H_a: \sigma_\gamma^2 > 0$.

Similarly, if $\sigma_\alpha^2 = 0$, then $E[MS(A)] = E[MS(B \text{ in } A)]$ and

$$F = \frac{MS(A)}{MS(B \text{ in } A)}$$

has an F distribution with $\nu_1 = n_1 - 1$ and $\nu_2 = n_1(n_2 - 1)$ degrees of freedom. This statistic is used to test $H_0: \sigma_\alpha^2 = 0$ against $H_a: \sigma_\alpha^2 > 0$.

The notation used in the analysis of variance for a three-stage nested sampling design, the computational formulas, and statistical tests are shown in the accompanying boxes.

EXAMPLE 14.2

One job of computer scientists is to evaluate computer hardware and software systems. Computer performance evaluation for software involves monitoring the CPU times of processed jobs. In addition to job-to-job variability, the CPU time will vary depending on the day on which the job is submitted and the initiator (a hardware device that initiates job processing) on which the job runs. A three-stage nested sampling experiment was conducted to compare the three sources of variation. On each of five randomly selected days, two randomly selected initiators were monitored. Then four jobs of a particular type were randomly selected from each initiator. The CPU times (in seconds) are shown in Table 14.5. Perform an analysis of variance on the data and test the following hypotheses:

a. $H_0: \sigma_\alpha^2 = 0$ against $H_a: \sigma_\alpha^2 > 0$ (i.e., whether the day-to-day variation exceeds the initiators-within-days variation)

b. $H_0: \sigma_\gamma^2 = 0$ against $H_a: \sigma_\gamma^2 > 0$ (i.e., whether the initiators-within-days variation exceeds the jobs-within-initiators variation)

TABLE 14.5
CPU Times for
Example 14.2

		DAY				
		1	2	3	4	5
INITIATOR	1	5.61	1.22	.89	3.69	7.61
		3.44	1.86	1.26	10.84	6.02
		.66	.05	1.43	1.07	.52
		.29	2.11	1.90	2.46	1.98
	2	8.17	1.53	6.27	15.20	2.41
		.13	1.03	1.01	3.62	3.02
		4.22	3.67	2.55	10.22	1.77
		2.50	2.29	1.52	1.83	1.38

SOLUTION

We first make some preliminary calculations:

$$\sum_{i=1}^{5} \sum_{j=1}^{2} \sum_{k=1}^{4} y_{ijk} = \text{Total of all 40 observations} = 129.25$$

$$CM = \frac{(\text{Total of all } n \text{ observations})^2}{n} = \frac{(129.25)^2}{40} = 417.6391$$

$$SS(\text{Total}) = \sum_{i=1}^{5} \sum_{j=1}^{2} \sum_{k=1}^{4} y_{ijk}^2 - CM = 831.9165 - 417.6391 = 414.2774$$

$$SS(A) = \sum_{i=1}^{5} \frac{A_i^2}{(2)(4)} - CM$$

$$= \frac{(25.02)^2 + (13.76)^2 + (16.83)^2 + (48.93)^2 + (24.71)^2}{8}$$

$$-417.6391$$

$$= 95.2754$$

$$SS(B \text{ in } A) = \sum_{i=1}^{5} \sum_{j=1}^{2} \frac{B_{ij}^2}{4} - \sum_{i=1}^{5} \frac{A_i^2}{(2)(4)}$$

$$= \frac{(10.00)^2 + (15.02)^2 + \cdots + (8.58)^2}{4} - 512.9145$$

$$= 36.4393$$

$$SS(C \text{ in } B) = SS(\text{Total}) - SS(A) - SS(B \text{ in } A)$$

$$= 414.2774 - 95.2754 - 36.4393$$

$$= 282.5627$$

$$MS(A) = \frac{SS(A)}{5-1} = \frac{95.2754}{4} = 23.8188$$

$$MS(B \text{ in } A) = \frac{SS(B \text{ in } A)}{5(2-1)} = \frac{36.4393}{5} = 7.2879$$

$$MS(C \text{ in } B) = \frac{SS(C \text{ in } B)}{(5)(2)(4-1)} = \frac{282.5627}{30} = 9.4187$$

These quantities are entered in the ANOVA table shown in Table 14.6.

TABLE 14.6
Analysis of Variance Table
for Example 14.2

SOURCE	df	SS	MS	F
A(Days)	4	95.2754	23.8188	3.27
B in A (Initiators within days)	5	36.4393	7.2879	.77
C in B (Jobs within initiators)	30	282.5627	9.4187	
Total	39	414.2774		

a. The computed value of F for testing $H_0: \sigma_\alpha^2 = 0$ against $H_a: \sigma_\alpha^2 > 0$ is

$$F = \frac{MS(A)}{MS(B \text{ in } A)} = \frac{23.8188}{7.2879} = 3.27$$

The tabulated value of F_α for $\alpha = .05$ and $\nu_1 = n_1 - 1 = 4$ and $\nu_2 = n_1(n_2 - 1) = 5$ degrees of freedom (see Table 9 of Appendix II) is $F_{.05} = 5.19$. Since the computed value does not exceed this critical value, there is insufficient evidence to indicate that $\sigma_\alpha^2 > 0$; that is, we cannot conclude that the variation between days exceeds the variation of initiators within days.

b. The computed value of F for testing $H_0: \sigma_\gamma^2 = 0$ against $H_a: \sigma_\gamma^2 > 0$ is

$$F = \frac{MS(B \text{ in } A)}{MS(C \text{ in } B)} = \frac{7.2879}{9.4187} = .77$$

Since the tabulated value of F_α for $\alpha = .05$ with $\nu_1 = n_1(n_2 - 1) = 5$ and $\nu_2 = n_1 n_2(n_3 - 1) = 30$ degrees of freedom, $F_{.05} = 2.53$, is larger than the computed value, there is insufficient evidence to indicate that $\sigma_\gamma^2 > 0$. We cannot conclude that initiators-within-days variation exceeds the jobs-within-initiators variation. ∎

The SAS printout for the analysis of variance for the three-stage nested design of Example 14.2 is shown in Figure 14.4. The printout is similar to that for the two-stage nested design (see Figure 14.3). However, the breakdown of SS(Total) is presented in three tables. The first table (box 1) shows the partitioning of SS(Total) into sources of variation due to MODEL and ERROR. For a three-stage nested design, ERROR corresponds to C in B, the last-stage source. The two remaining sources of variation, A and B in A, are included in MODEL variation. The second table (box 2) gives the individual sums of squares and computed F values for these first- and second-stage sources of variation, A and B in A. [*Warning:* Both F values are computed using MS(ERROR)—that is, MS(C in B)—in the denominator of the test statistic. Thus, the F value for source A is incorrect if we want to test $H_0: \sigma_\alpha^2 = 0$. However, SAS provides an option whereby the programmer may specify the appropriate mean square to be used in computing an F statistic. The correct F value for source A, computed using MS(B in A) in the denominator, is shown in the third table (box 3).] See Section 14.8 for the SAS commands required for analyzing a three-stage nested design.

FIGURE 14.4 SAS Computer Printout for the CPU Times of Example 14.2

DEPENDENT VARIABLE: Y

Box 1

SOURCE	DF	SUM OF SQUARES	MEAN SQUARE	F VALUE	PR > F	R-SQUARE	C.V.
MODEL	9	131.71471250	14.63496806	1.55	0.1748	0.317938	94.9787
ERROR	30	282.56272500	9.41875750		ROOT MSE		Y MEAN
CORRECTED TOTAL	39	414.27743750			3.06899943		3.23125000

Box 2

SOURCE	DF	ANOVA SS	F VALUE	PR > F
A	4	95.27542500	2.53	0.0612
B(A)	5	36.43928750	0.77	0.5763

Box 3

TESTS OF HYPOTHESES USING THE ANOVA MS FOR B(A) AS AN ERROR TERM

SOURCE	DF	ANOVA SS	F VALUE	PR > F
A	4	95.27542500	3.27	0.1131

||||||||||||||

EXERCISES 14.7–14.11

14.7 Quality control engineers at DuPont utilize nested sampling schemes to determine the percentage of a product shipped that conforms to specifications.* First, a random sample of n_1 production lots is selected; then, a random sample of n_2 batches is selected from each production lot. Finally, n_3 shipping lots are randomly selected from each batch for inspection. Suppose $n_1 = 10$, $n_2 = 5$, and $n_3 = 20$. Give the sources and degrees of freedom for an analysis of variance for the nested sampling design.

14.8 A three-stage nested sampling design utilized $n_1 = 4$ first-stage units, $n_2 = 3$ second-stage units per first-stage unit, and $n_3 = 2$ third-stage units per second-stage unit. The data are shown in the accompanying table.

		FIRST-STAGE UNIT			
		1	2	3	4
SECOND-STAGE UNIT WITHIN FIRST-STAGE UNIT	1	7	9	5	10
		6	10	7	8
	2	8	9	6	7
		6	7	6	9
	3	7	8	7	10
		8	9	6	11

a. Perform an analysis of variance for the data and construct an analysis of variance table to display the results.

b. Do the data provide sufficient evidence to indicate that the variation between first-stage units exceeds the variation within first-stage units? Test using $\alpha = .05$.

c. Do the data provide sufficient evidence to indicate that the variation between second-stage units exceeds the variation within second-stage units? Test using $\alpha = .05$.

14.9 A three-stage nested sampling experiment was conducted using $n_1 = 10$ first-stage units. A random sample of $n_2 = 4$ second-stage units was selected from within each first-stage unit and $n_3 = 3$ third-stage units were randomly selected from within each second-stage unit.

a. Use an analysis of variance table to display the sources and degrees of freedom for an analysis of variance.

b. Suppose that MS(A) = 14.7, MS(B in A) = 3, and SS(Total) = 225.3. Find MS(C in B).

c. Refer to the model given in this section for a three-stage nested sampling design. Do the data provide sufficient evidence to indicate that σ_γ^2 is greater than 0? Test at the .05 level of significance.

d. Do the data provide sufficient evidence to indicate that σ_α^2 exceeds 0? Test at the .05 level of significance.

---

*Henderson, R. K. "On Making the Transition from Inspection to Process Control." Paper presented at Joint Statistical Meetings, American Statistical Association and Biometric Society, August 1986, Chicago, Ill.

14.10 An experiment was conducted to estimate the mean level of sulfur content in coal produced by a particular mine. Five days were randomly selected and identified as coal sampling days. On each day, five coal cars were randomly selected and portions of coal were removed from each. Two specimens were prepared from each portion and analyzed for sulfur content. The data are shown in the accompanying table.

		DAY				
		1	2	3	4	5
	1	.107	.091	.110	.088	.089
		.105	.089	.113	.092	.088
	2	.104	.093	.108	.091	.087
		.103	.090	.110	.093	.089
COAL CARS WITHIN DAYS	3	.101	.092	.111	.092	.092
		.099	.093	.108	.089	.090
	4	.106	.091	.106	.088	.091
		.105	.091	.108	.087	.090
	5	.108	.092	.106	.091	.086
		.104	.090	.109	.088	.089

a. Perform an analysis of variance for the data and construct an analysis of variance table to display the results.

b. Do the data provide sufficient evidence to indicate that the variation in sulfur content between days exceeds the variation within days? Test using $\alpha = .05$.

c. Do the data provide sufficient evidence to indicate that the variation of sulfur content between cars within a day exceeds the variation within the coal specimens? Test using $\alpha = .05$.

14.11 The strength of paper depends upon the length and other characteristics of the wood fiber stock entering the paper machine. Consequently, as the source of the fiber stock varies over time, we expect the strength of the produced paper to vary also. To test this theory, 6 days were randomly selected from within a 4-month period of time and an end-of-the-roll paper patch was selected from each of three randomly selected rolls. Two strength tests were conducted on each of the 18 patches of paper. The strength measurements (pounds per square inch) are shown in the table.

		DAY					
		1	2	3	4	5	6
	1	20.7	22.1	19.0	20.6	23.2	20.7
		19.3	20.4	19.9	18.9	22.5	18.5
ROLLS WITHIN DAYS	2	21.2	21.6	18.8	19.8	24.2	19.6
		20.1	22.5	19.3	20.1	22.9	21.3
	3	19.9	20.9	20.2	20.7	23.4	20.0
		20.5	22.1	19.4	19.2	24.6	18.6

a. Perform an analysis of variance for the data and construct an analysis of variance table to display the results.
b. Do the data provide sufficient evidence to indicate that the variation in paper strength between days exceeds the variation within days? Test using $\alpha = .05$.
c. Do the data provide sufficient evidence to indicate that the variation in strength from roll to roll exceeds the variation between strength tests within a roll? Test using $\alpha = .05$.

| | | | | | | | | | | | | |

SECTION 14.4

ESTIMATING A POPULATION MEAN BASED ON NESTED SAMPLING

Regardless of the number of stages in a nested sampling design, the overall mean $\bar{y}$ of all n measurements in the experiment is a good estimator of the population mean μ. A $(1 - \alpha)100\%$ confidence interval for μ is shown in the box.

A $(1 - \alpha)100\%$ CONFIDENCE INTERVAL FOR μ

$$\bar{y} + t_{\alpha/2} \sqrt{\frac{MS(A)}{n}}$$

where $t_{\alpha/2}$ is the tabulated value of Student's t with $\nu = n_1 - 1$ degrees of freedom.

EXAMPLE 14.3

Refer to the nested sampling of a batch of antibiotic, as described in Section 14.1. Suppose we want to estimate the mean potency of a batch based on ten specimens selected within the vat and five equipment measurements on each specimen. (The *potency* is the concentration of the drug measured in micrograms per milliliter.) The sample mean for the $n = n_1 n_2 = 10(5) = 50$ measurements is 10.1, and the mean squares are given in Table 14.7. Find a 95% confidence interval for the mean potency of antibiotic in the batch.

TABLE 14.7
Analysis of Variance Table for the Antibiotic Sampling of Example 14.3

SOURCE	df	MS
A (Specimens)	9	.1309
B in A (Measurements within specimens)	40	.0141
Total	49	

SOLUTION

The tabulated value of $t_{.025}$ for $\nu = n_1 - 1 = 9$ degrees of freedom is 2.262. Substituting $\bar{y}$, MS(A), and $t_{.025}$ into the formula for the confidence interval for μ, we obtain

$$\bar{y} \pm t_{.025} \sqrt{\frac{MS(A)}{n}}$$

$$10.10 \pm 2.262 \sqrt{\frac{.1309}{50}}$$

or $10.10 \pm .12$. Therefore, we estimate that the mean potency of the batch is enclosed by the confidence interval from 9.98 to 10.22 micrograms per milliliter.

■

The formula for the variance of $\bar{y}$ can almost be deduced intuitively by considering the estimation problem in Example 14.3. The larger the number n_1 of specimens selected from within the batch, the more you will "average out" the error from specimen to specimen. Similarly, increasing the number n_2 of measurements per specimen will average out the equipment measurement error. In fact, it can be shown that the variance of $\bar{y}$ (to be derived in Example 14.5 in Section 14.5) is

$$V(\bar{y}) = \frac{\sigma_\alpha^2}{n_1} + \frac{\sigma^2}{n_1 n_2}$$

By examining this formula, you can see that $V(\bar{y})$ can be reduced by increasing n_1, or n_2, or both.

Suppose it costs c_1 dollars to select one first-stage unit and c_2 dollars to select one second-stage unit. Then, for a fixed value of $V(\bar{y})$, it can be shown that the values of n_1 and n_2 that minimize the total cost of sampling are

$$n_2 = \sqrt{\frac{\sigma^2 c_1}{\sigma_\alpha^2 c_2}}$$

and

$$n_1 = \frac{1}{B}\left(\sigma_\alpha^2 + \frac{\sigma^2}{\sqrt{\frac{\sigma^2 c_1}{\sigma_\alpha^2 c_2}}}\right) = \frac{1}{B}\left(\sigma_\alpha^2 + \frac{\sigma^2}{n_2}\right)$$

where B is the desired bound on $V(\bar{y})$.

EXAMPLE 14.4

Refer to Example 14.3. Suppose you want to reduce the standard deviation of $\bar{y}$ to approximately .03. Also assume that it costs \$10 to select a specimen from the vat and \$3 for each specimen measurement. What are the sample sizes, n_1 and n_2, that will minimize the cost of sampling?

SOLUTION

We first need to obtain estimates of σ_α^2 and σ^2. Since $E[MS(B \text{ within } A)] = \sigma^2$, we can use $MS(B \text{ within } A) = .0141$ (see Table 14.7) as an estimate of σ^2. Also, since $E[MS(A)] = n_2\sigma_\alpha^2 + \sigma^2$, it follows that

$$\hat{\sigma}_\alpha^2 = \frac{MS(A) - MS(B \text{ within } A)}{n_2}$$

is an unbiased estimator of σ_α^2. Therefore,

$$\hat{\sigma}^2 = MS(B \text{ within } A) = .0141$$

and

$$\hat{\sigma}_\alpha^2 = \frac{MS(A) - MS(B \text{ within } A)}{n_2} = \frac{.1309 - .0141}{5} = .0234$$

We want $\sigma_{\bar{y}} = .03$, or $V(\bar{y}) = \sigma_{\bar{y}}^2 = B = (.03)^2 = .0009$.

Substituting B, the estimates of σ_α^2 and σ^2, and the sampling unit costs into the formula for the optimal sample sizes, we obtain

$$n_2 = \sqrt{\frac{\sigma^2 c_1}{\sigma_\alpha^2 c_2}} = \sqrt{\frac{(.0141)(10)}{(.0234)(3)}} = \sqrt{2.008} = 1.4 \approx 2$$

and

$$n_1 = \frac{1}{B}\left(\sigma_\alpha^2 + \frac{\sigma^2}{n_2}\right) = \frac{1}{.0009}\left(.0234 + \frac{.0141}{2}\right)$$
$$= 33.83 \approx 34$$

Therefore, we would select $n_1 = 34$ specimens per batch and take $n_2 = 2$ measurements per specimen to minimize the cost of estimating μ where $\sigma_{\bar{y}} \approx .03$. ∎

EXERCISES 14.12–14.15

14.12 Estimate the mean sulfur content of coal produced by the mine in Exercise 14.10, using a 95% confidence interval.

14.13 A producer of an antibiotic uses a two-stage nested sampling process to monitor the potency of its product. The potency of the antibiotic varies from one point in the vat to another and the analytical procedure used to measure potency produces variable measurements on the same antibiotic specimen. For this reason, specimens of the antibiotic are periodically collected at ten randomly selected spots in the vat and two portions are measured by chemical analysis within each specimen. In one such monitoring process, the sum of squares between specimens was equal to 32.4 and the total sum of squares for all $n = 20$ measurements was 37.7.
  a. Construct an analysis of variance table for the experiment.
  b. Do the data provide sufficient evidence to indicate that the variation of potency between antibiotic specimens within the vat is greater than the variation of potency measurements within the same specimen? Test using $\alpha = .05$.
  c. If the mean of the 20 potency measurements is 78.3, find a 95% confidence interval for the mean potency of antibiotic within the vat.

14.14 In Exercise 9.22, we examined data on the compressive and tensile strength of mortar used to coat the inside of water pipeline. The thickness of the mortar layer for one segment of the pipeline was specified to be $\frac{7}{16}$ inch. To ensure that the contractor was conforming to specifications, the pipeline coated each day was checked by selecting a point in the mortared pipeline and drilling eight holes in the mortar at equidistant points around the circumference of the pipe. The mortar thickness was then measured at each point. On the basis of the means, ranges, and standard deviations provided by the researchers, we have constructed data for the mortar thickness measurements for 5 days, as shown in the table (page 748).

		DAY		
1	2	3	4	5
.52	.39	.72	.61	.59
.58	.43	.65	.63	.55
.55	.41	.69	.55	.64
.50	.46	.62	.52	.58
.55	.48	.71	.60	.55

Source: Based on data supplied in Aroni, S. and Fletcher, G. "Observations on Mortar Lining of Steel Pipelines." *Transportation Engineering Journal*, Nov. 1979.

a. Do the data provide sufficient evidence to indicate a variance in mean thickness from one day to another? Test using $\alpha = .05$.

b. Find a 95% confidence interval for the mean mortar thickness on day 5.

c. Find a 95% confidence interval for the mean mortar thickness for all days.

14.15 An experiment was performed to compare the mean percentages of iron in two shiploads of Chilean lumpy iron ore. The ore was sampled using a manual sampling procedure—stopping the ore conveyor belt at a random point in time and removing the ore on a 1-meter segment (increment) of belt. Twelve 1-meter increments of ore were randomly selected from each shipload. The percentage of iron in each increment of ore is shown in the table.

SHIPLOAD A	SHIPLOAD B
62.66	64.05
62.87	62.44
63.22	63.18
63.01	62.23
62.10	63.91
63.43	62.27
63.22	62.86
63.57	62.61
61.75	64.12
63.15	61.95
63.08	62.72
63.22	62.77

Source: Sato, T., Ito, K., Chujo, S., and Takahashi, U. "Examples of Experiments on Systematic Sampling of Iron Ore." *Reports of Statistical Application Research*, Union of Japanese Scientists and Engineers, Vol. 18, No. 1, 1971, pp. 33–37.

a. The shiploads A and B, consignments of 35,323 and 52,520 long tons of ore, respectively, represent randomly selected shiploads of ore from the Chilean source. Do the data present sufficient evidence to indicate that the variation in the percentage of iron in the ore between shiploads exceeds the variation in the percentage of iron in increments taken within shiploads? Test using $\alpha = .05$.

b. Find a 90% confidence interval for the mean percentage of iron in the ore at the Chilean mine.

SECTION 14.5

SOME COMMENTS ON THE THEORY UNDERLYING A NESTED SAMPLING DESIGN

The overall mean $\bar{y}$ in a nested sampling experiment possesses a normal sampling distribution with

$$E(\bar{y}) = \mu$$

and

$$V(\bar{y}) = \frac{\sigma_\alpha^2}{n_1} + \frac{\sigma^2}{n_1 n_2} \quad \text{(for a two-stage design)}$$

$$V(\bar{y}) = \frac{\sigma_\alpha^2}{n_1} + \frac{\sigma_\gamma^2}{n_1 n_2} + \frac{\sigma^2}{n_1 n_2 n_3} \quad \text{(for a three-stage design)}$$

The source mean squares have familiar sampling distributions. If MS denotes a source mean square and ν denotes its degrees of freedom, then it can be shown that

$$\chi^2 = \frac{\nu MS}{E(MS)}$$

possesses a chi-square distribution with ν degrees of freedom. Furthermore, it can be shown that $\bar{y}$ and the mean squares are independent random variables. We will prove some of these statements in the following examples; others will be left to you to prove in the Optional Exercises. The proof of the independence of $\bar{y}$ and the mean squares is beyond the scope of this text.

EXAMPLE 14.5

Find the sampling distribution of the overall sample mean, $\bar{y}$, for a two-stage nested sampling design.

SOLUTION

As we stated in Section 14.2, y_{ij} ($i = 1, 2, \ldots, n_1$; $j = 1, 2, \ldots, n_2$) is normally distributed with $E(y_{ij}) = \mu$ and $V(y_{ij}) = \sigma_\alpha^2 + \sigma^2$. It follows that $\bar{y}$, a linear function of the sample observations, is also normally distributed. To find the mean and variance of its normal sampling distribution, we first need to express $\bar{y}$ in terms of the models for the individual observations. Thus,

$$\bar{y} = \frac{1}{n_1 n_2}(y_{11} + y_{12} + \cdots + y_{1n_2} + y_{21} + \cdots + y_{2n_2} + \cdots + y_{n_1 n_2})$$

$$= \frac{1}{n_1 n_2}[(\mu + \alpha_1 + \varepsilon_{11}) + (\mu + \alpha_1 + \varepsilon_{12}) + \cdots + (\mu + \alpha_{n_1} + \varepsilon_{n_1 n_2})]$$

$$= \frac{1}{n_1 n_2}[n_1 n_2 \mu + n_2(\alpha_1 + \alpha_2 + \cdots + \alpha_{n_1}) + \varepsilon_{11} + \varepsilon_{12} + \cdots + \varepsilon_{n_1 n_2}]$$

$$= \mu + \bar{\alpha} + \bar{\varepsilon}$$

where $\bar{\alpha}$ is the mean of the n_1 first-stage random components $(\alpha_1, \alpha_2, \ldots, \alpha_{n_1})$ and $\bar{\varepsilon}$ is the mean of the $n = n_1 n_2$ second-stage random components $(\varepsilon_{11}, \varepsilon_{12}, \ldots, \varepsilon_{n_1 n_2})$.

Applying Theorem 6.7 for obtaining the mean and variance of a linear function of random variables, we have

$$E(\bar{y}) = E(\mu + \bar{\alpha} + \bar{\varepsilon}) = \mu + 0 + 0 = \mu$$
$$V(\bar{y}) = V(\mu + \bar{\alpha} + \bar{\varepsilon})$$

Since $V(\mu) = 0$ and since $\bar{\alpha}$ and $\bar{\varepsilon}$ are uncorrelated, it follows that

$$V(\bar{y}) = V(\mu) + V(\bar{\alpha}) + V(\bar{\varepsilon}) = 0 + \frac{\sigma_{\alpha}^2}{n_1} + \frac{\sigma^2}{n_1 n_2}$$

$$= \frac{\sigma_{\alpha}^2}{n_1} + \frac{\sigma^2}{n_1 n_2}$$

Therefore, for a two-stage nested sampling design, $\bar{y}$ is normally distributed with

$$E(\bar{y}) = \mu$$

and

$$V(\bar{y}) = \frac{\sigma_{\alpha}^2}{n_1} + \frac{\sigma^2}{n_1 n_2}$$ ∎

EXAMPLE 14.6

Find the sampling distribution of

$$\frac{(n_1 - 1)MS(A)}{E[MS(A)]}$$

for a two-stage nested sampling design.

SOLUTION

We first write

$$MS(A) = \frac{SS(A)}{n_1 - 1} = n_2 \frac{\sum_{i=1}^{n_1} (\bar{A}_i - \bar{y})^2}{n_1 - 1}$$

Note that $\bar{A}_i$, the mean of the n_2 observations in the ith first-stage unit, is equal to

$$\bar{A}_i = \mu + \alpha_i + \bar{\varepsilon}_i \quad (i = 1, 2, \ldots, n_1)$$

where $\bar{\varepsilon}_i$ is equal to the mean of the n_2 second-stage random components in the ith first-stage unit.

Since $\bar{A}_i$ is a linear function of normally distributed random variables, it is normally distributed with

$$E(\bar{A}_i) = \mu$$

and

$$V(\bar{A}_i) = \sigma_\alpha^2 + \frac{\sigma^2}{n_2}$$

For any i and j ($i \neq j$), $\bar{A}_i$ and $\bar{A}_j$ are independent because the random elements that appear in $\bar{A}_i$ differ from those contained in $\bar{A}_j$. Therefore, because $\bar{A}_1, \bar{A}_2, \ldots, \bar{A}_{n_1}$ are identically distributed random variables, it follows that

$$s_{\bar{A}}^2 = \frac{\sum_{i=1}^{n_1} (\bar{A}_i - \bar{y})^2}{n_1 - 1} = \frac{MS(A)}{n_2}$$

is an unbiased estimator of $V(\bar{A}_i)$ and, from Theorem 7.4,

$$\chi^2 = \frac{(n_1 - 1)s_{\bar{A}}^2}{V(\bar{A}_i)} = \frac{(n_1 - 1)MS(A)/n_2}{E[MS(A)/n_2]} = \frac{(n_1 - 1)MS(A)}{E[MS(A)]} = \frac{(n_1 - 1)MS(A)}{n_2\sigma_\alpha^2 + \sigma^2}$$

possesses a chi-square sampling distribution with $\nu = n_1 - 1$ degrees of freedom.

∎

EXAMPLE 14.7

Show that

$$t = \frac{\bar{y} - \mu}{\sqrt{\dfrac{MS(A)}{n}}}$$

has a Student's t distribution with $\nu = n_1 - 1$ degrees of freedom.

SOLUTION

Since $E(\bar{y}) = \mu$ and $V(\bar{y}) = \dfrac{\sigma_\alpha^2}{n_1} + \dfrac{\sigma^2}{n_1 n_2}$, it follows that

$$z = \frac{\bar{y} - \mu}{\sqrt{\dfrac{\sigma_\alpha^2}{n_1} + \dfrac{\sigma^2}{n_1 n_2}}}$$

is a normal random variable with mean equal to 0 and variance equal to 1. From Example 14.6, we know that

$$\chi^2 = \frac{(n_1 - 1)MS(A)}{n_2\sigma_\alpha^2 + \sigma^2}$$

is a chi-square random variable with $\nu = n_1 - 1$ degrees of freedom. Then, using the fact (proof omitted) that $\bar{y}$ and $MS(A)$ are independent, we conclude that

$$t = \frac{z}{\sqrt{\dfrac{\chi^2}{\nu}}} = \frac{(\bar{y} - \mu) \Big/ \sqrt{\dfrac{\sigma_\alpha^2}{n_1} + \dfrac{\sigma^2}{n_1 n_2}}}{\sqrt{\left[\dfrac{(n_1 - 1)\text{MS}(A)}{n_2 \sigma_\alpha^2 + \sigma^2}\right] \Big/ (n_1 - 1)}}$$

possesses a Student's t distribution with $\nu = n_1 - 1$ degrees of freedom. This quantity reduces algebraically to

$$t = \frac{\bar{y} - \mu}{\sqrt{\dfrac{\text{MS}(A)}{n}}}$$

where $n = n_1 n_2$. ∎

EXAMPLE 14.8

Show that when $\sigma_\alpha^2 = 0$,

$$F = \frac{\text{MS}(A)}{\text{MS}(B \text{ in } A)}$$

possesses an F distribution with $\nu_1 = n_1 - 1$ and $\nu_2 = n_1(n_2 - 1)$ degrees of freedom. (Recall that this F statistic was used as a test statistic in Section 14.2 to test $H_0 : \sigma_\alpha^2 = 0$.)

SOLUTION

We have already shown that

$$\chi_1^2 = \frac{(n_1 - 1)\text{MS}(A)}{E[\text{MS}(A)]}$$

is a chi-square random variable with $\nu_1 = n_1 - 1$ degrees of freedom. In a similar manner, we can show that

$$\chi_2^2 = \frac{n_1(n_2 - 1)\text{MS}(B \text{ in } A)}{E[\text{MS}(B \text{ in } A)]}$$

is a chi-square random variable with $\nu_2 = n_1(n_2 - 1)$ degrees of freedom. Then, since we have stated that $\text{MS}(A)$ and $\text{MS}(B \text{ in } A)$ are independent (proof omitted), it follows from Definition 7.3 that

$$F = \frac{\chi_1^2}{\chi_2^2} \cdot \frac{\nu_2}{\nu_1} = \left\{ \frac{(n_1 - 1)\text{MS}(A)}{E[\text{MS}(A)]} \Big/ \frac{n_1(n_2 - 1)\text{MS}(B \text{ in } A)}{E[\text{MS}(B \text{ in } A)]} \right\} \left\{ \frac{n_1(n_2 - 1)}{n_1 - 1} \right\}$$

$$= \frac{\text{MS}(A)}{\text{MS}(B \text{ in } A)} \left\{ \frac{E[\text{MS}(B \text{ in } A)]}{E[\text{MS}(A)]} \right\} = \left[\frac{\text{MS}(A)}{\text{MS}(B \text{ in } A)} \right] \left(\frac{\sigma^2}{\sigma^2 + n_2 \sigma_\alpha^2} \right)$$

possesses an F distribution with $\nu_1 = n_1 - 1$ and $\nu_2 = n_1(n_2 - 1)$ degrees of freedom. When $\sigma_\alpha^2 = 0$, this F statistic reduces to

$$F = \frac{\text{MS}(A)}{\text{MS}(B \text{ in } A)}$$ ∎

**EXERCISES
14.16–14.20**

OPTIONAL EXERCISES

14.16 We have shown in Example 14.6 that

$$\chi^2 = \frac{(n_1 - 1)MS(A)/n_2}{V(\bar{A}_i)} = \frac{(n_1 - 1)MS(A)/n_2}{\sigma_\alpha^2 + \sigma^2/n_2}$$

possesses a chi-square distribution with $\nu = n_1 - 1$ degrees of freedom. Use the fact that $E(\chi^2) = \nu$ to show that $E[MS(A)] = \sigma^2 + n_2\sigma_\alpha^2$.

14.17 Use Theorem 7.4 and the logic of Example 14.6 to show that

$$\chi^2 = \frac{n_1(n_2 - 1)MS(B \text{ in } A)}{\sigma^2}$$

possesses a chi-square distribution with $\nu = n_1(n_2 - 1)$ degrees of freedom.

14.18 Use the result of Optional Exercise 14.17 to show that

$$E[MS(B \text{ in } A)] = \sigma^2$$

14.19 Refer to the model for a three-stage nested sampling design given in Section 14.3. Find the mean and variance of $\bar{y}$.

14.20 Suppose that the cost of selecting a single first-stage sampling unit is c_1, the corresponding cost of a second-stage unit is c_2, and the total cost C of sampling for the experiment is $C = c_1n_1 + c_2n_2$. If we want the variance of $\bar{y}$ to equal B, i.e.,

$$V(\bar{y}) = \frac{\sigma_\alpha^2}{n_1} + \frac{\sigma^2}{n_1n_2} = B$$

show that the cost C is minimized when

$$n_2 = \sqrt{\left(\frac{\sigma^2}{\sigma_\alpha^2}\right)\left(\frac{c_1}{c_2}\right)}$$

and

$$n_1 = \frac{1}{B}\left(\sigma_\alpha^2 + \frac{\sigma^2}{n_2}\right)$$

S E C T I O N 14.6

COMPARING TWO OR MORE POPULATION MEANS USING NESTED SAMPLING

Nested sampling is often used for experiments conducted to compare population means, particularly experiments involving materials or product testing where the response may be subject to a large measurement error.

For example, suppose we want to compare the mean strength $E(y)$ for paper produced on $n_1 = 3$ different paper machines. Four ($n_2 = 4$) sheets of paper are randomly selected from rolls of paper produced by each of the three machines. Then, because the strength measurements of the paper vary substantially from one measurement to another, the strength y of the paper is measured at $n_3 = 3$ locations within each sheet of paper. This experiment involves a single factor—machines—at three levels. The four sheets of paper randomly selected from rolls produced by each of the three machines are first-stage sampling units; the three locations within each sheet are second-stage sampling units. The reason for selecting more than one second-stage unit per first-stage unit is that we are attempting to average out the large amount of variation that occurs when repeated strength measurements are made on the same sheet of paper. The sources of variation and the analysis of variance would appear as shown in Table 14.8.

TABLE 14.8

Analysis of Variance for Comparing Means for a Single Factor Using Two-Stage Nested Sampling

SOURCE	df	SS
Machines (A)	$n_1 - 1 = 2$	SS(A)
Sheets within machines (B in A)	$n_1(n_2 - 1) = 9$	SS(B in A)
Repeated measurements within sheets (C in B)	$n_1 n_2(n_3 - 1) = 24$	SS(C in B)
Total	$n_1 n_2 n_3 - 1 = 35$	SS(Total)

The sums of squares for the analysis of variance shown in Table 14.8 can be computed using the formulas for a three-stage nested sampling design. SS(A) is computed as if the $\bar{A}_i$ were the means of first-stage units rather than factor levels.

Note from Table 14.8 that n_1 represents the number of levels of the factor or, equivalently, the number of population means that we want to compare. These n_1 population means could also be the n_1 means corresponding to the factor–level combinations of a factorial experiment. For example, suppose we were to conduct a 3×4 factorial experiment by randomly selecting $n_2 = 4$ first-stage units for each of the $n_1 = (3)(4) = 12$ factor–level combinations and then randomly selecting $n_3 = 2$ second-stage units from within each first-stage sampling unit. If we denote the two factors as A and B, respectively, and the first- and second-stage sampling units as C and D, respectively, then the sources of variation and degrees of freedom for the analysis of variance would appear as shown in Table 14.9.

TABLE 14.9

Analysis of Variance for a 3×4 Factorial Experiment Using Two-Stage Nested Sampling

SOURCE	df	SS
Main effect A	$\left. \begin{array}{l} 2 \\ \end{array} \right.$	SS(A)
Main effect B	$n_1 - 1 = 12 - 1 = 11 \begin{cases} 2 \\ 3 \\ 6 \end{cases}$	SS(B)
AB interaction		SS(AB)
C in AB	$n_1(n_2 - 1) = 36$	SS(C in AB)
D in C	$n_1 n_2(n_3 - 1) = 48$	SS(D in C)
Total	95	SS(Total)

You will observe that Table 14.9 is similar to Table 14.8. The only difference is that we have partitioned the sum of squares for the $n_1 = 12$ treatments (corresponding to the 3×4 factor–level combinations) into components corresponding to main effect A, main effect B, and the AB interaction.

The formulas for the sums of squares for a factorial experiment are similar to those given in Chapter 13 for a replicated factorial experiment and for the nested sources of variation given in this chapter. However, because of the complexity of the calculations, we advise you to perform the calculations on a computer using one of the standard computer program packages. The commands for calling these packages are given in Section 14.8. The next example will familiarize you with the SAS printout for a 2 × 3 factorial experiment using nested sampling.

EXAMPLE 14.9

Refer to the strength of concrete experiment described in Example 14.1. The state highway department also conducted an experiment to compare the mean compressive strengths of two types of Portland cement. (Cement type is determined by the length of time that the concrete mix must set before hardening.) Type 1 cement, the most popular brand, requires a 28-day concrete mix setting, while type 3 cement needs only a 7-day setting. Because concrete strength depends on the proportion of water mixed with cement, three different water/cement ratios—.50 (dry cement), .65, and .80 (wet cement)—were applied to each type. (The water/cement ratio is computed by dividing the weight of water used in the mix by the weight of the cement.) Thus, a 2 × 3 factorial experiment was conducted. Three batches of concrete were prepared for each of the 2 × 3 = 6 factor–level combinations of cement type and water mix, and then two specimens were randomly selected from each batch. The compressive strength measurements (in thousands of pounds per square inch) of the specimens are shown in Table 14.10. The SAS printout for the nested 2 × 3 factorial experiment is provided in Figure 14.5. Identify and interpret the key elements of the printout.

TABLE 14.10
Compressive Strength Measurements for Concrete in Example 14.9

		TYPE 1 CEMENT			TYPE 3 CEMENT		
		WATER/CEMENT RATIO			WATER/CEMENT RATIO		
		.50	.65	.80	.50	.65	.80
BATCH	1	4.55	5.17	4.20	4.50	4.30	3.21
		4.26	5.39	4.15	4.67	4.27	3.43
	2	5.01	4.93	4.02	4.91	4.18	3.07
		4.81	5.02	4.26	5.16	4.01	3.15
	3	4.80	4.95	4.37	4.82	4.19	2.96
		4.93	4.92	4.28	4.75	3.98	2.80

FIGURE 14.5 SAS Computer Printout for the Nested 2 × 3 Factorial Experiment of Example 14.9

```
DEPENDENT VARIABLE: Y
    SOURCE          DF    SUM OF SQUARES    MEAN SQUARE    F VALUE      PR > F     R-SQUARE      C.V.
1   MODEL           17    15.73365556       0.92550915     64.42        0.0001     0.983830      2.7593
    ERROR           18     0.25860000       0.01436667                  ROOT MSE                 Y MEAN
    CORRECTED TOTAL 35    15.99225556                                   0.11986103               4.34388889

    SOURCE          DF      ANOVA SS      F VALUE    PR > F
2   A                1     3.77654444     262.87     0.0001
    B                2     8.60390556     299.44     0.0001
    A*B              2     2.41183889      83.94     0.0001
    C(A*B)          12     0.94136667       5.46     0.0007

    TESTS OF HYPOTHESES USING THE ANOVA MS FOR C(A*B) AS AN ERROR TERM
    SOURCE          DF      ANOVA SS      F VALUE    PR > F
3   A                1     3.77654444      48.14     0.0001
    B                2     8.60390556      54.84     0.0001
    A*B              2     2.41183889      15.37     0.0005
```

SOLUTION   The SAS printout presents the results of the ANOVA in three tables. The first table (box 1) partitions (SS)Total into two sources of variation, MODEL and ERROR. For all nested designs analyzed using SAS, ERROR corresponds to the last-stage source—in this case, D in C, where C represents batches and D represents specimens. All remaining sources of variation are included in MODEL. The key element from the first table is

$$MS(ERROR) = MS(D \text{ in } C) = .01436667$$

This value is used in computing the F values shown in Table 14.11.

TABLE 14.11 ANOVA Summary for Example 14.9

SOURCE	df	SS	MS	F	p-value
Type (A)	1 ⎫	3.776 ⎫	3.776 ⎫	48.14 ⎫	.0001 ⎫
Water/cement ratio (B)	2 ⎬(box 2)	8.604 ⎬(box 2)	4.302 ⎬(not shown on printout)	54.84 ⎬(box 3)	.0001 ⎬(box 3)
Type–ratio interaction (AB)	2 ⎪	2.412 ⎪	1.206 ⎪	15.37 ⎭	.0005 ⎭
Batches within treatments (C in AB)	12 ⎭	.941 ⎭	.078 ⎭	5.46 (box 2)	.0007 (box 2)
Specimens within batches (D in C)	18 (box 1)	.259 (box 1)	.014 (box 1)		
Total	35 (box 1)	15.992 (box 1)			

The second table (box 2) breaks down SS(MODEL) into four sources of variation: main effect for factor A (type), main effect for factor B (water/cement ratio), AB interaction, and C in AB. As we mentioned above, SAS computes all F values shown in this table using MS(ERROR)—that is, MS(D in C)—in the denominator. Ignore the F values for the three fixed effects, A, B, and A*B. The correct F ratios for these terms are computed using MS(C in AB) in the denominator. However, the F ratio corresponding to source C in AB, F = 5.46, is computed correctly. The observed significance level of the test, .0007 (shown under the PR > F column), indicates that batches-within-treatments (C in AB) variation exceeds specimens-within-batches (D in C) variation.

The proper F values for testing the contribution of the fixed-effect terms (main effects and interaction) are shown in the third table (box 3). The corresponding observed significance levels are all less than .01, indicating that differences exist among the mean compressive strengths of the six treatments.

An ANOVA table summarizing these results is illustrated in Table 14.11. The box from which each result was obtained from the SAS printout is indicated in parentheses. ∎

The formula for the confidence interval for the mean of a specific factor level combination is identical to the formula given in Section 14.4, except that MS(A) is replaced by the mean square associated with the first-stage sampling units. For example, if $\bar{y}_i$ is the mean for one of the $n_1 = 12$ factor–level combinations of Table 14.9, then the confidence interval for the corresponding population mean is

$$\bar{y}_i \pm t_{\alpha/2} \sqrt{\frac{MS(C \text{ in } AB)}{n_2 n_3}}$$

Note that MS(C in AB) is the mean square associated with the first-stage sampling units and $n_2 n_3$ is equal to the number of measurements used to compute $\bar{y}_i$. The degrees of freedom associated with $t_{\alpha/2}$ will equal the number for MS(C in AB)—namely, 36.

EXAMPLE 14.10

Refer to the nested 2×3 factorial experiment of Example 14.9. Most construction companies use a 28-day setting concrete mix (type 1 cement) with a water/cement ratio of .50. Use Tables 14.10 and 14.11 to construct a 95% confidence interval for the mean compressive strength of a batch of type 1 cement mixed using a water/cement ratio of .50.

SOLUTION

The tabulated value of $t_{.025}$ for $\nu = $ df(C in AB) $= 12$ is 2.179. The sample mean of the $n_2 n_3 = 6$ compressive strength measurements corresponding to type 1 cement mixed with a water/cement ratio of .50 is

$$\bar{y} = \frac{4.55 + 4.26 + \cdots + 4.93}{6} = \frac{28.36}{6} = 4.727$$

Substituting $\bar{y}$, MS(C in AB), and $t_{\alpha/2}$ into the equation, we have

$$\bar{y} \pm t_{\alpha/2} \sqrt{\frac{MS(C \text{ in } AB)}{n_2 n_3}}$$

$$4.727 \pm 2.179 \sqrt{\frac{.078}{6}}$$

$$4.727 \pm .248$$

or (4.479, 4.975). Therefore, we estimate that the mean compressive strength of the batch will fall between 4,479 and 4,975 pounds per square inch with 95% confidence. ∎

A confidence interval for the difference between a pair of treatment means is shown in the box (page 758).

A $(1 - \alpha)100\%$ CONFIDENCE INTERVAL FOR THE DIFFERENCE BETWEEN A PAIR OF TREATMENT MEANS BASED ON NESTED SAMPLING

$$(\bar{y}_i - \bar{y}_j) \pm t_{\alpha/2} \sqrt{\frac{2[\text{MS(First-stage sampling units)}]}{n}}$$

where $\bar{y}_i$ and $\bar{y}_j$ are the sample means corresponding to treatments i and j, n is the number of observations in each mean, and the degrees of freedom for $t_{\alpha/2}$ are the same as those associated with the sum of squares for the first-stage sampling units.

Assumptions: The nested sampling was conducted according to the nested sampling designs described in Sections 14.1–14.3. For these designs, each treatment will receive the same number of observations.

EXAMPLE 14.11

Refer to Examples 14.9 and 14.10. Suppose we are interested in comparing the most popular cement mix, type 1 cement with a .50 water/cement ratio, to type 3 cement mix with a .65 water/cement ratio. Find a 95% confidence interval for the difference in the mean compressive strengths of the two cement mixes.

SOLUTION

For this nested factorial experiment, the first-stage sampling units are batches within treatments—that is, C in AB. Hence, we use MS(C in AB) in the formula for the confidence interval. As in Example 14.10, the number of degrees of freedom associated with $t_{.025}$ is $\nu = \text{df}(C \text{ in } AB) = 12$ and $t_{.025} = 2.179$. The sample mean for the six measurements corresponding to type 3 cement mixed at a .65 water/cement ratio is

$$\bar{y}_2 = \frac{4.30 + 4.27 + \cdots + 3.98}{6} = \frac{24.93}{6} = 4.155$$

while the sample mean for type 1 cement mixed at a water/cement ratio of .50 is, from Example 14.10, $\bar{y}_1 = 4.727$. Substituting these values into the formula, we have

$$(\bar{y}_1 - \bar{y}_2) \pm t_{\alpha/2} \sqrt{\frac{2\text{MS}(C \text{ in } AB)}{6}}$$

$$(4.727 - 4.155) \pm 2.179 \sqrt{\frac{2(.078)}{6}}$$

$$.572 \pm .351$$

or (.221, .923). Thus, we estimate that the mean compressive strength of a batch of type 1 cement mixed using a .50 water/cement ratio will exceed the corresponding mean of a batch of type 3 cement mixed using a .65 water/cement ratio by anywhere from 221 to 923 pounds per square inch. ∎

14.21 An experiment was conducted to compare the rates of corrosion for four types of steel. Five specimens of each type of steel were subjected to a corrosive environment for a 1-month period of time and two measurements of the amount of corrosion were made on each specimen. The analysis of variance for the data is shown in the table.

SOURCE	df	SS	MS
Between steels		42.7	
Specimens within steels		58.1	
Measurements within specimens			
Total		133.4	

a. Complete the analysis of variance table.

b. Suppose that the sample mean measure of corrosion for steel type A is 7.4. Find a 95% confidence interval for the mean corrosion, μ_A.

c. Suppose that the sample means for steel types A and B were 7.4 and 6.2, respectively. Find a 95% confidence interval for the difference in mean corrosion between the two types of steel.

14.22 Carbon-fiber-reinforced thermoplastics (CFRTPs) are used by design engineers to replace materials such as wood, metals, and fiberglass-reinforced thermoplastics (FRTPs). Although the processing of CFRTPs closely parallels that of FRTPs, the properties of the two types of thermoplastic vary as a result of molding conditions. A nested factorial experiment was conducted to compare the "glassiness" of CFRTPs and FRTPs. Three factors were involved in the experiment: fiber type (glass or carbon), fiber concentration (20% or 30%), and four molding temperatures (180°, 200°, 220°, and 240°). For each of these $2 \times 2 \times 4 = 16$ factor–level combinations, two pieces of plastic were prepared and two glassmeter readings were made on each piece of thermoplastic. The two pairs of observations for each of the 16 factor–level combinations are recorded in the accompanying table. (High glassmeter readings are associated with a high degree of "glassiness.")

		FIBER TYPE							
		Glass				Carbon			
		FIBER CONCENTRATION				FIBER CONCENTRATION			
		20%		30%		20%		30%	
MOLDING TEMPERATURE	180°	12	10	15	16	10	9	7	9
		9	10	14	13	11	12	11	10
	200°	14	15	16	18	9	10	10	9
		15	14	17	18	8	8	7	8
	220°	20	21	20	19	11	13	12	12
		23	21	19	21	10	11	9	10
	240°	26	24	21	20	12	14	16	17
		26	26	22	21	10	9	20	18

a. Construct an analysis of variance table showing all sources of variation and their respective degrees of freedom.

b. Refer to the SAS computer printout shown here for this $2 \times 2 \times 4$ factorial experiment with nesting. Do the data provide sufficient evidence to indicate that the variation in glassiness measurements between pieces of thermoplastic is greater than the variation of measurements taken within pieces of thermoplastic? Test using $\alpha = .05$.

SAS Computer Printout for Exercise 14.22

DEPENDENT VARIABLE: GLASS

SOURCE	DF	SUM OF SQUARES	MEAN SQUARE	F VALUE	PR > F	R-SQUARE	C.V.
MODEL	31	1736.00000000	56.00000000	64.00	0.0001	0.984127	6.4511
ERROR	32	28.00000000	0.87500000			ROOT MSE	GLASS MEAN
CORRECTED TOTAL	63	1764.00000000				0.93541435	14.50000000

SOURCE	DF	ANOVA SS	F VALUE	PR > F
TEMP	3	594.50000000	226.48	0.0001
FIBER	1	784.00000000	896.00	0.0001
FIBER*TEMP	3	118.50000000	45.14	0.0001
CONC	1	7.56250000	8.64	0.0061
CONC*TEMP	3	15.68750000	5.98	0.0024
FIBER*CONC	1	3.06250000	3.50	0.0705
FIBER*CONC*TEMP	3	158.18750000	60.26	0.0001
PIEC(FIBE*CONC*TEMP)	16	54.50000000	3.89	0.0005

TESTS OF HYPOTHESES USING THE ANOVA MS FOR PIEC(FIBE*CONC*TEMP) AS AN ERROR TERM

SOURCE	DF	ANOVA SS	F VALUE	PR > F
TEMP	3	594.50000000	58.18	0.0001
FIBER	1	784.00000000	230.17	0.0001
FIBER*TEMP	3	118.50000000	11.60	0.0003
CONC	1	7.56250000	2.22	0.1557
CONC*TEMP	3	15.68750000	1.54	0.2439
FIBER*CONC	1	3.06250000	0.90	0.3571
FIBER*CONC*TEMP	3	158.18750000	15.48	0.0001

c. The sum of squares for subsamples is equal to the sum of the sums of squares of deviations of pairs of observations within the 32 pieces of thermoplastic about their respective means. Let y_{ij} denote the jth observation ($j = 1, 2$) within the ith piece of thermoplastic ($i = 1, 2, \ldots, 32$) and let T_i denote the sum of the two observations on thermoplastic piece i ($i = 1, 2, \ldots, 32$). Then the sum of squares for subsamples is equal to

$$SS(B \text{ within } A) = \sum_{j=1}^{2} \sum_{i=1}^{32} y_{ij}^2 - \frac{\sum_{i=1}^{32} T_i^2}{2}$$

Use this formula to calculate the sum of squares for subsamples and confirm that it is equal to the corresponding quantity shown in the SAS computer printout.

d. Do the data provide sufficient evidence to indicate any interactions among the three factors? Test using $\alpha = .05$.

e. Find a 95% confidence interval for the mean strength of glass-reinforced, 30% fiber concentration thermoplastic molded at 240°.

f. Find a 95% confidence interval for the difference between the mean strengths of glass-reinforced, 30% fiber concentration plastic molded at 200° and the same plastic molded at 240°.

OPTIONAL EXERCISES

14.23 Suppose you were to conduct a 3×4 factorial experiment using a two-stage nested sampling design, replicated with n_1 first-stage units per treatment and n_2 second-stage units per first-stage unit. Let

$$y_{ijk} = \mu_i + \alpha_j + \varepsilon_{ijk}$$

represent a response on the kth second-stage unit in the jth first-stage unit for the ith treatment ($i = 1, 2, \ldots, 12; j = 1, 2, \ldots, n_1; k = 1, 2, \ldots, n_2$). Find the variance of $\bar{y}_i$, the sample mean of all observations receiving treatment i.

14.24 Refer to Optional Exercise 14.23. If μ_i and μ_j represent the means for two different treatments, and $\bar{y}_i$ and $\bar{y}_j$ represent the corresponding sample means, find the variance of $(\bar{y}_i - \bar{y}_j)$. Use this result to justify the formula for the confidence interval for the difference between a pair of treatment means given in this section.

| | | | | | | | | | |

S E C T I O N 14.7

COMPUTER PRINTOUTS FOR A NESTED SAMPLING DESIGN

The computer printouts for the BMDP and SPSS^X statistical program packages are similar to the SAS computer printouts shown in the previous sections. With minor differences (rounding errors, etc.) all show the analysis of variance table for the experiment. Figure 14.6 shows the SAS, SPSS^X, and BMDP printouts for the two-stage nested sampling design of Example 14.1. Instructions on the use of the three packages listed above are contained in Section 14.8.

FIGURE 14.6   Computer Printouts for Two-Stage Nested Sampling Design of Example 14.1

(a) SAS

DEPENDENT VARIABLE: Y

SOURCE	DF	SUM OF SQUARES	MEAN SQUARE	F VALUE	PR > F	R-SQUARE	C.V.
MODEL	5	2.46974667	0.49394933	5.21	0.0022	0.520386	6.1356
ERROR	24	2.27624000	0.09484333		ROOT MSE		Y MEAN
CORRECTED TOTAL	29	4.74598667			0.30796645		5.01933333

SOURCE	DF	ANOVA SS	F VALUE	PR > F
A	5	2.46974667	5.21	0.0022

(b) SSPS^X

TESTS OF SIGNIFICANCE FOR Y USING SEQUENTIAL SUMS OF SQUARES

SOURCE OF VARIATION	SUM OF SQUARES	DF	MEAN SQUARE	F	SIG. OF F
WITHIN CELLS	2.27624	24	.09484		
CONSTANT	755.81106	1	755.81106	7969.04426	.000
A	2.46975	5	.49395	5.20805	.002

(c) BMDP

ANALYSIS OF VARIANCE FOR DEPENDENT VARIABLE   1

	SOURCE	ERROR TERM	SUM OF SQUARES	D.F.	MEAN SQUARE	F	PROB.	EXPECTED MEAN SQUARE
1	MEAN	A	755.8111	1	755.811059	1530.14	0.0000	30(1) + 5(2) + (3)
2	A	B(A)	2.4697	5	0.493949	5.21	0.0022	5(2) + (3)
3	B(A)		2.2762	24	0.094843			(3)

ESTIMATES OF VARIANCE COMPONENTS

(1)	25.17724
(2)	0.07982
(3)	0.09484

COMPUTER
PROGRAMS FOR A
NESTED SAMPLING
DESIGN (OPTIONAL)

The SAS, SPSS^X, and BMDP statistical program packages have nested sampling design options in their respective analysis of variance procedures. These options are fairly easy to use once you have attained a fundamental understanding of nested models and their ANOVA F tests. Minitab is currently without an ANOVA procedure for nested designs, and will be omitted from this discussion.*

The ANOVA procedures for random-effect models are accessed similarly to those for fixed-effect models (see Section 13.13). In all three computer packages, you must specify the sources of variation of the random-effect model. The nested sampling option is called whenever you include a nested term (such as B in A) in the model. However, the packages differ regarding which nested terms must necessarily be specified. For example, in a two-stage nested design, BMDP requires that you specify both the first-stage term, A, and the second-stage term, B in A. In contrast, only the first-stage term, A, must be specified in SAS. However, BMDP *automatically* conducts the appropriate F tests on the variance components of the random-effect model, while SAS and SPSS^X may not. That is, for some nested designs, you must specify the appropriate error terms for testing variance components with the SAS and SPSS^X procedures. In the following examples, we give the programming statements necessary to call the ANOVA procedures for nested sampling designs in all three packages.

EXAMPLE 14.12

Refer to the two-stage nested sampling design of Example 14.1, where we analyzed the compressive strengths of $n_2 = 5$ concrete specimens in each of $n_1 = 6$ batches. The $n = n_1 n_2 = 30$ measurements are reproduced in Table 14.12. Write the computer commands that will produce an ANOVA summary table similar to Table 14.3 for this design.

TABLE 14.12

		BATCH			
1	2	3	4	5	6
5.01	4.74	4.99	5.64	5.07	5.90
4.61	4.41	4.55	5.02	4.93	5.27
5.22	4.98	4.87	4.89	4.81	5.65
4.93	4.26	4.19	5.51	5.19	4.96
5.37	4.80	4.77	5.17	5.48	5.39

SOLUTION

The appropriate ANOVA commands for each of the three program packages, SAS, SPSS^X, and BMDP, are listed in Programs 14.1–14.3, respectively.

Consider first the SAS program in Program 14.1. The commands that call the ANOVA nested design procedure are given in lines 34–36. Note that only a single source of variation is specified in the CLASSES and MODEL statements—namely, the first-stage effect for batches (called A). The second-stage source of variation, specimens within batches (i.e., B in A), will appear on the SAS printout as ERROR

*The analysis of experimental designs more complex than a two-way factorial (such as nested designs) is accomplished in Minitab using a regression approach. This requires specifying the full design model, including dummy variables and interactions. Consult the Minitab reference manual for details on how to program these more complicated designs.

and need not be specified. The SAS default option is to conduct ANOVA F tests using MS(Error) in the denominator of the F ratio. Thus, for a two-stage nested sampling design, SAS will conduct the correct variance component F test since

$$F = \frac{MS(A)}{MS(Error)} = \frac{MS(A)}{MS(B\ in\ A)}$$

PROGRAM 14.1
SAS ANOVA Commands for a Two-Stage Nested Sampling Design

Command
line

```
 1    DATA CONCRETE;  ⎫
 2    INPUT A B Y;     ⎬  Data entry instructions
 3    CARDS;          ⎭
 4    1  1  5.01      ⎫
 5    1  2  4.61      ⎪
 6    1  3  5.22      ⎬  Input data values
 .    .  .   .        ⎪  (one observation per line)
 .    .  .   .        ⎪
 .    .  .   .        ⎪
33    6  5  5.39      ⎭
34    PROC ANOVA;     ⎫
35    CLASSES A;      ⎬  Statistical analysis instructions
36    MODEL Y = A;    ⎭
```

PROGRAM 14.2
SPSSˣ ANOVA Commands for a Two-Stage Nested Sampling Design

Command
line

```
 1    DATA LIST FREE/A,B,Y}  Data entry instructions
 2    MANOVA      Y BY A(1,6)/  ⎫  Statistical analysis instructions
 3                DESIGN=A/      ⎭
 4    BEGIN DATA
 5    1  1  5.01      ⎫
 6    1  2  4.61      ⎪
 7    1  3  5.22      ⎬  Input data values
 .    .  .   .        ⎪  (one observation per line)
 .    .  .   .        ⎪
 .    .  .   .        ⎪
34    6  5  5.39      ⎭
35    END DATA
```

PROGRAM 14.3
BMDP ANOVA Commands for a Two-Stage Nested Sampling Design

Command
line

```
 1    BMDP 8V}  Statistical analysis instruction
 2    / PROGRAM        TITLE IS 'NESTED DESIGN'.  ⎫
 3    / INPUT          VARIABLES=6.                ⎬  Data entry
 4                     FORMAT IS FREE.             ⎪  instructions
 5    / VARIABLE       NAMES=A1,A2,A3,A4,A5,A6.    ⎭
 6    / DESIGN         LEVELS=5,6.                 ⎫
 7                     NAMES=B,A.                  ⎬  Statistical analysis instructions
 8                     RANDOM=B,A.                 ⎪
 9                     MODEL='A,B(A)'.             ⎭
10    / END
11    5.01 4.74 4.99 5.64 5.07 5.90   ⎫
12    4.61 4.41 4.55 5.02 4.93 5.27   ⎪
 .      .    .    .    .    .    .    ⎬  Input data values
 .      .    .    .    .    .    .    ⎪  (six measurements per line)
 .      .    .    .    .    .    .    ⎪
15    5.37 4.80 4.77 5.17 5.48 5.39   ⎭
16    /END
```

The commands that call the SPSS^X ANOVA nested design procedure are given in lines 2–3 of Program 14.2. As with SAS, only the first-stage effect for batches (A) needs to be specified in the MANOVA (line 2) and DESIGN (line 3) statements. Note that the range of the levels for the effect A is given in parentheses in the MANOVA statement. The second-stage source of variation, specimens within batches (B in A), will appear on the SPSS^X ANOVA summary table as WITHIN CELLS (i.e., error). The SPSS^X default is to conduct ANOVA F tests using MS(Within cells) in the denominator of the F ratio. Since MS(Within cells) = MS(B in A) for this two-stage design, the appropriate F test will be conducted.

Program 14.3 lists the BMDP commands for analyzing a two-stage nested sampling design. The BMDP 8V procedure is used when analyzing mixed models with equal cell sizes, of which nested designs are a special case.* This BMDP program differs from the SAS and SPSS^X programs in two ways. First, the input data values (lines 11–15) for the BMDP program are entered in a slightly different form. SAS and SPSS^X read in one observation per line, where each observation consists of a value or level of A, a level of B, and a strength measurement y for that particular combination of A and B. In contrast, only the strength measurements (y's) are included in the BMDP data lines. The BMDP data lines take the form of a matrix, where each column of the matrix represents a different level of A, and each row represents a different level of B. For this example, we have a 5 × 6 matrix—five values (rows) of B for each of the six levels (columns) of A—where each entry in the matrix represents the strength measurement corresponding to a particular A–B combination. BMDP requires us to give a variable name to each column of the matrix. The number of columns or variables (6) is specified in the INPUT paragraph (line 3), while the variable names are specified (A1, A2, . . . , A6) in the VARIABLE paragraph (line 5).

The DESIGN paragraph (lines 6–9) includes instructions on how to read the data cards and specifies the sources of variation to be included in the ANOVA. The LEVELS statement (line 6) gives the number of levels of each of the two factors or sources of variation in the experiment, A and B. The first level value corresponds to the number of rows of the data input matrix and the second value corresponds to the number of columns. Since we sampled $n_2 = 5$ second-stage (B) units for each of $n_1 = 6$ first-stage (A) units, we write

```
LEVELS=5,6,
```

The NAMES statement (line 7) gives names to the sources of variation that appear on the ANOVA summary table. Since only the first character of each name is printed, this character must be unique. The order of the names in the NAMES statement must be consistent with the order of the sources of variation in the LEVELS statement.

---

*For users of BMDP on a mainframe computer, the appropriate program is called with the JCL statement (line 1)

```
//EXEC BIMED,PROGRAM=BMDP8V
```

The second major difference between the BMDP and the SAS and SPSS^X programs is that BMDP requires us to classify and list all sources of variation in the model, including the last-stage effect. Each source must be classified as either FIXED or RANDOM in the DESIGN paragraph. Since both A and B are random effects for this nested design, we include both names in a RANDOM statement (line 8). The different sources of variation, including any nesting or crossing (i.e., interaction) relationships, are then specified between single quotes in a MODEL statement (line 9). For this two-stage nested design, we list two sources of variation, A and B(A), where B(A) is the BMDP notation for "B in A". BMDP will automatically compute the appropriate ANOVA F ratio for the first-stage variance component:

$$F = \frac{\text{MS}(A)}{\text{MS}(B \text{ in } A)}$$ ∎

ANOVA commands for the more complex three-stage nested sampling designs and nested factorial designs follow the same general format as those given above. The source of variation for the units in the last stage of the nested design is *not* specified in the SAS and SPSS^X packages, while all sources of variation are identified in the BMDP program. For any type of nested design, including factorials, BMDP will compute the proper ANOVA F statistics for testing the variance components of the model. For designs more complex than two-stage nested designs, SAS and SPSS^X require you to specify the appropriate source of variation to be used in the denominator of the F ratio.

Sample programs for analyzing a three-stage nested sampling design are listed for each of the three packages in Programs 14.4(a)–14.4(c) on page 766. Here, we consider an experiment where we sample $n_1 = 4$ first-stage (A) units, $n_2 = 2$ second-stage (B) units, and $n_3 = 5$ third-stage (C) units. Note that in the SAS and SPSS^X programs [Programs 14.4(a)–(b), respectively], only the sources of variation corresponding to the first- and second-stage units, A and B within A, are specified. SAS uses the same notation as BMDP for nested effects [for example, B(A) represents B within A]. In SPSS^X, nested effects are written as, for example, B WITHIN A.

The proper F ratio for testing the first-stage effect A is

$$F = \frac{\text{MS}(A)}{\text{MS}(B \text{ in } A)}$$

This test is requested in the TEST statement (line 8) of the SAS program. In the SAS TEST statement, the source of variation that appears in the numerator of the F statistic follows H=, and the source of variation that appears in the denominator follows E=. Thus, for this example, the SAS TEST statement is written

```
TEST H=A E=B(A);
```

The results of this test appear below the default ANOVA F tests on the SAS printout. Ignore the default F test corresponding to source A, since it is computed

PROGRAM 14.4

ANOVA Commands for
a Three-Stage Nested
Sampling Design: First-
Stage Effect = A,
Second-Stage Effect =
B in A, Third-Stage
Effect = C in B

(a) SAS

Command
line

```
1    DATA NESTED ;          ⎫
2    INPUT A B C Y ;        ⎬  Data entry instructions
3    CARDS;                 ⎭
4    ⎡ 40 data lines      ⎤
     ⎢ (one observation   ⎥
     ⎣ per line)          ⎦
44   PROC ANOVA;            ⎫
45   CLASSES A B;           ⎬
46   MODEL Y=A B(A);        ⎭  Statistical  analysis  cards
47   TEST H=A E=B(A);
```

(b) SPSS^X

Command
line

```
1    DATA LIST    FREE/A,B,C,Y} Data entry instruction
2    MANOVA       Y BY A(1,4) B(1,2)/                  ⎫  Statistical
3                 DESIGN=A VS 1, B WITHIN A=1 VS W/   ⎬  analysis
                                                       ⎭  instructions
4    BEGIN DATA
     ⎡ 40 data lines     ⎤
     ⎢ (one observation  ⎥
     ⎣ per line)         ⎦
45   END DATA
```

(c) BMDP

Command
line

```
1    BMDP 8V} Statistical analysis instruction
2    / PROBLEM      TITLE IS 'NESTED DESIGN',  ⎫
3    / INPUT        VARIABLES=4,                ⎬  Data entry
4                   FORMAT IS FREE,             ⎭  instructions
5    / VARIABLE     NAMES=A1,A2,A3,A4,
6    / DESIGN       LEVELS=5,2,4,               ⎫
7                   NAMES=C,B,A,                ⎬  Statistical
8                   RANDOM=C,B,A,               ⎥  analysis
9                   MODEL='A,B(A),C(AB)',      ⎭  instructions
10   / END
     ⎡ 10 data lines              ⎤
     ⎢ (four measurements         ⎥
     ⎢ per line)                  ⎥
     ⎢ Row 1:   B = 1, C = 1      ⎥
     ⎢ Row 2:   B = 2, C = 1      ⎥
     ⎢ Row 3:   B = 1, C = 2      ⎥
     ⎢    .          .            ⎥
     ⎢    .          .            ⎥
     ⎢    .          .            ⎥
     ⎣ Row 10:  B = 2, C = 5      ⎦
21   /END
```

using MS(Error) in the denominator. However, the default F ratio for testing the second-stage effect

$$F = \frac{MS(B \text{ in } A)}{MS(\text{Error})}$$

is correct since $MS(C \text{ in } B) = MS(\text{Error})$ for this design.

Tests for the first-stage and second-stage effects are specified in the DESIGN statement (line 3) of the SPSS^X program. In the DESIGN statement, each source of variation and its corresponding error term for the ANOVA F test are listed. The source of variation to be tested (i.e., the source appearing in the numerator of the F ratio) appears to the left of VS, while the corresponding error term (i.e., the source appearing in the denominator of the F ratio) appears to the right. Error terms are designated by either a letter or a number. The letter W represents the within-cells variation (i.e., experimental error). Any other sources of variation can be designated error terms by setting the source equal to a number (for example, B WITHIN A=1). Thus, the statement

```
DESIGN=A VS 1,   B WITHIN A=1 VS W/
```

requests that in addition to experimental error, two sources of variation be included in the model: A and B WITHIN A. The first-stage source A will be tested using the error term designated by 1, i.e., B WITHIN A, while the second-stage source B WITHIN A will be tested using the within-cell variation denoted by W.

Program 14.4(c) lists the BMDP commands for the three-stage nested sampling design. As with the two-stage case, the data lines are in the form of a matrix; the columns of the matrix correspond to the $n_1 = 4$ first-stage A units. Each row of the matrix (a single data line) represents a combination of one of the $n_2 = 2$ second-stage B units and one of the $n_3 = 5$ third-stage C units. The LEVELS statement (line 6) identifies the number of levels (or sample size) at each stage; the stage with the slowest moving index is stated first. For example, the command

```
LEVELS=5,2,4,
```

implies that the subscripts for C change slowest, and those for A the fastest. The measurements appear on the data lines as follows, where $a_i b_j c_k$ represents the measurement for the ith level of A, the jth level of B, and the kth level of C:

Data line 1: $a_1b_1c_1$ $a_2b_1c_1$ $a_3b_1c_1$ $a_4b_1c_1$
Data line 2: $a_1b_2c_1$ $a_2b_2c_1$ $a_3b_2c_1$ $a_4b_2c_1$
Data line 3: $a_1b_1c_2$ $a_2b_1c_2$ $a_3b_1c_2$ $a_4b_1c_2$
Data line 4: $a_1b_2c_2$ $a_2b_2c_2$ $a_3b_2c_2$ $a_4b_2c_2$

        .
        .
        .

Data line 10: $a_1b_2c_5$ $a_2b_2c_5$ $a_3b_2c_5$ $a_4b_2c_5$

After identifying all effects as RANDOM (line 8), we list the three sources of variation in the MODEL statement (line 9). Note that the source C within B is

PROGRAM 14.5 ANOVA Commands for a Two-Stage Nested 2×3 Factorial Design: Fixed Effects = A, B; Random Effects = C in AB, D in C

(a) SAS

Command
line

```
 1    DATA FACTORS;        ⎫
                           ⎬  Data entry instructions
 2    INPUT A  B  C  D  Y; ⎪
 3    CARDS;               ⎭
      ⎡ 60 data lines
      ⎢ (one observation
      ⎣ per line)
64    PROC ANOVA;          ⎫
65    CLASSES A  B  C;     ⎪
66    MODEL Y=A  B  A*B  C(A  B);   ⎬  Statistical analysis instructions
67    TEST H=A  A*B  E=C(A  B);     ⎭
```

(b) SPSSˣ

Command
line

```
 1    DATA LIST  FREE/A,B,C,D,Y} Data entry instruction
 2    MANOVA     Y BY A(1,2) B(1,3) C(1,5)/      ⎫
 3               DESIGN=A VS 1, B VS 1, A BY B VS 1, ⎬  Statistical
 4                     C WITHIN A BY B=1 VS W/   ⎭  analysis
 5    BEGIN DATA                                    instructions
      ⎡ 60 data lines
      ⎢ (one observation
      ⎣ per line)
66    END DATA
```

(c) BMDP

Command
line

```
 1    BMDP 8V} Statistical analysis identification instruction
 2    / PROBLEM       TITLE IS 'NESTED FACTORIAL',⎫
 3    / INPUT         VARIABLES=6,                ⎬ Data entry
 4                    FORMAT IS FREE,             ⎭ instructions
 5    / VARIABLE      NAMES=A1B1,A1B2,A1B3,A2B1,A2B2,A2B3, ⎫
 6    / DESIGN        LEVELS=5,2,2,3,                      ⎪
 7                    NAMES=C,D,A,B,              ⎬ Statistical
 8                    RANDOM=C,D,                    analysis
 9                    FIXED=A,B                      instructions
10                    MODEL='A,B,C(AB),D(ABC)',   ⎪
11    / END                                       ⎭
      ⎡ 10 data lines
      ⎢ (six measurements
      ⎢ per line)
      ⎢ Row 1:   C = 1, D = 1
      ⎢ Row 2:   C = 1, D = 2
      ⎢ Row 3:   C = 2, D = 1
      ⎢    .          .
      ⎢    .          .
      ⎢    .          .
      ⎣ Row 10:  C = 5, D = 2
22    /END
```

denoted by C(AB), since in fact C is nested within both A and B. BMDP automatically conducts the appropriate ANOVA F tests on the variance components of the nested model.

As we have illustrated in this discussion and in Chapter 13, the ANOVA procedures of the SAS, SPSS^X, and BMDP packages are very flexible in terms of the types of sampling designs that they can accommodate. However, the more complex the design, the more complex the ANOVA commands become. Thus, we recommend that you consult a reference manual whenever you want to use a computer to analyze one of these more involved designs. In closing, we give the ANOVA commands for such a sampling design in Programs 14.5(a)–(c). The experimental design is a two-stage nested 2×3 factorial, where A and B represent the fixed effects (A at two levels, B at three levels), C represents the first-stage random effect ($n_1 = 5$ units sampled), and D represents the second-stage random effect ($n_2 = 2$ units sampled).

| | | | | | | | | | | |

SECTION 14.9

SUMMARY

Nested sampling involves the selection of a random sample of **first-stage sampling units**. **Second-stage** random samples are then selected from within each first-stage unit; **third-stage** random samples are selected from within each second-stage unit, and so on. Observations are then made on the sampling units in the final stage of the design.

Nested sampling designs provide a useful way of partitioning the sum of squares for error, SSE, into portions attributable to two or more sources of random error. This permits us to test hypotheses about these variance components to determine whether they, in fact, exist.

Recognition of the presence of two or more sources of error and the use of a nested sampling design can often lead to a reduction in the cost of estimating a population mean μ. Knowing the relative costs of obtaining sampling units from each stage, we can select optimal sample sizes required at each stage to minimize the cost of a fixed amount of information about μ.

| | | | | | | | | | | |

SUPPLEMENTARY EXERCISES
14.25–14.31

14.25 An experiment was conducted to investigate the abrasion resistance of a plastic imitation leather upholstery material. Patches of the material were randomly selected from within 15 batches of the material. Each patch was divided into four test specimens and each specimen was then subjected to a test for abrasion. The following results were obtained:

$$SS(Batch) = 78.9 \qquad SS(Total) = 121.6$$

a. Construct an analysis of variance table for this experiment.

b. Do the data provide sufficient evidence to indicate that the variation in resistance to abrasion between batches of material exceeds the variation within batches? Test using $\alpha = .05$.

c. Suppose the overall sample mean of the abrasion resistance measurements was 37.4. Find a 95% confidence interval for the mean measure of abrasion of the plastic upholstery material.

14.26 An experiment was conducted to determine whether batch-to-batch variation in the copper content of bronze castings exceeds the variation within batches. Samples of four specimens were prepared from each of five batches and the percentage of copper was measured for each specimen. The data are shown in the accompanying table.

		BATCH		
1	2	3	4	5
81	85	87	94	88
77	91	82	90	86
83	88	89	86	91
84	90	87	91	90

a. Perform an analysis of variance for the data and construct an analysis of variance table to display the results.

b. Do the data provide sufficient evidence to indicate that the variation in the percentage of copper between batches exceeds the variation within batches? Test using $\alpha = .05$.

14.27 *Sintering* is the process of producing a solid, but porous, metal by subjecting powdered metal to high temperature. At high temperatures, the particles of powder become bonded, one to another, and form a porous solid. An experiment was conducted to investigate the porosity of sintered copper. Six batches of sintered copper were produced and the porosity (the percentage of total volume occupied by voids) was estimated for three specimens from each batch. The results are shown in the table.

		BATCH			
1	2	3	4	5	6
21	24	19	27	22	27
23	25	22	24	20	26
21	23	20	25	20	23

a. Perfom an analysis of variance for the data and construct an analysis of variance table to display the results.

b. Estimate σ_α^2 and σ^2.

c. Do the data provide sufficient evidence to indicate that the porosity between batches exceeds the porosity within batches? Test using $\alpha = .05$.

14.28 Refer to Exercise 14.27. Find a 95% confidence interval for the mean porosity of the sintered copper.

14.29 A study was conducted to estimate the mean phosphorus concentration (in milligrams per liter) in a large northern lake. Five widely separated locations were randomly selected within the lake and three water specimens were collected at each location. Two analyses of phosphorus content were obtained for each specimen, with the results listed in the accompanying table.

		LOCATION				
		1	2	3	4	5
	1	.010	.013	.009	.011	.014
		.008	.017	.015	.015	.006
SPECIMEN	2	.009	.008	.010	.008	.018
WITHIN		.012	.010	.014	.013	.010
LOCATION	3	.011	.012	.017	.010	.005
		.006	.011	.011	.014	.013

a. Do the data provide sufficient evidence to indicate a variation in phosphorus content from one location to another? Test using $\alpha = .05$.
b. Do the data provide sufficient evidence to indicate variation in phosphorus content within locations? Test using $\alpha = .05$.
c. Find a 95% confidence interval for the mean phosphorus content in the lake.

14.30 In Exercise 14.13, we described the procedure used by a producer of antibiotics to measure the potency of the antibiotic at a particular point in time within the vat. Suppose the company has two different instruments, A and B, for measuring the potency of the antibiotic and wants to determine whether a difference exists between the mean readings obtained by two different analytical methods—that is, whether $\mu_A - \mu_B \neq 0$. The company conducted an unpaired experiment by collecting 20 specimens; ten specimens were randomly assigned to analytical method A, and ten to method B. Each specimen was divided into two portions and one measurement was taken on each.

a. Construct an analysis of variance table showing the sources of variation and their respective degrees of freedom.
b. Suppose that SS(Methods) = 2.5, SS(Between specimens within methods) = 66.7, and SS(Total) = 79.2. Find all sums of squares and mean squares for the analysis of variance table constructed in part **a**.
c. Do the data provide sufficient evidence to indicate a difference between the mean readings produced by the two analytical methods? Test using $\alpha = .05$.
d. Suppose that $\bar{y}_A = 78.3$ and $\bar{y}_B = 77.8$. Find a 95% confidence interval for the difference, $\mu_A - \mu_B$.

14.31 Refer to Exercises 14.13 and 14.30. Suppose that instead of taking independent random samples of ten specimens to be used for each analytical method, the company collected only ten specimens. Each specimen was divided into two portions, with one measured by analytical method A and the other by method B.

a. What type of design was used for this experiment?
b. Suppose that SS(Methods) = 2.47, SS(Specimens) = 1.58, and SS(Total) = 13.05. Construct an analysis of variance table showing sources, degrees of freedom, sums of squares, and mean squares.

c. Do the data provide sufficient evidence to indicate that there is a difference between the mean readings provided by the two analytical methods? Test using $\alpha = .05$.

d. Suppose that $\bar{y}_A = 78.3$ and $\bar{y}_B = 77.8$. Find a 95% confidence interval for the difference, $\mu_A - \mu_B$.

REFERENCES

BMDP User's Digest, 2nd ed. MaryAnn Hill, ed. Los Angeles: BMDP Statistical Software, 1982.

BMDPC: User's Guide to BMDP on the IBM PC. BMDP Statistical Software, Inc. Los Angeles, Calif. 90025.

Cochran, W. G. and Cox, G. M. *Experimental Designs*, 2nd ed. New York: John Wiley & Sons, Inc., 1957.

Cox, D. R. *Planning of Experiments.* New York: John Wiley & Sons, Inc., 1958.

Davies, O. L. *Statistical Methods in Research and Production*, 3rd ed. London: Oliver and Boyd, 1958.

Davies, O. L. *The Design and Analysis of Industrial Experiments*, 2nd ed. New York: Hafner Publishing Co., 1956.

Dixon, W. J., Brown, M. B., Engelman, L., Frane, J. W., Hill, M. A., Jennrich, R. I., and Toporek, J. D. *BMDP Statistical Software*, 1985 ed. Berkeley: University of California Press.

Graybill, F. A. *An Introduction to Linear Statistical Models.* New York: McGraw-Hill, 1961.

Mendenhall, W. *Introduction to Linear Models and the Design and Analysis of Experiments.* Belmont, Calif.: Wadsworth, 1968.

Norusis, M. J. *SPSS/PC+: SPSS for the IBM PC/XT/AT*, 1986 ed. SPSS, Inc., Suite 3000, 444 N. Michigan Avenue, Chicago, Ill. 60611.

Norusis, M. J. *The SPSS Guide to Data Analysis*, 1986 ed. SPSS, Inc., Suite 3000, 444 N. Michigan Avenue, Chicago, Ill. 60611.

Ott, L. *An Introduction to Statistical Methods and Data Analysis.* North Scituate, Mass.: Duxbury Press, 1978.

SAS Procedures Guide for Personal Computers, Version 6 ed., 1986. SAS Institute, Inc., Box 8000, Cary, N.C. 27511.

SAS Statistics Guide for Personal Computers, Version 6 ed., 1986. SAS Institute, Inc., Box 8000, Cary, N.C. 27511.

SAS User's Guide: Statistics, Version 5 ed., 1985. A. A. Ray, ed. SAS Institute, Inc., Box 8000, Cary, N.C. 27511.

SPSS^X User's Guide, 1983 ed. SPSS, Inc., Suite 3000, 444 N. Michigan Ave., Chicago, Ill. 60611.

Steel, R. G. D. and Torrie, J. H. *Principles and Procedures of Statistics.* New York: McGraw-Hill Book Company, 1960.

Walpole, R. E. and Myers, R. H. *Probability and Statistics for Engineers and Scientists*, 3rd ed. New York: Macmillan, 1985.

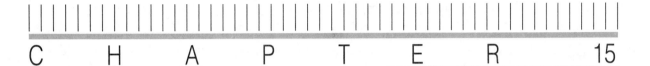

C H A P T E R 15

OBJECTIVE

To show how to analyze enumerative (count) data obtained by the classification of experimental observations.

CONTENTS

THE ANALYSIS OF ENUMERATIVE DATA

SECTION 15.1

COUNT DATA AND THE MULTINOMIAL EXPERIMENT

Many types of experiments result in observations on qualitative variables. For example, suppose that a particular personal computer (PC) is manufactured on one of five different production lines, A, B, C, D, or E. In order to compare the proportions of defective PCs that can be attributed to the five production lines, all defective computers located by an inspection program are classified each day according to the production line. Each PC is an experimental unit and the observation is a letter that identifies the production line on which it was produced. Production line is clearly a qualitative variable.

Suppose that $n = 103$ computers are found to be defective in a given week. The $n = 103$ qualitative observations, each resulting in an A, B, C, D, or E, produce counts giving the numbers of defectives emerging from the five production lines. For example, if there were $n_1 = 15$ A's, $n_2 = 27$ B's, $n_3 = 31$ C's, $n_4 = 19$ D's, and $n_5 = 11$ E's, the classified data would appear as shown in Table 15.1, which shows the counts in each category of the classification. Note that the sum of the numbers of defective PCs produced by the five lines must equal the total number of defectives:

$$n = n_1 + n_2 + n_3 + n_4 + n_5 = 15 + 27 + 31 + 19 + 11 = 103$$

TABLE 15.1

Classification of the $n = 103$ Defective Personal Computers According to Production Line

	PRODUCTION LINE			
A	B	C	D	E
15	27	31	19	11

The classification experiment that we have just described is called a **multinomial experiment** and represents an extension of the binomial experiment discussed in Section 4.5. Such an experiment consists of n identical trials—that is, observations on n experimental units. Each trial must result in one and only one of k outcomes, the k classification categories (for the binomial experiment, $k = 2$). The probability that the outcome of a single trial will fall in category i is p_i ($i = 1, 2, \ldots, k$). Finally, the trials are independent and we are interested in the numbers of observations, $n_1, n_2, \ldots, n_k$, falling in the k classification categories.

PROPERTIES OF THE MULTINOMIAL EXPERIMENT

1. The experiment consists of n identical trials.
2. There are k possible outcomes to each trial.
3. The probabilities of the k outcomes, denoted by $p_1, p_2 \ldots, p_k$, remain the same from trial to trial, where $p_1 + p_2 + \cdots + p_k = 1$.
4. The trials are independent.
5. The random variables of interest are the counts $n_1, n_2, \ldots, n_k$ in each of the k classification categories.

Multinomial experiments are numerous in engineering, business, and the social sciences. The ultimate objective of these experiments is to make inferences about the unknown category probabilities, $p_1, p_2, \ldots, p_k$. The practical question to be answered in the study of the defective personal computers is whether the proportions of defective PCs differ among the five production lines. Do the data provide evidence to contradict the null hypothesis H_0: $p_1 = p_2 = \cdots = p_5$? If the data contradict this hypothesis, the manufacturer would want to know why the rate of production of defectives is greater on some production lines than others and would take countermeasures to reduce the production of defectives.

This chapter is concerned with the analysis of count or enumerative data—specifically, data generated by a multinomial experiment. In the following sections we will learn how to make inferences about the category probabilities for data classified according to one or two qualitative variables. The statistic used for most of these inferences is one that possesses, approximately, the familiar chi-square distribution. Although the proof of the adequacy of this approximation is beyond the scope of this text, some aspects of the theory can be deduced from what we have learned in earlier chapters.

| | | | | | | | | | | | | |

SECTION 15.2

THE MULTINOMIAL PROBABILITY DISTRIBUTION

The procedure for deriving the **multinomial probability distribution** $p(n_1, n_2, \ldots, n_k)$ for the category counts, $n_1, n_2, \ldots, n_k$, is identical to the procedure employed for a binomial experiment. To simplify our notation, we will illustrate the procedure for $k = 3$ categories. The derivation of $p(n_1, n_2, \ldots, n_k)$ for k categories is similar.

Let the three outcomes corresponding to the $k = 3$ categories be denoted as A, B, and C, with respective category probabilities p_1, p_2, and p_3. Then any observation of the outcome of n trials will result in a simple event of the type shown in Table 15.2. The outcome of each trial is indicated by the letter that was observed. Thus, the simple event in Table 15.2 is one that results in C on the first trial, A on the second, A on the third, ..., and B on the last.

TABLE 15.2
A Typical Simple Event for a Multinomial Experiment ($k = 3$)

			TRIAL				
1	2	3	4	5	6	$\cdots$	n
C	A	A	B	A	C	$\cdots$	B

Now consider a simple event that will result in n_1 A outcomes, n_2 B outcomes, and n_3 C outcomes, where $n_1 + n_2 + n_3 = n$. One of these simple events is shown in Figure 15.1. The probability of the simple event shown in the figure, which results in n_1 A outcomes, n_2 B outcomes, and n_3 C outcomes, is

$$(p_1)^{n_1}(p_2)^{n_2}(p_3)^{n_3}$$

FIGURE 15.1
A Simple Event Containing n_1 A, n_2 B, and n_3 C Outcomes

$$\underbrace{AAA \cdots A}_{n_1} \underbrace{BBB \cdots B}_{n_2} \underbrace{CCC \cdots C}_{n_3}$$

How many simple events will there be in the sample space S that will imply n_1 A's, n_2 B's, and n_3 C's? This number is equal to the number of different ways that we can arrange the n_1 A's, n_2 B's, and n_3 C's in the n distinct positions of Figure 15.1. The number of ways that we could assign n_1 positions to A, n_2 to B, and n_3 to C is given by Theorem 3.3 as

$$\frac{n!}{n_1!n_2!n_3!}$$

Therefore, there are $n!/(n_1!n_2!n_3!)$ simple events resulting in n_1 A's, n_2 B's, and n_3 C's, each with probability $(p_1)^{n_1}(p_2)^{n_2}(p_3)^{n_3}$. It then follows that the probability of observing n_1 A's, n_2 B's, and n_3 C's in n trials is equal to the sum of the probabilities of these simple events:

$$p(n_1, n_2, n_3) = \frac{n!}{n_1!n_2!n_3!}(p_1)^{n_1}(p_2)^{n_2}(p_3)^{n_3}$$

The preceding derivation produced the multinomial probability distribution for $k = 3$ categories. The multinomial probability distribution for the general case of k categories is shown in the box.

THE MULTINOMIAL PROBABILITY DISTRIBUTION

$$p(n_1, n_2, \ldots, n_k) = \frac{n!}{n_1!n_2!\cdots n_k!}(p_1)^{n_1}(p_2)^{n_2}(p_3)^{n_3}\cdots(p_k)^{n_k}$$

where

$$p_1 + p_2 + \cdots + p_k = 1$$

and

$$n_1 + n_2 + \cdots + n_k = n$$

The expected value of the number of counts for a particular category, say category i, follows directly from our knowledge of the properties of a binomial random variable. If we combine all categories other than category i into a single category, then the multinomial classification becomes a binomial classification with n_i observations in category i and $(n - n_i)$ observations in the combined category. Then, from our knowledge of the expected value and variance of a binomial random variable, it follows that

$$E(n_i) = np_i$$
$$V(n_i) = np_i(1 - p_i)$$

We leave it as an optional exercise for you to show that the covariance between any pair of category counts, say n_i and n_j, is

$$\text{Cov}(n_i, n_j) = -np_ip_j$$

The multinomial probability distribution, along with the expected values, variances, and covariances associated with the category counts, provide background information that will aid you in understanding the workings of the chi-square statistic used in the following sections.

15.1 An electrical current traveling through a resistor may take one of three different paths, with probabilities $p_1 = .25$, $p_2 = .30$, and $p_3 = .45$, respectively. Suppose we monitor the path taken in $n = 10$ consecutive trials.
  a. Find the probability that the electrical current will travel the first path $n_1 = 2$ times, the second path $n_2 = 4$ times, and the third path $n_3 = 4$ times.
  b. Find $E(n_2)$ and $V(n_2)$.
  c. Find $\text{Cov}(n_1, n_3)$.

15.2 Jobs submitted at a university computer center may run under one of four different priority classes: urgent, normal priority, low priority, and stand-by. The computer center estimates that 10% of the jobs are submitted as urgent, 50% as normal priority, 20% as low priority, and 20% as stand-by. Suppose that $n = 20$ jobs are submitted simultaneously.
  a. Find the probability that $n_1 = 2$ jobs will be submitted as urgent, $n_2 = 12$ as normal priority, $n_3 = 5$ as low priority, and $n_4 = 1$ as stand-by.
  b. Find $E(n_3)$ and $V(n_3)$.
  c. Find $\text{Cov}(n_2, n_3)$.

15.3 A sample of size n is selected from a large lot of shear drill bits. Suppose that a proportion p_1 contains exactly one defect and a proportion p_2 contains more than one defect (with $p_1 + p_2 < 1$). The cost of replacing or repairing the defective drill bits is $C = 4n_1 + n_2$, where n_1 denotes the number of bits with one defect and n_2 denotes the number with two or more defects. Find the expected value and variance of C.

OPTIONAL EXERCISE

15.4 For the multinomial probability distribution, show that

$$\text{Cov}(n_i, n_j) = -np_ip_j$$

[*Hint:* First show that $E(n_in_j) = n(n-1)p_ip_j$.]

ESTIMATING CATEGORY PROBABILITIES

A multinomial experiment can always be reduced to a binomial experiment by isolating one category, say category i, and then combining all others. Since we know that $\hat{p} = y/n$ is a good estimator of the binomial parameter p, it follows that

$$\hat{p}_i = \frac{n_i}{n}$$

is a good estimator of p_i, the probability associated with category i in a multinomial experiment. It also follows that $\hat{p}_i$ will possess the same properties as $\hat{p}$—namely, that when n is large, $\hat{p}_i$ will be approximately normally distributed (by the central limit theorem) with

$$E(\hat{p}_i) = p_i$$

and

$$V(\hat{p}_i) = \frac{p_i(1 - p_i)}{n}$$

Consequently, a large-sample confidence interval for p_i may be constructed as shown in the box.

A LARGE-SAMPLE $(1 - \alpha)100\%$ CONFIDENCE INTERVAL FOR p_i

$$\hat{p}_i \pm z_{\alpha/2} \sqrt{\frac{\hat{p}_i(1 - \hat{p}_i)}{n}}$$

Values of $z_{\alpha/2}$ can be found in Table 4 of Appendix II.

We will estimate the difference between a pair of category probabilities, say categories i and j $(i \neq j)$, using $(\hat{p}_i - \hat{p}_j)$. This linear function of $\hat{p}_i$ and $\hat{p}_j$ will be approximately normally distributed with

$$E(\hat{p}_i - \hat{p}_j) = p_i - p_j$$

and

$$V(\hat{p}_i - \hat{p}_j) = V(\hat{p}_i) + V(\hat{p}_j) - 2\,\text{Cov}(\hat{p}_i, \hat{p}_j)$$

Since the covariance of two category counts, say n_i and n_j $(i \neq j)$, was given in Section 15.2 as

$$\text{Cov}(n_i, n_j) = -np_i p_j$$

it follows that the covariance between the corresponding estimators, $\hat{p}_i$ and $\hat{p}_j$, is

$$\text{Cov}(\hat{p}_i, \hat{p}_j) = E[(\hat{p}_i - p_i)(\hat{p}_j - p_j)] = E\left[\left(\frac{n_i}{n} - \frac{np_i}{n}\right)\left(\frac{n_j}{n} - \frac{np_j}{n}\right)\right]$$

$$= E\left[\frac{1}{n^2}(n_i - np_i)(n_j - np_j)\right] = \frac{1}{n^2}E\left[(n_i - np_i)(n_j - np_j)\right]$$

$$= \frac{1}{n^2}\text{Cov}(n_i, n_j) = \frac{1}{n^2}(-np_i p_j)$$

$$= \frac{-p_i p_j}{n}$$

Therefore,

$$V(\hat{p}_i - \hat{p}_j) = V(\hat{p}_i) + V(\hat{p}_j) - 2\,\text{Cov}(\hat{p}_i, \hat{p}_j)$$

$$= \frac{p_i(1 - p_i)}{n} + \frac{p_j(1 - p_j)}{n} + \frac{2p_i p_j}{n}$$

and a large-sample $(1 - \alpha)100\%$ confidence interval for $(p_i - p_j)$ is as indicated in the box.

A LARGE-SAMPLE $(1 - \alpha)100\%$ CONFIDENCE INTERVAL FOR $(p_i - p_j)$

$$(\hat{p}_i - \hat{p}_j) \pm z_{\alpha/2} \sqrt{\frac{\hat{p}_i(1 - \hat{p}_i) + \hat{p}_j(1 - \hat{p}_j) + 2\hat{p}_i\hat{p}_j}{n}}$$

Values of $z_{\alpha/2}$ can be found in Table 4 of Appendix II.

EXAMPLE 15.1

Refer to Table 15.1 and find a 95% confidence interval for the proportion p_1 of all defective personal computers that can be attributed to production line A. Note that p_1 is *not* the proportion of PCs produced by production line A that are defective. Rather, it is the proportion of all defective PCs that are produced by production line A.

SOLUTION

From Table 15.1, we have $n_1 = 15$ and $n = 103$. Therefore, a 95% confidence interval for p_1 is

$$\hat{p}_1 \pm z_{.025} \sqrt{\frac{\hat{p}_1(1 - \hat{p}_1)}{n}} \quad \text{where} \quad \hat{p}_1 = \frac{n_1}{n} = \frac{15}{103} = .146$$

$$.146 \pm 1.96 \sqrt{\frac{(.146)(.854)}{103}}$$

or $.146 \pm .068$. Therefore, our interval estimate for p_1 is from .078 to .214. ∎

EXAMPLE 15.2

Refer to Example 15.1 and find a 95% confidence interval for $(p_1 - p_2)$, the difference between the proportions of defective PCs attributable to production lines A and B, respectively.

SOLUTION

From Table 15.1, we have $n_2 = 27$ and $\hat{p}_2 = n_2/n = 27/103 = .262$. Then a 95% confidence interval for $(p_1 - p_2)$ is

$$(\hat{p}_1 - \hat{p}_2) \pm z_{\alpha/2} \sqrt{\frac{\hat{p}_1(1 - \hat{p}_1) + \hat{p}_2(1 - \hat{p}_2) + 2\hat{p}_1\hat{p}_2}{n}}$$

$$(.146 - .262) \pm 1.96 \sqrt{\frac{(.146)(.854) + (.262)(.738) + 2(.146)(.262)}{103}}$$

$$-.116 \pm .121$$

Therefore, our interval estimate of $(p_1 - p_2)$, the difference in the proportions of the defective PCs attributable to production lines A and B is $-.237$ to .005. ∎

| | | | | | | | | | | | | |

EXERCISES 15.5–15.9

15.5 A multinomial experiment with five possible outcomes and 90 trials produced the data shown in the table.

		OUTCOME			
	1	2	3	4	5
n_i	11	17	24	28	10

Find a 99% confidence interval for each of the following:
a. p_1 **b.** p_3 **c.** $(p_5 - p_2)$ **d.** $(p_3 - p_4)$

15.6 A traffic study found that of 972 automobiles entering a busy intersection during the period from 4 P.M. to 7 P.M., 357 turned left, 321 turned right, and 294 drove straight through the intersection.
a. Construct a 95% confidence interval for the true proportion of automobiles that drive straight through the intersection during this period.
b. Construct a 95% confidence interval for the difference between the proportions of automobiles that turn left and turn right, respectively, during this period. Interpret the interval.

15.7 In March 1981, a waterborne nonbacterial gastroenteritis outbreak occurred in Colorado due to a long-standing filter deficiency and malfunction of a sewage treatment plant. A study was conducted to determine whether the incidence of gastrointestinal disease during the epidemic was related to water consumption (*American Water Works Journal*, Jan. 1986). A telephone survey of households yielded the accompanying information on daily consumption of 8-ounce glasses of water for a sample of 40 residents who exhibited gastroenteritis symptoms during the epidemic.

	DAILY CONSUMPTION OF 8-OUNCE GLASSES OF WATER				
	0	1–2	3–4	5 or more	Total
Number of respondents with symptoms	6	11	13	10	40

Source: Hopkins, R. S. et al. "Gastroenteritis: Case Study of a Colorado Outbreak." *American Water Works Journal*, Vol. 78, No. 1, Jan. 1986, p. 42 (Table 1). Copyright © 1986, American Water Works Association. Reprinted by permission.

a. Use a 99% confidence interval to estimate the percentage of gastroenteritis cases who drink 1–2 glasses of water per day.
b. Use a 99% confidence interval to estimate the difference between the percentages of gastroenteritis cases who drink 1–2 and 0 glasses of water per day.

15.8 A survey was conducted to investigate the popularity of four statistical computer program packages, SAS, SPSS^X, BMDP, and Minitab, among statistics professors. The accompanying table summarizes the responses for a random sample of 100 statistics professors, each of whom was asked to state which of the four packages he or she prefers.

Package Preferred	SAS	SPSS$^{\times}$	BMDP	MINITAB
Number of Professors	37	22	15	26

a. Estimate the true proportion of statistics professors who prefer Minitab, using a 90% confidence interval.

b. Use a 90% confidence interval to estimate the difference between the proportions of statistics professors who prefer SAS and who prefer BMDP.

15.9 Because of erratic rainfall patterns and low water-holding capacities of soils in Florida, supplemental irrigation is required for producing most crops. A research team has developed five alternative water management strategies for irrigating crop land in central Florida. A random sample of 100 agricultural engineers were interviewed and asked which of the strategies he or she believes would yield maximum productivity. A summary of their responses is shown in the table.

Strategy	A	B	C	D	E
Frequency	17	27	22	15	19

a. Find a 90% confidence interval for the true proportion of agricultural engineers who recommend strategy C.

b. Find a 90% confidence interval for the difference between the true proportions of agricultural engineers who recommend strategies E and B.

c. Find a 90% confidence interval for the true difference between the percentages of agricultural engineers who recommend strategies A and D.

S E C T I O N   15.4

TESTING HYPOTHESES ABOUT THE CATEGORY PROBABILITIES

Suppose we want to test a hypothesis about the category probabilities for the home computer experiment using the data given in Table 15.1. Specifically, we might want to test the (null) hypothesis that the proportions of defectives attributable to the five production lines are equal, i.e., H_0: $p_1 = p_2 = \cdots = p_5 = .2$, against the alternative hypothesis that at least two of the probabilities are unequal. Intuitively, we would choose a test statistic based on the deviations of the observed category counts, $n_1, n_2, \ldots, n_5$, from their expected values

$$E(n_i) = np_i = (103)(.2) = 20.6 \quad (i = 1, 2, \ldots, 5)$$

Large deviations between the observed and expected category counts would provide evidence to indicate that the hypothesized category probabilities are incorrect.

The statistic used to test hypotheses about the category probabilities of a k-category multinomial experiment, one based on the deviations between observed and expected cell counts, is

$$X^2 = \sum_{i=1}^{k} \frac{[n_i - E(n_i)]^2}{E(n_i)} = \sum_{i=1}^{k} \frac{(n_i - np_i)^2}{np_i}$$

When the number n of trials is large enough so that $E(n_i) \geq 5$ for $i = 1, 2,$ $\ldots, k$, the statistic X^2 will possess (proof omitted) approximately a chi-square sampling distribution.* The value of X^2 will be larger than expected if the deviations $[n_i - E(n_i)]$ are large. Therefore, the rejection region for the test is $X^2 > \chi_\alpha^2$, where χ_α^2 is the value of χ^2 that locates an area α in the upper tail of the chi-square distribution (see Figure 15.2).

FIGURE 15.2
Rejection Region for the
Chi-square Test

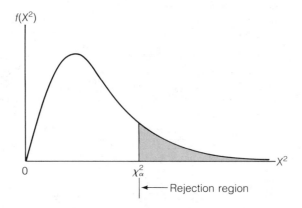

The number of degrees of freedom for the approximating chi-square distribution will always equal k less 1 degree of freedom for every linearly independent restriction placed on the category counts. For example, we always have at least one linear restriction on the category counts because their sum must equal the sample size, n:

$$n_1 + n_2 + \cdots + n_k = n$$

A TEST OF A HYPOTHESIS ABOUT MULTINOMIAL PROBABILITIES

H_0: $p_1 = p_{1,0}, \; p_2 = p_{2,0}, \; \ldots, \; p_k = p_{k,0}$, where $p_{1,0}, p_{2,0}, \; \ldots, \; p_{k,0}$ represent the hypothesized values of the multinomial probabilities

H_a: At least one of the multinomial probabilities does not equal its hypothesized value

Test statistic: $X^2 = \sum_{i=1}^{k} \dfrac{[n_i - E(n_i)]^2}{E(n_i)}$

where $E(n_i) = np_{i,0}$, the expected number of outcomes of type i assuming H_0 is true. The total sample size is n.

Rejection region: $X^2 > \chi_\alpha^2$, where χ_α^2 has $(k - 1)$ df

Assumption: $E(n_i) \geq 5$ for all n_i

---

*For some applications, the expected cell counts can be less than 5. More on this subject can be found in the paper by Cochran (1952) listed in the references at the end of this chapter.

Other restrictions arise if we must estimate the category probabilities. Since each estimate will involve a linear function of the category counts, the degrees of freedom for chi-square will be reduced by 1 for each category parameter that must be estimated.

A test of a hypothesis that the category probabilities assume specified values results in only a single linear restriction on the category counts—namely, $n_1 + n_2 + \cdots + n_k = n$. No category probabilities need to be estimated because their values are specified in H_0. The test procedure is described in the preceding box. We will illustrate this simple application of the chi-square test in Example 15.3.

EXAMPLE 15.3

Refer to the data provided in Table 15.1. Test the hypothesis that the proportions of all defective computers attributable to the five production lines are equal. Test using $\alpha = .05$.

SOLUTION

We want to test $H_0: p_1 = p_2 = \cdots = p_5 = .2$ against the alternative hypothesis, H_a: At least two of the category probabilities are unequal. We have already calculated

$$E(n_i) = np_i = (103)(.2) = 20.6 \quad (i = 1, 2, \ldots, 5)$$

The observed and the expected category counts (in parentheses) are shown in Table 15.3. Substituting the observed and expected values of the category counts into the formula for X^2, we obtain

$$X^2 = \sum_{i=1}^{k} \frac{(n_i - np_i)^2}{np_i} = \frac{(15 - 20.6)^2}{20.6} + \frac{(27 - 20.6)^2}{20.6} + \cdots + \frac{(11 - 20.6)^2}{20.6}$$

$$= 13.36$$

The rejection region for the test is $X^2 > \chi^2_{.05}$, where $\chi^2_{.05}$ is based on $k - 1 = 5 - 1 = 4$ degrees of freedom. This value, found in Table 7 of Appendix II, is $\chi^2_{.05} = 9.48773$. Since the observed value of X^2 exceeds this value, there is sufficient evidence to reject H_0. It appears that at least one production line is responsible for a higher proportion of defective computers than the other lines.

TABLE 15.3
Observed and Expected Category Counts for the Data of Table 15.1

Observed	15	27	31	19	11
Expected	(20.6)	(20.6)	(20.6)	(20.6)	(20.6)

■

EXERCISES 15.10–15.16

15.10 Refer to Exercise 15.5. Test the hypothesis that the five proportions are equal. Use $\alpha = .05$.

15.11 Refer to Exercise 15.6.
   a. Do the data disagree with the hypothesis that the traffic is equally divided among the three directions? Test using $\alpha = .05$.

b. Do the data provide sufficient evidence to indicate that more than one-third of all automobiles entering the intersection turn left? Test using $\alpha = .05$.

15.12 Refer to the gastroenteritis case study described in Exercise 15.7. Conduct a test to determine whether the incidence of gastrointestinal disease during the epidemic is related to water consumption. Use $\alpha = .01$.

15.13 Refer to Exercise 15.8. Test the hypothesis that the proportions of statistics professors who prefer SAS, SPSS^X, BMDP, and Minitab are equal. Use $\alpha = .05$.

15.14 Refer to Exercise 15.9. Do the data present sufficient evidence to indicate a preference for one or more of the five water management strategies? Test using $\alpha = .05$.

15.15 J. A. Breaugh investigated employees' reactions to compressed work weeks (*Personnel Psychology*, Summer 1983). *Compressed work weeks* are defined as "alternative work schedules in which a trade is made between the number of hours worked per day, and the number of days worked per week, in order to work the standard number of weekly hours in less than 5 days." A field study was conducted at a large midwestern continuous-processing (7 days/24 hours) chemical plant which had experimented with four different work schedules, two of which were compressed:

Three 8-hour fixed shifts (day, evening, midnight)

Three 8-hour rotating shifts

Two 12-hour fixed shifts (12 A.M.–12 P.M., 12 P.M.–12 A.M.)

Two 12-hour rotating shifts

Six hundred seventy-one hourly employees were asked to rank the four work schedules in order of preference. The accompanying table gives the number of first-place rankings for each schedule. Is there sufficient evidence to indicate that the hourly employees have a preference for one of the work schedules? Test using $\alpha = .01$.

8-hour fixed	8-hour rotating	12-hour fixed	12-hour rotating
389	54	208	20

OPTIONAL EXERCISE

15.16 A general proof of the fact that X^2 possesses approximately a chi-square sampling distribution when n is large is beyond the scope of this text. However, it can be justified for the binomial case ($k = 2$). In Optional Exercise 7.46, we stated that if z is a standard normal random variable, then z^2 is a chi-square random variable with 1 degree of freedom. Denote the two cell counts for a binomial experiment as $n_1 = y$ and $n_2 = (n - y)$. Then, for large n,

$$z = \frac{y - np}{\sqrt{npq}}$$

has approximately a standard normal distribution and z^2 will be approximately distributed as a chi-square random variable with 1 degree of freedom. Show algebraically that for $k = 2$, $X^2 = z^2$.

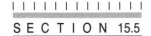

CONTINGENCY TABLES

Many applications of the test of hypothesis concerning the category probabilities presented in Section 15.4 imply a one-directional classification of the data. For example, the categories for the defective computer data of Example 15.3 correspond to the "values" assumed by the qualitative variable, production line. Sometimes we may want to classify data according to two directions of classification—that is, according to two qualitative variables. The objective of such a classification usually is to determine whether the two directions of classification are dependent.

For example, the past energy shortages have made many consumers more aware of the size of the automobiles they purchase. Suppose an automobile manufacturer is interested in determining if there is a relationship between the size and manufacturer of newly purchased automobiles. One thousand recent buyers of American-made cars are randomly sampled and each purchase is classified with respect to the size and manufacturer of the purchased automobile. The data are summarized in the **two-way table** shown in Table 15.4. This table is called a **contingency table**; it presents multinomial count data classified on two scales, or dimensions, of classification, namely, automobile size and manufacturer.

TABLE 15.4

Contingency Table for Automobile Size Example

	MANUFACTURER				TOTALS
	A	B	C	D	
Small	157	65	181	10	413
Medium	126	82	142	46	396
Large	58	45	60	28	191
TOTALS	341	192	383	84	1,000

Each cell of Table 15.4, located in a specific row and column, represents one of the $k = (3)(4) = 12$ categories of a two-directional classification of the $n = 1,000$ observations. The symbols representing the cell counts for the experiment in Table 15.4 are shown in Table 15.5a (page 786) and the corresponding cell, row, and column probabilities are shown in Table 15.5b. Thus, n_{11} represents the number of buyers who purchase a small car of manufacturer A and p_{11} represents the corresponding cell probability. The row and column totals, designated as r_1, r_2, r_3 and c_1, c_2, c_3, and c_4, respectively, are shown in Table 15.5a. The corresponding row and column probability totals are shown in Table 15.5b. The probability totals for the rows and columns are called **marginal probabilities**. The marginal probability p_1 is the probability that a small car is purchased, and the marginal probability p_A is the probability that a car by manufacturer A is purchased. Thus,

$$p_1 = p_{11} + p_{12} + p_{13} + p_{14}$$

and

$$p_A = p_{11} + p_{21} + p_{31}$$

TABLE 15.5

a. Observed Counts for Contingency Table

	MANUFACTURER				TOTALS
	A	B	C	D	
Small	n_{11}	n_{12}	n_{13}	n_{14}	r_1
Medium	n_{21}	n_{22}	n_{23}	n_{24}	r_2
Large	n_{31}	n_{32}	n_{33}	n_{34}	r_3
TOTALS	c_1	c_2	c_3	c_4	n

b. Probabilities for Contingency Table

	MANUFACTURER				TOTALS
	A	B	C	D	
Small	p_{11}	p_{12}	p_{13}	p_{14}	p_1
Medium	p_{21}	p_{22}	p_{23}	p_{24}	p_2
Large	p_{31}	p_{32}	p_{33}	p_{34}	p_3
TOTALS	p_A	p_B	p_C	p_D	1

You can see that the experiment we have described is a multinomial experiment with a total of 1,000 trials and $(3)(4) = 12$ categories. If the 1,000 recent buyers are randomly chosen, the trials are considered independent and the probabilities are viewed as remaining constant from trial to trial.

Suppose we want to know whether the two classifications, manufacturer and size, are dependent. That is, if we know which size car a buyer will choose, does that information give us a clue about the manufacturer of the car the buyer will choose? In a probabilistic sense we know (Chapter 3) that independence of events A and B implies $P(A \cap B) = P(A)P(B)$. Similarly, in the contingency table analysis, if the two classifications are independent, the probability that an item is classified in any particular cell of the table is the product of the corresponding marginal probabilities. Thus, under the hypothesis of independence, in Table 15.5b we must have

$$p_{11} = p_1 p_A \qquad p_{12} = p_1 p_B$$

and so forth. Therefore, the null hypothesis that the directions of classification are independent is equivalent to the hypothesis that every cell probability in the contingency table is equal to the product of its respective row and column marginal probabilities. If the data disagree with the expected cell counts computed from these probabilities, there is evidence to indicate that the two directions of classification are dependent.

If we were to calculate the expected cell counts for our example, you would immediately perceive a difficulty. The marginal probabilities are unknown and must be estimated. The best estimate of the ith row marginal probability, call it p_i, is

$$\hat{p}_i = \frac{r_i}{n} = \frac{\text{Row } i \text{ total}}{n}$$

Similarly, the best estimate of the jth marginal column probability is

$$\hat{p}_j = \frac{c_j}{n} = \frac{\text{Column } j \text{ total}}{n}$$

Therefore, the estimated expected cell count for the cell in the ith row and jth column of the contingency table is

$$\hat{E}(n_{ij}) = n\hat{p}_i\hat{p}_j$$
$$= n\left(\frac{r_i}{n}\right)\left(\frac{c_j}{n}\right) = \frac{r_i c_j}{n}$$
$$= \frac{(\text{Row } i \text{ total})(\text{Column } j \text{ total})}{n}$$

The general form of an $r \times c$ contingency table (one containing r rows and c columns) is shown in Table 15.6. When n is large, the test statistic

$$X^2 = \sum_{j=1}^{c}\sum_{i=1}^{r}\frac{[n_{ij} - \hat{E}(n_{ij})]^2}{\hat{E}(n_{ij})}$$
$$= \sum_{j=1}^{c}\sum_{i=1}^{r}\frac{\left(n_{ij} - \frac{r_i c_j}{n}\right)^2}{\left(\frac{r_i c_j}{n}\right)}$$

will possess approximately a chi-square distribution. The rejection region for the test will be $X^2 > \chi_\alpha^2$ (see Figure 15.3, page 788).

TABLE 15.6
General $r \times c$ Contingency Table

		COLUMN 1	2	$\cdots$	c	ROW TOTALS
ROW	1	n_{11}	n_{12}	$\cdots$	n_{1c}	r_1
	2	n_{21}	n_{22}	$\cdots$	n_{2c}	r_2
	$\vdots$	$\vdots$	$\vdots$		$\vdots$	$\vdots$
	r	n_{r1}	n_{r2}	$\cdots$	n_{rc}	r_r
COLUMN TOTALS		c_1	c_2	$\cdots$	c_c	n

To determine the number of degrees of freedom for the approximating chi-square distribution, note that $k = rc$. From this we must subtract 1 degree of freedom because the sum of all rc cell counts must equal n. We also subtract $(r - 1)$ because we must estimate the $(r - 1)$ row marginal probabilities. (The last row probability will then be determined because the sum of the row probabilities must equal 1.) Similarly, we must subtract $(c - 1)$ because we must estimate

FIGURE 15.3

Rejection Region for
the Chi-square Test
for Dependence

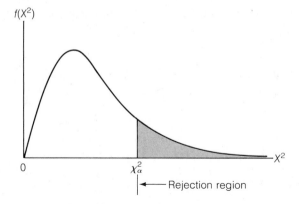

$(c - 1)$ column marginal probabilities. Therefore, the degrees of freedom for chi-square will be

$$df = k - \left(\begin{array}{c}\text{the number of linearly independent}\\ \text{restrictions on the cell counts}\end{array}\right)$$

$$= rc - (1) - (r - 1) - (c - 1)$$

$$= rc - r - c + 1$$

$$= (r - 1)(c - 1)$$

The chi-square test is summarized in the box and its use is illustrated in Example 15.4.

GENERAL FORM OF A CONTINGENCY TABLE ANALYSIS: A TEST FOR INDEPENDENCE

H_0:   The two classifications are independent

H_a: The two classifications are dependent

Test statistic: $X^2 = \sum\limits_{j=1}^{c} \sum\limits_{i=1}^{r} \dfrac{[n_{ij} - \hat{E}(n_{ij})]^2}{\hat{E}(n_{ij})}$

where

$$\hat{E}(n_{ij}) = \frac{r_i c_j}{n}$$

Rejection region: $X^2 > \chi_{\alpha}^2$, where χ_{α}^2 has $(r - 1)(c - 1)$ df.

Assumptions: 1. The n observed counts are a random sample from the population of interest. We may then consider this to be a multinomial experiment with $r \times c$ possible outcomes.

2. For the χ^2 approximation to be valid, we require that the estimated expected counts be greater than or equal to 5 in all cells.

EXAMPLE 15.4

Use the data in Table 15.4 to decide whether a buyer's choice of manufacturer depends on the size of the automobile that the buyer wants to purchase. Test using $\alpha = .05$.

SOLUTION

The first step in the analysis of a contingency table is to calculate the estimated expected cell counts. For example,

$$E(n_{11}) = \frac{r_1 c_1}{n} = \frac{(413)(341)}{1,000} = 140.833$$

$$E(n_{12}) = \frac{r_1 c_2}{n} = \frac{(413)(192)}{1,000} = 79.296$$

$$\vdots \qquad \qquad \vdots$$

$$E(n_{34}) = \frac{r_3 c_4}{n} = \frac{(191)(84)}{1,000} = 16.044$$

The cell counts and the corresponding estimated expected values (in parentheses) are shown in Table 15.7.

TABLE 15.7
Observed and Estimated Expected (in Parentheses) Counts

	MANUFACTURER			
	A	B	C	D
SMALL	157 (140.833)	65 (79.296)	181 (158.179)	10 (34.692)
MEDIUM	126 (135.036)	82 (76.032)	142 (151.668)	46 (33.264)
LARGE	58 (65.131)	45 (36.672)	60 (73.153)	28 (16.044)

We next compute the X^2 test statistic:

$$X^2 = \frac{[n_{11} - \hat{E}(n_{11})]^2}{\hat{E}(n_{11})} + \frac{[n_{12} - \hat{E}(n_{12})]^2}{\hat{E}(n_{12})} + \cdots + \frac{[n_{34} - \hat{E}(n_{34})]^2}{\hat{E}(n_{34})}$$

$$= \sum_{j=1}^{4} \sum_{i=1}^{3} \frac{[n_{ij} - \hat{E}(n_{ij})]^2}{\hat{E}(n_{ij})}$$

Substituting the data of Table 15.7 into this expression, we obtain

$$X^2 = \frac{(157 - 140.833)^2}{140.833} + \frac{(65 - 79.296)^2}{79.296} + \cdots + \frac{(28 - 16.044)^2}{16.044} = 45.81$$

The rejection region for the test is $X^2 > \chi^2_{.05} = 12.5916$, where $\chi^2_{.05}$ is based on $(r - 1)(c - 1) = (2)(3) = 6$ degrees of freedom. Since the computed value of X^2, 45.81, exceeds this value, we reject H_0 and conclude that the two directions of data classification are dependent. It appears that the choice of automobile manufacturer is dependent on the size of automobile that the buyer wants to purchase. ∎

15.17 Find the rejection region for a test of independence of two directions of classification if the contingency table contains r rows and c columns and:
a. $r = 2$, $c = 2$, $\alpha = .05$
b. $r = 3$, $c = 6$, $\alpha = .10$
c. $r = 3$, $c = 4$, $\alpha = .01$

15.18 Test the null hypothesis that the rows and columns for the 4×3 contingency table shown here are independent. Test using $\alpha = .01$.

		COLUMN		
		1	2	3
ROW	1	20	30	50
	2	40	20	40
	3	100	50	50
	4	40	0	60

15.19 Researchers at the Oak Ridge (Tennessee) National Laboratory have developed a computer program to estimate the numbers of expected and excess cases of thyroid cancer occurring in the lifetime of those exposed to atomic weapons tests at the Nevada Test Site in the 1950's (*Health Physics*, Jan. 1986). Of the approximately 23,000 people exposed to the weapons testing fallout, 58 were expected to develop thyroid cancer in their remaining lifetimes. According to the computer program, the 58 cases can be categorized by level of radiation (dose) at the time of exposure and sex as shown in the table.

Suppose that the data represent a random sample of 58 thyroid cancer patients selected from the target population. Conduct a test to determine whether the two directions of classification, dose at time of exposure and sex, are independent. Use $\alpha = .01$.

		SEX		TOTALS
		Males	Females	
DOSE (rad)	Less than 1	6	13	19
	1–10	8	18	26
	11 or more	3	10	13
TOTALS		17	41	58

Source: Zeighami, E. A. and Morris, M. D. "Thyroid Cancer Risk in the Population Around the Nevada Test Site." *Health Physics*, Vol. 50, No. 1, Jan. 1986, p. 26 (Table 2).

15.20 One criterion used to evaluate employees in the assembly section of a large factory is the number of defective pieces per 1,000 parts produced. The quality control department wants to find out whether there is a relationship between years of experience and defect rate. Since the job is rather repetitive, after the initial training period, any improvement due to a learning effect might be offset by a decrease in the motivation of a worker. A defect rate is calculated for each worker for a yearly evaluation. The results for 100 workers are given in the table. Is there evidence of a relationship between defect rate and years of experience? Use $\alpha = .05$.

		YEARS OF EXPERIENCE (AFTER TRAINING PERIOD)		
		1	2–5	6–10
	High	6	9	9
DEFECT RATE	Average	9	19	23
	Low	7	8	10

15.21 The nuclear mishap on Three Mile Island near Harrisburg, Pennyslvania, on March 28, 1979, forced many local residents to evacuate their homes. In order to assess the impact of the accident on the area population, a questionnaire was designed and mailed to a sample of 150 households within 2 weeks after the accident occurred. One question concerned residents' attitudes toward a full evacuation: "Should there have been a full evacuation of the immediate area?" Respondents were grouped according to the distance (in miles) of the community in which they reside from Three Mile Island and their responses were recorded. A summary of the results, adapted from the survey report, is shown in the table. Conduct a test to determine if local residents' attitudes toward a full evacuation is independent of distance of residence from Three Mile Island. Use $\alpha = .10$.

		DISTANCE FROM THREE MILE ISLAND (MILES)						TOTALS
		1–3	4–6	7–9	10–12	13–15	15+	
FULL	Yes	7	11	10	5	4	29	66
EVACUATION	No	9	11	13	6	6	39	84
TOTALS		16	22	23	11	10	68	150

Source: Brown, S. et al. "Final Report on a Survey of Three Mile Island Area Residents." Department of Geography, Michigan State University, Aug. 1979.

15.22 A 4-year study of computer abuse was recently completed by researchers. The numbers of four types of abuse reported and verified by year are shown in the table. Do the data provide sufficient evidence to indicate that the proportions of different types of abuse are changing over time? Test using $\alpha = .05$.

		TYPE OF ABUSE				TOTAL
		Financial Fraud	Theft of Information or Property	Unauthorized Use of Information	Vandalism	
	1	7	5	9	8	29
YEAR	2	22	18	6	6	52
	3	12	15	6	12	45
	4	21	15	16	9	61
TOTAL		62	53	37	35	187

15.23 The advances of medical technology have led to an increased survival rate of infants born with genetic diseases. Since no true cure for genetic disease is currently available, some physicians provide genetic counseling for their patients in order to prevent the birth of genetically defective infants. However, genetic counseling has faced a certain amount of resistance from both physicians and patients. In one part of a survey of general and family practitioners, pediatricians, and obstetrician-gynecologists in the cities of Phoenix and Tucson, Arizona, each physician was classified according to religion and opinion on genetic counseling. A summary of the responses for Jewish, Protestant, and Catholic physicians is shown in the table. Is there evidence of a difference in the proportions of physicians who strongly support genetic counseling among the three religions? Test using $\alpha = .01$.

	RELIGION			TOTALS
	Jewish	Protestant	Catholic	
Strongly support genetic counseling	21	36	10	67
Do not strongly support genetic counseling	26	142	52	220
TOTALS	47	178	62	287

Source: Weitz, R. "Barriers to Acceptance of Genetic Counseling Among Primary Care Physicians." *Social Biology*, Fall 1979, *26*, p. 192.

15.24 In recent years, many companies have converted to the metric system of measurement. To investigate this phenomenon, B. D. Phillips, H. A. G. Lakhani, and S. L. George analyzed data collected on 757 small manufacturers for a recently completed U.S. Metric Board study (*Technological Forecasting and Social Change*, Apr. 1984). The firms were cross-classified according to metric conversion (converters or nonconverters) and level of technology (high or non–high technology). The contingency table for the data is shown here.

	HIGH TECH	NON–HIGH TECH
Metric converters	81	296
Nonconverters	80	300

Reprinted by permission of the publisher. Copyright 1984 by Elsevier Science Publishing Co., Inc.

a. Calculate the estimated expected number of firms in each of the four cells of the table.
b. Calculate the X^2 statistic.
c. Is there sufficient evidence to indicate that the distributions of percentages of metric converters and nonconverters differ for high tech and non–high tech firms? Test using $\alpha = .01$.

CONTINGENCY TABLES WITH FIXED MARGINAL TOTALS

A situation that can occur in the analysis of contingency table data is that one or more of the categories may contain an insufficient number of observations. To illustrate, we will consider the study of the relationship between the manufacturer of a purchased automobile and its size, described in Section 15.5. If one of the manufacturers produces only a small number of automobiles, it is conceivable that very few of these automobiles would appear in the random sample of $n = 1,000$. This would cause the expected cell counts for the manufacturer to be small, perhaps less than the required 5. To guard against this possibility, experimenters often fix either the row or column totals. For our example, we would fix the column totals by randomly and independently sampling a fixed number of automobiles sold by each manufacturer. This would increase the likelihood that the estimated expected cell counts would be of adequate size.

For example, suppose we take random samples of 250 automobiles produced by each manufacturer. The results might appear as shown in Table 15.8. Note the difference between this sampling procedure and the one described in Section 15.5, where we assumed that a *single* random sample of $n = 1,000$ automobiles was selected from among the population of all newly purchased automobiles. In this section we have randomly and independently selected *four* samples, 250 automobiles for each manufacturer. Therefore, the data of Table 15.8 result from four multinomial experiments, each with $k = 3$ cells, corresponding to the four manufacturers.

TABLE 15.8

A Size–Manufacturer Contingency Table with Column Totals Fixed

| | MANUFACTURER | | | | TOTALS |
	A	B	C	D	
Small	122	64	107	34	327
Medium	101	95	92	119	407
Large	27	91	51	97	266
TOTALS	250	250	250	250	1,000

A chi-square test to detect dependence between row and column classifications, when either the column or the row totals are fixed, is conducted in exactly the same way as the test of Section 15.5. It can be shown (proof omitted) that the X^2 statistic will possess a sampling distribution that is approximately a chi-square distribution with $(r - 1)(c - 1)$ degrees of freedom. The test procedure is summarized in the box on page 794. An application of the test to the comparison of two or more binomial proportions is illustrated in Example 15.5.

EXAMPLE 15.5

In order to compare the proportions of defective dishwashers produced by three production lines, an engineer randomly sampled 500 washers from each line. The numbers of defectives for the three lines were found to be 12, 17, and 7, respectively. Do the data provide sufficient evidence to indicate differences in the

---

GENERAL FORM OF CONTINGENCY TABLE ANALYSIS: A TEST FOR INDEPENDENCE WITH ROW* TOTALS FIXED

If row totals are fixed:

H_0:   The row proportions in each cell do not depend on the row; that is, the distributions of observations in the column categories are the same for each row

H_a: The row proportions in some (or all) of the cells depend on the row; that is, the distributions of observations in the column categories differ for at least two of the rows

Test statistic: $X^2 = \sum\limits_{j=1}^{c} \sum\limits_{i=1}^{r} \dfrac{[n_{ij} - \hat{E}(n_{ij})]^2}{\hat{E}(n_{ij})}$

where

$$\hat{E}(n_{ij}) = \frac{r_i c_j}{n}$$

Rejection region: $X^2 > \chi_\alpha^2$, where χ_α^2 has $(r-1)(c-1)$ df

Assumptions: 1. A random sample is selected from each population for which the row totals are fixed.
2. The samples are independently selected.
3. We require the estimated expected value of each cell to be at least 5 in order to use the χ^2 approximation.

---

proportions of defective washers produced by the three production lines? Test using $\alpha = .05$.

SOLUTION

The data are presented as a contingency table in Table 15.9. The objective of this experiment is to compare three binomial proportions of defectives, p_1, p_2, and p_3, based on three independent binomial experiments, each containing 500 observations.

TABLE 15.9

Contingency Table for Example 15.5

	PRODUCTION LINE			TOTALS
	1	2	3	
Number of defectives	12	17	7	36
Number of nondefectives	488	483	493	1,464
TOTALS	500	500	500	1,500

*Note that to obtain the procedure for conducting a χ^2 analysis for fixed column totals, it is necessary only to interchange the words *column* and *row* in the box.

The null hypothesis is that the proportions of defectives for the three production lines are identical, i.e.,

$$H_0: \quad p_1 = p_2 = p_3$$

against the alternative hypothesis

H_a: At least two of the proportions, p_1, p_2, and p_3, differ.

Note that the null hypothesis we have specified implies that the numbers of defectives and nondefectives are independent of the production line. Therefore, we test H_0: $p_1 = p_2 = p_3$ using the chi-square test for a contingency table analysis.

The estimated expected cell counts are computed using the formula

$$\hat{E}(n_{ij}) = \frac{r_i c_j}{n}$$

Therefore,

$$\hat{E}(n_{11}) = \frac{r_1 c_1}{n} = \frac{(36)(500)}{1,500} = 12$$

and

$$\hat{E}(n_{12}) = \frac{r_1 c_2}{n} = \frac{(36)(500)}{1,500} = 12$$

These, along with the remaining estimated expected cell counts (in parentheses), are shown in Table 15.10.

TABLE 15.10

Observed and Estimated Expected Cell Counts for Example 15.5

	PRODUCTION LINE		
	1	2	3
Number of defectives	12 (12)	17 (12)	7 (12)
Number of nondefectives	488 (488)	483 (488)	493 (488)

The computed value of X^2 is

$$X^2 = \sum_{j=1}^{c} \sum_{i=1}^{r} \frac{[n_{ij} - \hat{E}(n_{ij})]^2}{\hat{E}(n_{ij})} = \frac{(12 - 12)^2}{12} + \frac{(17 - 12)^2}{12} + \cdots + \frac{(493 - 488)^2}{488}$$

$$= 4.269$$

The rejection region for the test is $X^2 > \chi^2_{.05}$ where $\chi^2_{.05} = 5.99147$ is based on $(r - 1)(c - 1) = (1)(2) = 2$ degrees of freedom. Since the computed value of X^2 does not exceed $\chi^2_{.05}$, there is insufficient evidence to indicate differences in the proportions of defective washers produced by the three production lines. Note that we do not accept H_0—that is, we do not conclude that $p_1 = p_2 = p_3$—because we would be concerned about the possibility of making a Type II error, failing to detect differences in the proportions of defectives if, in fact, differences exist. The test conclusion simply means that if differences exist, they were too small to detect using samples of 500 washers from each production line.   ∎

15.25 Four independent random samples of 100 observations each were classified according to how they fall in one of three categories. The contingency table, with columns corresponding to the four samples and rows corresponding to the three categories, is shown here.

		COLUMN (SAMPLES)			
		1	2	3	4
ROW (CATEGORIES)	1	20	30	20	10
	2	30	40	40	60
	3	50	30	40	30

a. Do the data provide sufficient evidence to indicate that the distributions of observations within the rows are dependent on the columns? Test using $\alpha = .05$.
b. Form a 95% confidence interval for the probability of observing a response in the second row and third column.
c. Form a 95% confidence interval for the difference between the probability of observing a response in the first row, first column, and the probability of observing a response in the first row, fourth column.

15.26 The scarce and essential metal, manganese, has been found in abundance in nodules on the deep sea floor. In order to investigate the relationship between the magnetic age of the earth's crust on the ocean floor and the probability of finding manganese nodules in that location, crust specimens were selected from seven magnetic age locations and the percentage of specimens containing manganese nodules was recorded for each. The data are shown in the accompanying table. Is there sufficient evidence to indicate that the probability of finding manganese nodules in the deep sea earth's crust is dependent on the magnetic age of the crust? Test using $\alpha = .05$.

AGE	NUMBER OF SPECIMENS	PERCENTAGE WITH MANGANESE NODULES
Miocene–recent	389	5.9
Oligocene	140	17.9
Eocene	214	16.4
Paleocene	84	21.4
Late Cretaceous	247	21.1
Early and Middle Cretaceous	1,120	14.2
Jurassic	99	11.0

Source: Menard, H. W. "Time, Chance, and the Origin of Manganese Nodules." *American Scientist*, Sept.–Oct. 1976.

15.27 A bank conducted a survey to compare the attitudes of young married customers (those having recently opened their first bank accounts) with the attitudes of older, established customers to their new automated teller system. Two hundred customers were randomly selected from each of these categories and asked whether they preferred the automated teller system to the personal service obtained at the bank. A summary of the customers' responses is shown in the table. Do the data provide sufficient evidence to indicate a difference between the proportions of new and established customers who favor use of the automatic teller? Test using $\alpha = .05$. Interpret the results of the test.

	New Customer	Established Customer	TOTALS
Favor the automatic teller	87	65	152
Favor personal service	113	135	248
TOTALS	200	200	400

15.28 An electric utilities company must choose between two technology options for generating electricity for its customers in the future: coal or nuclear energy. In order to assess the attitudes of local citizens, the power company conducted a public opinion poll. Four sectors were considered: news media, coal miners' union, environmentalists and conservationists, and local groups. Fifty persons were randomly sampled from each sector and asked to give their opinions. The results of the survey are shown in the table.

	News Media	Coal Miners' Union	Environmentalist and Conservationists	Local Groups	TOTALS
Support Coal Option	21	42	11	25	99
Support Nuclear Option	18	2	16	13	49
Neutral	11	6	23	12	52
TOTALS	50	50	50	50	200

a. Does public opinion regarding the choice of future technology options for generating electricity differ among the four groups? Test using $\alpha = .10$.

b. Does there appear to be more overall support for the coal option rather than the nuclear option? Test using $\alpha = .10$.

c. Construct a 90% confidence interval for the percentage of environmentalists and conservationists who support the nuclear option.

| | | | | | | | | | | | | |
S E C T I O N   15.7

COMPUTER PRINTOUTS FOR A CONTINGENCY TABLE ANALYSIS

All four of the statistical computer packages discussed throughout this text have procedures for contingency table analysis. The SAS, SPSS[X], Minitab, and BMDP computer printouts for the contingency table analysis of the automobile size–manufacturer data of Table 15.4 are shown in Figure 15.4. You can see that the outputs of the four packages are very similar. All four printouts show the computed value of the χ^2 statistic for testing independence of classifications below the contingency table; in addition, the SAS, SPSS[X], and BMDP printouts report the observed significance level (*p*-value) of the test. SAS and SPSS[X] are the only packages that give the expected cell frequencies in the contingency table.

FIGURE 15.4

Computer Printouts for the Contingency Table Analysis of the Data in Table 15.4

a. SAS

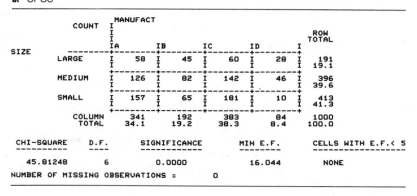

```
                    TABLE OF SIZE BY MANUFACT

     SIZE          MANUFACT

     FREQUENCY:
     EXPECTED  :
      PERCENT  :
      ROW PCT  :
      COL PCT  :A        :B       :C       :D      :   TOTAL
     ---------+--------+--------+--------+--------+
     LARGE    :    58  :    45  :    60  :    28  :    191
              :  65.1  :  36.7  :  73.2  :  16.0  :  19.10
              :  5.80  :  4.50  :  6.00  :  2.80  :
              : 30.37  : 23.56  : 31.41  : 14.66  :
              : 17.01  : 23.44  : 15.67  : 33.33  :
     ---------+--------+--------+--------+--------+
     MEDIUM   :   126  :    82  :   142  :    46  :    396
              : 135.0  :  76.0  : 151.7  :  33.3  :  39.60
              : 12.60  :  8.20  : 14.20  :  4.60  :
              : 31.82  : 20.71  : 35.86  : 11.62  :
              : 36.95  : 42.71  : 37.08  : 54.76  :
     ---------+--------+--------+--------+--------+
     SMALL    :   157  :    65  :   181  :    10  :    413
              : 140.8  :  79.3  : 158.2  :  34.7  :  41.30
              : 15.70  :  6.50  : 18.10  :  1.00  :
              : 38.01  : 15.74  : 43.83  :  2.42  :
              : 46.04  : 33.85  : 47.26  : 11.90  :
     ---------+--------+--------+--------+--------+
     TOTAL        341      192      383       84      1000
                34.10    19.20    38.30     8.40   100.00

               STATISTICS FOR 2-WAY TABLES

     CHI-SQUARE                          45.812   DF=   6   PROB=0.0001
     PHI                                  0.214
     CONTINGENCY COEFFICIENT              0.209
     CRAMER'S V                           0.151
     LIKELIHOOD RATIO CHISQUARE          50.610   DF=   6   PROB=0.0001
```

b. SPSS[X]

```
                        MANUFACT
                 COUNT  I
                        I                                      ROW
                        I                                     TOTAL
                        IA        IB        IC        ID      I
     SIZE        -------+---------+---------+---------+---------+
            LARGE    I    58  I    45  I    60  I    28  I     191
                     I        I        I        I        I     19.1
                     +--------+--------+--------+--------+
            MEDIUM   I   126  I    82  I   142  I    46  I     396
                     I        I        I        I        I     39.6
                     +--------+--------+--------+--------+
            SMALL    I   157  I    65  I   181  I    10  I     413
                     I        I        I        I        I     41.3
                     +--------+--------+--------+--------+
            COLUMN       341      192      383       84      1000
            TOTAL       34.1     19.2     38.3      8.4     100.0

     CHI-SQUARE      D.F.       SIGNIFICANCE      MIN E.F.      CELLS WITH E.F.< 5
     ----------      ----       ------------      --------      ------------------

       45.81248        6           0.0000          16.044             NONE
     NUMBER OF MISSING OBSERVATIONS =          0
```

FIGURE 15.4
(Continued)

c. Minitab

```
-- CHISQUARE FOR DATA IN C1-C4
    EXPECTED FREQUENCIES ARE PRINTED BELOW OBSERVED FREQUENCIES
            C1          C2          C3          C4      TOTALS
    ----------------------------------------------------------
     1        157          65         181          10       413
            140.8        79.3       158.2        34.7
    ----------------------------------------------------------
     2        126          82         142          46       396
            135.0        76.0       151.7        33.3
    ----------------------------------------------------------
     3         58          45          60          28       191
             65.1        36.7        73.2        16.0
    ----------------------------------------------------------
    TOTALS    341         192         383          84      1000

    TOTAL CHI SQUARE =
                  1.86  +   2.58  +   3.28  +  17.58  +
                  0.60  +   0.47  +   0.62  +   4.84  +
                  0.77  +   1.88  +   2.38  +   9.00  +
                                  =  45.86

    DEGREES OF FREEDOM = (3-1) X (4-1) = 6
```

d. BMDP

```
SIZE                    MANUFACT
------                  -------
                A           B           C           D       TOTAL
    -------------------------------------------------------------
    SMALL       157          65         181          10       413
    MEDIUM      126          82         142          46       396
    LARGE        58          45          60          28       191
    -------------------------------------------------------------
    TOTAL       341         192         383          84      1000
            ALL CASES HAD COMPLETE DATA FOR THIS TABLE.


    MINIMUM ESTIMATED EXPECTED VALUE IS      16.04

    STATISTIC                          VALUE    D.F.    PROB.
    PEARSON CHISQUARE                 45.812       6   0.0000
```

In optional Section 15.8 we give you appropriate commands for calling the contingency table analysis procedure of each computer package.

| | | | | | | | | | | | | |

SECTION 15.8

COMPUTER PROGRAMS FOR A CONTINGENCY TABLE ANALYSIS (OPTIONAL)

In this section we give the computer programs that produced the contingency table analysis printouts shown in Figure 15.4 for the automobile size–manufacturer data. For convenience, the data are reproduced in Table 15.11.

TABLE 15.11
Contingency Table for
Automobile Size Example

	MANUFACTURER				TOTALS
	A	B	C	D	
Small	157	65	181	10	413
Medium	126	82	142	46	396
Large	58	45	60	28	191
TOTALS	341	192	383	84	1,000

The SAS commands that produced the printout shown in Figure 15.4a are given in Program 15.1. Command lines 1–15 create a SAS data set named AUTO, which contains three variables. Two of the variables—size of the automobile sold (SIZE) and manufacturer (MANUFACT)—are qualitative. Note that both qualitative variables are followed by a dollar sign ($) in the INPUT statement. The third variable (NUMBER) is quantitative and represents the number of newly purchased automobiles for each size–manufacturer combination.

PROGRAM 15.1

SAS Commands for Analyzing the Contingency Table Shown in Table 15.11

```
Command
line
  1   DATA AUTO;                                        ⎫
  2   INPUT SIZE $ MANUFACT $ NUMBER;                   ⎬ Data entry instructions
  3   CARDS;                                            ⎭
  4   SMALL A 157    ⎫
  5   SMALL B 65     ⎪
  6   SMALL C 181    ⎪
  7   SMALL D 10     ⎪
  8   MEDIUM A 126   ⎪
  9   MEDIUM B 82    ⎬ Input data values: One
 10   MEDIUM C 142   ⎪ observation per line
 11   MEDIUM D 46    ⎪
 12   LARGE A 58     ⎪
 13   LARGE B 45     ⎪
 14   LARGE C 60     ⎪
 15   LARGE D 28     ⎭
 16   PROC FREQ;                                        ⎫ Statistical
 17   TABLES SIZE*MANUFACT/EXPECTED CHISQ;              ⎬ analysis
 18   WEIGHT NUMBER;                                    ⎭ instructions
```

The PROC FREQ statement (line 16) commands SAS to run its frequency (or contingency) tables procedure. The TABLES statement (line 17) defines the classification variables of the table. Variable names are separated by an asterisk (*). The values of the first variable, SIZE, will form the rows of the table and the values of the second variable, MANUFACT, form the columns.

The SAS printout shown in Figure 15.4a includes expected cell frequencies as well as the computed value of the χ^2 statistic for testing independence of classifications. The expected cell frequencies and χ^2 statistic are options that the programmer may call by including a slash (/) and the key words EXPECTED and CHISQ after the table request.*

The last SAS statement (line 18) of Program 15.1 defines the weighting variable of the contingency table. The weighting variable is necessary when the cell counts have already been tabulated, as is the case in Table 15.11. Since the variable NUMBER represents this cell count, NUMBER is entered after the key word

*When no options are included, SAS prints only cell frequencies and row and column percentages. Omit the slash if no options are specified.

WEIGHT. As a result, SAS produces a table showing the NUMBER of new car purchases in each of the SIZE–MANUFACT cells.

In most practical situations, you will not have the benefit of pretabulated cell counts. That is, if the data are *raw*, you will simply have a list of observations. In this example, for each of the 1,000 observations (new car purchases), you will have recorded the size and manufacturer of the car. You will then want to use the computer both to tabulate and to analyze the frequencies of each size–manufacturer combination. The SAS program statements required to accomplish this are nearly identical to those of Program 15.1, and are listed in Program 15.2. Note that when the data are raw (i.e., when the cell counts have not been tabulated), a WEIGHT command is not necessary and is omitted from the SAS program. Each of the 1,000 observations on new car purchase will contribute a value of 1 to the frequency counts, thus producing a contingency table identical to the one shown in Figure 15.4a.

PROGRAM 15.2

SAS Commands for Analyzing the Raw Data of Table 15.11 (Cell Counts Not Yet Tabulated)

```
Command
line
  1    DATA AUTO;                                   ⎫
  2    INPUT SIZE $ MANUFACT $ @@;                  ⎬ Data entry instructions
  3    CARDS;                                       ⎭
  4    SMALL A SMALL D LARGE B MEDIUM B             ⎫
  5    LARGE C SMALL B MEDIUM A MEDIUM A            ⎪ Input data values:
       .       .       .       .                   ⎬ One thousand observations
       .       .       .       .                   ⎪ (four observations per line)
       .       .       .       .                   ⎪
253    MEDIUM A SMALL A MEDIUM C LARGE B            ⎭
254    PROC FREQ;                                   ⎫ Statistical analysis
255    TABLES SIZE*MANUFACT/EXPECTED CHISQ;         ⎬ instructions
```

The SPSS[X], Minitab, and BMDP program statements that produced the contingency table printouts shown in Figures 15.4b–15.4d vary slightly from those required by SAS. Some of these packages do not allow nonnumeric data to be read onto a data set and, consequently, the programmer must define dummy numeric variables to represent a classification variable (e.g., 0 for small, 1 for medium, and 2 for large cars).

The commands necessary for analyzing the data in the contingency table (Table 15.11) for each of four computer packages—SAS, SPSS[X], Minitab, and BMDP— are given in Programs 15.3a–15.3d, respectively (pages 802–803). In the program statements, we are assuming that the cell counts have already been tabulated and that you want to calculate the chi-square statistic for testing independence of classifications. Program statements for creating and analyzing a contingency table when the data are raw are presented in Programs 15.4a–15.4d (pages 804–805).

PROGRAM 15.3 Program Statements for Analyzing a Contingency Table (Table 15.11)

PACKAGE	COMMAND LINE	PROGRAM STATEMENTS	DESCRIPTION
a. SAS	1	DATA AUTO;	
	2	INPUT SIZE $ MANUFACT $ NUMBER;	
	3	CARDS;	
	⋮	SMALL A 157 SMALL B 65 ⎤ Input data values LARGE D 28 ⎦	The values of the character variable SIZE are SMALL, MEDIUM, LARGE. The values of the character variable MANUFACT are A, B, C, D. NUMBER represents the freqency of each cell of the table. Twelve observations (cells) are read (one observation per line).
	4	PROC FREQ;	
	5	TABLES SIZE*MANUFACT/EXPECTED CHISQ;	The value of the chi-square statistic for testing independence of classifications is calculated.
	6	WEIGHT NUMBER;	
b. SSPS^X	1	DATA LIST FREE/NUMBER*SIZE (A5) MANUFACT (A1)	The values of the alphanumeric variable SIZE are SMALL, MEDIUM, LARGE.
	2	WEIGHT BY NUMBER	The values of the alphanumeric variable MANUFACT are A, B, C, D.
	3	CROSSTABS TABLES=SIZE BY MANUFACT	
	4	OPTIONS 14	
	5	STATISTICS 1	
	6	BEGIN DATA	NUMBER represents the frequency of each cell of the table.
		157 SMALL A ⎤ 65 SMALL B ⎥ ⋮ ⎥ Input data values 28 LARGE D ⎦	Twelve observations (cells) are read (one observation per line).
	7	END DATA	The value of the chi-square statistic for testing independence of classifications is calculated.

PROGRAM 15.3 (Continued)

c. Minitab

```
1   READ THE TABLE INTO C1, C2, C3, C4
    157 65 181 10  ⎤
    126 82 142 46  ⎥ Input data values
    58 45 60 28    ⎦
2   CHISQUARE ANALYSIS ON TABLE IN C1, C2, C3, C4
```

Contingency table is read into columns C1–C4 of the worksheet.

Columns represent values of MANUFACTURER (A,B,C,D).

Rows represent values of SIZE (SMALL, MEDIUM, LARGE).

The value of the chi-square statistic for testing independence of classifications is calculated.

d. BMDP

```
1   BMDP 4F
2   / PROGRAM TITLE IS 'AUTO EXAMPLE'.
3   / INPUT     VARIABLES=3.
4              FORMAT IS FREE.
5   / VARIABLE  NAMES ARE SIZE,MANUFACT,NUMBER.
6   / CATEGORY  CODES(1) ARE 0,1,2.
7              NAMES(1) ARE SMALL,MEDIUM,LARGE.
8              CODES(2) ARE 1,2,3,4.
9              NAMES(2) ARE A,B,C,D.
10  / TABLE    COLUMNS ARE MANUFACT.
11             ROWS ARE SIZE.
12             COUNT IS NUMBER.
13  / STATISTICS CHISQUARE.
14  / END
    0 1 157  ⎤
    0 2 65   ⎥ Input data values
    . . .    ⎥
    2 4 28   ⎦
15  /END
```

The values of the variable SIZE are coded 0 = SMALL, 1 = MEDIUM, 2 = LARGE.

The values of the variable MANUFACT are coded 1 = A, 2 = B, 3 = C, 4 = D.

NUMBER represents the frequency of each cell.

Twelve observations (cells) are read (one observation per line).

The value of the chi-square statistic for testing independence of classifications is calculated.

PROGRAM 15.4 Program Statements for Creating and Analyzing a Contingency Table (Table 15.11) from the Raw Data

PACKAGE	COMMAND LINE	PROGRAM STATEMENTS	DESCRIPTION
a. SAS	1	`DATA AUTO;`	
	2	`INPUT SIZE $ MANUFACT $ @@;`	The values of the character variable SIZE are SMALL, MEDIUM, LARGE.
	3	`CARDS;`	The values of the character variable MANUFACT are A, B, C, D.
		`SMALL A SMALL D LARGE B MEDIUM B` `LARGE C SMALL B MEDIUM A MEDIUM A` ⎤ Input data values	One thousand observations are read (four observations per line).
		`. . .` `MEDIUM A SMALL A MEDIUM C LARGE B` ⎦	
	4	`PROC FREQ;`	A frequency table is produced.
	5	`TABLES SIZE*MANUFACT/EXPECTED CHISQ;`	The value of the chi-square statistic for testing independence of classifications is calculated.
b. SSPS^X	1	`DATA LIST FREE/SIZE (A5) MANUFACT (A1)`	The values of the alphanumeric variable SIZE are SMALL, MEDIUM, LARGE.
	2	`CROSSTABS TABLES=SIZE BY MANUFACT`	The values of the alphanumeric variable MANUFACT are A, B, C, D.
	3	`OPTIONS 14`	
	4	`STATISTICS 1`	
	5	`BEGIN DATA`	
		`SMALL A SMALL D LARGE B MEDIUM B` `LARGE C SMALL B MEDIUM A MEDIUM A` ⎤ Input data values	One thousand observations are read (four observations per line).
		`. . .` `MEDIUM A SMALL A MEDIUM C LARGE B` ⎦	A contingency table is produced.
	6	`END DATA`	The value of the chi-square statistic for testing independence of classifications is calculated.

PROGRAM 15.4 (Continued)

c. Minitab

```
1    DIMENSION WORKSHEET TO 100 ROWS
2    SET SIZE CODE IN C1
       0 0 2 1 2 0 1 1 0 2
       1 0 0 0 2 1 2 2 0 1          Input data values
       . . .
3    SET MANUFACTURER CODE IN C2
       0 2 1 1 1 0 1 0 1 2
       1 4 2 2 3 2 1 1 4 2          Input data values
       2 1 4 4 1 3 2 1 2 3
       . . .
4    NAME C1 'SIZE', C2 'MANUFACT'
       4 1 3 1 1 1 1 3 2
5    CONTINGENCY TABLE ANALYSIS ON DATA IN C1 C2
```

One thousand coded values of SIZE are read in C1, where 0 = SMALL, 1 = MEDIUM, 2 = LARGE.

One thousand coded values of MANUFACT are read in C2, where 1 = A, 2 = B, 3 = C, 4 = D.

The rows of C1 and C2 correspond to the 1,000 observations. For example, the first element in the first row of C1 is 0 and the first element in the first row of C2 is 1; thus, the first observation represents a small car purchase from manufacturer A. Similarly, the second observation represents a small car purchase (0 is the second element in the first row of C1) from manufacturer D (4 is the second element in the first row of C2).

A contingency table is produced.

The value of the chi-square statistic for testing independence of classifications is calculated.

d. BMDP

```
1    BMDP 4F
2    / PROBLEM      TITLE IS 'AUTO EXAMPLE'.
3    / INPUT        VARIABLES=2.
4                   FORMAT IS SLASH.
5    / VARIABLE     NAMES ARE SIZE, MANUFACT.
6    / CATEGORY     CODES(1) ARE 0, 1, 2.
7                   NAMES(1) ARE SMALL, MEDIUM, LARGE.
8                   CODES(2) ARE 1, 2, 3, 4.
9                   NAMES(2) ARE A, B, C, D.
10   / TABLE        COLUMNS ARE MANUFACT.
11                  ROWS ARE SIZE.
12   / STATISTICS   CHISQUARE.
13   / END
       0 1/0 4/2 2/1 2/2 3/0 2/1 1/1 1/1 3/0 2/
       . . .                                        Input data values
       2 4/0 1/1 1/1 3/2 2/2 4/1 1/0 1/1 3/2 2
14   /END
```

The values of the variable SIZE are coded 0 = SMALL, 1 = MEDIUM, 2 = LARGE.

The values of the variable MANUFACT are coded 1 = A, 2 = B, 3 = C, 4 = D.

Observations on the data cards must be separated by a slash (/) when FORMAT IS SLASH (line 4) is used.

A contingency table is produced.

The value of the chi-square statistic for testing independence of classifications is calculated.

SUMMARY

The use of **count data** to test hypotheses about **multinomial probabilities** represents a very useful statistical technique. In a **one-dimensional table** we can use count data to test the hypothesis that the multinomial probabilities are equal to specified values. In the **two-dimensional contingency table**, we can test the independence of the two classifications. And these by no means exhaust the uses of the X^2 statistic. Many other applications can be found in the references at the end of this chapter.

Caution should be exercised to avoid misuse of the χ^2 procedure. The experiment must be multinomial.* Expected cell counts should be larger than or equal to 5 in order for the chi-square distribution to provide an adequate approximation to the sampling distribution of X^2.

If the X^2 value does not exceed the tabled critical value of χ^2, *do not accept* the hypothesis of independence. You would be risking a Type II error (accepting H_0 if it is false), and the probability β of committing such an error is unknown. The usual alternative hypothesis is that the classifications are dependent. Because there is literally an infinite number of ways two classifications can be dependent, it is difficult to calculate one or even several values of β to represent such a broad research hypothesis. Therefore, we avoid concluding that two classifications are independent, even when X^2 is small.

Finally, if a contingency table X^2 value *does* exceed the critical value, we must be careful to avoid inferring that a causal relationship exists between the classifications. Our alternative hypothesis states that the two classifications are statistically dependent, and **statistical dependence does not imply causality**. Therefore, the existence of a causal relationship cannot be established by a contingency table analysis.

15.29 A random sample of 250 observations was classified according to the row and column categories shown in the accompanying table.

		COLUMN		
		1	2	3
	1	20	20	10
ROW	2	10	20	70
	3	20	50	30

a. Do the data provide sufficient evidence to conclude that the row and column classifications are dependent? Test using $\alpha = .05$.

b. Would the analysis change if the row totals were fixed before the data were collected?

*When the row (or column) totals are fixed, each row (or column) represents a separate multinomial experiment.

c. Do the assumptions required for the analysis to be valid differ depending on whether the row (or column) totals are fixed? Explain.

15.30 A random sample of 150 observations was classified into the categories shown below:

	CATEGORY				
	1	2	3	4	5
n_i	28	35	33	25	29

a. Do the data provide sufficient evidence to indicate that the categories are not equally likely? Use $\alpha = .10$.
b. Form a 90% confidence interval for p_2, the probability that an observation will fall in category 2.

15.31 A computer used by a 24-hour banking service is supposed to assign each transaction to one of five memory locations at random. A check at the end of a day's transactions gave the following counts to each of the five memory locations:

Memory Location	1	2	3	4	5
Number of Transactions	90	78	100	72	85

Is there evidence to indicate a difference among the proportions of transactions assigned to the five memory locations? Test using $\alpha = .025$.

15.32 Does the propensity for worker injuries depend on the length of time that a worker has been on the job? An analysis of 714 worker injuries by one manufacturer gave the results shown in the table for the distribution of injuries over the eight 1-hour time periods per shift.

Hour of Shift	1	2	3	4	5	6	7	8
Number of Accidents	93	71	79	72	98	89	102	110

a. Do the data imply that the probabilities of worker accidents are higher in some time periods than in others? Test using $\alpha = .10$.
b. Do the data provide sufficient evidence to indicate that the probability of an accident during the last 4 hours of a shift is greater than during the first 4 hours? Test using $\alpha = .10$. [*Hint:* Test $H_0: p_1 = .5$, where p_1 is the probability of an accident during the last 4 hours.]

15.33 One of the major problems confronting urban planners is crime prevention. How should the buildings, sites, and neighborhoods in an urban (or urban renewal) area be physically designed in order to minimize crime levels? Over the past two decades, two planning theories have emerged—the *defensible space* approach and the *opportunity* approach. The defensible space theory suggests that land use characteristics and the design of streets, buildings, and building sites affect crime through informal social (i.e., nonpolice) control. Urban areas are designed to maximize the use of public areas, both day and night. The more people that use the streets, the more "eyes" there are for informal surveillance, which, in turn, discourages criminal activity. The opportunity theory, on the other hand, suggests that proximity to criminal opportunities is the most important determinant of crime. The physical characteristics of the urban area are designed so that the degree of access, ease of entrance and exit by potential offenders, and the supply of potential targets are minimized.

To assess the validity of the two different planning perspectives, Greenberg and Rohe (*Journal of the American Planning Association*, Winter 1984) examined differences in physical characteristics and various dimensions of informal social control in six neighborhoods in Atlanta, Georgia. The neighborhoods were grouped into three pairs—white middle-income pair, black lower-middle income pair, and black lower-income pair—where one member of each pair had a relatively high crime rate, and the other had a relatively low crime rate. The number of census blocks in each neighborhood pair were then classified according to crime rate (high or low) and several physical characteristics. The classification for street front characteristics is shown in the table. [*Note:* Greenberg and Rohe reported the results in terms of percentages. For the reader's convenience, we have converted all percentages to cell counts.]

STREET CHARACTERISTICS	WHITE MIDDLE-INCOME PAIR		BLACK LOWER-MIDDLE INCOME PAIR		BLACK LOWER-INCOME PAIR	
	High	Low	High	Low	High	Low
Major Thoroughfare	20	9	25	1	22	30
Small Neighborhood Street	7	9	25	27	8	42
Other	21	15	36	14	3	23
TOTALS	48	33	86	42	33	95

Source: Greenberg, S. W. and Rohe, W. M. "Neighborhood Design and Crime: A Test of Two Perspectives." *Journal of the American Planning Association*, Vol. 50, No. 1, 1984, pp. 48–60.

a. For each neighborhood pair, conduct a chi-square test for independence of classifications. Use $\alpha = .05$ and interpret your results.

b. For each neighborhood pair, construct a 95% confidence interval for the difference between the percentage of high-crime and low-crime neighborhood blocks that front a major thoroughfare.

15.34 A recent study showed that there is a basic resistance on the part of managers to use information produced by a computer. At one firm, over 40% of all computer-generated reports provided to managerial personnel were not used in any manner. A breakdown of the actual number of reports received and not used by each of four groups of employees is provided in the table. Is there sufficient evidence to conclude that the percentages of computer-generated reports not used differ among the four receiver groups? Test at $\alpha = .05$.

RECEIVER GROUP	COPIES OF REPORTS GENERATED	COPIES NOT USED
Directors and general managers	67	29
Middle managers	42	20
Supervisors	9	0
Clerks	180	42

Source: Edwards, C. "Encouraging Usage of Computer-Produced Management Information." *The Journal of the Operational Research Society*, Vol. 34, No. 3, Mar. 1983, p. 201. Copyright 1983, Pergamon Press, Ltd. Reprinted with permission.

15.35 Refer to Exercise 15.34. Models were devised to encourage the usage of computer-generated reports by management. A flexible model, in which either the receiver or the producer can undertake the task of encouraging use, was tested at the firm. Forty-one receivers and 41 producers accepted the responsibility of encouraging the use of computer-generated reports. The number of times each group satisfied the conditions specified in the model is shown in the accompanying table. Test the hypothesis that the proportion of times the model conditions are satisfied is identical for the two groups. Use $\alpha = .01$.

	RECEIVERS	PRODUCERS
Number of times model conditions satisfied	20	0
Number of times model conditions not satisfied	21	41

Source: Edwards, C. "Encouraging Usage of Computer-Produced Management Information." *The Journal of the Operational Research Society*, Vol. 34, No. 3, Mar. 1983, p. 201. Copyright 1983, Pergamon Press, Ltd. Reprinted with permission.

15.36 Along with the technological age comes the problem of workers being replaced by machines. A labor management organization wants to study the problem of workers displaced by automation within three industries. Case reports for 100 workers whose loss of job is directly attributable to technological advances are selected within each industry. For each worker selected, it is determined whether he or she was given another job within the same company, found a job with another company in the same industry, found a job in a new industry, or has been unemployed for longer than 6 months. The results are given in the accompanying table.

		SAME COMPANY	NEW COMPANY (SAME INDUSTRY)	NEW INDUSTRY	UNEMPLOYED
	A	62	11	20	7
INDUSTRY	B	45	8	38	9
	C	68	19	8	5

a. Does the plight of workers displaced by automation depend on the industry? Test using $\alpha = .01$.

b. Estimate the difference between the proportions of displaced workers who find work in another industry for industries A and C. Use a 95% confidence interval.

15.37 Video engineers have developed a new method of shortening the time required for broadcasting a television commercial. This technique, called *time compression*, has enabled television advertisers to cut the high costs of television commercials. But can shorter commercials be effective? In order to answer this question, 200 college students were randomly divided into three groups. The first group (57 students) was shown a videotape of a television program which included a 30-second commercial; the second group (74 students) was shown the same videotape but with the 24-second time-compressed version of the commerical; and the third group (69 students) was shown a 20-second time-compressed version of the commercial. Two days after viewing the tape, the three groups of students were asked to name the brand that was advertised. The numbers of students recalling the brand name for each of the three groups are given in the table.

		TYPE OF COMMERICAL			TOTALS
		Normal version (30 seconds)	Time-compressed version 1 (24-seconds)	Time-compressed version 2 (20 seconds)	
RECALL OF	Yes	15	32	10	57
BRAND NAME	No	42	42	59	143
TOTALS		57	74	69	200

a. Do the data provide sufficient evidence (at $\alpha = .05$) that the two directions of classification, type of commercial and recall of brand name, are dependent? Interpret your results.

b. Construct a 95% confidence interval for the difference between the proportions recalling brand name for viewers of normal and 24-second time-compressed commercials.

15.38 A survey conducted by the *Wall Street Journal* (Jan. 11, 1983) provides a comparison of the expectations of corporate managers for firms of different sizes. Random samples of 310 top executives at large companies, 305 from medium-sized companies, and 207 from small companies were asked, "Compared with 1982, what is your company's profit outlook for 1983?" The percentages of those executives expecting a much brighter 1983 are shown in the table for the three samples. Do the data provide sufficient evidence to indicate that the percentages expecting a much brighter 1983 depend on company size? Test using $\alpha = .10$.

		SIZE OF COMPANY		
		Large	Medium	Small
EXPECTATION OF	Much Brighter	17%	20%	8%
CORPORATE HEAD	Sample Size	310	305	207

15.39 A decision support system (DSS) is a computerized system designed to aid in the management and analysis of large data sets. Ideally, a DSS should include four components: (1) a data extraction system, (2) a relational data base organization, (3) analysis models, and (4) a user-friendly interactive dialogue between the user and the system. A state highway agency recently installed a DSS to help monitor data on road construction contract bids. As part of a self-examination, the agency selected 151 of the most recently encountered problems that could be traced directly to the DSS and classified each according to the component of origination. Can it be concluded from the data in the table that the proportions of problems are different for at least two of the four DSS components? Test using $\alpha = .05$.

Component	1	2	3	4
Number of Problems	31	28	45	47

15.40 An experiment was conducted to compare the fidelity and selectivity of radio receivers. One-hundred fifty receivers were tested and classified as low, medium, or high in each of the two categories. Do the data in the table provide sufficient evidence to indicate a dependence between fidelity and selectivity? Test using $\alpha = .025$.

		SELECTIVITY		
		Low	Medium	High
	Low	10	11	6
FIDELITY	Medium	30	52	19
	High	12	8	2

15.41 Vira-A is one of the few existing antiviral drugs on the market. Manufactured by Warner-Lambert Co., it is used primarily to treat herpes virus infections of the eye and brain in adults. However, the *Wall Street Journal* (Oct. 8, 1980) reports that Warner-Lambert will seek approval from the Food and Drug Administration (FDA) to use the drug in treating a rare, but usually fatal, infant illness.

Warner-Lambert claims that Vira-A is useful in treating babies suffering from herpes simplex virus infection. The illness, often transmitted from an infected mother, can leave surviving infants with permanent brain damage. In one portion of a study conducted in 18 health centers throughout the country, the drug was given to 24 babies suffering from the disease. Included in the study was a control group of 19 infected babies who were left untreated. The number of infants surviving the illness in each group is given in the accompanying table.

	SURVIVORS	DEATHS	TOTALS
Treated group (drug)	15	9	24
Control group (no drug)	5	14	19
TOTALS	20	23	43

a. Conduct a test to determine if the distributions of the percentages of survivors and deaths are different for infected babies treated with the drug and those left untreated.

b. You can perform the identical analysis of part **a** by comparing the binomial proportions p_1 and p_2 using the techniques of Chapter 9, where p_1 is the true survival rate of infected babies treated with the drug, and p_2 is the true survival rate of infected babies who are left untreated. Test the hypothesis that p_1 is larger than p_2, using a significance level of $\alpha = .05$.

c. Based *only* on the results of the survivor-rate study shown here, would you back Warner-Lambert's claim that the drug Vira-A is useful in treating babies suffering from herpes simplex virus infection?

15.42 A machine shop wants to compare the quality of work of its four machinists. A sample of 300 machined parts is selected from a 1-week output; the parts are inspected to determine whether they are defective and are categorized according to which machinist did the work. The results are given in the table. Is there evidence that the quality of work differs among the four machinists? Test using $\alpha = .01$.

	MACHINIST			
	A	B	C	D
Defective	18	7	12	6
Nondefective	62	81	58	56

15.43 The use of high-level computer programming languages (for example, Fortran, Cobol, Algol, and Pascal) with microprocessors and minicomputers has increased dramatically over the past few years. This has increased the need for new and better methods of performance evaluation. In one study, a researcher developed a measurement system for evaluating two high-level programming languages, Algol and Pascal. The reported results include a distribution of the relative frequency of occurrence of the different types of statements used in typical Algol and Pascal programs of approximately the same size. The reported percentages were used to tabulate the information given in the table.

		ALGOL	PASCAL
	IF	125	2,045
	FOR	968	350
TYPE OF	IO	135	1,847
STATEMENT	Assignment	8,923	4,763
	Other	261	465
TOTALS		10,412	9,470

Source: Adapted from De Prycker, M. "On the Development of a Measurement System for High-Level Language Program Statistics." *IEEE Transactions on Computers*, Vol. C-31, No. 9, Sept. 1982, pp. 888–890.

a. Assuming fixed marginals for the two programming languages, conduct a test to determine if the percentages of the different types of programming statements differ for the two languages. Test using $\alpha = .05$.

b. Construct a 95% confidence interval for the difference in the percentages of assignment statements used in the two languages.

15.44 A survey was conducted to determine whether a relationship exists between a new college graduate's expectations of acquiring rewarding employment and the graduate's college major. Three hundred graduates were randomly selected from among prospective graduates in the social, biological, and physical sciences, and from the arts and humanities. A summary of their responses is shown in the accompanying table. Do the data provide sufficient evidence to indicate differences in the patterns of response for the four types of majors? Test using $\alpha = .05$.

	COLLEGE MAJOR				TOTALS
	Social Sciences	Biological Sciences	Physical Sciences	Arts and Humanities	
High expectations for employment	12	27	43	16	98
Modest expectations for employment	36	45	38	27	146
Poor expectations for employment or no opinion	14	6	3	33	56
TOTALS	62	78	84	76	300

REFERENCES

Agresti, A. and Finlay, B. *Statistical Methods for the Social Sciences*, 2nd ed. San Francisco: Dellen Publishing Company, 1986.

BMDP User's Digest, 2nd ed. MaryAnn Hill, ed. Los Angeles: BMDP Statistical Software, 1982.

BMDPC: User's Guide to BMDP on the IBM PC. BMDP Statistical Software, Inc. Los Angeles, Calif. 90025.

Bowker, A. and Lieberman, G. *Engineering Statistics*, 2nd ed. Englewood Cliffs, N. J.: Prentice-Hall, 1972.

Cochran, W. G. "The χ^2 Test of Goodness of Fit," *Annals of Mathematical Statistics*, 23, 1952.

Cochran, W. G. "Some Methods for Strengthening the Common χ^2 Tests," *Biometrics*, Vol. 10.

Davies, O. L. *The Design and Analysis of Industrial Experiments*, 2nd ed. New York: Hafner Publishing Co., 1956.

Devore, J. L. *Probability and Statistics for Engineering and the Sciences*. Monterey, Calif.: Brooks/Cole Publishing, 1982.

Dixon, W. J., Brown, M. B., Engelman, L., Frane, J. W., Hill, M. A., Jennrich, R. I., and Toporek, J. D. *BMDP Statistical Software Manual*, 1985 ed. Berkeley: University of California Press.

Johnson, N. and Leone, F. *Statistics and Experimental Design in Engineering and the Physical Sciences*, Vol. II, 2nd ed. New York: John Wiley, 1977.

Norusis, M. J. *The SPSS Guide to Data Analysis*, 1986 ed. SPSS, Inc., Suite 3000, 444 N. Michigan Avenue, Chicago, Ill. 60611.

Norusis, M. J. *SPSS/PC+: SPSS for the IBM PC/XT/AT*, 1986 ed. SPSS, Inc., Suite 3000, 444 N. Michigan Avenue, Chicago, Ill. 60611.

Ryan, T. A., Joiner, B. L., and Ryan, B. F. *Minitab Reference Manual*. Minitab Project, University Park, Pa., 1985.

Ryan, T. A., Joiner, B. L., and Ryan, B. F. *Minitab Student Handbook*, 2nd ed. Boston: Duxbury, 1985.

SAS Procedures Guide for Personal Computers, Version 6 ed., 1986. SAS Institute, Inc., Box 8000, Cary, N.C. 27511.

SAS Statistics Guide for Personal Computers, Version 6 ed., 1986. SAS Institute, Inc., Box 8000, Cary, N.C. 27511.

SAS User's Guide: Statistics, Version 5 ed., 1985. A. A. Ray, ed. SAS Institute, Inc., Box 8000, Cary, N.C. 27511.

Savage, I. R. "Bibliography of Nonparametric Statistics and Related Topics," *Journal of the American Statistical Association*, 48, 1953, pp. 844–906.

Siegel, S. *Nonparametric Statistics for the Behavioral Sciences*. New York: McGraw-Hill, 1956.

SPSS^X User's Guide, 1983 ed. SPSS, Inc., Suite 3000, 444 N. Michigan Ave., Chicago, Ill. 60611.

Walpole, R. and Myers, R. *Probability and Statistics for Engineers and Scientists*, 3rd ed. New York: Macmillan, 1985.

C H A P T E R 16

OBJECTIVE

To present some statistical tests that do not require the normality assumptions necessary for the Student's t and F tests of Chapters 9, 10, and 13.

CONTENTS

NONPARAMETRIC STATISTICS

INTRODUCTION

The t and F tests for comparing two or more population means (discussed in Chapters 9 and 13) are unsuitable for some types of data that fall into one of two categories. The first are data sets that do not satisfy the assumptions upon which the t and F tests are based. For both tests, we assume that the random variables being measured have normal probability distributions with equal variances. Yet in practice, the observations from one population may exhibit much greater variability than those from another, or the probability distributions may be decidedly nonnormal. For example, the distribution might be very flat, peaked, or strongly skewed to the right or left. When any of the assumptions required for the t and F tests are seriously violated, the computed t and F statistics may not follow the standard t and F distributions. If this is true, the tabulated values of t and F given in Tables 6, 8, 9, 10, and 11 of Appendix II are not applicable, the correct value of α for the test is unknown, and the t and F tests are of dubious value.

The second type of data for which the t and F tests are inappropriate are responses that are not susceptible to measurement but that can be *ranked in order of magnitude*. For example, suppose we want to compare the ease of operation of two types of computer software based on subjective evaluations of trained observers. Although we cannot give an exact value to the variable Ease of operation of the software package, we may be able to decide that package A is better than package B. If packages A and B are evaluated by each of ten observers, we have the standard problem of comparing the probability distributions for two populations of ratings—one for package A and one for package B. But the t test of Chapter 9 would be inappropriate, because the only data that can be recorded are preferences; that is, each observer decides either that A is better than B or vice versa.

Consider another example of this type of data. Most firms that plan to purchase a new product first test the product to determine its acceptability. For a computer, an automobile, or some other piece of equipment, this may involve tests in which engineers rank the new product in order of preference with respect to one or more currently popular product types or makes. An engineer probably has a preference for each product, but the strength of the preference is difficult, if not impossible, to measure. Consequently, the best we can do is to have each engineer examine the new product along with a few established products and rank them according to preference: 1 for the product that is most preferred, 2 for the product with the second greatest preference, and so on.

The **nonparametric** counterparts of the t and F tests compare the relative locations of the probability distributions of the sampled populations, rather than specific parameters of these populations (such as the means or variances). If it can be inferred that the distribution of preferences for a new product lies above (to the right of) the others, as illustrated in Figure 16.1, the implication is that the engineers tend to prefer the new product to the other products.

Many nonparametric methods use the **relative ranks** of the sample observations rather than their actual numerical values. These tests are particularly valuable when we are unable to obtain numerical measurements of some phenomenon

FIGURE 16.1

Probability Distributions of Strengths of Preference Measurements (New Product Is Preferred)

Old product　　New product

Strength of preference measurements

but are able to rank the measurements relative to each other. Statistics based on the ranks of measurements are called **rank statistics**.

In Section 16.2, we discuss a simple nonparametric test for the location of a single population. In Sections 16.3 and 16.5, we will present rank statistics for comparing two probability distributions using independent random samples. In Sections 16.4 and 16.6, we will use the matched-pairs and randomized block designs to make nonparametric comparisons of populations. Finally, in Section 16.7, we present a nonparametric measure of correlation between two variables— **Spearman's rank correlation coefficient**. For a more complete discussion of tests based on rank statistics, see Lehmann (1975).

| | | | | | | | | | | | | |

S E C T I O N 16.2

THE SIGN TEST FOR A SINGLE POPULATION

Recall from Section 9.5 that small-sample procedures for testing a hypothesis about a population mean require that the population have an approximately normal distribution. For situations in which we collect a small sample ($n < 30$) from a nonnormal population, the t test is not valid and we must resort to a nonparametric procedure. The simplest nonparametric technique to apply in this situation is the **sign test**. The sign test is specifically designed for testing hypotheses about the median of any continuous population. Like the mean, the median is a measure of the center, or location, of the distribution; consequently, the sign test is sometimes referred to as a **test for location**.

Let $y_1, y_2, \ldots, y_n$ be a random sample from a population with unknown median M. Suppose we want to test the null hypothesis $H_0: M = 100$ against the one-sided alternative $H_a: M > 100$. From Definition 1.8 we know that the median is a number such that half the area under the probability distribution lies to the left of M and half lies to the right (see Figure 16.2). Therefore, the probability

FIGURE 16.2

Location of the Population Median, M

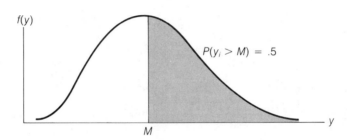

$f(y)$

$P(y_i > M) = .5$

M　　　　　　y

that a y-value selected from the population is larger than M is .5, i.e., $P(y_i > M) = .5$. If, in fact, the null hypothesis is true, then we should expect to observe approximately half the sample y-values greater than $M = 100$.

The sign test utilizes the test statistic, S, where

S = Number of y_i's that exceed 100

Notice that S depends only on the *sign* (positive or negative) of the difference between each sample value y_i and 100. That is, we are simply counting the number of positive $(+)$ signs among the sample differences $(y_i - 100)$. If S is "too large" (i.e., if we observe an unusually large number of y_i's exceeding 100), then we will reject H_0 in favor of the alternative H_a: $M > 100$.

The rejection region for the sign test is derived as follows. Let each sample difference $(y_i - 100)$ denote the outcome of a single trial in an experiment consisting of n identical trials. If we call a positive difference a "success" and a negative difference a "failure," then S is the number of successes in n trials. Under H_0, the probability of observing a success on any one trial is

$$p = P(\text{Success}) = P(y_i - 100 > 0) = P(y_i > 100) = .5$$

Since the trials are independent, the properties of a binomial experiment, listed in Chapter 4, are satisfied. Therefore, S has a binomial distribution with parameters n and $p = .5$. We can use this fact to calculate the observed significance level (p-value) of the sign test, as illustrated in the following example.

EXAMPLE 16.1

Bacteria are a most important component of microbial ecosystems in sewage treatment plants. Water management engineers have determined that the percentages of active bacteria in sewage specimens collected at a particular plant have a distribution with a median of 40 percent. If the median percentage is larger than 40, then adjustments in the sewage treatment process must be made. The percentages of active bacteria in a random sample of ten sewage specimens are given below:

41   33   43   52   46   37   44   49   53   30

Do the data provide sufficient evidence to indicate that the median percentage of active bacteria in sewage specimens is greater than 40? Test using $\alpha = .05$.

SOLUTION

We want to test

H_0: $M = 40$

H_a: $M > 40$

using the sign test. The test statistic is

S = Number of y_i's in the sample that exceed 40

   $= 7$

where S has a binomial distribution with parameters $n = 10$ and $p = .5$.

From Definition 9.4, the observed significance level (p-value) of the test is the probability that we observe a value of the test statistic S that is at least as contradictory to the null hypothesis as the computed value. For this one-sided case, the p-value is the probability that we observe a value of S greater than or equal to 7. We find this probability using the binomial table for $n = 10$ and $p = .5$ in Table 2 of Appendix II:

$$p\text{-value} = P(S \geq 7) = 1 - P(S \leq 6) = 1 - .828 = .172$$

Since the p-value, .172, is larger than $\alpha = .05$, we cannot reject the null hypothesis. That is, there is insufficient evidence to indicate that the median percentage of active bacteria in sewage specimens exceeds 40. ∎

A summary of the sign test for both one-sided and two-sided alternatives is provided in the box.

SIGN TEST FOR A POPULATION MEDIAN, M

ONE-TAILED TEST

H_0: $M = M_0$

H_a: $M > M_0$
    [or, H_a: $M < M_0$]

Test statistic:

S = Number of sample observations greater than M_0

[or, S = Number of sample observations less than M_0]

TWO-TAILED TEST

H_0: $M = M_0$

H_a: $M \neq M_0$

Test statistic:

S = larger of S_1 and S_2

where

S_1 = Number of sample observations greater than M_0

and

S_2 = Number of sample observations less than M_0

[*Note:* By definition, $S_2 = n - S_1$.]

Observed significance level:

p-value $= P(S \geq S_c)$

Observed significance level:

p-value $= 2P(S \geq S_c)$

where S_c is the computed value of the test statistic and S has a binomial distribution with parameters n and $p = .5$.

Rejection region: Reject H_0 if $\alpha > p$-value.

Assumption: The sample is randomly selected from a continuous probability distribution. [*Note:* No assumptions have to be made about the shape of the probability distribution.]

For a two-tailed test, you may calculate the test statistic as either

S_1 = Number of y_i's greater than M_0
  = Number of successes in n trials

or

S_2 = Number of y_i's less than M_0
  = Number of failures in n trials

Note that $S_1 + S_2 = n$; therefore, $S_2 = n - S_1$. In either case, the p-value of the test is double the corresponding one-sided p-value. To simplify matters, we suggest using the larger of S_1 and S_2 as the test statistic and calculating the p-value as shown in the box.

Recall from Section 7.5 that a normal distribution with mean $\mu = np$ and $\sigma = \sqrt{npq}$ can be used to approximate the binomial distribution for large n. When $p = .5$, the normal approximation performs reasonably well even for n as small as 10 (see Example 7.8). Thus, for $n \geq 10$, we can conduct the sign test using the familiar standard normal z statistic of Chapter 9. This large-sample sign test is summarized in the next box. Review Chapter 9 for examples of how to apply this test.

SIGN TEST BASED ON A LARGE SAMPLE ($n \geq 10$)

ONE-TAILED TEST	TWO-TAILED TEST
H_0: $M = M_0$	H_0: $M = M_0$
H_a: $M > M_0$	H_a: $M \neq M_0$
[or, H_a: $M < M_0$]	

Test statistic: Test statistic:

$$z = \frac{S - E(S)}{\sqrt{V(S)}} = \frac{S - .5n}{\sqrt{(.5)(.5)n}} = \frac{S - .5n}{.5\sqrt{n}}$$

[*Note:*   The value of S is calculated as shown in the previous box.]

Rejection region:	Rejection region:
$z > z_\alpha$	$z < -z_{\alpha/2}$ or $z > z_{\alpha/2}$
[or, $z < -z_\alpha$]	

where tabulated values of z_α and $z_{\alpha/2}$ are given in Table 4 of Appendix II.

| | | | | | | | | | | | | |

EXERCISES 16.1–16.7

16.1 Suppose you want to use the sign test to test the null hypothesis that the population median equals 75, i.e., H_0: $M = 75$. Use Table 2 of Appendix II to find the observed significance level (p-value) of the test for each of the following situations:

a. H_a: $M > 75$, $n = 5$, $S = 2$
b. H_a: $M \neq 75$, $n = 20$, $S = 16$
c. H_a: $M < 75$, $n = 10$, $S = 8$

16.2 A random sample of six observations from a continuous population resulted in the following:

18.2   21.3   20.5   19.4   19.6   17.7

Is there sufficient evidence to indicate that the population median differs from 20? Test using $\alpha = .05$. [*Hint:* Use the binomial formula for $p(x)$ given in Section 4.5 to calculate the p-value of the test.]

16.3 The building specifications in a certain city require that the sewer pipe used in residential areas have a median breaking strength of more than 2,500 pounds per lineal foot. A manufacturer who would like to supply the city with sewer pipe has submitted a bid and provided the following additional information. An independent contractor randomly selected seven sections of the manufacturer's pipe and tested each for breaking strength. The results (pounds per lineal foot) are shown below:

2,610   2,750   2,420   2,510   2,540   2,490   2,680

Is there sufficient evidence to conclude that the manufacturer's sewer pipe meets the required specifications? Use a significance level of $\alpha = .10$.

16.4 *Scram* is the term used by nuclear engineers to describe a rapid emergency shutdown of a nuclear reactor. The nuclear industry has made a concerted effort to reduce significantly the number of unplanned scrams each year. The number of unplanned scrams at each of a random sample of 20 nuclear reactor units in 1984 are given below (*Transactions of the American Nuclear Society*, Vol. 50, 1985):

1	8	0	3	3	9	1	2	4	5
4	3	1	2	7	10	2	6	3	0

Test the hypothesis that the median number of unplanned scrams at nuclear reactor plants in 1984 is less than 5. Use $\alpha = .10$.

16.5 An important property of certain products that are in powder or granular form is their particle size distribution. For example, refractory cements are adversely affected by too high a proportion of coarse granules, which can lead to weakness due to poor packing. The data below represent the percentages of coarse granules for a random sample of eight refractory cement specimens from a large lot.

1.7   .9   3.4   2.5   3.1   .6   1.0   2.1

Is there sufficient evidence to indicate that fewer than half of the cement specimens in the lot have more than 2% coarse granules? Test using $\alpha = .05$.

OPTIONAL EXERCISES

16.6 Suppose we want to test $H_0: M = M_0$ against $H_a: M \neq M_0$ using the sign test, where

S_1 = Number of sample observations greater than M_0

and

S_2 = Number of sample observations less than M_0

Show that $P(S_1 \geq c) = P(S_2 \leq n - c)$, where $0 \leq c \leq n$.

16.7 Refer to the two-tailed test of Exercise 16.6. Use the result of that exercise to show that the observed significance level for the test is

$$p\text{-value} = 2P(S_1 \geq c)$$

S E C T I O N   16.3

COMPARING TWO POPULATIONS: THE WILCOXON RANK SUM TEST FOR INDEPENDENT RANDOM SAMPLES

Suppose two independent random samples are to be used to compare two populations and the t test of Chapter 9 is inappropriate for making the comparison. Either we are unwilling to make assumptions about the form of the underlying probability distributions, or we are unable to obtain exact values of the sample measurements. For either of these situations, if the data can be ordered, the **Wilcoxon rank sum test** can be used to test the null hypothesis that the probability distributions associated with the two populations are equivalent against the alternative hypothesis that one population probability distribution is shifted to the right (or left) of the other.

For example, suppose that an experiment is conducted to compare the ratings of a computer word-processing package by two groups of people—word-processing operators who must use the package and computer programming specialists who are trained to develop computer software. Independent random samples of $n_1 = 7$ word-processing operators and $n_2 = 7$ programming specialists were selected for the experiment. Each was asked to rate the package on a scale from 1 to 100, with 100 denoting the best rating. After the data were recorded, the fourteen ratings were ranked in order of magnitude, 1 for the smallest and 14 for the largest. Tied observations (if they occur) are assigned ranks equal to the average of the ranks of the tied observations. For example, if the second and third ranked observations were tied, each would be assigned the rank 2.5. The data for the experiment and their ranks are shown in Table 16.1.

TABLE 16.1

Ratings of a New Word-Processing Package by Word-Processing Operators and Programming Specialists

WORD-PROCESSING OPERATORS		PROGRAMMING SPECIALISTS	
Rating	Rank	Rating	Rank
35	5	45	7
50	8	60	10
25	3	40	6
55	9	90	13
10	1	65	11
30	4	85	12
20	2	95	14
$n_1 = 7$	$T_1 = 32$	$n_2 = 7$	$T_2 = 73$

WILCOXON RANK SUM TEST: INDEPENDENT RANDOM SAMPLES*

ONE-TAILED TEST

H_0: The sampled populations have identical probability distributions

H_a: The probability distribution for population 2 is shifted to the right of that for population 1

Test statistic: Either rank sum can be used as the test statistic. We will use T_2.

Rejection region: $T_2 \geq T_U$ where T_U is the upper value given by Table 15 of Appendix II, for the chosen *one-tailed* α value.

TWO-TAILED TEST

H_0: The sampled populations have identical probability distributions

H_a: The probability distribution for population 2 is shifted to the left *or* to the right of that for population 1

Test statistic: Either rank sum can be used. We will denote this rank sum as T.

Rejection region: $T \leq T_L$ or $T \geq T_U$ where T_L is the lower value given by Table 15 of Appendix II for the chosen *two-tailed* α value, and T_U is the upper value from Table 15.

[*Note:* If the one-sided alternative is that the probability distribution for population 2 is shifted to the *left* of the distribution for population 1, we reject H_0 if $T_2 \leq T_L$.]

Assumptions:
1. The two samples are random and independent.
2. The observations obtained can be ranked in order of magnitude. [*Note:* No assumptions have to be made about the shapes of the population probability distributions.]
3. Tied observations are assigned ranks equal to the average of the ranks of the tied observations. (For example, if the second and third largest observations are tied, we would assign each the rank 2.5.)

The Wilcoxon rank sum test is based on the sums of the ranks (called **rank sums**) for the two samples. The logic is that if the null hypothesis

H_0: The two population probability distributions are identical

is true, then any one ranking of the $n = n_1 + n_2$ observations is just as likely as any other. Then, for equal sample sizes, we would expect the rank sums, T_1 and T_2, to be nearly equal.

---

*Another statistic used for comparing two populations based on independent random samples is the *Mann–Whitney U statistic.* The U statistic is a simple function of the rank sums. It can be shown that the Wilcoxon rank sum test and the Mann–Whitney U test are equivalent.

In contrast, if the one-sided alternative hypothesis

H_a: Probability distribution for population 2 is shifted to the right of that for population 1

is true, then, for equal sample sizes, we would expect the rank sum T_2 to be larger than the rank sum T_1. In fact, it can be shown (proof omitted) that, regardless of the sample sizes n_1 and n_2,

$$T_1 + T_2 = \frac{n(n + 1)}{2}$$

where $n = n_1 + n_2$. Therefore, as T_1 becomes smaller, T_2 will become larger and we would reject H_0 and accept H_a for large values of T_2.

In Example 16.3, we will illustrate the procedure for finding the rejection region for a specified value of α. For the moment, we will use the rejection regions provided in Table 15 of Appendix II to compare the programmer and word-processor operator ratings of Table 16.1. A summary of the Wilcoxon rank sum test for independent random samples is shown in the preceding box.

EXAMPLE 16.2

Refer to Table 16.1 and test the null hypothesis that the probability distributions of the operator and programmer ratings are identical against the alternative hypothesis that one of the distributions is shifted to the right of the other. Test using $\alpha = .05$.

SOLUTION

We can use either rank sum as the test statistic for this two-tailed test and we will reject H_0 if that rank sum, say T_1, is very small or very large—that is, if $T_1 \le T_L$ or $T_1 \ge T_U$. The tabulated values of T_L and T_U, the lower- and upper-tail values of the rank sum distribution, are given in Table 15 of Appendix II. The critical values of the rank sum for a one-tailed test with $\alpha = .025$ and for a two-tailed test with $\alpha = .05$ are given in Table 15a, which is reproduced in Table 16.2. Table 15b of Appendix II gives the critical values, T_L and T_U, for a

TABLE 16.2   A Partial Reproduction of Table 15 of Appendix II

n_2 \ n_1	3 T_L	3 T_U	4 T_L	4 T_U	5 T_L	5 T_U	6 T_L	6 T_U	7 T_L	7 T_U	8 T_L	8 T_U	9 T_L	9 T_U
3	5	16	6	18	6	21	7	23	7	26	8	28	8	31
4	6	18	11	25	12	28	12	32	13	35	14	38	15	41
5	6	21	12	28	18	37	19	41	20	45	21	49	22	53
6	7	23	12	32	19	41	26	52	28	56	29	61	31	65
7	7	26	13	35	20	45	28	56	37	68	39	73	41	78
8	8	28	14	38	21	49	29	61	39	73	49	87	51	93
9	8	31	15	41	22	53	31	65	41	78	51	93	63	108
10	9	33	16	44	24	56	32	70	43	83	54	98	66	114

one-tailed test with $\alpha = .05$ and for a two-tailed test with $\alpha = .10$. Examining Table 16.2, you will find that the critical values (shaded) corresponding to $n_1 = n_2 = 7$ are $T_L = 37$ and $T_U = 68$. Therefore, for $\alpha = .05$, we will reject H_0 if

$$T_1 \leq 37 \quad \text{or} \quad T_1 \geq 68$$

Since the observed value of the test statistic, $T_1 = 32$ (calculated in Table 16.1), is less than 37, we reject the hypothesis that the distributions of ratings are identical. There is sufficient evidence to indicate that one of the distributions is shifted to the right of the other.   ∎

EXAMPLE 16.3

Suppose that the alternative hypothesis in Example 16.2 had implied a one-tailed test. For example, suppose that we wanted to test H_0 against the alternative

H_a: Distribution 2 is shifted to the right of distribution 1

Locate the rejection region for the test using $\alpha = .025$.

SOLUTION

We can use either T_1 or T_2 as the test statistic; small values of T_1 and large values of T_2 support the alternative hypothesis. If we use T_1 as the test statistic, we will reject H_0 if $T \leq T_L$ where T_L is the lower-tail value of the rank sum, given in Table 16.2, for $n_1 = n_2 = 7$. This value is 37. Therefore, the rejection region for the one-tailed test with $\alpha = .025$ is $T_1 \leq 37$.

If we choose T_2 as the test statistic, we would reject H_0 if T_2 is large, say $T_2 \geq T_U$. The value of T_U given in Table 16.2 for $n_1 = n_2 = 7$ is $T_U = 68$. The two tests are equivalent.   ∎

EXAMPLE 16.4

Consider a Wilcoxon rank sum test for $n_1 = n_2 = 4$. Find the value of T_L such that $P(T_1 \leq T_L) \approx .05$. This value of T_L would be appropriate for a one-tailed test with $\alpha = .05$.

SOLUTION

To solve this problem, we use the probability methods of Chapter 3. If H_0 is true—i.e., if the two population probability distributions are identical—then any one ranking of the $n_1 + n_2 = 8$ observations is as likely as any other and each would represent a simple event for the experiment. For example, suppose that the four observations associated with samples 1 and 2 are denoted as y_{11}, y_{12}, y_{13}, y_{14}, and y_{21}, y_{22}, y_{23}, y_{24}, respectively. One ranking of the data that will produce the smallest possible value for T_1 is shown in Table 16.3.

TABLE 16.3

One Ranking of the $n_1 + n_2 = 8$ Observations of Example 16.4

SAMPLE 1		SAMPLE 2	
Observation	Rank	Observation	Rank
y_{11}	4	y_{21}	6
y_{12}	1	y_{22}	5
y_{13}	3	y_{23}	7
y_{14}	2	y_{24}	8
	$T_1 = 10$		$T_2 = 26$

To find the value T_L such that $P(T_1 \le T_L) \approx .05$, we find, $P(T_1 = 10)$, $P(T_1 = 11), \ldots$, and sum these probabilities until

$$P(T_1 = 10) + P(T_1 = 11) + \cdots + P(T_1 = T_L) \approx .05$$

The number of simple events in the sample space S is equal to the number of ways that you can arrange the integers, $1, 2, \ldots, 8$—namely, $8!$. Since the simple events are equiprobable, the probability of each simple event E_i in the sample space is

$$P(E_i) = \frac{1}{8!}$$

The number of rankings that will result in $T_1 = 10$ is equal to the number of ways that you can arrange the four ranks for sample 1 and the four ranks for sample 2. The number of distinctly different arrangements of one sample of four ranks is $4!$. Therefore, the number of ways that the two samples, each containing four ranks, can be arranged is

$$(4!)(4!)$$

Therefore, there will be $(4!)(4!)$ simple events in the event $T_1 = 10$, each with probability $P(E_i) = 1/8!$. Then,

$$P(T_1 = 10) = \frac{4!4!}{8!} = \frac{1}{70} = .0143$$

Next, consider the rank sum $T_1 = 11$. The only way that T_1 can equal 11 is if the ranks assigned to sample 1 are 1, 2, 3, and 5. Then

$$P(T_1 = 11) = \frac{4!4!}{8!} = \frac{1}{70} = .0143$$

and

$$P(T_1 \le 11) = P(T_1 = 10) + P(T_1 = 11) = 2(.0143) = .0286$$

Since this value is less than $\alpha = .05$, we will calculate the probability of observing the next larger rank sum for T_1—namely, $T_1 = 12$. We can obtain a rank sum $T_1 = 12$ if either the ranks 1, 2, 3, and 6 or the ranks 1, 2, 4, and 5 are assigned to sample 1. The probability of each of these occurrences is $1/70$. Therefore,

$$P(T_1 = 12) = P\{1, 2, 3, 6\} + P\{1, 2, 4, 5\} = \frac{1}{70} + \frac{1}{70} = .0286$$

and

$$P(T_1 \le 12) = P(T_1 = 10) + P(T_1 = 11) + P(T_1 = 12) = .0572$$

Since we want T_L to be the value such that $P(T_1 \le T_L)$ is close to $\alpha = .05$, it follows that $T_L = 12$. This is the tabulated value for T_L given in Table 15b of Appendix II ($\alpha = .05$).   ∎

Like the sign test, the Wilcoxon rank sum test can be conducted using the familiar z test statistic of Section 9.4 when the samples are large. The following (which we state without proof) leads to a large-sample Wilcoxon rank sum test. It can be shown that the mean and variance of the rank sum T_2 are

$$E(T_2) = \frac{n_1 n_2 + n_2(n_2 + 1)}{2}$$

and

$$V(T_2) = \frac{n_1 n_2 (n_1 + n_2 + 1)}{12}$$

Then, when n_1 and n_2 are large (say, $n_1 > 10$ and $n_2 > 10$), the sampling distribution of

$$z = \frac{T_2 - E(T_2)}{\sqrt{V(T_2)}} = \frac{T_2 - \left[\dfrac{n_1 n_2 + n_2(n_2 + 1)}{2}\right]}{\sqrt{\dfrac{n_1 n_2 (n_1 + n_2 + 1)}{12}}}$$

will have, approximately, a standard normal distribution.

THE WILCOXON RANK SUM TEST FOR LARGE SAMPLES ($n_1 > 10$ AND $n_2 > 10$)

H_0: The probability distributions for populations 1 and 2 are identical

H_a: The two population probability distributions are not identical (a two-tailed test)

or

H_a: The probability distribution for population 1 is shifted to the right (or left) of the distribution for population 2 (a one-tailed test)

Test statistic:

$$z = \frac{T_2 - E(T_2)}{\sqrt{V(T_2)}} = \frac{T_2 - \left[\dfrac{n_1 n_2 + n_2(n_2 + 1)}{2}\right]}{\sqrt{\dfrac{n_1 n_2 (n_1 + n_2 + 1)}{12}}}$$

Rejection region:
Reject H_0 if $z > z_{\alpha/2}$ or $z < -z_{\alpha/2}$ for a two-tailed test. For a one-tailed test, place all of α in one tail of the z distribution. To detect a shift in the distribution of the population 1 observations to the right of the distribution of the population 2 observations, reject H_0 if $z < -z_\alpha$. To detect a shift in the opposite direction, reject H_0 if $z > z_\alpha$. Tabulated values of z are given in Table 4 of Appendix II.

We summarize this large-sample test in the preceding box. The mechanics of the test can be seen by reconsidering one or more of the examples provided in Section 9.4.

16.8 Suppose you want to compare two treatments, A and B, and you want to determine whether the distribution of the population of B measurements is shifted to the right of the distribution of the population of A measurements.
a. If $n_A = 7$, $n_B = 5$, and $\alpha = .05$, give the rejection region for the test.
b. Suppose you want to detect a shift in the distributions, either A to the right of B or vice versa. Locate the rejection region for the test (assume $n_A = 7$, $n_B = 5$, and $\alpha = .05$).

16.9 Independent random samples were selected from two populations. The data are shown in the table.

SAMPLE FROM POPULATION 1	SAMPLE FROM POPULATION 2
15	6
16	13
13	8
14	9
12	7
17	5
	4
	10

a. Use the Wilcoxon rank sum test to determine whether the data provide sufficient evidence to indicate a shift in the locations of the probability distributions of the sampled populations. Test using $\alpha = .05$.
b. Do the data provide sufficient evidence to indicate that the probability distribution for population 1 is shifted to the right of the probability distribution for population 2? Use the Wilcoxon rank sum test with $\alpha = .05$.

16.10 The data in the table represent daily accumulated streamflow and precipitation (in inches) for two U.S. Geological Survey stations in Colorado. Conduct a test to determine whether the distributions of daily accumulated streamflow and precipitation for the two stations differ in location. Use $\alpha = .10$.

STATION 1		STATION 2	
127.96	178.21	114.79	117.64
210.07	285.37	109.11	302.74
203.24	100.85	330.33	280.55
108.91	85.89	85.54	145.11
			95.36

Source: Gastwirth, J. L. and Mahmoud, H. "An Efficient Robust Nonparametric Test for Scale Change for Data from a Gamma Distribution." *Technometrics*, Vol. 28, No. 1, Feb. 1986, p. 83 (Table 2).

16.11 An experiment was conducted to compare the strengths of two types of paper—a special bond paper that cannot be photocopied and a standard paper that can be copied. Ten pieces of each type of paper were randomly selected from the production line and produced the strength measurements shown in the accompanying table. Higher measurements indicate greater strength. Test the hypothesis that there is no difference in the distributions of strengths for the two types of paper against the alternative hypothesis that the special paper that cannot be photocopied tends to be of greater strength. Use $\alpha = .05$.

STANDARD A		SPECIAL B	
1.19	1.18	1.48	1.22
1.40	1.47	1.30	1.53
1.29	1.39	1.62	1.30
1.53	1.20	1.49	1.47
1.31	1.35	1.24	1.55

16.12 The presence of lead in drinking water is cause for alarm in some older cities, many of which used lead pipes for water service lines. In a recent study, researchers attempted to document the level of lead, copper, and iron in the Boston water supply for areas supplied by lead service lines. The data shown in the table are the mean concentrations of lead (milligrams per liter) for samples collected over the period from 1976 to 1981. In May 1977, the city began treating the water with sodium hydroxide to reduce traces of metal in the water. Compare the mean concentration levels of the samples before and after May 1977 using a Wilcoxon rank sum test. Do the data provide sufficient evidence to indicate a reduction in lead content in the city's water supply after the initiation of the sodium hydroxide water treatment? Test using $\alpha = .05$.

DATE	MEAN CONCENTRATION OF LEAD	DATE	MEAN CONCENTRATION OF LEAD
Feb. 1976	.074	June 1977	.035
Mar. 1976	.064	July 1977	.060
Apr. 1976	.069	Aug. 1977	.055
May 1976	.063	Oct. 1977	.035
July 1976	.077	Nov. 1977	.031
Sept. 1976	.095	Dec. 1977	.039
Oct. 1976	.092	Jan. 1978	.038
Nov. 1976	.091	Mar. 1978	.049
Dec. 1976	.067	Apr. 1978	.073
		May 1978	.047

Source: Karalekas, P. C., Jr., Ryan, C. R., and Taylor, F. B. "Control of Lead, Copper, and Iron Pipe Corrosion in Boston." *American Water Works Journal*, Feb. 1983, pp. 92–95. Copyright © 1983, American Water Works Association. Reprinted by permission.

16.13 Oil producers are interested in finding high strength nickel alloys that are corrosion-resistant. Nickel alloys are especially susceptible to hydrogen embrittlement, a process that results when the alloy is cathodically charged in a sulfuric acid solution.

In order to rate the performance of two incoloy alloys, 800 and 902, hydrogen-charged tensile specimens of each alloy were measured for the amount of ductility loss (recorded as a percentage reduction of area). The measurements for eight tensile specimens of each type are given in the table. Conduct a test to determine whether the probability distributions of ductility losses differ for the two nickel alloys. Use $\alpha = .05$.

ALLOY 800	ALLOY 902
59.2	67.2
78.8	46.8
79.2	50.2
75.0	44.5
66.3	61.3
69.8	58.7
66.2	40.9
70.7	55.4

16.14 The proliferation of new computer software systems has brought the need for readable manuals documenting their use. An experiment was conducted to compare the usability of two instructional manuals for a construction bid analysis and monitoring software system. Thirty users were randomly divided into two groups of equal size. Each group was assigned a manual and asked to rate its readability on a scale from 1 to 10 (where 1 is poor and 10 is excellent). Do the data listed in the accompanying table provide sufficient evidence to indicate a difference in the readability of the two manuals? Test at $\alpha = .01$ using the Wilcoxon rank sum test for large samples.

MANUAL 1		MANUAL 2	
6	5	8	5
5	2	7	6
8	6	10	9
1	4	3	8
3	6	7	8
9	5	9	4
8	9	6	8
4		8	

16.15 A preliminary study was conducted to obtain information on the background levels of the toxic substance polychlorinated biphenyl (PCB) in soil samples in the United Kingdom (*Chemosphere*, Feb. 1986). Such information could then be used as a benchmark against which PCB levels at waste disposal facilities in the United Kingdom can be compared. The accompanying table contains the measured PCB levels of soil samples taken at 14 rural and 15 urban locations in the United Kingdom (PCB concentration is measured in .0001 gram per kilogram of soil). From these preliminary results, the researchers reported "a significant difference between (the PCB levels) for rural areas . . . and for urban areas." Do the data support the researchers' conclusions? Test using $\alpha = .05$.

RURAL		URBAN	
3.5	5.3	24.0	11.0
8.1	9.8	29.0	49.0
1.8	15.0	16.0	22.0
9.0	12.0	21.0	13.0
1.6	8.2	107.0	18.0
23.0	9.7	94.0	12.0
1.5	1.0	141.0	18.0
		11.0	

Source: Badsha, K. and Eduljee, G. "PCB in the U.K. Environment—A Preliminary Survey." *Chemosphere*, Vol. 15, No. 2, Feb. 1986, p. 213 (Table 1). Copyright 1986, Pergamon Press, Ltd. Reprinted with permission.

OPTIONAL EXERCISES

16.16 Use the formula for the sum of an arithmetic progression to show that

$$T_1 + T_2 = \frac{n(n + 1)}{2}$$

for the Wilcoxon rank sum test.

16.17 Show that for the special case where $n_1 = 2$ and $n_2 = 2$, the formula for the expected value of the Wilcoxon rank sum T_2 given in this section holds. [*Hint:* List the $(n_1 + n_2)! = 4!$ ways that the ranks can be assigned, and compute T_2 for each assignment. Then use the fact that the probability of any assignment is equally likely.]

16.18 Consider the Wilcoxon rank sum T_1 for the case where $n_1 = 3$ and $n_2 = 3$. Use the technique outlined in this section to find T_L such that $P(T_1 \leq T_L) \approx .05$.

SECTION 16.4

COMPARING TWO POPULATIONS: THE WILCOXON SIGNED RANKS TEST FOR A MATCHED-PAIRS DESIGN

Nonparametric techniques can also be used to compare two probability distributions when a matched-pairs design (Section 8.7) is used. A **matched-pairs design** is a randomized block design with $k = 2$ treatments. In this section we will show how a rank sum test can be used to test the hypothesis that two population probability distributions are identical against the alternative hypothesis that one is shifted to the right (or left) of the other.

For example, for some paper products, the softness of the paper is an important consideration in determining consumer acceptance. One method of assessing softness is to have judges give softness ratings to samples of the products. Suppose each of ten judges is given a sample of two products that a company wants to compare. Each judge rates the softness of each product on a scale from 1 to 10, with higher ratings implying a softer product. The results of the experiment are shown in Table 16.4.

TABLE 16.4
Paper Softness Ratings

JUDGE	PRODUCT A	PRODUCT B	DIFFERENCE (A − B)	ABSOLUTE VALUE OF DIFFERENCE	RANK OF ABSOLUTE VALUE
1	6	4	2	2	5
2	8	5	3	3	7.5
3	4	5	−1	1	2
4	9	8	1	1	2
5	4	1	3	3	7.5
6	7	9	−2	2	5
7	6	2	4	4	9
8	5	3	2	2	5
9	6	7	−1	1	2
10	8	2	6	6	10

$$T_+ = \text{Sum of positive ranks} = 46$$
$$T_- = \text{Sum of negative ranks} = 9$$

Since this is a matched-pairs design, we analyze the differences between measurements within each pair. If almost all of the differences are positive (or negative), we have evidence to indicate that the population probability distributions differ in location—that is, one is shifted to the right or to the left of the other. The nonparametric approach requires us to calculate the ranks of the absolute values of the differences between the measurements (the ranks of the differences after removing any minus signs). Note that tied absolute differences are assigned the average of the ranks they would receive if they were unequal but successive measurements. After the absolute differences are ranked, the sum of the ranks of the positive differences, T_+, and the sum of the ranks of the negative differences, T_-, are computed.

To test the null hypothesis

H_0:   The probability distributions of the ratings for products A and B are identical

against the alternative hypothesis

H_a: The probability distribution of the ratings for product A is shifted to the right or left of the probability distribution for the ratings for product B

we use the test statistic

$T = $ Smaller of the positive and negative rank sums T_+ and T_-

The rejection region for the test includes the smallest values of T and is located so that $P(T \leq T_0) = \alpha$ for a one-tailed statistical test and $P(T \leq T_0) = \alpha/2$ for a two-tailed test. Values of T_0 for $n = 5$ to $n = 50$ pairs are presented in Table 16 of Appendix II. The Wilcoxon signed ranks test is summarized in the box and demonstrated in Example 16.5.

THE WILCOXON SIGNED RANKS TEST FOR A MATCHED-PAIRS EXPERIMENT

ONE-TAILED TEST

H_0: Two sampled populations have identical probability distributions

H_a: The probability distribution for population 1 is shifted to the right of that for population 2

Test statistic: T_-, the negative rank sum (We assume the differences are computed as $y_1 - y_2$.)

TWO-TAILED TEST

H_0: Two sampled populations have identical probability distributions

H_a: The probability distribution for population 1 is shifted to the right *or* to the left of that for population 2

Test statistic: T, the smaller of the positive and negative rank sums, T_+ and T_- (We assume the differences are computed as $y_1 - y_2$.)

[*Note:* Eliminate zero differences from the calculation of the test statistic, since these values contribute to neither the positive nor the negative rank sum.]

Rejection region: $T_- \leq T_0$ where T_0 is found in Table 16 of Appendix II.

Rejection region: $T \leq T_0$ where T_0 is found in Table 16 of Appendix II.

[*Note:* If the alternative hypothesis is that the probability distribution for population 1 is shifted to the left of the distribution for population 2, we use T_+ as the test statistic and reject H_0 if $T_+ \leq T_0$.]

Assumptions: 1. A random sample of pairs of observations has been selected from the two populations.
2. The absolute differences in the paired observations can be ranked. [*Note:* No assumptions must be made about the form of the population probability distributions.]
3. Differences equal to 0 are eliminated and n is reduced accordingly. Tied differences are assigned ranks equal to the average of the ranks of the tied observations.

EXAMPLE 16.5

Refer to the data shown in Table 16.4. Compare the judges' ratings of products 1 and 2, using a Wilcoxon signed ranks test. For $\alpha = .05$, test

H_0: The distributions of product ratings are identical for products 1 and 2

against the alternative hypothesis

H_a: The distribution of ratings for one of the products is shifted to the left (or right) of the other distribution—that is, one of the products is rated higher than the other

SOLUTION

The test statistic for this two-tailed test is the smaller rank sum, namely, $T_- = 9$. The rejection region is $T \leq T_0$ where values of T_0 are given in Table 16 of Appendix II. A portion of this table is reproduced in Table 16.5. Examining Table 16.5 in the column corresponding to a two-tailed test, the row corresponding to $\alpha = .05$, and the column for $n = 10$ pairs, we read $T_0 = 8$. Therefore, we will reject H_0 if T is less than or equal to 8. Since the smaller rank sum, $T_- = 9$, is not less than or equal to 8, we cannot reject H_0. There is insufficient evidence to indicate a shift in the distributions of ratings for the two products.

TABLE 16.5

A Partial Reproduction of Table 16 of Appendix II

ONE-TAILED	TWO-TAILED	$n = 5$	$n = 6$	$n = 7$	$n = 8$	$n = 9$	$n = 10$
$\alpha = .05$	$\alpha = .10$	1	2	4	6	8	11
$\alpha = .025$	$\alpha = .05$		1	2	4	6	8
$\alpha = .01$	$\alpha = .02$			0	2	3	5
$\alpha = .005$	$\alpha = .01$				0	2	3
		$n = 11$	$n = 12$	$n = 13$	$n = 14$	$n = 15$	$n = 16$
$\alpha = .05$	$\alpha = .10$	14	17	21	26	30	36
$\alpha = .025$	$\alpha = .05$	11	14	17	21	25	30
$\alpha = .01$	$\alpha = .02$	7	10	13	16	20	24
$\alpha = .005$	$\alpha = .01$	5	7	10	13	16	19
		$n = 17$	$n = 18$	$n = 19$	$n = 20$	$n = 21$	$n = 22$
$\alpha = .05$	$\alpha = .10$	41	47	54	60	68	75
$\alpha = .025$	$\alpha = .05$	35	40	46	52	59	66
$\alpha = .01$	$\alpha = .02$	28	33	38	43	49	56
$\alpha = .005$	$\alpha = .01$	23	28	32	37	43	49
		$n = 23$	$n = 24$	$n = 25$	$n = 26$	$n = 27$	$n = 28$
$\alpha = .05$	$\alpha = .10$	83	92	101	110	120	130
$\alpha = .025$	$\alpha = .05$	73	81	90	98	107	117
$\alpha = .01$	$\alpha = .02$	62	69	77	85	93	102
$\alpha = .005$	$\alpha = .01$	55	61	68	76	84	92

The Wilcoxon signed ranks procedure can also be used to test the location of a *single* population. That is, the Wilcoxon signed ranks test can be used as an alternative to the sign test of Section 16.2. For example, suppose we want to test the following hypotheses about a population median:

H_0: $M = 100$

H_a: $M > 100$

To conduct the test we calculate the differences $(y_i - 100)$ for the sample. Recall that the sign test depends only on the number of positive differences in the sample. The signed ranks test, on the other hand, requires that we first rank the differences, then sum the ranks of the positive differences. Thus, the Wilcoxon signed ranks test for a single sample is conducted exactly as the signed ranks procedure for matched pairs, except that the differences are calculated by subtracting the hypothesized value of the median from each sample observation. We summarize the procedure in the box.

THE WILCOXON SIGNED RANKS TEST FOR THE MEDIAN, *M*, OF A SINGLE POPULATION

ONE-TAILED TEST	TWO-TAILED TEST
H_0: $M = M_0$	H_0: $M = M_0$
H_a: $M > M_0$	H_a: $M \neq M_0$
[or, H_a: $M < M_0$]	

Test statistic: | *Test statistic:*

T_-, the negative rank sum [or, T_+, the positive rank sum] | T, the smaller of the positive and negative rank sums, T_+ and T_-

[*Note:* The sample differences are computed as $(y_i - M_0)$.]

Rejection region: | *Rejection region:*

$T_- \leq T_0$ [or, $T_+ \leq T_0$] | $T \leq T_0$

where T_0 is found in Table 16 of Appendix II.

Assumptions: 1. A random sample of observations has been selected from the population.
2. The absolute differences $y_i - M_0$ can be ranked. [No assumptions must be made about the form of the population probability distribution.]
3. Differences equal to 0 are eliminated and n is reduced accordingly. Tied differences are assigned ranks equal to the average of the ranks of the tied observations.

| | | | | | | | | | | | |

EXERCISES
16.19–16.30

16.19 Suppose you want to test a hypothesis that two treatments, A and B, are equivalent against the alternative that the responses for A tend to be larger than those for B.
   a. If $n = 8$ and $\alpha = .01$, give the rejection region for a Wilcoxon signed ranks test.
   b. Suppose you want to detect a difference in the locations of the distributions of the responses for A and B if such a difference exists. If $n = 7$ and $\alpha = .10$, give the rejection region for the Wilcoxon signed ranks test.

16.20 A random sample of nine pairs of measurements is shown in the table.

PAIR	SAMPLE DATA FROM POPULATION 1	SAMPLE DATA FROM POPULATION 2
1	8	7
2	10	1
3	6	4
4	10	10
5	7	4
6	8	3
7	4	6
8	9	2
9	8	4

   a. Use the Wilcoxon signed ranks test to determine whether the data provide sufficient evidence to indicate that the probability distribution for population 1 is shifted to the right of the probability distribution for population 2. Test using $\alpha = .05$.
   b. Use the Wilcoxon signed ranks test to determine whether the data provide sufficient evidence to indicate that the probability distribution for population 1 is shifted either to the right or to the left of the probability distribution for population 2. Test using $\alpha = .05$.

16.21 Refer to Exercise 16.3. Conduct the test for location using the Wilcoxon signed ranks procedure. Use $\alpha = .05$.

16.22 Dental researchers have developed a new material for preventing cavities, a plastic sealant, which is applied to the chewing surfaces of teeth. To determine whether the sealant is effective, it was applied to half of the teeth of each of 12 school-age children. After 5 years, the numbers of cavities in the sealant-coated teeth and untreated teeth were counted. The results are given in the table. Is there sufficient evidence to indicate that sealant-coated teeth are less prone to cavities than are untreated teeth? Test using $\alpha = .05$.

CHILD	SEALANT-COATED	UNTREATED	CHILD	SEALANT-COATED	UNTREATED
1	3	3	7	1	5
2	1	3	8	2	0
3	0	2	9	1	6
4	4	5	10	0	0
5	1	0	11	0	3
6	0	1	12	4	3

16.23 A team of research engineers has isolated a chemical that will speed the pulse of the corn earworm. The goal is to control these crop-damaging insects by inducing fatal heart attacks with the chemical. Tests of the new compound have been conducted in ten different corn-producing states. Two fields were randomly selected from each state—one field was sprayed with the new chemical and the other field acted as the control. The crop yield per acre was recorded for all 20 farms, with the results shown in the accompanying table. Is there evidence that the new chemical will increase crop yield per acre? Test using $\alpha = .05$.

STATE	YIELD PER ACRE	
	New Chemical	Control
1	50.0	35.0
2	50.0	25.0
3	70.0	57.0
4	76.0	60.0
5	58.9	34.0
6	60.0	35.0
7	50.0	55.0
8	33.0	30.0
9	34.0	33.0
10	61.7	51.1

16.24 Refer to Exercise 16.5. Conduct the test for location using the Wilcoxon signed ranks procedure.

16.25 Tetrachlorodibenzo-p-dioxin (TCDD) is a highly toxic substance found in industrial wastes. A study was conducted to determine the amount of TCDD present in the tissues of bullfrogs inhabiting the Rocky Branch Creek in central Arkansas, an area known to be contaminated by TCDD. The level of TCDD (in parts per trillion) was measured in several specific tissues of four female bullfrogs and the ratio of TCDD in the tissue to TCDD in the leg muscle of the frog was recorded for each. The relative ratios of contaminant for two tissues, the liver and the ovaries, are given for each of the four frogs in the accompanying table. According to the researchers, "the data set suggests that the relative level of TCDD in the ovaries of female frogs is higher than the level in the liver of the frogs." Test this claim using $\alpha = .05$. [Hint: Find the approximate rejection region by using the value of T_0 given in Table 16 of Appendix II for $n = 5$.]

	FROG			
	A	B	C	D
Liver	11.0	14.6	14.3	12.2
Ovaries	34.2	41.2	32.5	26.2

Source: Korfmacher, W. A., Hansen, E. B., and Rowland, K. L. "Tissue Distribution of 2,3,7,8-TCDD in Bullfrogs Obtained from a 2,3,7,8-TCDD-Contaminated Area." *Chemosphere*, Vol. 15, No. 2, Feb. 1986, p. 125. Copyright 1986, Pergamon Press, Ltd. Reprinted with permission.

16.26 Synthetic fibers (such as rayon, nylon, and polyester) account for approximately 70% of all fibers used by American mills in their production of textile products. An experiment was conducted to compare the breaking tenacity of synthetic fibers produced using two methods of spinning: wet spinning and dry spinning. Specimens of ten different synthetic fibers were selected and each was split into two filaments. One filament was processed using the wet spinning method, and the other using the dry spinning method; the breaking tenacity (grams per denier) of each filament was then measured. Do the data shown in the table provide sufficient evidence to indicate a difference in the breaking tenacity of synthetic fibers produced by the two methods? Test using $\alpha = .05$.

FIBER	DRY SPINNING	WET SPINNING
Acetate	1.3	1.0
Acrylic	2.7	2.5
Aramid	4.8	4.7
Modacrylic	2.6	2.8
Nylon	4.5	4.2
Olefin	5.9	5.8
Polyester	4.5	4.3
Rayon	1.6	1.1
Spandex	.7	.9
Triacetate	1.3	.9

16.27 Wall and Peterson (1986) developed a heat transfer model for predicting winter heat loss in wastewater treatment clarifiers. Part of their analysis involved a comparison of clear-sky solar irradiation for horizontal surfaces at different sites in the Midwest. The day-long solar irradiation levels (in BTU/sq.ft.) at two midwestern locations of different latitudes—St. Joseph, Missouri and Iowa Great Lakes—were recorded on each of seven clear-sky winter days. The data are given in the table. Conduct a nonparametric test to compare the distributions of daily irradiation levels of the two locations. Test using $\alpha = .02$.

DATE	ST. JOSEPH, MISSOURI	IOWA GREAT LAKES
December 21	782	593
January 6	965	672
January 21	948	750
February 6	1,181	988
February 21	1,414	1,226
March 7	1,633	1,462
March 21	1,852	1,698

Source: Wall, D. J. and Peterson, G. "Model for Winter Heat Loss in Uncovered Clarifiers." *Journal of Environmental Engineering*, Vol. 112, No. 1, Feb. 1986, p. 128.

OPTIONAL EXERCISES

16.28 For the Wilcoxon signed ranks test, show that

$$T_+ + T_- = \frac{n(n + 1)}{2}$$

where n is the number of nonzero differences that are ranked.

16.29 For the special case $n = 2$, with no ties in the data (that is, no differences of zero), list the eight different ways in which the two absolute differences can be ranked. [*Note:* The number of arrangements, 8, results from the general formula $2^n \cdot n!$.]

16.30 For the special case described in Optional Exercise 16.29, show that

$$E(T_+) = \frac{n(n + 1)}{4}$$

[*Hint:* Find T_+ for each of the eight arrangements of the ranks listed in Optional Exercise 16.29, and use the fact that any particular arrangement will occur with probability $\frac{1}{8}$.]

S E C T I O N   16.5

THE KRUSKAL–WALLIS *H* TEST FOR A COMPLETELY RANDOMIZED DESIGN

In Section 13.5 we compare the means of k populations based on data collected according to a completely randomized design. The analysis of variance F test, used to test the null hypothesis of equality of means, is based on the assumption that the populations are normally distributed with common variance σ^2.

The **Kruskal–Wallis *H* test** is the nonparametric equivalent of the analysis of variance F test. It tests the null hypothesis that all k populations possess the same probability distribution against the alternative hypothesis that the distributions differ in location—that is, one or more of the distributions are shifted to the right or left of each other. The advantage of the Kruskal–Wallis *H* test over the F test is that we need make no assumptions about the nature of the sampled populations.

A completely randomized design specifies that we select independent random samples of $n_1, n_2, \ldots, n_k$ observations from the k populations. To conduct the test, we first rank all $n = n_1 + n_2 + \cdots + n_k$ observations and compute the rank sums, $R_1, R_2, \ldots, R_k$, for the k samples. The ranks of tied observations are averaged in the same manner as for the Wilcoxon rank sum test. Then, if H_0 is true, and if the sample sizes, $n_1, n_2, \ldots, n_k$, each equal 5 or more, then the test statistic

$$H = \frac{12}{n(n + 1)} \sum_{i=1}^{k} \frac{R_i^2}{n_i} - 3(n + 1)$$

will have a sampling distribution that can be approximated by a chi-square distribution with $(k - 1)$ degrees of freedom. Large values of H imply rejection of H_0. Therefore, the rejection region for the test is $H > \chi_\alpha^2$ where χ_α^2 is the value that locates α in the upper tail of the chi-square distribution.

The test is summarized in the box and its use is illustrated in Example 16.6.

KRUSKAL–WALLIS *H* TEST FOR COMPARING *k* POPULATION PROBABILITY DISTRIBUTIONS

H_0: The k population probability distributions are identical

H_a: At least two of the k population probability distributions differ in location

Test statistic: $$H = \frac{12}{n(n+1)} \sum_{i=1}^{k} \frac{R_i^2}{n_i} - 3(n+1)$$

where

n_i = Number of measurements in sample i

R_i = Rank sum for sample i, where the rank of each measurement is computed according to its relative magnitude in the totality of data for the k samples

n = Total sample size = $n_1 + n_2 + \cdots + n_k$

Rejection region: $H > \chi_\alpha^2$ with $(k-1)$ degrees of freedom

Assumptions: 1. The k samples are random and independent.
2. There are 5 or more measurements in each sample.
3. The observations can be ranked.

[*Note*: No assumptions have to be made about the shape of the population probability distributions.]

EXAMPLE 16.6

Independent random samples of three different brands of magnetron tubes (the key components in microwave ovens) were subjected to stress testing, and the number of hours each operated without repair was recorded. Although these times do not represent typical lifelengths, they do indicate how well the tubes can withstand extreme stress. The data are shown in Table 16.6. Experience has shown that the distributions of lifelengths for manufactured products are often nonnormal, thus violating the assumptions required for the proper use of an analysis of variance F test. Use the Kruskal–Wallis H test to determine whether evidence exists to conclude that the brands of magnetron tubes tend to differ in length of life under stress. Test using $\alpha = .05$.

TABLE 16.6

Lengths of Life for Magnetron Tubes in Example 16.6

BRAND		
A	B	C
36	49	71
48	33	31
5	60	140
67	2	59
53	55	42

SOLUTION

The first step in performing the Kruskal–Wallis H test is to rank the $n = 15$ observations in the complete data set. The ranks and rank sums for the three samples are shown in Table 16.7.

TABLE 16.7
Ranks and Rank Sums
for Example 16.6

A	RANK	B	RANK	C	RANK
36	5	49	8	71	14
48	7	33	4	31	3
5	2	60	12	140	15
67	13	2	1	59	11
53	9	55	10	42	6
	$R_1 = 36$		$R_2 = 35$		$R_3 = 49$

We want to test the null hypothesis

H_0: The population probability distributions of length of life under stress are identical for the three brands of magnetron tubes

against the alternative hypothesis

H_a: At least two of the population probability distributions differ in location

using the test statistic

$$H = \frac{12}{n(n + 1)} \sum_{i=1}^{k} \frac{R_i^2}{n_i} - 3(n + 1)$$

$$= \frac{12}{(15)(16)} \left[\frac{(36)^2}{5} + \frac{(35)^2}{5} + \frac{(49)^2}{5} \right] - 3(16)$$

$$= 1.22$$

FIGURE 16.3
Rejection Region for the
Comparison of Three
Probability Distributions

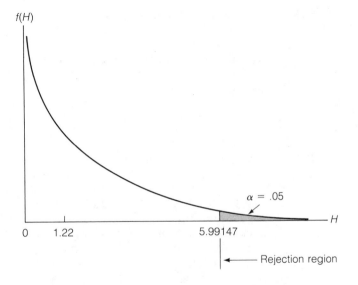

The rejection region for the H test is $H > \chi_\alpha^2$ where χ_α^2 is based on $(k - 2)$ degrees of freedom and the tabulated values of χ_α^2 are given in Table 7 of Appendix II. For $\alpha = .05$ and $(k - 1) = 2$ degrees of freedom, $\chi_{.05}^2 = 5.99147$. The rejection region for the test is $H > 5.99147$, as shown in Figure 16.3 (page 841). Since the computed value of H, $H = 1.22$, is less than $\chi_{.05}^2$, we cannot reject H_0. There is insufficient evidence to indicate a difference in location among the distributions of lifelengths for the three brands of magnetron tubes.   ∎

EXERCISES
16.31–16.38

16.31 Independent random samples were selected from four populations. The data are shown in the table.

SAMPLE 1	SAMPLE 2	SAMPLE 3	SAMPLE 4
8.1	7.8	2.1	4.8
2.3	3.5	3.3	3.9
3.4	5.2	1.0	2.6
5.1	5.0	4.5	6.7
3.6	4.6	.8	5.2
	7.7		

a. Do the data provide sufficient evidence to indicate a difference in location between at least two of the four probability distributions? Use the Kruskal–Wallis H test with $\alpha = .05$.

b. What assumptions must be satisfied in order for the test results in part **a** to be valid?

16.32 The EPA wants to determine whether temperature changes in the ocean's water caused by a nuclear power plant will have a significant effect on the animal life in the region. Recently hatched specimens of a certain species of fish are randomly divided into four groups. The groups are placed in separate simulated ocean environments that are identical in every way except for water temperature. Six months later, the specimens are weighed. The results (in ounces) are given in the table. Do the data provide sufficient evidence to indicate that one (or more) of the temperatures tend(s) to produce larger weight increases than the other temperatures? Test using $\alpha = .10$.

WATER TEMPERATURE			
38°F	42°F	46°F	50°F
22	15	14	17
24	21	28	18
16	26	21	13
18	16	19	20
19	25	24	21
	17	23	

16.33 B. L. Davis and M. K. Mount conducted a study to evaluate the effectiveness of performance appraisal training in an organizational setting (*Personnel Psychology*, Vol. 37, Aug. 1984, pp. 439–452). Middle-level managers were randomly selected and assigned to one of three training conditions: no training, computer-assisted training, or computer-assisted training plus a behavior modeling workshop. After the formal training, the managers were administered a 25-question multiple-choice test of managerial knowledge and the number of correct answers was recorded for each. The data in the table are adapted from summary information provided in the article. Is there sufficient evidence to indicate that the relative frequency distributions of scores differ in location for the three types of performance appraisal training? Test using $\alpha = .01$.

NO TRAINING	COMPUTER-ASSISTED TRAINING	COMPUTER TRAINING PLUS WORKSHOP
16	19	12
18	22	19
11	13	18
14	15	22
23	20	16
	18	25
	21	

16.34 As oil drilling costs rise at unprecedented rates, the task of measuring drilling performance becomes essential to a successful oil company. One method of lowering drilling costs is to increase drilling speed. Researchers at Cities Service Co. have developed a drill bit, called the PD-1, which they believe penetrates rock at a faster rate than any other bit on the market. It is decided to compare the speed of the PD-1 with the two fastest drill bits known, the IADC 1-2-6 and the IADC 5-1-7, at 15 drilling locations in Texas. Five drilling sites were randomly assigned to each bit, and the rate of penetration (RoP) in feet per hour was recorded after drilling 3,000 feet at each site. Based on the information given in the table, can Cities Service Co. conclude that the RoP probability distributions differ for at least two of the three drill bits? Test at the $\alpha = .05$ level of significance.

PD-1	IADC 1-2-6	IADC 5-1-7
35.2	25.8	14.7
30.1	29.7	28.9
37.6	26.6	23.3
34.3	30.1	16.2
31.5	28.8	20.1

16.35 A *modulator/demodulator*, or *modem*, is a device that converts electrical impulses sent from a computer into audio tones that travel over telephone lines to a remote terminal. The performance of a modem varies, depending on the speed with which it can send and receive signals (called the *baud rate*) and whether it transmits and receives data at the same time (*full duplex*) or must take turns with the computer transmitting data (*half duplex*). A new type of modem, called a *smart modem*, has been developed. Smart modems have built-in microprocessors with advanced features that improve overall modem performance and efficiency. Four new modems with self-contained microprocessors are currently on the market: Bizcomp 1012, Cermetek 212A, Hayes Smartmodem 1200, and Vadic 3451 Auto-Dial. Suppose that five users of each type of modem are randomly selected and asked to rate modem performance (measured on a scale from 1 to 100). Based on the data shown in the table (page 844), is there sufficient evidence to indicate a difference among the performance ratings of the four smart modems? Test using $\alpha = .10$.

BIZCOMP 1012	CERMETEK 212A	SMARTMODEM 1200	VADIC 3451
87	81	69	98
80	66	72	78
91	52	70	94
63	90	83	90
72	75	80	86

16.36 Vanadium (V) is a recently recognized essential trace element. An experiment was conducted to compare the concentrations of V in biological materials using isotope dilution mass spectrometry (*Analytical Chemistry*, Nov. 1985).

The accompanying table gives the quantities of V (measured in nanograms per gram) in dried samples of oyster tissue, citrus leaves, bovine liver, and human serum. Conduct a nonparametric test to determine whether the distributions of V concentrations for the four biological materials differ in location. Test using $\alpha = .05$.

OYSTER TISSUE	CITRUS LEAVES	BOVINE LIVER	HUMAN SERUM
2.35	2.32	.39	.10
1.30	3.07	.54	.17
.34	4.09	.30	.14
			.16
			.16

Source: Fassett, J. D. and Kingston, H. M. "Determination of Nanogram Quantities of Vanadium in Biological Material by Isotope Dilution Thermal Ionization Mass Spectrometry with Ion Counting Detection." *Analytical Chemistry*, Vol. 57, No. 13, Nov. 1985, p. 2475 (Table II). Copyright 1985, American Chemical Society. Reprinted with permission.

16.37 Phosphoric acid is chemically produced by reacting phosphate rock with sulfuric acid. An important consideration in the chemical process is the length of time required for the chemical reaction to reach a specified temperature. The shorter the length of time, the higher the reactivity of the phosphate rock. An experiment was conducted to compare the reactivity of phosphate rock mined in north, central, and south Florida. Rock samples were collected from each location and placed in vacuum bottles with a 56% strength sulfuric acid solution. The time (in seconds) for the chemical reaction to reach 200°F was recorded for each sample. Do the data provide sufficient evidence to indicate a difference in the reactivity of phosphoric rock mined at the three locations? Test using $\alpha = .05$.

SOUTH	CENTRAL	NORTH
40.6	41.1	25.6
42.0	38.3	36.4
37.5	40.2	28.2
38.1	33.5	31.3
41.9	35.7	29.5
		22.8
		27.5

OPTIONAL EXERCISE

16.38 Use the sum of an arithmetic progression to show that for the Kruskal–Wallis H test,

$$R_1 + R_2 + \cdots + R_k = \frac{n(n + 1)}{2}$$

where k is the number of probability distributions being compared and n is the total sample size.

S E C T I O N   16.6

THE FRIEDMAN F_r TEST FOR A RANDOMIZED BLOCK DESIGN

In this section we present the nonparametric equivalent of the analysis of variance F test for a randomized block design given in Section 13.6. The test, proposed by Milton Friedman (a Nobel prize winner in economics), is particularly appropriate for comparing the relative locations of k or more population probability distributions when the normality and common variance assumptions required for an analysis of variance are not (or may not be) satisfied.

To conduct the F_r test, we first rank the observations within each block and then compute the rank sums, $R_1, R_2, \ldots, R_k$, for the k treatments. If H_0 is true—

FRIEDMAN F_r TEST FOR A RANDOMIZED BLOCK DESIGN

H_0:   The probability distributions for the k treatments are identical

H_a: At least two of the probability distributions differ in location

Test statistic: $F_r = \dfrac{12}{bk(k + 1)} \displaystyle\sum_{i=1}^{k} R_i^2 - 3b(k + 1)$

where

   b = Number of blocks

   k = Number of treatments

   R_i = Rank sum of the ith treatment, where the rank of each measurement is computed relative to its position *within its own block*

Rejection region: $F_r > \chi_\alpha^2$ with $(k - 1)$ degrees of freedom

Assumptions: 1. The treatments are randomly assigned to experimental units within the blocks.
   2. The measurements can be ranked within blocks.
   3. Either the number of blocks (b) or the number of treatments (k) should exceed 5 for the χ^2 approximation to be adequate.

[*Note:*   No assumptions have to be made about the shape of the population probability distributions.]

that is, if the population probability distributions are identical—and if the number n of observations is large, then the F_r statistic

$$F_r = \frac{12}{bk(k+1)} \sum_{i=1}^{k} R_i^2 - 3b(k+1)$$

will possess a sampling distribution that can be approximated by a chi-square distribution with $(k-1)$ degrees of freedom. In order for the approximation to be reasonably good, we require that either b, the number of blocks, or k, the number of treatments, exceed 5. The rejection region for the test consists of large values of F_r. Therefore, we reject H_0 if $F_r > \chi_\alpha^2$.

The **Friedman F_r test** is summarized in the preceding box and its use is illustrated in Example 16.7.

EXAMPLE 16.7

The corrosion of different metals is a problem in many mechanical devices. Three sealers used to help retard corrosion were tested to determine whether there were any differences among them. Samples of ten different metal compositions were treated with each of the three sealers and the amount of corrosion was measured after exposure to the same environmental conditions for 1 month. The data and their associated ranks are shown in Table 16.8. Is there any evidence of a difference in the probability distributions of the amounts of corrosion among the three types of sealers? Test using $\alpha = .05$.

TABLE 16.8
Data and Ranks for the Randomized Block Design of Example 16.7

METAL	SEALER					
	1	Rank	2	Rank	3	Rank
1	21	2	23	3	15	1
2	29	2	30	3	21	1
3	16	1	19	3	18	2
4	20	3	19	2	18	1
5	13	2	10	1	14	3
6	5	1	12	3	6	2
7	18	2.5	18	2.5	12	1
8	26	2	32	3	21	1
9	17	2	20	3	9	1
10	4	2	10	3	2	1
	$R_1 = \overline{19.5}$		$R_2 = \overline{26.5}$		$R_3 = \overline{14}$	

SOLUTION

We want to test the null hypothesis

H_0: The probability distributions of the amounts of corrosion are identical for the three sealers

against the alternative hypothesis

H_a: At least two of the probability distributions differ in location

The ranks of the three treatments within each block and the treatment rank sums are shown in Table 16.8. Therefore, the calculated value of the F_r statistic is

$$F_r = \frac{12}{bk(k+1)} \sum_{i=1}^{k} R_i^2 - 3b(k+1)$$

$$= \frac{12}{10(3)(4)} [(19.5)^2 + (26.5)^2 + (14)^2] - 3(10)(4)$$

$$= 7.85$$

The rejection region for the test is $F_r > \chi^2_{.05}$, where the tabulated value (given in Table 7 of Appendix II) of $\chi^2_{.05}$, based on $k - 1 = 2$ degrees of freedom, is 5.99147. Thus, we will reject H_0 if $F_r > 5.99147$.

Since the computed value of the test statistic, $F_r = 7.85$, exceeds $\chi^2_{.05} = 5.99147$, there is sufficient evidence to reject H_0 and conclude that differences exist in the locations of two or more of the corrosion probability distributions. The practical conclusion is that there is evidence to indicate a difference among the sealing abilities of the three sealers. ∎

EXERCISES 16.39–16.44

16.39 An experiment was conducted using a randomized block design with four treatments and six blocks. The ranks of the measurements within each block are shown in the table. Use the Friedman F_r test for a randomized block design to determine whether the data provide sufficient evidence to indicate that at least two of the treatment probability distributions differ in location. Test using $\alpha = .05$.

TREATMENT	BLOCK					
	1	2	3	4	5	6
1	3	3	2	3	2	3
2	1	1	1	2	1	1
3	4	4	3	4	4	4
4	2	2	4	1	3	2

16.40 A serious drought-related problem for farmers is the spread of aflatoxin, a highly toxic substance caused by mold, which contaminates field corn. In higher levels of contamination, aflatoxin is potentially hazardous to animal and possibly human health. (Officials of the FDA have set a maximum limit of 20 parts per billion aflatoxin as safe for interstate marketing.) Three sprays, A, B, and C, have been developed to control aflatoxin in field corn. To determine whether differences exist among the sprays, ten ears of corn are randomly chosen from a contaminated corn field and each is divided into three pieces of equal size. The sprays are then randomly assigned to the pieces for each ear of corn, thus setting up a randomized block design. The table gives the amount (in parts per billion) of aflatoxin present in the corn samples after spraying. Use the Friedman test to determine whether there are differences among the probability distributions of the amounts of aflatoxin present for the three sprays. Test at the $\alpha = .05$ level of significance.

EAR	SPRAY		
	A	B	C
1	21	23	15
2	29	30	21
3	16	19	18
4	20	19	18
5	13	10	14
6	5	12	6
7	18	18	12
8	26	32	21
9	17	20	9
10	4	10	2

16.41 An *optical mark reader* (OMR) is a machine that is able to "read" pencil marks that have been entered on a special form. When connected to a computer, such systems are able to read and analyze data in one step. As a result, the keypunching of data into a machine-readable code can be eliminated. Eliminating this step reduces the possibility that the data will be contaminated by human error. OMRs are used by schools to grade exams and by survey research organizations to compile data from questionnaires. A manufacturer of OMRs believes its product can operate equally well in a variety of temperature and humidity environments. To determine whether operating data contradict this belief, the manufacturer asks a well-known industrial testing laboratory to test its product. Five recently produced OMRs were randomly selected and each was operated in six different environments. The number of forms each was able to process in an hour was recorded and used as a measure of the OMR's operating efficiency. These data appear in the table. Use the Friedman F_r test to determine whether evidence exists to indicate that the probability distributions for the number of forms processed per hour differ in location for at least two of the environments. Test using $\alpha = .10$.

MACHINE NUMBER	ENVIRONMENT					
	1	2	3	4	5	6
1	8,001	8,025	8,100	8,055	7,991	8,007
2	7,910	7,932	7,900	7,990	7,892	7,922
3	8,111	8,101	8,201	8,175	8,102	8,235
4	7,802	7,820	7,904	7,850	7,819	8,100
5	7,500	7,601	7,702	7,633	7,600	7,561

16.42 An experiment is conducted to investigate the toxic effect of three chemicals, A, B, and C, on the skin of rats. Three adjacent 1-inch squares are marked on the backs of eight rats and each of the three chemicals is applied to each rat. The squares of skin are then scored from 0 to 10, depending on the degree of irritation. The data are given in the accompanying table. Is there sufficient evidence to support the alternative hypothesis that at least two of the probability distributions of skin irritation scores corresponding to the three chemicals differ in location? Use $\alpha = .01$.

RAT	CHEMICAL		
	A	B	C
1	6	5	3
2	9	8	4
3	6	9	3
4	5	8	6
5	7	8	9
6	5	7	6
7	6	7	5
8	6	5	7

16.43 Research into the human-engineering aspect of computing has grown tremendously due to the ever-increasing number of computer users. The end goal is to create "user-friendly" hardware and software that reduces as much as possible any form of human strain or stress resulting from computer usage. A recent topic of concern to computer terminal users is the color of the characters displayed on the video display screens. Early full-screen video display terminals presented the viewer with white characters on a black background. Initially, viewers found the high degree of contrast easy on the eyes. However, after an extended period of use, black and white displays were frequently found to cause temporary eye irritations. In the past few years, experimentation with other colors revealed that yellow/amber displays may be the easiest on the eyes. In one German study, video display terminals were produced with white/black and six different symbol colors. (*Computers & Electronics*, Apr. 1983). Thirty test subjects were asked to specify which color combination they preferred by ranking each of the seven color combinations on a scale from 0 (no preference) to 10. Although the raw data were not revealed, the mean preference scores for each color were provided by the researchers. Based on these means, we have simulated the individual preference scores for ten subjects in the accompanying table.

SUBJECT	GREEN/ BLACK	WHITE/ BLACK	YELLOW/ WHITE	ORANGE/ WHITE	YELLOW	YELLOW/ AMBER	YELLOW/ ORANGE
1	7	6	7	2	8	9	3
2	8	6	9	4	9	8	1
3	5	5	7	1	6	8	2
4	3	4	2	0	2	6	0
5	9	8	8	3	9	9	2
6	7	5	6	2	7	7	1
7	6	7	8	4	6	9	5
8	6	5	8	1	8	9	1
9	9	9	8	2	9	8	0
10	9	8	8	3	9	10	1

Source: Adapted from Solomon, L. and Burawa, A. "Maximize Your Computing Comfort & Efficiency." *Computers & Electronics*, Apr. 1983, pp. 35–40.

a. Do the data provide sufficient evidence to indicate a difference among the preference scores for the seven video display color combinations? Test using $\alpha = .05$.

b. Do the data provide sufficient evidence to indicate that video display terminal users prefer a yellow/amber color combination over a green/black color combination? Test using $\alpha = .05$.

c. Do the data provide sufficient evidence to indicate that video display terminal users prefer a yellow/amber color combination over yellow? Test using $\alpha = .05$.

16.44 Nuclear power continues to be one of mankind's biggest fears. However, research has indicated that antismoking campaigns can save more lives each year than nuclear power plant safety programs (*Dun's Review*, Sept. 1979). In an effort to quantify the public's perception of risk, a research organization asked three groups of people to rank 30 products or activities from most risky to least risky. The three groups who took part in the survey were the League of Women Voters, college students, and business and professional club members. The accompanying table gives the rankings for the three groups and, in parentheses, the estimated number of deaths per year attributed to the 30 activities. Use the Friedman test to determine whether the three groups differ regarding their concepts of risk involved with these thirty activities. Test using $\alpha = .05$.

ACTIVITY AND DEATHS PER YEAR	LEAGUE OF WOMEN VOTERS	COLLEGE STUDENTS	BUSINESS AND PROFESSIONAL CLUB MEMBERS
Smoking (150,000)	4	3	4
Alcoholic beverages (100,000)	6	7	5
Motor vehicles (50,000)	2	5	3
Handguns (17,000)	3	2	1
Electric power (14,000)	18	19	19
Motorcycles (3,000)	5	6	2
Swimming (3,000)	19	30	17
Surgery (2,800)	10	11	9
X-rays (2,300)	22	17	24
Railroads (1,950)	24	23	20
General (private) aviation (1,300)	7	15	11
Large construction (1,000)	12	14	13
Bicycles (1,000)	16	24	14
Hunting (800)	13	18	10
Home appliances (200)	29	27	27
Fire fighting (195)	11	10	6
Police work (160)	8	8	7
Contraceptives (150)	20	9	22
Commercial aviation (130)	17	16	18
Nuclear power (100)	1	1	8
Mountain climbing (30)	15	22	12
Power mowers (24)	27	28	25
High school & college football (23)	23	26	21
Skiing (18)	21	25	16
Vaccinations (10)	30	29	29
Food coloring[a]	26	20	30
Food preservatives[a]	25	12	28
Pesticides[a]	9	4	15
Prescription antibiotics[a]	28	21	26
Spray cans[a]	14	13	23

[a]Not available

Source: "What Price Safety? The 'Zero-Risk' Debate." *Dun's Review*, Sept. 1979. Reprinted with permission, *Business Month* magazine. Copyright © 1979 by Business Magazine Corporation, 38 Commercial Wharf, Boston, Massachusetts 02110.

| | | | | | | | | | | | | |

S E C T I O N 16.7

**SPEARMAN'S RANK
CORRELATION
COEFFICIENT**

Several different nonparametric statistics have been developed to measure and to test for correlation between two random variables. The one we discuss in this section is known as **Spearman's rank correlation coefficient,** r_s.

SPEARMAN'S RANK CORRELATION COEFFICIENT

$$r_s = \frac{SS_{uv}}{\sqrt{SS_{uu}SS_{vv}}}$$

where

$$SS_{uv} = \sum(u_i - \bar{u})(v_i - \bar{v}) = \sum u_i v_i - \frac{\left(\sum u_i\right)\left(\sum v_i\right)}{n}$$

$$SS_{uu} = \sum(u_i - \bar{u})^2 = \sum u_i^2 - \frac{\left(\sum u_i\right)^2}{n}$$

$$SS_{vv} = \sum(v_i - \bar{v})^2 = \sum v_i^2 - \frac{\left(\sum v_i\right)^2}{n}$$

u_i = Rank of the ith observation on x

v_i = Rank of the ith observation on y

n = Number of pairs of observations (number of observations in each sample)

SHORTCUT FORMULA FOR r_s

$$r_s = 1 - \frac{6\sum d_i^2}{n(n^2 - 1)}$$

where

$d_i = u_i - v_i$ (difference in the ranks of the ith observations on x and y)

The easiest way to understand why Spearman's r_s is a good measure of the relationship between two random variables is to review the method for calculating the simple coefficient of correlation r, discussed in Chapter 10. We will assume that we have collected a random sample of n data points, consisting of paired values of two random variables (x, y). The simple correlation coefficient,

$$r = \frac{SS_{xy}}{\sqrt{SS_{xx}SS_{yy}}}$$

always assumes a value in the interval $-1 \leq r \leq 1$. Positive values of r indicate that y increases as x increases; negative values imply that y decreases as x increases.

Spearman's r_s is calculated in the same way as the simple coefficient of correlation, except that the original values of y and x are replaced by their ranks. The n values of y are ranked, with the smallest value of y receiving rank 1 and the largest receiving rank n. The values of x are similarly ranked. Then, letting u_i be the rank of x_i ($i = 1, 2, \ldots, n$) and v_i be the rank of y_i ($i = 1, 2, \ldots, n$), we can write Spearman's rank correlation coefficient as

$$r_s = \frac{\sum (u_i - \bar{u})(v_i - \bar{v})}{\sqrt{\sum (u_i - \bar{u})^2 \sum (v_i - \bar{v})^2}}$$

Or, using the notation of Chapter 10, we have

$$r_s = \frac{SS_{uv}}{\sqrt{SS_{uu}SS_{vv}}}$$

When there are no ties in the rankings, the formula for r_s reduces algebraically to the simple formula

$$r_s = 1 - \frac{6 \sum d_i^2}{n(n^2 - 1)}$$

where $d_i = u_i - v_i$, the difference in the ranks for the ith pair of observations. This formula for calculating r_s can be used when the number of tied ranks is small relative to n.*

SPEARMAN'S NONPARAMETRIC TEST FOR RANK CORRELATION

ONE-TAILED TEST

H_0: $\rho_s = 0$

H_a: $\rho_s > 0$
(or H_a: $\rho_s < 0$)

Test statistic: r_s, the sample Spearman's rank correlation

Rejection region: $r_s > r_{s,\alpha}$
(or $r_s < -r_{s,\alpha}$ when
H_a: $\rho_s < 0$)
where $r_{s,\alpha}$ is the value from Table 17 of Appendix II corresponding to the upper-tail area α and n pairs of observations

TWO-TAILED TEST

H_0: $\rho_s = 0$

H_a: $\rho_s \neq 0$

Test statistic: r_s, the sample Spearman's rank correlation

Rejection region:
$r_{s,\alpha} < -r_{s,\alpha/2}$ or $r_s > r_{s,\alpha/2}$

where $r_{s,\alpha/2}$ is the value from Table 17 of Appendix II corresponding to the upper-tail area $\alpha/2$ and n pairs of observations

---

*The shortcut formula is not exact when there are tied measurements, but it is a good approximation when the total number of ties is not large relative to n.

The interpretation of r_s is similar to the interpretation of the simple coefficient of correlation r. Spearman's r_s will assume a value in the interval $-1 \leq r_s \leq 1$. Values of r_s near 0 imply little or no correlation between the rankings; values near $+1$ or -1 imply strong positive and negative correlations, respectively.

The null hypothesis that the population value of Spearman's rank correlation coefficient, ρ_s, equals 0 implies that any one matching of the ranks, u_i and v_i ($i = 1, 2, \ldots, n$), is just as likely as any other. The rejection region for the test is determined by identifying the rankings for which $r_s > r_{s,\alpha}$. These values have been tabulated and are presented in Table 17 of Appendix II.

The test for rank correlation between two random variables is summarized in the preceding box and its use is illustrated in Example 16.8.

EXAMPLE 16.8

A large manufacturing firm wants to determine whether a relationship exists between the number of work-hours an employee misses per year and the employee's annual wages. A sample of 15 employees produced the data shown in Table 16.9. Is there sufficient evidence to indicate that the work-hours missed are correlated with annual wages? Test using $\alpha = .05$.

TABLE 16.9
Work-Hours Missed and
Annual Wages for
Example 16.8

EMPLOYEE	WORK-HOURS MISSED	ANNUAL WAGES (Thousands of dollars)
1	49	15.8
2	36	17.5
3	127	11.3
4	91	13.2
5	72	13.0
6	34	14.5
7	155	11.8
8	11	20.2
9	191	10.8
10	6	18.8
11	63	13.8
12	79	12.7
13	43	15.1
14	57	24.2
15	82	13.9

SOLUTION

We first rank the $n = 15$ values of work-hours missed and then repeat the process for the annual salaries. These rankings are shown in Table 16.10 (page 854). The next step is to calculate the differences, d_i ($i = 1, 2, \ldots, 15$), in the pairs of ranks for the $n = 15$ employees. These differences and their squares are shown in the d_i and d_i^2 columns of Table 16.10. Substituting Σd_i^2 into the shortcut formula for r_s, we obtain

$$r_s = 1 - \frac{6 \sum d_i^2}{n(n^2 - 1)} = 1 - \frac{6(1,038)}{15(224)} = -.854$$

TABLE 16.10

Calculation Table for
Example 16.8

EMPLOYEE	HOURS MISSED	RANK	ANNUAL WAGES	RANK	d_i	d_i^2
1	49	6	15.8	11	−5	25
2	36	4	17.5	12	−8	64
3	127	13	11.3	2	11	121
4	91	12	13.2	6	6	36
5	72	9	13.0	5	4	16
6	34	3	14.5	9	−6	36
7	155	14	11.8	3	11	121
8	11	2	20.2	14	−12	144
9	191	15	10.8	1	14	196
10	6	1	18.8	13	−12	144
11	63	8	13.8	7	1	1
12	79	10	12.7	4	6	36
13	43	5	15.1	10	−5	25
14	57	7	24.2	15	−8	64
15	82	11	13.9	8	3	9

$$\Sigma d_i^2 = 1{,}038$$

Since we want to detect either a positive or a negative correlation between work-hours missed and annual salary, we will test the null hypothesis

$$H_0: \quad \rho_s = 0$$

against the two-sided alternative hypothesis

$$H_a: \quad \rho_s \neq 0$$

The rejection region for the test, using $\alpha = .05$, can be found by examining Table 17 of Appendix II. A partial reproduction is shown in Table 16.11. The critical values of r_s given in Table 16.11 are those for a one-tailed test. For example, if $n = 10$ and we want to detect positive rank correlation, i.e., $H_a: \rho_s > 0$, we would reject H_0 if $r_s > .564$ for $\alpha = .05$. If we want to conduct a test for negative correlation for the same sample size and for $\alpha = .05$, we would reject H_0 if $r_s < -.564$. The critical values for a two-tailed test are obtained by choosing the column headed by $\alpha/2$—that is, half of your desired α value. Since we want $\alpha = .05$, we will look for the critical values for our test statistic in the .025 column opposite $n = 15$. Since this value of r_s is .525 (shaded), we will reject H_0 if

$$r_s > .525 \quad \text{or} \quad r_s < -.525$$

You can see that the calculated value, $r_s = -.854$, lies in the rejection region. Therefore, we reject H_0 and conclude that the population rank correlation coefficient, ρ_s, between work-hours missed and annual salary differs from 0.

TABLE 16.11

A Partial Reproduction of
Table 17 of Appendix II

The α values correspond to a one-tailed test of H_0: $\rho_s = 0$. The tabled value of α should be doubled for two-tailed tests.

n	$\alpha = .05$	$\alpha = .025$	$\alpha = .01$	$\alpha = .005$
5	.900	—	—	—
6	.829	.886	.943	—
7	.714	.786	.893	—
8	.643	.738	.833	.881
9	.600	.683	.783	.833
10	.564	.648	.745	.794
11	.523	.623	.736	.818
12	.497	.591	.703	.780
13	.475	.566	.673	.745
14	.457	.545	.646	.716
15	.441	.525	.623	.689
16	.425	.507	.601	.666

**EXERCISES
16.45–16.53**

16.45 A random sample of seven pairs of observations are recorded on two variables, x and y. The data are shown in the table. Use Spearman's nonparametric test for rank correlation to answer the following:

a. Do the data provide sufficient evidence to conclude that ρ_s, the rank correlation between x and y, is greater than 0? Test using $\alpha = .05$.

b. Do the data provide sufficient evidence to conclude that $\rho_s \neq 0$? Test using $\alpha = .05$.

PAIR	x	y
1	65	58
2	57	61
3	55	58
4	38	23
5	29	34
6	43	38
7	49	37

16.46 The Federal Communications Commission (FCC) specifies that radiated electromagnetic emissions from digital devices are to be measured in an open-field test site. In order to verify test-site acceptability, the site attenuation (i.e., the transmission loss from the input of one half-wave dipole to the output of another when both dipoles are positioned over the ground plane) must be evaluated. A study conducted at a test site in Fort Collins, Colorado, yielded the accompanying data on site attenuation (in decibels) and transmission frequency (in megahertz) for dipoles at a distance of 3 meters.

TRANSMISSION FREQUENCY x, MHz	SITE ATTENUATION y, db
50	11.5
100	15.8
200	18.2
300	22.6
400	26.2
500	27.1
600	29.5
700	30.7
800	31.3
900	32.6
1,000	34.9

Source: Bennett, W. S. "An Error Analysis of the FCC Site-Attenuation Approximation." *IEEE Transactions on Electromagnetic Compatibility*, Vol. EMC-27, No. 3, Aug. 1985, p. 113 (Table IV). © 1985, IEEE.

a. Find the Spearman coefficient of correlation between transmission frequency (x) and site attenuation level (y).

b. Conduct a test to determine if site attenuation level (y) is positively correlated with transmission frequency (x). Test using $\alpha = .10$.

16.47 Refer to Exercise 16.44, which discusses the *Dun's Review* survey on ranking 30 activities from most risky to least risky. Is the public's perception of risk consistent with actual risk (based on estimated number of deaths per year)? To answer this question, we can apply Spearman's nonparametric test for rank correlation.

a. Compute the Spearman coefficient of correlation between the rankings of the League of Women Voters and the actual rankings. Is there evidence of positive correlation between the rankings? Test using $\alpha = .05$.

b. Is there evidence of positive correlation between the rankings of college students and the actual rankings? Test using $\alpha = .05$.

c. Is there evidence of positive correlation between the rankings of business and professional club members and the actual rankings? Test using $\alpha = .05$.

16.48 Refer to Exercises 16.44 and 16.47. Conduct a test to determine if there is evidence of positive correlation between the rankings of:

a. The League of Women Voters and college students

b. The League of Women Voters and business and professional club members

c. College students and business and professional club members

16.49 A new computer software query package has been designed to achieve more efficient access and maintenance of large-scale data sets. Efficiency is measured in terms of the number of disk I/O's (called *storage blocks*) required to access and maintain the data set. The smaller the number of blocks that are read, the faster the operation takes place. To evaluate the performance of the new software system, the number of disk I/O's required to access a large-scale data set was recorded for each of a sample of 15 data sets of various sizes (where size is measured as the number of records in the data set). The results are shown in the table.

DATA SET	NUMBER OF RECORDS x, thousands	NUMBER OF DISK I/O'S y, thousands
1	350	36
2	200	20
3	450	45
4	50	5
5	400	40
6	150	18
7	350	38
8	300	32
9	150	21
10	500	54
11	100	11
12	400	43
13	200	19
14	50	7
15	250	26

a. Compute Spearman's rank correlation coefficient to measure the strength of the relationship between the number of records in a data set and the number of disk I/O's required to access the data set.

b. Test the null hypothesis that the number of records and the number of disk I/O's are not correlated against the alternative that these variables are positively correlated. Use $\alpha = .01$.

16.50 The accompanying table shows a portion of the experimental data obtained in a study of the radial tension strength of concrete pipe. The concrete pipe used for the experiment had an inside diameter of 84 inches and a wall thickness of approximately 8.75 inches. In addition, it was reinforced with cold drawn wire. The table shows the load (in pounds per foot) until the first crack was observed and the age (in days) for each of nine pipe specimens. Do the data present sufficient evidence to indicate correlation between load and age? Test using $\alpha = .05$.

LOAD	AGE	LOAD	AGE
11,450	20	10,540	25
10,420	20	9,470	31
11,142	20	9,190	31
10,840	25	9,540	31
11,170	25		

Source: Heger, F. J. and McGrath, T. J. "Radial Tension Strength of Pipe and Other Curved Flexural Members." *Journal of the American Concrete Institute*, Vol. 80, No. 1, 1983, pp. 33–39.

16.51 An automated system for marking large numbers of student computer programs, called AUTOMARK, has been used successfully at McMaster University in Ontario,

Canada. AUTOMARK takes into account both program correctness and program style when marking student assignments. To evaluate the effectiveness of the automated system, AUTOMARK was used to grade the FORTRAN77 assignments of a class of 33 students. These grades were then compared to the grades assigned by the instructor. The results are shown in the accompanying table. Is there evidence to indicate that the AUTOMARK and instructor grade assignments are correlated? Use Spearman's test for rank correlation at $\alpha = .05$.

AUTOMARK GRADE	INSTRUCTOR GRADE	AUTOMARK GRADE	INSTRUCTOR GRADE
12.2	10	17.8	17
10.6	11	18.0	17
15.1	12	18.2	17
16.2	12	18.4	17
16.6	12	18.6	17
16.6	13	19.0	17
17.2	14	19.3	17
17.6	14	19.5	17
18.2	14	19.7	17
16.5	15	18.6	18
17.2	15	19.0	18
18.2	15	19.2	18
15.1	16	19.4	18
17.2	16	19.6	18
17.5	16	20.1	18
18.6	16	19.2	19
18.8	16		

Source: Redish, K. A. and Smyth, W. F. "Program Style Analysis: A Natural By-product of Program Compilation." *Communications of the Association for Computing Machinery*, Vol. 29, No. 2, Feb. 1986, p. 132 (Figure 4). Copyright 1986, Association for Computing Machinery, Inc.

OPTIONAL EXERCISES

16.52 Show that for the special case where $n = 3$, $-1 \le r_s \le 1$. [*Hint:* List each of the $3! \times 3! = 36$ different arrangements of the x and y rankings and compute r_s for each.]

16.53 Show that for the special case where $n = 3$, $E(r_s) = 0$. (This fact is also true in general.) [*Hint:* Use the results of Optional Exercise 16.52 and the fact that any of the arrangements has a probability of $\frac{1}{36}$ of occurring.]

SECTION 16.8

SUMMARY

We have presented several useful **nonparametric techniques** for testing the location of a single population, or for comparing two or more populations. Nonparametric techniques are useful when the underlying assumptions for their parametric counterparts are not justified or when it is impossible to assign specific values to the observations. Nonparametric methods provide more general comparisons of populations than parametric methods, because they compare the probability distributions of the populations rather than specific parameters.

Rank sums are the primary tools of nonparametric statistics. The **Wilcoxon rank sum test** can be used to compare two populations based on an independent sampling experiment, and the **Wilcoxon signed ranks test** can be used for a **matched-pairs experiment**. The **Kruskal–Wallis H test** is applied when comparing k populations using a **completely randomized design**. The **Friedman F_r test** is used to compare k populations when a randomized block design is conducted.

The strength of nonparametric statistics lies in their general applicability. Few restrictive assumptions are required, and they may be used for observations that can be ranked but not exactly measured. Therefore, nonparametric methods provide useful alternatives to the parametric tests of Chapters 9, 10, and 13.

| | | | | | | | | | | | | | |

SUPPLEMENTARY EXERCISES 16.54–16.77

16.54 The data for three independent random samples are shown in the accompanying table. It is known that the sampled populations are *not* normally distributed. Use an appropriate test to determine whether the data provide sufficient evidence to indicate that at least two of the populations differ in location. Test using $\alpha = .05$.

SAMPLE FROM POPULATION 1	SAMPLE FROM POPULATION 2	SAMPLE FROM POPULATION 3
18	12	87
32	33	53
43	10	65
15	34	50
63	18	64
		77

16.55 A random sample of nine pairs of observations are recorded on two variables, x and y. The data are shown in the table.

PAIR	x	y
1	19	12
2	27	19
3	15	7
4	35	25
5	13	11
6	29	10
7	16	16
8	22	10
9	16	18

a. Do the data provide sufficient evidence to indicate that ρ_s, the Spearman rank correlation between x and y, differs from 0? Test using $\alpha = .05$.

b. Do the data provide sufficient evidence to indicate that the probability distribution for x is shifted to the right of that for y? Test using $\alpha = .05$.

16.56 Two independent random samples produced the measurements listed in the table. Do the data provide sufficient evidence to conclude that there is a difference between the locations of the probability distributions for the sampled populations? Test using $\alpha = .05$.

SAMPLE FROM POPULATION 1	SAMPLE FROM POPULATION 2
1.2	1.5
1.9	1.3
.7	2.9
2.5	1.9
1.0	2.7
1.8	3.5
1.1	

16.57 An experiment was conducted using a randomized block design with six treatments and four blocks. The data are shown in the table. Do the data provide sufficient evidence to conclude that at least two of the treatment probability distributions differ in location? Test using $\alpha = .05$.

TREATMENT	BLOCK			
	1	2	3	4
1	75	77	70	80
2	65	69	63	69
3	74	78	69	80
4	80	80	75	86
5	69	72	63	77
6	78	79	72	81

16.58 A random sample of ten measurements from a continuous population produced the following measurements:

286	416	333	307	115
197	225	108	345	279

Conduct a test to determine whether the median of the population exceeds 200. Test using $\alpha = .01$.

16.59 Medical researchers believe that exposure to dust from cotton bract induces respiratory disease in susceptible field workers. An experiment was conducted to determine the effect of air-dried green cotton bract extract (GBE) on the cells of non-dust exposed mill workers (*Environmental Research*, Feb. 1986). Blood samples taken on six workers were incubated with varying concentrations of GBE. After a short period of time, the cyclic AMP level (a measure of cell activity expressed in picomoles per million cells) of each blood sample was measured. The data for two GBE concentrations, 0 mg/ml (salt solution) and .2 mg/ml, are reproduced in the table. [Note that one blood sample was taken from each worker, with one aliquot exposed to the salt buffer solution and the other to the GBE.] Conduct a test to detect a shift in the locations of the cyclic AMP level distributions for the two GBE concentrations. Test using $\alpha = .10$.

WORKER	GBE CONCENTRATION (mg/ml)	
	0	.2
A	8.8	4.4
B	13.0	5.7
C	9.2	4.4
D	6.5	4.1
F	9.1	4.4
H	17.0	7.9

Source: Butcher, B. T., Reed, M. A., and O'Neil, C. E. "Biochemical and Immunologic Characterization of Cotton Bract Extract and Its Effect on *in vitro* Cyclic AMP Production." *Environmental Research*, Vol. 39, No. 1, Feb. 1986, p. 119.

BEFORE	AFTER
12	4
5	2
10	7
9	3
14	8
6	

16.60 A study was conducted to determine whether the installation of a traffic light was effective in reducing the number of accidents at a busy intersection. Samples of 6 months prior to installation and 5 months after installation of the light yielded the numbers of accidents per month indicated in the table.
a. Is there sufficient evidence to conclude that the probability distribution of the number of monthly accidents before installation of the traffic light is shifted to the right of the probability distribution of the number of accidents after installation of the light? Test using $\alpha = .025$.
b. Explain why this type of data might or might not be suitable for analysis using the t test of Chapter 9.

16.61 Weevils cause millions of dollars worth of damage each year to cotton crops. Three chemicals designed to control weevil populations were applied, one to each of three fields of cotton. After 3 months, ten plots of equal size were randomly selected within each field and the percentage of cotton plants with weevil damage was recorded for each. Do the data in the table provide sufficient evidence to indicate a difference in location among the distributions of damage rates corresponding to the three treatments? Use $\alpha = .05$.

A	B	C
10.8	22.3	9.8
15.6	19.5	12.3
19.2	18.6	16.2
17.9	24.3	14.1
18.3	19.9	15.3
9.8	20.4	10.8
16.7	23.6	12.2
19.0	21.2	17.3
20.3	19.8	15.1
19.4	22.6	11.3

16.62 An experiment was conducted to study the effect of reinforced flanges on the torsional capacity of reinforced concrete T-beams (*Journal of the American Concrete Institute*, Jan.–Feb. 1986). Several different types of T-beams were used in the experiment, each type having a different flange width. The beams were tested under combined torsion and bending until failure (i.e., cracking). One variable of interest is the cracking torsion moment at the top of the flange of the T-beam. Cracking torsion moments for eight beams with 70-cm slab widths and eight beams with 100-cm slab widths are recorded below:

70-cm slab width: 6.00, 7.20, 10.20, 13.20, 11.40, 13.60, 9.20, 11.20
100-cm slab width: 6.80, 9.20, 8.80, 13.20, 11.20, 14.90, 10.20, 11.80

Is there evidence of a difference in the locations of the cracking torsion moment distributions for the two types of T-beams? Test using $\alpha = .10$.

16.63 *Acid rain* is considered by some environmentalists to be the nation's most serious environmental problem. It is formed by the combination of water vapor in clouds with nitrogen oxide and sulfuric dioxide emissions from the burning of coal, oil, and natural gas. The acidity of rain in central and northern Florida consistently ranges from 4.5 to 5 on the pH scale, a decidedly acid condition. To determine the effects of acid rain on the acidity of soils in a natural ecosystem, engineers at the University of Florida's Institute of Food and Agricultural Sciences irrigated experimental plots near Gainesville, Florida, with acid rain at two pH levels: 3.7 and 4.5. The acidity of the soil was then measured at three different depths: 0–15, 15–30, and 30–46 centimeters. Tests were conducted during three different time periods in 1981. The resulting soil pH values are shown in the table. Suppose the main objective of the experiment is to compare the acidity of soil irrigated with pH 4.5 acid rain to the acidity of soil irrigated with pH 3.7 acid rain, and that the different soil depths are to be treated as blocks.

		APRIL 3, 1981		JUNE 16, 1981		JUNE 30, 1981	
		ACID RAIN pH		ACID RAIN pH		ACID RAIN pH	
		3.7	4.5	3.7	4.5	3.7	4.5
	0–15 cm	5.33	5.33	5.47	5.47	5.20	5.13
SOIL DEPTH	15–30 cm	5.27	5.03	5.50	5.53	5.33	5.20
	30–46 cm	5.37	5.40	5.80	5.60	5.33	5.17

Source: "Acid Rain Linked to Growth of Coal-Fired Power." *Florida Agricultural Research 83*, Vol. 2, No. 1, Winter 1983.

a. Use a nonparametric test to compare the soil pH values of the two treatments on April 3, 1981.
b. Use a nonparametric test to compare the soil pH values of the two treatments on June 16, 1981.
c. Use a nonparametric test to compare the soil pH values of the two treatments on June 30, 1981.
d. Comment on the validity of the tests in parts a–c.

16.64 The thermogravimetric balance (TG) is a new technique developed to evaluate the thermal behavior of chemical compounds. Abou El Naga and Salem (1986) compared the TG technique to the standard method of evaluating the thermooxidation stability of base oils and their additive blends (e.g., transformer oils, turbine oils, transmission oils, and so forth). For each of a sample of ten base oils, the amount of oxidative compounds formed at the oxidation point was determined using the TG technique and the total percentage of oxidation products was determined by the standard method. The results of the experiment are shown in the accompanying table. Do the data provide sufficient evidence to indicate that the oxidation measurements of the two methods are positively correlated? Test using $\alpha = .01$.

BASE OIL	TG TECHNIQUE Amount of Oxidative Compounds, % weight	STANDARD METHOD Total Oxidation Products, %
1	25.4	2.3
2	27.11	2.5
3	28.0	2.65
4	17.9	1.3
5	18.9	1.45
6	22.9	1.9
7	30.8	3.3
8	18.6	1.4
9	24.4	2.1
10	29.8	2.9

Source: Abou El Naga, H. H. and Salem, A. E. M. "Base Oils Thermooxidation." *Lubrication Engineering*, Vol. 24, No. 4, Apr. 1986, p. 213. Reprinted by permission of the American Society of Lubrication Engineers. All rights reserved.

16.65 Governmental agencies periodically monitor nuclear-powered electrical generating plants for the purpose of establishing radiation guidelines. These guidelines then permit the detection of any changes resulting from operation that may endanger the surrounding environment. In 1978–1979, the Department of Health and Rehabilitative Services (DHRS) monitored three nuclear power plants in Florida for radiation in air particulates. The data shown in the table represent mean gross beta values (pCi/m^3)—a measure of radioactive air particulates—recorded at each plant and an Orlando control site for the first 10 weeks in 1979.

WEEK	ORLANDO	TURKEY POINT	ST. LUCIE	CRYSTAL RIVER
1	.048	.023	.023	.041
2	.019	.025	.020	.032
3	.022	.026	.022	.025
4	.015	.026	.028	.027
5	.027	.034	.031	.030
6	.122	.035	.025	.033
7	.013	.033	.020	.080
8	.007	.022	.022	.026
9	.025	.021	.013	.015
10	.025	.027	.026	.042

Source: "Monitoring of Nuclear Power Plant Environs in Florida: 1978–79." Dept. of Health and Rehabilitative Services, Health & Technical Support Services, Radiological Health Services.

a. Do the data provide sufficient evidence to indicate a difference in the radioactivity of air particulates at the four Florida sites? Test using $\alpha = .05$.

b. Do the data provide sufficient evidence to indicate a difference in the radioactivity of air particulates at the Crystal River plant site and the Orlando control site? Test using $\alpha = .05$.

16.66 Bulk specimens of Chilean lumpy iron ore (95% particle size, 150 millimeters) were randomly sampled from a 35,325-long-ton shipload of ore and the percentage of iron in each ore specimen was determined (*Reports of Statistical Application Research, Union of Japanese Scientists and Engineers*, Vol. 18, 1971). The data for ten bulk specimens are given below:

> 63.01   61.75   63.22   62.38   62.80
> 63.92   62.94   63.71   62.10   64.34

Is there sufficient evidence to indicate that the median percentage of iron in bulk specimens from the shipload of ore differs from 63? Test using $\alpha = .05$.

16.67 In the early 1960's, the air-conditioning systems of a fleet of Boeing 720 jet airplanes came under investigation. The accompanying table presents the lifelengths (in hours) of the air-conditioning systems in two different Boeing 720 planes. Assuming the data represent random samples from the respective populations, is there evidence of a shift in the location of the lifelength distributions of air-conditioning systems for the two Boeing 720 planes? Test using $\alpha = .05$.

PLANE 1		PLANE 2	
23	49	59	230
118	10	32	54
90	310	14	152
29	76	102	67
156	62	66	34

Source: Hollander, M., Park, D. H., and Proschan, F. "Testing Whether *F* Is 'More NBU' Than Is *G*." *Microelectronics and Reliability*, Vol. 26, No. 1, 1986, p. 43 (Table 1). Copyright 1986, Pergamon Press, Ltd. Reprinted with permission.

16.68 An experiment was conducted to compare three calculators, A, B, and C, according to ease of operation. To make the comparison, six randomly selected students were assigned to perform the same sequence of arithmetic operations on each of the three calculators. The order of use of the calculators varied in a random manner from student to student. The times necessary for the completion of the sequence of tasks (in seconds) are recorded in the table. Do the data provide sufficient evidence to indicate that the task completion times tend to be lower for at least one of the calculator types?

STUDENT	CALCULATOR TYPE		
	A	B	C
1	306	330	300
2	260	265	285
3	281	290	277
4	288	301	305
5	301	309	319
6	262	245	240

16.69 The oxygen supply of coastal marine sediments is of great importance to plant and animal communities in the ocean. An engineer wants to compare the mean depths of penetration of oxygen into coastal marine sediments at five different water depths: 5, 10, 15, 20, and 40 meters. Independent samples of marine sediments at each water depth were randomly selected and measured for depth of oxygen penetration by polarographic oxygen microelectrodes. The data are recorded (in millimeters) in the accompanying table. Is there evidence of a difference among the probability distributions of the depths of penetration of oxygen into marine sediments for the five water depths? Test using $\alpha = .05$.

WATER DEPTH (METERS)				
5	10	15	20	40
1.2	3.7	1.7	2.8	4.4
3.6	4.6	2.2	4.6	5.3
1.3	3.1	1.6	3.7	3.6
1.8	2.0	1.5	3.2	4.8
2.4	3.0	2.0	3.1	3.9

16.70 Airplane pilots are trained to act quickly and decisively in the face of an emergency. Aeronautical specialists believe that by finding the most efficient arrangement of instruments on the control panel of a plane they can save the pilots precious seconds in an emergency. Two different arrangements were compared by simulating an emergency condition and then measuring the reaction time required to correct the problem. Twenty pilots were selected and randomly assigned to the two different arrangements. The following reaction times were measured (in seconds):

Arrangement 1:   9, 4, 9, 5, 11, 4, 12, 5, 7, 6
Arrangement 2:   6, 13, 8, 11, 16, 9, 8, 10, 9, 12

Use a nonparametric test to check the specialist's claim that arrangement 1 leads to faster reaction times. Use $\alpha = .05$.

16.71 As part of a computer system performance evaluation, a system manager is interested in predicting the response time for computer terminals. *Terminal response time* is defined as the length of time (in seconds) required for the computer to respond to a command sent from a computer terminal by pressing one of the terminal's program function keys. One variable that influences terminal response time is the number of simultaneous users (that is, the number of users accessing the computer's central processing unit at the same time the command was sent). Refer to the sample data given in the table.

NUMBER OF SIMULTANEOUS USERS x	TERMINAL RESPONSE TIME y, seconds
1	.22
2	.59
3	1.36
4	1.01
5	1.42

a. Compute Spearman's rank coefficient of correlation between the number of simultaneous users and terminal response time.

b. Do the data provide sufficient evidence to indicate that there is a positive correlation between the two variables? Test using $\alpha = .10$.

16.72 An experiment was conducted to compare the compressive strengths of two types of plastic. Eight specimens of each type of plastic were tested and the compressive strengths (in thousands of pounds per square inch) were recorded as follows:

Type 1:　14.1, 14.3, 13.8, 14.2, 14.0, 14.5, 13.9, 13.7
Type 2:　14.2, 13.9, 13.8, 14.3, 14.1, 13.4, 13.8, 14.0

Do the data provide evidence of a difference in the probability distributions of the compressive strength measurements of the two plastics? Test using $\alpha = .05$.

16.73 A state highway patrol was interested in knowing whether frequent patrolling of highways substantially reduced the number of speeders. Two similar interstate highways were selected for the study—one very heavily patrolled and the other only occasionally patrolled. After 1 month, random samples of 100 cars were chosen on each highway and the number of cars exceeding the speed limit was recorded. This process was repeated on 5 randomly selected days. The data are shown in the table.

DAY	HIGHWAY 1 (Heavily patrolled)	HIGHWAY 2 (Occasionally patrolled)
1	35	60
2	40	36
3	25	48
4	38	54
5	47	63

a. Do the data provide evidence to indicate that the heavily patrolled highway tends to have fewer speeders per 100 cars than the occasionally patrolled highway? Test using $\alpha = .05$.

b. Use the paired t test with $\alpha = .05$ to compare the population mean number of speeders per 100 cars for the two highways. What assumptions are necessary for the validity of this procedure?

16.74 The number of machine breakdowns per month was recorded for 9 months on two identical machines used to fill 16-ounce bottles of cough syrup. The data are shown in the table. Do the data provide sufficient evidence to indicate a difference in the monthly breakdown rates for the two machines? Use $\alpha = .05$.

MONTH	MACHINE A	MACHINE B
March	5	9
April	16	14
May	9	11
June	12	17
July	11	13
August	8	8
September	15	14
October	8	7
November	9	15

16.75 A study was conducted to explore the sources of occupational stress for engineers (*IEEE Transactions on Engineering Management*, Feb. 1986). One of the objectives was to determine "if there are consistent and significant differences among engineers at different levels of the organizational hierarchy in the degree to which they consider the different factors as sources of stress." A sample of male engineers from different types of organizations in Ontario, Canada, was administered the Stress Diagnostic Survey (SDS). The SDS provides stress ratings for each of 15 categories of work stressors. The researchers ranked 15 stress categories from 1 (highest stress) to 15 (lowest stress) for each of four groups of engineers—nonsupervisors, first-level supervisors, second-level supervisors, and third-level supervisors—as shown in the table. Conduct a test to determine whether the rank orderings of the stress categories differ among the four groups of engineers. Test using $\alpha = .01$.

STRESS CATEGORY	NONSUPERVISORS	1ST LEVEL	2ND LEVEL	3RD LEVEL
Politics	5	6	4	7
Underutilization	3	5	7	5
Human resources development	4	3	2	3
Supervisory style	6	7	8	8
Rewards	1	1	1	4
Organizational structure	9	9	10	12
Participation	2	4	5	6
Role ambiguity	10	13	12	14
Overload/quantitative	15	15	15	15
Overload/qualitative	12	8	6	2
Time pressure	8	2	3	1
Role conflict	13	10	13	10
Career progression	7	10	9	13
Job scope	11	12	14	11
Responsibility for people	14	14	11	9

Source: Saleh, S. D. and Desai, K. "Occupational Stress for Engineers." *IEEE Transactions on Engineering Management*, Vol. EM-33, No. 1, Feb. 1986, p. 8 (Table II). © 1986, IEEE.

OPTIONAL SUPPLEMENTARY EXERCISES

[*Note:* The following exercises require the use of a computer and computer simulation techniques.]

16.76 Throughout this chapter we have omitted the theoretical derivations of the null distributions of the various nonparametric test statistics. However, we can use computer simulation to derive approximate rejection regions for the tests. Consider the problem of finding the approximate sampling distribution of the Wilcoxon rank sum statistic for the case $n_1 = n_2 = 10$.

  a. Write a computer program that will randomly order the $n = n_1 + n_2 = 20$ ranks and compute the corresponding Wilcoxon rank sum T_1. This can be accomplished using a random number generator (see Section 7.6).

  b. Write a computer program that will repeat the instructions of part **a** $N = 1,000$ times.

  c. Construct a relative frequency distribution for the $N = 1,000$ computer-generated values of T_1 (refer to Chapter 2). This simulated distribution represents an approximation to the sampling distribution of T_1.

  d. Use the simulated sampling distribution to determine the value $T_{.05}$, such that $P(T_1 \leq T_{.05}) = .05$. This value represents the one-tailed critical value of the Wilcoxon rank sum test for $\alpha = .05$.

16.77 Follow the steps outlined in Optional Exercise 16.76 to find the approximate critical value (at $\alpha = .05$) of Spearman's test for rank correlation for the case $n = 10$. [*Hint:* In part **a** you will need to randomly order the $n = 10$ y ranks and $n = 10$ x ranks.]

REFERENCES

Friedman, M. "The Use of Ranks to Avoid the Assumption of Normality Implicit in the Analysis of Variance," *Journal of the American Statistical Association*, Vol. 32, 1937.

Gibbons, J. D. *Nonparametric Statistical Inference*. New York: McGraw-Hill, 1971.

Hollander, M. and Wolfe, D. A. *Nonparametric Statistical Methods*. New York: Wiley, 1973.

Kruskal, W. H. and Wallis, W. A. "Use of Ranks in One-Criterion Variance Analysis," *Journal of the American Statistical Association*, Vol. 47, 1952.

Lehmann, E. L. *Nonparametrics: Statistical Methods Based on Ranks*. San Francisco: Holden-Day, 1975.

Siegel, S. *Nonparametric Statistics for the Behavioral Sciences*. New York: McGraw-Hill, 1956.

Wilcoxon, F. and Wilcox, R. A. "Some Rapid Approximate Statistical Procedures," The American Cyanamid Co., 1964.

OBJECTIVE

To present some statistical methods for estimating the probability that a manufactured product or a system will perform satisfactorily for a specified period of time.

CONTENTS

APPLICATIONS: PRODUCT AND SYSTEM RELIABILITY

SECTION 17.1

INTRODUCTION

Do your stereo system and your automobile perform well for a reasonably long period of time? If they do, we would say that these products are *reliable*.

The **reliability** of a product is the probability that the product will meet certain specifications for a given period of time. For example, suppose we want a new automobile to perform without malfunction for a period of 2 years or for 20,000 miles. The probability that an automobile will meet these specifications is the *reliability* of the automobile.

DEFINITION 17.1

The **reliability** of a product is the probability that the product will meet a set of specifications for a given period of time.

Some products need to function on a one-time basis. Others repeat a function over and over until they eventually fail. For example, a fuse either works or does not work when an electrical circuit is overloaded. The reliability of a fuse is the probability that it will work when subjected to a specific overload. In contrast, an automobile is used over and over again; its reliability is the probability that the automobile will perform without a major malfunction for some specified period of time.

SECTION 17.2

FAILURE TIME DISTRIBUTIONS

The *length of life* of a product is the length of time until the product fails to perform according to specifications. When the product fails to perform according to specifications, it is said to have *failed*.

The time at which a single product item fails is called the **failure time** for the item. For example, the length of life of an abrasive grinding wheel is the length of time until the wheel fails to perform according to specifications. The specifications may have been determined by the manufacturer or the user may have written his or her own specifications. The length of time until failure is called the failure time of the wheel.

DEFINITION 17.2

The **failure time** of a product is the length of time that the product performs according to specifications.

The failure time t for any product varies from one item to another and is, in fact, a random variable. The density function for a product failure time is called a **failure time distribution**. A typical failure time distribution might appear as shown in Figure 17.1.

FIGURE 17.1
A Failure Time Distribution

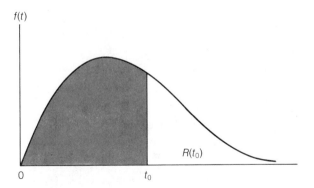

$f(t)$

$R(t_0)$

0 t_0

DEFINITION 17.3

The **failure time distribution** for a product is the density function of the failure time t.

If we denote the failure time density function by the symbol $f(t)$, then the probability that the product will fail before time t_0 is

$$F(t_0) = \int_0^{t_0} f(t)\, dt$$

This probability is the shaded area under the density function shown in Figure 17.1.

Suppose that a product is said to be *reliable* if it survives until time t_0. Then the reliability of the product—that is, the probability that it will survive until time t_0—is

$$R(t_0) = 1 - F(t_0)$$

This probability, $R(t_0)$, is the unshaded area under the density function to the right of t_0 in Figure 17.1.

Realistically, the failure time distribution is a conceptual relative frequency distribution of the lengths of life of some group of product items of specific interest—say, those manufactured in a given week, month, or year. Based on an analysis of sample data, we may select one of the density functions described in Chapter 5 to model this distribution. The family of density functions represented by the Weibull distribution (discussed in Section 5.7) is often used for this purpose.

S E C T I O N 17.3

HAZARD RATES

The failure time distribution for a product enables us to calculate the probability $F(t_0)$ that an item will fail before time t_0 and the probability $R(t_0) = 1 - F(t_0)$ that the item will survive until time t_0. The probability that an item will fail in the interval $(t, t + dt)$ is the shaded area shown in Figure 17.2 (page 872). The density $f(t)$, the height of the shaded rectangle, is proportional to this probability.

FIGURE 17.2

A Failure Time Distribution
Showing the Approximate
Probability of Failure
During the Interval
$(t, t + dt)$

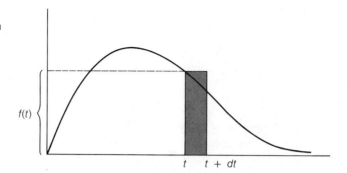

Another way to describe the life characteristics of a product is to use a measure of the probability of failure as the product gets older—that is, the probability that the product will fail in the interval $(t, t + dt)$, given that the item has survived to time t.

If we define the events

A: Item fails in the interval $(t, t + dt)$

B: Item survives until time t

then the probability of failure in the time interval $(t, t + dt)$, given that the item has survived to time t, is

$$P(A \mid B) = \frac{P(A \cap B)}{P(B)}$$

But, the event $A \cap B$ is equivalent to the event A—that is, an item must have survived to time t in order for it to be able to fail in the interval $(t, t + dt)$. Therefore,

$$P(A \cap B) = P(A)$$

This probability is approximately equal to the shaded area in Figure 17.2. Then the probability of failure in the interval $(t, t + dt)$, given that the item has survived to time t, is

$$P(A \mid B) = \frac{P(A \cap B)}{P(B)} \approx \frac{f(t) \, dt}{1 - F(t)} = \frac{f(t) \, dt}{R(t)}$$

The quantity

$$z(t) = \frac{f(t)}{R(t)}$$

is proportional to this conditional probability and is called the **hazard rate** for the product.

Knowledge about a product's hazard rate often helps us to select the appropriate failure time density function for the product. The following example illustrates the point.

DEFINITION 17.4

The **hazard rate** for a product is defined to be

$$z(t) = \frac{f(t)}{1 - F(t)} = \frac{f(t)}{R(t)}$$

where $f(t)$ is the density function of the product's failure time distribution.

EXAMPLE 17.1

The exponential distribution (discussed in Section 5.6) is often used in industry to model the failure time distribution of a product. Find the hazard rate for the exponential distribution.

SOLUTION

The exponential density function and cumulative distribution function are, respectively,

$$f(t) = \frac{e^{-t/\beta}}{\beta} \qquad 0 \le t < \infty, \quad \beta > 0$$

and

$$F(t) = \int_{-\infty}^{t} f(y) \, dy = \int_{0}^{t} \frac{e^{-y/\beta}}{\beta} \, dy = 1 - e^{-t/\beta}$$

Then the hazard rate for the exponential distribution is

$$z(t) = \frac{f(t)}{1 - F(t)} = \frac{\dfrac{e^{-t/\beta}}{\beta}}{1 - (1 - e^{-t/\beta})} = \frac{1}{\beta}$$

Since $\beta = E(t)$ is the mean life of the product, it follows that the hazard rate is constant (see Figure 17.3). Therefore, a product that has an exponential failure time distribution never becomes fatigued. It is just as likely to survive any one unit of time as it is any other.

FIGURE 17.3
Hazard Rate for the
Exponential Failure Time
Distribution

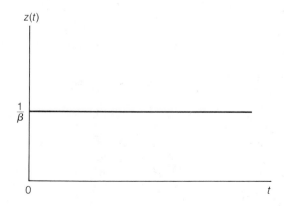

∎

Clearly, the exponential distribution would not provide a good model for the failure time distribution of humans or for industrial products that become fatigued and more prone to failure as they get older. But it does provide a good model for some products, particularly for complex systems whose parts are replaced as they fail. After such systems have been in operation for a while, the probability of failure tends to be as likely in any one unit of time as in any other. Failure time distributions that exhibit this property (i.e., a constant hazard rate) are often called **memoryless distributions**.

The Weibull distribution density function (discussed in Section 5.7) and cumulative distribution function are, respectively,

$$f(t) = \frac{\alpha}{\beta} t^{\alpha - 1} e^{-t^{\alpha}/\beta} \qquad 0 \le t < \infty; \quad \alpha > 0; \quad \beta > 0$$

and

$$F(t) = 1 - e^{-t^{\alpha}/\beta}$$

By changing the shape parameter α and the scale parameter β, we obtain a variety of density functions useful for modeling failure time distributions for many industrial products. For $\alpha = 1$, we obtain the exponential distribution.

EXAMPLE 17.2

Find the hazard rate for the Weibull distribution and graph $z(t)$ versus time for $\alpha = 1, 2,$ and 3.

SOLUTION

Using the density function and cumulative distribution functions given above, we determine the hazard rate for the Weibull distribution:

$$z(t) = \frac{f(t)}{1 - F(t)} = \frac{\left(\frac{\alpha}{\beta}\right) t^{\alpha - 1} e^{-t^{\alpha}/\beta}}{1 - (1 - e^{-t^{\alpha}/\beta})} = \frac{\alpha}{\beta} t^{\alpha - 1}$$

When the shape parameter α is equal to 1, we obtain

$$z(t) = \frac{1}{\beta}$$

which is the constant hazard rate for the exponential distribution. For $\alpha = 2$,

$$z(t) = \frac{2}{\beta} t$$

the equation of a straight line passing through the origin. For $\alpha = 3$,

$$z(t) = \frac{3}{\beta} t^2$$

a second-order function of time t. Graphs of these hazard rates are shown in Figure 17.4. Note that the hazard rate increases more rapidly with time for larger values of the shape parameter α.

FIGURE 17.4
Graphs of the Hazard
Rate for Weibull
Distributions with
$\alpha = 1, 2, 3$

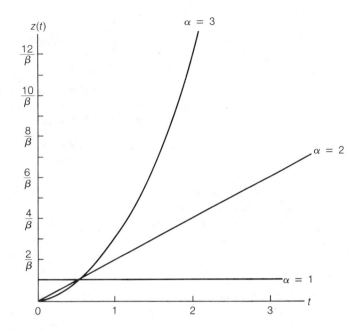

EXERCISES 17.1–17.6

17.1 Find and graph the hazard rate $z(t)$ for a Weibull distribution with:
  a. $\alpha = 1$, $\beta = 4$
  b. $\alpha = 2$, $\beta = 2$
  c. $\alpha = 3$, $\beta = 6$

17.2 Graph the hazard rate $z(t)$ for Weibull distributions with $\beta = 1$ and $\alpha = 1, 2, 3$, and 4.

17.3 Suppose the failure time distribution for a product can be approximated by a normal distribution with $\mu = 3$ and $\sigma = 1$.
  a. Find $f(t)$, $F(t)$, and $z(t)$ for $t = 0, 1, 2, \ldots, 6$.
  b. Plot the values of $z(t)$ for corresponding values of t and obtain a graph of the hazard rate for this normal failure time distribution.

17.4 The lifetime t (in hours) of a certain electronic component is a random variable with density function

$$f(t) = \begin{cases} \dfrac{1}{100}e^{-t/100} & t > 0 \\ 0 & \text{elsewhere} \end{cases}$$

  a. Find $F(t)$ and $R(t)$.
  b. What is the reliability of the component at $t = 25$ hours?
  c. Find $z(t)$ and interpret the result.

17.5 A drill bit has a failure time distribution given by the density

$$f(t) = \begin{cases} \dfrac{2te^{-t^2/100}}{100} & 0 \le t < \infty \\ 0 & \text{elsewhere} \end{cases}$$

a. Find $F(t)$.

b. Find expressions for the reliability $R(t)$ and the hazard rate $z(t)$ of the drill bit at time t.

c. Use the results of part **b** to find $R(8)$ and $z(8)$.

17.6 The failure of a computer disk pack is considered to be an *initial failure* if it occurs prior to time $t = \alpha$ and a *wear-out failure* if it occurs after time $t = \beta$. Suppose the failure time distribution during the useful life of the disk pack is given by

$$f(t) = \frac{1}{\beta - \alpha} \qquad \alpha \leq t \leq \beta$$

a. Find $F(t)$ and $R(t)$.

b. Find the hazard rate $z(t)$.

c. Graph the hazard rate of the disk pack for $\alpha = 100$ hours and $\beta = 1{,}500$ hours.

d. For $\alpha = 100$ and $\beta = 1{,}500$, what is the reliability of the disk pack at time $t = 500$ hours? What is the hazard rate?

SECTION 17.4

LIFE TESTING: CENSORED SAMPLING

A **life test** is an experiment conducted to obtain sample values of the lengths of life of some product items. Typically, a random sample of n items is placed on test under specified environmental conditions and left on test until they fail. The recorded times to failure, $t_1, t_2, \ldots, t_n$, provide a random sample of observations on the length of life t of the product. If for convenience we let t_1 represent the smallest failure time, t_2 the second smallest, . . . , and t_n the largest, then the times might appear as points on a time line, as shown in Figure 17.5.

FIGURE 17.5

Failure Times of n Items of Some Product

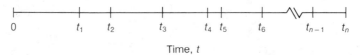

Time, t

In many situations, life tests are conducted to determine the quality of a manufactured product prior to sale. Waiting for the last few items in a sample to fail can be time-consuming and expensive. To reduce the cost of waiting for some long-life items, tests are often concluded after a specified length T of test time. When we do this, we say that the life test is **censored** at time T. A second type of censored sampling occurs when we conclude the testing after a fixed number r of items have failed.

If a life test is censored at a fixed time T, the length of testing time is fixed. This makes it easier to schedule the life-testing equipment, but the number r of failures observed prior to time T is a random variable. Thus, r could assume any integer value in the interval $0 \leq r \leq n$, and it is possible that no failure times would be observed. If the test is censored after a fixed number r of failures have been observed, we know that we will always acquire the values of r failure times, but the length of the life test will be variable and equal to the length of time t_r until failure of the rth item.

There are many other types of life-testing procedures. In **life testing with replacement**, product items are replaced on the test equipment as soon as an item fails, a procedure that makes maximum use of the test equipment. Other tests are designed to investigate the effect of various stresses on a product by testing the items under varying stresses. Tests of this type are called **accelerated life tests**. Descriptions of these and other life test procedures, as well as methods for using the sample data to estimate the parameters of failure time distributions, are described in the references at the end of the chapter.

| | | | | | | | | | | | |

SECTION 17.5

ESTIMATING THE PARAMETERS OF AN EXPONENTIAL FAILURE TIME DISTRIBUTION

The methods for finding estimators of the parameters of failure time distributions are those described in the preceding chapters. We can use the method of moments (Chapter 8), method of maximum likelihood (Chapter 8), or the method of least squares (Chapter 10). Depending on the failure time distribution and the number of parameters involved, finding the estimator may be easy or difficult. For example, finding the maximum likelihood estimator of the parameter of an exponential distribution based on simple random sampling is easy (see Example 8.5), but solving the maximum likelihood equations obtained for estimating the parameters of a Weibull distribution is relatively difficult.

It is also difficult to obtain estimators and their sampling distributions based on certain types of sampling, especially when the sampling has been censored at a fixed time T. Consequently, in this and the following sections, we will present estimation procedures for the exponential and the Weibull failure time distributions. Estimation procedures for these and other failure time distributions are discussed in the literature or in texts on product and system reliability.

We will first consider estimators for the mean failure time β for the exponential distribution. Estimators for β are the same regardless of whether the life test is censored or uncensored; the estimator is always equal to the total observed life divided by the number r of failures observed. For example, if a random sample of n items is selected from the population and the life test is concluded after the rth failure is observed ($r > 0$), then

$$\hat{\beta} = \frac{\sum_{i=1}^{r} t_i + (n-r)t_r}{r} = \frac{\text{Total observed life}}{r}$$

If we wait until all n items fail, then $r = n$ and the estimator is the sample mean failure time:

$$\hat{\beta} = \frac{\sum_{i=1}^{n} t_i}{n} = \bar{t}$$

Note that for both the censored and uncensored sampling situations, the numerator in the above expressions is equal to the total length of life observed for the n items during the length of the life test.

For censored sampling with a fixed time of testing T,

$$\hat{\beta} = \frac{\sum\limits_{i=1}^{r} t_i + (n - r)T}{r} = \frac{\text{Total observed life}}{r} \qquad \text{for } r \geq 1$$

Again, note that the numerator in this expression is the total length of life recorded for the n items until the life test is concluded at time T.

POINT ESTIMATORS OF THE MEAN LIFE β FOR AN EXPONENTIAL DISTRIBUTION

For uncensored life testing:

$$\hat{\beta} = \frac{\sum\limits_{i=1}^{n} t_i}{n}$$

For censored sampling with r fixed:

$$\hat{\beta} = \frac{\sum\limits_{i=1}^{r} t_i + (n - r)t_r}{r}$$

For censored sampling with test time T fixed:

$$\hat{\beta} = \frac{\sum\limits_{i=1}^{r} t_i + (n - r)T}{r}$$

The formulas for $(1 - \alpha)100\%$ confidence intervals for β are shown in the following boxes. The confidence interval based on sampling censored at a fixed point in time is only approximate.

A $(1 - \alpha)100\%$ CONFIDENCE INTERVAL FOR β
BASED ON CENSORED SAMPLING WITH r FIXED

$$\frac{2(\text{Total life})}{\chi^2_{\alpha/2}} \leq \beta \leq \frac{2(\text{Total life})}{\chi^2_{(1-\alpha/2)}}$$

where

$$\text{Total life} = \sum_{i=1}^{r} t_i + (n - r)t_r$$

and $\chi^2_{\alpha/2}$ and $\chi^2_{(1-\alpha/2)}$ are the tabulated values of a chi-square statistic, based on $2r$ degrees of freedom, that locate $\alpha/2$ in the upper and lower tails, respectively, of the chi-square distribution.

AN APPROXIMATE $(1 - \alpha)100\%$ CONFIDENCE INTERVAL FOR β BASED ON CENSORED SAMPLING WITH T FIXED

$$\frac{2(\text{Total life})}{\chi^2_{\alpha/2}} \leq \beta \leq \frac{2(\text{Total life})}{\chi^2_{(1-\alpha/2)}}$$

where

$$\text{Total life} = \sum_{i=1}^{r} t_i + (n - r)T$$

and $\chi^2_{\alpha/2}$ and $\chi^2_{(1-\alpha/2)}$ are the tabulated upper- and lower-tail values of a chi-square distribution based on $(2r + 2)$ degrees of freedom.

EXAMPLE 17.3

Suppose that the length of time between malfunctions for a particular type of aircraft engine has an exponential failure time distribution. Ten of the engines were tested until six of the engines malfunctioned. The times to malfunction were 48, 35, 91, 62, 59, and 77 hours, respectively. Find a 95% confidence interval for the mean time β between malfunctions for the engines.

SOLUTION

Since this life test was concluded after the sixth failure was observed, it represents censored sampling with $r = 6$. The total observed life for the test was

$$\text{Total life} = \sum_{i=1}^{r} t_i + (n - r)t_r$$

$$= 372 + 364 = 736 \text{ hours}$$

The tabulated values of $\chi^2_{.025}$ and $\chi^2_{.975}$, based on $2r = 2(6) = 12$ degrees of freedom, are 23.3367 and 4.40379, respectively. Then the 95% confidence interval for β is

$$\frac{2(736)}{23.3367} \leq \beta \leq \frac{2(736)}{4.40379}$$

or $63.08 \leq \beta \leq 334.26$. Our interpretation is that the true mean time β between malfunctions of this particular type of aircraft engine falls between 63.08 hours and 334.26 hours, with 95% confidence.   ∎

EXAMPLE 17.4

Refer to Example 17.3.

a. Find a 95% confidence interval for the hazard rate of the aircraft engine.
b. Find a 95% confidence interval for the reliability of the system at time 50 hours.

SOLUTION

a. Recall from Section 17.3 that the hazard rate for the exponential distribution is $1/\beta$. We therefore begin with the 95% confidence interval for β derived in

Example 17.3 and transform it to a confidence interval for $1/\beta$:

$$63.08 \le \beta \le 334.26$$

$$\frac{1}{334.26} \le \frac{1}{\beta} \le \frac{1}{63.08}$$

$$.003 \le \frac{1}{\beta} \le .016$$

Thus, the hazard rate for the aircraft engine at time t (which is proportional to the probability that the engine will fail during a fixed small interval of time, given that the engine has survived to time t) falls between .003 and .016 with 95% confidence.

b. From Example 17.1, the cumulative distribution function for the exponential distribution is

$$F(t) = 1 - e^{-t/\beta}$$

By definition, the reliability of the aircraft engine at time t_0 is

$$R(t_0) = 1 - F(t_0)$$
$$= 1 - (1 - e^{-t_0/\beta}) = e^{-t_0/\beta}$$

or, for $t_0 = 50$ hours, $R(50) = e^{-50/\beta}$.
　　Then $63.08 \le \beta \le 334.26$ is equivalent to

$$e^{-50/63.08} \le e^{-50/\beta} \le e^{-50/334.26}$$

$$.453 \le e^{-50/\beta} \le .861$$

Therefore, the probability that the engine survives at least 50 hours may be as low as .453 or as high as .861, with 95% confidence.　　　　　　■

EXERCISES 17.7–17.12

17.7 A wet-mix, steel-fiber reinforced microsilica concrete (called *shotcrete*), used extensively in Scandinavia, is now being marketed in the United States. The material is said to have a minimum 28-day breaking strength of 9,000 pounds per square inch (psi) of compression. To investigate the breaking strength of the new product, seven pieces of shotcrete were subjected to 9,000 psi of compression daily until they failed. The times to failure were 33, 35, 61, 38, 21, 41, and 52 days. Assuming that the shotcrete has an exponential failure time distribution when subjected to 9,000 psi of compression, find a 90% confidence interval for the mean time β until the shotcrete fails.

17.8 Refer to Exercise 17.7.
　　a. Find a 90% confidence interval for the probability that the shotcrete will not fail before the 28-day specified minimum.
　　b. Find a 90% confidence interval for the hazard rate of the shotcrete.

17.9 A study was conducted to estimate the mean life (in miles) of a certain type of locomotive using censored sampling (*Technometrics*, May 1985). Ninety-six locomotives were operated for either 135 thousand miles or until failure. Of these, 37 failed before the 135-thousand-mile period. The accompanying table contains the miles to failure for these locomotives. Assuming an exponential failure time distribution, construct a 95% confidence interval for the mean miles to failure of the locomotives.

MILES TO FAILURE					
22.5	57.5	78.5	91.5	113.5	122.5
37.5	66.5	80.0	93.5	116.0	123.0
46.0	68.0	81.5	102.5	117.0	127.5
48.5	69.5	82.0	107.0	118.5	131.0
51.5	76.5	83.0	108.5	119.0	132.5
53.0	77.0	84.0	112.5	120.0	134.0
54.5					

Source: Schmee, J., Gladstein, D., and Nelson, W. "Confidence Limits for Parameters of a Normal Distribution from Singly-Censored Samples, Using Maximum Likelihood." *Technometrics*, Vol. 27, No. 2, May 1985, p. 119.

17.10 Suppose that an integrated circuit chip possesses an exponential failure time distribution. Fifteen chips were put on accelerated life test until five of the chips failed. The first five failures occurred at 18.2, 19.5, 24.8, 31.0, and 45.6 (in thousands of hours).
  a. Find a 95% confidence interval for the mean time between failures of the circuit chips.
  b. Find a 95% confidence interval for the reliability of the circuit chips at 20,000 hours.

17.11 A sample of 100 high-reliability capacitors was placed on test for 2,000 hours. At the end of this period only three capacitors had malfunctioned, with failure times of 810, 1,422, and 1,816 hours. Assuming an exponential failure time distribution, construct a 99% confidence interval for the mean time between failures of the capacitors. Interpret the interval.

17.12 Refer to Exercise 17.11.
  a. Find a 99% confidence interval for the hazard rate of the capacitors.
  b. Find a 99% confidence interval for the reliability of the capacitors at 3,000 hours.
  c. Find a 99% confidence interval for the probability that a capacitor will fail before 2,000 hours.

SECTION 17.6

ESTIMATING THE PARAMETERS OF A WEIBULL FAILURE TIME DISTRIBUTION

The method of maximum likelihood (discussed in Section 8.3) can be used to obtain estimates of the shape and scale parameters of the Weibull distribution, but the procedure is difficult and beyond the scope of this text. The interested reader should consult the references listed at the end of the chapter. The disadvantage of the method of maximum likelihood is that the estimates of α and β are obtained by solving a complicated pair of simultaneous nonlinear equations. The advantage of the method is that when the sample size n is large, maximum likelihood estimators possess sampling distributions that are approximately normal with known means and variances. This fact can be used to form large-sample confidence intervals using the method described in Section 8.4.

Instead of using the method of maximum likelihood to estimate α and β, we will use the method of least squares. You will recall that the cumulative distribution function for the Weibull distribution is

$$F(t) = 1 - e^{-t^{\alpha}/\beta}$$

Then the probability of survival to time t is

$$R(t) = 1 - F(t) = e^{-t^{\alpha}/\beta}$$

and

$$\frac{1}{R(t)} = e^{t^{\alpha}/\beta}$$

Taking the natural logarithms of both sides of this equation, we obtain

$$\ln\left[\frac{1}{R(t)}\right] = \frac{t^{\alpha}}{\beta}$$

$$-\ln R(t) = \frac{t^{\alpha}}{\beta}$$

$$\ln[-\ln R(t)] = -\ln \beta + \alpha \ln t$$

To use the method of least squares, we need to estimate the survival function based on life test data. One way to do this is to place a random sample of n items on life test and count the number of survivors at the end of one unit of time (for example, a week, or a month), after two units of time, and, in general, after i units of time, $i = 1, 2, \ldots$. The intervals of time are shown in Figure 17.6. An estimate of the proportion of survivors at time i is

$$\hat{R}(i) = \frac{n_i}{n}$$

where

n_i = Number of survivors at the end of the ith time unit

n = Total number of items placed on test

We would calculate $\hat{R}(i)$ for $i = 1, 2, \ldots$, and then fit the least squares line

$$\underbrace{\ln[-\ln \hat{R}(i)]}_{y} = \underbrace{-\ln \beta}_{\beta_0} + \underbrace{\alpha}_{\beta_1} \underbrace{\ln i}_{x}$$

to the data points (x_i, y_i), $i = 1, 2, \ldots$, where the ith data point is

$$y_i = \ln[-\ln \hat{R}(i)] \quad \text{and} \quad x_i = \ln i$$

FIGURE 17.6

EXAMPLE 17.5

A manufacturer of hydraulic seals conducted a life test during which the seals were subjected to a fluid pressure that was 200% of the pressure normally maintained in hydraulic systems in which the seal is used. One hundred seals were placed on test and the number of survivors was recorded at the end of each day for a period of 7 days, as listed in the following table.

Day	1	2	3	4	5	6	7
Number of Survivors	69	48	33	21	13	7	4

Use the data to estimate the parameters α and β for a Weibull distribution.

SOLUTION

The first step is to calculate $\hat{R}(i)$ and $\ln[-\ln \hat{R}(i)]$ for each of the seven time intervals. These calculations are shown in Table 17.1. The SAS printout for a simple linear regression for the data is shown in Figure 17.7.

TABLE 17.1

TIME i	$x_i = \ln i$	NUMBER OF SURVIVORS	$\hat{R}(i)$	$-\ln \hat{R}(i)$	$y_i = \ln[-\ln \hat{R}(i)]$
1	0	69	.69	.37106	−.99138
2	.69315	48	.48	.73397	−.30929
3	1.09861	33	.33	1.10866	.10315
4	1.38629	21	.21	1.56065	.44510
5	1.60944	13	.13	2.04022	.71306
6	1.79176	7	.07	2.65926	.97805
7	1.94591	4	.04	3.21888	1.16903

FIGURE 17.7 SAS Printout for Example 17.5

```
DEPENDENT VARIABLE: Y
SOURCE              DF      SUM OF SQUARES      MEAN SQUARE      F VALUE       PR > F      R-SQUARE         C.V.
MODEL               1       3.46814159          3.46814159       1074.81       0.0001      0.995370         18.8654
ERROR               5       0.01613374          0.00322675                     ROOT MSE                     Y MEAN
CORRECTED TOTAL     6       3.48427533                                         0.05680447                   0.30110339

SOURCE              DF      TYPE I SS       F VALUE      PR > F        DF        TYPE III SS      F VALUE      PR > F
X                   1       3.46814159      1074.81      0.0001        1         3.46814159       1074.81      0.0001

                                    T FOR HO:        PR > :T:      STD ERROR OF
PARAMETER           ESTIMATE        PARAMETER=0                    ESTIMATE
INTERCEPT           -1.05097757     -22.60           0.0001        0.04649563
X                   1.11019209      32.78            0.0001        0.03386353
```

From the printout, you can see that the least squares estimates are

$$\hat{\beta}_0 = -1.05098 \quad \text{and} \quad \hat{\beta}_1 = 1.11019$$

Since $\beta_0 = -\ln \beta$ and $\beta_1 = \alpha$, we have

$$\hat{\alpha} = \hat{\beta}_1 = 1.11019$$

and

$$\hat{\beta}_0 = -\ln \hat{\beta} \quad \text{or} \quad \hat{\beta} = e^{-\hat{\beta}_0} = 2.86045$$

Therefore, based on the method of least squares, we would use a Weibull distribution with parameters $\alpha = 1.11019$ and $\beta = 2.86045$ to model the failure time distribution of the hydraulic seals.

Note that we could form confidence intervals for α and β using the confidence limits for β_0 and β_1. The confidence interval for α would be the usual regression confidence interval for β_1 because $\alpha = \beta_1$. The confidence limits for β would be computed by substituting the upper and lower confidence limits for β_0 into the relationship $\beta = e^{-\beta_0}$.

To illustrate, the formula for a $(1 - \alpha)100\%$ confidence interval for β_0 given in Section 11.7 is

$$\hat{\beta}_0 \pm (t_{\alpha/2})s_{\hat{\beta}_0}$$

where $t_{\alpha/2}$ depends on $(n - 2)$ degrees of freedom and $s_{\hat{\beta}_0}$ is the estimated standard error of the estimate $\hat{\beta}_0$. From the SAS printout (Figure 17.7), $\hat{\beta}_0 = -1.05098$ and $s_{\hat{\beta}_0} = .46496$, and for $n - 2 = 7 - 2 = 5$ degrees of freedom and $\alpha = .05$, $t_{\alpha/2} = t_{.025} = 2.571$. Then a 95% confidence interval for β_0 is given by

$$-1.05098 \pm (2.571)(.046496)$$

$$-1.05098 \pm .11954$$

or $(-1.17052, -.93144)$. Consequently, a 95% confidence interval for the Weibull parameter β is

$$(e^{.93144}, e^{1.17052})$$

or $(2.5382, 3.2237)$. ∎

EXAMPLE 17.6

Use the estimates of α and β derived in Example 17.5 to find the probability that a hydraulic seal placed on test will survive at least 3 days.

SOLUTION

Recall that the probability of survival to time t_0 under a Weibull distribution is given by

$$R(t_0) = 1 - F(t_0) = e^{-t_0^{\alpha}/\beta}$$

Substituting the estimates $\hat{\alpha} = 1.11019$ and $\hat{\beta} = 2.86045$ into the equation for $t_0 = 3$ days, we have

$$\hat{R}(3) = e^{-31.11019/2.86045} = e^{-1.18375} = .30613$$

Therefore, the probability that the hydraulic seal will survive at least 3 days is estimated to be .30613.   ■

When n is small, we can still use the method of least squares to estimate the parameters of a Weibull distribution, but the preferred procedure is to estimate the probability of survival to time t, $R(t) = 1 - F(t)$, after each failure time has been observed. The data points used for the least squares methods are $[t_1, \hat{R}(t_1)]$, $[t_2, \hat{R}(t_2)], \ldots, [t_r, \hat{R}(t_r)]$, where t_1 is the first observed failure, t_2 the second, and so on. When this method of defining the data points is used, the estimator of the survival rates used in Example 17.5 is modified to

$$\hat{R}(t_i) = \frac{n_i + 1}{n + 1}$$

where n_i is the number of survivors when the ith failure time t_i is observed and n is the sample size.*

In concluding, note that $\hat{\alpha}$ and $\hat{\beta}$ will not possess the properties of the usual least squares estimators of β_0 and β_1. The response variable

$$y = \ln[-\ln \hat{R}(t)]$$

is not a normally distributed random variable and, in addition, the observed values of y are correlated. This is because the number of survivors at one point in time is dependent on the number observed at some previous point in time. The extent to which these violations of the regression analysis assumptions affect the properties of the estimators is unknown, but it is probably slight when the sample size n and the number r of observed failures are large.

| | | | | | | | | | | | | |

**EXERCISES
17.13–17.18**

17.13 Suppose the lifelength (in years) of a memory chip in a mainframe computer has a Weibull failure time distribution. In order to estimate the Weibull parameters, α and β, 50 chips were placed on test and the number of survivors was recorded at the end of each year, for a period of 8 years. The data are shown in the accompanying table.

Year	1	2	3	4	5	6	7	8
Number of Survivors	47	39	29	18	11	5	3	1

a. Use the method of least squares to derive estimates of α and β.
b. Construct a 95% confidence interval for α.
c. If you have access to a linear regression computer package, construct a 95% confidence interval for β.

*Some statisticians use $\hat{R}(t_i) = (n_i + 1/2)/n$. See Miller and Freund (1977).

17.14 Refer to Exercise 17.13.
   a. Use the estimates of α and β to find the probability that a memory chip will fail before 5 years.
   b. Estimate the reliability of the memory chips at time $t = 7$ years.

17.15 Refer to Exercise 17.13.
   a. Using the least squares estimates of α and β, find and graph the hazard rate, $z(t)$.
   b. Compute the hazard rate at time $t = 4$ years and interpret its value.

17.16 Engineers often use a Weibull failure time distribution for a "weakest link" product, i.e., a product consisting of multiple parts (e.g., roller bearings) that fails when the first part (or weakest link) fails. Nelson (*Journal of Quality Technology*, July 1985) applied the Weibull distribution to the lifelengths of a sample of $n = 138$ roller bearings. The accompanying table gives the number of bearings still in operation at the end of each 100-hour period until all bearings failed.

Hours (hundreds)	1	2	3	4	5	6	7	8	12	13	17	19	24	51
Number of Bearings	138	114	104	64	37	29	20	10	8	6	4	3	2	1

Source: Nelson, W. "Weibull Analysis of Reliability Data with Few or No Failures." *Journal of Quality Technology*, Vol. 17, No. 3, July 1985, p. 141 (Table 1). © 1985 American Society for Quality Control. Reprinted by permission.

   a. Use the method of least squares to estimate the Weibull parameters α and β.
   b. Construct a 99% confidence interval for α. If you have access to a regression computer package, obtain a 99% confidence interval for β.
   c. Estimate the reliability of the roller bearings at $t = 300$ hours.
   d. Estimate the probability that a roller bearing will fail before 200 hours.

17.17 A manufacturer of washing machines conducted a life test during which he monitored 12 new machines for a period of 3 years and recorded the time to a major repair for each. At the end of the 3-year testing period, two machines had not yet required a major repair. The failure times (in months) of the remaining ten washing machines were 14, 28, 9, 13, 6, 20, 10, 17, 30, and 20. Assume the lifelength (in years) of the machines has a Weibull failure time distribution with unknown parameters α and β.
   a. Construct a table for the data listing the number of machines surviving (that is, without major repair) at the end of each year.
   b. Apply the method of least squares to the data in the table of part **a** in order to derive estimates of α and β.
   c. Find a 95% confidence interval for α. If you have access to a regression computer package, find a 95% confidence interval for β.
   d. The manufacturer guarantees all machines against a major repair for 2 years. Using the least squares estimates of α and β, find the probability that a new washer will have to be repaired under the guarantee.

17.18 In order to evaluate the performance of rebuilt hydraulic pumps at an aircraft rework facility, 20 pumps were placed on test and the number of pumps still running at the end of each week was recorded for a period of 6 weeks, as listed in the accompanying table.

Week	1	2	3	4	5	6
Number of Pumps	14	11	9	7	5	4

a. Use the data to estimate the parameters, α and β, for a Weibull failure time distribution.

b. Construct a 90% confidence interval for α. If you have access to a regression computer package, find a 90% confidence interval for β.

c. Find the reliability of the rebuilt hydraulic pumps at time $t = 2$ weeks.

SECTION 17.7

SYSTEM RELIABILITY

Systems—electronic, mechanical, or a combination of both—are composed of components, some of which are combined to form smaller subsystems. We will identify a component of a system by a capital letter and portray it graphically as a box. Two systems, each composed of three components, A, B, and C, are shown in Figure 17.8.

FIGURE 17.8
Two Systems Each
Composed of Three
Components, A, B, and C

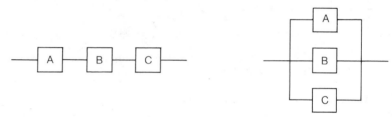

a. Series System   **b.** Parallel System

Suppose that a system is composed of k components. If the system fails when any one of the components fails, it is called a **series system**. A three-component series system is represented graphically in Figure 17.8a. If a system fails only when *all* of its components fail, it is called a **parallel system**. A three-component parallel system is represented graphically in Figure 17.8b.

Figure 17.9a shows a system composed of five components, A, B, C, D, and E. Components D and E form a two-component parallel subsystem. This subsystem is connected in series with components A, B, and C. Figure 17.9b represents a system containing two parallel subsystems connected in series. The first parallel subsystem contains three components, A, B, and C. The second contains two series subsystems—the first composed of components D and E, and the second composed of components F and G.

FIGURE 17.9
Two Systems

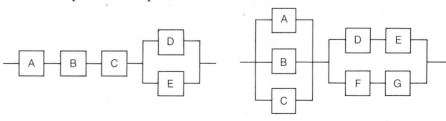

a.   **b.**

DEFINITION 17.5

A **series system** is one that fails if any one of its components fails.

DEFINITION 17.6

A **parallel system** is one that fails only if all of its components fail.

Suppose that the reliability of component i—the probability that it will function properly under specified conditions—is p_i and that the k components of a system are mutually independent. That is, we assume that the operation of one component does not affect the operation of any of the others. Then the reliability of a system can be calculated using the multiplicative rule of probability.

Since a series system will function only if all of its components function, the reliability of a series system is

$P(\text{Series system functions}) = P(\text{All components function})$

Then, because the components operate independently of each other, we can apply the multiplicative rule of probability:

$P(\text{Series system functions})$

$\qquad = P(\text{A functions}) \; P(\text{B functions}) \; \cdots \; P(\text{K functions})$

$\qquad = p_A p_B p_C \cdots p_K$

THEOREM 17.1

The reliability of a series system consisting of k independently operating components, A, B, . . . , K, is

$\qquad P(\text{Series system functions}) = p_A p_B \cdots p_K$

where p_i is the probability that the ith component functions, $i = $ A, B, . . . , K.

The reliability of a parallel system containing k components can be calculated in a similar manner. Since a parallel system will fail only if all components fail,

$P(\text{Parallel system fails}) = (1 - p_A)(1 - p_B) \cdots (1 - p_K)$

and

$P(\text{Parallel system functions}) = 1 - P(\text{Parallel system fails})$

$\qquad\qquad\qquad\qquad\qquad = 1 - (1 - p_A)(1 - p_B) \cdots (1 - p_K)$

> **THEOREM 17.2**
>
> The reliability of a parallel system consisting of k independently operating components is
>
> $$P(\text{Parallel system functions}) = 1 - (1 - p_A)(1 - p_B) \cdots (1 - p_K)$$
>
> where p_i is the probability that the ith component functions, $i = $ A, B, ..., K.

Theorems 17.1 and 17.2 can be used to calculate the reliability of series systems, parallel systems, or any combinations thereof, as long as the systems satisfy the assumption that the components operate independently. The following examples illustrate the procedure.

EXAMPLE 17.7

Given that $p_A = .90$, $p_B = .95$, and $p_C = .90$, find the reliability of the series system shown in Figure 17.8a.

SOLUTION

Since this is a series system consisting of three components, A, B, and C, it follows from Theorem 17.1 that the reliability of this system is

$$P(\text{System functions}) = p_A p_B p_C = (.90)(.95)(.90) = .7695 \qquad \blacksquare$$

EXAMPLE 17.8

Suppose that the components in Example 17.7 were connected in parallel, as shown in Figure 17.8b. Find the reliability of the system.

SOLUTION

To find the reliability of this parallel system, we apply Theorem 17.2:

$$
\begin{aligned}
P(\text{System functions}) &= 1 - (1 - p_A)(1 - p_B)(1 - p_C) \\
&= 1 - (.10)(.05)(.10) \\
&= 1 - .0005 \\
&= .9995 \qquad \blacksquare
\end{aligned}
$$

Examples 17.7 and 17.8 demonstrate that the *reliability of a series system is always less than the reliability of its least reliable component. In contrast, the reliability of a parallel system is always greater than the reliability of its most reliable component.*

To find the reliability of a system containing subsystems, we first find the reliability of the smallest subsystems. Then we find the reliability of the systems in which they are contained.

EXAMPLE 17.9

Find the reliability of the system shown in Figure 17.9a, given the following component reliabilities: $p_A = .95$, $p_B = .99$, $p_C = .97$, $p_D = .90$, and $p_E = .90$.

SOLUTION

The complete system is composed of three components, A, B, and C, and a subsystem connected in series. The parallel subsystem, comprised of components D and E, is shown here:

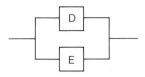

The reliability of this subsystem is

$$P(\text{Subsystem D and E functions}) = p_{DE}$$
$$= 1 - (1 - p_D)(1 - p_E)$$
$$= 1 - (.1)(.1) = .99$$

We now view the complete system as one consisting of four components: components A, B, and C and the subsystem (D, E), connected in series. To find its reliability, we apply Theorem 17.1. The reliability of the complete system is

$$P(\text{System functions}) = p_A p_B p_C p_{DE} = (.95)(.99)(.97)(.99)$$
$$= .9031622 \qquad \blacksquare$$

EXAMPLE 17.10

Find the reliability of the system shown in Figure 17.9b, given that $p_A = .90$, $p_B = .95$, $p_C = .95$, $p_D = .92$, $p_E = .97$, $p_F = .92$, and $p_G = .97$.

SOLUTION

An examination of Figure 17.9b shows that the system is a series of two parallel subsystems. The first parallel subsystem contains components A, B, and C. The second is a parallel subsystem of two series subsystems, the first containing components D and E, and the second containing components F and G.

Since the reliabilities of the pairs of components (D, E) and (F, G) are identical, it follows that the reliabilities of these two series subsystems are equal:

By Theorem 17.1, the reliability of these series subsystems is

$$p_{DE} = p_{FG} = p_D p_E = p_F p_G = (.92)(.97) = .8924$$

We now consider the reliability of the parallel subsystem containing these two series subsystems:

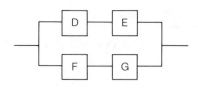

By Theorem 17.2, we have

$$
\begin{aligned}
p_{\text{DEFG}} &= 1 - (1 - p_{\text{DE}})(1 - p_{\text{FG}}) \\
&= 1 - (1 - .8924)(1 - .8924) \\
&= 1 - .0115778 \\
&= .9884222
\end{aligned}
$$

Next, we compute the reliability of the parallel subsystem consisting of components A, B, and C:

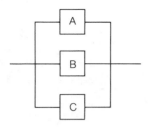

By Theorem 17.2,

$$
\begin{aligned}
p_{\text{ABC}} &= 1 - (1 - p_{\text{A}})(1 - p_{\text{B}})(1 - p_{\text{C}}) \\
&= 1 - (1 - .90)(1 - .95)(1 - .95) \\
&= 1 - .00025 \\
&= .99975
\end{aligned}
$$

We have calculated the reliabilities of the two parallel subsystems. These two subsystems are connected in series, as shown here:

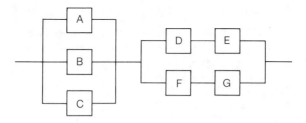

Thus, the reliability of the complete system is

$$
\begin{aligned}
P(\text{System functions}) &= p_{\text{ABC}}p_{\text{DEFG}} \\
&= (.99975)(.9884222) \\
&= .9881751
\end{aligned}
$$

17.19 Consider a series system consisting of four components, A, B, C, and D, with probabilities of functioning given by $p_A = .88$, $p_B = .95$, $p_C = .90$, and $p_D = .80$. Find the reliability of the system.

17.20 Consider a parallel system consisting of four components, A, B, C, and D, with probabilities of functioning given by $p_A = .90$, $p_B = .99$, $p_C = .92$, and $p_D = .85$. Find the reliability of the system.

17.21 A system consists of eight components, as shown in the accompanying diagram. Find the reliability of the system, given that the individual probabilities of functioning are $p_A = .90$, $p_B = .95$, $p_C = .85$, $p_D = .85$, $p_E = .98$, $p_F = .80$, $p_G = .95$, and $p_H = .95$.

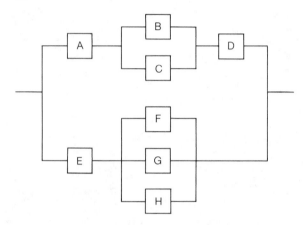

17.22 Consider an electrical circuit consisting of two subcircuits, the first of which involves components A, B, and C in parallel and the second of which involves components D and E in parallel. Suppose that the individual reliabilities of the components are given by $p_A = .95$, $p_B = .95$, $p_C = .90$, $p_D = .90$, and $p_E = .98$.
a. Find the reliability of the system if the two subcircuits are connected in series.
b. Find the reliability of the system if the two subcircuits are connected in parallel.

17.23 The reliability of a system consisting of three identical components is .95. What must be the probability of functioning for each component if:
a. The components are connected in series?
b. The components are connected in parallel?

This chapter presents statistical methods for evaluating the life characteristics of a manufactured product—whether a simple device such as a light bulb or a complex system composed of a number of components or subsystems.

The **reliability** of a product is the probability that the product will function a specified length of time under specified environmental conditions. The length of time that a unit of product performs satisfactorily is called its **length of life**. The time at which it fails is called its **failure time**.

If a population of product items was placed on test, they would fail at varying times and the relative frequency distribution of failure times would be called a **failure time distribution**. This distribution can often be modeled by one of the probability density functions presented in Chapter 5. The exponential distribution and the Weibull distribution are widely used as models for product failure time distributions. The normal and gamma distributions are also used for some products.

Another way to describe the failure characteristics of a product is in terms of its hazard rate. The **hazard rate** $z(t)$ at time t is proportional to the probability that an item will fail during a fixed small interval of time, given that the item has survived to time t:

$$z(t) = \frac{f(t)}{1 - F(t)}$$

The hazard rate provides a measure of the propensity of the product to fail as it gets older.

To find an appropriate model for a product, we conduct a **life test**. That is, we place a random sample of product items on test and record the observed times to failure.

The length of time for a life test is often shortened by **censoring** the sample— that is, stopping the life test either after a specified number r of failures have been observed or after a specified amount of time T has elapsed. Regardless of how the sample data are collected, by censored or uncensored sampling, we employ the methods of Chapter 8 to find estimators of the parameters of the failure time distribution. Parameter estimates can then be substituted into the formula for the failure time distribution:

$$P(\text{Survival to time } t_0) = 1 - F(t_0)$$

In concluding, we learned how to calculate the reliability of systems of components, assuming that the components operate independently of each other and that the component reliabilities are known. We considered systems that can be viewed as a **series system**, a **parallel system**, or some combination of the two.

Because of its importance, much work has been done on life-testing methodology and on methods for evaluating system reliability. This chapter provides only an introduction to the topic. If you seek more information on the subject, you may consult the references at the end of this chapter.

SUPPLEMENTARY EXERCISES 17.24–17.38

17.24 A certain component has an exponential failure time distribution with mean $\beta = 3$ hours.
  a. Find the probability that the component will fail before time $t = 2$ hours.
  b. What is the reliability of the component at time $t = 5$ hours? Interpret this value.
  c. Find the hazard rate for the component and interpret its value.

17.25 Suppose the lifelength (in hours) of a fluorescent light has a Weibull failure time distribution with parameters $\alpha = .05$ and $\beta = .70$.
  a. Find the probability that the fluorescent light will fail before time $t = 1,000$ hours.
  b. Find the reliability of the fluorescent light at time $t = 500$ hours and interpret its value.
  c. Find the hazard rate for the fluorescent light at time $t = 500$ hours and interpret its value.

17.26 Consider the gamma failure time distribution with $\alpha = 2$ and $\beta = 1$ given by the density function

$$f(t) = \begin{cases} te^{-t} & 0 \le t < \infty \\ 0 & \text{elsewhere} \end{cases}$$

  a. Find $F(t)$. [*Hint:* $\int te^{-t}\,dt = -te^{-t} + \int e^{-t}\,dt$]
  b. Find expressions for the reliability $R(t)$ and the hazard rate $z(t)$.
  c. Use the results of part **b** to find $R(3)$ and $z(3)$. Interpret these values.

17.27 Consider the uniform failure time distribution given by the density function

$$f(t) = \begin{cases} \dfrac{1}{\beta} & 0 \le t \le \beta \\ 0 & \text{elsewhere} \end{cases}$$

  a. Find $F(t)$, $R(t)$, and $z(t)$.
  b. Graph the hazard rate $z(t)$ for $t = 0, 1, 2, \dots, 5$ when $\beta = 10$.
  c. Compute the reliability of the system at $t = 4$ when $\beta = 10$.

17.28 To investigate the performance of the central processing unit (CPU) of a certain type of microcomputer, 20 CPUs were placed on test for a period of 5,000 hours. When the test was terminated, four CPUs had failed with failure times of 1,850, 2,090, 3,440, and 3,970 hours. Assume a negative exponential failure time distribution.
  a. Find a 90% confidence interval for the mean time β until failure of the microcomputer's CPU.
  b. Find a 90% confidence interval for the reliability of the CPU at time $t = 2,000$ hours.
  c. Find a 90% confidence interval for the hazard rate of the CPU.

17.29 A certain type of coating for pipes is designed to resist corrosion. Five hundred pieces of coated pipe were placed on test and subjected to a 90% solution of hydrochloric acid. At the end of each hour, for a period of 5 hours, the number of pipe specimens that had resisted corrosion was recorded, as shown in the accompanying table.

Hour	1	2	3	4	5
Number Resisting Corrosion	438	280	146	51	15

  a. Use the data in the table to estimate the parameters α and β of a Weibull failure time distribution.

b. Use the estimates obtained in part **a** to find expressions for the hazard rate $z(t)$ and the reliability $R(t)$ of the coated pipes.

c. Find the probability that a piece of coated pipe will resist corrosion under similar experimental conditions for at least 1 hour.

17.30 A piece of equipment consists of seven tubes connected as shown in the diagram. Find the reliability of the system if the tubes have probabilities of functioning given by $p_A = .80$, $p_B = .90$, $p_C = .85$, $p_D = .85$, $p_E = .75$, $p_F = .75$, and $p_G = .95$.

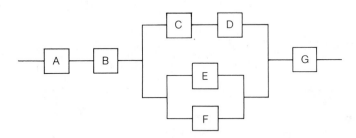

17.31 Two resistors connected in series have exponential failure time distributions with mean $\beta = 1,000$ hours. At time $t = 1,400$ hours, what is the reliability of the system?

17.32 Four components, A, B, C, and D, are connected in parallel. Suppose that components A and B have normal failure time distributions with parameters $\mu = 500$ hours and $\sigma = 100$ hours, while components C and D have Weibull failure time distributions with parameters $\alpha = .5$ and $\beta = 100$. Find the reliability of the system at time $t = 300$ hours.

17.33 The service life (in hours) of a semiconductor has an approximate exponential failure time distribution. Ten semiconductors are placed on life test until four fail. The failure times for these four semiconductors are 585, 972, 1,460, and 2,266 hours.

a. Construct a 95% confidence interval for the mean time β until failure for the semiconductors.

b. What is the probability that a semiconductor will still be in operation after 4,000 hours? Find a 95% confidence interval for this probability.

c. Compute and interpret the hazard rate for the semiconductors. Construct a 95% confidence interval for this hazard rate.

17.34 The failure times (in hours) of electronic components in a guidance system for a missile have a Weibull distribution with unknown parameters α and β. In order to derive estimates of these parameters, 1,000 components were placed on life test and every 50 hours the number of components still in operation was recorded. The data are provided in the table.

Hours	50	100	150	200	250	300	350
Number in Operation	611	362	231	136	84	53	17

a. Find estimates of α and β. If you have access to a regression computer package, find 99% confidence intervals for both α and β.

b. Calculate the reliability of the electronic components at $t = 200$ hours.

c. Find and graph the hazard rate for $t = 50, 100, 150, \ldots$.

17.35 Consider the product system shown in the diagram. Given the individual component reliabilities $p_A = .85$, $p_B = .75$, $p_C = .75$, $p_D = .90$, and $p_E = .95$, find the overall reliability of the system.

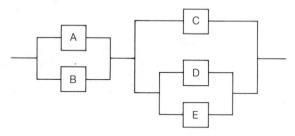

OPTIONAL SUPPLEMENTARY EXERCISES

17.36 Show that the hazard rate $z(t)$ can be expressed as

$$z(t) = \frac{-d[\ln R(t)]}{dt}$$

[*Hint:* Make use of the fact that $R(t) = 1 - F(t)$ and, hence, that

$$f(t) = \frac{dF(t)}{dt} = \frac{-dR(t)}{dt}$$

Then substitute these expressions into the formula given in Definition 17.4.]

17.37 Use the result of Exercise 17.36 and the relation $f(t) = z(t)R(t)$ to show that the failure time density can be expressed as

$$f(t) = z(t)e^{-\int_0^t z(y)\,dy}$$

[*Hint:* The differential equation

$$z(t) = \frac{-d[\ln R(t)]}{dt}$$

has

$$R(t) = e^{-\int_0^t z(y)\,dy}$$

as its solution.]

17.38 Suppose we are concerned only with the initial failure of a component. That is, once the component has survived past a certain time $t = \alpha$, we treat the component (for all practical purposes) as if it never failed. In this situation it is reasonable to use the hazard rate

$$z(t) = \begin{cases} \beta(1 - t/\alpha) & 0 < t < \alpha \\ 0 & \text{elsewhere} \end{cases}$$

a. Use the result of Exercise 17.37 to find expressions for $f(t)$, $F(t)$, and $R(t)$.
b. Show that the probability of initial failure, i.e., the probability that the component fails before time $t = \alpha$, is $1 - e^{-\alpha\beta/2}$.

REFERENCES

Barlow, R. E. and Proschan, F. *The Mathematical Theory of Reliability.* New York: John Wiley and Sons, 1965.

Box, G. E. P. "Problems in the Analysis of Growth and Wear Curves." *Biometrics,* 6, 1950.

Cohen, A. C., Jr. "On Estimating the Mean and Standard Deviation of Truncated Normal Distribution." *Journal of the American Statistical Association, 44,* 1949, pp. 518–525.

Cohen, A. C., Jr. "A Note on Truncated Distributions." *Industrial Quality Control,* 6, 1949, p. 22.

Davis, D. J. "An Analysis of Some Failure Data." *Journal of the American Statistical Association, 47,* 1952, pp. 113–150.

Epstein, B. "Statistical Problems in Life Testing." *Seventh Annual Quality Control Conference Papers,* 1953, pp. 385–398.

Epstein, B. and Sobel, M. "Life Testing." *Journal of the American Statistical Association, 48,* 1953, pp. 486–502.

Miller, I. and Freund, J. E. *Probability and Statistics for Engineers,* 2nd ed. Englewood Cliffs, N. J.: Prentice-Hall, 1977.

Weibull, W. "A Statistical Distribution Function of Wide Applicability." *Journal of Applied Mechanics,* 18, 1951, pp. 293–297.

Zelen, M. *Statistical Theory of Reliability.* Madison, WI: University of Wisconsin Press, 1963.

OBJECTIVE

To present some statistical procedures for monitoring the quality of a manufactured product and for controlling the quality of products shipped to consumers.

CONTENTS

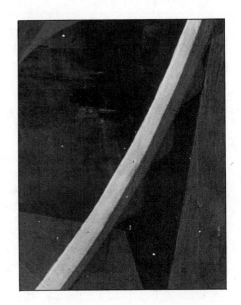

APPLICATIONS: QUALITY CONTROL

SECTION 18.1

INTRODUCTION

When we think of product quality, we think of a set of characteristics that we expect a product to possess. We want light bulbs to have a long length of life, toweling paper to be strong and absorbent, and a quarter-pound hamburger to weigh at least a quarter pound. But producing a quality product is not an easy job. Variation in the characteristics of raw materials and workmanship tend to produce variation in product quality. The length of life of a light bulb produced in an automated production line may differ markedly from the length of life of a bulb produced seconds later. Similarly, the strength of paper produced by a paper machine may vary from one point in time to another due to variations in the characteristics of pulp fed into the machine.

Manufacturers monitor the quality of a product by measuring the product's quality characteristics periodically over time. For example, the manufacturer of shafts for an electrical motor might select one shaft every 10 minutes and measure its diameter. These measurements, plotted against time, provide visual evidence of the ability of the process to produce shafts with diameters that meet a customer's specifications. For example, the diameters of ten shafts, all specified to be 1.500 inches, plus or minus .010 inch, might appear as shown in Figure 18.1. Although the diameters of these ten shafts vary from one point in time to another, all fall within the limits specified by the manufacturer's customer.

FIGURE 18.1

A Plot of the Diameters of Ten Motor Shafts

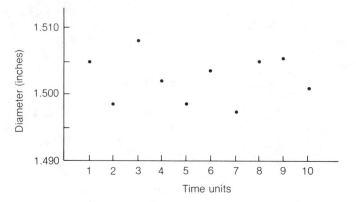

It is common to classify the variation in a variable that measures a product quality characteristic into one of two categories. The first category is the variation in the variable that is due to an **assignable cause**. Variations due to assignable causes are produced by such things as the wear in a metallic cutting machine, the wear in an abrasive wheel, changes in the humidity and temperature in the production area, worker fatigue, and so on. The effects of wear in cutting edges, abrasive surfaces, or changes in the environment are usually evidenced by gradual trends in a characteristic over time (see Figure 18.2a). In contrast, the sudden failure of a part in a machine or the accidental pollution of a source of raw material will often produce an abrupt change in the level of a quality characteristic (see Figure 18.2b). Quality control and production engineers attempt to identify trends or abrupt changes in a quality characteristic when they occur and to modify the process to reduce or eliminate this type of variation.

FIGURE 18.2

Plots of a Quality
Characteristic That
Suggest Variation Due
to Assignable Causes

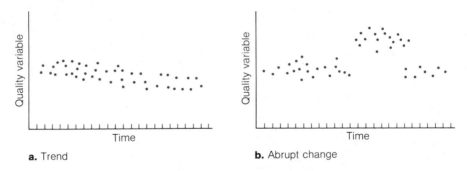

a. Trend   **b.** Abrupt change

Even when variation due to assignable causes is accounted for, measurements
taken on a product quality characteristic tend to vary in a random manner from
one point in time to another. This second category of variation—**random (or
chance) variation**—is caused by minute and random changes in raw materials,
worker behavior, and so on. Since some stable system of chance causes is inherent
in any production process, this type of variation is accepted as the normal variation of the process. When the quality characteristics of a product are subject
only to random variation, the process is said to be **in control**.

The subject of quality control encompasses a broad area of knowledge concerned with techniques, both managerial and statistical, to control the quality of
a product. One of the most powerful of these tools is the **control chart**, developed
by W. A. Shewhart in 1924. Control charts are constructed to monitor either a
quantitative characteristic or a qualitative attribute of a manufactured product.
The power of the control chart lies in its ability to separate out the two types of
variation—variation due to assignable causes and random variation. This permits
the quality control engineer to make better decisions regarding the quality level
of the production process.

In this chapter, we will introduce control charts and a few other statistical
methods useful in quality control. For additional information, you may consult
the references listed at the end of the chapter.

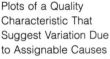

SECTION 18.2

CONTROL CHART FOR MEANS: $\bar{x}$-CHART

A control chart constructed to monitor a quantitative quality characteristic is
usually based on random samples of several units of the product rather than on
the characteristics of individual industrial units as shown in Figure 18.1. For
example, the manufacturer of electrical shafts in Section 18.1 might select a
sample of five shafts at the end of each hour. A plot showing the mean diameters
of the samples, one mean corresponding to each point in time, is called a **control
chart for means** or an **$\bar{x}$-chart**.*

Control charts are constructed after a process has been adjusted to correct for
assignable causes of variation and the process is deemed to be in control. Such

*In order to be consistent with the symbols used in quality control literature, we will use x (rather
than y) to denote a quantitative quality characteristic variable.

a chart would show only random variation in the sample mean over time. Theoretically, $\bar{x}$ should vary about the process mean, $E(x) = \mu$, and fall within the limits $\mu \pm 3\sigma_{\bar{x}}$ or $\mu \pm 3\sigma/\sqrt{n}$ with a high probability. A control chart constructed for the means of samples of $n = 5$ motor shafts taken each hour might appear as shown in Figure 18.3.

FIGURE 18.3

$\bar{x}$-Chart for Samples of $n = 5$ Shaft Diameters

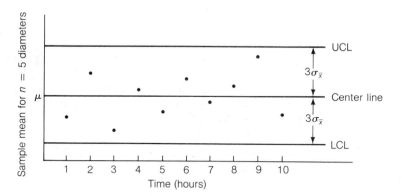

An $\bar{x}$-chart, such as that shown in Figure 18.3, contains three horizontal lines. The **center line** establishes the mean value μ of the process. Although this value is usually unknown, it can be estimated by averaging a large number (for example, 20) sample means obtained when the process is in control. For example, if we average the values of k sample means, then

$$\text{Center line} = \bar{\bar{x}} = \frac{\sum_{i=1}^{k} \bar{x}_i}{k}$$

The two lines located above and below the center line establish the **upper control limit (UCL)** and the **lower control limit (LCL)**, between which we would expect the sample means to fall if the process is in control. They are located a distance of $3\sigma_{\bar{x}} = 3\sigma/\sqrt{n}$ above and below the center line.

The process standard deviation σ is usually unknown, but it can be estimated from a large sample of data collected while the process is in control. Prior to the advent of computers, it was common to estimate σ by first computing the sample range R, the difference between the largest and smallest sample measurements. The process standard deviation σ was then estimated by dividing the average $\bar{R}$ of k sample ranges by a constant d_2, the value of which depended on the sample size n:

$$\hat{\sigma} = \frac{\bar{R}}{d_2} = \frac{\sum_{i=1}^{k} R_i/k}{d_2}$$

Since the control limits are located a distance of $3\sigma_{\bar{x}} = 3\sigma/\sqrt{n}$ above and below the center line, this distance was estimated to be

$$3\hat{\sigma}_{\bar{x}} = \frac{3(\bar{R}/d_2)}{\sqrt{n}} = \frac{3}{d_2\sqrt{n}}\bar{R} = A_2\bar{R}$$

where

$$A_2 = \frac{3}{d_2\sqrt{n}}$$

Values of A_2 and d_2 for sample sizes $n = 2$ to $n = 25$ are given in Table 18 of Appendix II.

Today, the sample measurements for quality control processes can be entered into a computer which is programmed to compute the means and standard deviations of the individual samples, as well as the means and standard deviations of the data contained in any set of k samples. For large samples, the best estimate of σ is then the standard deviation s of the data contained in the k sets of data.* The computer calculates $\bar{\bar{x}}$ and s and provides a printout of the control chart.

LOCATION OF CENTER LINE AND CONTROL LIMITS FOR AN x̄-CHART

Center line: $\bar{\bar{x}} = \dfrac{\sum_{i=1}^{k} \bar{x}_i}{k}$

UCL: $\bar{\bar{x}} + A_2\bar{R}$

LCL: $\bar{\bar{x}} - A_2\bar{R}$

where

$$\bar{R} = \frac{\sum_{i=1}^{k} R_i}{k}$$

and A_2 is given in Table 18 of Appendix II.

[*Note:* For large samples (say, $n > 15$), the upper and lower control limits may be computed as follows:

UCL: $\bar{\bar{x}} + 3s/\sqrt{n}$

LCL: $\bar{\bar{x}} - 3s/\sqrt{n}$

where s is the standard deviation of all nk sample measurements.]

EXAMPLE 18.1

Suppose the process for manufacturing electrical shafts is in control. At the end of each hour, for a period of 20 hours, the manufacturer selected a random sample of four shafts and measured the diameter of each. The measurements (in inches)

---

*Grant and Leavenworth (1980) suggest using s to estimate σ when the sample size n is greater than 15. For smaller samples, $\bar{R}/d_2$ will usually provide a better estimate.

for the 20 samples are recorded in Table 18.1. Construct a control chart for the sample means and interpret the results.

SOLUTION

The first step in constructing an $\bar{x}$-chart is to compute the sample mean, $\bar{x}$, and range, R, for each of the 20 samples. These values are shown in the last two columns of Table 18.1.

TABLE 18.1
Samples of $n = 4$ Shaft Diameters, Example 18.1

SAMPLE NUMBER	SAMPLE MEASUREMENTS (inches)				SAMPLE MEAN $\bar{x}$	RANGE R
1	1.505	1.499	1.501	1.488	1.4983	.017
2	1.496	1.513	1.512	1.501	1.5055	.017
3	1.516	1.485	1.492	1.503	1.4990	.031
4	1.507	1.492	1.511	1.491	1.5003	.020
5	1.502	1.491	1.501	1.502	1.4990	.011
6	1.502	1.488	1.506	1.483	1.4948	.023
7	1.489	1.512	1.496	1.501	1.4995	.023
8	1.485	1.518	1.494	1.513	1.5025	.033
9	1.503	1.495	1.503	1.496	1.4993	.008
10	1.485	1.519	1.503	1.507	1.5035	.034
11	1.491	1.516	1.497	1.493	1.4993	.025
12	1.486	1.505	1.487	1.492	1.4925	.019
13	1.510	1.502	1.515	1.499	1.5065	.016
14	1.495	1.485	1.493	1.503	1.4940	.018
15	1.504	1.499	1.504	1.500	1.5018	.005
16	1.499	1.503	1.508	1.497	1.5018	.011
17	1.501	1.493	1.509	1.491	1.4985	.018
18	1.497	1.510	1.496	1.500	1.5008	.014
19	1.503	1.526	1.497	1.500	1.5065	.029
20	1.494	1.501	1.508	1.519	1.5055	.025

Next, we calculate $\bar{\bar{x}}$, the average of the 20 sample means, and $\bar{R}$, the average of the 20 sample ranges:

$$\bar{\bar{x}} = \frac{\sum_{i=1}^{20} \bar{x}_i}{20} = \frac{(1.4983 + 1.5055 + 1.4990 + \cdots + 1.5055)}{20}$$
$$= 1.50045$$

$$\bar{R} = \frac{\sum_{i=1}^{20} \bar{R}_i}{20} = \frac{(.017 + .017 + .031 + \cdots + .025)}{20}$$
$$= .01985$$

The value of $\bar{\bar{x}} = 1.50045$ locates the center line on the control chart. To find upper and lower control limits, we need the value of the control limit factor

A_2, found in Table 18 of Appendix II. For $n = 4$ measurements in each sample, $A_2 = .729$. Then

$$\text{UCL} = \bar{\bar{x}} + A_2\bar{R} = 1.50045 + (.729)(.01985)$$
$$= 1.51492$$
$$\text{LCL} = \bar{\bar{x}} - A_2\bar{R} = 1.50045 - (.729)(.01985)$$
$$= 1.48598$$

Using these limits, we construct the control chart for the sample means shown in Figure 18.4. Note that all 20 sample means fall within the control limits.

FIGURE 18.4

$\bar{x}$-Chart for Shaft Diameters, Example 18.1

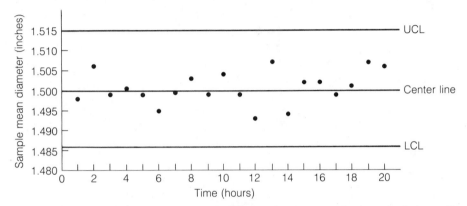

A computer-generated (SAS) control chart for the sample means of Example 18.1 is shown in Figure 18.5 (page 906). The upper and lower control limits on the SAS printout differ slightly from those in Figure 18.4 since the computer uses the standard deviation s of all 80 measurements to establish these limits. From a portion of the SAS printout (not shown here), $s = .009244$. Then

$$3\hat{\sigma}_{\bar{x}} = \frac{3s}{\sqrt{n}} = \frac{3(.009244)}{\sqrt{4}} = .013866$$

Thus, the computer-generated control limits are given by

$$\text{UCL} = \bar{\bar{x}} + 3\hat{\sigma}_{\bar{x}} = 1.50045 + .013866$$
$$= 1.51432$$
$$\text{LCL} = \bar{\bar{x}} - 3\hat{\sigma}_{\bar{x}} = 1.50045 - .013866$$
$$= 1.48658$$ ∎

The purpose of the $\bar{x}$-chart is to detect departures from process control. If the process is in control, the probability that a sample mean will fall within the control limits is very high. This result is due to the Central Limit Theorem, which guarantees that the sampling distribution of $\bar{x}$ will be approximately normal for large samples. Consequently, the probability that $\bar{x}$ will fall within the control limits, i.e., $\pm 3\hat{\sigma}_{\bar{x}}$, is approximately .997. Therefore, a sample mean falling outside the control limits is taken as an indication of possible trouble in the production

FIGURE 18.5 Computer-Generated $\bar{x}$-Chart for Example 18.1

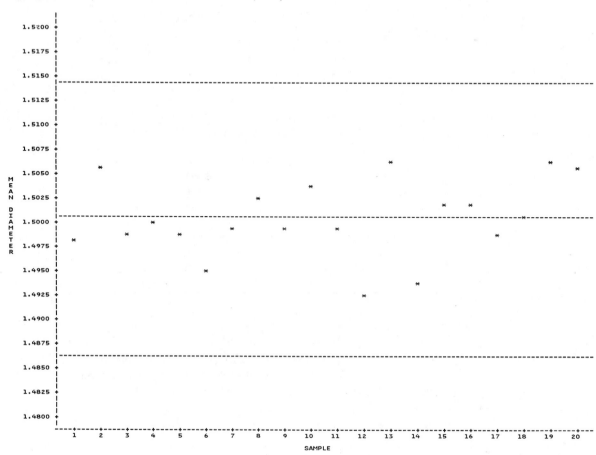

process. When this occurs, we say with a high degree of confidence that the process is "out of control," and process engineers are usually assigned to determine the cause of the unusually large (or small) value of $\bar{x}$.

On the other hand, when all the sample means fall within the control limits (as in Figure 18.4) we say that the process is "in control." However, we do not have the same degree of confidence in this statement as with the "out of control" conclusion above. In one sense, we are using the control chart to test the null hypothesis H_0: Process in control (i.e., no assignable causes of variation are present). As you recall from Chapter 9, we must be careful not to accept H_0 since the probability of a Type II error is unknown. In practice, when quality control engineers say "the process is in control," they really mean that "it pays to act as if no assignable causes of variation are present." In this situation, it is better to leave the process alone than to spend a great deal of time and money looking for trouble that may not exist.

Control charts are also examined for trends or cyclical behavior in the values of $\bar{x}$ collected over time. To engineers familiar with the process, this type of behavior may suggest the presence of one or more assignable causes that can be determined and controlled. A simple test for trend is presented in Section 18.4.

In concluding our discussion of the x̄-chart, we emphasize that the chart describes the process as it is, not the way we want it to be. The process mean and control limits may differ markedly from the specifications set by the manufacturer of the product. For example, although a manfacturer may want to produce electrical motor shafts with a diameter of 1.500 inches, the actual process mean will usually differ from 1.500, at least by some small amount. Also, the control limits obtained from the control chart are appropriate only for analyzing "past" data, i.e., the data that were used in their calculation. Thus, they may require modification before applying them to future production data. For example, in cases where the process is found to be out of control, the control limits and center line are often modified by recalculating their values using only the sample means that fall within the control limits. If the cause of the problem has been corrected, these new values serve as control limits for future data.

EXERCISES 18.1–18.5

18.1 A corporation that manufactures field rifles for the Department of Defense operates a production line that turns out finished firing pins. In order to monitor the process, an inspector randomly selects five firing pins from the production line, measures their lengths (in inches), and repeats this process at 30-minute intervals over a 5-hour period.

30-MINUTE INTERVAL	FIRING PIN LENGTHS				
1	1.05	1.03	.99	1.00	1.03
2	.93	.96	1.01	.98	.97
3	1.02	.99	.99	1.00	.98
4	.98	1.01	1.02	.99	.97
5	1.02	.99	1.04	1.07	.98
6	1.05	.98	.96	.91	1.02
7	.92	.95	1.00	.99	1.01
8	1.06	.98	.98	1.04	1.00
9	.97	.99	.99	.98	1.01
10	1.00	.96	1.02	1.03	.99

a. Use the data for the ten time periods listed in the table to calculate the center line for an x̄-chart.

b. Compute upper and lower control limits.

c. Calculate and plot the ten sample means to form an x̄-chart for the firing pin lengths.

d. Suppose the Defense Department's specification for the firing pins is that they be 1.00 inch plus or minus .08 inch in length. Does the manufacturing process appear to be in control?

18.2 Molded-rubber expansion joints, used in heating and air conditioning systems, are designed to have internal diameters of 5 inches. To monitor the manufacturing process, eight joints were randomly selected from the production line and their diameters (in inches) measured each hour, for a period of 12 hours, as shown in the table. The data for the 12 samples will be used to construct an $\bar{x}$-chart.

HOUR	MOLDED-RUBBER EXPANSION JOINT DIAMETERS							
1	5.08	5.01	4.99	4.93	4.98	5.00	5.04	4.97
2	4.88	5.10	4.93	5.02	5.06	4.99	4.92	4.91
3	4.99	5.00	5.02	5.01	5.03	4.92	4.97	5.01
4	5.04	4.96	5.01	5.00	5.00	4.98	4.91	4.96
5	5.00	4.93	4.94	5.02	5.01	4.97	5.08	5.11
6	4.83	4.92	4.96	4.91	5.01	5.03	4.93	5.00
7	5.02	5.01	4.96	4.98	5.00	5.07	4.94	5.01
8	4.91	5.00	4.97	5.03	5.02	4.99	4.98	4.99
9	5.06	5.04	4.99	5.02	4.97	5.00	5.01	5.01
10	4.92	4.98	5.01	5.01	4.97	5.00	5.02	4.93
11	5.01	5.00	5.02	4.98	4.99	5.00	5.01	5.01
12	4.92	5.12	5.06	4.93	4.98	5.02	5.04	4.97

a. Locate the center line for the $\bar{x}$-chart.
b. Locate upper and lower control limits.
c. Calculate and plot the 12 sample means to produce an $\bar{x}$-chart for the joint diameters. Does the process appear to be in control?

18.3 Each month, the quality control engineer at a bottle manufacturing company randomly samples three finished bottles from the production process at 20 points in time (days) and records the weight of each bottle (in ounces). The data for last month's inspection are provided in the table.

DAY	BOTTLE WEIGHTS			DAY	BOTTLE WEIGHTS		
1	5.6	5.8	5.8	11	6.2	5.6	5.8
2	5.7	6.3	6.0	12	5.9	5.7	5.9
3	6.1	5.3	6.0	13	5.2	5.5	5.7
4	6.3	5.8	5.9	14	6.0	6.1	6.0
5	5.2	5.9	6.3	15	6.3	5.7	5.9
6	6.0	6.7	5.2	16	5.8	6.2	6.1
7	5.8	5.7	6.1	17	6.1	6.4	6.6
8	5.8	6.0	6.2	18	6.2	5.7	5.7
9	6.4	5.6	5.9	19	5.3	5.5	5.4
10	6.0	5.7	6.1	20	6.0	6.1	6.0

a. Construct an $\bar{x}$-chart for the weights of the finished bottles.
b. Does the process appear to be in control for this particular month?

18.4 One of the operations in a plant consists of thread grinding a fitting for an aircraft hydraulic system. In order to monitor the process, a production supervisor randomly sampled five fittings for each hour, for a period of 20 hours, and measured the pitch diameters of the threads. The measurements, expressed in units of .0001 inch in excess of .4000 inch, are shown in the table. (For example, the value 36 represents .4036 inch.)

HOUR	PITCH DIAMETERS OF THREADS					HOUR	PITCH DIAMETERS OF THREADS				
1	36	35	34	33	32	11	34	38	35	34	38
2	31	31	34	32	30	12	36	38	39	39	40
3	30	30	32	30	32	13	36	40	35	26	33
4	32	33	33	32	35	14	36	35	37	34	33
5	32	34	37	37	35	15	30	37	33	34	35
6	32	32	31	33	33	16	28	31	33	33	33
7	33	33	36	32	31	17	33	30	34	33	35
8	23	33	36	35	36	18	27	28	29	27	30
9	43	36	35	24	31	19	35	36	29	27	32
10	36	35	36	41	41	20	33	35	35	39	36

Source: Grant, E. L. and Leavenworth, R. S. *Statistical Quality Control*, 5th ed. New York: McGraw-Hill, 1980 (Table 1-1). Reprinted with permission.

a. Construct an x̄-chart for the process.
b. Locate the center line, upper control limit, and lower control limit on the x̄-chart.
c. Does the process appear to be in control?
d. Eliminate the points that fall outside the control limts and recalculate their values. Would you recommend using these modified control limits for future data?

18.5 A rheostat knob, produced by plastic molding, contains a metal insert. The fit of this knob into its assembly is determined by the distance from the back of the knob to the far side of a pin hole. In order to monitor the molding operation, five knobs from each hour's production were randomly sampled and the dimension measured on each. The accompanying table gives the distance measurements (in inches) for the first 27 hours the process was in operation.

HOUR	DISTANCE MEASUREMENTS	HOUR	DISTANCE MEASUREMENTS
1	.140, .143, .137, .134, .135	15	.144, .142, .143, .135, .145
2	.138, .143, .143, .145, .146	16	.140, .132, .144, .145, .141
3	.139, .133, .147, .148, .139	17	.137, .137, .142, .143, .141
4	.143, .141, .137, .138, .140	18	.137, .142, .142, .145, .143
5	.142, .142, .145, .135, .136	19	.142, .142, .143, .140, .135
6	.136, .144, .143, .136, .137	20	.136, .142, .140, .139, .137
7	.142, .147, .137, .142, .138	21	.142, .144, .140, .138, .143
8	.143, .137, .145, .137, .138	22	.139, .146, .143, .140, .139
9	.141, .142, .147, .140, .140	23	.140, .145, .142, .139, .137
10	.142, .137, .145, .140, .132	24	.134, .147, .143, .141, .142
11	.137, .147, .142, .137, .135	25	.138, .145, .141, .137, .141
12	.137, .146, .142, .142, .140	26	.140, .145, .143, .144, .138
13	.142, .142, .139, .141, .142	27	.145, .145, .137, .138, .140
14	.137, .145, .144, .137, .140		

Source: Grant, E. L. and Leavenworth, R. S. *Statistical Quality Control*, 5th ed. New York: McGraw-Hill, 1980 (Table 1–2). Reprinted with permission.

a. Construct an x̄-chart for the process.
b. Locate the center line, upper control limit, and lower control limit on the x̄-chart.
c. Does the process appear to be in control?

CONTROL CHART FOR PROCESS VARIATION: R-CHART

We want to control not only the mean value of some quality characteristic, but also its variability. An increase in the process standard deviation σ means that the quality characteristic variable will vary over a wider range, thereby increasing the probability of producing a product that falls outside of the manufacturer's specifications. Consequently, a process that is in control generates data with a relatively constant process mean μ and standard deviation σ.

The variation in a quality characteristic is monitored using a **range chart** or **R-chart**. Thus, in addition to calculating the mean $\bar{x}$ for each sample, we also calculate and plot the sample range R. As with an $\bar{x}$-chart, an R-chart also contains a center line and lines corresponding to the upper and lower control limits. An R-chart for the 20 sample ranges of Table 18.1 is shown in Figure 18.6.

FIGURE 18.6

R-Chart for the $k = 20$ Sample Ranges of Table 18.1

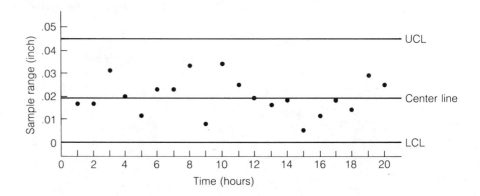

The expected value and standard deviation of the sample range are

$$E(R) = d_2\sigma \quad \text{and} \quad \sigma_R = d_3\sigma$$

where d_2 and d_3 are constants (see Table 18 of Appendix II) that depend on the sample size n. Therefore, we would locate the center line of the R-chart at $d_2\sigma$ where, if σ is unknown, $E(R)$ is estimated by the mean $\bar{R}$ of the ranges of k samples.*

The upper and lower control limits are located a distance $3\sigma_R = 3d_3\sigma$ above and below the center line. Using $\bar{R}/d_2$ to estimate σ, we locate the upper and lower control limits as follows:

$$\text{UCL:} \quad \bar{R} + 3\frac{d_3}{d_2}\bar{R} = \bar{R}\left(1 + 3\frac{d_3}{d_2}\right) = \bar{R}D_4$$

where

$$D_4 = 1 + 3\frac{d_3}{d_2}$$

*As an alternative procedure, we could estimate σ using the standard deviation of all the data contained in the k samples.

and

$$\text{LCL:} \quad \bar{R} - 3\frac{d_3}{d_2}\bar{R} = \bar{R}\left(1 - 3\frac{d_3}{d_2}\right) = \bar{R}D_3$$

where

$$D_3 = 1 - 3\frac{d_3}{d_2}$$

Values of D_3 and D_4 have been computed for sample sizes of $n = 2$ to $n = 25$, and appear in Table 18 of Appendix II.

LOCATION OF CENTER LINE AND CONTROL LIMITS FOR AN *R*-CHART

Center line: $\bar{R}$

UCL: $\bar{R}D_4$

LCL: $\bar{R}D_3$

Values of D_3 and D_4 are given in Table 18 of Appendix II for $n = 2$ to $n = 25$.

EXAMPLE 18.2

Find the location of the center line and the upper and lower control limits for the *R*-chart shown in Figure 18.6.

SOLUTION

In Example 18.1, we calculated the mean of the 20 sample ranges (the center line) to be $\bar{R} = .01985$. For $n = 4$, the values of D_3 and D_4 given in Table 18 of Appendix II are $D_3 = 0$ and $D_4 = 2.282$. Then the upper and lower control limits for the *R*-chart are:

$$\text{UCL} = \bar{R}D_4 = (.01985)(2.282) = .045298$$

$$\text{LCL} = \bar{R}D_3 = (.01985)(0) = 0$$

To monitor the variation in shaft diameters produced by the manufacturing process, a quality control engineer would check to determine that the sample range does not exceed .045298 inch. ■

The practical implications to be derived from an *R*-chart are similar to those associated with an $\bar{x}$-chart. Values of *R* that fall outside of the control limits are suspect and suggest a possible change in the process. Trends in the sample range may also indicate problems, such as wear within a machine. (We investigate this type of problem in the next section.) As in the case of the $\bar{x}$-chart, the *R*-chart can provide an indication of possible trouble in a process. A process engineer then attempts to locate the difficulty, if in fact it exists.

| | | | | | | | | | | |

EXERCISES 18.6–18.11

18.6 Refer to Exercise 18.1. Suppose the inspector wants to monitor the variation in firing pin lengths with an *R*-chart.
  a. Locate the center line for the *R*-chart.
  b. Locate upper and lower control limits for the *R*-chart.
  c. Calculate and plot the ten sample ranges in an *R*-chart. Does the process variation appear to be in control?

18.7 Construct an *R*-chart for the data of Exercise 18.2 to monitor the variation in the diameters of the molded-rubber expansion joints produced by the manufacturing process. Does the process appear to be in control?

18.8 Construct an *R*-chart for the data of Exercise 18.3 to monitor the variation in the weights of the finished bottles. Does the process appear to be in control?

18.9 Construct an *R*-chart for the data of Exercise 18.4 to monitor the variation in pitch diameters of the threaded fittings. Does the process appear to be in control?

18.10 Refer to Exercise 18.9. Modify the control limits on the *R*-chart so it can be applied to future data.

18.11 Construct an *R*-chart for the data of Exercise 18.5. Does the process variation appear to be in control?

| | | | | | | | | | | |

S E C T I O N 18.4

DETECTING TRENDS IN A CONTROL CHART: RUNS ANALYSIS

As mentioned in the previous two sections, control charts are also examined for trends in the values of $\bar{x}$ or R collected over time. Even when the sample values fall within the control limits, such a trend may indicate the presence of one or more assignable causes of variation. For example, the true process mean may have shifted slightly due to wear in the machine.

Trends in the process can be detected by observing runs of points above or below the center line of a control chart. Here a **run** is defined as a sequence of one or more consecutive points which all fall above (or all fall below) the center line. The runs (indicated in brackets) for the *R*-chart of Figure 18.6 are shown in Figure 18.7. Sample ranges that fall above the center line are denoted by a "+" symbol, and ranges that fall below the center line by a "−" symbol.

FIGURE 18.7
Runs for the *k* = 20
Sample Ranges in the
R-Chart, Figure 18.6

Note that the sequence of 20 points consists of a total of eight runs, starting with a run of two "−", followed by a run of two "+", and so on. Considerable work has been done by researchers on the development of statistical tests based on the **theory of runs**. Many of these techniques are useful for testing whether the sample observations have been drawn at random from the target population. These tests require that the total number of runs, long and short alike, be determined. In quality control, however, a few simple rules have been developed for detecting trend that are based on only the *extreme (or longest) runs*, in the control chart.

To illustrate, consider the sequence of runs in Figure 18.7. The extreme run in the sequence is composed of seven "−" symbols. These represent the seven consecutive sample ranges that all fell below the center line during hours 12, 13, . . . , 18. How likely is it to observe seven consecutive points on the control chart, all on the same side of the center line, if in fact no assignable causes of variation are present? To answer this question, we use the laws of probability learned in Chapter 3.

First, note that the probability of any one point falling above (or below) the center line is $\frac{1}{2}$ when the process is in control. Then, from the Multiplicative Law of Probability for independent events (see Chapter 3), the probability of seven consecutive points falling, say, *above* the center line is

$$\left(\frac{1}{2}\right)\left(\frac{1}{2}\right)\left(\frac{1}{2}\right)\left(\frac{1}{2}\right)\left(\frac{1}{2}\right)\left(\frac{1}{2}\right)\left(\frac{1}{2}\right) = \left(\frac{1}{2}\right)^7 = \frac{1}{128}$$

Likewise, the probability of seven consecutive points falling *below* the center line is $(\frac{1}{2})^7 = \frac{1}{128}$. Therefore, the probability of seven consecutive points falling on the same side of the center line is, by the Additive Law of Probability,

$P(7 \text{ consecutive points on the same side of the center line})$

$\quad = P(7 \text{ consecutive points above the center line})$
$\quad\quad + P(7 \text{ consecutive points below the center line})$

$$= \frac{1}{128} + \frac{1}{128} = \frac{2}{128} = \frac{1}{64}$$

or .0156. Since it is very unlikely to observe such a pattern if the process is in control, the trend in the control chart is taken as a signal of possible trouble in the production process.

A probability such as the one above can be calculated for any run in the control chart, and, based on its value, a decision made about whether to look for trouble in the process. Grant and Leavenworth (1980) recommend looking for assignable causes of variation if any one of the following sequences of points occurs in the control chart:

Seven or more consecutive points on the same side of the center line

At least 10 out of 11 consecutive points on the same side of the center line

At least 12 out of 14 consecutive points on the same side of the center line

At least 14 out of 17 consecutive points on the same side of the center line

---

These rules are easy to apply in practice since they simply require one to count consecutive points in the control chart. In each case, the probability of observing that sequence of points when the process is in control is approximately .01. (We leave proof of this result to you as an exercise.) Consequently, if one of these sequences occurs, we are highly confident that some problem in the production process, possibly a shift in the process mean, exists.

More formal statistical tests of runs are available. Consult the references at the end of this chapter if you want to learn more about these techniques.

**EXERCISES
18.12–18.16**

18.12 Refer to the $\bar{x}$- and R-charts, Exercises 18.1 and 18.6. Conduct a runs analysis to detect any trend in the process.

18.13 Refer to the $\bar{x}$- and R-charts, Exercises 18.2 and 18.7. Conduct a runs analysis to detect any trend in the process.

18.14 Refer to the $\bar{x}$- and R-charts, Exercises 18.3 and 18.8. Conduct a runs analysis to detect any trend in the process.

18.15 Refer to the $\bar{x}$- and R-charts, Exercises 18.4 and 18.9. Conduct a runs analysis to detect any trend in the process.

18.16 Refer to the $\bar{x}$- and R-charts, Exercises 18.5 and 18.11. Conduct a runs analysis to detect any trend in the process.

S E C T I O N   18.5

**CONTROL CHART
FOR PERCENT
DEFECTIVES: *p*-CHART**

In addition to measuring quantitative quality characteristics, we are also interested in monitoring the proportion p of the items produced that are defective. As in the case of the $\bar{x}$-chart, random samples of n items are selected from the production line at the end of some specified interval of time. For each sample, we compute the sample proportion

$$\hat{p} = \frac{y}{n}$$

where y is the number of defective items in the sample. The sample proportions are then plotted against time and displayed in a ***p*-chart**.

The center line for a p-chart is determined by combining the data contained in a large number k of samples. The estimate of the process proportion defective p is

$$\bar{p} = \frac{\text{Total number of defectives}}{\text{Total number inspected}} = \frac{n \sum_{i=1}^{k} \hat{p}_i}{nk} = \frac{\sum_{i=1}^{k} \hat{p}_i}{k}$$

The upper and lower control limits are located a distance of

$$3\sigma_{\hat{p}} = 3 \sqrt{\frac{p(1 - p)}{n}}$$

above and below the center line. Using $\bar{p}$ to estimate the process proportion defective p, we find

$$\text{UCL} = \bar{p} + 3 \sqrt{\frac{\bar{p}(1 - \bar{p})}{n}}$$

$$\text{LCL} = \bar{p} - 3 \sqrt{\frac{\bar{p}(1 - \bar{p})}{n}}$$

The interpretation of a p-chart is similar to the interpretations of $\bar{x}$- and R-charts. We expect the sample proportions defective to fall within the control limits. Failure to do so suggests difficulties with the production process and should be investigated.

LOCATION OF CENTER LINE AND CONTROL LIMITS FOR *p*-CHART

$$\text{Center line:} \quad \bar{p} = \frac{\text{Total number of defectives in } k \text{ samples}}{\text{Total number of items inspected}}$$

$$= \frac{\sum_{i=1}^{k} \hat{p}_i}{k}$$

$$\text{UCL:} \quad \bar{p} + 3 \sqrt{\frac{\bar{p}(1 - \bar{p})}{n}}$$

$$\text{LCL:} \quad \bar{p} - 3 \sqrt{\frac{\bar{p}(1 - \bar{p})}{n}}$$

EXAMPLE 18.3

To monitor the manufacturing process of rubber support bearings used between the superstructure and foundation pads of nuclear power plants, a quality control engineer randomly samples 100 bearings from the production line each day over a 15-day period. The bearings were inspected for defects and the number of defectives found each day are recorded in Table 18.2. Construct a *p*-chart for the fraction of defective bearings.

TABLE 18.2 Defective Bearings in 15 Samples of $n = 100$, Example 18.3

DAY	1	2	3	4	5	6	7	8	9	10	11	12	13	14	15
Number of Defectives	2	6	3	4	4	2	3	4	3	2	5	3	3	2	2
Proportion of Defectives	.02	.06	.03	.04	.04	.02	.03	.04	.03	.02	.05	.03	.03	.02	.02

SOLUTION

The center line for the *p*-chart is the proportion of defective bearings in the combined sample of $n = 1,500$ bearings:

$$\bar{p} = \frac{\text{Total number of defective bearings}}{\text{Total number inspected}} = \frac{48}{1,500} = .032$$

Upper and lower control limits are then computed as follows:

$$\text{UCL} = \bar{p} + 3 \sqrt{\frac{\bar{p}(1 - \bar{p})}{n}} = .032 + 3 \sqrt{\frac{(.032)(.968)}{1,500}}$$

$$= .032 + .0136 = .0456$$

$$\text{LCL} = \bar{p} - 3 \sqrt{\frac{\bar{p}(1 - \bar{p})}{n}} = .032 - 3 \sqrt{\frac{(.032)(.968)}{1,500}}$$

$$= .032 - .0136 = .0184$$

Thus, if the process is in control, we expect the sample proportion of defective rubber bearings to fall between .0184 and .0456 with a high probability.

A control chart for the percentage of defective bearings is shown in Figure 18.8. Note that on days 2 and 11, the sample proportion fell outside the control limits. This suggests possible problems with the manufacturing process and warrants further investigation.

FIGURE 18.8
p-Chart for the Percentage of Defective Bearings, Example 18.3

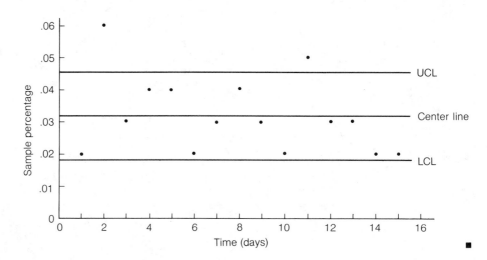

Once the problem that caused the two unusually large percentages of defectives in Example 18.3 has been identified and corrected, the control limits should be modified so that they can be applied to future data. As mentioned in Section 18.2, one method of adjusting is to recalculate their values based on only the sample points that fall within the control limits of Figure 18.8. Omitting the data for days 2 and 11, we obtain the modified values

$$\bar{p} = \frac{\text{Total number of defective bearings (excluding days 2 and 11)}}{\text{Total number of inspected (excluding days 2 and 11)}} = \frac{37}{1,400} = .026$$

$$\text{UCL} = \bar{p} + 3\sqrt{\frac{\bar{p}(1-\bar{p})}{n}} = .026 + 3\sqrt{\frac{(.026)(.974)}{1,400}}$$

$$= .026 + .013 = .039$$

$$\text{LCL} = \bar{p} - 3\sqrt{\frac{\bar{p}(1-\bar{p})}{n}} = .026 - 3\sqrt{\frac{(.026)(.974)}{1,400}}$$

$$= .026 - .013 = .013$$

Now a control chart with center line $\bar{p} = .026$, UCL $= .039$, and LCL $= .013$ can be used to monitor the percentage defective produced in future days of the process.

EXERCISES
18.17–18.20

18.17 Prestressed concrete cylinder pipe (PCCP) is a rigid pipe designed to take optimum advantage of the tensile strength of steel and the compressive strength and corrosive inhibiting properties of concrete. PCCP, produced in laying lengths of 24 feet, is susceptible to major stress cracks during the manufacturing process. To monitor the process, 20 sections of PCCP were sampled each week for a 6-week period. The number of defective sections (i.e., sections with major stress cracks) in each sample is recorded in the table.

Week	1	2	3	4	5	6
Number of Defectives	1	0	2	2	3	1

a. Construct a *p*-chart for the sample percentage of defective PCCP sections manufactured.
b. Locate the center line on the *p*-chart.
c. Locate upper and lower control limits on the *p*-chart. Does the process appear to be in control?

18.18 A manufacturer of computer terminal fuses wants to establish a control chart to monitor the production process. Each hour, for a period of 25 hours, during a time when the process is known to be in control, a quality control engineer randomly selected and tested 100 fuses from the production line. The number of defective fuses found each hour is recorded in the table.

Hour	1	2	3	4	5	6	7	8	9	10	11	12	13
Number Defective	6	4	9	3	0	6	4	2	1	2	1	3	4

Hour	14	15	16	17	18	19	20	21	22	23	24	25
Number Defective	5	5	2	1	1	0	3	7	9	2	10	3

a. Construct a *p*-chart for the sample percentage of defective terminal fuses.
b. Locate the center line on the *p*-chart.
c. Locate upper and lower control limits on the *p*-chart. Does the process appear to be in control?
d. Conduct a runs analysis on the points on the *p*-chart. What does this imply?

18.19 Refer to Exercise 18.18. Suppose the next sample of 100 terminal fuses selected from the production line contains 11 defectives. Is the process now "out of control"? Explain.

18.20 An electronics company manufactures several types of cathode ray tubes on a mass production basis. To monitor the process, 50 tubes of a certain type were randomly sampled from the production line and inspected each day over a 1-month period. The number of defectives found each day is provided in the table (page 918).

DAY	NUMBER DEFECTIVE	DAY	NUMBER DEFECTIVE
1	11	12	23
2	15	13	15
3	12	14	12
4	10	15	11
5	9	16	11
6	12	17	16
7	12	18	15
8	14	19	10
9	9	20	13
10	13	21	12
11	15		

a. Construct a *p*-chart for the sample fraction of defective cathode ray tubes.
b. Locate the center line on the *p*-chart.
c. Locate the upper and lower control limits on the *p*-chart.
d. Does the process appear to be in control? If not, modify the control limits for future data.
e. Conduct a runs analysis to detect a trend in the production process.

SECTION 18.6

CONTROL CHART FOR THE NUMBER OF DEFECTS PER ITEM: *c*-CHART

In addition to various other quality characteristics, we may be interested in the number of defects or blemishes contained in each single item of the product. For example, a manufacturer of office furniture might randomly sample one piece of furniture from the production line every 15 minutes and record the number of blemishes on the finish. Similarly, a textile manufacturer might inspect a randomly selected 1-square-foot piece of material each hour and count the number of minor defects that it contains. The objective of this procedure is to monitor the number of defects per item and to detect situations where this variable is out of control. In the notation used in quality control, the number of defects per item is denoted by the symbol *c* and a control chart used to monitor this variable over time is called a *c*-**chart**.

The Poisson probability distribution (Section 4.8) provides a good model for the probability distribution for the number *c* of defects contained in some manufactured product. From Section 4.8, we recall that if *c* possesses a Poisson probability distribution with parameter λ, then

$$E(c) = \lambda \quad \text{and} \quad \sigma_c = \sqrt{\lambda}$$

To construct a *c*-chart, we observe *c* over a reasonably large number, *k*, of equally spaced points in time and use the average value of *c*, $\bar{c}$, to estimate λ. Then since $E(c) = \lambda$, we would locate the center line of the *c*-chart at

$$\text{Center line:} \quad \bar{c} = \frac{\sum_{i=1}^{k} c_i}{k}$$

The upper and lower control limits are located a distance of $3\sigma_c$ (estimated to be $3\sqrt{\bar{c}}$) above and below the center line. Thus, the upper and lower control limits are located at

$UCL: \quad \bar{c} + 3\sqrt{\bar{c}}$

$LCL: \quad \bar{c} - 3\sqrt{\bar{c}}$

LOCATION OF CENTER LINE AND CONTROL LIMITS FOR A c-CHART

Center line: $\bar{c}$

$UCL: \quad \bar{c} + 3\sqrt{\bar{c}}$

$LCL: \quad \bar{c} - 3\sqrt{\bar{c}}$

where

$$\bar{c} = \frac{\sum_{i=1}^{k} c_i}{k}$$

EXAMPLE 18.4

The number of noticeable defects found by quality control inspectors in a randomly selected 1-square-meter specimen of woolen fabric from a certain loom is recorded each hour for a period of 20 hours. The results are shown in Table 18.3. Assuming that the number of defects per square meter has an approximate Poisson probability distribution, construct a c-chart to monitor the textile production process.

TABLE 18.3
Number of Defects Observed in Specimens of Woolen Fabric over 20 Consecutive Hours, Example 18.4

Hour	1	2	3	4	5	6	7	8	9	10
Number of Defects	11	14	10	8	3	9	10	2	5	6
Hour	11	12	13	14	15	16	17	18	19	20
Number of Defects	12	3	4	5	6	8	11	8	7	9

SOLUTION

The first step is to estimate λ, the mean number of defects per square meter of woolen fabric. This value, $\bar{c}$, also represents the center line for the control chart:

$$\bar{c} = \frac{\sum c_i}{n} = \frac{151}{20} = 7.55$$

Upper and lower control limits are then calculated as follows:

$UCL = \bar{c} + 3\sqrt{\bar{c}} = 7.55 + 3\sqrt{7.55} = 15.79$

$LCL = \bar{c} - 3\sqrt{\bar{c}} = 7.55 - 3\sqrt{7.55} = -.69$

Since a negative number of defects cannot be observed, the LCL is adjusted up to 0.

The control chart for the data appears in Figure 18.9. According to current standards, the textile process produces an allowable number of defects in woolen fabric if the number of defects per square meter does not exceed 15. At no time during the 20-hour period did the process appear to be out of control.

FIGURE 18.9
c-Chart for the Number of Defects per Square Meter of Woolen Fabric, Example 18.4

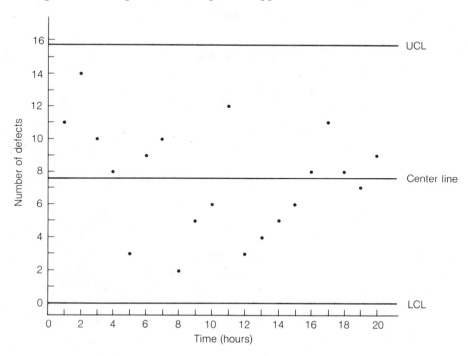

However, before we conclude that the process is in control, we should check for trends on the number of defects over time, i.e., we should perform a runs analysis as described in Section 18.4. Using the symbols "+" and "−" to denote points above and below the center line, respectively, we obtain the sequence of runs shown in Figure 18.10. Note that the extreme runs in the sequence (runs 1 and 6) include only four points. Also, none of the other unlikely sequences given in the box in Section 18.4 occurs. Therefore, it does not appear that any trend exists in the data. At this point in time, the process appears to be in control.

FIGURE 18.10
Runs for the $k = 20$ Numbers of Defects in the c-Chart, Figure 18.9

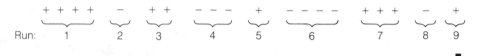

■

EXERCISES 18.21–18.24

18.21 The number of imperfections (scratches, chips, cracks, and blisters) in manufactured custom wood cabinet panels is important both to the customer and to the custom builder. To monitor the manufacturing process, a finished panel 4 feet by 8 feet was selected and inspected for imperfections each hour for 15 consecutive hours. The number of imperfections per panel is recorded in the table.

Panel	1	2	3	4	5	6	7	8	9	10	11	12	13	14	15
Number of Defects	4	2	3	3	9	4	5	3	8	7	3	6	5	7	3

a. Plot the number of defects per panel in a c-chart.
b. Locate the center line for the c-chart.
c. Locate upper and lower control limits for the c-chart. Is the process in control?
d. Conduct a runs analysis for the c-chart. What does this imply?

18.22 A quality control study was undertaken by the supervisor of a computer card key-punch operation. For each of the last 20 days' output, 25 punched cards were randomly selected and inspected for keypunching errors. The numbers of errors observed per day are recorded in the table.

Day	1	2	3	4	5	6	7	8	9	10
Number of Errors	7	3	9	8	2	5	10	5	7	6

Day	11	12	13	14	15	16	17	18	19	20
Number of Errors	1	4	11	8	6	6	9	2	12	9

a. Plot the number of keypunching errors per day in a c-chart.
b. Locate the center line for the c-chart.
c. Locate upper and lower control limits for the c-chart. Is the process in control?
d. Conduct a runs analysis for the c-chart. What does this imply?

18.23 A certain airplane model is susceptible to alignment errors in the manufacturing process. To monitor this process, the total number of alignment errors observed at final inspection for each of the first 25 aircraft produced were recorded, as shown in the accompanying table.

AIRPLANE	NUMBER OF ALIGNMENT ERRORS	AIRPLANE	NUMBER OF ALIGNMENT ERRORS
1	7	14	9
2	6	15	8
3	6	16	15
4	7	17	6
5	4	18	4
6	7	19	13
7	8	20	7
8	12	21	8
9	9	22	15
10	9	23	6
11	8	24	6
12	5	25	10
13	5		

Source: Grant, E. L. and Leavenworth, R. S. *Statistical Quality Control*, 5th ed. New York: McGraw-Hill, 1980 (Table 8–1). Reprinted with permission.

a. Construct a c-chart for the number of alignment errors per aircraft.
b. Locate the center and upper and lower control limits on the c-chart.
c. Does the process appear to be in control? Would you recommend using these control limits for future data?

18.24 Refer to Exercise 18.23. The numbers of alignment errors observed for each of the next 25 aircraft produced are shown in the table.

AIRPLANE	NUMBER OF ALIGNMENT ERRORS	AIRPLANE	NUMBER OF ALIGNMENT ERRORS
26	7	39	11
27	13	40	8
28	4	41	10
29	5	42	8
30	9	43	7
31	3	44	16
32	4	45	13
33	6	46	12
34	7	47	9
35	14	48	11
36	18	49	11
37	11	50	8
38	11		

Source: Grant, E. L. and Leavenworth, R. S. *Statistical Quality Control*, 5th ed. New York: McGraw-Hill, 1980 (Table 8–1). Reprinted with permission.

a. Add these 25 points to the c-chart of Exercise 18.23. Does the process still appear to be in control?
b. Conduct a runs analysis for the revised c-chart. What do you detect?

SECTION 18.7

TOLERANCE LIMITS

The Shewhart control charts described in the previous sections provide valuable information on the quality of the production process as a whole. Even if the process is deemed to be "in control," however, an individual manufactured item may not always meet specifications. Therefore, in addition to process control, it is often important to know that a large proportion of the individual quality measurements fall within certain limits with a high degree of confidence. An interval that includes a certain percentage of measurements with a known probability is called a **tolerance interval** and the endpoints of the interval are called **tolerance limits**.

Tolerance intervals are identical to the confidence intervals of Chapter 8, except that we are attempting to capture a proportion γ of measurements in a population rather than a population parameter (e.g., the population mean μ). For example, a production supervisor may want to establish tolerance limits for 99% of the length measurements of eyescrews manufactured on the production line, using a 95% tolerance interval. Here, the confidence coefficient is $1 - \alpha = .95$ and the proportion of measurements the supervisor wants to capture is $\gamma = .99$. The confidence coefficient, .95, has the same meaning as in Chapter 8. That is, approximately 95 out of every 100 similarly constructed tolerance intervals will contain 99% of the length measurements in the population.

When the population of measurements that characterize the product is normally distributed with known mean μ and known standard deviation σ, tolerance limits are easily constructed. In fact, such an interval is a 100% tolerance interval, i.e., the confidence coefficient is 1.0. For example, suppose the lengths of the eyescrews above have a normal distribution with $\mu = .50$ inch and $\sigma = .01$ inch. From our knowledge of the standard normal (z) distribution, we know with certainty (i.e., with probability $1 - \alpha = 1.0$) that 99% of the measurements will fall within $z = 2.58$ standard deviations of the mean. Thus, a 100% tolerance interval for 99% of the length measurements is

$$\mu \pm 2.58\sigma = .50 \pm 2.58(.01)$$
$$= .50 \pm .0258$$

or (.4742, .5258).

In practice, quality control engineers will rarely know the true values of μ and σ. Fortunately, tolerance intervals can be constructed by substituting the sample estimates $\bar{x}$ and s for μ and σ, respectively. Due to the errors introduced by the sample estimators, however, the confidence coefficient for the tolerance interval will no longer equal 1.0. The procedure for constructing tolerance limits for a normal population of measurements is described in the box.

A TOLERANCE INTERVAL FOR THE MEASUREMENTS IN A NORMAL POPULATION

A $100(1 - \alpha)$% tolerance interval for 100γ% of the measurements in a normal population is given by

$\bar{x} \pm Ks$

where

$\bar{x}$ = Mean of a sample of n measurements

s = Sample standard deviation

and K is found from Table 19 of Appendix II, based on the values of the confidence coefficient $(1 - \alpha)$, γ, and the sample size n.

Assumption: The population of measurements is approximately normal.

EXAMPLE 18.5

Refer to Example 8.1. Use the sample information provided in Table 18.1 to find a 95% tolerance interval for 99% of the shaft diameters produced by the manufacturing process. Assume that the distribution of shaft diameters is approximately normal.

SOLUTION

Table 18.1 contains diameters for 20 samples of four shafts each, or a total of $n = 80$ shaft diameters. The mean diameter of the entire sample is

$$\bar{x} = \frac{\sum\limits_{i=1}^{80} x_i}{80} = \frac{(1.505 + 1.499 + 1.501 + 1.488 + \cdots + 1.519)}{80} = 1.50045$$

(Note that this is the same value as the center line, $\bar{\bar{x}}$, computed in Example 18.1.) The sample standard deviation (given in Example 18.1) is $s = .009244$.

Since we desire a tolerance interval for 99% of the shaft diameters, $\gamma = .99$. Also, the confidence coefficient is $1 - \alpha = .95$. Table 19 of Appendix II gives the values of K for several values of γ and $1 - \alpha$. For $\gamma = .99$, $1 - \alpha = .95$, and $n = 80$, Table 19 gives $K = 2.986$. Then, the 95% tolerance interval is

$$\bar{x} \pm 2.986s = 1.50045 \pm (2.986)(.009244)$$

$$= 1.50045 \pm .02760$$

or $(1.47285, 1.52805)$. Thus, the lower and upper 95% tolerance limits for 99% of the shaft diameters are 1.47285 inches and 1.52805 inches, respectively. Our confidence in the procedure is based on the premise that approximately 95 out of every 100 similarly constructed tolerance intervals will contain 99% of the shaft diameters in the population. ∎

The technique applied in Example 18.5 gives tolerance limits for a normal distribution of measurements. If we are unwilling or unable to make the normality assumption, we must resort to a nonparametric method. Nonparametric tolerance limits are based on only the smallest and largest measurements in the sample data, as shown in the box. These tolerance intervals can be applied to any distribution of measurements.

A NONPARAMETRIC TOLERANCE INTERVAL

Let y_{min} and y_{max} be the smallest and largest observations, respectively, in a sample of size n from any distribution of measurements. Then we can select n so that

(y_{min}, y_{max})

forms a $100(1 - \alpha)\%$ tolerance interval for at least $100\gamma\%$ of the population. Values of n for several values of the confidence coefficient $(1 - \alpha)$ and γ are given in Table 20 of Appendix II.

EXAMPLE 18.6

Refer to Example 18.5. Find the sample size required so that the interval (y_{min}, y_{max}) forms a 95% tolerance interval for at least 90% of the shaft diameters produced by the manufacturing process.

SOLUTION

Here, the confidence coefficient is $1 - \alpha = .95$ and the proportion of measurements we want to capture is $\gamma = .90$. From Table 20 of Appendix II, the sample size corresponding to $1 - \alpha = .95$ and $\gamma = .90$ is $n = 46$. Therefore, if we randomly sample $n = 46$ shafts, the smallest and largest diameters in the sample will represent the lower and upper tolerance limits, respectively, for at least 90% of the shaft diameters with confidence coefficient .95. ∎

The information provided by tolerance intervals is often used to determine whether specifications are being satisfied. For example, in many production processes a design engineer will set specification limits on the manufactured product. To determine whether the specifications are realistic, the specification limits are compared to the "natural" tolerance limits of the process, i.e., the tolerance limits obtained from sampling. If the tolerance limits do not fall within the specification limits, a review of the production process is strongly recommended. An investigation may find that the specifications are tighter than necessary for the functioning of the production and, consequently, should be widened. Or, if the specifications cannot be changed, a fundamental change in the production process may be necessary in order to reduce product variability.

EXERCISES 18.25–18.29

18.25 Refer to Exercise 18.3. Use all the sample information to find a 95% tolerance interval for 90% of the finished bottle weights. Assume the distribution of bottle weights is approximately normal.

18.26 Refer to Exercise 18.4.
 a. Use all the sample information to find a 95% tolerance interval for 99% of all the pitch diameters. Assume the distribution of pitch diameters is approximately normal.
 b. Specifications require the pitch diameter of the thread to fall within .4037 ± .0013 inch. Based on the "natural" tolerance limits of the process (i.e., the tolerance limits of part **a**), does it appear that the specifications are being met?
 c. How large a sample is required to construct a nonparametric 95% tolerance interval for at least 95% of the pitch diameters? If n is large enough for this case, give the nonparametric tolerance limits.

18.27 Refer to Exercise 18.5. Find a 99% tolerance interval for at least 95% of the distance measurements assuming each of the following:
 a. A normal distribution.
 b. A nonnormal distribution.

18.28 J. Namias used the techniques of statistical quality control to determine when to conduct a search for specific causes of consumer complaints at a beverage company (*Journal of Marketing Research*, Aug. 1964). Namias discovered that when the process was in control, the biweekly complaint rate of a bottled product (i.e., the number of customer complaints per 10,000 bottles sold in a 2-week period) had an approximately normal distribution with $\mu = 26$ and $\sigma = 11.3$. Customer complaints primarily concerned chipped bottles that looked dangerous.

a. Find a tolerance interval for 99% of the complaint rates when the bottling process is assumed to be in control. What is the confidence coefficient for the interval? Explain.

b. In one 2-week period, the observed complaint rate was 93.12 complaints per 10,000 bottles sold. Based on your knowledge of statistical quality control, do you think the observed rate is due to chance or some specific cause? (In actuality, a search for a possible problem in the bottling process led to a discovery of rough handling of the bottled beverage in the warehouse by newly hired workers. As a result, a training program for new workers was instituted.)

18.29 Many hand tools used by mechanics involve attachments that fit into sockets (e.g., a socket wrench). In the manufacturing of the tools, specifications require that the inside diameter of the socket be larger than the outside diameter of the extension. That is, there must be enough clearance so that the extensions actually fit in the sockets. In order to establish tolerances for the tools, independent random samples of 50 sockets and 50 attachments were selected from the production process and the diameters (inside for sockets and outside for extensions) were measured. An analysis revealed that the distributions for both dimensions were approximately normal. The means and standard deviations (in inches) for the two samples are given in the accompanying table.

	SOCKETS (1)	ATTACHMENTS (2)
Sample Mean	.5120	.5005
Standard Deviation	.0010	.0015

a. Find a 95% tolerance interval for 99% of the socket diameters.

b. Find a 95% tolerance interval for 99% of the attachment diameters.

c. Specifications require that the clearance between attachment and socket (i.e., the difference between the inside socket diameter and outside attachment diameter) be at least .004 inch. Based on the tolerance limits from parts **a** and **b**, is it likely to find an extension and socket with less than the desired minimum clearance of .004 inch?

d. Specifications also require a maximum of .015 inch clearance between attachment and socket, to prevent fits that are too loose. Based on the tolerance intervals from parts **a** and **b**, would you expect to find some attachment and socket pairs that fit too loosely?

e. Refer to part **d**. Calculate the approximate probability of observing a loose fit. [*Hint:* Use the fact that the difference between the inside socket diameter and outside attachment diameter is approximately normal (since the two distributions are normal) with mean $\mu_1 - \mu_2$ and variance $\sigma_1^2 + \sigma_2^2$ (from Theorem 6.7).]

| | | | | | | | | | | | | |

SECTION 18.8

ACCEPTANCE SAMPLING FOR DEFECTIVES

In the preceding sections, we have learned how control charts can be used during the manufacturing process to monitor and improve the quality of a product. After manufacturing, items of the product are stored (and packaged) in *lots* containing anywhere from two to many thousands of items per lot, the *lot size* depending on the nature of the product. At this point, just prior to shipment, a second statistical tool—an **acceptance sampling plan**—is often employed to reduce the proportion of defective items shipped to customers.

An acceptance sampling plan works in the following way. A fixed number n of items is sampled from each lot, carefully inspected, and each item is judged to be either defective or nondefective. If the number y of defectives in the sample is less than or equal to a prespecified **acceptance number** a, the lot is accepted. If the number of defectives exceeds a, the lot is rejected and withheld for either a second sampling, a complete inspection, or some other procedure (see Figure 18.11). The objectives of the sampling plan are to accept and ship lots containing a small fraction p of defectives, to reject and withhold lots containing a high fraction of defectives, and to do both with a high probability.

FIGURE 18.11

Accepting or Rejecting Lots Based on the Number of Defectives in a Sample of n Items

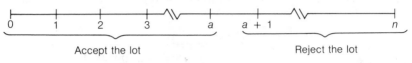

Number y of defectives

At this point you may wonder why quality control engineers resort to sampling rather than an inspection of all items in the lot. That is, why not 100% inspection? First, 100% inspection often turns out to be impractical or uneconomical. Second, studies have shown that the quality of the product shipped is often better with acceptance sampling than with 100% inspection, especially when there are a great many similar items of a product to be inspected. With 100% inspection, inspectors' fatigue on repetitive operations is always a danger. Also, psychologically, laborers have more of a tendency to make a quality product when only a few items are inspected.

Upon reflection, you can see that the decision procedure for accepting or rejecting a lot with acceptance sampling is simply a test of a hypothesis about the lot fraction defective p. The manufacturer (or customer) has in mind some lot fraction defective, say p_0, called the **acceptable quality level (AQL)**. If the lot fraction p is below $p_0 = $ AQL, the lot is deemed acceptable. The probability α of rejecting

$$H_0: \quad p = p_0$$

if in fact $p = p_0$ (that is, if the lot is actually acceptable) is called the **producer's risk**. In other words, even if $p = p_0$, the manufacturer (the producer) will withhold $100\alpha\%$ of the acceptable lots from shipment and be subjected to the cost of resampling, and so on.

DEFINITION 18.1

The **acceptable quality level (AQL)** is an upper limit, p_0, on the fraction defective that a producer is willing to tolerate.

DEFINITION 18.2

The **producer's risk** is the probability α of rejecting lots if in fact the lot fraction defective is equal to p_0, the acceptable quality level. In the terminology of hypothesis testing, the producer's risk is the probability of a Type I error.

The consumer, the purchaser of the product, is also subject to a risk—namely, the risk of accepting lots containing a high fraction defective p. The consumer will usually have in mind a lot fraction defective p_1 which is the largest lot fraction defective that he or she will tolerate. The probability β of accepting lots containing fraction defective p_1 is called the **consumer's risk**.

DEFINITION 18.3

The **consumer's risk** is the probability β of accepting lots containing fraction defective p_1 where p_1 is the upper limit in lot fraction defective acceptable to the consumer. In the terminology of hypothesis testing, the consumer's risk is the probability of a Type II error.

An **operating characteristic curve** is a graph of the probability of lot acceptance $P(A)$ versus lot fraction defective p. A typical operating characteristic curve, shown in Figure 18.12, completely characterizes a sampling plan and shows the probability of lot acceptance equal to 1 when $p = 0$ and equal to 0 when $p = 1$. As the lot fraction defective p increases, the probability $P(A)$ of lot acceptance decreases until it reaches 0. The producer's risk α is equal to $1 - P(A)$ when $p = p_0$. The consumer's risk β is equal to $P(A)$ when $p = p_1$.

FIGURE 18.12

A Typical Operating Characteristic Curve

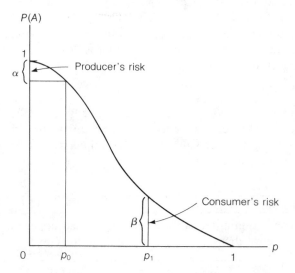

DEFINITION 18.4

The **operating characteristic curve** for a sampling plan is a graph of the probability of lot acceptance, $P(A)$, versus the lot fraction defective, p.

The operating characteristic curve for a sampling plan can be constructed by calculating $P(A)$ for various values of the lot fraction defective p. As explained in Sections 4.5 and 4.7, the probability distribution for the number y of defectives in a sample of n items from a lot will depend on the lot size N. If N is large and n is small relative to N, then the probability distribution for y can be approximated by a **binomial probability distribution** (Section 4.5):

$$p(y) = \binom{n}{y} p^y q^{n-y} \qquad y = 0, 1, 2, \ldots, n$$

where

$$q = 1 - p$$

If N is small or n is large relative to N, then y will have a **hypergeometric probability distribution** (Section 4.7):

$$p(y) = \frac{\binom{r}{y}\binom{N-r}{n-y}}{\binom{N}{n}}$$

where

N = Lot size

r = Number of defectives in the lot

$p = \dfrac{r}{N}$ = Lot fraction defective

n = Sample size

y = Number of defectives in the sample

Using the appropriate probability distribution for a sampling plan with sample size n and acceptance number a, we can compute the probability of accepting a lot with lot fraction defective p:

$$P(A) = P(y \le a) = p(0) + p(1) + \cdots + p(a)$$

We will illustrate the procedure with the next example.

EXAMPLE 18.7

A manufacturer of metal gaskets ships a particular gasket in lots of 500 each. The acceptance sampling plan used prior to shipment is based on a sample size $n = 10$ and acceptance number $a = 1$.

a. Find the producer's risk if the AQL is .05.
b. Find the consumer's risk if the lot fraction defective is $p_1 = .20$.
c. Draw a rough sketch of the operating characteristic curve for the sampling plan.

SOLUTION

a. The producer's risk is $\alpha = 1 - P(A)$ when $p = p_0 = .05$. For $N = 500$ and $n = 10$, y will possess approximately a binomial probability distribution. Then, if in fact $p = .05$,

$$P(A) = p(0) + p(1)$$

$$= \binom{10}{0}(.05)^0(.95)^{10} + \binom{10}{1}(.05)^1(.95)^9 = .914$$

and the producer's risk is

$$\alpha = 1 - P(A) = 1 - .914 = .086$$

This means that the producer will reject 8.6% of the lots, even if the lot fraction defective is as small as .05.

b. The consumer's risk is $\beta = P(A)$ when $p = .20$:

$$\beta = P(A) = p(0) + p(1)$$

$$= \binom{10}{0}(.2)^0(.8)^{10} + \binom{10}{1}(.2)^1(.8)^9 = .376$$

Thus, the consumer risks accepting lots containing a lot fraction defective equal to $p_1 = .20$ approximately 37.6% of the time. The fact that β is so large for $p_1 = .20$ indicates that this sampling plan would be of little value in practice. The plan needs to be based on a larger sample size.

c. A rough sketch of the operating characteristic curve for the sampling plan can be obtained using the two points calculated in parts **a** and **b** and the fact that $P(A) = 1$ when $p = 0$ and $P(A) = 0$ when $p = 1$. The sketch is shown in Figure 18.13.

FIGURE 18.13
A Rough Sketch of the Operating Characteristic Curve of $n = 10$ and $a = 1$

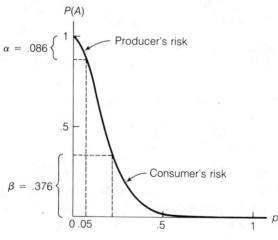

In practice, engineers do not construct sampling plans for specific lot sizes and AQLs because they have been constructed and have been in use for years. One of the most widely used collections of sampling plans is known as the **Military Standard 105D (MIL-STD-105D)**. The sampling plans contained in MIL-STD-105D employ a sample size n that varies with the lot size N. The sample sizes specified in the plans were chosen in order to give reasonable values of consumer risk. In addition, the plans have been constructed so that each falls into one of three levels of inspection categories: reduced (I), normal (II), or tightened (III). Lower consumer risks are associated with tighter plans.

Two of the MIL-STD-105D tables are reproduced in Tables 21 and 22 of Appendix II. The following example illustrates their use.

EXAMPLE 18.8

Find the appropriate MIL-STD-105D normal (level) general inspection sampling plan for a lot size of 500 items and an acceptable quality level of .065.

SOLUTION

The first step in selecting the sampling plan is to identify the MIL-STD-105D code corresponding to a lot size of 500 and a normal inspection level—that is, level II. This code letter, H, is found in Table 21 of Appendix II, in the row corresponding to lot size 281–500 and in the column labeled II.

The second step in selecting the plan is to determine the sample size and acceptance number from Table 22 of Appendix II. The sample size code letters appear in the first column of the table. The recommended sample sizes are shown in the second column. Moving down column 1 to code letter H, we see that the recommended sample size (column 2) is $n = 50$. To find the acceptance number, move across the top row to 6.5%, or, equivalently, AQL = .065. The acceptance (Ac) number, $a = 7$, is shown at the intersection of the 6.5 column and the H row. The number 8 that also appears at this intersection is the rejection number for the sampling plan—that is, we reject a lot if y is greater than or equal to 8.

You can see that this MIL-STD-105D sampling plan uses a much larger sample ($n = 50$) than the plan of Example 18.7. Because of this larger sample size, the probability of lot acceptance, $P(A)$, calculated for a given lot fraction defective p, would be much smaller than for the plan of Example 18.7. We would say that the MIL-STD-105D plan is *tighter* than the plan of Example 18.7. The consumer risk is less or, equivalently, it allows fewer bad lots to be shipped.   ∎

The probability of acceptance $P(A)$ for the MIL-STD-105D sampling plan can be calculated as described earlier in this section. For example, for a lot fraction defective $p = .10$ in Example 18.8, we have

$$P(A) = P(y \leq 7)$$
$$= \sum_{y=0}^{7} p(y)$$

where $p(y)$ is a hypergeometric probability distribution with $N = 500$, $n = 50$, and the number r of defectives in the lot is $Np = (500)(.1) = 50$. The actual calculation of $P(A)$ is tedious and is best accomplished by using a computer.

EXERCISES
18.30–18.34

18.30 Consider a sampling plan with sample size $n = 5$ and acceptance number $a = 0$.
    a. Calculate the probability of lot acceptance for fractions defective $p = .1$, $.3$, and $.5$. Sketch the operating characteristic curve for the plan.
    b. Find the producer's risk if AQL $= .01$.
    c. Find the consumer's risk if $p_1 = .10$.

18.31 Consider a sampling plan with sample size $n = 15$ and acceptance number $a = 1$.
    a. Calculate the probability of lot acceptance for fractions defective $p = .1$, $.2$, $.3$, $.4$, and $.5$. Sketch the operating characteristic curve for the plan.
    b. Find the producer's risk if AQL $= .05$.
    c. Find the consumer's risk if $p_1 = .20$.

18.32 Find the appropriate MIL-STD-105D general inspection sampling plan for a lot size of 5,000 items and an AQL of 4% under each of the following inspection categories:
    a. Reduced (I) inspection level
    b. Normal (II) inspection level
    c. Tightened (III) inspection level

18.33 The tensile strengths of wires in a certain lot of size 400 are specified to exceed 5 kilograms. Consider an acceptance sampling plan based on a sample of $n = 10$ wires and acceptance number $a = 1$.
    a. Find the producer's risk if the AQL is 2.5%.
    b. Find the consumer's risk if the lot fraction failing to meet specifications is $p_1 = .15$.
    c. Draw a rough sketch of the operating characteristic curve for the sampling plan. Do you think the sampling plan is acceptable? Explain.

18.34 Refer to Exercise 18.33. Find the appropriate MIL-STD-105D normal (level) general inspection sampling plan for a lot size of 400 wires and an AQL of 2.5%.

S E C T I O N 18.9

OTHER SAMPLING PLANS

In Section 18.8 we presented a sampling plan based on the number of defectives contained in a single sample. A second type of acceptance sampling plan is one based on double or multiple sampling. A **double sampling plan** involves the selection of n_1 items from the lot. The lot is accepted if the number y_1 of defectives in the sample is $y_1 \leq a_1$ and rejected if $y_1 \geq r_1$ (where $r_1 > a_1$), as shown in Figure 18.14. If y_1 falls *between* a_1 and r_1, then a second sample of n_2 items is

FIGURE 18.14
Location of the Acceptance Number a_1 and Rejection Number r_1 for the First Sample in a Double Sampling Plan

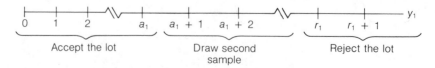

selected from the lot and the total number y of defectives in the $(n_1 + n_2)$ sampled items is recorded. If y is less than or equal to a second acceptance number a_2, the lot is accepted; otherwise, it is rejected.

The ultimate in multiple sampling is **sequential sampling**. In a sequential sampling plan, the items are selected from the lot, one by one. As each item is selected, a decision is made to accept the lot, to reject the lot, or to sample the next item from the lot. With this type of sampling, the decision to accept (or to reject) the lot might occur as early as the first, second, or third items sampled. It is also possible that the decision to accept or to reject the lot might require a very large sample. Thus, in sequential sampling, the sample size n is a random variable.

In addition to single, multiple, and sequential sampling plans based on the number y of defects observed, similar plans have been developed to utilize measurements on quantitative variables. Thus, instead of examining each item in a sample and rating it as defective or nondefective, we make our decision to reject or to accept the lot based on a quantitative measurement taken on each of the items. For example, a purchaser of 50-gallon barrels of acetone might be primarily concerned that each barrel contain at least 50 gallons. A typical sampling plan might involve sampling ten barrels from each lot and measuring the exact number y of gallons in each barrel. We could classify each barrel that contains less than 50 gallons as defective and base our decision to reject or to accept the lot on the number of defective barrels in the sample. Alternatively, we could base our decision on the sample mean, $\bar{y}$, the average amount of acetone in the ten barrels. A sampling plan based on the mean of a sample of quantitative measurements is called **acceptance sampling by variables**. One of the most widely used collections of such sampling plans is **Military Standard 414 (MIL-STD-414)**.

The literature on acceptance sampling plans is extensive. For collections of sampling plans and for more information on the subject, we refer you to the references at the end of the chapter. Before leaving this discussion, however, we leave you with this thought: *It does not always pay to sample.* There may be certain situations where the cost of sampling is so prohibitive that the only alternatives are either 100% inspection or no inspection at all. Thus, total cost plays an important role in the acceptance sampling plan selection process.

SECTION 18.10
EVOLUTIONARY OPERATIONS

An **evolutionary operation** is a technique designed to improve the yield and/or the quality of an industrial product by extracting information from an operating process. To illustrate the procedure, suppose that some quality characteristic of a chemical product—say, viscosity—is dependent on a number of variables, including the temperature of the raw materials and the pressure maintained within the vat in which they are mixed. To investigate the effect of these variables on the viscosity of a batch, we could simulate the process in a laboratory and conduct a multivariable experiment (for example, a factorial experiment) as described in Chapter 13. But this process would be costly and it is possible that the simulation would behave differently from the production process.

A second and less costly procedure is to concentrate on only two or three of the independent variables and to vary the settings of these variables according to a designed experiment. The key is to make the changes in the independent variables so small that there is no *observable* change in the quality of the product. In order to detect the effect of these small changes, we repeat the experiment over and over again until the sample sizes are so large that even small changes in the mean value of the quality variable are significant when tested statistically.

For example, suppose we know that a number of controllable process variables, including the temperature and pressure of raw materials, affect the viscosity of a batch-produced chemical. We are afraid to make experimental changes in these variables out of fear that we might produce a bad product and an accompanying financial loss. However, we know that very slight changes in temperature and pressure—say, changes of 2°F and 2 pounds per square inch (psi)—would have a negligible effect on product quality.

To investigate the effects of temperature and pressure, we will conduct an experiment in the operating process using the experimental design shown in Figure 18.15. The four temperature—pressure combinations at the corners of the design are the four factor–level combinations of a 2×2 factorial experiment (Section 13.7). The pressure–temperature combination (50°F, 130 psi) was added at the center of the design region to enable us to detect a relatively high (or low) mean viscosity in the center of the experimental region, in case it exists.

FIGURE 18.15

An Experimental Design for an Evolutionary Operation

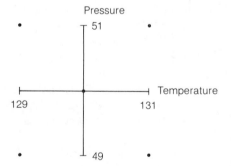

To conduct the evolutionary operation, we would assign one of the five pressure–temperature combinations to each batch of chemical and measure the viscosity y for each. If the manufacturer produces ten batches per day, we would obtain two replications of the five treatments contained in the design shown in Figure 18.15. If we were to conduct statistical tests to detect changes in the mean viscosity based on the data for 1 day, or perhaps even for 100 days, it is conceivable that no changes in mean viscosity would be evident. However, if we continue to collect data over a long period of time, obtaining two replications of the experiment each day, we would eventually detect changes in mean viscosity (if they exist). Thus, the logic of an evolutionary operation is that a production process produces data at the same time that it generates a product. Why not utilize the information that is free (except for the cost of collection)? Although the individual observations contain very little information on the effect that pressure and temperature have on mean viscosity, the weight of huge amounts of data eventually will show us how to change these variables to produce desirable changes in mean viscosity. Thus, repeated experimentation over time enables the process to evolve to a higher level of quality and/or yield.

| | | | | | | | | | | |

S E C T I O N 18.11

SUMMARY

Chapter 18 introduced three statistical techniques useful in quality control: statistical control charts, acceptance sampling plans, and evolutionary operations. Although these methods involve relatively simple applications of statistical theory, they represent very powerful techniques for controlling and improving the quality of a manufactured product.

Control charts are used to monitor the quality characteristics of a product during the manufacturing process. Control charts indicate when something may be wrong with the manufacturing process. Although they do not specify what is wrong, the behavior of the plotted points on a control chart often provides clues that will lead a production engineer to the source of the trouble. The $\bar{x}$-charts and **R-charts** monitor the mean and variability of a quantitative quality characteristic using the mean and range of samples selected at fixed intervals of time. A **p-chart** is used to monitor the fraction defective produced by a manufacturing process and a chart showing the **number of defects** tracks the mean number of defects per item.

Quality assurance sampling plans are utilized to prevent bad lots of product from being shipped to a manufacturer's customers. Thus, a sampling plan is intended to provide a screen at the door of the shipping department. The screen is supposed to pass good lots and withhold bad lots, both with a high probability. We described a single sampling plan for the number of defectives in a sample. We stated that a plan is completely described by its **operating characteristic curve**, a graph of the probability $P(A)$ of accepting a lot versus the lot fraction defective p ($0 \leq p \leq 1$). **Acceptance sampling by variables, double and sequential sampling plans** can also be used to withhold lots that contain an overly large fraction defective.

The third statistical technique presented was the concept of an **evolutionary operation**, the idea of experimenting and improving the quality of an industrial product during an ongoing manufacturing operation. Very small changes are made in the controllable process variables—changes so small that their effects would be undetectable using small to moderate sample sizes. The effects of these changes are detected by replicating the treatments of a simple experimental design, over and over again, until the effects of the small variable changes are clear. We then move the process control variable settings to the values that provide the best mean value for the quality characteristic and then repeat the experimentation using a new selection of design points. If the control variables really affect the quality characteristic, this process will lead to higher and higher levels of quality.

18.35 Specifications require the nickel content of manufactured stainless steel hydraulic valves to be 13% by weight. To monitor the production process, four valves were selected from the production line each hour over an 8-hour period and the percentage nickel content was measured for each, with the results recorded in the table.

HOUR	NICKEL CONTENT			
1	13.1	12.8	12.7	12.9
2	12.5	13.0	13.6	13.1
3	12.9	12.9	13.2	13.3
4	12.4	13.0	12.1	12.6
5	12.8	11.9	12.7	12.4
6	13.0	13.6	13.2	12.9
7	13.5	13.5	13.1	12.7
8	12.6	13.9	13.3	12.8

  a. Construct a control chart for the mean nickel content of the hydraulic valves.
  b. Establish control limits for the mean using Table 18 of Appendix II.
  c. Establish control limits for the mean using the standard deviation of the overall sample. Compare to the limits obtained in part **b**.
  d. Do all observed sample means lie within the control limits? What are the consequences of this?
  e. Find a 99% tolerance interval for 99% of the nickel contents in the hydraulic valves. Assume that the distribution of nickel contents is approximately normal.

18.36 Refer to Exercise 18.35. Construct a control chart with control limits for the variability in the nickel contents of the hydraulic valves. Interpret your results.

18.37 A quality control inspector is studying the alternative sampling plans $(n = 5, a = 1)$ and $(n = 25, a = 5)$.
  a. Sketch the operating characteristic curves for both plans, using lot fractions defective .05, .10, .20, .30, and .40.
  b. As a seller producing lots with AQL = .10, which of the two sampling plans would you prefer? Why?
  c. As a buyer wanting to protect against accepting lots with fraction defective exceeding $p_1 = .30$, which of the two sampling plans would you prefer? Why?

18.38 A company manufactures rolled steel for nuclear submarines. To monitor the production process, a quality control inspector sampled finished rolls of steel from the production line, one each hour for 12 consecutive hours. The number of imperfections discovered on each roll is recorded in the table.

Hour	1	2	3	4	5	6	7	8	9	10	11	12
Number of Imperfections	14	10	8	7	11	12	6	15	13	4	9	10

  a. Construct a control chart for the number of imperfections per finished roll of steel.

b. Locate the center line and upper and lower control limits on the chart.

c. Does the manufacturing process appear to be in control?

18.39 For a lot of 250 electron tubes with an acceptance quality level of 10%, find the appropriate MIL-STD-105D general inspection sampling plan under each of the following inspection categories:

a. Normal inspection level

b. Tightened inspection level

18.40 High-level computer technology has developed bit-sized microprocessors for use in operating industrial "robots." To monitor the fraction of defective microprocessors produced by a manufacturing process, 50 microprocessors are sampled each hour. The results for 20 hours of sampling are provided in the table.

Sample	1	2	3	4	5	6	7	8	9	10
Number of Defectives	5	6	4	7	1	3	6	5	4	5

Sample	11	12	13	14	15	16	17	18	19	20
Number of Defectives	8	3	2	1	0	1	1	2	3	3

a. Construct a control chart for the proportion of defective microprocessors.

b. Locate the center line and upper and lower control limits on the chart. Does the process appear to be in control?

c. Conduct a runs analysis for the control chart. Interpret the result.

18.41 A construction engineer buys steel cable in large rolls to use in supporting equipment and temporary structures during the process of erecting permanent structures. Specifications require the breaking strength of the steel cable to exceed 200 pounds. For a lot size of 1,500 large rolls of steel cable, consider an acceptance sampling plan based on a sample of $n = 20$ rolls and acceptance number $a = 2$.

a. Find the producer's risk if the AQL is .05.

b. Find the consumer's risk if the lot fraction failing to meet breaking strength specifications is $p_1 = .10$.

c. Draw a rough sketch of the operating characteristic curve for the sampling plan. Is the sampling plan reasonable?

18.42 Refer to Exercise 18.41. Find the appropriate MIL-STD-105D normal (level) general inspection sampling plan for a lot size of 1,500 large rolls of steel cable and an AQL of .05.

18.43 Refer to Exercise 18.42.

a. Find the producer's risk under the inspection sampling plan. [*Hint:* Use the normal approximation to the binomial.]

b. Find the consumer's risk if the lot fraction failing to meet breaking strength specifications is $p_1 = .08$. [*Hint:* Use the normal approximation to the binomial.]

18.44 Refer to the stress analysis on epoxy-repaired truss joints described in Exercise 8.31. Tests were conducted on epoxy-bonded truss joints made of wood in order to determine tolerances for actual glue line shear stress (*Journal of Structural Engineering*, Feb. 1986). The mean and standard deviation of the shear strengths (pounds per square inch) for a random sample of 100 Southern Pine truss joints are

$$\bar{x} = 1,312 \qquad s = 422$$

a. Assuming the distribution of strength measurements is approximately normal, construct a 95% tolerance interval for 99% of the shear strengths.

b. Interpret the interval obtained in part **b**.

c. Explain how you could obtain a tolerance interval when the normality assumption is not satisfied.

REFERENCES

Box, G. E. P. "Evolutionary Operation: A Method for Increasing Industrial Productivity," *Applied Statistics, 6*, 1957, pp. 3–23.

Box, G. E. P. and Hunter, J. S. "Condensed Calculations for Evolutionary Operation Programs," *Technometrics, 1*, 1959, pp. 77–95.

Deming, W. E. *Quality, Productivity, and Competitive Position.* Cambridge, Mass.: MIT Press, 1982.

Grant, E. L. and Leavenworth, R. S. *Statistical Quality Control*, 5th ed. New York: McGraw-Hill, 1980.

Hald, A. *Statistical Theory of Sampling Inspection of Attributes.* New York: Academic Press, 1981.

Juran, J. M. and Gryna, F. M. *Quality Planning and Analysis.* New York: McGraw-Hill, 1970.

Mendenhall, W. *The Design and Analysis of Experiments.* Belmont, California: Wadsworth Publishing Co., 1968.

Military Standard 105D. Washington, D. C.: U. S. Government Printing Office, 1963.

Miller, I. and Freund, J. E. *Probability and Statistics for Engineers*, 2nd ed. Englewood Cliffs, N. J.: Prentice-Hall, 1977.

National Bureau of Standards, *Tables of the Binomial Distribution.* Washington, D. C.: U. S. Government Printing Office, 1950.

Ott, E. R. *Process Quality Control: Trouble-shooting and Interpretation of Data.* New York: McGraw-Hill, 1975.

Romig, H. G. *50–100 Binomial Tables.* New York: John Wiley, 1953.

Scheaffer, R. L. and McClave, J. T. *Statistics for Engineers*, 2nd ed. Boston: Duxbury Press, 1986.

Shewhart, W. A. *Economic Control of Quality of Manufactured Product.* Princeton, N. J.: Van Nostrand Reinhold, 1931.

| | | | | | | | | | | | | | | | | | | | | | | | | | | | | | | | | | | | | | | | | | | | | | | | | | | | | |

A P P E N D I X   I

CONTENTS

MATRIX ALGEBRA

SECTION I.1

**MATRICES
AND MATRIX
MULTIPLICATION**

For some statistical procedures (e.g., multiple regression), the formulas for conducting the analysis are more easily given using **matrix algebra** instead of ordinary algebra. By arranging the data in particular rectangular patterns called **matrices** and performing various operations with them, we can obtain the results of the analyses much more quickly. In this appendix, we will define a matrix and explain various operations that can be performed with them. (We explained how to use this information to conduct a regression analysis in Section 11.4.)

Three matrices, **A**, **B**, and **C**, are shown below. Note that each matrix is a rectangular arrangement of numbers with one number in every row–column position.

$$\mathbf{A} = \begin{bmatrix} 2 & 3 \\ 0 & 1 \\ -1 & 6 \end{bmatrix} \qquad \mathbf{B} = \begin{bmatrix} 3 & 0 & 1 \\ -1 & 0 & 1 \\ 4 & 2 & 0 \end{bmatrix} \qquad \mathbf{C} = \begin{bmatrix} 1 \\ 2 \\ 1 \end{bmatrix}$$

DEFINITION I.1

A **matrix** is a rectangular array of numbers.*

The numbers that appear in a matrix are called **elements** of the matrix. If a matrix contains r rows and c columns, there will be an element in each of the row–column positions of the matrix, and the matrix will have $r \times c$ elements. For example, the matrix **A** shown above contains $r = 3$ rows, $c = 2$ columns, and $rc = (3)(2) = 6$ elements, one in each of the 6 row–column positions.

DEFINITION I.2

A number in a particular row–column position is called an **element** of the matrix.

Notice that the matrices **A**, **B**, and **C** contain different numbers of rows and columns. The numbers of rows and columns give the **dimensions** of a matrix.

DEFINITION I.3

A matrix containing r rows and c columns is said to be an $r \times c$ **matrix**, where r and c are the **dimensions** of the matrix.

DEFINITION I.4

If $r = c$, a matrix is said to be a **square matrix**.

---

*For our purpose, we assume that the numbers are real.

When we give a formula in matrix notation, the elements of a matrix will be represented by symbols. For example, if we have a matrix

$$A = \begin{bmatrix} a_{11} & a_{12} & a_{13} \\ a_{21} & a_{22} & a_{23} \end{bmatrix}$$

the symbol a_{ij} will denote the element in the ith row and jth column of the matrix. The first subscript always identifies the row and the second identifies the column in which the element is located. For example, the element a_{12} is in the first row and second column of matrix A. The rows are always numbered from top to bottom, and the columns are always numbered from left to right.

Matrices are usually identified by capital letters, such as A, B, C, corresponding to the letters of the alphabet employed in ordinary algebra. The difference is that in ordinary algebra, a letter is used to denote a single real number, while in matrix algebra, a letter denotes a rectangular array of numbers. The operations of matrix algebra are very similar to those of ordinary algebra—you can add matrices, subtract them, multiply them, and so on. However, there are a few operations that are unique to matrices, such as the **transpose of a matrix**. For example, if

$$A = \begin{bmatrix} 5 \\ 1 \\ 0 \\ 4 \\ 2 \end{bmatrix} \quad \text{and} \quad B = \begin{bmatrix} 1 & 0 \\ 1 & 1 \\ 1 & 4 \\ 1 & 2 \\ 1 & 6 \end{bmatrix}$$

then the transpose matrices of the A and B matrices, denoted as A' and B', respectively, are

$$A' = \begin{bmatrix} 5 & 1 & 0 & 4 & 2 \end{bmatrix} \quad \text{and} \quad B' = \begin{bmatrix} 1 & 1 & 1 & 1 & 1 \\ 0 & 1 & 4 & 2 & 6 \end{bmatrix}$$

DEFINITION I.5

The **transpose of a matrix** A, denoted as A', is obtained by interchanging corresponding rows and columns of the A matrix. That is, the ith row of the A matrix becomes the ith column of the A' matrix.

Since we are concerned mainly with the applications of matrix algebra to the solution of the least squares equations in multiple regression (See Chapter 11), we will define only the operations and types of matrices that are pertinent to that subject.

The most important operation for us is matrix multiplication, which requires **row–column multiplication**. To illustrate this process, suppose we wish to find the product **AB**, where

$$A = \begin{bmatrix} 2 & 1 \\ 4 & -1 \end{bmatrix} \quad \text{and} \quad B = \begin{bmatrix} 2 & 0 & 3 \\ -1 & 4 & 0 \end{bmatrix}$$

We will always multiply the rows of **A** (the matrix on the left) by the columns of **B** (the matrix on the right). The product formed by the first row of **A** times the first column of **B** is obtained by multiplying the elements in corresponding positions and summing these products. Thus, the first row, first column product, shown diagrammatically below, is

$$(2)(2) + (1)(-1) = 4 - 1 = 3$$

$$AB = \begin{bmatrix} 2 & 1 \\ 4 & -1 \end{bmatrix} \begin{bmatrix} 2 & 0 & 3 \\ -1 & 4 & 0 \end{bmatrix} = \begin{bmatrix} 3 & & \\ & & \end{bmatrix}$$

Similarly, the first row, second column product is

$$(2)(0) + (1)(4) = 0 + 4 = 4$$

So far we have

$$AB = \begin{bmatrix} 3 & 4 & \\ & & \end{bmatrix}$$

To find the complete matrix product **AB**, all we need to do is find each element in the **AB** matrix. Thus, we will define an element in the ith row, jth column of **AB** as the product of the ith row of **A** and the jth column of **B**. We complete the process in Example I.1.

EXAMPLE I.1

Find the product **AB**, where

$$A = \begin{bmatrix} 2 & 1 \\ 4 & -1 \end{bmatrix} \quad \text{and} \quad B = \begin{bmatrix} 2 & 0 & 3 \\ -1 & 4 & 0 \end{bmatrix}$$

SOLUTION

If we represent the product **AB** as

$$C = \begin{bmatrix} c_{11} & c_{12} & c_{13} \\ c_{21} & c_{22} & c_{23} \end{bmatrix}$$

we have already found $c_{11} = 3$ and $c_{12} = 4$. Similarly, the element c_{21}, the element in the second row, first column of **AB**, is the product of the second row of **A** and the first column of **B**:

$$(4)(2) + (-1)(-1) = 8 + 1 = 9$$

Proceeding in a similar manner to find the remaining elements of **AB**, we have

$$AB = \begin{bmatrix} 2 & 1 \\ 4 & -1 \end{bmatrix} \begin{bmatrix} 2 & 0 & 3 \\ -1 & 4 & 0 \end{bmatrix} = \begin{bmatrix} 3 & 4 & 6 \\ 9 & -4 & 12 \end{bmatrix}$$

∎

Now, try to find the product **BA**, using matrices **A** and **B** from Example I.1. You will observe two very important differences between multiplication in matrix algebra and multiplication in ordinary algebra:

1. You cannot find the product **BA**, because you cannot perform row–column multiplication. You can see that the dimensions do not match by placing the matrices side-by-side:

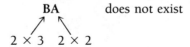

$$\text{BA} \qquad \text{does not exist}$$
$$2 \times 3 \quad 2 \times 2$$

The number of elements (3) in a row of **B** (the matrix on the left) does not match the number of elements (2) in a column of **A** (the matrix on the right). Therefore, you cannot perform row–column multiplication, and the matrix product **BA** does not exist. The point is, not all matrices can be multiplied. You can find products for matrices **AB**, only where **A** is $r \times d$ and **B** is $d \times c$. That is:

REQUIREMENT FOR MULTIPLICATION

$$\text{AB}$$
$$r \times d \quad d \times c$$

The two inner dimension numbers must be equal. The dimensions of the product will always be given by the outer dimension numbers:

DIMENSIONS OF AB ARE $r \times c$

$$\text{AB}$$
$$r \times d \quad d \times c$$

---

2. The second difference between ordinary and matrix multiplication is that in ordinary algebra, $ab = ba$. In matrix algebra, **AB** usually does not equal **BA**. In fact, as noted in item 1 above, it may not even exist.

DEFINITION I.6

The product **AB** of an $r \times d$ matrix **A** and a $d \times c$ matrix **B** is an $r \times c$ matrix **C**, where the element c_{ij} ($i = 1, 2, \ldots, r; j = 1, 2, \ldots, c$) of **C** is the product of the ith row of **A** and the jth column of **B**.

EXAMPLE I.2

Given the matrices below, find $\mathbf{IA}$ and $\mathbf{IB}$.

$$\mathbf{A} = \begin{bmatrix} 2 \\ 1 \\ 3 \end{bmatrix} \qquad \mathbf{B} = \begin{bmatrix} 3 & 0 \\ 1 & 2 \\ 4 & -1 \end{bmatrix} \qquad \mathbf{I} = \begin{bmatrix} 1 & 0 & 0 \\ 0 & 1 & 0 \\ 0 & 0 & 1 \end{bmatrix}$$

SOLUTION

Notice that the product

exists and that it will be of dimensions 3×1:

$$\mathbf{IA} = \begin{bmatrix} 1 & 0 & 0 \\ 0 & 1 & 0 \\ 0 & 0 & 1 \end{bmatrix} \begin{bmatrix} 2 \\ 1 \\ 3 \end{bmatrix} = \begin{bmatrix} 2 \\ 1 \\ 3 \end{bmatrix}$$

Similarly,

exists and is of dimensions 3×2:

$$\mathbf{IB} = \begin{bmatrix} 1 & 0 & 0 \\ 0 & 1 & 0 \\ 0 & 0 & 1 \end{bmatrix} \begin{bmatrix} 3 & 0 \\ 1 & 2 \\ 4 & -1 \end{bmatrix} = \begin{bmatrix} 3 & 0 \\ 1 & 2 \\ 4 & -1 \end{bmatrix}$$

Notice that the $\mathbf{I}$ matrix possesses a special property. We have $\mathbf{IA} = \mathbf{A}$ and $\mathbf{IB} = \mathbf{B}$. We will comment further on this property in Section I.2. ∎

EXERCISES I.1–I.6

I.1 Consider the matrices A, B, and C:

$$\mathbf{A} = \begin{bmatrix} 3 & 0 \\ -1 & 4 \end{bmatrix} \qquad \mathbf{B} = \begin{bmatrix} 2 & 1 \\ 0 & -1 \end{bmatrix} \qquad \mathbf{C} = \begin{bmatrix} 1 & 0 & 3 \\ -2 & 1 & 2 \end{bmatrix}$$

a. Find **AB**.   **b.** Find **AC**.   **c.** Find **BA**.

I.2 Consider the matrices A, B, and C:

$$\mathbf{A} = \begin{bmatrix} 3 & 1 & 3 \\ 2 & 0 & 4 \\ -4 & 1 & 2 \end{bmatrix} \qquad \mathbf{B} = \begin{bmatrix} 1 & 0 & 2 \end{bmatrix} \qquad \mathbf{C} = \begin{bmatrix} 3 \\ 0 \\ 2 \end{bmatrix}$$

a. Find **AC**.   **b.** Find **BC**.
c. Is it possible to find **AB**? Explain.

I.3 Assume that **A** is a 3 × 2 matrix and **B** is a 2 × 4 matrix.
a. What are the dimensions of **AB**?
b. Is it possible to find the product **BA**? Explain.

I.4 Assume that matrices **B** and **C** are of dimensions 1 × 3 and 3 × 1, respectively.
a. What are the dimensions of the product **BC**?
b. What are the dimensions of **CB**?
c. If **B** and **C** are the matrices shown in Exercise I.2, find **CB**.

I.5 Consider the matrices **A**, **B**, and **C**:

$$
A = \begin{bmatrix} 1 & 0 & 0 \\ 0 & 3 & 0 \\ 0 & 0 & 2 \end{bmatrix} \qquad
B = \begin{bmatrix} 2 & 3 \\ -3 & 0 \\ 4 & -1 \end{bmatrix} \qquad
C = \begin{bmatrix} 3 & 0 & 2 \end{bmatrix}
$$

a. Find **AB**.   **b.** Find **CA**.   **c.** Find **CB**.

I.6 Consider the matrices:

$$
A = \begin{bmatrix} 3 & 0 & -1 & 2 \end{bmatrix} \qquad
B = \begin{bmatrix} 2 \\ -1 \\ 0 \\ 3 \end{bmatrix}
$$

a. Find **AB**.   **b.** Find **BA**.

SECTION I.2

IDENTITY MATRICES AND MATRIX INVERSION

In ordinary algebra, the number 1 is the identity element for the multiplication operation. That is, 1 is the element such that any other number, say c, multiplied by the identity element is equal to c. Thus, $4(1) = 4$, $(-5)(1) = -5$, etc.

The corresponding identity element for multiplication in matrix algebra, identified by the symbol **I**, is a matrix such that

$$
\mathbf{AI} = \mathbf{IA} = \mathbf{A} \quad \text{for any matrix } \mathbf{A}
$$

The difference between identity elements in ordinary algebra and matrix algebra is that in ordinary algebra, there is only one identity element, the number 1. In matrix algebra, the identity matrix must possess the correct dimensions in order for the product **IA** to exist. Consequently, there is an infinitely large number of identity matrices—all square and all possessing the same pattern. The 1 × 1, 2 × 2, and 3 × 3 identity matrices are

$$
\underset{1 \times 1}{\mathbf{I}} = [1] \qquad
\underset{2 \times 2}{\mathbf{I}} = \begin{bmatrix} 1 & 0 \\ 0 & 1 \end{bmatrix} \qquad
\underset{3 \times 3}{\mathbf{I}} = \begin{bmatrix} 1 & 0 & 0 \\ 0 & 1 & 0 \\ 0 & 0 & 1 \end{bmatrix}
$$

In Example I.2, we demonstrated the fact that this matrix satisfies the property

$$
\mathbf{IA} = \mathbf{A}
$$

EXAMPLE I.3

If **A** is the matrix shown below, find **IA** and **AI**.

$$A = \begin{bmatrix} 3 & 4 & -1 \\ 1 & 0 & 2 \end{bmatrix}$$

SOLUTION

$$\underset{2 \times 2 \quad 2 \times 3}{\mathbf{IA}} = \begin{bmatrix} 1 & 0 \\ 0 & 1 \end{bmatrix} \begin{bmatrix} 3 & 4 & -1 \\ 1 & 0 & 2 \end{bmatrix} = \begin{bmatrix} 3 & 4 & -1 \\ 1 & 0 & 2 \end{bmatrix} = \mathbf{A}$$

$$\underset{2 \times 3 \quad 3 \times 3}{\mathbf{AI}} = \begin{bmatrix} 3 & 4 & -1 \\ 1 & 0 & 2 \end{bmatrix} \begin{bmatrix} 1 & 0 & 0 \\ 0 & 1 & 0 \\ 0 & 0 & 1 \end{bmatrix} = \begin{bmatrix} 3 & 4 & -1 \\ 1 & 0 & 2 \end{bmatrix} = \mathbf{A}$$

Notice that the identity matrices used to find the products **IA** and **AI** were of different dimensions. This was necessary in order for the products to exist. ∎

DEFINITION I.7

If **A** is any matrix, then a matrix **I** is defined to be an **identity matrix** if **AI** = **IA** = **A**. The matrices that satisfy this definition possess the pattern

$$I = \begin{bmatrix} 1 & 0 & 0 & \cdots & 0 \\ 0 & 1 & 0 & \cdots & 0 \\ 0 & 0 & 1 & \cdots & 0 \\ \cdot & \cdot & \cdot & \cdots & \cdot \\ \cdot & \cdot & \cdot & \cdots & \cdot \\ 0 & 0 & 0 & \cdots & 1 \end{bmatrix}$$

The identity element assumes importance when we consider the process of division and its role in the solution of equations. In ordinary algebra, division is essentially multiplication using the reciprocals of elements. For example, the equation

$$2X = 6$$

can be solved by dividing both sides of the equation by 2, *or* it can be solved by *multiplying* both sides of the equation by $\frac{1}{2}$, which is the reciprocal of 2. Thus,

$$\left(\frac{1}{2}\right)2X = \frac{1}{2}(6)$$
$$X = 3$$

What is the reciprocal of an element? It is the element such that the reciprocal times the element is equal to the identity element. Thus, the reciprocal of 3 is $\frac{1}{3}$ because

$$3\left(\frac{1}{3}\right) = 1$$

The identity matrix plays the same role in matrix algebra. Thus, the reciprocal of a matrix A, called **A-inverse** and denoted by the symbol A^{-1}, is a matrix such that $AA^{-1} = A^{-1}A = I$.

Inverses are defined only for square matrices, but not all square matrices possess inverses. (Those that do play an important role in solving the least squares equations and in other aspects of a regression analysis.) We will show you one important application of the inverse matrix in Section I.3. The procedure for finding the inverse of a matrix is demonstrated in Section I.4.

DEFINITION I.8

The square matrix A^{-1} is said to be the **inverse** of the square matrix A if

$$A^{-1}A = AA^{-1} = I$$

The procedure for finding an inverse matrix is computationally quite tedious and is performed most often using a computer. There is one exception. Finding the inverse of one type of matrix, called a **diagonal matrix**, is easy. A diagonal matrix is one that has nonzero elements down the **main diagonal** (running top left of the matrix to bottom right) and 0 elements elsewhere. For example, the identity matrix is a diagonal matrix (with 1's along the main diagonal), as are the following matrices:

$$A = \begin{bmatrix} 3 & 0 & 0 \\ 0 & 1 & 0 \\ 0 & 0 & 2 \end{bmatrix} \qquad B = \begin{bmatrix} 5 & 0 & 0 & 0 \\ 0 & 2 & 0 & 0 \\ 0 & 0 & 1 & 0 \\ 0 & 0 & 0 & 5 \end{bmatrix}$$

DEFINITION I.9

A **diagonal matrix** is one that contains nonzero elements on the main diagonal and 0 elements elsewhere.

You can verify that the inverse of

$$A = \begin{bmatrix} 3 & 0 & 0 \\ 0 & 1 & 0 \\ 0 & 0 & 2 \end{bmatrix} \quad \text{is} \quad A^{-1} = \begin{bmatrix} \frac{1}{3} & 0 & 0 \\ 0 & 1 & 0 \\ 0 & 0 & \frac{1}{2} \end{bmatrix}$$

i.e., $AA^{-1} = I$. In general, the inverse of a diagonal matrix is given by the following theorem:

THEOREM I.1

The inverse of a diagonal matrix

$$D = \begin{bmatrix} d_{11} & 0 & 0 & \cdots & 0 \\ 0 & d_{22} & 0 & \cdots & 0 \\ 0 & 0 & d_{33} & \cdots & 0 \\ \vdots & \vdots & \vdots & \cdots & \vdots \\ 0 & 0 & 0 & \cdots & d_{nn} \end{bmatrix} \quad \text{is} \quad D^{-1} = \begin{bmatrix} 1/d_{11} & 0 & 0 & \cdots & 0 \\ 0 & 1/d_{22} & 0 & \cdots & 0 \\ 0 & 0 & 1/d_{33} & \cdots & 0 \\ \vdots & \vdots & \vdots & \cdots & \vdots \\ 0 & 0 & 0 & \cdots & 1/d_{nn} \end{bmatrix}$$

EXERCISES I.7–I.11

I.7 Consider the following matrix:

$$A = \begin{bmatrix} 3 & 0 & 2 \\ -1 & 1 & 4 \end{bmatrix}$$

a. Give the identity matrix that will be used to obtain the product IA.
b. Show that $IA = A$.
c. Give the identity matrix that will be used to find the product AI.
d. Show that $AI = A$.

I.8 For the matrices A and B given here, show that $AB = I$, that $BA = I$, and, consequently, verify that $B = A^{-1}$.

$$A = \begin{bmatrix} 1 & 0 & 0 \\ 0 & 2 & 0 \\ 0 & 0 & 3 \end{bmatrix} \qquad B = \begin{bmatrix} 1 & 0 & 0 \\ 0 & \frac{1}{2} & 0 \\ 0 & 0 & \frac{1}{3} \end{bmatrix}$$

I.9 If

$$A = \begin{bmatrix} 12 & 0 & 0 & 8 \\ 0 & 12 & 0 & 0 \\ 0 & 0 & 8 & 0 \\ 8 & 0 & 0 & 8 \end{bmatrix} \quad \text{verify that} \quad A^{-1} = \begin{bmatrix} \frac{1}{4} & 0 & 0 & -\frac{1}{4} \\ 0 & \frac{1}{12} & 0 & 0 \\ 0 & 0 & \frac{1}{8} & 0 \\ -\frac{1}{4} & 0 & 0 & \frac{3}{8} \end{bmatrix}$$

I.10 If

$$A = \begin{bmatrix} 3 & 0 & 0 \\ 0 & 5 & 0 \\ 0 & 0 & 7 \end{bmatrix} \quad \text{show that} \quad A^{-1} = \begin{bmatrix} \frac{1}{3} & 0 & 0 \\ 0 & \frac{1}{5} & 0 \\ 0 & 0 & \frac{1}{7} \end{bmatrix}$$

I.11 Verify Theorem I.1.

||||||||||||||

S E C T I O N I.3

SOLVING SYSTEMS OF SIMULTANEOUS LINEAR EQUATIONS

Consider the following set of simultaneous linear equations in two unknowns:

$$2v_1 + v_2 = 7$$
$$v_1 - v_2 = 2$$

Note that the solution for these equations is $v_1 = 3$, $v_2 = 1$.

Now define the matrices

$$A = \begin{bmatrix} 2 & 1 \\ 1 & -1 \end{bmatrix} \qquad V = \begin{bmatrix} v_1 \\ v_2 \end{bmatrix} \qquad G = \begin{bmatrix} 7 \\ 2 \end{bmatrix}$$

Thus, A is the matrix of coefficients of v_1 and v_2, V is a column matrix containing the unknowns (written in order, top to bottom), and G is a column matrix containing the numbers on the right-hand side of the equal signs.

Now, the system of simultaneous equations shown above can be rewritten as a **matrix equation**:

$$AV = G$$

By a matrix equation, we mean that the product matrix, AV, is equal to the matrix G. *Equality of matrices means that corresponding elements are equal.* You can see that this is true for the expression $AV = G$, since

$$AV = \begin{bmatrix} 2 & 1 \\ 1 & -1 \end{bmatrix} \begin{bmatrix} v_1 \\ v_2 \end{bmatrix} = \begin{bmatrix} (2v_1 + v_2) \\ (v_1 - v_2) \end{bmatrix} = G$$

$$2 \times 2 \quad 2 \times 1 \qquad\qquad\qquad\qquad 2 \times 1$$

The matrix procedure for expressing a system of two simultaneous linear equations in two unknowns can be extended to express a set of k simultaneous equations in k unknowns. If the equations are written in the orderly pattern

$$a_{11}v_1 + a_{12}v_2 + \cdots + a_{1k}v_k = g_1$$
$$a_{21}v_1 + a_{22}v_2 + \cdots + a_{2k}v_k = g_2$$
$$\vdots \qquad \vdots \qquad\qquad \vdots \qquad \vdots$$
$$a_{k1}v_1 + a_{k2}v_2 + \cdots + a_{kk}v_k = g_k$$

then the set of simultaneous linear equations can be expressed as the matrix equation $\mathbf{AV} = \mathbf{G}$, where

$$\mathbf{A} = \begin{bmatrix} a_{11} & a_{12} & \cdots & a_{1k} \\ a_{21} & & \cdots & a_{2k} \\ \vdots & & & \vdots \\ a_{k1} & & \cdots & a_{kk} \end{bmatrix} \qquad \mathbf{V} = \begin{bmatrix} v_1 \\ v_2 \\ \vdots \\ v_k \end{bmatrix} \qquad \mathbf{G} = \begin{bmatrix} g_1 \\ g_2 \\ \vdots \\ g_k \end{bmatrix}$$

Now let use solve this system of simultaneous equations. (If they are uniquely solvable, it can be shown that $\mathbf{A}^{-1}$ exists.) Multiplying both sides of the matrix equation by $\mathbf{A}^{-1}$, we have

$$(\mathbf{A}^{-1})\mathbf{AV} = (\mathbf{A}^{-1})\mathbf{G}$$

But since $\mathbf{A}^{-1}\mathbf{A} = \mathbf{I}$, we have

$$(\mathbf{I})\mathbf{V} = \mathbf{A}^{-1}\mathbf{G}$$
$$\mathbf{V} = \mathbf{A}^{-1}\mathbf{G}$$

In other words, if we know $\mathbf{A}^{-1}$, we can find the solution to the set of simultaneous linear equations by obtaining the product $\mathbf{A}^{-1}\mathbf{G}$.

MATRIX SOLUTION TO A SET OF SIMULTANEOUS LINEAR EQUATIONS, AV = G

Solution: $\mathbf{V} = \mathbf{A}^{-1}\mathbf{G}$

EXAMPLE I.4

Apply the boxed result to find the solution to the set of simultaneous linear equations

$$2v_1 + v_2 = 7$$
$$v_1 - v_2 = 2$$

SOLUTION

The first step is to obtain the inverse of the coefficient matrix,

$$\mathbf{A} = \begin{bmatrix} 2 & 1 \\ 1 & -1 \end{bmatrix}$$

namely,

$$\mathbf{A}^{-1} = \begin{bmatrix} \frac{1}{3} & \frac{1}{3} \\ \frac{1}{3} & -\frac{2}{3} \end{bmatrix}$$

(This matrix can be found using a packaged computer program for matrix inversion or, for this simple case, you could use the procedure explained in Section I.4.) As a check, note that

$$\mathbf{A}^{-1}\mathbf{A} = \begin{bmatrix} \frac{1}{3} & \frac{1}{3} \\ \frac{1}{3} & -\frac{2}{3} \end{bmatrix} \begin{bmatrix} 2 & 1 \\ 1 & -1 \end{bmatrix} = \begin{bmatrix} 1 & 0 \\ 0 & 1 \end{bmatrix} = \mathbf{I}$$

The second step is to obtain the product $\mathbf{A}^{-1}\mathbf{G}$. Thus,

$$\mathbf{V} = \mathbf{A}^{-1}\mathbf{G} = \begin{bmatrix} \frac{1}{3} & \frac{1}{3} \\ \frac{1}{3} & -\frac{2}{3} \end{bmatrix} \begin{bmatrix} 7 \\ 2 \end{bmatrix} = \begin{bmatrix} 3 \\ 1 \end{bmatrix}$$

Since

$$\mathbf{V} = \begin{bmatrix} v_1 \\ v_2 \end{bmatrix} = \begin{bmatrix} 3 \\ 1 \end{bmatrix}$$

it follows that $v_1 = 3$ and $v_2 = 1$. You can see that these values of v_1 and v_2 satisfy the simultaneous linear equations and are the values that we specified as a solution at the beginning of this section.   ■

EXERCISES I.12–I.13

I.12 Suppose the simultaneous linear equations

$$3v_1 + v_2 = 5$$
$$v_1 - v_2 = 3$$

are expressed as a matrix equation,

$$\mathbf{AV} = \mathbf{G}$$

a. Find the matrices $\mathbf{A}$, $\mathbf{V}$, and $\mathbf{G}$.
b. Verify that

$$\mathbf{A}^{-1} = \begin{bmatrix} \frac{1}{4} & \frac{1}{4} \\ \frac{1}{4} & -\frac{3}{4} \end{bmatrix}$$

[*Note:*   A procedure for finding $\mathbf{A}^{-1}$ is given in Section I.4.]
c. Solve the equations by finding $\mathbf{V} = \mathbf{A}^{-1}\mathbf{G}$.

I.13 For the simultaneous linear equations

$$10v_1 + 20v_3 - 60 = 0$$
$$20v_2 - 60 = 0$$
$$20v_1 + 68v_3 - 176 = 0$$

a. Find the matrices $\mathbf{A}$, $\mathbf{V}$, and $\mathbf{G}$.
b. Verify that

$$\mathbf{A}^{-1} = \begin{bmatrix} \frac{17}{70} & 0 & -\frac{1}{14} \\ 0 & \frac{1}{20} & 0 \\ -\frac{1}{14} & 0 & \frac{1}{28} \end{bmatrix}$$

c. Solve the equations by finding $\mathbf{V} = \mathbf{A}^{-1}\mathbf{G}$.

SECTION I.4

A PROCEDURE FOR INVERTING A MATRIX

There are several different methods for inverting matrices. All are tedious and time-consuming. Consequently, in practice, you will invert almost all matrices using an electronic computer. The purpose of this section is to present one method so that you will be able to invert small (2×2 or 3×3) matrices manually and so that you will appreciate the enormous computing problem involved in inverting large matrices (and, consequently, in fitting linear models containing many terms to a set of data). Particularly, you will be able to understand why rounding errors creep into the inversion process and, consequently, why two different computer programs might invert the same matrix and produce inverse matrices with slightly different corresponding elements.

The procedure we will demonstrate to invert a matrix $\mathbf{A}$ requires that we perform a series of operations on the rows of the $\mathbf{A}$ matrix. For example, suppose

$$\mathbf{A} = \begin{bmatrix} 1 & -2 \\ -2 & 6 \end{bmatrix}$$

We will identify two different ways to operate on a row of a matrix:*

1. We can multiply every element in one particular row by a constant, c. For example, we could operate on the first row of the $\mathbf{A}$ matrix by multiplying every element in the row by a constant, say 2. Then the resulting row would be $[2 \quad -4]$.

2. We can operate on a row by multiplying another row of the matrix by a constant and then adding (or subtracting) the elements of that row to elements in corresponding positions in the row operated upon. For example, we could operate on the first row of the $\mathbf{A}$ matrix by multiplying the second row by a constant, say 2:

$$2[-2 \quad 6] = [-4 \quad 12]$$

Then we add this row to row 1:

$$[(1 - 4) \quad (-2 + 12)] = [-3 \quad 10]$$

Note one important point. We operated on the *first* row of the $\mathbf{A}$ matrix. Although we used the second row of the matrix to perform the operation, *the second row would remain unchanged*. Therefore, the row operation on the $\mathbf{A}$ matrix that we have just described would produce the new matrix,

$$\begin{bmatrix} -3 & 10 \\ -2 & 6 \end{bmatrix}$$

Matrix inversion using row operations is based on an elementary result from matrix algebra. It can be shown (proof omitted) that performing a series of row operations on a matrix $\mathbf{A}$ is equivalent to multiplying $\mathbf{A}$ by a matrix $\mathbf{B}$, i.e., row operations produce a new matrix, $\mathbf{BA}$. This result is used as follows: Place the $\mathbf{A}$ matrix and an identity matrix $\mathbf{I}$ of the same dimensions, side-by-side. Then

---

*We omit a third row operation, because it would add little and could be confusing.

perform the same series of row operations on both **A** and **I** until the **A** matrix has been changed into the identity matrix **I**. This means that you have multiplied both **A** and **I** by some matrix **B** such that:

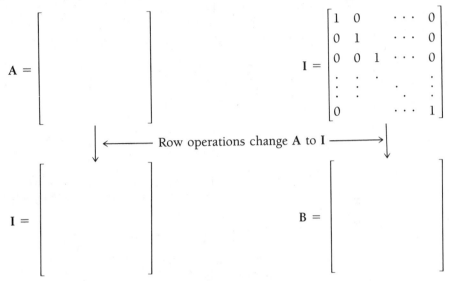

$$\mathbf{BA} = \mathbf{I} \quad \text{and} \quad \mathbf{BI} = \mathbf{B}$$

Since **BA** = **I**, it follows that **B** = **A**$^{-1}$. Therefore, as the **A** matrix is transformed by row operations into the identity matrix **I**, the identity matrix **I** is transformed into **A**$^{-1}$, i.e.,

$$\mathbf{BI} = \mathbf{B} = \mathbf{A}^{-1}$$

We will show you how this procedure works with two examples.

EXAMPLE I.5

Find the inverse of the matrix,

$$\mathbf{A} = \begin{bmatrix} 1 & -2 \\ -2 & 6 \end{bmatrix}$$

SOLUTION

Place the **A** matrix and a 2×2 identity matrix side-by-side and then perform the following series of row operations (we will indicate by arrow the row operated upon in each operation):

$$\mathbf{A} = \begin{bmatrix} 1 & -2 \\ -2 & 6 \end{bmatrix} \qquad \mathbf{I} = \begin{bmatrix} 1 & 0 \\ 0 & 1 \end{bmatrix}$$

OPERATION 1   Multiply the first row by 2 and add to the second row:

$$\rightarrow \begin{bmatrix} 1 & -2 \\ 0 & 2 \end{bmatrix} \qquad \begin{bmatrix} 1 & 0 \\ 2 & 1 \end{bmatrix}$$

OPERATION 2   Multiply the second row by $\frac{1}{2}$:

$$\rightarrow \begin{bmatrix} 1 & -2 \\ 0 & 1 \end{bmatrix} \qquad \begin{bmatrix} 1 & 0 \\ 1 & \frac{1}{2} \end{bmatrix}$$

OPERATION 3   Multiply the second row by 2 and add it to the first row:

$$\rightarrow \begin{bmatrix} 1 & 0 \\ 0 & 1 \end{bmatrix} \qquad \begin{bmatrix} 3 & 1 \\ 1 & \frac{1}{2} \end{bmatrix}$$

Thus,

$$\mathbf{A}^{-1} = \begin{bmatrix} 3 & 1 \\ 1 & \frac{1}{2} \end{bmatrix}$$

The final step in finding an inverse is to check your solution by finding the product $\mathbf{A}^{-1}\mathbf{A}$ to see if it equals the identity matrix $\mathbf{I}$. To check:

$$\mathbf{A}^{-1}\mathbf{A} = \begin{bmatrix} 3 & 1 \\ 1 & \frac{1}{2} \end{bmatrix} \begin{bmatrix} 1 & -2 \\ -2 & 6 \end{bmatrix} = \begin{bmatrix} 1 & 0 \\ 0 & 1 \end{bmatrix}$$

Since this product is equal to the identity matrix, it follows that our solution for $\mathbf{A}^{-1}$ is correct.   ■

EXAMPLE I.6

Find the inverse of the matrix,

$$\mathbf{A} = \begin{bmatrix} 2 & 0 & 3 \\ 0 & 4 & 1 \\ 3 & 1 & 2 \end{bmatrix}$$

SOLUTION

Place an identity matrix alongside the $\mathbf{A}$ matrix and perform the row operations:

OPERATION 1   Multiply row 1 by $\frac{1}{2}$:

$$\rightarrow \begin{bmatrix} 1 & 0 & \frac{3}{2} \\ 0 & 4 & 1 \\ 3 & 1 & 2 \end{bmatrix} \qquad \begin{bmatrix} \frac{1}{2} & 0 & 0 \\ 0 & 1 & 0 \\ 0 & 0 & 1 \end{bmatrix}$$

OPERATION 2   Multiply row 1 by 3 and subtract from row 3:

$$\begin{bmatrix} 1 & 0 & \frac{3}{2} \\ 0 & 4 & 1 \\ \rightarrow 0 & 1 & -\frac{5}{2} \end{bmatrix} \qquad \begin{bmatrix} \frac{1}{2} & 0 & 0 \\ 0 & 1 & 0 \\ -\frac{3}{2} & 0 & 1 \end{bmatrix}$$

OPERATION 3   Multiply row 2 by $\frac{1}{4}$:

$$\rightarrow \begin{bmatrix} 1 & 0 & \frac{3}{2} \\ 0 & 1 & \frac{1}{4} \\ 0 & 1 & -\frac{5}{2} \end{bmatrix} \qquad \begin{bmatrix} \frac{1}{2} & 0 & 0 \\ 0 & \frac{1}{4} & 0 \\ -\frac{3}{2} & 0 & 1 \end{bmatrix}$$

OPERATION 4   Subtract row 2 from row 3:

$$
\rightarrow
\begin{bmatrix}
1 & 0 & \frac{3}{2} \\
0 & 1 & \frac{1}{4} \\
0 & 0 & -\frac{11}{4}
\end{bmatrix}
\quad
\begin{bmatrix}
\frac{1}{2} & 0 & 0 \\
0 & \frac{1}{4} & 0 \\
-\frac{3}{2} & -\frac{1}{4} & 1
\end{bmatrix}
$$

OPERATION 5   Multiply row 3 by $-\frac{4}{11}$:

$$
\rightarrow
\begin{bmatrix}
1 & 0 & \frac{3}{2} \\
0 & 1 & \frac{1}{4} \\
0 & 0 & 1
\end{bmatrix}
\quad
\begin{bmatrix}
\frac{1}{2} & 0 & 0 \\
0 & \frac{1}{4} & 0 \\
\frac{12}{22} & \frac{1}{11} & -\frac{4}{11}
\end{bmatrix}
$$

OPERATION 6   Operate on row 2 by subtracting $\frac{1}{4}$ of row 3:

$$
\rightarrow
\begin{bmatrix}
1 & 0 & \frac{3}{2} \\
0 & 1 & 0 \\
0 & 0 & 1
\end{bmatrix}
\quad
\begin{bmatrix}
\frac{1}{2} & 0 & 0 \\
-\frac{3}{22} & \frac{5}{22} & \frac{1}{11} \\
\frac{12}{22} & \frac{1}{11} & -\frac{4}{11}
\end{bmatrix}
$$

OPERATION 7   Operate on row 1 by subtracting $\frac{3}{2}$ of row 3:

$$
\rightarrow
\begin{bmatrix}
1 & 0 & 0 \\
0 & 1 & 0 \\
0 & 0 & 1
\end{bmatrix}
\quad
\begin{bmatrix}
-\frac{7}{22} & -\frac{3}{22} & \frac{6}{11} \\
-\frac{3}{22} & \frac{5}{22} & \frac{1}{11} \\
\frac{6}{11} & \frac{1}{11} & -\frac{4}{11}
\end{bmatrix}
= \mathbf{A}^{-1}
$$

To check the solution, we find the product,

$$
\mathbf{A}^{-1}\mathbf{A} =
\begin{bmatrix}
-\frac{7}{22} & -\frac{3}{22} & \frac{6}{11} \\
-\frac{3}{22} & \frac{5}{22} & \frac{1}{11} \\
\frac{6}{11} & \frac{1}{11} & -\frac{4}{11}
\end{bmatrix}
\begin{bmatrix}
2 & 0 & 3 \\
0 & 4 & 1 \\
3 & 1 & 2
\end{bmatrix}
$$

$$
=
\begin{bmatrix}
1 & 0 & 0 \\
0 & 1 & 0 \\
0 & 0 & 1
\end{bmatrix}
$$

Since the product $\mathbf{A}^{-1}\mathbf{A}$ is equal to the identity matrix, it follows that our solution for $\mathbf{A}^{-1}$ is correct.   ∎

Examples I.5 and I.6 indicate the strategy employed when performing row operations on the **A** matrix to change it into an identity matrix. Multiply the first row by a constant to change the element in the top left row into a 1. Then perform operations to change all elements in the first column into 0's. Then operate on the second row and change the second diagonal element into a 1. Then operate to change all elements in the second column beneath row 2 into 0's. Then operate on the diagonal element in row 3, etc. When all elements on the main diagonal are 1's and all below the main diagonal are 0's, perform row operations to change the last column to 0; then the next-to-last, etc., until you get back to the first row. The procedure for changing the off-diagonal elements to 0's is indicated diagrammatically as shown:

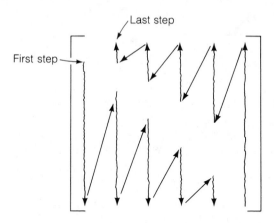

The preceding instructions on how to invert a matrix using row operations suggest that the inversion of a large matrix would involve many multiplications, subtractions, and additions and, consequently, could produce large rounding errors in the calculations unless you carry a large number of significant figures in the calculations. This explains why two different multiple regression analysis computer programs may produce different estimates of the same β parameters, and it emphasizes the importance of carrying a large number of significant figures in all computations when inverting a matrix.

EXERCISE I.14

I.14 Invert the following matrices and check your answers to make certain that $A^{-1}A = AA^{-1} = I$:

a. $A = \begin{bmatrix} 3 & 2 \\ 4 & 5 \end{bmatrix}$ **b.** $A = \begin{bmatrix} 3 & 0 & -2 \\ 1 & 4 & 2 \\ 5 & 1 & 1 \end{bmatrix}$

c. $A = \begin{bmatrix} 1 & 0 & 1 \\ 0 & 2 & 1 \\ 1 & 1 & 3 \end{bmatrix}$ **d.** $A = \begin{bmatrix} 4 & 0 & 10 \\ 0 & 10 & 0 \\ 10 & 0 & 5 \end{bmatrix}$

[*Note:* No answers are given to these exercises. You will know your answers are correct if $A^{-1}A = I$.]

APPENDIX II

CONTENTS

USEFUL STATISTICAL TABLES

TABLE 1 Random Numbers

ROW \ COLUMN	1	2	3	4	5	6	7	8	9	10	11	12	13	14
1	10480	15011	01536	02011	81647	91646	69179	14194	62590	36207	20969	99570	91291	90700
2	22368	46573	25595	85393	30995	89198	27982	53402	93965	34095	52666	19174	39615	99505
3	24130	48360	22527	97265	76393	64809	15179	24830	49340	32081	30680	19655	63348	58629
4	42167	93093	06243	61680	07856	16376	39440	53537	71341	57004	00849	74917	97758	16379
5	37570	39975	81837	16656	06121	91782	60468	81305	49684	60672	14110	06927	01263	54613
6	77921	06907	11008	42751	27756	53498	18602	70659	90655	15053	21916	81825	44394	42880
7	99562	72905	56420	69994	98872	31016	71194	18738	44013	48840	63213	21069	10634	12952
8	96301	91977	05463	07972	18876	20922	94595	56869	69014	60045	18425	84903	42508	32307
9	89579	14342	63661	10281	17453	18103	57740	84378	25331	12566	58678	44947	05585	56941
10	85475	36857	53342	53988	53060	59533	38867	62300	08158	17983	16439	11458	18593	64952
11	28918	69578	88231	33276	70997	79936	56865	05859	90106	31595	01547	85590	91610	78188
12	63553	40961	48235	03427	49626	69445	18663	72695	52180	20847	12234	90511	33703	90322
13	09429	93969	52636	92737	88974	33488	36320	17617	30015	08272	84115	27156	30613	74952
14	10365	61129	87529	85689	48237	52267	67689	93394	01511	26358	85104	20285	29975	89868
15	07119	97336	71048	08178	77233	13916	47564	81056	97735	85977	29372	74461	28551	90707
16	51085	12765	51821	51259	77452	16308	60756	92144	49442	53900	70960	63990	75601	40719
17	02368	21382	52404	60268	89368	19885	55322	44819	01188	65255	64835	44919	05944	55157
18	01011	54092	33362	94904	31273	04146	18594	29852	71585	85030	51132	01915	92747	64951
19	52162	53916	46369	58586	23216	14513	83149	98736	23495	64350	94738	17752	35156	35749
20	07056	97628	33787	09998	42698	06691	76988	13602	51851	46104	88916	19509	25625	58104
21	48663	91245	85828	14346	09172	30168	90229	04734	59193	22178	30421	61666	99904	32812
22	54164	58492	22421	74103	47070	25306	76468	26384	58151	06646	21524	15227	96909	44592
23	32639	32363	05597	24200	13363	38005	94342	28728	35806	06912	17012	64161	18296	22851
24	29334	27001	87637	87308	58731	00256	45834	15398	46557	41135	10367	07684	36188	18510
25	02488	33062	28834	07351	19731	92420	60952	61280	50001	67658	32586	86679	50720	94953
26	81525	72295	04839	96423	24878	82651	66566	14778	76797	14780	13300	87074	79666	95725
27	29676	20591	68086	26432	46901	20849	89768	81536	86645	12659	92259	57102	80428	25280

(continued)

TABLE 1 Continued

ROW \ COLUMN	1	2	3	4	5	6	7	8	9	10	11	12	13	14
28	00742	57392	39064	66432	84673	40027	32832	61362	98947	96067	64760	64584	96096	98253
29	05366	04213	25669	26422	44407	44048	37937	63904	45766	66134	75470	66520	34693	90449
30	91921	26418	64117	94305	26766	25940	39972	22209	71500	64568	91402	42416	07844	69618
31	00582	04711	87917	77341	42206	35126	74087	99547	81817	42607	43808	76655	62028	76630
32	00725	69884	62797	56170	86324	88072	76222	36086	84637	93161	76038	65855	77919	88006
33	69011	65795	95876	55293	18988	27354	26575	08625	40801	59920	29841	80150	12777	48501
34	25976	57948	29888	88604	67917	48708	18912	82271	65424	69774	33611	54262	85963	03547
35	09763	83473	73577	12908	30883	18317	28290	35797	05998	41688	34952	37888	38917	88050
36	91576	42595	27958	30134	04024	86385	29880	99730	55536	84855	29080	09250	79656	73211
37	17955	56349	90999	49127	20044	59931	06115	20542	18059	02008	73708	83517	36103	42791
38	46503	18584	18845	49618	02304	51038	20655	58727	28168	15475	56942	53389	20562	87338
39	92157	89634	94824	78171	84610	82834	09922	25417	44137	48413	25555	21246	35509	20468
40	14577	62765	35605	81263	39667	47358	56873	56307	61607	49518	89656	20103	77490	18062
41	98427	07523	33362	64270	01638	92477	66969	98420	04880	45585	46565	04102	46880	45709
42	34914	63976	88720	82765	34476	17032	87589	40836	32427	70002	70663	88863	77775	69348
43	70060	28277	39475	46473	23219	53416	94970	25832	69975	94884	19661	72828	00102	66794
44	53976	54914	06990	67245	68350	82948	11398	42878	80287	88267	47363	46634	06541	97809
45	76072	29515	40980	07391	58745	25774	22987	80059	39911	96189	41151	14222	60697	59583
46	90725	52210	83974	29992	65831	38857	50490	83765	55657	14361	31720	57375	56228	41546
47	64364	67412	33339	31926	14883	24413	59744	92351	97473	89286	35931	04110	23726	51900
48	08962	00358	31662	25388	61642	34072	81249	35648	56891	69352	48373	45578	78547	81788
49	95012	68379	93526	70765	10592	04542	76463	54328	02349	17247	28865	14777	62730	92277
50	15664	10493	20492	38391	91132	21999	59516	81652	27195	48223	46751	22923	32261	85653
51	16408	81899	04153	53381	79401	21438	83035	92350	36693	31238	59649	91754	72772	02338
52	18629	81953	05520	91962	04739	13092	97662	24822	94730	06496	35090	04822	86774	98289
53	73115	35101	47498	87637	99016	71060	88824	71013	18735	20286	23153	72924	35165	43040
54	57491	16703	23167	49323	45021	33132	12544	41035	80780	45393	44812	12515	98931	91202
55	30405	83946	23792	14422	15059	45799	22716	19792	09983	74353	68668	30429	70735	25499
56	16631	35006	85900	98275	32388	52390	16815	69298	82732	38480	73817	32523	41961	44437
57	96773	20206	42559	78985	05300	22164	24369	54224	35083	19687	11052	91491	60383	19746
58	38935	64202	14349	82674	66523	44133	00697	35552	35970	19124	63318	29686	03387	59846
59	31624	76384	17403	53363	44167	64486	64758	75366	76554	31601	12614	33072	60332	92325
60	78919	19474	23632	27889	47914	02584	37680	20801	72152	39339	34806	08930	85001	87820
61	03931	33309	57047	74211	63445	17361	62825	39908	05607	91284	68833	25570	38818	46920
62	74426	33278	43972	10119	89917	15665	52872	73823	73144	88662	88970	74492	51805	99378
63	09066	00903	20795	95452	92648	45454	09552	88815	16553	51125	79375	97596	16296	66092

64	42238	12426	87025	14267	20979	04508	64535	31355	86064	29472	47689	05974	52468	16834
65	16153	08002	26504	41744	81959	65642	74240	56302	00033	67107	77510	70625	28725	34191
66	21457	40742	29820	96783	29400	21840	15035	34537	33310	06116	95240	15957	16572	06004
67	21581	57802	02050	89728	17937	37621	47075	42080	97403	48626	68995	43805	33386	21597
68	55612	78095	83197	33732	05810	24813	86902	60397	16489	03264	88525	42786	05269	92532
69	44657	66999	99324	51281	84463	60563	79312	93454	68876	25471	93911	25650	12682	73572
70	91340	84979	46949	81973	37949	61023	43997	15263	80644	43942	89203	71795	99533	50501
71	91227	21199	31935	27022	84067	05462	35216	14486	29891	68607	41867	14951	91696	85065
72	50001	38140	66321	19924	72163	09538	12151	06878	91903	18749	34405	56087	82790	70925
73	65390	05224	72958	28609	81406	39147	25549	48542	42627	45233	57202	94617	23772	07896
74	27504	96131	83944	41575	10573	08619	64482	73923	36152	05184	94142	25299	84387	34925
75	37169	94851	39117	89632	00959	16487	65536	49071	39782	17095	02330	74301	00275	48280
76	11508	70225	51111	38351	19444	66499	71945	05422	13442	78675	84081	66938	93654	59894
77	37449	30362	06694	54690	04052	53115	62757	95348	78662	11163	81651	50245	34971	52924
78	46515	70331	85922	38329	57015	15765	97161	17869	45349	61796	66345	81073	49106	79860
79	30986	81223	42416	58353	21532	30502	32305	86482	05174	07901	54339	58861	74818	46942
80	63798	64995	46583	09785	44160	78128	83991	42865	92520	83531	80377	35909	81250	54238
81	82486	84846	99254	67632	43218	50076	21361	64816	51202	88124	41870	52689	51275	83556
82	21885	32906	92431	09060	64297	51674	64126	62570	26123	05155	59194	52799	28225	85762
83	60336	98782	07408	53458	13564	59089	26445	29789	85205	41001	12535	12133	14645	23541
84	43937	46891	24010	25560	86355	33941	25786	54990	71899	15475	95434	98227	21824	19585
85	97656	63175	89303	16275	07100	92063	21942	18611	47348	20203	18534	03862	78095	50136
86	03299	01221	05418	38982	55758	92237	26759	86367	21216	98442	08303	56613	91511	75928
87	79626	06486	03574	17668	07785	76020	79924	25651	83325	88428	85076	72811	22717	50585
88	85636	68335	47539	03129	65651	11977	02510	26113	99447	68645	34327	15152	55230	93448
89	18039	14367	61337	06177	12143	46609	32989	74014	64708	00533	35398	58408	13261	47908
90	08362	15656	60627	36478	65648	16764	53412	09013	07832	41574	17639	82163	60859	75567
91	79556	29068	04142	16268	15387	12856	66227	38358	22478	73373	88732	09443	82558	05250
92	92608	82674	27072	32534	17075	27698	98204	63863	11951	34648	88022	56148	34925	57031
93	23982	25835	40055	67006	12293	02753	14827	23235	35071	99704	37543	11601	35503	85171
94	09915	96306	05908	97901	28395	14186	00821	80703	70426	75647	76310	88717	37890	40129
95	59037	33300	26695	62247	69927	76123	50842	43834	86654	70959	79725	93872	28117	19233
96	42488	78077	69882	61657	34136	79180	97526	43092	04098	73571	80799	76536	71255	64239
97	46764	86273	63003	93017	31204	36692	40202	35275	57306	55543	53203	18098	47625	88684
98	03237	45430	55417	63282	90816	17349	88298	90183	36600	78406	06216	95787	42579	90730
99	86591	81482	52667	61582	14972	90053	89534	76036	49199	43716	97548	04379	46370	28672
100	38534	01715	94964	87288	65680	43772	39560	12918	86537	62738	19636	51132	25739	56947

Source: Abridged from W. H. Beyer, Ed., CRC Standard Mathematical Tables, 24th ed. (Cleveland: The Chemical Rubber Company), 1976. Reproduced by permission of the publisher. Copyright The Chemical Rubber Co., CRC Press, Inc., Boca Raton, Florida.

TABLE 2 Binomial Probabilities

Tabulated values are $\sum_{y=0}^{k} p(y)$. (Computations are rounded at the third decimal place.)

a. $n = 5$

k \ p	0.01	0.05	0.10	0.20	0.30	0.40	0.50	0.60	0.70	0.80	0.90	0.95	0.99
0	.951	.774	.590	.328	.168	.078	.031	.010	.002	.000	.000	.000	.000
1	.999	.977	.919	.737	.528	.337	.188	.087	.031	.007	.000	.000	.000
2	1.000	.999	.991	.942	.837	.683	.500	.317	.163	.058	.009	.001	.000
3	1.000	1.000	1.000	.993	.969	.913	.812	.663	.472	.263	.081	.023	.001
4	1.000	1.000	1.000	1.000	.998	.990	.969	.922	.832	.672	.410	.226	.049

b. $n = 10$

k \ p	0.01	0.05	0.10	0.20	0.30	0.40	0.50	0.60	0.70	0.80	0.90	0.95	0.99
0	.904	.599	.349	.107	.028	.006	.001	.000	.000	.000	.000	.000	.000
1	.996	.914	.736	.376	.149	.046	.011	.002	.000	.000	.000	.000	.000
2	1.000	.988	.930	.678	.383	.167	.055	.012	.002	.000	.000	.000	.000
3	1.000	.999	.987	.879	.650	.382	.172	.055	.011	.001	.000	.000	.000
4	1.000	1.000	.998	.967	.850	.633	.377	.166	.047	.006	.000	.000	.000
5	1.000	1.000	1.000	.994	.953	.834	.623	.367	.150	.033	.002	.000	.000
6	1.000	1.000	1.000	.999	.989	.945	.828	.618	.350	.121	.013	.001	.000
7	1.000	1.000	1.000	1.000	.998	.988	.945	.833	.617	.322	.070	.012	.000
8	1.000	1.000	1.000	1.000	1.000	.998	.989	.954	.851	.624	.264	.086	.004
9	1.000	1.000	1.000	1.000	1.000	1.000	.999	.994	.972	.893	.651	.401	.096

c. $n = 15$

k \ p	0.01	0.05	0.10	0.20	0.30	0.40	0.50	0.60	0.70	0.80	0.90	0.95	0.99
0	.860	.463	.206	.035	.005	.000	.000	.000	.000	.000	.000	.000	.000
1	.990	.829	.549	.167	.035	.005	.000	.000	.000	.000	.000	.000	.000
2	1.000	.964	.816	.398	.127	.027	.004	.000	.000	.000	.000	.000	.000
3	1.000	.995	.944	.648	.297	.091	.018	.002	.000	.000	.000	.000	.000
4	1.000	.999	.987	.836	.515	.217	.059	.009	.001	.000	.000	.000	.000
5	1.000	1.000	.998	.939	.722	.403	.151	.034	.004	.000	.000	.000	.000
6	1.000	1.000	1.000	.982	.869	.610	.304	.095	.015	.001	.000	.000	.000
7	1.000	1.000	1.000	.996	.950	.787	.500	.213	.050	.004	.000	.000	.000
8	1.000	1.000	1.000	.999	.985	.905	.696	.390	.131	.018	.000	.000	.000
9	1.000	1.000	1.000	1.000	.996	.966	.849	.597	.278	.061	.002	.000	.000
10	1.000	1.000	1.000	1.000	.999	.991	.941	.783	.485	.164	.013	.001	.000
11	1.000	1.000	1.000	1.000	1.000	.998	.982	.909	.703	.352	.056	.005	.000
12	1.000	1.000	1.000	1.000	1.000	1.000	.996	.973	.873	.602	.184	.036	.000
13	1.000	1.000	1.000	1.000	1.000	1.000	1.000	.995	.965	.833	.451	.171	.010
14	1.000	1.000	1.000	1.000	1.000	1.000	1.000	1.000	.995	.965	.794	.537	.140

TABLE 2   Continued

d. $n = 20$

k	0.01	0.05	0.10	0.20	0.30	0.40	0.50	0.60	0.70	0.80	0.90	0.95	0.99
0	.818	.358	.122	.012	.001	.000	.000	.000	.000	.000	.000	.000	.000
1	.983	.736	.392	.069	.008	.001	.000	.000	.000	.000	.000	.000	.000
2	.999	.925	.677	.206	.035	.004	.000	.000	.000	.000	.000	.000	.000
3	1.000	.984	.867	.411	.107	.016	.001	.000	.000	.000	.000	.000	.000
4	1.000	.997	.957	.630	.238	.051	.006	.000	.000	.000	.000	.000	.000
5	1.000	1.000	.989	.804	.416	.126	.021	.002	.000	.000	.000	.000	.000
6	1.000	1.000	.998	.913	.608	.250	.058	.006	.000	.000	.000	.000	.000
7	1.000	1.000	1.000	.968	.772	.416	.132	.021	.001	.000	.000	.000	.000
8	1.000	1.000	1.000	.990	.887	.596	.252	.057	.005	.000	.000	.000	.000
9	1.000	1.000	1.000	.997	.952	.755	.412	.128	.017	.001	.000	.000	.000
10	1.000	1.000	1.000	.999	.983	.872	.588	.245	.048	.003	.000	.000	.000
11	1.000	1.000	1.000	1.000	.995	.943	.748	.404	.113	.010	.000	.000	.000
12	1.000	1.000	1.000	1.000	.999	.979	.868	.584	.228	.032	.000	.000	.000
13	1.000	1.000	1.000	1.000	1.000	.994	.942	.750	.392	.087	.002	.000	.000
14	1.000	1.000	1.000	1.000	1.000	.998	.979	.874	.584	.196	.011	.000	.000
15	1.000	1.000	1.000	1.000	1.000	1.000	.994	.949	.762	.370	.043	.003	.000
16	1.000	1.000	1.000	1.000	1.000	1.000	.999	.984	.893	.589	.133	.016	.000
17	1.000	1.000	1.000	1.000	1.000	1.000	1.000	.996	.965	.794	.323	.075	.001
18	1.000	1.000	1.000	1.000	1.000	1.000	1.000	.999	.992	.931	.608	.264	.017
19	1.000	1.000	1.000	1.000	1.000	1.000	1.000	1.000	.999	.988	.878	.642	.182

(continued)

TABLE 2 Continued

e. $n = 25$

k \ p	0.01	0.05	0.10	0.20	0.30	0.40	0.50	0.60	0.70	0.80	0.90	0.95	0.99
0	.778	.277	.072	.004	.000	.000	.000	.000	.000	.000	.000	.000	.000
1	.974	.642	.271	.027	.002	.000	.000	.000	.000	.000	.000	.000	.000
2	.998	.873	.537	.098	.009	.000	.000	.000	.000	.000	.000	.000	.000
3	1.000	.966	.764	.234	.033	.002	.000	.000	.000	.000	.000	.000	.000
4	1.000	.993	.902	.421	.090	.009	.000	.000	.000	.000	.000	.000	.000
5	1.000	.999	.967	.617	.193	.029	.002	.000	.000	.000	.000	.000	.000
6	1.000	1.000	.991	.780	.341	.074	.007	.000	.000	.000	.000	.000	.000
7	1.000	1.000	.998	.891	.512	.154	.022	.001	.000	.000	.000	.000	.000
8	1.000	1.000	1.000	.953	.677	.274	.054	.004	.000	.000	.000	.000	.000
9	1.000	1.000	1.000	.983	.811	.425	.115	.013	.000	.000	.000	.000	.000
10	1.000	1.000	1.000	.994	.902	.586	.212	.034	.002	.000	.000	.000	.000
11	1.000	1.000	1.000	.998	.956	.732	.345	.078	.006	.000	.000	.000	.000
12	1.000	1.000	1.000	1.000	.983	.846	.500	.154	.017	.000	.000	.000	.000
13	1.000	1.000	1.000	1.000	.994	.922	.655	.268	.044	.002	.000	.000	.000
14	1.000	1.000	1.000	1.000	.998	.966	.788	.414	.098	.006	.000	.000	.000
15	1.000	1.000	1.000	1.000	1.000	.987	.885	.575	.189	.017	.000	.000	.000
16	1.000	1.000	1.000	1.000	1.000	.996	.946	.726	.323	.047	.000	.000	.000
17	1.000	1.000	1.000	1.000	1.000	.999	.978	.846	.488	.109	.002	.000	.000
18	1.000	1.000	1.000	1.000	1.000	1.000	.993	.926	.659	.220	.009	.000	.000
19	1.000	1.000	1.000	1.000	1.000	1.000	.998	.971	.807	.383	.033	.001	.000
20	1.000	1.000	1.000	1.000	1.000	1.000	1.000	.991	.910	.579	.098	.007	.000
21	1.000	1.000	1.000	1.000	1.000	1.000	1.000	.998	.967	.766	.236	.034	.000
22	1.000	1.000	1.000	1.000	1.000	1.000	1.000	1.000	.991	.902	.463	.127	.002
23	1.000	1.000	1.000	1.000	1.000	1.000	1.000	1.000	.998	.973	.729	.358	.026
24	1.000	1.000	1.000	1.000	1.000	1.000	1.000	1.000	1.000	.996	.928	.723	.222

TABLE 3
Exponentials

c	e^{-c}	c	e^{-c}	c	e^{-c}
.00	1.000000	2.35	.095369	4.70	.009095
.05	.951229	2.40	.090718	4.75	.008652
.10	.904837	2.45	.086294	4.80	.008230
.15	.860708	2.50	.082085	4.85	.007828
.20	.818731	2.55	.078082	4.90	.007447
.25	.778801	2.60	.074274	4.95	.007083
.30	.740818	2.65	.070651	5.00	.006738
.35	.704688	2.70	.067206	5.05	.006409
.40	.670320	2.75	.063928	5.10	.006097
.45	.637628	2.80	.060810	5.15	.005799
.50	.606531	2.85	.057844	5.20	.005517
.55	.576950	2.90	.055023	5.25	.005248
.60	.548812	2.95	.052340	5.30	.004992
.65	.522046	3.00	.049787	5.35	.004748
.70	.496585	3.05	.047359	5.40	.004517
.75	.472367	3.10	.045049	5.45	.004296
.80	.449329	3.15	.042852	5.50	.004087
.85	.427415	3.20	.040762	5.55	.003887
.90	.406570	3.25	.038774	5.60	.003698
.95	.386741	3.30	.036883	5.65	.003518
1.00	.367879	3.35	.035084	5.70	.003346
1.05	.349938	3.40	.033373	5.75	.003183
1.10	.332871	3.45	.031746	5.80	.003028
1.15	.316637	3.50	.030197	5.85	.002880
1.20	.301194	3.55	.028725	5.90	.002739
1.25	.286505	3.60	.027324	5.95	.002606
1.30	.272532	3.65	.025991	6.00	.002479
1.35	.259240	3.70	.024724	6.05	.002358
1.40	.246597	3.75	.023518	6.10	.002243
1.45	.234570	3.80	.022371	6.15	.002133
1.50	.223130	3.85	.021280	6.20	.002029
1.55	.212248	3.90	.020242	6.25	.001930
1.60	.201897	3.95	.019255	6.30	.001836
1.65	.192050	4.00	.018316	6.35	.001747
1.70	.182684	4.05	.017422	6.40	.001661
1.75	.173774	4.10	.016573	6.45	.001581
1.80	.165299	4.15	.015764	6.50	.001503
1.85	.157237	4.20	.014996	6.55	.001430
1.90	.149569	4.25	.014264	6.60	.001360
1.95	.142274	4.30	.013569	6.65	.001294
2.00	.135335	4.35	.012907	6.70	.001231
2.05	.128735	4.40	.012277	6.75	.001171
2.10	.122456	4.45	.011679	6.80	.001114
2.15	.116484	4.50	.011109	6.85	.001059
2.20	.110803	4.55	.010567	6.90	.001008
2.25	.105399	4.60	.010052	6.95	.000959
2.30	.100259	4.65	.009562	7.00	.000912

(continued)

TABLE 3
Continued

c	e^{-c}	c	e^{-c}	c	e^{-c}
7.05	.000867	8.05	.000319	9.05	.000117
7.10	.000825	8.10	.000304	9.10	.000112
7.15	.000785	8.15	.000289	9.15	.000106
7.20	.000747	8.20	.000275	9.20	.000101
7.25	.000710	8.25	.000261	9.25	.000096
7.30	.000676	8.30	.000249	9.30	.000091
7.35	.000643	8.35	.000236	9.35	.000087
7.40	.000611	8.40	.000225	9.40	.000083
7.45	.000581	8.45	.000214	9.45	.000079
7.50	.000553	8.50	.000204	9.50	.000075
7.55	.000526	8.55	.000194	9.55	.000071
7.60	.000501	8.60	.000184	9.60	.000068
7.65	.000476	8.65	.000175	9.65	.000064
7.70	.000453	8.70	.000167	9.70	.000061
7.75	.000431	8.75	.000158	9.75	.000058
7.80	.000410	8.80	.000151	9.80	.000056
7.85	.000390	8.85	.000143	9.85	.000053
7.90	.000371	8.90	.000136	9.90	.000050
7.95	.000353	8.95	.000130	9.95	.000048
8.00	.000336	9.00	.000123	10.00	.000045

TABLE 4

Normal Curve Areas

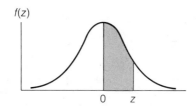

z	.00	.01	.02	.03	.04	.05	.06	.07	.08	.09
.0	.0000	.0040	.0080	.0120	.0160	.0199	.0239	.0279	.0319	.0359
.1	.0398	.0438	.0478	.0517	.0557	.0596	.0636	.0675	.0714	.0753
.2	.0793	.0832	.0871	.0910	.0948	.0987	.1026	.1064	.1103	.1141
.3	.1179	.1217	.1255	.1293	.1331	.1368	.1406	.1443	.1480	.1517
.4	.1554	.1591	.1628	.1664	.1700	.1736	.1772	.1808	.1844	.1879
.5	.1915	.1950	.1985	.2019	.2054	.2088	.2123	.2157	.2190	.2224
.6	.2257	.2291	.2324	.2357	.2389	.2422	.2454	.2486	.2517	.2549
.7	.2580	.2611	.2642	.2673	.2704	.2734	.2764	.2794	.2823	.2852
.8	.2881	.2910	.2939	.2967	.2995	.3023	.3051	.3078	.3106	.3133
.9	.3159	.3186	.3212	.3238	.3264	.3289	.3315	.3340	.3365	.3389
1.0	.3413	.3438	.3461	.3485	.3508	.3531	.3554	.3577	.3599	.3621
1.1	.3643	.3665	.3686	.3708	.3729	.3749	.3770	.3790	.3810	.3830
1.2	.3849	.3869	.3888	.3907	.3925	.3944	.3962	.3980	.3997	.4015
1.3	.4032	.4049	.4066	.4082	.4099	.4115	.4131	.4147	.4162	.4177
1.4	.4192	.4207	.4222	.4236	.4251	.4265	.4279	.4292	.4306	.4319
1.5	.4332	.4345	.4357	.4370	.4382	.4394	.4406	.4418	.4429	.4441
1.6	.4452	.4463	.4474	.4484	.4495	.4505	.4515	.4525	.4535	.4545
1.7	.4554	.4564	.4573	.4582	.4591	.4599	.4608	.4616	.4625	.4633
1.8	.4641	.4649	.4656	.4664	.4671	.4678	.4686	.4693	.4699	.4706
1.9	.4713	.4719	.4726	.4732	.4738	.4744	.4750	.4756	.4761	.4767
2.0	.4772	.4778	.4783	.4788	.4793	.4798	.4803	.4808	.4812	.4817
2.1	.4821	.4826	.4830	.4834	.4838	.4842	.4846	.4850	.4854	.4857
2.2	.4861	.4864	.4868	.4871	.4875	.4878	.4881	.4884	.4887	.4890
2.3	.4893	.4896	.4898	.4901	.4904	.4906	.4909	.4911	.4913	.4916
2.4	.4918	.4920	.4922	.4925	.4927	.4929	.4931	.4932	.4934	.4936
2.5	.4938	.4940	.4941	.4943	.4945	.4946	.4948	.4949	.4951	.4952
2.6	.4953	.4955	.4956	.4957	.4959	.4960	.4961	.4962	.4963	.4964
2.7	.4965	.4966	.4967	.4968	.4969	.4970	.4971	.4972	.4973	.4974
2.8	.4974	.4975	.4976	.4977	.4977	.4978	.4979	.4979	.4980	.4981
2.9	.4981	.4982	.4982	.4983	.4984	.4984	.4985	.4985	.4986	.4986
3.0	.4987	.4987	.4987	.4988	.4988	.4989	.4989	.4989	.4990	.4990

Source: Abridged from Table I of A. Hald, *Statistical Tables and Formulas* (New York: John Wiley & Sons, Inc.), 1952. ©1952 by John Wiley & Sons, Inc. Reproduced by permission.

TABLE 5
Gamma Function

Values of $\Gamma(n) = \int_0^\infty e^{-x}x^{n-1}dx$; $\Gamma(n + 1) = n\Gamma(n)$

n	$\Gamma(n)$	n	$\Gamma(n)$	n	$\Gamma(n)$	n	$\Gamma(n)$
1.00	1.00000	1.25	.90640	1.50	.88623	1.75	.91906
1.01	.99433	1.26	.90440	1.51	.88659	1.76	.92137
1.02	.98884	1.27	.90250	1.52	.88704	1.77	.92376
1.03	.98355	1.28	.90072	1.53	.88757	1.78	.92623
1.04	.97844	1.29	.89904	1.54	.88818	1.79	.92877
1.05	.97350	1.30	.89747	1.55	.88887	1.80	.93138
1.06	.96874	1.31	.89600	1.56	.88964	1.81	.93408
1.07	.96415	1.32	.89464	1.57	.89049	1.82	.93685
1.08	.95973	1.33	.89338	1.58	.89142	1.83	.93969
1.09	.95546	1.34	.89222	1.59	.89243	1.84	.94261
1.10	.95135	1.35	.89115	1.60	.89352	1.85	.94561
1.11	.94739	1.36	.89018	1.61	.89468	1.86	.94869
1.12	.94359	1.37	.88931	1.62	.89592	1.87	.95184
1.13	.93993	1.38	.88854	1.63	.89724	1.88	.95507
1.14	.93642	1.39	.88785	1.64	.89864	1.89	.95838
1.15	.93304	1.40	.88726	1.65	.90012	1.90	.96177
1.16	.92980	1.41	.88676	1.66	.90167	1.91	.96523
1.17	.92670	1.42	.88636	1.67	.90330	1.92	.96878
1.18	.92373	1.43	.88604	1.68	.90500	1.93	.97240
1.19	.92088	1.44	.88580	1.69	.90678	1.94	.97610
1.20	.91817	1.45	.88565	1.70	.90864	1.95	.97988
1.21	.91558	1.46	.88560	1.71	.91057	1.96	.98374
1.22	.91311	1.47	.88563	1.72	.91258	1.97	.98768
1.23	.91075	1.48	.88575	1.73	.91466	1.98	.99171
1.24	.90852	1.49	.88595	1.74	.91683	1.99	.99581
						2.00	1.00000

Source: Abridged from W. H. Beyer, ed., *Handbook of Tables for Probability and Statistics*, 1966. Reproduced by permission of the publisher, The Chemical Rubber Company.

TABLE 6

Critical Values of *t*

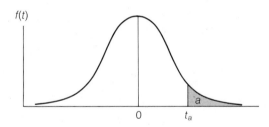

ν	$t_{.100}$	$t_{.050}$	$t_{.025}$	$t_{.010}$	$t_{.005}$
1	3.078	6.314	12.706	31.821	63.657
2	1.886	2.920	4.303	6.965	9.925
3	1.638	2.353	3.182	4.541	5.841
4	1.533	2.132	2.776	3.747	4.604
5	1.476	2.015	2.571	3.365	4.032
6	1.440	1.943	2.447	3.143	3.707
7	1.415	1.895	2.365	2.998	3.499
8	1.397	1.860	2.306	2.896	3.355
9	1.383	1.833	2.262	2.821	3.250
10	1.372	1.812	2.228	2.764	3.169
11	1.363	1.796	2.201	2.718	3.106
12	1.356	1.782	2.179	2.681	3.055
13	1.350	1.771	2.160	2.650	3.012
14	1.345	1.761	2.145	2.624	2.977
15	1.341	1.753	2.131	2.602	2.947
16	1.337	1.746	2.120	2.583	2.921
17	1.333	1.740	2.110	2.567	2.898
18	1.330	1.734	2.101	2.552	2.878
19	1.328	1.729	2.093	2.539	2.861
20	1.325	1.725	2.086	2.528	2.845
21	1.323	1.721	2.080	2.518	2.831
22	1.321	1.717	2.074	2.508	2.819
23	1.319	1.714	2.069	2.500	2.807
24	1.318	1.711	2.064	2.492	2.797
25	1.316	1.708	2.060	2.485	2.787
26	1.315	1.706	2.056	2.479	2.779
27	1.314	1.703	2.052	2.473	2.771
28	1.313	1.701	2.048	2.467	2.763
29	1.311	1.699	2.045	2.462	2.756
∞	1.282	1.645	1.960	2.326	2.576

Source: From M. Merrington, "Table of Percentage Points of the *t*-Distribution," *Biometrika*, 1941, *32*, 300. Reproduced by permission of the *Biometrika* Trustees.

TABLE 7

Critical Values of χ^2

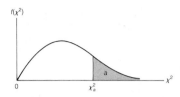

DEGREES OF FREEDOM	$\chi^2_{.995}$	$\chi^2_{.990}$	$\chi^2_{.975}$	$\chi^2_{.950}$	$\chi^2_{.900}$
1	0.0000393	0.0001571	0.0009821	0.0039321	0.0157908
2	0.0100251	0.0201007	0.0506356	0.102587	0.210720
3	0.0717212	0.114832	0.215795	0.351846	0.584375
4	0.206990	0.297110	0.484419	0.710721	1.063623
5	0.411740	0.554300	0.831211	1.145476	1.61031
6	.675727	0.872085	1.237347	1.63539	2.20413
7	.989265	1.239043	1.68987	2.16735	2.83311
8	1.344419	1.646482	2.17973	2.73264	3.48954
9	1.734926	2.087912	2.70039	3.32511	4.16816
10	2.15585	2.55821	3.24697	3.94030	4.86518
11	2.60321	3.05347	3.81575	4.57481	5.57779
12	3.07382	3.57056	4.40379	5.22603	6.30380
13	3.56503	4.10691	5.00874	5.89186	7.04150
14	4.07468	4.66043	5.62872	6.57063	7.78953
15	4.60094	5.22935	6.26214	7.26094	8.54675
16	5.14224	5.81221	6.90766	7.96164	9.31223
17	5.69724	6.40776	7.56418	8.67176	10.0852
18	6.26481	7.01491	8.23075	9.39046	10.8649
19	6.84398	7.63273	8.90655	10.1170	11.6509
20	7.43386	8.26040	9.59083	10.8508	12.4426
21	8.03366	8.89720	10.28293	11.5913	13.2396
22	8.64272	9.54249	10.9823	12.3380	14.0415
23	9.26042	10.19567	11.6885	13.0905	14.8479
24	9.88623	10.8564	12.4011	13.8484	15.6587
25	10.5197	11.5240	13.1197	14.6114	16.4734
26	11.1603	12.1981	13.8439	15.3791	17.2919
27	11.8076	12.8786	14.5733	16.1513	18.1138
28	12.4613	13.5648	15.3079	16.9279	18.9392
29	13.1211	14.2565	16.0471	17.7083	19.7677
30	13.7867	14.9535	16.7908	18.4926	20.5992
40	20.7065	22.1643	24.4331	26.5093	29.0505
50	27.9907	29.7067	32.3574	34.7642	37.6886
60	35.5346	37.4848	40.4817	43.1879	46.4589
70	43.2752	45.4418	48.7576	51.7393	55.3290
80	51.1720	53.5400	57.1532	60.3915	64.2778
90	59.1963	61.7541	65.6466	69.1260	73.2912
100	67.3276	70.0648	74.2219	77.9295	82.3581

DEGREES OF FREEDOM	$\chi^2_{.100}$	$\chi^2_{.050}$	$\chi^2_{.025}$	$\chi^2_{.010}$	$\chi^2_{.005}$
1	2.70554	3.84146	5.02389	6.63490	7.87944
2	4.60517	5.99147	7.37776	9.21034	10.5966
3	6.25139	7.81473	9.34840	11.3449	12.8381
4	7.77944	9.48773	11.1433	13.2767	14.8602
5	9.23635	11.0705	12.8325	15.0863	16.7496
6	10.6446	12.5916	14.4494	16.8119	18.5476
7	12.0170	14.0671	16.0128	18.4753	20.2777
8	13.3616	15.5073	17.5346	20.0902	21.9550
9	14.6837	16.9190	19.0228	21.6660	23.5893
10	15.9871	18.3070	20.4831	23.2093	25.1882
11	17.2750	19.6751	21.9200	24.7250	26.7569
12	18.5494	21.0261	23.3367	26.2170	28.2995
13	19.8119	22.3621	24.7356	27.6883	29.8194
14	21.0642	23.6848	26.1190	29.1413	31.3193
15	22.3072	24.9958	27.4884	30.5779	32.8013
16	23.5418	26.2962	28.8454	31.9999	34.2672
17	24.7690	27.5871	30.1910	33.4087	35.7185
18	25.9894	28.8693	31.5264	34.8053	37.1564
19	27.2036	30.1435	32.8523	36.1908	38.5822
20	28.4120	31.4104	34.1696	37.5662	39.9968
21	29.6151	32.6705	35.4789	38.9321	41.4010
22	30.8133	33.9244	36.7807	40.2894	42.7956
23	32.0069	35.1725	38.0757	41.6384	44.1813
24	33.1963	36.4151	39.3641	42.9798	45.5585
25	34.3816	37.6525	40.6465	44.3141	46.9278
26	35.5631	38.8852	41.9232	45.6417	48.2899
27	36.7412	40.1133	43.1944	46.9630	49.6449
28	37.9159	41.3372	44.4607	48.2782	50.9933
29	39.0875	42.5569	45.7222	49.5879	52.3356
30	40.2560	43.7729	46.9792	50.8922	53.6720
40	51.8050	55.7585	59.3417	63.6907	66.7659
50	63.1671	67.5048	71.4202	76.1539	79.4900
60	74.3970	79.0819	83.2976	88.3794	91.9517
70	85.5271	90.5312	95.0231	100.425	104.215
80	96.5782	101.879	106.629	112.329	116.321
90	107.565	113.145	118.136	124.116	128.299
100	118.498	124.342	129.561	135.807	140.169

TABLE 8

Percentage Points of the
F Distribution, $\alpha = .10$

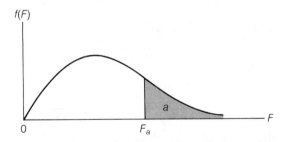

f(F)

ν_2 \ ν_1	NUMERATOR DEGREES OF FREEDOM								
	1	2	3	4	5	6	7	8	9
1	39.86	49.50	53.59	55.83	57.24	58.20	58.91	59.44	59.86
2	8.53	9.00	9.16	9.24	9.29	9.33	9.35	9.37	9.38
3	5.54	5.46	5.39	5.34	5.31	5.28	5.27	5.25	5.24
4	4.54	4.32	4.19	4.11	4.05	4.01	3.98	3.95	3.94
5	4.06	3.78	3.62	3.52	3.45	3.40	3.37	3.34	3.32
6	3.78	3.46	3.29	3.18	3.11	3.05	3.01	2.98	2.96
7	3.59	3.26	3.07	2.96	2.88	2.83	2.78	2.75	2.72
8	3.46	3.11	2.92	2.81	2.73	2.67	2.62	2.59	2.56
9	3.36	3.01	2.81	2.69	2.61	2.55	2.51	2.47	2.44
10	3.29	2.92	2.73	2.61	2.52	2.46	2.41	2.38	2.35
11	3.23	2.86	2.66	2.54	2.45	2.39	2.34	2.30	2.27
12	3.18	2.81	2.61	2.48	2.39	2.33	2.28	2.24	2.21
13	3.14	2.76	2.56	2.43	2.35	2.28	2.23	2.20	2.16
14	3.10	2.73	2.52	2.39	2.31	2.24	2.19	2.15	2.12
15	3.07	2.70	2.49	2.36	2.27	2.21	2.16	2.12	2.09
16	3.05	2.67	2.46	2.33	2.24	2.18	2.13	2.09	2.06
17	3.03	2.64	2.44	2.31	2.22	2.15	2.10	2.06	2.03
18	3.01	2.62	2.42	2.29	2.20	2.13	2.08	2.04	2.00
19	2.99	2.61	2.40	2.27	2.18	2.11	2.06	2.02	1.98
20	2.97	2.59	2.38	2.25	2.16	2.09	2.04	2.00	1.96
21	2.96	2.57	2.36	2.23	2.14	2.08	2.02	1.98	1.95
22	2.95	2.56	2.35	2.22	2.13	2.06	2.01	1.97	1.93
23	2.94	2.55	2.34	2.21	2.11	2.05	1.99	1.95	1.92
24	2.93	2.54	2.33	2.19	2.10	2.04	1.98	1.94	1.91
25	2.92	2.53	2.32	2.18	2.09	2.02	1.97	1.93	1.89
26	2.91	2.52	2.31	2.17	2.08	2.01	1.96	1.92	1.88
27	2.90	2.51	2.30	2.17	2.07	2.00	1.95	1.91	1.87
28	2.89	2.50	2.29	2.16	2.06	2.00	1.94	1.90	1.87
29	2.89	2.50	2.28	2.15	2.06	1.99	1.93	1.89	1.86
30	2.88	2.49	2.28	2.14	2.05	1.98	1.93	1.88	1.85
40	2.84	2.44	2.23	2.09	2.00	1.93	1.87	1.83	1.79
60	2.79	2.39	2.18	2.04	1.95	1.87	1.82	1.77	1.74
120	2.75	2.35	2.13	1.99	1.90	1.82	1.77	1.72	1.68
∞	2.71	2.30	2.08	1.94	1.85	1.77	1.72	1.67	1.63

(Left axis label: DENOMINATOR DEGREES OF FREEDOM)

Source: From M. Merrington and C. M. Thompson, "Tables of Percentage Points of the Inverted Beta (F)-Distribution," *Biometrika*, 1943, 33, 73–88. Reproduced by permission of the *Biometrika* Trustees.

TABLE 8   Continued

DENOMINATOR DEGREES OF FREEDOM

ν_2 \ ν_1	NUMERATOR DEGREES OF FREEDOM									
	10	12	15	20	24	30	40	60	120	∞
1	60.19	60.71	61.22	61.74	62.00	62.26	62.53	62.79	63.06	63.33
2	9.39	9.41	9.42	9.44	9.45	9.46	9.47	9.47	9.48	9.49
3	5.23	5.22	5.20	5.18	5.18	5.17	5.16	5.15	5.14	5.13
4	3.92	3.90	3.87	3.84	3.83	3.82	3.80	3.79	3.78	3.76
5	3.30	3.27	3.24	3.21	3.19	3.17	3.16	3.14	3.12	3.10
6	2.94	2.90	2.87	2.84	2.82	2.80	2.78	2.76	2.74	2.72
7	2.70	2.67	2.63	2.59	2.58	2.56	2.54	2.51	2.49	2.47
8	2.54	2.50	2.46	2.42	2.40	2.38	2.36	2.34	2.32	2.29
9	2.42	2.38	2.34	2.30	2.28	2.25	2.23	2.21	2.18	2.16
10	2.32	2.28	2.24	2.20	2.18	2.16	2.13	2.11	2.08	2.06
11	2.25	2.21	2.17	2.12	2.10	2.08	2.05	2.03	2.00	1.97
12	2.19	2.15	2.10	2.06	2.04	2.01	1.99	1.96	1.93	1.90
13	2.14	2.10	2.05	2.01	1.98	1.96	1.93	1.90	1.88	1.85
14	2.10	2.05	2.01	1.96	1.94	1.91	1.89	1.86	1.83	1.80
15	2.06	2.02	1.97	1.92	1.90	1.87	1.85	1.82	1.79	1.76
16	2.03	1.99	1.94	1.89	1.87	1.84	1.81	1.78	1.75	1.72
17	2.00	1.96	1.91	1.86	1.84	1.81	1.78	1.75	1.72	1.69
18	1.98	1.93	1.89	1.84	1.81	1.78	1.75	1.72	1.69	1.66
19	1.96	1.91	1.86	1.81	1.79	1.76	1.73	1.70	1.67	1.63
20	1.94	1.89	1.84	1.79	1.77	1.74	1.71	1.68	1.64	1.61
21	1.92	1.87	1.83	1.78	1.75	1.72	1.69	1.66	1.62	1.59
22	1.90	1.86	1.81	1.76	1.73	1.70	1.67	1.64	1.60	1.57
23	1.89	1.84	1.80	1.74	1.72	1.69	1.66	1.62	1.59	1.55
24	1.88	1.83	1.78	1.73	1.70	1.67	1.64	1.61	1.57	1.53
25	1.87	1.82	1.77	1.72	1.69	1.66	1.63	1.59	1.56	1.52
26	1.86	1.81	1.76	1.71	1.68	1.65	1.61	1.58	1.54	1.50
27	1.85	1.80	1.75	1.70	1.67	1.64	1.60	1.57	1.53	1.49
28	1.84	1.79	1.74	1.69	1.66	1.63	1.59	1.56	1.52	1.48
29	1.83	1.78	1.73	1.68	1.65	1.62	1.58	1.55	1.51	1.47
30	1.82	1.77	1.72	1.67	1.64	1.61	1.57	1.54	1.50	1.46
40	1.76	1.71	1.66	1.61	1.57	1.54	1.51	1.47	1.42	1.38
60	1.71	1.66	1.60	1.54	1.51	1.48	1.44	1.40	1.35	1.29
120	1.65	1.60	1.55	1.48	1.45	1.41	1.37	1.32	1.26	1.19
∞	1.60	1.55	1.49	1.42	1.38	1.34	1.30	1.24	1.17	1.00

TABLE 9

Percentage Points of the
F Distribution, $\alpha = .05$

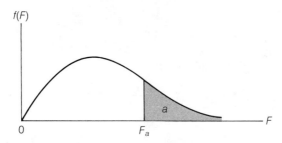

ν_1	NUMERATOR DEGREES OF FREEDOM								
ν_2	1	2	3	4	5	6	7	8	9
1	161.4	199.5	215.7	224.6	230.2	234.0	236.8	238.9	240.5
2	18.51	19.00	19.16	19.25	19.30	19.33	19.35	19.37	19.38
3	10.13	9.55	9.28	9.12	9.01	8.94	8.89	8.85	8.81
4	7.71	6.94	6.59	6.39	6.26	6.16	6.09	6.04	6.00
5	6.61	5.79	5.41	5.19	5.05	4.95	4.88	4.82	4.77
6	5.99	5.14	4.76	4.53	4.39	4.28	4.21	4.15	4.10
7	5.59	4.74	4.35	4.12	3.97	3.87	3.79	3.73	3.68
8	5.32	4.46	4.07	3.84	3.69	3.58	3.50	3.44	3.39
9	5.12	4.26	3.86	3.63	3.48	3.37	3.29	3.23	3.18
10	4.96	4.10	3.71	3.48	3.33	3.22	3.14	3.07	3.02
11	4.84	3.98	3.59	3.36	3.20	3.09	3.01	2.95	2.90
12	4.75	3.89	3.49	3.26	3.11	3.00	2.91	2.85	2.80
13	4.67	3.81	3.41	3.18	3.03	2.92	2.83	2.77	2.71
14	4.60	3.74	3.34	3.11	2.96	2.85	2.76	2.70	2.65
15	4.54	3.68	3.29	3.06	2.90	2.79	2.71	2.64	2.59
16	4.49	3.63	3.24	3.01	2.85	2.74	2.66	2.59	2.54
17	4.45	3.59	3.20	2.96	2.81	2.70	2.61	2.55	2.49
18	4.41	3.55	3.16	2.93	2.77	2.66	2.58	2.51	2.46
19	4.38	3.52	3.13	2.90	2.74	2.63	2.54	2.48	2.42
20	4.35	3.49	3.10	2.87	2.71	2.60	2.51	2.45	2.39
21	4.32	3.47	3.07	2.84	2.68	2.57	2.49	2.42	2.37
22	4.30	3.44	3.05	2.82	2.66	2.55	2.46	2.40	2.34
23	4.28	3.42	3.03	2.80	2.64	2.53	2.44	2.37	2.32
24	4.26	3.40	3.01	2.78	2.62	2.51	2.42	2.36	2.30
25	4.24	3.39	2.99	2.76	2.60	2.49	2.40	2.34	2.28
26	4.23	3.37	2.98	2.74	2.59	2.47	2.39	2.32	2.27
27	4.21	3.35	2.96	2.73	2.57	2.46	2.37	2.31	2.25
28	4.20	3.34	2.95	2.71	2.56	2.45	2.36	2.29	2.24
29	4.18	3.33	2.93	2.70	2.55	2.43	2.35	2.28	2.22
30	4.17	3.32	2.92	2.69	2.53	2.42	2.33	2.27	2.21
40	4.08	3.23	2.84	2.61	2.45	2.34	2.25	2.18	2.12
60	4.00	3.15	2.76	2.53	2.37	2.25	2.17	2.10	2.04
120	3.92	3.07	2.68	2.45	2.29	2.17	2.09	2.02	1.96
∞	3.84	3.00	2.60	2.37	2.21	2.10	2.01	1.94	1.88

Source: From M. Merrington and C. M. Thompson, "Tables of Percentage Points of the Inverted Beta (F)-Distribution," *Biometrika*, 1943, 33, 73–88. Reproduced by permission of the *Biometrika* Trustees.

ν_1 ν_2	NUMERATOR DEGREES OF FREEDOM									
	10	12	15	20	24	30	40	60	120	∞
1	241.9	243.9	245.9	248.0	249.1	250.1	251.1	252.2	253.3	254.3
2	19.40	19.41	19.43	19.45	19.45	19.46	19.47	19.48	19.49	19.50
3	8.79	8.74	8.70	8.66	8.64	8.62	8.59	8.57	8.55	8.53
4	5.96	5.91	5.86	5.80	5.77	5.75	5.72	5.69	5.66	5.63
5	4.74	4.68	4.62	4.56	4.53	4.50	4.46	4.43	4.40	4.36
6	4.06	4.00	3.94	3.87	3.84	3.81	3.77	3.74	3.70	3.67
7	3.64	3.57	3.51	3.44	3.41	3.38	3.34	3.30	3.27	3.23
8	3.35	3.28	3.22	3.15	3.12	3.08	3.04	3.01	2.97	2.93
9	3.14	3.07	3.01	2.94	2.90	2.86	2.83	2.79	2.75	2.71
10	2.98	2.91	2.85	2.77	2.74	2.70	2.66	2.62	2.58	2.54
11	2.85	2.79	2.72	2.65	2.61	2.57	2.53	2.49	2.45	2.40
12	2.75	2.69	2.62	2.54	2.51	2.47	2.43	2.38	2.34	2.30
13	2.67	2.60	2.53	2.46	2.42	2.38	2.34	2.30	2.25	2.21
14	2.60	2.53	2.46	2.39	2.35	2.31	2.27	2.22	2.18	2.13
15	2.54	2.48	2.40	2.33	2.29	2.25	2.20	2.16	2.11	2.07
16	2.49	2.42	2.35	2.28	2.24	2.19	2.15	2.11	2.06	2.01
17	2.45	2.38	2.31	2.23	2.19	2.15	2.10	2.06	2.01	1.96
18	2.41	2.34	2.27	2.19	2.15	2.11	2.06	2.02	1.97	1.92
19	2.38	2.31	2.23	2.16	2.11	2.07	2.03	1.98	1.93	1.88
20	2.35	2.28	2.20	2.12	2.08	2.04	1.99	1.95	1.90	1.84
21	2.32	2.25	2.18	2.10	2.05	2.01	1.96	1.92	1.87	1.81
22	2.30	2.23	2.15	2.07	2.03	1.98	1.94	1.89	1.84	1.78
23	2.27	2.20	2.13	2.05	2.01	1.96	1.91	1.86	1.81	1.76
24	2.25	2.18	2.11	2.03	1.98	1.94	1.89	1.84	1.79	1.73
25	2.24	2.16	2.09	2.01	1.96	1.92	1.87	1.82	1.77	1.71
26	2.22	2.15	2.07	1.99	1.95	1.90	1.85	1.80	1.75	1.69
27	2.20	2.13	2.06	1.97	1.93	1.88	1.84	1.79	1.73	1.67
28	2.19	2.12	2.04	1.96	1.91	1.87	1.82	1.77	1.71	1.65
29	2.18	2.10	2.03	1.94	1.90	1.85	1.81	1.75	1.70	1.64
30	2.16	2.09	2.01	1.93	1.89	1.84	1.79	1.74	1.68	1.62
40	2.08	2.00	1.92	1.84	1.79	1.74	1.69	1.64	1.58	1.51
60	1.99	1.92	1.84	1.75	1.70	1.65	1.59	1.53	1.47	1.39
120	1.91	1.83	1.75	1.66	1.61	1.55	1.50	1.43	1.35	1.25
∞	1.83	1.75	1.67	1.57	1.52	1.46	1.39	1.32	1.22	1.00

DENOMINATOR DEGREES OF FREEDOM

TABLE 10

Percentage Points of the
F Distribution, $\alpha = .025$

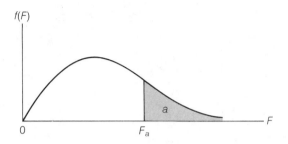

ν_1 ν_2	NUMERATOR DEGREES OF FREEDOM								
	1	2	3	4	5	6	7	8	9
1	647.8	799.5	864.2	899.6	921.8	937.1	948.2	956.7	963.3
2	38.51	39.00	39.17	39.25	39.30	39.33	39.36	39.37	39.39
3	17.44	16.04	15.44	15.10	14.88	14.73	14.62	14.54	14.47
4	12.22	10.65	9.98	9.60	9.36	9.20	9.07	8.98	8.90
5	10.01	8.43	7.76	7.39	7.15	6.98	6.85	6.76	6.68
6	8.81	7.26	6.60	6.23	5.99	5.82	5.70	5.60	5.52
7	8.07	6.54	5.89	5.52	5.29	5.12	4.99	4.90	4.82
8	7.57	6.06	5.42	5.05	4.82	4.65	4.53	4.43	4.36
9	7.21	5.71	5.08	4.72	4.48	4.32	4.20	4.10	4.03
10	6.94	5.46	4.83	4.47	4.24	4.07	3.95	3.85	3.78
11	6.72	5.26	4.63	4.28	4.04	3.88	3.76	3.66	3.59
12	6.55	5.10	4.47	4.12	3.89	3.73	3.61	3.51	3.44
13	6.41	4.97	4.35	4.00	3.77	3.60	3.48	3.39	3.31
14	6.30	4.86	4.24	3.89	3.66	3.50	3.38	3.29	3.21
15	6.20	4.77	4.15	3.80	3.58	3.41	3.29	3.20	3.12
16	6.12	4.69	4.08	3.73	3.50	3.34	3.22	3.12	3.05
17	6.04	4.62	4.01	3.66	3.44	3.28	3.16	3.06	2.98
18	5.98	4.56	3.95	3.61	3.38	3.22	3.10	3.01	2.93
19	5.92	4.51	3.90	3.56	3.33	3.17	3.05	2.96	2.88
20	5.87	4.46	3.86	3.51	3.29	3.13	3.01	2.91	2.84
21	5.83	4.42	3.82	3.48	3.25	3.09	2.97	2.87	2.80
22	5.79	4.38	3.78	3.44	3.22	3.05	2.93	2.84	2.76
23	5.75	4.35	3.75	3.41	3.18	3.02	2.90	2.81	2.73
24	5.72	4.32	3.72	3.38	3.15	2.99	2.87	2.78	2.70
25	5.69	4.29	3.69	3.35	3.13	2.97	2.85	2.75	2.68
26	5.66	4.27	3.67	3.33	3.10	2.94	2.82	2.73	2.65
27	5.63	4.24	3.65	3.31	3.08	2.92	2.80	2.71	2.63
28	5.61	4.22	3.63	3.29	3.06	2.90	2.78	2.69	2.61
29	5.59	4.20	3.61	3.27	3.04	2.88	2.76	2.67	2.59
30	5.57	4.18	3.59	3.25	3.03	2.87	2.75	2.65	2.57
40	5.42	4.05	3.46	3.13	2.90	2.74	2.62	2.53	2.45
60	5.29	3.93	3.34	3.01	2.79	2.63	2.51	2.41	2.33
120	5.15	3.80	3.23	2.89	2.67	2.52	2.39	2.30	2.22
∞	5.02	3.69	3.12	2.79	2.57	2.41	2.29	2.19	2.11

DENOMINATOR DEGREES OF FREEDOM

ν_2 \ ν_1	NUMERATOR DEGREES OF FREEDOM									
	10	12	15	20	24	30	40	60	120	∞
1	968.6	976.7	984.9	993.1	997.2	1001	1006	1010	1014	1018
2	39.40	39.41	39.43	39.45	39.46	39.46	39.47	39.48	39.49	39.50
3	14.42	14.34	14.25	14.17	14.12	14.08	14.04	13.99	13.95	13.90
4	8.84	8.75	8.66	8.56	8.51	8.46	8.41	8.36	8.31	8.26
5	6.62	6.52	6.43	6.33	6.28	6.23	6.18	6.12	6.07	6.02
6	5.46	5.37	5.27	5.17	5.12	5.07	5.01	4.96	4.90	4.85
7	4.76	4.67	4.57	4.47	4.42	4.36	4.31	4.25	4.20	4.14
8	4.30	4.20	4.10	4.00	3.95	3.89	3.84	3.78	3.73	3.67
9	3.96	3.87	3.77	3.67	3.61	3.56	3.51	3.45	3.39	3.33
10	3.72	3.62	3.52	3.42	3.37	3.31	3.26	3.20	3.14	3.08
11	3.53	3.43	3.33	3.23	3.17	3.12	3.06	3.00	2.94	2.88
12	3.37	3.28	3.18	3.07	3.02	2.96	2.91	2.85	2.79	2.72
13	3.25	3.15	3.05	2.95	2.89	2.84	2.78	2.72	2.66	2.60
14	3.15	3.05	2.95	2.84	2.79	2.73	2.67	2.61	2.55	2.49
15	3.06	2.96	2.86	2.76	2.70	2.64	2.59	2.52	2.46	2.40
16	2.99	2.89	2.79	2.68	2.63	2.57	2.51	2.45	2.38	2.32
17	2.92	2.82	2.72	2.62	2.56	2.50	2.44	2.38	2.32	2.25
18	2.87	2.77	2.67	2.56	2.50	2.44	2.38	2.32	2.26	2.19
19	2.82	2.72	2.62	2.51	2.45	2.39	2.33	2.27	2.20	2.13
20	2.77	2.68	2.57	2.46	2.41	2.35	2.29	2.22	2.16	2.09
21	2.73	2.64	2.53	2.42	2.37	2.31	2.25	2.18	2.11	2.04
22	2.70	2.60	2.50	2.39	2.33	2.27	2.21	2.14	2.08	2.00
23	2.67	2.57	2.47	2.36	2.30	2.24	2.18	2.11	2.04	1.97
24	2.64	2.54	2.44	2.33	2.27	2.21	2.15	2.08	2.01	1.94
25	2.61	2.51	2.41	2.30	2.24	2.18	2.12	2.05	1.98	1.91
26	2.59	2.49	2.39	2.28	2.22	2.16	2.09	2.03	1.95	1.88
27	2.57	2.47	2.36	2.25	2.19	2.13	2.07	2.00	1.93	1.85
28	2.55	2.45	2.34	2.23	2.17	2.11	2.05	1.98	1.91	1.83
29	2.53	2.43	2.32	2.21	2.15	2.09	2.03	1.96	1.89	1.81
30	2.51	2.41	2.31	2.20	2.14	2.07	2.01	1.94	1.87	1.79
40	2.39	2.29	2.18	2.07	2.01	1.94	1.88	1.80	1.72	1.64
60	2.27	2.17	2.06	1.94	1.88	1.82	1.74	1.67	1.58	1.48
120	2.16	2.05	1.94	1.82	1.76	1.69	1.61	1.53	1.43	1.31
∞	2.05	1.94	1.83	1.71	1.64	1.57	1.48	1.39	1.27	1.00

DENOMINATOR DEGREES OF FREEDOM

TABLE 11

Percentage Points of the
F Distribution, $\alpha = .01$

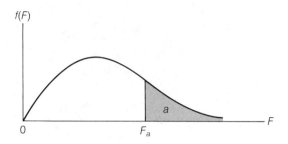

ν_1 ν_2	NUMERATOR DEGREES OF FREEDOM								
	1	2	3	4	5	6	7	8	9
1	4,052	4,999.5	5,403	5,625	5,764	5,859	5,928	5,982	6,022
2	98.50	99.00	99.17	99.25	99.30	99.33	99.36	99.37	99.39
3	34.12	30.82	29.46	28.71	28.24	27.91	27.67	27.49	27.35
4	21.20	18.00	16.69	15.98	15.52	15.21	14.98	14.80	14.66
5	16.26	13.27	12.06	11.39	10.97	10.67	10.46	10.29	10.16
6	13.75	10.92	9.78	9.15	8.75	8.47	8.26	8.10	7.98
7	12.25	9.55	8.45	7.85	7.46	7.19	6.99	6.84	6.72
8	11.26	8.65	7.59	7.01	6.63	6.37	6.18	6.03	5.91
9	10.56	8.02	6.99	6.42	6.06	5.80	5.61	5.47	5.35
10	10.04	7.56	6.55	5.99	5.64	5.39	5.20	5.06	4.94
11	9.65	7.21	6.22	5.67	5.32	5.07	4.89	4.74	4.63
12	9.33	6.93	5.95	5.41	5.06	4.82	4.64	4.50	4.39
13	9.07	6.70	5.74	5.21	4.86	4.62	4.44	4.30	4.19
14	8.86	6.51	5.56	5.04	4.69	4.46	4.28	4.14	4.03
15	8.68	6.36	5.42	4.89	4.56	4.32	4.14	4.00	3.89
16	8.53	6.23	5.29	4.77	4.44	4.20	4.03	3.89	3.78
17	8.40	6.11	5.18	4.67	4.34	4.10	3.93	3.79	3.68
18	8.29	6.01	5.09	4.58	4.25	4.01	3.84	3.71	3.60
19	8.18	5.93	5.01	4.50	4.17	3.94	3.77	3.63	3.52
20	8.10	5.85	4.94	4.43	4.10	3.87	3.70	3.56	3.46
21	8.02	5.78	4.87	4.37	4.04	3.81	3.64	3.51	3.40
22	7.95	5.72	4.82	4.31	3.99	3.76	3.59	3.45	3.35
23	7.88	5.66	4.76	4.26	3.94	3.71	3.54	3.41	3.30
24	7.82	5.61	4.72	4.22	3.90	3.67	3.50	3.36	3.26
25	7.77	5.57	4.68	4.18	3.85	3.63	3.46	3.32	3.22
26	7.72	5.53	4.64	4.14	3.82	3.59	3.42	3.29	3.18
27	7.68	5.49	4.60	4.11	3.78	3.56	3.39	3.26	3.15
28	7.64	5.45	4.57	4.07	3.75	3.53	3.36	3.23	3.12
29	7.60	5.42	4.54	4.04	3.73	3.50	3.33	3.20	3.09
30	7.56	5.39	4.51	4.02	3.70	3.47	3.30	3.17	3.07
40	7.31	5.18	4.31	3.83	3.51	3.29	3.12	2.99	2.89
60	7.08	4.98	4.13	3.65	3.34	3.12	2.95	2.82	2.72
120	6.85	4.79	3.95	3.48	3.17	2.96	2.79	2.66	2.56
∞	6.63	4.61	3.78	3.32	3.02	2.80	2.64	2.51	2.41

Source: From M. Merrington and C. M. Thompson, "Tables of Percentage Points of the Inverted Beta (F)-Distribution," *Biometrika*, 1943, 33, 73–88. Reproduced by permission of the *Biometrika* Trustees.

ν_2 \ ν_1	NUMERATOR DEGREES OF FREEDOM									
	10	12	15	20	24	30	40	60	120	∞
1	6,056	6,106	6,157	6,209	6,235	6,261	6,287	6,313	6,339	6,366
2	99.40	99.42	99.43	99.45	99.46	99.47	99.47	99.48	99.49	99.50
3	27.23	27.05	26.87	26.69	26.60	26.50	26.41	26.32	26.22	26.13
4	14.55	14.37	14.20	14.02	13.93	13.84	13.75	13.65	13.56	13.46
5	10.05	9.89	9.72	9.55	9.47	9.38	9.29	9.20	9.11	9.02
6	7.87	7.72	7.56	7.40	7.31	7.23	7.14	7.06	6.97	6.88
7	6.62	6.47	6.31	6.16	6.07	5.99	5.91	5.82	5.74	5.65
8	5.81	5.67	5.52	5.36	5.28	5.20	5.12	5.03	4.95	4.86
9	5.26	5.11	4.96	4.81	4.73	4.65	4.57	4.48	4.40	4.31
10	4.85	4.71	4.56	4.41	4.33	4.25	4.17	4.08	4.00	3.91
11	4.54	4.40	4.25	4.10	4.02	3.94	3.86	3.78	3.69	3.60
12	4.30	4.16	4.01	3.86	3.78	3.70	3.62	3.54	3.45	3.36
13	4.10	3.96	3.82	3.66	3.59	3.51	3.43	3.34	3.25	3.17
14	3.94	3.80	3.66	3.51	3.43	3.35	3.27	3.18	3.09	3.00
15	3.80	3.67	3.52	3.37	3.29	3.21	3.13	3.05	2.96	2.87
16	3.69	3.55	3.41	3.26	3.18	3.10	3.02	2.93	2.84	2.75
17	3.59	3.46	3.31	3.16	3.08	3.00	2.92	2.83	2.75	2.65
18	3.51	3.37	3.23	3.08	3.00	2.92	2.84	2.75	2.66	2.57
19	3.43	3.30	3.15	3.00	2.92	2.84	2.76	2.67	2.58	2.49
20	3.37	3.23	3.09	2.94	2.86	2.78	2.69	2.61	2.52	2.42
21	3.31	3.17	3.03	2.88	2.80	2.72	2.64	2.55	2.46	2.36
22	3.26	3.12	2.98	2.83	2.75	2.67	2.58	2.50	2.40	2.31
23	3.21	3.07	2.93	2.78	2.70	2.62	2.54	2.45	2.35	2.26
24	3.17	3.03	2.89	2.74	2.66	2.58	2.49	2.40	2.31	2.21
25	3.13	2.99	2.85	2.70	2.62	2.54	2.45	2.36	2.27	2.17
26	3.09	2.96	2.81	2.66	2.58	2.50	2.42	2.33	2.23	2.13
27	3.06	2.93	2.78	2.63	2.55	2.47	2.38	2.29	2.20	2.10
28	3.03	2.90	2.75	2.60	2.52	2.44	2.35	2.26	2.17	2.06
29	3.00	2.87	2.73	2.57	2.49	2.41	2.33	2.23	2.14	2.03
30	2.98	2.84	2.70	2.55	2.47	2.39	2.30	2.21	2.11	2.01
40	2.80	2.66	2.52	2.37	2.29	2.20	2.11	2.02	1.92	1.80
60	2.63	2.50	2.35	2.20	2.12	2.03	1.94	1.84	1.73	1.60
120	2.47	2.34	2.19	2.03	1.95	1.86	1.76	1.66	1.53	1.38
∞	2.32	2.18	2.04	1.88	1.79	1.70	1.59	1.47	1.32	1.00

DENOMINATOR DEGREES OF FREEDOM

TABLE 12
Critical Values of the Sample Coefficient of Correlation, r

SAMPLE SIZE n	$r_{.050}$	$r_{.025}$	$r_{.010}$	$r_{.005}$
3	.988	.9969	.99951	.99988
4	.900	.950	.980	.99000
5	.805	.878	.934	.959
6	.729	.811	.882	.917
7	.669	.754	.833	.875
8	.621	.707	.789	.834
9	.582	.666	.750	.798
10	.549	.632	.715	.765
11	.521	.602	.685	.735
12	.497	.576	.658	.708
13	.476	.553	.634	.684
14	.457	.532	.612	.661
15	.441	.514	.592	.641
16	.426	.497	.574	.623
17	.412	.482	.558	.606
18	.400	.468	.543	.590
19	.389	.456	.529	.575
20	.378	.444	.516	.561
21	.369	.433	.503	.549
22	.360	.423	.492	.537
27	.323	.381	.445	.487
32	.296	.349	.409	.449
37	.275	.325	.381	.418
42	.257	.304	.358	.393
47	.243	.288	.338	.372
52	.231	.273	.322	.354
62	.211	.250	.295	.325
72	.195	.232	.274	.302
82	.183	.217	.257	.283
92	.173	.205	.242	.267
102	.164	.195	.230	.254

Source: From E. S. Pearson and H. O. Hartley (eds.), "Percentage Points for the Distribution of the Correlation Coefficient, r, when $\rho = 0$," *Biometrika Tables for Statisticians*, Vol. I, 1966, p. 146. Reproduced by permission of the *Biometrika* Trustees.

TABLE 13   Percentage Points of the Studentized Range $q(p, \nu)$, $\alpha = .05$

ν \\ p	2	3	4	5	6	7	8	9	10	11
1	17.97	26.98	32.82	37.08	40.41	43.12	45.40	47.36	49.07	50.59
2	6.08	8.33	9.80	10.88	11.74	12.44	13.03	13.54	13.99	14.39
3	4.50	5.91	6.82	7.50	8.04	8.48	8.85	9.18	9.46	9.72
4	3.93	5.04	5.76	6.29	6.71	7.05	7.35	7.60	7.83	8.03
5	3.64	4.60	5.22	5.67	6.03	6.33	6.58	6.80	6.99	7.17
6	3.46	4.34	4.90	5.30	5.63	5.90	6.12	6.32	6.49	6.65
7	3.34	4.16	4.68	5.06	5.36	5.61	5.82	6.00	6.16	6.30
8	3.26	4.04	4.53	4.89	5.17	5.40	5.60	5.77	5.92	6.05
9	3.20	3.95	4.41	4.76	5.02	5.24	5.43	5.59	5.74	5.87
10	3.15	3.88	4.33	4.65	4.91	5.12	5.30	5.46	5.60	5.72
11	3.11	3.82	4.26	4.57	4.82	5.03	5.20	5.35	5.49	5.61
12	3.08	3.77	4.20	4.51	4.75	4.95	5.12	5.27	5.39	5.51
13	3.06	3.73	4.15	4.45	4.69	4.88	5.05	5.19	5.32	5.43
14	3.03	3.70	4.11	4.41	4.64	4.83	4.99	5.13	5.25	5.36
15	3.01	3.67	4.08	4.37	4.60	4.78	4.94	5.08	5.20	5.31
16	3.00	3.65	4.05	4.33	4.56	4.74	4.90	5.03	5.15	5.26
17	2.98	3.63	4.02	4.30	4.52	4.70	4.86	4.99	5.11	5.21
18	2.97	3.61	4.00	4.28	4.49	4.67	4.82	4.96	5.07	5.17
19	2.96	3.59	3.98	4.25	4.47	4.65	4.79	4.92	5.04	5.14
20	2.95	3.58	3.96	4.23	4.45	4.62	4.77	4.90	5.01	5.11
24	2.92	3.53	3.90	4.17	4.37	4.54	4.68	4.81	4.92	5.01
30	2.89	3.49	3.85	4.10	4.30	4.46	4.60	4.72	4.82	4.92
40	2.86	3.44	3.79	4.04	4.23	4.39	4.52	4.63	4.73	4.82
60	2.83	3.40	3.74	3.98	4.16	4.31	4.44	4.55	4.65	4.73
120	2.80	3.36	3.68	3.92	4.10	4.24	4.36	4.47	4.56	4.64
∞	2.77	3.31	3.63	3.86	4.03	4.17	4.29	4.39	4.47	4.55

(continued)

TABLE 13 Continued

ν \\ p	12	13	14	15	16	17	18	19	20
1	51.96	53.20	54.33	55.36	56.32	57.22	58.04	58.83	59.56
2	14.75	15.08	15.38	15.65	15.91	16.14	16.37	16.57	16.77
3	9.95	10.15	10.35	10.52	10.69	10.84	10.98	11.11	11.24
4	8.21	8.37	8.52	8.66	8.79	8.91	9.03	9.13	9.23
5	7.32	7.47	7.60	7.72	7.83	7.93	8.03	8.12	8.21
6	6.79	6.92	7.03	7.14	7.24	7.34	7.43	7.51	7.59
7	6.43	6.55	6.66	6.76	6.85	6.94	7.02	7.10	7.17
8	6.18	6.29	6.39	6.48	6.57	6.65	6.73	6.80	6.87
9	5.98	6.09	6.19	6.28	6.36	6.44	6.51	6.58	6.64
10	5.83	5.93	6.03	6.11	6.19	6.27	6.34	6.40	6.47
11	5.71	5.81	5.90	5.98	6.06	6.13	6.20	6.27	6.33
12	5.61	5.71	5.80	5.88	5.95	6.02	6.09	6.15	6.21
13	5.53	5.63	5.71	5.79	5.86	5.93	5.99	6.05	6.11
14	5.46	5.55	5.64	5.71	5.79	5.85	5.91	5.97	6.03
15	5.40	5.49	5.57	5.65	5.72	5.78	5.85	5.90	5.96
16	5.35	5.44	5.52	5.59	5.66	5.73	5.79	5.84	5.90
17	5.31	5.39	5.47	5.54	5.61	5.67	5.73	5.79	5.84
18	5.27	5.35	5.43	5.50	5.57	5.63	5.69	5.74	5.79
19	5.23	5.31	5.39	5.46	5.53	5.59	5.65	5.70	5.75
20	5.20	5.28	5.36	5.43	5.49	5.55	5.61	5.66	5.71
24	5.10	5.18	5.25	5.32	5.38	5.44	5.49	5.55	5.59
30	5.00	5.08	5.15	5.21	5.27	5.33	5.38	5.43	5.47
40	4.90	4.98	5.04	5.11	5.16	5.22	5.27	5.31	5.36
60	4.81	4.88	4.94	5.00	5.06	5.11	5.15	5.20	5.24
120	4.71	4.78	4.84	4.90	4.95	5.00	5.04	5.09	5.13
∞	4.62	4.68	4.74	4.80	4.85	4.89	4.93	4.97	5.01

Source: *Biometrika Tables for Statisticians*, Vol. I, 3d ed., edited by E. S. Pearson and H. O. Hartley (Cambridge University Press, 1966). Reproduced by permission of Professor E. S. Pearson and the *Biometrika* Trustees.

TABLE 14   Percentage Points of the Studentized Range $q(p, \nu)$, $\alpha = .01$

ν \ p	2	3	4	5	6	7	8	9	10	11
1	90.03	135.0	164.3	185.6	202.2	215.8	227.2	237.0	245.6	253.2
2	14.04	19.02	22.29	24.72	26.63	28.20	29.53	30.68	31.69	32.59
3	8.26	10.62	12.17	13.33	14.24	15.00	15.64	16.20	16.69	17.13
4	6.51	8.12	9.17	9.96	10.58	11.10	11.55	11.93	12.27	12.57
5	5.70	6.98	7.80	8.42	8.91	9.32	9.67	9.97	10.24	10.48
6	5.24	6.33	7.03	7.56	7.97	8.32	8.61	8.87	9.10	9.30
7	4.95	5.92	6.54	7.01	7.37	7.68	7.94	8.17	8.37	8.55
8	4.75	5.64	6.20	6.62	6.96	7.24	7.47	7.68	7.86	8.03
9	4.60	5.43	5.96	6.35	6.66	6.91	7.13	7.33	7.49	7.65
10	4.48	5.27	5.77	6.14	6.43	6.67	6.87	7.05	7.21	7.36
11	4.39	5.15	5.62	5.97	6.25	6.48	6.67	6.84	6.99	7.13
12	4.32	5.05	5.50	5.84	6.10	6.32	6.51	6.67	6.81	6.94
13	4.26	4.96	5.40	5.73	5.98	6.19	6.37	6.53	6.67	6.79
14	4.21	4.89	5.32	5.63	5.88	6.08	6.26	6.41	6.54	6.66
15	4.17	4.84	5.25	5.56	5.80	5.99	6.16	6.31	6.44	6.55
16	4.13	4.79	5.19	5.49	5.72	5.92	6.08	6.22	6.35	6.46
17	4.10	4.74	5.14	5.43	5.66	5.85	6.01	6.15	6.27	6.38
18	4.07	4.70	5.09	5.38	5.60	5.79	5.94	6.08	6.20	6.31
19	4.05	4.67	5.05	5.33	5.55	5.73	5.89	6.02	6.14	6.25
20	4.02	4.64	5.02	5.29	5.51	5.69	5.84	5.97	6.09	6.19
24	3.96	4.55	4.91	5.17	5.37	5.54	5.69	5.81	5.92	6.02
30	3.89	4.45	4.80	5.05	5.24	5.40	5.54	5.65	5.76	5.85
40	3.82	4.37	4.70	4.93	5.11	5.26	5.39	5.50	5.60	5.69
60	3.76	4.28	4.59	4.82	4.99	5.13	5.25	5.36	5.45	5.53
120	3.70	4.20	4.50	4.71	4.87	5.01	5.12	5.21	5.30	5.37
∞	3.64	4.12	4.40	4.60	4.76	4.88	4.99	5.08	5.16	5.23

(continued)

TABLE 14   Continued

ν \ p	12	13	14	15	16	17	18	19	20
1	260.0	266.2	271.8	277.0	281.8	286.3	290.0	294.3	298.0
2	33.40	34.13	34.81	35.43	36.00	36.53	37.03	37.50	37.95
3	17.53	17.89	18.22	18.52	18.81	19.07	19.32	19.55	19.77
4	12.84	13.09	13.32	13.53	13.73	13.91	14.08	14.24	14.40
5	10.70	10.89	11.08	11.24	11.40	11.55	11.68	11.81	11.93
6	9.48	9.65	9.81	9.95	10.08	10.21	10.32	10.43	10.54
7	8.71	8.86	9.00	9.12	9.24	9.35	9.46	9.55	9.65
8	8.18	8.31	8.44	8.55	8.66	8.76	8.85	8.94	9.03
9	7.78	7.91	8.03	8.13	8.23	8.33	8.41	8.49	8.57
10	7.49	7.60	7.71	7.81	7.91	7.99	8.08	8.15	8.23
11	7.25	7.36	7.46	7.56	7.65	7.73	7.81	7.88	7.95
12	7.06	7.17	7.26	7.36	7.44	7.52	7.59	7.66	7.73
13	6.90	7.01	7.10	7.19	7.27	7.35	7.42	7.48	7.55
14	6.77	6.87	6.96	7.05	7.13	7.20	7.27	7.33	7.39
15	6.66	6.76	6.84	6.93	7.00	7.07	7.14	7.20	7.26
16	6.56	6.66	6.74	6.82	6.90	6.97	7.03	7.09	7.15
17	6.48	6.57	6.66	6.73	6.81	6.87	6.94	7.00	7.05
18	6.41	6.50	6.58	6.65	6.72	6.79	6.85	6.91	6.97
19	6.34	6.43	6.51	6.58	6.65	6.72	6.78	6.84	6.89
20	6.28	6.37	6.45	6.52	6.59	6.65	6.71	6.77	6.82
24	6.11	6.19	6.26	6.33	6.39	6.45	6.51	6.56	6.61
30	5.93	6.01	6.08	6.14	6.20	6.26	6.31	6.36	6.41
40	5.76	5.83	5.90	5.96	6.02	6.07	6.12	6.16	6.21
60	5.60	5.67	5.73	5.78	5.84	5.89	5.93	5.97	6.01
120	5.44	5.50	5.56	5.61	5.66	5.71	5.75	5.79	5.83
∞	5.29	5.35	5.40	5.45	5.49	5.54	5.57	5.61	5.65

Source: *Biometrika Tables for Statisticians,* Vol. I, 3d ed., edited by E. S. Pearson and H. O. Hartley (Cambridge University Press, 1966). Reproduced by permission of Professor E. S. Pearson and the *Biometrika* Trustees.

TABLE 15 Critical Values of T_L and T_U for the Wilcoxon Rank Sum Test: Independent Samples

a. $\alpha = .025$ one-tailed; $\alpha = .05$ two-tailed

n_2 \ n_1	3		4		5		6		7		8		9		10	
	T_L	T_U	T_L	T_U	T_L	T_U	T_L	T_U	T_L	T_U	T_L	T_U	T_L	T_U	T_L	T_U
3	5	16	6	18	6	21	7	23	7	26	8	28	8	31	9	33
4	6	18	11	25	12	28	12	32	13	35	14	38	15	41	16	44
5	6	21	12	28	18	37	19	41	20	45	21	49	22	53	24	56
6	7	23	12	32	19	41	26	52	28	56	29	61	31	65	32	70
7	7	26	13	35	20	45	28	56	37	68	39	73	41	78	43	83
8	8	28	14	38	21	49	29	61	39	73	49	87	51	93	54	98
9	8	31	15	41	22	53	31	65	41	78	51	93	63	108	66	114
10	9	33	16	44	24	56	32	70	43	83	54	98	66	114	79	131

b. $\alpha = .05$ one-tailed; $\alpha = .10$ two-tailed

n_2 \ n_1	3		4		5		6		7		8		9		10	
	T_L	T_U	T_L	T_U	T_L	T_U	T_L	T_U	T_L	T_U	T_L	T_U	T_L	T_U	T_L	T_U
3	6	15	7	17	7	20	8	22	9	24	9	27	10	29	11	31
4	7	17	12	24	13	27	14	30	15	33	16	36	17	39	18	42
5	7	20	13	27	19	36	20	40	22	43	24	46	25	50	26	54
6	8	22	14	30	20	40	28	50	30	54	32	58	33	63	35	67
7	9	24	15	33	22	43	30	54	39	66	41	71	43	76	46	80
8	9	27	16	36	24	46	32	58	41	71	52	84	54	90	57	95
9	10	29	17	39	25	50	33	63	43	76	54	90	66	105	69	111
10	11	31	18	42	26	54	35	67	46	80	57	95	69	111	83	127

Source: From F. Wilcoxon and R. A. Wilcox, "Some Rapid Approximate Statistical Procedures," 1964, pp. 20–23. Reproduced with the permission of American Cyanamid Company.

TABLE 16

Critical Values of T_0 in the Wilcoxon Matched Pairs Signed Ranks Test

ONE-TAILED	TWO-TAILED	$n = 5$	$n = 6$	$n = 7$	$n = 8$	$n = 9$	$n = 10$
$\alpha = .05$	$\alpha = .10$	1	2	4	6	8	11
$\alpha = .025$	$\alpha = .05$		1	2	4	6	8
$\alpha = .01$	$\alpha = .02$			0	2	3	5
$\alpha = .005$	$\alpha = .01$				0	2	3
		$n = 11$	$n = 12$	$n = 13$	$n = 14$	$n = 15$	$n = 16$
$\alpha = .05$	$\alpha = .10$	14	17	21	26	30	36
$\alpha = .025$	$\alpha = .05$	11	14	17	21	25	30
$\alpha = .01$	$\alpha = .02$	7	10	13	16	20	24
$\alpha = .005$	$\alpha = .01$	5	7	10	13	16	19
		$n = 17$	$n = 18$	$n = 19$	$n = 20$	$n = 21$	$n = 22$
$\alpha = .05$	$\alpha = .10$	41	47	54	60	68	75
$\alpha = .025$	$\alpha = .05$	35	40	46	52	59	66
$\alpha = .01$	$\alpha = .02$	28	33	38	43	49	56
$\alpha = .005$	$\alpha = .01$	23	28	32	37	43	49
		$n = 23$	$n = 24$	$n = 25$	$n = 26$	$n = 27$	$n = 28$
$\alpha = .05$	$\alpha = .10$	83	92	101	110	120	130
$\alpha = .025$	$\alpha = .05$	73	81	90	98	107	117
$\alpha = .01$	$\alpha = .02$	62	69	77	85	93	102
$\alpha = .005$	$\alpha = .01$	55	61	68	76	84	92
		$n = 29$	$n = 30$	$n = 31$	$n = 32$	$n = 33$	$n = 34$
$\alpha = .05$	$\alpha = .10$	141	152	163	175	188	201
$\alpha = .025$	$\alpha = .05$	127	137	148	159	171	183
$\alpha = .01$	$\alpha = .02$	111	120	130	141	151	162
$\alpha = .005$	$\alpha = .01$	100	109	118	128	138	149
		$n = 35$	$n = 36$	$n = 37$	$n = 38$	$n = 39$	
$\alpha = .05$	$\alpha = .10$	214	228	242	256	271	
$\alpha = .025$	$\alpha = .05$	195	208	222	235	250	
$\alpha = .01$	$\alpha = .02$	174	186	198	211	224	
$\alpha = .005$	$\alpha = .01$	160	171	183	195	208	
		$n = 40$	$n = 41$	$n = 42$	$n = 43$	$n = 44$	$n = 45$
$\alpha = .05$	$\alpha = .10$	287	303	319	336	353	371
$\alpha = .025$	$\alpha = .05$	264	279	295	311	327	344
$\alpha = .01$	$\alpha = .02$	238	252	267	281	297	313
$\alpha = .005$	$\alpha = .01$	221	234	248	262	277	292
		$n = 46$	$n = 47$	$n = 48$	$n = 49$	$n = 50$	
$\alpha = .05$	$\alpha = .10$	389	408	427	446	466	
$\alpha = .025$	$\alpha = .05$	361	379	397	415	434	
$\alpha = .01$	$\alpha = .02$	329	345	362	380	398	
$\alpha = .005$	$\alpha = .01$	307	323	339	356	373	

Source: From F. Wilcoxon and R. A. Wilcox, "Some Rapid Approximate Statistical Procedures," 1964, p. 28. Reproduced with the permission of American Cyanamid Company.

TABLE 17
Critical Values of
Spearman's Rank
Correlation Coefficient

The α-values correspond to a one-tailed test of H_0: $\rho_S = 0$. The value should be doubled for two-tailed tests.

n	$\alpha = .05$	$\alpha = .025$	$\alpha = .01$	$\alpha = .005$
5	.900	—	—	—
6	.829	.886	.943	—
7	.714	.786	.893	—
8	.643	.738	.833	.881
9	.600	.683	.783	.833
10	.564	.648	.745	.794
11	.523	.623	.736	.818
12	.497	.591	.703	.780
13	.475	.566	.673	.745
14	.457	.545	.646	.716
15	.441	.525	.623	.689
16	.425	.507	.601	.666
17	.412	.490	.582	.645
18	.399	.476	.564	.625
19	.388	.462	.549	.608
20	.377	.450	.534	.591
21	.368	.438	.521	.576
22	.359	.428	.508	.562
23	.351	.418	.496	.549
24	.343	.409	.485	.537
25	.336	.400	.475	.526
26	.329	.392	.465	.515
27	.323	.385	.456	.505
28	.317	.377	.448	.496
29	.311	.370	.440	.487
30	.305	.364	.432	.478

Source: From E. G. Olds, "Distribution of Sums of Squares of Rank Differences for Small Samples," *Annals of Mathematical Statistics*, 1938, 9. Reproduced with the permission of the Institute of Mathematical Statistics.

TABLE 18 Factors Used When Constructing Control Charts

NUMBER OF OBSERVATIONS IN SAMPLE	CHART FOR AVERAGES FACTORS FOR CONTROL LIMITS			CHART FOR STANDARD DEVIATIONS FACTORS FOR CENTRAL LINE		FACTORS FOR CONTROL LIMITS				CHART FOR RANGES FACTORS FOR CENTRAL LINE		FACTORS FOR CONTROL LIMITS				
n	A	A_1	A_2	c_2	$1/c_2$	B_1	B_2	B_3	B_4	d_2	$1/d_2$	d_3	D_1	D_2	D_3	D_4
2	2.121	3.760	1.880	.5642	1.7725	0	1.843	0	3.267	1.128	.8865	.853	0	3.686	0	3.276
3	1.732	2.394	1.023	.7236	1.3820	0	1.858	0	2.568	1.693	.5907	.888	0	4.358	0	2.575
4	1.501	1.880	.729	.7979	1.2533	0	1.808	0	2.266	2.059	.4857	.880	0	4.698	0	2.282
5	1.342	1.596	.577	.8407	1.1894	0	1.756	0	2.089	2.326	.4299	.864	0	4.918	0	2.115
6	1.225	1.410	.483	.8686	1.1512	.026	1.711	.030	1.970	2.534	.3946	.848	0	5.078	0	2.004
7	1.134	1.277	.419	.8882	1.1259	.105	1.672	.118	1.882	2.704	.3698	.833	.205	5.203	.076	1.924
8	1.061	1.175	.373	.9027	1.1078	.167	1.638	.185	1.815	2.847	.3512	.820	.387	5.307	.136	1.864
9	1.000	1.094	.337	.9139	1.0942	.219	1.609	.239	1.761	2.970	.3367	.808	.546	5.394	.184	1.816
10	.949	1.028	.308	.9227	1.0837	.262	1.584	.284	1.716	3.078	.3249	.797	.687	5.469	.223	1.777
11	.905	.973	.285	.9300	1.0753	.299	1.561	.321	1.679	3.173	.3152	.787	.812	5.534	.256	1.744
12	.866	.925	.266	.9359	1.0684	.331	1.541	.354	1.646	3.258	.3069	.778	.924	5.592	.284	1.719
13	.832	.884	.249	.9410	1.0627	.359	1.523	.382	1.618	3.336	.2998	.770	1.026	5.646	.308	1.692
14	.802	.848	.235	.9453	1.0579	.384	1.507	.406	1.594	3.407	.2935	.762	1.121	5.693	.329	1.671
15	.775	.816	.223	.9490	1.0537	.406	1.492	.428	1.572	3.472	.2880	.755	1.207	5.737	.348	1.652
16	.750	.788	.212	.9523	1.0501	.427	1.478	.448	1.552	3.532	.2831	.749	1.285	5.779	.364	1.636
17	.728	.762	.203	.9551	1.0470	.445	1.465	.466	1.534	3.588	.2787	.743	1.359	5.817	.379	1.621
18	.707	.738	.194	.9576	1.0442	.461	1.454	.482	1.518	3.640	.2747	.738	1.426	5.854	.392	1.608
19	.688	.717	.187	.9599	1.0418	.477	1.443	.497	1.503	3.689	.2711	.733	1.490	5.888	.404	1.596
20	.671	.697	.180	.9619	1.0396	.491	1.433	.510	1.490	3.735	.2677	.729	1.548	5.922	.414	1.586
21	.655	.679	.173	.9638	1.0376	.504	1.424	.523	1.477	3.778	.2647	.724	1.606	5.950	.425	1.575
22	.640	.662	.167	.9655	1.0358	.516	1.415	.534	1.466	3.819	.2618	.720	1.659	5.979	.434	1.566
23	.626	.647	.162	.9670	1.0342	.527	1.407	.545	1.455	3.858	.2592	.716	1.710	6.006	.443	1.557
24	.612	.632	.157	.9684	1.0327	.538	1.399	.555	1.445	3.895	.2567	.712	1.759	6.031	.452	1.548
25	.600	.619	.153	.9696	1.0313	.548	1.392	.565	1.435	3.931	.2544	.709	1.804	6.058	.459	1.541
Over 25	$\dfrac{3}{\sqrt{n}}$	$\dfrac{3}{\sqrt{n}}$	—	—	—	a	b	a	b	—	—	—	—	—	—	—

[a] $1 - \dfrac{3}{\sqrt{2n}}$

[b] $1 + \dfrac{3}{\sqrt{2n}}$

Source: *ASTM Manual on Quality Control of Materials*, American Society for Testing Materials, Philadelphia, Pa., 1951. Copyright ASTM. Reprinted with permission.

TABLE 19

Values of K for Tolerance
Limits for Normal
Distributions

n	$1 - \alpha = .95$			$1 - \alpha = .99$		
γ	.90	.95	.99	.90	.95	.99
2	32.019	37.674	48.430	160.193	188.491	242.300
3	8.380	9.916	12.861	18.930	22.401	29.055
4	5.369	6.370	8.299	9.398	11.150	14.527
5	4.275	5.079	6.634	6.612	7.855	10.260
6	3.712	4.414	5.775	5.337	6.345	8.301
7	3.369	4.007	5.248	4.613	5.488	7.187
8	3.136	3.732	4.891	4.147	4.936	6.468
9	2.967	3.532	4.631	3.822	4.550	5.966
10	2.839	3.379	4.433	3.582	4.265	5.594
11	2.737	3.259	4.277	3.397	4.045	5.308
12	2.655	3.162	4.150	3.250	3.870	5.079
13	2.587	3.081	4.044	3.130	3.727	4.893
14	2.529	3.012	3.955	3.029	3.608	4.737
15	2.480	2.954	3.878	2.945	3.507	4.605
16	2.437	2.903	3.812	2.872	3.421	4.492
17	2.400	2.858	3.754	2.808	3.345	4.393
18	2.366	2.819	3.702	2.753	3.279	4.307
19	2.337	2.784	3.656	2.703	3.221	4.230
20	2.310	2.752	3.615	2.659	3.168	4.161
25	2.208	2.631	3.457	2.494	2.972	3.904
30	2.140	2.549	3.350	2.385	2.841	3.733
35	2.090	2.490	3.272	2.306	2.748	3.611
40	2.052	2.445	3.213	2.247	2.677	3.518
45	2.021	2.408	3.165	2.200	2.621	3.444
50	1.996	2.379	3.126	2.162	2.576	3.385
55	1.976	2.354	3.094	2.130	2.538	3.335
60	1.958	2.333	3.066	2.103	2.506	3.293
65	1.943	2.315	3.042	2.080	2.478	3.257
70	1.929	2.299	3.021	2.060	2.454	3.225
75	1.917	2.285	3.002	2.042	2.433	3.197
80	1.907	2.272	2.986	2.026	2.414	3.173
85	1.897	2.261	2.971	2.012	2.397	3.150
90	1.889	2.251	2.958	1.999	2.382	3.130
95	1.881	2.241	2.945	1.987	2.368	3.112
100	1.874	2.233	2.934	1.977	2.355	3.096
150	1.825	2.175	2.859	1.905	2.270	2.983
200	1.798	2.143	2.816	1.865	2.222	2.921
250	1.780	2.121	2.788	1.839	2.191	2.880
300	1.767	2.106	2.767	1.820	2.169	2.850
400	1.749	2.084	2.739	1.794	2.138	2.809
500	1.737	2.070	2.721	1.777	2.117	2.783
600	1.729	2.060	2.707	1.764	2.102	2.763
700	1.722	2.052	2.697	1.755	2.091	2.748
800	1.717	2.046	2.688	1.747	2.082	2.736
900	1.712	2.040	2.682	1.741	2.075	2.726
1000	1.709	2.036	2.676	1.736	2.068	2.718
∞	1.645	1.960	2.576	1.645	1.960	2.576

Source: From *Techniques of Statistical Analysis* by C. Eisenhart, M. W. Hastay, and W. A. Wallis. Copyright 1947, McGraw-Hill Book Company, Inc. Reproduced with permission of McGraw-Hill.

TABLE 20

Sample Size *n* for Nonparametric Tolerance Limits

γ	$1 - \alpha$					
	.50	.70	.90	.95	.99	.995
.995	336	488	777	947	1,325	1,483
.99	168	244	388	473	662	740
.95	34	49	77	93	130	146
.90	17	24	38	46	64	72
.85	11	16	25	30	42	47
.80	9	12	18	22	31	34
.75	7	10	15	18	24	27
.70	6	8	12	14	20	22
.60	4	6	9	10	14	16
.50	3	5	7	8	11	12

Source: Tables A-25d of Wilfrid J. Dixon and Frank J. Massey, Jr., *Introduction to Statistical Analysis*, 3rd ed., McGraw-Hill Book Company, New York, 1969. Used with permission of McGraw-Hill Book Company.

TABLE 21

Sample Size Code Letters: MIL-STD-105D

LOT OR BATCH SIZE	SPECIAL INSPECTION LEVELS				GENERAL INSPECTION LEVELS		
	S-1	S-2	S-3	S-4	I	II	III
2–8	A	A	A	A	A	A	B
9–15	A	A	A	A	A	B	C
16–25	A	A	B	B	B	C	D
26–50	A	B	B	C	C	D	E
51–90	B	B	C	C	C	E	F
91–150	B	B	C	D	D	F	G
151–280	B	C	D	E	E	G	H
281–500	B	C	D	E	F	H	J
501–1,200	C	C	E	F	G	J	K
1,201–3,200	C	D	E	G	H	K	L
3,201–10,000	C	D	F	G	J	L	M
10,001–35,000	C	D	F	H	K	M	N
35,001–150,000	D	E	G	J	L	N	P
150,001–500,000	D	E	G	J	M	P	Q
500,001 and over	D	E	H	K	N	Q	R

TABLE 22   A Portion of the Master Table for Normal Inspection (Single Sampling): MIL–STD–105D

ACCEPTABLE QUALITY LEVELS (NORMAL INSPECTION)

Sample Size Code Letter	Sample Size	.010	.015	.025	.040	.065	.10	.15	.25	.40	.65	1.0	1.5	2.5	4.0	6.5	10	15	25	40	65
A	2	↓	↓	↓	↓	↓	↓	↓	↓	↓	↓	↓	↓	↓	↓	↓	↓	0 1	1 2	2 3	3 4
B	3	↓	↓	↓	↓	↓	↓	↓	↓	↓	↓	↓	↓	↓	↓	↓	0 1	1 2	2 3	3 4	5 6
C	5	↓	↓	↓	↓	↓	↓	↓	↓	↓	↓	↓	↓	↓	↓	0 1	1 2	2 3	3 4	5 6	7 8
D	8	↓	↓	↓	↓	↓	↓	↓	↓	↓	↓	↓	↓	↓	0 1	1 2	2 3	3 4	5 6	7 8	10 11
E	13	↓	↓	↓	↓	↓	↓	↓	↓	↓	↓	↓	↓	0 1	1 2	2 3	3 4	5 6	7 8	10 11	14 15
F	20	↓	↓	↓	↓	↓	↓	↓	↓	↓	↓	↓	0 1	1 2	2 3	3 4	5 6	7 8	10 11	14 15	21 22
G	32	↓	↓	↓	↓	↓	↓	↓	↓	↓	↓	0 1	1 2	2 3	3 4	5 6	7 8	10 11	14 15	21 22	↑
H	50	↓	↓	↓	↓	↓	↓	↓	↓	↓	0 1	1 2	2 3	3 4	5 6	7 8	10 11	14 15	21 22	↑	↑
J	80	↓	↓	↓	↓	↓	↓	↓	↓	0 1	1 2	2 3	3 4	5 6	7 8	10 11	14 15	21 22	↑	↑	↑
K	125	↓	↓	↓	↓	↓	↓	↓	0 1	1 2	2 3	3 4	5 6	7 8	10 11	14 15	21 22	↑	↑	↑	↑
L	200	↓	↓	↓	↓	↓	↓	0 1	1 2	2 3	3 4	5 6	7 8	10 11	14 15	21 22	↑	↑	↑	↑	↑
M	315	↓	↓	↓	↓	↓	0 1	1 2	2 3	3 4	5 6	7 8	10 11	14 15	21 22	↑	↑	↑	↑	↑	↑
N	500	↓	↓	↓	↓	0 1	1 2	2 3	3 4	5 6	7 8	10 11	14 15	21 22	↑	↑	↑	↑	↑	↑	↑
P	800	↓	↓	↓	0 1	1 2	2 3	3 4	5 6	7 8	10 11	14 15	21 22	↑	↑	↑	↑	↑	↑	↑	↑
Q	1,250	↓	↓	0 1	1 2	2 3	3 4	5 6	7 8	10 11	14 15	21 22	↑	↑	↑	↑	↑	↑	↑	↑	↑
R	2,000	↓	0 1	1 2	2 3	3 4	5 6	7 8	10 11	14 15	21 22	↑	↑	↑	↑	↑	↑	↑	↑	↑	↑

↓ = Use first sampling plan below arrow. If sample size equals or exceeds lot or batch size, do 100% inspection.
↑ = Use first sampling plan above arrow.
Ac = Acceptance number
Re = Rejection number

APPENDIX III

DDT ANALYSES ON FISH SAMPLES, TENNESSEE RIVER, ALABAMA

DDT ANALYSES
ON FISH SAMPLES COLLECTED SUMMER 1980
TENNESSEE RIVER, ALABAMA

OBS	LOCATION	SPECIES	LENGTH	WEIGHT	DDT
1	FCM5	CHANNEL CATFISH	42.5	732	10.00
2	FCM5	CHANNEL CATFISH	44.0	795	16.00
3	FCM5	CHANNEL CATFISH	41.5	547	23.00
4	FCM5	CHANNEL CATFISH	39.0	465	21.00
5	FCM5	CHANNEL CATFISH	50.5	1252	50.00
6	FCM5	CHANNEL CATFISH	52.0	1255	150.00
7	LCM3	CHANNEL CATFISH	40.5	741	28.00
8	LCM3	CHANNEL CATFISH	48.0	1151	7.70
9	LCM3	CHANNEL CATFISH	48.0	1186	2.00
10	LCM3	CHANNEL CATFISH	43.5	754	19.00
11	LCM3	CHANNEL CATFISH	40.5	679	16.00
12	LCM3	CHANNEL CATFISH	47.5	985	5.40
13	SCM1	CHANNEL CATFISH	44.5	1133	2.60
14	SCM1	CHANNEL CATFISH	46.0	1139	3.10
15	SCM1	CHANNEL CATFISH	48.0	1186	3.50
16	SCM1	CHANNEL CATFISH	45.0	984	9.10
17	SCM1	CHANNEL CATFISH	43.0	965	7.80
18	SCM1	CHANNEL CATFISH	45.0	1084	4.10
19	TRM275	CHANNEL CATFISH	48.0	986	8.40
20	TRM275	CHANNEL CATFISH	45.0	1023	15.00
21	TRM275	CHANNEL CATFISH	49.0	1266	25.00
22	TRM275	CHANNEL CATFISH	50.0	1086	5.60
23	TRM275	CHANNEL CATFISH	46.0	1044	4.60
24	TRM275	CHANNEL CATFISH	52.0	1770	8.20
25	TRM280	CHANNEL CATFISH	48.0	1048	6.10
26	TRM280	CHANNEL CATFISH	51.0	1641	13.00
27	TRM280	CHANNEL CATFISH	48.5	1331	6.00
28	TRM280	CHANNEL CATFISH	51.0	1728	6.60
29	TRM280	CHANNEL CATFISH	44.0	917	5.50
30	TRM280	CHANNEL CATFISH	51.0	1398	11.00
31	TRM280	SMALL MOUTH BUFFALO	49.0	1763	4.50
32	TRM280	SMALL MOUTH BUFFALO	46.0	1459	4.20
33	TRM280	SMALL MOUTH BUFFALO	52.0	2302	3.00
34	TRM280	SMALL MOUTH BUFFALO	46.0	1614	2.30
35	TRM280	SMALL MOUTH BUFFALO	46.0	1444	2.50
36	TRM280	SMALL MOUTH BUFFALO	48.0	2006	6.80
37	TRM285	CHANNEL CATFISH	44.0	936	19.00
38	TRM285	CHANNEL CATFISH	42.0	1058	7.20
39	TRM285	CHANNEL CATFISH	42.5	800	6.00
40	TRM285	CHANNEL CATFISH	45.5	1087	10.00
41	TRM285	CHANNEL CATFISH	48.0	1329	12.00
42	TRM285	CHANNEL CATFISH	44.0	897	2.80
43	TRM285	LARGE MOUTH BASS	28.5	778	0.48
44	TRM285	LARGE MOUTH BASS	26.0	532	0.18
45	TRM285	LARGE MOUTH BASS	25.5	441	0.34
46	TRM285	LARGE MOUTH BASS	25.0	544	0.11
47	TRM285	LARGE MOUTH BASS	23.0	393	0.22
48	TRM285	LARGE MOUTH BASS	28.0	733	0.80
49	TRM290	CHANNEL CATFISH	41.0	961	8.70
50	TRM290	CHANNEL CATFISH	44.0	886	22.00
51	TRM290	CHANNEL CATFISH	41.0	678	13.00
52	TRM290	CHANNEL CATFISH	42.0	1011	3.50
53	TRM290	CHANNEL CATFISH	42.5	947	9.30
54	TRM290	CHANNEL CATFISH	44.0	989	21.00
55	TRM290	SMALL MOUTH BUFFALO	43.5	1291	3.40
56	TRM290	SMALL MOUTH BUFFALO	46.5	1186	13.00
57	TRM290	SMALL MOUTH BUFFALO	43.0	1293	5.60
58	TRM290	SMALL MOUTH BUFFALO	47.0	1709	12.00
59	TRM290	SMALL MOUTH BUFFALO	46.0	1425	21.00
60	TRM290	SMALL MOUTH BUFFALO	41.0	1176	8.00
61	TRM295	CHANNEL CATFISH	36.0	980	12.00
62	TRM295	CHANNEL CATFISH	47.5	1176	6.00
63	TRM295	CHANNEL CATFISH	41.5	989	4.70
64	TRM295	CHANNEL CATFISH	49.5	1084	31.00
65	TRM295	CHANNEL CATFISH	46.0	1115	5.20
66	TRM295	CHANNEL CATFISH	46.5	724	27.00
67	TRM300	CHANNEL CATFISH	36.0	847	18.00
68	TRM300	CHANNEL CATFISH	37.0	876	7.50
69	TRM300	CHANNEL CATFISH	35.0	844	3.00
70	TRM300	CHANNEL CATFISH	36.0	908	13.00
71	TRM300	CHANNEL CATFISH	48.0	1358	7.30
72	TRM300	CHANNEL CATFISH	49.0	1019	15.00
73	TRM300	SMALL MOUTH BUFFALO	35.5	1300	1.30
74	TRM300	SMALL MOUTH BUFFALO	46.0	1365	4.80
75	TRM300	SMALL MOUTH BUFFALO	45.0	1437	5.10
76	TRM300	SMALL MOUTH BUFFALO	44.5	1460	5.10
77	TRM300	SMALL MOUTH BUFFALO	49.0	1671	4.00
78	TRM300	SMALL MOUTH BUFFALO	47.5	1717	10.00
79	TRM305	CHANNEL CATFISH	35.0	613	12.00
80	TRM305	CHANNEL CATFISH	51.0	353	22.00
81	TRM305	CHANNEL CATFISH	42.5	909	10.00
82	TRM305	CHANNEL CATFISH	38.0	886	11.00
83	TRM305	CHANNEL CATFISH	41.0	890	17.00
84	TRM305	CHANNEL CATFISH	47.0	1031	9.70
85	TRM310	CHANNEL CATFISH	45.0	1083	12.00
86	TRM310	CHANNEL CATFISH	45.5	864	4.70
87	TRM310	CHANNEL CATFISH	45.0	886	6.00
88	TRM310	CHANNEL CATFISH	45.0	965	3.80
89	TRM310	CHANNEL CATFISH	39.0	537	17.00
90	TRM310	CHANNEL CATFISH	40.5	630	12.00
91	TRM310	SMALL MOUTH BUFFALO	46.0	1486	1.40
92	TRM310	SMALL MOUTH BUFFALO	47.0	1743	6.10
93	TRM310	SMALL MOUTH BUFFALO	48.5	2061	2.80
94	TRM310	SMALL MOUTH BUFFALO	48.0	1707	4.80
95	TRM310	SMALL MOUTH BUFFALO	38.0	862	5.70
96	TRM310	SMALL MOUTH BUFFALO	38.5	911	3.30
97	TRM315	CHANNEL CATFISH	29.5	476	3.30
98	TRM315	CHANNEL CATFISH	42.0	743	3.70
99	TRM315	CHANNEL CATFISH	47.5	1128	9.90
100	TRM315	CHANNEL CATFISH	43.5	848	6.80
101	TRM315	CHANNEL CATFISH	47.5	1091	13.00

DDT ANALYSES
ON FISH SAMPLES COLLECTED SUMMER 1980
TENNESSEE RIVER, ALABAMA

OBS	LOCATION	SPECIES	LENGTH	WEIGHT	DDT
102	TRM315	CHANNEL CATFISH	43.5	715	8.80
103	TRM320	CHANNEL CATFISH	47.5	983	57.00
104	TRM320	CHANNEL CATFISH	51.5	1251	96.00
105	TRM320	CHANNEL CATFISH	49.5	1255	360.00
106	TRM320	CHANNEL CATFISH	47.0	1152	130.00
107	TRM320	CHANNEL CATFISH	47.5	1085	13.00
108	TRM320	CHANNEL CATFISH	47.0	1118	61.00
109	TRM320	SMALL MOUTH BUFFALO	36.0	1285	12.00
110	TRM320	SMALL MOUTH BUFFALO	34.5	1178	33.00
111	TRM320	SMALL MOUTH BUFFALO	44.5	1492	48.00
112	TRM320	SMALL MOUTH BUFFALO	46.0	1524	10.00
113	TRM320	SMALL MOUTH BUFFALO	46.0	1473	44.00
114	TRM320	SMALL MOUTH BUFFALO	32.5	520	0.43
115	TRM325	CHANNEL CATFISH	46.0	863	1100.00
116	TRM325	CHANNEL CATFISH	40.0	549	9.40
117	TRM325	CHANNEL CATFISH	43.5	810	4.10
118	TRM325	CHANNEL CATFISH	46.5	908	2.80
119	TRM325	CHANNEL CATFISH	43.0	804	0.74
120	TRM325	CHANNEL CATFISH	47.5	1179	14.00
121	TRM330	CHANNEL CATFISH	32.0	556	22.00
122	TRM330	CHANNEL CATFISH	40.5	659	9.10
123	TRM330	CHANNEL CATFISH	51.5	1229	140.00
124	TRM330	CHANNEL CATFISH	48.0	1050	4.20
125	TRM330	CHANNEL CATFISH	47.0	952	12.00
126	TRM330	CHANNEL CATFISH	41.0	826	2.00
127	TRM330	SMALL MOUTH BUFFALO	33.5	599	0.30
128	TRM330	SMALL MOUTH BUFFALO	47.0	1704	1.20
129	TRM340	CHANNEL CATFISH	50.0	1207	7.10
130	TRM340	CHANNEL CATFISH	45.0	911	180.00
131	TRM340	CHANNEL CATFISH	49.0	1498	1.50
132	TRM340	CHANNEL CATFISH	49.5	1496	2.40
133	TRM340	CHANNEL CATFISH	50.0	1142	4.30
134	TRM340	CHANNEL CATFISH	45.0	879	3.90
135	TRM340	SMALL MOUTH BUFFALO	32.5	525	0.99
136	TRM340	SMALL MOUTH BUFFALO	38.0	806	0.45
137	TRM340	SMALL MOUTH BUFFALO	38.5	694	2.50
138	TRM340	SMALL MOUTH BUFFALO	36.0	643	0.25
139	TRM345	LARGE MOUTH BASS	26.5	514	0.58
140	TRM345	LARGE MOUTH BASS	23.5	358	2.00
141	TRM345	LARGE MOUTH BASS	30.0	856	2.20
142	TRM345	LARGE MOUTH BASS	29.0	793	7.40
143	TRM345	LARGE MOUTH BASS	17.5	173	0.35
144	TRM345	LARGE MOUTH BASS	36.0	1433	1.90

CENTRAL PROCESSING UNIT (CPU) TIMES OF 1,000 COMPUTER JOBS

APPENDIX IV   CENTRAL PROCESSING UNIT (CPU) TIMES OF 1,000 COMPUTER JOBS

OBS	CPUTIME	OBS	CPUTIME	OBS	CPUTIME	OBS	CPUTIME	OBS	CPUTIME	OBS	CPUTIME	OBS	CPUTIME	OBS	CPUTIME	OBS	CPUTIME	OBS	CPUTIME
1	1.86	2	3.49	3	2.63	4	3.49	5	1.69	6	1.83	7	0.81	8	4.70	9	0.85	10	4.24
11	3.49	12	2.75	13	1.65	14	0.92	15	0.62	16	0.41	17	3.23	18	4.13	19	3.23	20	1.89
21	2.66	22	3.52	23	2.39	24	1.60	25	1.88	26	0.36	27	11.85	28	0.87	29	3.10	30	0.70
31	3.23	32	2.64	33	1.69	34	0.41	35	3.29	36	1.81	37	1.45	38	0.79	39	0.38	40	0.45
41	3.49	42	2.59	43	1.78	44	4.92	45	0.48	46	1.66	47	1.86	48	3.23	49	0.43	50	1.27
51	1.29	52	1.29	53	3.20	54	4.92	55	3.01	56	2.40	57	4.85	58	11.72	59	1.71	60	15.99
61	16.46	62	16.90	63	16.87	64	17.28	65	16.87	66	17.65	67	17.38	68	17.55	69	16.25	70	3.30
71	15.47	72	3.55	73	2.79	74	4.12	75	4.50	76	2.56	77	4.31	78	4.14	79	1.17	80	3.75
81	3.80	82	3.18	83	1.59	84	5.05	85	5.19	86	5.02	87	5.02	88	5.02	89	4.99	90	3.11
91	3.31	92	3.30	93	4.91	94	1.46	95	2.53	96	2.08	97	4.38	98	2.30	99	3.34	100	4.09
101	3.00	102	2.63	103	5.10	104	5.07	105	3.20	106	3.32	107	5.02	108	3.21	109	3.36	110	3.40
111	2.61	112	3.73	113	4.50	114	1.35	115	4.47	116	2.78	117	2.85	118	6.64	119	2.69	120	3.02
121	5.10	122	5.53	123	8.13	124	7.51	125	7.43	126	8.89	127	6.34	128	8.84	129	0.40	130	0.66
131	4.98	132	13.19	133	2.41	134	1.66	135	1.68	136	5.01	137	3.55	138	14.00	139	18.90	140	5.02
141	6.67	142	2.54	143	3.24	144	5.83	145	9.23	146	6.57	147	1.23	148	8.43	149	2.82	150	3.91
151	3.19	152	4.42	153	15.11	154	3.63	155	2.86	156	7.52	157	2.03	158	1.95	159	6.87	160	2.83
161	6.34	162	6.98	163	2.97	164	6.47	165	7.64	166	8.94	167	4.49	168	4.55	169	0.75	170	4.22
171	1.79	172	13.27	173	16.02	174	1.96	175	0.75	176	4.22	177	0.80	178	0.82	179	19.51	180	5.24
181	10.07	182	6.02	183	3.64	184	17.56	185	9.57	186	3.19	187	6.33	188	10.88	189	3.94	190	11.60
191	18.39	192	18.17	193	11.59	194	10.21	195	10.55	196	11.56	197	4.23	198	11.52	199	2.96	200	7.52
201	11.34	202	2.04	203	6.51	204	11.34	205	3.70	206	2.24	207	5.94	208	5.66	209	5.65	210	6.27
211	5.71	212	6.09	213	7.86	214	3.59	215	5.27	216	3.72	217	10.24	218	11.09	219	13.23	220	16.64
221	6.19	222	4.01	223	4.00	224	10.57	225	13.35	226	19.96	227	1.58	228	16.83	229	6.24	230	8.13
231	3.49	232	1.57	233	5.86	234	3.17	235	9.02	236	7.06	237	2.49	238	6.32	239	2.59	240	19.25
241	3.81	242	10.46	243	1.00	244	1.77	245	10.56	246	7.16	247	2.18	248	1.10	249	1.87	250	7.90
251	11.51	252	1.78	253	1.78	254	2.59	255	3.04	256	2.53	257	3.28	258	4.25	259	4.24	260	4.16
261	3.93	262	12.64	263	20.78	264	20.79	265	5.10	266	5.24	267	5.25	268	8.66	269	11.00	270	8.48
271	8.02	272	16.94	273	15.97	274	3.19	275	3.24	276	1.81	277	0.64	278	4.67	279	1.02	280	7.73
281	3.34	282	11.77	283	15.97	284	3.76	285	2.41	286	1.61	287	1.82	288	4.25	289	3.17	290	2.23
291	12.19	292	3.19	293	0.62	294	2.16	295	1.94	296	1.71	297	0.47	298	1.65	299	6.54	300	4.20
301	0.84	302	2.59	303	3.13	304	0.48	305	2.63	306	2.96	307	0.99	308	0.85	309	5.33	310	1.83
311	2.02	312	1.01	313	3.68	314	1.88	315	0.97	316	3.28	317	0.52	318	0.68	319	2.15	320	3.17
321	3.43	322	9.45	323	3.19	324	17.44	325	4.23	326	0.55	327	3.12	328	0.51	329	0.60	330	1.92
331	1.97	332	2.67	333	3.16	334	0.45	335	1.59	336	1.74	337	1.22	338	0.61	339	0.59	340	1.53
341	0.96	342	0.97	343	3.21	344	0.23	345	5.27	346	3.14	347	0.75	348	0.68	349	2.07	350	13.38
351	0.18	352	3.23	353	2.20	354	1.70	355	6.23	356	0.68	357	10.67	358	3.15	359	2.11	360	2.22
361	0.24	362	3.21	363	3.15	364	11.39	365	0.48	366	3.00	367	17.30	368	3.45	369	7.51	370	8.95
371	7.82	372	8.74	373	11.35	374	12.18	375	13.10	376	3.17	377	14.19	378	10.55	379	14.43	380	14.35
381	11.81	382	3.27	383	15.51	384	14.66	385	16.33	386	16.75	387	15.55	388	8.01	389	3.16	390	2.17
391	15.57	392	14.85	393	15.88	394	1.86	395	3.17	396	15.76	397	0.33	398	20.09	399	9.25	400	0.55
401	2.79	402	2.26	403	4.41	404	3.16	405	2.14	406	0.51	407	0.53	408	5.31	409	0.10	410	3.41
411	3.45	412	7.22	413	2.77	414	2.00	415	0.89	416	1.69	417	2.73	418	4.08	419	6.29	420	12.87
421	3.12	422	0.64	423	2.01	424	2.69	425	2.70	426	3.15	427	1.57	428	1.54	429	0.46	430	1.55
431	2.45	432	0.46	433	0.47	434	4.17	435	4.09	436	0.56	437	0.15	438	0.81	439	2.72	440	4.08
441	6.45	442	11.60	443	12.96	444	2.78	445	2.74	446	4.06	447	6.30	448	12.86	449	12.86	450	3.26
451	2.98	452	0.11	453	1.10	454	1.42	455	1.74	456	2.93	457	0.84	458	2.41	459	2.51	460	2.16
461	1.67	462	3.29	463	1.14	464	1.15	465	1.22	466	1.54	467	0.54	468	1.49	469	2.67	470	2.46
471	1.46	472	2.61	473	2.63	474	2.63	475	3.53	476	5.04	477	3.50	478	4.98	479	3.65	480	3.54
481	1.17	482	3.49	483	3.50	484	3.54	485	3.52	486	3.55	487	3.12	488	3.09	489	4.61	490	2.70
491	3.80	492	7.10	493	3.35	494	3.01	495	3.07	496	3.19	497	4.75	498	2.62	499	4.17	500	4.98

OBS	CPUTIME	OBS	CPUTIME	OBS	CPUTIME	OBS	CPUTIME	OBS	CPUTIME	OBS	CPUTIME	OBS	CPUTIME	OBS	CPUTIME	OBS	CPUTIME	OBS	CPUTIME
501	5.11	502	4.05	503	3.92	504	1.52	505	4.53	506	3.28	507	3.63	508	1.63	509	1.59	510	1.56
511	3.61	512	3.53	513	4.31	514	4.83	515	1.97	516	5.09	517	5.48	518	5.39	519	5.43	520	5.27
521	5.35	522	2.01	523	2.00	524	2.02	525	5.09	526	5.08	527	5.08	528	7.28	529	1.16	530	3.08
531	2.22	532	5.38	533	19.61	534	0.63	535	3.01	536	9.98	537	15.92	538	17.45	539	7.64	540	17.22
541	5.00	542	18.97	543	6.51	544	1.82	545	0.54	546	2.68	547	9.79	548	10.68	549	10.41	550	20.47
551	10.86	552	9.89	553	10.43	554	9.39	555	3.63	556	3.61	557	10.59	558	9.15	559	9.48	560	1.12
561	2.19	562	3.13	563	2.15	564	1.55	565	1.81	566	1.77	567	1.79	568	1.74	569	2.59	570	2.61
571	3.93	572	3.89	573	4.26	574	6.94	575	2.11	576	1.43	577	2.63	578	2.85	579	1.52	580	1.40
581	2.85	582	2.08	583	3.22	584	2.65	585	4.08	586	5.52	587	5.04	588	6.73	589	3.47	590	1.90
591	4.70	592	5.57	593	7.19	594	7.26	595	12.04	596	5.77	597	2.84	598	2.78	599	1.05	600	0.88
601	3.15	602	1.78	603	1.93	604	1.83	605	3.11	606	3.75	607	11.93	608	0.43	609	2.33	610	3.28
611	3.86	612	2.52	613	8.12	614	17.61	615	4.29	616	2.20	617	4.43	618	5.44	619	4.64	620	6.30
621	7.55	622	11.75	623	0.34	624	7.87	625	2.05	626	5.28	627	0.28	628	8.09	629	1.79	630	1.08
631	2.70	632	0.33	633	1.86	634	6.88	635	5.90	636	2.67	637	1.48	638	8.70	639	3.78	640	2.73
641	3.44	642	7.69	643	8.61	644	0.68	645	8.53	646	8.60	647	3.69	648	3.46	649	3.38	650	8.73
651	0.18	652	9.16	653	9.39	654	10.88	655	11.03	656	11.16	657	11.18	658	5.50	659	12.22	660	0.25
661	3.11	662	10.82	663	11.62	664	3.43	665	11.68	666	16.39	667	0.29	668	3.94	669	2.00	670	0.28
671	4.08	672	3.26	673	3.15	674	0.26	675	2.28	676	2.49	677	2.00	678	2.26	679	0.30	680	0.28
681	2.35	682	3.73	683	2.02	684	1.28	685	3.83	686	0.29	687	2.64	688	0.29	689	0.27	690	2.74
691	0.10	692	0.19	693	1.05	694	5.54	695	2.00	696	3.88	697	3.14	698	0.38	699	0.96	700	9.95
701	0.39	702	1.78	703	0.61	704	6.66	705	0.85	706	1.67	707	2.03	708	3.89	709	4.34	710	4.88
711	4.09	712	11.05	713	5.11	714	1.75	715	0.06	716	3.12	717	1.57	718	0.12	719	2.72	720	1.76
721	1.32	722	0.62	723	1.60	724	6.79	725	2.28	726	1.67	727	1.38	728	1.08	729	10.09	730	3.31
731	3.36	732	3.36	733	3.36	734	7.28	735	2.38	736	2.40	737	4.88	738	0.28	739	3.49	740	0.27
741	1.99	742	0.27	743	12.38	744	1.06	745	2.28	746	1.75	747	2.01	748	4.88	749	0.53	750	0.28
751	0.27	752	1.99	753	0.36	754	6.54	755	2.39	756	2.07	757	18.60	758	0.67	759	8.90	760	4.07
761	0.65	762	2.87	763	7.71	764	7.62	765	2.45	766	14.23	767	10.12	768	5.24	769	8.72	770	13.75
771	6.32	772	16.15	773	6.57	774	16.26	775	8.89	776	16.27	777	10.71	778	5.14	779	4.51	780	6.03
781	2.74	782	3.16	783	3.25	784	8.17	785	15.82	786	18.98	787	5.06	788	12.97	789	19.55	790	5.77
791	16.97	792	0.34	793	4.88	794	7.03	795	15.30	796	19.53	797	7.45	798	12.26	799	19.37	800	7.18
801	5.28	802	2.24	803	17.70	804	8.39	805	8.96	806	3.76	807	1.40	808	3.04	809	3.35	810	2.67
811	3.16	812	4.67	813	6.69	814	4.89	815	10.56	816	6.90	817	6.23	818	1.81	819	1.85	820	2.22
821	2.23	822	4.21	823	4.68	824	2.73	825	2.12	826	1.10	827	1.09	828	1.11	829	1.33	830	1.29
831	4.93	832	2.76	833	2.85	834	2.75	835	0.77	836	5.55	837	0.54	838	3.55	839	2.01	840	1.58
841	0.55	842	10.54	843	0.54	844	1.57	845	1.00	846	1.81	847	2.05	848	4.83	849	1.41	850	0.64
851	0.61	852	2.37	853	1.46	854	0.60	855	4.07	856	0.61	857	3.05	858	3.96	859	0.99	860	0.81
861	1.49	862	1.53	863	0.65	864	2.05	865	0.46	866	1.79	867	1.55	868	0.82	869	0.10	870	0.34
871	0.97	872	0.41	873	0.86	874	13.85	875	0.93	876	1.65	877	1.88	878	1.63	879	0.67	880	1.14
881	2.77	882	0.76	883	0.64	884	1.96	885	7.57	886	1.52	887	3.44	888	0.81	889	1.80	890	4.80
891	2.45	892	1.89	893	7.64	894	2.92	895	4.76	896	5.72	897	3.89	898	4.16	899	3.97	900	4.59
901	7.41	902	7.52	903	7.59	904	8.72	905	4.52	906	0.50	907	2.08	908	0.52	909	2.04	910	0.47
911	9.35	912	7.52	913	7.58	914	3.21	915	1.15	916	1.65	917	1.52	918	3.48	919	6.41	920	0.84
921	1.41	922	2.65	923	4.54	924	7.11	925	1.69	926	2.40	927	2.19	928	0.18	929	0.14	930	2.20
931	5.21	932	2.25	933	0.18	934	0.22	935	14.23	936	18.20	937	0.20	938	2.06	939	0.19	940	9.91
941	0.16	942	0.18	943	2.19	944	0.40	945	3.19	946	0.47	947	9.17	948	0.26	949	9.95	950	0.23
951	2.34	952	0.21	953	0.24	954	2.75	955	0.21	956	0.44	957	2.19	958	3.15	959	3.18	960	0.47
961	3.20	962	11.43	963	0.20	964	0.20	965	0.45	966	2.14	967	2.49	968	0.24	969	3.14	970	0.19
971	0.45	972	5.03	973	2.06	974	0.36	975	6.19	976	0.45	977	2.17	978	3.18	979	0.20	980	0.19
981	0.22	982	11.90	983	8.44	984	2.24	985	2.47	986	12.33	987	3.84	988	8.52	989	6.27	990	2.19
991	2.02	992	2.82	993	2.19	994	5.13	995	2.43	996	0.23	997	0.43	998	6.22	999	3.98	1000	2.05

PERCENTAGE IRON CONTENT FOR 390 IRON ORE SPECIMENS

Number	% Iron	Number	% Iron	Number	% Iron	Number	% Iron	Number	% Iron
1	66.08	43	66.18	85	64.41	127	66.75	169	65.62
2	65.92	44	66.11	86	64.01	128	66.56	170	65.96
3	65.83	45	66.03	87	64.39	129	66.83	171	66.46
4	65.81	46	66.11	88	64.89	130	66.34	172	66.34
5	65.77	47	66.28	89	64.89	131	66.78	173	65.98
6	65.83	48	66.09	90	65.09	132	66.67	174	65.69
7	65.82	49	66.08	91	64.02	133	66.86	175	65.83
8	66.06	50	66.24	92	64.82	134	66.82	176	66.28
9	65.85	51	66.01	93	64.35	135	66.82	177	66.05
10	65.75	52	66.33	94	64.86	136	66.86	178	66.18
11	66.02	53	66.59	95	64.51	137	66.68	179	65.83
12	65.81	54	65.82	96	64.95	138	66.80	180	66.09
13	65.85	55	66.21	97	65.48	139	66.86	181	66.29
14	65.66	56	66.32	98	65.24	140	66.81	182	66.04
15	65.97	57	66.10	99	64.93	141	66.69	183	66.36
16	65.80	58	65.77	100	65.19	142	66.68	184	66.16
17	65.97	59	65.86	101	65.30	143	66.71	185	66.37
18	66.08	60	65.80	102	64.95	144	66.34	186	66.25
19	65.79	61	66.07	103	64.96	145	66.18	187	66.86
20	66.13	62	65.94	104	64.98	146	65.78	188	66.16
21	66.23	63	66.16	105	65.04	147	65.63	189	66.09
22	65.81	64	65.86	106	65.73	148	66.81	190	66.36
23	65.99	65	65.79	107	66.18	149	65.80	191	66.29
24	66.18	66	65.77	108	65.61	150	65.80	192	66.57
25	65.97	67	65.69	109	66.53	151	66.54	193	66.50
26	66.30	68	65.87	110	65.48	152	66.51	194	66.31
27	66.23	69	65.88	111	65.75	153	63.95	195	65.81
28	65.99	70	65.31	112	65.95	154	65.86	196	65.98
29	65.99	71	65.30	113	65.98	155	65.95	197	65.97
30	65.78	72	65.61	114	65.96	156	65.56	198	66.16
31	65.97	73	65.00	115	65.86	157	65.81	199	66.23
32	66.32	74	64.76	116	66.25	158	66.30	200	66.18
33	65.73	75	64.45	117	65.80	159	65.68	201	65.88
34	66.06	76	65.02	118	65.77	160	65.40	202	66.07
35	65.67	77	65.25	119	65.75	161	65.83	203	65.93
36	65.98	78	64.75	120	65.98	162	65.57	204	65.50
37	66.32	79	64.66	121	66.49	163	65.39	205	66.35
38	65.56	80	65.21	122	66.41	164	65.57	206	66.24
39	65.85	81	64.50	123	66.38	165	65.57	207	66.33
40	66.11	82	64.18	124	66.11	166	65.89	208	66.27
41	66.02	83	64.25	125	65.88	167	65.79	209	66.41
42	65.88	84	64.24	126	66.39	168	65.60	210	66.67

Number	% Iron	Number	% Iron	Number	% Iron	Number	% Iron	Number	% Iron
211	66.62	247	65.69	283	65.26	319	65.55	355	66.28
212	65.99	248	65.30	284	65.65	320	65.91	356	65.92
213	66.23	249	65.78	285	65.47	321	65.50	357	64.44
214	66.77	250	65.81	286	65.24	322	65.72	358	66.28
215	66.29	251	65.42	287	66.23	323	65.76	359	65.34
216	66.20	252	65.13	288	65.74	324	65.50	360	66.55
217	66.54	253	65.50	289	65.66	325	65.38	361	66.72
218	66.45	254	65.98	290	65.73	326	65.95	362	66.50
219	66.57	255	65.38	291	65.41	327	65.48	363	66.62
220	66.69	256	65.83	292	65.20	328	65.40	364	66.02
221	66.19	257	65.64	293	65.24	329	65.38	365	65.53
222	66.01	258	65.07	294	66.26	330	65.55	366	66.19
223	66.02	259	65.37	295	65.50	331	65.87	367	65.78
224	65.88	260	65.32	296	65.73	332	65.92	368	65.92
225	66.30	261	65.27	297	65.28	333	66.58	369	66.10
226	66.56	262	65.22	298	64.98	334	66.04	370	65.80
227	66.06	263	65.13	299	65.19	335	66.00	371	64.66
228	66.27	264	64.98	300	65.68	336	65.74	372	65.80
229	65.94	265	65.49	301	65.68	337	65.84	373	65.55
230	65.14	266	65.34	302	65.81	338	63.48	374	65.54
231	65.41	267	64.70	303	65.85	339	63.29	375	65.90
232	65.18	268	65.52	304	65.73	340	63.62	376	66.30
233	65.50	269	65.33	305	65.69	341	63.89	377	66.70
234	65.70	270	65.44	306	65.75	342	64.14	378	66.75
235	65.57	271	65.29	307	65.46	343	63.53	379	66.80
236	65.79	272	65.72	308	65.73	344	63.57	380	66.70
237	65.51	273	65.51	309	65.21	345	63.52	381	66.29
238	65.97	274	65.07	310	65.48	346	63.85	382	66.26
239	66.05	275	65.28	311	64.85	347	64.44	383	66.34
240	66.27	276	65.44	312	65.10	348	63.75	384	66.64
241	65.23	277	65.40	313	65.33	349	63.95	385	66.54
242	66.00	278	65.63	314	65.00	350	63.60	386	66.60
243	65.14	279	65.72	315	65.03	351	62.77	387	66.58
244	65.85	280	65.47	316	64.95	352	66.43	388	66.49
245	65.66	281	65.09	317	65.16	353	66.74	389	66.70
246	66.50	282	65.04	318	65.39	354	66.40	390	66.38

Source: Takahashi, U. and Imaizami, M. "Sampling Experiment of Fine Iron Ore," *Reports of Statistical Application Research*, Union of Japanese Scientists and Engineers. Vol. 18, No. 1, 1971.

CHAPTER 1

1.6a. Quantitative   **b.** Qualitative   **c.** Qualitative   **d.** Quantitative   **e.** Quantitative   **f.** Quantitative
g. Qualitative   **h.** Quantitative
1.7b. Plastic leaded chip carriers and small-outline transistors   **1.10b.** IBM   **1.11b.** Yes
1.12b. Minerals C and E   **1.13c.** Yes   **1.20** Mean = 6, median = 5, mode: all 5 measurements
1.21 Mean = 6, median = 5.5, modes: 4 and 6   **1.22** Mean = 9.50, median = 7.40, mode = 12.00
1.23 Mean = 4.04, median = 3, mode = 3   **1.24** Mean = 2,262.57, median = 2,200, mode = 2,130
1.25 Range = 14, variance = 14.19, standard deviation = 3.77

1.26

INTERVAL		NUMBER IN INTERVAL	PROPORTION	TCHEBYSHEFF'S THEOREM	EMPIRICAL RULE
$\bar{y} \pm s$	(2.99, 10.53)	19	.76	—	.68
$\bar{y} \pm 2s$	(−.77, 14.29)	24	.96	.75	.95
$\bar{y} \pm 3s$	(−4.54, 18.07)	25	1.00	.889	Almost 1.00

1.27d. $\bar{y} = 62.96$, $s = .61$ **e.** 96.97%, yes

1.29

INTERVAL		NUMBER IN INTERVAL	NUMBER EXPECTED BY TCHEBYSHEFF'S THEOREM	NUMBER EXPECTED BY EMPIRICAL RULE
$\bar{y} \pm s$	(1.01, 7.07)	39	—	38
$\bar{y} \pm 2s$	(−2.02, 10.10)	54	42	53
$\bar{y} \pm 3s$	(−5.05, 13.13)	56	50	Almost all

1.30b. Mean = .033, standard deviation = .016   **c.** 95.65% within interval (.00027, .06547)
1.31b. Mean = .080, standard deviation = .039   **c.** 91.30% within interval (.002, .158)
1.32b. Mean = .169, standard deviation = .052   **c.** 95.65% within interval (.066, .272)
1.33 25th percentile = 4, 50th percentile = 6, 75th percentile = 9
1.34a. Eighty percent of the operations analysts earn less than $2,470 monthly; 20% earn more.
b. Lower quartile = $2,120, upper quartile = $2,400   **c.** −1.39
1.35a. 7.15   **b.** 221.12   **c.** 4.83   **1.36a.** 99% of the rounding errors are less than .53.   **b.** 4.43
1.37a. 6   **b.** 1.64   **1.38a.** Yes; 13 is a suspect outlier.   **b.** No outliers
1.39a. Yes; 10.55 is a suspect outlier; 8.05, 8.72, 8.72, and 8.80 are highly suspect outliers.
b. 8.72, 8.72, and 8.05 are outliers.
1.40a. No   **b.** No outliers   **1.43c.** Yes
1.46a. Vendor 1: mean = 3.156, median = 3.1, modes: 2.9, 3.1, and 3.5; vendor 2: mean = 3.100, median = 3.2,
modes: 2.8 and 3.2; vendor 3: mean = 2.922, median = 2.8, mode = 2.6
b. Vendor 1: range = .8, variance = .083, standard deviation = .288; vendor 2: range = .8, variance = .075,
standard deviation = .274; vendor 3: range = 1.4, variance = .219, standard deviation = .468
c. Statement could be misleading; vendor 2 has smaller variation.
1.47 12%   **1.48a.** Approx. 18%   **b.** Approx. 17%   **c.** Approx. 55%   **d.** Approx. 14%
1.49a. Approx. 68% between 3.5 and 6.5 parts per million; approx. 95% between 2 and 8 parts per million; almost all between .5
and 9.5 parts per million   **b.** No
1.50c. Based on this information, it appears to be effective.
1.52b. Mean = 63.71, median = 4, mode = 1   **c.** Range = 1,602, variance = 48,341.4, standard deviation = 219.87
d. Lower quartile = 1, upper quartile = 13, IQR = 12   **e.** All measurements exceeding 49 are outliers.

1.53b. $\bar{y} = 3.386$, $s^2 = 10.508$, $s = 3.242$ **c.**

INTERVAL		NUMBER IN INTERVAL	NUMBER EXPECTED BY EMPIRICAL RULE
$\bar{y} \pm s$	(.144, 6.628)	31	24
$\bar{y} \pm 2s$	(−3.098, 9.870)	33	34
$\bar{y} \pm 3s$	(−6.772, 13.544)	35	Almost all

1.54a. Mean = 70, median = 60, modes: all measurements, range = 60, variance = 800, standard deviation = 28.28
b. Mean = 70, median = 70, modes: all measurements, range = 40, variance = 250, standard deviation = 15.81
c. Mean = 70, median = 70, mode = 70, range = 0, variance = 0, standard deviation = 0
d. Mean = 74, median = 90, modes: all measurements, range = 90, variance = 1,367.50, standard deviation = 36.98
1.55a. Secondary sulfate   **b.** 20%; 19%; 22%
1.57a. Mean = 117.82, median = 117.5, modes: 97, 112, 124, 128, and 131
b. Range = 62, variance = 225.33, standard deviation = 15.01

c.

INTERVAL		NUMBER IN INTERVAL	NUMBER EXPECTED BY EMPIRICAL RULE
$\bar{y} \pm s$	(102.81, 132.83)	31	34
$\bar{y} \pm 2s$	(87.80, 147.84)	49	48
$\bar{y} \pm 3s$	(72.79, 162.85)	50	Almost all

d. No outliers   **e.** 128

1.58a. Quantitative   **b.** Frequency distribution   **c.** Approx. .28
d. Yes, because the interval (.9995, 1.0005) has a large number of observations while the interval (.9985, .9995) contains no observations.   **1.60b.** 80%   **1.61c.** 1,013.85

1.62b.

SEX/LIFTS	INTERVAL	PROPORTION
Male/1	(13.13, 47.37)	Approx. .95
Male/4	(10.43, 37.23)	Approx. .95
Female/1	(13.57, 26.01)	Approx. .95
Female/4	(9.36, 22.28)	Approx. .95

c. Male: yes, $25 < \bar{y} + 2s = 37.23$; female: no, $25 > \bar{y} + 2s = 22.28$ **1.64a.** 29,600   **b.** 64%
1.65d. $\bar{y} = 65.94$, $s = .16$ **e.** 96.67%   **g.** Yes

CHAPTER 2

2.1 a, b, e, and g   **2.2** c and d
2.3a. Illegal data set name   **b.** Missing semicolon   **c.** Undefined division operator   **d.** Semicolons in data cards
2.4a.

```
DATA MICRO;
  INPUT TIME BRAND $ EXP;
  CARDS;
  2.0   A   2
   .    .   .
   .    .   .
  1.2   B   7
```

b.

```
DATA MICRO;
  INPUT TIME BRAND $ EXP;
  COST = TIME*200;
  CARDS;
  2.0   A   2
   .    .   .
   .    .   .
  1.2   B   7
```

2.5

```
DATA FROSH;
  INPUT VERBAL MATH GPA MAJOR $;
  EXAMTOT = VERBAL + MATH;
  CARDS;
  81   87   3.49   COMPSCI
   .    .    .       .
   .    .    .       .
   .    .    .       .
  51   75   2.48   MATH
```

2.6 a, b, e, and g

2.7a. Missing asterisk after DDT   **b.** Undefined division operator   **c.** Invalid keyword READ
d. Need blank between 100 and CATFISH   **e.** Command must begin in column 1

```
2.8 DATA LIST FREE/TIME EXP * BRAND (A1)
    COMPUTE COST = TIME * 200
    BEGIN DATA
    2.0  2  A
     .   .  .
     .   .  .
     .   .  .
    1.2  7  B
    END DATA
2.9 DATA LIST FREE/VERBAL MATH GPA * MAJOR (A7)
    COMPUTE EXAMTOT = VERBAL + MATH
    BEGIN DATA
    81  87  3.49  COMPSCI
     .   .    .      .
     .   .    .      .
     .   .    .      .
    51  75  2.48  MATH
    END DATA
```

2.10a. Undefined column indicator   **b.** Variable values not numeric   **c.** Variables not referenced by their column names
d. Undefined keyword END

```
2.11 READ C1 C2 C3        2.12 READ C1 C2 C3
     [ Data cards ]            2.0  1  2
     MULTIPLY C2 2 C4           .   .  .
     SUBTRACT C4 C3 C5          .   .  .
     STOP                       .   .  .
                              1.2  2  7
                              MULTIPLY C1 BY 200, PUT INTO C4
                              STOP
2.13 READ C1 C2 C3 C4       2.14 a, b, e, and g
     81  87  3.49  1
      .   .    .   .
      .   .    .   .
      .   .    .   .
     51  75  2.48  2
     ADD C1 C2, PUT INTO C5
     STOP
```

2.15a. Need slash at beginning of statement   **b.** Invalid keyword FREEFIELD
c. Illegal variable name: FISH SPECIES   **d.** Missing period   **e.** Undefined division operator
f. Missing slash   **g.** BASS and CATFISH should be coded numerically.   **h.** Use END instead of STOP.

```
2.16 / INPUT VARIABLES ARE 3, FORMAT IS FREE.
     / VARIABLE NAMES ARE TIME, BRAND, EXP, COST. ADD = NEW.
     / TRANSFORM COST = TIME * 200.
     / END
     2.0  0  2
      .   .  .
      .   .  .
      .   .  .
     1.2  1  7
     /END
2.17 / INPUT VARIABLES ARE 4, FORMAT IS FREE.
     / VARIABLE NAMES ARE VERBAL,MATH,GPA,MAJOR,EXAMTOT.
                ADD = NEW.
     / TRANSFORM EXAMTOT = VERBAL + MATH.
     / END
     81  87  3.49  1
      .   .    .   .
      .   .    .   .
     51  75  2.48  2
     /END
```

2.18 DATA ROCK;
INPUT POINTS @@;
CARDS;
196 82 400 12 310
PROC CHART;
VBAR POINTS/TYPE=PERCENT;
PROC UNIVARIATE;
VAR POINTS;

2.19 DATA FISH;
INPUT DDT @@;
CARDS;
13.00 3.50 9.30 21.00 3.40
 .   .   .   .   .
 .   .   .   .   .
 .   .   .   .   .
3.30 3.30 3.70 9.90 6.80
PROC CHART;
VBAR DDT/TYPE=PERCENT;
PROC UNIVARIATE;
VAR DDT;

2.20 DATA LIST FREE/POINTS
FREQUENCIES VARIABLES=POINTS/
          HISTOGRAM=PERCENT/
          STATISTICS=ALL
BEGIN DATA
196 82 400 12 310
END DATA

2.21 DATA LIST FREE/DDT
FREQUENCIES VARIABLES=DDT/
          HISTOGRAM=PERCENT/
          STATISTICS=ALL
BEGIN DATA
13.00 3.50 9.30 21.00 3.40
 .   .   .   .   .
 .   .   .   .   .
 .   .   .   .   .
3.30 3.30 3.70 9.90 6.80
END DATA

2.22 SET C1
196 82 400 12 310
NAME C1 = 'POINTS'
HISTOGRAM OF C1
DESCRIBE C1
STOP

2.23 SET C1
13.00 3.50 9.30 21.00 3.40
 .   .   .   .   .
 .   .   .   .   .
 .   .   .   .   .
3.30 3.30 3.70 9.90 6.80
NAME C1 = 'DDT'
HISTOGRAM OF C1
DESCRIBE C1
STOP

2.24a. BMDP 5D
/ PROBLEM TITLE IS 'HISTOGRAM OF ROCK DATA'.
/ INPUT VARIABLE IS 1. FORMAT IS STREAM.
/ VARIABLE NAME IS POINTS.
/ PLOT VARIABLE IS POINTS.
/ END
196 82 400 12 310
/END

b. BMDP 1D
/ PROBLEM TITLE IS 'DESCRIBE ROCK DATA'.
/ INPUT VARIABLE IS 1. FORMAT IS STREAM.
/ VARIABLE NAME IS POINTS.
/ END
196 82 400 12 310
/END

2.25a. BMDP 5D
/ PROBLEM TITLE IS 'HISTOGRAM OF FISH DATA'.
/ INPUT VARIABLE IS 1. FORMAT IS STREAM.
/ VARIABLE NAME IS DDT.
/ PLOT VARIABLE IS DDT.
/ END
13.00 3.50 9.30 21.00 3.40
 .   .   .   .   .
 .   .   .   .   .
 .   .   .   .   .
3.30 3.30 3.70 9.90 6.80
/END

b. BMDP 1D
```
  / PROBLEM TITLE IS 'DESCRIBE FISH DATA',
  / INPUT VARIABLE IS 1, FORMAT IS STREAM,
  / VARIABLE NAME IS DDT,
  / END
  13,00 3,50 9,30 21,00 3,40
     .     .     .      .     .
     .     .     .      .     .
     .     .     .      .     .
   3,30 3,30 3,70   9,90 6,80
  /END
```

CHAPTER 3

3.1a. $\frac{6}{16}$ **b.** $\frac{15}{16}$ **c.** No   **3.2a.** EO, EC, NO, NC   **b.** $\frac{5}{24}, 0, \frac{12}{24}, \frac{7}{24}$ **c.** $\frac{5}{24}$ **d.** $\frac{7}{24}$ **3.3a.** $\frac{6}{21}$ **b.** $\frac{3}{21}$ **c.** $\frac{18}{21}$

3.4a. (40, 300), (40, 350), (40, 400), (45, 300), (45, 350), (45, 400), (50, 300), (50, 350), (50, 400)

3.5a. The simple events are the eight action/queries listed in the table.   **b.** $\frac{139}{548}, \frac{104}{548}, \frac{68}{548}, \frac{87}{548}, \frac{25}{548}, \frac{52}{548}, \frac{41}{548}, \frac{32}{548}$ **c.** $\frac{155}{548}$ **d.** $\frac{507}{548}$

3.6 $6 \times 6 \times 6$ **3.7** 16   **3.8a.** 720   **b.** 1,000   **3.9a.** 16   **b.** 24   **3.10** 15,504

3.11a. 729   **b.** 120   **3.12** 63,063,000   **3.13** $\frac{64}{1,326}$ **3.14** $\frac{48}{2,598,960}$

3.15a. 1, 2, 3, 4, 5, 6, 7, 8, 9, 10, 11, 13, 15, 17, 19, 20, 21, 22, 23, 24, 25, 26, 27, 28, 29, 31, 33, 35

b. 19, 21, 23, 25, 27, 29, 31, 33, 35

c. 2, 4, 6, 8, 10, 11, 13, 15, 17, 19, 20, 21, 22, 23, 24, 25, 26, 27, 28, 29, 30, 31, 32, 33, 34, 35, 36

d. 0, 00, 1, 3, 5, 7, 9, 12, 14, 16, 18, 19, 21, 23, 25, 27, 30, 32, 34, 36   **e.** 29, 31, 33, 35

3.16a. $\frac{28}{38} = .737$ **b.** $\frac{9}{38} = .237$ **c.** $\frac{27}{38} = .711$ **d.** $\frac{20}{38} = .526$ **e.** $\frac{4}{38} = .105$ **3.17** .984

3.18b. The simple events are the seven steel sheet types listed in the table.   **c.** .27, .12, .30, .15, .08, .05, .03   **d.** .05

e. .84   **f.** .52

3.19a. $\frac{2}{9}$ **b.** $\frac{8}{9}$ **3.20a.** $\frac{4}{9} = .444$ **b.** $\frac{9}{18} = .500$ **c.** $\frac{9}{18} = .500$ **3.21a.** .111   **b.** .125   **3.22** .559   **3.23** .25

3.24a. .571   **b.** .143   **3.25a.** $\frac{6}{60} = .1$ **b.** $\frac{9}{40} = .225$ **3.26** $(\frac{48}{1,326})(\frac{1,192}{1,225}) \approx .035$ **3.27** .00001

3.28a. .75   **b.** .13   **c.** .85   **d.** .81   **3.29a.** .729   **b.** .001   **c.** Claim is probably false.   **3.30** .005

3.31a. .85   **b.** .40

c. (0, 0); (0, 50); (0, 100); (0, 500); (0, 1,000); (50, 0); (50, 50); (50, 100); (50, 500); (50, 1,000); (100, 0); (100, 50); (100, 100); (100, 500); (100, 1,000); (500, 0); (500, 50); (500, 100); (500, 500); (500, 1,000); (1,000, 0); (1,000, 50); (1,000, 100); (1,000, 500); (1,000, 1,000)

d. .36, .06, .09, .06, .03, .06, .01, .015, .01, .005, .09, .015, .0225, .015, .075, .06, .01, .015, .01, .005, .03, .005, .075, .005, .0025

e. .64

3.32a. (A, B) and (A, C) **b.** (B, C) **3.33a.** .0256   **b.** .418   **c.** Independent events   **3.34a.** .970   **b.** $1 - .99^k$

3.35a. .01099   **b.** .000000001   **3.37a.** .516   **b.** .385   **c.** .099   **3.38** .6982   **3.39** .236, .194, .570

3.40a. S_4 **b.** S_4 or S_6 **3.41** Probably understated, since P(At least one defective) $= .039$ if claim is true

3.42 Yes   **3.43a.** Probably not, since P(3 misses) $= .166$ if $p = .45$ **b.** Yes, since P(10 misses) $= .0025$ if $p = .45$

3.45 Perhaps because during the Depression, only the wealthy could afford telephone service; thus, sample was not random

3.47a. 1,440   **b.** 240   **3.48a.** $\frac{3}{10}$ **b.** No   **3.49a.** .14   **b.** .65   **c.** .065   **3.50a.** 16   **b.** $\frac{1}{16}$ **c.** Yes

3.51a. .15   **b.** .80   **c.** .60   **d.** None are mutually exclusive.   **3.52a.** 30   **b.** 20   **d.** $\frac{1}{20}$ **e.** $\frac{10}{20}$

3.53a. .20   **b.** .80   **c.** .40   **3.54** $\frac{1}{8,000} = .000125$

3.55a. .06   **b.** .94   **3.56a.** .122   **b.** .719   **c.** .682   **3.57** .891   **3.58** 336   **3.59** 26

3.60a. .53   **b.** .57   **c.** .12   **d.** .15   **e.** .29   **f.** .46   **g.** No   **h.** No   **3.61** Company A

3.62a. .0116   **b.** .87   **3.63a.** 60   **b.** $\frac{3}{5}$ **c.** $\frac{3}{10}$ **3.64a.** $\frac{24}{36} = .67$ **b.** .116   **c.** $\frac{6}{36} = .167$

3.65a. .0019808   **b.** .00394   **c.** .0000154   **3.66a.** 60   **b.** 54   **3.67** .01782

CHAPTER 4

4.1a. Yes   **c.** .15   **d.** .75   **4.2a.**

y	0	1	2	3
$p(y)$	$\frac{1}{8}$	$\frac{3}{8}$	$\frac{3}{8}$	$\frac{1}{8}$

b. $\frac{7}{8}$

4.3

y	0	1	2	3
p(y)	.004096	.064512	.338688	.592704

4.4

y	1	2	3
p(y)	$\frac{3}{5}$	$\frac{3}{10}$	$\frac{1}{10}$

4.5 Mean = 10.15, variance = 3.1275   **4.6** Mean = 1.5, variance = .75   **4.7** Mean = 2.52; variance = .4032
4.8a. $300   **b.** 18,000   **c.** $31.67 to $568.33   **4.9** $E(y) = 63,000$; $\sigma^2 = 61,321,000,000$ **4.10** $\mu = \$11,500$; yes
4.11a. $\mu = 15.14$, $\sigma = 1.11$ **b.** (12.92, 17.36)   **c.** .29   **4.12** $\mu = \$152,250$; $\sigma^2 = 703,687,500$
4.13 $\mu = \$350$; $\sigma^2 = 4,500$ **4.17** $p(0) = \frac{1}{16}$, $p(1) = \frac{1}{4}$, $p(2) = \frac{3}{8}$, $p(3) = \frac{1}{4}$, $p(4) = \frac{1}{16}$
4.19a. .358   **b.** .736   **c.** .736   **d.** .264   **e.** Improbable that $y \geq 3$ if claim is valid
4.20a. .4978714; .3793306   **b.** .0004267; .0003553
4.21a. .0005063   **b.** Suspect that the probability of a single failure exceeds .15
4.22a. Yes   **b.** .007   **c.** It probably differs from .5.   **4.23a.** $\mu = .0389$; $\sigma^2 = .0388985$ **b.** No
4.24a. .189   **b.** Approx. 0   **c.** Strong evidence that $p < .95$ **4.25a.** .001   **b.** .623   **c.** .999

4.30a. .064   **b.** .1875   **c.** .12288   **4.31a.**

y	6	7	8	9
p(y)	.13824	.08294	.04645	.02477

4.32a. $\mu = 2$, $\sigma = 1.83$ **b.** .9502   **4.33a.**

y	1	2	3	4	5
p(y)	.7	.21	.063	.0189	.00567

4.34a. $\mu = 1.43$, $\sigma = .78$ **b.** .91   **4.35** .8165
4.36a. .657   **b.** $\mu = 3.33$, $\sigma = 2.79$ **c.** No; $y = 10$ falls outside the interval $(\mu - 2\sigma, \mu + 2\sigma)$ **4.37** .671
4.38a. .0000374   **b.** $\mu = 25,706.9$, $\sigma^2 = 660,821,102.2$ **c.** No; $y = 100,000$ falls outside the interval $(\mu - 2\sigma, \mu + 2\sigma)$
4.40a. 0   **b.** $\frac{1}{12}$ **c.** $\frac{1}{3}$ **d.** $\frac{1}{6}$

4.41a.

y	3	4	5	6	7
p(y)	.071	.354	.424	.141	.010

b. $\mu = 4.67$, $\sigma = .841$ **c.** (2.99, 6.35)   **d.** .99

4.42a. 0   **b.** .354   **c.** .425   **d.** .575   **e.** 0   **f.** 0   **4.43a.** .399   **b.** .009   **c.** .601   **4.44a.** .6   **b.** .071
4.46a. .677   **b.** .034   **c.** .950   **d.** .000123   **4.47b.** $\mu = 5.5$, $\sigma = 2.35$; interval is (.80, 10.20)   **c.** .970
4.48a. .323   **b.** .036   **c.** .135   **4.49a.** .251   **b.** .558   **c.** .011
4.50a. $\mu = 1.57$, $\sigma = 1.25$ **b.** .2089   **4.51a.** .08   **b.** Yes, since $P(y > 3) = .019$
4.52a. $\mu = .75$, $\sigma^2 = .75$ **b.** .528   **c.** .472
4.53a. $\mu = 15$, $\sigma = 3.87$ **b.** Approx. 0, since $y = 27$ falls outside the interval $(\mu - 3\sigma, \mu + 3\sigma)$
c. No; probability will be smaller

4.54a. $\sigma = 2$ **b.** No; probability is .003   **4.55** $P(y \leq 30) \approx \sum_{y=0}^{30} \frac{6^y e^{-6}}{y!}$

4.61a. $E(\text{Loss}) = \$2,450$ for each firm   **b.** Firm A: $661.44, firm B: $701.78; firm B   **c.** Pure risk
4.62a. $\mu = .5$, $\sigma = .707$ **b.** No   **c.** Percent defective probably greater than .1%
4.63a. .7763   **b.** .1118   **4.64a.** .10   **b.** .70

4.65a.

x	−100,000	−70,000	−40,000	−20,000	10,000	60,000	380,000	410,000
p(x)	.36	.12	.01	.18	.03	.0225	.12	.02

x	460,000	860,000	900,000	930,000	980,000	1,380,000	1,900,000
p(x)	.03	.01	.06	.01	.015	.01	.0025

b. $E(x) = 126,000$; $\text{Var}(x) = 122,642,000,000$ **c.** .2775   **d.** .36

4.66a. .2399   **b.** .6801   **c.** .9999   **4.67a.** $p(y) = (.001)(.999)^{y-1}$ **b.** .995   **c.** Claim probably not valid
4.68a. $\mu = 23,850$, $\sigma = 7,087.14$ **b.** $\mu = \$1,431,000$, $\sigma = \$425,228.40$ **4.69a.** $\mu = 5$, $\sigma^2 = 4$ **b.** .617   **c.** .006
4.70a. .0778   **b.** .6826   **c.** .0102   **4.71a.** .402   **b.** .161   **4.72a.** .10   **b.** .074
4.73a. .016   **b.** .949   **c.** Doubt that p is as large as .40   **4.74a.** .6561   **b.** .9999   **c.** .0001
4.75 .263   **4.76** .049; .0956; .1399; .1821

CHAPTER 5

5.1a. $c = \frac{3}{8}$ **b.** $F(y) = y^3/8$ **c.** $\frac{1}{8}$ **d.** .0156   **5.2a.** $c = \frac{2}{3}$ **b.** $F(y) = \frac{1}{3}(4y - y^2)$ **c.** .48
5.3a. $c = 1$ **b.** $F(y) = 1 - e^{-y}$ **c.** .9257   **5.4a.** $F(y) = y^2/4$, $0 < y < 2$ **b.** $F(y)$ is NBU.
5.5 $\mu = 1.5$, $\sigma^2 = .15$, $\sigma = .387$; .952   **5.6** $\mu = .444$, $\sigma^2 = .080247$, $\sigma = .283$; 1.0
5.7 $\mu = 1$, $\sigma^2 = 1$, $\sigma = 1$; .950   **5.8** $\mu = 0$, $\sigma^2 = 5$ **b.** $\mu = 0$, $\sigma^2 = \frac{5}{3,600} = .0014$ **c.** $\mu = 0$, $\sigma^2 = 18,000$
5.13a. $\mu = 7$, $\sigma = .29$ **b.** .3   **5.14a.** $\mu = 1.375$, $\sigma^2 = .2552$, $\sigma = .51$ **b.** 1.0   **c.** .286
5.15a. $\mu = 1.2d$, $\sigma = .46d$ **b.** $\frac{3}{8}$ **5.19a.** .2417   **b.** .3496   **c.** .5336   **d.** .3936   **e.** .0918   **f.** .9564
5.20a. 1.645   **b.** 1.96   **c.** $-.84$ **d.** -3.01 **e.** 1.88   **f.** .15   **5.21** .0122   **5.22** .8%
5.23a. New compact tube   **b.** Standard compact tube   **5.24a.** .0139   **b.** .2417   **c.** 278
5.25a. .9671   **b.** .2611   **c.** No   **5.27** $c = \frac{1}{16}$
5.28a. $\mu = .21$, $\sigma^2 = .0147$
b. No, since $y = .60$ is larger than $\mu + 3\sigma$; either the values of α and β have changed or the 1982–1985 distribution no longer is approximated by the gamma
5.29a. $\mu = 1,000$, $\sigma^2 = 1,000,000$ **b.** .135   **c.** .777   **5.30a.** $\sigma = 10$ **b.** .753   **c.** .135
5.31a. .082   **b.** Suspect that $\beta > 24$ **5.32a.** $\mu_A = 4$, $\mu_B = 4$ **b.** $\sigma_A^2 = 8$, $\sigma_B^2 = 16$ **c.** Formula B
5.33a. .1353   **b.** .9817   **5.38** $\mu = \nu$, $\sigma^2 = 2\nu$

5.39 y	2	5	8	11	14	17
$f(y)$	.03843	.07788	.08437	.06560	.03944	.018896

5.40 $\sigma^2 = 21.46$; .9626   **5.41a.** .9981   **b.** .4448   **c.** $\mu = 2.866$, $\sigma = .8042$ **d.** .9571
5.42a. .6321   **b.** $\mu = 1.77246$, $\sigma = .92649$ **c.** .9626   **d.** No; $P(y > 6) = .0001234$
5.43a. $\mu = 5.65$, $\sigma^2 = 31.9225$ **b.** .6542   **c.** .1703   **5.44** 1.75 months
5.45a. $.88623\sqrt{\beta}$ **b.** $(.2146)\beta$ **c.** $e^{-c^2/\beta}$ **5.49** $c = 168$
5.50a. $\mu = \frac{2}{11}$, $\sigma^2 = \frac{3}{242}$ **b.** .046   **c.** .264   **5.51a.** $\mu = .0385$, $\sigma^2 = .00137$ **b.** .778
5.52a. $\mu = \frac{1}{2}$, $\sigma^2 = \frac{1}{20}$ **b.** .028   **5.53a.** .834   **b.** .006
5.60a. $(1 + t)^{-1}$ **b.** $\mu = -1$, $\sigma^2 = 1$ **5.61a.** $c = 2$ **b.** $F(y) = 1 - e^{-y^2}$ **c.** .00193
5.62a. $\mu = 5$, $\sigma^2 = \frac{25}{3}$ **b.** .3   **5.63a.** .3935   **b.** .0240   **c.** .0498
5.64a. .0918   **b.** 0   **c.** Lowered 4.87 decibels (to 95.13)   **5.65** .0721
5.66a. $\mu = .1923$, $\sigma^2 = .0058$ **b.** .090   **c.** .007   **5.67a.** .321   **b.** .105   **5.68a.** .0401   **b.** .1056
5.69 $\frac{1}{6}$ **5.70** .2231   **5.71a.** $\alpha = 9$, $\beta = 2$ **b.** $\mu = .818$, $\sigma^2 = .0124$ **c.** .624
5.72a. $\alpha = 2$, $\beta = 16$ **b.** $\mu = 3.545$, $\sigma^2 = 3.434$ **c.** .1054   **5.73** .3935   **5.74a.** Approx. 0   **b.** No

CHAPTER 6

6.1a.

y_1	0	1	2	3	4	5
$p_1(y_1)$	.300	.100	.025	.300	.125	.150

b.

y_2	0	1	2
$p_2(y_2)$	.10	.55	.35

c.

y_1	0	1	2	3	4	5
$p_1(y_1 \mid 0)$	0	.50	.25	0	.25	0

y_1	0	1	2	3	4	5
$p_1(y_1 \mid 1)$	.364	.091	0	.545	0	0

y_1	0	1	2	3	4	5
$p_1(y_1 \mid 2)$	.286	0	0	0	.286	.429

d.

y_2	0	1	2
$p_2(y_2 \mid 0)$	0	.667	.333

y_2	0	1	2
$p_2(y_2 \mid 2)$	1	0	0

y_2	0	1	2
$p_2(y_2 \mid 4)$	.2	0	.8

y_2	0	1	2
$p_2(y_2 \mid 1)$	.5	.5	0

y_2	0	1	2
$p_2(y_2 \mid 3)$	0	1	0

y_2	0	1	2
$p_2(y_2 \mid 5)$	0	0	1

6.2a. $p(y_1, y_2) = \frac{1}{36}$ $(y_1 = 1, 2, \ldots, 6; y_2 = 1, 2, \ldots, 6)$
b. $p_1(y_1) = \frac{1}{6}$ $(y_1 = 1, 2, \ldots, 6); p_2(y_2) = \frac{1}{6}$ $(y_2 = 1, 2, \ldots, 6)$
c. $p(y_1 \mid y_2) = \frac{1}{6}$ $(y_1 = 1, 2, \ldots, 6; y_2 = 1, 2, \ldots, 6); p(y_2 \mid y_1) = \frac{1}{6}$ $(y_1 = 1, 2, \ldots, 6; y_2 = 1, 2, \ldots, 6)$
d. y_1 and y_2 are independent

6.3a.

y_2	0	1	2	3
$p_2(y_2)$	.11	.25	.40	.24

b.

y_2	0	1	2	3
$p_2(y_2 \mid 2)$	.175	.250	.375	.200

6.4a.

	y_1 0	1	2	3
y_2 0	0	$\frac{3}{35}$	$\frac{6}{35}$	$\frac{1}{35}$
1	$\frac{2}{35}$	$\frac{12}{35}$	$\frac{6}{35}$	0
2	$\frac{2}{35}$	$\frac{3}{35}$	0	0

b.

y_1	0	1	2	3
$p_1(y_1)$	$\frac{4}{35}$	$\frac{18}{35}$	$\frac{12}{35}$	$\frac{1}{35}$

6.6a. $c = -1$ **b.** $f_2(y_2) = \frac{3}{2} - y_2$ **c.** $f_1(y_1 \mid y_2) = \dfrac{y_1 - y_2}{\frac{3}{2} - y_2}$

6.7a. $c = 4$ **b.** $f_1(y_1) = 2y_1; f_2(y_2) = 2y_2$ **c.** $f_1(y_1 \mid y_2) = 2y_1; f_2(y_2 \mid y_1) = 2y_2$
6.8a. $c = 2$ **b.** $f_1(y_1) = 2y_1 e^{-y_1^2}$ **c.** $f_2(y_2 \mid y_1) = 1/y_1$ $(0 \le y_2 \le y_1)$
6.10a. $c = 1$ **b.** $f_1(y_1) = e^{-y_1}$ **c.** $f_2(y_2) = e^{-y_2}$ **d.** $f_1(y_1 \mid y_2) = e^{-y_1}$ **e.** $f_2(y_2 \mid y_1) = e^{-y_2}$ **f.** .399
6.11a. 1.69 **b.** 3.46 **6.12a.** 0 **b.** 2 **6.13a.** $\frac{19}{12}$ **b.** $\frac{5}{12}$ **c.** 2 **d.** $\frac{2}{3}$
6.14a. 1 **b.** 1 **c.** 2 **d.** 1 **6.18** No **6.19** Yes **6.20** No **6.21** Yes
6.22a. $f(y_1, y_2) = \frac{1}{25} e^{-(y_1 + y_2)/5}$ **b.** 10 **6.25** .375 **6.26** 0
6.27 $\frac{1}{144}$ **6.28** 0 **6.29** $-.0366$ **6.30** $\frac{1}{11}$
6.31 0 **6.32** $\frac{1}{75}$ **6.36** 0 **6.37** $E(\ell) = 11, V(\ell) = 54.5$ **6.38** $E(\ell) = -4, V(\ell) = 256$
6.39 $E(y_1 + y_2) = 7, V(y_1 + y_2) = 5.83$ **6.40** 1.6484, 3.46 ± 3.2968 **6.41** $\frac{1}{9}$ **6.42** $E(\hat{p}) = p, V(\hat{p}) = pq/n$

6.44a.

y_1	0	10	20	30	40	50	60	70	80	90
$p_1(y_1)$	.006	.010	.019	.135	.220	.235	.130	.118	.071	.056

y_2	1	2	3	4	5
$p_2(y_2)$	.105	.405	.355	.058	.077

b.

y_1	0	10	20	30	40	50	60	70	80	90
$p_1(y_1 \mid 1)$	.0095	.0190	.0190	.2381	.3810	.2381	.0476	.0476	0	0

y_1	0	10	20	30	40	50	60	70	80	90
$p_1(y_1 \mid 2)$	.0123	.0123	.0247	.1852	.2469	.1852	.1234	.0741	.0741	.0617

y_1	0	10	20	30	40	50	60	70	80	90
$p_1(y_1 \mid 3)$	0	0	0	.0704	.1408	.2254	.1408	.2254	.1127	.0845

y_1	0	10	20	30	40	50	60	70	80	90
$p_1(y_1 \mid 4)$	0	.0172	.0345	.0862	.1724	.4310	.1724	.0517	.0172	.0172

y_1	0	10	20	30	40	50	60	70	80	90
$p_1(y_1 \mid 5)$	0	.0260	.0649	.0649	.2597	.3896	.1948	0	0	0

c. .9296　**d.** 2.597　**e.** Yes; no　**f.** Mean = \$1,296,500, standard deviation = \$2,891

6.45a. $f_1(y_1) = y_1 + \frac{1}{2}; f_2(y_2) = y_2 + \frac{1}{2}$　**c.** $f_1(y_1 \mid y_2) = \dfrac{y_1 + y_2}{y_2 + \frac{1}{2}}; f_2(y_2 \mid y_1) = \dfrac{y_1 + y_2}{y_1 + \frac{1}{2}}$

e. Yes; no　**f.** $E(d) = \frac{5}{12}$, $V(d) = \frac{5}{144}$; $.42 \pm .559$
6.46a. .17　**b.** $\frac{3}{17} = .1765$　**c.** Yes　**d.** No　**6.49a.** $c = \frac{1}{3}$　**b.** No

CHAPTER 7

7.1a. $f(g) = 1$　$(0 \le g \le 1)$　**b.** $f(g) = (g + 1)/2$　$(-1 \le g \le 1)$　**c.** $f(g) = 2/g^3$　$(1 \le g < \infty)$
7.2a. $f(g) = e^3$　$(0 \le g \le e^{-3})$　**b.** $f(g) = -e^{-g}$　$(0 \le g < \infty)$　**c.** $f(g) = 3e^{-3(g-1)}$　$(1 \le g < \infty)$
7.3 $f(g) = \frac{1}{15}e^{-(g-2)/15}$　$(g \ge 2)$　**7.4** $f(g) = \begin{cases} \frac{1}{5} & \text{if } 0 \le g \le \frac{5}{2} \\ \frac{1}{10} & \text{if } \frac{5}{2} \le g \le \frac{15}{2} \end{cases}$
7.5 $y = \sqrt{g}$, where g is uniform on the interval $0 \le g \le 1$
7.6 $y = \ln(g)$, where g is uniform on the interval $0 \le g \le 1$　**7.7** $f(g) = e^{-g}$　$(0 < g < \infty)$
7.8a. Approx. normal with $\mu_{\bar{y}_{25}} = 17$ and $\sigma_{\bar{y}_{25}} = 2$; approx. normal with $\mu_{\bar{y}_{100}} = 17$ and $\sigma_{\bar{y}_{100}} = 1$　**c.** .6826, .9544
7.9a. Approx. normal with $\mu_{\bar{y}} = 51$ and $\sigma_{\bar{y}} = 2.09$　**b.** .3156　**c.** .1694　**7.10a.** .0043　**b.** .6065
7.11a. .0233　**b.** Strong evidence that $\mu < 9.2$　**7.12a.** $\mu_{\bar{y}} = .000005$, $\sigma_{\bar{y}} = .00000028$　**b.** .1446
7.13a. .0062　**b.** Strong evidence that $\mu > 60$　**7.14** .9966　**7.15b.** .9544　**7.17a.** .985　**b.** .9878
7.18a. .0139　**b.** .4052　**c.** .017, .416; yes　**7.19a.** Approx. 0　**b.** .0094　**7.20** .8186
7.21a. .0262　**b.** .0125　**c.** .2266
7.28a. $f(g) = 6g^5$　$(0 < g < 1)$　**b.** $f(g) = 3(3 - g)^2$　$(2 < g < 3)$　**c.** $f(g) = -3e^{3g}$　$(0 < g < \infty)$
7.29 $f(g) = \begin{cases} (g + 2/200) & \text{if } -2 < g < 8 \\ \frac{1}{20} & \text{if } 8 < g < 23 \end{cases}$　**7.30** .4514　**7.31** .0082
7.32a. Approx. normal with $\mu_{\bar{y}} = 45$ and $\sigma_{\bar{y}} = .258$　**b.** .8508　**c.** Approx. 0
7.33a. Approx. normal with $\mu_{\bar{y}} = 121.74$ and $\sigma_{\bar{y}} = 4.86$　**b.** .7348
7.34 $y = 1 + \sqrt{g}$　**7.35** $y = \sqrt{-\ln(1 - g)}$
7.36a. Approx. normal with $\mu_{\bar{y}} = 1,312$ and $\sigma_{\bar{y}} = 42.2$　**b.** .006　**c.** Most likely $\mu > 1,312$
7.37 .2938　**7.38** .6950　**7.39** .008
7.40a. Approx. normal with $\mu_{\bar{y}} = 9.8$ and $\sigma_{\bar{y}} = .078$　**b.** .8997　**c.** No; $z = -3.21$
7.41 .9706　**7.42** .5211　**7.45** $f(g) = (1/\beta)e^{-g/\beta}$　$(g > 0)$, exponential

CHAPTER 8

8.1b. $\hat{\theta}_1$ **8.2a.** $\hat{\lambda}_1$ and $\hat{\lambda}_3$ **b.** $\hat{\lambda}_1$ **8.3b.** pq/n **8.4c.** $\beta^2/2n$ **8.6a.** $B = 1 - \theta/2$ **c.** $(\theta - 2)^2/3$
8.7a. $\bar{y}$ **b.** Yes **8.8a.** y/n **b.** Yes **c.** y/n **d.** Yes **8.9a.** $\bar{y}$ **b.** Yes **c.** β^2/n
8.10a. $\bar{y}/2$ **b.** $E(\hat{\beta}) = \beta$, $V(\hat{\beta}) = \beta^2/2n$ **8.11a.** $\bar{y}/2$ **b.** $E(\hat{\beta}) = \beta$, $V(\hat{\beta}) = \beta^2/2n$ **8.12** $\Sigma y_i^2/n$
8.13a. 2.898 **b.** 2.262 **c.** 1.761 **8.15** $\hat{p} \pm z_{\alpha/2}\sqrt{\dfrac{\hat{p}\hat{q}}{n}}$ **8.16** $\bar{y} \pm z_{\alpha/2}\sqrt{\bar{y}/n}$ **8.17** $\bar{y} \pm z_{\alpha/2}(\bar{y}/\sqrt{n})$
8.23 $4.985 \pm .0083$ **8.24** $\bar{y} = 239.2$, $s = 29.29$ **b.** 239.2 ± 20.08 **8.25** $4{,}001.46 \pm 387.21$
8.26a. 84.84 **b.** 84.84 ± 4.03 **c.** .95 **8.27a.** 9.9 ± 4.81 **b.** 6.7 ± 6.19 **8.28a.** 105.7 ± 7.776
8.29a. 35 ± 1.645 **8.30** -16 ± 11.58; disk drive 2 **8.31** -40 ± 95.12 **8.32a.** $.9162 \pm 1.5124$
8.33a. 4.23 ± 4.60 **b.** 4.23 ± 4.36 **8.34a.** 436.5 ± 47.61 **b.** $-1.09 \pm .506$
8.35a. $1.18 \pm .807$ **c.** Restrained mean exceeds unrestrained mean **8.36** 198 ± 41.2
8.37a. 1.865 ± 1.486 **b.** Yes, since the interval includes only positive differences **8.38a.** 5.45 ± 2.49 **b.** Yes
8.39 $.24 \pm .118$ **8.40** $.16 \pm .085$ **8.41a.** $.23 \pm .0165$ **b.** $.20 \pm .0156$ **8.42** $.75 \pm .1225$ **8.43** $.18 \pm .099$
8.44 New York: $.262 \pm .0084$, Wisconsin: $.15 \pm .026$, Maine: $.661 \pm .0833$, Florida: $.005 \pm .0048$, Virginia: $.224 \pm .0937$
8.45 $.58 \pm .057$ **8.46a.** $.067 \pm .144$ **b.** $-.247 \pm .138$ **8.47** $.078 \pm .035$
8.48 $-.227 \pm .166$; Japan **8.49** $.051 \pm .127$
8.50a. 14.0671 **b.** 23.5418 **c.** 23.2093 **d.** 17.5346 **e.** 16.7496 **8.51** $(.0028, .0105)$ **8.52** $(.085, 1.140)$
8.53 $(.0000457, .0003216)$ **8.54a.** $(.163, 29.141)$ **8.55a.** 2.40 **b.** 3.35 **c.** 1.65 **d.** 5.86
8.56a. 3.18 **b.** 2.62 , **c.** 2.10 **8.57** $(1.53, 3.64)$; yes **8.58a.** $(.1796, .8072)$ **c.** Assembly line 1
8.59 $(.28, 4.57)$; yes **8.60a.** $(.040, 1.63)$ **8.61a.** 722 **b.** 174 **8.62** $n_1 = n_2 = 195$ **8.63** 1,729
8.64 2,401 **8.66** $1{,}173.6 \pm 25.97$ **8.67a.** $(.436, 17.81)$ **b.** No evidence of a difference
8.68 $.619 \pm .0657$ **8.69** $(14{,}575.78, 91{,}381.97)$ **8.70** 1.375 ± 2.570 **8.71** 192
8.72a. $1{,}500 \pm 837.7$ **b.** South **8.73a.** $.60 \pm .10$ **b.** Increase sample sizes or decrease confidence coefficient
8.74 $12.4 \pm .6553$ **8.75a.** -4.8 ± 5.0 **c.** No **d.** 74 **8.76** $(.41, 4.32)$
8.77a. 14.1 ± .375 days **b.** $\$2{,}256 \pm \$60$ **8.78a.** $.22 \pm .142$ **b.** Yes
8.79a. $.24 \pm .084$ **c.** The 99% confidence interval would be wider. **8.80** $-2.7 \pm .61$

8.81a. Bias $= \frac{1}{2}$ **b.** $1/(12n)$ **c.** $\bar{y} - \frac{1}{2}$ **8.82b.** $\lambda_1/n_1 + \lambda_2/n_2$; $\bar{y}_1/n_1 + \bar{y}_2/n_2$ **c.** $(\bar{y}_1 - \bar{y}_2) \pm z_{\alpha/2}\sqrt{\dfrac{\bar{y}_1}{n_1} + \dfrac{\bar{y}_2}{n_2}}$

8.83c. $\dfrac{2y}{\chi^2_{\alpha/2}} < \beta < \dfrac{2y}{\chi^2_{(1-\alpha/2)}}$ **8.84b.** $\dfrac{y^2}{\chi^2_{\alpha/2}} < \sigma^2 < \dfrac{y^2}{\chi^2_{(1-\alpha/2)}}$ **8.86** $y \pm 1.96$ **8.87** LCL $= y/.95$

CHAPTER 9

9.3a. .032 **b.** .370 **c.** .132 **d.** .006 **e.** Type I: $y \geq 8$; Type II: $y \geq 5$
9.4a. $y \geq 9$ **b.** .596 **c.** .005 **9.5a.** .033 **b.** .617 **c.** .029 **9.10** Yes; $z = 3.54$ **9.11** Yes; $z = 9.90$
9.12a. No; $z = 1.41$ **b.** Small **9.13** Yes; $t = 1.64$ **9.14a.** Yes; $t = 2.89$ **9.15** .5871
9.16a. .4721 **b.** Approx. 1 **9.18** Yes; $z = 3.54$ **9.19** $z = 6.83$; reject H_0 **9.20** Yes; $t = 3.08$
9.21 $t = 3.02$; reject H_0 **9.22a.** Yes; $z = 11.00$ **b.** Yes; $z = 11.24$ **9.23** Yes; $z = 11.14$ **9.24** No; $t = -.33$
9.25 $t = -1.17$; do not reject H_0 **9.26** $t = -7.40$; reject H_0 **9.27** $t = 3.70$; reject H_0 **9.28a.** Yes; $t = 2.67$
9.29 $t = 1.93$; reject H_0 **9.30** Yes; $z = 2.84$ **9.31** $z = 1.83$; reject H_0 **9.32** No; $z = -1.58$
9.33 Yes; $z = -2.74$ **9.34** $z = .44$; do not reject H_0 **9.35** $z = 3.27$; reject H_0 **9.36** Yes; $z = 5.52$
9.37 No; $z = 1.19$ **9.38** Yes; $z = 1.96$ **9.39** Yes; $z = -2.40$ **9.40** $\chi^2 = 2.97$; no **9.41** Yes; $\chi^2 = 688$
9.42a. No; $\chi^2 = 6.912$ **9.43** Yes; $\chi^2 = 54.00$ **9.44a.** No; $F = 1.09$ **9.45** No; $F = 1.31$
9.46 $F = 2.03$; do not reject H_0 **9.47** Yes; $F = 1.75$ **9.50a.** .025 **b.** .05 **c.** .0038 **d.** .1056
9.51a. .3124 **b.** .0178 **c.** Approx. 0 **d.** .1470 **9.52** .0793
9.53a. $.05 < p\text{-value} < .10$ **b.** Approx. 0 **c.** $p\text{-value} < .01$ **d.** $.02 < p\text{-value} < .05$ **e.** .0023 **f.** .0164
9.54a. .066 **b.** .942; .058 **c.** .322; .678 **9.55** H_0: $\mu = 22$, H_a: $\mu < 22$ **9.56** H_0: $p = \frac{1}{6}$, H_a: $p \neq \frac{1}{6}$
9.57 H_0: $(\mu_1 - \mu_2) = 0$, H_a: $(\mu_1 - \mu_2) > 0$, where μ_1 is mean rating of vendor's product and μ_2 is mean rating of rival vendor's product
9.58 H_0: $p = .10$, H_a: $p > .10$ **9.59** H_0: $\mu = \frac{1}{2}$, H_a: $\mu \neq \frac{1}{2}$
9.60a. Yes; $z = -2.52$ **b.** $p\text{-value} = .0059$ **9.61** $F = 10.00$; yes
9.62a. $z = 1.99$; yes **b.** $z = .15$; no **9.63** $t = 1.54$; do not reject H_0
9.64a. No; $t = -2.20$ **c.** $.1 < \beta < .5$ **d.** $.01 < p\text{-value} < .025$ **9.65** $z = -.97$; no
9.66 $z = .74$; no **9.67** $\chi^2 = 133.9$; yes **9.68** $z = -1.78$; yes **9.69** Yes; $z = 2.79$ **9.70** $z = -6.29$; yes

9.71a. $t = .193$; no   **b.** p-value $> .2$ **9.72** No; $z = -.925$ **9.73** No; $F = 4.47$
9.74 No; $t = -1.29$ **9.75** $t = 2.54$; yes

CHAPTER 10

10.2 $\beta_0 = 1$, $\beta_1 = 1$; $y = 1 + x$ **10.3a.** $y = 2 + 2x$ **b.** $y = 4 + x$ **c.** $y = -2 + 4x$ **d.** $y = -4 - x$
10.5a. $\beta_1 = 2$, $\beta_0 = 3$ **b.** $\beta_1 = 1$, $\beta_0 = 1$ **c.** $\beta_1 = 3$, $\beta_0 = -2$ **d.** $\beta_1 = 5$, $\beta_0 = 0$ **e.** $\beta_1 = -2$, $\beta_0 = 4$
10.6a. $\hat{\beta}_0 = 0$, $\hat{\beta}_1 = .8571$ **10.7a.** $\hat{\beta}_0 = 2$, $\hat{\beta}_1 = -1.2$ **10.8a.** $\hat{\beta}_0 = 2,084,880$, $\hat{\beta}_1 = -1,050$
10.9a. Yes   **b.** $\hat{y} = 14.175 + .02243x$ **10.10b.** $\hat{\beta}_0 = 7.725$, $\hat{\beta}_1 = -.5625$ **d.** 5.475
10.11a. $\hat{y} = 1.40 + .1010x$ **c.** 23.62   **10.17** $s^2 = .03128571$
10.18a. SSE $= 1.143$, $s^2 = .2857$ **b.** SSE $= 1.6$, $s^2 = .5333$ **c.** SSE $= 203,000$, $s^2 = 67,666.67$
d. SSE $= 4.3675$, $s^2 = .727917$ **e.** SSE $= 49.03$, $s^2 = 3.77$
10.19a. $\hat{y} = -2.095 + .003693x$ **c.** SSE $= .02899$, $s^2 = .00414$
10.20a. $\hat{y} = 6.514 + 10.83x$ **c.** 12.90   **d.** SSE $= .8825$, $s^2 = .1103$
10.22a. $\hat{y} = 9.3134 + 6.91175x$ **c.** SSE $= 171.396$, $s^2 = 21.425$
10.25a. Yes; $t = 6.708$ **b.** Yes; $t = -5.196$ **c.** Yes; $t = -12.764$ **d.** Yes; $t = 10.79$ **e.** Yes; $t = -5.594$
f. Yes; $t = 28.30$
10.26 Yes; $t = 44.45$ **10.27** Yes; $t = 6.34$ **10.28** Yes; $t = 3.03$
10.29a. $\hat{y} = 10.39 + 90.92x$ **b.** $H_a: \beta_1 > 0$; reject H_0, $t = 3.182$ **c.** 90.92 ± 57.58
10.30 $\hat{\beta}_1 = 7.77$; yes, $t = 11.64$ **10.31a.** $\hat{y} = -.031 + .317x$ **b.** $.317 \pm .124$
10.32a. Yes; $t = 3.39$ **b.** $.3307 \pm .2079$ **10.36** $r = .913$, $r^2 = .8338$
10.37a. $r = .998$, $r^2 = .996$ **b.** Reject H_0; $r_{.05} = .582$; yes   **10.38a.** Yes; positive   **b.** $.9433$ **c.** Yes; $r_{.025} = .514$
10.39 $r_1 = -.97$, $r_2 = -.11$; x_1 **10.40a.** Yes; $-r_{.05} = -.549$ **b.** Yes; $-r_{.05} = -.549$ **c.** $.8464$ **d.** $.5041$
10.41a. $r^2 = .8089$ **b.** Yes; $r = .8994$ **10.42b.** Probably; however, we cannot conduct a test since n is not given.
10.43 $\hat{y} = -.1935 + 1.2206x$, $r^2 = .5921$ **10.47** $5.475 \pm .778$ **10.48a.** 78.58 ± 35.56 **b.** 78.58 ± 12.60
10.51 181.678 ± 9.413 **10.52a.** $\hat{y} = 2.815 + 6.249x$ **c.** 22.81 ± 5.05 **d.** 21.56 ± 1.81
10.53a. $\hat{y} = -.7995 + .93205x$ **b.** Yes; $t = 8.30$ **c.** 15.5113 ± 2.7125
10.58b. $\hat{y} = -13.622 - .053x$ **d.** Reject H_0; $t = -6.797$ **e.** $-.053 \pm .015$ **f.** $r = -.923$ **g.** $r^2 = .852$
h. $.779 \pm .474$ **i.** $.779 \pm .150$
10.59b. $\hat{y} = -2.209 + 1.308x$ **c.** $r^2 = .9868$ **d.** $s = 2.858$ **e.** Yes; $t = 21.187$ **f.** $1.308 \pm .229$
g. 37.029 ± 11.381
10.60a. $\hat{y} = -1.081 + .676x$ **b.** Yes; $t = 2.5$ **c.** $.676 \pm .860$ **d.** 15.811 ± 1.334 **e.** 14.459 ± 3.689
10.61a. $\hat{y} = 1,855.35 + 47.07x$ **c.** Yes, $t = 93.36$ **d.** $r = .9995$, $r^2 = .9991$ **e.** $36,169 \pm 296$ **f.** $36,169 \pm 415$
10.62a. Yes; $t = -4.36$ **b.** $10,020 \pm 422$ **10.63a.** No; $t = 2.609$ **b.** $r^2 = .531$ **c.** 33.38 ± 5.14
10.64a. Yes   **b.** Yes   **d.** 8.29   **e.** 21.15   **10.65a.** $\hat{y} = -.11238 + .09439x$ **b.** Yes; $t = 11.39$ **c.** $.926 \pm .197$
10.66a. $\hat{y} = .702 + .353x$ **c.** $r = .9836$, $r^2 = .9675$ **d.** Yes; $t = 10.92$ **e.** 13.05 ± 2.91
10.67b. $\hat{y} = 6.735 - .142x$ **c.** Yes; $t = -5.562$ **d.** 3.895 ± 1.036

CHAPTER 11

11.1a. $Y = \begin{bmatrix} 4 \\ 3 \\ 3 \\ 1 \\ -1 \end{bmatrix}$; $X = \begin{bmatrix} 1 & -2 \\ 1 & -1 \\ 1 & 0 \\ 1 & 1 \\ 1 & 2 \end{bmatrix}$ **b.** $X'X = \begin{bmatrix} 5 & 0 \\ 0 & 10 \end{bmatrix}$; $X'Y = \begin{bmatrix} 10 \\ -12 \end{bmatrix}$ **c.** $\hat{\beta} = \begin{bmatrix} 2 \\ -1.2 \end{bmatrix}$ **d.** $\hat{y} = 2 - 1.2x$

11.2a. $Y = \begin{bmatrix} 1 \\ 2 \\ 2 \\ 3 \\ 5 \\ 5 \end{bmatrix}$; $X = \begin{bmatrix} 1 & 1 \\ 1 & 2 \\ 1 & 3 \\ 1 & 4 \\ 1 & 5 \\ 1 & 6 \end{bmatrix}$ **b.** $X'X = \begin{bmatrix} 6 & 21 \\ 21 & 91 \end{bmatrix}$; $X'Y = \begin{bmatrix} 18 \\ 78 \end{bmatrix}$ **d.** $\hat{\beta} = \begin{bmatrix} 0 \\ \frac{6}{7} \end{bmatrix}$ **e.** $\hat{y} = \frac{6}{7}x$

11.3a. Y is 10×1; X is 10×3 **c.** $\hat{\beta} = \begin{bmatrix} 2.789 \\ .715 \\ -.039 \end{bmatrix}$; $\hat{y} = 2.789 + .715x - .039x^2$

11.4a. $Y = \begin{bmatrix} .24 \\ .38 \\ .44 \\ .61 \\ .75 \end{bmatrix}$; $X = \begin{bmatrix} 1 & 1.0 & 1.00 \\ 1 & 3.5 & 12.25 \\ 1 & 6.0 & 36.00 \\ 1 & 8.5 & 72.25 \\ 1 & 11.0 & 121.00 \end{bmatrix}$ **b.** $X'X = \begin{bmatrix} 5 & 30 & 242.5 \\ 30 & 242.5 & 2,205 \\ 242.5 & 2,205 & 21,308.125 \end{bmatrix}$; $X'Y = \begin{bmatrix} 2.42 \\ 17.645 \\ 155.5575 \end{bmatrix}$

c. $(X'X)^{-1} = \begin{bmatrix} 1.786 & -.612 & .043 \\ -.612 & .279 & -.022 \\ .043 & -.022 & .002 \end{bmatrix}$ **d.** $\hat{\beta} = \begin{bmatrix} .2135 \\ .0349 \\ .0013 \end{bmatrix}$; $\hat{y} = .2135 + .0349x + .0013x^2$

11.5 Yes; $t = -5.196$ **11.6** $-1.2 \pm .543$ **11.7** Yes; $t = -2.22$ **11.8** $t = .98$; do not reject H_0
11.9b. $\hat{y} = 1,717.4952 - 150.26x$; term appears to be useful ($t = -5.72$)   **c.** Yes, $t = -5.47$, p-value $= .0001$
11.10a. $\hat{y} = 22.925 - 3.525x_1 - .375x_2$ **b.** No; $t = -1.52$ **c.** No; $t = -.16$ **d.** -3.525 ± 14.680
e. $-.375 \pm 14.680$
11.11a. Quadratic   **b.** Reject H_0: $\beta_2 = 0$; $t = -15.78$ **c.** .00005   **d.** 188.89
11.14a. No; $F = 1.28$ **11.15** Yes; $F = 443.18$ **11.16a.** Yes   **b.** Yes; $F = 38.836$
11.17a. Yes; $F = 17.8$ **b.** Reject H_0: $\beta_1 = 0$; $t = -3.50$ **c.** -6.38 ± 4.723
11.18a. $R^2 = .262$; no   **b.** $F = 1.896$; do not reject H_0 **11.19a.** Yes; $F = 618.96$ **b.** Yes; $F = 1,540.32$
11.20a. Yes; $F = 120.65$ **b.** $t = 9.86$; reject H_0 **11.21** $F = 1.056$; do not reject H_0
11.22 $.8 \pm .941$ **11.23** $.8 \pm 1.96$ **11.24** $1.714 \pm .620$ **11.25** 1.714 ± 1.297 **11.26** $3.464 \pm .077$
11.27 $3.464 \pm .193$ **11.28** 26.075 ± 51.167 **11.29** 26.075 ± 78.159
11.39 Unable to test model adequacy since there are no degrees of freedom for estimating σ^2 (i.e., df $= n - 3 = 0$)
11.40 No; appears that multicollinearity exists
11.41a. $\hat{y} = 2.7433 + .800976x_1$; yes, $t = 15.92$ **b.** $\hat{y} = 1.6647 + 12.3954x_2$; yes, $t = 11.76$
c. $\hat{y} = -11.7953 + 25.0682x_3$; yes (at $\alpha = .05$), $t = 2.51$
11.42a. $-.406, -.206, -.047, .053, .112, .212, .271, .471, .330, .230, -.411, -.611$ **b.** Yes; needs curvature
c. No outliers   **d.** Yes; needs curvature
11.43a. $10.146, 2.065, -.412, 1.986, -8.529, -3.052, 3.164, -5.313, 1.616, .111, -1.737$ **c.** No
d. Assumptions appear to be satisfied
11.44a. $\hat{y} = 40.0714 - .214286x$.   **b.** $-4.643, -1.429, -3.857, 5.000, 4.214, .429, -6.357$ **c.** Yes; needs curvature
d. $\hat{y} = -1,199.9286 + 75.2143x - 1.142857x^2$; yes, $t = -9.37$
11.45a. 2.714   **b.** (3.8612, 4.5156)   **c.** No
11.46a. $\hat{y} = .94 - .214x$ **b.** $0, .02, -.026, .034, .088, -.112, -.058, .002, .036, .016$
c. Football shape; unequal variances   **d.** Use the transformation $y^* = \sin^{-1}\sqrt{y}$ and fit the model $y^* = \beta_0 + \beta_1 x + \varepsilon$
e. $\hat{y}^* = 1.307 - .2496x$; possibly

11.47a. $Y = \begin{bmatrix} 1.1 \\ 1.9 \\ 3.0 \\ 3.8 \\ 5.1 \\ 6.0 \end{bmatrix}$; $X = \begin{bmatrix} 1 & -5 \\ 1 & -3 \\ 1 & -1 \\ 1 & 1 \\ 1 & 3 \\ 1 & 5 \end{bmatrix}$ **b.** $X'X = \begin{bmatrix} 6 & 0 \\ 0 & 70 \end{bmatrix}$; $X'Y = \begin{bmatrix} 20.9 \\ 34.9 \end{bmatrix}$ **c.** $\hat{\beta} = \begin{bmatrix} 3.4833333 \\ .4985714 \end{bmatrix}$

d. $\hat{y} = 3.48333 + .49857x$ **e.** SSE $= .0681912$, $s^2 = .0170478$ **f.** Yes; $t = 31.95$ **g.** $r^2 = .9961$ **h.** $3.7326 \pm .1148$
11.48b. $\hat{y} = 2.46 + .41x_1 + 1.614x_2$ **c.** SSE $= 2.4363$, $s^2 = .34804$ **d.** Yes; $F = 109.65$ **e.** $R^2 = .969$
f. Yes; $t = 3.11$ **g.** $4.87 \pm .74$ **h.** 4.87 ± 1.34
11.49a. $\hat{w} = -.104 + .282\sqrt{t}$ **b.** Yes; $t = 83.93$
11.50a. $\hat{y} = .13202 - 9.30712x_1 + 1.55756x_2$ **b.** Reject H_0; $F = 35.84$ (p-value $= .0005$)
c. No; $t = -1.84$ (p-value $= .1153$)   **d.** Yes; $t = 8.47$ (p-value $= .0001$)   **e.** .922766   **f.** .152141

11.51e. $\hat{y} = -.093 + .917x$ **f.** SSE $= 1.556$, $s^2 = .1556$ **g.** Yes; $t = 13.75$ **h.** $r^2 = .9498$
11.52a. $4.03 \pm .24$ **b.** $4.03 \pm .76$
11.53a. Yes; $F = 24.412$ **b.** $t = -2.01$; reject H_0 **c.** $t = .3125$; do not reject H_0 **d.** $t = 2.385$; reject H_0
11.54a. $3.197, 3.215, -2.258, .269, -1.713, -7.186, -2.659, 3.359, -2.132, 5.904$ **b.** Yes; needs curvature
c. No outliers **d.** Yes; needs curvature
11.55a. $E(y) = \beta_0 + \beta_1 x + \beta_2 x^2$ **b.** $\hat{y} = 66.95 - 243.52x + 245.98x^2$ **c.** SSE $= 92.52$, $s^2 = 46.26$, $R^2 = .92$
11.56a. $\hat{y} = 90.1 - 1.84x_1 + .285x_2$ **b.** $R^2 = .916$ **c.** Yes; $F = 65.429$ **d.** Yes; $t = -5.01$
11.57a. $\hat{y} = -3.373 + .0036x_1 + .9476x_2$ **b.** Yes; $F = 20.90$ **c.** $.718 \pm .971$
11.58a. $\hat{y} = .04564705 + .00078505x_1 + .23737262x_2 - .00003809x_1x_2$ **b.** SSE $= 2.71515$, $s^2 = .16970$
11.59a. Yes; $F = 84.96$, p-value $= .0001$ **11.60** $(7.32285, 9.44865)$
11.61a. $R^2 = .90$ **b.** $s = 4.5485$ **c.** $F = 33.08$ (p-value $= .0003$); reject H_0
d. Yes; $t = 2.13$ (p-value $= .0706/2 = .0353$)
11.62a. 13.6812 **b.** Yes
11.63b. $-50.76, -47.44, -15.96, -33.72, -54.64, -78.24, 16.84, -46.52, 43.24, -64.96, -55.96, -50.52, 436.52, 37.88,$
-36.00; almost all residuals are negative, one large positive residual **c.** Omit data for this employee from analysis
d. $\hat{y} = 191.26 - 9.58x$; $r^2 = .483$; model is adequate ($t = -3.35$)
11.64 Yes; $t = 4.44$
11.65b. Clone 5263: $F = 338.55$, reject H_0; clone 5319: $F = 297.88$, reject H_0; clone 5331: $F = 175.91$, reject H_0;
clone 5271: $F = 331.50$, reject H_0 **c.** Clone 5263: yes, $t = 12$; clone 5319: yes, $t = 8$; clone 5331: yes, $t = 7$;
clone 5271: yes, $t = 7$
11.66a. $E(S_v) = \beta_0 + \beta_1 x + \beta_2 x^2$ **b.** $E(V_v) = \beta_0 + \beta_1 x + \beta_2 x^2$
11.67b. No; $F = 4.82$ (p-value $= .1155$) **11.68b.** No, $t = -2.10$ (p-value $= .126$)
11.69b. Yes, $F = 28.72$ (p-value $= .0058$) **c.** $(138.35, 1,470.38)$

CHAPTER 12

12.1a. Quantitative **b.** Quantitative **c.** Qualitative **d.** Qualitative **e.** Qualitative **f.** Qualitative
g. Quantitative **h.** Qualitative
12.2a. Quantitative **b.** Qualitative **c.** Quantitative **d.** Quantitative **e.** Qualitative
12.3a. Quantitative **b.** Qualitative **c.** Quantitative **d.** Quantitative **e.** Qualitative
12.4a. First-order **b.** Second-order **c.** Third-order **d.** Third-order **e.** Second-order **f.** First-order
12.5a. (i) First-order (ii) Third-order (iii) First-order (iv) Second-order
b. (i) $E(y) = \beta_0 + \beta_1 x$ (ii) $E(y) = \beta_0 + \beta_1 x + \beta_2 x^2 + \beta_3 x^3$ (iii) $E(y) = \beta_0 + \beta_1 x$
(iv) $E(y) = \beta_0 + \beta_1 x + \beta_2 x^2$ **c.** (i) $\beta_1 > 0$ (ii) $\beta_3 > 0$ (iii) $\beta_1 < 0$ (iv) $\beta_2 < 0$
12.6a. Second-order **c.** Downward curvature **12.7a.** First-order **c.** Positive (upward) slope
12.8 $E(y) = \beta_0 + \beta_1 x + \beta_2 x^2$ **12.9b.** $E(y) = \beta_0 + \beta_1 x$; $E(y) = \beta_0 + \beta_1 x$; $E(y) = \beta_0 + \beta_1 x + \beta_2 x^2$
12.10 $E(y) = \beta_0 + \beta_1 x + \beta_2 x^2$ **12.11** $E(y) = \beta_0 + \beta_1 x$ **12.12** Reject H_0: $\beta_2 < 0$, $t < -1.717$
12.13a. $E(y) = \beta_0 + \beta_1 x_1 + \beta_2 x_2$ **b.** Include $\beta_3 x_1 x_2$ **c.** Include $\beta_4 x_1^2 + \beta_5 x_2^2$
12.16a. Both quantitative **b.** $E(y) = \beta_0 + \beta_1 x_1 + \beta_2 x_2$ **c.** Include $\beta_3 x_1 x_2 + \beta_4 x_1^2 + \beta_5 x_2^2$
d. H_0: $\beta_3 = 0$ against H_a: $\beta_3 \neq 0$
12.17 $E(y) = \beta_0 + \beta_1 x_1 + \beta_2 x_2 + \beta_3 x_1 x_2 + \beta_4 x_1^2 + \beta_5 x_2^2$
12.18a. Both quantitative **b.** $E(y) = \beta_0 + \beta_1 x_1 + \beta_2 x_2$ **c.** $E(y) = \beta_0 + \beta_1 x_1 + \beta_2 x_2 + \beta_3 x_1 x_2$
d. $E(y) = \beta_0 + \beta_1 x_1 + \beta_2 x_2 + \beta_3 x_1 x_2 + \beta_4 x_1^2 + \beta_5 x_2^2$
12.19a. $E(y) = \beta_0 + \beta_1 x_1 + \beta_2 x_2 + \beta_3 x_1 x_2 + \beta_4 x_1^2 + \beta_5 x_2^2$ **b.** $E(y) = \beta_0 + \beta_1 x_1 + \beta_2 x_2$
c. $E(y) = \beta_0 + \beta_1 x_1 + \beta_2 x_2 + \beta_3 x_1 x_2$ **d.** $\beta_1 + \beta_3 x_2$ **e.** $\beta_2 + \beta_3 x_1$
12.20a. $E(y) = \beta_0 + \beta_1 x_1 + \beta_2 x_2 + \beta_3 x_1 x_2 + \beta_4 x_1^2 + \beta_5 x_2^2$ **b.** $E(y) = \beta_0 + \beta_1 x_1 + \beta_2 x_2 + \beta_3 x_1 x_2$
12.21a. $u = (x - 85.1)/14.81$ **b.** $-.668, .446, 1.026, -1.411, -.223, 1.695, -.527, -.338$ **c.** $.997$ **d.** $.377$
e. $\hat{y} = 110.95 + 14.376u + 7.425u^2$
12.22a. $u = (x - 33)/2.16$ **b.** $-1.389, -.926, -.463, 0, .463, .926, 1.389$ **c.** $.99966$ **d.** 0
e. $\hat{y} = 37.5714 - .4629u - 5.3333u^2$
12.23a. $u = (x - 5.5)/1.74$ **b.** -1.433 ($x = 3$), $-.860$ ($x = 4$), $-.287$ ($x = 5$), $.287$ ($x = 6$), $.860$ ($x = 7$), 1.433 ($x = 8$)
c. $.9913$ **d.** 0 **e.** $\hat{y} = 1079.49 - 262.14u - 196.64u^2$
12.24a. $u_1 = (x_1 - 940)/200.12$, $u_2 = (x_2 - .68)/.34$ **b.** $.998$; $.621$ **c.** $.989$; $.362$
d. $\hat{y} = .4376 + .6068u_1 + .3151u_2 + .3373u_1u_2 + .2250u_1^2 + .0266u_2^2$

12.25a. H_a: At least one parameter (β_3, β_4, or β_5) is not 0   **c.** 3, 24   **12.26** $F = 1.15$; do not reject H_0

12.27a. H_0: $\beta_1 = \beta_2 = \beta_3 = \beta_4 = \beta_5 = 0$; H_a: At least one β is not 0   **b.** $F = 18.39$; reject H_0

c. H_0: $\beta_3 = \beta_4 = \beta_5 = 0$; H_a: At least one β is not 0   **d.** $F = 8.46$; reject H_0 **e.** Complete model

12.28a. $F = 31.55$; reject H_0 **b.** Complete model   **12.29** $F = 24.19$; yes

12.30 $\hat{y} = 10.2$ for level 1; $\hat{y} = 6.2$ for level 2; $\hat{y} = 22.2$ for level 3; $\hat{y} = 12.2$ for level 4

12.31 H_0: $\beta_1 = \beta_2 = \beta_3 = 0$; H_a: At least one β is not 0

12.32 $E(y) = \beta_0 + \beta_1 x_1 + \beta_2 x_2 + \beta_3 x_3 + \beta_4 x_4$, where $x_1 = \begin{cases} 1 & \text{if brand 1} \\ 0 & \text{otherwise} \end{cases}$, $x_2 = \begin{cases} 1 & \text{if brand 2} \\ 0 & \text{otherwise} \end{cases}$,

$x_3 = \begin{cases} 1 & \text{if brand 3} \\ 0 & \text{otherwise} \end{cases}$, $x_4 = \begin{cases} 1 & \text{if brand 4} \\ 0 & \text{otherwise} \end{cases}$

12.33a. Group is a qualitative variable.

b. $E(y) = \beta_0 + \beta_1 x_1 + \beta_2 x_2$, where $x_1 = \begin{cases} 1 & \text{if group 2} \\ 0 & \text{otherwise} \end{cases}$, $x_2 = \begin{cases} 1 & \text{if group 3} \\ 0 & \text{otherwise} \end{cases}$

12.34a. $E(y) = \beta_0 + \beta_1 x_1 + \beta_2 x_2 + \beta_3 x_3$, where $x_1 = \begin{cases} 1 & \text{if Trilogy} \\ 0 & \text{otherwise} \end{cases}$,

$x_2 = \begin{cases} 1 & \text{if Set the Hostages Free} \\ 0 & \text{otherwise} \end{cases}$, $x_3 = \begin{cases} 1 & \text{if Queen of Hearts} \\ 0 & \text{otherwise} \end{cases}$ **c.** $\beta_0 + \beta_3$

12.35 $E(y) = \beta_0 + \beta_1 x_1$ **12.36** Include $\beta_2 x_2 + \beta_3 x_3$ **12.37** Include $\beta_4 x_1 x_2 + \beta_5 x_1 x_3$

12.38 If $\beta_4 = \beta_5 = 0$ **12.39** If $\beta_2 = \beta_3 = \beta_4 = \beta_5 = 0$

12.40b. Yes; $F = 10.60$ **c.** No; $t = -1.56$ **d.** $E(y) = \beta_0 + \beta_1 x_1 + \beta_2 x_2 + \beta_3 x_1 x_2$ **e.** Lines will have different slopes

12.41a. $E(y) = \beta_0 + \beta_1 x_1 + \beta_2 x_2 + \beta_3 x_3$, where $x_2 = \begin{cases} 1 & \text{if program B} \\ 0 & \text{otherwise} \end{cases}$ and $x_3 = \begin{cases} 1 & \text{if program C} \\ 0 & \text{otherwise} \end{cases}$

b. H_0: $\beta_2 = \beta_3 = 0$; H_a: At least one β is not 0

12.42 $F = 2.60$, do not reject H_0

12.43a. H_0: $\beta_2 = \beta_3 = 0$; H_a: At least one β is not 0   **b.** $F = 6.99$; reject H_0

c. Fit the model $E(y) = \beta_0 + \beta_1 x_1 + \beta_2 x_2 + \beta_3 x_3 + \beta_4 x_1 x_2 + \beta_5 x_1 x_3$ and test H_0: $\beta_4 = \beta_5 = 0$.

12.44a. Brand of extractor is qualitative; size of orange is quantitative.

b. $E(y) = \beta_0 + \beta_1 x_1 + \beta_2 x_2$, where $x_2 = \begin{cases} 1 & \text{if extractor B} \\ 0 & \text{otherwise} \end{cases}$ **c.** Include $\beta_3 x_1 x_2$ **e.** H_0: $\beta_3 = 0$ against H_a: $\beta_3 \neq 0$

12.45a. $E(y) = \beta_0 + \beta_1 x_1 + \beta_2 x_2 + \beta_3 x_3$ **b.** $E(y) = \beta_0 + \beta_1 x_1 + \beta_2 x_2 + \beta_3 x_3 + \beta_4 x_1 x_2 + \beta_5 x_1 x_3$

c. TDS-3A: $\beta_1 + \beta_4$; FE: $\beta_1 + \beta_5$; AL: β_1 **d.** Test H_0: $\beta_4 = \beta_5 = 0$

12.46a. $E(y) = \beta_0 + \beta_1 x_1 + \beta_2 x_2 + \beta_3 x_3 + \beta_4 x_1 x_2 + \beta_5 x_1 x_3$,

where $x_2 = \begin{cases} 1 & \text{if algorithm B} \\ 0 & \text{otherwise} \end{cases}$ and $x_3 = \begin{cases} 1 & \text{if algorithm C} \\ 0 & \text{otherwise} \end{cases}$

b. $F = 4.90$, reject H_0 (p-value $= .0394$)   **c.** $E(y) = \beta_0 + \beta_1 x_1 + \beta_2 x_2 + \beta_3 x_3$ **d.** $F = 2.36$; no

12.47 $E(y) = \beta_0 + \beta_1 x_1 + \beta_2 x_1^2$ **12.48** Include $\beta_3 x_2 + \beta_4 x_3$ **12.49** Include $\beta_5 x_1 x_2 + \beta_6 x_1 x_3 + \beta_7 x_1^2 x_2 + \beta_8 x_1^2 x_3$

12.50 If $\beta_5 = \beta_6 = \beta_7 = \beta_8 = 0$ **12.51** If $\beta_2 = \beta_5 = \beta_6 = \beta_7 = \beta_8 = 0$

12.52 If $\beta_3 = \beta_4 = \beta_5 = \beta_6 = \beta_7 = \beta_8 = 0$ **12.53a.** H_0: $\beta_4 = \beta_5 = 0$ **b.** H_0: $\beta_3 = \beta_4 = \beta_5 = 0$ **12.54** $F = .93$; no

12.55a. $E(y) = \beta_0 + \beta_1 x_1 + \beta_2 x_1^2 + \beta_3 x_2 + \beta_4 x_3 + \beta_5 x_1 x_2 + \beta_6 x_1 x_3 + \beta_7 x_1^2 x_2 + \beta_8 x_1^2 x_3$ **b.** H_0: $\beta_5 = \beta_6 = \beta_7 = \beta_8 = 0$

12.56a. $E(y) = \beta_0 + \beta_1 x_1 + \beta_2 x_1^2 + \beta_3 x_2 + \beta_4 x_3 + \beta_5 x_1 x_2 + \beta_6 x_1 x_3 + \beta_7 x_1^2 x_2 + \beta_8 x_1^2 x_3$

b. H_0: $\beta_3 = \beta_4 = \beta_5 = \beta_6 = \beta_7 = \beta_8 = 0$ **c.** $F = .34$; do not reject H_0

12.57 $F = 6.71$; yes   **12.58a.** x_2 ($t = -90$ is the largest in magnitude)   **b.** Yes

12.59a. x_4, x_5, x_6 **b.** No; there may be other important variables, as yet unspecified.

c. $E(y) = \beta_0 + \beta_1 x_4 + \beta_2 x_5 + \beta_3 x_6 + \beta_4 x_4 x_5 + \beta_5 x_4 x_6 + \beta_6 x_5 x_6$

12.60a. At least one β is not 0.   **c.** 2, 29   **d.** Reject H_0 if $F > 3.33$.   **12.61** No estimate of σ^2

12.62a. Two   **b.** $E(y) = \beta_0 + \beta_1 x_1 + \beta_2 x_2$, where $x_1 = \begin{cases} 1 & \text{if full-service only} \\ 0 & \text{otherwise} \end{cases}$ and $x_2 = \begin{cases} 1 & \text{if both} \\ 0 & \text{otherwise} \end{cases}$

12.63a. $E(y) = \beta_0 + \beta_1 x_1 + \beta_2 x_2$, where $x_1 =$ total area and $x_2 = \begin{cases} 1 & \text{if central air conditioning} \\ 0 & \text{otherwise} \end{cases}$

b. $E(y) = \beta_0 + \beta_1 x_1 + \beta_2 x_1^2 + \beta_3 x_2 + \beta_4 x_1 x_2 + \beta_5 x_1^2 x_2$ **c.** H_0: $\beta_2 = \beta_4 = \beta_5 = 0$

12.64a. $F = 2.49$; no   **b.** $F = 209$; reject H_0 **12.65a.** H_0: $\beta_2 = \beta_5 = 0$ **b.** H_0: $\beta_3 = \beta_4 = \beta_5 = 0$ **12.66** $F = 2.15$; no

12.67a. $E(y) = \beta_0 + \beta_1 x_1 + \beta_2 x_1^2 + \beta_3 x_2 + \beta_4 x_1 x_2 + \beta_5 x_1^2 x_2$ **c.** $E(y) = \beta_0 + \beta_1 x_1 + \beta_2 x_2$

12.68a. $E(y) = \beta_0 + \beta_1 x_1 + \beta_2 x_2 + \beta_3 x_1 x_2 + \beta_4 x_1^2 + \beta_5 x_2^2$ **c.** $\hat{y} = 54.5 + .007697 x_1 + .554111 x_2 + .000113 x_1 x_2$

e. $F = 44.67$; reject H_0 (p-value $= .0005$)   **f.** $t = .68$; no

12.69a. $\hat{y} = 22.0189 - .1807x_1 - .2498x_2 - 4.6910x_3 + 3.6745x_4 + 22.5201x_5$ **b.** $R^2 = .60$ **c.** 8.66
d. $F = 87.45$; yes   **e.** $t = -4.65$; yes   **f.** (2.885, 4.464)
12.70a. $\hat{y} = 59.0717 - 1.7003x_1 - 1.5722x_2 - 1.6971x_3 - .5638x_1x_2 + .9463x_1x_3 - 1.1963x_2x_3 - 17.3858x_1^2$
$- 13.8459x_2^2 - 17.4512x_3^2$ **b.** $F = 1.78$; do not reject H_0
12.71a. $E(y) = \beta_0 + \beta_1x_1 + \beta_2x_2 + \beta_3x_3 + \beta_4x_4$ **b.** $\hat{y} = 25.10 + 3.85x_1 + 6.55x_2 - 2.65x_3 - 4.75x_4$ **c.** $F = 23.97$; yes
12.72a. $F = 149$; reject H_0

CHAPTER 13

13.1a.

SOURCE	df	SS	MS	F
Treatments	2	11.075	5.538	3.15
Error	7	12.301	1.757	
Total	9	23.376		

b. Do not reject H_0 **c.** 2.54 ± 1.92 **d.** Smaller   **e.** 3.975 ± 1.567 **f.** 43

13.2a.

SOURCE	df	SS	MS	F
Treatments	4	24.7	6.175	4.91
Error	30	37.7	1.257	
Total	34	62.4		

b. 5   **c.** $F = 4.91$; yes   **d.** p-value $< .01$ **e.** $t = -.67$; no   **f.** $-.4 \pm .99$ **g.** $3.7 \pm .70$

13.3a.

SOURCE	df	SS	MS	F
Treatments	3	18.868	6.289	17.31
Error	10	3.633	.363	
Total	13	22.501		

b. $F = 17.31$; yes   **c.** $.15 \pm .60$ **d.** -1.83 ± 1.10

13.4 $F = 9.50$; yes

13.5a.

SOURCE	df	SS	MS	F
Treatments	3	.00678	.00226	3.50
Error	20	.01292	.00065	
Total	23	.01970		

b. $F = 3.50$; yes   **c.** $.08167 \pm .0216$ **d.** $-.04 \pm .0306$ **e.** 50

13.6 No; do not reject H_0, $F = .10$

13.7a.

SOURCE	df	SS	MS	F
Treatments	4	215.85	53.96	12.05
Error	18	80.58	4.48	
Total	22	296.43		

b. $F = 12.05$; yes   **c.** -2.25 ± 2.98 **d.** 138

13.10a. CM = 33,235,588,432; SST = 1,937,808,567.8   **b.** 148,746,738.4   **c.** 2,086,555,306.2

d.

SOURCE	df	SS	MS	F
Treatments	4	1,937,808,567.8	484,452,141.95	2,533.86
Error	778	148,746,738.4	191,191.18	
Total	782	2,086,555,306.2		

e. $F = 2,533.86$; yes   **f.** $3,660 \pm 135.93$ **g.** $-5,210 \pm 148.31$

13.11a.

SOURCE	df	SS	MS	F
Treatments	4	357,986.87	89,496.718	5.67
Error	73	1,151,602.00	15,775.370	
Total	77	1,509,588.87		

b. Yes; reject H_0, $F = 5.67$

13.12a

SOURCE	df	SS	MS	F
Treatments	2	23.167	11.583	12.64
Blocks	3	14.250	4.75	5.18
Error	6	5.500	.917	
Total	11	42.917		

b. Yes; $F = 12.64$ **c.** Yes; $F = 5.18$ **d.** -2.5 ± 1.32

13.13a.

SOURCE	df	SS	MS	F
Treatments	3	27.1	9.03	4.05
Blocks	5	74.5	14.90	6.68
Error	15	33.4	2.23	
Total	23	135.0		

b. No   **c.** Yes   **d.** -2.4 ± 1.51

13.14a. $F = 6.36$; yes   **b.** -4.8 ± 3.75 **13.15a.** Reject H_0, $F = 57.99$ **b.** .58 ± .10

13.16a. $F = .3381$; no   **b.** p-value is larger than .10

c.

SOURCE	df	SS	MS	F
Engineers	2	.9411	.4706	.3381
Contracts	5	10,240.0	2,048.0	1,471.0
Error	10	13.92	1.392	
Total	17	10,254.9		

d. $-.1833 \pm 1.234$

13.17a. Randomized block design   **b.**

SOURCE	df	SS	MS	F
Methods	1	.505	.5050	1.71
Increments	15	12.146	.8097	2.75
Error	15	4.421	.2947	
Total	31	17.072		

c. $F = 1.71$; no   **d.** $-.251 \pm .409$ **e.** $t = -1.31$; do not reject H_0

13.18a.

SOURCE	df	SS	MS	F
Machines	2	27,875.05	13,937.52	5.06
Functions	4	25,150.16	6,287.54	2.28
Error	8	22,016.91	2,752.11	
Total	14	75,042.12		

b. $F = 5.06$; yes   **c.** Do not reject H_0; $F = 2.28$

13.19a. $F = 39.46$; yes   **b.** $F = 1.93$; no   **c.** -27 ± 6.85

13.20a.

SOURCE	df	SS	MS	F
A	2	100	50	25
B	1	559	559	279.5
AB	2	5	2.5	1.25
Error	18	36	2.0	
Total	23	700		

b. Do not reject H_0; $F = 1.25$ **c.** Reject H_0; $F = 25.0$ **d.** Reject H_0; $F = 279.5$

13.21a.

SOURCE	df	SS	MS	F
A	3	13.79167	4.5972	2.57
B	2	6.58333	3.2917	1.84
AB	6	35.08333	5.8472	3.26
Error	12	21.50000	1.7917	
Total	23	76.95833		

b. $F = 3.26$; yes   **c.** 4.5 ± 1.687 **d.** -3.5 ± 2.385

13.22a.

SOURCE	df	SS	MS	F
Method	1	.00008929	.00008929	6.16
Lab	6	.00824321	.00137387	94.75
Method–Lab	6	.00019121	.00003187	2.20
Error	14	.00020300	.00001450	
Total	27	.00872671		

c. $F = 2.20$; no   **d.** $F = 94.75$; yes   **e.** $F = 6.16$; yes   **f.** $-.0185 \pm .0082$

13.23a.

SOURCE	df	SS	MS	F
Machines (A)	1	.101	.101	22.95
Materials (B)	2	.812	.406	92.27
AB	2	.768	.384	87.27
Error	12	.053	.0044	
Total	17	1.734		

b. Yes; $F = 87.27$ **d.** No; interaction is present   **e.** $.413 \pm .118$

13.24a.

SOURCE	df	SS	MS	F
Metal	1	51.842	51.842	41.15
Time	1	76.832	76.832	60.99
Metal–Time	1	2.738	2.738	2.17
Error	16	20.156	1.25975	
Total	19	151.568		

c. $F = 2.17$; no   **d.** $16.08 \pm .876$ **e.** -3.18 ± 1.239

13.25a.

SOURCE	df	SS	MS	F
Amount (A)	3	104.19	34.73	20.12
Method (B)	3	28.63	9.54	5.53
AB	9	25.13	2.79	1.62
Error	32	55.25	1.73	
Total	47	213.20		

b. Do not reject H_0; $F = 1.62$ **d.** Amount (A): reject H_0, $F = 20.12$; Method (B): reject H_0, $F = 5.53$ **e.** 21.40 ± 1.95
f. $-.70 \pm 2.77$

13.26a.

	Light	Heavy
Female	146.40	116.00
Male	104.00	98.00

b. 6,739.605   **c.** SS(Sex) $= 114.005$, SS(Weight) $= 41.405$, SS(SW) $= 18.605$

d.

	Sample Variance	SS(Deviations Within)
Female, Light	46.3761	324.6327
Female, Heavy	8.5849	60.0943
Male, Light	25.4016	177.8112
Male, Heavy	32.4900	227.4300

e. 789.968   **f.** 963.983

g.

SOURCE	df	SS	MS	F
Sex (S)	1	114.005	114.005	4.04
Weight (W)	1	41.405	41.405	1.47
SW	1	18.605	18.605	0.66
Error	28	789.968	28.213	
Total	31	963.983		

h. No; $F = .66$ **i.** -5.30 ± 5.44 **j.** -2.25 ± 5.44

13.27a. Complete model: $E(y) = \beta_0 + \beta_1 x_1 + \beta_2 x_2 + \beta_3 x_3$,
where $x_1 = \begin{cases} 1 & \text{if citrus leaves} \\ 0 & \text{if not} \end{cases}$, $x_2 = \begin{cases} 1 & \text{if bovine liver} \\ 0 & \text{if not} \end{cases}$, $x_3 = \begin{cases} 1 & \text{if human serum} \\ 0 & \text{if not} \end{cases}$;
reduced model: $E(y) = \beta_0$
13.28a. Complete model: $E(y) = \beta_0 + \beta_1 x_1 + \beta_2 x_2 + \beta_3 x_3 + \beta_4 x_4 + \cdots + \beta_{14} x_{14}$,
where $x_1 = \begin{cases} 1 & \text{if 2 weeks} \\ 0 & \text{if not} \end{cases}$, $x_2 = \begin{cases} 1 & \text{if 6 weeks} \\ 0 & \text{if not} \end{cases}$, $x_3 = \begin{cases} 1 & \text{if 14 weeks} \\ 0 & \text{if not} \end{cases}$,
$x_4 = \begin{cases} 1 & \text{if Location 2} \\ 0 & \text{if not} \end{cases}$, $\ldots$, $x_{14} = \begin{cases} 1 & \text{if location 12} \\ 0 & \text{if not} \end{cases}$;
reduced model: $E(y) = \beta_0 + \beta_4 x_4 + \cdots + \beta_{14} x_{14}$
13.29a. Complete model: $E(y) = \beta_0 + \beta_1 x_1 + \beta_2 x_2 + \beta_3 x_3 + \beta_4 x_4 + \beta_5 x_5 + \beta_6 x_6 + \beta_7 x_7 + \beta_8 x_1 x_2 + \beta_9 x_1 x_3 + \beta_{10} x_1 x_4 + \beta_{11} x_1 x_5 + \beta_{12} x_1 x_6 + \beta_{13} x_1 x_7$,
where $x_1 = \begin{cases} 1 & \text{if } A_2 \\ 0 & \text{if not} \end{cases}$, $x_2 = \begin{cases} 1 & \text{if } L_2 \\ 0 & \text{if not} \end{cases}$, $x_3 = \begin{cases} 1 & \text{if } L_3 \\ 0 & \text{if not} \end{cases}$, $\ldots$, $x_7 = \begin{cases} 1 & \text{if } L_7 \\ 0 & \text{if not} \end{cases}$;
reduced model: $E(y) = \beta_0 + \beta_1 x_1 + \beta_2 x_2 + \cdots + \beta_7 x_7$
13.30a. $\hat{y} = 7.46 + .0466 x_1 + 5.44 x_2 - .0148 x_1 x_2$ **b.** $\hat{y} = 12.9 + .0318 x_1$ **c.** $\hat{y} = 7.46 + .0466 x_1$
f. $t = -1.47$; do not reject H_0
13.31b. $E(y) = \beta_0 + \beta_1 x_1 + \beta_2 x_2 + \beta_3 x_1 x_2 + \beta_4 x_1^2 + \beta_5 x_2^2$
c. SSE $= .00014989$, $s^2 = .00000714$, SS(Total) $= 24.85834074$ **d.** $R^2 = .999$
e. The relationship between $E(y)$ and x_1 is independent of x_2.   **f.** $t = 12.41$; yes
g. $\hat{y} = -.28015 + .001 x_1 - .2855 x_2 + .00073 x_1 x_2 + .000000959 x_1^2 + .5514 x_2^2$ **h.** 2.6515   **i.** (2.6494, 2.6538)

13.32a. Pay rate (quantitative); length of workday (quantitative)

b. Three pay rates: P_1, P_2, P_3; three workday lengths: L_1, L_2, L_3; treatments: $P_1L_1, P_1L_2, P_1L_3, P_2L_1, P_2L_2, P_2L_3, P_3L_1, P_3L_2, P_3L_3$

c. $E(y) = \beta_0 + \beta_1x_1 + \beta_2x_1^2 + \beta_3x_2 + \beta_4x_2^2 + \beta_5x_1x_2 + \beta_6x_1^2x_2 + \beta_7x_1x_2^2 + \beta_8x_1^2x_2^2$ **d.** Fourth-order

e. $E(y) = \beta_0 + \beta_1x_1 + \beta_2x_1^2 + \beta_3x_2 + \beta_4x_2^2 + \beta_5x_1x_2$ **f.** 0   **g.** 3   **h.** Complete, 9 df; second-order, 12 df

i. Second-order   **j.** Test $H_0: \beta_5 = \beta_6 = \beta_7 = \beta_8 = 0$ using the reduced model $E(y) = \beta_0 + \beta_1x_1 + \beta_2x_1^2 + \beta_3x_2 + \beta_4x_2^2$

k. Test $H_0: \beta_5 = 0$

13.33a. Yes; $F = 40.78$ is significant at $\alpha = .0001$ **b.** FOIL and SHIFT*OPERATOR

c. $E(y) = \beta_0 + \beta_1x_1 + \beta_2x_2 + \beta_3x_3 + \beta_4x_4 + \beta_5x_1x_2 + \beta_6x_1x_3 + \beta_7x_1x_4 + \beta_8x_2x_3 + \beta_9x_2x_4 + \beta_{10}x_3x_4 +$
$\beta_{11}x_1x_2x_3 + \beta_{12}x_1x_2x_4 + \beta_{13}x_1x_3x_4 + \beta_{14}x_2x_3x_4 + \beta_{15}x_1x_2x_3x_4$ **d.** 16

13.34a. $F = 74.16$; yes   **b.** Charging time–alloy type and material–alloy type

13.35a. $E(y) = \beta_0 + \beta_1x_1 + \beta_2x_1^2$ **b.** $E(y) = (\beta_0 + \beta_3) + (\beta_1 + \beta_6)x_1 + (\beta_2 + \beta_9)x_1^2$

c. $E(y) = (\beta_0 + \beta_3 + \beta_4 + \beta_5) + (\beta_1 + \beta_6 + \beta_7 + \beta_8)x_1 + (\beta_2 + \beta_9 + \beta_{10} + \beta_{11})x_1^2$

13.36a. $\hat{y} = 31.15 + .153x_1 - .00396x_1^2 + 17.05x_2 + 19.1x_3 - 14.3x_2x_3 + .151x_1x_2 + .017x_1x_3 - .08x_1x_2x_3 -$
$.00356x_1^2x_2 + .0006x_1^2x_3 + .0012x_1^2x_2x_3$

b. Rolled/inconel: $\hat{y} = 53 + .241x_1 - .00572x_1^2$; rolled/incoloy: $\hat{y} = 50.25 + .17x_1 - .00336x_1^2$;
drawn/inconel: $\hat{y} = 48.2 + .304x_1 - .00752x_1^2$; drawn/incoloy: $\hat{y} = 31.15 + .153x_1 - .00396x_1^2$

13.37 $F = 304.6$; reject $H_0: \beta_3 = \beta_4 = \cdots = \beta_{11} = 0$

13.38a.

SOURCE	df
A	2
B	3
C	1
AB	6
AC	2
BC	3
ABC	6
Error	120
Total	143

b. $F = 5.21$; yes   **c.** $F = 2.79$; yes   **d.** SS(Total) $= 34.99$, $R^2 = .52$

13.39a. $3 \times 4 \times 3 \times 3 = 108$

b. The complete model has 108 terms, including β_0, 9 main effect terms, 30 2-way interactions, 44 3-way interactions, and 24 4-way interactions.   **c.** $H_0: \beta_{10} = \beta_{11} = \cdots = \beta_{107} = 0$

13.40a. Randomized block design   **b.**

SOURCE	df
Temperature	1
Pressure	1
Temperature–Pressure	1
Week	2
Error	6
Total	11

13.41b. Yes; p-value $= .0072$ **c.** No (at $\alpha = .05$); p-value $= .0646$ **13.42** $\omega = 8.67$; $\mu_{PD-1} > \mu_{ADC\ 5-1-7}$

13.43 $\omega = .138$; only means for 6 and 14 weeks do not appear to differ

13.44 $\omega = .182$; means for the following treatment pairs appear to differ: (A_1B_1, A_1B_3), (A_1B_1, A_2B_1), (A_1B_1, A_2B_2), (A_1B_1, A_2B_3), (A_1B_1, A_1B_2), (A_1B_3, A_2B_3), (A_1B_3, A_1B_2), (A_2B_1, A_2B_3), (A_2B_1, A_1B_2), (A_2B_2, A_2B_3), (A_2B_2, A_1B_2)

13.45 $\omega = 4.53$. There is a significant difference between the two groups of means, (μ_5, μ_3, μ_4) and (μ_2, μ_1). No significant differences exist within groups.

13.46 $\omega = 1.82$. There is a significant difference between the two groups of means, (μ_5, μ_3, μ_0) and (μ_{10}). No significant differences exist within groups.

13.47a. $F = 13.00$; yes   **b.** $\omega = .618$; μ_2 is significantly larger than either μ_1 or μ_3.

13.48a. $F = 22.23$; yes   **b.** $F = 20.45$; yes   **c.** 1.59 ± 4.088 **13.49a.** $F = 4.81$; no

13.50a. Factorial   **b.** No replications   **c.** $E(y) = \beta_0 + \beta_1 x_1 + \beta_2 x_2 + \beta_3 x_1 x_2 + \beta_4 x_1^2 + \beta_5 x_2^2$ **d.** Test $H_0: \beta_3 = 0$
e. $\hat{y} = -384.75 + 3.73x_1 + 12.72x_2 - .05x_1x_2 - .009x_1^2 - .322x_2^2$; $t = -2.05$, reject H_0 (p-value $= .07$)
13.51a. Completely randomized design   **b.** $F = 7.02$; yes   **c.** $8.15 \pm .44$ **13.52a.** $F = 6.63$; yes   **b.** -2.1 ± 1.28

13.53a.

SOURCE	df	SS	MS	F
Treatments	1	.0625	.0625	.83
Error	14	1.0575	.0755	
Total	15	1.1200		

b. $F = .83$; no   **c.** $.125 \pm .295$ **d.** 59

13.54a. $t = .91$ **c.** $H_a: \mu_1 > \mu_2$ **d.** No, since test is one-tailed

13.55a.

SOURCE	df	SS	MS	F
Time	2	8,200.389	4,100.195	856.52
Temperature	2	4,376.722	2,188.361	457.14
Time–Temperature	4	103.278	25.820	5.39
Error	27	129.250	4.787	
Total	35	12,809.639		

b. $F = 5.39$; yes   **c.** 50 ± 1.86 **d.** -22.25 ± 2.63

13.56a. $\hat{y} = -12.3055 - .1875x_1 + .10x_2 + .01125x_1x_2 + .0171x_1^2 + .00146x_2^2$ **b.** 50.78   **c.** 151.528
d. (85.218, 89.338)
13.57a. $F = 2.32$; no   **b.** $F = 4.68$; no

13.58a.

SOURCE	df	SS	MS	F
Union	1	7,168.444	7,168.444	26.34
Plan	2	5,425.167	2,712.584	9.97
Union–Plan	2	159.389	79.695	0.29
Error	30	8,165.000	272.167	
Total	35	20,918.000		

c. $F = .29$; no   **d.** 354.67 ± 11.079 **e.** -25 ± 15.668
13.59a. $F = .39$; no   **b.** $F = 19.17$; reject H_0 **c.** $.023 \pm .181$

13.60a.

SOURCE	df
A	3
B	3
AB	9
C	3
AC	9
BC	9
ABC	27
Error	128
Total	191

b. $F = 1.69$; yes   **c.** $F = 10.53$; yes

d. B_1: $-3.585 \pm .1917$; B_2: $-3.3875 \pm .1917$; B_3: $-3.1242 \pm .1917$; B_4: $-3.3542 \pm .1917$
13.61a. CM $= 8,912,304,000$; SST $= 92,833,225$ **b.** $75,145,616$ **c.** $167,978,841$

d.

SOURCE	df	SS	MS	F
Treatments	1	92,833,225	92,833,225	121.07
Error	98	75,145,616	766,792	
Total	99	167,978,841		

13.62b. $\hat{y} = 7,834.67 + 91.76x$ **c.** $9,669.87 \pm 176.43$ **d.** .553
13.63a. CM = 46,104,100; SST = 336,400   **b.** 261,121   **13.64b.** $\hat{y} = 582.33 + 5.524x$ **c.** 692.81 ± 102.96

CHAPTER 14

14.1a.

SOURCE	df	SS	MS	F
A	3	74.9167	24.9722	19.98
B in A	8	10.000	1.2500	
Total	11	84.9167		

b. $\hat{\sigma}_\alpha^2 = 7.91$, $\hat{\sigma}^2 = 1.25$ **c.** $F = 19.98$; yes

14.2a. 6   **b.** 5   **c.** 30   **d.** $y_{ij} = \mu + \alpha_i + \varepsilon_{ij}$ ($i = 1, 2, \ldots, 6; j = 1, 2, \ldots, 5$)
e. $\hat{\sigma}_\alpha^2 = 7.14$, $\hat{\sigma}^2 = 1.933$ **f.** $F = 3.69$; yes

14.3a.

SOURCE	df	SS	MS	F
A (Site)	2	.002573	.00129	.02
B in A (Within site)	12	.669600	.05580	
Total	14	.672173		

b. Do not reject H_0; $F = .02$

14.4a.

SOURCE	df	SS	MS	F
A (Batch)	9	6,902.125	766.903	78.99
B in A (Within batch)	30	291.250	9.708	
Total	39	7,193.375		

b. $\hat{\sigma}_\alpha^2 = 766.903$, $\hat{\sigma}^2 = 9.708$ **c.** $F = 78.99$; yes

14.5a. $\hat{\sigma}_B^2 = .364$, $\hat{\sigma}_W^2 = .057$ **b.** $F = 6.34$; yes

14.6a.

SOURCE	df	SS	MS	F
A (Day)	2	115.267	57.633	6.60
B in A (Within days)	27	235.700	8.730	
Total	29	350.967		

b. $F = 6.60$; yes

14.7

SOURCE	df
A (Production lot)	9
B in A (Batch within lot)	40
C in B (Shipping lot within batch)	950
Total	999

14.8a.

SOURCE	df	SS	MS	F
A	3	35.5	11.833	9.47
B in A	8	10.0	1.250	1.15
C in B	12	13.0	1.083	
Total	23	58.5		

b $F = 9.47$; yes   **c.** $F = 1.15$; no

14.9a.

SOURCE	df
A	9
B in A	30
C in B	80
Total	119

b. .0375   **c.** $F = 80$; yes   **d.** $F = 4.9$; yes

14.10a.

SOURCE	df	SS	MS	F
A (Days)	4	.0033933	.00084833	135.30
B in A (Cars within days)	20	.0001254	.00000627	2.29
C in B (Specimens within cars)	25	.0000685	.00000274	
Total	49	.0035872		

b. $F = 135.30$; yes   **c.** $F = 2.29$; yes

14.11a.

SOURCE	df	SS	MS	F
A (Days)	5	70.435833	14.0872	37.59
B in A (Rolls within days)	12	4.496667	.3747	.47
C in B (Tests within rolls)	18	14.455000	.8031	
Total	35	89.387500		

b. $F = 37.59$; yes   **c.** $F = .47$; no

14.12 $.09666 \pm .01143$

14.13a.

SOURCE	df	SS	MS	F
A (Specimens)	9	32.4	3.60	6.79
B in A (Portions within specimens)	10	5.3	.53	
Total	19	37.7		

b. $F = 6.79$; yes   **c.** $78.3 \pm .96$

14.14a. $F = 25.97$; yes   **b.** $.582 \pm .245$ **c.** $.5632 \pm .1095$ **14.15a.** $F = .0029$; no   **b.** $62.93 \pm .0447$

14.21a.

SOURCE	df	SS	MS
Between steels	3	42.7	14.23
Specimens within steels	16	58.1	3.63
Measurements within specimens	20	32.6	1.63
Total	39	133.4	

b. 7.4 ± 3.80 **c.** 1.2 ± 5.37

14.22a.

SOURCE	df
Temperature (T)	3
Fiber (F)	1
TF	3
Concentration (C)	1
CT	3
CF	1
CTF	3
Piece (CTF)	16
Error	32
Total	63

b. $F = 3.89$; yes   **d.** $F = 15.48$; yes   **e.** 21 ± 1.96 **f.** 3.75 ± 2.77

14.23 $\dfrac{\sigma_\alpha^2}{n_1} + \dfrac{\sigma^2}{n_1 n_2}$

14.25a.

SOURCE	df	SS	MS	F
Batch	14	78.9	5.64	5.94
Specimens within batch	45	42.7	.95	
Total	59	121.6		

b. $F = 5.94$; yes   **c.** $37.4 \pm .66$

14.26a.

SOURCE	df	SS	MS	F
Batch	4	198	49.5	5.99
Specimens within batch	15	124	8.3	
Total	19	322		

b. $F = 5.99$; yes

14.27a.

SOURCE	df	SS	MS	F
Batch	5	78.444	15.689	7.43
Specimens within batch	12	25.333	2.111	
Total	17	103.777		

b. 4.526; 2.111   **c.** $F = 7.43$; yes

14.28 22.889 ± 2.400 **14.29a.** $F = 1.25$; no   **b.** $F = .61$; no   **c.** $.0113 \pm .0016$

14.30a.

SOURCE	df
Method	1
Specimens within method	18
Portion within specimen	20
Total	39

b. MS(Method) = 2.5; MS(Specimens within method) = 3.71; MS(Portions within specimen) = .5   **c.** $F = .67$; no
d. $.5 \pm .905$

14.31a. Randomized block design   **b.**

SOURCE	df	SS	MS
Methods (Treatments)	1	2.47	2.47
Specimens (Blocks)	9	1.58	.175
Error	9	9.00	1.00
Total	19	13.05	

c. $F = 2.47$; no   **d.** $.5 \pm 1.01$

CHAPTER 15

15.1a. .0654 **b.** 3.0; 2.1 **c.** -1.125 **15.2a.** .0033067 **b.** 4.0; 3.2 **c.** -2.0

15.3 $E(C) = n(4p_1 + p_2)$; $V(C) = 16np_1(1 - p_1) + np_2(1 - p_2) - 8np_1p_2$

15.5a. $.122 \pm .089$ **b.** $.267 \pm .120$ **c.** $-.078 \pm .147$ **d.** $-.044 \pm .206$

15.6a. $.302 \pm .029$ **b.** $.037 \pm .052$ **15.7a.** $.275 \pm .182$ **b.** $.125 \pm .261$ **15.8a.** $.260 \pm .072$ **b.** $.220 \pm .113$

15.9a. $.22 \pm .068$ **b.** $-.08 \pm .111$ **c.** $.02 \pm .093$ **15.10** $X^2 = 13.89$; reject H_0

15.11a. $X^2 = 6.167$; yes **b.** $z = 2.24$; yes **15.12** $X^2 = 2.6$; do not reject H_0 **15.13** $X^2 = 10.16$; reject H_0

15.14 $X^2 = 4.4$; no **15.15** $X^2 = 508.74$; yes **15.17a.** $X^2 > 3.84146$ **b.** $X^2 > 15.9871$ **c.** $X^2 > 16.8119$

15.18 $X^2 = 66.25$; reject H_0 **15.19** $X^2 = .32$; do not reject H_0 **15.20** $X^2 = 1.351$; no

15.21 $X^2 = .449$; no **15.22** $X^2 = 15.859$; no **15.23** $X^2 = 14.73$; yes

15.24a.

	High Tech	Non–High Tech
Converters	80.2	296.8
Nonconverters	80.8	299.2

b. .021 **c.** No

15.25a. $X^2 = 28.51$; yes **b.** $.4 \pm .096$ **c.** $.1 \pm .098$ **15.26** $X^2 = 38.4$; yes **15.27** $X^2 = 5.136$; yes

15.28a. $X^2 = 44.55$; yes **b.** $z = 8.06$; yes **c.** $.32 \pm .109$

15.29a. $X^2 = 54.14$; yes **b.** No **c.** Yes **15.30a.** $X^2 = 2.13$; no **b.** $.233 \pm .057$

15.31 $X^2 = 5.506$; no **15.32a.** $X^2 = 12.57$; yes **b.** $z = 3.14$; yes

15.33a. White middle: $X^2 = 2.74$, do not reject H_0; black lower-middle: $X^2 = 19.04$, reject H_0; black lower: $X^2 = 12.68$, reject H_0 **b.** White middle: $.144 \pm .206$; black lower-middle: $.267 \pm .106$; black lower: $.351 \pm .186$

15.34 $X^2 = 19.27$; yes **15.35** $X^2 = 26.452$; reject H_0 **15.36a.** $X^2 = 31.8$; yes **b.** $.12 \pm .0947$

15.37a. $X^2 = 14.67$; yes **b.** $-.169 \pm .161$ **15.38** $X^2 = 13.289$; yes **15.39** $X^2 = 7.384$; no **15.40** $X^2 = 5.719$; no

15.41a. $X^2 = 5.581$; reject H_0 **b.** $z = 2.56$; reject H_0 **c.** Yes **15.42** $X^2 = 8.806$; no

15.43a. $X^2 = 4,755$; reject H_0 **b.** $.354 \pm .012$ **15.44** $X^2 = 61.489$; yes

CHAPTER 16

16.1a. .812 **b.** .012 **c.** .055 **16.2** No; $S = 4$, $p\text{-value} = .6875$

16.3 No; $S = 5$, $p\text{-value} = .2266$

16.4 $S = 14$, reject H_0; $p\text{-value} = .058$ **16.5** No; $S = 4$, $p\text{-value} = .6367$

16.8a. Reject H_0 if $T \leq 22$ or $T \geq 43$. **b.** Reject H_0 if $T_A \leq 20$ or $T_A \geq 45$.

16.9a. $T_1 = 67.5$; yes **b.** $T_1 = 67.5$; yes **16.10** Do not reject H_0; $T_1 = 71$ **16.11** $T_A = 85$; no

16.12 $T_{1976} = 131$; yes **16.13** $T_{800} = 96$; yes **16.14** $z = 1.846$; no **16.15** $z = 3.91$; yes

16.18 $T_L = 6$ **16.19a.** Reject H_0 if $T_- \leq 2$ **b.** Reject H_0 if $T \leq 4$ **16.20a.** $T_- = 2.5$; yes **b.** $T_- = 2.5$; yes

16.21 $T_- = 5.5$; do not reject H_0 **16.22** $T_+ = 11$; no **16.23** $T_- = 3$; yes **16.24** $T_+ = 17$; no

16.25 $T_- = 0$; reject H_0 **16.26** $T_- = 9$; no **16.27** $T_- = 0$; reject H_0 **16.31a.** $H = 8.008$; yes **16.32** $H = 2.03$; no

16.33 $H = .98$; no **16.34** $H = 10.91$; yes **16.35** $H = 5.73$; no

16.36 $H = 11.17$; reject H_0 (test is approximate due to inadequate sample size) **16.37** $H = 11.03$; yes

16.39 $F_r = 13.0$; yes **16.40** $F_r = 7.85$; yes **16.41** $F_r = 12.89$; yes **16.42** $F_r = 1.75$; no

16.43a. $F_r = 43.56$; yes **b.** $F_r = 2.5$; no **c.** $F_r = 1.6$; no **16.44** $F_r = 1.82$; no

16.45a. $r_s = .866$; yes **b.** Yes **16.46a.** 1.00 **b.** Reject H_0

16.47a. $r_s = .53$; yes **b.** $r_s = .30$; no **c.** $r_s = .64$; yes **16.48a.** $r_s = .82$; yes **b.** $r_s = .92$; yes **c.** $r_s = .67$; yes

16.49a. $r_s = .9812$ **b.** Reject H_0 **16.50** $r_s = -.738$; yes **16.51** $r_s = .835$; yes

16.54 $H = 9.859$; yes **16.55a.** $r_s = .4$; no **b.** $T_- = 1.5$; yes **16.56** $T_2 = 55.5$; no

16.57 $F_r = 19.21$; yes **16.58** $S = 7$, do not reject H_0; $p\text{-value} = .172$ **16.59** $T_- = 0$; reject H_0

16.60a. $T_A = 19$; yes **16.61** $H = 19.4658$; yes **16.62** $T_A = 66$; no

16.63a. $F_r = 0$; do not reject H_0 **b.** $F_r = 0$; do not reject H_0 **c.** $F_r = 3.0$; do not reject H_0 ($\alpha = .05$)

16.64 $r_s = 1.0$; yes **16.65a.** $F_r = 6.15$; no **b.** $F_r = 1.6$; no **16.66** No, $S = 5$; $p\text{-value} = 1.00$

16.67 No; $T_A = 106$ **16.68** $F_r = 1.33$; no **16.69** $H = 15.68$; yes **16.70** $T_1 = 78.5$; reject H_0

16.71a. $r_s = .90$ **b.** Yes **16.72** $T = 61.5$; no **16.73a.** $T_+ = 1$; yes **b.** $t = -2.96$; reject H_0

16.74 $T_+ = 7$; no **16.75** $F_r = 1.18$; do not reject H_0

CHAPTER 17

17.1a. $z(t) = \frac{1}{4}$ **b.** $z(t) = t$ **c.** $z(t) = \frac{1}{2}t^2$

17.3a. $f(0) = .0044, F(0) = .0013, z(0) = .0044; f(1) = .0540, F(1) = .0228, z(1) = .0553; f(2) = .2420, F(2) = .1587,$
$z(2) = .2876; f(3) = .3989, F(3) = .5000, z(3) = .7979; f(4) = .2420, F(4) = .8413, z(4) = 1.5247; f(5) = .0540,$
$F(5) = .9772, z(5) = 2.3680; f(6) = .0044, F(6) = .9987, z(6) = 3.4091$

17.4a. $F(t) = 1 - e^{-t/100}; R(t) = e^{-t/100}$ **b.** .7788   **c.** $z(t) = .01$

17.5a. $F(t) = 1 - e^{-t^2/100}$ **b.** $R(t) = e^{-t^2/100}; z(t) = t/50$ **c.** $R(8) = .5273; z(8) = .16$

17.6a. $F(t) = (t - \alpha)/(\beta - \alpha); R(t) = (\beta - t)/(\beta - \alpha)$ **b.** $z(t) = 1/(\beta - t)$ **d.** .7143; .001

17.7 (23.728, 85.532)   **17.8a.** (.3073, .7208)   **b.** (.0117, .0421)   **17.9** (211.4, 394.5)

17.10a. (58.106, 366.557)   **b.** (.7088, .9469)   **17.11** (18,041.27, 294,622.44)

17.12a. (.0000034, .0000554)   **b.** (.8468, .9899)   **c.** (.0068, .1049)

17.13a. $\hat{\alpha} = 1.9879; \hat{\beta} = 16.029$ **b.** (1.9405, 2.0353)   **c.** (14.9423, 17.1947)

17.14a. .7834   **b.** .0505   **17.15a.** $z(t) = .124t^{.9879}$ **b.** $z(4) = .4877$

17.16a. $\hat{\alpha} = 1.033, \hat{\beta} = 5.641$ **b.** $1.033 \pm .440$; (1.996, 15.942)   **c.** .576   **d.** .304

17.17a.

Year	1	2	3
Number of Survivors	9	4	2

b. $\hat{\alpha} = 1.6938; \hat{\beta} = 3.326$ **c.** α: $(-.7335, 4.1211)$; β: (.5383, 20.5250)   **d.** .6219

17.18a. $\hat{\alpha} = .8426; \hat{\beta} = 2.9364$ **b.** α: (.7577, .9276); β: (2.6401, 3.2660)   **c.** .5430

17.19 .60192   **17.20** .999988   **17.21** .99507   **17.22a.** .99775   **b.** .9999995   **17.23a.** .983   **b.** .632

17.24a. .4866   **b.** .1889   **c.** .3333   **17.25a.** .8671   **b.** .1424   **c.** .000195

17.26a. $F(t) = 1 - e^{-t}(t + 1)$ **b.** $R(t) = e^{-t}(t + 1); z(t) = t/(t + 1)$ **c.** $R(3) = .199; z(3) = .75$

17.27a. $F(t) = t/\beta; R(t) = 1 - t/\beta; z(t) = 1/(\beta - t)$ **c.** .6

17.28a. (9,979.8, 46,367.0)   **b.** (.8184, .9578)   **c.** (.0000216, .0001002)

17.29a. $\hat{\alpha} = 2.0312; \hat{\beta} = 7.3942$ **b.** $z(t) = (.2747)t^{1.0312}, R(t) = e^{-t^{2.0312}/7.3942}$ **c.** .8735

17.30 .6721   **17.31** .0608   **17.32** .9758

17.33a. (2,153.3, 17,322.3)   **b.** .4284; (.1560, .7938)   **c.** .000212; (.0000577, .0004644)

17.34a. $\hat{\alpha} = 1.0380, \hat{\beta} = 119.8543; \alpha$: (.8889, 1.1871), β: (55.44, 259.09)   **b.** .1299   **c.** $z(t) = .0087t^{.0380}$

17.35 .9613

17.38a. $f(t) = \beta(1 - t/\alpha)e^{-\beta t(1-t/2\alpha)}, 0 < t < \alpha; F(t) = 1 - e^{-\beta t(1-t/2\alpha)}, 0 < t < \alpha; R(t) = \beta(1 - t/\alpha), 0 < t < \alpha$

CHAPTER 18

18.1a. .9958   **b.** LCL = .9531, UCL = 1.0385   **d.** Yes

18.2a. 4.99114   **b.** LCL = 4.93923, UCL = 5.04305   **c.** Yes

18.3a. $\bar{x} = 5.89667$, LCL = 5.36471, UCL = 6.42863   **b.** Yes

18.4b. $\bar{x} = .40336$, LCL = .40300, UCL = .40371

c. No; sample means for hours 10, 12, and 18 fall outside the control limits

d. $\bar{x} = .40333$, LCL = .40296, UCL = .40371; yes

18.5b. $\bar{x} = .14065$, LCL = .13565, UCL = .14565   **c.** Yes   **18.6a.** .074   **b.** LCL = 0, UCL = .15651   **c.** Yes

18.7 $\bar{R} = .13917$, LCL = .01893, UCL = .25941; yes   **18.8** $\bar{R} = .52$, LCL = 0, UCL = 1.339; no, range for day 6 is 1.5

18.9 $\bar{R} = .00062$, LCL = 0, UCL = .00131; no, ranges for hours 9 and 13 fall outside control limits

18.10 $\bar{R} = .00051$, LCL = 0, UCL = .00107   **18.11** $\bar{R} = .00867$, LCL = 0, UCL = .01833; yes

18.12 No trends   **18.13** No trends   **18.14** No trends

18.15 Sample means for seven consecutive hours (9–15) fall above the center line; evidence of trend

18.16 No trends   **18.17b.** .075   **c.** LCL = 0, UCL = .25169; yes

18.18b. .0372   **c.** LCL = 0, UCL = .09398   **18.19** Yes; $\hat{p} = .11$ for this sample lies outside the control limits

18.20b. .2571   **c.** LCL = .0717, UCL = .4426   **d.** No; $\bar{p} = .247$, LCL = .064, UCL = .430   **e.** No trends

18.21b. 4.8   **c.** LCL = 0, UCL = 11.37; yes   **d.** No evidence of trend

18.22b. 6.5   **c.** LCL = 0, UCL = 14.15; yes   **d.** No evidence of trend

18.23b. $\bar{c} = 8.00$, LCL $= 0$, UCL $= 16.49$ **c.** Yes; yes

18.24a. No, number of errors for plane 36 falls outside control limits   **b.** No trends

18.25 $5.8967 \pm .6459$ **18.26a.** $.40336 \pm .00104$ **b.** Yes   **c.** $n = 93$; $(.4023, .4043)$

18.27a. $.14065 \pm .00827$ **b.** $(.132, .148)$

18.28a. 26 ± 29.11; $1 - \alpha = 1.00$ **b.** Specific cause, since 93.12 falls outside the tolerance interval

18.29a. $.5120 \pm .0031$ **b.** $.5005 \pm .0047$ **c.** No   **d.** Yes   **e.** $.0262$

18.30a. $.5905, .1681, .0313$ **b.** $.049$ **c.** $.5905$

18.31a. $.5490, .1671, .0353, .0052, .000488$ **b.** $.1710$ **c.** $.1671$

18.32a. $n = 80$, $a = 7$ **b.** $n = 200$, $a = 14$ **c.** $n = 315$, $a = 21$

18.33a. $.0246$ **b.** $.5443$ **18.34** $n = 50$, $a = 3$

18.35b. LCL $= 12.3452$, UCL $= 13.5298$ **c.** LCL $= 12.2811$, UCL $= 13.5939$ **d.** Yes for both sets of limits
e. 12.9375 ± 1.6334

18.36 $\bar{R} = .8125$, LCL $= 0$, UCL $= 1.8541$

18.37b. Producer's risk for plan 1: $.0815$, producer's risk for plan 2: $.0334$; prefer plan 2
c. Consumer's risk for plan 1: $.5282$, consumer's risk for plan 2: $.1935$; prefer plan 2

18.38b. $\bar{c} = 9.92$, LCL $= .47$, UCL $= 19.37$ **c.** Yes   **18.39a.** $n = 32$, $a = 7$ **b.** $n = 50$, $a = 10$

18.40b. $\bar{p} = .07$, LCL $= 0$, UCL $= .178$; yes   **c.** Evidence of trend

18.41a. $.0755$ **b.** $.6769$ **18.42** $n = 125$, $a = 10$ **18.43a.** $.041$ **b.** $.564$ **18.44a.** $1,312 \pm 1,238.15$

APPENDIX I

I.1a. $\begin{bmatrix} 6 & 3 \\ -2 & -5 \end{bmatrix}$ **b.** $\begin{bmatrix} 3 & 0 & 9 \\ -9 & 4 & 5 \end{bmatrix}$ **c.** $\begin{bmatrix} 5 & 4 \\ 1 & -4 \end{bmatrix}$ **I.2a.** $\begin{bmatrix} 15 \\ 14 \\ -8 \end{bmatrix}$ **b.** $[7]$ **c.** No

I.3a. 3×4 **b.** No   **I.4a.** 1×1 **b.** 3×3 **c.** $\begin{bmatrix} 3 & 0 & 6 \\ 0 & 0 & 0 \\ 2 & 0 & 4 \end{bmatrix}$

I.5a. $\begin{bmatrix} 2 & 3 \\ -9 & 0 \\ 8 & -2 \end{bmatrix}$ **b.** $[3 \quad 0 \quad 4]$ **c.** $[14 \quad 7]$ **I.6a.** $[12]$ **b.** $\begin{bmatrix} 6 & 0 & -2 & 4 \\ -3 & 0 & 1 & -2 \\ 0 & 0 & 0 & 0 \\ 9 & 0 & -3 & 6 \end{bmatrix}$

I.7a. $\begin{bmatrix} 1 & 0 \\ 0 & 1 \end{bmatrix}$ **c.** $\begin{bmatrix} 1 & 0 & 0 \\ 0 & 1 & 0 \\ 0 & 0 & 1 \end{bmatrix}$ **I.12a.** $A = \begin{bmatrix} 3 & 1 \\ 1 & -1 \end{bmatrix}$; $V = \begin{bmatrix} v_1 \\ v_2 \end{bmatrix}$; $G = \begin{bmatrix} 5 \\ 3 \end{bmatrix}$ **c.** $V = \begin{bmatrix} 2 \\ -1 \end{bmatrix}$

I.13a. $A = \begin{bmatrix} 10 & 0 & 20 \\ 0 & 20 & 0 \\ 20 & 0 & 68 \end{bmatrix}$; $V = \begin{bmatrix} v_1 \\ v_2 \\ v_3 \end{bmatrix}$; $G = \begin{bmatrix} 60 \\ 60 \\ 176 \end{bmatrix}$ **c.** $V = \begin{bmatrix} 2 \\ 3 \\ 2 \end{bmatrix}$

I N D E X

Designer: Janet Bollow
Cover designer: John Williams
Technical artist: Reese Thornton
Proofreader: Ellen Z. Curtin
Production manager: Susan Reiland
Typesetter: Typeset in 10/12 Berkeley Old Style
          by Jonathan Peck Typographers, Ltd.